D1594852

CONTEMPORARY STATISTICAL MODELS

for the Plant and Soil Sciences

CONTEMPORARY STATISTICAL MODELS

for the Plant and Soil Sciences

Oliver Schabenberger
Francis J. Pierce

CRC PRESS

Boca Raton London New York Washington, D.C.

Library of Congress Cataloging-in-Publication Data

Schabenberger, Oliver.
Contemporary statistical models for the plant and soil sciences / Oliver Schabenberger and Francis J. Pierce.
p. cm.
Includes bibliographical references (p.).
ISBN 1-58488-111-9 (alk. paper)
1. Plants, Cultivated—Statistical methods. 2. Soil science—Statistical methods. I. Pierce, F. J. (Francis J.) II. Title.

SB91 .S36 2001
630′.727—dc21 2001043254

Direct all inquiries to CRC Press LLC, 2000 N.W. Corporate Blvd., Boca Raton, Florida 33431.

Visit the CRC Press Web site at www.crcpress.com

International Standard Book Number 1-58488-111-9
Library of Congress Card Number 2001043254
Printed in the United States of America 1 2 3 4 5 6 7 8 9 0
Printed on acid-free paper

To Lisa and Linda
for enduring support and patience

Contents

Preface

To the Reader

Statistics is essentially a discipline of the twentieth century, and for several decades it was keenly involved with problems of interpreting and analyzing empirical data that originate in agronomic investigations. The vernacular of experimental design in use today bears evidence of the agricultural connection and origin of this body of theory. Omnipresent terms, such as *block* or *split-plot,* emanated from descriptions of blocks of land and experimental plots in agronomic field designs. The theory of randomization in experimental work was developed by Fisher to neutralize in particular the spatial effects among experimental units he realized existed among field plots. Despite its many origins in agronomic problems, statistics today is often unrecognizable in this context. Numerous recent methodological approaches and advances originated in other subject-matter areas and agronomists frequently find it difficult to see their immediate relation to questions that their disciplines raise. On the other hand, statisticians often fail to recognize the riches of challenging data analytical problems contemporary plant and soil science provides. One could gain the impressions that

- statistical methods of concern to plant and soil scientists are completely developed and understood;
- the analytical tools of *classical* statistical analysis learned in a one- or two-semester course for non-statistics majors are sufficient to cope with data analytical problems;
- recent methodological work in statistics applies to other disciplines such as human health, sociology, or economics, and has no bearing on the work of the agronomist;
- there is no need to consider contemporary statistical methods and no gain in doing so.

These impressions are incorrect. Data collected in many investigations and the circumstances under which they are accrued often bear little resemblance to classically designed experiments. Much of the data analysis in the plant and soil sciences is nevertheless viewed in the experimental design framework. Ground and remote sensing technology, yield monitoring, and geographic information systems are but a few examples where analysis cannot necessarily be cast, nor should it be coerced, into a standard analysis of variance framework. As our understanding of the biological/physical/environmental/ecological mechanisms increases, we are more and more interested in what some have termed the space/time dynamics of the processes we observe or set into motion by experimentation. It is one thing to collect data in space and/or over time, it is another matter to apply the appropriate statistical tools to infer what the data are trying to tell us. While many of the advances in statistical methodologies in past decades have not explicitly focused on agronomic applications, it would be incorrect to assume that these methods are not fruitfully applied there. Geostatistical methods, mixed models for repeated measures and longitudinal data, generalized linear models for non-normal (= non-Gaussian) data, and nonlinear models are cases in point.

The dedication of time, funds, labor, and technology to study design and data accrual often outstrip the efforts devoted to the analysis of the data. Does it not behoove us to make the most of the data, extract the most information, and apply the most appropriate techniques? Data sets are becoming richer and richer and there is no end in sight to the opportunities for data collection. Continuous time monitoring of experimental conditions is already a reality in biomedical studies where wristwatch-like devices report patient responses in a continuous stream. Through sensing technologies, variables that would have been observed only occasionally and on a whole-field level can now be observed routinely and spatially explicit. As one colleague put it: "*What do you do the day you receive your first five million observations?*" We do not have (all) the answers for data analysis needs in the information technology age. We subscribe wholeheartedly, however, to its emerging philosophy : *Do not to be afraid to get started, do not to be afraid to stop, and apply the best available methods along the way.*

In the course of many consulting sessions with students and researchers from the life sciences, we realized that the statistical tools covered in a one- or two-semester statistical methods course are insufficient to cope successfully with the complexity of empirical research data. Correlated, clustered, and spatial data, non-Gaussian (non-Normal) data and nonlinear responses are common in practice. The complexity of these data structures tends to outpace the basic curriculum. Most studies do not collect just one data structure, however. Remotely sensed leaf area index, repeated measures of plant yield, ordinal responses of plant injury, the presence/absence of disease and random sampling of soil properties, for example, may all be part of one study and comprise the threads from which scientific conclusions must be woven. Diverse data structures call for diverse tools. This text is an attempt to *squeeze between two covers* many statistical methods pertinent to research in the life sciences. Any one of the main chapters (§4 to 9) could have easily been expanded to the size of the entire text, and there are several excellent textbooks and monographs that do so. Invariably, we are guilty of omission.

To the User

Contemporary statistical models cannot be *appreciated* to their full potential without a good understanding of theory. Hence, we place emphasis on that. They also cannot be *applied* to their full potential without the aid of statistical software. Hence, we place emphasis on that. The main chapters are roughly equally divided between coverage of essential theory and applications. Additional theoretical derivations and mathematical details needed to develop a deeper understanding of the models can be found on the companion CD-ROM. The choice to focus on The SAS® System for calculations was simple. It is, in our opinion, the most powerful statistical computing platform and the most widely available and accepted computing environment for statistical problems in academia, industry, and government. In areas where The SAS® System was not sufficient for our purposes, e.g., spatial data analysis, we employed the S-PLUS® package, in particular the S+SpatialStats® module. The important portions of the executed computer code are shown in the text along with the output. All data sets and SAS® or S-PLUS® codes are contained on the CD-ROM.

To the Instructor

This text is both a reference and textbook and was developed with a reader in mind who has had a first course in statistics, covering simple and multiple linear regression, analysis of variance, who is familiar with the principles of experimental design and is willing to absorb a few concepts from linear algebra necessary to discuss the theory. A graduate-level course in statistics may focus on the theory in the main text and the mathematical details appendix. A graduate-level service course in statistical methods may focus on the theory and applications in the main chapters. A graduate-level course in the life sciences can focus on the applications and through them develop an appreciation of the theory. Chapters 1 and 2 introduce statistical models and the key data structures covered in the text. The notion of clustering in data is a recurring theme of the text. Chapter 3 discusses requisite linear algebra tools, which are indispensable to the discussion of statistical models beyond simple analysis of variance and regression. Depending on the audiences previous exposure to basic linear algebra, this chapter can be skipped. Several possible course concentrations are possible. For example,

1. A course on linear models beyond the basic stats-methods course: §1, 2, (3), 4
2. A course on modeling nonlinear response: §1, 2, (3), 5, 6, parts of 8
3. A course on correlated data: §1, 2, (3), 7, 8, parts of 9
4. A course on mixed models: §1, 2, (3), parts of 4, 7, 8
5. A course on spatial data analysis: §1, 2, (3), 9

In a statistics curriculum the coverage of §4 to 9 should include the mathematical details and special topics sections §A4 to A9.

We did not include exercises in this text; the book can be used in various types of courses at different levels of technical difficulty. We did not want to suggest a particular type or level through exercises. Although the applications (case studies) in §5 to 9 are lengthy, they do not consitute the *final word* on any particular data. Some data sets, such as the Hessian fly experiment or the Poppy count data are visited repeatedly in different chapters and can be tackled with different tools. We encourage comparative analyses for other data sets. If the applications leave the reader wanting to try out a different approach, to tackle the data from a new angle, and to improve upon our analysis, we wronged enough to get that right.

This text would not have been possible without the help and support of others. Data were kindly made available by A.M. Blackmer, R.E. Byers, R. Calhoun, J.R. Craig, D. Gilstrap, C.A. Gotway Crawford, J.R. Harris, L.P. Hart, D. Holshouser, D.E. Karcher, J.J. Kells, J. Kelly, D. Loftis, R. Mead, G. A. Milliken, P. Mou, T.G. Mueller, N.L. Powell, R. Reed, J. D. Rimstidt, R. Witmer, J. Walters, and L.W. Zelazny. Dr. J.R. Davenport (Washington State University-IRAEC) kindly provided the aerial photo of the potato circle for the cover. Several graduate students at Virginia Tech reviewed the manuscript in various stages and provided valuable insights and corrections. We are grateful in particular to S.K. Clark, S. Dorai-Raj, and M.J. Waterman. Our thanks to C.E. Watson (Mississippi State University) for a detailed review and to Simon L. Smith for EXP®. Without drawing on the statistical expertise of J.B. Birch (Virginia Tech) and T.G. Gregoire (Yale University), this text would have been more difficult to finalize. Without the loving support of our families it would have been impossible. Finally, the fine editorial staff at CRC Press LLC, and in particular our editor, Mr. John Sulzycki, brought their skills to bear to make this project a reality. We thank all of these individuals for contributing to the parts of the book that are right. Its flaws are our responsibility.

Oliver Schabenberger
Virginia Polytechnic Institute and State University

Francis J. Pierce
Washington State University

About the Authors

Oliver Schabenberger is associate professor in the department of statistics at Virginia Polytechnic Institute and State University where he conducts research on parametric and non-parametric statistical methods for nonlinear response, non-normal data, longitudinal and spatial data. He is particularly interested in the application of statistics to agronomy and natural resource disciplines. He teaches statistical methods courses for non-majors and courses for majors in applied statistics, biological statistics, and spatial statistics. In addition to his teaching and research responsibilities, Dr. Schabenberger has served as statistical consultant to faculty and graduate students at Virginia Tech since 1999 and from 1996 to 1999 in the department of crop and soil sciences at Michigan State University. He holds degrees in forest engineering (Dipl.-Ing. F.H.) from the Fachhochschule für Forstwirtschaft in Rottenburg, Germany, forest science (Diplom) from the Albert-Ludwigs University in Freiburg, Germany, statistics (M.S.) and forestry (Ph.D.) from Virginia Polytechnic Institute and State University. He is a member of the American Statistical Association, the International Biometric Society (Eastern North American Region), the American Society of Agronomy, and the Crop Science Society of America. He was recently appointed research affiliate in the School of Forestry and Environmental Studies at Yale University.

Francis J. Pierce is the director of the center for precision agricultural systems at Washington State University, located at the WSU Irrigated Agriculture Research & Extension Center (IAREC) in Prosser, Washington. He is also a professor in the departments of crop and soil sciences and biological systems engineering and directs the WSU Public Agricultural Weather System. Dr. Pierce received his M.S. and Ph.D. degrees in soil science from the University of Minnesota in 1980 and 1984. He spent the next 16 years at Michigan State University, where he has served as professor of soil science in the department of crop and soil sciences since 1995. His expertise is in soil management and he has been involved in the development and evaluation of precision agriculture since 1991. The Center for Precision Agricultural Systems was funded by the Washington Legislature as part of the University's Advanced Technology Initiative in 1999. As center director, Dr. Pierce's mission is to advance the science and practice of precision agriculture in Washington. The center's efforts will support the competitive production of Washington's agricultural commodities, stimulate the state's economic development, and protect the region's environmental and natural resources. Dr. Pierce has edited three other books, *Soil Management for Sustainability*, *Advances in Soil Conservation*, and *The State of Site-Specific Management for Agriculture*.

The CD-ROM

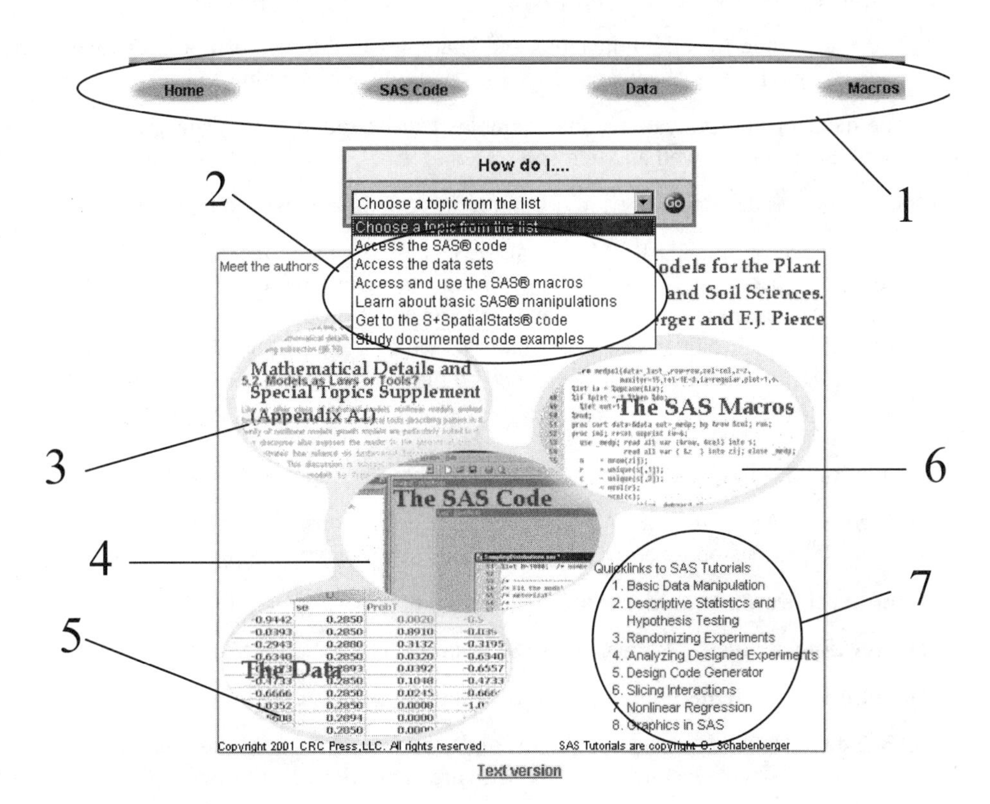

The CD-ROM is a key component of this book. It is organized so that its contents can be browsed with a standard browser such as Netscape® Navigator or Microsoft® Internet Explorer. To efficiently move about the various pages, the browser should have frame capabilities and be enabled to interpret JavaScript™ code. The graphic above shows the main page. It is opened as `default.htm` in the root directory of the CD-ROM with Windows® 95/98/NT/2000. Users of ISO9660-compliant operating systems (Macintosh, Solaris, OS2) can browse the files from the \iso9660\ directory of the CD-ROM.

1. The navigation bar remains in the top frame of many pages, enabling the user to quickly jump from one page to the next.

2. The drop-down box contains links to files that explain in more detail how the various pages function. For example, under "Access the SAS® Code", the reader finds explanations on how to copy SAS® programs from the browser window to The SAS® System and how to include programs directly from the CD-ROM into an SAS® session.

3. For each of the main chapters (§4 to 9) additional sections of texts that cover mathematical derivations, special topics, and supplementary applications are available. If the text refers to a chapter as §A5.9.3, for example, the "A" implies that the section can be found here on the CD-ROM. The supplementary sections have been compiled into one file in Adobe® portable document format (pdf). Alternatively, the contents can be browsed from the CD-ROM without the need for the Acrobat® Reader® software. This software is not supplied with this text and is available as a free download from `www.adobe.com/products/acrobat/readstep.html`.

4. The SAS® code for all applications and examples in this text can be found here. The code is organized by chapters and sections or applications within a chapter.

5. The data for all applications and examples can be found here. Data are available as SAS® data steps as well as Microsoft® Excel format.

6. SAS® macros developed by authors and referred to in the text can be found here.

7. We included on the CD-ROM several SAS® tutorials that cover elementary topics for the novice user, such as navigating through the windows of the user interface and basic data manipulation, as well as more advanced programming techniques such as publication quality graphics. These tutorials were developed by the first author. Additional SAS® related materials developed by him can be found at `www.stat.vt.edu/~oliver`.

The files on the CD-ROM can be accessed directly. The important directories are:

\Data\SAS: SAS® data steps for examples and applications. Files with *.txt and *.sas extensions. The *.txt files are displayed when browsing.

\Data\Excel: Data sets in Microsoft® Excel format.

\Data\SASpermanent6: Permanent SAS® data sets in Version 6.12 format.

\Data\SASpermanent8: Permanent SAS® data sets in Version 7.0 (8.0) format.

\SASCode: The SAS® programs for examples and applications. Files with *.txt and *.sas extensions. The *.txt files are displayed when browsing.

\SASMacros:The SAS® macros used in certain applications. Files with *.txt and *.sas extensions. The *.txt files are displayed when browsing.

\SPlus: The S-PLUS® code for certain applications.

\Text: Contains the pdf file for Appendix A, "Mathematical Details and Special Topics".

The JavaScript™, HTML, and SAS® code as well as all SAS® programs and macros were written by the first author. It has been tested under a fair number of circumstances, but we cannot guarantee that it always will. Please see disclaimer in `Readme.txt` file on CD-ROM.

Chapter 1

Statistical Models

"A theory has only the alternative of being right or wrong. A model has a third possibility: it may be right, but irrelevant." Manfred Eigen. In the Physicist's Conception of Nature *(Jagdish Mehra, Ed.) 1973.*

1.1 Mathematical and Statistical Models

Box 1.1 Statistical Models

- **A scientific model is the abstraction of a real phenomenon or process that isolates those aspects relevant to a particular inquiry.**
- **Inclusion of stochastic (random) elements in a mathematical model leads to more parsimonious and often more accurate abstractions than complex deterministic models.**
- **A special case of a stochastic model is the statistical model which contains unknown constants to be estimated from empirical data.**

The ability to represent phenomena and processes of the biological, physical, chemical, and social world through models is one of the great scientific achievements of humankind. Scientific models isolate and abstract the elementary facts and relationships of interest and provide the logical structure in which a system is studied and from which inferences are drawn. Identifying the important components of a system and isolating the facts of primary interest is necessary to focus on those aspects relevant to a particular inquiry. Abstraction is necessary to cast the facts in a logical system that is concise, deepens our insight, and is understood by others to foster communication, critique, and technology transfer. Mathematics is the most universal and powerful logical system, and it comes as no surprise that most scientific models in the life sciences or elsewhere are either developed as mathematical abstractions of real phenomena or can be expressed as such. A purely **mathematical model** is a mechanistic (= deterministic) device in that for a given set of inputs, it predicts the output with absolute certainty. It leaves nothing to chance. Beltrami (1998, p. 86), for example, develops the following mathematical model for the concentration α of a pollutant in a river at point s and time t:

$$\alpha(s,t) = \alpha_0(s - ct)\exp\{-\mu t\}. \qquad [1.1]$$

In this equation μ is a proportionality constant, measuring the efficiency of bacterial decomposition of the pollutant, c is the water velocity, and $\alpha_0(s)$ is the initial pollutant concentration at site s. Given the inputs c, μ, and α_0, the pollutant concentration at site s and time t is predicted with certainty. This would be appropriate if the model were correct, all its assumptions met, and the inputs measured or ascertained with certainty. Important assumptions of [1.1] are (i) a homogeneous pollutant concentration in all directions except for downstream flow, (ii) the absence of diffusive effects due to contour irregularities and turbulence, (iii) the decay of the pollutant due to bacterial action, (iv) the constancy of the bacterial efficiency, and (v) thorough mixing of the pollutant in the water. These assumptions are reasonable but not necessarily true. By *ignoring* diffusive effects, for example, it is really assumed that the positive and negative effects due to contour irregularities and turbulence will *average out*. The uncertainty of the effects at a particular location along the river and point in time can be incorporated by casting [1.1] as a **stochastic model**,

$$\alpha(s,t) = \alpha_0(s - ct)\exp\{-\mu t\} + e, \qquad [1.2]$$

where e is a random variable with mean 0, variance σ^2, and some probability distribution. Allowing for the random deviation e, model [1.2] now states explicitly that $\alpha(s,t)$ is a random variable and the expression $\alpha_0(s - ct)\exp\{-\mu t\}$ is the expected value or average pollutant concentration,

$$\mathrm{E}[\alpha(s,t)] = \alpha_0(s - ct)\exp\{-\mu t\}.$$

Olkin et al. (1978, p. 4) conclude: "*The assumption that chance phenomena exist and can be described, whether true or not, has proved valuable in almost every discipline.*" Of the many reasons for incorporating stochastic elements in scientific models, an incomplete list includes the following.

- The model is not correct for a particular observation, but correct on average.
- Omissions and assumptions are typically necessary to abstract a phenomenon.
- Even if the nature of all influences were known, it may be impossible to measure or even observe all the variables.
- Scientists do not develop models without validation and calibration with real data. The innate variability (nonconstancy) of empirical data stems from systematic and random effects. Random measurement errors, observational (sampling) errors due to sampling a population rather than measuring its entirety, experimental errors due to lack of homogeneity in the experimental material or the application of treatments, account for stochastic variation in the data even if all systematic effects are accounted for.
- Randomness is often introduced deliberately because it yields representative samples from which unbiased inferences can be drawn. A random sample from a population will represent the population (on average), regardless of the sample size. Treatments are assigned to experimental units by a random mechanism to neutralize the effects of unaccounted sources of variation which enables unbiased estimates of treatment means and their differences (Fisher 1935). Replication of treatments guarantees that experimental error variation can be estimated. Only in combination with randomization will this estimate be free of bias.
- Stochastic models are often more parsimonious than deterministic models and easier to study. A deterministic model for the germination of seeds from a large lot, for example, would incorporate a plethora of factors, their actions and interactions. The plant species and variety, storage conditions, the germination environment, amount of non-seed material in the lot, seed-to-seed differences in nutrient content, plant-to-plant interactions, competition, soil conditions, etc. must be accounted for. Alternatively, we can think of the germination of a particular seed from the lot as a Bernoulli random variable with success (germination) probability π. That is, if Y_i takes on the value 1 if seed i germinates and the value 0 otherwise, then the probability distribution of Y_i is simply

$$p(y_i) = \begin{cases} \pi & y_i = 1 \\ 1 - \pi & y_i = 0. \end{cases}$$

If seeds germinate independently and the germination probability is constant throughout the seed lot, this simple model permits important conclusions about the nature of the seed lot based on a sample of seeds. If n seeds are gathered from the lot for a germination test and $\widehat{\pi} = \sum_{i=1}^{n} Y_i/n$ is the sample proportion of germinated seeds, the stochastic behavior of the estimator $\widehat{\pi}$ is known, provided the seeds were selected at random. For all practical purposes it is not necessary to know the precise germination percentage in the lot. This would require either germinating all seeds or a deterministic model that can be applied to all seeds in the lot. It is entirely sufficient to be able to state with a desired level of confidence that the germination percentage is within certain narrow bounds.

Statistical models, in terminology that we adopt for this text, are stochastic models that contain unknown constants (**parameters**). In the river pollution example, the model

$$\alpha(s,t) = \alpha_0(s - ct)\exp\{-\mu t\} + e, \quad \mathrm{E}[e] = 0, \mathrm{Var}[e] = \sigma^2$$

is a stochastic model if all parameters $(\alpha_0, c, \mu, \sigma^2)$ are known. (Note that e is not a constant but a random variable. Its mean and variance are constants, however.) Otherwise it is a statistical model and those constants that are unknown must be estimated from data. In the seed germination example, the germination probability π is unknown, hence the model

$$p(y_i) = \begin{cases} \pi & y_i = 1 \\ 1 - \pi & y_i = 0 \end{cases}$$

is a statistical one. The parameter π is estimated based on a sample of n seeds from the lot. This usage of the term parameter is consistent with statistical theory but not necessarily with modeling practice. Any quantity that drives a model is often termed a *parameter* of the model. We will refer to parameters only if they are unknown constants. Variables that can be measured, such as plant density in the model of a yield-density relationship are, not parameters. The rate of change of plant yield as a function of plant density is a parameter.

1.2 Functional Aspects of Models

Box 1.2 What a Model Does

- **Statistical models describe the distributional properties of one or more response variables, thereby decomposing variability in known and unknown sources.**
- **Statistical models represent a mechanism from which data with the same statistical properties as the observed data can be generated.**
- **Statistical models are assumed to be correct on average. The quality of a model is not necessarily a function of its complexity or size, but is determined by its utility in a particular study or experiment to answer the questions of interest.**

A statistical model describes completely or incompletely the distributional properties of one or more variables, which we shall call the **response(s)**. If the description is complete and values for all parameters are given, the distributional properties of the response are known. A simple linear regression model for a random sample of $i = 1, \cdots, n$ observations on response Y_i and associated regressor x_i, for example, can be written as

$$Y_i = \beta_0 + \beta_1 x_i + e_i, \quad e_i \sim iid\, G\left(0, \sigma^2\right). \qquad [1.3]$$

The **model errors** e_i are assumed to be *i*ndependent and *i*dentically *d*istributed (iid) according to a Gaussian distribution (we use this denomination instead of *Normal* distribution throughout) with mean 0 and variance σ^2. As a consequence, Y_i is also distributed as a Gaussian random variable with mean $\mathrm{E}[Y_i] = \beta_0 + \beta_1 x_i$ and variance σ^2,

$$Y_i \sim G\left(\beta_0 + \beta_1 x_i, \sigma^2\right).$$

The Y_i are not identically distributed because their means are different, but they remain independent (a result of drawing a random sample). Since a Gaussian distribution is completely specified by its mean and variance, the distribution of the Y_i is completely known, once values for the parameters β_0, β_1, and σ^2 are known. For many statistical purposes the assumption of Gaussian errors is more than what is required. To derive unbiased estimators of the intercept β_0 and slope β_1, it is sufficient that the errors have zero mean. A simple linear regression model with lesser assumptions than [1.3] would be, for example,

$$Y_i = \beta_0 + \beta_1 x_i + e_i, \quad e_i \sim iid\, \left(0, \sigma^2\right).$$

The errors are assumed independent zero mean random variables with equal variance (homoscedastic), but their distribution is otherwise not specified. This is sometimes referred to as the **first-two-moments specification** of the model. Only the mean and variance of the Y_i can be inferred:

$$\begin{aligned} \mathrm{E}[Y_i] &= \beta_0 + \beta_1 x_i \\ \mathrm{Var}[Y_i] &= \sigma^2. \end{aligned}$$

If the parameters β_0, β_1, and σ^2 were known, this model would be an incomplete description of the distributional properties of the response Y_i. Implicit in the description of distributional properties is a separation of variability into known sources, e.g., the dependency of Y on x, and unknown sources (error) and a description of the form of the dependency. Here, Y is assumed to depend linearly on the regressor. Expressing which regressors Y depends on individually and simultaneously and how this dependency can be crafted mathematically is one important aspect of statistical modeling.

To conceptualize what constitutes a useful statistical model, we appeal to what we consider the most important functional aspect. A statistical model provides a mechanism to generate the essence of the data such that the properties of the data generated under the model are statistically equivalent to the observed data. In other words, the observed data can be considered as one particular realization of the stochastic process that is implied by the model. If the relevant features of the data cannot be realized under the assumed model, it is not useful.

The upper left panel in Figure 1.1 shows $n = 21$ yield observations as a function of the amount of nitrogen fertilization. Various candidate models exist to model the relationship between plant yield and fertilizer input. One class of models, the linear-plateau models

(§5.8.3, §5.8.4), are segmented models connecting a linear regression with a flat plateau yield. The upper right panel of Figure 1.1 shows the distributional specification of a linear plateau model. If α denotes the nitrogen concentration at which the two segments connect, the model for the average plant yield at concentration N can be written as

$$\mathrm{E}[Yield] = \begin{cases} \beta_0 + \beta_1 N & N \leq \alpha \\ \beta_0 + \beta_1 \alpha & N > \alpha. \end{cases}$$

If $I(x)$ is the indicator function returning value 1 if the condition x holds and 0 otherwise, the statistical model can also be expressed as

$$Yield = (\beta_0 + \beta_1 N)I(N \leq \alpha) + (\beta_0 + \beta_1 \alpha)I(N > \alpha) + e,\ e \sim G(0, \sigma^2). \quad [1.4]$$

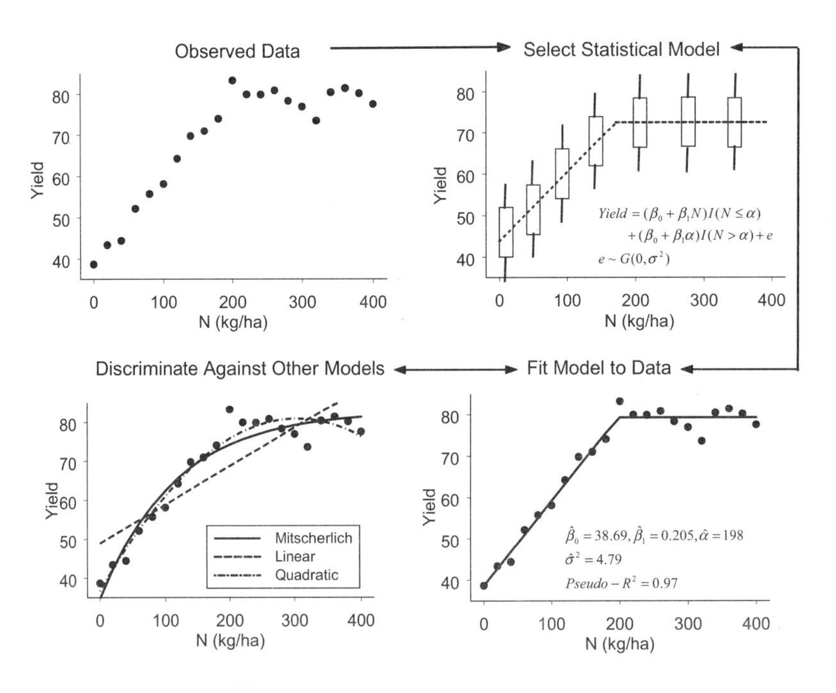

Figure 1.1. Yield data as a function of N input (upper left panel), linear-plateau model as an assumed data-generating mechanism (upper right panel). Fit of linear-plateau model and competing models are shown in the lower panels.

The constant variance assumption is shown in Figure 1.1 as box-plots at selected amounts of N with constant width. If [1.4] is viewed as the data-generating mechanism for the data plotted in the first panel, then for certain values of β_0, β_1, α, and σ^2, the statistical properties of the observed data should not be unusual compared to the data generated under the model. Rarely is there only a single model that could have generated the data, and statistical modeling invariably involves the comparison of competing models. The lower left panel of Figure 1.1 shows three alternatives to the linear-plateau model. A simple linear regression model without plateau, a quadratic polynomial response model, and a nonlinear model known as Mitscherlich's equation (§5.8.1, §5.8.2).

Choosing among competing models involves many factors and criteria, not all of which are statistical. Of the formal procedures, hypothesis tests for nested models and summary measures of model performance are the most frequently used. Two models are nested if the *smaller* (reduced) model can be obtained from the *larger* (full) model by placing restrictions on the parameters of the latter. A hypothesis test stating the restriction as the null hypothesis is performed. If the null hypothesis is rejected, the smaller model is deemed of lesser quality than the large model. This, of course, does not imply that the full model fits the data sufficiently well to be useful. Consider as an example the simple linear regression model discussed earlier and three competitors:

$$
\begin{aligned}
\boxed{\text{A}}&: \quad Y_i = \beta_0 + e_i \\
\boxed{\text{B}}&: \quad Y_i = \beta_0 + \beta_1 x_i + e_i \\
\boxed{\text{C}}&: \quad Y_i = \beta_0 + \beta_1 x_i + \beta_2 x_i^2 + e_i \\
\boxed{\text{D}}&: \quad Y_i = \beta_0 + \beta_1 x_i + \beta_2 z_i + e_i.
\end{aligned}
$$

Models $\boxed{\text{A}}$ through $\boxed{\text{C}}$ are **nested models**, so are $\boxed{\text{A}}$, $\boxed{\text{B}}$, and $\boxed{\text{D}}$. The restriction $\beta_2 = 0$ in model $\boxed{\text{C}}$ produces model $\boxed{\text{B}}$, the restriction $\beta_1 = 0$ in model $\boxed{\text{B}}$ produces the intercept-only model $\boxed{\text{A}}$, and $\beta_1 = \beta_2 = 0$ yields $\boxed{\text{A}}$ from $\boxed{\text{C}}$. To decide among these three models, one can commence by fitting model $\boxed{\text{C}}$ and perform hypothesis tests for the respective restrictions. Based on the results of these tests we are led to the *best* model among $\boxed{\text{A}}$, $\boxed{\text{B}}$, $\boxed{\text{C}}$. Should we have started with model $\boxed{\text{D}}$ instead? Model $\boxed{\text{D}}$ is a two-regressor multiple linear regression model, and a comparison between models $\boxed{\text{C}}$ and $\boxed{\text{D}}$ by means of a hypothesis test is not possible; the models are not nested. Other criteria must be employed to discriminate between them. The appropriate criteria will depend on the intended use of the statistical model. If it is important that the model fits well to the data at hand, one may rely on the coefficient of determination (R^2). To guard against overfitting, Mallow's C_p statistic or likelihood-based statistics such as Akaike's information criterion (AIC) can be used. If precise predictions are required then one can compare models based on cross-validation criteria or the PRESS statistic. Variance inflation factors and other collinearity diagnostics come to the fore if statistical properties of the parameter estimates are important (see, e.g., Myers 1990, and our §§4.4, A4.8.3). Depending on which criteria are chosen, different models might emerge as *best*.

Among the *informal* procedures of model critique are various graphical displays, such as plots of residuals against predicted values and regressors, partial residual plots, normal probability and Q-Q plots, and so forth. These are indispensable tools of statistical analysis but they are often overused and misused. As we will see in §4.4 the standard collection of residual measures in linear models are ill-suited to pass judgment about whether the unobservable model errors are Gaussian-distributed, or not. For most applications, it is not the Gaussian assumption whose violation is most damaging, but the homogeneous variance and the independence assumptions (§4.5). In nonlinear models the behavior of the residuals in a correctly chosen model can be very different from the *textbook* behavior of fitted residuals in a linear model (§5.7.1). Plotting studentized residuals against fitted values in a linear model, one expects a band of random scatter about zero. In a nonlinear model where intrinsic curvature is large, one should look for a negative trend between the residuals and fitted values as a sign of a well-fitting model.

Besides model discrimination based on statistical procedures or displays, the subject matter hopefully plays a substantive role in choosing among competing models. Interpretability and parsimony are critical assets of a useful statistical model. Nothing is gained by building models that are so large and complex that they are no longer interpretable as a whole or involve factors that are impossible to observe in practice. Adding variables to a regression model will necessarily increase R^2, but can also create conditions where the relationships among the regressor variables render estimation unstable, predictions imprecise, and interpretation increasingly difficult (see §§4.4.4, 4.4.5 on **collinearity**, its impact, diagnosis, and remedy). Medieval Franciscan monk William of Ockham (1285-1349) is credited with coining *pluralitas non est ponenda sine neccesitate* or *plurality should not be assumed (posited) without necessity*. Also known as *Ockham's Razor*, this tenet is often loosely phrased as "among competing explanations, pick the simplest one." When choosing among competing statistical models, *simple* does not just imply the smallest possible model. The selected model should be simple to fit, simple to interpret, simple to justify, and simple to apply. Nonlinear models, for example, have long been considered difficult to fit to data. Even recently, Black (1993, p. 65) refers to the "drudgery connected with the actual fitting" of nonlinear models. Although there are issues to be considered when modeling nonlinear relationships that do not come to bear with linear models, the actual process of fitting a nonlinear model with today's computing support is hardly more difficult than fitting a linear regression model (see §5.4).

Returning to the yield-response example in Figure 1.1, of the four models plotted in the lower panels, the straight-line regression model is ruled out because of poor fit. The other three models, however, have very similar goodness-of-fit statistics and we consider them competitors. Table 1.1 gives the formulas for the mean yield under these models. Each is a three-parameter model, the first two are nonlinear, the quadratic polynomial is a linear model. All three models are easy to fit with statistical software. The selection thus boils down to their interpretability and justifiability.

Each model contains one parameter measuring the average yield if no N is applied. The linear-plateau model achieves the yield maximum of $\beta_0 + \beta_1\alpha$ at precisely $N = \alpha$; the Mitscherlich equation approaches the yield maximum λ asymptotically (as $N \rightarrow \infty$). The quadratic polynomial does not have a yield plateau or asymptote, but achieves a yield maximum at $N = -\gamma_1/2\gamma_2$. Increasing yields are recorded for $N < -\gamma_1/2\gamma_2$ and decreasing yields for $N > -\gamma_1/2\gamma_2$. Since the yield increase is linear in the plateau model (up to $N = \alpha$), a single parameter describes the rate of change in yield. In the Mitscherlich model where the transition between yield minimum ξ and upper asymptote α is smooth, no single parameter measures the rate of change, but one parameter (κ) governs it:

$$\frac{\partial \mathrm{E}[Yield]}{\partial N} = \kappa(\lambda - \xi)\exp\{-\kappa N\}.$$

The standard interpretation of regression coefficients in a multiple linear regression equation is to measure the change in the mean response if the associated regressor increases by one unit while all other regressors are held constant. In the quadratic polynomial the linear and quadratic coefficients (γ_1, γ_2) cannot be interpreted this way. Changing N while holding N^2 constant is not possible. The quadratic polynomial has a linear rate of change,

$$\frac{\partial \mathrm{E}[Yield]}{\partial N} = \gamma_1 + 2\gamma_2 N,$$

so that γ_1 can be interpreted as the rate of change at $N = 0$ and $2\gamma_2$ as the rate of change of the rate of change.

Table 1.1. Three-parameter yield response models and the interpretation of their parameters

Model	E[$Yield$]	No. of params.	Interpretation
Linear-plateau	$\begin{cases} \beta_0 + \beta_1 N & N \leq \alpha \\ \beta_0 + \beta_1 \alpha & N > \alpha \end{cases}$	3	β_0: E[$Yield$] at $N = 0$ β_1: Change in E[$Yield$] per 1 kg/ha additional N prior to reaching plateau α: N amount where plateau is reached
Mitscherlich	$\lambda + (\xi - \lambda)\exp\{-\kappa N\}$	3	λ: Upper E[$Yield$] asymptote ξ: E[$Yield$] at $N = 0$ κ: Governs rate of change
Quadr. polynomial	$\gamma_0 + \gamma_1 N + \gamma_2 N^2$	3	γ_0: E[$Yield$] at $N = 0$ γ_1: ∂E[$Yield$]$/\partial N$ at $N = 0$ γ_2: $\frac{1}{2}\partial^2$E[$Yield$]$/\partial N^2$

Interpretability of the model parameters clearly favors the nonlinear models. Biological relationships rarely exhibit sharp transitions and kinks. Smooth, gradual transitions are more likely. The Mitscherlich model may be more easily justifiable than the linear-plateau model. If no decline of yields was observed over the range of N applied, resorting to a model that will invariably have a maximum at some N amount, be it within the observed range or outside of it, is tenuous.

One appeal of the linear-plateau model is to estimate the amount of N at which the two segments connect, a special case of what is known in dose-response studies as an effective (or critical) dosage. To compute an effective dosage in the Mitscherlich model the user must specify the response the dosage is supposed to achieve. For example, the nitrogen fertilizer amount N_K that produces $K\%$ of the asymptotic yield in the Mitscherlich model is obtained by solving $\lambda K/100 = \lambda + (\xi - \lambda)\exp\{-\kappa N_K\}$ for N_K,

$$N_K = -\frac{1}{\kappa}\ln\left\{\frac{\lambda}{\lambda - \xi}\left(\frac{100 - K}{100}\right)\right\}.$$

In the linear-plateau model, the plateau yield is estimated as $\widehat{\beta}_0 + \widehat{\beta}_1\widehat{\alpha} = 38.69 + 0.205*198 = 79.28$. The parameter estimates obtained by fitting the Mitscherlich equation to the data are $\widehat{\lambda} = 82.953$, $\widehat{\xi} = 35.6771$, and $\widehat{\kappa} = 0.00857$. The estimated plateau yield is 95.57% of the estimated asymptotic yield, and the N amount that achieves the plateau yield in the Mitscherlich model is

$$\widehat{N}_k = -\frac{1}{0.00857}\ln\left\{\frac{82.953}{82.953 - 35.6771}\left(\frac{100 - 95.57}{100}\right)\right\} = 298.0 \text{ kg/ha},$$

considerably larger than the critical dosage in the linear-plateau model.

1.3 The Inferential Steps — Estimation and Testing

Box 1.3 Inference

- **Inference in a statistical model involves the estimation of model parameters, the determination of the precision of the estimates, predictions and their precision, and the drawing of conclusions about the true values of the parameters.**
- **The two most important statistical estimation principles are the principles of least squares and the maximum likelihood principle. Each appears in many different flavors.**

After selecting a statistical model its parameters must be estimated. If the fitted model is accepted as a useful abstraction of the phenomenon under study, further inferential steps involve the calculation of confidence bounds for the parameters, the testing of hypotheses about the parameters, and the calculation of predicted values. Of the many estimation principles at our disposal, the most important ones are the least squares and the maximum likelihood principles. Almost all of the estimation methods that are discussed and applied in subsequent chapters are applications of these basic principles. Least squares (LS) was advanced and maximum likelihood (ML) proposed by Carl Friedrich Gauss (1777-1855) in the early nineteenth century. R.A. Fisher is usually credited with the (re-)discovery of the likelihood principle.

Least Squares

Assume that a statistical model for the observed data Y_i can be expressed as

$$Y_i = f_i(\theta_1, \cdots, \theta_p) + e_i, \qquad [1.5]$$

where the θ_j $(j = 1, \cdots, p)$ are parameters of the mean function $f_i()$, and the e_i are zero mean random variables. The distribution of the e_i can depend on other parameters, but not on the θ_j. LS is a semi-parametric principle, in that only the mean and variance of the e_i as well as their covariances (correlations) are needed to derive estimates. The distribution of the e_i can be otherwise unspecified. In fact, least squares can be motivated as a geometric rather than a statistical principle (§4.2.1). The assertion found in many places that the e_i are Gaussian random variables is not needed to derive the estimators of $\theta_1, \cdots, \theta_p$.

Different flavors of the LS principle are distinguished according to the variances and covariances of the error terms. In ordinary least squares (OLS) estimation the e_i are uncorrelated and homoscedastic ($\text{Var}[e_i] = \sigma^2$). Weighted least squares (WLS) assumes uncorrelated errors but allows their variances to differ ($\text{Var}[e_i] = \sigma_i^2$). Generalized least squares (GLS) accommodates correlations among the errors and estimated generalized least squares (EGLS) allows these correlations to be unknown. There are other varieties of the least squares principles, but these four are of primary concern in this text. If the mean function $f_i(\theta_1, \cdots, \theta_p)$ is nonlinear in the parameters $\theta_1, \cdots, \theta_p$, the respective methods are referred to as nonlinear OLS, nonlinear WLS, and so forth (see §5).

The general philosophy of least squares estimation is most easily demonstrated for the case of OLS. The principle seeks to find those values $\widehat{\theta}_1, \cdots, \widehat{\theta}_p$ that minimize the sum of squares

$$S(\theta_1, \cdots, \theta_p) = \sum_{i=1}^{n} (y_i - f_i(\theta_1, \cdots, \theta_p))^2 = \sum_{i=1}^{n} e_i^2.$$

This is typically accomplished by taking partial derivatives of $S(\theta_1, \cdots, \theta_p)$ with respect to the θ_j and setting them to zero. The system of equations

$$\begin{aligned} \partial S(\theta_1, \cdots, \theta_p)/\partial\theta_1 &= 0 \\ &\vdots \\ \partial S(\theta_1, \cdots, \theta_p)/\partial\theta_p &= 0 \end{aligned}$$

is known as the *normal* equations. For linear and nonlinear models a solution is best described in terms of matrices and vectors; it is deferred until necessary linear algebra tools have been discussed in §3. The solutions $\widehat{\theta}_1, \cdots, \widehat{\theta}_p$ to this minimization problem are called the ordinary least squares estimators (OLSE). The residual sum of squares SSR is obtained by evaluating $S(\theta_1, \cdots, \theta_p)$ at the least squares estimate.

Least squares estimators have many appealing properties. In linear models, for example,

- the $\widehat{\theta}_j$ are linear functions of the observations $y_1, \cdots, y_n$, which makes it easy to establish statistical properties of the $\widehat{\theta}_j$ and to test hypotheses about the unknown θ_j.
- The linear combination $a_1\widehat{\theta}_1 + \cdots + a_p\widehat{\theta}_p$ is the best linear unbiased estimator (BLUE) of $a_1\theta_1 + \cdots + a_p\theta_p$ (Gauss-Markov theorem). If, in addition, the e_i are Gaussian-distributed, then $a_1\widehat{\theta}_1 + \cdots + a_p\widehat{\theta}_p$ is the minimum variance unbiased estimator of $a_1\theta_1 + \cdots + a_p\theta_p$. No other unbiased estimator can beat its performance, linear or not.
- If the e_i are Gaussian, then two nested models can be compared with a sum of squares reduction test. If SSR_f and SSR_r denote the residual sums of squares in the full and reduced model, respectively, and MSR_f the residual mean square in the full model, then the statistic

  $$F_{obs} = \frac{(SSR_r - SSR_f)/q}{MSR_f}$$

  has an F distribution with q numerator and dfR_f denominator degrees of freedom under the null hypothesis. Here, q denotes the number of restrictions imposed (usually the number of parameters eliminated from the full model) and dfR_f denotes the residual degrees of freedom in the full model. F_{obs} is compared against the F_{α,q,dfR_f} and the null hypothesis is rejected if F_{obs} exceeds the cutoff. The distributional properties of F_{obs} are established in §4.2.3 and §A4.8.2. The sum of squares reduction test can be shown to be equivalent to many standard procedures. For the one-sample and pooled t-tests, we demonstrate the equivalency in §1.4.

A *downside* of least squares estimation is its focus on parameters of the mean function $f_i()$. The principle does not lend itself to estimation of parameters associated with the distribution of the errors e_i, for example, the variance of the model errors. In least squares estimation, these parameters must be obtained by other principles.

Maximum Likelihood

Maximum likelihood is a parametric principle; it requires that the joint distribution of the observations $Y_1, \cdots, Y_n$ is known except for the parameters to be estimated. For example, one may assume that $Y_1, \cdots, Y_n$ follow a multivariate Gaussian distribution (§3.7) and base ML estimation on this fact. If the observations are statistically independent the joint density (y continuous) or mass (y discrete) function is the product of the marginal distributions of the Y_i, and the likelihood is calculated as the product of individual contributions, one for each sample. Consider the case where

$$Y_i = \begin{cases} 1 & \text{if a seed germinates} \\ 0 & \text{otherwise,} \end{cases}$$

a binary response variable. If a random sample of n seeds is obtained from a seed lot then the probability mass function of y_i is

$$p(y_i; \pi) = \pi^{y_i}(1 - \pi)^{1-y_i},$$

where the parameter π denotes the probability of germination in the seed lot. The joint mass function of the random sample becomes

$$p(y_1, \cdots, y_n; \pi) = \prod_{i=1}^{n} \pi^{y_i}(1 - \pi)^{1-y_i} = \pi^r(1 - \pi)^{n-r}, \qquad [1.6]$$

with $r = \sum_{i=1}^{n} y_i$, the number of germinated seeds. For any given value $\widetilde{\pi}$, the probability $p(y_1, \cdots, y_n; \widetilde{\pi})$ can be thought of as the probability of observing the sample $y_1, \cdots, y_n$ if the germination probability is $\widetilde{\pi}$. The maximum likelihood principle estimates π by that value which maximizes $p(y_1, \cdots, y_n; \pi)$, because this is the value most likely to have generated the data.

Since $p(y_1, \cdots, y_n; \pi)$ is now considered a function of π for a given sample $y_1, \cdots, y_n$, we write $\mathcal{L}(\pi; y_1, \cdots, y_n)$ for the function to be maximized and call it the **likelihood** function. Whatever technical device is necessary, maximum likelihood estimators (MLEs) are found as those values that maximize $\mathcal{L}(\pi; y_1, \cdots, y_n)$ or, equivalently, maximize the **log-likelihood** function $\ln\{\mathcal{L}(\pi; y_1, \cdots, y_n)\} = l(\pi; y_1, \cdots, y_n)$. Direct maximization is often possible. Such is the case in the seed germination example. From [1.6] the log-likelihood is computed as $l(\pi; y_1, \cdots, y_n) = r\ln\{\pi\} + (n - r)\ln\{1 - \pi\}$, and taking the derivative with respect to π yields

$$\partial l(\pi; y_1, \cdots, y_n)/\partial\pi = r/\pi - (n - r)/(1 - \pi).$$

The MLE $\widehat{\pi}$ of π is the solution of $r/\widehat{\pi} = (n - r)/(1 - \widehat{\pi})$, or $\widehat{\pi} = r/n = \overline{y}$, the sample proportion of germinated seeds.

The maximum likelihood principle is intuitive and maximum likelihood estimators have many appealing properties. For example,

- ML produces estimates for all parameters, not only for those in the mean function;

- If the data are Gaussian, MLEs of mean parameters are identical to least squares estimates;
- MLEs are functionally invariant. If $\widehat{\pi} = \overline{y}$ is the MLE of π, then $\ln\{\widehat{\pi}/(1-\widehat{\pi})\}$ is the MLE of $\ln\{\pi/(1-\pi)\}$, the logit of π, for example.
- MLEs are usually asymptotically efficient. With increasing sample size their distribution tends to a Gaussian distribution, they are asymptotically unbiased and the most efficient estimators.

On the *downside* we note that MLEs do not necessarily exist, are not necessarily unique, and are often biased estimators. Variations of the likelihood idea of particular importance for the discussion in this text are restricted maximum likelihood (REML, §7), quasi-likelihood (QL, §8), and composite likelihood (CL, §9).

To compare two nested models in the least squares framework, the sum of squares reduction test is a convenient and powerful device. It is intuitive in that a restriction imposed on a statistical model necessarily will result in an increase of the residual sum of squares. Whether that increase is statistically significant can be assessed by comparing the F_{obs} statistic against appropriate cutoff values or by calculating the p-value of the F_{obs} statistic under the null distribution (see §1.6). If $\mathfrak{L}_f$ is the likelihood in a statistical model and $\mathfrak{L}_r$ is the likelihood if the model is reduced according to a restriction imposed on the full model, then $\mathfrak{L}_r$ cannot exceed $\mathfrak{L}_f$. In the discrete case where likelihoods have interpretation of true probabilities, the ratio $\mathfrak{L}_f/\mathfrak{L}_r$ expresses how much more likely it is that the full model generated the data compared to the reduced model. A similar interpretation applies in the case where Y is continuous, although the likelihood ratio then does not measure a ratio of probabilities but a ratio of densities. If the ratio is sufficiently large, then the reduced model should be rejected. For many important cases, e.g., when the data are Gaussian-distributed, the distribution of $\mathfrak{L}_f/\mathfrak{L}_r$ or a function thereof is known. In general, the **likelihood ratio** statistic

$$\Lambda = 2\ln\{\mathfrak{L}_f/\mathfrak{L}_r\} = 2\{l_f - l_r\} \qquad [1.7]$$

has an asymptotic Chi-squared distribution with q degrees of freedom, where q equals the number of restrictions imposed on the full model. In other words, q is equal to the number of parameters in the full model minus the number of parameters in the reduced model. The reduced model is rejected in favor of the full model at the $\alpha \times 100\%$ significance level if Λ exceeds $\chi^2_{\alpha,q}$, the α right-tail probability cutoff of a χ^2_q distribution. In cases where an exact likelihood-ratio test is possible, it is preferred over the asymptotic test, which is exact only as sample size tends to infinity.

As in the sum of squares reduction test, the likelihood ratio test requires that the models being compared are nested. In the seed germination example the full model

$$p(y_i;\pi) = \pi^{y_i}(1-\pi)^{1-y_i}$$

leaves unspecified the germination probability π. To test whether the germination probability in the seed lot takes on a given value, π_0, the model reduces to

$$p(y_i;\pi_0) = \pi_0^{y_i}(1-\pi_0)^{1-y_i}$$

under the hypothesis $H_0{:}\,\pi = \pi_0$. The log-likelihood in the full model is evaluated at the maximum likelihood estimate $\widehat{\pi} = r/n$. There are no unknowns in the reduced model and its log-likelihood is evaluated at the hypothesized value $\pi = \pi_0$. The two log-likelihoods become

$$l_f = l(\widehat{\pi}; y_1, \cdots, y_n) = r\ln\{r/n\} + (n-r)\ln\{1 - r/n\}$$
$$l_r = l(\pi_0; y_1, \cdots, y_n) = r\ln\{\pi_0\} + (n-r)\ln\{1 - \pi_0\}.$$

After some minor manipulations the likelihood-ratio test statistic can be written as

$$\Lambda = 2\{l_f - l_r\} = 2r\ln\left\{\frac{r}{n\pi_0}\right\} + 2(n-r)\ln\left\{\frac{n-r}{n-n\pi_0}\right\}.$$

Note that $n\pi_0$ is the expected number of seeds germinating if the null hypothesis is true. Similarly, $n - n\pi_0$ is the expected number of seeds that fail to germinate under H_0. The quantities r and $n - r$ are the observed numbers of seeds in the two categories. The likelihood-ratio statistic in this case takes on the familiar form (Agresti 1990),

$$\Lambda = \sum_{\text{all categories}} \text{observed count} \times \ln\left\{\frac{\text{observed count}}{\text{expected count}}\right\}.$$

1.4 *t*-Tests in Terms of Statistical Models

The idea of testing hypotheses by comparing nested models is extremely powerful and general. As noted previously, we encounter such tests as sum of squares reduction tests, which compare the residual sum of squares of a full and reduced model, and likelihood-ratio tests that rest on the difference in the log-likelihoods. In linear regression models with Gaussian errors, the test statistic

$$t_{obs} = \frac{\widehat{\beta}_j}{\text{ese}\left(\widehat{\beta}_j\right)},$$

which is the ratio of a parameter estimate and its estimated standard error, is appropriate to test the hypothesis $H_0: \beta_j = 0$. This test turns out to be equivalent to a comparison of two models. The full model containing the regressor associated with β_j and a reduced model from which the regressor has been removed. The comparison of nested models lurks in other procedures, too, which on the surface do not appear to have much in common with statistical models. In this section we formulate the well-known one- and two-sample (pooled) t-tests in terms of statistical models and show how the comparison of two nested models is equivalent to the standard tests.

One-Sample t-Test

The one-sample t-test of the hypothesis that the mean μ of a population takes on a particular value μ_0, $H_0: \mu = \mu_0$, is appropriate if the data are a random sample from a Gaussian population with unknown mean μ and unknown variance σ^2. The general setup of the test as discussed in an introductory statistics course is as follows. Let $Y_1, \cdots, Y_n$ denote a random sample from a $G(\mu, \sigma^2)$ distribution. To test $H_0: \mu = \mu_0$ against the alternative $H_1: \mu \neq \mu_0$, compare the test statistic

$$t_{obs} = \frac{|\overline{y} - \mu_0|}{s/\sqrt{n}},$$

where s is the sample standard deviation, against the $\alpha/2$ (right-tailed) cutoff of a t_{n-1} distribution. If $t_{obs} > t_{\alpha/2,n-1}$, reject H_0 at the α significance level. We note in passing that all cutoff values in this text are understood as cutoffs for right-tailed probabilities, e.g., $\Pr(t_n > t_{\alpha,n}) = \alpha$.

First notice that a two-sided t-test is equivalent to a one-sided F-test where the critical value is obtained as the α cutoff from an F distribution with one numerator and $n-1$ denominator degrees of freedom and the test statistic is the square of t_{obs}. An equivalent test of $H_0\colon \mu = \mu_0$ against $H_1\colon \mu \neq \mu_0$ thus rejects H_0 at the $\alpha \times 100\%$ significance level if

$$F_{obs} = t_{obs}^2 > F_{\alpha,1,n-1} = t_{\alpha/2,n-1}^2.$$

The statistical models reflecting the null and alternative hypothesis are

$$\begin{array}{ll} H_0 \text{ true:} & \boxed{A}\colon Y_i = \mu_0 + e_i, \quad e_i \sim G(0, \sigma^2) \\ H_1 \text{ true:} & \boxed{B}\colon Y_i = \mu + e_i, \quad e_i \sim G(0, \sigma^2) \end{array}$$

Model $\boxed{B}$ is the full model because μ is not specified under the alternative and by imposing the constraint $\mu = \mu_0$, model $\boxed{A}$ is obtained from model $\boxed{B}$. The two models are thus nested and we can compare how well they fit the data by calculating their respective residual sum of squares $SSR = \sum_{i=1}^{n}(y_i - \widehat{\mathrm{E}}[Y_i])^2$. Here, $\widehat{\mathrm{E}}[Y_i]$ denotes the mean of Y_i evaluated at the least squares estimates. Under the null hypothesis there are no parameters since μ_0 is a **known** constant, so that $SSR_{\boxed{A}} = \sum_{i=1}^{n}(y_i - \mu_0)^2$. Under the alternative the least squares estimate of the unknown mean μ is the sample mean $\overline{y}$. The residual sum of squares thus takes on the familiar form $SSR_{\boxed{B}} = \sum_{i=1}^{n}(y_i - \overline{y})^2$. Some simple manipulations yield

$$SSR_{\boxed{A}} - SSR_{\boxed{B}} = n(\overline{y} - \mu_0)^2.$$

The residual mean square in the full model, MSR_f, is the sample variance $s^2 = (n-1)^{-1}\sum_{i=1}^{n}(y_i - \overline{y})^2$ and the sum of squares reduction test statistic becomes

$$F_{obs} = \frac{n(\overline{y} - \mu_0)^2}{s^2} = t_{obs}^2.$$

The critical value for an $\alpha \times 100\%$ level test is $F_{\alpha,1,n-1}$ and the sum of squares reduction test is thereby shown to be equivalent to the standard one-sample t-test.

A likelihood-ratio test comparing models $\boxed{A}$ and $\boxed{B}$ can also be developed. The probability density function of a $G(\mu, \sigma^2)$ random variable is

$$f(y; \mu, \sigma^2) = \frac{1}{\sqrt{2\pi\sigma^2}} \exp\left\{ -\frac{1}{2}\frac{(y-\mu)^2}{\sigma^2} \right\}$$

and the log-likelihood function in a random sample of size n is

$$l(\mu, \sigma^2; y_1, \cdots, y_n) = -\frac{n}{2}\ln\{2\pi\} - \frac{n}{2}\ln\{\sigma^2\} - \frac{1}{2\sigma^2}\sum_{i=1}^{n}(y_i - \mu)^2. \qquad [1.8]$$

The derivatives with respect to μ and σ^2 are

$$\partial l(\mu, \sigma^2; y_1, \cdots, y_n)/\partial\mu = \frac{1}{\sigma^2}\sum_{i=1}^{n}(y_i - \mu) \equiv 0$$

$$\partial l(\mu, \sigma^2; y_1, \cdots, y_n)/\partial\sigma^2 = -\frac{n}{2\sigma^2} + \frac{1}{2\sigma^4}\sum_{i=1}^{n}(y_i - \mu)^2 \equiv 0.$$

In the full model where both μ and σ^2 are unknown, the respective MLEs are the solutions to these equations, namely $\widehat{\mu} = \overline{y}$ and $\widehat{\sigma}^2 = n^{-1}\sum_{i=n}^{n}(y_i - \overline{y})^2$. Notice that the MLE of the error variance is not the sample variance s^2. It is a biased estimator related to the sample variance by $\widehat{\sigma}^2 = s^2(n-1)/n$. In the reduced model the mean is fixed at μ_0 and only σ^2 is a parameter of the model. The MLE in model A becomes $\widehat{\sigma}_0^2 = n^{-1}\sum_{i=1}^{n}(y_i - \mu_0)^2$. The likelihood ratio test statistic is obtained by evaluating the log-likelihoods at $\widehat{\mu}$, $\widehat{\sigma}^2$ in model B and at μ_0, $\widehat{\sigma}_0^2$ in model A. Perhaps surprisingly, Λ reduces to

$$\Lambda = n\ln\left\{\frac{\widehat{\sigma}_0^2}{\widehat{\sigma}^2}\right\} = n\ln\left\{\frac{\widehat{\sigma}^2 + (\overline{y} - \mu_0)^2}{\widehat{\sigma}^2}\right\}.$$

The second expression uses the fact that $\widehat{\sigma}_0^2 = \widehat{\sigma}^2 + (\overline{y} - \mu_0)^2$. If the sample mean is far from the hypothesized value, the variance estimate in the reduced model will be considerably larger than that in the full model. That is the case if μ is far removed from μ_0, because $\overline{y}$ is an unbiased estimators of the true mean μ. Consequently, we reject H_0: $\mu = \mu_0$ for large values of Λ. Based on the fact that Λ has an approximate χ_1^2 distribution, the decision rule can be formulated to reject H_0 if $\Lambda > \chi_{\alpha,1}^2$. However, we may be able to determine a function of the data in which Λ increases monotonically. If this function has a known rather than an approximate distribution, an exact test is possible. It is sufficient to concentrate on $\widehat{\sigma}_0^2/\widehat{\sigma}^2$ to this end since Λ is increasing in $\widehat{\sigma}_0^2/\widehat{\sigma}^2$. Writing

$$\widehat{\sigma}_0^2/\widehat{\sigma}^2 = 1 + \frac{(\overline{y} - \mu_0)^2}{\widehat{\sigma}^2}$$

and using the fact that $\widehat{\sigma}^2 = s^2(n-1)/n$, we obtain

$$\widehat{\sigma}_0^2/\widehat{\sigma}^2 = 1 + \frac{(\overline{y} - \mu_0)^2}{\widehat{\sigma}^2} = 1 + \frac{1}{n-1}\frac{n(\overline{y} - \mu_0)^2}{s^2} = 1 + \frac{1}{n-1}F_{obs}.$$

Instead of rejecting for large values of Λ we can also reject for large values of F_{obs}. Since the distribution of F_{obs} under the null hypothesis is $F_{1,(n-1)}$, an exact test is possible, and this test is the same as the sum of squares reduction test.

Pooled T-Test

In the two-sample case the hypothesis that two populations have the same mean, H_0: $\mu_1 = \mu_2$, can be tested with the pooled t-test under the following assumptions. Y_{1j}, $j = 1, \cdots, n_1$ are a random sample from a $G(\mu_1, \sigma^2)$ distribution and Y_{2j}, $j = 1, \cdots, n_2$ are a random sample from a $G(\mu_2, \sigma^2)$ distribution, drawn independently of the first sample. The common variance σ^2 is unknown and can be estimated as the pooled sample variance from which the procedure derives its name:

$$s_p^2 = \frac{(n_1 - 1)s_1^2 + (n_2 - 1)s_2^2}{n_1 + n_2 - 2}.$$

The procedure for testing $H_0\colon \mu_1 = \mu_2$ against $H_1\colon \mu_1 \neq \mu_2$ is to compare the value of the test statistic

$$t_{obs} = \frac{|\overline{y}_1 - \overline{y}_2|}{\sqrt{s_p^2\left(\frac{1}{n_1} + \frac{1}{n_2}\right)}}$$

against the $t_{\alpha/2, n_1+n_2-2}$ cutoff.

The distributional properties of the data under the null and alternative hypothesis can again be formulated as nested statistical models. Of the various mathematical constructions that can accomplish this, we prefer the one relying on a dummy regressor variable. Let z_{ij} denote a binary variable that takes on value 1 for all observations sampled from population 2, and the value 0 for all observations from population 1. That is,

$$z_{ij} = \begin{cases} 0 & i = 1\ (= \text{group } 1) \\ 1 & i = 2\ (= \text{group } 2) \end{cases}$$

The statistical model for the two-group data can be written as

$$Y_{ij} = \mu_1 + z_{ij}\beta + e_{ij}, \quad e_{ij} \sim iid\, G\left(0, \sigma^2\right).$$

The distributional assumptions for the errors reflect the independence of samples from a group, among the groups, and the equal variance assumption.

For the two possible values of z_{ij}, the model can be expressed as

$$Y_{ij} = \begin{cases} \mu_1 + e_{ij} & i = 1\ (= \text{group } 1) \\ \mu_1 + \beta + e_{ij} & i = 2\ (= \text{group } 2). \end{cases}$$

The parameter β measures the difference between the means in the two groups, $\mathrm{E}[Y_{1j}] - \mathrm{E}[Y_{2j}] = \mu_1 - \mu_2 = \beta$. The hypothesis $H_0\colon \mu_1 - \mu_2 = 0$ is the same as $H_0\colon \beta = 0$. The reduced and full models to be compared are

$$\begin{aligned} H_0 \text{ true:}\quad & \boxed{A}\colon Y_{ij} = \mu_1 + e_{ij}, && e_{ij} \sim G\left(0, \sigma^2\right) \\ H_1 \text{ true:}\quad & \boxed{B}\colon Y_{ij} = \mu_1 + z_{ij}\beta + e_{ij}, && e_{ij} \sim G\left(0, \sigma^2\right). \end{aligned}$$

It is a nice exercise to derive the least squares estimators of μ_1 and β in the full and reduced models and to calculate the residual sums of squares from it. Briefly, for the full model, one obtains $\widehat{\mu}_1 = \overline{y}_1$, $\widehat{\beta} = \overline{y}_2 - \overline{y}_1$, and $SSR_{\boxed{B}} = (n_1 - 1)s_1^2 + (n_2 - 1)s_2^2$, $MSR_{\boxed{B}} = s_p^2(1/n_1 + 1/n_2)$. In the reduced model the least squares estimate of μ_1 becomes $\widehat{\mu}_1 = (n_1\overline{y}_1 + n_2\overline{y}_2)/(n_1 + n_2)$ and

$$SSR_{\boxed{A}} - SSR_{\boxed{B}} = \frac{n_1 n_2}{n_1 + n_2}(\overline{y}_1 - \overline{y}_2)^2.$$

The test statistic for the sum of squares reduction test,

$$F_{obs} = \frac{(SSR_{\boxed{A}} - SSR_{\boxed{B}})/1}{MSR_{\boxed{B}}} = \frac{(\overline{y}_1 - \overline{y}_2)^2}{s_p^2\left(\frac{1}{n_1} + \frac{1}{n_2}\right)}$$

is again the square of the t_{obs} statistic.

1.5 Embedding Hypotheses

Box 1.4 Embedded Hypothesis

- **A hypothesis is said to be embedded in a model if invoking the hypothesis creates a reduced model that can be compared to the full model with a sum of squares reduction or likelihood ratio test.**

- **Reparameterization is often necessary to reformulate a model so that embedding is possible.**

The idea of the sum of squares reduction test is intuitive. Impose a restriction on a model and determine whether the resulting increase in an uncertainty measure is statistically significant. If the change in the residual sum of squares is significant, we conclude that the restriction does not hold. One could call the procedure a sum of squares increment test, but we prefer to view it in terms of the reduction that is observed when the restriction is lifted from the reduced model. The restriction is the null hypothesis, and its rejection leads to the rejection of the reduced model. It is advantageous to formulate statistical models so that hypotheses of interest can be tested through comparisons of nested models. We then say that the hypotheses of interest can be **embedded** in the model.

As an example, consider the comparison of two simple linear regression lines, one for each of two groups (control and treated group, for example). The possible scenarios are (i) the same trend in both groups, (ii) different intercepts but the same slopes, (iii) same intercepts but different slopes, and (iv) different intercepts and different slopes (Figure 1.2). Which of the four scenarios best describes the mechanism that generated the observed data can be determined by specifying a full model representing case (iv) in which the other three scenarios are nested. We choose case (iv) as the full model because it has the most unknowns. Let Y_{ij} denote the jth observation from group i $(i = 1, 2)$ and define a dummy variable

$$z_{ij} = \begin{cases} 1 & \text{if observation from group 1} \\ 0 & \text{if observation from group 2.} \end{cases}$$

The model representing case (iv) can be written as

$$Y_{ij} = \beta_0 + \beta_1 z_{ij} + \beta_2 x_{ij} + \beta_3 x_{ij} z_{ij} + e_{ij},$$

where x_{ij} is the value of the continuous regressor for observation j from group i. To see how the dummy variable z_{ij} creates two separate trends in the groups, consider

$$\mathrm{E}[Y_{ij}|z_{ij} = 1] = \beta_0 + \beta_1 + (\beta_2 + \beta_3)x_{ij}$$
$$\mathrm{E}[Y_{ij}|x_{ij} = 0] = \beta_0 + \beta_2 x_{ij}.$$

The intercepts are $(\beta_0 + \beta_1)$ in group 1 and β_0 in group 2. Similarly, the slopes are $(\beta_2 + \beta_3)$ in group 1 and β_2 in group 2. The restrictions (null hypotheses) that reduce the full model to the other three cases are

$$\text{(i): } H_0 : \beta_1 = \beta_3 = 0 \quad \text{(ii): } H_0 : \beta_3 = 0 \quad \text{(iii): } H_0 : \beta_1 = 0.$$

Notice that the term $x_{ij}z_{ij}$ has the form of an interaction between the regressor and the dummy variable that identifies group membership. If $\beta_3 = 0$, the lines are parallel. This is the very meaning of the absence of an interaction: the comparison of groups no longer depends on the value of x.

The hypotheses can be tested by fitting the full and the three reduced models and performing the sum of squares reduction tests. For linear models, the results of reduction tests involving only one regressor or effect are given by standard regression packages, and a good package is capable of testing more complicated constraints such as (i) based on a fit of the full model only (§4.2.3).

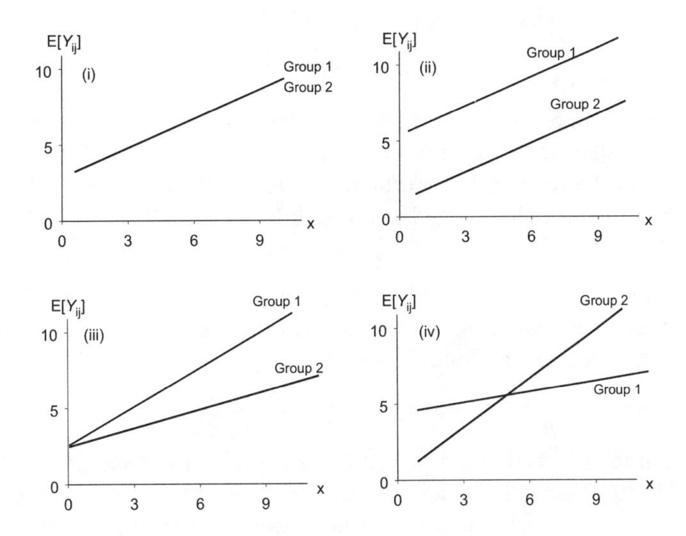

Figure 1.2. Comparison of simple linear regressions among two groups.

Many statistical models can be expressed in alternative ways and this can change the formulation of the hypothesis. Consider a completely randomized experiment with r replications of t treatments. The linear statistical model for this experiment can be written in at least two ways, known as the means and effects models (§4.3.1)

$$\begin{array}{lll} \text{Means model:} & Y_{ij} = \mu_i + e_{ij} & e_{ij} \sim iid\left(0, \sigma^2\right) \\ \text{Effects model:} & Y_{ij} = \mu + \tau_i + e_{ij} & e_{ij} \sim iid\left(0, \sigma^2\right) \\ & i = 1, \cdots, t;\ j = 1, \cdots, r. & \end{array}$$

The treatment effects τ_i are simply $\mu_i - \mu$ and μ is the average of the treatment means μ_i. Under the hypothesis of equal treatment means, H_0:$\mu_1 = \mu_2 = \cdots = \mu_t$, the means model reduces to $Y_{ij} = \mu + e_{ij}$, where μ is the unknown mean common to all treatments. The equivalent hypothesis in the effects model is H_0: $\tau_1 = \tau_2 = \cdots = \tau_t$. Since $\sum_{i=1}^{t}\tau_i = 0$ by construction, one can also state the hypothesis as H_0: all $\tau_i = 0$. Notice that the two-sample t-test problem in §1.4 is a special case of this problem with $t = 2, r = n_1 = n_2$.

In particular for nonlinear models, it may not be obvious how to embed a hypothesis in a model. This is the case when the model is not expressed in terms of the quantities of interest. Recall the Mitscherlich yield equation

$$\mathrm{E}[Y] = \lambda + (\xi - \lambda)\exp\{-\kappa x\}, \qquad [1.9]$$

where λ is the upper yield asymptote, ξ is the yield at $x = 0$, and κ governs the rate of change. Imagine that x is the amount of a nutrient applied and we are interested in estimating and testing hypotheses about the amount of the nutrient already in the soil. Call this parameter α. Black (1993, p. 273) terms $-1 \times \alpha$ the availability index of the nutrient in the soil. It turns out that α is related to the three parameters in [1.9],

$$\alpha = \ln\{(\lambda - \xi)/\lambda\}/\kappa.$$

Once estimates of λ, ξ, and κ have been obtained, this quantity can be estimated by plugging in the estimates. The standard error of this estimate of α will be very difficult to obtain owing to the nonlinearity of the relationship. Furthermore, to test the restriction that $\alpha = -20$, for example, requires fitting a reduced model in which $\ln\{(\lambda - \xi)/\lambda\}/\kappa = -20$. This is not a well-defined problem.

To enable estimation and testing of hypotheses in nonlinear models, the model should be rewritten to contain the quantities of interest. This process, termed **reparameterization** (§5.7), yields for the Mitscherlich equation

$$\mathrm{E}[Y] = \lambda(1 - \exp\{-\kappa(x - \alpha)\}). \qquad [1.10]$$

The parameter ξ in model [1.9] was replaced by $\lambda(1 - \exp\{\kappa\alpha\})$ and after collecting terms, one arrives at [1.10]. The sum of squares decomposition, residuals, and fit statistics are identical when the two models are fit to data. Testing the hypothesis $\alpha = -20$ is now straightforward. Obtain the estimate of α and its estimated standard error from a statistical package (see §§5.4, 5.8.1) and calculate a confidence interval for α. If it does not contain the value -20, reject H_0:$\alpha = -20$. Alternatively, fit the reduced model $\mathrm{E}[Y] = \lambda(1 - \exp\{-\kappa(x + 20)\})$ and perform a sum of squares reduction test.

1.6 Hypothesis and Significance Testing — Interpretation of the *p*-Value

Box 1.5 *p*-Value

- **p-values are probabilities calculated under the assumption that a null hypothesis is true. They measure the probability to observe an experimental outcome at least as extreme as the observed outcome.**

- **p-values are frequently misinterpreted in terms of an error probability of rejecting the null hypothesis or, even worse, as a probability that the null hypothesis is true.**

A distinction is made in statistical theory between hypothesis and significance testing. The former relies on comparing the observed value of a test statistic with a critical value and to reject the null hypothesis if the observed value is more extreme than the critical value. Most statistical computing packages apply the significance testing approach because it does not involve critical values. In order to derive a critical value, one must decide first on the Type-I error probability α to reject a null hypothesis that is true. The significance approach relies on calculating the p-value of a test statistic, the probability to obtain a value of the test statistic at least as extreme as the observed one, provided that the null hypothesis is true. The connection between the two approaches lies in the Type-I error rate α. If one rejects the null hypothesis when the p-value is less than α, and fails to reject otherwise, significance and hypothesis testing lead to the same decisions. Statistical tests done by hand are almost always performed as hypothesis tests, and the results of tests carried out with computers are usually reported as p-values. We will not make a formal distinction between the two approaches here and note that p-values are more informative than decisions based on critical values. To attach *, **, ***, or some notation to the results of tests that are significant at the $\alpha = 0.05$, 0.01, and 0.001 level is commonplace but arbitrary. When the p-value is reported each reader can draw his/her own conclusion about the fate of the null hypothesis.

Even if results are reported with notations such as *, **, *** or by attaching lettering to an ordered list of treatment means, these displays are often obtained by converting p-values from statistical output. The ubiquitous p-values are probably the most misunderstood and misinterpreted quantities in applied statistical work. To draw correct conclusions from output of statistical packages it is imperative to interpret them properly. Common misconceptions are that (i) the p-value measures an error probability for the rejection of the hypothesis, (ii) the p-value measures the probability that the null hypothesis is true, (iii) small p-values imply that the alternative hypothesis is correct. To rectify these misconceptions we briefly discuss the rationale of hypothesis testing from a probably unfamiliar angle, Monte-Carlo testing, and demonstrate the calculation of p-values with a spatial point pattern example.

The frequentist approach to measuring model-data agreement is based on the notion of comparing a model against different data sets. Assume a particular model holds (is true) for the time being. In the test of two nested models we assume that the restriction imposed on the full model holds and we *accept* the reduced model unless we can find evidence to the contrary. That is, we are working under the assumption that the null hypothesis holds until it is rejected. Because there is uncertainty in the outcome of the experiment we do not expect the observed data and the postulated model to agree perfectly. But if chance is the only explanation for the disagreement between data and model, there is no reason to reject the model from a statistical point of view. It may fit the data poorly because of large variability in the data, but it remains correct on average.

The problem, of course, is that we observe only one experimental outcome and do not see other sets of data that have been generated by the model under investigation. If that were the case we could devise the following test procedure. Calculate a test statistic from the observed data. Generate all possible data sets consistent with the null hypothesis if this number is finite or generate a sufficiently large number of data sets if there are infinitely many experimental outcomes. Denote the number of data sets so generated by k. Calculate the test statistic in each of the k realizations. Since we assume the null hypothesis to be true, the value of the test statistic calculated from the observed data is added to the test statistics calculated from the generated data sets and the $k+1$ values are ranked. If the data were generated by a mechanism that does not agree with the model under investigation, the observed value of the test statistic should be unusual, and its rank should be extreme. At this point we need to invoke a decision rule according to which values of the test statistic are deemed sufficiently rare or unusual to reject H_0. If the observed value is among those values considered rare enough to reject H_0, this is the decision that follows. The critical rank is a measure of the acceptability of the model (McPherson 1990). The decision rule cannot alone be the attained rank of a test statistic, for example, "reject H_0 if the observed test statistic ranks fifth." If k is large the probability of a particular value can be very small. As k tends to infinity, the probability to observe a particular rank tends to zero. Instead we define cases deemed inconsistent with H_0 by a range of ranks. Outcomes **at least as extreme** as the critical rank lead to the rejection of H_0. This approach of testing hypotheses is known under several names. In the design of experiment it is termed the **randomization** approach. If the number of possible data sets under H_0 is finite it is also known as **permutation** testing. If a random sample of the possible data sets is drawn it is referred to as **Monte-Carlo** testing (see, e.g. Kempthorne 1952, 1955; Kempthorne and Doerfler 1969; Rubinstein 1981; Diggle 1983, Hinkelmann and Kempthorne 1994, our §A4.8.2 and §9.7.3)

The reader is most likely familiar with procedures that calculate an observed value of a test statistic and then (i) compare the value against a cutoff from a tabulated probability distribution or (ii) calculate the p-value of the test statistic. The procedure based on generating data sets under the null hypothesis as outlined above is no different from this *classical* approach. The distribution table from which cutoffs are obtained or the distribution from which p-values are calculated reflect the probability distribution of the test statistic if the null hypothesis is true. The list of k test statistic obtained from data sets generated under the null hypothesis also reflects the distribution of the test statistic under H_0. The critical value (cutoff) in a test corresponds to the critical rank in the permutation/Monte-Carlo/randomization procedure.

To illustrate the calculation of p-values and the Monte-Carlo approach, we consider the data shown in Figure 1.3, which might represent the locations on a field at which 150 weeds emerged. Chapter 9.7 provides an in-depth discussion of spatial point patterns and their analysis. We wish to test whether a process that places weeds completely at random (uniformly and independently) in the field could give rise to the distribution shown in Figure 1.3 or whether the data generating process is clustered. In a clustered process, events exhibit more grouping than in a spatially completely random process. If we calculate the average distance between an event and its nearest neighbor, we expect this distance to be smaller in a clustered process. For the observed pattern (Figure 1.3) the average nearest neighbor distance is $\overline{y} = 0.03964$.

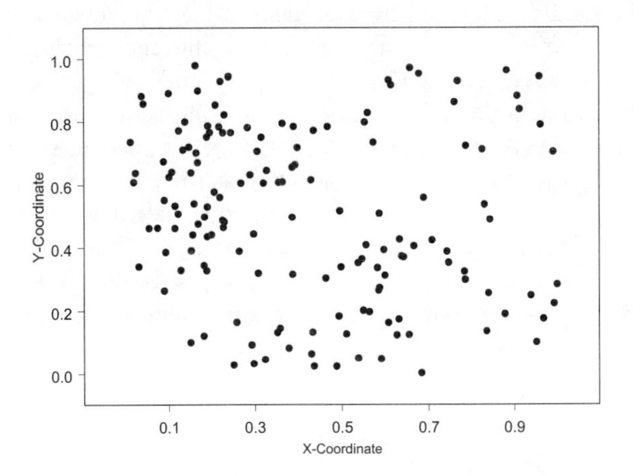

Figure 1.3. Distribution of 150 weed plants on a simulated field.

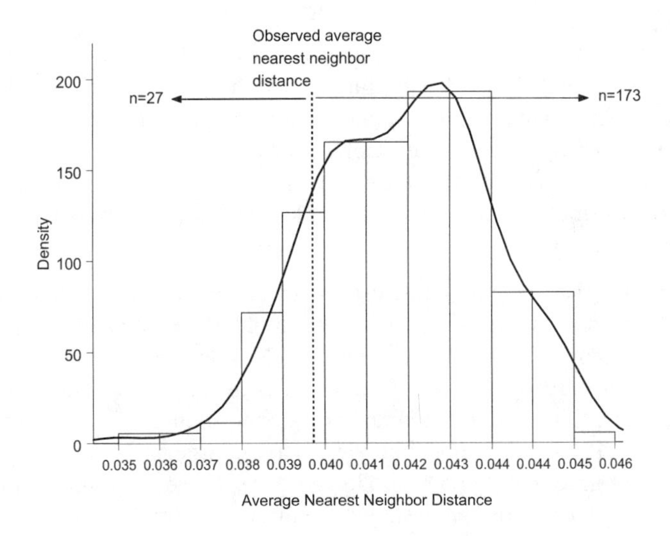

Figure 1.4. Distribution of average nearest neighbor distances in $k = 200$ simulations of a spatial process that places 150 plants completely at random on the $(0,1) \times (0,1)$ square.

Two hundred point patterns were then generated from a process that is spatially completely neutral. Events are placed uniformly and independently of each other. This process represents the stochastic model consistent with the null hypothesis. Each of the $k = 200$ simulated processes has the same boundary box as the observed pattern and the same number of points (150). Figure 1.4 shows the histogram of the $k + 1 = 201$ average nearest-neighbor distances. Note that the observed distance ($\overline{y} = 0.03964$) is part of the histogram.

The line in the figure is a nonparametric estimate of the probability density of the average nearest-neighbor statistics. This density is only an estimate of the true distribution function of the test statistic since infinitely many realizations are possible under the null hypothesis.

The observed statistic is the 28th smallest among the 201 values. Under the alternative hypothesis of a clustered process, the average nearest-neighbor distance should be smaller than under the null hypothesis. Plants that appear in clusters are closer to each other on average than plants distributed completely at random. Small average nearest-neighbor distances are thus extreme under the null hypothesis. For a 5% significance test the critical rank would be 10.05. If the observed value ranks 10th or lower, H_0 is rejected. This is not the case and we fail to reject the hypothesis that the plants are distributed completely at random. There is insufficient evidence at the 5% significance level to conclude that a clustered process gives rise to the spatial distribution in Figure 1.3. The p-value is calculated as the proportion of values at least as extreme as the observed value, hence $p = 28/201 = 0.139$. Had we tested whether the observed point distribution is more regular than a completely random pattern, large average nearest neighbor distances would be consistent with the alternative hypothesis and the p-value would be $174/201 = 0.866$.

Because the null hypothesis is true when the $k = 200$ patterns are generated and for the purpose of the statistical decision, the observed value is judged as if the null hypothesis is true; the p-value is **not** a measure for the probability that H_0 is wrong. The p-value is a conditional probability under the assumption that H_0 is correct. As this probability becomes smaller and smaller, we will eventually distrust the condition.

In a test that is not based on Monte-Carlo arguments, the estimated density in Figure 1.4 is replaced by the exact probability density function of the test statistic. This distribution can be obtained by complete enumeration in randomization or permutation tests, by deriving the distribution of a test statistic from first principles, or by approximation. In the t-test examples of §1.4, the distribution of F_{obs} under H_0 is known to be that of an F random variable and t_{obs} is known to be distributed as a t random variable. The likelihood ratio test statistic [1.7] is approximately distributed as a Chi-squared random variable.

In practice, rejection of the null hypothesis is tantamount to the acceptance of the alternative hypothesis. Implicit in this step is the assumption that if an outcome is extreme under the null hypothesis, it is not so under the alternative. Consider the case of a simple null and alternative hypotheses, e.g., $H_0{:}\mu = \mu_0$, $H_1{:}\mu = \mu_1$. If μ_1 and μ_0 are close, an experimental outcome extreme under H_0 is also extreme under H_1. Rejection of H_0 should then not prompt the acceptance of H_1. But this test most likely will have a large probability of a Type-II error, to fail to reject an incorrect null hypothesis (low power). These situations can be avoided by controlling the Type-II error of the test, which ultimately implies collecting samples of sufficient size (sufficient to achieve a desired power for a stipulated difference $\mu_1 - \mu_0$). Finally, we remind the reader of the difference between statistical and practical significance. Just because statistical significance is attached to a result does not imply that it is a meaningful result from a practical point of view. If it takes $n = 5,000$ samples to detect a significant difference between two treatments, their actual difference is probably so small that hardly anyone will be interested in knowing it.

1.7 Classes of Statistical Models

1.7.1 The Basic Component Equation

Box 1.6 Some Terminology

A statistical model consists of at least the following components: Response, error, systematic part, parameters.

- **Response:** **The outcome of interest being measured, counted, or classified. Notation:** Y
- **Parameter:** **Any unknown constant in the mean function or distribution of random variables.**
- **Systematic part: The mean function of a model.**
- **Model errors:** **The difference between observations and the mean function.**
- **Prediction:** **Evaluation of the mean function at the estimated values of the parameters.**
- **Fitted residual:** **The difference between observed and fitted values:** $\widehat{e}_i = Y_i - \widehat{Y}_i$

Most statistical models are supported by the decomposition

$$\text{response} = \text{structure} + \text{error}$$

of observations into a component associated with identifiable sources of variability and an error component that McPherson (1990) calls the component equation. The equation is of course not a panacea for all statistical problems, errors may be multiplicative, for example. The attribute being modeled is termed the **response** or **outcome**. An observation in the narrow sense is a recorded value of the response. In a broader sense, observations also incorporate information about the variables in the structural part to which the response is related.

The component equation is eventually expressed in mathematical terms. A very general expression for the systematic part is

$$f(x_{0i}, x_{1i}, x_{2i}, ..., x_{ki}, \theta_0, \theta_1, ..., \theta_p)$$

where $x_{0i}, \cdots, x_{ki}$ are measured variables and $\theta_0, \theta_1, \cdots, \theta_p$ are parameters. The response is typically denoted Y, and an appropriate number of subscripts must be added to associate a single response with the parts of the model structure. If a single subscript is sufficient the basic component equation of a statistical model becomes

$$Y_i = f(x_{0i}, x_{1i}, x_{2i}, ..., x_{ki}, \theta_0, \theta_1, ..., \theta_p) + e_i. \qquad [1.11]$$

The specification of the component equation is not complete without the means, variances, and covariances of all random variables involved and if possible, their distribution laws. If these are unknown, they add additional parameters to the model. The assumption that the user's model is correct is reflected in the zero mean assumption of the errors $(\mathrm{E}[e_i] = 0)$. Since then $\mathrm{E}[Y_i] = f(x_{0i}, x_{1i}, x_{2i}, ..., x_{ki}, \theta_0, \theta_1, ..., \theta_p)$, the function $f()$ is often called the **mean function** of the model.

The process of fitting the model to the observed responses involves estimation of the unknown quantities in the systematic part and the parameters of the error distribution. Once the parameters are estimated the **fitted** values can be calculated:

$$\widehat{Y}_i = f\left(x_{0i}, x_{1i}, x_{2i}, ..., x_{ki}, \widehat{\theta}_0, \widehat{\theta}_1, ..., \widehat{\theta}_p\right) = \widehat{f}_i$$

and a second decomposition of the observations has been attained:

$$\text{response} = \text{fitted value} + \text{fitted residual} = \widehat{f}_i + \widehat{e}_i.$$

A caret placed over a symbol denotes an estimated quantity. Fitted values are calculated for the observed values of $x_{0i}, \cdots, x_{ki}$. Values calculated for any combination of the x variables, whether part of the data set or not, are termed **predicted** values. It is usually assumed that the **fitted residual** $\widehat{e}_i$ is an estimate of the unobservable model error e_i which justifies model diagnostics based on residuals. But unless the fitted values estimate the systematic part of the model without bias *and* the model is correct $(\mathrm{E}[e_i] = 0)$, the fitted residuals will not even have a zero mean. And the fitted values $\widehat{f}_i$ may be biased estimators of f_i, even if the model is correctly specified. This is common when f is a nonlinear function of the parameters.

The measured variables contributing to the systematic part of the model are termed here **covariates**. In regression applications, they are also referred to as regressors or independent variables, while in analysis of variance models the term covariate is sometimes reserved for those variables which are measured on a continuous scale. The term independent variable should be avoided, since it is not clear what the variable is independent of. The label is popular though, since the response is often referred to as the **dependent** variable. In many regression models the covariates are in fact very highly dependent **on each other**; therefore, the term *independent variable* is misleading. Covariates that can take on only two values, 0 or 1, are also called **design variables**, dummy variables, or binary variables. They are typical in analysis of variance models. In observational studies (see §2.3) covariates are also called *explanatory variables*. We prefer the term covariate to encompass all of the above. The precise nature of a covariate will be clear from context.

1.7.2 Linear and Nonlinear Models

Box 1.7 Nonlinearity

- **Nonlinearity does not refer to curvature of the mean function as a function of covariates.**

- **A model is nonlinear if at least one derivative of the mean function with respect to the parameters depends on at least one parameter.**

The distinction between linear and nonlinear models is often obstructed by references to graphs of the predicted values. If a graph of the predicted values appears to have curvature, the underlying statistical model may still be linear. The polynomial

$$Y_i = \beta_0 + \beta_1 x_i + \beta_2 x_i^2 + e_i$$

is a linear model, but when $\widehat{y}$ is graphed vs. x, the predicted values exhibit curvature. The acid test for linearity is as follows: if the derivatives of the model's systematic part with respect to the parameters do not depend on any of the parameters, the model is linear. Otherwise, the model is nonlinear. For example,

$$Y_i = \beta_0 + \beta_1 x_i + e_i$$

has a linear mean function, since

$$\begin{aligned} \partial(\beta_0 + \beta_1 x_i)/\partial\beta_0 &= 1 \\ \partial(\beta_0 + \beta_1 x_i)/\partial\beta_1 &= x_i \end{aligned}$$

and neither of the derivatives depends on any parameters. The quadratic polynomial $Y_i = \beta_0 + \beta_1 x_i + \beta_2 x_i^2 + e_i$ is also a linear model, since

$$\begin{aligned} \partial\left(\beta_0 + \beta_1 x_i + \beta_2 x_i^2\right)/\partial\beta_0 &= 1 \\ \partial\left(\beta_0 + \beta_1 x_i + \beta_2 x_i^2\right)/\partial\beta_1 &= x_i \\ \partial\left(\beta_0 + \beta_1 x_i + \beta_2 x_i^2\right)/\partial\beta_2 &= x_i^2. \end{aligned}$$

Linear models with curved mean function are termed **curvilinear**. The model

$$Y_i = \beta_0\left(1 + e^{\beta_1 x_i}\right) + e_i,$$

on the other hand, is nonlinear, since the derivatives

$$\begin{aligned} \partial\left(\beta_0\left(1 + e^{\beta_1 x_i}\right)\right)/\partial\beta_0 &= 1 + e^{\beta_1 x_i} \\ \partial\left(\beta_0\left(1 + e^{\beta_1 x_i}\right)\right)/\partial\beta_1 &= \beta_0 x_i e^{\beta_1 x_i} \end{aligned}$$

depend on the model parameters.

A model can be linear in some and nonlinear in other parameters. $\mathrm{E}[Y_i] = \beta_0 + e^{\beta_1 x_i}$, for example, is linear in β_0 and nonlinear in β_1, since

$$\begin{aligned} \partial\left(\beta_0 + e^{\beta_1 x_i}\right)/\partial\beta_0 &= 1 \\ \partial\left(\beta_0 + e^{\beta_1 x_i}\right)/\partial\beta_1 &= x_i e^{\beta_1 x_i}. \end{aligned}$$

If a model is nonlinear in at least one parameter, the entire model is considered nonlinear.

Linearity refers to linearity in the parameters, not the covariates. Transformations of the covariates such as e^x, $\ln(x)$, $1/x$, $\sqrt{x}$ do not change the linearity of the model, although they will affect the degree of curvature seen in a plot of y against x. Polynomial models which raise a covariate to successively increasing powers are always linear models.

1.7.3 Regression and Analysis of Variance Models

Box 1.8 Regression vs. ANOVA

- **Regression model:** **Covariates are continuous**
- **ANOVA model:** **Covariates are classification variables**
- **ANCOVA model:** **Covariates are a mixture of continuous and classification variables.**

A parametric regression model in general is a linear or nonlinear statistical model in which the covariates are continuous, while in an analysis of variance model covariates represent **classification variables**. Assume that a plant growth regulator is applied at four different rates, 0.5, $1.0, 1.3$, and $2.4\,\text{kg} \times \text{ha}^{-1}$. A simple linear regression model relating plant yield on a per-plot basis to the amount of growth regulator applied associates the ith plot's response to the applied rates directly:

$$Y_i = \theta_0 + \theta_1 x_i + e_i.$$

If the first four observations received 0.5, 1.0, 1.3, and 1.3 kg $\times$ ha^{-1}, respectively, the model for these observations becomes

$$\begin{aligned}
Y_1 &= \theta_0 + 0.5{*}\theta_1 + e_1 \\
Y_2 &= \theta_0 + 1.0{*}\theta_1 + e_2 \\
Y_3 &= \theta_0 + 1.3{*}\theta_1 + e_3 \\
Y_4 &= \theta_0 + 1.3{*}\theta_1 + e_4.
\end{aligned}$$

If rate of application is a classification variable, information about the actual amounts of growth regulator applied is not taken into account. Only information about which of the four levels of application rate an observation is associated with is considered. The corresponding ANOVA (classification) model becomes

$$Y_{ij} = \mu_i + e_{ij},$$

where μ_i is the mean yield if the ith level of the growth regulator is applied. The double subscript is used to emphasize that multiple observations can share the same growth regulator level (replications). This model can be expanded using a series of dummy covariates, $z_{1j}, \cdots, z_{4j}$, say. Let z_{ij} take on the value 1 if the jth observation received the ith level of the growth regulator, and 0 otherwise. The expanded ANOVA model then becomes

$$\begin{aligned}
Y_{ij} &= \mu_i + e_{ij} \\
&= \mu_1 z_{1j} + \mu_2 z_{2j} + \mu_3 z_{3j} + \mu_4 z_{4j} + e_{ij}.
\end{aligned}$$

In this form, the ANOVA model is a multiple regression model with four covariates and no intercept.

The role of the dummy covariates in ANOVA models is to select the parameters (effects) associated with a particular response. The relationship between plot yield and growth regulation is described more parsimoniously in the regression model, which contains only two parameters, the intercept θ_0 and the slope θ_1. The ANOVA model allots four parameters $(\mu_1, \cdots, \mu_4)$ to describe the systematic part of the model. The downside of the regression model is that if the relationship between y and x is not linear, the model will not apply and inferences based on the model may be incorrect.

ANOVA and linear regression models can be cast in the same framework; they are both linear statistical models. Classification models may contain continuous covariates in addition to design variables, and regression models may contain binary covariates in addition to continuous ones. An example of the first type of model arises when adjustments are made for known systematic differences in initial conditions among experimental units to which the treatments are applied. Assume, for example, that the soils of the plots on which growth regulators were applied had different lime requirements. Let u_{ij} be the lime requirement of the jth plot receiving the ith rate of the regulator; then the systematic effect of lime requirement on plot yield can be accounted for by incorporating u_{ij} as a continuous covariate in the classification model:

$$Y_{ij} = \mu_i + \theta u_{ij} + e_{ij} \qquad [1.12]$$

This is often termed an **analysis of covariance** (ANCOVA) model. The parameter θ measures the change in plant yield if lime requirement changes by one unit. This change is the same for all levels of growth regulator. If the lime requirement effect depends on the particular growth regulator, interactions can be incorporated:

$$Y_{ij} = \mu_i + \theta_i u_{ij} + e_{ij}.$$

The presence of these interactions is easily tested with a sum of squares reduction test since the previous two models are nested $(H_0\colon \theta_1 = \theta_2 = \theta_3 = \theta_4)$.

The same problem can be approached from a regression standpoint. Consider the initial model, $Y_{ij} = \theta_0 + \theta u_{ij} + e_{ij}$, linking plot yield to lime requirement. Because a distinct number of rates were applied on the plot, the simple linear regression can be extended to accommodate separate intercepts for the growth regulators. Replace the common intercept θ_0 by

$$\mu_i = \mu_1 z_{1j} + \mu_2 z_{2j} + \mu_3 z_{3j} + \mu_4 z_{4j}$$

and model [1.12] results. Whether a regression model is enlarged to accommodate a classification variable or a classification model is enlarged to accommodate a continuous covariate, the same models result.

1.7.4 Univariate and Multivariate Models

Box 1.9 Univariate vs. Multivariate Models

- **Univariate statistical models analyze one response at a time while multivariate models analyze several responses simultaneously.**

- **Multivariate outcomes can be measurements of different responses, e.g., canning quality, biomass, yield, maturity, or measurements of the same response at multiple points in time, space, or time and space.**

Most experiments produce more than just a single response. Statistical models that model one response independently of other experimental outcomes are called **univariate** models, whereas **multivariate** models simultaneously model several response variables. Models with more than one covariate are sometimes incorrectly termed *multivariate* models. The **multiple** linear regression model $Y_i = \beta_0 + \beta_1 x_{1i} + \beta_2 x_{2i} + e_i$ is a univariate model.

The advantage of multivariate over univariate models is that multivariate models incorporate the relationships between experimental outcomes into the analysis. This is particularly meaningful if the multivariate responses are observations of the *same* attribute at different locations or time points. When data are collected as longitudinal, repeated measures, or spatial data, the temporal and spatial dependencies among the observations must be taken into account (§2.5). In a repeated measures study, for example, this requires modeling the observations jointly, rather than through separate analyses by time points. By separately analyzing the outcomes by year in a multi-year study, little insight is gained into the time-dependency of the system.

Multivariate responses in this text are confined to the special case where the same response variable is measured repeatedly, that is, longitudinal, repeated measures, and spatial data. Developing statistical models for such data (§§ 7, 8, 9) requires a good understanding of the notion and consequences of clustering in data (discussed in §2.4 and §7.1).

1.7.5 Fixed, Random, and Mixed Effects Models

Box 1.10 Fixed and Random Effects

- **The distinction of fixed and random effects applies to the unknown model components:**
  - **— a fixed effect is an unknown constant (does not vary),**
  - **— a random effect is a random variable.**
- **Random effects arise from subsampling, random selection of treatment levels, and hierarchical random processes, e.g., in clustered data (§2.4).**
- **Fixed effects model: All effects are fixed (apart from the error)**
- **Random effects model: All effects are random (apart from intercept)**
- **Mixed effects model: Some effects are fixed, others are random (not counting an intercept and the model error)**

The distinction between fixed, random, and mixed effects models is not related to the nature of the covariates, but the unknown quantities of the statistical model. In this text we assume that covariate values are not associated with error. A **fixed effects** model contains only constants in its systematic part and one random variable (**the** error term). The variance of the error term measures residual variability. Most traditional regression models are of this type. In designed experiments, fixed effects models arise when the levels of the treatments are chosen deliberately by the researcher as the only levels of interest. A fixed effects model for a randomized complete block design with one treatment factor implies that the blocks are predetermined as well as the factor levels.

A random effects model consists of random variables only, apart from a possible grand mean. These arise when multiple random processes are in operation. Consider sampling two hundred bags of seeds from a large seed lot. Fifty laboratories are randomly selected to receive four bags each for analysis of germination percentage and seed purity from a list of laboratories. Upon repetition of this experiment a different set of laboratories would be selected to receive different bags of seeds. Two random processes are at work. One source of variability is due to selecting laboratories at random from the population of all possible laboratories that could have performed the analysis. A second source of variability stems from randomly determining which particular four bags of seeds are sent to a laboratory. This variability is a measure for the heterogeneity within the seed lot, and the first source represents variability among laboratories. If the two random processes are independent, the variance of a single germination test result is the sum of two variance components,

$$\text{Var}[Y_{ij}] = \sigma_l^2 + \sigma_b^2.$$

Here, σ_l^2 measures lab-to-lab variability and σ_b^2 variability in test results within a lab (seed lot heterogeneity). A statistical model for this experiment is

$$Y_{ij} = \mu + \alpha_i + e_{ij}$$
$$i = 1, \cdots, 50;\ j = 1, \cdots, 4,$$

where α_i is a random variable with mean 0 and variance σ_l^2, e_{ij} is a random variable (independent of the α_i) with mean 0 and variance σ_b^2, and Y_{ij} is the germination percentage reported by the ith lab for the jth bag. The grand mean is expressed by μ, the true germination percentage of the lot. A fixed grand mean should always be included in random effects models unless the response has zero average.

Mixed effects models arise when some of the model components are fixed, while others are random. A mixed model contains at least two random variables (counting the model errors e) and two unknown constants in the systematic part (counting the grand mean). Mixed models can be found in multifactor experiments where levels of some factors are predetermined while levels of other factors are chosen at random. If two levels of water stress (irrigated, not irrigated) are combined with six genotypes selected from a list of 30 possible genotypes, a two-factor mixed model results:

$$Y_{ijk} = \mu + \alpha_i + \gamma_j + (\alpha\gamma)_{ij} + e_{ijk}.$$

Here, Y_{ijk} is the response of genotype j under water stress level i in replicate k. The α_i's denote the fixed effects of water stress, the γ_j's the random effects of genotype with mean 0 and variance σ_γ^2. Interaction terms such as $(\alpha\gamma)_{ij}$ are random effects, if at least one of the factors involved in the interaction is a random factor. Here, $(\alpha\gamma)_{ij}$ is a random effect with mean 0 and variance $\sigma_{\alpha\gamma}^2$. The e_{ijk} finally denote the experimental errors. Mixed model structures also result when treatments are allocated to experimental units by separate randomizations. A split-plot design randomly allocates levels of the whole-plot factor to large experimental units (whole-plots) and independently thereof randomly allocates levels of one or more other factors within the whole-plots. The two randomizations generate two types of experimental errors, one associated with the whole-plots, one associated with the sub-plots. If the levels of the whole- and sub-plot treatment factor are selected at random, the resulting model is a random model. If the levels of at least one of the factors were predetermined, a mixed model results.

In observational studies (see §2.3), mixed effects models have gained considerable popularity for longitudinal data structures (Jones 1993; Longford 1993; Diggle, Liang, and Zeger 1994; Vonesh and Chinchilli 1997; Gregoire et al. 1997, Verbeke and Molenberghs 1997; Littell et al. 1996). Longitudinal data are measurements taken repeatedly on observational units without the creation of experimental conditions by the experimenter. These units are often termed subjects or clusters. In the absence of randomization, mixed effects arise because some "parameters" of the model (slopes, intercept) are assumed to vary at random from subject to subject while other parameters remain constant across subjects. Chapter 7 discusses mixed models for longitudinal and repeated measures data in great detail. The distinction between fixed and random effects and its bearing on data analysis and data interpretation are discussed there in more detail.

1.7.6 Generalized Linear Models

Box 1.11 Generalized Linear Models (GLM)

- **GLMs provide a unified framework in which Gaussian and non-Gaussian data can be modeled. They combine aspects of linear and nonlinear models. Covariate effects are linear on a transformed scale of the mean response. The transformations are usually nonlinear.**

- **Classical linear regression and classification models with Gaussian errors are special cases of generalized linear models.**

Generalized linear models (GLMs) are statistical models for a large family of probability distributions known as the exponential family (§6.2.1). This family includes such important distributions as the Gaussian, Gamma, Chi-squared, Beta, Bernoulli, Binomial, and Poisson distributions. We consider generalized linear models among the most important statistical models today. The frequency (or lack thereof) with which they are applied in problems in the plant and soil sciences belies their importance. They are based on work by Nelder and Wedderburn (1972) and Wedderburn (1974), subsequently popularized in the monograph by McCullagh and Nelder (1989). If responses are continuous, modelers typically resort to linear or nonlinear statistical models of the kind

$$Y_i = f(x_{0i}, x_{1i}, x_{2i}, ..., x_{ki}, \theta_0, \theta_1, ..., \theta_p) + e_i,$$

where it is assumed that the model residuals have zero mean, are independent, and have some common variance $\text{Var}[e_i] = \sigma^2$. For purposes of parameter estimation, these assumptions are usually sufficient. For purposes of statistical inference, such as the test of hypotheses or the calculation of confidence intervals, distributional assumptions about the model residuals are added. All too often, the errors are assumed to follow a Gaussian distribution. There are many instances in which the Gaussian assumption is not tenable. For example, if the response is not a continuous characteristic, but a frequency count, or when the error distribution is clearly skewed. Generalized linear models allow the modeling of such data when the response distribution is a member of the exponential family (§6.2.1). Since the Gaussian distribution is a member of the exponential family, linear regression and analysis of variance methods are special cases of generalized linear models.

Besides non-Gaussian error distributions, generalized linear models utilize a model component known as the **link function**. This is a transformation which maps the expected values of the response onto a scale where covariate effects are additive. For a simple linear regression model with Gaussian error,

$$Y_i = \beta_0 + \beta_1 x_i + e_i;\; e_i \sim G\left(0, \sigma^2\right),$$

the expectation of the response is already linear. The applicable link function is the identity function. Assume that we are concerned with a binary response, for example, whether a particular plant disease is present or absent. The mean (expected value) of the response is the probability π that the disease occurs and a model is sought that relates the mean to some environmental factor x. It would be unreasonable to model this probability as a linear function of x, $\pi = \beta_0 + \beta_1 x$. There is no guarantee that the predicted values are between 0 and 1, the only acceptable range for probabilities. A monotone function that maps values between 0 and 1 onto the real line is the **logit** function

$$\eta = \ln\left\{\frac{\pi}{1-\pi}\right\}.$$

Rather then modeling π as a linear function of x, it is the transformed value η that is modeled as a linear function of x,

$$\eta = \ln\left\{\frac{\pi}{1-\pi}\right\} = \beta_0 + \beta_1 x.$$

Since the logit function links the mean π to the covariate, it is called the link function of the model. For any given value of η, the mean response is calculated by inverting the relationship,

$$\pi = \frac{1}{1+\exp\{-\eta\}} = \frac{1}{1+\exp\{-\beta_0 - \beta_1 x\}}.$$

In a generalized linear model, a linear function of covariates is selected in the same way as in a regression or classification model. Under a distributional assumption for the responses and after selecting a link function, the unknown parameters can be estimated. There are important differences between applying a link function and transformations such as the arcsine, square root, logarithmic transform that are frequently applied in statistical work. The latter transformations are applied to the individual responses Y in order to achieve greater variance homogeneity and/or symmetry, usually followed by a standard linear model analysis on the transformed scale assuming Gaussian errors. In a generalized linear model the link function transforms the mean response $\mathrm{E}[Y]$ and the distributional properties of the response are not changed. A Binomial random variable is analyzed as a Binomial random variable, a Poisson random variable as a Poisson random variable.

Because η is a linear function of the covariates, statistical inference about the model parameters is straightforward. Tests for treatment main effects and interactions, for example, are simple if the outcomes of an experiment with factorial treatment structure are analyzed as a generalized linear model. They are much more involved if the model is a general nonlinear model. Chapter 6 provides a thorough discussion of generalized linear models and numerous applications. We mention in passing that statistical models appropriate for ordinal outcomes such as visual ratings of plant quality or injury can be derived as extensions of generalized linear models (see §6.5)

1.7.7 Errors in Variable Models

The distinction between response variable and covariates applied in this text implies that the response is subject to variability and the covariates are not. When measuring the height and biomass of a plant it is obvious, however, that both variables are subject to variability, even if uncertainty stems from measurement error alone. If the focus of the investigation is to describe the association or strength of dependency between the two variables, methods of correlation analysis can be employed. If, however, one wishes to describe or model biomass as a function of height, treating one variable as fixed (the covariate) and the other as random (the response) ignores the variability in the covariate.

We will treat covariates as fixed throughout, consistent with classical statistical models. If covariate values are random, this approach remains intact in the following scenarios (Seber and Wild 1989):

- Response Y and covariate X are both random and linked by a relationship $f(x, \theta_1, \cdots, \theta_p)$. If the values of X can be measured accurately and the measured value is a realization of X, the systematic part of the model is interpreted **conditionally** on the observed values of X. We can write this as

  $$\mathrm{E}[Y|X = x] = f(x, \theta_1, \cdots, \theta_p),$$

  read as *conditional on observing covariate value x, the mean of Y is $f(x, \theta_1, \cdots, \theta_p)$.*
- Response Y and covariate X are both random and $f()$ is an empirical relationship developed by the modeler. Even if X can only be measured with error, modeling proceeds conditional on the observed values of X and the same conditioning argument applies: $\mathrm{E}[Y|X = x] = f(x, \theta_1, \cdots, \theta_p)$.

To shorten notation we usually drop the *condition on X* argument in expectation operations. If the conditioning argument is not applied, an error-in-variable model is called for. These models are beyond the scope of this text. The interested reader is directed to the literature: Berkson (1950); Kendall and Stuart (1961); Moran (1971); Fedorov (1974); Seber and Wild (1989); Bunke and Bunke (1989); Longford (1993); Carroll, Ruppert, and Stefanski (1995).

Chapter 2

Data Structures

"Modern statisticians are familiar with the notions that any finite body of data contains only a limited amount of information, on any point under examination; that this limit is set by the nature of the data themselves, and cannot be increased by any amount of ingenuity expended in their statistical examination: that the statistician's task, in fact, is limited to the extraction of the whole of the available information on any particular issue." Fisher, R.A., The Design of Experiments, 4th ed., *Edinburgh: Oliver and Boyd, 1947, p. 39)*

2.1 Introduction

The statistical model is an abstraction of a data-generating mechanism that captures those features of the experimental process that are pertinent to a particular inquiry. The data structure is part of the generating mechanism that a model must recognize. The pooled t-test of §1.4, for example, is appropriate when independent random samples are obtained from two groups (populations) and the response variable is Gaussian distributed. Two data structures fit this scenario: (i) two treatments are assigned completely at random to n_1 and n_2 experimental units, respectively, and a Gaussian distributed response is measured once on each unit; (ii) two existing populations are identified which have a Gaussian distribution with the same variance; n_1 individuals are randomly selected from the first and – independent thereof – n_2 individuals are randomly selected from the second population. Situation (i) describes a designed experiment and situation (ii) an observational study. Either situation can be analyzed with the same statistical model

$$Y_{ij} = \begin{cases} \mu_1 + e_{ij} & i = 1 \text{ (=group 1)} \\ \mu_1 + \beta + e_{ij} & i = 2 \text{ (=group 2)} \end{cases}, \quad e_{ij} \sim iid\, G\left(0, \sigma^2\right).$$

In other instances the statistical model must be altered to accommodate a different data structure. Consider the Mitscherlich or linear-plateau model of §1.2. We tacitly assumed there that observations corresponding to different levels of N input were independent. If the 21 N levels are randomly assigned to some experimental units, this assumption is reasonable. Even if there are replications of each N level the models fitted in §1.2 still apply if the replicate observations for a particular N level are averaged. Now imagine the following alteration of the experimental protocol. Each experimental unit receives 0 kg/ha N at the beginning of the study, and every few days 5 kg/ha are added until eventually all experimental units have received all of the 21 N levels. Whether the statistical model remains valid and if not, how it needs to be altered, depends on the changes in the data structure that have been incurred by the protocol alteration.

A data structure comprises three key aspects. The **response type** (e.g., continuous or discrete, §2.2), the **study type** (e.g., designed experiment vs. observational study, §2.3), and the degree of **data clustering** (the hierarchical nature of the data, §2.4). Agronomists are mostly familiar with statistical models for continuous response from designed experiments. The statistical models underpinning the analyses are directed by the treatment, error control, and observational design and require comparatively little interaction between user and model. The temptation to apply the same types of analyses, i.e., the same types of statistical models, in other situations, is understandable and may explain why analysis of proportions or ordinal data by analysis of variance methods is common. But if the statistical model reflects a mechanism that cannot generate data with the same pertinent features as the data at hand, if it generates a different *kind* of data, how can inferences based on these models be reliable?

The analytical task is to construct models from the classes in §1.7 that represent appropriate generating mechanisms. Discrete response data, for example, will lead to generalized linear models, continuous responses with nonlinear mean function will lead to nonlinear models. Hierarchical structures in the data, for example, from splitting experimental units, often call for mixed model structures. The powerful array of tools with which many (most?) data analytic problems in the plant and soil sciences can be tackled is attained by combining the modeling elements. A discrete outcome such as the number of infected plants observed in a split-plot design will automatically lead to generalized linear models of the mixed variety (§8). A continuous outcome for georeferenced data will lead to a spatial random field model (§9).

The main chapters of this text are organized according to response types and levels of clustering. Applications within each chapter reflect different study types. Models appropriate for continuous responses are discussed in §4, §5, §7 to 9. Models for discrete and other non-normal responses appear in §§6, 8. The statistical models in §§4 to 6 apply to uncorrelated (nonclustered) data. Chapters 7 and 8 apply to single- or multiple-level hierarchical data. Chapter 9 applies to a very special case of clustered data structures, spatial data.

2.2 Classification by Response Type

There are numerous classifications of random variables and almost any introductory statistics book presents its own variation. The variation we find most useful is given below. For the purpose of this text, outcomes (random variables) are classified as **discrete** or **continuous**, depending on whether the possible values of the variable (its support) are countable or not (Figure 2.1). The support of a continuous random variable Y is not countable, which implies that if a and b $(a < b)$ are two points in the support of Y, an infinite number of possible values can be placed between them.

The support of a discrete random variable is always countable; it can be enumerated, even if it is infinitely large. The number of earthworms under a square meter of soil does not have a theoretical upper limit, although one could claim that worms cannot be packed arbitrarily dense. The support can nevertheless be enumerated: $\{0, 1, 2, \cdots, \}$. A discrete variable is observed (*measured*) by counting the number of times a particular value in the support occurs. If the support itself consists of counts, the discrete variable is called a **count variable**. The number of earthworms per square meter or the number of seeds germinating out of 100 seeds are examples of count variables. Among count variables a further distinction is helpful. Some counts can be converted into proportions because they have a natural denominator. An outcome in the seed count experiment can be reported as "Y seeds out of n germinated" or as a proportion: "The proportion of germinated seeds is Y/n." We term such count variables **counts with a natural denominator**. Counts without a natural denominator cannot be converted into proportions. The number of poppies per square meter or the number of chocolate chips in a cookie are examples. The distinction between the two types of count variables is helpful because a reasonable probability model for counts with a natural denominator is often the Binomial distribution and counts without a natural denominator are often modeled as Poisson variables.

If the discrete support consists of labels that indicate membership in a category, the variable is termed **categorical**. The presence or absence of a disease, the names of nematode species, and plant quality ratings are categorical outcomes. The support of a categorical variable may consist of numbers but continuity of the variable is not implied. The visual assessment of turfgrass quality on a scale from 1 to 9 creates an ordinal rather than a continuous variable. Instead of the numbers 1 through 9, the labels "a" through "i" may be used. The two labeling systems represent the same ordering. Using numbers to denote category labels, so-called **scoring**, is common and gives a false sense of continuity that may result in the application of statistical models to ordered data which are designed for continuous data. A case in point is the fitting of analysis of variance models to ordinal scores.

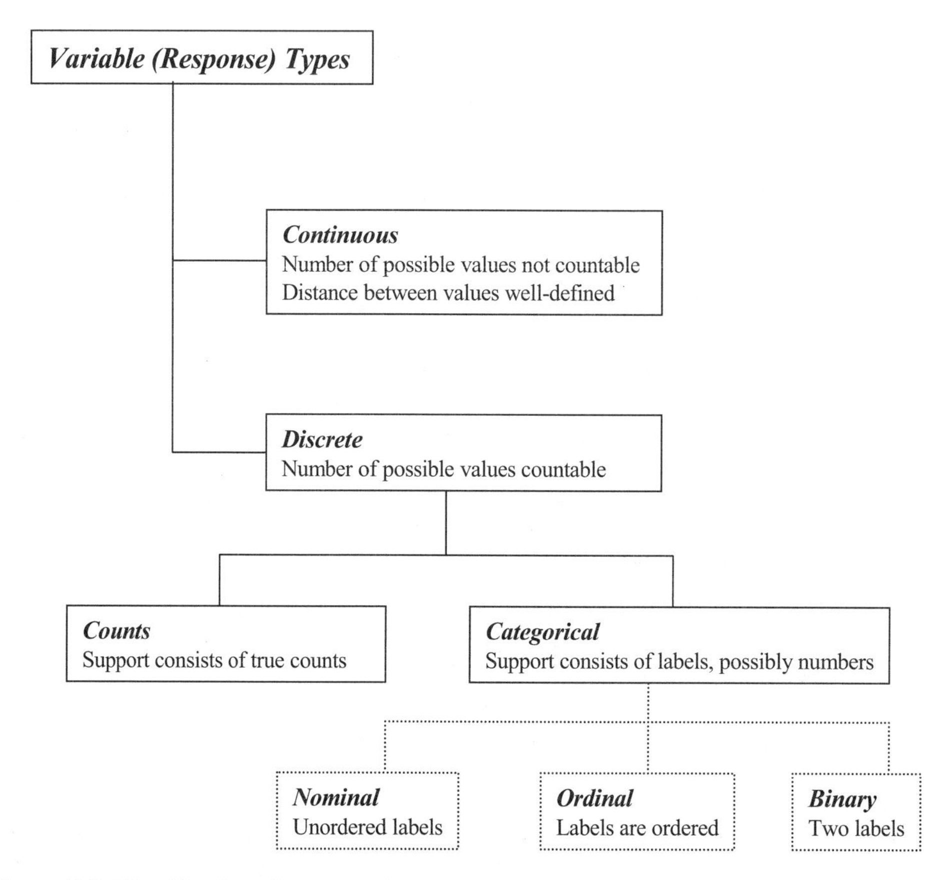

Figure 2.1. Classification of response types.

There is a transition from discrete to continuous variables, which is best illustrated using proportions. Consider counting the number of plants X out of a total of k plants that die after application of an herbicide. Since both X and k are integers, the support of Y, the proportion of dead plants, is discrete:

$$\left\{0, \frac{1}{k}, \frac{2}{k}, \ldots, \frac{k-1}{k}, 1\right\}.$$

As k increases so does the number of elements in the support and provided k is sufficiently large, it can be justified to consider the support infinitely large and no longer countable. The discrete proportion is then treated for analytic purposes as a continuous variable.

2.3 Classification by Study Type

Box 2.1 Study Types

- **Designed experiment: Conditions (treatments) are applied by the experimenter *and* the principles of experimental design (replication, randomization, blocking) are observed.**

- **Comparative experiment: A designed experiment where changes in the conditions (treatments) are examined as the cause of a change in the response.**

- **Observational study: Values of the covariates (conditions) are merely observed, not applied. If conditions are applied, design principles are not or cannot be followed. Inferences are associative rather than causal because of the absence of experimental control.**

- **Comparative study: An observational study examining whether changes in the conditions can be associated with changes in the response.**

- **Validity of inference is derived from *random allocation* of treatments to experimental units in designed experiments and *random sampling* of the population in observational studies.**

The two fundamental situations in which data are gathered are the designed experiment and the observational study. Control over experimental conditions and deliberate varying of these conditions is occasionally cited as the defining feature of a designed experiment (e.g., McPherson 1990, Neter et al. 1990). Observational studies then are experiments where conditions are beyond the control of the experimenter and covariates are merely observed. We do not fully agree with this delineation of designed and observational experiments. Application of treatments is an insufficient criterion for *design* and the existence of factors not controlled by the investigator does not rule out a designed experiment, provided uncontrolled effects can be properly neutralized via randomization. Unless the principles of experimental design, **randomization**, **replication**, and across-unit homogeneity (**blocking**) are observed, data should not be considered generated by a designed experiment. This narrow definition is necessary since designed experiments are understood to lead to cause-and-effect conclusions rather than associative interpretations. Experiments are usually designed as comparative experiments where a change in treatment levels is to be shown to be the cause of changes in the response.

Experimental control must be exercised properly, which implies that (i) treatments are randomly allocated to experimental units to neutralize the effects of uncontrolled factors; (ii) treatments are replicated to allow the estimation of experimental error variance; and (iii) experimental units are grouped into homogeneous blocks prior to treatment application to eliminate controllable factors that are related to the response. The only negotiable of the three principles is that of blocking. If a variable by which the experimental units should be blocked is not taken into account, the experimental design will lead to unbiased estimates of treatment differences and error variances provided randomization is applied. The resulting design may be inefficient with a large experimental error component and statistical tests may be lacking power. Inferences remain valid, however. The completely randomized design (CRD) which does not involve any blocking factors can indeed be more efficient than a randomized complete block design if the experimental material is homogeneous. The other principles are not negotiable. If treatments are replicated but not randomly assigned to experimental units, the basis for causal inferences has been withdrawn and the data must not be viewed as if they had come from a designed experiment. The data should be treated as observational.

An observational study in the narrow sense produces data where the values of covariates are merely observed, not assigned. Typical examples are studies where a population is sampled and along with a response, covariates of immediate or potential interest are observed. In the broader sense we include experiments where the factors of interest have not been randomized and experiments without proper replication. The validity of inferences in observational studies derives from random sampling of the population, not random assignment of treatments. Conclusions in observational studies are narrower than in designed experiments, since a dependence of the response on covariates does not imply that a change in the value of the covariate will **cause** a corresponding change in the value of the response. The variables are simply **associated** with each other in the particular data set. It cannot be ruled out that other effects caused the changes in the response and confound inferences.

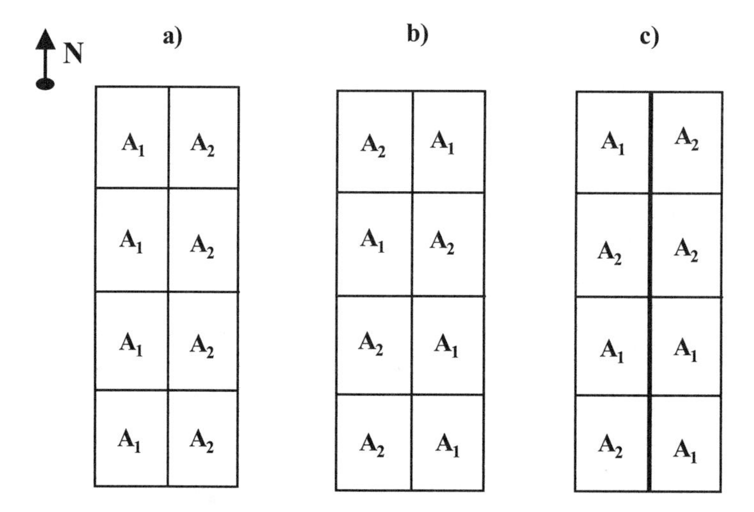

Figure 2.2. Three randomizations of two treatments (A_1, A_2) in four replications. (a) and (b) are complete random assignments, whereas (c) restricts each treatment to appear exactly twice in the east and twice in the west strip of the experimental field.

In a designed experiment, a single randomization is obtained, based on which the experiment is performed. It is sometimes remarked that in the particular randomization obtained, treatment differences may be confounded with nontreatment effects. For example, in a completely randomized design, it is possible, although unlikely, that the r replications of treatment A_1 come to lie in the westernmost part of the experimental area and the replications of treatment A_2 are found in the easternmost areas (Figure 2.2a). If wind or snowdrift induce an east-west effect on the outcome, treatment differences will be confounded with the effects of snow accumulation. One may be more comfortable with a more *balanced* arrangement (Figure 2.2b). If the experimenter feels uncomfortable with the result of a randomization, then most likely the chosen experimental design is inappropriate. If the experimental units are truly homogeneous, treatments can be assigned completely at random and the actual allocation obtained should not matter. If one's discomfort is the reflection of an anticipated east-west effect the complete random assignment is not appropriate, the east-west effect can be controlled by blocking (Figure 2.2c).

The effect of randomization cannot be gleaned from the experimental layout obtained. One must envision the design under repetition of the assignment procedure. The neutralizing effect of randomization does not apply to a single experiment, but is an average effect across all possible arrangements of treatments to experimental units under the particular design. While blocking **eliminates** the effects of systematic factors in any given design, randomization **neutralizes** the unknown effects in the population of all possible designs. Randomization is the means to estimate treatment differences and variance components without bias. Replication, the independent assignment of treatments, enables estimation of experimental error variance. Replication alone does not lead to unbiased estimates of treatment effects or experimental error, a fact that is often overlooked.

2.4 Clustered Data

Box 2.2 Clusters

- **A cluster is a collection of observations that share a stochastic, temporal, spatial, or other association that suggests to treat them as a group.**
- **While dependencies and interrelations may exist among the units within a cluster, it is often reasonable to treat observations from different clusters as independent.**
- **Clusters commonly arise from**
  - **— hierarchical random sampling (nesting of sampling units)**
  - **— hierarchical random assignment (splitting),**
  - **— repeated observations taken on experimental or sampling units.**

Clustering of data refers to the hierarchical structure in data. It is an important feature of the data-generating mechanism and as such it plays a critical role in formulating statistical models, in particular mixed models and models for spatial data. In general, a cluster represents a collection of observations that are somehow stochastically related, whereas observations from different clusters are typically independent (stochastically unrelated). The two primary situations that lead to clustered data structures are (i) hierarchical random processes and (ii) repeated measurements (Figure 2.3). Grouping of observations into sequences of repeated observations collected on the same entity or subject has long been recognized as a clustered data structure. Clustering through hierarchical random processes such as subsampling or splitting of experimental units also gives rise to hierarchical data structures.

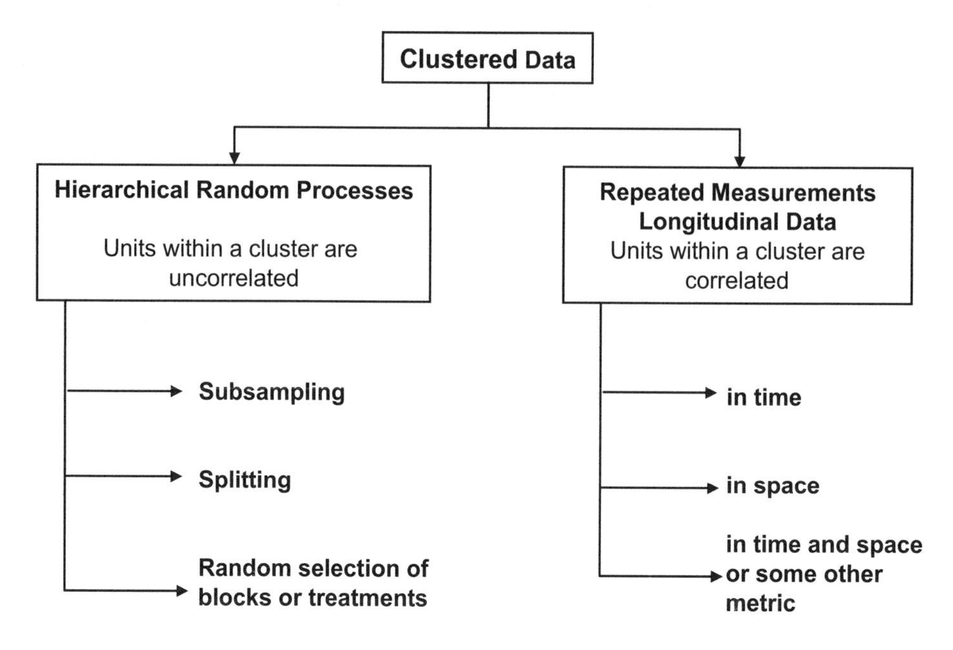

Figure 2.3. Frequently encountered situations that give rise to clustered data.

2.4.1 Clustering through Hierarchical Random Processes

Observational studies as well as designed experiments can involve subsampling. In the former, the level-one units are sampled from a population and contain multiple elements. A random sample of these elements, the subsamples, constitute the observational data. For example, forest stands are selected at random from a list of stands and within each stand a given number of trees are selected. The forest stand is the cluster and the trees selected within are the cluster elements. These share a commonality, they belong to the same stand (cluster). Double-tubed steel probes are inserted in a truckload of wheat kernels and the kernels trapped in the probe are extracted. The contents of the probe are well-mixed and two 100-g subsamples are selected and submitted for assay to determine deoxynivalenol concentration. The probes are the clusters and the two 100-g samples are the within-cluster observational units. A subsampling structure can also be cast as a nested structure, in which the subsamples are nested within the level-one sampling units. In designed experiments subsampling is common when experimental units cannot be measured in their entirety. An experimental unit may contain eight rows of crop but only a subsample thereof is harvested and analyzed. The experimental unit is the cluster. In contrast to the observational study, the first random process is not the random selection of the experimental unit, but the random allocation of the treatments to the experimental unit.

Splitting, the process of assigning levels of one factor within experimental units of another factor, also gives rise to clustered data structures. Consider two treatment factors A and B with a and b levels, respectively. The a levels of factor A are assigned to experimental units according to some standard experimental design, e.g., a completely randomized design, a randomized complete block design, etc. Each experimental unit for factor A, the whole-plots, is then split (divided) into b experimental units (the sub-plots) to which the levels of factor B are randomly assigned. This process of splitting and assigning the B levels is carried out independently for each whole-plot. Each whole-plot constitutes a cluster and the sub-plot observations for factor B are the within-cluster units. This process can be carried out to another level of clustering when the sub-plots are split to accommodate a third factor, i.e., as a split-split-plot design.

The third type of clustering through hierarchical random processes, the random selection of block or treatment levels, is not as obvious, but it shares important features with the subsampling and splitting procedures. When treatment levels are randomly selected from a list of possible treatments, the treatment effects in the statistical model become random variables. The experimental errors, also random variables, then are nested within another random effect. To compare the three hierarchical approaches, consider the case of a designed experiment. Experiment [A] is a completely randomized design (CRD) with subsampling, experiment [B] is a split-plot design with the whole-plot factor arranged in a CRD, and experiment [C] is a CRD with the treatments selected at random. The statistical models for the three designs are

[A] $$Y_{ijk} = \mu + \tau_i + e_{ij} + d_{ijk}$$
$$e_{ij} \sim iid\left(0, \sigma_e^2\right);\ d_{ijk} \sim iid\left(0, \sigma_d^2\right)$$

[B] $$Y_{ijk} = \mu + \alpha_i + e_{ij} + \beta_k + (\alpha\beta)_{ik} + d_{ijk}$$
$$e_{ij} \sim iid\left(0, \sigma_e^2\right);\ d_{ijk} \sim iid\left(0, \sigma_d^2\right)$$

[C] $$Y_{ij} = \mu + \tau_i + e_{ij}$$
$$\tau_i \sim iid\left(0, \sigma_\tau^2\right);\ e_{ij} \sim iid\left(0, \sigma_e^2\right).$$

In all models the e_{ij} denote experimental errors. In [B] the e_{ij} are the whole-plot experimental errors and the d_{ijk} are the sub-plot experimental errors. In [A] the d_{ijk} are subsampling (observational) errors. The τ_i in [C] are the random treatment effects. Regardless of the type of design, the error terms in all three models are independent by virtue of the random selection of observational units or the random assignment of treatments. Every model contains two random effects where the second effect has one more subscript. The clusters are formed by the e_{ij} in [A] and [B] and by the τ_i in [C]. While the within-cluster units are uncorrelated, it turns out that the responses Y_{ijk} and Y_{ij} are not necessarily independent of each other. For two subsamples from the same experimental unit in [A] and two observations from the same whole-plot in [B], we have $\text{Cov}[Y_{ijk}, Y_{ijk'}] = \sigma_e^2$. For two replicates of the same treatment in [C], we obtain $\text{Cov}[Y_{ij}, Y_{ij'}] = \sigma_\tau^2$. This is an important feature of statistical models with multiple, nested random effects. They **induce** correlations of the responses (§7.5.1).

2.4.2 Clustering through Repeated Measurements

A different type of clustered data structure arises if the same observational or experimental unit is measured repeatedly. An experiment where one randomization of treatments to experimental units occurs at the beginning of the study and the units are observed repeatedly through time without re-randomization is termed a **repeated measures** study. If the data are observational in nature, we prefer the term **longitudinal data**. The cluster is then formed by the collection of repeated measurements. Characteristic of both is the absence of a random assignment or selection within the cluster. Furthermore, the repeat observations from a given cluster usually are **autocorrelated** (§2.5) It is not necessary that the repeated measurements are collected over time, although this is by far the most frequent case. It is only required that the cluster elements in longitudinal and repeated measures data be ordered along some metric. This metric may be temporal, spatial, or spatiotemporal. The cluster elements in the following example are arranged spatially along transects.

Example 2.1. The phosphate load of pastures under rotational grazing is investigated for two forage species, alfalfa and birdsfoot trefoil. Since each pasture contains a watering hole where heifers may concentrate, Bray-1 soil P (Bray-P1) is measured at various distances from the watering hole. The layout of the observational units for one replicate of the alfalfa treatment is shown in Figure 2.4.

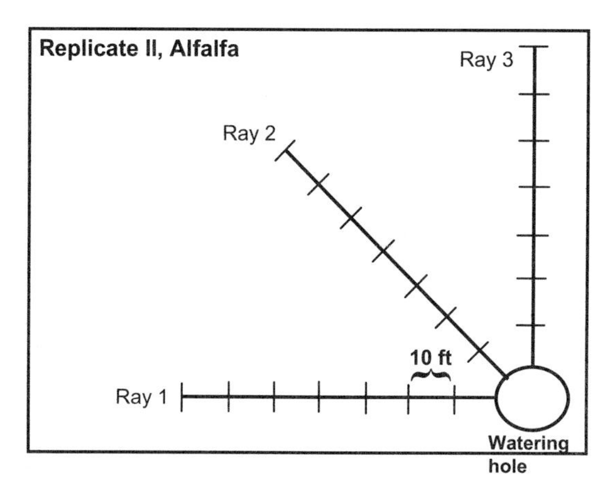

Figure 2.4. Alfalfa replicate in soil P study. The replicate serves as the cluster for the three rays along which soil samples are collected in 10-ft spacing. Each ray is a cluster of seven soil samples.

The measurements along each ray are ordered along a spatial metric. Similarly, the distances between measurements on different rays within a replicate are defined by the Euclidean distance between any two points of soil sampling.

A critical difference between this design and a subsampling or split-plot design is the systematic arrangement of elements within a cluster. This lack of randomization is particularly apparent in longitudinal or repeated measures structures in time. A common practice is to analyze repeated measures data as if it arises from a split-plot experiment. The repeated measurements made on the experimental units of a basic design such as a randomized block design are assumed to constitute a split of these units. This practice is not appropriate unless the temporal measurements are as exchangeable (permutable) as treatment randomizations. Since the flow of time is unidirectional, this is typically not the case.

Longitudinal studies where the cluster elements are arranged by time are frequent in the study of growth processes. For example, plants are randomly selected and their growth is recorded repeatedly over time. An alternative data collection scheme would be to randomly select plants at different stages of development and record their current yield, a so-called **cross-sectional** study. If subjects of different ages are observed at one point in time, the differences in yield between these subjects do not necessarily represent true growth due to the entanglement of age and **cohort** effect. The term cohort was coined in the social sciences and describes a group of individuals which differs from other individuals in a systematic fashion. For example, the cohort of people in their twenties today is not necessarily comparable to the cohort of twenty year olds in 1950. The problem of confounding age and cohort effects applies to data in the plant and soil sciences similarly.

In a long-term ecological research project one may be interested in the long-term effects of manure management according to guidelines on soil phosphorus levels. The hypothesis is that management according to guidelines will reduce the Bray-P1 levels over time. A cross-sectional study would identify n sites that have been managed according to guidelines for a varied number of years and analyze (model) the Bray-P1 levels as a function of time in management. A longitudinal study identifies n sites managed according to guidelines and follows them for several years. The sequence of observations over time on a given site in the longitudinal study is that sites change in Bray-P1 phosphorus. In order to conclude that changes over time in Bray-P1 are due to adherence to management guidelines in the cross-sectional study, cohort effects must be eliminated or assumed absent. That is, the cohort of sites that were in continued manure management for five years at the beginning of the study are assumed to develop in the next seven years to the Bray-P1 levels exhibited by those sites that were in continued manure management for twelve years at the onset of the study.

The most meaningful approach to measuring the growth of a subject is to observe the subject at different points in time. All things being equal (*ceteris paribus*), the error-free change will represent true growth. The power of longitudinal and repeated measures data lies in the efficient estimation of growth without confounding of cohort effects. In the vernacular of longitudinal data analysis this is phrased as "each cluster [subject] serves as its own control." Deviations are not measured relative to the overall trend across all clusters, the population average, but relative to the trend specific to each cluster.

The within-cluster elements of longitudinal and repeated measures data can be ordered in a nontemporal fashion, for example spatially (e.g., Example 2.1). On occasion, spatial and temporal metrics are intertwined. Following Gregoire and Schabenberger (1996a, 1996b) we analyze data from tree stem analysis where the diameter of a tree is recorded in three-foot intervals in §8.4.1. Tree volume is modeled as a function of the diameter to which the tree has tapered at a given measurement interval. The diameter can be viewed as a proxy for a spatial metric, the distance above ground, or a temporal metric, the age of the tree at which a certain measurement height was achieved.

A repeated measured data structure with spatiotemporal metric is discussed in the following example.

Example 2.2. Pierce et al. (1994) discuss a tillage experiment where four tillage strategies (Moldboard plowing; Plowing Spring 1987 and Fall 1996; Plowing Spring 1986, Spring 1991, Fall 1995; No-Tillage) were arranged in a randomized complete block design with four blocks. Soil characteristics were obtained April 10, 1987, April 22, 1991, July 26, 1996, and November 20, 1997. At each time point, soil samples were extracted from four depths: 0 to 2, 2 to 4, 4 to 6, and 6 to 8 inches. At a given time point the data have a repeated measures structure with spatial metric. For a given sampling depth, the data have a repeated measures structure with temporal metric. The combined data have a repeated measures structure with spatiotemporal metric. Figure 2.5 depicts the sample mean pH across the spatiotemporal metric for the No-Tillage treatment.

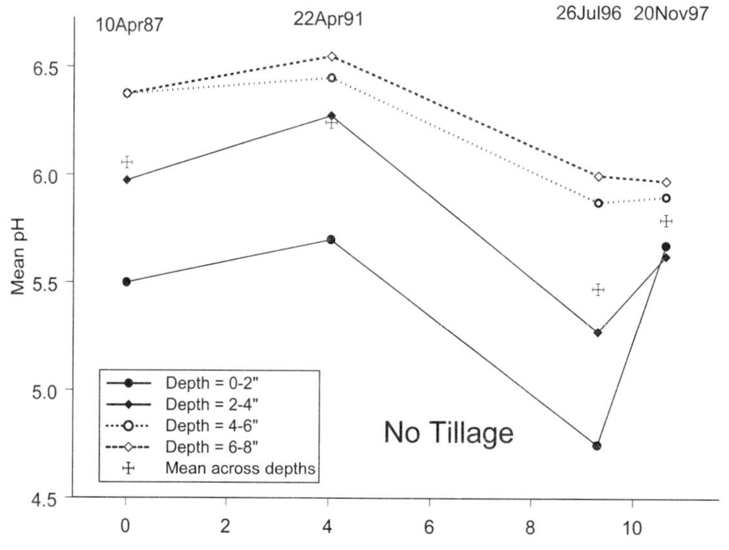

Figure 2.5. Depth × Time sample mean pH for the No-Tillage treatment. Cross-hairs depict sample mean pH at a given depth.

While in longitudinal studies, focus is on modeling the population-average or cluster-specific trends as a function of time, space, etc., in repeated measures studies, interest shifts to examining whether treatment comparisons depend on time and/or space. If the trends in treatment response over time are complex, or the number of re-measurements large, modeling the trends over time similar to a longitudinal study and comparing those among treatments may be a viable approach to analysis (see Rasse, Smucker, and Schabenberger 1999 for an example).

In studies where data are collected over long periods of time, it invariably happens that data become unbalanced. At certain measurement times not all sampling or experimental units can be observed. Data are considered **missing** if observations intended for collection are inaccessible, lost, or unavailable. Experimental units are lost to real estate sales, animals die, patients drop out of a study, a lysimeter is destroyed, etc. Missing data implies unplanned imbalance but the reverse is not necessarily true. Unbalanced data contain missing values only if one intended to measure the absent observations. If the process that produced the missing observation is related to the unobserved outcome, conclusions based on data that ignore the missing value process are biased. Consider the study of soil properties under bare soil and a cover crop in a repeated measures design. Due to drought the bare soil cannot be drilled at certain sampling times. The absence of the observations is caused by soil properties we intended to measure. The missing values mask the effect we are hoping to detect, and ignoring them will bias the comparison between cover crop and bare soil treatments. Now assume that a technician applying pesticides on an adjacent area accidentally drives through one of the bare soil replicates. Investigators are forced to ignore the data from the particular replicate because soil properties have been affected by trafficking. In this scenario the unavailability of the data from the trafficked replicate is not related to the characteristic we intended to record. Ignoring data from the replicate will not bias conclusions about the treatments. It will only lower the power of the comparisons due to reduced sample size.

Little and Rubin (1987) classify missing value mechanisms into three categories. If the probability of a value being missing is unrelated to both the observed and unobserved data, missingness is termed completely at random. If missingness does not relate to the unobserved data (but possibly to the observed data) it is termed random, and if it is dependent on the unobserved data as in the first scenario above, it is informative. Informative missingness is troublesome because it biases the results. Laird (1988) calls random missingness ignorable missingness because it does not negatively affect certain statistical estimation methods (see also Rubin 1976, Diggle et al. 1994). When analyzing repeated measures or longitudinal data containing missing values we hope that the missing value mechanism is at least random. Missingness is then ignorable.

The underlying metric of a longitudinal or repeated measures data set can be continuous or discrete and measurements can be equally or unequally spaced. Some processes are inherently discrete when measurements cannot be taken at arbitrary points in time. Jones (1993) cites the recording of the daily maximum temperature which yields only one observation per day. If plant leaf temperatures are measured daily for four weeks at 12:00 p.m., however, the underlying time scale is continuous since the choice of measurement time was arbitrary. One could have observed the temperatures at any other time of day. The measurements are also equally spaced. Equal or unequal spacing of measurements does not imply discreteness or continuity of the underlying metric. If unequal spacing is deliberate (planned), the metric is usually continuous. There is some ambiguity, however. In Example 2.2 the temporal metric has four entries. One could treat this as unequally spaced yearly measurements or equally spaced daily measurements with many missing values ($3,874$ missing values between April 10, 1987 and November 20, 1997). The former representation clearly is more useful. In the study of plant growth, measurements will be denser during the growing season compared to periods of dormancy or lesser biological activity. Unequal spacing is neither good nor bad. Measurements should be collected when it is most meaningful. The statistical models for analyzing repeated measures and longitudinal data should follow suit. The mixed models for clustered data discussed in §§7 and 8 are models of this type. They can accommodate unequal spacing without difficulty as well as (random) missingness of observations. This is in sharp contrast to more traditional models for repeated measures analysis, e.g., repeated measures analysis of variance, where the absence of a single temporal observation for an experimental unit results in the loss of all repeated measurements for that unit, a wasteful proposition.

2.5 Autocorrelated Data

Box 2.3 Autocorrelation

The correlation ρ_{xy} between two random variables measures the strength (and direction) of the (linear) dependency between X and Y.

- **If X and Y are *two different* variables ρ_{xy} is called the product-moment correlation coefficient. If X and Y are observations of the *same* variable at different times or locations, ρ_{xy} is called the autocorrelation coefficient.**

- **Autocorrelation is the correlation of a variable in time or space *with itself*. While estimating the product-moment correlation requires n *pairs* of observations $(x_1, y_1), ..., (x_n, y_n)$, autocorrelation is determined from a single (time) sequence $(x_1, x_2, ..., x_t)$.**

- **Ignoring autocorrelation (treating data as if they were independent) distorts statistical inferences.**

- **Autocorrelations are usually positive, which implies that an above-average value is likely to be followed by another above average value.**

2.5.1 The Autocorrelation Function

Unless elements of a cluster are selected or assigned at random, the responses within a cluster are likely to be correlated. This type of correlation does not measure the (linear) dependency between two different attributes, but between different values of the same attribute. Hence the name **autocorrelation**. When measurements are collected over time it is also referred to as **serial correlation**, a term that originated in the study of time series data (Box, Jenkins, and Reinsel 1994). The (product-moment) correlation coefficient between random variables X and Y is defined as

$$\text{Corr}[X, Y] = \rho_{xy} = \frac{\text{Cov}[X, Y]}{\sqrt{\text{Var}[X]\text{Var}[Y]}}, \qquad [2.1]$$

where $\text{Cov}[X, Y]$ denotes the covariance between the two random variables. The coefficient ranges from –1 to 1 and measures the strength of the linear dependency between X and Y. It is related to the coefficient of determination (R^2) in a linear regression of Y on X (or X on Y) by $R^2 = \rho_{xy}^2$. A positive value of ρ_{xy} implies that an above-average value of X is likely to be paired with an above-average value of Y. A negative correlation coefficient implies that above-average values of X are paired with below-average values of Y. Autocorrelation coefficients are defined in the same fashion, a covariance divided by the square root of a variance product. Instead of two different variables X and Y, the covariance and variances pertain to the same attribute measured at two different points in time, two different points in space, and so forth.

Focus on the temporal case for the time being and let $Y(t_1), \cdots, Y(t_n)$ denote a sequence of observations collected at times t_1 through t_n. Denote the mean at time t_i as $\mathrm{E}[Y(t_i)] = \mu(t_i)$. The covariance between two values in the sequence is given by

$$\mathrm{Cov}[Y(t_i), Y(t_j)] = \mathrm{E}[(Y(t_i) - \mu(t_i))(Y(t_j) - \mu(t_j))] = \mathrm{E}[Y(t_i)Y(t_j)] - \mu(t_i)\mu(t_j).$$

The variance of the sequence at time t_i is $\mathrm{Var}[Y(t_i)] = \mathrm{E}\left[(Y(t_i) - \mu(t_i))^2\right]$ and the autocorrelation between observations at times t_i and t_j is measured by the correlation

$$\mathrm{Corr}[Y(t_i), Y(t_j)] = \frac{\mathrm{Cov}[Y(t_i), Y(t_j)]}{\sqrt{\mathrm{Var}[Y(t_i)]\mathrm{Var}[Y(t_j)]}}. \qquad [2.2]$$

If data measured repeatedly are uncorrelated, they should scatter around the trend over time in an unpredictable, nonsystematic fashion (open circles in Figure 2.6). Data with positive autocorrelation show long runs of positive or negative residuals. If an observation is likely to be below (above) average at some time point, it was likely to be below (above) average in the immediate past. While negative product-moment correlations between random variables are common, negative autocorrelation is fairly rare. It would imply that above (below) average values are likely to be preceded by below (above) average values. This is usually indication of an incorrectly specified mean function. For example, a circadian rhythm or seasonal fluctuation was omitted.

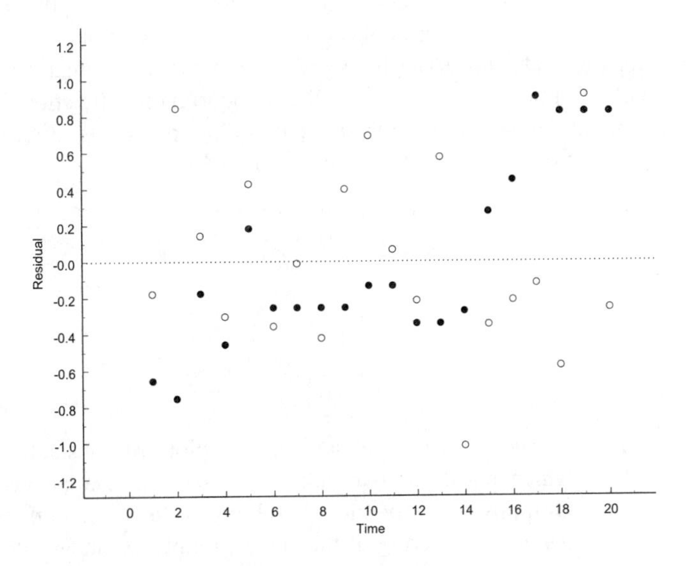

Figure 2.6. Autocorrelated and independent data with the same dispersion, $\sigma^2 = 0.3$. Open circles depict independent observations, closed circles observations with positive autocorrelation. Data are shown as deviations from the mean (residuals). A run of negative residuals for the autocorrelated data occurs between times 6 and 14.

In a temporal sequence with n observations there are possibly n different variances and $n(n-1)/2$ different covariances. The number of unknowns in the autocorrelation structure is larger than the number of observations. To facilitate estimation of the autocorrelation structure stationarity assumptions are common in time series. Similar assumptions are made for longitudinal and repeated measures data (§7) and for spatial data (§9). Details of stationarity properties will be discussed at length in the pertinent chapters; here we note only that a (second-order) stationary time series has reached a form of equilibrium. The variances of the observations are constant and do not depend on the time point. The strength of the autocorrelation does not depend on the origin of time, only on elapsed time. In a stationary time series the correlation between observations two days apart is the same, whether the first day is a Monday or a Thursday. Under these conditions the correlation [2.2] can be expressed as

$$\mathrm{Corr}[Y(t_i), Y(t_j)] = \frac{C(|t_i - t_j|)}{\sigma^2}, \qquad [2.3]$$

where $\sigma^2 = \mathrm{Var}[Y(t_i)]$. The function $C()$ is called the (auto-)**covariance function** of the process and $h_{ij} = |t_i - t_j|$ is the time **lag** between $Y(t_i)$ and $Y(t_j)$. Since $\mathrm{Var}[Y(t_i)] = \mathrm{Cov}[Y(t_i), Y(t_i)] = C(0)$, the (auto-)**correlation function** in a series can be expressed for any time lag h as

$$R(h) = \frac{C(h)}{C(0)}. \qquad [2.4]$$

The autocorrelation function is an important device in analyzing clustered data. In some applications it is modeled implicitly (indirectly), that is, the formulation of the model implies a particular correlation function. In other applications it is modeled explicitly (directly), that is, the analyst develops a model for $R(h)$ (or $C(h)$) in the same way that a model for the mean structure is built. Implicit specification of the autocorrelation function is typical for clustered data structures that arise through hierarchical random processes (Figure 2.3). In the case of the split-plot design of §2.4 we have for observations from the same whole-plot

$$\mathrm{Cov}[Y_{ijk}, Y_{ijk'}] = \sigma_e^2$$
$$\mathrm{Var}[Y_{ijk}] = \sigma_e^2 + \sigma_d^2,$$

so that the autocorrelation function is

$$R(h) = \frac{\sigma_e^2}{\sigma_e^2 + \sigma_d^2} = \rho.$$

The correlations among observations from the same whole-plot are constant. The function $R()$ does not depend on which particular two sub-plot observations are considered. This structure is known as the **compound-symmetric** or **exchangeable** correlation structure. The process of randomization makes the ordering of the sub-plot units exchangeable. If repeated measures experiments are analyzed as split-plot designs it is implicitly assumed that the autocorrelation structure of the repeated measure is exchangeable.

With temporal data it is usually more appropriate to assume that correlations (covariances) decrease with increasing temporal separation. The autocorrelation function $R(h)$ approaches 0 as the time lag h increases. When modeling $R(h)$ or $C(h)$ directly, we rely on models for the covariance function that behave in a way the user deems reasonable. The following models are some popular candidates of covariance functions where correlations decrease with lag ($\rho > 0, \alpha > 0$),

$$C(h) = C(0)\rho^{|i-j|}$$
$$C(h) = C(0)\rho^{|t_i-t_j|}$$
$$C(h) = C(0)\exp\{-|t_i - t_j|^2/\alpha^2\}.$$

2.5.2 Consequences of Ignoring Autocorrelation

Analyzing data that exhibit positive autocorrelation as if they were uncorrelated has serious implications. In short, estimators are typically inefficient although they may remain unbiased. Estimates of the precision of the estimators (the estimated standard errors) can be severely biased. The precision of the estimators is usually overstated, resulting in test statistics that are too large and p-values that are too small. These are ramifications of "pretending to have more information than there really is." We demonstrate this effect with a simple example. Assume that two treatments (groups) are to be compared. For each group a series of positively autocorrelated observations is collected, for example, a repeated measure. The observations are thus correlated within a group but are uncorrelated between the groups. For simplicity we assume that both groups are observed at the same points in time $t_1, \cdots, t_n$, that both groups have the same (known) variances, and that the observations within a group are equicorrelated and Gaussian-distributed. In essence this setup is a generalization of the two-sample z-test for Gaussian data with correlations among the observations from a group. In parlance of §2.4 there are two clusters with n observations each.

Let $Y_1(t_1), \cdots, Y_1(t_n)$ denote the n observations from group 1 and $Y_2(t_1), \cdots, Y_2(t_n)$ the observations from group 2. The distributional assumptions can be expressed as

$$Y_1(t_i) \sim G(\mu_1, \sigma^2)$$
$$Y_2(t_i) \sim G(\mu_2, \sigma^2)$$
$$\text{Cov}[Y_1(t_i), Y_1(t_j)] = \begin{cases} \sigma^2\rho & i \neq j \\ \sigma^2 & i = j \end{cases}$$
$$\text{Cov}[Y_1(t_i), Y_2(t_j)] = 0.$$

We are interested in comparing the means of the two groups, $H_0: \mu_1 - \mu_2 = 0$. First assume that the correlations are ignored. Then one would estimate μ_1 and μ_2 by the respective sample means

$$\overline{Y}_1 = \frac{1}{n}\sum_{i=1}^{n} Y_1(t_i) \qquad \overline{Y}_2 = \frac{1}{n}\sum_{i=1}^{n} Y_2(t_i)$$

and determine their variances to be $\text{Var}[\overline{Y}_1] = \text{Var}[\overline{Y}_2] = \sigma^2/n$. The test statistic (assuming σ^2 known) for $H_0: \mu_1 - \mu_2 = 0$ is

$$Z^*_{obs} = \frac{\overline{Y}_1 - \overline{Y}_2}{\sigma\sqrt{2/n}}.$$

If the within-group correlations are taken into account, $\overline{Y}_1$ and $\overline{Y}_2$ remain unbiased estimators of μ_1 and μ_2, but their variance is no longer σ^2/n. Instead,

$$\begin{aligned}\operatorname{Var}\left[\overline{Y}_1\right] &= \frac{1}{n^2}\operatorname{Var}\left[\sum_{i=1}^{n} Y_1(t_i)\right] = \frac{1}{n^2}\sum_{i=1}^{n}\sum_{j=1}^{n}\operatorname{Cov}[Y_1(t_i), Y_1(t_j)] \\ &= \frac{1}{n^2}\left\{\sum_{i=1}^{n}\operatorname{Var}[Y_1(t_i)] + \sum_{i=1}^{n}\sum_{\substack{j=1\\ j\neq i}}^{n}\operatorname{Cov}[Y_1(t_i), Y_1(t_j)]\right\} \\ &= \frac{1}{n^2}\left\{n\sigma^2 + n(n-1)\sigma^2\rho\right\} = \frac{\sigma^2}{n}\{1 + (n-1)\rho\}.\end{aligned}$$

The same expression is derived for $\operatorname{Var}[\overline{Y}_2]$. The variance of the sample means is larger (if $\rho > 0$) than σ^2/n, the variance of the sample mean in the absence of correlations. The precision of the estimate of the group difference, $\overline{Y}_1 - \overline{Y}_2$, is overestimated by the multiplicative factor $1 + (n-1)\rho$. The correct test statistic is

$$Z_{obs} = \frac{\overline{y}_1 - \overline{y}_2}{\sigma\sqrt{\frac{2}{n}(1 + (n-1)\rho)}} < Z^*_{obs}.$$

The incorrect test statistic will be too large, and the p-value of that test will be too small. One may declare the two groups as significantly different, whereas the correct test may fail to find significant differences in the group means. The evidence against the null hypothesis has been overstated by ignoring the correlation.

Another way of approaching this issue is in terms of the **effective sample size**. One can ask: "How many samples of the uncorrelated kind provide the same precision as a sample of correlated observations?" Cressie (1993, p. 15) calls this the equivalent number of independent observations. Let n denote the number of samples of the correlated kind and n' the equivalent number of independent observations. The effective sample size is calculated as

$$n' = n\frac{\operatorname{Var}\left[\overline{Y}\text{ assuming independence}\right]}{\operatorname{Var}\left[\overline{Y}\text{ under autocorrelation}\right]} = \frac{n}{1 + (n-1)\rho}.$$

Ten observations equicorrelated with $\rho = 0.3$ provide as much information as 2.7 independent observations. This seems like a hefty penalty for having correlated data. Recall that the group mean difference was estimated as the difference of the arithmetic sample means, $\overline{Y}_1 - \overline{Y}_2$. Although this difference is an unbiased estimate of the group mean difference $\mu_1 - \mu_2$, it is an efficient estimate only in the case of uncorrelated data. If the correlations are taken into account, more efficient estimators of $\mu_1 - \mu_2$ are available and the increase in efficiency works to offset the smaller effective sample size.

When predicting new observations based on a statistical model with correlated errors, it turns out that the correlations enhance the predictive ability. Geostatistical kriging methods, for example, utilize the spatial autocorrelation among observations to predict an attribute of interest at a location where no observation has been collected (§9.4). The autocorrelation between the unobserved attributed and the observed values allows one to glean information about the unknown value from surrounding values. If data were uncorrelated, surrounding values would not carry any stochastic information.

2.5.3 Autocorrelation in Designed Experiments

Serial, spatial, or other autocorrelation in designed experiments is typically due to an absence of randomization. Time *can* be a randomized treatment factor in an experiment, for example, when only parts of an experimental unit are harvested at certain points in time and the order in which the parts are harvested is randomized. In this case *time* is a true sub-plot treatment factor in a split-plot design. More often, however, time effects refer to repeated measurements of the same experimental or observational units and cannot be randomized. The second measurement is collected after the first and before the third measurement. The obstacle to randomization is the absence of a conveyance for time-travel. Randomization restrictions can also stem from mechanical or technical limitations in the experimental setup. In strip-planter trials the order of the strips is often not randomized. In line sprinkler experiments, water is emitted from an immobile source through regularly spaced nozzles and treatments are arranged in strips perpendicular to the line source. Hanks et al. (1980), for example, describe an experiment where three cultivars are arranged in strips on either side of the sprinkler. The irrigation level, a factor of interest in the study, decreases with increasing distance from the sprinkler and cannot be randomized. Along the strip of a particular cultivar one would expect spatial autocorrelations.

A contemporary approach to analyzing designed experiments derives validity of inference not from randomization alone, but from information about the spatial, temporal, and other dependencies that exist among the experimental units. Randomization neutralizes spatial dependencies at the scale of the experimental unit, but not on smaller or larger scales. Fertility trends that run across an experimental area and are not eliminated by blocking are neutralized by randomization if one considers averages across all possible experimental arrangements. In a particular experiment, these spatial effects may still be present and increase error variance. Local trends in soil fertility, competition among adjacent plots, and imprecision in treatment application (drifting spray, subsurface movement of nutrients, etc.) create effects which increase variability and can bias unadjusted treatment contrasts. Stroup et al. (1994) note that combining adjacent experimental units into blocks in agricultural variety trials can be at odds with an assumption of homogeneity within blocks when more than eight to twelve experimental units are grouped. Spatial trends will then be removed only incompletely. An analysis which takes explicitly into account the spatial dependencies of experimental units may be beneficial compared to an analysis which relies on conceptual assignments that did not occur. A simple approach to account for spatial within-block effects is based on analyzing differences of nearest-neighbors in the field layout. These methods originated with Papadakis (1937) and were subsequently modified and improved upon (see, for example, Wilkinson et al. 1983, Besag and Kempton 1986). Gilmour et al. (1997) discuss some of the difficulties and pitfalls when differencing data to remove effects of spatial variation. Grondona and Cressie (1991) developed spatial analysis of variance derived from expected mean squares where expectations are taken over the distribution induced by treatment randomization *and* a spatial process. We prefer modeling techniques rooted in random field theory (§9.5) to analyze experimental data in which spatial dependencies were not fully removed or neutralized (Zimmerman and Harville 1991, Brownie et al. 1993, Stroup et al.

1994, Brownie and Gumpertz 1997, Gilmour et al. 1997, Gotway and Stroup 1997, Verbyla et al. 1999). This is a modeling approach that considers the spatial process as the driving force behind the data. Whether a spatial model will provide a more efficient analysis than a traditional analysis of variance will depend to what extent the components of the spatial process are conducive to modeling. We agree with Besag and Kempton (1986) that many agronomic experiments are not carried out in a sophisticated manner. The reasons may be convenience, unfamiliarity of the experimenter with more complex design choices, or tradition. Bartlett (1938, 1978a) views analyses that emphasize the spatial context over the design context as ancillary devices to salvage efficiency in experiments that could have been designed more appropriately. We do not agree with this view of spatial analyses as *salvage tools*. One might then adopt the viewpoint that randomization and replication are superfluous since treatment comparisons can be recovered through spatial analysis tools. By switching from the classical design-based analysis to an analysis based on spatial modeling, inferences have become conditional on the correctness of the model. The ability to draw causal conclusions has been lost. There is no free lunch.

2.6 From Independent to Spatial Data — a Progression of Clustering

Box 2.4 Progression of Clustering

- **Clustered data can be decomposed into k clusters of size n_k where $n = \sum_{i=1}^{k} n_i$ is the total number of observations and n_i is the size of the ith cluster.**
- **Spatial data is a special case of clustered data where the entire data comprises a single cluster.**
  - **— Unclustered data: $k = n, n_i = 1$**
  - **— Longitudinal data: $k < n, n_i > 1$**
  - **— Spatial data: $k = 1, n_1 = n$.**

As far as the correlation in data is concerned this text considers three prominent types of data structures. Models for independent (uncorrelated) data that do not exhibit clustering are covered in Chapters 4 to 6, statistical models for clustered data where observations from different clusters are uncorrelated but correlations within a cluster are possible are considered in Chapters 7 and 8. Models for spatial data are discussed in §9. Although the statistical tools can differ greatly from chapter to chapter, there is a natural progression in these three data structures. Before we can make this progression more precise, a few introductory comments about spatial data are in order. More detailed coverage of spatial data types and their underlying stochastic processes is deferred until §9.1.

A data set is termed spatial if, along with the attribute of interest, Y, the spatial locations of the attributes are recorded. Let $\mathbf{s}$ denote the vector of coordinates at which Y was observed. If we restrict discussion for the time being to observations collected in the plane, then $\mathbf{s}$ is a two-dimensional vector containing the longitude and latitude. In a time series we can represent the observation collected at time t as $Y(t)$. Similarly for spatial data we represent the observation as $Y(\mathbf{s})$. A univariate data set of n spatial observations thus consists of $Y(\mathbf{s}_1),\ Y(\mathbf{s}_2), \cdots, Y(\mathbf{s}_n)$. The three main types of spatial data, geostatistical data, lattice data, and point pattern data, are distinguished according to characteristics of the set of locations. If the underlying space is continuous and samples can be gathered at arbitrary points, the data are termed **geostatistical.** Yield monitoring a field or collecting soil samples produces geostatistical data. The soil samples are collected at a finite number of sites but could have been gathered anywhere within the field. If the underlying space is made up of discrete units, the spatial data type is termed **lattice data**. Recording events by city block, county, region, soil type, or experimental unit in a field trial, yields lattice data. Lattices can be regular or irregular. Satellite pixel images also give rise to spatial lattice data as pixels are discrete units. No observations can be gathered between two pixels. Finally, if the set of locations is itself random, the data are termed a **point pattern**. For geostatistical and lattice data the locations at which data are collected are not considered random, even if they are randomly placed. Whether data are collected on a systematic grid or by random sampling has no bearing on the nature of the spatial data type. With point patterns we consider the locations at which events occur as random; for example, when the location of disease-resistant plants is recorded.

Modeling techniques for spatial data are concerned with separating the variability in $Y(\mathbf{s})$ into components due to large-scale trends, smooth-scale, microscale variation and measurement error (see §9.3). The smooth- and microscale variation in spatial data is stochastic variation, hence random. Furthermore, we expect observations at two spatial locations $\mathbf{s}_i$ and $\mathbf{s}_j$ to exhibit more similarity if the locations are close and less similarity if $\mathbf{s}_i$ is far from $\mathbf{s}_j$. This phenomenon is often cited as Tobler's first law of geography: "Everything is related to everything else, but near things are more related than distant things" (Tobler 1970). Spatial data are autocorrelated and the degree of spatial dependency typically decreases with increasing separation. For a set of spatial data $Y(\mathbf{s}_1),\ Y(\mathbf{s}_2), \cdots, Y(\mathbf{s}_n)$ Tobler's law implies that all covariances $\text{Cov}[Y(\mathbf{s}_i), Y(\mathbf{s}_j)]$ are potentially nonzero. In contrast to clustered data one cannot identify a group of observations where members of the group are correlated but are independent of members of other groups. This is where the progression of clustering comes in. Let k denote the number of clusters in a data set and n_i the number of observations for the ith cluster. The total size of the data is then

$$n = \sum_{i=1}^{k} n_i.$$

Consider a square field with sixteen equally spaced grid points. Soil samples are collected at the sixteen points and analyzed for soil organic matter (SOM) content (Figure 2.7a). Although the grid points are regularly spaced, the observations are a realization of geostatistical data, not lattice data. Soil samples could have been collected anywhere in the field. If, however, a grid point is considered the centroid of a rectangular area represented by the point, these would be lattice data. Because the data are spatial, every SOM measurement might be correlated with every other measurement. If a cluster is the collection of observations that are potentially correlated, the SOM data comprise a single cluster ($k = 1$) of size $n = 16$.

Panels (b) and (c) of Figure 2.7 show two experimental design choices to compare four treatments. A completely randomized design with four replications is arranged in panel (b) and a completely randomized design with two replications and two subsamples per experimental unit in panel (c). The data in panel (b) represent uncorrelated observations or $k = 16$ clusters of size $n_i = 1$ each. The subsampling design is a clustered design in the narrow sense, because it involves two hierarchical random processes. Treatments are assigned completely at random to the experimental units such that each treatment is replicated twice. Then two subsamples are selected from each unit. These data comprise $k = 8$ clusters of size $n_i = 2$. There will be correlations among the observations from the same experimental unit because they share the same experimental error term (see §2.4.1).

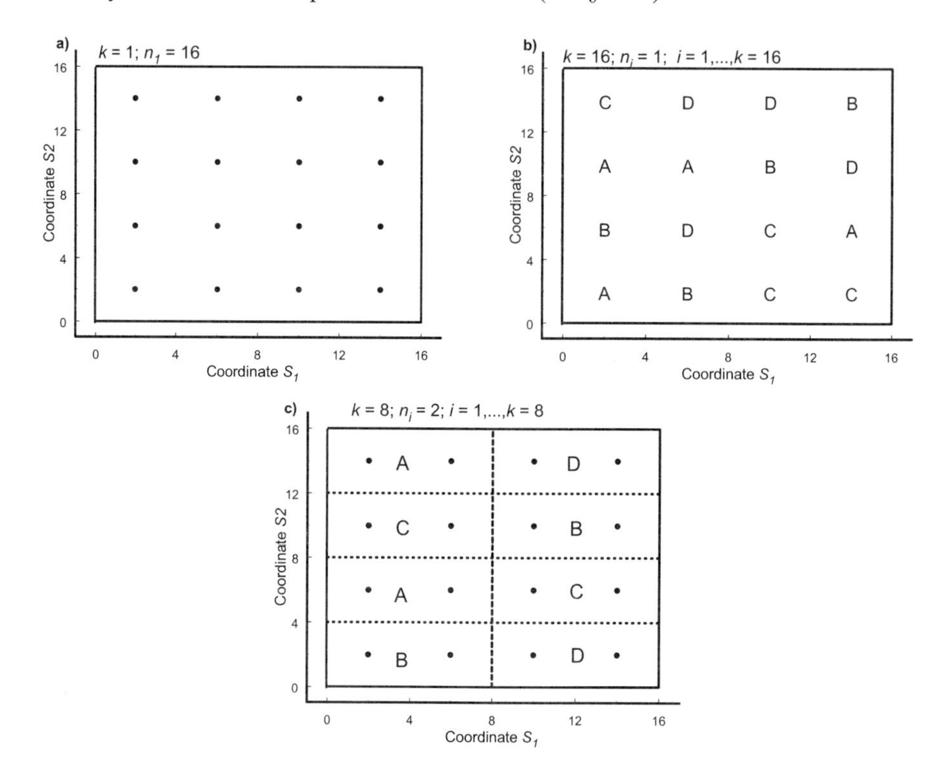

Figure 2.7. Relationship between clustered and spatial data. Panel a) shows the grid points at which a spatial data set for soil organic matter is collected. A completely randomized design (CRD) with four treatments and four replicates is shown in (b), and a CRD with two replicates of four treatments and two subsamples per experimental unit is shown in (c).

Statistical models that can accommodate all three levels of *clustering* are particularly appealing (to us). The mixed models of §7 and §8 have this property. They reduce to standard models for uncorrelated data when each observation represents a *cluster* by itself and to (certain) models for spatial data when the entire data is considered a single cluster. It is reasonable to assume that the SOM data in Figure 2.7a are correlated while the CRD data in Figure 2.7b are independent, because they appeal to different data generating mechanisms; randomization in Figure 2.7b and a stochastic process (a random field) in Figure 2.7a. More and more agronomists are not just faced with a single data structure in a given experiment. Some responses may be continuous, others discrete. Some data have a design context, other a spatial context. Some data are longitudinal, some are cross-sectional. The organization of the remainder of this book by response type and level of clustering is shown in Figure 2.8.

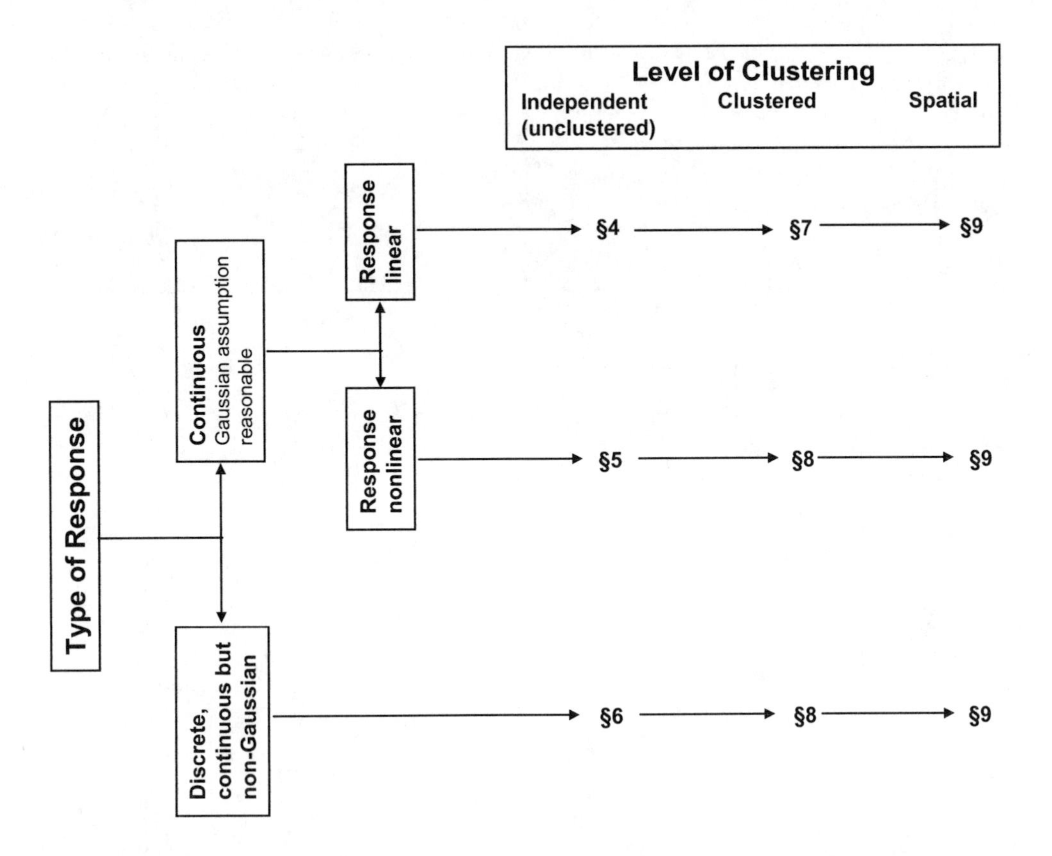

Figure 2.8. Organization of chapters by response type and level of clustering in data.

Chapter 3

Linear Algebra Tools

"The bottom line for mathematicians is that the architecture has to be right. In all the mathematics that I did, the essential point was to find the right architecture. It's like building a bridge. Once the main lines of the structure are right, then the details miraculously fit. The problem is the overall design." Freeman Dyson. In Interview with Donald J. Albers ("Freeman Dyson: Mathematician, Physicist, and Writer"). The College Mathematics Journal, *vol. 25, no. 1, January 1994.*

"It is easier to square the circle than to get round a mathematician." Augustus De Morgan. In Eves, H., Mathematical Circles, *Boston: Prindle, Weber and Schmidt, 1969.*

3.1 Introduction

The discussion of the various statistical models in the following chapters requires linear algebra tools. The reader familiar with the basic concepts and operations such as matrix addition, subtraction, multiplication, transposition, inversion and the expectation and variance of a random vector may skip this chapter. For the reader unfamiliar with these concepts or in need of a refresher, this chapter provides the necessary background and tools. We have compiled the most important rules and results needed to follow the notation and mathematical operations in subsequent chapters. Texts by Graybill (1969), Rao and Mitra (1971), Searle (1982), Healy (1986), Magnus (1988), and others provide many more details about matrix algebra useful in statistics. Specific techniques such as Taylor series expansions of vector-valued functions are introduced as needed.

Without using matrices and vectors, mathematical expressions in statistical model inference quickly become unwieldy. Consider a two-way $(a \times b)$ factorial treatment structure in a randomized complete block design with r blocks. The experimental error (residual) sum of squares in terms of scalar quantities is obtained as

$$SSR = \sum_{i=1}^{a}\sum_{j=1}^{b}\sum_{k=1}^{r}\left(y_{ijk} - \overline{y}_{i..} - \overline{y}_{.j.} - \overline{y}_{..k} + \overline{y}_{ij.} + \overline{y}_{i.k} + \overline{y}_{.jk} - \overline{y}_{...}\right)^2,$$

where y_{ijk} denotes an observation for level i of factor A, level j of factor B in block k. $\overline{y}_{i..}$ is the sample mean of all observations for level i of A and so forth. If only one treatment factor is involved, the experimental error sum of squares fomula becomes

$$SSR = \sum_{i=1}^{t}\sum_{j=1}^{r}\left(y_{ij} - \overline{y}_{i.} - \overline{y}_{.j} + \overline{y}_{..}\right)^2.$$

Using matrices and vectors, the residual sum of squares can be written in either case as $SSR = \mathbf{y}'(\mathbf{I} - \mathbf{H})\mathbf{y}$ for a properly defined vector $\mathbf{y}$ and matrix $\mathbf{H}$.

Consider the following three linear regression models:

$$\begin{aligned} Y_i &= \beta_0 + \beta_1 x_i + e_i \\ Y_i &= \beta_1 x_i + e_i \\ Y_i &= \beta_0 + \beta_1 x_i + \beta_2 x_i^2 + e_i, \end{aligned}$$

a simple linear regression, a straight line regression through the origin, and a quadratic polynomial. If the least squares estimates are expressed in terms of scalar quantities, we get for the simple linear regression model the formulas

$$\widehat{\beta}_0 = \overline{y} - \widehat{\beta}_1\overline{x} \qquad \widehat{\beta}_1 = \left(\sum_{i=1}^{n}(y_i - \overline{y})(x_i - \overline{x})\right) \Big/ \sum_{i=1}^{n}(x_i - \overline{x})^2,$$

in the case of the straight line through the origin,

$$\widehat{\beta}_1 = \left(\sum_{i=1}^{n} y_i x_i\right) \Big/ \left(\sum_{i=1}^{n} x_i^2\right),$$

and

$$\widehat{\beta}_0 = \overline{y} - \widehat{\beta}_1\overline{x} - \widehat{\beta}_2\overline{z}$$

$$\widehat{\beta}_1 = \frac{\left\{\sum_{i=1}^{n}(z_i - \overline{z})y_i\right\}\sum_{i=1}^{n}(x_i - \overline{x})(z_i - \overline{z}) - \left\{\sum_{i=1}^{n}(x_i - \overline{x})y_i\right\}\sum_{i=1}^{n}(z_i - \overline{z})^2}{\left\{\sum_{i=1}^{n}(x_i - \overline{x})(z_i - \overline{z})\right\}^2 - \sum_{i=1}^{n}(x_i - \overline{x})^2\sum_{i=1}^{n}(z_i - \overline{z})^2}$$

for the quadratic polynomial ($z_i = x_i^2$). By properly defining matrices **X**, **Y**, **e**, and $\boldsymbol{\beta}$, we can write either of these models as $\mathbf{Y} = \mathbf{X}\boldsymbol{\beta} + \mathbf{e}$ and the least squares estimates are

$$\widehat{\boldsymbol{\beta}} = (\mathbf{X}'\mathbf{X})^{-1}\mathbf{X}'\mathbf{y}.$$

Matrix algebra allows us to efficiently develop and discuss theory and methods of statistical models.

3.2 Matrices and Vectors

Box 3.1 Matrices and Vectors

- **A matrix is a two-dimensional array of real numbers.**
- **The size of a matrix (its order) is specified as (rows $\times$ columns).**
- **A vector is a one-dimensional array of real numbers. Vectors are special cases of matrices with either a single row (row vector) or a single column (column vector).**
- **Matrices are denoted by uppercase bold-faced letters, vectors with lowercase bold-faced letters. Vectors of random variables are an exception. The random vectors are denoted by boldface uppercase letters and the realized values by boldface lowercase letters.**

In this text matrices are rectangular arrays of real numbers (we do not consider complex numbers). The size of a matrix is called its **order** and written as (row $\times$ column). A (3×4) matrix has three rows and four columns. When referring to individual elements of a matrix, a double subscript denotes the row and column of the matrix in which the element is located. For example, if **A** is a $(n \times k)$ matrix, a_{35} is the element positioned in the third row, fifth column. We sometimes write $\mathbf{A} = [a_{ij}]$ to show that the individual elements of **A** are the a_{ij}. If it is necessary to explicitly identify the order of **A**, a subscript is used, for example, $\mathbf{A}_{(n \times k)}$.

Matrices (and vectors) are distinguished from scalars with bold-face lettering. Matrices with a single row or column dimension are called **vectors**. A vector is usually denoted by a lowercase boldface letter, e.g., **z**, **y**. If uppercase lettering such as **Z** or **Y** is used for vectors, it is implied that the elements of the vector are **random** variables. This can cause a little confusion. Is $\mathbf{X}$ a vector of random variables or a matrix of constants? To reduce confusion as much as possible the following rules (and exceptions) are adopted:

- $\mathbf{X}$ always refers to a matrix of constants, never to a random vector.
- $\mathbf{Z}$ also refers to a matrix of constants, except when we consider spatial data, where the vector $\mathbf{Z}$ has a special meaning in terms of $\mathbf{Z}(\mathbf{s})$. If $\mathbf{Z}$ appears by itself without the argument $(\mathbf{s})$, it denotes a matrix of constants.
- An exception from the uppercase/lowercase notation is made for the vector $\mathbf{b}$, which is a vector of random variables in mixed models (§7).
- If $\mathbf{Y}$ denotes a vector of random variables, then $\mathbf{y}$ denotes its realized values. $\mathbf{y}$ is thus a vector of constants.
- Boldface Greek lettering is used to denote vectors of parameters, e.g., $\boldsymbol{\theta}$ or $\boldsymbol{\beta}$. When placing a caret ($\widehat{}$) or tilde ($\sim$) over the Greek letter, we refer to an estimator or estimate of the parameter vector. An **estimator** is a random variable; for example, $(\mathbf{X}'\mathbf{X})^{-1}\mathbf{X}'\mathbf{Y}$ is random since it depends on the random vector $\mathbf{Y}$. An **estimate** is a constant; the estimate $(\mathbf{X}'\mathbf{X})^{-1}\mathbf{X}'\mathbf{y}$ is the realized value of the estimator $(\mathbf{X}'\mathbf{X})^{-1}\mathbf{X}'\mathbf{Y}$.

An $(n \times 1)$ vector is also called a **column vector**, $(1 \times k)$ vectors are referred to as **row vectors**. By convention all vectors in this text are column vectors unless explicitly stated otherwise. Any matrix can be partitioned into a series of column vectors. $\mathbf{A}_{(n\times k)}$ can be thought of as the horizontal concatenation of k $(n \times 1)$ column vectors: $\mathbf{A} = [\mathbf{a}_1, \mathbf{a}_2, ..., \mathbf{a}_k]$. For example,

$$\mathbf{A}_{(5\times 4)} = \begin{bmatrix} 1 & 2.3 & 0 & 1 \\ 1 & 9.0 & 1 & 0 \\ 1 & 3.1 & 0 & 1 \\ 1 & 4.9 & 1 & 0 \\ 1 & 3.2 & 1 & 0 \end{bmatrix} \qquad [3.1]$$

consists of 4 column vectors

$$\mathbf{a}_1 = \begin{bmatrix} 1 \\ 1 \\ 1 \\ 1 \\ 1 \end{bmatrix}, \mathbf{a}_2 = \begin{bmatrix} 2.3 \\ 9.0 \\ 3.1 \\ 4.9 \\ 3.2 \end{bmatrix}, \mathbf{a}_3 = \begin{bmatrix} 0 \\ 1 \\ 0 \\ 1 \\ 1 \end{bmatrix}, \mathbf{a}_4 = \begin{bmatrix} 1 \\ 0 \\ 1 \\ 0 \\ 0 \end{bmatrix}.$$

$\mathbf{A}$ can also be viewed as a vertical concatenation of row vectors: $\mathbf{A} = \begin{bmatrix} \boldsymbol{\alpha}_1' \\ \boldsymbol{\alpha}_2' \\ \boldsymbol{\alpha}_3' \\ \boldsymbol{\alpha}_4' \\ \boldsymbol{\alpha}_5' \end{bmatrix}$, where $\boldsymbol{\alpha}_1' = [1, 2.3, 0, 1]$.

We now define some special matrices encountered frequently throughout the text.

- $\mathbf{1}_n$: the **unit** vector; an $(n \times 1)$ vector whose elements are 1.

- $\mathbf{J}_n$: the **unit** matrix of size n; an $(n \times n)$ matrix consisting of ones everywhere, e.g.,

$$\mathbf{J}_5 = [\mathbf{1}_5, \mathbf{1}_5, \mathbf{1}_5, \mathbf{1}_5, \mathbf{1}_5] = \begin{bmatrix} 1 & 1 & 1 & 1 & 1 \\ 1 & 1 & 1 & 1 & 1 \\ 1 & 1 & 1 & 1 & 1 \\ 1 & 1 & 1 & 1 & 1 \\ 1 & 1 & 1 & 1 & 1 \end{bmatrix}.$$

- $\mathbf{I}_n$: the **identity** matrix of size n; an $(n \times n)$ matrix with ones on the diagonal and zeros elsewhere,

$$\mathbf{I}_5 = \begin{bmatrix} 1 & 0 & 0 & 0 & 0 \\ 0 & 1 & 0 & 0 & 0 \\ 0 & 0 & 1 & 0 & 0 \\ 0 & 0 & 0 & 1 & 0 \\ 0 & 0 & 0 & 0 & 1 \end{bmatrix}.$$

- $\mathbf{0}_n$: the $(n \times 1)$ zero vector; a column vector consisting of zeros only.
- $\mathbf{0}_{(n \times k)}$: an $(n \times k)$ matrix of zeros.

If the order of these matrices is obvious from the context, the subscripts are omitted.

3.3 Basic Matrix Operations

Box 3.2 Basic Operations

- **Basic matrix operations are addition, subtraction, transposition, and inversion.**
- **The transpose of a matrix $\mathbf{A}_{(n \times k)}$ is obtained by exchanging its rows and columns (row 1 becomes column 1 etc.). It is denoted with a single quote (the transpose of A is denoted A′) and has order $(k \times n)$.**
- **The sum of two matrices is the elementwise sum of its elements. The difference between two matrices is the elementwise difference between its elements.**
- **Two matrices A and B are conformable for addition and subtraction if they have the same order and for multiplication A*B if the number of columns in A equals the number of rows in B.**

Basic matrix operations are addition, subtraction, multiplication, transposition, and inversion. Matrix inversion is discussed in §3.4, since it requires some additional comments. The **transpose** of a matrix **A** is obtained by exchanging its rows and columns. The first row becomes the first column, the second row the second column, and so forth. It is denoted by attaching a single quote (′) to the matrix symbol. Symbolically, $\mathbf{A}' = [a_{ji}]$. The (5×4) matrix **A** in [3.1] has transpose

$$\mathbf{A}' = \begin{bmatrix} 1 & 1 & 1 & 1 & 1 \\ 2.3 & 9.0 & 3.1 & 4.9 & 3.2 \\ 0 & 1 & 0 & 1 & 1 \\ 1 & 0 & 1 & 0 & 0 \end{bmatrix}$$

The transpose of a column vector **a** is a row vector, and vice versa:

$$\mathbf{a}' = \begin{bmatrix} a_1 \\ a_2 \\ \vdots \\ a_n \end{bmatrix}' = [a_1, a_2, ..., a_n].$$

Transposing a matrix changes its order from $(n \times k)$ to $(k \times n)$ and transposing a transpose produces the original matrix, $(\mathbf{A}')' = \mathbf{A}$.

The **sum** of two matrices is the elementwise sum of its elements. The difference between two matrices is the elementwise difference of its elements. For this operation to be successful, the matrices being added or subtracted must be of identical order, and they must be **conformable** for addition. Symbolically,

$$\mathbf{A}_{(n\times k)} + \mathbf{B}_{(n\times k)} = [a_{ij} + b_{ij}], \qquad \mathbf{A}_{(n\times k)} - \mathbf{B}_{(n\times k)} = [a_{ij} - b_{ij}]. \qquad [3.2]$$

Define the following matrices

$$\mathbf{A} = \begin{bmatrix} 1 & 2 & 0 \\ 3 & 1 & -3 \\ 4 & 1 & 2 \end{bmatrix}, \mathbf{B} = \begin{bmatrix} -2 & 1 & 1 \\ 4 & 1 & 8 \\ 1 & 3 & 4 \end{bmatrix}, \mathbf{C} = \begin{bmatrix} 1 & 0 \\ 2 & 3 \\ 2 & 1 \end{bmatrix}, \mathbf{D} = \begin{bmatrix} 3 & 9 \\ 0 & 1 \\ -2 & 3 \end{bmatrix}.$$

A and **B** are both of order (3×3), while **C** and **D** are of order (3×2). The sum

$$\mathbf{A} + \mathbf{B} = \begin{bmatrix} 1-2 & 2+1 & 0+1 \\ 3+4 & 1+1 & -3+8 \\ 4+1 & 1+3 & 2+4 \end{bmatrix} = \begin{bmatrix} -1 & 3 & 1 \\ 7 & 2 & 5 \\ 5 & 4 & 6 \end{bmatrix}$$

is defined as well as the difference

$$\mathbf{C} - \mathbf{D} = \begin{bmatrix} 1-3 & 0-9 \\ 2-0 & 3-1 \\ 2-(-2) & 1-3 \end{bmatrix} \begin{bmatrix} -2 & -9 \\ 2 & 2 \\ 4 & -2 \end{bmatrix}.$$

The operations $\mathbf{A} + \mathbf{C}, \mathbf{A} - \mathbf{C}, \mathbf{B} + \mathbf{D}$, for example, are not possible since the matrices do not conform. Addition and subtraction are commutative, i.e., $\mathbf{A} + \mathbf{B} = \mathbf{B} + \mathbf{A}$ and can be combined with transposition,

$$(\mathbf{A} + \mathbf{B})' = \mathbf{A}' + \mathbf{B}'. \qquad [3.3]$$

Multiplication of two matrices requires a different kind of conformity. The product **A*B** is possible if the number of columns in **A** equals the number of rows in **B**. For example, **A*C** is defined if **A** has order (3×3) and **C** has order (3×2). **C*A** is not possible, however. The order of the matrix product equals the number of rows of the first and the number of columns of the second matrix. The product of an $(n \times k)$ and a $(k \times p)$ matrix is an $(n \times p)$ matrix. The product of a row vector $\mathbf{a}'$ and a column vector **b** is termed the **inner product** of **a** and **b**. Because a $(1 \times k)$ matrix is multiplied with a $(k \times 1)$ matrix, the inner product is a scalar.

$$\mathbf{a}'\mathbf{b} = [a_1, ..., a_k] \begin{bmatrix} b_1 \\ b_2 \\ \vdots \\ b_k \end{bmatrix} = a_1 b_1 + a_2 b_2 + ... + a_k b_k = \sum_{i=1}^{k} a_i b_i. \quad [3.4]$$

The square root of the inner product of a vector with itself is the (Euclidean) **norm** of the vector, denoted $||\mathbf{a}||$:

$$||\mathbf{a}|| = \sqrt{\mathbf{a}'\mathbf{a}} = \sqrt{\sum_{i=1}^{k} a_i^2}. \quad [3.5]$$

The norm of the difference of two vectors **a** and **b** measures their Euclidean distance:

$$||\mathbf{a} - \mathbf{b}|| = \sqrt{(\mathbf{a} - \mathbf{b})'(\mathbf{a} - \mathbf{b})} = \sqrt{\sum_{i=1}^{k} (a_i - b_i)^2}. \quad [3.6]$$

It plays an important role in statistics for spatial data (§9) where **a** and **b** are vectors of spatial coordinates (longitude and latitude).

Multiplication of the matrices $\mathbf{A}_{(n \times k)}$ and $\mathbf{B}_{(k \times p)}$ can be expressed as a series of inner products. Partition $\mathbf{A}'$ as $\mathbf{A}' = [\boldsymbol{\alpha}_1, \boldsymbol{\alpha}_2, ..., \boldsymbol{\alpha}_n]$ and $\mathbf{B}_{(k*p)} = [\mathbf{b}_1, \mathbf{b}_2, ..., \mathbf{b}_p]$. Here, $\boldsymbol{\alpha}_i$ is the ith row of **A** and $\mathbf{b}_i$ is the ith column of **B**. The elements of the matrix product **A*B** can be written as a matrix whose typical elements are the inner products of the rows of **A** with the columns of **B**

$$\mathbf{A}_{(n \times k)} \mathbf{B}_{(k \times p)} = [\boldsymbol{\alpha}_i' \mathbf{b}_j]_{(n \times p)}. \quad [3.7]$$

For example, define two matrices

$$\mathbf{A} = \begin{bmatrix} 1 & 2 & 0 \\ 3 & 1 & -3 \\ 4 & 1 & 2 \end{bmatrix}, \mathbf{B} = \begin{bmatrix} 1 & 0 \\ 2 & 3 \\ 2 & 1 \end{bmatrix}.$$

Then matrix can be partitioned into vectors corresponding to the rows of **A** and the columns of **B** as

$$\boldsymbol{\alpha}_1 = \begin{bmatrix} 1 \\ 2 \\ 0 \end{bmatrix} \quad \boldsymbol{\alpha}_2 = \begin{bmatrix} 3 \\ 1 \\ -3 \end{bmatrix} \quad \boldsymbol{\alpha}_3 = \begin{bmatrix} 4 \\ 1 \\ 2 \end{bmatrix} \quad \mathbf{b}_1 = \begin{bmatrix} 1 \\ 2 \\ 2 \end{bmatrix} \quad \mathbf{b}_2 = \begin{bmatrix} 0 \\ 3 \\ 1 \end{bmatrix}.$$

The product **AB** can now be written as

$$\mathbf{A}_{(n\times k)}\mathbf{B}_{(k\times p)} = [\boldsymbol{\alpha}_i'\mathbf{b}_j] = \begin{bmatrix} \boldsymbol{\alpha}_1'\mathbf{b}_1 & \boldsymbol{\alpha}_1'\mathbf{b}_2 \\ \boldsymbol{\alpha}_2'\mathbf{b}_1 & \boldsymbol{\alpha}_2'\mathbf{b}_2 \\ \boldsymbol{\alpha}_3'\mathbf{b}_1 & \boldsymbol{\alpha}_3'\mathbf{b}_2 \end{bmatrix}$$

$$= \begin{bmatrix} 1*1+2*2+0*2 & 1*0+2*3+0*1 \\ 3*1+1*2-3*2 & 3*0+1*3-3*1 \\ 4*1+1*2+2*2 & 4*0+1*3+2*1 \end{bmatrix} = \begin{bmatrix} 5 & 6 \\ -1 & 0 \\ 10 & 5 \end{bmatrix}$$

Matrix multiplication is not a symmetric operation. **AB** does not yield the same product as **BA**. First, for both products to be defined, **A** and **B** must have the same row and column order (must be **square**). Even then, the outcome of the multiplication might differ. The operation **AB** is called postmultiplication of **A** by **B**, **BA** is called premultiplication of **A** by **B**.

To multiply a matrix by a scalar, simply multiply every element of the matrix by the scalar:

$$c\mathbf{A} = [ca_{ij}]. \quad [3.8]$$

Here are more basic results about matrix multiplication.

$$\begin{aligned} \mathbf{C}(\mathbf{A}+\mathbf{B}) &= \mathbf{CA}+\mathbf{CB} \\ (\mathbf{AB})\mathbf{C} &= \mathbf{A}(\mathbf{BC}) \\ c(\mathbf{A}+\mathbf{B}) &= c\mathbf{A}+c\mathbf{B} \\ (\mathbf{A}+\mathbf{B})(\mathbf{C}+\mathbf{D}) &= \mathbf{AC}+\mathbf{AD}+\mathbf{BC}+\mathbf{BD}. \end{aligned} \quad [3.9]$$

The following results combine matrix multiplication and transposition.

$$\begin{aligned} (\mathbf{AB})' &= \mathbf{B}'\mathbf{A}' \\ (\mathbf{ABC})' &= \mathbf{C}'\mathbf{B}'\mathbf{A}' \\ (c\mathbf{A})' &= c\mathbf{A}' \\ (a\mathbf{A}+b\mathbf{B})' &= a\mathbf{A}'+b\mathbf{B}' \end{aligned} \quad [3.10]$$

Matrices are called **square**, when their row and column dimensions are identical. If, furthermore, $a_{ij} = a_{ji}$ for all $j \neq i$, the matrix is called **symmetric**. Symmetric matrices are encountered frequently in the form of variance-covariance or correlation matrices (see §3.5). A symmetric matrix can be constructed from any matrix **A** with the operations $\mathbf{A}'\mathbf{A}$ or $\mathbf{AA}'$. If all off-diagonal cells of a matrix are zero, i.e. $a_{ij} = 0$ if $j \neq i$, the matrix is called **diagonal** and is obviously symmetric. The identity matrix is a diagonal matrix with 1's on the diagonal. On occasion, diagonal matrices are written as Diag(**a**), where **a** is a vector of diagonal elements. If, for example, $\mathbf{a}' = [1,2,3,4,5]$, then

$$\text{Diag}(\mathbf{a}) = \begin{bmatrix} 1 & 0 & 0 & 0 & 0 \\ 0 & 2 & 0 & 0 & 0 \\ 0 & 0 & 3 & 0 & 0 \\ 0 & 0 & 0 & 4 & 0 \\ 0 & 0 & 0 & 0 & 5 \end{bmatrix}.$$

3.4 Matrix Inversion — Regular and Generalized Inverse

Box 3.3 Matrix Inversion

- **The matrix B for which AB = I is called the inverse of A. It is denoted as $\mathbf{A}^{-1}$.**
- **The rank of a matrix equals the number of its linearly independent columns.**
- **Only square matrices of full rank have an inverse. These are called non-singular matrices. If the inverse of A exists, it is unique.**
- **A singular matrix does not have a regular inverse. However, a generalized inverse $\mathbf{A}^{-}$ for which $\mathbf{AA}^{-}\mathbf{A} = \mathbf{A}$ holds, can be found for any matrix.**

Multiplying a scalar with its reciprocal produces the multiplicative identity, $c \times (1/c) = 1$, provided $c \neq 0$. For matrices an operation such as this simple division is not so straightforward. The multiplicative identity for matrices should be the identity matrix **I**. The matrix **B** which yields $\mathbf{AB} = \mathbf{I}$ is called the **inverse of A**, denoted $\mathbf{A}^{-1}$. Unfortunately, an inverse does not exist for all matrices. Matrices which are not square do not have regular inverses. If $\mathbf{A}^{-1}$ exists, **A** is called **nonsingular** and we have

$$\mathbf{A}^{-1}\mathbf{A} = \mathbf{AA}^{-1} = \mathbf{I}.$$

To see how important it is to have access to the inverse of a matrix, consider the following equation which expresses the vector **y** as a linear function of **c**,

$$\mathbf{y}_{(n\times 1)} = \mathbf{X}_{(n\times k)}\mathbf{c}_{(k\times 1)}.$$

To solve for **c** we need to eliminate the matrix **X** from the right-hand side of the equation. But **X** is not square and thus cannot be inverted. However, $\mathbf{X}'\mathbf{X}$ is a square, symmetric matrix. Premultiply both sides of the equation with $\mathbf{X}'$

$$\mathbf{X}'\mathbf{y} = \mathbf{X}'\mathbf{X}\mathbf{c}. \qquad [3.11]$$

If the inverse of $\mathbf{X}'\mathbf{X}$ exists, premultiply both sides of the equation with it to isolate **c**:

$$(\mathbf{X}'\mathbf{X})^{-1}\mathbf{X}'\mathbf{y} = (\mathbf{X}'\mathbf{X})^{-1}\mathbf{X}'\mathbf{X}\mathbf{c} = \mathbf{I}\mathbf{c} = \mathbf{c}.$$

For the inverse of a matrix to exist, the matrix must be of **full rank**. The rank of a matrix **A**, denoted $r(\mathbf{A})$, is equal to the number of its linearly independent columns. For example,

$$\mathbf{A} = \begin{bmatrix} 1 & 1 & 0 \\ 1 & 1 & 0 \\ 1 & 0 & 1 \\ 1 & 0 & 1 \end{bmatrix}$$

has three columns, $\mathbf{a}_1$, $\mathbf{a}_2$, and $\mathbf{a}_3$. But because $\mathbf{a}_1 = \mathbf{a}_2 + \mathbf{a}_3$, the columns of $\mathbf{A}$ are not linearly independent. In general, let $\mathbf{a}_1$, $\mathbf{a}_2$, ..., $\mathbf{a}_k$ be a set of column vectors. If a set of scalars $c_1, ..., c_k$ can be found, not all of which are zero, such that

$$c_1\mathbf{a}_1 + c_2\mathbf{a}_2 + ... + c_k\mathbf{a}_k = 0, \quad [3.12]$$

the k column vectors are said to be linearly dependent. If the only set of constants for which [3.12] holds is $c_1 = c_2 = ... = c_k = 0$, the vectors are linearly independent. An $(n \times k)$ matrix $\mathbf{X}$ whose rank is less than k is called **rank-deficient** (or **singular**) and $\mathbf{X}'\mathbf{X}$ does not have a regular inverse. A few important results about the rank of a matrix and the rank of matrix products follow:

$$\begin{aligned} r(\mathbf{A}) &= r(\mathbf{A}') = r(\mathbf{A}'\mathbf{A}) = r(\mathbf{A}\mathbf{A}') \\ r(\mathbf{AB}) &\leq \min\{r(\mathbf{A}), r(\mathbf{B})\} \\ r(\mathbf{A}+\mathbf{B}) &\leq r(\mathbf{A}) + r(\mathbf{B}) \end{aligned} \quad [3.13]$$

If $\mathbf{X}$ is rank-deficient, $\mathbf{X}'\mathbf{X}$ will still be symmetric, but its inverse does not exist. How can we then isolate $\mathbf{c}$ in [3.11]? It can be shown that for any matrix $\mathbf{A}$, a matrix $\mathbf{A}^-$ can be found which satisfies

$$\mathbf{A}\mathbf{A}^-\mathbf{A} = \mathbf{A}. \quad [3.14]$$

$\mathbf{A}^-$ is called the generalized inverse or pseudo-inverse of $\mathbf{A}$. The terms g-inverse or conditional inverse are also in use. It can be shown that a solution of $\mathbf{X}'\mathbf{y} = \mathbf{X}'\mathbf{X}\mathbf{c}$ is obtained with a generalized inverse as $\mathbf{c} = (\mathbf{X}'\mathbf{X})^-\mathbf{X}'\mathbf{y}$. Apparently, if $\mathbf{X}$ is not of full rank, all we have to do is substitute the generalized inverse for the regular inverse. Unfortunately, generalized inverses are not unique. The condition [3.14] is satisfied by (infinitely) many matrices. The solution $\mathbf{c}$ is hence not unique either. Assume that $\mathbf{G}$ is a generalized inverse of $\mathbf{X}'\mathbf{X}$. Then any vector

$$\mathbf{c} = \mathbf{G}\mathbf{X}'\mathbf{y} + (\mathbf{G}\mathbf{X}'\mathbf{X} - \mathbf{I})\mathbf{d}$$

is a solution where $\mathbf{d}$ is a conformable but otherwise arbitrary vector (Searle 1971, p. 9; Rao and Mitra 1971, p. 44). In analysis of variance models, $\mathbf{X}$ contains dummy variables coding the treatment and design effects and is typically rank-deficient (see §3.9.1). Statistical packages that use different generalized inverses will return different estimates of these effects. This would pose a considerable problem, but fortunately, several important properties of generalized inverses come to our rescue in statistical inference. If $\mathbf{G}$ is a generalized inverse of $\mathbf{X}'\mathbf{X}$, then

(i) $\mathbf{G}'$ is also a generalized inverse of $\mathbf{X}'\mathbf{X}$;

(ii) $\mathbf{G}'\mathbf{X}$ is a generalized inverse of $\mathbf{X}$;

(iii) $\mathbf{X}\mathbf{G}\mathbf{X}'$ is invariant to the choice of $\mathbf{G}$;

(iv) $\mathbf{X}'\mathbf{X}\mathbf{G}\mathbf{X}' = \mathbf{X}'$ and $\mathbf{X}\mathbf{G}\mathbf{X}'\mathbf{X} = \mathbf{X}$;

(v) $\mathbf{GX'X}$ and $\mathbf{X'XG}$ are symmetric.

Consider the third result, for example. If $\mathbf{XGX'}$ is invariant to the choice of the particular generalized inverse, then $\mathbf{XGX'y}$ is also invariant. But $\mathbf{GX'y} = (\mathbf{X'X})^{-}\mathbf{X'y}$ is the solution derived above and so $\mathbf{Xc}$ is invariant. In statistical models, $\mathbf{c}$ is often the least squares estimate of the model parameters and $\mathbf{X}$ a regressor or design matrix. While the estimates $\mathbf{c}$ will depend on the choice of the generalized inverse, the predicted values will not. The two statistical packages should report the same fitted values.

For any matrix $\mathbf{A}$ there is one unique matrix $\mathbf{B}$ that satisfies the following conditions:

$$\text{(i) } \mathbf{ABA} = \mathbf{A} \qquad \text{(ii) } \mathbf{BAB} = \mathbf{B} \qquad \text{(iii) } (\mathbf{BA})' = \mathbf{BA} \qquad \text{(iv) } (\mathbf{AB})' = \mathbf{AB}. \qquad [3.15]$$

Because of (i), $\mathbf{B}$ is a generalized inverse of $\mathbf{A}$. The matrix satisfying (i) to (iv) of [3.15] is called the **Moore-Penrose** inverse, named after work by Penrose (1955) and Moore (1920). Different classes of generalized inverses have been defined depending on subsets of the four conditions. A matrix $\mathbf{B}$ satisfying (i) is the *standard* generalized inverse. If $\mathbf{B}$ satisfies (i) and (ii), it is termed the **reflexive** generalized inverse according to Urquhart (1968). Special cases of generalized inverses satisfying (i), (ii), (iii) or (i), (ii), (iv) are the **left** and **right** inverses. Let $\mathbf{A}_{(p \times k)}$ be of rank k. Then $\mathbf{A'A}$ is a $(k \times k)$ matrix of rank k by [3.13] and its inverse exists. The matrix $(\mathbf{A'A})^{-1}\mathbf{A'}$ is called the **left inverse** of $\mathbf{A}$, since left multiplication of $\mathbf{A}$ by $(\mathbf{A'A})^{-1}\mathbf{A'}$ produces the identiy matrix:

$$(\mathbf{A'A})^{-1}\mathbf{A'A} = \mathbf{I}.$$

Similarly, if $\mathbf{A}_{(p \times k)}$ is of rank p, then $\mathbf{A'}(\mathbf{AA'})^{-1}$ is its **right inverse**, since $\mathbf{AA'}(\mathbf{AA'})^{-1} = \mathbf{I}$. Left and right inverses are not regular inverses, but generalized inverses. Let $\mathbf{C}_{(k \times p)} = \mathbf{A'}(\mathbf{AA'})^{-1}$ then $\mathbf{AC} = \mathbf{I}$, but the product $\mathbf{CA}$ cannot be computed since the matrices do not conform to multiplication. It is easy to verify, however, that $\mathbf{ACA} = \mathbf{A}$, hence $\mathbf{C}$ is a generalized inverse of $\mathbf{A}$ by [3.14]. Left and right inverses are sometimes called **normalized** generalized inverse matrices (Rohde 1966, Morris and Odell 1968, Urquhart 1968). Searle (1971, pp. 1-3) explains how to construct a generalized inverse that satisfies [3.15]. Given a matrix $\mathbf{A}$ and arbitrary generalized inverses $\mathbf{B}$ of $(\mathbf{AA'})$ and $\mathbf{C}$ of $(\mathbf{A'A})$, the unique Moore-Penrose inverse can be constructed as $\mathbf{A'BACA'}$.

For inverses and all generalized inverses, we note the following results (here $\mathbf{B}$ is rank-deficient and $\mathbf{A}$ is of full rank):

$$\begin{aligned}
(\mathbf{B}^{-})' &= (\mathbf{B}')^{-} \\
(\mathbf{A}^{-1})' &= (\mathbf{A}')^{-1} \\
(\mathbf{A}^{-1})^{-1} &= \mathbf{A} \\
(\mathbf{AB})^{-1} &= \mathbf{B}^{-1}\mathbf{A}^{-1} \\
(\mathbf{B}^{-})^{-} &= \mathbf{B} \\
r(\mathbf{B}^{-}) &= r(\mathbf{B}).
\end{aligned} \qquad [3.16]$$

Finding the inverse of a matrix is simple for certain patterned matrices such as full-rank (2×2) matrices, diagonal matrices, and block-diagonal matrices. The inverse of a full-rank

(2×2) matrix

$$\mathbf{A} = \begin{bmatrix} a_{11} & a_{12} \\ a_{21} & a_{22} \end{bmatrix}$$

is

$$\mathbf{A}^{-1} = \frac{1}{a_{22}a_{11} - a_{12}a_{21}} \begin{bmatrix} a_{22} & -a_{12} \\ -a_{21} & a_{11} \end{bmatrix}. \qquad [3.17]$$

The inverse of a full-rank diagonal matrix $\mathbf{D}_{(k*k)} = \text{Diag}(\mathbf{a})$ is obtained by replacing the diagonal elements by their reciprocals

$$\mathbf{D}^{-1} = \text{Diag}\left(\left[\frac{1}{a_1}, \frac{1}{a_2}, \ldots, \frac{1}{a_k}\right]\right). \qquad [3.18]$$

A **block-diagonal** matrix is akin to a diagonal matrix, where matrices instead of scalars form the diagonal. For example,

$$\mathbf{B} = \begin{bmatrix} \mathbf{B}_1 & \mathbf{0} & \mathbf{0} & \mathbf{0} \\ \mathbf{0} & \mathbf{B}_2 & \mathbf{0} & \mathbf{0} \\ \mathbf{0} & \mathbf{0} & \mathbf{B}_3 & \mathbf{0} \\ \mathbf{0} & \mathbf{0} & \mathbf{0} & \mathbf{B}_4 \end{bmatrix},$$

is a block-diagonal matrix where the matrices $\mathbf{B}_1, \cdots, \mathbf{B}_4$ form the blocks. The inverse of a block-diagonal matrix is obtained by separately inverting the matrices on the diagonal, provided these inverses exist:

$$\mathbf{B}^{-1} = \begin{bmatrix} \mathbf{B}_1^{-1} & \mathbf{0} & \mathbf{0} & \mathbf{0} \\ \mathbf{0} & \mathbf{B}_2^{-1} & \mathbf{0} & \mathbf{0} \\ \mathbf{0} & \mathbf{0} & \mathbf{B}_3^{-1} & \mathbf{0} \\ \mathbf{0} & \mathbf{0} & \mathbf{0} & \mathbf{B}_4^{-1} \end{bmatrix}.$$

3.5 Mean, Variance, and Covariance of Random Vectors

Box 3.4 Moments

- **A random vector is a vector whose elements are random variables.**
- **The expectation of a random vector is the vector of the expected values of its elements.**
- **The variance-covariance matrix of a random vector $\mathbf{Y}_{(n\times 1)}$ is a square, symmetric matrix of order $(n \times n)$. It contains the variances of the elements of Y on the diagonal and their covariances in the off-diagonal cells.**

If the elements of a vector are random variables, it is called a random vector. If Y_i, $(i = 1, \cdots, n)$, denotes the ith element of the random vector $\mathbf{Y}$, then Y_i has mean (expected value) $\mathrm{E}[Y_i]$ and variance $\mathrm{Var}[Y_i] = \mathrm{E}[Y_i^2] - \mathrm{E}[Y_i]^2$. The expected value of a random vector is the vector of the expected values of its elements,

$$\mathrm{E}[\mathbf{Y}] = [\mathrm{E}[Y_i]] = \begin{bmatrix} \mathrm{E}[Y_1] \\ \mathrm{E}[Y_2] \\ \mathrm{E}[Y_3] \\ \vdots \\ \mathrm{E}[Y_n] \end{bmatrix}. \qquad [3.19]$$

In the following, let $\mathbf{A}$, $\mathbf{B}$, and $\mathbf{C}$ be matrices of constants (not containing random variables), $\mathbf{a}$, $\mathbf{b}$, and $\mathbf{c}$ vectors of constants and a, b, and c scalar constants. Then

$$\begin{aligned} &\mathrm{E}[\mathbf{A}] = \mathbf{A} \\ &\mathrm{E}[\mathbf{AYB} + \mathbf{C}] = \mathbf{A}\mathrm{E}[\mathbf{Y}]\mathbf{B} + \mathbf{C} \\ &\mathrm{E}[\mathbf{AY} + \mathbf{c}] = \mathbf{A}\mathrm{E}[\mathbf{Y}] + \mathbf{c}. \end{aligned} \qquad [3.20]$$

If $\mathbf{Y}$ and $\mathbf{U}$ are random vectors, then

$$\begin{aligned} &\mathrm{E}[\mathbf{AY} + \mathbf{BU}] = \mathbf{A}\mathrm{E}[\mathbf{Y}] + \mathbf{B}\mathrm{E}[\mathbf{U}] \\ &\mathrm{E}[a\mathbf{Y} + b\mathbf{U}] = a\mathrm{E}[\mathbf{Y}] + b\mathrm{E}[\mathbf{U}]. \end{aligned} \qquad [3.21]$$

The **covariance matrix** between two random vectors $\mathbf{Y}_{(k\times 1)}$ and $\mathbf{U}_{(p\times 1)}$ is a $(k \times p)$ matrix, its ij^{th} element is the covariance between Y_i and U_j.

$$\mathrm{Cov}[\mathbf{Y}, \mathbf{U}] = [\mathrm{Cov}[Y_i, U_j]].$$

It can be expressed in terms of expectation operations as

$$\mathrm{Cov}[\mathbf{Y}, \mathbf{U}] = \mathrm{E}\big[(\mathbf{Y} - \mathrm{E}[\mathbf{Y}])(\mathbf{U} - \mathrm{E}[\mathbf{U}])'\big] = \mathrm{E}[\mathbf{YU}'] - \mathrm{E}[\mathbf{Y}]\mathrm{E}[\mathbf{U}]'. \qquad [3.22]$$

The covariance of linear combinations of random vectors are evaluated similarly to the scalar case (assuming $\mathbf{W}$ and $\mathbf{V}$ are also random vectors):

$$\begin{aligned} \mathrm{Cov}[\mathbf{AY}, \mathbf{U}] &= \mathbf{A}\mathrm{Cov}[\mathbf{Y}, \mathbf{U}] \\ \mathrm{Cov}[\mathbf{Y}, \mathbf{BU}] &= \mathrm{Cov}[\mathbf{Y}, \mathbf{U}]\mathbf{B}' \\ \mathrm{Cov}[\mathbf{AY}, \mathbf{BU}] &= \mathbf{A}\mathrm{Cov}[\mathbf{Y}, \mathbf{U}]\mathbf{B}' \\ \mathrm{Cov}[a\mathbf{Y} + b\mathbf{U}, c\mathbf{W} + d\mathbf{V}] &= ac\mathrm{Cov}[\mathbf{Y}, \mathbf{W}] + bc\mathrm{Cov}[\mathbf{U}, \mathbf{W}] \\ &\quad + ad\mathrm{Cov}[\mathbf{Y}, \mathbf{V}] + bd\mathrm{Cov}[\mathbf{U}, \mathbf{V}]. \end{aligned} \qquad [3.23]$$

The **variance-covariance matrix** is the covariance of a random vector with itself

$$\begin{aligned} \mathrm{Var}[\mathbf{Y}] = \mathrm{Cov}[\mathbf{Y}, \mathbf{Y}] &= [\mathrm{Cov}[Y_i, Y_j]] \\ &= \mathrm{E}\big[(\mathbf{Y} - \mathrm{E}[\mathbf{Y}])\,\mathrm{E}(\mathbf{Y} - \mathrm{E}[\mathbf{Y}])'\big] \\ &= \mathrm{E}[\mathbf{YY}'] - \mathrm{E}[\mathbf{Y}]\mathrm{E}[\mathbf{Y}]'. \end{aligned}$$

The variance-covariance matrix contains on its diagonal the variances of the observations and the covariances among the random vector's elements in the off-diagonal cells. Variance-covariance matrices are square and symmetric, since $\text{Cov}[Y_i, Y_j] = \text{Cov}[Y_j, Y_i]$:

$$\text{Var}\left[\mathbf{Y}_{(k\times 1)}\right] = \begin{bmatrix} \text{Var}[Y_1] & \text{Cov}[Y_1,Y_2] & \text{Cov}[Y_1,Y_3] & \dots & \text{Cov}[Y_1,Y_k] \\ \text{Cov}[Y_2,Y_1] & \text{Var}[Y_2] & \text{Cov}[Y_2,Y_3] & \dots & \text{Cov}[Y_2,Y_k] \\ \text{Cov}[Y_3,Y_1] & \text{Cov}[Y_3,Y_2] & \text{Var}[Y_3] & \dots & \text{Cov}[Y_3,Y_k] \\ \vdots & \vdots & \vdots & \ddots & \vdots \\ \text{Cov}[Y_k,Y_1] & \text{Cov}[Y_k,Y_2] & \text{Cov}[Y_k,Y_3] & \dots & \text{Var}[Y_k] \end{bmatrix}.$$

To designate the mean and variance of a random vector $\mathbf{Y}$, we use the notation $\mathbf{Y} \sim (\text{E}[\mathbf{Y}], \text{Var}[\mathbf{Y}])$. For example, homoscedastic zero mean errors $\mathbf{e}$ are designated as $\mathbf{e} \sim (\mathbf{0}, \sigma^2\mathbf{I})$.

The elements of a random vector $\mathbf{Y}$ are said to be uncorrelated if the variance-covariance matrix of $\mathbf{Y}$ is a diagonal matrix

$$\text{Var}\left[\mathbf{Y}_{(k\times 1)}\right] = \begin{bmatrix} \text{Var}[Y_1] & 0 & 0 & \dots & 0 \\ 0 & \text{Var}[Y_2] & 0 & \dots & 0 \\ 0 & 0 & \text{Var}[Y_3] & \dots & 0 \\ \vdots & \vdots & \vdots & \ddots & \vdots \\ 0 & 0 & 0 & \dots & \text{Var}[Y_k] \end{bmatrix}.$$

Two random vectors $\mathbf{Y}_1$ and $\mathbf{Y}_2$ are said to be uncorrelated if their variance-covariance matrix is block-diagonal

$$\text{Var}\begin{bmatrix} \mathbf{Y}_1 \\ \mathbf{Y}_2 \end{bmatrix} = \begin{bmatrix} \text{Var}[\mathbf{Y}_1] & \mathbf{0} \\ \mathbf{0} & \text{Var}[\mathbf{Y}_2] \end{bmatrix}.$$

The rules above for working with the covariance of two (or more) random vectors can be readily extended to variance-covariance matrices:

$$\begin{aligned} \text{Var}[\mathbf{AY}] &= \mathbf{A}\text{Var}[\mathbf{Y}]\mathbf{A}' \\ \text{Var}[\mathbf{Y}+\mathbf{a}] &= \text{Var}[\mathbf{Y}] \\ \text{Var}[\mathbf{a}'\mathbf{Y}] &= \mathbf{a}'\text{Var}[\mathbf{Y}]\mathbf{a} \\ \text{Var}[a\mathbf{Y}] &= a^2\text{Var}[\mathbf{Y}] \\ \text{Var}[a\mathbf{Y}+b\mathbf{U}] &= a^2\text{Var}[\mathbf{Y}] + b^2\text{Var}[\mathbf{U}] + 2ab\text{Cov}[\mathbf{Y},\mathbf{U}]. \end{aligned} \qquad [3.24]$$

3.6 The Trace and Expectation of Quadratic Forms

Box 3.5 Trace and Quadratic Forms

- **The trace of a matrix A, denoted tr(A), equals the sum of its diagonal elements.**

- **The expression y′Ay is called a quadratic form. It is a scalar quantity. Sums of squares are quadratic forms. If Y is a random vector and A a matrix of**

constants, the expected value of $\mathbf{Y}'\mathbf{AY}$ is

$$\mathrm{E}[\mathbf{Y}'\mathbf{AY}] = \mathrm{tr}(\mathbf{A}\mathrm{Var}[\mathbf{Y}]) + \mathrm{E}[\mathbf{Y}]'\mathbf{A}\mathrm{E}[\mathbf{Y}].$$

The trace of a matrix $\mathbf{A}$ is the sum of its diagonal elements, denoted $\mathrm{tr}(\mathbf{A})$ and plays an important role in determining the expected value of quadratic forms in random vectors. If $\mathbf{y}$ is an $(n \times 1)$ vector and $\mathbf{A}$ an $(n \times n)$ matrix, $\mathbf{y}'\mathbf{Ay}$ is a **quadratic form** in $\mathbf{y}$. Notice that quadratic forms are scalars. Consider a regression model $\mathbf{Y} = \mathbf{X}\boldsymbol{\beta} + \mathbf{e}$, where the elements of $\mathbf{e}$ have 0 mean and constant variance σ^2, $\mathbf{e} \sim (\mathbf{0}, \sigma^2\mathbf{I})$. The total sum of squares corrected for the mean

$$SST_m = \mathbf{Y}'(\mathbf{I} - \mathbf{11}'/n)\mathbf{Y}$$

is a quadratic form as well as the regression (model) and residual sums of squares:

$$\begin{aligned} SSM_m &= \mathbf{Y}'(\mathbf{H} - \mathbf{11}'/n)\mathbf{Y} \\ SSR &= \mathbf{Y}'(\mathbf{I} - \mathbf{H})\mathbf{Y} \\ \mathbf{H} &= \mathbf{X}(\mathbf{X}'\mathbf{X})^{-1}\mathbf{X}'. \end{aligned}$$

To determine distributional properties of these sums of squares and expected mean squares, we need to know, for example, $\mathrm{E}[SSR]$.

If $\mathbf{Y}$ has mean $\boldsymbol{\mu}$ and variance-covariance matrix $\mathbf{V}$, the quadratic form $\mathbf{Y}'\mathbf{AY}$ has expected value

$$\mathrm{E}[\mathbf{Y}'\mathbf{AY}] = \mathrm{tr}(\mathbf{AV}) + \boldsymbol{\mu}'\mathbf{A}\boldsymbol{\mu}. \qquad [3.25]$$

In evaluating such expectations, several properties of the trace operator $\mathrm{tr}(\bullet)$ are helpful:

$$\begin{aligned} &\text{(i) } \mathrm{tr}(\mathbf{ABC}) = \mathrm{tr}(\mathbf{BCA}) = \mathrm{tr}(\mathbf{CAB}) \\ &\text{(ii) } \mathrm{tr}(\mathbf{A} + \mathbf{B}) = \mathrm{tr}(\mathbf{A}) + \mathrm{tr}(\mathbf{B}) \\ &\text{(iii) } \mathbf{y}'\mathbf{Ay} = \mathrm{tr}(\mathbf{y}'\mathbf{Ay}) \\ &\text{(iv) } \mathrm{tr}(\mathbf{A}) = \mathrm{tr}(\mathbf{A}') \\ &\text{(v) } \mathrm{tr}(c\mathbf{A}) = c\,\mathrm{tr}(\mathbf{A}) \\ &\text{(vi) } \mathrm{tr}(\mathbf{A}) = r(\mathbf{A}) \text{ if } \mathbf{AA} = \mathbf{A} \text{ and } \mathbf{A}' = \mathbf{A}. \end{aligned} \qquad [3.26]$$

Property (i) states that the trace is invariant under cyclic permutations and (ii) that the trace of the sum of two matrices is identical to the sum of their traces. Property (iii) emphasizes that quadratic forms are scalars and any scalar is of course equal to its trace (a scalar is a (1×1) matrix). We can now apply [3.25] with $\mathbf{A} = (\mathbf{I} - \mathbf{H})$, $\mathbf{V} = \sigma^2\mathbf{I}$, $\boldsymbol{\mu} = \mathbf{X}\boldsymbol{\beta}$ to find the expected value of SSR in the linear regression model:

$$\begin{aligned} \mathrm{E}[SSR] &= \mathrm{E}[\mathbf{Y}'(\mathbf{I} - \mathbf{H})\mathbf{Y}] \\ &= \mathrm{tr}\big((\mathbf{I} - \mathbf{H})\sigma^2\mathbf{I}\big) + \boldsymbol{\beta}'\mathbf{X}'(\mathbf{I} - \mathbf{H})\mathbf{X}\boldsymbol{\beta} \\ &= \sigma^2\mathrm{tr}(\mathbf{I} - \mathbf{H}) + \boldsymbol{\beta}'\mathbf{X}'\mathbf{X}\boldsymbol{\beta} - \boldsymbol{\beta}'\mathbf{X}'\mathbf{HX}\boldsymbol{\beta}. \end{aligned}$$

At this point we notice that $(\mathbf{I} - \mathbf{H})$ is symmetric and that $(\mathbf{I} - \mathbf{H})(\mathbf{I} - \mathbf{H}) = (\mathbf{I} - \mathbf{H})$. We can apply rule (vi) and find $\mathrm{tr}(\mathbf{I} - \mathbf{H}) = n - r(\mathbf{H})$. Furthermore,

$$\boldsymbol{\beta}'\mathbf{X}'\mathbf{H}\mathbf{X}\boldsymbol{\beta} = \boldsymbol{\beta}'\mathbf{X}'\mathbf{X}(\mathbf{X}'\mathbf{X})^{-1}\mathbf{X}'\mathbf{X}\boldsymbol{\beta} = \boldsymbol{\beta}'\mathbf{X}'\mathbf{X}\boldsymbol{\beta},$$

which leads to $\mathrm{E}[SSR] = \sigma^2\mathrm{tr}(\mathbf{I} - \mathbf{H}) = \sigma^2(n - r(\mathbf{H}))$. We have established that $S^2 = SSR/(n - r(\mathbf{H}))$ is an unbiased estimator of the residual variability in the classical linear model. Applying the rules about the rank of a matrix in [3.13] we see that the rank of the matrix **H** is identical to the rank of the regressor matrix **X** and so $S^2 = SSR/[n - r(\mathbf{X})]$.

3.7 The Multivariate Gaussian Distribution

Box 3.6 Multivariate Gaussian

- **A random vector Y with mean E[Y] $= \boldsymbol{\mu}$ and variance-covariance matrix Var[Y] $= \Sigma$ is said to be multivariate Gaussian-distributed if its probability density function is given by [3.28]. Notation: Y $\sim G(\boldsymbol{\mu}, \Sigma)$.**

- **The elements of a multivariate Gaussian random vector have (univariate) Gaussian distributions.**

- **Independent random variables (vectors) are also uncorrelated. The reverse is not necessarily true, but holds for Gaussian random variables (vectors). That is, Gaussian random variables (vectors) that are uncorrelated are necessarily stochastically independent.**

The Gaussian (*Normal*) distributions are arguably the most important family of distributions in all of statistics. This does not stem from the fact that many attributes are Gaussian-distributed. Most outcomes are not Gaussian. We thus prefer the label Gaussian distribution over *Normal* distribution. First, statistical methods are usually more simple and mathematically straightforward if data are Gaussian-distributed. Second, the **Central Limit Theorem** (CLT) permits approximating the distribution of averages in random samples by a Gaussian distribution regardless of the distribution from which the sample was drawn, provided sample size is sufficiently large.

A scalar random variable Y is said to be (univariate) Gaussian-distributed with mean μ and variance σ^2 if its probability density function is

$$f(y) = \frac{1}{\sqrt{2\pi\sigma^2}}\exp\left\{ -\frac{1}{2\sigma^2}(y-\mu)^2 \right\}, \qquad [3.27]$$

a fact denoted by $Y \sim G(\mu, \sigma^2)$. The distribution of a random vector $\mathbf{Y}_{(n \times 1)}$ is said to be multivariate Gaussian with mean $\mathrm{E}[\mathbf{Y}] = \boldsymbol{\mu}$ and variance-covariance matrix $\mathrm{Var}[\mathbf{Y}] = \boldsymbol{\Sigma}$ if its density is given by

$$f(\mathbf{y}) = \frac{|\boldsymbol{\Sigma}|^{-1/2}}{(2\pi)^{n/2}} \exp\left\{ -\frac{1}{2}(\mathbf{y}-\boldsymbol{\mu})'\boldsymbol{\Sigma}^{-1}(\mathbf{y}-\boldsymbol{\mu}) \right\}. \qquad [3.28]$$

The term $|\boldsymbol{\Sigma}|$ is called the **determinant** of $\boldsymbol{\Sigma}$. We express the fact that $\mathbf{Y}$ is multivariate Gaussian with the shortcut notation

$$\mathbf{Y} \sim G_n(\boldsymbol{\mu}, \boldsymbol{\Sigma}).$$

The subscript n identifies the dimensionality of the distribution and thereby the order of $\boldsymbol{\mu}$ and $\boldsymbol{\Sigma}$. It can be omitted if the dimension is clear from the context. Some important properties of Gaussian-distributed random variables follow:

(i) $\mathrm{E}[\mathbf{Y}] = \boldsymbol{\mu}$, $\mathrm{Var}[\mathbf{Y}] = \boldsymbol{\Sigma}$.

(ii) If $\mathbf{Y} \sim G_n(\boldsymbol{\mu}, \boldsymbol{\Sigma})$, then $\mathbf{Y} - \boldsymbol{\mu} \sim G_n(\mathbf{0}, \boldsymbol{\Sigma})$.

(iii) $(\mathbf{Y}-\boldsymbol{\mu})'\boldsymbol{\Sigma}^{-1}(\mathbf{Y}-\boldsymbol{\mu}) \sim \chi^2_n$, a Chi-squared variable with n degrees of freedom.

(iv) If $\boldsymbol{\mu} = \mathbf{0}$ and $\boldsymbol{\Sigma} = \sigma^2\mathbf{I}$, then $G(\mathbf{0}, \sigma^2\mathbf{I})$ is called the **standard** multivariate Gaussian distribution.

(v) If $\mathbf{Y} \sim G_n(\boldsymbol{\mu}, \boldsymbol{\Sigma})$, then $\mathbf{U} = \mathbf{A}_{(k \times n)}\mathbf{Y} + \mathbf{b}_{(k \times 1)}$ is Gaussian-distributed with mean $\mathbf{A}\boldsymbol{\mu} + \mathbf{b}$ and variance-covariance matrix $\mathbf{A}\boldsymbol{\Sigma}\mathbf{A}'$ (linear combinations of Gaussian variables are also Gaussian).

A direct relationship between absence of correlations and stochastic independence exists only for Gaussian-distributed random variables. In general, $\mathrm{Cov}[U_1, U_2] = 0$ implies only that U_1 and U_2 are not correlated. It is not a sufficient condition for their stochastic independence which requires that the joint density of U_1 and U_2 factors into a product of marginal densities. In the case of Gaussian-distributed random variables zero covariance *does* imply stochastic independence. Let $\mathbf{Y} \sim G_n(\boldsymbol{\mu}, \boldsymbol{\Sigma})$ and partition $\mathbf{Y}$ into two vectors

$$\mathbf{Y}_{(n \times 1)} = \begin{bmatrix} \mathbf{Y}_{1(s \times 1)} \\ \mathbf{Y}_{2(t \times 1)} \end{bmatrix} \text{ with } n = s + t \text{ and partition further}$$

$$\mathrm{E}[\mathbf{Y}] = \boldsymbol{\mu} = \begin{bmatrix} \mathrm{E}[\mathbf{Y}_1] = \boldsymbol{\mu}_1 \\ \mathrm{E}[\mathbf{Y}_2] = \boldsymbol{\mu}_2 \end{bmatrix}, \mathrm{Var}[\mathbf{Y}] = \boldsymbol{\Sigma} = \begin{bmatrix} \boldsymbol{\Sigma}_{11} & \boldsymbol{\Sigma}_{12} \\ \boldsymbol{\Sigma}_{21} & \boldsymbol{\Sigma}_{22} \end{bmatrix}$$

If $\boldsymbol{\Sigma}_{12(s \times t)} = \mathbf{0}$, then $\mathbf{Y}_1$ and $\mathbf{Y}_2$ are uncorrelated *and* independent. As a corollary note that the elements of $\mathbf{Y}$ are mutually independent if and only if $\boldsymbol{\Sigma}$ is a diagonal matrix. We can learn more from this partitioning of $\mathbf{Y}$. The distribution of any subset of $\mathbf{Y}$ is itself Gaussian-distributed, for example,

$$\begin{aligned} \mathbf{Y}_1 &\sim G_s(\boldsymbol{\mu}_s, \boldsymbol{\Sigma}_{11}) \\ Y_i &\sim G(\mu_i, \sigma_{ii}) \end{aligned} \quad .$$

Another important result pertaining to the multivariate Gaussian distribution tells us that if $\mathbf{Y} = [\mathbf{Y}_1', \mathbf{Y}_2']'$ is Gaussian-distributed, the conditional distribution of $\mathbf{Y}_1$ given $\mathbf{Y}_2 = \mathbf{y}_2$ is also Gaussian. Furthermore, the conditional mean $\mathrm{E}[\mathbf{Y}_1|\mathbf{y}_2]$ is a linear function of $\mathbf{y}_2$. The general result is as follows (see, e.g. Searle 1971, Ch. 2.4). If

$$\begin{bmatrix}\mathbf{Y}_1\\ \mathbf{Y}_2\end{bmatrix} \sim G\left(\begin{bmatrix}\boldsymbol{\mu}_1\\ \boldsymbol{\mu}_2\end{bmatrix}, \begin{bmatrix}\mathbf{V}_{11} & \mathbf{V}_{12}\\ \mathbf{V}_{12}' & \mathbf{V}_{22}\end{bmatrix}\right),$$

then $\mathbf{Y}_1|\mathbf{y}_2 \sim G(\boldsymbol{\mu}_1 + \mathbf{V}_{12}\mathbf{V}_{22}^{-1}(\mathbf{y}_2 - \boldsymbol{\mu}_2), \mathbf{V}_{11} - \mathbf{V}_{12}\mathbf{V}_{22}^{-1}\mathbf{V}_{12}')$. This result is important when predictors for random variables are derived, for example in linear mixed models (§7), and to evaluate the optimality of kriging methods for geostatistical data (§9).

3.8 Matrix and Vector Differentiation

Estimation of the parameters of a statistical model commonly involves the minimization or maximization of a function. Since the parameters are elements of vectors and matrices, we need tools to perform basic calculus with vectors and matrices. It is sufficient to discuss matrix differentiation. Vector differentiation then follows immediately as a special case. Assume that the elements of matrix $\mathbf{A}$ depend on a scalar parameter θ, i.e.,

$$\mathbf{A} = [a_{ij}(\theta)].$$

The derivative of $\mathbf{A}$ with respect to θ is defined as the matrix of derivatives of its elements,

$$\partial\mathbf{A}/\partial\theta = \left[\frac{\partial a_{ij}(\theta)}{\partial\theta}\right]. \qquad [3.29]$$

Many statistical applications involve derivatives of the logarithm of the determinant of a matrix, $\ln(|\mathbf{A}|)$. Although we have not discussed the determinant of a matrix in detail, we note that

$$\partial\ln(|\mathbf{A}|)/\partial\theta = \frac{1}{|\mathbf{A}|}\frac{\partial|\mathbf{A}|}{\partial\theta} = tr\left(\mathbf{A}^{-1}\partial\mathbf{A}/\partial\theta\right). \qquad [3.30]$$

The derivatives of the inverse of $\mathbf{A}$ with respect to θ are calculated as

$$\partial\mathbf{A}^{-1}/\partial\theta = -\mathbf{A}^{-1}(\partial\mathbf{A}/\partial\theta)\mathbf{A}^{-1}. \qquad [3.31]$$

Occasionally the derivative of a function with respect to an entire vector is needed. For example, let

$$f(\mathbf{x}) = x_1 + 3x_2 - 6x_2x_3 + 4x_3^2.$$

The derivative of $f(\mathbf{x})$ with respect to $\mathbf{x}$ is the vector of partial derivatives

$$\partial f(\mathbf{x})/\partial\mathbf{x} = \begin{bmatrix}\partial f(\mathbf{x})/\partial x_1\\ \partial f(\mathbf{x})/\partial x_2\\ \partial f(\mathbf{x})/\partial x_3\end{bmatrix} = \begin{bmatrix}1\\ 3 - 6x_3\\ -6x_2 + 8x_3\end{bmatrix}.$$

Some additional results for vector and matrix calculus follow. Again, $\mathbf{A}$ is a matrix whose elements depend on θ. Then,

(i) $\partial\mathrm{tr}(\mathbf{AB})/\partial\theta = \mathrm{tr}((\partial\mathbf{A}/\partial\theta)\mathbf{B}) + \mathrm{tr}(\mathbf{A}\partial\mathbf{B}/\partial\theta)$

(ii) $\partial\mathbf{x}'\mathbf{A}^{-1}\mathbf{x}/\partial\theta = -\mathbf{x}'\mathbf{A}^{-1}(\partial\mathbf{A}/\partial\theta)\mathbf{A}^{-1}\mathbf{x}$

(iii) $\partial \mathbf{x}'\mathbf{a}/\partial \mathbf{x} = \partial \mathbf{a}'\mathbf{x}/\partial \mathbf{x} = \mathbf{a}$

(iv) $\partial \mathbf{x}'\mathbf{A}\mathbf{x}/\partial \mathbf{x} = 2\mathbf{A}\mathbf{x}$. [3.32]

We can put these rules to the test to find the maximum likelihood estimator of $\boldsymbol{\beta}$ in the Gaussian linear model $\mathbf{Y} = \mathbf{X}\boldsymbol{\beta} + \mathbf{e}$, $\mathbf{e} \sim G_n(\mathbf{0}, \boldsymbol{\Sigma})$. To this end we need to find the solution $\widehat{\boldsymbol{\beta}}$, which maximizes the likelihood function. From [3.28] the density function is given by

$$f(\mathbf{y}, \boldsymbol{\beta}) = \frac{|\boldsymbol{\Sigma}|^{-1/2}}{(2\pi)^{n/2}} \exp\left\{ -\frac{1}{2}(\mathbf{y} - \mathbf{X}\boldsymbol{\beta})'\boldsymbol{\Sigma}^{-1}(\mathbf{y} - \mathbf{X}\boldsymbol{\beta})\right\}.$$

Consider this a function of $\boldsymbol{\beta}$ for a given set of data $\mathbf{y}$, and call it the likelihood function $\mathfrak{L}(\boldsymbol{\beta}; \mathbf{y})$. Maximizing $\mathfrak{L}(\boldsymbol{\beta}; \mathbf{y})$ is equivalent to maximizing its logarithm. The log-likelihood function in the Gaussian linear model becomes

$$\ln\{\mathfrak{L}(\boldsymbol{\beta}; \mathbf{y})\} = l(\boldsymbol{\beta};\mathbf{y}) = -\frac{1}{2}\ln(|\boldsymbol{\Sigma}|) - \frac{n}{2}\ln(2\pi) - \frac{1}{2}(\mathbf{y} - \mathbf{X}\boldsymbol{\beta})'\boldsymbol{\Sigma}^{-1}(\mathbf{y} - \mathbf{X}\boldsymbol{\beta}). \quad [3.33]$$

To find $\widehat{\boldsymbol{\beta}}$, find the solution to $\partial l(\boldsymbol{\beta};\mathbf{y})/\partial\boldsymbol{\beta} = \mathbf{0}$. First we derive the derivative of the log-likelihood:

$$\begin{aligned}
\partial l(\boldsymbol{\beta};\mathbf{y})/\partial\boldsymbol{\beta} &= -\frac{1}{2}\frac{\partial(\mathbf{y} - \mathbf{X}\boldsymbol{\beta})'\boldsymbol{\Sigma}^{-1}(\mathbf{y} - \mathbf{X}\boldsymbol{\beta})}{\partial\boldsymbol{\beta}} \\
&= -\frac{1}{2}\frac{\partial}{\partial\boldsymbol{\beta}}\{\mathbf{y}'\boldsymbol{\Sigma}^{-1}\mathbf{y} - \mathbf{y}'\boldsymbol{\Sigma}^{-1}\mathbf{X}\boldsymbol{\beta} - \boldsymbol{\beta}'\mathbf{X}'\boldsymbol{\Sigma}^{-1}\mathbf{y} + \boldsymbol{\beta}'\mathbf{X}'\boldsymbol{\Sigma}^{-1}\mathbf{X}\boldsymbol{\beta}\} \\
&= -\frac{1}{2}\frac{\partial}{\partial\boldsymbol{\beta}}\{\mathbf{y}'\boldsymbol{\Sigma}^{-1}\mathbf{y} - 2\mathbf{y}'\boldsymbol{\Sigma}^{-1}\mathbf{X}\boldsymbol{\beta} + \boldsymbol{\beta}'\mathbf{X}'\boldsymbol{\Sigma}^{-1}\mathbf{X}\boldsymbol{\beta}\} \\
&= -\frac{1}{2}\{-2\mathbf{X}'\boldsymbol{\Sigma}^{-1}\mathbf{y} + 2\mathbf{X}'\boldsymbol{\Sigma}^{-1}\mathbf{X}\boldsymbol{\beta}\}.
\end{aligned}$$

Setting the derivative to zero yields the maximum likelihood equations for $\boldsymbol{\beta}$:

$$\mathbf{X}'\boldsymbol{\Sigma}^{-1}\mathbf{y} = \mathbf{X}'\boldsymbol{\Sigma}^{-1}\mathbf{X}\boldsymbol{\beta}.$$

If $\mathbf{X}$ is of full rank, the maximum likelihood estimator becomes $\widehat{\boldsymbol{\beta}} = (\mathbf{X}'\boldsymbol{\Sigma}^{-1}\mathbf{X})^{-1}\mathbf{X}'\boldsymbol{\Sigma}^{-1}\mathbf{y}$, which is also the generalized least squares estimator of $\boldsymbol{\beta}$ (see §4.2 and §A4.8.1).

3.9 Using Matrix Algebra to Specify Models

3.9.1 Linear Models

Box 3.7 Linear Model

- **The linear regression or classification model in matrix/vector notation is**

$$\mathbf{Y} = \mathbf{X}\boldsymbol{\beta} + \mathbf{e},$$

where Y is the $(n \times 1)$ vector of observations, X is an $(n \times k)$ matrix of regressor or dummy variables, and e is a vector of random disturbances (errors). The first column of X usually consists of ones, modeling an intercept in regression or a grand mean in classification models. The vector $\boldsymbol{\beta}$ contains the parameters of the mean structure.

The notation for the parameters is arbitrary; we only require that parameters are denoted with Greek letters. In regression models the symbols $\beta_1, \beta_2, \cdots, \beta_k$ are traditionally used, while α, ρ, τ, etc., are common in analysis of variance models. When parameters are combined into a parameter vector, the generic symbols $\boldsymbol{\beta}$ or $\boldsymbol{\theta}$ are entertained. As an example, consider the linear regression model

$$Y_i = \beta_0 + \beta_1 x_{1i} + \beta_2 x_{2i} + \cdots + \beta_k x_{ki} + e_i, i = 1, \cdots, n. \qquad [3.34]$$

Here, x_{ki} denotes the value of the kth covariate (regressor) for the ith observation. The model for n observations is

$$\begin{aligned}
Y_1 &= \beta_0 + \beta_1 x_{11} + \beta_2 x_{21} + \cdots + \beta_k x_{k1} + e_1 \\
Y_2 &= \beta_0 + \beta_1 x_{12} + \beta_2 x_{22} + \cdots + \beta_k x_{k2} + e_2 \\
Y_3 &= \beta_0 + \beta_1 x_{13} + \beta_2 x_{23} + \cdots + \beta_k x_{k3} + e_3 \\
Y_4 &= \beta_0 + \beta_1 x_{14} + \beta_2 x_{24} + \cdots + \beta_k x_{k4} + e_4 \\
&\vdots \\
Y_n &= \beta_0 + \beta_1 x_{1n} + \beta_2 x_{2n} + \cdots + \beta_k x_{kn} + e_n
\end{aligned}$$

To express the linear regression model for this set of data in matrix/vector notation define:

$$\mathbf{Y}_{(n\times 1)} = \begin{bmatrix} Y_1 \\ Y_2 \\ Y_3 \\ \vdots \\ Y_n \end{bmatrix}, \ \boldsymbol{\beta}_{(k+1\times 1)} = \begin{bmatrix} \beta_0 \\ \beta_1 \\ \beta_2 \\ \vdots \\ \beta_k \end{bmatrix}, \ \mathbf{e}_{(n\times 1)} = \begin{bmatrix} e_1 \\ e_2 \\ e_3 \\ \vdots \\ e_n \end{bmatrix},$$

and finally collect the regressors in the matrix

$$\mathbf{X}_{(n\times(k+1))} = \begin{bmatrix} 1 & x_{11} & x_{21} & \dots & x_{k1} \\ 1 & x_{12} & x_{22} & \dots & x_{k2} \\ \vdots & \vdots & \vdots & \ddots & \vdots \\ 1 & x_{1n} & x_{2n} & \dots & x_{kn} \end{bmatrix}.$$

Combining terms, model [3.34] can be written as

$$\mathbf{Y} = \mathbf{X}\boldsymbol{\beta} + \mathbf{e}. \qquad [3.35]$$

Notice that $\mathbf{X}\boldsymbol{\beta}$ is evaluated as

$$\mathbf{X}\boldsymbol{\beta} = \begin{bmatrix} 1 & x_{11} & x_{21} & \dots & x_{k1} \\ 1 & x_{12} & x_{22} & \dots & x_{k2} \\ \vdots & \vdots & \vdots & \ddots & \vdots \\ 1 & x_{1n} & x_{2n} & \dots & x_{kn} \end{bmatrix} \begin{bmatrix} \beta_0 \\ \beta_1 \\ \beta_2 \\ \vdots \\ \beta_k \end{bmatrix} = \begin{bmatrix} \beta_0 + \beta_1 x_{11} + \beta_2 x_{21} + \dots + \beta_k x_{k1} \\ \beta_0 + \beta_1 x_{12} + \beta_2 x_{22} + \dots + \beta_k x_{k2} \\ \vdots \\ \beta_0 + \beta_1 x_{1n} + \beta_2 x_{2n} + \dots + \beta_k x_{kn} \end{bmatrix}.$$

$\mathbf{X}$ is often called the **regressor matrix** of the model.

The specification of the model is not complete without specifying means, variances, and covariances of all random components. The assumption that the model is correct leads naturally to a zero mean assumption for the errors, $\mathrm{E}[\mathbf{e}] = \mathbf{0}$. In the model $\mathbf{Y} = \mathbf{X}\boldsymbol{\beta} + \mathbf{e}$, where $\boldsymbol{\beta}$ contains fixed effects only, $\mathrm{Var}[\mathbf{e}] = \mathrm{Var}[\mathbf{Y}]$. Denote this variance-covariance matrix as $\mathbf{V}$. If the residuals are uncorrelated, $\mathbf{V}$ is a diagonal matrix and can be written as

$$\mathbf{V} = \mathrm{Diag}\left(\sigma_1^2, \sigma_2^2, \dots, \sigma_n^2\right) = \begin{bmatrix} \sigma_1^2 & 0 & \cdots & 0 \\ 0 & \sigma_2^2 & \cdots & 0 \\ \vdots & & \ddots & \vdots \\ 0 & 0 & \cdots & \sigma_n^2 \end{bmatrix}.$$

If the variances of the residuals are in addition homogeneous,

$$\mathbf{V} = \sigma^2 \begin{bmatrix} 1 & 0 & \cdots & 0 \\ 0 & 1 & \cdots & 0 \\ \vdots & & \ddots & \vdots \\ 0 & 0 & \cdots & 1 \end{bmatrix} = \sigma^2 \mathbf{I}_n,$$

where σ^2 is the common variance of the model disturbances. The classical linear model with homoscedastic, uncorrelated errors is finally

$$\mathbf{Y} = \mathbf{X}\boldsymbol{\beta} + \mathbf{e}, \quad \mathbf{e} \sim (\mathbf{0}, \sigma^2\mathbf{I}). \qquad [3.36]$$

The assumption of Gaussian error distribution is deliberately omitted. To estimate the parameter vector $\boldsymbol{\beta}$ by least squares does not require a Gaussian distribution. Only if hypotheses about $\boldsymbol{\beta}$ are tested does a distributional assumption for the errors come into play.

Model [3.34] is a regression model consisting of fixed coefficients only. How would the notation change if the model incorporates effects (classification variables) rather than coefficients? In §1.7.3 it was shown that ANOVA models can be expressed as regression models by constructing appropriate dummy regressors, which associate an observation with elements of the parameter vector. Consider a randomized complete block design with four treatments in three blocks. Written as an effects model (§4.3.1), the linear model for the block design is

$$Y_{ij} = \mu + \rho_j + \tau_i + e_{ij}, \qquad [3.37]$$

where Y_{ij} is the response in the experimental unit receiving treatment i in block j, ρ_j is the effect of block $i = 1, \cdots, 3$, τ_i is the effect of treatment $i = 1, \cdots, 4$, and the e_{ij} are the experimental errors assumed uncorrelated with mean 0 and variance σ^2. Define the column vector of parameters

$$\boldsymbol{\theta} = [\mu, \rho_1, \rho_2, \rho_3, \tau_1, \tau_2, \tau_3, \tau_4]',$$

and the response vector $\mathbf{Y}$, design matrix $\mathbf{P}$, and vector of experimental errors as

$$\mathbf{Y} = \begin{bmatrix} Y_{11} \\ Y_{21} \\ Y_{31} \\ Y_{41} \\ \hline Y_{12} \\ Y_{22} \\ Y_{32} \\ Y_{42} \\ \hline Y_{13} \\ Y_{23} \\ Y_{33} \\ Y_{43} \end{bmatrix}, \quad \mathbf{P} = \begin{bmatrix} 1 & 1 & 0 & 0 & 1 & 0 & 0 & 0 \\ 1 & 1 & 0 & 0 & 0 & 1 & 0 & 0 \\ 1 & 1 & 0 & 0 & 0 & 0 & 1 & 0 \\ 1 & 1 & 0 & 0 & 0 & 0 & 0 & 1 \\ \hline 1 & 0 & 1 & 0 & 1 & 0 & 0 & 0 \\ 1 & 0 & 1 & 0 & 0 & 1 & 0 & 0 \\ 1 & 0 & 1 & 0 & 0 & 0 & 1 & 0 \\ 1 & 0 & 1 & 0 & 0 & 0 & 0 & 1 \\ \hline 1 & 0 & 0 & 1 & 1 & 0 & 0 & 0 \\ 1 & 0 & 0 & 1 & 0 & 1 & 0 & 0 \\ 1 & 0 & 0 & 1 & 0 & 0 & 1 & 0 \\ 1 & 0 & 0 & 1 & 0 & 0 & 0 & 1 \end{bmatrix}, \quad \mathbf{e} = \begin{bmatrix} e_{11} \\ e_{21} \\ e_{31} \\ e_{41} \\ \hline e_{12} \\ e_{22} \\ e_{32} \\ e_{42} \\ \hline e_{13} \\ e_{23} \\ e_{33} \\ e_{43} \end{bmatrix},$$

and model [3.37] in matrix/vector notation becomes

$$\mathbf{Y} = \mathbf{P}\boldsymbol{\theta} + \mathbf{e}. \qquad [3.38]$$

The first four rows of $\mathbf{Y}$, $\mathbf{P}$, and $\mathbf{e}$ correspond to the four observations from block 1, observations five through eight are associated with block 2, and so forth. Comparing [3.35] and [3.38] the models look remarkably similar. The matrix $\mathbf{P}$ in [3.38] contains only dummy variables, however, and is termed a **design matrix** whereas $\mathbf{X}$ in [3.35] contains continuous regressors. The parameter vector $\boldsymbol{\theta}$ in [3.38] contains the block and treatment effects, $\boldsymbol{\beta}$ in [3.35] contains the slopes (gradients) of the response in the covariates. It can easily be verified that the design matrix $\mathbf{P}$ is rank-deficient. The sum of columns two through four is identical to the first column. Also, the sum of the last four columns equals the first column. Generalized inverses must be used to calculate parameter estimates in analysis of variance models. The ramifications with regard to the uniqueness of the estimates have been discussed in §3.4. The regressor matrix $\mathbf{X}$ in regression models is usually of full (column) rank. A rank deficiency in regression models occurs only if one covariate is a linear combination of one or more of the other covariates. For example, if $\mathbf{Y}$ is regressed on the height of a plant in inches and meters simultaneously. While the inverse matrix of $\mathbf{X}'\mathbf{X}$ usually exists in linear regression models, the problem of **near-linear** dependencies among the columns of $\mathbf{X}$ can arise if covariates are closely interrelated. This condition is known as multicollinearity. A popular method for combating collinearity is ridge regression (§4.4.5).

The variance-covariance matrix of $\mathbf{e}$ in [3.38] is a diagonal matrix by virtue of the random assignment of treatments to experimental units within blocks and independent randomizations between blocks. Heterogeneity of the variances across blocks could still exist. If, for example, the homogeneity of experimental units differs between blocks 1 and 2, but not between blocks 2 and 3, Var[$\mathbf{e}$] would become

$$\text{Var}[\mathbf{e}] = \text{Diag}\left(\sigma_1^2, \sigma_1^2, \sigma_1^2, \sigma_1^2, \sigma_2^2, \sigma_2^2, \sigma_2^2, \sigma_2^2, \sigma_2^2, \sigma_2^2, \sigma_2^2, \sigma_2^2\right). \qquad [3.39]$$

3.9.2 Nonlinear Models

A nonlinear mean function is somewhat tricky to express in matrix/vector notation. Building on the basic component equation in §1.7.1, the systematic part for the ith observation can be expressed as $f(x_{1i}, x_{2i}, \cdots, x_{ki}, \theta_1, \theta_2, \cdots, \theta_p)$. In §3.9.1 we included x_{0i} to allow for an intercept term. Nonlinear models do not necessarily possess intercepts and the number of parameters usually does not equal the number of regressors. The x's and θ's can be collected into two vectors to depict the systematic component for the ith observation as a function of covariates and parameters as $f(\mathbf{x}_i, \boldsymbol{\theta})$, where

$$\mathbf{x}_i = [x_{1i}, x_{2i}, \cdots, x_{ki}]', \quad \boldsymbol{\theta} = [\theta_1, \theta_2, \cdots, \theta_p]'.$$

As an example, consider

$$Y_i = x_{1i}^{\alpha}(\beta_1 + \beta_2 x_{2i}),$$

a nonlinear model used by Cole (1975) to model forced expiratory volume of humans (Y_i) as a function of height (x_1) and age (x_2). Put $\mathbf{x}_i = [x_{1i}, x_{2i}]'$ and $\boldsymbol{\theta} = [\alpha, \beta_1, \beta_2]'$ and add a stochastic element to the model:

$$Y_i = f(\mathbf{x}_i, \boldsymbol{\theta}) + e_i. \qquad [3.40]$$

To express model [3.40] for the vector of responses $\mathbf{Y} = [Y_1, Y_2, \cdots, Y_n]'$, replace the function $f()$ with vector notation, and remove the index i from its arguments,

$$\mathbf{Y} = \mathbf{f}(\mathbf{x}, \boldsymbol{\theta}) + \mathbf{e}. \qquad [3.41]$$

[3.41] is somewhat careless notation, since $\mathbf{f}()$ is not a vector function. We think of it as the function $f()$ applied to the arguments $\mathbf{x}_i, \boldsymbol{\theta}$ in turn:

$$\mathbf{Y} = \begin{bmatrix} Y_1 \\ Y_2 \\ Y_3 \\ \vdots \\ Y_n \end{bmatrix}, \quad \mathbf{f}(\mathbf{x}, \boldsymbol{\theta}) = \begin{bmatrix} f(\mathbf{x}_1, \boldsymbol{\theta}) \\ f(\mathbf{x}_2, \boldsymbol{\theta}) \\ f(\mathbf{x}_3, \boldsymbol{\theta}) \\ \vdots \\ f(\mathbf{x}_n, \boldsymbol{\theta}) \end{bmatrix}, \quad \mathbf{e} = \begin{bmatrix} e_1 \\ e_2 \\ e_3 \\ \vdots \\ e_n \end{bmatrix}.$$

3.9.3 Variance-Covariance Matrices and Clustering

In §2.6 a progression of clustering was introduced leading from unclustered data (all data points uncorrelated) to clustered and spatial data depending on the number and size of the clusters in the data set. This progression corresponds to particular structures of the variance-covariance matrix of the observations $\mathbf{Y}$. Consider Figure 2.7 (page 56), which shows spatial data in panel (a), unclustered data in panel (b), and clustered data in panel (c). Denote the responses associated with the panels as $\mathbf{Y}_a$, $\mathbf{Y}_b$, and $\mathbf{Y}_c$, respectively. Possible models for the three cases are

$$\begin{aligned}&\text{a)} \quad Y_{ai} = \beta_0 + \beta_1 s_{1i} + \beta_2 s_{2i} + e_i\\&\text{b)} \quad Y_{bij} = \mu + \tau_j + e_{ij}\\&\text{c)} \quad Y_{cijk} = \mu + \tau_j + e_{ij} + d_{ijk},\end{aligned}$$

where e_{ij} denotes the experimental error for replicate i of treatment j, and d_{ijk} the subsampling error for sample k of replicate i of treatment j. The indices range as follows.

Table 3.1. Indices for the models corresponding to panels (a) to (c) of Figure 2.7

Panel of Figure 2.7	$i=$	$j=$	$k=$	No. of obs.
(a)	$1,\cdots,16$			16
(b)	$1,\cdots,4$	$1,\cdots,4$		16
(c)	$1,\cdots,2$	$1,\cdots,4$	$1,\cdots,2$	16

To complete the specification of the error structure we put $\mathrm{E}[e_i] = \mathrm{E}[e_{ij}] = \mathrm{E}[d_{ijk}] = 0$,

$$\begin{aligned}&\text{(a)} \quad \mathrm{Cov}[e_i, e_k] = \sigma^2\exp\{-3h_{ij}/\alpha\}\\&\text{(b), (c)} \quad \mathrm{Cov}[e_{ij}, e_{kl}] = \begin{cases}\sigma_e^2 & i=k, j=l\\ 0 & \text{otherwise}\end{cases}\\&\text{(d)} \quad \mathrm{Cov}[d_{ijk}, d_{lmn}] = \begin{cases}\sigma_d^2 & i=l, j=m, k=n\\ 0 & \text{otherwise}\end{cases}\\&\qquad \mathrm{Cov}[e_{ij}, d_{ijk}] = 0.\end{aligned}$$

The covariance model for two spatial observations in model (a) is called the exponential model, where h_{ij} is the Euclidean distance between Y_{ai} and Y_{aj}. The parameter α measures the range at which observations are (practically) uncorrelated (see §7.5.2 and §9.2.2 for details). In models (b) and (c), the error structure states that experimental and subsampling errors are uncorrelated with variances σ_e^2 and σ_d^2, respectively. The variance-covariance matrix in each model is a (16×16) matrix. In the case of model a) we have

$$\begin{aligned}\mathrm{Var}[Y_{ai}] &= \mathrm{Cov}[e_i, e_i] = \sigma^2\exp\{-3h_{ii}/\alpha\} = \sigma^2\exp\{-0/\alpha\} = \sigma^2\\ \mathrm{Cov}[Y_{ai}, Y_{aj}] &= \mathrm{Cov}[e_i, e_j] = \sigma^2\exp\{-3h_{ij}/\alpha\}\end{aligned}$$

and the variance-covariance matrix is

$$\mathrm{Var}[\mathbf{Y}_a] = \sigma^2\begin{bmatrix}1 & \exp\{-3h_{12}/\alpha\} & \exp\{-3h_{13}/\alpha\} & \cdots & \exp\{-3h_{1\,16}/\alpha\}\\ \exp\{-3h_{21}/\alpha\} & 1 & \exp\{-3h_{23}/\alpha\} & \cdots & \exp\{-3h_{2\,16}/\alpha\}\\ \exp\{-3h_{31}/\alpha\} & \exp\{-3h_{32}/\alpha\} & 1 & \cdots & \exp\{-3h_{3\,16}/\alpha\}\\ \vdots & \vdots & & \ddots & \vdots\\ \exp\{-3h_{16\,1}/\alpha\} & \exp\{-3h_{16\,2}/\alpha\} & \exp\{-3h_{16\,3}/\alpha\} & \cdots & 1\end{bmatrix}.$$

In model (b) we have

$$\mathrm{Var}[Y_{bij}] = \sigma_e^2,\ \mathrm{Cov}[Y_{bij}, Y_{bkl}] = 0 \text{ (if } i \neq k \text{ or } j \neq l)$$

and the variance-covariance matrix is diagonal:

$$\text{Var}[\mathbf{Y}_b] = \text{Var}\left[\begin{bmatrix} Y_{b11} \\ Y_{b12} \\ Y_{b13} \\ Y_{b14} \\ Y_{b21} \\ Y_{b22} \\ \vdots \\ Y_{b44} \end{bmatrix}\right] = \sigma^2 \begin{bmatrix} 1 & 0 & 0 & 0 & 0 & 0 & \cdots & 0 \\ 0 & 1 & 0 & 0 & 0 & 0 & \cdots & 0 \\ 0 & 0 & 1 & 0 & 0 & 0 & \cdots & 0 \\ 0 & 0 & 0 & 1 & 0 & 0 & \cdots & 0 \\ 0 & 0 & 0 & 0 & 1 & 0 & \cdots & 0 \\ 0 & 0 & 0 & 0 & 0 & 1 & \cdots & 0 \\ \vdots & \vdots & \vdots & \vdots & \vdots & \vdots & \ddots & \vdots \\ 0 & 0 & 0 & 0 & 0 & 0 & \cdots & 1 \end{bmatrix}.$$

The first four entries of $\mathbf{Y}_b$ correspond to the first replicates of treatments 1 to 4 and so forth. To derive the variance-covariance matrix for model (c), we need to separately investigate the covariances among observations from the same experimental unit *and* from different units. For the former we have

$$\text{Cov}[Y_{cijk}, Y_{cijk}] = \text{Var}[e_{ij} + d_{ijk}] = \sigma_e^2 + \sigma_d^2$$

and

$$\begin{aligned} \text{Cov}[Y_{cijk}, Y_{cijl}] &= \text{Cov}[e_{ij} + d_{ijk}, e_{ij} + d_{ijl}] \\ &= \text{Cov}[e_{ij}, e_{ij}] + \text{Cov}[e_{ij}, d_{ijl}] + \text{Cov}[d_{ijk}, e_{ij}] + \text{Cov}[d_{ijk}, d_{ijl}] \\ &= \sigma_e^2 + 0 + 0 + 0. \end{aligned}$$

For observations from different experimental units we have

$$\begin{aligned} \text{Cov}[Y_{cijk}, Y_{clmn}] &= \text{Cov}[e_{ij} + d_{ijk}, e_{lm} + d_{lmn}] \\ &= \text{Cov}[e_{ij}, e_{lm}] + \text{Cov}[e_{ij}, d_{lmn}] + \text{Cov}[d_{ijk}, e_{lm}] + \text{Cov}[d_{ijk}, d_{lmn}] \\ &= 0 + 0 + 0 + 0. \end{aligned}$$

If the elements of $\mathbf{Y}_c$ are arranged by grouping observations from the same cluster together, the variance-covariance matrix can be written as

$$\text{Var}[\mathbf{Y}_c] = \left(\sigma_e^2 + \sigma_d^2\right) \begin{bmatrix} 1 & \phi & 0 & 0 & 0 & 0 & 0 & 0 & 0 \\ \phi & 1 & 0 & 0 & 0 & 0 & 0 & 0 & 0 \\ 0 & 0 & 1 & \phi & 0 & 0 & 0 & 0 & 0 \\ 0 & 0 & \phi & 1 & 0 & 0 & 0 & 0 & 0 \\ 0 & 0 & 0 & 0 & 1 & \phi & 0 & 0 & 0 \\ 0 & 0 & 0 & 0 & \phi & 1 & & \vdots & \vdots \\ 0 & 0 & 0 & 0 & 0 & 0 & \ddots & & \\ 0 & 0 & 0 & 0 & 0 & 0 & \cdots & 1 & \phi \\ 0 & 0 & 0 & 0 & 0 & 0 & \cdots & \phi & 1 \end{bmatrix}$$

where $\phi = \sigma_e^2/(\sigma_e^2 + \sigma_d^2)$. The observations from a single cluster are shaded.

When data are clustered and clusters are uncorrelated but observations within a cluster are correlated, $\text{Var}[\mathbf{Y}]$ has a **block-diagonal** structure, and each block corresponds to a different cluster. If data are unclustered (cluster size 1) and uncorrelated, $\text{Var}[\mathbf{Y}]$ is diagonal. If data consists of a single cluster of size n (spatial data), the variance-covariance matrix consists of a single block.

Chapter 4

The Classical Linear Model: Least Squares and Some Alternatives

"The ususal criticism is that the formulae ... can tell us nothing new, and nothing worth knowing of the biology of the phenomenon. This appears to me to be very ill-founded. In the first place, quantitative expression in place of a vague idea ... is not merely a mild convenience. It may even be a very great convenience, and it may even be indispensable in making certain systematic and biological deductions. But further, it may suggest important ideas to the underlying processes involved." Huxley, J.S., Problems of Relative Growth. *New York: Dial Press, 1932.*

Appendix A on CD-ROM

4.1 Introduction

Contemporary statistical models are the main concern of this text and the reader may rightfully challenge us to define what this means. This is no simple task since we hope that our notion of what is contemporary will be obsolete shortly. In fact, if the title of this text is outdated by the time it goes to press, we would be immensely satisfied. Past experience has led us to believe that most of the statistical models utilized by plant and soil scientists can be cast as classical regression or analysis of variance models to which a student will be introduced in a one- or two-semester sequence of introductory statistics courses. A randomized complete block design with fixed block and treatment effects, for example, is such a standard linear model since it comprises a single error term, a linear mean structure of block and treatment effects, and its parameters are best estimated by ordinary least squares. In matrix/vector notation this classical model can be written as

$$\mathbf{Y} = \mathbf{X}\boldsymbol{\beta} + \mathbf{e}, \qquad [4.1]$$

where $\mathrm{E}[\mathbf{e}] = \mathbf{0}$, $\mathrm{Var}[\mathbf{e}] = \sigma^2\mathbf{I}$, and $\mathbf{X}$ is an $(n \times k)$ matrix.

We do not consider such models as contemporary and draw the line between *classical* and *contemporary* where the linear fixed effects model with a single error term is no longer appropriate because

- such a simple model is no longer a good description of the data generating mechanism

or

- least squares estimators of the model parameters are no longer best suited although the model may be correct.

It is the many ways in which model [4.1] breaks down that we discuss here. If the model does not hold we are led to alternative model formulations; in the second case we are led to alternative methods of parameter estimation. Figure 4.1 is an attempt to roadmap where these breakdowns will take us. Before engaging nonlinear, generalized linear, linear mixed, nonlinear mixed, and spatial models, this chapter is intended to reacquaint the reader with the basic concepts of statistical estimation and inference in the classical linear model and to introduce some methods that have gone largely unnoticed in the plant and soil sciences. Sections §4.2 and §4.3 are largely a review of the analysis of variance and regression methods. The important sum of squares reduction test that will be used frequently throughout this text is discussed in §4.2.3. Standard diagnostics for performance of regression models are discussed in §4.4 along with some remedies for model breakdowns such as ridge regression to combat collinearity of the regressors (§4.4.5). §4.5 concentrates on diagnosing classification models with special emphasis on the homogeneous variance assumption. In the sections that follow we highlight some alternative approaches to statistical estimation that (in our opinion) have not received the attention they deserve, specifically L_1- and M-Estimation (§4.6) and nonparametric regression (§4.7). Mathematical details on these topics which reach beyond the coverage in the main text can be found in Appendix A on the CD-ROM (§A4.8).

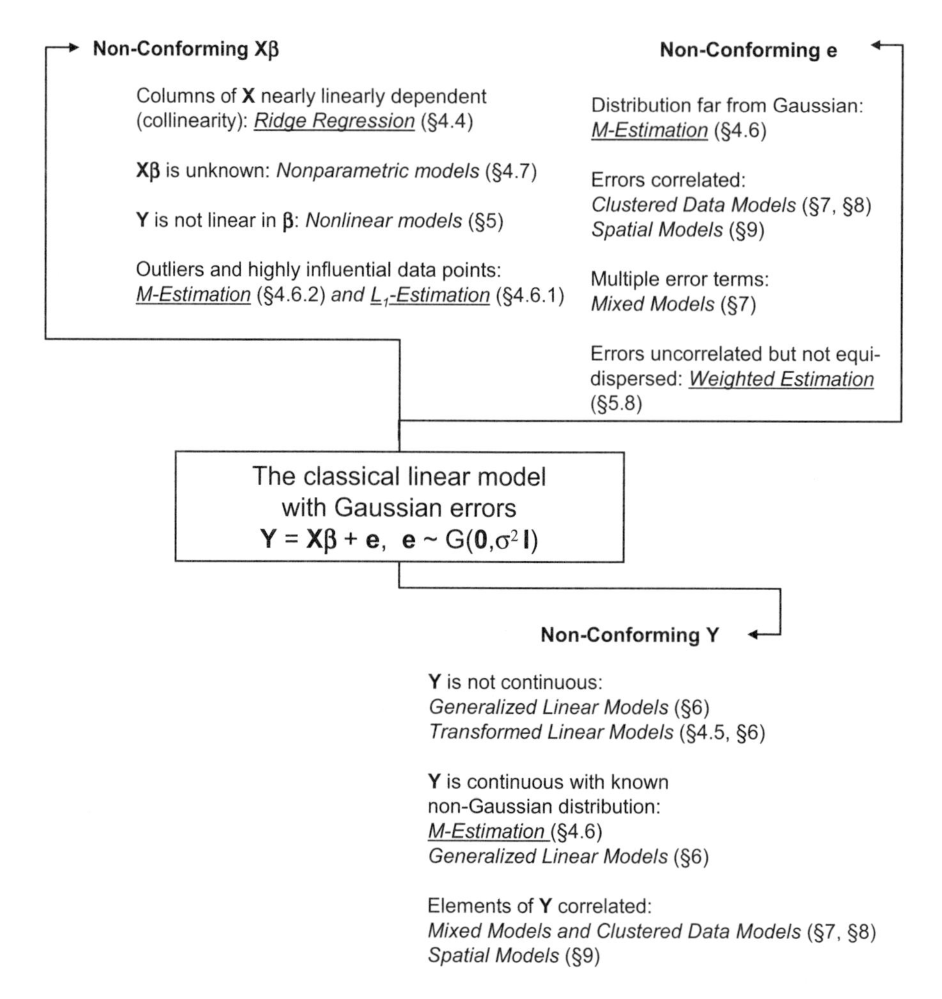

Figure 4.1. The classical linear Gaussian model and breakdowns of its components that lead to alternative methods of estimation (underlined) or alternative statistical models (italicized).

The linear model $\mathbf{Y} = \mathbf{X}\boldsymbol{\beta} + \mathbf{e}$, $\mathbf{e} \sim (\mathbf{0}, \sigma^2\mathbf{I})$ encompasses a plethora of important situations. Standard analysis of variance and simple and multiple linear regression are among them. In this chapter we focus by way of example on a small subset of situations and those model breakdowns that lead to alternative estimation methods. The applications to which we will return frequently are now introduced.

Example 4.1. Lime Application.
Each of two lime types (agricultural lime (AL) and granulated lime (GL)) were applied at each of five rates (0, 1, 2, 4, 8 tons) independently on five replicate plots (Pierce and Warncke 2000). Hence, a total of $5 \times 5 \times 2$ experimental units were involved in the experiment. The pH in soil samples obtained one week after lime application is shown in Table 4.1. If the treatments are assigned completely at random to the experimental units, the observations in Table 4.1 can be thought of as realizations of random devia-

tions around a mean value common to a particular Lime $\times$ Rate combination. We use uppercase lettering when referring to a factor and lowercase lettering when referring to one or more levels of a factor. For example, "Lime Type" designates the factor and "lime types" the two levels.

By virtue of this random assignment of treatments to experimental units, the fifty responses are also independent and if error variability is homogeneous across treatments, the responses are homoscedastic. Under these conditions, model [4.1] applies and can be expressed, for example, as

$$Y_{ijk} = \mu_{ij} + e_{ijk}$$
$$i = 1, 2;\ j = 1, \cdots, 5;\ k = 1, \cdots, 5.$$

Here, μ_{ij} is the mean pH on plots receiving Lime Type i at Rate of Application j. The e_{ijk} are experimental errors associated with the k replicates of the ij^{th} treatment combination.

Table 4.1. pH in 5×2 factorial arrangement with five replicates

	Agricultural Lime					Granulated Lime				
Rep.	**0**	**1**	**2**	**4**	**8**	**0**	**1**	**2**	**4**	**8**
1	5.735	5.845	5.980	6.180	6.415	5.660	5.925	5.800	6.005	6.060
2	5.770	5.880	5.965	6.060	6.475	5.770	4.740	5.760	5.940	5.985
3	5.730	5.865	5.975	6.135	6.495	5.760	5.785	5.805	5.865	5.920
4	5.735	5.875	5.975	6.205	6.455	5.720	5.725	5.805	5.910	5.845
5	5.750	5.865	6.075	6.095	6.410	5.700	5.765	5.670	5.850	6.065

Of interest to the modeler is to uncover the relationship among the 10 treatment means $\mu_{11}, \cdots, \mu_{25}$. Figure 4.2 displays the sample averages of the five replicates for each treatment and a sharp increase of pH with increasing rate of application for agricultural lime and a weaker increase for granulated lime is apparent.

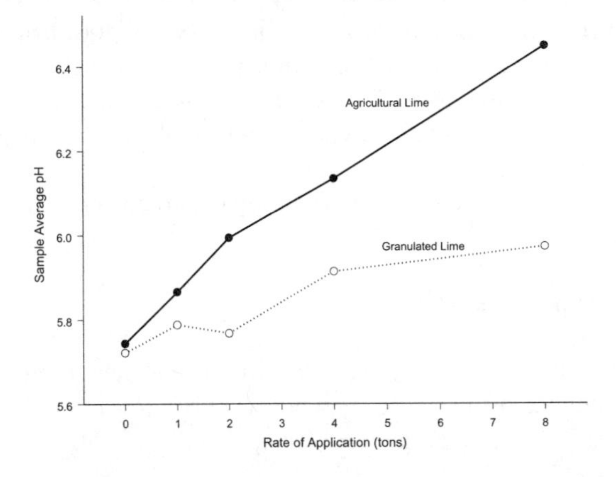

Figure 4.2. Sample average pH in soil samples one week after lime application.

Treatment comparisons can be approached by first partitioning the variability in $\mathbf{Y}$ into a source due to experimental error and a source due to treatments and subsequent testing of linear combinations of the μ_{ij}. $H_0: \mu_{12} - 2\mu_{13} + \mu_{14} = 0$, for example, posits that the average of rate 1 ton and 4 tons is equal to the mean pH when applying 2 tons for AL. These comparisons can be structured efficiently, since the factors Lime Type and Rate of Application are crossed; each level of Lime Type appears in combination with each level of factor Rate. The treatment variability is thus partitioned into main effects and interactions and the linear model $\mathbf{Y} = \mathbf{X}\boldsymbol{\mu} + \mathbf{e}$ is similarly expanded. In §4.2.2 we discuss variability partitioning in general and in §4.3.3 specifically for classification models.

The linear model in Example 4.1 is determined by the experimental design. Randomization ensures independence of the errors, the treatment structure determines which treatment effects are to be included in the model and if design effects (blocking, replication at different locations, or time points) were present they, too, would be included in the mean structure. Analysis focuses on the predetermined questions of interest. For example,

- Do factors Lime Type and Rate of Application interact?
- Are there significant main effects of Lime Type and Rate of Application?
- How can the trend between pH and application rates be modeled and does this trend depend on which type of lime is applied?
- At which rate of application do the lime types differ significantly in pH?

In the next example, developing an appropriate mean structure is the focal point of the analysis. The modeler must apply a series of hypothesis tests and diagnostic procedures to arrive at a final model on which inference and conclusions can be based with confidence.

Example 4.2. Turnip Greens. Draper and Smith (1981, p. 406) list data from a study of Vitamin B_2 content in the leaves of turnip plants (Wakeley, 1949). For each of $n = 27$ plants, the concentration of B_2 vitamin (milligram per gram) was measured as the response of interest. Along with the vitamin content the explanatory variables

$$\begin{aligned} X_1 &= \text{Amount of radiation (Sunlight) during the preceding half-day} \\ &\qquad (g \text{ cal } \times cm^{-2} \times min^{-1}) \\ X_2 &= \text{Soil Moisture tension} \\ X_3 &= \text{Air Temperature } (°F) \end{aligned}$$

were measured (Table 4.2). Only three levels of soil moisture were observed for X_2 $(2.0, 7.0, 47.4)$ with nine plants per level, whereas only a few or no duplicate values are available for the variables Sunlight and Air Temperature.

Table 4.2. Partial Turnip Green Data from Draper and Smith (1981)

Plant #	Vitamin B_2 (y_i)	Sunlight (x_1)	Soil Moisture (x_2)	Air Temp. (x_3)
1	110.4	176	7.0	78
2	102.8	155	7.0	89
3	101.0	273	7.0	89
4	108.4	273	7.0	72
5	100.7	256	7.0	84
6	100.3	280	7.0	87
7	102.0	280	7.0	74
8	93.7	184	7.0	87
⋮	⋮	⋮	⋮	⋮
23	61.0	76	47.4	74
24	53.2	213	47.4	76
25	59.4	213	47.4	69
26	58.7	151	47.4	75
27	58.0	205	47.4	76

Draper, N.R. and Smith, H. (1981) *Applied Regression Analysis. 2nd ed.* Wiley and Sons, New York.

Any one of the explanatory variables does not seem very closely related to the vitamin content of the turnip leaves (Figure 4.3). Running separate linear regressions between Y and each of the three explanatory variables, only Soil Moisture seems to explain a significant amount of Vitamin B_2 variation. Should we conclude based on this finding that the amount of sunlight and the air temperature have no effect on the vitamin content of turnip plant leaves? Is it possible that Air Temperature is an important predictor of vitamin B_2 content if we simultaneously adjust for Soil Moisture? Is a linear trend in Soil Moisture reasonable even if it appears to be significant?

Analysis of the Turnip Greens data does not utilize a linear model suggested by the processes of randomization and experimental control. It is the modeler's task to discover the importance of the explanatory variables on the response and their interaction with each other in building a model for these data. We use methods of multiple linear regression (MLR) to that end. The purposes of an MLR model can be any or all of the following:

- to determine a (small) set of explanatory variables from which the response can be predicted with reasonable confidence and to discover the interrelationships among them;

- to develop a mathematical model that describes how the mean response changes with explanatory variables;

- to predict the outcome of interest for values of the explanatory variables not in the data set.

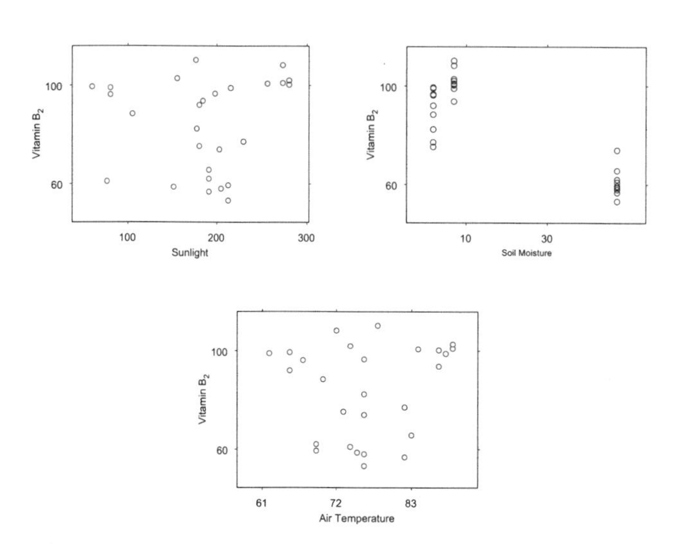

Figure 4.3. Vitamin B_2 in leaves of 27 turnip plants plotted against explanatory variables.

Examples 4.1 and 4.2 are analyzed with standard analysis of variance and multiple linear regression techniques. The parameters of the respective models will be estimated by ordinary least squares or one of its variations (§4.2.1) because of the efficiency of least squares estimates under standard conditions. Least squares estimates are not necessarily the *best* estimates. They are easily distorted in a variety of situations. Strong dependencies among the columns in the regressor matrix **X**, for example, can lead to numerical instabilities producing least squares estimates of inappropriate sign, inappropriate magnitude, and of low precision. Diagnosing and remedying this condition – known as multicollinearity – is discussed in §4.4.4, with additional details in §A4.8.3.

Outlying observations also have high (negative) influence on least squares analysis and a single outlier can substantially distort the analysis. Methods resistant and/or robust against outliers were developed decades ago but are being applied to agronomic data only infrequently. To delete *suspicious* observations from the analysis is a common course of action, but the fact that an observation is *outlying* does not warrant its removal. Outliers can be the most interesting observations in a set of data that should be investigated with extra thoroughness. Outliers can be due to a breakdown of an assumed model; therefore, the model needs to be changed, not the data. One such breakdown concerns the distribution of the model errors. Compared to a Gaussian distribution, *outliers* will be occurring more frequently if the distribution of the errors is heavy-tailed or skewed. Another model breakdown agronomists should be particularly aware of concerns the presence of block $\times$ treatment interactions in randomized complete block designs (RCBD). The standard analysis of an RCBD is not valid if treatment comparisons do not remain constant from block to block. A single observation – often an extreme observation – can induce a significant interaction.

Example 4.3. Dollar Spot Counts. A turfgrass experiment was conducted in a randomized complete block design with fourteen treatments in four blocks. The outcome of interest was the number of dollar spot infection centers on each experimental unit (Table 4.3).

Table 4.3. Dollar spot counts† in a randomized complete block design

	Treatment													
Block	**1**	**2**	**3**	**4**	**5**	**6**	**7**	**8**	**9**	**10**	**11**	**12**	**13**	**14**
1	16	40	29	30	31	42	36	47	97	110	181	63	40	14
2	37	81	19	68	71	99	44	48	81	108	85	19	33	16
3	24	74	38	25	66	40	45	79	42	88	105	21	20	31
4	34	88	27	34	42	40	45	39	92	278	152	46	23	39

†Data kindly provided by David Gilstrap, Department of Crop and Soil Sciences, Michigan State University. Used with permission.

Denote the entry for block (row) i and treatment (column) j as Y_{ij}. Since the data are counts (without natural demoninator, §2.2), one may consider the entry in each cell as a realization of a Poisson random variable with mean $\mathrm{E}[Y_{ij}]$. Poisson random variables with mean greater than 15, say, can be well approximated by Gaussian random variables. The entries in Table 4.3 appear sufficiently large to invoke the approximation. If one analyzes these data with a standard analysis of variance and performs hypothesis tests based on the Gaussian distribution of the model errors, will the observation for treatment 10 in block 4 negatively affect the inference? Are there any other unusual or influential observations? Are there interactions between treatments and blocks? If so, could they be induced by extreme observations? Will the answers to these important questions change if a transformation of the counts is employed?

In §4.5.3 we apply the outlier resistant method of Median Polishing to study the potential block $\times$ treatment interactions and in §4.6.4 we estimate treatment effects with an outlier robust method.

Many agronomic data sets are comparatively small, and estimates of residual variation (mean square errors) rest on only a few degrees of freedom. This is particularly true for designed experiments but analysis in observational studies may be hampered too. Losing additional degrees of freedom by removing *suspicious* observations is then particularly costly since it can reduce the power of the analysis considerably. An outlier robust method that retains all observations but reduces their negative influence on the analysis is then preferred.

Example 4.4. Prediction Efficiency. Mueller et al. (2001) investigate the accuracy and precision of mapping spatially variable soil attributes for site-specific fertility management. The efficiency of geostatistical prediction via kriging (§9.4) was expressed as the ratio of the kriging mean square error relative to a whole-field average prediction which does not take into account the spatial dependency of the data. Data were collected on a

20.4-ha field in Clinton County, Michigan, on a 30×30 meter grid. The field had been in a corn (*Zea mays* L.)-soybean (*Glycine max* L. [Merr.]) rotation for 22 years and a portion of the field had been subirrigated since 1988. Of the attributes for which spatial analyses were performed, we focus on

- pH = soil pH (1:1 soil water mixture)
- P = Bray P-1 extractable soil phosphorus
- Ca = Calcium 1M NH_3OAc extractable
- Mg = Magnesium 1M NH_3OAc extractable
- CEC = Cation exchange capacity
- Lime, Prec = Lime and P fertilizer recommendations according to tristate fertilizer recommendations (Vitosh et al. 1995) for corn with a uniform yield goal of 11.3 MG ha^{-1}.

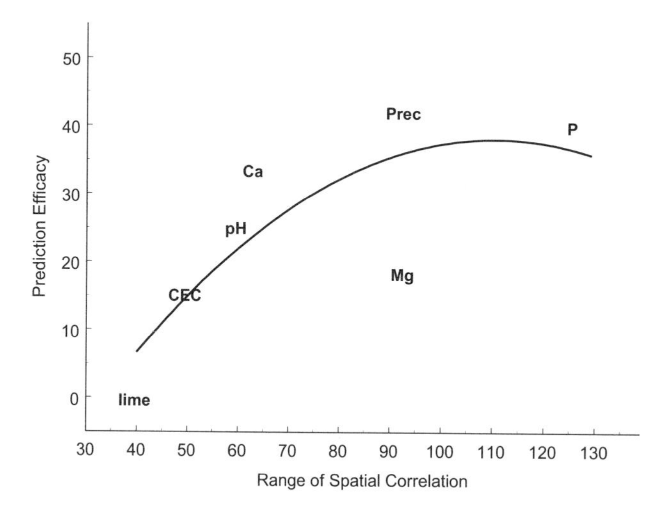

Figure 4.4. Prediction efficiency for kriging of various soil attributes as a function of the range of spatial correlation overlaid with predictions from a quadratic polynomial. Adapted from Figure 1.12 in Mueller (1998). Data kindly provided by Dr. Thomas G. Mueller, Department of Agronomy, University of Kentucky. Used with permission.

The precision with which observations at unobserved spatial locations can be predicted based on geostatistical methods (§9.4) is a function of the spatial autocorrelation among the observations. The stronger the autocorrelation, the greater the precision of spatially explicit methods compared to whole-field average prediction which does not utilize spatial information. The degree of autocorrelation is strongly related to the range of the spatial process. The range is defined as the spatial separation distance beyond which measurements of an attribute can be considered uncorrelated. It is expected that the geostatistical efficiency increases with the range of the attribute. This is clearly seen in Figure 4.4. Finding a model that captures (on average) the dependency between range and efficiency is of primary interest in this application. In contrast to the Turnip Greens study (Example 4.2), tests of hypotheses about the relationship between prediction efficiency and covariates are secondary in this study.

The seven observations plotted in Figure 4.4 suggest a quadratic trend between efficiency and the range of the spatial dependency but the Mg observation clearly stands out. It is considerably off the trend. Deleting this observation will result in *cleaner* least squares estimates of the relationship but also reduces the data set by 14%. In §4.6.3 we examine the impact of the Mg observation on the least squares estimates and fit a model to these data that is robust to outlying observations.

4.2 Least Squares Estimation and Partitioning of Variation

4.2.1 The Principle

Box 4.1 Least Squares

- **Least squares estimation rests on a geometric principle. The parameters are estimated by those values that minimize the sum of squared deviations between observations and the model: $(\mathbf{Y}-\mathbf{X}\boldsymbol{\beta})'(\mathbf{Y}-\mathbf{X}\boldsymbol{\beta})$. It does not require Gaussian errors.**

- **Ordinary least squares (OLS) leads to best linear unbiased estimators in the classical model and minimum variance unbiased estimators if $\mathbf{e} \sim G(0, \sigma^2\mathbf{I})$.**

Recall the classical linear model $\mathbf{Y} = \mathbf{X}\boldsymbol{\beta} + \mathbf{e}$ where the errors are uncorrelated and homoscedastic, $\mathrm{Var}[\mathbf{e}] = \sigma^2\mathbf{I}$. The least squares principle chooses as estimators of the parameters $\boldsymbol{\beta} = [\beta_0, \beta_1, \cdots, \beta_{k-1}]'$ those values that minimize the sum of the squared residuals

$$S(\boldsymbol{\beta}) = ||\mathbf{e}||^2 = \mathbf{e}'\mathbf{e} = (\mathbf{Y}-\mathbf{X}\boldsymbol{\beta})'(\mathbf{Y}-\mathbf{X}\boldsymbol{\beta}). \qquad [4.2]$$

One approach is to set derivatives of $S(\boldsymbol{\beta})$ with respect to $\boldsymbol{\beta}$ to zero and to solve. This leads to the **normal equations** $\mathbf{X}'\mathbf{Y} = \mathbf{X}'\mathbf{X}\widehat{\boldsymbol{\beta}}$ with solution $\widehat{\boldsymbol{\beta}} = (\mathbf{X}'\mathbf{X})^{-1}\mathbf{X}'\mathbf{Y}$ provided $\mathbf{X}$ is of full rank k. If $\mathbf{X}$ is rank deficient, a generalized inverse is used instead and the estimator becomes $\widehat{\boldsymbol{\beta}} = (\mathbf{X}'\mathbf{X})^{-}\mathbf{X}'\mathbf{Y}$.

The calculus approach disguises the geometric principle behind least squares somewhat. The simple identity

$$\mathbf{Y} = \mathbf{X}\widehat{\boldsymbol{\beta}} + (\mathbf{Y}-\mathbf{X}\widehat{\boldsymbol{\beta}}), \qquad [4.3]$$

that expresses the observations as the sum of a vector $\mathbf{X}\widehat{\boldsymbol{\beta}}$ of fitted values and a residual vector $(\mathbf{Y}-\mathbf{X}\widehat{\boldsymbol{\beta}})$ leads to the following argument. The $(n \times 1)$ vector $\mathbf{Y}$ is a point in an n-dimensional space $\mathbb{R}^n$. If the $(n \times k)$ matrix $\mathbf{X}$ is of full rank, its columns generate a k-

dimensional subspace of $\mathbb{R}^n$. In other words, the mean values $\mathbf{X}\boldsymbol{\beta}$ cannot be points anywhere in $\mathbb{R}^n$, but can only "live" in a subspace thereof. Whatever estimator $\widetilde{\boldsymbol{\beta}}$ we choose, $\mathbf{X}\widetilde{\boldsymbol{\beta}}$ will also be a point in this subspace. So why not choose $\mathbf{X}\widetilde{\boldsymbol{\beta}}$ so that its distance from $\mathbf{Y}$ is minimized by projecting $\mathbf{Y}$ perpendicularly onto the space generated by $\mathbf{X}\boldsymbol{\beta}$ (Figure 4.5)? This requires that

$$\left(\mathbf{Y}-\mathbf{X}\widehat{\boldsymbol{\beta}}\right)'\mathbf{X}\widehat{\boldsymbol{\beta}}=\mathbf{0} \Leftrightarrow \widehat{\boldsymbol{\beta}}'\mathbf{X}'\mathbf{Y}=\widehat{\boldsymbol{\beta}}'\mathbf{X}'\mathbf{X}\widehat{\boldsymbol{\beta}} \Leftrightarrow \mathbf{X}'\mathbf{Y}=\mathbf{X}'\mathbf{X}\widehat{\boldsymbol{\beta}} \qquad [4.4]$$

and the ordinary least squares (OLS) estimate follows.

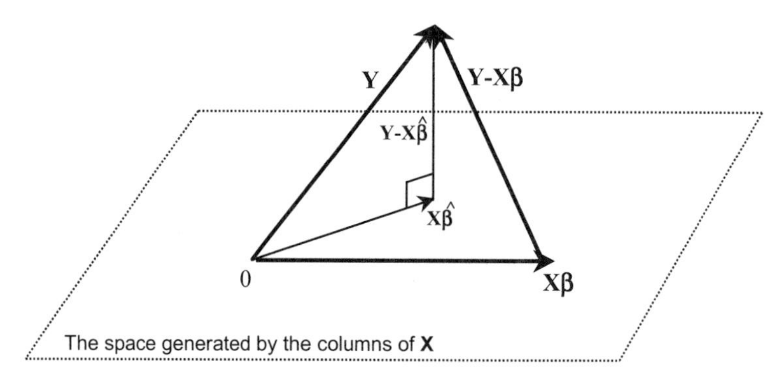

Figure 4.5. The geometry of least squares.

If the classical linear model holds, the least square estimator enjoys certain optimal properties. It is a **best linear unbiased estimator** (BLUE) since $\mathrm{E}[\widehat{\boldsymbol{\beta}}]=\boldsymbol{\beta}$ and no other unbiased estimator that is a linear function of $\mathbf{Y}$ has smaller variability. These appealing features do not require Gaussianity. It is possible, however, that some other, nonlinear estimator of $\boldsymbol{\beta}$ would have greater precision. If the model errors are Gaussian, i.e., $\mathbf{e}\sim G(\mathbf{0},\sigma^2\mathbf{I})$, the ordinary least squares estimator of $\boldsymbol{\beta}$ is a **minimum variance unbiased estimator** (MVUE), extending its optimality beyond those estimators which are linear in $\mathbf{Y}$. We will frequently denote the ordinary least squares estimator as $\widehat{\boldsymbol{\beta}}_{OLS}$, to distinguish it from the **generalized** least squares estimator $\widehat{\boldsymbol{\beta}}_{GLS}$ that arises if $\mathbf{e}\sim(\mathbf{0},\mathbf{V})$, where $\mathbf{V}$ is a general variance-covariance matrix. In this case we minimize $\mathbf{e}'\mathbf{V}^{-1}\mathbf{e}$ and obtain the estimator

$$\widehat{\boldsymbol{\beta}}_{GLS}=\left(\mathbf{X}'\mathbf{V}^{-1}\mathbf{X}\right)^{-1}\mathbf{X}'\mathbf{V}^{-1}\mathbf{Y}. \qquad [4.5]$$

Table 4.4 summarizes some properties of the ordinary and generalized least square estimators. We see that the ordinary least squares estimator remains unbiased, even if the error variance is not $\sigma^2\mathbf{I}$. However, $\mathrm{Var}[\mathbf{a}'\widehat{\boldsymbol{\beta}}_{OLS}]$ is typically larger than $\mathrm{Var}[\mathbf{a}'\widehat{\boldsymbol{\beta}}_{GLS}]$ in this case. The OLS estimator is less efficient. Additional details about the derivation and properties of $\widehat{\boldsymbol{\beta}}_{GLS}$ can be found in §A4.8.1. A third case positioned between ordinary and generalized least squares arises when $\mathbf{V}$ is a **diagonal** matrix and is termed **weighted least squares** estimation (WLSE). It is the appropriate estimation principle if the model errors are heteroscedastic but uncorrelated. If $\mathrm{Var}[\mathbf{e}]=\mathrm{Diag}(\boldsymbol{\sigma})=\mathbf{W}$, where $\boldsymbol{\sigma}$ is a vector containing the variances of the e_i, then the weighted least squares estimator is $\widehat{\boldsymbol{\beta}}_{WLS}=(\mathbf{X}'\mathbf{W}^{-1}\mathbf{X})^{-1}\mathbf{X}'\mathbf{W}^{-1}\mathbf{Y}$. If $\mathbf{V}=\sigma^2\mathbf{I}$, the

GLS estimator reduces to the OLS estimator, if $\mathbf{V} = \text{Diag}(\boldsymbol{\sigma})$ then GLS reduces to WLS, and if $\boldsymbol{\sigma} = \sigma^2\mathbf{1}$, then WLS reduces to OLS.

Table 4.4. Properties of ordinary and generalized least squares estimators

	$\mathbf{e} \sim (\mathbf{0}, \sigma^2\mathbf{I})$	$\mathbf{e} \sim (\mathbf{0}, \mathbf{V})$
$\text{E}\left[\widehat{\boldsymbol{\beta}}_{OLS}\right]$	$\boldsymbol{\beta}$	$\boldsymbol{\beta}$
$\text{E}\left[\widehat{\boldsymbol{\beta}}_{GLS}\right]$	$\boldsymbol{\beta}$	$\boldsymbol{\beta}$
$\text{Var}\left[\widehat{\boldsymbol{\beta}}_{OLS}\right]$	$\sigma^2(\mathbf{X}'\mathbf{X})^{-1}$	$(\mathbf{X}'\mathbf{X})^{-1}\mathbf{X}'\mathbf{V}\mathbf{X}(\mathbf{X}'\mathbf{X})^{-1}$
$\text{Var}\left[\widehat{\boldsymbol{\beta}}_{GLS}\right]$	$\sigma^2(\mathbf{X}'\mathbf{X})^{-1}$	$(\mathbf{X}'\mathbf{V}^{-1}\mathbf{X})^{-1}$
Notes	$\widehat{\boldsymbol{\beta}}_{OLS}$ is BLUE if $\mathbf{e}$ is Gaussian, $\widehat{\boldsymbol{\beta}}_{OLS}$ is MVUE	$\widehat{\boldsymbol{\beta}}_{GLS}$ is BLUE if $\mathbf{e}$ is Gaussian, $\widehat{\boldsymbol{\beta}}_{GLS}$ is MVUE

Because of the linearity of $\widehat{\boldsymbol{\beta}}$ in $\mathbf{Y}$ it is simple to derive the distribution of $\widehat{\boldsymbol{\beta}}$ if the errors of the linear model are Gaussian. In the model with general error variance $\mathbf{V}$, for example,

$$\widehat{\boldsymbol{\beta}}_{GLS} \sim G\left(\boldsymbol{\beta}, \left(\mathbf{X}'\mathbf{V}^{-1}\mathbf{X}\right)^{-1}\right).$$

Since OLS is a special case of GLS with $\mathbf{V} = \sigma^2\mathbf{I}$ we also have

$$\widehat{\boldsymbol{\beta}}_{OLS} \sim G\left(\boldsymbol{\beta}, \sigma^2(\mathbf{X}'\mathbf{X})^{-1}\right).$$

Standard hypothesis tests can be derived based on the Gaussian distribution of the estimator and usually lead to t- or F-tests (§A4.8.2). If the model errors are not Gaussian, the asymptotic distribution of $\widehat{\boldsymbol{\beta}}$ is Gaussian nevertheless. With sufficiently large sample size one can thus proceed as if Gaussianity of $\widehat{\boldsymbol{\beta}}$ holds.

4.2.2 Partitioning Variability through Sums of Squares

Recall that the norm $||\mathbf{a}||$ of a $(p \times 1)$ vector $\mathbf{a}$ is defined as

$$||\mathbf{a}|| = \sqrt{\mathbf{a}'\mathbf{a}} = \sqrt{\sum_{i=1}^{p} a_i^2}$$

and measures the length of the vector. By the orthogonality of the residual vector $\mathbf{Y} - \mathbf{X}\widehat{\boldsymbol{\beta}}$ and the vector of predicted values $\mathbf{X}\widehat{\boldsymbol{\beta}}$ (Figure 4.5), the length of the observed vector $\mathbf{Y}$ is related to the length of the predictions and residuals by the Pythagorean theorem:

$$||\mathbf{Y}||^2 = ||\mathbf{Y} - \mathbf{X}\widehat{\boldsymbol{\beta}}||^2 + ||\mathbf{X}\widehat{\boldsymbol{\beta}}||^2. \qquad [4.6]$$

The three terms in [4.6] correspond to the uncorrected total sum of squares $(SST = ||\mathbf{Y}||^2)$, the residual sum of squares $(SSR = ||\mathbf{Y} - \mathbf{X}\widehat{\boldsymbol{\beta}}||^2)$, and the model sum of squares

$(SSM = ||\mathbf{X}\widehat{\boldsymbol{\beta}}||^2)$. Some straightforward manipulations yield the simpler expressions shown in the analysis of variance table for [4.1]. For example,

$$SSR = \left(\mathbf{Y} - \mathbf{X}\widehat{\boldsymbol{\beta}}\right)'\left(\mathbf{Y} - \mathbf{X}\widehat{\boldsymbol{\beta}}\right) = \left(\mathbf{Y} - \mathbf{X}\widehat{\boldsymbol{\beta}}\right)'\mathbf{Y}$$
$$= \mathbf{Y}'\mathbf{Y} - \widehat{\boldsymbol{\beta}}'\mathbf{X}'\mathbf{Y}. \qquad [4.7]$$

Table 4.5. Analysis of variance table for the standard linear model ($r(\mathbf{X})$ denotes rank of $\mathbf{X}$)

Source	df	SS	MS
Model	$r(\mathbf{X})$	$\widehat{\boldsymbol{\beta}}'\mathbf{X}'\mathbf{Y} = SSM$	$SSM/r(\mathbf{X})$
Residual (Error)	$n - r(\mathbf{X})$	$\mathbf{Y}'\mathbf{Y} - \widehat{\boldsymbol{\beta}}'\mathbf{X}'\mathbf{Y} = SSR$	$SSR/(n - r(\mathbf{X}))$
Uncorrected Total	n	$\mathbf{Y}'\mathbf{Y} = SST$	

When the $\mathbf{X}$ matrix contains an intercept (regression models) or grand mean (classification models), it is common to correct the entries of the analysis of variance table for the mean with the term $n\overline{Y}^2$ (Table 4.6).

Table 4.6. Analysis of variance table corrected for the mean

Source	df	SS	MS
Model	$r(\mathbf{X}) - 1$	$\widehat{\boldsymbol{\beta}}'\mathbf{X}'\mathbf{Y} - n\overline{Y}^2 = SSM_m$	$SSM_m/(r(\mathbf{X}) - 1)$
Residual (Error)	$n - r(\mathbf{X})$	$\mathbf{Y}'\mathbf{Y} - \widehat{\boldsymbol{\beta}}'\mathbf{X}'\mathbf{Y} = SSR$	$SSR/(n - r(\mathbf{X}))$
Total	$n - 1$	$\mathbf{Y}'\mathbf{Y} - n\overline{Y}^2 = SST_m$	

The model sum of squares SSM measures the joint explanatory power of the variables (including the intercept). If the explanatory variables in $\mathbf{X}$ were unrelated to the response, we would use $\overline{Y}$ to predict the mean response, which is the ordinary least squares estimate in an intercept-only model. SSM_m, which measures variability explained beyond an intercept-only model, is thus the appropriate statistic for evaluating the predictive value of the explanatory variables and we use ANOVA tables in which the pertinent terms are corrected for the mean. The coefficient of determination (R^2) is defined as

$$R^2 = \frac{SSM_m}{SST_m} = \frac{\widehat{\boldsymbol{\beta}}'\mathbf{X}'\mathbf{Y} - n\overline{Y}^2}{\mathbf{Y}'\mathbf{Y} - n\overline{Y}^2}. \qquad [4.8]$$

4.2.3 Sequential and Partial Sums of Squares and the Sum of Squares Reduction Test

To assess the contribution of **individual** columns (variables) in $\mathbf{X}$, SSM_m is further decomposed into $r(\mathbf{X}) - 1$ single degree of freedom components. Consider the regression model

$$Y_i = \beta_0 + \beta_1 x_{1i} + \beta_2 x_{2i} + \cdots + \beta_{k-1} x_{k-1i} + e_i.$$

To underline that SSM is the joint contribution of β_0 through β_{k-1}, we use the expression

$$SSM = SS(\beta_0, \beta_1, \cdots, \beta_{k-1}).$$

Similarly, $SSM_m = SS(\beta_1, \cdots, \beta_{k-1}|\beta_0)$ is the joint contribution of $\beta_1, \cdots, \beta_{k-1}$ after adjustment for the intercept (correction for the mean, Table 4.6). A partitioning of SSM_m into **sequential** one degree of freedom sums of squares is

$$\begin{aligned} SSM_m &= SS(\beta_1|\beta_0) \\ &+ SS(\beta_2|\beta_0, \beta_1) \\ &+ SS(\beta_3|\beta_0, \beta_1, \beta_2) \\ &+ \cdots \\ &+ SS(\beta_{k-1}|\beta_0, \cdots, \beta_{k-2}). \end{aligned} \qquad [4.9]$$

$SS(\beta_2|\beta_0, \beta_1)$, for example, is the sum of squares contribution accounted for by adding the regressor X_2 to a model already containing an intercept and X_1. The test statistic $SS(\beta_2|\beta_0, \beta_1)/MSR$ can be used to test whether the addition of X_2 to a model containing X_1 and an intercept provides significant improvement of fit and hence is a gauge for the explanatory value of the regressor X_2. If the model errors are Gaussian, $SS(\beta_2|\beta_0, \beta_1)/MSR$ has an F distribution with one numerator and $n - r(\mathbf{X})$ denominator degrees of freedom. Since the sum of squares $SS(\beta_2|\beta_0, \beta_1)$ has a single degree of freedom, we can also express this test statistic as

$$F_{obs} = \frac{MS(\beta_2|\beta_0, \beta_1)}{MSR},$$

where $MS(\beta_2|\beta_0, \beta_1) = SS(\beta_2|\beta_0, \beta_1)/1$ is the sequential **mean** square. This is a special case of a **sum of squares reduction test**. Imagine we wish to test whether adding regressors X_2 and X_3 simultaneously to a model containing X_1 and an intercept improves the model fit. The change in the model sum of squares is calculated as

$$SS(\beta_2, \beta_3|\beta_0, \beta_1) = SS(\beta_0, \beta_1, \beta_2, \beta_3) - SS(\beta_0, \beta_1).$$

To obtain $SS(\beta_2, \beta_3|\beta_0, \beta_1)$ we fit a model containing four regressors and obtain its model sum of squares. Call it the **full model**. Then a **reduced model** is fit, containing only an intercept and X_1. The difference of the model sums of squares of the two models is the contribution of adding X_2 and X_3 simultaneously. The **mean** square associated with the addition of the two regressors is $MS(\beta_2, \beta_3|\beta_0, \beta_1) = SS(\beta_2, \beta_3|\beta_0, \beta_1)/2$. Since both models have the same (corrected or uncorrected) total sum of squares we can express the test mean square also in terms of a residual sum of squares difference. This leads us to the general version of the sum of squares reduction test.

Consider a **full** model M_f and a **reduced** model M_r where M_r is obtained from M_f by constraining some (or all) of its parameters. Usually the constraints mean setting one or more parameters to zero, but other constraints are possible, for example, $\beta_1 = \beta_2 = 6$. If (i) SSR_f is the residual sum of squares obtained from fitting the full model, (ii) SSR_r is the respective sum of square for the reduced model, and (iii) q is the number of parameters constrained in the full model to obtain M_r, then

$$F_{obs} = \frac{(SSR_r - SSR_f)/q}{SSR_f/dfR_f} = \frac{(SSR_r - SSR_f)/q}{MSR_f} \qquad [4.10]$$

is distributed as an F random variable on q numerator and dfR_f denominator degrees of freedom, provided that the model errors are Gaussian-distributed. Here, dfR_f are the residual degrees of freedom in the full model, i.e., the denominator of F_{obs} is the mean square error in the full model. [4.10] is called a **sum of squares reduction** statistic because the term $SSR_r - SSR_f$ in the numerator measures how much the residual sum of squares is reduced when the constraints on the reduced model are removed.

It is now easy to see that $F_{obs} = MSM_m/MSR$ is a special case of a sum of squares reduction test. In a model with k parameters (including the intercept), we have $SSM = SS(\beta_0, \beta_1, \beta_2, \cdots, \beta_{k-1})$ and can test $H_0: \beta_1 = \beta_2 = ... = \beta_{k-1} = 0$ by choosing a reduced model containing only an intercept. The model sum of squares of the reduced model is $SS(\beta_0)$ and their difference,

$$SS(\beta_0, \beta_1, \beta_2, \cdots, \beta_{k-1}) - SS(\beta_0)$$

is SSM_m with $k-1$ degrees of freedom. See §A4.8.2 for more details on the sum of squares reduction test.

The sequential sum of squares decomposition [4.9] is not unique; it depends on the order in which the regressors enter the model. Regardless of this order, the sequential sums of squares add up to the model sum of squares. This may appear to be an appealing feature, but in practice it is of secondary importance. A decomposition in one degree of freedom sum of squares that does not (necessarily) add up to anything useful but is much more relevant in practice is based on **partial** sums of squares. A partial sum of squares is the contribution made by one explanatory variable in the presence of **all other regressors**, not only the regressors preceding it. In a four regressor model (not counting the intercept), for example, the partial sums of squares are

$$\begin{array}{l} SS(\beta_1|\beta_0, \beta_2, \beta_3, \beta_4) \\ SS(\beta_2|\beta_0, \beta_1, \beta_3, \beta_4) \\ SS(\beta_3|\beta_0, \beta_1, \beta_2, \beta_4) \\ SS(\beta_4|\beta_0, \beta_1, \beta_2, \beta_3). \end{array}$$

Partial sums of squares do not depend on the order in which the regressors enter a model and are usually more informative for purposes of hypothesis testing.

Example 4.2 Turnip Greens (continued). Recall the Turnip Greens data on p. 91. We need to develop a model that relates the Viatmin B_2 content in turnip leaves to the explanatory variables Sunlight (X_1), Soil Moisture (X_2), and Air Temperature (X_3). Figure 4.3 on p. 92 suggests that the relationship between vitamin content and soil moisture is probably quadratic, rather than linear. We fit the following multiple regression model to the 27 observations

$$Y_i = \beta_0 + \beta_1 x_{1i} + \beta_2 x_{2i} + \beta_3 x_{3i} + \beta_4 x_{4i}^2 + e_i,$$

where $x_{4i} = x_{2i}^2$. Assume the following questions are of particular interest:

- Does the addition of Air Temperature (X_3) significantly improve the fit of the model in the presence of the other variables?
- Is the quadratic Soil Moisture term necessary?
- Is the simultaneous contribution of Air Temperature and Sunlight to the model significant?

In terms of the model parameters, the questions translate into the hypotheses

- H_0: $\beta_3 = 0$ (given X_1, X_2, X_4 are in the model)
- H_0: $\beta_4 = 0$ (given X_1, X_2, X_3 are in the model)
- H_0: $\beta_1 = \beta_3 = 0$ (given X_2 and X_4 are in the model).

The full model is the four regressor model and the reduced models are given in Table 4.7 along with their residual sums of squares. Notice that the reduced model for the third hypothesis is a quadratic polynomial in soil moisture.

Table 4.7. Residual sums of squares and degrees of freedom of various models along with test statistics for sum of squares reduction test

Hypothesis	Reduced Model Contains	Residual SS	Residual df	F_{obs}	P-value
$\beta_3 = 0$	X_1, X_2, X_4	$1,031.18$	23	5.677	0.0263
$\beta_4 = 0$	X_1, X_2, X_3	$2,243.18$	23	38.206	0.0001
$\beta_1 = \beta_3 = 0$	X_2, X_4	$1,180.60$	24	4.843	0.0181
Full Model		819.68	22		

The test statistic for the sum of squares reduction tests is

$$F_{obs} = \frac{(SSR_r - SSR_f)/q}{SSR_f/22},$$

where q is the difference in residual degrees of freedom between the full and a reduced model. For example, for the test of H_0: $\beta_3 = 0$ we have

$$F_{obs} = \frac{(1,031.18 - 819.68)/1}{819.68/22} = \frac{211.50}{37.259} = 5.677$$

and for $H_0 : \beta_4 = 0$

$$F_{obs} = \frac{(2,243.18 - 819.68)/1}{819.68/22} = \frac{1,423.50}{37.259} = 38.206.$$

P-values are calculated from an F distribution with q numerator and 22 denominator degrees of freedom.

Using `proc glm` of The SAS® System, the full model is analyzed with the statements

```
proc glm data=turnip;
  model vitamin = sunlight moisture airtemp x4 ;
run; quit;
```

Output 4.1.

```
                            The GLM Procedure

Dependent Variable: vitamin
                                   Sum of
Source                  DF        Squares    Mean Square   F Value   Pr > F
Model                    4    8330.845545    2082.711386     55.90   <.0001
Error                   22     819.684084      37.258367
Corrected Total         26    9150.529630

            R-Square      Coeff Var       Root MSE     vitamin Mean
            0.910422       7.249044       6.103963         84.20370

Source                  DF      Type I SS    Mean Square   F Value   Pr > F
sunlight                 1      97.749940      97.749940      2.62   0.1195
moisture                 1    6779.103952    6779.103952    181.95   <.0001
airtemp                  1      30.492162      30.492162      0.82   0.3754
X4                       1    1423.499492    1423.499492     38.21   <.0001

Source                  DF    Type III SS    Mean Square   F Value   Pr > F
sunlight                 1      71.084508      71.084508      1.91   0.1811
moisture                 1    1085.104925    1085.104925     29.12   <.0001
airtemp                  1     211.495902     211.495902      5.68   0.0263
X4                       1    1423.499492    1423.499492     38.21   <.0001

                                         Standard
    Parameter         Estimate              Error     t Value     Pr > |t|

    Intercept      119.5714052        13.67649483        8.74       <.0001
    sunlight        -0.0336716         0.02437748       -1.38       0.1811
    moisture         5.4250416         1.00526166        5.40       <.0001
    airtemp         -0.5025756         0.21094164       -2.38       0.0263
    X4              -0.1209047         0.01956034       -6.18       <.0001
```

The analysis of variance for the full model leads to $SSM_m = 8,330.845$, $SSR = 819.684$, and $SST_m = 9150.529$. The mean square error estimate is $MSR = 37.258$. The four regressors jointly account for 91% of the variability in the vitamin B_2 content of turnip leaves. The test statistic $F_{obs} = MSM_m/MSR = 55.90$ ($p < .0001$) is used to test the global hypothesis H_0: $\beta_1 = \beta_2 = \beta_3 = \beta_4 = 0$. Since it is rejected, we conclude that at least one of the regressors explains a significant amount of vitamin B_2 variability in the presence of the others.

Sequential sums of squares are labeled `Type I SS` by `proc glm` (Output 4.1). For example, $SS(\beta_1|\beta_0) = 97.749$, $SS(\beta_2|\beta_0, \beta_1) = 6779.10$. Also notice that the sequential sums of squares add up to SSM_m,

$$97.749 + 6,779.104 + 30.492 + 1,423.500 = 8,330.845 = SSM_m.$$

The partial sums of squares are listed as `Type III SS`. The partial hypothesis $H_0 : \beta_3 = 0$ can be answered directly from the output of the fitted full model without actually fitting a reduced model and calculating the sum of squares reduction test. The F_{obs} statistics shown as `F Value` and the p-values shown as `Pr > F` in the `Type III SS` table are partial tests of the individual parameters. Notice that the last sequential and the last partial sums of squares are always identical. The hypothesis $H_0 : \beta_4 = 0$ can be tested based on either sum of squares for `x4`.

The last table of Output 4.1 shows the parameter estimates and their standard errors. For example, $\widehat{\beta}_0 = 119.571$, $\widehat{\beta}_1 = -0.0337$, and so forth. The t_{obs} statistics shown in the column `t Value` are the ratios of a parameter estimate and its estimated standard error, $t_{obs} = \widehat{\beta}_j / \text{ese}(\widehat{\beta}_j)$. The two-sided t-tests are identical to the partial F-tests in the `Type III SS` table. See §A4.8.2 for the precise correspondence between partial F-tests for single variables and the t-tests.

In the Turnip Greens example, sequential and partial sums of squares are not identical. If they were, it would not matter in which order the regressors enter the model. When should we expect the two sets of sum of squares to be the same? For example, when the explanatory variables are **orthogonal** which requires that the inner product of pairs of columns of $\mathbf{X}$ is zero. The sum of square contribution of one regressor then does not depend on whether the other regressor is in the model. Sequential and partial sums of squares may coincide under conditions less stringent than orthogonality of the columns of $\mathbf{X}$. In classification (ANOVA) models where the columns of $\mathbf{X}$ consist of dummy (design) variables, the sums of squares coincide if the data exhibit a certain balance. Hinkelmann and Kempthorne (1994, pp. 87-88) show that for the two-way classification without interaction, a sufficient condition is equal frequencies (replication) of the factor-level combinations. An analysis of variance table where sequential and partial sums of squares are identical is termed an **orthogonal** ANOVA. There are other differences in the sum of squares partitioning between regression and analysis of variance models. Most notably, in ANOVA models we usually test subsets rather than individual parameters and the identity of the parameters in subsets reflects informative structural relationships among the factor levels. The single degree of freedom sum of squares partitioning in ANOVA classification models can be accomplished with sums of squares of orthogonal contrasts, which are linear combinations of the model parameters. Because of these subtleties and the importance of ANOVA models in agronomic data analysis, we devote §4.3 to classification models exclusively.

Sequential sums of squares are of relatively little interest unless there is a *natural* order in which the various explanatory variables or effects should enter the model. A case in point are polynomial regression models where regressors reflect successively higher powers of a single variable. Consider the cubic polynomial

$$Y_i = \beta_0 + \beta_1 x_i + \beta_2 x_i^2 + \beta_3 x_i^3 + e_i.$$

The model sum of squares is decomposed sequentially as

$$\begin{aligned} SSM_m &= SS(\beta_1|\beta_0) \\ &+ SS(\beta_2|\beta_0, \beta_1) \\ &+ SS(\beta_3|\beta_0, \beta_1, \beta_2). \end{aligned}$$

Read from the bottom, these sums of squares can be used to answer the questions

Is there a cubic trend beyond the linear and quadratic trends?
Is there a quadratic trend beyond the linear trend?
Is there a linear trend?

Statisticians and practitioners do not perfectly agree on how to build a final model based on the answers to these questions. One school of thought is that if an interaction is found significant the associated main effects (lower-order terms) should also be included in the model. Since we can think of the cubic term as the interaction between the linear and quadratic term ($x^3 = x \times x^2$), if x^3 is found to make a significant contribution in the presence of x and x^2, these terms would not be tested further and remain in the model. The second school of thought tests all terms individually and retains only the significant ones. A third-order term without the linear or quadratic term in the model is then possible. In observational studies where the regressors are rarely orthogonal, we adopt the first philosophy. In designed experiments where the treatments are levels on a continuous scale such as a rate of application, we prefer to adopt the second school of thougt. If the design matrix is orthogonal we can then easily test the significance of a cubic term independently from the linear or quadratic term using orthogonal polynomial contrasts.

4.3 Factorial Classification

A factorial classification model is a statistical model in which two or more classification variables (factors) are related to the response of interest and the levels of the factors are **crossed**. Two factors A and B are said to be crossed if each level of one factor is combined with each level of the other factor. The factors in Example 4.1 are **crossed**, for example, since each level of factor Lime Type (AL, GL) is combined with each level of the factor Rate of Application. If in addition there are replicate values for the factor level combination, we can study main effects and interactions of the factors. A randomized complete block design, for example, also involves two factors (a block and a treatment factor), but since each treatment appears exactly once in each block, the design does not provide replications of the block $\times$ treatment combinations. This is the reason why block $\times$ treatment interactions cannot be studied in a randomized block design but in the generalized randomized block design where treatments are replicated within the blocks.

If the levels of one factor are not identical across the levels of another factor, the factors are said to be **nested**. For example, if the rates of application for agricultural lime would be different from the rates of application of granulated lime, the Rate factor would be nested within the factor Lime Type. In multilocation trials where randomized block designs with the same treatments are performed at different locations, the block effects are nested within locations, because block number 1 at location 1 is not the same physical entity as block number 1 at location 2. If the same treatments are applied at either location, treatments are crossed with locations, however. In studies of heritability where a random sample of dams are mated to a particular sire, dams are nested within sires and the offspring are nested within dams. If factors are nested, a study of their interaction is not possible. This can be gleaned from the degree of freedom decomposition. Consider two factors A and B with a and b levels, respectively. If the factors are crossed the treatment source of variability is associated with $a \times b - 1$ degrees of freedom. It decomposes into $(a - 1)$ degrees of freedom for the A main effect, $(b-1)$ degrees of freedom for the B main effect, and $(a-1)(b-1)$ degrees of freedom for the interaction. If B is nested within A, the treatment degrees of freedom decompose into $(a-1)$ degrees of freedom for the A main effect and $a(b-1)$ degrees of freedom for the nested factor combining the degrees of freedom for the B main effect and the $A \times B$ interaction: $(b-1)+(a-1)(b-1)=a(b-1)$. In this section we focus on the two-factor crossed classification with replications in which main effects and interactions can be studied (see §4.3.2 for definitions). Rather than providing a comprehensive treatise of experimental data analysis this section intends to highlight the standard operations and techniques that apply in classification models where treatment comparisons are of primary concern.

4.3.1 The Means and Effects Model

Box 4.2 Means and Effects Model

- **The means model expresses observations as random deviations from the cell means; its design matrix is of full rank.**
- **The effects model decomposes cell means in grand mean, main effects, and interactions; its design matrix is deficient in rank.**

Two equivalent ways of representing a classification models are termed the **means** and the **effects** model. We prefer effects models in general, although on the surface means models are simpler. But the study of main effects and interactions, which are of great concern in classification models, is simpler in means models. In the two-way classification with replications (e.g., Example 4.1) an observation Y_{ijk} for the kth replicate of Lime Type i and Rate j can be expressed as a random deviation from the mean of that particular treatment combination:

$$Y_{ijk} = \mu_{ij} + e_{ijk}. \qquad [4.11]$$

Here, e_{ijk} denotes the experimental error associated with the kth replicate of the ith lime type and j^{th} rate of application $\{i=1, a=2; j=1,\cdots,b=5; k=1,\cdots,r=5\}$. μ_{ij} denotes the mean pH of an experimental unit receiving lime type i at rate j; hence the name means model. To finish the model formulation assume e_{ijk} are uncorrelated random variables with mean 0 and common variance σ^2. Model [4.11] can then be written in matrix vector notation as $\mathbf{Y} = \mathbf{X}\boldsymbol{\mu} + \mathbf{e}$, where

$$\mathbf{Y}_{(50\times 1)} = \begin{bmatrix} \mathbf{Y}_{11} \\ \mathbf{Y}_{12} \\ \mathbf{Y}_{13} \\ \mathbf{Y}_{14} \\ \mathbf{Y}_{15} \\ \mathbf{Y}_{21} \\ \mathbf{Y}_{22} \\ \mathbf{Y}_{23} \\ \mathbf{Y}_{24} \\ \mathbf{Y}_{25} \end{bmatrix} = \begin{bmatrix} \mathbf{1} & \mathbf{0} & \mathbf{0} & \mathbf{0} & \mathbf{0} & \mathbf{0} & \mathbf{0} & \mathbf{0} & \mathbf{0} & \mathbf{0} \\ \mathbf{0} & \mathbf{1} & \mathbf{0} & \mathbf{0} & \mathbf{0} & \mathbf{0} & \mathbf{0} & \mathbf{0} & \mathbf{0} & \mathbf{0} \\ \mathbf{0} & \mathbf{0} & \mathbf{1} & \mathbf{0} & \mathbf{0} & \mathbf{0} & \mathbf{0} & \mathbf{0} & \mathbf{0} & \mathbf{0} \\ \mathbf{0} & \mathbf{0} & \mathbf{0} & \mathbf{1} & \mathbf{0} & \mathbf{0} & \mathbf{0} & \mathbf{0} & \mathbf{0} & \mathbf{0} \\ \mathbf{0} & \mathbf{0} & \mathbf{0} & \mathbf{0} & \mathbf{1} & \mathbf{0} & \mathbf{0} & \mathbf{0} & \mathbf{0} & \mathbf{0} \\ \mathbf{0} & \mathbf{0} & \mathbf{0} & \mathbf{0} & \mathbf{0} & \mathbf{1} & \mathbf{0} & \mathbf{0} & \mathbf{0} & \mathbf{0} \\ \mathbf{0} & \mathbf{0} & \mathbf{0} & \mathbf{0} & \mathbf{0} & \mathbf{0} & \mathbf{1} & \mathbf{0} & \mathbf{0} & \mathbf{0} \\ \mathbf{0} & \mathbf{0} & \mathbf{0} & \mathbf{0} & \mathbf{0} & \mathbf{0} & \mathbf{0} & \mathbf{1} & \mathbf{0} & \mathbf{0} \\ \mathbf{0} & \mathbf{0} & \mathbf{0} & \mathbf{0} & \mathbf{0} & \mathbf{0} & \mathbf{0} & \mathbf{0} & \mathbf{1} & \mathbf{0} \\ \mathbf{0} & \mathbf{0} & \mathbf{0} & \mathbf{0} & \mathbf{0} & \mathbf{0} & \mathbf{0} & \mathbf{0} & \mathbf{0} & \mathbf{1} \end{bmatrix}_{(50\times 10)} \begin{bmatrix} \mu_{11} \\ \mu_{12} \\ \mu_{13} \\ \mu_{14} \\ \mu_{15} \\ \mu_{21} \\ \mu_{22} \\ \mu_{23} \\ \mu_{24} \\ \mu_{25} \end{bmatrix} + \begin{bmatrix} \mathbf{e}_{11} \\ \mathbf{e}_{12} \\ \mathbf{e}_{13} \\ \mathbf{e}_{14} \\ \mathbf{e}_{15} \\ \mathbf{e}_{21} \\ \mathbf{e}_{22} \\ \mathbf{e}_{23} \\ \mathbf{e}_{24} \\ \mathbf{e}_{25} \end{bmatrix}. \qquad [4.12]$$

The (5×1) vector $\mathbf{Y}_{11}$, for example, contains the replicate observations for lime type 1 and application rate 0 tons. Notice that $\mathbf{X}$ is not the identity matrix. Each $\mathbf{1}$ in $\mathbf{X}$ is a (5×1) vector of ones and each $\mathbf{0}$ is a (5×1) vector of zeros. The design matrix $\mathbf{X}$ is of full column rank and the inverse of $\mathbf{X}'\mathbf{X}$ exists:

$$(\mathbf{X}'\mathbf{X})^{-1} = \text{Diag}\left(\frac{1}{5} * \mathbf{1}_{10}\right) = \begin{bmatrix} 0.2 & 0 & 0 & 0 & \cdots & 0 \\ 0 & 0.2 & 0 & 0 & \cdots & 0 \\ 0 & 0 & 0.2 & 0 & \cdots & 0 \\ 0 & 0 & 0 & 0.2 & \cdots & 0 \\ \vdots & \vdots & \vdots & \vdots & \ddots & \vdots \\ 0 & 0 & 0 & 0 & \cdots & 0.2 \end{bmatrix}.$$

The ordinary least squares estimate of μ is thus simply the vector of sample means in the $a \times b$ groups.

$$\widehat{\boldsymbol{\mu}} = (\mathbf{X}'\mathbf{X})^{-1}\mathbf{X}'\mathbf{y} = \begin{bmatrix} \overline{y}_{11.} \\ \overline{y}_{12.} \\ \vdots \\ \overline{y}_{15.} \\ \overline{y}_{21.} \\ \overline{y}_{22.} \\ \vdots \\ \overline{y}_{25.} \end{bmatrix}.$$

A different parameterization of the two-way model can be derived if we think of the μ_{ij} as **cell means** in the body of a two-dimensional table in which thc factors are cross-classified (Table 4.8). The row and column averages of this table are denoted $\mu_{i.}$ and $\mu_{.j}$ where the dot replaces the index over which averaging is carried out. We call $\mu_{i.}$ and $\mu_{.j}$ the **marginal means** for factor levels i of Lime type and j of Rate of Application, respectively, since they occupy positions in the margin of the table (Table 4.8).

Table 4.8. Cell and marginal means in Lime Application (Example 4.1)

	Rate of Application					
	0	**1**	**2**	**4**	**8**	
Agricultural lime	μ_{11}	μ_{12}	μ_{13}	μ_{14}	μ_{15}	$\mu_{1.}$
Granulated lime	μ_{21}	μ_{22}	μ_{23}	μ_{24}	μ_{25}	$\mu_{2.}$
	$\mu_{.1}$	$\mu_{.2}$	$\mu_{.3}$	$\mu_{.4}$	$\mu_{.5}$	

Marginal means are arithmetic averages of cell means (Yandell, 1997, p. 109), even if the data are unbalanced:

$$\mu_{i.} = \frac{1}{b}\sum_{j=1}^{b}\mu_{ij} \qquad \mu_{.j} = \frac{1}{a}\sum_{i=1}^{a}\mu_{ij}.$$

The grand mean μ is defined as the average of all cell means, $\mu = \sum_i\sum_j\mu_{ij}/ab$. To construe marginal means as weighted means – where weighing would take into account the number of observations n_{ij} for a particular cell – would define population quantities as functions of sample quantities. The relationships of the means should not depend on how many observations are sampled or how many times treatments are replicated. See Chapters 4.6 and 4.7 in Searle (1987) for a comparison of the weighted and unweighted schemes.

We can now write the mathematical identity

$$\begin{aligned} \mu_{ij} = \mu &\quad + (\mu_{i.} - \mu) &\quad + (\mu_{.j} - \mu) &\quad + (\mu_{ij} - \mu_{i.} - \mu_{.j} + \mu) \\ = \mu &\quad + \alpha_i &\quad + \beta_j &\quad + (\alpha\beta)_{ij}. \end{aligned} \qquad [4.13]$$

Models based on this decomposition are termed **effects** models since $\alpha_i = (\mu_{i.} - \mu)$ measures the effect of the ith level of factor A, $\beta_j = (\mu_{.j} - \mu)$ the effect of the jth level of factor B and $(\alpha\beta)_{ij}$ their interaction. The nature and precise interpretation of main effects and interactions is studied in more detail in §4.3.2. For now we notice that the effects obey certain constraints by construction:

$$\begin{aligned} &\sum_{i=1}^{a} \alpha_i = 0 \\ &\sum_{j=1}^{b} \beta_j = 0 \\ &\sum_{i=1}^{a} (\alpha\beta)_{ij} = \sum_{j=1}^{b} (\alpha\beta)_{ij} = 0. \end{aligned} \qquad [4.14]$$

A two-way factorial layout coded as an effects model can be expressed as a sum of separate vectors. We obtain

$$\mathbf{Y} = \mathbf{1}\boldsymbol{\mu} + \mathbf{X}_{\alpha}\boldsymbol{\alpha} + \mathbf{X}_{\beta}\boldsymbol{\beta} + \mathbf{X}_{(\alpha\beta)}\boldsymbol{\phi} + \mathbf{e}, \qquad [4.15]$$

where

$$\begin{bmatrix} \mathbf{Y}_{11} \\ \mathbf{Y}_{12} \\ \mathbf{Y}_{13} \\ \mathbf{Y}_{14} \\ \mathbf{Y}_{15} \\ \mathbf{Y}_{21} \\ \mathbf{Y}_{22} \\ \mathbf{Y}_{23} \\ \mathbf{Y}_{24} \\ \mathbf{Y}_{25} \end{bmatrix} = \begin{bmatrix} \mathbf{1} \\ \mathbf{1} \\ \mathbf{1} \\ \mathbf{1} \\ \mathbf{1} \\ \mathbf{1} \\ \mathbf{1} \\ \mathbf{1} \\ \mathbf{1} \\ \mathbf{1} \end{bmatrix} \mu + \begin{bmatrix} \mathbf{1} & \mathbf{0} \\ \mathbf{1} & \mathbf{0} \\ \mathbf{1} & \mathbf{0} \\ \mathbf{1} & \mathbf{0} \\ \mathbf{1} & \mathbf{0} \\ \mathbf{0} & \mathbf{1} \\ \mathbf{0} & \mathbf{1} \\ \mathbf{0} & \mathbf{1} \\ \mathbf{0} & \mathbf{1} \\ \mathbf{0} & \mathbf{1} \end{bmatrix} \begin{bmatrix} \alpha_1 \\ \alpha_2 \end{bmatrix} + \begin{bmatrix} \mathbf{1} & \mathbf{0} & \mathbf{0} & \mathbf{0} & \mathbf{0} \\ \mathbf{0} & \mathbf{1} & \mathbf{0} & \mathbf{0} & \mathbf{0} \\ \mathbf{0} & \mathbf{0} & \mathbf{1} & \mathbf{0} & \mathbf{0} \\ \mathbf{0} & \mathbf{0} & \mathbf{0} & \mathbf{1} & \mathbf{0} \\ \mathbf{0} & \mathbf{0} & \mathbf{0} & \mathbf{0} & \mathbf{1} \\ \mathbf{1} & \mathbf{0} & \mathbf{0} & \mathbf{0} & \mathbf{0} \\ \mathbf{0} & \mathbf{1} & \mathbf{0} & \mathbf{0} & \mathbf{0} \\ \mathbf{0} & \mathbf{0} & \mathbf{1} & \mathbf{0} & \mathbf{0} \\ \mathbf{0} & \mathbf{0} & \mathbf{0} & \mathbf{1} & \mathbf{0} \\ \mathbf{0} & \mathbf{0} & \mathbf{0} & \mathbf{0} & \mathbf{1} \end{bmatrix} \begin{bmatrix} \beta_1 \\ \beta_2 \\ \beta_3 \\ \beta_4 \\ \beta_5 \end{bmatrix} + \mathbf{X}_{(\alpha\beta)} \begin{bmatrix} (\alpha\beta)_{11} \\ (\alpha\beta)_{12} \\ (\alpha\beta)_{13} \\ (\alpha\beta)_{14} \\ (\alpha\beta)_{15} \\ (\alpha\beta)_{21} \\ (\alpha\beta)_{22} \\ (\alpha\beta)_{23} \\ (\alpha\beta)_{24} \\ (\alpha\beta)_{25} \end{bmatrix} + \begin{bmatrix} \mathbf{e}_{11} \\ \mathbf{e}_{12} \\ \mathbf{e}_{13} \\ \mathbf{e}_{14} \\ \mathbf{e}_{15} \\ \mathbf{e}_{21} \\ \mathbf{e}_{22} \\ \mathbf{e}_{23} \\ \mathbf{e}_{24} \\ \mathbf{e}_{25} \end{bmatrix}.$$

The matrix $\mathbf{X}_{(\alpha\beta)}$ is the same as the $\mathbf{X}$ matrix in [4.12]. Although the latter is non-singular, it is clear that in the effects model the complete design matrix $\mathbf{P} = \left[\mathbf{1}, \mathbf{X}_{\alpha}, \mathbf{X}_{\beta}, \mathbf{X}_{(\alpha\beta)}\right]$ is rank-deficient. The columns of $\mathbf{X}_{\alpha}$, the columns of $\mathbf{X}_{\beta}$, and the columns of $\mathbf{X}_{(\alpha\beta)}$ sum to 1. This results from the linear constraints in [4.14].

We now proceed to define various effect types based on the effects model.

4.3.2 Effect Types in Factorial Designs

Box 4.3 Effect Types in Crossed Classifications

- **Simple effects are comparisons of the cell means where one factor is held fixed.**
- **Interaction effects are contrasts among simple effects.**
- **Main effects are contrasts among the marginal means and can be expressed as averages of simple effects.**
- **Simple main effects are differences of cell and marginal means.**
- **Slices are collections of simple main effects.**

Hypotheses can be expressed in terms of relationships among the cell means μ_{ij} or in terms of the effects α_i, β_j, and $(\alpha\beta)_{ij}$ in [4.13]. These relationships are classified into the following categories (effects):

- simple effects
- interaction effects
- main effects, and
- simple main effects

A **simple effect** is the most elementary comparison. It is a comparison of the μ_{ij} where one of the factors is held fixed. For example, $\mu_{13} - \mu_{23}$ is a comparison of lime types at rate 2 tons, and $\mu_{22} - \mu_{23}$ is a comparison of the 1 and 2 ton application rate for granulated lime. By **comparisons** we do not just have pairwise tests in mind, but more generally, **contrasts**. A contrast is a linear function of parameters in which the coefficients of the linear function sum to zero. A simple effect of application rate for agricultural lime $(i = 1)$ is a contrast among the μ_{1j},

$$\ell = \sum_{j=1}^{b} c_j \mu_{1j},$$

where $\sum_{j=1}^{b} c_j = 0$. The c_j are called the contrast coefficients. The simple effect $\mu_{22} - \mu_{23}$ has contrast coefficients $(0, 1, -1, 0, 0)$.

Interaction effects are contrasts among simple effects. Consider $\ell_1 = \mu_{11} - \mu_{21}$, $\ell_2 = \mu_{12} - \mu_{22}$, and $\ell_3 = \mu_{13} - \mu_{23}$ which are simple Lime Type effects at rates 0, 1, and 2 tons, respectively. The contrast

$$\ell_4 = 1 \times \ell_1 - 2 \times \ell_2 + 1 \times \ell_3$$

is an interaction effect. It tests whether the difference in pH between lime types changes linearly with application rates between 0 and 2 tons. Interactions are interpreted as the non-constancy of contrasts in one factor as the levels of the other factor are changed. In the absence of interactions, $\ell_4 = 0$. In this case it would be reasonable to disengage the Rate of Application in comparisons of Lime Types and vice versa. Instead, one should focus on contrasts among the marginal averages across rates of applications and averages across lime types.

Main effects are contrasts among the marginal means. A main effect of Lime Type is a contrast among the $\mu_{i.}$ and a main effect of Application Rate is a contrast among the $\mu_{.j}$. Since simple effects are the elementary comparisons, it is of interest to find out how main effects relate to these. Consider the contrast $\mu_{1.} - \mu_{2.}$, which is the main effect of Lime Type since the factor has only two levels. Write $\mu_{1.} = 0.2(\mu_{11} + \mu_{12} + \mu_{13} + \mu_{14} + \mu_{15})$ and similarly for $\mu_{2.}$. Thus,

$$\begin{aligned} \mu_{1.} - \mu_{2.} &= \frac{1}{5}(\mu_{11} + \mu_{12} + \mu_{13} + \mu_{14} + \mu_{15}) - \frac{1}{5}(\mu_{21} + \mu_{22} + \mu_{23} + \mu_{24} + \mu_{25}) \\ &= \frac{1}{5}(\mu_{11} - \mu_{21}) + \frac{1}{5}(\mu_{12} - \mu_{22}) + \cdots + \frac{1}{5}(\mu_{15} - \mu_{25}). \end{aligned} \qquad [4.16]$$

Each of the five terms in this expression is a Lime Type simple effect, one at each rate of application and we conclude that main effects are *averages* of simple effects. The marginal difference, $\mu_{1.} - \mu_{2.}$, will be identical to the simple effects in [4.16] if the lines in Figure 4.2 (p. 89) are parallel. This is the condition under which interactions are absent and immediately sheds some light on the proper interpretation and applicability of main effects. If interactions are present, some of the simple effects in [4.16] may be of positive sign, others may be of negative sign. The marginal difference, $\mu_{1.} - \mu_{2.}$, may be close to zero, masking the fact that lime types are effective at various rates of application. If the sign of the differences remains the same across application rates, as is the case here (see Figure 4.2), an interpretation of the main effects remains possible even in the light of interactions.

The presence of interactions implies that at least some simple effects change with the level of the factor held fixed in the simple effects. One may thus be tempted to perform separate one-way analyses at each rate of application and separate one-way analyses for each lime type. The drawback of this approach is a considerable loss of degrees of freedom and hence a loss of power. If, for example, an analysis is performed with data from 0 tons alone, the residual error will be associated with $2(5-1) = 8$ degrees of freedom, whereas $2{*}5(5-1) = 40$ degrees of freedom are availablein the two-way factorial analysis. It is more efficient to test the hypotheses in the analysis based on the full data.

A comparison of application rates for each lime type involves tests of the hypotheses

$$\begin{aligned} &\boxed{A} \quad H_0: \mu_{11} = \mu_{12} = \mu_{13} = \mu_{14} = \mu_{15} \\ &\boxed{B} \quad H_0: \mu_{21} = \mu_{22} = \mu_{23} = \mu_{24} = \mu_{25}. \end{aligned} \qquad [4.17]$$

$\boxed{A}$ is called the **slice** of the Lime × Rate interaction at lime type 1 and $\boxed{B}$ is called the **slice** of the Lime × Rate interaction at lime type 2. Similarly, slices at the various application rates are tests of the hypotheses

$$
\begin{array}{ll}
\boxed{A} & H_0: \mu_{11} = \mu_{21} \\
\boxed{B} & H_0: \mu_{12} = \mu_{22} \\
\boxed{C} & H_0: \mu_{13} = \mu_{23} \\
\boxed{D} & H_0: \mu_{14} = \mu_{24} \\
\boxed{E} & H_0: \mu_{15} = \mu_{25}.
\end{array}
$$

How do slices relate to simple and main effects? Kirk (1995, p. 377) defines as **simple main effects** comparisons of the type $\mu_{ik} - \mu_{i.}$ and $\mu_{kj} - \mu_{.j}$. Consider the first case. If all simple main effects are identical, then

$$\mu_{i1} - \mu_{i.} = \mu_{i2} - \mu_{i.} = \cdots = \mu_{ib} - \mu_{i.}$$

which implies $\mu_{i1} = \mu_{i2} = \mu_{i3} = \mu_{i4} = \mu_{ib}$. This is the slice at the ith level of factor A. Schabenberger, Gregoire, and Kong (2000) made this correspondence between the various effects in a factorial structure more precise in terms of matrices and vectors. They also proved the following result noted earlier by Winer (1971, p. 347). If you assemble all possible slices of $A \times B$ by the levels of A (for example, $\boxed{A}$ and $\boxed{B}$ in [4.17]), the sum of squares associated with this assembly is identical to the sum of squares for the B main effect plus that for the $A \times B$ interaction.

4.3.3 Sum of Squares Partitioning through Contrasts

The definition of main effects as contrasts among marginal means and interactions as contrasts among simple effects seems at odds with the concept of **the** main effect of factor A (or B) and **the** interaction of A and B. **The** main effect of a factor can be represented by a collection of contrasts among marginal means, all of which are **mutually orthogonal**. Two contrasts $\ell_1 = \sum_{i=1}^{a} c_{1i}\mu_{i.}$ and $\ell_2 = \sum_{i=1}^{a} c_{2i}\mu_{i.}$ are orthogonal if

$$\sum_{i=1}^{a} c_{1i}c_{2i} = 0,$$

We do not distinguish between balanced and unbalanced cases since the definition of population quantities ℓ_1 and ℓ_2 should be independent of the sample design. A **complete set** of orthogonal contrasts for a factor with p levels is any set of $(p-1)$ contrasts in which the members are mutually orthogonal. We can always fall back on the following complete set:

$$\boldsymbol{\Theta}^p = \begin{bmatrix} 1 & -1 & 0 & 0 & \cdots & 0 \\ 1 & 1 & -2 & 0 & \cdots & 0 \\ \vdots & & & & & \vdots \\ 1 & 1 & 1 & 1 & \cdots & -(p-1) \end{bmatrix}_{((p-1)\times p)} . \qquad [4.18]$$

For the factor Lime Type with only two levels a *complete set* contains only a single contrast with coefficients 1 and -1, for the factor Rate of Application the set would be

$$\mathbf{\Theta}^5 = \begin{bmatrix} 1 & -1 & 0 & 0 & 0 \\ 1 & 1 & -2 & 0 & 0 \\ 1 & 1 & 1 & -3 & 0 \\ 1 & 1 & 1 & 1 & -4 \end{bmatrix}.$$

The sum of squares for a contrast among the marginal means of application rate is calculated as

$$SS(\ell) = \frac{\widehat{\ell}^2}{\sum_{j=1}^{b} c_j^2/n_{.j}}$$

and if contrasts are orthogonal, their sum of squares contributions are additive. If the contrast sums of squares for any four orthogonal contrasts among the marginal Rate means $\mu_{.j}$ are added, the resulting sum of squares is that of **the** Rate main effect. Using the generic complete set above, let the contrasts be

$$\begin{aligned}
\ell_1 &= \mu_{.1} - \mu_{.2} \\
\ell_2 &= \mu_{.1} + \mu_{.2} - 2\mu_{.3} \\
\ell_3 &= \mu_{.1} + \mu_{.2} + \mu_{.3} - 3\mu_{.4} \\
\ell_4 &= \mu_{.1} + \mu_{.2} + \mu_{.3} + \mu_{.4} - 4\mu_{.5}.
\end{aligned}$$

The contrast sum of squares are calculated in Table 4.9. The only contrast defining **the** Lime Type main effect is $\mu_{1.} - \mu_{2.}$.

Table 4.9. Main effects contrast for Rate of Application and Lime Type ($n_{.j} = 10$, $n_{i.} = 25$)

			$\sum c^2/n$	$SS(\ell)$
Rate 0 tons	$\overline{y}_{.1} = 5.733$	$\widehat{\ell}_1 = -0.094$	0.2	0.044
Rate 1 ton	$\overline{y}_{.2} = 5.827$	$\widehat{\ell}_2 = -0.202$	0.6	0.068
Rate 2 tons	$\overline{y}_{.3} = 5.881$	$\widehat{\ell}_3 = -0.633$	1.2	0.334
Rate 4 tons	$\overline{y}_{.4} = 6.025$	$\widehat{\ell}_4 = -1.381$	2.0	0.954
Rate 8 tons	$\overline{y}_{.5} = 6.212$			
Agricultural lime	$\overline{y}_{1.} = 6.038$	$\widehat{\ell} = 0.205$	0.08	0.0525
Granulated lime	$\overline{y}_{2.} = 5.833$			

The sum of squares for the Rate main effect is

$$SS(\text{Rate}) = 0.044 + 0.068 + 0.334 + 0.954 = 1.40$$

and that for the Lime Type main effect is $SS(\text{Lime}) = 0.0525$.

The contrast set from which to obtain the interaction sum of squares is constructed by unfolding the main effects contrasts to correspond to the cell means and multiplying the contrast coefficients element by element. Since there are four contrasts defining the Rate main effect and one contrast for the Lime main effect, there will be $4 \times 1 = 4$ contrasts defining the interactions (Table 4.10).

Table 4.10. Construction of interaction contrasts by unfolding the design (empty cells correspond to zero coefficients)

	c_{11}	c_{12}	c_{13}	c_{14}	c_{15}	c_{21}	c_{22}	c_{23}	c_{24}	c_{25}
Rate Main Effects										
ℓ_1	1	-1				1	-1			
ℓ_2	1	1	-2			1	1	-2		
ℓ_3	1	1	1	-3		1	1	1	-3	
ℓ_4	1	1	1	1	-4	1	1	1	1	-4
Lime Main Effect										
ℓ	1	1	1	1	1	-1	-1	-1	-1	-1
Interaction Contrasts										
$\ell_1 \times \ell$	1	-1				-1	1			
$\ell_2 \times \ell$	1	1	-2			-1	-1	2		
$\ell_3 \times \ell$	1	1	1	-3		-1	-1	-1	3	
$\ell_4 \times \ell$	1	1	1	1	-4	-1	-1	-1	-1	4

Sum of squares of the interaction contrasts are calculated as linear functions among the cell means (μ_{ij}). For example,

$$SS(\ell_1 \times \ell) = \frac{(\overline{y}_{11.} - \overline{y}_{12.} - \overline{y}_{21.} + \overline{y}_{22.})^2}{\{1^2 + (-1)^2 + (-1)^2 + 1^2\}/5}.$$

The divisor of 5 stems from the fact that each cell mean is estimated as the arithmetic average of the five replications for that treatment combination. The interaction sum of squares is then finally

$$SS(\text{Lime} \times \text{Rate}) = SS(\ell_1 \times \ell) + SS(\ell_2 \times \ell) + SS(\ell_3 \times \ell) + SS(\ell_4 \times \ell).$$

This procedure of calculating main effects and interaction sums of squares is cumbersome. It demonstrates, however, that a main effect with $(a-1)$ degrees of freedom can be partitioned into $(a-1)$ mutually orthogonal single degree of freedom contrasts. It also highlights why there are $(a-1)(b-1)$ degrees of freedom in the interaction between two factors with a and b levels, respectively, because of the way they are obtained from crossing $(a-1)$ and $(b-1)$ main effects contrasts. In practice one obtains the sum of squares of contrasts and the sum of squares partitioning into main effects and interactions with a statistical computing package.

4.3.4 Effects and Contrasts in The SAS® System

Contrasts in The SAS® System are coded in the effects decomposition [4.13], not the means model. To test for main effects, interactions, simple effects, and slices, the contrasts need to be expressed in terms of the α_i, β_j, and $(\alpha\beta)_{ij}$. Consider the `proc glm` statements for the two-factor completely randomized design in Example 4.1. In the `model` statement of the code segment

```
proc glm data=limereq;
  class lime rate;
  model ApH = lime rate lime*rate;
run;
```

`lime` refers to the main effects α_i of factor Lime Type, `rate` to the main effects of factor Application Rate and `lime*rate` to the $(\alpha\beta)_{ij}$ interaction terms. In Output 4.2 we find sequential (`Type I`) and partial (`Type III`) sums of squares and tests for the three effects listed in the `model` statement. Because the design is orthogonal the two groups of sums of squares are identical. By default, `proc glm` produces these tests for any term listed on the right-hand side of the `model` statement. Also, we obtain $SSR = 0.12885$ and $MSR = 0.00322$. **The** main effects of Lime Type and Rate of Application are shown as sources `LIME` and `RATE` on the output. The difference between our calculation of $SS(\text{Rate}) = 1.40$ and the calculation by SAS® $(SS(\text{Rate}) = 1.398)$ is due to round-off errors. There is a significant interaction between the two factors ($F_{obs} = 24.12, p < 0.0001$). Neither of the main effects is masked. Since the trends in application rates do not cross (Figure 4.2, p. 89) this is to be expected.

Output 4.2.

```
                         The GLM Procedure

                      Class Level Information

              Class          Levels    Values
              LIME                2    AG lime Pell lime
              RATE                5    0 1 2 4 8

                   Number of observations     50

Dependent Variable: APH

                                       Sum of
Source                     DF         Squares    Mean Square  F Value   Pr > F
Model                       9      2.23349200     0.24816578    77.04   <.0001
Error                      40      0.12885000     0.00322125
Corrected Total            49      2.36234200

            R-Square     Coeff Var      Root MSE     APH Mean
            0.945457      0.956230      0.056756     5.935400

Source                     DF     Type I SS    Mean Square  F Value   Pr > F
LIME                        1    0.52428800     0.52428800   162.76   <.0001
RATE                        4    1.39845700     0.34961425   108.53   <.0001
LIME*RATE                   4    0.31074700     0.07768675    24.12   <.0001

Source                     DF   Type III SS    Mean Square  F Value   Pr > F
LIME                        1    0.52428800     0.52428800   162.76   <.0001
RATE                        4    1.39845700     0.34961425   108.53   <.0001
LIME*RATE                   4    0.31074700     0.07768675    24.12   <.0001
```

We now reconstruct the main effects and interaction tests with contrasts for expository purposes. Recall that an A factor main effect contrast is a contrast among the marginal means $\mu_{i.}$ and a B factor main effect is a contrast among the marginal means $\mu_{.j}$. We obtain

$$\sum_{i=1}^{a} c_i \mu_{i.} = \sum_{i=1}^{a} c_i \sum_{j=1}^{b} \frac{1}{b}\left(\mu + \alpha_i + \beta_j + (\alpha\beta)_{ij}\right)$$
$$= \sum_{i=1}^{a} c_i \alpha_i + \frac{1}{b}\sum_{i=1}^{a}\sum_{j=1}^{b} c_i (\alpha\beta)_{ij}$$

and

$$\sum_{j=1}^{b} c_j \mu_{.j} = \sum_{j=1}^{b} c_j \sum_{i=1}^{a} \frac{1}{a}\left(\mu + \alpha_i + \beta_j + (\alpha\beta)_{ij}\right)$$
$$= \sum_{j=1}^{b} c_j \beta_j + \frac{1}{a}\sum_{j=1}^{b}\sum_{i=1}^{a} c_j (\alpha\beta)_{ij}.$$

Fortunately, The SAS® System does not require that we specify coefficients for effects which contain other effects for which coefficients are given. For main effect contrasts it is sufficient to specify the contrast coefficients for `lime` or `rate`; the interaction coefficients c_i/b and c_j/a are assigned automatically. Using the generic contrast set [4.18] for the main effects and the unfolded contrast coefficients for the interaction in Table 4.10, the following `contrast` statements in `proc glm` add Output 4.3 to Output 4.2.

```
proc glm data=limereq;
  class lime rate;
  model ApH = lime rate lime*rate;
  contrast 'Lime main effect' lime 1 -1;
  contrast 'Rate main effect' rate 1 -1 ,
                              rate 1  1 -2 ,
                              rate 1  1  1 -3 ,
                              rate 1  1  1  1 -4;
  contrast 'Lime*Rate interaction' lime*rate 1 -1  0  0  0 -1  1  0  0  0,
                                   lime*rate 1  1 -2  0  0 -1 -1  2  0  0,
                                   lime*rate 1  1  1 -3  0 -1 -1 -1  3  0,
                                   lime*rate 1  1  1  1 -4 -1 -1 -1 -1  4;
run;
```

Output 4.3.

Contrast	DF	Contrast SS	Mean Square	F Value	Pr > F
Lime main effect	1	0.52428800	0.52428800	162.76	<.0001
Rate main effect	4	1.39845700	0.34961425	108.53	<.0001
Lime*Rate interaction	4	0.31074700	0.07768675	24.12	<.0001

For the simple effect $\sum_{j=1}^{b} c_j \mu_{ij}$ substitution of [4.13] yields

$$\sum_{j=1}^{b} c_j \mu_{ij} = \sum_{j=1}^{b} c_j\left(\mu + \alpha_i + \beta_j + (\alpha\beta)_{ij}\right)$$
$$= \sum_{j=1}^{b} c_j \beta_j + \sum_{j=1}^{b} c_j (\alpha\beta)_{ij}. \qquad [4.19]$$

In this case the user must specify the c_j coefficients for the interaction terms **and** the β_j. If only the latter are given, SAS® will apply the rules for testing main effects and assign the average coefficient c_j/a to the interaction terms, which does not produce a simple effect.

Recall that interaction effects are contrasts among simple effects. Consider the question whether the simple effect $\sum_{j=1}^{b} c_j\mu_{ij}$ is constant for the first two levels of factor A. The interaction effect becomes

$$\sum_{j=1}^{b} c_j\mu_{1j} - \sum_{j=1}^{b} c_j\mu_{2j} = \sum_{j=1}^{b} c_j\beta_j + \sum_{j=1}^{b} c_j(\alpha\beta)_{1j} - \sum_{j=1}^{b} c_j\beta_j - \sum_{j=1}^{b} c_j(\alpha\beta)_{2j}$$
$$= \sum_{j=1}^{b} c_j(\alpha\beta)_{1j} - \sum_{j=1}^{b} c_j(\alpha\beta)_{2j}. \qquad [4.20]$$

Genuine interaction effects will involve **only** the $(\alpha\beta)_{ij}$ terms.

For the lime requirement data we now return to the research questions raised in the introduction:

A: Do Lime Type and Application Rate interact?

B: Are there main effects of Lime Type and Application Rate?

C: Is there a difference between lime types at the 1 ton application rate?

D: Does the difference between lime types depend on whether 1 or 2 tons are applied?

E: How does the comparison of lime types change with application rate?

Questions A and B refer to **the** interactions and **the** main effects of the two factors. Although we have seen in Output 4.3 how to obtain the main effects with contrasts, it is of course much simpler to locate the particular sources in the `Type III` sum of squares table. C is a simple effect ([4.19]), D an interaction effect ([4.20]), and E slices of the Lime Type × Rate interaction by application rates. The `proc glm` statements are as follows (Output 4.4).

```
proc glm data=limereq;
  class lime rate ;
  model aph = lime rate lime*rate;                          /* A and B */
  contrast 'Lime at 1 ton              (C)'
            lime 1 -1 lime*rate 0 1  0 0 0 0 -1 0 0 0; /* C   */
  contrast 'Lime effect at 1 vs. 2 (D)'
                  lime*rate 0 1 -1 0 0 0 -1 1 0 0; /* D   */
  lsmeans lime*rate / slice=(rate);                         /* E   */
run; quit;
```

In specifying contrast coefficients for an effect in SAS®, one should pay attention to (i) the `Class Level Information` table printed at the top of the output (see Output 4.2) and (ii) the order in which the factors are listed in the `class` statement. The `Class Level Information` table depicts how the levels of the factors are ordered internally. If factor variables are character variables, the default ordering of the levels is alphabetical, which may be counterintuitive in the assignment of contrast coefficients (for example, "0 tons/acre" appears before "100 tons/acre" which appears before "50 tons/acre").

Output 4.4.

```
                              The GLM Procedure

                          Class Level Information
                Class           Levels     Values
                LIME                 2     AG lime Pell lime
                RATE                 5     0 1 2 4 8

                        Number of observations     50

Dependent Variable: APH
                                        Sum of
Source                      DF         Squares    Mean Square   F Value   Pr > F
Model                        9      2.23349200     0.24816578     77.04   <.0001
Error                       40      0.12885000     0.00322125
Corrected Total             49      2.36234200

Source                      DF     Type I SS      Mean Square   F Value   Pr > F
LIME                         1      0.52428800     0.52428800    162.76   <.0001
RATE                         4      1.39845700     0.34961425    108.53   <.0001
LIME*RATE                    4      0.31074700     0.07768675     24.12   <.0001

Source                      DF   Type III SS      Mean Square   F Value   Pr > F
LIME                         1      0.52428800     0.52428800    162.76   <.0001
RATE                         4      1.39845700     0.34961425    108.53   <.0001
LIME*RATE                    4      0.31074700     0.07768675     24.12   <.0001

 Contrast                        DF    Contrast SS     Mean Squ.  F Value    Pr > F

 Lime at 1 ton             (C)    1      0.015210      0.015210      4.72    0.0358
 Lime effect at 1 vs. 2    (D)    1      0.027380      0.027380      8.50    0.0058

                  LIME*RATE Effect Sliced by RATE for APH

                                Sum of
    RATE        DF             Squares       Mean Square      F Value      Pr > F

    0            1            0.001210          0.001210         0.38      0.5434
    1            1            0.015210          0.015210         4.72      0.0358
    2            1            0.127690          0.127690        39.64      <.0001
    4            1            0.122102          0.122102        37.91      <.0001
    8            1            0.568822          0.568822       176.58      <.0001
```

The order in which variables are listed in the `class` statement determines how SAS® organizes the cell means. The subscript of the factor listed first varies slower than the subscript of the factor listed second and so forth. The `class` statement

```
class lime rate;
```

results in cell means ordered $\mu_{11}, \mu_{12}, \mu_{13}, \mu_{14}, \mu_{15}, \mu_{21}, \mu_{22}, \mu_{23}, \mu_{24}, \mu_{25}$. Contrast coefficients are assigned in the same order to the `lime*rate` effect in the `contrast` statement. If the `class` statement were

```
class rate lime;
```

the cell means would be ordered $\mu_{11}, \mu_{21}, \mu_{12}, \mu_{22}, \mu_{13}, \mu_{23}, \mu_{14}, \mu_{24}, \mu_{15}, \mu_{25}$ and the arrangement of contrast coefficients for the `lime*rate` effect changes accordingly.

The `slice` option of the `lsmeans` statement makes it particularly simple to obtain comparisons of Lime Types separately at each level of the Rate factor. Slices by either factor can be obtained with the statement

```
lsmeans lime*rate / slice=(rate lime);
```

and slices of three-way interactions are best carried out as

```
lsmeans A*B*C / slice=(A*B A*C B*C);
```

so that only a single factor is being compared at each combination of two other factors.

The table of effect slices at the bottom of Output 4.4 conveys no significant difference among lime types at 0 tons ($F_{obs} = 0.38, p = 0.5434$), but the F statistics increase with increasing rate of application. This represents the increasing separation of the trends in Figure 4.2. Since factor Lime Type has only two levels the contrast comparing Lime Types at 1 ton is identical to the slice at that rate ($F_{obs} = 4.72, p = 0.0358$).

One could perform slices of Lime Type $\times$ Rate in the other direction, comparing the rates of application at each lime type. We notice, however, that the rate of application is a **quantitative** factor. It is thus more meaningful to test the nature of the trend between pH and Rate of Application with **regression contrasts** (orthogonal polynomials). Since five rates were applied we can test for quartic, cubic, quadratic, and linear trends. Published tables of orthogonal polynomial coefficients require that the levels of the factor are evenly spaced and that the data are balanced. In this example, the factors are unevenly spaced. The correct contrast coefficients can be calculated with the `OrPol()` function of SAS/IML®. The `%orpoly` macro contained on the CD-ROM finds coefficients up to the eighth degree for any particular factor level spacing. When the macro is executed for the factor level spacing $0, 1, 2, 4, 8$ the coefficients shown in Output 4.5 result.

```
data rates;
   input levels @@;
   datalines;
0 1 2 4 8
;;
run;
%orpoly(data=rates,var=levels);
```

Output 4.5.

	linear	quadratic	cubic	quartic
0	-.47434	0.54596	-.46999	0.23672
1	-.31623	0.03522	0.42226	-.72143
2	-.15811	-.33462	0.51435	0.63125
4	0.15811	-.65163	-.57050	-.15781
8	0.79057	0.40507	0.10388	0.01127

The coefficients are fractional numbers. Sometimes, they can be converted to values that are easier to code using the following trick. Since the contrast F statistic is unaffected by a re-scaling of the contrast coefficients, we can multiply the coefficients for a contrast with an arbitrary constant. Dividing the coefficients by the smallest coefficient for each trend yields the coefficients in Table 4.11.

Table 4.11. Contrast coefficients from Output 4.5 divided by the smallest coefficient in each column

	Order of Trend			
Level	**Linear**	**Quadratic**	**Cubic**	**Quartic**
0 tons	– 3	15.5	– 4.5	21
1 ton	– 2	1	4.05	– 64
2 tons	– 1	– 9.5	4.95	56
4 tons	1	– 18.5	– 5.5	– 14
8 tons	5	11.5	1	1

Since the factors interact we test the order of the trends for agricultural lime (AL) and granulated lime (GL) separately by adding to the `proc glm` code (Output 4.6) the statements

```
contrast 'rate quart (AL)'  rate  21   -64    56    -14    1
                       lime*rate  21   -64    56    -14    1;
contrast 'rate cubic (AL)'  rate  -4.5  4.05  4.95  -5.5  1
                       lime*rate  -4.5  4.05  4.95  -5.5  1;
contrast 'rate quadr.(AL)'  rate  15.5  1     -9.5  -18.5 11.5
                       lime*rate  15.5  1     -9.5  -18.5 11.5;
contrast 'rate linear(AL)'  rate  -3    -2    -1     1    5
                       lime*rate  -3    -2    -1     1    5;

contrast 'rate quart (GL)'    rate  21   -64    56    -14    1
                lime*rate 0 0 0 0 0  21   -64    56    -14    1;
contrast 'rate cubic (GL)'    rate  -4.5  4.05  4.95  -5.5  1
                lime*rate 0 0 0 0 0  -4.5  4.05  4.95  -5.5  1;
contrast 'rate quadr.(GL)'    rate  15.5  1     -9.5  -18.5 11.5
                 lime*rate 0 0 0 0 0 15.5  1     -9.5  -18.5 11.5;
contrast 'rate linear(GL)'    rate  -3    -2    -1     1    5
                lime*rate 0 0 0 0 0  -3    -2    -1     1    5;
```

Output 4.6.

```
Contrast              DF    Contrast SS    Mean Square   F Value   Pr > F

rate quart (AL)        1     0.00128829     0.00128829      0.40   0.5307
rate cubic (AL)        1     0.00446006     0.00446006      1.38   0.2463
rate quadr.(AL)        1     0.01160085     0.01160085      3.60   0.0650
rate linear(AL)        1     1.46804112     1.46804112    455.74   <.0001

rate quart (GL)        1     0.01060179     0.01060179      3.29   0.0772
rate cubic (GL)        1     0.00519994     0.00519994      1.61   0.2112
rate quadr.(GL)        1     0.00666459     0.00666459      2.07   0.1581
rate linear(GL)        1     0.20129513     0.20129513     62.49   <.0001
```

For either Lime Type, one concludes that the trend of pH is linear in application rate at the 5% significance level. For agricultural lime a slight quadratic effect is noticeable. The interaction between the two factors should be evident in a comparison of the linear trends among the Lime Types. This is accomplished with the `contrast` statement

```
contrast 'Linear(AL vs. GL)'
     lime*rate -3  -2  -1  1  5  3  2  1  -1  -5;
```

Output 4.7.

```
Contrast              DF   Contrast SS   Mean Square  F Value   Pr > F
Linear(AL vs. GL)      1    0.29106025    0.29106025    90.36   <.0001
```

The linear trends are significantly different ($F_{obs} = 90.36, p < 0.0001$). From Figure 4.2 this is evidently due to differences in the slopes of the two lime types.

4.4 Diagnosing Regression Models

Box 4.4 Model Diagnostics

- **Diagnosing (criticizing) a linear regression model utilizes**
  - **— residual analysis**
  - **— case deletion diagnostics**
  - **— collinearity diagnostics and**
  - **— other diagnostics concerned with assumptions of the linear model.**
- **If breakdowns of a correct model cannot be remedied because of outlying or influential observations, alternative estimation methods may be employed.**

Diagnosing the model and its agreement with a particular data set is an essential step in developing a good statistical model and sound inferences. Estimating model parameters and drawing statistical inferences must be accompanied by sufficient criticism of the model. This criticism should highlight whether the assumptions of the model are met and if not, to what degree they are violated. Key assumptions of the classical linear model [4.1] are

- correctness of the model ($\mathrm{E}[\mathbf{e}] = \mathbf{0}$) and
- homoscedastic, uncorrelated errors ($\mathrm{Var}[\mathbf{e}] = \sigma^2\mathbf{I}$).

Often the assumption of Gaussian errors is added and must be diagnosed, too, not because least squares estimation requires it, but to check the validity of exact inferences. The importance of these assumptions, in terms of complications introduced into the analysis by their violation, follows roughly the same order.

4.4.1 Residual Analysis

Box 4.5 Residuals

- **The raw residual $\widehat{e}_i = y_i - \widehat{y}_i$ is not a good diagnostic tool, since it does not mimic the behavior of the model disturbances e_i. The n residuals are correlated, heteroscedastic, and do not constitute n pieces of information.**

- **Studentized residuals are suitably standardized residuals that at least remedy the heteroscedasticity problem.**

Since the model residuals (errors) $\mathbf{e}$ are unobservable it seems natural to focus model criticism on the fitted residuals $\widehat{\mathbf{e}} = \mathbf{y} - \widehat{\mathbf{y}} = \mathbf{y} - \mathbf{X}\widehat{\boldsymbol{\beta}}$. The fitted residuals do not behave exactly as the model residuals. If $\widehat{\boldsymbol{\beta}}$ is an unbiased estimator of $\boldsymbol{\beta}$, then $\mathrm{E}[\widehat{\mathbf{e}}] = \mathrm{E}[\mathbf{e}] = \mathbf{0}$, but the fitted residuals are neither uncorrelated nor homoscedastic. We can write $\widehat{\mathbf{e}} = \mathbf{y} - \widehat{\mathbf{y}} = \mathbf{y} - \mathbf{H}\mathbf{y}$, where $\mathbf{H} = \mathbf{X}(\mathbf{X}'\mathbf{X})^{-1}\mathbf{X}'$ is called the "Hat" matrix since it produces the $\widehat{\mathbf{y}}$ values when post-multiplied by $\mathbf{y}$ (see §A4.8.3 for details). The hat matrix is a symmetric idempotent matrix which implies that

$$\mathbf{H}' = \mathbf{H}, \mathbf{H}\mathbf{H} = \mathbf{H} \text{ and } (\mathbf{I} - \mathbf{H})(\mathbf{I} - \mathbf{H}) = (\mathbf{I} - \mathbf{H}).$$

The variance of the fitted residuals now follows easily:

$$\mathrm{Var}[\widehat{\mathbf{e}}] = \mathrm{Var}[\mathbf{Y} - \mathbf{H}\mathbf{Y}] = \mathrm{Var}[(\mathbf{I} - \mathbf{H})\mathbf{Y}] = \sigma^2\mathbf{I}(\mathbf{I} - \mathbf{H})(\mathbf{I} - \mathbf{H}) = \sigma^2(\mathbf{I} - \mathbf{H}). \quad [4.21]$$

$\mathbf{H}$ is not a diagonal matrix and the entries of its diagonal are not equal. The fitted residuals thus are neither uncorrelated nor homoscedastic. Furthermore, $\mathbf{H}$ is a singular matrix of rank $n - r(\mathbf{X})$. If one fits a standard regression model with intercept and $k - 1$ regressors only $n - k$ least squares residuals carry information about the model disturbances, the remaining residuals are redundant. In classification models where the rank of $\mathbf{X}$ can be large relative to the sample size, only a few residuals are nonredundant.

Because of their heteroscedastic nature, we do not recommend using the raw residuals $\widehat{e}_i$ to diagnose model assumptions. First, the residual should be properly scaled. The ith diagonal value of $\mathbf{H}$ is called the **leverage** of the ith data point. Denote it as h_{ii} and it follows that

$$\mathrm{Var}[\widehat{e}_i] = \sigma^2(1 - h_{ii}). \quad [4.22]$$

Standardized residuals have mean 0 and variance 1 and are obtained as $\widehat{e}_i^* = \widehat{e}_i/(\sigma\sqrt{1 - h_{ii}})$. Since the variance σ^2 is unknown, σ is replaced by its estimate $\widehat{\sigma}$, the square root of the model mean square error. The residual

$$r_i = \frac{\widehat{e}_i}{\widehat{\sigma}\sqrt{1 - h_{ii}}} \quad [4.23]$$

is called the **studentized** residual and is a more appropriate diagnostic measure. If the model errors e_i are Gaussian, the scale-free studentized residuals are akin to t random variables which justifies — to some extent — their use in diagnosing outliers in regression models (see below and Myers 1990, Ch. 5.3). Plots of r_i against the regressor variables can be used to diagnose the equal variance assumption and the need to transform regressor variables. A graph of r_i against the fitted values $\widehat{y}_i$ highlights the correctness of the model. In either type of plot the residuals should appear as a *stable* band of random scatter around a horizontal line at zero.

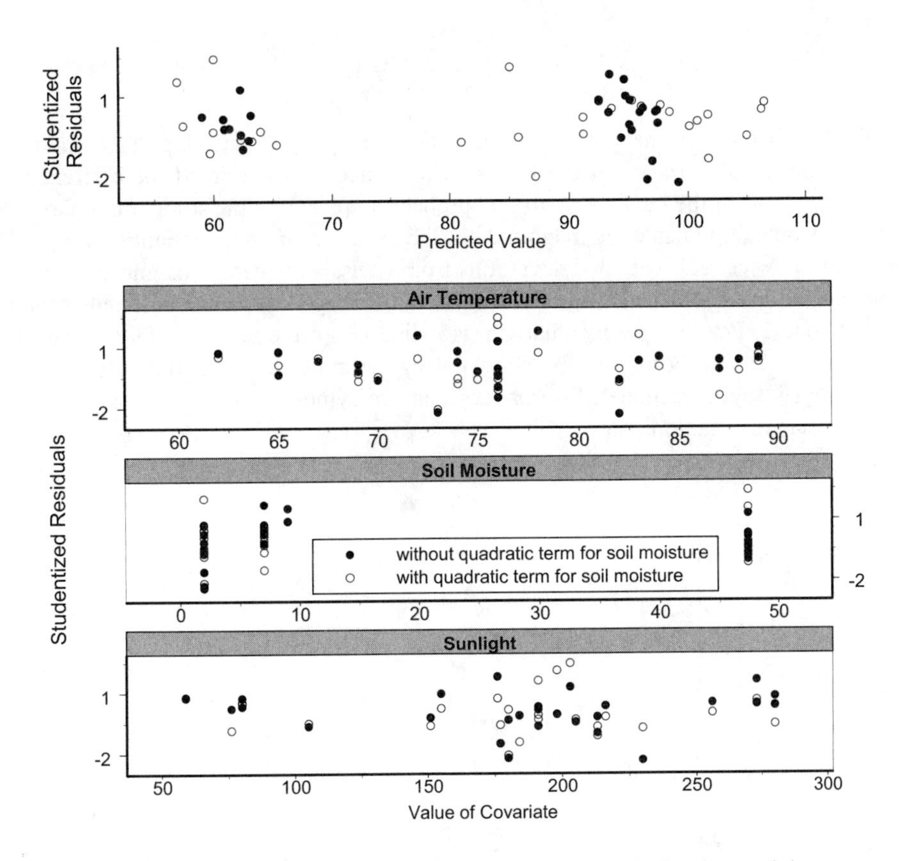

Figure 4.6. Studentized residuals in the turnip green data analysis for the model $y_i = \beta_0 + \beta_1 x_{1i} + \beta_2 x_{2i} + \beta_3 x_{3i}$ (full circles) and $y_i = \beta_0 + \beta_1 x_{1i} + \beta_2 x_{2i} + \beta_3 x_{3i} + \beta_4 x_{2i}^2$ (open circles).

For the Turnip Greens data (Example 4.2) studentized residuals for the model with and without the squared term for soil moisture are shown in Figure 4.6. The quadratic trend in the residuals for soil moisture is obvious if $X_4 = X_2^2$ is omitted.

A problem of the residual by covariate plot is the interdependency of the covariates. The plot of r_i vs. Air Temperature, for example, assumes that the other variables are held constant. But changing the amount of sunlight obviously changes air temperature. How this **collinearity** of the regressor variables impacts not only the interpretation but also the stability of the least squares estimates is addressed in §4.4.4.

To diagnose the assumption of Gaussian model errors one can resort to graphical tools such as normal quantile and normal probability plots, to formal tests for normality, for example, the tests of Shapiro and Wilk (1965) and Anderson and Darling (1954) or to goodness-of-fit tests based on the empirical distribution function. The normal probability plot is a plot of the ranked studentized residuals (ordinate) against the expected value of the ith smallest value in a random sample of size n from a standard Gaussian distribution (abscissa). This expected value of the ith smallest value can be approximated as the $(p \times 100)^{\text{th}}$ $G(0,1)$ percentile z_p,

$$p = \left(\frac{i - 0.375}{n + 0.25} \right).$$

The reference line in a normal probability plot has intercept $-\mu/\sigma$ and slope $1/\sigma$ where μ and σ^2 are the mean and variance of the Gaussian reference distribution. If the reference is the $G(0,1)$, deviations of the points in a normal probability plot from the straight line through the origin with slope 1.0 indicate the magnitude of the departure from Gaussianity. Myers (1990, p. 64) shows how certain patterned deviations from Gaussianity can be diagnosed in this plot. For studentized residuals the 45°-line is the correct reference since these residuals have mean 0 and variance 1. For the raw residuals the 45°-line is not the correct reference since their variance is not 1. The normal probability plot of the studentized residuals for the Turnip Greens analysis suggests model disturbances that are symmetric but less heavy in the tails than a Gaussian distribution.

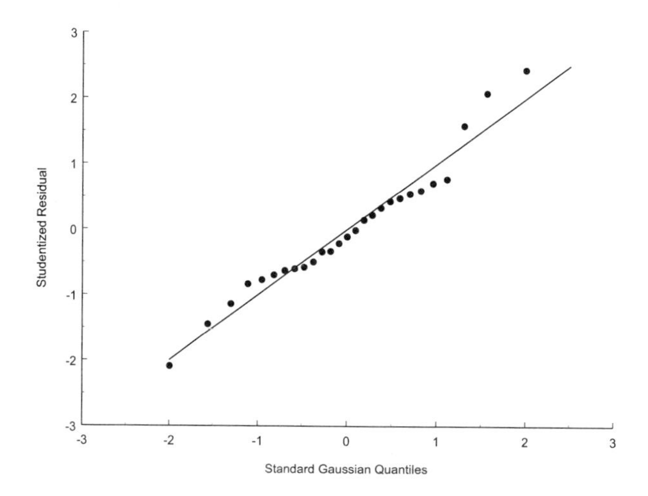

Figure 4.7. Normal probability plot of studentized residuals for full model in Turnip Greens analysis.

Plots of raw or studentized residuals against regressor or fitted values are helpful to visually assess the quality of a model without attaching quantitative performance measures. Our objection concerns using these types of residuals for methods of residual analysis that predicate a random sample, i.e., independent observations. These methods can be graphical (such as the normal probability plot) or quantitative (such as tests of Gaussianity). Such diagnostic tools should be based on $n - k$ residuals that are homoscedastic and uncorrelated. These can be constructed as **recursive** or **linearly recovered** errors.

4.4.2 Recursive and Linearly Recovered Errors

The three goals of error recovery are (i) to remove the correlation among the fitted residuals, (ii) to recover homoscedastic variables, and (iii) to avoid the illusion that there are actually n *observations*. From the previous discussion it is clear that if $\mathbf{X}$ contains k linearly independent columns, only $n - k$ errors can be recovered after a least squares fit, since the ordinary least square residuals satisfy k constraints (the vector $\widehat{\mathbf{e}}$ is orthogonal to each column of $\mathbf{X}$). P-values in tests for normality that depend on sample size will be reported erroneously if the set of n raw or studentized residuals is used for testing. The correlations among the $n - k$ nonredundant residuals will further distort these p-values since the tests assume independence of the observations. Although studentized residuals have unit variance, they still are correlated and for the purpose of validating the Gaussianity assumption, they suffer from similar problems as the raw residuals.

Error recovery is based either on employing the projection properties of the "Hat" matrix, $\mathbf{H} = \mathbf{X}(\mathbf{X}'\mathbf{X})^{-1}\mathbf{X}'$, or on sequentially forecasting observations based on a fit of the model to preceding observations. Recovered errors of the first type are called Linear Unbiased Scaled (LUS) residuals (Theil 1971), errors of the second type are called **recursive** or sequential residuals. Kianifard and Swallow (1996) provide a review of the development of recursive residuals. We will report only the most important details here.

Consider the standard linear model with Gaussian errors,

$$\mathbf{Y}_{(n\times 1)} = \mathbf{X}_{(n\times k)}\boldsymbol{\beta} + \mathbf{e},\ \mathbf{e} \sim G\left(\mathbf{0}, \sigma^2\mathbf{I}\right),$$

and fit the model to k data points. The remaining $n - k$ data points are then entered sequentially and the jth recursive residual is the scaled difference of predicting the next observation from the model fit to the previous observations. More formally, let $\mathbf{X}_{j-1}$ be the matrix consisting of the first $j - 1$ rows of $\mathbf{X}$. If $\mathbf{X}'_{j-1}\mathbf{X}_{j-1}$ is nonsingular and $j \geq k + 1$ the parameter vector $\boldsymbol{\beta}$ can be estimated as

$$\widehat{\boldsymbol{\beta}}_{j-1} = \left(\mathbf{X}'_{j-1}\mathbf{X}_{j-1}\right)^{-1}\mathbf{X}'_{j-1}\mathbf{y}_{j-1}. \qquad [4.24]$$

Now consider adding the next observation y_j. Define as the (unstandardized) recursive residual w_i^* the difference between y_j and the predicted value based on fiting the model to the preceding observations,

$$w_i^* = y_j - \mathbf{x}'_j\widehat{\boldsymbol{\beta}}_{j-1}.$$

Finally, scale w_i^* and define the recursive residual (Brown et al. 1975) as

$$w_i = \frac{w_i^*}{\sqrt{1 + \mathbf{x}'_j\left(\mathbf{X}'_{j-1}\mathbf{X}_{j-1}\right)^{-1}\mathbf{x}_j}},\ j = k + 1, \cdots, n. \qquad [4.25]$$

The w_i are independent random variables with mean 0 and variance $\text{Var}[w_i] = \sigma^2$, just as the model disturbances. However, only $n - k$ of them are available. Recursive residuals are unfortunately not unique. They depend on the set of k points chosen initially and on the order of the remaining data. It is not at all clear how to compute the *best* set of recursive residuals. Because data points with high leverage can have negative impact on the analysis, one possibility is to order the data by leverage to circumvent calculation of recursive residuals for potentially influential data points. In other circumstances, for example, when the detection of outliers is paramount, one may want to produce recursive residuals for precisely these observations. On occasion a particular ordering of the data suggests itself, for example, if a covariate relates to time or distance. A second complication of initially fitting the model to k observations is the need for $\mathbf{X}'_{j-1}\mathbf{X}_{j-1}$ to be nonsingular. When columns of $\mathbf{X}$ contain repeat values, for example, treatment variables or soil moisture in the Turnip Greens data (Table 4.2), the data must be rearranged or the number of data points in the initial fit must be enlarged. The latter leads to the recovery of fewer errors.

Recursive residuals in regression models can be calculated with `proc autoreg` of The SAS® System (Release 7.0 or higher). The following code simulates a data set with fifteen observation and a systematic quadratic trend. The error distribution was chosen as t on seven degrees of freedom and is thus slightly heavier in the tails than the standard Gaussian distribution. The normal probability plots of the raw and recursive residuals are shown in Figure 4.8.

```
data simu;
  do i = 1 to 15;
    x   = ranuni(4355)*30;
    u   = ranuni(6573);
    e   = tinv(u,7);
    x2  = x*x;
    fn  = 0.1 + 0.2*x - 0.04*x2;
    y = fn + e;
    output;
  end; drop i;
run;

proc reg data=simu noprint;
  model y = x x2;
  output out=regout student=student rstudent=rstudent residual=res h=h;
run; quit;
/* Sort by descending leverage to eliminate high leverage points first */
proc sort data=regout; by descending h; run;

proc autoreg data=regout noprint;
  model y = x x2;
  output out=residuals residual=res blus=blus recres=recres;
run;
proc univariate data=residuals normal plot; var student recres; run;
```

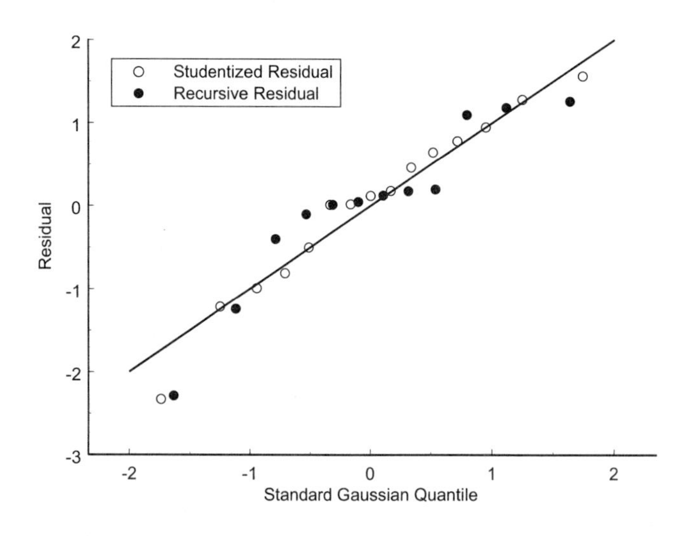

Figure 4.8. Normal probability plots of studentized and recursive residuals for simulated data. Solid line represents standard Gaussian distribution.

The recursive residuals show the deviation from Gaussianity more clearly than the studentized residuals which tend to cluster around the reference line. The p-values for the Shapiro-Wilk test for Gaussianity were 0.7979 and 0.1275 for the studentized and recursive residuals, respectively. Applying the Shapiro-Wilk test incorrectly to the correlated raw residuals leads to considerably less evidence for detecting a departure from Gaussianity. The p-values for the Anderson-Darling test for Gaussianity were > 0.25 and 0.1047, respectively. In all instances the recursive residuals show more evidence against Gaussianity.

We emphasized recursive residuals for diagnosing the assumption of normal disturbances. This is certainly not the only possible application. Recursive residuals are also key in detecting outliers, detecting changes in the regression coefficients, testing for serial correlation, and testing for heteroscedasticity. For details and further references, see Galpin and Hawkins (1984) and Kianifard and Swallow (1996).

The second method of error recovery leads to Linearly Unbiased Scaled (LUS) estimates of e_i and is due to Theil (1971). It exploits the projection properties in the standard linear model. Denote the $(t = n - k \times 1)$ vector of LUS residuals as

$$\mathbf{R}_{(t)} = [R_1, R_2, \cdots, R_{n-k}]'.$$

The name reflects that $\mathbf{R}_{(t)}$ is a linear function of $\mathbf{Y}$, has the same expectation as the model disturbances ($\mathrm{E}[\mathbf{R}] = \mathrm{E}[\mathbf{e}] = \mathbf{0}$), and has a scalar covariance matrix ($\mathrm{Var}[\mathbf{R}_{(t)}] = \sigma^2\mathbf{I}$). In contrast to recursive residuals it is not necessary to fit the model to an initial set of k observations, since the process is not sequential. This allows the recovery of $t = n - k$ uncorrelated, homoscedastic residuals in the presence of classification variables. The error recovery proceeds as follows. Let $\mathbf{M} = \mathbf{I} - \mathbf{H}$ and recall that $\mathrm{Var}[\widehat{\mathbf{e}}] = \sigma^2\mathbf{M}$. If one premultiplies $\widehat{\mathbf{e}}$ with a matrix $\mathbf{Q}'$ such that

$$\mathrm{Var}[\mathbf{Q}'\widehat{\mathbf{e}}] = \sigma^2 \begin{bmatrix} \mathbf{I}_t & \mathbf{0} \\ \mathbf{0} & \mathbf{0} \end{bmatrix},$$

then the first t elements of $\mathbf{Q}'\widehat{\mathbf{e}}$ are the LUS estimates of $\mathbf{e}$. Does such a matrix $\mathbf{Q}$ exist? This is indeed the case and unfortunately there are many such matrices. The spectral decomposition of a real symmetric matrix $\mathbf{A}$ is $\mathbf{PDP}' = \mathbf{A}$, where $\mathbf{P}$ is an orthogonal matrix containing the ordered eigenvectors of $\mathbf{A}$, and $\mathbf{D}$ is a diagonal matrix containing the ordered eigenvalues of $\mathbf{A}$. Since $\mathbf{M}$ is symmetric idempotent (a projector) it has a spectral decomposition. Furthermore, since the eigenvalues of a projector are either 1 or 0, and the number of nonzero eigenvalues equals the rank of a matrix, there are $t = n - k$ eigenvalues of value 1 and the remaining values are 0. $\mathbf{D}$ thus has precisely the structure

$$\mathbf{D} = \begin{bmatrix} \mathbf{I}_t & \mathbf{0} \\ \mathbf{0} & \mathbf{0} \end{bmatrix}$$

we are looking for. So, if the spectral decomposition of $\mathbf{M}$ is $\mathbf{M} = \mathbf{QDQ}'$, then $\mathbf{Q}'\mathbf{MQ} = \mathbf{D}$, since $\mathbf{Q}$ is an orthogonal matrix ($\mathbf{Q}'\mathbf{Q} = \mathbf{I}$). Jensen and Ramirez (1999) call the first t elements of $\mathbf{Q}'\widehat{\mathbf{e}}$ the **linearly recovered errors**. Their stochastic properties in t-dimensional space are identical to those of the model disturbances $\mathbf{e}$ in n-dimensional space. The nonuniqueness of the LUS estimates stems from the nondistinctness of the eigenvalues of $\mathbf{M}$. Any orthogonal rotation $\mathbf{BQ}'$ will also produce a LUS estimator for $\mathbf{e}$. Theil (1971) settled the problem of how to make the LUS unique by defining as *best* the set of LUS residuals that minimize the expected sum of squares of estimation errors and termed those the BLUS residuals. While this is a sensible criterion, Kianifard and Swallow (1996) point out that there is no reason why recovered errors being *best* in this sense should necessarily lead to tests with high power. To diagnose normality of the model disturbances, Jensen and Ramirez (1999) observe that the skewness and kurtosis of the least square residuals are closer to the values expected under Gaussianity than are those of nonnormal model errors. The diagnostic tests are thus slanted toward Gaussianity. These authors proceed to settle the nonuniqueness of the LUS residuals by finding that orthogonal rotation $\mathbf{BQ}'$ which maximizes the kurtosis (4th moment) of the recovered errors. With respect to the fourth moment, this recovers (rotated) errors that are *as ugly as possible*, slanting diagnostic tests away from Gaussianity. Particularly with small sample sizes this may be of importance to retain power. Linearly recovered errors and recursive residuals are no longer in a direct correspondence with the data points. The jth recursive residual does not represent the residual for the jth observation. Their use as diagnostic tools is thus restricted to those inquiries where the individual observation is immaterial, e.g., tests for normality.

4.4.3 Case Deletion Diagnostics

Box 4.6 Influential Data Point

- **A data point is called highly influential if an important aspect of the fitted model changes considerably if the data point is removed.**
- **A high leverage point is unusual with respect to other x values; it has the potential to be an influential data point.**
- **An outlying observation is unusual with respect to other Y values.**

Case deletion diagnostics assess the influence of individual observations on the overall analysis. The idea is to remove a data point and refit the model without it. The change in a particular aspect of the fitted model, e.g., the residual for the deleted data point or the least squares estimates, is a measure for its influence on the analysis. Problematic are **highly influential** points (hips), since they have a tendency to dominate the analysis. Fortunately, in linear regression models, these diagnostics can be calculated without actually fitting a regression model n times but in a single fit of the entire data set. This is made possible by the Sherman-Morrison-Woodbury theorem which is given in §A4.8.3. For a data point to be a hip, it must be either an outlier or a high leverage data point. Outliers are data points that are unusual relative to the other Y values. The Magnesium observation in the prediction efficiency data set (Example 4.4, p. 93) might be an outlier. The attribute outlier does not have a negative connotation. It designates a data point as unusual and does not automatically warrant deletion. An outlying data point is only outlying with respect to a particular statistical model or criterion. If, for example, the trend in y is quadratic in x, but a simple linear regression model, $y_i = \beta_0 + \beta_1 x_i + e_i$, is fit to the data, many data points may be classified as outliers because they do not agree with the model. The reason therefore is an incorrect model, not erroneous observations. According to the commonly applied definition of what constitutes outliers based on the box-plot of a set of data, outliers are those values that are more than 1.5

times the interquartile-range above the third or below the first quartile of the data. When randomly sampling $1,000$ observations from a Gaussian distribution, one should expect to see on average about seven outliers according to this definition.

To understand the concept of leverage, we concentrate on the "Hat" matrix

$$\mathbf{H} = \mathbf{X}(\mathbf{X}'\mathbf{X})^{-1}\mathbf{X}' \qquad [4.26]$$

introduced in the previous section. This matrix does not depend on the observed responses $\mathbf{y}$, only on the information in $\mathbf{X}$. Its ith diagonal element, h_{ii}, measures the **leverage** of the ith observation and expresses how unusual or extreme the covariate record of this observation is relative to the other observations. In a simple linear regression, a single x value far removed from the bulk of the x-data is typically a data point with high leverage. If $\mathbf{X}$ has full rank k, a point is considered a high leverage point if

$$h_{ii} > 2k/n.$$

High leverage points deserve special attention because they *may* be influential. They have the **potential** to pull the fitted regression toward it. A high leverage point that follows the regression trend implied by the remaining observations will not exert undue influence on the least squares estimates and is of no concern. The decision whether a high leverage point is influential thus rests on combining information about leverage with the magnitude of the residual. As leverage increases, smaller and smaller residual values are needed to declare a point as influential. Two statistics are particularly useful in this regard. The **RStudent** residual is a studentized residual that combines leverage and the fitted residual similar to r_i in [4.23],

$$\text{RStudent}_i = \widehat{e}_i / \left(\widehat{\sigma}_{-i}\sqrt{1-h_{ii}}\right). \qquad [4.27]$$

Here, $\widehat{\sigma}^2_{-i}$ is the mean square error estimate obtained after removal of the ith data point. The **DFFITS** (**dif**ference in **fit**, **s**tandardized) statistic measures the change in fit in terms of standard error units when the ith observation is deleted:

$$\text{DFFITS}_i = \text{RStudent}_i\sqrt{h_{ii}/(1-h_{ii})}. \qquad [4.28]$$

A DFFITS_i value of 2.0, for example, implies that the fit at y_i will change by two standard error units if the ith data point is removed. DFFITs are useful to asses whether a data point is highly influential and RStudent residuals are good at determining outlying observations. According to [4.27] a data point may be a hip if it has a moderate residual $\widehat{e}_i$ and high leverage or if it is not a high leverage point (h_{ii} small), but unusual in the $\mathbf{y}$ space ($\widehat{e}_i$ large).

RStudent residuals and DFFITS measure changes in residuals or fitted values as the ith observation is deleted. The influence of removing the ith observation on the least squares estimate $\widehat{\boldsymbol{\beta}}$ is measured by Cook's Distance D_i (Cook 1977):

$$D_i = r_i^2 h_{ii}/((k+1)(1-h_{ii})) \qquad [4.29]$$

When diagnosing (criticizing) a particular model, one or more of these statistics may be important. If the purpose of the model is mainly predictive, the change in the least squares estimates (D_i) is secondary to RStudent residuals and DFFITS. If the purpose of the model lies in testing hypotheses about β, Cook's distance will gain importance. Rule of thumb (rot) values for the various statistics are shown in Table 4.12, but caution should be exercised in their application:

- If a decision is made to delete a data point because of its influence, the data set changes and the leverage, RStudent values, etc. of the remaining observations also change. Data points not influential before may be designated influential now and continuance in this mode leads to the elimination of too many data points.
- Data points often are unusual in groups. This is especially true for clustered data where entire clusters may be unusual. The statistics discussed here apply to single-case deletion only.
- The fact that a data point is influential does not imply that the data point must be deleted. It should prompt the investigator to question the validity of the observation *and* the validity of the model.
- Several of the case deletion diagnostics have known distributional properties under Gaussianity of the model errors. When these properties are tractable, relying on rules of thumbs instead of exact p-values is difficult to justify. For important results about the distributional properties of these diagnostics see, for example, Jensen and Ramirez (1998) and Dunkl and Ramirez (2001).
- Inexperienced modelers tend to delete more observations based on these criteria than warranted.

Table 4.12. Rules of thumbs (rots) for leverage and case deletion diagnostics: k denotes the number of parameters (regressors plus intercept), n the number of observations

Statistic	Formula	Rule of thumb	Conclusion
Leverage	h_{ii}	$h_{ii} > 2k/n$	high leverage point
RStudent$_i$	$\widehat{e}_i/(\widehat{\sigma}_{-i}\sqrt{1-h_{ii}})$	$\lvert\text{RStudent}_i\rvert > 2$	outlier (hip)
DFFITS$_i$	$r_i^2 h_{ii}/(k(1-h_{ii}))$	$\lvert\text{DFFITS}_i\rvert > 2\sqrt{k/n}$	hip
Cook's Distance	$D_i = r_i^2 h_{ii}/(k(1-h_{ii}))$	$D_i > 1$	hip

Example 4.4. Prediction Efficiency (continued). Recall the Prediction Efficiency example introduced on p. 93 and the data shown in Figure 4.4. Although Mg is obviously a *strange* observation in this data set, the overall trend between prediction efficiency and the range of the spatial correlation could be quadratic.

Let X_i denote the spatial range of the ith attribute and Y_i its prediction efficiency, we commence by fitting a quadratic polynomial

$$Y_i = \beta_0 + \beta_1 x_i + \beta_2 x_i^2 + e_i \qquad [4.30]$$

with `proc reg` in The SAS® System:

```
proc reg data=range;
  model eff30 = range range2;
run; quit;
```

Output 4.8.

```
                    The REG Procedure
                     Model: MODEL1
               Dependent Variable: eff30

                  Analysis of Variance

                              Sum of           Mean
Source               DF      Squares         Square    F Value    Pr > F

Model                 2    854.45840      427.22920       3.34    0.1402
Error                 4    511.52342      127.88086
Corrected Total       6   1365.98182


      Root MSE              11.30844    R-Square     0.6255
      Dependent Mean        25.04246    Adj R-Sq     0.4383
      Coeff Var             45.15707


                  Parameter Estimates

                        Parameter       Standard
   Variable     DF       Estimate          Error    t Value    Pr > |t|

   Intercept     1      -39.12105       37.76100      -1.04      0.3587
   range         1        1.39866        1.00851       1.39      0.2378
   range2        1       -0.00633        0.00609      -1.04      0.3574
```

Ordinary least squares estimates are $\widehat{\beta}_0 = -39.121$, $\widehat{\beta}_1 = 1.398$, and $\widehat{\beta}_2 = -0.00633$. The coefficient of determination is

$$R^2 = 1 - \frac{SSR}{SST_m} = 0.6255.$$

Removing the Mg observation and refitting the quadratic polynomial leads to $\widehat{\beta}_0 = -75.102$, $\widehat{\beta}_1 = 2.414$, and $\widehat{\beta}_2 = -0.0119$ and the coefficient of determination for the quadratic model increases from 0.6255 to 0.972 (output not shown). Obviously, this data point exerts large influence on the analysis.

The various case deletion diagnostics are listed in Table 4.13. The purpose of their calculation is to determine whether the Mg observation is an influential data point. The rule of thumb value for a high leverage point is $2k/n = 2*3/7 = 0.857$ and for a DFFITS is $2\sqrt{3/7} = 1.309$. The diagnostics were calculated with the `/influence` option of the `model` statement in `proc reg`. This option produces raw residuals, RStudent residuals, leverages, DFFITS, and other diagnostics:

```
proc reg data=range;
  model eff30 = range range2 / influence;
  output out=cookd cookd=cookd;
run; quit;
proc print data=cookd; run;
```

The two most extreme regressor values, for P and lime, have the largest leverages. P is a borderline high leverage point (Table 4.13). The residual of the P observations is not too large so that it is not considered an outlying observation as judged by RStudent. It has considerable influence on the fitted values (DFFITS_2) and the least squares estimates (D_2). Removing the P observation will change the fit dramatically. Since it is not a candidate observation for deletion, it remains in the data set. The Mg data point is not a high leverage point, since it is near the center of the observed range (Figure 4.4). Its unusual behavior in the **y**-space is evidenced by a very large RStudent residual and a large DFFITs value. It is an outlier with considerable influence on fitted values. Its influence on the least squares estimates as judged by Cook's Distance is not critical. Since the purpose of modeling these data is the derivation of a predictive equation of efficiency as a function of spatial range, influence on fitted (predicted) values is more important than influence on the actual estimates.

Table 4.13. Case deletion diagnostics for quadratic polynomial in Prediction Efficiency application (*rot* denotes a rule-of-thumb value)

Obs.	Attribute	$\widehat{e}_i$	h_{ii} (rot $= 0.857$)	RStudent_i (rot $= 2$)	DFFITS_i (rot $= 1.309$)	D_i (rot $= 1$)
1	pH	3.6184	0.2157	0.3181	0.1668	0.0119
2	P	3.3750	0.9467	1.4673	6.1842	9.8956
3	Ca	9.9978	0.2309	1.0109	0.5539	0.1017
4	Mg	– 17.5303	0.3525	– 6.2102	– 4.5820	0.6734
5	CEC	0.7828	0.2711	0.0703	0.0429	0.0007
6	Lime	– 6.4119	0.6307	– 0.9135	– 1.1936	0.4954
7	Prec	6.1682	0.3525	0.6240	0.4604	0.0834

4.4.4 Collinearity Diagnostics

Case deletion diagnostics focus on the relationship among the rows of **Y** (outliers) and rows of **X** (leverage) and their impact on the analysis. Collinearity is a condition among the **columns** of **X**. The two extreme cases are (i) complete independence of its columns (orthogonality) and (ii) one or more exact linear dependencies among the columns. In the first case, the inner product (§3.3) between any columns of **X** is 0 and $\mathbf{X}'\mathbf{X}$ is a diagonal matrix. Since $\text{Var}[\widehat{\boldsymbol{\beta}}] = \sigma^2(\mathbf{X}'\mathbf{X})^{-1}$, the variance-covariance matrix is diagonal and the least squares estimates are uncorrelated. Removing a covariate from **X** or adding a covariate to **X** that is orthogonal to all of the other columns does not affect the least squares estimates.

If one or more exact linear dependencies exist, **X** is rank-deficient, the inverse of $(\mathbf{X}'\mathbf{X})$ does not exist, and ordinary least squares estimates $\widehat{\boldsymbol{\beta}} = (\mathbf{X}'\mathbf{X})^{-1}\mathbf{X}'\mathbf{y}$ cannot be calculated. Instead, a generalized inverse (§3.4) is used and the solution $\widehat{\boldsymbol{\beta}} = (\mathbf{X}'\mathbf{X})^{-}\mathbf{X}'\mathbf{y}$ is known to be nonunique.

The **collinearity** problem falls between these two extremes. In a multiple regression model the columns of **X** are almost never orthogonal and exact dependencies among the columns also rarely exist. Instead, the columns of **X** are *somewhat* interrelated. One can calculate pairwise correlation coefficients among the columns to detect which columns relate to one another (linearly) in pairs. For the Turnip Greens data the pairwise correlations among the columns of $\mathbf{X} = [\mathbf{x}_1, \mathbf{x}_2, \mathbf{x}_3, \mathbf{x}_4]$ are shown in Table 4.14. Substantial correlations exist between Sunlight and Air Temperature (X_1, X_3) and Soil Moisture and its square (X_2, X_4).

Table 4.14. Pairwise Pearson correlation coefficients for covariates in Turnip Greens data.

	Sunlight X_1	**Soil Moisture** X_2	**Air Temp.** X_3	**Soil Moisture**^{2} X_4
X_1		0.0112	0.5373	– 0.0380
X_2	0.0112		– 0.0149	0.9965
X_3	0.5373	– 0.0149		– 0.0702
X_4	– 0.0380	0.9965	– 0.0702	

Large pairwise correlations are a sufficient but not a necessary condition for collinearity. A collinearity problem can exist even if pairwise correlations among the regressors are small and a near-linear dependency involves more than two regressors. For example, if

$$\sum_{j=1}^{k} c_j \mathbf{x}_j = c_1\mathbf{x}_1 + c_2\mathbf{x}_2 + \cdots + c_k\mathbf{x}_k \approx \mathbf{0}$$

holds, $\mathbf{X}$ will be almost rank-deficient although the pairwise correlations among the $\mathbf{x}_i$ may be small. Collinearity negatively affects all regression calculations that involve the $(\mathbf{X}'\mathbf{X})^{-1}$ matrix: the least squares estimates, their standard error estimates, the precision of predictions, test statistics, and so forth. Least squares estimates tend to be unstable (large standard errors), large in magnitude, and have signs at odds with the subject matter.

Example 4.2. Turnip Greens (continued). Collinearity can be diagnosed in a first step by removing columns from $\mathbf{X}$. The estimates of the remaining parameters will change unless the columns of $\mathbf{X}$ are orthogonal. The table that follows shows the coefficient estimates and their standard errors when X_4, the quadratic term for Soil Moisture, is in the model (full model) and when it is removed.

Table 4.15. Effect on the least squares estimates of removing X_4 = Soil Moisture^2 from the full model

	Full Model		**Reduced Model**	
Parameter	**Estimate**	**Est. Std. Error**	**Estimate**	**Est. Std. Error**
β_0	119.571	13.676	82.0694	19.8309
β_1 (sunlight)	– 0.0336	0.0244	0.0227	0.0365
β_2 (soil moist.)	5.4250	1.0053	– 0.7783	0.0935
β_3 (air temp.)	– 0.5025	0.2109	0.1640	0.2933
β_4 (soil moist^2)	– 0.1209	0.0195	.	.
R^2	0.9104		0.7549	

The estimates of β_0 through β_3 change considerably in size and sign depending on whether X_4 is in the model. While for a given soil moisture an increase in air temperature of 1° Fahrenheit reduces vitamin B_2 concentration by 0.5025 milligrams $\times$ gram^{-1} in the full model, it increases it by 0.164 milligrams $\times$ gram^{-1} in the reduced model.

The effect of radiation (X_1) also changes sign. The coefficients appear to be unstable. If theory suggests that vitamin B_2 concentration should increase with additional sunlight exposure and increases in air temperature, the coefficients are not meaningful from a biological point of view in the full model, although this model provides significantly better fit of the data than the model without X_4.

A simple but efficient collinearity diagnostic is based on variance inflation factors (VIFs) of the regression coefficients. The VIF of the jth regression coefficient is obtained from the coefficient of determination (R^2) in a regression model where the jth covariate is the response and all other covariates form the **X** matrix (Table 4.16). If the coefficient of determination from this regression is R_j^2, then

$$VIF_j = 1/(1 - R_j^2). \qquad [4.31]$$

Table 4.16. Variance inflation factors for the full model in the Turnip Greens example

Response	Regressors	j	R_j^2	VIF_j
X_1	X_2, X_3, X_4	1	0.3888	1.636
X_2	X_1, X_3, X_4	2	0.9967	303.0
X_3	X_1, X_2, X_4	3	0.4749	1.904
X_4	X_1, X_2, X_3	4	0.9966	294.1

If a covariate is orthogonal to all other columns, its variance inflation factor is 1. With increasing linear dependence, the VIFs increase. As a rule of thumb, variance inflation factors greater than 10 are an indication of a collinearity problem. VIFs greater than 30 indicate a severe problem.

Other collinearity diagnostics are based on the principal value decomposition of a centered and scaled regressor matrix (see §A4.8.3 for details). A variable is centered and scaled by subtracting its sample mean and dividing by its sample standard deviation:

$$x_{ij}^* = (x_{ij} - \overline{x}_j)/s_j,$$

where $\overline{x}_j$ is the sample mean of the x_{ij} and s_j is the standard deviation of the jth column of **X**. Let $\mathbf{X}^*$ be the regressor matrix

$$\mathbf{X}^* = [\mathbf{x}_1^*, \cdots, \mathbf{x}_k^*]$$

and rewrite the model as

$$\mathbf{Y} = \beta_0^* + \mathbf{X}^*\boldsymbol{\beta}^* + \mathbf{e}. \qquad [4.32]$$

This is called the **centered** regression model and $\boldsymbol{\beta}^*$ is termed the vector of standardized coefficients. Collinearity diagnostics examine the conditioning of $\mathbf{X}^{*\prime}\mathbf{X}^*$. If **X** is an orthogonal matrix, then $\mathbf{X}^{*\prime}\mathbf{X}^*$ is orthogonal and its eigenvalues λ_j are all unity. If an exact linear dependency exists, at least one eigenvalue is zero (the rank of a matrix is equal to the number of nonzero eigenvalues). If eigenvalues of $\mathbf{X}^{*\prime}\mathbf{X}^*$ are close to zero, a collinearity problem exists. The jth **condition index** of $\mathbf{X}^{*\prime}\mathbf{X}^*$ is the square root of the ratio between the largest and the jth eigenvalue,

$$\psi_j = \sqrt{\frac{\max\{\lambda_j\}}{\lambda_j}}. \qquad [4.33]$$

A condition index in excess of 30 indicates a near linear dependency.

Variance inflation factors and condition indices are obtained with the `vif` and `collinoint` options of the `model` statement in `proc reg`. For example,

```
proc reg data=turnip;
  model vitamin = sun moist airtemp x4 / vif collinoint;
run; quit;
```

The `collin` option of the `model` statement also calculates eigenvalues and condition indices, but does not adjust them for the intercept as the `collinoint` option does. We prefer the latter. With the intercept-adjusted collinearity diagnostics, one obtains as many eigenvalues and condition indices as there are explanatory variables in **X**. They do not stand in a one-to-one correspondence with the regressors, however. If the first condition index is larger than 30, it does not imply that the first regressor needs to be removed. It means that there is at least one near-linear dependency among the regressors that may or may not involve X_1.

Table 4.17. Eigenvalues and condition indices for the full model in Turnip Greens data and for the reduced model without X_4

	Eigenvalues				**Condition Indices**			
	λ_1	λ_2	λ_3	λ_4	ψ_1	ψ_2	ψ_3	ψ_4
Full Model	2.00	1.53	0.46	0.001	1.00	1.14	2.08	34.86
Reduced Model	1.54	1.00	0.46		1.00	1.24	1.82	

For the Turnip Greens data the (intercept-adjusted) eigenvalues of $\mathbf{X}^{*\prime}\mathbf{X}^{*}$ and the condition indices are shown in Table 4.17. The full model containing X_4 has one eigenvalue that is close to zero and the associated condition index is greater than the rule of thumb value of 30. Removing X_4 leads to a model without a collinearity problem, but early on we found that the model without X_4 does not fit the data well. To keep the four regressor model and reduce the negative impact of collinearity on the least squares estimates, we now employ ridge regression.

4.4.5 Ridge Regression to Combat Collinearity

Box 4.7 Ridge Regression

- **Ridge regression is an ad-hoc regression method to combat collinearity. The ridge regression estimator allows for some bias in order to reduce the mean square error compared to OLS.**

- **The method has an ad-hoc character because the user must choose the ridge factor, a small number by which to *shrink* the least squares estimates.**

Least squares estimates are best linear unbiased estimators in the model

$$\mathbf{Y} = \mathbf{X}\boldsymbol{\beta} + \mathbf{e},\ \mathbf{e} \sim \left(\mathbf{0}, \sigma^2\mathbf{I}\right),$$

where *best* implies that the variance of the linear combination $\mathbf{a}'\widehat{\boldsymbol{\beta}}_{OLS}$ is smaller than the variance of $\mathbf{a}'\widetilde{\boldsymbol{\beta}}$, where $\widetilde{\boldsymbol{\beta}}$ is any other estimator that is linear in $\mathbf{Y}$. In general, the **mean square error** (MSE) of an estimator $f(\mathbf{Y})$ for a target parameter θ is defined as

$$\begin{aligned} MSE[f(\mathbf{Y}),\theta] &= \mathrm{E}\left[(f(\mathbf{Y})-\theta)^2\right] = \mathrm{E}\left[(f(\mathbf{Y})-\mathrm{E}[f(\mathbf{Y})])^2\right] + \mathrm{E}[f(\mathbf{Y})-\theta]^2 \\ &= \mathrm{Var}[f(\mathbf{Y})] + \mathrm{Bias}[f(\mathbf{Y}),\theta]^2. \end{aligned} \qquad [4.34]$$

The MSE has a variance and a (squared) bias component that reflect the estimator's **precision** and **accuracy**, respectively (low variance = high precision). The mean square error is a more appropriate measure for the performance of an estimator than the variance alone since it takes both precision and accuracy into account. If $f(\mathbf{Y})$ is unbiased for θ, then $MSE[f(\mathbf{Y}),\theta] = \mathrm{Var}[f(\mathbf{Y})]$ and choosing among unbiased estimators on the basis of their variances is reasonable. But unbiasedness is not necessarily the most desirable property. An estimator that is highly variable and unbiased may be less preferable than an estimator which has a small bias and high precision. Being slightly off-target if one is always close to the target should not bother us as much as being on target **on average** but frequently far from it.

The **realtive efficiency** of $f(\mathbf{Y})$ compared to $g(\mathbf{Y})$ as an estimator for θ is measured by the ratio of their respective mean square errors:

$$\mathrm{RE}[f(\mathbf{Y}), g(\mathbf{Y})|\theta] = \frac{MSE[g(\mathbf{Y}),\theta]}{MSE[f(\mathbf{Y}),\theta]} = \frac{\mathrm{Var}[g(\mathbf{Y})] + \mathrm{Bias}[g(\mathbf{Y}),\theta]^2}{\mathrm{Var}[f(\mathbf{Y})] + \mathrm{Bias}[f(\mathbf{Y}),\theta]^2}. \qquad [4.35]$$

If $\mathrm{RE}[f(\mathbf{Y}), g(\mathbf{Y})|\theta] > 1$ then $f(\mathbf{Y})$ is preferred, and $g(\mathbf{Y})$ is preferred if the efficiency is less than 1. Assume that $f(\mathbf{Y})$ is an unbiased estimator and that we choose $g(\mathbf{Y})$ by shrinking $f(\mathbf{Y})$ by some multiplicative factor c $(0 < c < 1)$. Then

$$\begin{aligned} MSE[f(\mathbf{Y}),\theta] &= \mathrm{Var}[f(\mathbf{Y})] \\ MSE[g(\mathbf{Y}),\theta] &= c^2\mathrm{Var}[f(\mathbf{Y})] + (c-1)^2\theta^2. \end{aligned}$$

The efficiency of the unbiased estimator $f(\mathbf{Y})$ relative to the **shrinkage** estimator $g(\mathbf{Y}) = cf(\mathbf{Y})$ is

$$\mathrm{RE}[f(\mathbf{Y}), cf(\mathbf{Y})|\theta] = \frac{c^2\mathrm{Var}[f(\mathbf{Y})] + (c-1)^2\theta^2}{\mathrm{Var}[f(\mathbf{Y})]} = c^2 + \theta^2\frac{(c-1)^2}{\mathrm{Var}[f(\mathbf{Y})]}.$$

If c is chosen such that $c^2 + \theta^2(c-1)^2/\mathrm{Var}[f(\mathbf{Y})]$ is less than 1, the biased estimator is more efficient than the unbiased estimtator and should be preferred.

How does this relate to least squares estimation in the classical linear model? The optimality of the least squares estimator of $\boldsymbol{\beta}$, since it is restricted to the class of unbiased estimators, implies that no other unbiased estimator will have a smaller mean square error. If the regressor matrix $\mathbf{X}$ exhibits near linear dependencies (collinearity), the variance of $\widehat{\beta}_j$ may be inflated, and sign and magnitude of $\widehat{\beta}_j$ may be distorted. The least squares estimator, albeit unbiased, has become unstable and imprecise. A slightly biased estimator with greater stability can be more efficient because it has a smaller mean square error. To motivate this estimator, consider the centered model [4.32]. The unbiased least squares estimator is

$$\widehat{\boldsymbol{\beta}}^{*}_{OLS} = \left(\mathbf{X}^{*\prime}\mathbf{X}^{*}\right)^{-1}\mathbf{X}^{*\prime}\mathbf{Y}.$$

The ill-conditioning of the $\mathbf{X}^{*\prime}\mathbf{X}^{*}$ matrix due to collinearity causes the instability of the estimator. To improve the conditioning of $\mathbf{X}^{*\prime}\mathbf{X}^{*}$, add a small positive amount δ to its diagonal values,

$$\widehat{\boldsymbol{\beta}}^{*}_{R} = \left(\mathbf{X}^{*\prime}\mathbf{X}^{*} + \delta\mathbf{I}\right)^{-1}\mathbf{X}^{*\prime}\mathbf{Y}. \qquad [4.36]$$

This estimator is known as the **ridge** regression estimator of $\boldsymbol{\beta}$ (Hoerl and Kennard 1970a, 1970b). It is applied in the centered and not the original model to remove the scale effects of the $\mathbf{X}$ columns. Whether a covariate is measured in inches or meters, the same correction should apply since only one ridge factor δ is added to all diagonal elements. Centering the model removes scaling effects, every column of $\mathbf{X}^{*}$ has a sample mean of 0 and a sample variance of 1. The value δ, some small positive number chosen by the user, is called the **ridge factor**.

The ridge regression estimator is not unbiased. Its bias (see §A4.8.4) is

$$\mathrm{E}\left[\widehat{\boldsymbol{\beta}}^{*}_{R} - \boldsymbol{\beta}^{*}\right] = -\,\delta\left(\mathbf{X}^{*\prime}\mathbf{X}^{*} + \delta\mathbf{I}\right)^{-1}\boldsymbol{\beta}^{*}.$$

If $\delta = 0$ the unbiased ordinary least squares estimator results and with increasing ridge factor δ, the ridge estimator applies more shrinkage.

To see that the ridge estimator is a shrinkage estimator consider a simple linear regression model in centered form,

$$\mathbf{Y} = \beta_0^{*} + x_i^{*}\beta_1^{*} + e_i,$$

where $x_i^{*} = (x_i - \overline{x})/s_x$. The least squares estimate of the standardized slope is

$$\widehat{\beta}_1^{*} = \frac{\sum_{i=1}^{n} x_i^{*} y_i}{\sum_{i=1}^{n} x_i^{*2}}$$

while the ridge estimate is

$$\widehat{\beta}_{1R}^{*} = \frac{\sum_{i=1}^{n} x_i^{*} y_i}{\delta + \sum_{i=1}^{n} x_i^{*2}}.$$

Since δ is positive $\widehat{\beta}_{1R}^{*} < \widehat{\beta}_1^{*}$. The standardization of the coefficients is removed by dividing the coefficient by $\sqrt{s_x}$, and we also have $\widehat{\beta}_{1R} < \widehat{\beta}_1$.

The choice of δ is made by the user which has led to some disapproval of ridge regression among statisticians. Although numerical methods exist to estimate the best value of the ridge factor from data, the most frequently used technique relies on graphs of the ridge trace. The **ridge trace** is a plot of $\widehat{\beta}_{jR}$ for various values of δ. The ridge factor is chosen as that value for which the estimates stabilize (Figure 4.9).

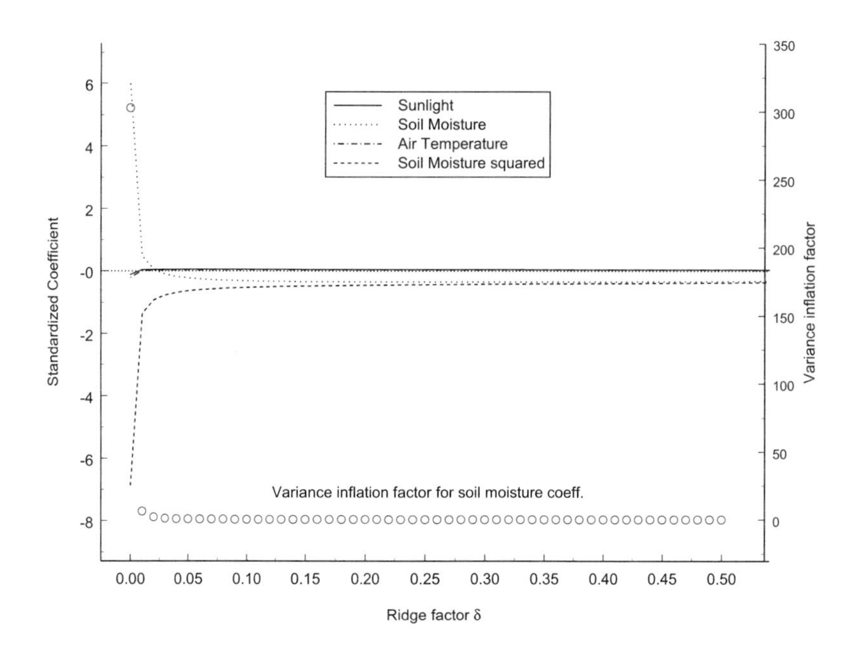

Figure 4.9. Ridge traces for full model in the Turnip Greens example. Also shown (empty circles) is the variance inflation factor for the Soil Moisture coefficient.

For the full model in the Turnip Greens example the two coefficients affected by the collinearity are $\widehat{\beta}_2$ and $\widehat{\beta}_4$. Adding only a small amount $\delta = 0.01$ to the diagonals of the $\mathbf{X}^{*\prime}\mathbf{X}^*$ matrix already stabilizes the coefficients (Figure 4.9). The variance inflation factor of the Soil Moisture coefficient is also reduced dramatically. When studying a ridge trace such as Figure 4.9 we look for various indications of having remedied collinearity.

- The smallest value of δ at which the parameter estimates are stabilized (do not change rapidly). This is not necessarily the value where the ridge trace becomes a flat line. Inexperienced users tend to choose a δ value that is too large, taking the notion of stability too literally.

- Changes in sign. One of the effects of collinearity is that signs of the coefficients are at odds with theory. The ridge factor δ should be chosen in such a way that the signs of the ridge estimates are meaningful on subject matter grounds.

- Coefficients that change the most when the ridge factor is altered are associated with variables involved in a near-linear dependency. The more the ridge traces swoop, the higher the degree of collinearity.

- If the degree of collinearity is high, a small value for δ will suffice, although this result may seem counterintuitive.

Table 4.18 shows the standardized ordinary least squares coefficients and the ridge estimates for $\delta = 0.01$ along with their estimated standard errors. Sign, size, and standard errors of the coefficient change drastically with only a small adjustment to the diagonal of $\mathbf{X}^{*\prime}\mathbf{X}^{*}$, in particular for β_2 and β_4.

Table 4.18. Ordinary least squares and ridge regression estimates ($\delta = 0.01$) along with their estimated standard errors

	Ordinary Least Squares		**Ridge Regression**	
Coefficient	**Estimate**	**Est. Std. Error**	**Estimate**	**Est. Std. Error**
β_1	– 0.11274	0.02438	0.03891	0.03240
β_2	5.98982	1.00526	0.47700	0.21146
β_3	– 0.20980	0.21094	0.01394	0.26045
β_4	– 6.87490	0.01956	– 1.34614	0.00411

The ridge regression estimates can be calculated in SAS® with the `ridge=` option of the `reg` procedure. The `outest=` option is required to identify a data set in which SAS® collects the ridge estimates. By adding `outstb` and `outvif` options, SAS® will also save the standardized coefficients along with the variance inflation factors in the `outest` data set. In the Turnip Greens example, the following code calculates ridge regression estimates for ridge factors $\delta = 0, 0.01, 0.02, \cdots, 0.5, 0.75, 1, 1.25, 1.5, 1.75,$ and 2. The subsequent `proc print` step displays the results.

```
proc reg data=turnip outest=regout outstb outvif
         ridge=0 to 0.5 by 0.01 0.5 to 2 by 0.25;
   model vitamin = sun moist airtemp x4 ;
run; quit;
proc print data=regout; run;
```

For more details on ridge regression and its applications the reader is referred to Myers (1990, Ch. 8).

4.5 Diagnosing Classification Models

4.5.1 What Matters?

Diagnosis of a classification model such as an analysis of variance model in an experimental design does not focus on the mean function as much as is the case for a regression model. In the classification model $\mathbf{Y} = \mathbf{X}\boldsymbol{\theta} + \mathbf{e}$, where $\mathbf{X}$ is a matrix of dummy (design) variables, collinearity issues do not arise. While there are no near-linear dependencies among the columns of $\mathbf{X}$, the columns exhibit perfect linear dependencies and $\mathbf{X}$ is not a full rank matrix. The implications of this rank-deficiency have been explored in §3.4 and are further discussed in §A4.8.1 and §A4.8.2. Essentially, we replace $(\mathbf{X}'\mathbf{X})^{-1}$ in all formulas with a generalized inverse $(\mathbf{X}'\mathbf{X})^{-}$, the least squares estimates are biased and not unique, but estimable linear combinations $\mathbf{a}'\boldsymbol{\theta}$ are unique and unbiased. Also, which effects are included in the $\mathbf{X}$ matrix, that is, the structure of the mean function, is typically not of great concern to the modeler. In the Lime Application example (Example 4.1), the mean structure consists of Lime Type and Rate of Application main effects and the Lime Type × Rate interactions. Even if the interactions were not significant, one would not remove them from the model as compared to a regression context where a model for the mean response is built from scratch.

Diagnostics in classification models focus primarily on the stochastic part, the model errors. The traditional assumptions that **e** is a vector of homoscedastic, independent, Gaussian disturbances are of concern. Performance evaluations of the analysis of variance F test, multiple comparison procedures, and other ANOVA components under departures from these assumptions have a long history. Based on results of these evaluations we rank the three error assumptions by increasing importance as:

- Gaussian distribution of **e**;
- homogeneous variances;
- independence of (lack of correlation among) error terms.

The analysis of variance F test is generally robust against mild and moderate departures of the error distribution from the Gaussian model. This result goes back to Pearson (1931), Geary (1947), and Gayen (1950). Box and Andersen (1955) note that robustness of the test increases with the sample size per group (number of replications per treatment). The reason for the robustness of the F test is that the group sample means are unbiased estimates of the group means and for sufficient sample size are Gaussian-distributed by the Central Limit Theorem regardless of the parent distribution from which the data are drawn. Finally, it must be noted that in designed experiments under randomization one can (and should) apply significance tests derived from randomization theory (§A4.8.2). These tests do not require Gaussian-distributed error terms. As shown by, e.g., Box and Andersen (1955), the analysis of variance F test is a good approximation to the randomization test.

A departure from the equal variance assumption is more troublesome. If the data are Gaussian-distributed and the sample sizes in the groups are (nearly) equal, the ANOVA F test retains a surprising robustness against moderate violations of this assumption (Box 1954a, 1954b, Welch 1937). This is not true for one-degree-of-freedom contrasts which are very much affected by unequal variances even if the group sizes are the same. As group sizes become more unbalanced, small departures from the equal variance assumption can have very negative effects. Box (1954a) studied the signficance level of the F test in group comparisons as a function of the ratio of the smallest group variance to the largest group variance. With equal replication and a variance ratio of $1 : 3$, the nominal significance level of 0.05 was only slightly exceeded ($0.056 - 0.074$), but with unequal group sizes the actual significance level was as high as 0.12. The test becomes liberal, rejecting more often than it should.

Finally, the independence assumption is the most critical since it affects the F test most severely. The essential problem has been outlined before. If correlations among observations are positive, the p-values of the standard tests that treat data as if they were uncorrelated are too small as are the standard error estimates.

There is no remedy for lack of independence except for observing proper experimental procedure. For example, randomizing treatments to experimental units and ensuring that experimental units not only receive treatments independently but also respond independently. Even then, data may be correlated, for example, when measurements are taken repeatedly over time. There is no univariate data transformation to uncorrelate observations but there are of course methods for diagnosing and correcting departures from the Gaussian and the homoscedasticity assumption. The remedies are often sought in transformations of the data. Hopefully, the transformation that stabilizes the variance also achieves greater symmetry in the data so that a Gaussian assumption is more reasonable. In what follows we discuss various transformations and illustrate corrective measures for unequal group variances by example. Because of the robustness of the F test against moderate departures from Gaussianity, we focus on transformations that stabilize variability.

For two-way layouts without replication (a randomized block design, for example), an exploratory technique called Median-Polishing is introduced in §4.5.3. We find it particularly useful to diagnose outliers and interactions that are problematic in the absence of replications.

4.5.2 Diagnosing and Combating Heteroscedasticity

Table 4.19 contains simulated data from a one-way classification with five groups (treatments) and ten samples per group (replications per treatment). A look at the sample means and sample standard deviations at the bottom of the table suggests that the variability increases with the treatment means. Are the discrepancies among the treatment sample standard deviations large enough to worry about the equal variance assumption? To answer this question, numerous statistical tests have been developed, e.g., Bartlett's test (Bartlett 1937a,b), Cochran's test (Cochran 1941), and Hartley's F-Max test (Hartley 1950).

These tests have in common a discouraging sensitivity to departures from the Gaussian assumption. Depending on whether the data have kurtosis higher or lower than that of the Gaussian distribution they tend to exaggerate or mask differences among the variances. Sahai and Ageel (2000, p. 107) argue that when these tests are used as a check before the analysis of variance and lead to rejection, that one can conclude that either the Gaussian error or the homogeneous variance assumption or both are violated. We discourage the practice of using a test designed for detecting heterogeneous variances to draw conclusions about departures from the Gaussian distribution simply because the test performs poorly in this circumstance. If the equal variance assumption is to be tested, we prefer a modification by Brown and Forsythe (1974) of a test by Levene (1960). Let

$$U_{ij} = |Y_{ij} - \text{median}\{Y_{ij}\}|$$

denote the absolute deviations of observations in group i from the group median. Then calculate an analysis of variance for the U_{ij} and reject the hypothesis of equal variances when significant group differences among the U_{ij} exist.

Originally, the Levene test was based on $U_{ij} = (Y_{ij} - \overline{Y}_{i.})^2$, rather than absolute deviations from the median. Other variations of the test include analysis of variance based on

$$\begin{aligned} U_{ij} &= |Y_{ij} - \overline{Y}_{i.}| \\ U_{ij} &= \ln\{|Y_{ij} - \overline{Y}_{i.}|\} \\ U_{ij} &= \sqrt{|Y_{ij} - \overline{Y}_{i.}|}. \end{aligned}$$

The first of these deviations as well as $U_{ij} = (Y_{ij} - \overline{Y}_{i.})^2$ are implemented through the `hovtest=` option of the `means` statement in `proc glm` of The SAS® System.

Table 4.19. Data set from a one-way classification with 10 replicates

Replicate	Treatment $i = 1$	2	3	4	5
1	1.375	2.503	5.278	12.301	7.219
2	2.143	0.431	4.728	11.139	24.166
3	1.419	16.493	6.833	19.349	37.613
4	0.295	11.096	9.609	7.166	9.910
5	2.572	2.436	10.428	10.372	8.509
6	0.312	7.774	7.221	8.956	17.380
7	2.272	6.520	12.753	9.362	19.566
8	1.806	6.386	10.694	9.237	15.954
9	1.559	5.927	2.534	12.929	11.878
10	1.409	8.883	7.093	6.485	15.941
$\overline{y}_{i.}$	1.514	6.845	7.717	10.729	16.814
s_i	0.757	4.676	3.133	3.649	9.000

For the data in Table 4.19 we fit a simple one-way model:

$$Y_{ij} = \mu + \tau_i + e_{ij},\ i = 1, \cdots, 5;\ j = 1, \cdots, 10,$$

where Y_{ij} denotes the jth replicate value observed for the ith treatment and obtain the Levene test with the statements

```
proc glm data=hetero;
   class tx;
   model y = tx /ss3;
   means tx / hovtest=levene(type=abs);
run; quit;
```

The Levene test clearly rejects the hypothesis of equal variances ($p = 0.0081$, Output 4.9). Even with equal sample sizes per group the test for equality of the treatment means ($F_{obs} = 12.44$, $p < 0.0001$) is not to be trusted if variance heterogeneity is that strong.

Output 4.9.

```
                         The GLM Procedure

                     Class Level Information
                 Class         Levels     Values
                 tx                 5     1 2 3 4 5

                    Number of observations    50

Dependent Variable: y
                                    Sum of
Source                      DF     Squares    Mean Square  F Value  Pr > F
Model                        4  1260.019647    315.004912    12.44  <.0001
Error                       45  1139.207853     25.315730
Corrected Total             49  2399.227500

            R-Square      Coeff Var      Root MSE        y Mean
            0.525177       57.67514      5.031474      8.723818

Source                      DF  Type III SS    Mean Square  F Value  Pr > F
tx                           4  1260.019647     315.004912    12.44  <.0001
```

Output 4.9 (continued).

```
            Levene's Test for Homogeneity of y Variance
            ANOVA of Absolute Deviations from Group Means

                           Sum of          Mean
Source          DF        Squares        Square      F Value     Pr > F
tx               4          173.6       43.4057         3.93     0.0081
Error           45          496.9       11.0431

      Level of                 --------------y--------------
      tx              N              Mean            Std Dev
      1              10         1.5135500         0.75679658
      2              10         6.8449800         4.67585820
      3              10         7.7172000         3.13257166
      4              10        10.7295700         3.64932521
      5              10        16.8137900         9.00064885
```

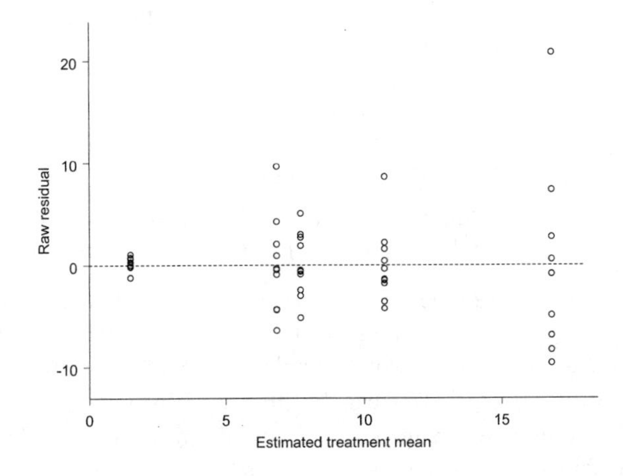

Figure 4.10. Raw residuals of one-way classification model fit to data in Table 4.19.

A graphical means of diagnosing heterogeneous variances is based on plotting the residuals against the treatment means. As is often the case with biological data, variability increases with the means and this should be obvious from the residual plot. Figure 4.10 shows a plot of the raw residuals $\widehat{e}_{ij} = y_{ij} - \widehat{y}_{ij}$ against the treatment means $\overline{y}_{i.}$. Not only is it obvious that the variability of the residuals is very heterogeneous in the groups, it also appears that the distribution of the residuals is right-skewed and not a Gaussian distribution.

Can the data be transformed to make the residual distribution more symmetric and its variability homogeneous across the treatments? Transformations to stabilize the variance exploit a functional relationship between group means and group variances. Let $\mu_i = \mu + \tau_i$ denote the mean in the ith group and σ_i^2 the error variance in the group. If $\mu_i = \sigma_i^2$, the variance stabilizing transform is the square root transformation $U_{ij} = \sqrt{Y_{ij}}$. The Poisson distribution (see §6.2) for counts has this property of mean-variance identity. For Binomial counts, such as the number of successes out of n independent trials, each of which can result in either a failure or success, the relationship between the mean and variance is $\mathrm{E}[Y_{ij}] = \pi_i$,

$\text{Var}[Y_{ij}] = n_{ij}\pi_i(1-\pi_i)$ where π_i is the probability of a success in any one trial. The variance stabilizing transform for this random variable is the arc-sine transformation $U_{ij} = \sin^{-1}\left(\sqrt{Y_{ij}/n_{ij}}\right)$. The transformed variable has variance $1/(4n_{ij})$. Anscombe (1948) has recommended modifications of these basic transformations. In the Poisson case when the average count is small, $U_{ij} = \sqrt{Y_{ij}+3/8}$ has been found to stabilize the variances better and in the Binomial case

$$U_{ij} = \sin^{-1}\left(\sqrt{\frac{Y_{ij}+3/8}{n_{ij}+3/4}}\right)$$

appears to be superior.

For continuous data, the relationship between mean and variance is not known unless the distribution of the responses is known (see §6.2 on how mean variance relationships for continuous data are used in formulating generalized linear models). If, however, it can be established that variances are proportional to some power of the mean, a transformation can be derived. Box and Cox (1964) assume that

$$\sqrt{\text{Var}[Y_{ij}]} = \sigma_i \propto \mu_i^{\beta}. \qquad [4.37]$$

If one takes a transformation $U_{ij} = Y_{ij}^{\lambda}$, then $\sqrt{\text{Var}[U_{ij}]} \propto \mu_i^{\beta+\lambda-1}$. Since variance homogeneity implies $\sigma_i \propto 1$, the transformation with $\lambda = 1-\beta$ stabilizes the variance. If $\lambda = 0$, the transformation is not defined and the logarithmic transformation is chosen.

For the data in Table 4.19 sample means and sample standard deviations are available for each group. We can empirically determine which transformation will result in a variance stabilizing transformation. For data in which standard deviations are proportional to a power of the mean the relationship $\sigma_i = \alpha\mu_i^{\beta}$ can be linearized by taking logarithms: $\ln\{\sigma_i\} = \ln\{\alpha\} + \beta\ln\{\mu_i\}$. Substituting estimates s_i for σ_i and $\overline{y}_{i.}$ for μ_i, this is a linear regression of the log standard deviations on the log sample means: $\ln\{s_i\} = \beta_0 + \beta\ln\{\overline{y}_{i.}\} + e_i^*$. Figure 4.11 shows a plot of the five sample standard deviation/sample mean points after taking logarithms. The trend is linear and the assumption that σ_i is proportional to a power of the mean is reasonable. The statements

```
proc means data=hetero noprint;
   by tx;
   var y;
   output out=meanstd(drop=_type_) mean=mean std=std;
run;
data meanstd; set meanstd;
  logstd  = log(std);
  logmn   = log(mean);
run;
proc reg data=meanstd;
  model logstd = logmn;
run; quit;
```

calculate the treatment specific sample means and standard deviations, take their logarithms and fit the simple linear regression model (Output 4.10). The estimate of the slope is $\widehat{\beta} = 0.958$ which suggests that $1 - 0.958 = 0.042$ is the power for the variance stabilizing transform. Since $1 - \widehat{\beta}$ is close to zero, one could also opt for a logarithmic transformation of the data. The analysis of variance and the Levene homogeneity of variance test for the transformed values is obtained with

```
data hetero; set hetero;
  powery = y**0.042;
run;
proc glm data=hetero;
  class tx;
  model powery = tx / ss3;
  means tx / hovtest=levene(type=abs);
run; quit;
```

and results are shown in Output 4.11.

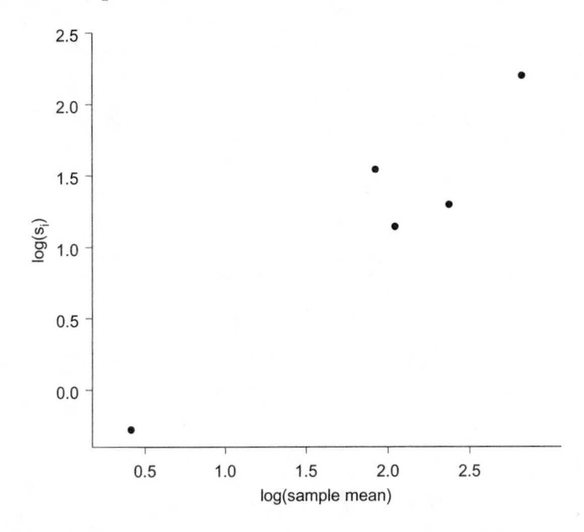

Figure 4.11. Logarithm of sample standard deviations vs. logarithm of sample means for data in Table 4.19 suggests a linear trend.

Output 4.10.

```
                    The REG Procedure
                 Dependent Variable: logstd

                   Analysis of Variance

                             Sum of           Mean
Source              DF      Squares         Square    F Value    Pr > F
Model                1      3.02949        3.02949      32.57    0.0107
Error                3      0.27901        0.09300
Corrected Total      4      3.30850

       Root MSE              0.30497    R-Square     0.9157
       Dependent Mean        1.17949    Adj R-Sq     0.8876
       Coeff Var            25.85570

                   Parameter Estimates

                     Parameter       Standard
  Variable     DF     Estimate          Error    t Value    Pr > |t|
  Intercept     1     -0.65538        0.34923      -1.88      0.1572
  logmn         1      0.95800        0.16785       5.71      0.0107
```

Output 4.11.

```
                              The GLM Procedure

                          Class Level Information
                     Class          Levels    Values
                     tx                  5    1 2 3 4 5

                        Number of observations    50

Dependent Variable: powery
                                       Sum of
Source                   DF          Squares    Mean Square   F Value   Pr > F
Model                     4       0.07225884     0.01806471     20.76   <.0001
Error                    45       0.03915176     0.00087004
Corrected Total          49       0.11141060

           R-Square      Coeff Var      Root MSE     powery Mean
           0.648581       2.736865      0.029496        1.077745

Source                   DF      Type III SS    Mean Square   F Value   Pr > F
tx                        4       0.07225884     0.01806471     20.76   <.0001

            Levene's Test for Homogeneity of powery Variance
              ANOVA of Absolute Deviations from Group Means

                              Sum of         Mean
      Source          DF     Squares       Square      F Value      Pr > F
      tx               4     0.00302     0.000755         2.19      0.0852
      Error           45      0.0155     0.000345

           Level of            ------------powery-----------
           tx            N             Mean              Std Dev

           1            10       1.00962741           0.03223304
           2            10       1.07004861           0.04554636
           3            10       1.08566961           0.02167296
           4            10       1.10276544           0.01446950
           5            10       1.12061483           0.02361509
```

The Levene test is (marginally) nonsignificant at the 5% level. Since we are interested in *accepting* the null hypothesis of equal variances in the five groups this result is not too convincing. We would like the p-value for the Levene test to be larger. The Levene test based on squared deviations $(Z_{ij} - \overline{Z}_{i.})^2$ has a p-value of 0.189 which is more appropriate. The group-specific standard deviations are noticeably less heterogeneous than for the untransformed data. The ratio of the largest to the smallest group sample variance for the original data was $9.0^2/0.7568^2 = 141.42$ whereas this ratio is only $0.0322^2/0.0145^2 = 4.93$ for the transformed data. The 5% critical value for a Hartley's F-Max test is 7.11 and we fail to reject the hypothesis of equal variances with this test too.

4.5.3 Median Polishing of Two-Way Layouts

A two-way layout is a cross-tabulation of two discrete variables. While common in the analysis of categorical data where the body of the table depicts absolute or relative frequencies, two-way layouts also arise in field experiments and the study of spatial data. For example, a randomized complete block design can be thought of as a two-way layout with blocks as the row variable, treatments as the column variable, and the observed outcomes in the body of the table. In this subsection we consider only nonreplicated two-way layouts with at most one value in each cell. If replicate values are available it is assumed that the data were preprocessed to reduce the replicates to a summary value (the total, median or average). Because of the importance of randomized complete block designs (RCBDs) in the plant and soil sciences we focus on these in particular.

An important issue that is often overlooked in RCBDs is the presence of block $\times$ treatment interactions. The *crux* of the standard RCBD with t treatments in b blocks is that each treatment appears only once in each block. To estimate the experimental error variance one cannot rely on replicate values of treatments within each block, as for example, in the generalized randomized block design (Hinkelmann and Kempthorne 1994, Ch. 9.7). Instead, the experimental error (residual) sum of squares in a RCBD is calculated as

$$SSR = \sum_{i=1}^{b}\sum_{j=1}^{t}\left(y_{ij} - \overline{y}_{i.} - \overline{y}_{.j} + \overline{y}_{..}\right)^2, \qquad [4.38]$$

which *is* an interaction sum of squares. If treatment comparisons depend on the block in which the comparison is made, $MSR = SSR/\{(b-1)(t-1)\}$ is no longer an unbiased estimator of the experimental error variance and test statistics having MSR in the denominator no longer have the needed distributional properties. It is the *absence* of the block $\times$ treatment interactions that ensures that MSR is an unbiased estimator of the experimental error variance. One can also not estimate the full block $\times$ treatment interactions as this would use up $(b-1)(t-1)$ degrees of freedom, leaving nothing for the estimation of the experimental error variance. Tukey (1949) developed a one-degree of freedom test for nonadditivity that tests for a particularly structured type of interaction in nonreplicated two-way layouts.

The interaction of blocks and treatment in a two-way layout can be caused by only a few extreme observations in some blocks. Hence it is important to determine whether outliers are present in the data and whether they induce a block $\times$ treatment interaction. In that case reducing their influence on the analysis by using robust estimation methods (see §4.6.4) can lead to a valid analysis of the block design. If the interaction is not due to outlying observations, then it must be incorporated in the model in such a way that degrees of freedom for the estimation of experimental error variance remain. To examine the possibility of block $\times$ treatment interaction and the presence of outliers we reach into the toolbox of exploratory data analysis. The method we prefer is termed **median polishing** (Tukey 1977, Emerson and Hoaglin 1983, Emerson and Wong 1985). Median polishing is also a good device to check on the Gaussianity of the model errors prior to an analysis of variance. In the analysis of spatial data median polishing also plays an important role as an exploratory tool and has been successfully applied as an alternative to *traditional* methods of geostatistical prediction (Cressie 1986).

To test nonreplicated two-way data for Gaussianity presents a considerable problem. The assumption of Gaussianity in the analysis of variance for a randomized complete block design requires that each of the treatment populations is Gaussian-distributed. It is thus not sufficient to simply calculate a normal probability plot or test of normality for the entire two-way data, since this assumes that the treatment means (and variances) are equal. Normal probability (and similar) plots of least squares residuals have their own problems. Although they have zero mean they are not homoscedastic and not uncorrelated (§4.4.1). Since least squares estimates are not outlier-resistant, the residuals may not reflect **how** extreme observations really are compared to a method of fitting that accounts for the mean structure in an outlier-resistant fashion. One can then expect to see the full force of the outlying observations in the residuals. Error recovery as an alternative (§4.4.2) produces only a few residuals because error degrees of freedom are often small.

To fix ideas, let Y_{ij} be the observation in row i and column j of a two-way table. In the following 3×3 table, for example, $Y_{13} = 42$.

	Variety		
Fertilizer	V_1	V_2	V_3
B_1	38	60	42
B_2	13	32	29
B_3	27	43	38

To decompose the variability in this table we apply the simple decomposition

$$\text{Data} = \text{All} + \text{Row} + \text{Column} + \text{Residual},$$

where *All* refers to an overall effect, *Row* to the row effects, and so forth. In a block design *Row* could denote the blocks, *Column* the treatments, and *All* an overall effect (intercept). With familiar notation the decomposition can be expressed mathematically as

$$\begin{aligned} &Y_{ij} = \mu + \alpha_i + \beta_j + e_{ij} \\ &i = 1, \cdots, I; \; j = 1, \cdots, J, \end{aligned}$$

where I and J denote the number of rows and columns, respectively. If there are no missing entries in the table, the least squares estimates of the effects are simple linear combinations of the various row and column averages:

$$\begin{aligned} \widehat{\mu} &= \frac{1}{IJ}\sum_{i=1}^{I}\sum_{j=1}^{J} y_{ij} = \overline{y}_{..} \\ \widehat{\alpha}_i &= \overline{y}_{i.} - \overline{y}_{..} \\ \widehat{\beta}_j &= \overline{y}_{.j} - \overline{y}_{..} \\ \widehat{e}_{ij} &= y_{ij} - \overline{y}_{i.} - \overline{y}_{.j} + \overline{y}_{..} \end{aligned}$$

Notice that the $\widehat{e}_{ij}$ are the terms in the residual sum of squares [4.38]. The estimates have the desirable property that $\sum_{i=1}^{I}\sum_{j=1}^{J}\widehat{e}_{ij} = 0$, mimicking a property of the model disturbances e_{ij} provided the decomposition is correct. The sample means unfortunately are sensitive to extreme observations (outliers). In least squares analysis based on means, the residual in any cell of the two-way table affects the fitted values in all other cells and the effect of an outlier is to leak its contribution across the estimates of the overall, row, and column effects (Emerson and Hoaglin 1983). Tukey (1977) suggested basing estimation of the effects on medians which are less affected by extreme observations. Leakage of outliers into estimates of overall, row, and column effects is then minimized. The method — termed median polishing — is iterative and sweeps medians out of rows, then columns out of the row-adjusted residuals, then rows out of column-adjusted residuals, and so forth. The process stops when the following conditions are met:

$$\begin{aligned} &\text{median}_i\{\widetilde{\alpha}_i\} = 0 \\ &\text{median}_j\left\{\widetilde{\beta}_j\right\} = 0 \\ &\text{median}_i\{\widetilde{e_{ij}}\} = \text{median}_j\{\widetilde{e_{ij}}\} = 0. \end{aligned} \qquad [4.39]$$

In [4.39] $\widetilde{\alpha_i}$ denotes the median-polished effect of the ith row, $\widetilde{\beta}_j$ the median-polished effect of the jth column and the residual is calculated as $\widetilde{e_{ij}} = y_{ij} - \widetilde{\mu} - \widetilde{\alpha_i} - \widetilde{\beta_j}$.

A short example of median-polishing follows, kindly made available by Dr. Jeffrey B. Birch[1] at Virginia Tech. The 3×3 layout represents the yield per plot as a function of type of fertilizer (B_1, B_2, B_3) and wheat variety (V_1, V_2, V_3).

	Variety		
Fertilizer	V_1	V_2	V_3
B_1	38	60	42
B_2	13	32	29
B_3	27	43	38

Step 1. Obtain row medians and subtract them from the y_{ij}.

0	0	0	0	(0)
0	38	60	42	(42)
0	13	32	29	(29)
0	27	43	38	(38)

$\Leftrightarrow$

0	0	0	0
42	− 4	18	0
29	− 16	3	0
38	− 11	5	0

Some housekeeping is done in the first row and column, storing $\widetilde{\mu}$ in the first cell, the $\widetilde{\alpha_i}$ in the first column, and the $\widetilde{\beta_j}$ in the first row of the table. The table on the right-hand side is obtained from the one on the left by adding the row medians to the first column and subtracting them from the remainder of the table.

Step 2. Obtain column medians and subtract them from the residuals (right-hand table from step 1).

0	0	0	0
42	− 4	18	0
29	− 16	3	0
38	− 11	5	0
(38)	(− 11)	(5)	(0)

$\Leftrightarrow$

38	− 11	5	0
4	7	13	0
− 9	− 5	− 2	0
0	0	0	0

The right-hand table in this step is obtained by adding the column medians to the first row and subtracting them from the remainder of the table.

Step 3. Repeat sweeping the row medians.

38	− 11	5	0	(0)
4	7	13	0	(7)
− 9	− 5	− 2	0	(− 2)
0	0	0	0	(0)

$\Leftrightarrow$

38	− 11	5	0
11	0	6	− 7
− 11	− 3	0	2
0	0	0	0

[1] Birch, J.B. Unpublished manuscript: *Contemporary Applied Statistics: An Exploratory and Robust Data Analysis Approach.*

The right-hand table is obtained by adding the row medians to the first column and subtracting them from the rest of the table.

Step 4. Repeat sweeping column medians.

38	– 11	5	0
11	0	6	– 7
– 11	– 3	0	2
0	0	0	0
(0)	(0)	(0)	(0)

At this point the iterative sweep stops. Nothing is to be added to the first row or subtracted from the rest of the table. Notice that the residuals in the body of the cell do **not** sum to zero.

The sweeped row and column effects represent the large-scale variation (the mean structure) of the data. The median-polished residuals represent the errors and are used to test for Gaussianity, outliers, and interactions. When row and column effects are not additive, Tukey (1977) suggests augmenting the model with an interaction term

$$Y_{ij} = \mu + \alpha_i + \beta_j + \theta\left(\frac{\alpha_i\beta_j}{\mu}\right) + e^*_{ij}. \qquad [4.40]$$

If this type of patterned interaction is present, then θ should differ from zero. After median polishing the no-interaction model, it is recommended to plot the residuals $\widetilde{e}_i$ against the **comparison values** $\widetilde{\alpha}_i\,\widetilde{\beta}_j\,/\,\widetilde{\mu}$. A trend in this plot indicates the presence of row $\times$ column interaction. To decide whether the interaction is induced by outlying observations or a more general phenomenon, fit a simple linear regression between $\widetilde{e}_{ij}$ and the comparison values by an outlier-robust method such as M-Estimation (§4.6.2) and test the slope against zero. Also, fit a simple regression by least squares. When the robust M-estimate of the slope is not different from zero but the least squares estimate is, the *interaction* between rows and columns is caused by outlying observations. Transforming the data or removing the outlier should then eliminate the interaction. If the M-estimate of the slope differs significantly from zero the interaction is not caused just by outliers. The slope of the diagnostic plot is helpful in determining the transformation that can reduce the nonadditivity of row and column effects. Power transformations of approximately $1 - \theta$ are in order (Emerson and Hoaglin 1983). In the interaction test a failure to reject $H_0\colon \theta = 0$ is the result of interest since one can then proceed with a simpler analysis without interaction. Even if the data are balanced (all cells of the table are filled), the analysis of variance based on the interaction model [4.40] is non-orthogonal and treatment means are not estimated by arithmetic averages. Because of the possibility of a Type-II error one should not accept H_0 when the p-value of the test is barely larger than 0.05 but require p-values in excess of $0.25 - 0.3$.

Example 4.3. Dollar Spot Counts (continued). Recall the turfgrass experiment in a randomized complete block design with $t = 14$ treatments in $b = 4$ blocks (p. 93). The outcome of interest was the number of leaves infected with dollar spot in a reference area of each experimental unit. One count was obtained from each unit.

In the chapter introduction we noted that count data are often modeled as Poisson variates which can be approximated well by Gaussian variates if the mean count is sufficiently large (>15, say). This does not eliminate the heterogeneous variance problem since for Poisson variates the mean equals the variance. The Gaussian distribution approximating the distribution of counts for treatments with low dollar spot incidence thus has a smaller variance than that for treatments with high incidence. As an alternative, transformations of the counts are frequently employed to stabilize the variance. In particular the square-root and logarithmic transformation are popular. In §6, we discuss generalized linear models that utilize the actual distribution of the count data, rather than relying on Gaussian approximations or transformations.

Here we perform median polishing for the dollar spot count data for the original counts and log-transformed counts. Particular attention will be paid to the observation for treatment 10 in block 4 that appears extreme compared to the remainder of the data (see Table 4.3, p. 93). Median polishing with The SAS® System is possible with the `%MedPol()` macro contained on the CD-ROM (`\SASMacros\MedianPolish.sas`). It requires an installation of the SAS/IML® module. The macro does not produce any output apart from the residual interaction plot if the `plot=1` option is active (which is the default). The results of median polishing are stored in a SAS® data set termed `_medpol` with the following structure.

Variable	Value
VALUE	contains the estimates of Grand, Row, Col, and Residual
Type	0 = grand, 1=row effect, 2=col effect, 3 = residual
ROW	indicator of row number
COL	indicator of col number
IA	interaction term $\alpha_i\beta_j/\mu$
Signal	$y_{ij} - \widetilde{e_{ij}}$

The median polish of the original counts and a printout of the first 25 observations of the result data set is produced by the statements

```
%include 'DriveLetterOfCDROM:\SASMacros\Median Polish.sas';

title1 'Median Polishing of dollar spot counts';
%MedPol(data=dollarspot,row=block,col=tmt,y=count,plot=1);
proc print data=_medpol(obs=25); run;
```

Median Polishing of dollar spot counts

Obs	Value	Row	Col	_Type_	IA	Signal
1	43.3888	0	0	0	.	.
2	-0.6667	1	0	1	.	.
3	0.6667	2	0	1	.	.
4	-6.0000	3	0	1	.	.
5	3.4445	4	0	1	.	.
6	-13.1111	0	1	2	.	.
7	36.7778	0	2	2	.	.
8	-16.7777	0	3	2	.	.
9	-12.5555	0	4	2	.	.
10	11.0556	0	5	2	.	.
11	0.9445	0	6	2	.	.

```
  12       -0.9444     0      7     2         .          .
  13        4.1112     0      8     2         .          .
  14       41.0556     0      9     2         .          .
  15       65.6112     0     10     2         .          .
  16       86.3889     0     11     2         .          .
  17       -8.6111     0     12     2         .          .
  18      -14.2222     0     13     2         .          .
  19      -17.9444     0     14     2         .          .
  20      -13.6111     1      1     3     0.20145    29.6111
  21      -39.5000     1      2     3    -0.56509    79.5000
  22        3.0556     1      3     3     0.25779    25.9444
  23       -0.1667     1      4     3     0.19292    30.1667
  24      -22.7777     1      5     3    -0.16987    53.7777
  25       -1.6667     1      6     3    -0.01451    43.6667
```

A graph of the column (treatment) effects against column index gives an indication of treatment differences to be expected when a formal comparison procedure is invoked (Figure 4.12).

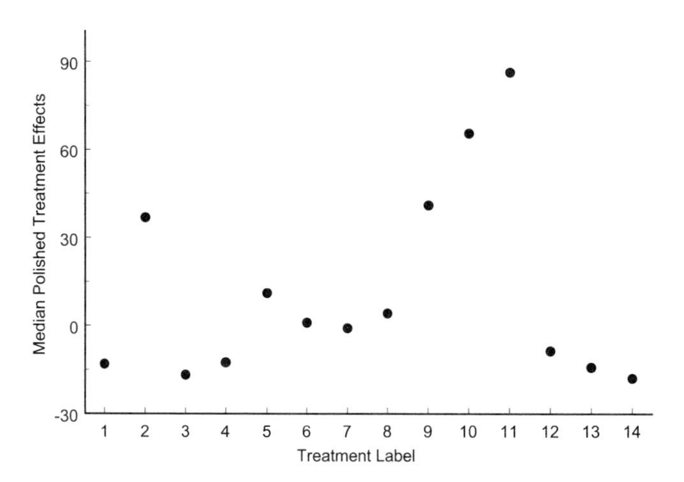

Figure 4.12. Column (treatment) effects $\widetilde{\beta}_j$ for dollar spot counts.

Whether a formal analysis can proceed without accounting for block × treatment interaction cannot be gleaned from a plot of the column effects alone. To this end, plot the median polished residuals against the interaction term (Figure 4.13) and calculate least squares and robust M-estimates (§4.6) of the simple linear regression after the call to `%MedPol()`:

```
%include 'DriveLetterOfCDROM:\SASMacros\MEstimation.sas';

title1 'M-Regression for median polished residuals of
        dollar spot counts';
%MEstim(data=_medpol(where=(_type_=3)),
          stmts=%str(model value = ia /s;)
       )
```

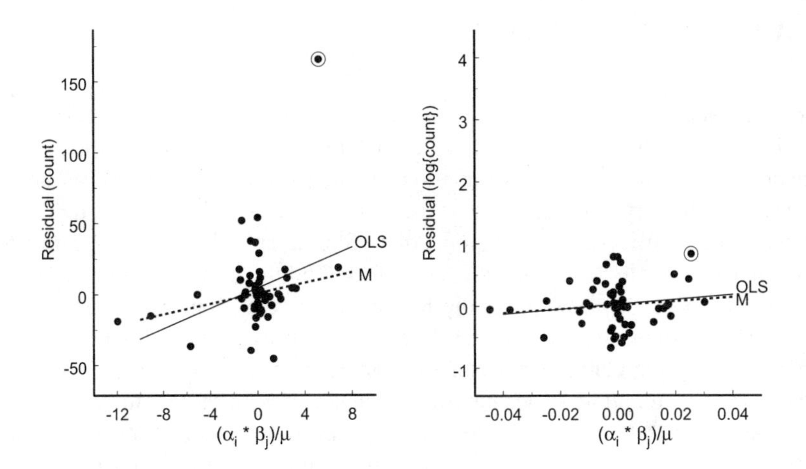

Figure 4.13. Median polished residual plots vs. interaction terms for original counts (left panel) and log-transformed counts (right panel). The outlier (treatment 10 in block 4) is circled.

The outlying observation for treatment 10 in block 4 pulls the least squares regression line toward it (Figure 4.13). The interaction is significant ($p = 0.0094$, Table 4.20) and remains significant when the influence of the outlier is reduced by robust M-Estimation ($p = 0.0081$). The interaction is thus not induced by the outlier alone but a general block $\times$ treatment interaction exists for these data. For the log-transformed counts (Figure 4.13, right panel) the interaction is not significant ($p = 0.2831$ for OLS and $p = 0.3898$ for M-Estimation) and the least squares and robust regression line are almost indistinguishable from each other. The observation for treatment 10 in block 4 is certainly not as extreme relative to the remainder of the data as is the case for the original counts.

Table 4.20. P-values for testing the the block $\times$ treatment interaction, $H_0: \theta = 0$

Response	Least Squares	M-Estimation
Count	0.0094	0.0081
log{Count}	0.2831	0.3898

Based on these findings a log transformation appears reasonable. It reduces the influence of extreme observations and allows analysis of the data without interaction. It also improves the symmetry of the data. The Shapiro-Wilk test for normality has p-values of < 0.0001 for the original data and 0.1069 for the log-transformed counts. The p-values of the interaction test for the log-transformed data are sufficiently large to proceed with confidence with an analysis of the transformed data without an interaction term. The larger p-value for the M-estimate of θ suggests to use a robustified analysis of variance rather than the standard, least squares-based analysis. This analysis is carried out in §4.6.4 after the details of robust parameter estimation have been covered.

4.6 Robust Estimation

Robustness of a statistical inference procedure refers to its stability when model assumptions are violated. For example, a procedure may be robust to an incorrectly specified mean model, heterogeneous variances, or the contamination of the error distribution. **Resistance**, on the other hand, is the property of a statistic or estimator to be stable when the data are grossly contaminated (many outliers). The sample median, for example, is a resistant estimator of the central tendency of sample data, since it is not affected by the absolute value of observations, only whether they occur to the right or the left of the middle. Statistical procedures that are based on sample medians instead of sample means such as Median Polishing (§4.5.3) are resistant procedures. The claim of robustness must be made with care since robustness with respect to one model breakdown does not imply robustness with respect to other model breakdowns. M-Estimation (§4.6.2, §4.6.3), for example, are robust against moderate outlier contamination of the data, but not against high leverage values. A procedure robust against mild departures from Gaussianity may be much less robust against violations of the homogeneous variance assumption.

In this section we focus in particular on mild-to-moderate outlier contamination of the data in regression and analysis of variance models. Rather than deleting outlying observations, we wish to keep them in the analysis to prevent a loss of degrees of freedom but reduce their negative influence. As mentioned earlier, deletion of diagnosed outliers is not an appropriate strategy, unless the observation can be clearly identified as in error. Anscombe (1960) concludes that "if we could be sure that an outlier was caused by a large measurement or execution error which could not be rectified (and if we had no interest in studying such errors for their own sake), we should be justified in entirely discarding the observation and all memory of it." Barnett and Lewis (1994) distinguish three origins for outlying observations:

- *The variability of the phenomenon being studied.* An unusual observation may well be within the range of natural variability; its appearance reflects the distributional properties of the response.
- *Erroneous measurements.* Mistakes in reading, recording, or calculating data.
- *Imperfect data collection (execution errors).* Collecting a biased sample or observing individuals not representing or belonging to the population under study.

If the variability of the phenomenon being studied is the cause of the outliers there is no justification for their deletion. Two alternatives to least squares estimation, L_1- and M-Estimation, are introduced in §4.6.1 and §4.6.2. The prediction efficiency data is revisited in §4.6.3 where the quadratic response model is fit by robust methods to reduce the negative influence of the outlying Mg observation without deleting it from the data set. In §4.6.4 we apply the M-Estimation principle to classification models.

4.6.1 L_1-Estimation

Box 4.8 Least Absolute Deviation (LAD) Regression

- **L_1-Regression minimizes the sum of absolute residuals and not the sum of squared residuals as does ordinary least squares estimation. An alternative name is hence *Least Absolute Deviation (LAD)* Regression.**
- **The L_1-norm of a vector $\mathbf{a}_{(k\times 1)}$ is defined as $\sum_{i=1}^{k}|a_i|$ and the L_2-norm is $\sqrt{\sum_{i=1}^{k} a_i^2}$. Least squares is an L_2-norm method.**
- **The fitted trend of an L_1-Regression with $k-1$ regressors and an intercept passes through k data points.**

The Euclidean norm of a $(k \times 1)$ vector $\mathbf{a}$,

$$||\mathbf{a}|| = \sqrt{\mathbf{a}'\mathbf{a}} = \sqrt{\sum_{i=1}^{k} a_i^2},$$

is also called its L_2-norm and the L_1-norm of a vector is the sum of the absolute values of its elements: $\sum_{i=1}^{k}|a_i|$. In L_1-Estimation the objective is not to find the estimates of the model parameters that minimize the sum of squared residuals (the L_2-norm) $\sum_{i=1}^{n} e_i^2$, but the sum of the absolute values of the residuals, $\sum_{i=1}^{n}|e_i|$. Hence the alternative name of Least Absolute Deviation (LAD) estimation. In terms of a linear model $\mathbf{Y} = \mathbf{X}\boldsymbol{\beta} + \mathbf{e}$, the L_1-estimates $\widehat{\boldsymbol{\beta}}_L$ are the values of $\boldsymbol{\beta}$ that minimize

$$\sum_{i=1}^{n} | \, y_i - \mathbf{x}_i'\boldsymbol{\beta} \, |. \qquad [4.41]$$

Squaring of the residuals in least squares estimation gives more weight to large residuals than taking their absolute value (see Figure 4.15 below).

A feature of L_1-Regression is that the model passes exactly through some of the data points. In the case of a simple linear regression, $y_i = \beta_0 + \beta_1 x_i + e_i$, the estimates $\widehat{\beta}_{0L} + \widehat{\beta}_{1L} x$ pass through two data points and in general if $\mathbf{X}$ is a $(n \times k)$ matrix of full rank k the model passes through k data points. This fact can be used to devise a brute-force method of finding the LAD estimates. Force the model through k data points and calculate its sum of absolute deviations $SAD = \sum_{i=1}^{n}|\widehat{e}_i|$. Repeat this for all possible sets of k points and choose the set which has the smallest SAD.

Example 4.5. Galton (1886), who introduced the concept of regression in studies of inheritance noted that the diameter of offspring peas was linearly increasing with the diameter of the parent peas. Seven of the data points from his 1886 publication are shown in Figure 4.14. Fitting a simple linear regression model to these data, there are

$7*6/2 = 21$ possible least absolute deviation regressions. Ten of them are shown along with the sum of their absolute deviations. SAD achieves a minimum at 1.80 for the line that passes through the first and last data point.

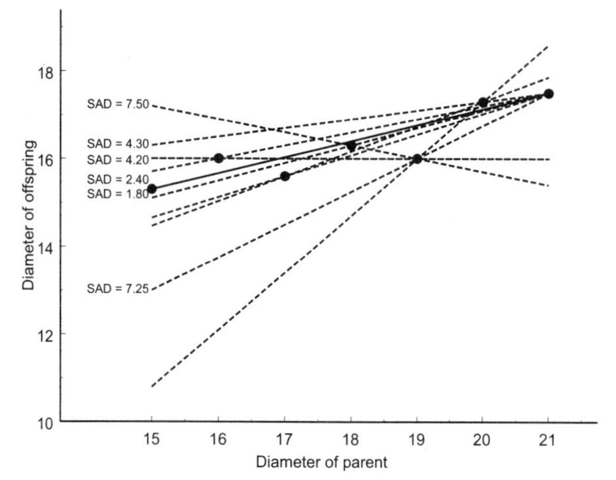

Figure 4.14. Ten of the 21 possible least absolute deviation lines for seven observations from Galton's pea diameter data. The line that minimizes $\sum_{i=1}^{n} |\widehat{e}_i|$ passes through the first and last data point: $\widehat{y}_i = 9.8 + 0.366x$.

This brute-force approach is computationally expensive since the total number of models being fit is large unless either n or k is small. The number of models that must be evaluated with this method is

$$\binom{n}{k} = \frac{n!}{k!(n-k)!}.$$

Fitting a four regressor model with intercept to a data set with $n = 30$ observations requires evaluation of $142,506$ models. In practice we rely on iterative algorithms to reduce the number of evaluations. One such algorithm is discussed in §A4.8.5 and implemented in the SAS® macro `%LAD()` (`\SASMacro\L1_Regression.sas` on CD-ROM). One can also fit least absolute deviation regression with the SAS/IML® function `lav()`.

To test the hypothesis H_0: $\mathbf{A}\boldsymbol{\beta} = \mathbf{d}$ in L_1-Regression we use an approximate test analogous to the sum of squares reduction test in ordinary least squares (§4.2.3 and §A4.8.2). If SAD_f is the sum of absolute deviations for the full and SAD_r is the sum of absolute deviations for the model reduced under the hypothesis, then

$$F_{obs} = \frac{(SAD_r - SAD_f)/q}{\widehat{\lambda}/2} \qquad [4.42]$$

is an approximate F statistic with q numerator and $n-k$ denominator degrees of freedom. Here, q is the rank of $\mathbf{A}$ and k is the rank of $\mathbf{X}$. An estimator of λ suggested by McKean and Schrader (1987) is described in §A4.8.5. Since the asymptotic variance-covariance matrix of $\widehat{\boldsymbol{\beta}}_L$ is $\text{Var}\left[\widehat{\boldsymbol{\beta}}_L\right] = \lambda(\mathbf{X}'\mathbf{X})^{-1}$, one can construct an approximate test of H_0:$\beta_j = 0$ as

$$t_{obs} = \frac{\widehat{\beta}_j}{c_{jj}}$$

where c_{jj} is the square root of the jth diagonal element of $\widehat{\lambda}(\mathbf{X}'\mathbf{X})^{-1}$, the estimated standard error of $\widehat{\beta}_j$. L_1-Regression is applied to the Prediction Efficiency data in §4.6.3.

4.6.2 M-Estimation

Box 4.9 M-Estimation

- **M-Estimation is an estimation technique robust to outlying observations.**
- **The idea is to curtail residuals that exceed a certain threshold, restricting the influence of observations with large residuals on the parameter estimates.**

M-Estimation was introduced by Huber (1964, 1973) as a robust technique for estimating location parameters (means) in data sets containing outliers. It can also be applied to the estimation of parameters in the mean function of a regression model. The idea of M-Estimation is simple (additional details in §A4.8.6). In least squares the objective function can be written as

$$Q = \sum_{i=1}^{n} \frac{1}{\sigma^2}(y_i - \mathbf{x}_i'\boldsymbol{\beta})^2 = \sum_{i=1}^{n} (e_i/\sigma)^2, \qquad [4.43]$$

a function of the squared residuals. If the contribution of large residuals e_i to the objective function Q can be reduced, the estimates should be more robust to extreme deviations. In M-Estimation we minimize the sum of a function $\rho(\bullet)$ of the residuals, the function being chosen so that large residuals are properly weighted. The objective function for M-Estimation can be written as

$$Q^* = \sum_{i=1}^{n} \rho\left(\frac{y_i - \mathbf{x}_i'\boldsymbol{\beta}}{\sigma}\right) = \sum_{i=1}^{n} \rho(e_i/\sigma) = \sum_{i=1}^{n} \rho(u). \qquad [4.44]$$

Least squares estimation is a special case of M-Estimation with $\rho(u) = u^2$ and maximum likelihood estimators are obtained for $\rho(u) = -\ln\{f(u)\}$ where $f(u)$ is the probability density function of the model errors. The name M-Estimation is derived from this relationship with maximum likelihood estimation.

The estimating equations of the minimization problem can be written as

$$\sum_{i=1}^{n} \psi(\widehat{u})\mathbf{x}_i = \mathbf{0}, \qquad [4.45]$$

where $\psi(u)$ is the first derivative of $\rho(u)$. An iterative algorithm for this problem is discussed in our §A4.8.6 and in Holland and Welsch (1977), Coleman et al. (1980), and Birch and Agard (1993). Briefly, the algorithm rests on rewriting the psi-function as $\psi(u) =$

$(\psi(u)/u)u$, and to let $w = \psi(u)/u$. The estimation problem is then expressed as a weighted least squares problem with weights w. In order for the estimates to be unique, consistent, and asymptotically Gaussian, $\rho(u)$ must be a convex function, so that its derivative $\psi(u)$ is monotonic. One of the classical truncation functions is

$$\rho(e_i) = \begin{cases} e_i^2 & |e_i| \le k \\ 2k|e_i| - k^2 & |e_i| > k. \end{cases} \qquad [4.46]$$

If a residual is less than some number k in absolute value it is retained; otherwise it is curtailed to $2k|e_i| - k^2$. This truncation function is a compromise between using squared residuals if they are small, and a value close to $|e_i|$ if the residual is large. It is a compromise between least squares and L_1-Estimation (Figure 4.15) combining the efficiency of the former with the robustness of the latter. Large residuals are not curtailed to $|e_i|$ but $2k|e_i| - k^2$ to ensure that $\rho(e)$ is convex (Huber 1964). Huber (1981) suggested to choose $k = 1.5 \times \widehat{\sigma}$ where $\widehat{\sigma}$ is an estimate of the standard deviation. In keeping with the robust/resistant theme one can choose the median absolute deviation $\widehat{\sigma} = 1.4826 \times \text{median}(|\widehat{e}_i|)$ or its rescaled version (Birch and Agard 1993)

$$\widehat{\sigma} = 1.4826 \times \text{median}(|\widehat{e}_i - \text{median}(\widehat{e}_i)|).$$

The value $k = 1.345\widehat{\sigma}$ was suggested by Holland and Welsch (1977). If the data are Gaussian-distributed this leads to M-estimators with relative efficiency of 95% compared to least squares estimators. For non-Gaussian data prone to outliers, $1.345\widehat{\sigma}$ yields more efficient estimators than $1.5\widehat{\sigma}$.

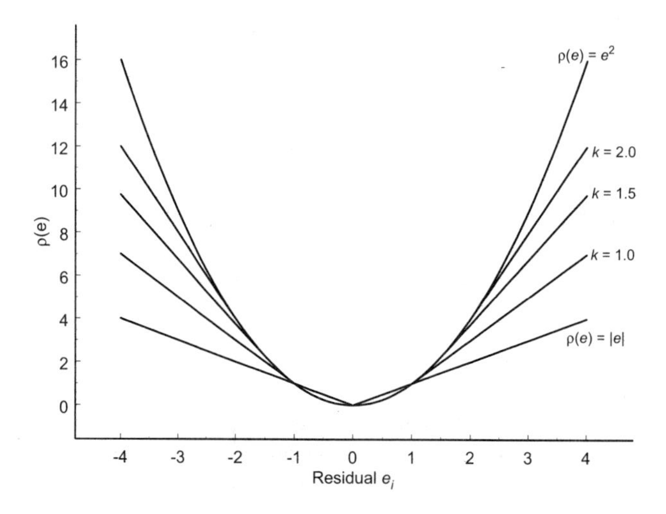

Figure 4.15. Residual transformations in least squares ($\rho(e) = e^2$), L_1-Estimation ($\rho(e) = |e|$), and M-Estimation ([4.46] for $k = 1.0, 1.5, 2.0$).

Since $\psi(\widehat{u})$ serves as a weight of a residual, it is sometimes termed the **residual weighing function** and presents an alternative way of defining the residual transformation in M-Estimation. For Huber's transformation with tuning constant $k = 1.345\widehat{\sigma}$ the weighing function is

$$\psi(e_i) = \begin{cases} e_i & |e_i| \leq k \\ -k & e_i < -k \\ k & e_i > k. \end{cases}$$

A weighing function due to Andrews et al. (1972) is

$$\psi(e_i) = \begin{cases} \sin(e_i/k) & |e_i| \leq k\pi \\ 0 & |e_i| > k\pi \end{cases}, \quad k = 2.1,$$

while Hampel (1974) suggested a step function

$$\psi(e_i) = \begin{cases} e_i & |e_i| \leq a \\ a \text{ sign}(e_i) & a < |e_i| \leq b \\ a(c \text{ sign}(e_i) - e_i) & b < |e_i| \leq c \\ 0 & |e_i| > c. \end{cases}$$

The user must set values for the tuning constants a, b, and c, similar to the choice of k in other weighing functions. Beaton and Tukey (1974) define the biweight M-estimate through the weight function

$$\psi(e_i) = \begin{cases} e_i\left(1 - \frac{e_i^2}{\sigma^2 k^2}\right)^2 & |e_i| \leq k \\ 0 & |e_i| > k. \end{cases}$$

and suggest the tuning constant $k = 4.685$. Whereas Huber's weighing function is an example of a monotonic function, the Beaton and Tukey ψ function is a redescending function which is preferred if the data are contaminated with gross outliers. A weight function that produces maximum likelihood estimates for Cauchy distributed data (a t distribution with a single degree of freedom) is

$$\psi(e_i) = 2e_i/\left(1 + e_i^2\right).$$

The Cauchy distribution is symmetric but much heavier in the tails than the Gaussian distribution and permits more extreme observations than the Gaussian model. A host of other weighing functions has been proposed in the literature. See Hampel et al. (1986) and Barnett and Lewis (1994) for a more detailed description.

Hypothesis tests and confidence intervals for the elements of $\boldsymbol{\beta}$ in M-Estimation rely on the asymptotic Gaussian distribution of $\widehat{\boldsymbol{\beta}}_M$. For finite sample sizes the p-values of the tests are only approximate. Several tests have been suggested in the literature. To test the general linear hypothesis, H_0: $\mathbf{A}\boldsymbol{\beta} = \mathbf{d}$, an analog of the sum of squares reduction test can be used. The test rests on comparing the sum of transformed residuals ($STR = \sum_{i=1}^{n} \rho(\widehat{e}_i)$) in a full and reduced model. If STR_r denotes this sum in the reduced and STR_f in the full model, the test statistic is

$$F_{obs} = \frac{(STR_r - STR_f)/q}{\widehat{\varphi}}, \quad [4.47]$$

where q is the rank of $\mathbf{A}$ and $\widehat{\varphi}$ is an estimate of error variability playing a similar role in M-Estimation as $\widehat{\sigma}^2$ in least squares estimation. If $s_i = \max\{-k, \min\{e_i, k\}\}$ then

$$\widehat{\varphi} = \frac{n}{m(n-k)} \sum_{i=1}^{n} s_i^2. \quad [4.48]$$

This variation of the sum of squares reduction test is due to Schrader and McKean (1977) and Schrader and Hettmansberger (1980). P-values are approximated as $\Pr(F_{obs} > F_{q,n-k})$. The test of Schrader and Hettmansberger (1980) does not adjust the variance of the M-estimates for the fact that they are weighted unequally as one would in weighted least squares. This is justifiable since the weights are random, whereas they are considered fixed in weighed least squares. To test the general linear hypothesis H_0: $\mathbf{A}\boldsymbol{\beta} = \mathbf{d}$, we prefer a test proposed by Birch and Agard (1993) which is a direct analog of the F test in the Gaussian linear models with unequal variances. The test statistic is

$$F_{obs} = (\mathbf{A}\boldsymbol{\beta} - \mathbf{d})'\left[\mathbf{A}(\mathbf{X}'\mathbf{W}\mathbf{X})^{-1}\mathbf{A}'\right]^{-1}(\mathbf{A}\boldsymbol{\beta} - \mathbf{d})/\left(qs^2\right),$$

where q is the rank of $\mathbf{A}$ and s^2 is an estimator of σ^2 proposed as

$$s^2 = \left(\frac{n^2}{n-p}\right)\widehat{\sigma}^2 \frac{\sum_{i=1}^{n}\psi(\widehat{u})^2}{\left(\sum_{i=1}^{n}\psi'(\widehat{u})\right)^2}. \quad [4.49]$$

Here, $\psi'(u)$ is the first derivative of u and the diagonal weight matrix $\mathbf{W}$ has entries $\psi(\widehat{u})/\widehat{u}$. Through simulation studies it was shown that this test has very appealing properties with respect to size (significance level) and power. Furthermore, if $\psi(u) = u$ which results in least squares estimates, s^2 is the traditional residual mean square error estimate.

4.6.3 Robust Regression for Prediction Efficiency Data

When fitting a quadratic polynomial to the Prediction Efficiency data by least squares and invoking appropriate model diagnostics, it was noted that the data point corresponding to the Magnesium observation was an outlier (see Table 4.13, p. 130). There is no evidence suggesting that the particular observation is due to a measurement or execution error and we want to retain it in the data set. The case deletion diagnostics in Table 4.13 suggest that the model fit may change considerably, depending on whether the data point is included or excluded. We fit the quadratic polynomial to these data with several methods. Ordinary least squares with all data points (OLS_{all}), ordinary least squares with the Mg observation deleted (OLS_{-Mg}), L_1-Regression, and M-Regression based on Huber's residual transformation [4.46].

M- and L_1-Estimation are implemented with various SAS® macros contained on the companion CD-ROM. Three of the macros require an installation of the SAS/IML® module. These are `%MRegress()`, `%MTwoWay()`, and `%LAD()`. A third, general-purpose macro `%MEstim()` performs M-Estimation in any linear model, and does not require SAS/IML®. `%MRegress()` in file `\SASMacros\M_Regression.sas` is specifically designed for regression models. It uses either the Huber weighing function or Beaton and Tukey's biweight function. The macro `%MTwoWay()` contained in file `\SASMacros\M_TwoWayAnalysis.sas` implements M-Estimation for the analysis of two-way layouts without replications (a randomized block design, for example). It also uses either the Huber or Tukey weighing functions. Finally, least absolute deviation regression is implemented in macro `%LAD()` (`\SASMacros\L1_Regression.sas`). The most general macro to fit linear models by M-Estimation is macro `%MEstim()` (`\SASMacros\MEstimation.sas`). This macro is considerably more involved than `%MRegress()` or `%MTwoWay()` and thus executes slightly slower. On the upside it can fit regression and classification models alike, does not require SAS/IML®, allows choosing from nine different residual weighing functions, user-modified tuning constants, tests of main effects and interactions, multiple comparisons, interaction slices, contrasts, etc. The call to the macro and its arguments mimics statements and options of the `mixed` procedure of The SAS® System. The macro output is also formatted similar to that produced by the `mixed` procedure.

To fit the quadratic model to the Prediction Efficiency data by least squares (with and without outlier), M-Estimation and L_1-Estimation, we use the following statements.

```
title 'Least Squares regression line with outlier';
proc reg data=range;
  model eff30 = range range2;
run; quit;

title 'Least Squares regression line without outlier';
proc reg data=range(where=(covar ne 'Mg'));
  model eff30 = range range2;
run; quit;

%include 'DriveLetterOfCDROM:\SASMacros\MEstimation.sas';
/* M-Estimation */
%MEstim(data=range,
        stmts=%str(model eff30 = range range2 /s;) )
proc print data=_predm;
  var eff30 _wght Pred Resid StdErrPred Lower Upper;
run;

%include 'DriveLetterOfCDROM:\SASMacros\L1_Regression.sas';
title 'L1-Regression for Prediction Efficacy Data';
%LAD(data=range,y=eff30,x=range range2);
```

L_1-estimates can also be calculated with the `LAV()` function call of the SAS/IML® module which provides greater flexibility in choosing the method for determining standard errors and tailoring of output than the `%LAD()` macro. The code segment

```
proc iml;
  use range; read all var {eff30} into y;
             read all var {range range2} into x;
  close range;
  x = J(nrow(x),1,1) || x;
  opt = {. 3 0 . };
  call lav(rc,xr,X,y,,opt);
quit;
```

produces the same analysis as the previous call to the `%LAD()` macro.

Output 4.12 reiterates the strong negative influence of the Mg observation. When it is contained in the model, the residual sum of squares is $SSR = 536.29$, the error mean square estimate is $MSR = 89.383$, and neither the linear nor the quadratic term are (partially) significant at the 5% level. Removing the Mg observation changes things dramatically. The mean square error estimate is now $MSR = 7.459$ and the linear and quadratic terms are (partially) significant.

Output 4.12.

```
                 Least Squares regression  with outlier

                          The REG Procedure
                            Model: MODEL1
                      Dependent Variable: eff30

                        Analysis of Variance

                                  Sum of           Mean
Source                  DF       Squares         Square    F Value    Pr > F
Model                    2     854.45840      427.22920       3.34    0.1402
Error                    4     511.52342      127.88086
Corrected Total          6    1365.98182

           Root MSE             11.30844    R-Square      0.6255
           Dependent Mean       25.04246    Adj R-Sq      0.4383
           Coeff Var            45.15707

                         Parameter Estimates

                        Parameter       Standard
   Variable      DF      Estimate          Error    t Value    Pr > |t|

   Intercept      1     -39.12105       37.76100      -1.04      0.3587
   range          1       1.39866        1.00851       1.39      0.2378
   range2         1      -0.00633        0.00609      -1.04      0.3574


                Least Squares regression without outlier

                          The REG Procedure
                            Model: MODEL1
                      Dependent Variable: eff30

                        Analysis of Variance

                                  Sum of           Mean
Source                  DF       Squares         Square    F Value    Pr > F
Model                    2    1280.23575      640.11788      52.02    0.0047
Error                    3      36.91880       12.30627
Corrected Total          5    1317.15455

           Root MSE              3.50803    R-Square      0.9720
           Dependent Mean       26.12068    Adj R-Sq      0.9533
           Coeff Var            13.43008

                         Parameter Estimates

                        Parameter       Standard
   Variable      DF      Estimate          Error    t Value    Pr > |t|

   Intercept      1     -75.10250       13.06855      -5.75      0.0105
   range          1       2.41407        0.35300       6.84      0.0064
   range2         1      -0.01198        0.00210      -5.72      0.0106
```

The M-estimates of the regression coefficients are $\widehat{\beta}_{0M} = -65.65$, $\widehat{\beta}_{1M} = 2.126$, and $\widehat{\beta}_{2M} = -0.0103$ (Output 4.13), values similar to the OLS estimates after the Mg observation has been deleted. The estimate of the residual variability (14.256) is similar to the mean square error estimate based on OLS estimates in the absence of the outlying observation. Of particular interest is the printout of the data set `_predm` that is generated automatically by the macro (Output 4.13). It contains the raw fitted residuals $\widehat{e}_i$ and the weights of the observations in the weighted least squares algorithm that underpins estimation. The Mg observation with a large residual of $\widehat{e}_4 = -24.3204$ receives the smallest weight in the analysis (_wght=0.12808).

Output 4.13.

```
Results - The MEstim Macro - Author: Oliver Schabenberger
                       Model Information

  Data Set                          WORK.RANGE
  Dependent Variable                EFF30
  Weighing Function                 HUBER
  Covariance Structure              Variance Components
  Estimation Method                 REML
  Residual Variance Method          Parameter
  Fixed Effects SE Method           Model-Based
  Degrees of Freedom Method         Between-Within
  Weighing constant                 1.345

                          Dimensions

           Covariance Parameters             1
           Columns in X                      3
           Columns in Z                      0
           Subjects                          7
           Max Obs Per Subject               1
           Observations Used                 7
           Observations Not Used             0
           Total Observations                7

                        Fit Statistics

     OLS Residual variance                     127.8809
     Rescaled MAD                               2.31598
     Birch and Agard estimate                  14.25655
     Observations used                                7
     Sum of       weights (M)                  5.605905
     Sum of       residuals (M)                -17.8012
     Sum of abs.  residuals (M)                37.02239
     Sum of squ.  residuals (M)                643.6672
     Sum of       residuals (OLS)              -142E-15
     Sum of abs.  residuals (OLS)              47.88448
     Sum of squ.  residuals (OLS)              511.5234


                   Covariance Parameters

                   Parameter     Estimate
                   Residual       14.2566

               Type 3 Tests of Fixed Effects

                     Num      Den
      Effect          DF       DF     F Value     Pr > F
      range            1        4       30.93     0.0051
      range2           1        4       20.31     0.0108
```

Output 4.13 (continued).

```
                         Solution for Fixed Effects

                                 Standard
    Effect          Estimate        Error       DF     t Value     Pr > |t|

    Intercept       -65.6491      14.0202        4       -4.68       0.0094
    range             2.1263       0.3823        4        5.56       0.0051
    range2          -0.01030     0.002285        4       -4.51       0.0108


                                                    StdErr
 Obs      eff30     _wght      Pred       Resid      Pred      Lower     Upper

  1     25.2566   1.00000   24.3163     0.9403   2.03354   18.6703   29.9623
  2     40.1652   1.00000   39.0396     1.1256   3.70197   28.7613   49.3179
  3     33.7026   0.47782   27.1834     6.5191   2.17049   21.1572   33.2097
  4     18.5732   0.12808   42.8935   -24.3204   2.82683   35.0450   50.7421
  5     15.3946   1.00000   14.3691     1.0256   2.03142    8.7289   20.0092
  6     -0.0666   1.00000    2.4030    -2.4696   3.08830   -6.1714   10.9775
  7     42.2717   1.00000   42.8935    -0.6218   2.82683   35.0450   50.742
```

Notice that the raw residuals sum to zero in the ordinary least squares analysis but not in M-Estimation due to the fact that residuals are weighted unequally. The results of L_1-Estimation of the model parameters are shown in Output 4.14. The estimates of the parameters are very similar to the M-estimates.

Output 4.14.

```
              Least Absolute Deviation = L1-Norm Regression
                     Author: Oliver Schabenberger
                   Model: eff30 = intcpt range range2

                  Data Set                      range
                  Number of observations           7
                  Response                      eff30
                  Covariates                    range range2


                          LAD Regression Results

               Median of response                  1366.302766
               Sum abs. residuals full model         35.363442
               Sum abs. residuals null model         82.238193
               Sum squared residuals full m.        624.274909
               Tau estimate                          15.085416


                 Parameter Estimates and Standard Errors

                   Estimate          Std.Err       T-Value     Pr(> |T|)

         INTERCPT    -57.217612      50.373024          .            .
         RANGE         1.917134       1.345349        1.425     0.227280
         RANGE2       -0.009095       0.008121       -1.120     0.325427
```

The estimates of the various fitting procedures are summarized in Table 4.21. The coefficients for L_1-, M-Regression, and OLS_{-Mg} are quite similar, but the former methods retain all observations in the analysis.

Table 4.21. Estimated coefficients for quadratic polynomial $y_i = \beta_0 + \beta_1 x_i + \beta_2 x_i^2 + e_i$ for Prediction Efficacy data

	OLS$_{all}$	**OLS**$_{-Mg}$	L_1**-Regression**	**M-Regression**
$\widehat{\beta}_0$	– 39.121	– 75.103	– 57.217	– 65.649
$\widehat{\beta}_1$	1.398	2.414	1.917	2.126
$\widehat{\beta}_2$	– 0.0063	– 0.0119	– 0.009	– 0.010
n	7	6	7	7
SSR	511.523	36.919	624.275	643.667
SAD	47.884		35.363	37.022

Among the fits based on all observations ($n = 7$), the residual sum of squares is minimized for ordinary least squares regression, as it should be. By the same token, L_1-Estimation yields the smallest sum of absolute deviations (SAD). M-Regression with Huber's weighing function, has a residual sum of squares slightly larger than the least absolute deviation regression. Notice that M-estimates are obtained by weighted least squares, and SSR is not the criterion being minimized. The sum of absolute deviations of M-Estimation is smaller than that of ordinary least squares and only slightly larger than that of the L_1 fit.

The similarity of the predicted trends for OLS_{-Mg}, L_1-, and M-Regression is apparent in Figure 4.16. While the Mg observation pulls the least squares regression line toward it, L_1- and M-estimates are not greatly affected by it. The predictions for L_1- and M-Regression are hard to distinguish; they are close to least squares predictions obtained *after* outlier deletion but do not require removal of the offending data point. The L_1-Regression passes through three data points since the model contains two regressors and an intercept. The data points are CEC, Prec and P.

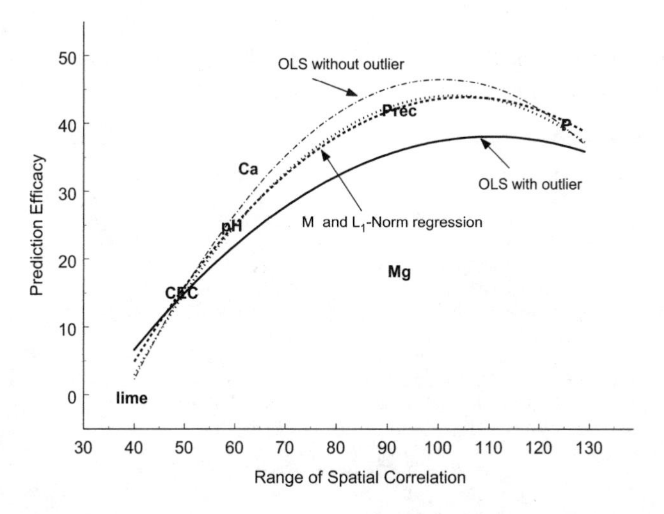

Figure 4.16. Fitted and predicted values in prediction efficacy example for ordinary least squares with and without outlying Mg observation, L_1- and M-Regression.

4.6.4 M-Estimation in Classification Models

Analysis of variance (classification) models that arise in designed experiments often have relatively few observations. An efficiently designed experiment will use as few replications as necessary to maintain a desirable level of test power, given a particular error-control and treatment design. As a consequence, error degrees of freedom in experimental designs are often small. A latin square design with four treatments, for example, has only six degrees of freedom for the estimation of experimental error variability. Removal of outliers from the data further reduces the degrees of freedom and the loss of observations is particularly damaging if only few degrees of freedom are available to begin with. It is thus not only with an eye toward stabilizing the estimates of block, treatment, and other effects that robust analysis of variance in the presence of outliers is important. Allowing outliers to remain in the data but appropriately downweighing their influence retains error degrees of freedom and inferential power. In this subsection, outlier-resistant M-Estimation is demonstrated for analysis of variance models using the `%MEstim()` macro written for The SAS® System. No further theoretical details are needed in this subsection beyond the discussion in §4.6.2 and the mathematical details in §A4.8.6 except for taking into account that the design matrix in analysis of variance models is rank-deficient. We discuss M-Estimation in classification models by way of example.

Example 4.3 Dollar Spot Counts (continued). Median polishing of the dollar spot count data suggested to analyze the log-transformed counts. We apply M-Estimation here with Huber's weighing function and the tuning constant 1.345. These are default settings of `%MEstim()`. The statements

```
%include 'DriveLetter:\SASMacros\MEstimation.sas';
%MEstim(data=dollarspot,
        stmts=%str(
           class block tmt;
           model lgcnt = block tmt;
           lsmeans tmt / diff;)     );
```

ask for a robust analysis of the log-transformed dollar spot counts in the block design and pairwise treatment comparisons (`lsmeans tmt / diff`).

Output 4.15. The MEstim Macro - Author: Oliver Schabenberger

```
                        Model Information
          Data Set                          WORK.DOLLSPOT
          Dependent Variable                LGCNT
          Weighing Function                 HUBER
          Covariance Structure              Variance Components
          Estimation Method                 REML
          Residual Variance Method          Parameter
          Fixed Effects SE Method           Model-Based
          Degrees of Freedom Method         Between-Within
          Weighing constant                 1.345

                     Class Level Information
            Class      Levels    Values
            BLOCK           4    1 2 3 4
            TMT            14    1 2 3 4 5 6 7 8 9 10 11 12 13 14
```

Output 4.15 (continued).

Dimensions

Covariance Parameters	1
Columns in X	19
Columns in Z	0
Subjects	56
Max Obs Per Subject	1
Observations Used	56
Observations Not Used	0
Total Observations	56

Fit Statistics

OLS Residual variance	0.161468
Rescaled MAD	0.38555
Birch and Agard estimate	0.162434
Observations used	56
Sum of weights (M)	55.01569
Sum of residuals (M)	0.704231
Sum of abs. residuals (M)	15.68694
Sum of squ. residuals (M)	6.336998
Sum of residuals (OLS)	-862E-16
Sum of abs. residuals (OLS)	15.89041
Sum of squ. residuals (OLS)	6.297243

Covariance Parameters

Parameter	Estimate
Residual	0.1624

Type 3 Tests of Fixed Effects

Effect	Num DF	Den DF	F Value	Pr > F
BLOCK	3	39	0.78	0.5143
TMT	13	39	7.38	<.0001

M-Estimated Least Square Means

Obs	Effect	TMT	Estimate	Standard Error	DF	t Value	Pr > |t|
1	TMT	1	3.2720	0.2015	39	16.24	<.0001
2	TMT	2	4.2162	0.2015	39	20.92	<.0001
3	TMT	3	3.3113	0.2015	39	16.43	<.0001
4	TMT	4	3.5663	0.2058	39	17.33	<.0001
5	TMT	5	3.9060	0.2015	39	19.38	<.0001
6	TMT	6	3.8893	0.2076	39	18.73	<.0001
7	TMT	7	3.7453	0.2015	39	18.59	<.0001
8	TMT	8	3.9386	0.2015	39	19.54	<.0001

... and so forth ...

Differences of M-Estimated Least Square Means

Effect	TMT	_TMT	Estimate	Standard Error	DF	t Value	Pr > |t|
TMT	1	2	-0.9442	0.2850	39	-3.31	0.0020
TMT	1	3	-0.03931	0.2850	39	-0.14	0.8910
TMT	1	4	-0.2943	0.2880	39	-1.02	0.3132
TMT	1	5	-0.6340	0.2850	39	-2.22	0.0320
TMT	1	6	-0.6173	0.2893	39	-2.13	0.0392
TMT	1	7	-0.4733	0.2850	39	-1.66	0.1048
TMT	1	8	-0.6666	0.2850	39	-2.34	0.0245
TMT	1	9	-1.0352	0.2850	39	-3.63	0.0008
TMT	1	10	-1.5608	0.2894	39	-5.39	<.0001
TMT	1	11	-1.5578	0.2850	39	-5.47	<.0001
TMT	1	12	-0.1448	0.2921	39	-0.50	0.6229
TMT	1	13	-0.05717	0.2850	39	-0.20	0.8420

Output 4.15 (continued).

```
TMT        1      14        0.1447      0.2850      39       0.51    0.6145
TMT        2       3        0.9049      0.2850      39       3.18    0.0029
TMT        2       4        0.6499      0.2880      39       2.26    0.0297
TMT        2       5        0.3102      0.2850      39       1.09    0.2831
TMT        2       6        0.3269      0.2893      39       1.13    0.2654
TMT        2       7        0.4709      0.2850      39       1.65    0.1065
TMT        2       8        0.2776      0.2850      39       0.97    0.3360
TMT        2       9      -0.09097      0.2850      39      -0.32    0.7513
TMT        2      10       -0.6166      0.2894      39      -2.13    0.0395
TMT        2      11       -0.6136      0.2850      39      -2.15    0.0376
TMT        2      12        0.7994      0.2921      39       2.74    0.0093
TMT        2      13        0.8870      0.2850      39       3.11    0.0035
TMT        2      14        1.0889      0.2850      39       3.82    0.0005
TMT        3       4       -0.2550      0.2880      39      -0.89    0.3814
... and so forth ...
```

The least squares estimates minimize the sum of squared residuals (6.297, Table 4.22). The corresponding value in the robust analysis is very close as are the F statistics for the treatment effects (7.60 and 7.38).

The treatment estimates are also quite close. Because of the orthogonality and the constant weights in the least squares analysis, the standard errors of the treatment effects are identical. In the robust analysis, the standard errors depend on the weights which differ from observation to observation depending on the size of their residuals. The subtle difference in precision is exhibited in the standard error of the estimate for treatment 6. Four observations are downweighed substantially in the robust analysis. In particular, treatment 6 in block 2 ($y_{26} = 99$, Table 4.3) received weight 0.77 and treatment 12 in block 1 ($y_{1,12} = 63$) receives weight 0.64. These observations have not been identified as potential outliers before. Combined with treatment 10 in block 4 these observations accounted for three of the four largest median polished residuals.

Table 4.22. Analysis of variance for Dollar Spot log(counts) by least squares and M-Estimation (Huber's weighing function)

	Least Squares	**M-Estimation**
Sum of squared residuals	6.29	6.34
Sum of absolute residuals	15.89	15.69
F_{obs} for treatment effect	7.60	7.38
p-value for treatment effect	0.0001	0.0001
Treatment Estimates (Std.Err)		
1	3.27 (0.20)	3.27 (0.20)
2	4.21 (0.20)	4.21 (0.20)
3	3.31 (0.20)	3.31 (0.20)
4	3.59 (0.20)	3.56 (0.20)
5	3.90 (0.20)	3.90 (0.20)
6	3.93 (0.20)	3.88 (0.21)
⋮	⋮	⋮

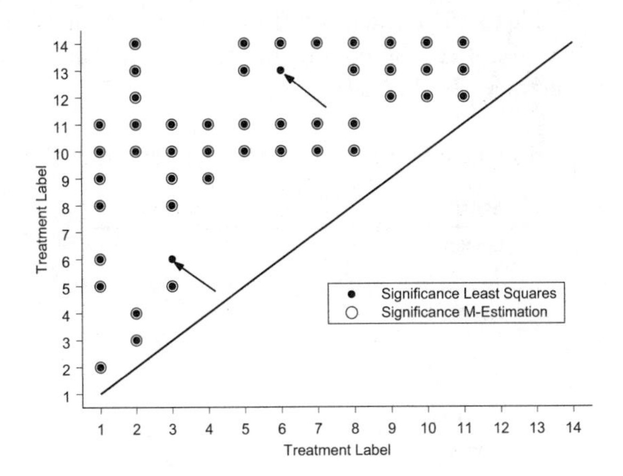

Figure 4.17. Results of pairwise treatment comparisons in robust ANOVA and ordinary least squares estimation. Dots reflect significance of treatment comparison at the 5% level in ANOVA, circles significance at the 5% level in M-Estimation.

The robust analysis downweighs the influence of $y_{26} = 99$ whereas the least squares analysis weighs all observations equally. Such differences, although small, can have a measurable impact on treatment comparisons. At the 5% significance level the least squares and robust analysis agree closely. However, treatments 3 and 6 as well as 6 and 13 are significantly different in the least squares analysis and not so in the robust analysis (Figure 4.17).

The detrimental effect of outlier deletion in experimental designs with few degrees of freedom to spare can be demonstrated with the following split-plot design. The whole-plot design is a randomized complete block design with four treatments in two blocks. Each whole-plot is then subdivided into three sub-plots to which the levels of the sub-plot treatment factor are randomly assigned. The analysis of variance of this design contains separate error terms for whole- and sub-plot factors. Tests of whole-plot effects use the whole-plot error which has only 3 degrees of freedom (Table 4.23).

Table 4.23. Analysis of variance for split-plot design

Source	Degrees of freedom
Block	1
A	3
Error(A)	3
B	2
$A \times B$	6
Error(B)	8
Total	23

The sub-plot error is associated with 8 degrees of freedom. Now assume that the observations from one of the eight whole plots appear errant. It is surmised that either the experimental unit is not comparable to other units, that the treatment was applied incorrectly, or that measurement errors were comitted. Removing the three observations for the whole-plot changes the analysis of variance (Table 4.24).

Table 4.24. Analysis of variance for split-plot design after removal of one whole-plot

Source	Degrees of freedom
Block	1
A	3
Error(A)	2
B	2
A × B	6
Error(B)	6
Total	20

At the 5% significance level the critical value in the F-test for the whole-plot (A) main effect is $F_{0.05,3,3} = 9.28$ in the complete design and $F_{0.05,3,2} = 19.16$ in the design with a lost whole-plot. The test statistic $F_{obs} = MS(A)/MS(\text{Error}(A))$ must double in value to find a significant difference among the whole-plot treatments. An analysis which retains extreme observations and reduces their impact is to be preferred.

Example 4.6. The data from a split-plot design with four whole-plot treatments arranged in a randomized complete block design with three blocks and three sub-plot treatments is given in the table below.

Table 4.25. Observations from split-plot design

Block	Whole-Plot Treatment	Sub-Plot Treatment B_1	B_2	B_3
1	A_1	3.8	5.3	6.2
	A_2	5.2	5.6	5.4
	A_3	6.0	5.6	7.6
	A_4	6.5	7.1	7.7
2	A_1	3.9	5.4	4.5
	A_2	6.0	6.1	6.2
	A_3	7.0	6.4	7.4
	A_4	7.4	8.3	6.9
3	A_1	4.9	6.4	3.5
	A_2	4.6	5.0	6.4
	A_3	6.8	6.2	7.7
	A_4	4.7	7.3	7.4

It is assumed that the levels of factor A are quantitative and equally spaced. Apart from tests for main effects and interactions, we are interested in testing for trends of the response with levels of A. We present a least squares analysis of variance and a robust analysis of variance using all the data. The linear statistical model underlying this analysis is

$$Y_{ijk} = \mu + \rho_i + \alpha_j + d_{ij} + \beta_k + (\alpha\beta)_{jk} + e_{ijk},$$

where ρ_i denotes the block effect $(i = 1,\cdots,3)$, α_j the main effects of factor A $(j = 1,\cdots,4)$, d_{ij} is the random whole-plot experimental error with mean 0 and variance σ_d^2, β_k are the main effects of factor B $(k = 1,2,3)$, $(\alpha\beta)_{jk}$ are the interaction effects, and e_{ijk} is the random sub-plot experimental error with mean 0 and variance σ_e^2.

The tabulated data do not suggest any data points as problematic. A graphical display adds more insight. The value 4.7 for the fourth level of factor A and the first level of B in the third block appears suspiciously small compared to the remainder of the data for the whole-plot factor level (Figure 4.18).

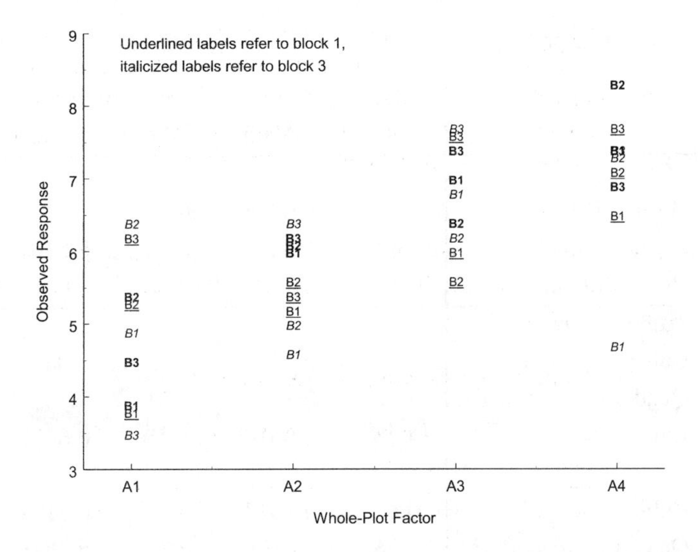

Figure 4.18. Data in split-plot design. Labels $B_1,\cdots,B_3$ in the graph area are drawn at the value of the response for the particular combination of factors A (horizontal axis) and B. Values from block 1 are underlined, appear in regular type for block 2, and are italicized for block 3.

Split-plot designs are special cases of mixed models (§7) that are best fit with the `mixed` procedure of The SAS® System. The standard and robust analysis using Huber's weight function with tuning constant 1.345 are produced with the statements

```
proc mixed data=spd;
 class block a b;
 model y = block a b a*b ;
 random block*a;
 lsmeans a b a*b / diff;
 lsmeans a*b / slice=(a b);
 contrast 'A cubic @ B1' a -1  3 -3  1 a*b -1 0 0  3 0 0 -3 0 0 1 0 0;
 contrast 'A quadr.@ B1' a  1 -1 -1  1 a*b  1 0 0 -1 0 0 -1 0 0 1 0 0;
 contrast 'A linear@ B1' a -3 -1  1  3 a*b -3 0 0 -1 0 0  1 0 0 3 0 0;
```

```
   /* and so forth for contrasts @ B2 and @B3 */
run;

/* Robust Analysis */
%include 'DriveLetter:\SASMacros\MEstimation.sas';
%MEstim(data=spd,
         stmts=%str(class block a b;
              model y = block a b a*b ;
              random block*a;
              parms / nobound;
              lsmeans a b a*b / diff;
              lsmeans a*b / slice=(a b);
              contrast 'A cubic @ B1' a -1  3 -3  1
                                   a*b -1 0 0 3 0 0 -3 0 0 1 0 0;
              contrast 'A quadr.@ B1' a  1 -1 -1  1
                                   a*b 1 0 0 -1 0 0 -1 0 0 1 0 0;
              contrast 'A linear@ B1' a -3 -1  1  3
                                   a*b -3 0 0 -1 0 0 1 0 0 3 0 0;
                /* and so forth for contrasts @ B2 and @B3 */
         ),converge=1E-4,fcn=huber    );
```

Table 4.26. Analysis of variance results in split-plot design

	Least Squares		**M-Estimation**	
	F Value	p-Value	F Value	p-Value
Model Effects				
A Main Effect	16.20	0.0028	26.68	0.3312
B Main Effect	4.21	0.0340	3.93	0.0408
$A \times B$ Interaction	1.87	0.1488	3.03	0.0355
Trend Contrasts				
Cubic trend @ B_1	1.07	0.3153	1.12	0.3051
Quadratic trend @ B_1	2.89	0.1085	2.51	0.1325
Linear trend @ B_1	14.45	0.0016	24.75	0.0001
Cubic trend @ B_2	0.04	0.8517	0.05	0.8198
Quadratic trend @ B_2	3.58	0.0766	5.32	0.0348
Linear trend @ B_2	10.00	0.0060	14.84	0.0014
Cubic trend @ B_3	1.18	0.2925	1.12	0.3055
Quadratic trend @ B_3	3.02	0.1013	6.84	0.0187
Linear trend @ B_3	23.57	0.0002	40.07	0.0001

The interaction is not significant in the least squares analysis ($p = 0.1488$, Table 4.26) and both main effects are. The robust analysis reaches a different conclusion. It indicates a significant $A \times B$ interaction ($p = 0.0355$) and a **masked** A main effect ($p = 0.3312$). At the 5% significance level the least squares results suggest linear trends of the response in A for all levels of B. A marginal quadratic trend can be noted for B_2 ($p = 0.0766$, Table 4.26). The robust analysis concludes a stronger linear effect at B_1 and quadratic effects at B_2 and B_3.

Can these discrepancies between the two analyses be explained by a single observation? Studying the weights w_i in M-Estimation for these data reveals that there are in fact two extreme observations. All weights are one except for A_4B_1 in block 3 $(y = 4.7)$ with $w = 0.4611$ and A_1B_3 in block 1 $(y = 6.2)$ with $w = 0.3934$. While the first observation was identified as suspicious in Figure 4.18, most analysts would have probably failed to recognize the large influence of the second observation. As can be seen from the magnitude of the weight, its residual is even larger than that of A_4B_1 in block 3. The weights can be viewed by printing the data set `_predm` which is generated by the `%MEstim()` macro (not shown here).

The estimated treatment cell means are very much the same for both analyses with the exception of $\widehat{\mu}_{13}$ and $\widehat{\mu}_{41}$ (Figure 4.19). The least squares estimate of μ_{13} is too large and the estimate of μ_{41} is too small. For the other treatment combinations the estimates are identical (because of the unequal weights the precision of the treatment estimates is not the same in the two analyses, even if the estimates agree). Subtle differences in estimates of treatment effects contribute to the disagreement in conclusions from the two analyses in Table 4.26. A second source of disagreement is the estimate of experimental error variance. The sub-plot error variance σ_e^2 was estimated as $\widehat{\sigma}^2 = 0.5584$ in the least squares analysis and as 0.3761 in the robust analysis. Comparisons of the treatments whose estimates are not affected by the outliers will be more precise and powerful in the robust analysis.

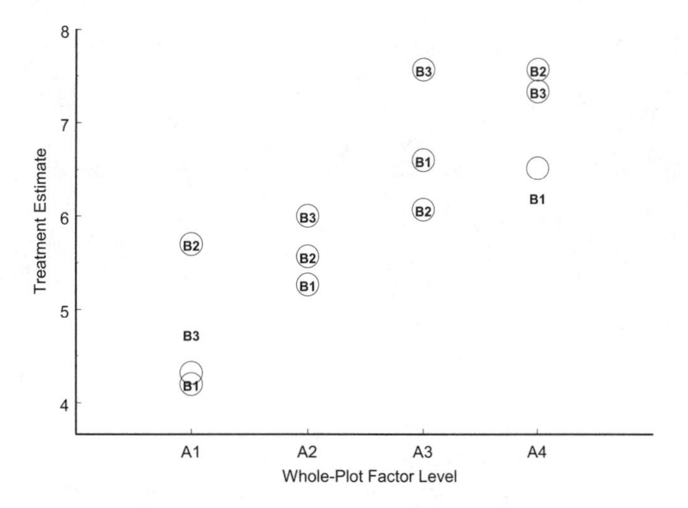

Figure 4.19. Estimated treatment means $\widehat{\mu}_{jk}$ in split-plot design. The center of the circles denotes the estimates in the robust analysis, the labels the location of the estimates in the least squares analysis.

The disagreement does not stop here. Comparing the levels of B at each level of A via slicing (see §4.3.2), the least squares analysis fails to find significant differences among the levels of B at the 5% level for any level of A. The robust analysis detects B effects at A_1 and A_3 (Table 4.27). The marginally significant slice at level A_4 in the least squares analysis ($p = 0.086$) is due to a greater separation of the means compared to the robust analysis (Figure 4.19).

Table 4.27. P-values for slices in split-plot design

Factor being compared	**at level**	**Least Squares** F Value	p-value	**M-Estimation** F Value	p-value
B	A_1	3.11	0.0725	5.38	0.0163
	A_2	0.73	0.4972	1.08	0.3618
	A_3	3.11	0.0725	4.61	0.0262
	A_4	2.87	0.0860	2.14	0.1505

4.7 Nonparametric Regression

Box 4.10 Nonparametric Regression

- **Nonparametric regression estimates the conditional expectation $f(x) = \mathrm{E}[Y|X = x]$ by a smooth but otherwise unspecified function.**
- **Nonparametric regression is based on estimating $f(x)$ within a local neighborhood of x by weighted averages or polynomials. This process is termed smoothing.**
- **The trade-off between bias and variance of a smoother is resolved by selecting a smoothing parameter that optimizes some goodness-of-fit criterion.**

Consider the case of a single response variable Y and a single covariate X. The **regression** of Y on X is the conditional expectation

$$f(x) = \mathrm{E}[Y|X = x]$$

and our analysis of the relationship between the two variables so far has revolved around a parametric model for $f(x)$, for example a quadratic polynomial $f(x) = \beta_0 + \beta_1 x + \beta_2 x^2$. Inferences drawn from the analysis are dependent on the model for $f(x)$ being correct. How are we to proceed in situations where the data do not suggest a particular class of parametric models? What can be gleaned about the conditional expectation $\mathrm{E}[Y|x]$ in an exploratory fashion that can aid in the development of a parametric model?

A starting point is to avoid any parametric specification of the mean function and to consider the general model

$$Y_i = f(x_i) + e_i. \qquad [4.50]$$

Rather than placing the onus on the user to select a parametric model for $f(x_i)$, we let the data guide us to a nonparametric estimate of $f(x_i)$.

Example 4.7. Paclobutrazol Growth Response. During the 1995 growing season the growth regulator Paclobutrazol was applied May 1, May 29, June 29, and July 24 on turf plots. If turfgrass growth is expressed relative to the growth of untreated plots we expect a decline of growth shortly after each application of the regulator and increasing growth as the regulator's effect wears off. Figure 4.20 shows the clipping percentages removed from Paclobutrazol-treated turf by regular mowing. The amount of clippings removed relative to the control is a surrogate measure of growth in this application.

The data points show the general trend that is expected. Decreased growth shortly after application with growth recovery before the next application. It is not obvious, however, how to parametrically model the clipping percentages over time. A single polynomial function would require trends of high order to pick up the fluctuations in growth response. One could also fit separate quadratic or cubic polynomials to the intervals $[0,2]$, $[2,3]$, and $[3,4]$ months. Before examining complicated parametric models, a nonparametric smooth of the data can (i) highlight pertinent features of the data, (ii) provide guidance for the specification of possible parametric structures and (iii) may answer the questions of interest.

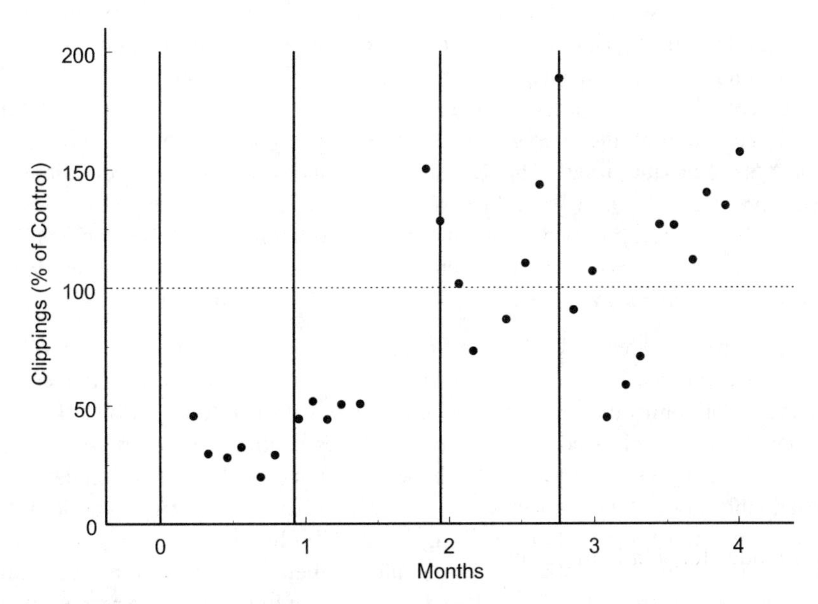

Figure 4.20. Clipping percentages of Paclobutrazol-treated turf plots relative to an untreated control. Vertical lines show times of treatment applications. Data kindly provided by Mr. Ronald Calhoun, Department of Crop and Soil Sciences, Michigan State University. Used with permission.

The result of a nonparametric regression analysis is a smoothed function $\widehat{f}(x)$ of the data and the terminology *data smoothing* is sometimes used in this context. We point out, however, that parametric regression analysis is also a form of data smoothing. The least smooth description of the data is obtained through interpolation the data points by connecting the dots. The parametric regression line, on the other hand, is a very smooth representation of the data. It passes through the center of the data scatter, but not necessarily any particular data pair (y_i, x_i). Part of the difficulty in modeling data nonparametrically lies in determining the appropriate degree of smoothness so that $\widehat{f}(x)$ retains the pertinent features of the data while filtering its random fluctuations.

Nonparametric statistical methods are not assumption-free as is sometimes asserted. $f(x)$ is assumed to belong to some collection of functions, possibly of infinite dimension, that share certain properties. It may be required that $f(x)$ is differentiable, for example. Secondly, the methods discussed in what follows also require that the errors e_i are uncorrelated and homoscedastic ($\text{Var}[e_i] = \sigma^2$).

4.7.1 Local Averaging and Local Regression

The unknown target function $f(x)$ is the conditional mean of the response if the covariate value is x. It is thus a small step to consider as an estimate of $f(x_i)$ some form of average calculated from the observations at x_i. If data are replicated such that r responses are observed for each value x_i, one technique of estimating $f(x_i)$ is to average the r response values and to interpolate the means. Since this procedure breaks down if $r = 1$, we assume now that each covariate value x_i is unique in the data set and if replicate measurements were made, that we operate with the average values at x_i. In this setup, estimation will be possible whether the x_i's are unique or not. The data set now comprises as many observations as there are unique covariate values, n, say. Instead of averaging y_i's at a given x_i we can also estimate $f(x_i)$ by averaging observations in the neighborhood of x_i. Also, the value of X at which a prediction is desired can be any value, whether it is part of the observed data or not. We simply refer to x_0 as the value at which the function $f()$ is to be estimated.

If Y and X are unrelated and $f(x)$ is a flat line, it does not matter how close to the target x_0 we select observations. But as $\text{E}[Y|x]$ changes with x, observations far removed from the point x_0 should not contribute much or anything to the estimate to avoid bias. If we denote the set of points that are allowed to contribute to the estimation of f at x_0 with $N(x_0)$, we need to decide (i) how to select the neighborhood $N(x_0)$ and (ii) how to weigh the points that are in the neighborhood. Points not in the neighborhood have zero weight and do not contribute to the estimation of $f(x_0)$, but points in the neighborhood can be weighted unequally. To assign larger weights to points close to x_0 and smaller weights to points far from x_0 is reasonable. The local estimate can be written as a linear combination of the responses,

$$\widehat{f}(x_0) = \sum_{i=1}^{n} W_0(x_i; \lambda) y_i. \qquad [4.51]$$

$W_0()$ is a function that assigns a weight to y_i based on the distance of x_i from x_0. The weights depend on the form of the weight function and the parameter λ, called the **smoothing parameter**. λ essentially determines the width of the neighborhood.

The simplest smoother is the **moving average** where $\widehat{f}(x_0)$ is calculated as the average of the λ points to the right or left of x_0. If x_0 is an observed value, then points within λ points of x_0 receive weight $W_0(x_i;\lambda) = 1/(2\lambda+1)$; all other points receive zero weight. In the interior of the data, each predicted value is thus the average of $100 \times (2\lambda+1)/n$ percent of the data. Instead of this symmetric nearest neighborhood the 2λ points closest to x_0 (the nearest neighborhood) may be selected. Because of the bias incurred by the moving average near the ends of the X data, the symmetric nearest neighborhood is usually preferred; see Hastie and Tibshirani (1990, p. 32). A moving average is not very smooth and has a jagged look if λ is chosen small. Figure 4.21a shows 819 observations of the light transmittance (PPDF) in the understory of a longleaf pine stand collected during a single day. Moving averages were calculated with symmetric nearest neighborhoods containing 1%, 5%, and 30% of the data corresponding roughly to smoothing parameters $\lambda = 4, \lambda = 20$, and $\lambda = 123$. With increasing λ the fitted profile becomes more smooth. At the same time, the bias in the profile increases. Near the center of the data, the large neighborhood $\lambda = 123$ in Figure 4.21d averages values to the left and right of the peak. The predicted values thus are considerably below what appears to be the maximum PPDF. For the small smoothing neighborhood $\lambda = 4$ in Figure 4.21b, the resulting profile is not smooth at all. It nearly interpolates the data.

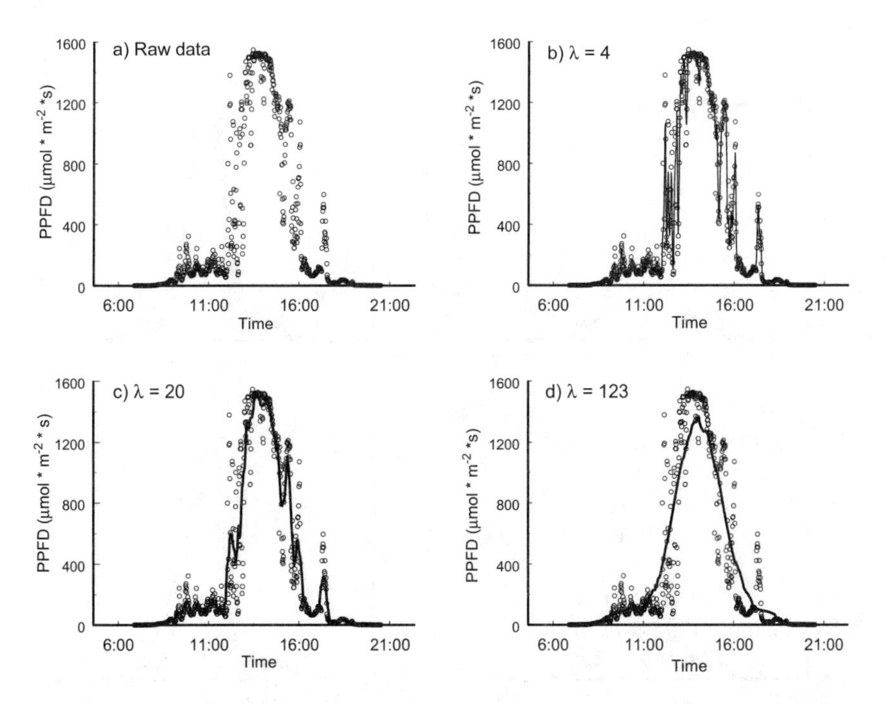

Figure 4.21. Light transmittance expressed as photosynthetic photon flux density (PPDF $\mu mol/m^2/s$) in understory of longleaf pine stand. (a): raw data measured in 10-second intervals. (b) to (d): moving average smoothers with symmetric nearest neighborhoods of 1% (b), 5% (c), and 30% of the $n = 819$ data points. Data kindly provided by Dr. Paul Mou, Department of Biology, University of North Carolina at Greensboro. Used with permission.

An improvement over the moving average is to fit linear regression lines to the data in each neighborhood, usually simple linear regressions or quadratic polynomials. The smoothing parameter λ must be chosen so that each neighborhood contains a sufficient number of points to estimate the regression parameters. As in the case of the moving average, the **running-line smoother** becomes smoother as λ increases. If the neighborhood includes all n points, it is identical to a parametric least squares polynomial fit.

As one moves through the data from point to point to estimate $f(x_0)$, the weight assigned to the point x_i changes. In case of the moving average or running-line smoother, the weight of x_i makes a discrete jump as the point enters and leaves the neighborhood. This accounts for their sometimes jagged look. Cleveland (1979) implemented a running-line smoother that eliminates this problem by varying the weights within the neighborhood in a smooth manner. The target point x_0 is given the largest weight and weights decrease with increasing distance $|x_i - x_0|$. The weight is exactly zero at the point in the neighborhood farthest from x_0. Cleveland's tri-cube weights are calculated as

$$W_0(x_i;\lambda) = \begin{cases} \left\{1 - \left(\frac{|x_i - x_0|}{\lambda}\right)^3\right\}^3 & 0 \le |x_j - x_i| < \lambda \\ 0 & \text{otherwise} \end{cases}.$$

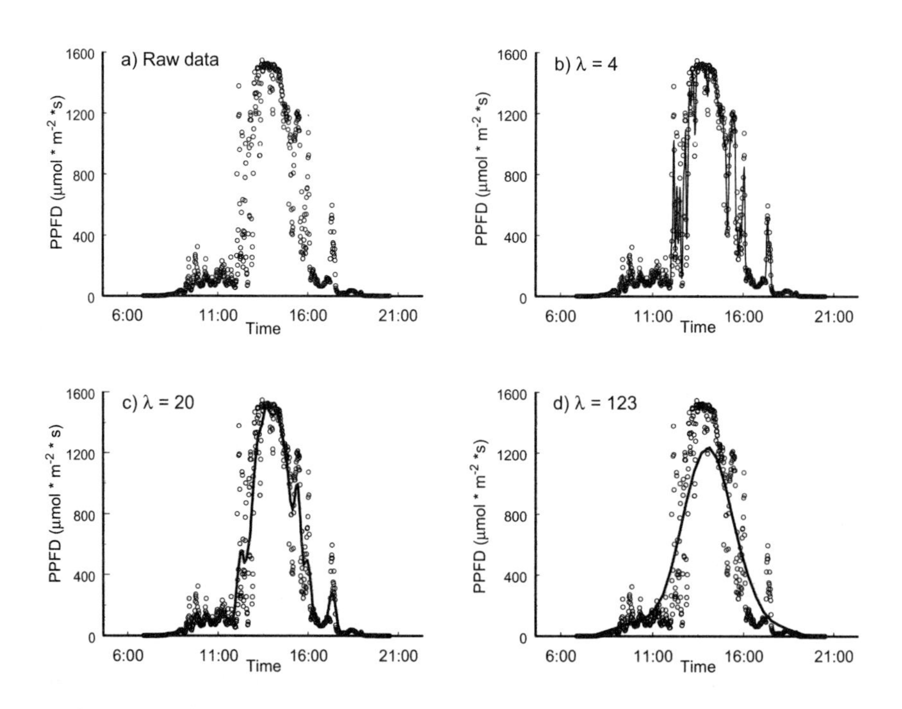

Figure 4.22. Loess fit of light transmittance data with smoothing parameters identical to those in Figure 4.21.

Cleveland's proposal was to initially fit a dth degree polynomial with weights $W_i(x_i;\lambda)$ and to obtain the residual $y_i - \widehat{f}(x_i)$ at each point. Then a second set of weights δ_i is defined based on the size of the residual with δ_i being small for large residuals. The local polynomial fit is repeated with $W_i(x_i;\lambda)$ replaced by $\delta_i W_i(x_j;\lambda)$ and the residual weights are again updated based on the size of the residuals after the second fit. This procedure is repeated until some convergence criterion is met. This robustified local polynomial fit was called robust locally weighted regression by Cleveland (1997) and is also known as **loess** regression or **lowess** regression (see also Cleveland et al. 1988).

Beginning with Release 8.0 of The SAS® System the `loess` procedure has been available to fit local polynomial models with and without robustified reweighing. Figure 4.22 shows loess fits for the light transmittance data with smoothing parameters identical to those in Figure 4.21 produced with `proc loess`. For the same smoothing parameter the estimated trend $\widehat{f}(x_i)$ is less jagged and erratic compared to a simple moving average. Small neighborhoods ($\lambda = 4$) still produce highly undersmoothed estimates with large variance.

A closely related class of smoothers uses a weight function that is a symmetric density function with a scale parameter λ, called the **bandwidth**, and calculates weighted averages of points in the neighborhood. The weight functions trail off with increasing distance of x_i from x_0 but do not have to reach zero. These weight functions are called **kernel functions** or simply **kernels** and the resulting estimators are called **kernel estimators**. We require the kernels to have certain properties. Their support should be bounded so that the function can be rescaled to the $[-1,1]$ interval and the function should be symmetric and integrate to one. Because of these requirements, the Gaussian probability density function

$$K(t) = \frac{1}{\sqrt{2\pi}}\exp\{-½t^2\}$$

is a natural choice. Other frequently used kernels are the quadratic kernel

$$K(t) = \begin{cases} 0.75(1-t^2) & |t| \le 1 \\ 0 & \text{otherwise} \end{cases}$$

due to Epanechnikov (1969), the triangular kernel $K(t) = 1 - |t| \times I(|t| \le 1)$ and the minimum variance kernel

$$K(t) = \begin{cases} \frac{3}{8}(3-5t^2) & |t| \le 1 \\ 0 & \text{otherwise} \end{cases}.$$

For a discussion of these kernels see Hastie and Tibshirani (1990), Härdle (1990), and Eubank (1988). The weight assigned to the point x_i in the estimation of $\widehat{f}(x_0)$ is now

$$W_0(x_i;\lambda) = \frac{c}{\lambda}K\left(\frac{|x_i - x_0|}{\lambda}\right).$$

If the constant c is chosen so that the weights sum to one, i.e.,

$$c^{-1} = \frac{1}{\lambda}\sum_{i=1}^{n} K\left(\frac{|x_i - x_0|}{\lambda}\right)$$

one arrives at the popular Nadaraya-Watson kernel estimator (Nadaraya 1964, Watson 1964)

$$\widehat{f}(x_0) = \frac{\sum_i K(|x_i - x_0|/\lambda) y_i}{\sum_i K(|x_0 - x_i|/\lambda)}. \quad [4.52]$$

Compared to the choice of bandwidth λ, the choice of kernel function is usually of lesser consequence for the resulting estimate $\widehat{f}(x)$.

The Nadaraya-Watson kernel estimator is a weighted average of the observations where the weights depend on the kernel function, bandwidth, and placement of the design points $x_1, \cdots, x_n$. Rather than estimating an average locally, one can also estimate a local mean function that depends on X. This leads to **kernel regression**. A local linear kernel regression estimate models the mean at x_0 as $\mathrm{E}[Y] = \beta_0^{(0)} + \beta_1^{(0)} x_0$ by weighted least squares where the weights for the sum of squares are given by the kernel weights. Once estimates are obtained, the mean at x_0 is estimated as

$$\widehat{\beta}_0^{(0)} + \widehat{\beta}_1^{(0)} x_0.$$

As x_0 is changed to the next location at which a mean prediction is desired, the kernel weights are recomputed and the weighted least squares problems is solved again, yielding new estimates $\widehat{\beta}_0^{(0)}$ and $\widehat{\beta}_1^{(0)}$.

4.7.2 Choosing the Smoothing Parameter

The smoothing parameter λ, the size of the nearest neighborhood in the moving average, running-line, and loess estimators, and the scaling parameter of the kernel in kernel smoothing has considerable influence on the shape of the estimated mean function $f(x)$. Undersmoothed fits, resulting from choosing λ too small are jagged and erratic, and do not allow the modeler to discern the pertinent features of the data. Oversmoothed fits result from choosing λ too large and hide important features of the data. To facilitate the choice of λ a customary approach is to consider the trade-off between the bias of an oversmoothed function and the variability of an undersmoothed function. The bias of a smoother is caused by the fact that the true expectation function $f(x)$ in the neighborhood around x_0 does not follow the trend assumed by the smoother in the local neighborhood. Kernel estimates and moving averages assume that $f(x)$ is flat in the vicinity of x_0 and local polynomials assume a linear or quadratic trend. A second source of bias is incurred at the endpoints of the x range, because the neighborhood is one-sided. Ignoring this latter effect and focusing on a moving average we estimate at point x_0,

$$\widehat{f}(x_0) = \frac{1}{2\lambda + 1} \sum_{i \in N(x_0)} y_i \quad [4.53]$$

with expectation $\mathrm{E}[\widehat{f}(x_0)] = 1/(2\lambda+1)\sum_{i \in N(x_0)} f(x_i)$ and variance $\mathrm{Var}[\widehat{f}(x_0)] = \sigma^2/(2\lambda+1)$. As λ is increased ($\widehat{f}(x)$ is less smooth), the bias $\mathrm{E}[\widehat{f}(x_0) - f(x_0)]$ increases, and the variance decreases. To balance bias and variance, we minimize for a given bandwidth a criterion combining both. Attempting to select the bandwidth that minimizes the residual variance alone is not meaningful since this invariably will lead to a small bandwidth that connects the data point. If the x_i are not replicated in the data the residual sum of squares would be exactly zero in this case. Among the criteria that combine accuracy and precision are the average mean square error $AMSE(\lambda)$ and the average prediction error $APSE(\lambda)$ at the observed values:

$$AMSE(\lambda) = \frac{1}{n}\sum_{i=1}^{n} \mathrm{E}\left[\left\{\widehat{f}(x_i) - f(x_i)\right\}^2\right]$$

$$APSE(\lambda) = \frac{1}{n}\sum_{i=1}^{n} \mathrm{E}\left[\{Y_i - f(x_i)\}^2\right] = AMSE + \sigma^2.$$

The prediction error focuses on the prediction of a new observation and thus has an additional term (σ^2). The bandwidth which minimizes one criterion also minimizes the other. It can be shown that

$$AMSE(\lambda) = \frac{1}{n}\sum_i \mathrm{Var}\left[\widehat{f}(x_i)\right] + \frac{1}{n}\sum_i \left(f(x_i) - \widehat{f}(x_i)\right)^2.$$

The term $f(x_i) - \widehat{f}(x_i)$ is the bias of the smooth at x_i and the average mean square error is the sum of the average variance and the average squared bias. The cross-validation statistic

$$CV = \frac{1}{n}\sum_{i=1}^{n}\left(y_i - \widehat{f}_{-i}(x_i)\right)^2 \qquad [4.54]$$

is an estimate of $APSE(\lambda)$, where $\widehat{f}_{-i}(x_i)$ is the *leave-one-out* prediction of $f()$ at x_i. That is, the nonparametric estimate of $f(x_i)$ is obtained after removing the ith data point from the data set. The cross-validation statistic is obviously related to the PRESS (**pr**ediction **e**rror **s**um of **s**quares) statistic in parametric regression models (Allen, 1974),

$$PRESS = \sum_{i=1}^{n}\left(y_i - \widehat{y}_{i,-i}\right)^2, \qquad [4.55]$$

where $\widehat{y}_{i,-i}$ is the predicted mean of the ith observation if that observation is left out in the estimation of the regression coefficients. Bandwidths that minimize the cross-validation statistic [4.54] are often too small, creating fitted values with too much variability. Various adjustments of the basic CV statistic have been proposed. The generalized cross-validation statistic of Craven and Wahba (1979),

$$GCV = \frac{1}{n}\sum_{i=1}^{n}\left(\frac{y_i - \widehat{f}(x_i)}{n - \nu}\right)^2, \qquad [4.56]$$

simplifies the calculation of CV and penalizes it at the same time. If the vector of fitted values at the observed data points is written as $\widehat{\mathbf{y}} = \mathbf{H}\mathbf{y}$, then the degrees of freedom $n - \nu$ are $n - \mathrm{tr}(\mathbf{H})$. Notice that the difference in the numerator term is no longer the leave-one-out residual, but the residual where $\widehat{f}$ is based on all n data points. If the penalty $n - \nu$ is applied directly to CV, a statistic results that Mays, Birch, and Starnes (2001) term

$$PRESS^* = \frac{PRESS}{n - \nu}.$$

Whereas bandwidths selected on the basis of CV are often too small, those selected based on $PRESS^*$ tend to be large. A penalized $PRESS$ statistic that is a compromise between CV

and GCV and appears to be *just right* (Mays et al. 2001) is

$$PRESS^{**} = \frac{PRESS}{n - \nu + (n-1)\{SSR_{max} - SSR(\lambda)\}/SSR_{max}}.$$

SSR_{max} is the largest residual sum of squares over all possible values of λ and $SSR(\lambda)$ is the residual sum of squares for the value of λ investigated.

CV and related statistics select the bandwidth based on the ability to predict a new observation. One can also concentrate on the ability to estimate the mean $f(x_i)$ which leads to consideration of $Q = \sum_{i=1}^{n}(f(x_i) - \widehat{f}(_i))^2$ as a selection criterion under squared error loss. Mallows' C_p statistic

$$C_p(\lambda) = n^{-1}SSR(\lambda) + 2\widehat{\sigma}^2\text{tr}(\mathbf{H})/n, \qquad [4.57]$$

is an estimate of E$[Q]$. In parametric regression models C_p is used as a model-building tool to develop a model that fits well and balances the variability of the coefficients in an over-fit model with the bias of an under-fit model (Mallows 1973). An estimate of σ^2 is needed to calculate the C_p statistic. Hastie and Tibshirani (1990, p. 48) recommend estimating σ^2 from a nonparametric fit with small smoothing parameter λ^* as

$$\widehat{\sigma}^2 = SSR(\lambda^*)/\{n - \text{tr}(2\mathbf{H}^* - \mathbf{H}^*\mathbf{H}^{*\prime})\}.$$

A different group of bandwidth selection statistics is based on information-theoretical measures which play an important role in likelihood inference. Among them are Akaike's information criterion $AIC(\lambda) = n\log\{AMSE(\lambda)\} + 2$ and variations thereof (see Eubank 1988, pp. 38-41; Härdle 1990, Ch. 5; Hurvich and Simonoff 1998).

Example 4.7 Paclobutrazol Growth Response (continued). A local quadratic polynomial (loess) smooth was obtained for a number of smoothing parameters ranging from 0.2 to 0.8 and $AIC(\lambda)$, $C_p(\lambda)$, and $GCV(\lambda)$ were calculated. Since the measures have different scales, we rescaled them to range between zero and one in Figure 4.23. $AIC(\lambda)$ is minimized for $\lambda = 0.4$, Mallows $C_p(\lambda)$ and the generalized cross-validation statistic are minimized for $\lambda = 0.35$. The smooth with $\lambda = 0.35$ is shown in Figure 4.24.

The estimate $\widehat{f}(x)$ traces the reversal in response trend after treatment application. It appears that approximately two weeks after the third and fourth treatment application the growth regulating effect of Paclobutrazol has disappeared, and there appears to be a growth stimulation relative to untreated plots.

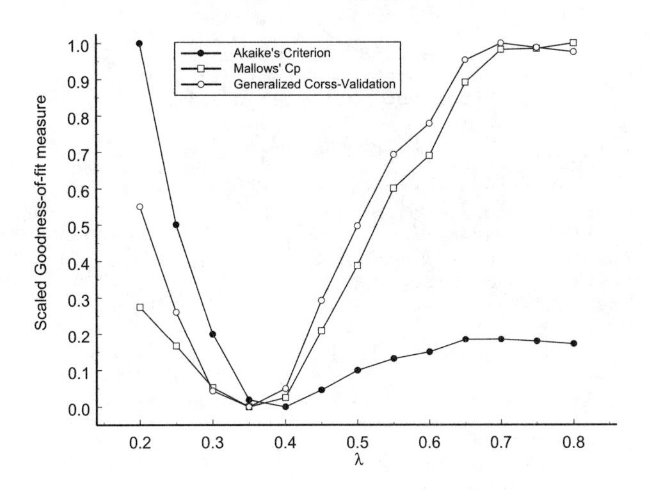

Figure 4.23. $AIC(\lambda)$, $CV(\lambda)$, and $GCV(\lambda)$ for Paclobutrazol response data in Figure 4.20. The goodness of fit measures were rescaled to range from 0 to 1.

The quadratic loess fit was obtained in The SAS® System with `proc loess`. Starting with Release 8.1 the `select=` option of the `model` statement in that procedure enables automatic selection of the smoothing parameter by $AIC(\lambda)$ or $GCV(\lambda)$ criteria. For the Paclobutrazol data, the following statements fit local quadratic polynomials and select the smoothing parameter based on the general cross-validation criteria ($GCV(\lambda)$, Output 4.16).

```
proc loess data=paclobutrazol;
  model clippct = time / degree=2 dfmethod=exact direct select=GCV;
  ods output OutputStatistics=loessFit;
run;
proc print data=loessFit; var Time DepVar Pred Residual LowerCl UpperCl;
run;
```

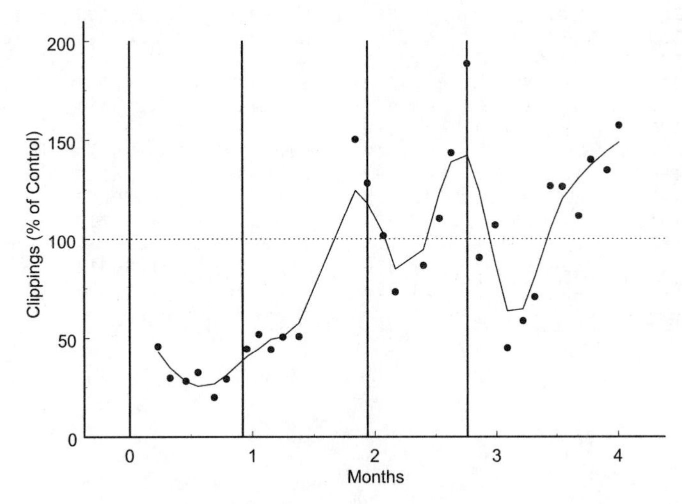

Figure 4.24. Local quadratic polynomial fit (loess) with $\lambda = 0.35$.

Output 4.16.

The LOESS Procedure
Selected Smoothing Parameter: 0.35
Dependent Variable: CLIPPCT

Fit Summary

Fit Method	Direct
Number of Observations	30
Degree of Local Polynomials	2
Smoothing Parameter	0.35000
Points in Local Neighborhood	10
Residual Sum of Squares	6613.29600
Trace[L]	10.44893
GCV	17.30123
AICC	7.70028
AICC1	233.02349
Delta1	18.62863
Delta2	18.28875
Equivalent Number of Parameters	9.52649
Lookup Degrees of Freedom	18.97483
Residual Standard Error	18.84163

Obs	TIME	DepVar	Pred	Residual	LowerCL	UpperCL
1	0.22998	45.6134	43.0768	2.53664	10.472	75.6815
2	0.32854	29.7255	34.9396	-5.2141	13.4187	56.4605
3	0.45996	28.0945	27.8805	0.21407	9.05546	46.7055
4	0.55852	32.5046	25.5702	6.9344	5.21265	45.9277
5	0.68994	19.9031	26.7627	-6.8596	5.06392	48.4615
6	0.7885	29.1638	31.1516	-1.9878	8.98544	53.3177
7	0.95277	44.3525	40.5377	3.81474	19.8085	61.267
8	1.05133	51.7729	44.6647	7.10817	24.9249	64.4045
9	1.1499	44.1395	49.392	-5.2525	29.4788	69.3052
10	1.24846	50.4982	50.6073	-0.1091	29.9399	71.2747
11	1.37988	50.6601	57.6029	-6.9428	31.5453	83.6605
12	1.83984	150.228	124.482	25.7452	101.39	147.575
13	1.9384	128.217	117.851	10.3662	96.7892	138.912
14	2.06982	101.588	102.656	-1.0678	82.8018	122.509
15	2.16838	73.1811	84.7463	-11.565	63.0274	106.465
16	2.39836	86.4965	94.4899	-7.9934	70.7593	118.221
17	2.52977	110.224	123.13	-12.906	101.038	145.222
18	2.62834	143.489	138.821	4.66873	118.077	159.564
19	2.75975	188.535	142.252	46.2833	120.837	163.666
20	2.85832	90.5399	124.276	-33.736	103.48	145.071
21	2.98973	106.795	87.7587	19.0363	66.5058	109.012
22	3.0883	44.9255	63.638	-18.713	42.3851	84.891
23	3.21971	58.5588	64.5384	-5.9796	43.2855	85.7913
24	3.31828	70.728	80.7667	-10.039	59.5138	102.02
25	3.44969	126.722	104.887	21.8345	83.6342	126.14
26	3.54825	126.386	120.239	6.14743	98.9857	141.492
27	3.67967	111.562	130.674	-19.113	110.239	151.11
28	3.77823	140.074	137.368	2.70567	118.34	156.396
29	3.90965	134.654	144.436	-9.7821	122.914	165.958
30	4.00821	157.226	148.854	8.37196	116.131	181.576

Chapter 5

Nonlinear Models

"Given for one instant an intelligence which could comprehend all the forces by which nature is animated and the respective situation of the beings who compose it — an intelligence sufficiently vast to submit these data to analysis — it would embrace in the same formula the movements of the greatest bodies of the universe and those of the lightest atom; for it, nothing would be uncertain and the future, as the past, would be present to its eyes." Pierre de LaPlace, Concerning Probability. In Newman, J.R., The World of Mathematics. *New York: Simon and Schuster, 1965, p. 1325.*

5.1 Introduction

Box 5.1 Nonlinear Models

- **Nonlinear models have advantages over linear models in that**
  - **their origin lies in biological/physical/chemical theory and principles;**
  - **their parameters reflect quantities important to the user;**
  - **they typically require fewer parameters than linear models;**
  - **they require substantial insight into the studied phenomenon.**

- **Nonlinear models have disadvantages over linear models in that**
  - **they require iterative fitting algorithms;**
  - **they require user-supplied starting values (initial estimates) for the parameters;**
  - **they permit only approximate (rather than exact) inference;**
  - **they require substantial insight into the studied phenomenon.**

Recall from §1.7.2 that nonlinear statistical models are defined as models in which the derivatives of the mean function with respect to the parameters depend on one or more of the parameters. A growing number of researchers in the biological sciences share our sentiment that relationships among biological variables are best described by nonlinear functions. Processes such as growth, decay, birth, mortality, abundance, and yield, rarely relate linearly to explanatory variables. Even the most basic relationships between plant yield and nutrient supply, for example, are nonlinear. Liebig's famous *law of the minimum* or *law of constant returns* has been interpreted to imply that for a single deficient nutrient crop yield, Y is proportional to the addition of a fertilizer X until a point is reached where another nutrient is in the minimum and yield is limited. At this point further additions of the fertilizer show no effect and the yield stays constant unless the deficiency of the limiting nutrient is removed. The proportionality between Y and X prior to reaching the yield limit implies a straight-line relationship that can be modeled with a linear model. As soon as the linear increase is combined with a plateau, the corresponding model is nonlinear. Such models are termed linear-plateau models (Anderson and Nelson 1975), linear response-and-plateau models (Waugh et al. 1973, Black 1993), or broken-stick models (Colwell et al. 1988). The data in Figure 5.1 show relative corn (*Zea mays* L.) yield percentages as a function of late-spring test nitrate concentrations in the top 30 cm of the soil. The data are a portion of a larger data set discussed and analyzed by Binford et al. (1992). A linear-plateau model has been fitted to these data and is shown as a solid line. Let Y denote the yield percent and x the soil nitrogen concentration. The linear-plateau model can be written as

$$\mathrm{E}[Y] = \begin{cases} \beta_0 + \beta_1 x & x \leq \alpha \\ \beta_0 + \beta_1 \alpha & x > \alpha, \end{cases} \qquad [5.1]$$

where α is the nitrogen concentration at which the two linear segments join. An alternative expression for model [5.1] is

$$\begin{aligned}\mathrm{E}[Y] &= (\beta_0 + \beta_1 x)I(x \leq \alpha) + (\beta_0 + \beta_1 \alpha)I(x > \alpha) \\ &= \beta_0 + \beta_1(xI(x \leq \alpha) + \alpha I(x > \alpha)).\end{aligned}$$

Here, $I(x \leq \alpha)$ is the indicator function that returns 1 if $x \leq \alpha$ and 0 otherwise. Similarly, $I(x > \alpha)$ returns 1 if $x > \alpha$ and 0 otherwise. If the concentration α at which the lines intersect is known, the term $z = (xI(x \leq \alpha) + \alpha I(x > \alpha))$ is known and one can set up an appropriate regressor variable by replacing the concentrations in excess of α with the value of α. The resulting model is a linear regression model $\mathrm{E}[Y] = \beta_0 + \beta_1 z$ with parameters β_0 and β_1. If α is not known and must be estimated from the data – as will usually be the case – this is a nonlinear model since the derivatives

$$\begin{aligned}\partial\mathrm{E}[Y]/\partial\beta_0 &= 1 \\ \partial\mathrm{E}[Y]/\partial\beta_1 &= xI(x \leq \alpha) + \alpha I(x > \alpha) \\ \partial\mathrm{E}[Y]/\partial\alpha &= \beta_1 I(x > \alpha)\end{aligned}$$

depend on model parameters.

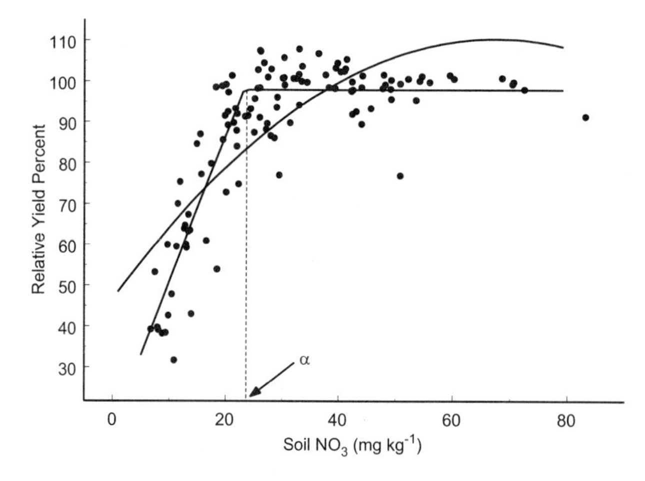

Figure 5.1. Relative corn yield percent as a function of late-spring test soil nitrogen concentration in top 30 cm of soil. Solid line is the fitted linear-plateau model. Dashed line is the fitted quadratic polynomial model. Data kindly made available by Dr. A. Blackmer, Department of Agronomy, Iowa State University. Used with permission. See also Binford, Blackmer, and Cerrato (1992) and the application in §5.8.4.

Should we guess a value for α from a graph of the data, assume it is the true value (without variability) and fit a simple linear regression model or should we let the data guide us to a best possible estimate of α and fit the model as a nonlinear regression model? As an alternative we can abandon the linear-plateau philosophy and fit a quadratic polynomial to the data, since a polynomial $\mathrm{E}[Y] = \beta_0 + \beta_1 x + \beta_2 x^2$ has curvature. That this polynomial fails to fit the data is easily seen from Figure 5.1. It breaks down in numerous places. The initial increase of yield with soil NO_3 is steeper than what a quadratic polynomial can accommodate. The maximum yield for the polynomial model occurs at a nitrate concentration that is upwardly biased. Anderson and Nelson (1975) have noticed that these two model breakdowns are rather typical when polynomials are fit to data for which a linear-plateau model is appropriate. In addition, the quadratic polynomial has a maximum at $x_{\text{max}} = -\beta_1/2\beta_2$ ($x_{\text{max}} = 67.4$ in Figure 5.1). The data certainly do not support the conclusion that the maximum yield is achieved at a nitrate concentration that high, nor do they support the idea of decreasing yields with increasing concentration.

This application serves to show that even in rather simple situations such as a linear-plateau model we are led to nonlinear statistical models for which linear models are a poor substitute. The two workarounds that result in a linear model are not satisfactory. To fix a guessed value for α in the analysis ignores the fact that the "guesstimate" is not free of uncertainty. It depends on the observed data and upon repetition of the experiment we are likely to arrive at a (slightly?) different value for α. This uncertainty must be incorporated into the analysis when determining the precision of the slope and intercept estimators, since the three parameters in the plateau model are not independent. Secondly, a visual guesstimate does not compare in accuracy (or precision) to a statistical estimate. Would you have guessed, based **only** on the data points in Figure 5.1, α to be 23.13? The second *workaround*, that of abandoning the linear-plateau model in favor of a polynomial model is even worse. Not only does the model not fit the data, polynomial models do not incorporate behavior one would expect of the yield percentages. For example, they do not provide a plateau. The linear-plateau model does; it is constructed to exhibit that behavior. If there is theoretical and/or empirical evidence that the response follows a certain characteristic trend, one should resort to statistical models that guarantee that the fitted model shares these characteristics. Cases in point are sigmoidal, convex, hyperbolic, asymptotic, plateau, and other relationships. Almost always, such models will be nonlinear.

In our experience the majority of statistical models fitted by researchers and practitioners to empirical data are nevertheless linear in the parameters. How can this discrepancy be explained? Straightforward inference in linear models contributes to it as does lack of familiarity with nonlinear fitting methods and software as does the exaggeration of the following perceived disadvantages of nonlinear models.

- Linear models are simple to fit and parameter estimation is straightforward. In nonlinear models, parameters are estimated iteratively. Initial estimates are successively improved until some convergence criterion is met. These initial estimates, also termed **starting values**, are supplied by the user. There is no guarantee that the iterative algorithm converges to a unique solution or converges at all. A model may apply to a set of data but because of poorly chosen starting values one may not obtain any parameter estimates at all.

- Curved trends can be modeled by curvilinear models (see §1.7.2), e.g., polynomial models of the form $Y_i = \beta_0 + \beta_1 x_i + \beta_2 x_i^2 + \cdots + \beta_k x_i^k + e_i$. One can motivate a polynomial as an approximation to a nonlinear model (see §5.3), but this approximation may be poor (see Figure 5.1).

- Statistical inference for linear models is well-established and when data are Gaussian-distributed, is exact. Even for Gaussian-distributed data, inference in nonlinear models is only approximate and relies on asymptotic results.

- Some nonlinear models can be transformed into linear models. $\mathrm{E}[Y_i] = \beta_0 \exp\{x\beta_1\}$, for example, can be linearized by taking logarithms: $\ln\{\mathrm{E}[Y_i]\} = \ln\{\beta_0\} + x\beta_1$, which is a linear regression with intercept $\ln\{\beta_0\}$ and slope β_1. In §4.5.2 a nonlinear relationship between group standard deviations and group means was linearized to determine a variance-stabilizing transform. Often, non-negligible transformation bias is incurred in this process, and interpretability of the parameters is sacrificed (§5.6).

- Treatment comparisons in linear models are simple. Set up a linear hypothesis H_0: $\mathbf{A}\boldsymbol{\beta} = \mathbf{0}$ and invoke a sum of squares reduction test. In the worst-case scenario this requires fitting of a full and a reduced model and constructing the sum of squares reduction test statistic by hand (§4.2.3). But usually, the test statistics can be obtained from a fit of the full model alone. Tests of hypotheses in nonlinear models require more often the fitting of a full and a reduced model or ingenious ways of model parameterization (§5.7).

Problems with iterative algorithms usually can be overcome by choosing an optimization method suited to the problem at hand (§5.4) and choosing starting values carefully (§5.4.3). Starting values can often be found by graphical examination of the data, fits of linearized or approximate models, or by simple mathematical techniques (§5.4.3). If the user has limited understanding of the properties of a nonlinear model and the interpretation of the parameters is unclear, choosing starting values can be difficult. Unless sample sizes are very small, nonlinear inference, albeit approximate, is reliable. It is our opinion that approximate inference in a properly specified nonlinear model outgains exact inference in a less applicable linear model. If a response is truly nonlinear, transformations to linearity are not without problems (see §5.6) and treatment comparisons are possible, if the model is parameterized properly.

The advantages of nonlinear modeling outweigh its disadvantages:

- Nonlinear models are more **parsimonious** than linear models. To invoke curvature with inflection with a polynomial model requires at least a cubic term. The linear polynomial $Y_i = \beta_0 + \beta_1 x_i + \beta_2 x_i^2 + \beta_3 x_i^3 + e_i$ has four parameters in the mean function. Nonlinear models can accommodate inflection points with fewer parameters. For example, the model $Y_i = 1 - \exp\{-\beta x^\alpha\} + e_i$ has only two parameters in the mean function.

- Many outcomes do not develop without bounds but reach upper/lower asymptotes and plateaus. It is difficult to incorporate **limiting** behavior into linear models. Many classes of nonlinear models have been studied and constructed to exhibit behavior such as asymptotic limits, inflections, and symmetry. Many models are mentioned throughout the text and §A5.9 lists many more members of the classes of sigmoidal, concave, and convex models.

- Many nonlinear models are derived from **elementary biological**, **physical**, or **chemical principles**. For example, if y is the size of an organism at time t and α is its maximum size, assuming that the rate of growth $\partial y/\partial t$ is proportional to the remaining growth $(a - y)$ leads to the differential equation: $\partial y/\partial t = \beta(\alpha - y)$. Upon integration one obtains a nonlinear three-parameter model for growth: $y(t) = \alpha + (\gamma - \alpha)\exp\{-\beta t\}$. The parameter γ denotes the initial size $y(0)$.

- Parameters of nonlinear models are typically meaningful quantities and have a direct interpretation applicable to the problem being studied. In the growth model $y(t) = \alpha + (\gamma - \alpha)\exp\{-\beta t\}$, α is the final size, γ is the initial size and β governs a rate of change that determines how quickly the organism grows from γ to α. In the soil sciences nitrogen mineralization potential is often modeled as a function of time by an exponential model of form $\mathrm{E}[Y] = \beta_0(1 - \exp\{-\beta_1 t\})$ where β_0 is the maximum amount mineralized and β_1 is the rate at which mineralization occurs. In the yield-density model $\mathrm{E}[Y] = (\alpha + \beta x)^{-1}$ where Y denotes yield per plant and x is the plant density per unit area, $1/\alpha$ measures the genetic potential of the species and $1/\beta$ the environmental potential (see §5.8.7 for an application).

This chapter is concerned with nonlinear statistical models with a single covariate. In §5.2 we investigate **growth models** as a particularly important family of nonlinear models to demonstrate how theoretical considerations give rise to nonlinear models through deterministic generating equations, but also to examine how nonlinear models evolved from mathematical equivalents of *laws of nature* to **empirical tools** for data summary and analysis. In §5.3 a relationship between linear polynomial and nonlinear models is drawn with the help of Taylor series expansions and the basic process of fitting a nonlinear model to data is discussed in §5.4. The test of hypotheses and inference about the parameters is covered in §5.5. Even if models can be transformed to a linear scale, we prefer to fit them in their nonlinear form to retain interpretability of the parameters and to avoid **transformation bias**. Transformations to stabilize the variance have already been discussed in §4.5.2 for linear models. Transformations to linearity (§5.6) are of concern if the modeler does not want to resort to nonlinear fitting methods. **Parameterization**, the process of changing the mathematical form of a nonlinear model by re-expressing the model in terms of different parameters greatly impacts the statistical properties of the parameter estimates and the convergence properties of the fitting algorithms. Problems in fitting a particular model can often be overcome by changing its parameterization (§5.7). Through reparameterization one can also make the model depend on parameters it did not contain originally, thereby facilitating statistical inference about these quantities. In §5.8 we discuss various analyses of nonlinear models from a standard textbook example to complex factorial treatment structures involving a nonlinear response. Since the selection of an appropriate model family is key in successful nonlinear modeling we present numerous concave, convex, and sigmoidal nonlinear models in §A5.9 (on CD-ROM). Additional mathematical details extending the discussion in the text can be found as Appendix A on the CD-ROM (§A5.10).

5.2. Models as Laws or Tools

Like no other class of statistical models, nonlinear models evolved from mathematical formulations of laws of nature to empirical tools describing pattern in data. Within the large family of nonlinear models, growth models are particularly suited to discuss this evolution. This discourse also exposes the reader to the genesis of some popular models and demonstrates how reliance on fundamental biological relationships naturally leads to nonlinearity. This discussion is adapted in part from a wonderful review article on the history of growth models by Zeger and Harlow (1987).

The origin of many nonlinear models in use today can be traced to scholarly efforts to discover laws of nature, to reveal scales of being, and to understand the forces of life. Assumptions were made about how elementary chemical, anatomical, and physical relationships perpetuate to form a living and growing organism and by extension, populations (collections of organisms). Robertson (1923), for example, saw a fundamental law of growth in the chemistry of cells. The equation on which he built describes a chemical reaction in which the product y is also a catalyst (an autocatalytic reaction). If α is the initial rate of growth and β the upper limit of growth, this relationship can be expressed in form of the differential equation

$$\partial\log\{y\}/\partial t = (\partial y/\partial t)/y = \alpha(1 - y/\beta), \qquad [5.2]$$

which is termed the **generating equation** of the process. Robertson viewed this relationship as fundamental to describe the increase in size (y) over time (t) for (all) biological entities. The solution to this differential equation is known as the logistic or autocatalytic model:

$$y(t) = \frac{\beta}{1 + \exp\{-\alpha(x - \gamma)\}}. \qquad [5.3]$$

Pearl and Reed (1924) promoted the autocatalytic concept not only for individual but also for population growth.

The term $\partial\log\{y\}/\partial t = (\partial y/\partial t)/y$ in [5.2] is known as the **specific growth rate**, a measure of the rate of change relative to size. Minot (1908) called it the power of growth and defined senescence as a loss in specific growth rate. He argued that $\partial\log\{y\}/\partial t$ is a concave decreasing function of time, since the rate of senescence decreases from birth. A mathematical example of a relationship satisfying Minot's assumptions about aging and death is the differential equation

$$\partial\log\{y\}/\partial t = \alpha\{\log\{\beta\} - \log\{y\}\}, \qquad [5.4]$$

where α is the intrinsic growth rate and β is a rate of decay. This model is due to Gompertz (1825) who posited it as a law of human mortality. It assumes that specific growth declines linearly with the logarithm of size. Gompertz (1825) reasoned that

> *"the average exhaustions of a man's power to avoid death were such that at the end of equal infinitely small intervals of time, he lost equal portions of his remaining power to oppose destruction."*

The Gompertz model, one of the more common growth models and named after him, is the solution to this differential equation:

$$y(t) = \beta\exp\{-\exp\{-\alpha(t - \gamma)\}\}. \qquad [5.5]$$

Like the logistic model it has upper and lower asymptotes and is sigmoidal in shape. Whereas the logistic model is symmetric about the inflection point $t = \gamma$, the Gompertz model is asymmetric.

Whether growth is autocatalytic or captured by the Gompertz model has been the focus of much debate. The key was whether one believed that specific growth rate is a linear or concave function of size. Courtis (1937) felt so strongly about the adequacy of the Gompertz model that he argued any biological growth, whether of an individual organism, its parts, or populations, can be described by the Gompertz model provided that for the duration of the study conditions (environments) remained constant.

The Gompertz and logistic models were developed for Size-vs.-Time relationships. A second developmental track focused on models where the size of one part (y_1) is related to the size of another (y_2) (so-called size-vs.-size models). Huxley (1932) proposed that specific growth rates of y_1 and y_2 should be proportional:

$$(\partial y_1/\partial t)/y_1 = \beta(\partial y_2/\partial t)/y_2. \qquad [5.6]$$

The parameter β measures the ratio of the specific growth rates of y_1 and y_2. The **isometric** case $\beta = 1$ was of special interest because it implies independence of size and shape. Quiring (1941) felt strongly that **allometry**, the proportionality of sizes, was a fundamental biological law. The study of its regularities, in his words,

"should lead to a knowledge of the fundamental laws of organic growth and explain the scale of being."

Integrating the differential equation, one obtains the basic allometric equation $\log\{y_1\} = \alpha + \beta\log\{y_2\}$ or in exponentiated form,

$$y_1 = \alpha y_2^{\beta}. \qquad [5.7]$$

We notice at this point that the models [5.3] – [5.7] are of course nonlinear. Nonlinearity is a result of integrating the underlying differential equations. The allometric model, however, can be linearized by taking logarithms on both sides of [5.7]. Pázman (1993, p.36) refers to such models as **intrinsically linear**. Models which cannot be transformed to linearity are then **intrinsically nonlinear**.

Allometric relationships can be embedded in more complicated models. Von Bertalanffy (1957) postulated that growth is the sum of positive (anabolic) forces that synthesize material and negative (metabolic) forces that reduce material in an organism. Studying the weight of animals, he found that the power $2/3$ for the metabolic rate describes the anabolic forces well. The model derived from the differential equation,

$$\partial y/\partial t = \alpha y^{2/3} - \beta y \qquad [5.8]$$

is known as the Von Bertalanffy model. Notice that the first term on the right-hand side is of the allometric form [5.7].

The paradigm shift from nonlinear models as mathematical expressions of laws to nonlinear models as empirical tools for data summary had numerous reasons. Cases that did not seem to fit any of the classical models could only be explained as aberrations in measurement protocol or environment or as new processes for which laws needed to be found. At the same time evidence mounted that the various laws could not necessarily coexist. Zeger and Harlow (1987) elaborate how Lumer (1937) showed that sigmoidal growth (Logistic or Gompertz) in different parts of an organism can disable allometry by permitting only certain parameter values in the allometry equation. The laws could not hold simultaneously. Laird (1965) argued that allometric analyses were consistent with sigmoidal growth provided certain conditions about specific growth rates are met. Despite the inconsistencies between sigmoidal and allometric growth, Laird highlighted the utility in both types of models. Finally, advances in computing technology made fitting of nonlinear models less time demanding and allowed examination of competing models for the same data set. Rather than adopting a single model family as the law to which a set of data must comply, the empirical nature of the data could be emphasized and different model families could be tested against a set of data to determine which described the observations best. Whether one adopts an underlying biological or chemical relationship as true, there is much to be learned from a model that fits the data well. Today, we are selecting nonlinear models because they offer certain patterns. If the data suggest a sigmoidal trend with limiting values, we will turn to the Logistic, Gompertz, and other families of models that exhibit the desired behavior. If empirical data suggest monotonic increasing or decreasing relationships, families of concave or convex models are to be considered.

One can argue whether the empiricism in modeling has been carried too far. Study of certain disciplines shows a prevalence of narrow classes of models. In (herbicide) dose-response experiments the logistic model (or log-logistic model if the regressor is log-transformed) is undoubtedly the most frequently used model. This is not the case because the underlying linearity of specific growth (decay) rates is widely adopted as the mechanism of herbicide response, but because in numerous works it was found that logistic functions fit herbicide dose-response data well (e.g., Streibig 1980, Streibig 1981, Lærke and Streibig 1995, Seefeldt et al. 1995, Hsiao et al. 1996, Sandral et al. 1997). As a result, analysts may resist the urge to thoroughly investigate alternative model families. Empirical models are not panaceas and examples where the logistic family does not describe herbicide dose-response behavior well can be found easily (see for example, Brain and Cousens 1989, Schabenberger et al. 1999). Sandland and McGilchrist (1979) and Sandland (1983) criticize the widespread application of the Von Bertalanffy model in the fisheries literature. The model's status according to Sandland (1983), goes "far beyond that accorded to purely empirical models." Cousens (1985) criticizes the categorical assumption of many weed-crop competition studies that crop yield is related to weed density in sigmoidal fashion (Zimdahl 1980, Utomo 1981, Roberts et al. 1982, Radosevich and Holt 1984). Models for yield loss as a function of weed density are more reasonably related to hyperbolic shapes according to Cousens (1985). The appropriateness of sigmoidal vs. hyperbolic models for yield loss depends on biological assumptions. If it is assumed that there is no competition between weeds and crop at low densities, a sigmoidal model suggests itself. On the other hand, if one assumes that at low weed densities weed plants interact with the crop but not each other and that a weed's influence increases with its size, hyperbolic models with a linear increase of yield loss at low weed densities of the type advocated by Cousens (1985) arise rather naturally. Because the biological explanations for the two model types are different, Cousens concludes that one must be rejected. We believe that much is to be gained from using nonlinear models that differ in their physical, biological, and chemical underpinnings. If a sigmoidal model fits a set of yield data better than the hyperbolic contrary to the experimenter's expectation, one is led to rethink the nature of the biological process, a most healthy exercise in any circumstance. If one adopts an attitude that models are selected that describe the data well, not because they comply with a narrow set of biological assumptions, any one of which may be violated in a particular case, the modeler gains considerable freedom. Swinton and Lyford (1996), for example, entertain a reparameterized form of Cousen's rectangular hyperbola to model yield loss as a function of weed density. Their model permits a test whether the yield loss function is indeed hyperbolic or sigmoidal and the question can be resolved via a statistical test if one is not willing to choose between the two model families on biological grounds alone. Cousens (1985) advocates **semi-empirical** model building. A biological process is divided into stages and likely properties of each stage are combined to formulate a resulting model "*based on biologically sound premises.*" His rectangular hyperbola mentioned above is derived on these grounds: (i) yield loss percentage ranges between 0% and 100% as weed density tends towards 0 or infinity, respectively; (ii) effects of individual weed plants on crop at low density are additive; (iii) the rate at which yield loss increases with increasing density is proportional to the squared yield loss per weed plant. Developing mathematical models in this fashion is highly recommended. Regarding the resulting equation as a biological law and to reject other models in its favor equates assumptions with knowledge.

At the other end of the modeling spectrum is analysis without any underlying generating equation or mechanism by fitting linear polynomial functions to the data. An early two-stage method for analyzing growth data from various individuals (clusters) was to fit (orthogonal) polynomials separately to the data from each individual in the first stage and to compare the polynomial coefficients in the second stage with analysis of variance methods (Wishart 1938). This approach is referred to by Sandland and McGilchrist (1979) as "statistical" modeling of growth while relying on nonlinear models derived from deterministic generating equations is termed "biological" modeling. Since all models examined in this text are statistical/stochastic in nature, we do not abide by this distinction. Using polynomials gives the modeler freedom, since it eliminates the need to develop or justify the theory behind a nonlinear relationship. On the other hand, it is well-documented that polynomials are not well-suited to describe growth data. Sandland and McGilchrist (1979) expressed desiderata for growth models, that strike a balance between adhering to deterministic biological laws and empiricism. In our opinion, these are desiderata for all models applied to biological data:

"Growth models should be flexible and able to fit a range of different shapes. They should be based on biological considerations bearing in mind the approximate nature of our knowledge of growth. A biologist should be able to draw some meaningful conclusions from the analysis. The biological considerations should cover not only the intrinsic growth process but also the random environment in which it is embedded."

5.3 Linear Polynomials Approximate Nonlinear Models

The connection between polynomial and nonlinear models is closer than one may think. Polynomial models can be considered approximations to (unknown) nonlinear models. Assume that two variables Y and x are functionally related. For the time being we ignore the possibly stochastic nature of their relationship. If the function $y = f(x)$ is known, y could be predicted for every value of x. Expanding $f(x)$ into a Taylor series around some other value x^*, and assuming that $f(x)$ is continuous, $f(x)$ can be expressed as a sum:

$$f(x) = f(x^*) + zf'(x^*) + z^2\frac{1}{2!}f''(x^*) + ... + z^r\frac{1}{r!}f'^{...'}(x^*) + R. \qquad [5.9]$$

Equation [5.9] is the Taylor series expansion of $f(x)$ around x^*. Here, $f'(x^*)$ denotes the first derivative of $f(x)$ with respect to x, evaluated at the point x^* and $z = (x - x^*)$. R is the remainder term of the expansion and measures the accuracy of the approximation of $f(x)$ by the series of order r. Replace $f(x^*)$ with β_0, $f'(x^*)$ with β_1, $f''(x^*)/2!$ with β_2 and so forth and [5.9] reveals itself as a polynomial in z:

$$y = \beta_0 + \beta_1 z + \beta_2 z^2 + ... + \beta_r z^r + R = \mathbf{z}'\boldsymbol{\beta} + R. \qquad [5.10]$$

The term $\beta_0 + \beta_1 z + \beta_2 z^2 + ... + \beta_r z_r$ is a linear approximation to $f(x)$ and, depending on the number of terms, can be made arbitrarily close to $f(x)$. If there are n distinct data points of x, a polynomial with degree $r = n - 1$ will fit the data perfectly and the remainder term will be exactly zero. If the degree of the polynomial is less than $n - 1$, $R \neq 0$ and R is a measure for the discrepancy between the true function $f(x)$ and its linear approximation $\mathbf{z}'\boldsymbol{\beta}$.

When fitting a linear polynomial, the appropriate degree r is of concern. While the flexibility of the polynomial increases with r, complexity of the model must be traded against quality of fit and poorer statistical properties of estimated coefficients in high-order, overfit polynomials. Single-covariate nonlinear statistical models, the topic of this chapter, target $f(x)$ directly, rather than its linear approximation.

Example 5.1. The data plotted in Figure 5.2 suggest a curved trend between y and x with inflection point. To incorporate the inflection, a model must be found for which the second derivative of the mean function depends on x. A linear polynomial in x must be carried at least to the third order. The four parameter linear model

$$Y_i = \beta_0 + \beta_1 x_i + \beta_2 x_i^2 + \beta_3 x_i^3 + e_i.$$

is a candidate. An alternative one-parameter nonlinear model could be

$$Y_i = 1 - \exp\left\{ - x_i^{\beta} \right\} + e_i.$$

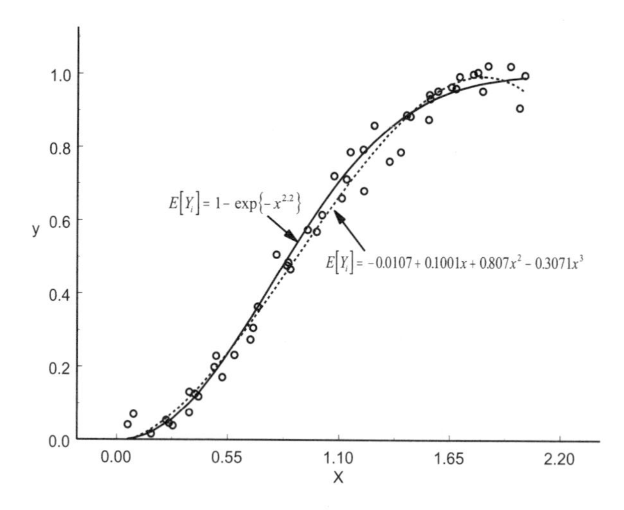

Figure 5.2. Data suggesting mean function with inflection summarized by nonlinear and linear polynomial models.

The nonlinear model is more parsimonious and also restricts $\mathrm{E}[Y_i]$ between zero and one. If the response is a true proportion the linear model does not guarantee predicted values inside the permissible range, whereas the nonlinear model does. The nonlinear function approaches the upper limit of 1.0 asymptotically as x grows. The fitted polynomial, because of its curvature, does not have an asymptote but achieves extrema at

$$x = \frac{-2\widehat{\beta}_2 \pm \sqrt{4\widehat{\beta}_2^2 - 12\widehat{\beta}_3\widehat{\beta}_1}}{6\widehat{\beta}_3} = \{0.0644, 1.687\}.$$

If a decrease in E$[Y]$ is not reasonable on biological grounds, the polynomial is a deficient model.

Linear polynomials are flexible modeling tools that do not appeal to a generating equation and for short data series, may be the only possible modeling choice. They are less parsimonious, poor at fitting asymptotic approaches to limiting values, and do not provide a biologically meaningful parameter interpretation. From the scientists' point of view, nonlinear models are certainly superior to polynomials. As an exploratory tool that points the modeler into the direction of appropriate nonlinear models, polynomials are valuable. For complex processes with changes of phase and temporal fluctuations, they may be the only models offering sufficient flexibility unless one resorts to nonparametric methods (see §4.7).

5.4 Fitting a Nonlinear Model to Data

Box 5.2. Model Fitting

- **An algorithm for fitting a nonlinear model is iterative and comprises three components:**
  - **a numerical rule for successively updating iterates (§5.4.1),**
  - **a method for deciding when to stop the process (§5.4.2),**
  - **starting values to get the iterative process under way (§5.4.3).**

- **Commonly used iterative algorithms are the Gauss-Newton and Newton-Raphson methods. Neither should be used in their original, unmodified form.**

- **Stop criteria for the iterative process should be true convergence, not termination, criteria to distinguish convergence to a global minimum from lack of progress of the iterative algorithm.**

- **Little constructive theory is available to select starting values. Some ad-hoc procedures have proven particularly useful in practice.**

5.4.1 Estimating the Parameters

The least squares principle of parameter estimation has equal importance for nonlinear models as for linear ones. The idea is to minimize the sum of squared deviations between observations and their mean. In the linear model $Y_i = \mathbf{x}_i'\boldsymbol{\beta} + e_i$, where E$[Y_i] = \mathbf{x}_i'\boldsymbol{\beta}$, $e_i \sim iid$

$(0, \sigma^2)$, this requires minimization of the residual sum of squares

$$S(\boldsymbol{\beta}) = \sum_{i=1}^{n} \left(y_i - \mathbf{x}_i'\boldsymbol{\beta}\right)^2 = (\mathbf{y} - \mathbf{X}\boldsymbol{\beta})'(\mathbf{y} - \mathbf{X}\boldsymbol{\beta}).$$

If $\mathbf{X}$ is of full rank, this problem has a closed-form unique solution, the OLS estimator $\widehat{\boldsymbol{\beta}} = (\mathbf{X}'\mathbf{X})^{-1}\mathbf{X}'\mathbf{y}$ (see §4.2.1). If the mean function is nonlinear, the basic model equation is

$$Y_i = f(\mathbf{x}_i, \boldsymbol{\theta}) + e_i, \; e_i \sim iid\left(0, \sigma^2\right), \; i = 1, \cdots, n, \qquad [5.11]$$

where $\boldsymbol{\theta}$ is the $(p \times 1)$ vector of parameters to be estimated and $f(\mathbf{x}_i, \boldsymbol{\theta})$ is the mean of Y_i. The residual sum of squares to be minimized now can be written as

$$\begin{aligned} S(\boldsymbol{\theta}) &= \sum_{i=1}^{n} \left(y_i - f(\mathbf{x}_i, \boldsymbol{\theta})\right)^2 \\ &= (\mathbf{y} - \mathbf{f}(\mathbf{x}, \boldsymbol{\theta}))'(\mathbf{y} - \mathbf{f}(\mathbf{x}, \boldsymbol{\theta})), \end{aligned} \qquad [5.12]$$

with

$$\mathbf{f}(\mathbf{x}, \boldsymbol{\theta}) = \begin{bmatrix} f(\mathbf{x}_1, \boldsymbol{\theta}) \\ f(\mathbf{x}_2, \boldsymbol{\theta}) \\ \vdots \\ f(\mathbf{x}_n, \boldsymbol{\theta}) \end{bmatrix}.$$

This minimization problem is not as straightforward as in the linear case since $\mathbf{f}(\mathbf{x}, \boldsymbol{\theta})$ is a nonlinear function of $\boldsymbol{\theta}$. The derivatives of $S(\boldsymbol{\theta})$ depend on the particular structure of the model, whereas in the linear case with $\mathbf{f}(\mathbf{x}, \boldsymbol{\theta}) = \mathbf{X}\boldsymbol{\beta}$ finding derivatives is easy. One method of minimizing [5.12] is to replace $\mathbf{f}(\mathbf{x}, \boldsymbol{\theta})$ with a linear model that approximates $\mathbf{f}(\mathbf{x}, \boldsymbol{\theta})$. In §5.3, a nonlinear function $f(x)$ was expanded into a Taylor series of order r. Since $\mathbf{f}(\mathbf{x}, \boldsymbol{\theta})$ has p unknowns in the parameter vector we expand it into a Taylor series of first order about each element of $\boldsymbol{\theta}$. Denote by $\boldsymbol{\theta}^0$ a vector of initial guesses of the parameters (a vector of starting values). The first-order Taylor series (see §A5.10.1 for details) of $\mathbf{f}(\mathbf{x}, \boldsymbol{\theta})$ around $\boldsymbol{\theta}^0$ is

$$\mathbf{f}(\mathbf{x}, \boldsymbol{\theta}) \doteq \mathbf{f}\left(\mathbf{x}, \boldsymbol{\theta}^0\right) + \frac{\partial \mathbf{f}(\mathbf{x}, \boldsymbol{\theta})}{\partial \boldsymbol{\theta}'}\bigg|_{\boldsymbol{\theta}^0} \left(\boldsymbol{\theta} - \boldsymbol{\theta}^0\right) = \mathbf{f}\left(\mathbf{x}, \boldsymbol{\theta}^0\right) + \mathbf{F}^0\left(\boldsymbol{\theta} - \boldsymbol{\theta}^0\right), \qquad [5.13]$$

where $\mathbf{F}^0$ is the $(n \times p)$ matrix of first derivatives of $\mathbf{f}(\mathbf{x}, \boldsymbol{\theta})$ with respect to the parameters, evaluated at the initial guess value $\boldsymbol{\theta}^0$. The residual $\mathbf{y} - \mathbf{f}(\mathbf{x}, \boldsymbol{\theta})$ in [5.12] is then approximated by the residual

$$\mathbf{y} - \mathbf{f}\left(\mathbf{x}, \boldsymbol{\theta}^0\right) - \mathbf{F}^0\left(\boldsymbol{\theta} - \boldsymbol{\theta}^0\right) = \mathbf{y} - \mathbf{f}\left(\mathbf{x}, \boldsymbol{\theta}^0\right) + \mathbf{F}^0\boldsymbol{\theta}^0 - \mathbf{F}^0\boldsymbol{\theta},$$

which is linear in $\boldsymbol{\theta}$ and minimizing [5.12] can be accomplished by standard linear least squares where the response $\mathbf{y}$ is replaced by the pseudo-response $\mathbf{y} - \mathbf{f}(\mathbf{x}, \boldsymbol{\theta}^0) + \mathbf{F}^0\boldsymbol{\theta}^0$ and the regressor matrix is given by $\mathbf{F}^0$. Since the estimates we obtain from this approximated linear least squares problem depend on our choice of starting values $\boldsymbol{\theta}^0$, the process cannot stop after just one update of the estimates. Call the estimates of this first fit $\boldsymbol{\theta}^1$. Then we recalculate the new pseudo-response as $\mathbf{y} - \mathbf{f}(\mathbf{x}, \boldsymbol{\theta}^1) + \mathbf{F}^1\boldsymbol{\theta}^1$ and the new regressor matrix is $\mathbf{F}^1$. This process continues until some convergence criterion is met, for example, until the relative change in residual sums of squares between two updates is minor. This approach to least squares fitting of the nonlinear model is termed the **Gauss-Newton** (GN) method of nonlinear least squares (see §A5.10.2 for more details).

A second, popular method of finding the minimum of [5.12], is the **Newton-Raphson** (NR) method. It is a generic method in the sense that it can be used to find the minimum of any function, not just a least squares objective function. Applied to the nonlinear least squares problem, the Newton-Raphson method differs from the Gauss-Newton method in the following way. Rather than approximating the model itself and substituting the approximation into the objective function [5.12], we approximate $S(\boldsymbol{\theta})$ directly by a second-order Taylor series and find the minimum of the resulting approximation (see §A5.10.4 for details). The NR method also requires initial guesses (starting values) of the parameters and is hence also iterative. Successive iterates $\widehat{\boldsymbol{\theta}}^u$ are calculated as

$$\begin{aligned} \text{Gauss-Newton:} \quad & \widehat{\boldsymbol{\theta}}^{u+1} = \widehat{\boldsymbol{\theta}}^u + \boldsymbol{\delta}^u_{GN} = \widehat{\boldsymbol{\theta}}^u + \left(\mathbf{F}^{u\prime}\mathbf{F}^u\right)^{-1}\mathbf{F}^{u\prime}\mathbf{r}\left(\widehat{\boldsymbol{\theta}}^u\right) \\ \text{Newton-Raphson:} \quad & \widehat{\boldsymbol{\theta}}^{u+1} = \widehat{\boldsymbol{\theta}}^u + \boldsymbol{\delta}^u_{NR} = \widehat{\boldsymbol{\theta}}^u + \left(\mathbf{F}^{u\prime}\mathbf{F}^u + \mathbf{A}^u\right)^{-1}\mathbf{F}^{u\prime}\mathbf{r}\left(\widehat{\boldsymbol{\theta}}^u\right). \end{aligned} \qquad [5.14]$$

The matrix $\mathbf{A}$ is defined in §A5.10.4 and $\mathbf{r}(\widehat{\boldsymbol{\theta}}^u) = \mathbf{y} - \mathbf{f}(\mathbf{x}, \widehat{\boldsymbol{\theta}}^u)$ is the vector of fitted residuals after the uth iteration. When the process has successfully converged we call the converged iterate the nonlinear least squares estimates $\widehat{\boldsymbol{\theta}}$ of $\boldsymbol{\theta}$.

The vector of starting values $\boldsymbol{\theta}^0$ is supplied by the user and their determination is important (§5.4.3). The closer the starting values are to the least squares estimate that minimizes [5.12], the faster and more reliable the iterative algorithm will converge. There is no guarantee that the GN and NR algorithms converge to the same estimates. They may, in fact, not converge at all. The GN method in particular is notorious for failing to converge if the starting values are chosen poorly, the residuals are large, and the $\mathbf{F}'\mathbf{F}$ matrix is ill-conditioned (close to singular).

We described the GN and NR method in their most basic form. Usually they are not implemented without some modifications. The GN algorithm, for example, does not guarantee that residual sums of squares between successive iterations decrease. Hartley (1961) proposed a modification of the basic Gauss-Newton step where the next iterate is calculated as

$$\widehat{\boldsymbol{\theta}}^{u+1} = \widehat{\boldsymbol{\theta}}^u + k\boldsymbol{\delta}^u_{GN}, \quad k \in (0,1) \qquad [5.15]$$

and k is chosen to ensure that the residual sum of square decreases between iterations. This is known as step-halving or step-shrinking. The GN method is also not a stable estimation method if the columns of $\mathbf{F}$ are highly collinear for the same reasons that ordinary least squares estimates are unstable if the columns of the regressor matrix $\mathbf{X}$ are collinear (see §4.4.4 on collinearity) and hence $\mathbf{X}'\mathbf{X}$ is ill-conditioned. Nonlinear models are notorious for ill-conditioning of the $\mathbf{F}'\mathbf{F}$ matrix which plays the role of the $\mathbf{X}'\mathbf{X}$ matrix in the approximate linear model of the GN algorithm. In particular when parameters appear in exponents, derivatives with respect to different parameters contain similar functions. Consider the simple two-parameter nonlinear model

$$\mathrm{E}[Y] = 1 - \beta\exp\{-x^{\theta}\} \qquad [5.16]$$

with $\beta = 0.5$, $\theta = 0.9$ (Figure 5.3). The derivatives are given by

$$\partial \mathrm{E}[Y]/\partial\beta = -\exp\{-x^{\theta}\}$$
$$\partial \mathrm{E}[Y]/\partial\theta = \beta\ln\{x\}x^{\theta}\exp\{-x^{\theta}\}.$$

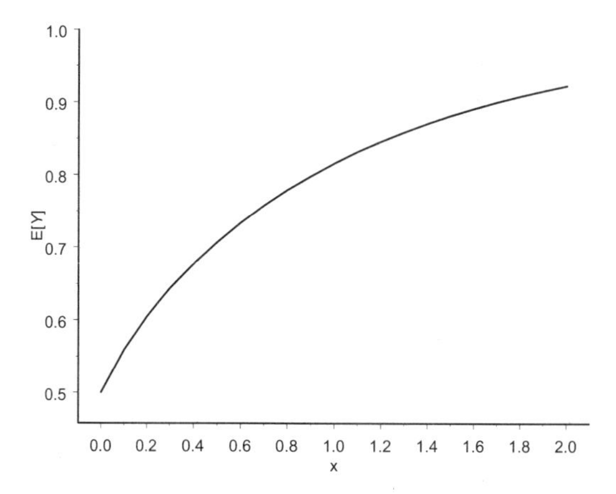

Figure 5.3. Nonlinear response function $\mathrm{E}[Y] = 1 - \beta\exp\{-x^{\theta}\}$ with $\beta = 0.5$, $\theta = 0.9$.

Assume the covariate vector is $\mathbf{x} = [0.2, 0.5, 0.7, 1.8]'$ and the $\mathbf{F}$ matrix becomes

$$\mathbf{F} = \begin{bmatrix} -0.7906 & -0.1495 \\ -0.5815 & -0.1087 \\ -0.4841 & -0.0626 \\ -0.1832 & 0.0914 \end{bmatrix}.$$

The correlation coefficient between the two columns of $\mathbf{F}$ is 0.9785. Ridging (§4.4.5) the $\mathbf{F}'\mathbf{F}$ matrix is one approach to modifying the basic GN method to obtain more stable estimates. This modification is known as the Levenberg-Marquardt method (Levenberg 1944, Marquardt 1963).

Calculating nonlinear parameter estimates by hand is a tedious exercise. Black (1993, p. 65) refers to it as the "drudgery connected with the actual fitting." Fortunately, we can rely on statistical computing packages to perform the necessary calculations and manipulations. We do caution the user, however, that simply because a software package claims to be able to fit nonlinear models does not imply that it can fit the models well. Among the features of a good package we expect suitable modifications of several basic algorithms, grid searches over sets of starting values, efficient step-halving procedures, the ability to apply ridge estimation when the $\mathbf{F}$ matrix is poorly conditioned, explicit control over the type and strictness of the convergence criterion, and automatic differentiation to free the user from having to specify derivatives. These are just some of the features found in the `nlin` procedure of The SAS® System. We now go through the "drudgery" of fitting a very simple, one-parameter nonlinear model by hand, and then show how to apply the `nlin` procedure.

Example 5.2. The nonlinear model we are fitting by the GN method is a special case of model [5.16] with $\beta = 1$:

$$Y_i = 1 - \exp\{-x_i^\theta\} + e_i,\, e_i \sim iid(0, \sigma^2),\, i = 1, \cdots, 4$$

The data set consists of the response vector $\mathbf{y} = [0.1, 0.4, 0.6, 0.9]'$ and the covariate vector $\mathbf{x} = [0.2, 0.5, 0.7, 1.8]'$. The mean vector and the matrix (vector) of derivatives are given by the following:

$$\mathbf{f}(\mathbf{x}, \theta) = \begin{bmatrix} 1 - \exp\{-0.2^\theta\} \\ 1 - \exp\{-0.5^\theta\} \\ 1 - \exp\{-0.7^\theta\} \\ 1 - \exp\{-1.8^\theta\} \end{bmatrix}; \quad \mathbf{F} = \begin{bmatrix} \ln\{0.2\}0.2^\theta \exp\{-0.2^\theta\} \\ \ln\{0.5\}0.5^\theta \exp\{-0.5^\theta\} \\ \ln\{0.7\}0.7^\theta \exp\{-0.7^\theta\} \\ \ln\{1.8\}1.8^\theta \exp\{-1.8^\theta\} \end{bmatrix}.$$

As a starting value we select $\theta^0 = 1.3$. From [5.14] the first evaluation of the derivative matrix and the residual vector gives

$$\mathbf{r}(\theta^0) = \mathbf{y} - \mathbf{f}(\mathbf{x}, \theta^0) = \begin{bmatrix} 0.1 - 0.1161 = & -0.0161 \\ 0.4 - 0.3338 = & 0.0662 \\ 0.6 - 0.4669 = & 0.1331 \\ 0.9 - 0.8832 = & 0.0168 \end{bmatrix}; \quad \mathbf{F}^0 = \begin{bmatrix} -0.1756 \\ -0.1875 \\ -0.1196 \\ 0.1474 \end{bmatrix}.$$

The first correction term is then $(\mathbf{F}^{0\prime}\mathbf{F}^0)^{-1}\mathbf{F}'^0\mathbf{r}(\theta^0) = 9.800 \times -0.023 = -0.2258$ and the next iterate is $\widehat{\theta}^1 = \theta^0 - 0.2258 = 1.0742$. Table 5.1 shows results of successive iterations with the GN method.

Table 5.1. Gauss-Newton iterations, $\theta^0 = 1.3$

Iteration u	θ^u	$\mathbf{F}^u$	$\mathbf{r}^u$	$(\mathbf{F}^{u\prime}\mathbf{F}^u)^{-1}$	$\mathbf{F}^{u\prime}\mathbf{r}^u$	δ^u	$S(\widehat{\theta}^u)$
0	1.3	$\begin{bmatrix} -0.1756 \\ -0.1875 \\ -0.1196 \\ 0.1474 \end{bmatrix}$	$\begin{bmatrix} -0.0161 \\ 0.0662 \\ 0.1331 \\ 0.0168 \end{bmatrix}$	9.8005	−0.0230	−0.2258	0.0226
1	1.0742	$\begin{bmatrix} -0.2392 \\ -0.2047 \\ -0.1230 \\ 0.1686 \end{bmatrix}$	$\begin{bmatrix} -0.0626 \\ 0.0219 \\ 0.1057 \\ 0.0526 \end{bmatrix}$	7.0086	0.0064	0.0445	0.0184
2	1.1187	$\begin{bmatrix} -0.2254 \\ -0.2014 \\ -0.1223 \\ 0.1647 \end{bmatrix}$	$\begin{bmatrix} -0.0523 \\ 0.0309 \\ 0.1112 \\ 0.0451 \end{bmatrix}$	7.4932	−0.0006	−0.0047	0.0181
3	1.1140	$\begin{bmatrix} -0.2268 \\ -0.2018 \\ -0.1224 \\ 0.1651 \end{bmatrix}$	$\begin{bmatrix} -0.0534 \\ 0.0300 \\ 0.1106 \\ 0.0459 \end{bmatrix}$	7.4406	0.0001	0.0006	0.0181
4	1.1146						

After one iteration the derivative matrix **F** and the residual vector **r** have stabilized and exhibit little change in successive iterations. The initial residual sum of squares $S(\theta^0) = 0.0226$ decreases by 18% in the first iteration and does not change after the third iteration. If convergence is measured as the (relative) change in $S(\theta)$, the algorithm is then considered converged.

To fit this model using The SAS® System, we employ `proc nlin`. Prior to Release 6.12 of SAS®, `proc nlin` required the user to supply first derivatives for the Gauss-Newton method and first and second derivatives for the Newton-Raphson method. Since Release 6.12 of The SAS® System, an automatic differentiator is provided by the procedure. The user supplies only the starting values and the model expression. The following statements read the data set and fit the model using the default Gauss-Newton algorithm. More sophisticated applications of the `nlin` procedure can be found in the example applications (§5.8) and a more in-depth discussion of its capabilities and options in §5.8.1. The statements

```
data Ex_51;
  input y x @@;
  datalines;
0.1 0.2 0.4 0.5 0.6 0.7 0.9 1.8
;;
run;
proc nlin data=Ex_51;
  parameters theta=1.3;
  model y = 1 - exp(-x**theta);
run;
```

produce Output 5.1. The Gauss-Newton method converged in six iterations to a residual sum of squares of $S(\widehat{\theta}) = 0.018095$ from the starting value $\theta^0 = 1.3$. The converged iterate is $\widehat{\theta} = 1.1146$ with an estimated asymptotic standard error $\text{ese}(\widehat{\theta}) = 0.2119$.

Output 5.1.

```
                    The NLIN Procedure
                      Iterative Phase
                   Dependent Variable y
                   Method: Gauss-Newton
                                          Sum of
                Iter        theta        Squares
                   0       1.3000         0.0227
                   1       1.0742         0.0183
                   2       1.1187         0.0181
                   3       1.1140         0.0181
                   4       1.1146         0.0181
                   5       1.1146         0.0181
                   6       1.1146         0.0181
         NOTE: Convergence criterion met.

                    Estimation Summary
            Method                        Gauss-Newton
            Iterations                               6
            R                                 2.887E-6
            PPC(theta)                        9.507E-7
            RPC(theta)                        7.761E-6
            Object                            4.87E-10
            Objective                         0.018095
            Observations Read                        4
            Observations Used                        4
            Observations Missing                     0
```

Output 5.1 (continued).

```
             NOTE: An intercept was not specified for this model

                                    Sum of          Mean    Asymptotic      Approx
Source                      DF     Squares        Square       F Value      Pr > F
Regression                   1      1.3219        1.3219        219.16      0.0007
Residual                     3      0.0181       0.00603
Uncorrected Total            4      1.3400
Corrected Total              3      0.3400

                                    Asymptotic
                                      Standard      Asymptotic 95% Confidence
  Parameter          Estimate            Error                 Limits
  theta                1.1146           0.2119         0.4401          1.7890
```

The `parameters` statement defines which quantities are parameters to be estimated and assigns starting values. The `model` statement defines the mean function $f(\mathbf{x}_i, \boldsymbol{\theta})$ to be fitted to the response variable (`y` in this example). All quantities not defined in the `parameters` statement must be either constants defined through SAS® programming statements or variables to be found in the data set. Since `x` is neither defined as a parameter nor a constant, SAS® will look for a variable by that name in the data set. If, for example, one may fit the same model where `x` is square-root transformed, one can simply put

```
proc nlin data=Ex_51;
  parameters theta=1.3;
  z = sqrt(x);
  model y = 1 - exp(-z**theta);
run;
```

As is the case for a linear model, the method of least squares provides estimates for the parameters of the mean function but not for the residual variability. In the model $\mathbf{Y} = \mathbf{f}(\mathbf{x}, \boldsymbol{\theta}) + \mathbf{e}$ with $\mathbf{e} \sim (\mathbf{0}, \sigma^2\mathbf{I})$ an estimate of σ^2 is required for evaluating confidence intervals and test statistics. Appealing to linear model theory it is reasonable to utilize the residual sum of squares obtained at convergence. Specifically,

$$\widehat{\sigma}^2 = \frac{1}{n-p} S(\widehat{\boldsymbol{\theta}}) = \frac{1}{n-p}\left(\mathbf{y} - \mathbf{f}\left(\mathbf{x}, \widehat{\boldsymbol{\theta}}\right)\right)'\left(\mathbf{y} - \mathbf{f}\left(\mathbf{x}, \widehat{\boldsymbol{\theta}}\right)\right) = \frac{1}{n-p}\mathbf{r}(\widehat{\boldsymbol{\theta}})'\mathbf{r}(\widehat{\boldsymbol{\theta}}). \quad [5.17]$$

Here, p is the number of parameters and $\widehat{\boldsymbol{\theta}}$ is the converged iterate of $\boldsymbol{\theta}$. If the model errors are Gaussian, $(n-p)\widehat{\sigma}^2/\sigma^2$ is approximately Chi-square distributed with $n-p$ degrees of freedom. The approximation improves with sample size n and is critical in the formulation of test statistics and confidence intervals. In Output 5.1 this estimate is shown in the analysis of variance table as the `Mean Square` of the `Residual` source, $\widehat{\sigma}^2 = 0.00603$.

5.4.2 Tracking Convergence

Since fitting algorithms for nonlinear models are iterative, some criterion must be employed to determine when iterations can be halted. The objective of nonlinear least squares estimation is to minimize a sum of squares criterion and one can, for example, monitor the residual sum of squares $S(\widehat{\boldsymbol{\theta}}^u)$ between iterations. In the unmodified Gauss-Newton algorithm it is not guaranteed that the residual sum of square in the uth iteration is less than the residual sum of squares in the previous iteration; therefore, this criterion is dangerous. Furthermore, we note that one should not use absolute convergence criteria such as the change in the parameter estimates between iterations, since they depend on the scale of the estimates. When the largest absolute change in a parameter estimates from one iteration to the next is only 0.0001 does not imply that changes are sufficiently small. If the current estimate of that parameter is 0.0002 the parameter estimate has changed by 50% between iterations.

Tracking changes in the residual sum of squares and changes in the parameter estimates monitors different aspects of the algorithm. A small relative change in $S(\widehat{\boldsymbol{\theta}})$ indicates that the sum of squares surface near the current iterate is relatively flat. A small relative change in the parameter estimates implies that a small increment of the estimates can be tolerated (Himmelblau 1972). Bates and Watts (1981) drew attention to the fact that convergence criteria are not just termination criteria. They should indicate that a global minimum of the sum of squares surface has been found and not that the iterative algorithm is lacking progress. Bates and Watts (1981) also point out that computation should not be halted when the relative **accuracy** of the parameter estimates seems adequate. The variability of the estimates should also be considered. A true measure of convergence according to Bates and Watts (1981) is based on the projection properties of the residual vector. Their criterion is

$$\sqrt{\frac{(n-p)}{pS\left(\widehat{\boldsymbol{\theta}}^u\right)}\mathbf{r}\left(\widehat{\boldsymbol{\theta}}\right)'\mathbf{F}(\mathbf{F}'\mathbf{F})^{-1}\mathbf{F}'\mathbf{r}\left(\widehat{\boldsymbol{\theta}}\right)}, \qquad [5.18]$$

and iterations are halted when this measure is less than some number ξ. The `nlin` procedure of The SAS® System implements the Bates and Watts criterion as the default convergence criterion with $\xi = 10^{-5}$.

The sum of squares surface in linear models with a full rank **X** matrix has a unique minimum, the values at the minimum being the least squares estimates (Figure 5.4). In nonlinear models the surface can be considerably more complicated with long, elongated valleys (Figure 5.5) or multiple local minima (Figure 5.6). When the sum of square surface has multiple extrema the iterative algorithm may be trapped in a region from which it cannot escape. If the surface has long, elongated valleys, it may require a large number of iterations to locate the minimum. The sum of squares surface is a function of the model and the data and reparameterization of the model (§5.7) can have tremendous impact on its shape. Well-chosen starting values (§5.4.3) help in the resolution of convergence problems.

To protect against the possibility that a local rather than a global minimum has been found we recommend starting the nonlinear algorithm with sufficiently different sets of starting values. If they converge to the same estimates it is reasonable to assume that the sum of squares surface has a global minimum at these values. Good implementations of modified algorithms, such as Hartley's modified Gauss-Newton method (Hartley 1961) can improve the convergence behavior if starting values are chosen far from the solution but cannot guarantee convergence to a global minimum.

Example 5.3. A simple linear regression model

$$\mathrm{E}[Y_i] = \beta_0 + \beta_1 x_i^3,\ i = 1,..,3$$

is fitted to responses $\mathbf{y} = [-0.1, 5, -0.2]'$ and data matrix

$$\mathbf{X} = \begin{bmatrix} 1 & -1.1447 \\ 1 & 0.0 \\ 1 & 1.13388 \end{bmatrix}.$$

A surface contour of the least squares objective function $S(\boldsymbol{\beta}) = \sum_i^3 (y_i - \beta_0 - \beta_1 x_i^3)^2$ shows elliptical contours with a single minimum $S(\widehat{\boldsymbol{\beta}}) = 17.295$ achieved by $\widehat{\boldsymbol{\beta}} = [1.63408, -0.22478]'$ (Figure 5.4). If a Gauss-Newton or Newton-Raphson algorithm is used to estimate the parameters of this linear model, either method will find the least squares estimates with a single update, regardless of the choice of starting values.

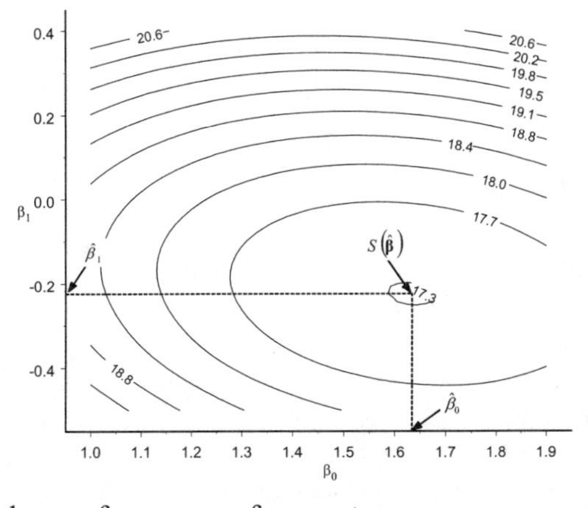

Figure 5.4. Residual sum of squares surface contours.

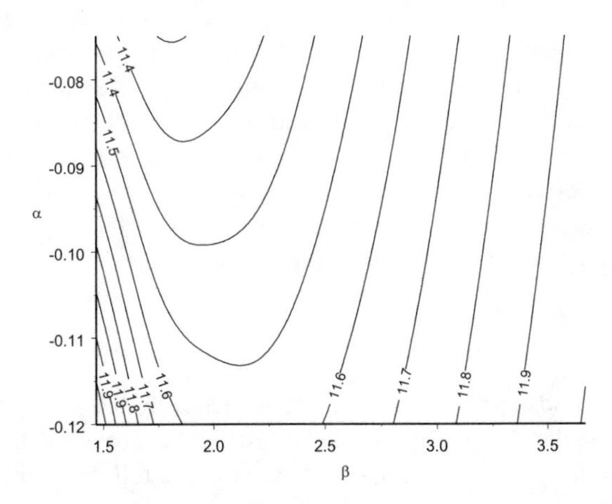

Figure 5.5. Sum of squares contour of model $1/(\alpha + \beta x)$ with elongated valley.

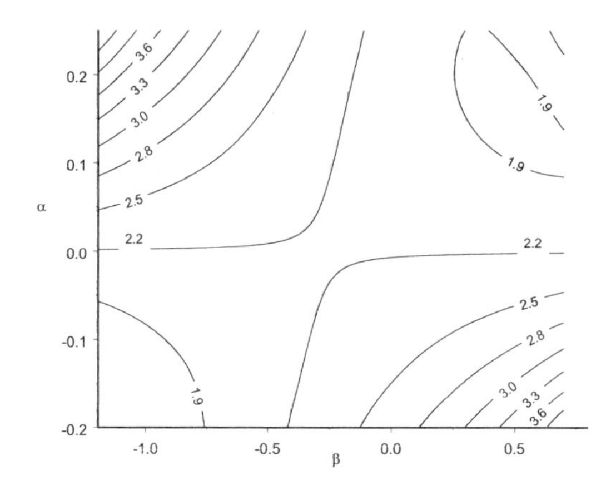

Figure 5.6. Residual sum of square contour for model $\alpha\exp\{-\beta x\}$ with two local minima. Adapted from Figure 3.1 in Seber, G.A.F. and Wild, C.J. (1989) *Nonlinear Regression*. Wiley and Sons, New York. Copyright © 1989 John Wiley and Sons, Inc. Reprinted by permission of John Wiley and Sons, Inc.

5.4.3 Starting Values

A complete algorithm for fitting a nonlinear model to data involves starting values to get the iterative process under way, numerical rules for obtaining a new iterate from previous ones, and a stopping rule indicating convergence of the iterative process. Although starting values initiate the process, we discuss the importance and methods for finding good starting values at this point, after the reader has gained appreciation for the difficulties one may encounter during iterations. These problems are amplified by poorly chosen starting values. When initial values are far from the solution, iterations may diverge, in particular with unmodified Gauss-Newton and Newton-Raphson algorithms. A nonlinear problem may have multiple roots (solutions) and poorly chosen starting values may lead to a local instead of a global minimum. Seber and Wild (1989, p. 665) convey that "the optimization methods themselves tend to be far better and more efficient at seeking a minimum than the various ad hoc procedures that are often suggested for finding starting values." This having been said, well-chosen starting values will improve convergence, reduce the need for numerical manipulations and speed up fitting of nonlinear models. While at times only wild guesses can be mustered, several techniques are available to determine starting values. Surprisingly, there are only few constructive theoretical results about choosing initial values. Most methods have an ad-hoc character; those discussed here have been found to work well in practice.

Graphing Data

One of the simplest methods to determine starting values is to discern reasonable values for $\boldsymbol{\theta}$ from a scatterplot of the data. A popular candidate for fitting growth data, for example, is the four-parameter logistic model. It can be parameterized in the following form,

$$\mathrm{E}[Y_i] = \delta + \frac{\alpha}{1 + \exp\{\beta - \gamma x_i\}}, \qquad [5.19]$$

where δ and $(\alpha + \delta)$ are lower and upper asymptotes, respectively, and the inflection point is located at $x^* = \beta/\gamma$ (Figure 5.7). Furthermore, the slope of the logistic function at the inflection point is a function of α and γ, $\partial f/\partial x_{|x^*} = \alpha\gamma/2$.

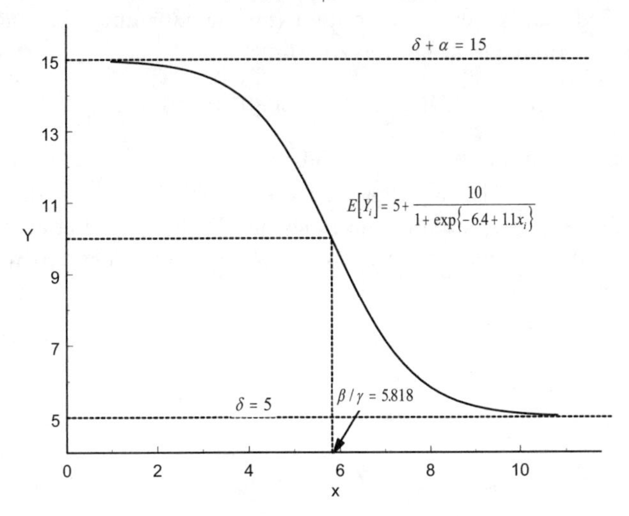

Figure 5.7. Four-parameter logistic model with parameters $\delta = 5$, $\alpha = 10$, $\beta = -6.4$, $\gamma = -1.1$.

Consider having to determine starting values for a logistic model with the data shown in Figure 5.8. The starting values for the lower and upper asymptote could be $\delta^0 = 5$, $\alpha^0 = 9$. The inflection point occurs approximately at $x = 6$, hence $\widehat{\beta}^0/\widehat{\gamma}^0 = 6$ and the slope at the inflection point is about –3. Solving the equations $\alpha^0\gamma^0/2 = -3$ and $\beta^0/\gamma^0 = 6$ for γ^0 and β^0 yields the starting values $\gamma^0 = -0.66$ and $\beta^0 = -4.0$.

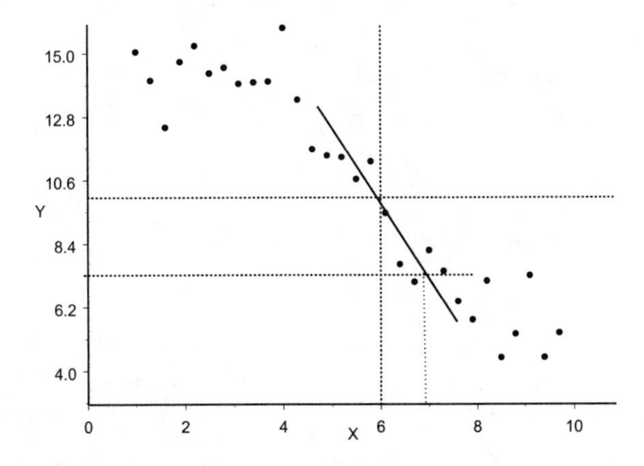

Figure 5.8. Observed data points in growth study.

With `proc nlin` of The SAS® System starting values are assigned in the `parameters` statement. The code

```
proc nlin data=Fig5_8;
  parameters delta=5 alpha=9 beta=-4.0 gamma=-0.66;
  model y = delta + alpha/(1+exp(beta-gamma*x));
run;
```

invokes the modified Gauss-Newton algorithm (the default) and convergence is achieved after seven iterations. Although the converged estimates

$$\widehat{\boldsymbol{\theta}} = [5.3746, 9.1688, -6.8014, -1.1621]'$$

are not too far from the starting values $\boldsymbol{\theta}^0 = [5.0, 9.0, -4.0, -0.66]'$ the initial residual sum of squares $S(\boldsymbol{\theta}^0) = 51.6116$ is more than twice the final sum of squares $S(\widehat{\boldsymbol{\theta}}) = 24.3406$ (Output 5.2). The `nlin` procedure uses the Bates and Watts (1981) criterion [5.18] to track convergence with a default benchmark of 10^{-5}. The criterion achieved when convergence was halted is shown as `R` in the `Estimation Summary`.

Output 5.2.

```
                       The NLIN Procedure
                        Iterative Phase
                     Dependent Variable y
                     Method: Gauss-Newton

                                                                Sum of
 Iter       delta        alpha         beta        gamma       Squares
    0      5.0000       9.0000      -4.0000      -0.6600       51.6116
    1      6.7051       7.0023      -7.3311      -1.2605       43.0488
    2      5.3846       9.1680      -6.5457      -1.1179       24.4282
    3      5.3810       9.1519      -6.8363      -1.1675       24.3412
    4      5.3736       9.1713      -6.7944      -1.1610       24.3403
    5      5.3748       9.1684      -6.8026      -1.1623       24.3403
    6      5.3746       9.1689      -6.8012      -1.1621       24.3403
    7      5.3746       9.1688      -6.8014      -1.1621       24.3403
 NOTE: Convergence criterion met.

                     Estimation Summary
             Method                      Gauss-Newton
             Iterations                             7
             R                               6.353E-6
             PPC(beta)                       6.722E-6
             RPC(beta)                       0.000038
             Object                           1.04E-9
             Objective                       24.34029
             Observations Read                     30
             Observations Used                     30
             Observations Missing                   0

                                 Sum of         Mean                  Approx
Source                  DF      Squares       Square    F Value     Pr > F
Regression               4       3670.4        917.6     136.97     <.0001
Residual                26      24.3403       0.9362
Uncorrected Total       30       3694.7
Corrected Total         29        409.0

                                   Approx     Approximate 95% Confidence
 Parameter      Estimate       Std Error                Limits
 delta            5.3746          0.5304       4.2845          6.4648
 alpha            9.1688          0.7566       7.6137         10.7240
 beta            -6.8014          1.4984      -9.8813         -3.7215
 gamma           -1.1621          0.2592      -1.6950         -0.6293
```

Grid Search

A grid search allows the numerical optimization method to evaluate $S(\boldsymbol{\theta}^0)$ for more than one set of starting values. The actual iterative part of model fitting then commences with the set of starting values that produced the smallest sum of squares among the grid values. Initial grid searches are **no substitute** for restarting the iterations with different sets of starting values. The latter procedure is recommended to ensure that the converged iterate is indeed a global, not a local minimum.

The SAS® statements

```
proc nlin data=Fig5_8;
  parameters delta=3 to 6 by 1             /* 4 grid values for δ */
             alpha=7 to 9 by 1             /* 3 grid values for α */
             beta=-6 to -3.0 by 1          /* 4 grid values for β */
             gamma=-1.5 to -0.5 by 0.5;/* 3 grid values for γ */
  model y = delta + alpha/(1+exp(beta-gamma*x));
run;
```

fit the four-parameter logistic model to the data in Figure 5.8. The residual sum of squares is evaluated at $4 \times 3 \times 4 \times 3 = 144$ parameter combinations. The best initial combination is the set of values that produces the smallest residual sum of squares. This turns out to be $\boldsymbol{\theta}^0 = [5, 9, -6, -1]$ (Output 5.3). The algorithm converges to the same estimates as above but this time requires only five iterations.

Output 5.3.

```
                           The NLIN Procedure
                              Grid Search
                         Dependent Variable y

                                                                 Sum of
          delta        alpha         beta        gamma          Squares
         3.0000       7.0000      -6.0000      -1.5000            867.2
         4.0000       7.0000      -6.0000      -1.5000            596.6
         5.0000       7.0000      -6.0000      -1.5000            385.9
         6.0000       7.0000      -6.0000      -1.5000            235.3
         3.0000       8.0000      -6.0000      -1.5000            760.4
... and so forth ...
         6.0000       8.0000      -6.0000      -1.0000          36.7745
         3.0000       9.0000      -6.0000      -1.0000            189.8
         4.0000       9.0000      -6.0000      -1.0000          79.9745
         5.0000       9.0000      -6.0000      -1.0000          30.1425
         6.0000       9.0000      -6.0000      -1.0000          40.3106
... and so forth ...
         5.0000       9.0000      -3.0000      -0.5000          82.4638
         6.0000       9.0000      -3.0000      -0.5000          86.0169

                           The NLIN Procedure
                            Iterative Phase
                         Dependent Variable y
                         Method: Gauss-Newton
                                                                        Sum of
    Iter       delta        alpha         beta        gamma            Squares
       0      5.0000       9.0000      -6.0000      -1.0000            30.1425
       1      5.4329       9.0811      -6.8267      -1.1651            24.3816
       2      5.3729       9.1719      -6.7932      -1.1607            24.3403
       3      5.3748       9.1684      -6.8026      -1.1623            24.3403
       4      5.3746       9.1689      -6.8011      -1.1621            24.3403
       5      5.3746       9.1688      -6.8014      -1.1621            24.3403
    NOTE: Convergence criterion met.
Remainder of Output as in Output 5.2.
```

Elimination of Linear Parameters

Some statistical models contain both linear and nonlinear parameters, for example

$$\mathrm{E}[Y_i] = \beta_0 + \beta_1 x_i + \beta_2 z_i^{\theta}.$$

Once θ is fixed, the model is linear in $\boldsymbol{\beta} = [\beta_0, \beta_1, \beta_2]'$. A common model to relate yield per plant Y to plant density x is due to Bleasdale and Nelder (1960),

$$\mathrm{E}[Y] = (\alpha + \beta x)^{-1/\theta}.$$

A special case of this model is the Shinozaki-Kira model with $\theta = 1$ (Shinozaki and Kira 1956). Starting values for the Bleasdale-Nelder model can be found by setting $\theta = 1$ and obtaining initial values for α and β from a simple linear regression $1/Y_i = \alpha + \beta x_i$. Once starting values for all parameters have been found the model is fit in nonlinear form.

The Mitscherlich model is popular in agronomy to express crop yield Y as a function of the availability of a nutrient x. One of the many parameterizations of the Mitscherlich model (see §5.7, §A5.9.1 on parameterizations of the Mitscherlich model and §5.8.1 for an application) is

$$\mathrm{E}[Y] = \alpha(1 - \exp\{-\kappa(x - x_0)\})$$

where α is the upper yield asymptote. κ is related to the rate of change and x_0 is the nutrient concentration at which mean yield is 0. A starting value α^0 can be found from a graph of the data as the plateau yield. Then the relationship can be re-expressed by taking logarithms as

$$\ln\{\alpha^0 - Y\} = \ln\{\alpha\} + \kappa x_0 - \kappa x = \beta_0 + \beta_1 x,$$

which is a simple linear regression with response $\ln\{\alpha^0 - Y\}$, intercept $\beta_0 = \ln\{\alpha\} + \kappa x_0$ and slope $\beta_1 = -\kappa$. The ordinary least squares estimate of κ serves as the starting value κ^0. Finally, if the yield without any addition of nutrient (e.g., the zero fertilizer control) is y^*, a starting value for x_0 is $(1/\kappa^0)\ln\{1 - y^*/\alpha^0\}$.

Reparameterization

It can be difficult to find starting values for parameters that have an unrestricted range ($-\infty$, ∞). On occasion, only the sign of the parameter value can be discerned. In these cases it helps to modify the parameterization of the model. Instead of the unrestricted parameter θ one can fit, for example, $\alpha = 1/(1 + \exp\{-\theta\})$ which is constrained to range from zero to one. Specifying a parameter in this range may be simpler. Once a reasonable estimate for α has been obtained one can change the parameterization back to the original state and use $\theta^0 = \ln\{\alpha^0/(1 - \alpha^0\}$ as initial value.

A reparameterization technique advocated by Ratkowsky (1990, Sec. 2.3.1) makes finding starting values particularly simple. The idea is to rewrite a given model in terms of its **expected value parameters** (see §5.7.1). They correspond to predicted values at selected values (x^*) of the regressor. From a scatterplot of the data one can then estimate $\mathrm{E}[Y|x^*]$ by visual inspection. Denote this expected value by μ^*. Set the expectation equal to $f(x^*, \boldsymbol{\theta})$ and replace one of the elements of $\boldsymbol{\theta}$. We illustrate with an example.

In biochemical applications the Michaelis-Menten model is popular to describe chemical reactions in enzyme systems. It can be written in the form

$$\mathrm{E}[Y] = \frac{Vx}{x+K}, \qquad [5.20]$$

where Y is the velocity of the chemical reaction and x is the substrate concentration. V and K are parameters of the model, measuring the theoretical maximum velocity (V) and the substrate concentration at which velocity $V/2$ is attained (K). Assume that no prior knowledge of the potential maximum velocity is available. From a scatterplot of Y vs. X the average velocity is estimated to be μ^* at substrate concentration x^*. Hence, if $X = x^*$, the expected reaction time is

$$\mu^* = \frac{Vx^*}{x^*+K}.$$

Solving this expression for V, the parameter which is difficult to specify initially, leads to

$$V = \mu^* \frac{x^*+K}{x^*}.$$

After substituting this expression back into [5.20] and some algebraic manipulations the reparameterized model becomes

$$\mathrm{E}[Y] = \frac{x\mu^*\left(\frac{x^*+K}{x^*}\right)}{x+K} = \mu^* \frac{x + \frac{x}{x^*}K}{x^* + \frac{x}{x^*}K}. \qquad [5.21]$$

This is a two-parameter model (μ^*, K) as the original model. The process can be repeated by choosing another value x^{**}, its expected value parameter μ^{**}, and replacing the parameter K.

Changing the parameterization of a nonlinear model to expected value parameters not only simplifies finding starting values, but also improves the statistical properties of the estimators (see §5.7.1 and the monographs by Ratkowsky 1983, 1990). A drawback of working with expected value parameters is that not all parameters can be replaced with their expected value equivalents, since the resulting system of equations may not have analytic solutions.

Finding Numeric Solutions

This method is similar in spirit to the method of expected value parameters and was suggested by Gallant (1975). For each of the p parameters of the model choose a point (x_j, y_j). This can be an observed data point or an estimate of the average of Y at x_j. Equate $y_j = f(x_j, \theta); j = 1, \cdots, p$ and solve the system of nonlinear equations. For the Michaelis-Menten model [5.20] the system comprises two equations:

$$y_1 = \frac{Vx_1}{x_1+K} \Leftrightarrow V = y_1(x_1+K)/x_1$$
$$y_2 = \frac{Vx_2}{x_2+K} \Leftrightarrow K = \frac{Vx_2}{y_2} - x_2.$$

Substituting the expression for K into the first equation and simplifying yields

$$V = y_1 \frac{1 - x_2/x_1}{1 - y_1 x_2/y_2}$$

and substituting this expression into $K = Vx_2/y_2 - x_2$ yields

$$K = \frac{y_1}{y_2} \frac{1 - x_2/x_1}{(1/x_2 - y_1/y_2)} - x_2.$$

This technique requires that the system of nonlinear equations can be solved and is a special case of a more general method proposed by Hartley and Booker (1965). They divide the n observations into pm sets, where p is the number of parameters and x_{hk} $(h = 1, \cdots, p; k = 1, \cdots, m)$ are the covariate values. Then the system of nonlinear equations

$$\overline{y}_h = \frac{1}{m} \sum_{k=1}^{m} f(x_{hk}, \theta)$$

is solved where $\overline{y}_h = \frac{1}{m}\sum_{k=1}^{m} y_{hk}$. Gallant's method is a special case where $m = 1$ and one selects *p representative* points from the data.

In practical applications, the various techniques for finding starting values are often combined. Initial values for some parameters are determined graphically, others are derived from expected value parameterization or subject-matter considerations, yet others are entirely guessed.

Example 5.4. Gregoire and Schabenberger (1996b) model stem volume in 336 yellow poplar (*Liriodendron tulipifera* L.) trees as a function of the relative diameter

$$t_{ij} = \frac{d_{ij}}{D_j},$$

where D_j is the stump diameter of the jth tree and d_{ij} is the diameter of tree j measured at the ith location along the bole (these data are visited in §8.4). At the tip of the tree $t_{ij} = 0$ and directly above ground $t_{ij} = 1$. The measurements along the bole were spaced 1.2 meters apart. The authors selected a volume-ratio model to describe the accumulation of volume with decreasing diameter (increasing height above ground):

$$V_{ij} = (\beta_0 + \beta_1 x_j)\exp\left\{ -\beta_2 \frac{t_{ij}}{1000} e^{\beta_3 t_{ij}} \right\} + e_{ij}. \qquad [5.22]$$

Here x_j = diameter at breast height squared times total tree height for the jth tree. This model consists of a linear part $(\beta_0 + \beta_1 x_j)$ representing the total volume of a tree and a multiplicative reduction term

$$R(\beta_2, \beta_3, t_{ij}) = \exp\left\{ -\beta_2 \frac{t_{ij}}{1000} e^{\beta_3 t_{ij}} \right\},$$

which by virtue of the parameterization and $0 \le t_{ij} \le 1$ is constrained between

$$\exp\left\{-\beta_2 \frac{1}{1000} e^{\beta_3}\right\} \leq R(\beta_2, \beta_3, t) \leq 1.$$

To find starting values for $\boldsymbol{\beta} = [\beta_0, \beta_1, \beta_2, \beta_3]'$, the linear component was fit to the total tree volumes of the 336 trees,

$$V_j = \beta_0 + \beta_1 x_j + e_j \ (j = 1, ..., 336)$$

and the linear least squares estimates $\widehat{\beta}_0$ and $\widehat{\beta}_1$ were used as starting values for β_0 and β_1. For the reduction term $R(\beta_2, \beta_3, t)$ one must have $\beta_2 > 0, \beta_3 > 0$; otherwise the term does not shrink toward 0 with increasing relative diameter t. Plotting $R(\beta_2, \beta_3, t)$ for a variety of parameter values indicated that especially β_2 is driving the shape of the reduction term and that values of β_3 between 5 and 8 had little effect on the reduction term. To derive a starting value for β_2, its expected value parameter can be computed. Solving

$$\mu^* = \exp\left\{-\beta_2 \frac{t^*}{1000} e^{\beta_3 t^*}\right\},$$

for β_2 yields

$$\beta_2 = -\ln\{\mu^*\} e^{-\beta_3 t^*} \frac{1000}{t^*}$$

Examining graphs of the tree profiles, it appeared that about 90% of the total tree volume were accrued on average at a height where the trunk diameter had decreased by 50%, i.e., at $t = 0.5$. With a guesstimate of $\beta_3 = 7.5$, the starting value for β_2 with $t^* = 0.5$ and $\mu^* = 0.9$ is $\beta_2^0 = 4.96$.

5.4.4 Goodness-of-Fit

The most frequently used goodness-of-fit (g-o-f) measure in the classical linear model

$$Y_i = \beta_0 + \sum_{j=1}^{k-1} \beta_j x_{ji} + e_i$$

is the simple (or multiple) coefficient of determination,

$$R^2 = \frac{SSM_m}{SST_m} = \frac{\sum_{i=1}^{n} (\widehat{y}_i - \overline{y})^2}{\sum_{i=1}^{n} (y_i - \overline{y})^2}. \qquad [5.23]$$

The appeal of the R^2 statistic is that it ranges between 0 and 1 and has an immediate interpretation as the proportion of variability in Y jointly explained by the regressor variables. Notice that this is a proportion of variability in Y about its mean $\overline{Y}$ since in the absence of any regressor information one would naturally predict $\mathrm{E}[Y_i]$ with the sample mean. Kvålseth

(1985) reviews alternative definitions of R^2. For example,

$$R^2 = 1 - \frac{\sum_{i=1}^{n}(y_i - \widehat{y}_i)^2}{\sum_{i=1}^{n}(y_i - \overline{y})^2} = 1 - \frac{SSR}{SST_m} = \frac{SST_m - SSR}{SST_m}. \qquad [5.24]$$

In the linear model with intercept term, the two R^2 statistics [5.23] and [5.24] are identical. [5.23] contains mean adjusted quantities and for models not containing an intercept other R^2-type measures have been proposed. Kvålseth (1985) mentions

$$R^2_{noint} = 1 - \frac{\sum_{i=1}^{n}(y_i - \widehat{y}_i)^2}{\sum_{i=1}^{n} y_i^2} \text{ and } R^{2*}_{noint} = \frac{\sum_{i=1}^{n}\widehat{y}_i^2}{\sum_{i=1}^{n} y_i^2}.$$

The R^2_{noint} statistics are appropriate only if the model does not contain an intercept **and** $\overline{y}$ is zero. Otherwise one may obtain misleading results. Nonlinear models do not contain an intercept in the typical sense and care must be exercised to select an appropriate goodness-of-fit measure. Also, [5.23] and [5.24] do not give identical results in the nonlinear case. [5.23] can easily exceed 1 and [5.24] can conceivably be negative. The key difficulty is that the decomposition

$$\sum_{i=1}^{n}(y_i - \overline{y})^2 = \sum_{i=1}^{n}(\widehat{y}_i - \overline{y})^2 + \sum_{i=1}^{n}(y_i - \widehat{y}_i)^2$$

no longer holds in nonlinear models. Ratkowsky (1990, p. 44) feels strongly that the danger of misinterpreting R^2 in nonlinear models is too great to rely on such measures and recommends basing goodness-of-fit decisions in a nonlinear model with p parameters on the mean square error

$$s^2 = \frac{1}{n-p}\sum_{i=1}^{n}\left(y_i - \widehat{y}_i^2\right) = \frac{1}{n-p}SSR. \qquad [5.25]$$

The mean square error is a useful goodness-of-fit statistic because it combines a measure of closeness between data and fit (SSR) with a penalty term $(n-p)$ to prevent overfitting. Including additional parameters will decrease SSR and the denominator. s^2 may increase if the added parameter does not improve the model fit. But a decrease of s^2 is not necessarily indicative of a statistically significant improvement of the model. If the models with and without the additional parameter(s) are nested, the sum of squares reduction test addresses the level of significance of the improvement. The usefulness of s^2 notwithstanding, we feel that a reasonable R^2-type statistic can be applied and recommend the statistic [5.24] as a goodness-of-fit measure in linear and nonlinear models. In the former, it yields the coefficient of determination provided the model contains an intercept term, and is also meaningful in the no-intercept model. In nonlinear models it cannot exceed 1 and avoids a serious pitfall of [5.23]. Although this statistic can take on negative values in nonlinear models, this has never happened in our experience and usually the statistic is bounded between 0 and 1. A negative value would indicate a serious problem with the considered model. Since the possibility of negative values exists theoretically, we do not term it a R^2 statistic in nonlinear models to avoid interpretation in terms of the proportion of total variation of Y about its mean accounted for by the model. We term it instead

$$\text{Pseudo-}R^2 = 1 - \frac{\sum_{i=1}^{n}(y_i - \widehat{y}_i)^2}{\sum_{i=1}^{n}(y_i - \overline{y})^2} = 1 - \frac{SSR}{SST_m}. \qquad [5.26]$$

Because the additive sum of squares decomposition does not hold in nonlinear models, Pseudo-R^2 should not be interpreted as the proportion of variability explained by the model.

5.5 Hypothesis Tests and Confidence Intervals

5.5.1 Testing the Linear Hypothesis

Statistical inference in nonlinear models is based on the asymptotic distribution of the parameter estimators $\widehat{\boldsymbol{\theta}}$ and $\widehat{\sigma}^2$. For any fixed sample size inferences are thus only approximate, even if the model errors are perfectly Gaussian-distributed. Exact nonlinear inferences (e.g., Hartley 1964) are typically not implemented in statistical software packages. In this section we discuss various approaches to the testing of hypotheses about $\boldsymbol{\theta}$ and the construction of confidence intervals for parameters, as well as confidence and prediction intervals for the mean response and new data points. The tests are straightforward extensions of similar tests and procedures for linear models (§A4.8.2) but in contrast to the linear case where the same test can be formulated in different ways, in nonlinear models the equivalent expressions will lead to tests with different properties. For example a Wald test of $H_0 : \mathbf{A}\boldsymbol{\beta} = \mathbf{d}$ in the linear model and the sum of squares reduction test of this hypothesis are equivalent. The sum of squares reduction test equivalent in the nonlinear model does not produce the same test statistic as the Wald test.

Consider the general nonlinear model $\mathbf{Y} = \mathbf{f}(\mathbf{x}, \boldsymbol{\theta}) + \mathbf{e}$, where errors are uncorrelated with constant variance, $\mathbf{e} \sim (\mathbf{0}, \sigma^2\mathbf{I})$. The nonlinear least squares estimator $\widehat{\boldsymbol{\theta}}$ has an asymptotic Gaussian distribution with mean $\boldsymbol{\theta}$ and variance-covariance matrix $\sigma^2(\mathbf{F}'\mathbf{F})^{-1}$. Notice, that even if the errors $\mathbf{e}$ are Gaussian, the least squares estimator is not. Contrast this with the linear model $\mathbf{Y} = \mathbf{X}\boldsymbol{\beta} + \mathbf{e}$, $\mathbf{e} \sim G(\mathbf{0}, \sigma^2\mathbf{I})$, where the ordinary least squares estimator

$$\widehat{\boldsymbol{\beta}} = (\mathbf{X}'\mathbf{X})^{-1}\mathbf{X}'\mathbf{Y}$$

is Gaussian with mean $\boldsymbol{\beta}$ and variance-covariance matrix $\sigma^2(\mathbf{X}'\mathbf{X})^{-1}$. The derivative matrix $\mathbf{F}$ in the nonlinear model plays a role akin to the $\mathbf{X}$ matrix in the linear model, but in contrast to the linear model, the $\mathbf{F}$ matrix is not known, unless $\boldsymbol{\theta}$ is known. The derivatives of a nonlinear model depend, by definition, on the unknown parameters. To estimate the standard error of the nonlinear parameter θ_j, we extract c^j, the jth diagonal element of the $(\mathbf{F}'\mathbf{F})^{-1}$ matrix. Similarly, in the linear model we extract d^j, the jth diagonal element of $(\mathbf{X}'\mathbf{X})^{-1}$. The estimated standard errors for θ_j and β_j, respectively, are

$$\textbf{Linear model:} \quad \text{se}\left(\widehat{\beta}_j\right) = \sqrt{\sigma^2 d^j} \quad \text{ese}\left(\widehat{\beta}_j\right) = \sqrt{\widehat{\sigma}^2 d^j} \quad d^j \text{ known}$$

$$\textbf{Nonlinear model:} \quad \text{ase}\left(\widehat{\theta}_j\right) = \sqrt{\sigma^2 c^j} \quad \text{ease}\left(\widehat{\theta}_j\right) = \sqrt{\widehat{\sigma}^2 \widehat{c}^j} \quad c^j \text{ unknown.}$$

The differences between the two cases are subtle. $\sqrt{\sigma^2 d^j}$ is the standard error of β_j in the linear model, but $\sqrt{\sigma^2 c^j}$ is only the **asymptotic** standard error (ase) of θ_j in the nonlinear case. Calculating estimates of these quantities requires substituting an estimate of σ^2 in the linear model whereas in the nonlinear case we also need to estimate the unknown c^j. This estimate is found by evaluating $\mathbf{F}$ at the converged iterate and extracting $\widehat{c}^j$ as the jth diagonal element of the $(\widehat{\mathbf{F}}'\widehat{\mathbf{F}})^{-1}$ matrix. We use ease$(\widehat{\theta}_j)$ to denote the **e**stimated **a**symptotic **s**tandard **e**rror of the parameter estimator.

Nonlinearity also affects the distributional properties of $\widehat{\sigma}^2$, not only those of the θ_js. In the linear model with Gaussian errors where $\mathbf{X}$ has rank p, $(n-p)\widehat{\sigma}^2/\sigma^2$ is a Chi-squared random variable with $n-p$ degrees of freedom. In a nonlinear model with p parameters, $(n-p)\widehat{\sigma}^2/\sigma^2$ is only approximately a Chi-squared random variable.

If sample size is sufficiently large we can rely on the asymptotic results and an approximate $(1-\alpha)100\%$ confidence interval for θ_j can be calculated as

$$\widehat{\theta}_j \pm z_{\alpha/2}\text{ease}(\widehat{\theta}_j) = \widehat{\theta}_j \pm z_{\alpha/2}\widehat{\sigma}\sqrt{\widehat{c}^j}.$$

Because we do not use ase$(\widehat{\theta}_j)$, but the estimated ase$(\widehat{\theta}_j)$, it is reasonable to use instead confidence intervals based on a t-distribution with $n-p$ degrees of freedom, rather than intervals based on the Gaussian distribution. If

$$\left(\widehat{\theta}_j - \theta_j\right)/\text{ese}\left(\widehat{\theta}_j\right)$$

is a standard Gaussian variable, then

$$\left(\widehat{\theta}_j - \theta_j\right)/\text{ease}\left(\widehat{\theta}_j\right)$$

can be treated as a t_{n-p} variable. As a consequence, an α-level test for $H_0{:}\theta_j = d$ vs. $H_1{:}\theta_j \neq d$ compares

$$t_{obs} = \frac{\widehat{\theta}_j - d}{\text{ease}(\widehat{\theta}_j)} \qquad [5.27]$$

against the $\alpha/2$ cutoff of a t_{n-p} distribution. H_0 is rejected if $|t_{obs}| > t_{\alpha/2,n-p}$ or, equivalently, if the $(1-\alpha)100\%$ confidence interval

$$\widehat{\theta}_j \pm t_{\alpha/2,n-p}\text{ease}(\widehat{\theta}_j) = \widehat{\theta}_j \pm t_{\alpha/2,n-p}\widehat{\sigma}\sqrt{\widehat{c}^j}. \qquad [5.28]$$

does not contain d. These intervals are calculated by the `nlin` procedure with $\alpha = 0.05$ for each parameter of the model by default.

The simple hypothesis $H_0{:}\theta_j = d$ is a special case of a **linear hypothesis** (see also §A4.8.2). Consider we wish to test whether

$$\mathbf{A}\boldsymbol{\theta} = \mathbf{d}.$$

In this constraint on the $(p \times 1)$ parameter vector $\boldsymbol{\theta}$, $\mathbf{A}$ is a $(q \times p)$ matrix of rank q. The nonlinear equivalent of the sum of squares reduction test consists of fitting the full model and to obtain its residual sum of squares $S(\widehat{\boldsymbol{\theta}})_f$. Then the constraint $\mathbf{A}\boldsymbol{\theta} = \mathbf{d}$ is imposed on the model and the sum of squares of this reduced model, $S(\widehat{\boldsymbol{\theta}})_r$, is obtained. The test statistic

$$F_{obs}^{(2)} = \frac{\left\{S(\widehat{\boldsymbol{\theta}})_r - S(\widehat{\boldsymbol{\theta}})_f\right\}/q}{S(\widehat{\boldsymbol{\theta}})_f/(n-p)} \qquad [5.29]$$

has an approximate F distribution with q numerator and $n - p$ denominator degrees of freedom. If the model errors are Gaussian-distributed, $F^{(2)}$ is also a likelihood ratio test statistic. We designate the test statistic as $F^{(2)}$ since there is another statistic that could be used to test the same hypothesis. It is based on the Wald statistic

$$F_{obs}^{(1)} = \left(\mathbf{A}\widehat{\boldsymbol{\theta}} - \mathbf{d}\right)'\left[\mathbf{A}(\mathbf{F}'\mathbf{F})^{-1}\mathbf{A}'\right]^{-1}\left(\mathbf{A}\widehat{\boldsymbol{\theta}} - \mathbf{d}\right)/\left(q\widehat{\sigma}^2\right). \qquad [5.30]$$

The asymptotic distribution of [5.30] is that of a Chi-squared random variable since asymptotically $\widehat{\sigma}^2$ converges to a constant, not a random variable. For fixed sample size it has been found that the distribution of [5.30] is better approximated by an F distribution with q numerator and $n - p$ denominator degrees of freedom, however. We thus use the same approximate distribution for [5.29] and [5.30]. The fact that the F distribution is a better approximate distribution than the asymptotic Chi-squared is also justification for using t-based confidence intervals rather than Gaussian confidence intervals. In contrast to the linear model where $F_{obs}^{(1)}$ and $F_{obs}^{(2)}$ are identical, the statistics differ in a nonlinear model. Their relative merits are discussed in §A5.10.6. In the special case $H_0\colon \theta_j = 0$, where $\mathbf{A}$ is a $(1 \times p)$ vector of zeros with a 1 in the jth position and $\mathbf{d} = 0$, $F_{obs}^{(1)}$ reduces to

$$F_{obs}^{(1)} = \frac{\widehat{\theta}_j^2}{\widehat{\sigma}^2 \widehat{c}^j} = t_{obs}^2.$$

Any demerits of the Wald statistic are also demerits of the t-test and t-based confidence intervals shown earlier.

Example 5.5. Velvetleaf Growth Response. Two herbicides (H_1, H_2) are applied at seven different rates and the dry weight percentages (relative to a no-herbicide control treatment) of velvetleaf (*Abutilon theophrasti* Medikus) plants are recorded (Table 5.2).

A graph of the data shows no clear inflection point in the dose response (Figure 5.9). The logistic model, although popular for modeling dose-response data, is not appropriate in this instance. Instead, we select a hyperbolic function, the three-parameter extended Langmuir model (Ratkowsky 1990).

Table 5.2. Velvetleaf herbicide dose response data
(Y_{ij} is the average across the eight replications for rate x_{ij})

	Herbicide 1		Herbicide 2	
i	**Rate** x_{i1} (lbs ae/acre)	Y_{i1}	**Rate** x_{i2} (lbs ae/acre)	Y_{i2}
1	$1E-8$	100.00	$1E-8$	100.000
2	0.0180	95.988	0.0468	75.800
3	0.0360	91.750	0.0938	54.925
4	0.0710	82.288	0.1875	33.725
5	0.1430	55.688	0.3750	16.250
6	0.2860	18.163	0.7500	10.738
7	0.5720	7.350	1.5000	8.713

Data kindly provided by Dr. James J. Kells, Department of Crop and Soil Sciences, Michigan State University. Used with permission.

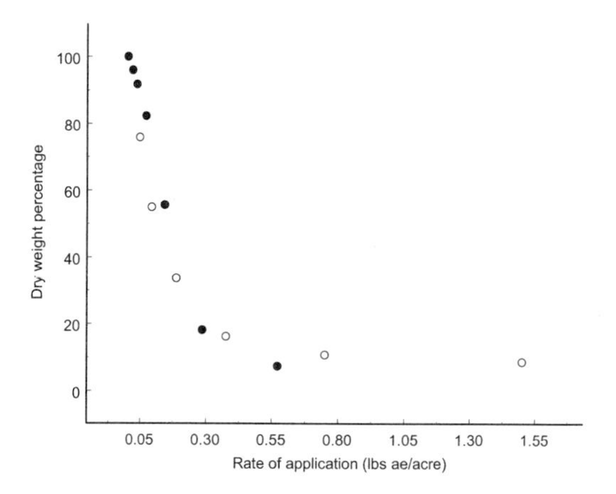

Figure 5.9. Velvetleaf dry weight percentages as a function of the amount of active ingredient applied. Closed circles correspond to herbicide 1, open circles to herbicide 2.

The fullest version of this model for the two-herbicide comparison is

$$\mathrm{E}[Y_{ij}] = \alpha_j \frac{\beta_j x_{ij}^{\gamma_j}}{1 + \beta_j x_{ij}^{\gamma_j}}. \qquad [5.31]$$

Y_{ij} denotes the observation made at rate x_{ij} for herbicide $j = 1, 2$. α_1 is the asymptote for herbicide 1 and α_2 the asymptote for herbicide 2. The model states that the two herbicides differ in the parameters; hence $\boldsymbol{\theta}$ is a (6×1) vector, $\boldsymbol{\theta} = [\alpha_1, \alpha_2, \beta_1, \beta_2, \gamma_1, \gamma_2]'$. To test the hypothesis that the herbicides have the same mean response function, we let

$$\mathbf{A} = \begin{bmatrix} 1 & -1 & 0 & 0 & 0 & 0 \\ 0 & 0 & 1 & -1 & 0 & 0 \\ 0 & 0 & 0 & 0 & 1 & -1 \end{bmatrix}.$$

The constraint $\mathbf{A}\boldsymbol{\theta} = \mathbf{0}$ implies that $[\alpha_1 - \alpha_2, \beta_1 - \beta_2, \gamma_1 - \gamma_2]' = [0,0,0]'$ and the reduced model becomes

$$\mathrm{E}[Y_{ij}] = \alpha\beta x_{ij}^{\gamma} \Big/ \left(1 + \beta x_{ij}^{\gamma}\right).$$

To test whether the herbicides share the same β parameter, $H_0\colon \beta_1 = \beta_2$, we let $\mathbf{A} = [0,0,1,-1,0,0]'$. The reduced model corresponding to this hypothesis is

$$\mathrm{E}[Y_{ij}] = \alpha_j\beta x_{ij}^{\gamma_j} \Big/ \left(1 + \beta x_{ij}^{\gamma_j}\right).$$

Since the dry weights are expressed relative to a no-treatment control we consider as the full model for analysis

$$Y_{ij} = 100\frac{\beta_j x_{ij}^{\gamma_j}}{1 + \beta_j x_{ij}^{\gamma_j}} + e_{ij} \qquad [5.32]$$

instead of [5.31]. To fit this model with the `nlin` procedure, it is helpful to rewrite it as

$$Y_{ij} = 100\left\{\frac{\beta_1 x_{i1}^{\gamma_1}}{1 + \beta_1 x_{i1}^{\gamma_1}}\right\} I\{j=1\} + 100\left\{\frac{\beta_2 x_{i2}^{\gamma_2}}{1 + \beta_2 x_{i2}^{\gamma_2}}\right\} I\{j=2\} + e_{ij}. \qquad [5.33]$$

$I\{\}$ is the indicator function that returns the value 1 if the condition inside the curly braces is true, and 0 otherwise. $I\{j=1\}$ for example, takes on value 1 for observations receiving herbicide 1 and 0 for observations receiving herbicide 2. The SAS® statements to accomplish the fit (Output 5.4) are

```
proc nlin data=herbicide noitprint;
  parameters beta1=0.049 gamma1=-1.570 /* starting values for j=1 */
             beta2=0.049 gamma2= 1.570;/* starting values for j=2 */
  alpha = 100;                         /* constrain α to 100      */
  term1 = beta1*(rate**gamma1);
  term2 = beta2*(rate**gamma2);
  model drypct = alpha * (term1 / (1 + term1))*(herb=1) +
                 alpha * (term2 / (1 + term2))*(herb=2);
run;
```

Output 5.4.

```
                    The NLIN Procedure

                   Estimation Summary
          Method                          Gauss-Newton
          Iterations                                10
          Subiterations                              5
          Average Subiterations                    0.5
          R                                   6.518E-6
          Objective                           71.39743
          Observations Read                         14
          Observations Used                         14
          Observations Missing                       0
   NOTE: An intercept was not specified for this model.

                           Sum of        Mean    Asymptotic      Approx
Source              DF    Squares      Square       F Value      Pr > F
Regression           4    58171.5     14542.9       2036.89      <.0001
Residual            10    71.3974      7.1397
Uncorrected Total   14    58242.9
Corrected Total     13    17916.9
```

Output 5.4 (continued).

Parameter	Estimate	Asymptotic Standard Error	Asymptotic 95% Confidence Limits	
beta1	0.0213	0.00596	0.00799	0.0346
gamma1	-2.0394	0.1422	-2.3563	-1.7225
beta2	0.0685	0.0122	0.0412	0.0958
gamma	-1.2218	0.0821	-1.4047	-1.0389

We note $S(\widehat{\boldsymbol{\theta}}) = 71.3974, n - p = 10, \widehat{\boldsymbol{\theta}} = [0.0213, \; -2.0394, 0.0685, \; -1.2218]'$ and the estimated asymptotic standard errors of 0.00596, 0.1422, 0.0122, and 0.0821. Asymptotic 95% confidence limits are calculated from [5.28]. For example, a 95% confidence interval for $\theta_2 = \gamma_1$ is

$$\widehat{\gamma}_1 \pm t_{0.025,10}\text{ese}(\widehat{\gamma}_1) = \; -2.0394 \pm 2.228{*}0.1422 = [\,-2.3563\,, \; -1.7225].$$

To test the hypothesis

$$H_0\colon \begin{bmatrix} \beta_1 - \beta_2 \\ \gamma_1 - \gamma_2 \end{bmatrix} = \begin{bmatrix} 0 \\ 0 \end{bmatrix},$$

that the two herbicides coincide in growth response, we fit the constrained model

$$Y_{ij} = 100\frac{\beta x_{ij}^{\gamma}}{1 + \beta x_{ij}^{\gamma}} + e_i. \tag{5.34}$$

with the statements

```
proc nlin data=herbicide noitprint  ;
  parameters beta=0.049 gamma=-1.570;
  alpha = 100;
  term1 = beta*(rate**gamma);
  model drypct = alpha*term1 / (1 + term1);
run;
```

The abridged output follows.

Output 5.5.

The NLIN Procedure

NOTE: Convergence criterion met.

Source	DF	Sum of Squares	Mean Square	Asymptotic F Value	Approx Pr > F
Regression	2	57835.3	28917.6	851.27	<.0001
Residual	12	407.6	33.9699		
Uncorrected Total	14	58242.9			
Corrected Total	13	17916.9			

Parameter	Estimate	Asymptotic Standard Error	Asymptotic 95% Confidence Limits	
beta	0.0419	0.0139	0.0117	0.0722
gamma	-1.5706	0.1564	-1.9114	-1.2297

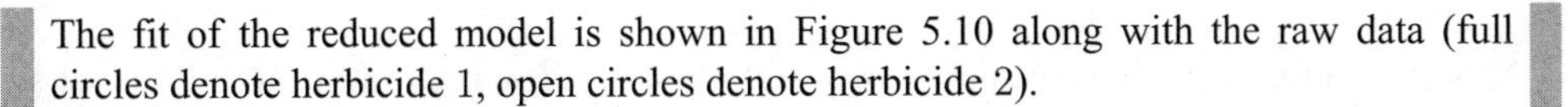

The fit of the reduced model is shown in Figure 5.10 along with the raw data (full circles denote herbicide 1, open circles denote herbicide 2).

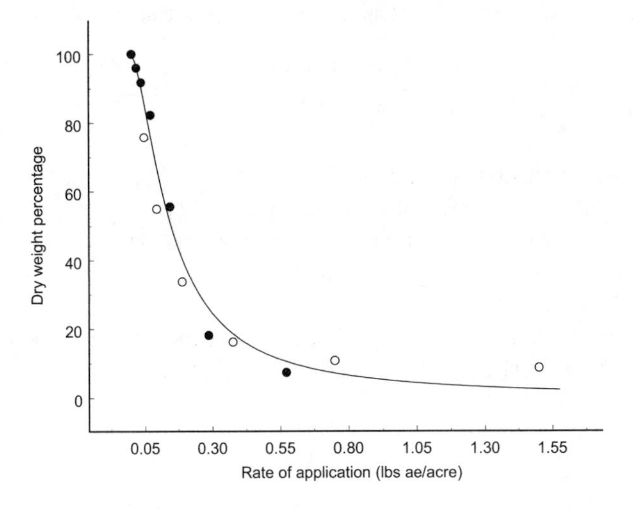

Figure 5.10. Observed dry weight percentages for velvetleaf weed plants treated with two herbicides at various rates. Rates are measured in pounds of acid equivalent per acre. Solid line is model [5.34] fit to the combined data in Table 5.2.

The sum of squares reduction test yields

$$F_{obs}^{(2)} = \frac{\left\{S(\widehat{\boldsymbol{\theta}})_r - S(\widehat{\boldsymbol{\theta}})_f\right\}/q}{S(\widehat{\boldsymbol{\theta}})_f/(n-p)} = \frac{\{407.6 - 71.3974\}/2}{71.3974/10} = 5.0 \;\; (p = 0.03125).$$

There is sufficient evidence at the 5% level to conclude that the growth responses of the two herbicides are different. At this point it is interesting to find out whether the herbicides differ in the parameter β, γ, or both. It is here, in the comparison of **individual** parameters across groups, that the Wald-type test can be implemented easily. To test H_0:$\beta_1 = \beta_2$, for example, we can fit the reduced model

$$Y_{ij} = 100\frac{\beta x_{ij}^{\gamma_j}}{1 + \beta x_{ij}^{\gamma_j}} + e_{ij} \qquad [5.35]$$

and use the sum of squares reduction test to compare with the full model [5.32]. Using the statements (output not shown)

```
proc nlin data=herbicide noitprint  ;
  parameters beta=0.049 gamma1=-1.570 gamma2=1.570;
  alpha = 100;
  term1 = beta*(rate**gamma1);
  term2 = beta*(rate**gamma2);
  model drypct = alpha * (term1 / (1 + term1))*(herb=1) +
                 alpha * (term2 / (1 + term2))*(herb=2);
run;
```

one obtains $S(\widehat{\boldsymbol{\theta}}_r) = 159.8$ and $F_{obs}^{(2)} = (159.8 - 71.3974)/7.13974 = 12.38$ with p-value 0.0055. At the 5% level the hypothesis of equal β parameters is rejected.

This test could have been calculated as a Wald-type test from the fit of the full model by observing that $\widehat{\beta}_1$ and $\widehat{\beta}_2$ are uncorrelated and the appropriate t statistic is

$$t_{obs} = \frac{\widehat{\beta}_1 - \widehat{\beta}_2}{\text{ease}\left(\widehat{\beta}_1 - \widehat{\beta}_2\right)} = \frac{0.0213 - 0.0685}{\sqrt{0.00596^2 + 0.0122^2}} = \frac{-0.04720}{0.01358} = -3.47622$$

Since the square of a t random variable with v degrees of freedom is identical to a F random variable with $1, v$ degrees of freedom, we can compare $t^2_{obs} = 3.47622^2 = 12.08$ to the F statistic of the sum of squares reduction test. The two tests are asymptotically equivalent, but differ for any finite sample size.

An alternative implementation of the t test that estimates the difference between β_1 and β_2 directly, is as follows. Let $\beta_2 = \beta_1 + \delta$ and write [5.32] as

$$Y_{ij} = 100\left\{\frac{\beta_1 x_{i1}^{\gamma_1}}{1 + \beta_1 x_{i1}^{\gamma_1}}\right\} I\{j = 1\} + 100\left\{\frac{(\beta_1 + \delta)x_{i2}^{\gamma_2}}{1 + (\beta_1 + \delta)x_{i2}^{\gamma_2}}\right\} I\{j = 2\} + e_{ij},$$

so that δ measures the difference between β_1 and β_2. The SAS® statements fitting this model are

```
proc nlin data=herbicide noitprint;
  parameters beta1=0.049 gamma1=-1.570 gamma2=-1.570 delta=0 ;
  alpha   = 100;
  beta    = beta1 + delta*(herb=2);
  term1   = beta*(rate**gamma1);
  term2   = beta*(rate**gamma2);
  model drypct = (alpha*term1 / (1 + term1))*(herb=1) +
                 (alpha*term2 / (1 + term2))*(herb=2);
run;
```

The estimate $\widehat{\delta} = 0.0472$ (Output 5.6, abridged) agrees with the numerator of t_{obs} above (apart from sign) and ease$(\widehat{\delta}) = 0.0136$ is identical to the denominator of t_{obs}.

Output 5.6. The NLIN Procedure

Source	DF	Sum of Squares	Mean Square	Asymptotic F Value	Approx Pr > F
Regression	4	58171.5	14542.9	2036.89	<.0001
Residual	10	71.3974	7.1397		
Uncorrected Total	14	58242.9			
Corrected Total	13	17916.9			

Parameter	Estimate	Asymptotic Standard Error	Asymptotic 95% Confidence Limits	
beta1	0.0213	0.00596	0.00799	0.0346
gamma1	-2.0394	0.1422	-2.3563	-1.7225
gamma2	-1.2218	0.0821	-1.4047	-1.0389
delta	0.0472	0.0136	0.0169	0.0776

This parameterization allows a convenient method to test for differences in parameters among groups. If the asymptotic 95% confidence interval for δ does not contain the value zero, the hypothesis $H_0{:}\delta = 0 \Leftrightarrow \beta_1 = \beta_2$ is rejected. The sign of $\widehat{\delta}$ informs us then which parameter is significantly larger. In this example $H_0{:}\,\delta = 0$ is rejected at the 5% level since the asymptotic 95% confidence interval $[0.0169, 0.0776]$ does not contain 0.

5.5.2 Confidence and Prediction Intervals

Once estimates of the model parameters have been obtained, the mean response at $\mathbf{x}_i$ is predicted by substituting $\widehat{\boldsymbol{\theta}}$ for $\boldsymbol{\theta}$. To calculate a $(1-\alpha)100\%$ confidence interval for the prediction $f(\mathbf{x}_i,\widehat{\boldsymbol{\theta}})$ we rely again on the asymptotic Gaussian distribution of $\widehat{\boldsymbol{\theta}}$. Details can be found in §A5.10.1. Briefly, the asymptotic standard error of $f(\mathbf{x}_i,\widehat{\boldsymbol{\theta}})$ is

$$\text{ase}\Big(f(\mathbf{x}_i,\widehat{\boldsymbol{\theta}})\Big) = \sqrt{\sigma^2\mathbf{F}_i'(\mathbf{F}'\mathbf{F})^{-1}\mathbf{F}_i'},$$

where $\mathbf{F}_i$ denotes the ith row of $\mathbf{F}$. To estimate this quantity, σ^2 is replaced by its estimate $\widehat{\sigma}^2$ and $\mathbf{F}$, $\mathbf{F}_i$ are evaluated at the converged iterate. The t-based confidence interval for $\mathrm{E}[Y]$ at $\mathbf{x}$ is

$$f(\mathbf{x}_i,\widehat{\boldsymbol{\theta}}) \pm t_{\alpha/2,n-p}\text{ease}\Big(f(\mathbf{x}_i,\widehat{\boldsymbol{\theta}})\Big).$$

These confidence intervals can be calculated with `proc nlin` of The SAS® System as we now illustrate with the small example visited earlier.

Example 5.2 Continued. Recall the simple one parameter model $Y_i = 1 - \exp\{-x_i^\theta\} + e_i,\ i = 1,\cdots,4$ that was fit to a small data set on page 199. The converged iterate was $\widehat{\theta} = 1.1146$ and the estimate of residual variability $\widehat{\sigma}^2 = 0.006$. The gradient matrix at $\widehat{\theta}$ evaluates to

$$\widehat{\mathbf{F}} = \begin{bmatrix} \ln\{0.2\}0.2^{\widehat{\theta}}\exp\Big\{-0.2^{\widehat{\theta}}\Big\} \\ \ln\{0.5\}0.5^{\widehat{\theta}}\exp\Big\{-0.5^{\widehat{\theta}}\Big\} \\ \ln\{0.7\}0.7^{\widehat{\theta}}\exp\Big\{-0.7^{\widehat{\theta}}\Big\} \\ \ln\{1.8\}1.8^{\widehat{\theta}}\exp\Big\{-1.8^{\widehat{\theta}}\Big\} \end{bmatrix} = \begin{bmatrix} -0.2267 \\ -0.2017 \\ -0.1224 \\ 0.1650 \end{bmatrix}$$

and $\widehat{\mathbf{F}}'\widehat{\mathbf{F}} = 0.13428$. The confidence intervals for $f(x,\theta)$ at the four observed data points are shown in the following table.

Table 5.3. Asymptotic 95% confidence intervals for $f(x,\theta)$; $t_{0.025,3} = 3.1824$, $\widehat{\sigma}^2 = 0.006$

x_i	$\widehat{\mathbf{F}}_i$	$\widehat{\mathbf{F}}_i'(\widehat{\mathbf{F}}'\widehat{\mathbf{F}})^{-1}\widehat{\mathbf{F}}_i'$	$\text{ease}\left(f(x_i,\widehat{\boldsymbol{\theta}})\right)$	**Asymptotic Conf. Interval**
0.2	– 0.2267	0.3826	0.0480	0.0004, 0.3061
0.5	– 0.2017	0.3030	0.0427	0.2338, 0.5092
0.7	– 0.1224	0.1116	0.0259	0.4067, 0.5718
1.8	0.1650	0.2028	0.0349	0.7428, 0.9655

Asymptotic 95% confidence intervals are obtained with the `l95m` and `u95m` options of the `output` statement in `proc nlin`. The `output` statement in the code segment that follows creates a new data set (here termed `nlinout`) that contains the predicted values (variable `pred`), the estimated asymptotic standard errors of the predicted values ($\text{ease}(f(x_i,\widehat{\theta}))$, variable `stdp`), and the lower and upper 95% confidence limits for $f(x_i,\theta)$ (variables `l95m` and `u95m`).

```
ods listing close;
proc nlin data=Ex_52;
  parameters theta=1.3;
  model y = 1 - exp(-x**theta);
  output out=nlinout pred=pred stdp=stdp l95m=l95m u95m=u95m;
run;
ods listing;
proc print data=nlinout; run;
```

Output 5.7.

```
Obs     y      x       PRED        STDP        L95M        U95M

 1     0.1    0.2    0.15323    0.048040    0.00035    0.30611
 2     0.4    0.5    0.36987    0.042751    0.23382    0.50592
 3     0.6    0.7    0.48930    0.025941    0.40674    0.57186
 4     0.9    1.8    0.85418    0.034975    0.74287    0.96549
```

Confidence intervals are intervals for the mean response $f(\mathbf{x}_i,\boldsymbol{\theta})$ at $\mathbf{x}_i$. **Prediction** intervals, on the contrary, have $(1-\alpha)100\%$ coverage probability for Y_i, which is a random variable. Prediction intervals are wider than confidence intervals for the mean because the estimated standard error is that of the difference $Y_i - f(\mathbf{x}_i,\widehat{\boldsymbol{\theta}})$ rather than that of $f(\mathbf{x}_i,\widehat{\boldsymbol{\theta}})$:

$$f(\mathbf{x}_i,\widehat{\boldsymbol{\theta}}) \pm t_{\alpha/2,n-p}\text{ease}\left(Y_i - f(\mathbf{x}_i,\widehat{\boldsymbol{\theta}})\right). \qquad [5.36]$$

In SAS®, prediction intervals are also obtained with the `output` statement of the `nlin` procedure, but instead of `l95m=` and `u95m=` use `l95=` and `u95=` to save prediction limits to the output data set.

5.6 Transf ormations

Box 5.3 Transformations

- **Transformations of the model are supposed to remedy a model breakdown such as**
  - **— nonhomogeneity of the variance;**
  - **— non-Gaussian error distribution;**
  - **— nonlinearity.**
- **Some nonlinear models can be transformed into linear models. However, de-transforming parameter estimates and predictions leads to bias (transformation bias).**
- **The transformation that linearizes the model may be different from the transformation that stabilizes the variance or makes the errors more Gaussian-like.**

5.6.1 Transformation to Linearity

Since linear models are arguably easier to work with, transforming a nonlinear model into a linear model is a frequently used device in data analysis. The apparent advantages are unbiased minimum variance estimation in the linearized form of the model and simplified calculations. Transformation to linearity obviously applies only to nonlinear models which can be linearized (intrinsically linear models in the sense of Pázman 1993, p. 36).

Example 5.6. Studies of allometry relate differences in anatomical shape to differences in size. Relating two size measures under the assumption of a constant ratio of their relative growth rates gives rise to a nonlinear model of form

$$\mathrm{E}[Y_i] = \beta_0 x_i^{\beta_1}, \qquad [5.37]$$

where Y is one size measurement (e.g., length of fibula) and x is the other size measure (e.g., length of sternum). β_1 is the ratio of the relative growth rates $\beta_1 = (x/y)\partial y/\partial x$. This model can be transformed to linearity by taking logarithms on both sides

$$\ln\{\mathrm{E}[Y_i]\} = U_i = \ln\{\beta_0\} + \beta_1 \ln\{x_i\}$$

which is a straight-line regression of U_i on the logarithm of x. If, however the underlying allometric relationship were given by

$$\mathrm{E}[Y_i] = \alpha + \beta_0 x_i^{\beta_1}$$

taking logarithms would not transform the model to a linear one.

Whether a nonlinear model which can be linearized should be linearized depends on a number of issues. We consider the nature of the model residuals first. A model popular in forestry is Schumacher's height-age relationship which predicts the height of the socially dominant trees in an even-aged stand (Y) from the reciprocal of the stand age ($x = 1/age$) (Schumacher 1939, Clutter et al. 1992). One parameterization of the mean function is

$$\mathrm{E}[Y] = \alpha\exp\{\beta x\}, \qquad [5.38]$$

and $\mathrm{E}[Y]$ is linearized by taking logarithms,

$$\ln\{\mathrm{E}[Y]\} = \ln\{\alpha\} + \beta x = \gamma + \beta x.$$

To obtain an observational model we need to add stochastic error terms. Residuals proportional to the expected value of the response give rise to

$$Y = \alpha\exp\{\beta x\}(1 + e^*) = \alpha\exp\{\beta x\} + \alpha\exp\{\beta x\}e^* = \alpha\exp\{\beta x\} + e^{**}. \qquad [5.39]$$

This is called a **constant relative error** model (Seber and Wild 1989, p. 15). Additive residuals, on the other hand, lead to a **constant absolute error** model

$$Y = \alpha\exp\{\beta x\} + \epsilon^*. \qquad [5.40]$$

A linearizing transform is not applied to the mean values but to the observables Y, so that the two error assumptions will lead to different properties for the transformed residuals. In the case of [5.39] the transformed observational model becomes

$$\ln\{Y\} = \gamma + \beta x + \ln\{1 + e^*\} = \gamma + \beta x + e.$$

If e^* has zero mean and constant variance, then $\ln\{1 + e^*\}$ will have approximately zero mean and constant variance (which can be seen easily from a Taylor series of $\ln\{1 + e^*\}$). For constant absolute error the logarithm of the observational model leads to

$$\begin{aligned}
\ln\{Y\} &= \ln\{\alpha e^{\beta x} + \epsilon^*\} \\
&= \ln\left\{\alpha e^{\beta x} + \epsilon^* \frac{\alpha\exp\{\beta x\}}{\alpha\exp\{\beta x\}}\right\} \\
&= \ln\left\{\alpha e^{\beta x}\left(1 + \frac{\epsilon^*}{\alpha\exp\{\beta x\}}\right)\right\} \\
&= \gamma + \beta x + \ln\left\{1 + \frac{\epsilon^*}{\alpha\exp\{\beta x\}}\right\} = \gamma + \beta x + \epsilon.
\end{aligned}$$

The error term of this model should have expectation close to zero expectation if ϵ^* is small on average compared to $\mathrm{E}[Y]$ (Seber and Wild, 1989, p. 15). $\mathrm{Var}[\epsilon]$ depends on the mean of the model, however. In case of [5.39] the linearization transform is also a variance-stabilizing transform but in the constant absolute error model linearization has created heterogeneous error variances.

Nonlinear models are parameterized so that the parameters represent important physical and biological measures (see §5.7 on parameterizations) such as rates of change, survival and mortality, upper and lower yield and growth asymptotes, densities, and so forth. Linearization destroys the natural interpretation of the parameters. Sums of squares and variability estimates are not reckoned on the original scale of measurement, but on the transformed scale. The variability of plant yield is best understood in yield units (bushels/acre, e.g.), not in square roots or logarithms of bushels/acre. Another problematic aspect of linearizing transformations relates to their **transformation bias** (also called **prediction bias**). Taking expectations of random variables is a linear operation and does not apply to nonlinear transforms. To obtain predicted values on the original scale after a linearized version of the model has been fit, the predictions obtained in the linearized model need to be detransformed. This process introduces bias.

Example 5.7. Consider a height-age equation of Schumacher form

$$\mathrm{E}[H_i] = \alpha\exp\{\beta x_i\},$$

where H_i is the height of the ith plant (or stand of plants) and x_i is the reciprocal of plant (stand) age. As before, the mean function is linearized by taking logarithms on both sides:

$$\ln\{\mathrm{E}[H_i]\} = \ln\{\alpha\} + \beta x_i = \mathbf{x}_i'\boldsymbol{\beta}.$$

Here $\mathbf{x}_i' = [1, x_i]$, $\boldsymbol{\beta} = [\ln\{\alpha\}, \ -\beta]'$ and $\exp\{\mathbf{x}_i'\boldsymbol{\beta}\} = \mathrm{E}[H_i]$. We can fit the linearized model

$$U_i = \mathbf{x}_i'\boldsymbol{\beta} + e_i \qquad [5.41]$$

where the e_i are assumed independent Gaussian with mean 0 and variance σ^2. A predicted value in this model on the linear scale is $\widehat{u} = \mathbf{x}'\widehat{\boldsymbol{\beta}}$ which has expectation

$$\mathrm{E}[\widehat{u}] = \mathrm{E}\left[\mathbf{x}'\widehat{\boldsymbol{\beta}}\right] = \mathbf{x}'\boldsymbol{\beta} = \ln\{\mathrm{E}[H]\}.$$

If the linearized model is correct, the predictions are unbiased for the logarithm of height. Are they also unbiased for height? To this end we detransform the linear predictions and evaluate $\mathrm{E}[\exp\{\widehat{u}\}]$. In general, if $\ln\{Y\}$ is Gaussian with mean μ and variance ξ, then Y has a Log-Gaussian distribution with

$$\mathrm{E}[Y] = \exp\{\mu + \xi/2\}$$
$$\mathrm{Var}[Y] = \exp\{2\mu + \xi\}(\exp\{\xi\} - 1).$$

We note that $\mathrm{E}[Y]$ is $\exp\{\xi/2\}$ times larger than $\exp\{\mu\}$, since ξ is a positive quantity. Under the assumption about the distribution of e_i made earlier, the predicted value $\widehat{u}$ is also Gaussian with mean $\mathbf{x}'\boldsymbol{\beta}$ and variance $\sigma^2\mathbf{x}'(\mathbf{X}'\mathbf{X})^{-1}\mathbf{x}$. The mean of a predicted height value is then

$$\mathrm{E}[\exp\{\widehat{u}\}] = \exp\left\{\mathbf{x}'\boldsymbol{\beta} + \frac{1}{2}\sigma^2\mathbf{x}'(\mathbf{X}'\mathbf{X})^{-1}\mathbf{x}\right\} = \mathrm{E}[H]\exp\left\{\frac{1}{2}\sigma^2\mathbf{x}'(\mathbf{X}'\mathbf{X})^{-1}\mathbf{x}\right\}.$$

This is an overestimate of the average height.

Transformation to achieve linearity can negatively affect other desirable properties of the model. The most desirable transformation is one which linearizes the model, stabilizes the variance, and makes the residuals Gaussian-distributed. The class of Box-Cox transformations (after Box and Cox 1964, see §5.6.2 below for Box-Cox transformations in the context of variance heterogeneity) is heralded to accomplish these multiple goals but is not without controversy. Transformations are typically more suited to rectify a particular problematic aspect of the model, such as variance heterogeneity or residuals which are far from a Gaussian distribution. The transformation that stabilizes the variance may be different from the transformation that linearizes the model.

Example 5.8. Yield density models are used in agronomy to describe agricultural output (yield) as a function of plant density (see §5.8.7 for an application). Two particularly simple representatives of yield-density models are due to Shinozaki and Kira (1956)

$$Y_i = \frac{1}{\beta_0 + \beta_1 x_i} + e_i$$

and Holliday (1960)

$$Y_i = \frac{1}{\beta_0 + \beta_1 x_i + \beta_2 x_i^2} + e_i.$$

Obviously the linearizing transform is the reciprocal $1/Y$. It turns out, however, that for many data sets to which these models are applied the appropriate transform to stabilize the variance is the logarithmic transform.

Nonlinearity of the model, given the reliability and speed of today's computer algorithms, is not considered a shortcoming of a statistical model. If estimation and inferences can be carried out on the original scale, there is no need to transform a model to linearity simply to invoke a linear regression routine and then to incur transformation bias upon detransformation of parameter estimates and predictions. The lack of Gaussianity of the model residuals is less critical for nonlinear models than it is for linear ones, since less is lost if the errors are non-Gaussian. Statistical inference in nonlinear models requires asymptotic results and the nonlinear least squares estimates are asymptotically Gaussian-distributed regardless of the distribution of the model errors. In linear models the difference between Gaussian and non-Gaussian errors is the difference between exact and approximate inference. In nonlinear models, inference is approximate anyway.

5.6.2 Transformation to Stabilize the Variance

Variance heterogeneity (**heteroscedasticity**), the case when the variance-covariance matrix of the model residuals is not given by

$$\text{Var}[\mathbf{e}] = \sigma^2\mathbf{I},$$

but a diagonal matrix whose entries are of different magnitude, is quite common in biological data. It is related to the intuitive observation that large entities vary more than small entities.

If error variance is a function of the regressor x two approaches can be used to remedy heteroscedasticity. One can apply a power transformation of the model or fit the model by nonlinear **weighted** least squares. The former approach relies on a transformation that stabilizes the variance. The latter approach accounts for changes in variability with x in the process of model fitting. Box and Cox (1964) made popular the family of power transformations defined by

$$U = \begin{cases} (Y^{\lambda} - 1)/\lambda & \lambda \neq 0 \\ \ln\{Y\} & \lambda = 0, \end{cases} \quad [5.42]$$

which apply when the response is non-negative ($Y > 0$). Expanding U into a first-order Taylor series around $\mathrm{E}[Y]$ we find the variance of the Box-Cox transformed variable to be approximately

$$\mathrm{Var}[U] = \mathrm{Var}[Y]\{\mathrm{E}[Y]\}^{2(\lambda-1)}. \quad [5.43]$$

By choosing λ properly the variance of U can be made constant which was the motivation behind finding a variance-stabilizing transform in §4.5.2 for linear models. There we were concerned with the comparison of groups where replicate values were available for each group. As a consequence we could estimate the mean and variance in each group, linearize the relationship between variances and means, and find a numerical estimate for the parameter λ. If replicate values are not available, λ can be determined by trial and error, choosing a value for λ, fitting the model and examining the fitted residuals $\widehat{e}_i$ until a suitable value of λ has been found. With some additional programming effort, λ can be estimated from the data. Seber and Wild (1989) discuss maximum likelihood estimation of the parameters of the mean function and λ jointly. To combat variance heterogeneity [5.43] suggests two approaches:

- transform both sides of the model according to [5.42] and fit a nonlinear model with response U_i;
- leave the response Y_i unchanged and allow for variance heterogeneity. The model is fit by nonlinear weighted least squares where the variance of the response is proportional to $\mathrm{E}[Y_i]^{2(\lambda-1)}$.

Carroll and Ruppert (1984) call these approaches *power-transform both sides* and *power-transformed weighted least squares* models. If the original, untransformed model is

$$Y_i = f(\mathbf{x}_i, \boldsymbol{\theta}) + e_i,$$

then the former is

$$U_i = f^*(\mathbf{x}_i, \boldsymbol{\theta}) + e_i^*$$

where

$$f^*(\mathbf{x}_i, \boldsymbol{\theta}) = \begin{cases} (f(\mathbf{x}_i, \boldsymbol{\theta})^{\lambda} - 1)/\lambda & \lambda \neq 0 \\ \ln\{f(\mathbf{x}_i, \boldsymbol{\theta})\} & \lambda = 0 \end{cases}$$

and $e_i^* \sim iid\, G(0, \sigma^2)$. The second approach uses the original response and

$$Y_i = f(\mathbf{x}_i, \boldsymbol{\theta}) + e_i,\ e_i \sim iid\ \ G\Big(0, \sigma^2 f(\mathbf{x}_i, \boldsymbol{\theta})^{\lambda}\Big).$$

For extensions of the Box-Cox method such as power transformations for negative responses see Box and Cox (1964), Carroll and Ruppert (1984), Seber and Wild (1989, Ch. 2.8) and references therein. Weighted nonlinear least squares is applied in §5.8.8.

5.7 Parameterization of Nonlinear Models

Box 5.4 Parameterization

- **Parameterization is the process of expressing the mean function of a statistical model in terms of parameters to be estimated.**
- **A nonlinear model can be parameterized in different ways. A particular parameterization is chosen so that the parameter estimators have a certain interpretation, desirable statistical properties, and allow embedding of hypotheses (see §1.5).**
- **Nonlinear models have two curvature components, called *intrinsic* and *parameter-effects* curvature. Changing the parameterization alters the degree of parameter-effects curvature.**
- **The smaller the curvature of a model, the more estimators behave like the efficient estimators in a linear model and the more reliable are inferential and diagnostic procedures.**

Parameterization is the process of expressing the mean function of a statistical model in terms of unknown constants (parameters) to be estimated. With nonlinear models the same model can usually be expressed (parameterized) in a number of ways.

Consider the basic differential equation $\partial y/\partial t = \kappa(\alpha - y)$ where $y(t)$ is the size of an organism at time t. According to this model, growth is proportional to the remaining size of the organism, α is the final size (total growth), and κ is the proportionality constant. Upon integration of this generating equation one obtains

$$y(t) = \alpha + (\xi - \alpha)\exp\{-\kappa t\} \quad [5.44]$$

where $\alpha, \kappa, \xi, x > 0$ and ξ is the initial size, $y(0)$. A simple reparameterization is obtained by setting $\theta = \xi - \alpha$ in [5.44]. The equation now becomes

$$y(t) = \alpha + \theta\exp\{-\kappa t\}, \quad [5.45]$$

a form in which it is known as the **monomolecular growth** model. If furthermore $\exp\{-\kappa\}$ is replaced by ψ, the resulting equation

$$y(t) = \alpha + \theta\psi^t \quad [5.46]$$

is known as the **asymptotic regression model**. Finally, one can put $\xi = \alpha(1 - \exp\{\kappa t_0\})$, where t_0 is the time at which the size (yield) is zero, $(y(t_0) = 0)$ to obtain the equation

$$y(t) = \alpha(1 - \exp\{-\kappa(t - t_0)\}). \qquad [5.47]$$

In this form the equation is known as the Mitscherlich *law* (or **Mitscherlich equation**), popular in agronomy to model crop yield as a function of fertilizer input (t) (see §5.8.1 for a basic application of the Mitscherlich model). In fisheries and wildlife research [5.46] is known as the Von Bertalanffy model and it finds application as Newton's law of cooling a body over time in physics. Ratkowsky (1990) discusses it as Mitscherlich's law and the Von Bertalanffy model but Seber and Wild (1989) argue that the Von Bertalanffy model is derived from a different generating equation; see §5.2. Equations [5.44] through [5.47] are four parameterizations of the same basic relationship. When fitting either parameterization to data, they yield the same goodness of fit, the same residual error sum of squares, and the same vector of fitted residuals. The interpretation of their parameters is different, however, as are the statistical properties of the parameter estimators. For example, the correlations between the parameter estimates can be quite different. In §4.4.4 it was discussed how correlations and dependencies can negatively affect the least squares estimate. This problem is compounded in nonlinear models which usually exhibit considerable correlations among the columns in the regressor matrix. A parameterization that reduces these correlations will lead to more reliable convergence of the iterative algorithm. To understand the effects of the parameterization on statistical inference, we need to consider the concept of the **curvature** of a nonlinear model.

5.7.1 Intrinsic and Parameter-Effects Curvature

The statistical properties of nonlinear models, their numerical behavior in the estimation process (convergence behavior), and the reliability of asymptotic inference for finite sample size are functions of a model's **curvature**. The curvature consists of two components, termed **intrinsic** curvature and **parameter-effects** curvature (Ratkowsky 1983, 1990). Intrinsic curvature measures how much the nonlinear model bends if the value of the *parameters* are changed by a small amount. This is not the same notion as the bending of the mean function the regressor x is changed. Models with curved mean function can have low intrinsic curvature. A **curvilinear** model such as $Y = \beta_0 + \beta_2 x^2 + e$ has a mean function that curves when $\mathrm{E}[Y]$ is plotted against x but has no intrinsic curvature. To understand these curvature components better, we consider how the mean function $f(\mathbf{x}, \boldsymbol{\theta})$ varies for different values of $\boldsymbol{\theta}$ for given values of the regressor. The result is called the **expectation surface** of the model. With only two data points x_1 and x_2, the expectation *surface* can be displayed in a two-dimensional coordinate system.

Example 5.9. Consider the linear mean function $\mathrm{E}[Y_i] = \theta x_i^2$ and design points at $x_1 = 1$ and $x_1 = 2$. The expectation surface is obtained by varying

$$\mathrm{E}[\mathbf{Y}] = \begin{bmatrix} \theta \\ 4\theta \end{bmatrix} = \begin{bmatrix} f_1(\theta) \\ f_2(\theta) \end{bmatrix}$$

for different values of θ. This *surface* can be plotted in a two-dimensional coordinate system as a straight line (Figure 5.11).

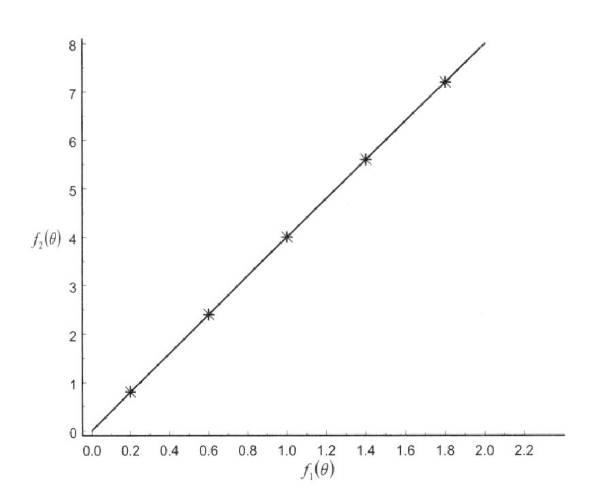

Figure 5.11. Expectation surface of $\mathrm{E}[Y_i] = \theta x_i^2$. Asterisks mark points on the surface for equally spaced values of $\theta = 0.2, 0.6, 1.0, 1.4,$ and 1.8.

Since the expectation surface does not bend as θ is changed, the model has no intrinsic curvature.

Nonlinear estimation rests on linearization techniques. The linear approximation

$$\mathbf{f}(\mathbf{x}, \boldsymbol{\theta}) - \mathbf{f}(\mathbf{x}, \widehat{\boldsymbol{\theta}}) \doteq \widehat{\mathbf{F}}\left(\boldsymbol{\theta} - \widehat{\boldsymbol{\theta}}\right)$$

amounts to approximating the expectation surface in the vicinity of $\widehat{\boldsymbol{\theta}}$ by a tangent plane. To locally replace the expectation surface with a plane is not justifiable if the intrinsic curvature is large, since then changes in $\boldsymbol{\theta}$ will cause considerable bending in $\mathbf{f}(\mathbf{x}, \boldsymbol{\theta})$ while the linear approximation assumes that the surface is flat in the neighborhood of $\widehat{\boldsymbol{\theta}}$. Seber and Wild (1989, p. 137) call this the *planar assumption* and point out that it is invalidated when the intrinsic curvature component is large. A second assumption plays into the quality of the linear approximation; when a regular grid of $\boldsymbol{\theta}$ values centered at the estimate $\widehat{\boldsymbol{\theta}}$ is distorted upon projection onto $\mathbf{f}(\mathbf{x}, \boldsymbol{\theta})$, the parameter-effects curvature is large. Straight, parallel, equispaced lines in the parameter space then do not map into straight, parallel, equispaced lines on the expectation surface. What Seber and Wild (1989, p. 137) call the *uniform-coordinate assumption* is invalid if the parameter-effects curvature is large. The linear model whose expectation surface is shown in Figure 5.11 has no parameter-effects curvature. Equally spaced values of θ, for which the corresponding points on the expectation surface are marked with asterisks, are equally spaced on the expectation surface.

Ratkowsky (1990) argues that intrinsic curvature is typically the smaller component and advocates focusing on parameter-effects curvature because it can be influenced by model parameterization. No matter how a model is parameterized, the intrinsic curvature remains the same. But the effect of parameterization is generally data-dependent. A parameterization with low parameter-effects curvature for one data set may produce high curvature for another data set. After these rather abstract definitions of curvature components and the sobering finding that it is hard to say how strong these components are for a particular model/data combination without knowing precisely the design points, one can rightfully ask, why worry? The answer lies in two facts. First, the effects of strong curvature on many aspects of statistical inference can be so damaging that one should be aware of the liabilities when choosing a parameterization that produces high curvature. Second, parameterizations are known for many popular models that typically result in models with low parameter-effects curvature and desirable statistical properties of the estimators. These are expected value parameterizations (Ratkowsky 1983, 1989).

Before we examine some details of the effects of curvature, we demonstrate the relationship of the two curvature components in nonlinear models for a simple case that is discussed in Seber and Wild (1989, pp. 98-102) and was inspired by Ratkowsky (1983).

Example 5.10. The nonlinear mean function $\mathrm{E}[Y_i] = \exp\{\theta x_i\}$ is investigated, where the only design points are $x_1 = 1$ and $x_2 = 2$. The expectation surface is then

$$\mathrm{E}[\mathbf{Y}] = \begin{bmatrix} \exp\{\theta\} \\ \exp\{2\theta\} \end{bmatrix} = \begin{bmatrix} f_1(\theta) \\ f_2(\theta) \end{bmatrix}.$$

A reparameterization of the model is achieved by letting $\psi^{x_i} = \exp\{\theta x_i\}$. In this form the expectation surface is

$$\mathrm{E}[\mathbf{Y}] = \begin{bmatrix} \psi \\ \psi^2 \end{bmatrix} = \begin{bmatrix} g_1(\psi) = f_1(\theta) \\ g_2(\psi) = f_2(\theta) \end{bmatrix}.$$

This surface is graphed in Figure 5.12 along with points on the surface corresponding to equally spaced sets of θ and ψ. Asterisks mark points on the surface for equally spaced values of $\theta = 0, 0.5, 1.0, 1.5, 2.0, 2.5, 3.0,$ and 3.5, and circles mark points for equally spaced values of $\psi = 2, 4, ..., 32$.

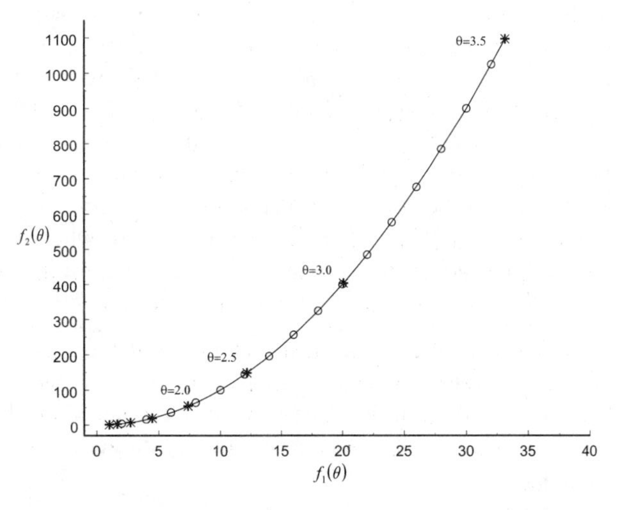

Figure 5.12. Expectation surface of $\mathrm{E}[Y_i] = \exp\{\theta x_i\} = \psi^{x_i}$ for $x_1 = 1, x_2 = 2$. Adapted from Figures 3.5a) and b) in Seber, G.A.F. and Wild, C.J. (1989) *Nonlinear Regression*. Wiley and Sons, New York. Copyright © 1989 John Wiley and Sons, Inc. Reprinted by permission of John Wiley and Sons, Inc.

The two parameterizations of the model produce identical expectation surfaces; hence their intrinsic curvatures are the same. The parameter-effects curvatures are quite different, however. Equally spaced values of θ are no longer equally spaced on the surface as in the linear case (Figure 5.11). Mapping onto the expectation surfaces results in considerable distortion, and parameter-effects curvature is large. But equally spaced values of ψ create almost equally spaced values on the surface. The parameter-effects curvature of model $\mathrm{E}[Y_i] = \psi^{x_i}$ is small.

What are the effects of strong curvature in a nonlinear model? Consider using the fitted residuals

$$\widehat{e}_i = y_i - f(\mathbf{x}_i, \widehat{\boldsymbol{\theta}})$$

as a diagnostic tool similar to diagnostics applicable in linear models (§4.4.1). There we can calculate studentized residuals (see §4.8.3) ,

$$r_i = \widehat{e}_i / \left(\widehat{\sigma}\sqrt{1 - h_{ii}}\right),$$

for example, that behave similar to the unobservable model errors. They have mean zero and constant variance. Here, h_{ii} is the ith diagonal element of the projection matrix $\mathbf{X}(\mathbf{X}'\mathbf{X})^{-1}\mathbf{X}'$. If the intrinsic curvature component is large, a similar residual obtained from a nonlinear model fit will not behave as expected. Let

$$r_i^* = \frac{y_i - f(\mathbf{x}_i, \widehat{\boldsymbol{\theta}})}{\sqrt{\widehat{\sigma}^2(1 - h_{ii}^*)}},$$

where h_{ii}^* is the ith diagonal element of $\widehat{\mathbf{F}}(\widehat{\mathbf{F}}'\widehat{\mathbf{F}})^{-1}\widehat{\mathbf{F}}'$. If intrinsic curvature is pronounced, the r_i^* will **not** have mean 0, will **not** have constant variance, and as shown by Seber and Wild (1989, p. 178) r_i^* is negatively correlated with the predicted values $\widehat{y}_i$. The diagnostic plot of r_i^* versus $\widehat{y}_i$, a tool commonly borrowed from linear model analysis will show a negative slope rather than a band of random scatter about 0. Seber and Wild (1989, p. 179) give expressions of what these authors call **projected** residuals that do behave as their counterparts in the linear model (see also Cook and Tsai, 1985), but none of the statistical packages we are aware of can calculate these. The upshot is that if intrinsic curvature is large, one may diagnose a nonlinear model as deficient based on a residual plot when in fact the model is adequate, or find a model to be adequate when in fact it is not. It is for this reason that we shy away from *standard* residual plots in nonlinear regression models in this chapter.

The residuals are not affected by the parameter-effects curvature, only by intrinsic curvature. Fitting a model in different parameterizations will produce the same set of fitted residuals (see §5.8.2 for a demonstration). The parameter estimates, on the other hand, are very much affected by parameter-effects curvature. The key properties of the nonlinear least squares estimators $\widehat{\boldsymbol{\theta}}$, namely

- asymptotic unbiasedness
- asymptotic minimum variance
- asymptotic Gaussianity

all depend on the validity of the tangent-plane approximation (the planar assumption). With increasing curvature this assumption becomes less tenable. The estimators will be increasingly biased; their sample distribution will deviate from a Gaussian distribution and, perhaps most damagingly, their variance will exceed $\sigma^2(\mathbf{F}'\mathbf{F})^{-1}$. Since this is the expression on which nonlinear statistical routines base the estimated asymptotic standard errors of the estimators, these packages will underestimate the uncertainty in $\widehat{\boldsymbol{\theta}}$, test statistics will be inflated and p-values will be too small. In §5.8.2 we examine how the sampling distribution of the parameter estimates depends on the curvature and how one can diagnose a potential problem.

Having outlined the damage incurred by pronounced curvature of the nonlinear model, we are challenged to remedy the problem. In Example 5.10 the parameterization $\mathrm{E}[Y_i] = \psi^{x_i}$ was seen to be preferable because it leads to smaller parameter-effects curvature. But as the design points are changed so changes the expectation surface and the curvature components are altered. One approach of reducing curvature is thus to choose the design points (regressor values) accordingly. But the design points which minimize the parameter-effects curvature may not be the levels of interest to the experimenter. In a rate application trial, fertilizers are often applied at equally spaced levels, e.g., 0 lbs/acre, 30 lbs/acre, 60 lbs/acre. If the response (crop yield, for example) is modeled nonlinearly the fertilizer levels 4.3 lbs/acre, 10.3 lbs/acre, 34.9 lbs/acre may minimize the parameter-effects curvature but are of little interest to the experimenter. In observational studies the regressor values cannot be controlled and parameter-effects curvature cannot be influenced by choosing design points. In addition, the appropriate nonlinear model may not be known in advance. The same set of regressor values will produce different degrees of curvature depending on the model. If alternative models are to be examined, fixing the design points to reduce curvature effects is not a useful solution in our opinion.

Ratkowsky (1983) termed nonlinear models with low curvature **close-to-linear** models, meaning that their estimators behave similarly to estimators in linear models. They should be close to Gaussian-distributed, almost unbiased, and the estimates of their precision based on the linearization should be reliable. Reducing the parameter-effects curvature component as much as possible will make nonlinear parameter estimates behave as closely to linear as possible for a given model and data set. Also, the convergence properties of the iterative fitting process should thereby be improved because models with high parameter-effects curvature have a shallow residual sum of squares surface. Subsequent reductions in $S(\widehat{\boldsymbol{\theta}})$ between iterations will be small and the algorithm requires many iterations to achieve convergence.

For the two parameterizations in Figure 5.12 the residual sum of squares *surface* can be shown as a line (Figure 5.13), since there is only one parameter (θ or ψ). The sum of squares surface of the model with higher parameter-effects curvature (θ) is more shallow over a wide range of θ values. Only if $\widehat{\theta}$ is close to about 2.2, a more rapid dip occurs, leading to the final estimate of $\widehat{\theta} = 2.699$. The minimum of the surface for $\mathrm{E}[Y_i] = \psi^{x_i}$ is more pronounced and regardless of the starting value, convergence will occur more quickly. The final estimate in this parameterization is $\widehat{\psi} = 14.879$. In either case, the residual sum of squares finally achieved is 8.30.

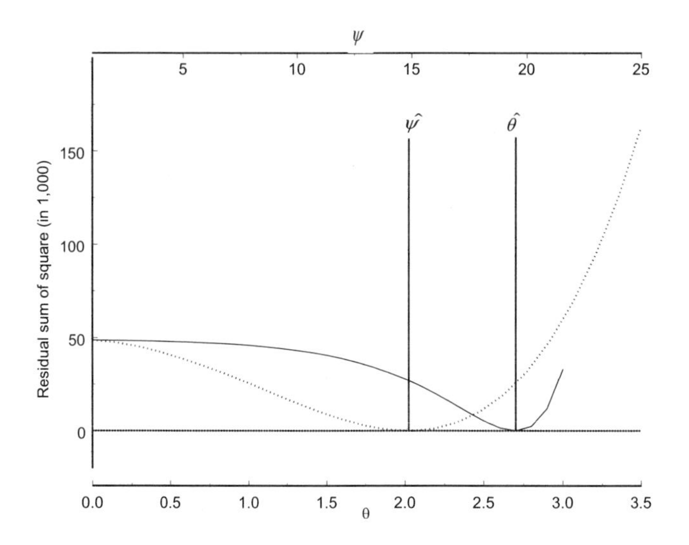

Figure 5.13. Residual sum of squares surfaces for $\mathrm{E}[Y_i] = \exp\{\theta x_i\} = \psi^{x_i}, y_1 = 12$, $y_2 = 221.5$. Other values as shown in Figure 5.12. Solid line: sum of squares surface for $\mathrm{E}[Y_i] = \exp\{\theta x_i\}$. Dashed line: sum of squares surface for $\mathrm{E}[Y_i] = \psi^{x_i}$.

Ratkowsky's technique for reducing the parameter-effects curvature relies on rewriting the model in its expected value parameterization. This method was introduced in §5.4.3 as a vehicle for determining starting values. In §5.8.2 we show how the parameters of the Mitscherlich equation have much improved properties in its expected value parameterization compared to some of the *standard* parameterizations of the model. Not only are they more Gaussian-distributed, they also are less biased and their standard error estimates are more reliable. Furthermore, the model in expected value parameterization converges more quickly and the estimators are less correlated (§5.8.1). As a consequence, the $(\mathbf{F}'\mathbf{F})$ matrix is better-conditioned and the iterative algorithm is more stable. Expected value parameterizations can also be used to our advantage to rewrite a nonlinear model in terms of parameters that are of more immediate interest to the researcher. Schabenberger et al. (1999) used this method to reparameterize log-logistic herbicide dose response models and term the technique reparameterization through defining relationships.

5.7.2 Reparameterization through Defining Relationships

An expected value parameter corresponds to a predicted value at a given value x^* of the regressor. Consider, for example, the log-logistic model

$$\mathrm{E}[Y] = \delta + \frac{\alpha - \delta}{1 + \psi \exp\{\beta \ln(x)\}}, \qquad [5.48]$$

popular in modeling the relationship between a dosage x and a response Y. The response changes in a sigmoidal fashion between the asymptotes α and δ. The rate of change and whether response increases or decreases in x depend on the parameters ψ and β. Assume we wish to replace a parameter, ψ say, by its expected value equivalent:

$$\mu^* = \delta + \frac{\alpha - \delta}{1 + \psi\exp\{\beta\ln(x^*)\}}.$$

Solving for ψ leads to

$$\psi = \left(\frac{\alpha - \delta}{\mu^* - \delta} - 1\right)\exp\{-\beta\ln(x^*)\}.$$

After substituting back into [5.48] and simplifying, one arrives at

$$\mathrm{E}[Y] = \delta + \frac{\alpha - \delta}{1 + \left(\frac{\alpha - \mu^*}{\mu^* - \delta} - 1\right)\exp\{\beta\ln(x/x^*)\}}.$$

This model should have less parameter-effects curvature than [5.48], since ψ was replaced by its expected value parameter. It has not become more interpretable, however. Expected value parameterization rests on estimating the unknown μ^* for a known value of x^*. This process can be reversed if we want to express the model as a function of an unknown value x^* for a known value μ^*. A common task in dose-response studies is to find that dosage which reduces/increases the response by a certain amount or percentage. In a study of insect mortality as a function of insecticide concentration, for example, we might be interested in the dosage at which 50% of the treated insects die (the so-called LD_{50} value). In biomedical studies, one is often interested in the dosage that cures 90% of the subjects. The idea is as follows. Consider the model $\mathrm{E}[Y|x] = f(x, \boldsymbol{\theta})$ with parameter vector $\boldsymbol{\theta} = [\theta_1, \theta_2, \cdots, \theta_p]'$. Find a value x^* which is of interest to the investigation, for example LD_{50} or GR_{50}, the dosage that reduces/increases growth by 50%. Our goal is to estimate x^*. Now set

$$\mathrm{E}[Y|x^*] = f(x^*, \boldsymbol{\theta}),$$

termed the **defining relationship**. Solve for one of the original parameters, θ_1 say. Substitute the expression obtained for θ_1 into the original model $f(x, \boldsymbol{\theta})$ and one obtains a model where θ_1 has been replaced by x^*. Schabenberger et al. (1999) apply these ideas in the study of herbicide dose response where the log-logistic function has negative slope and represents the growth of treated plants relative to an untreated control. Let λ_K be the value which reduces growth by $(K{*}100)$%. In a model with lower and upper asymptotes such as model [5.48] one needs to carefully define whether this is a reduction of the maximum response α or of the difference between the maximum and minimum response. Here, we define λ_K as the value for which

$$\mathrm{E}[Y|\lambda_K] = \delta + \left(\frac{100 - K}{100}\right)(\alpha - \delta).$$

The defining relationship to parameterize a log-logistic model in terms of λ_K is then

$$\delta + \left(\frac{100 - K}{K}\right)(\alpha - \delta) = \delta + \frac{(\alpha - \delta)}{1 + \psi\exp\{\beta\ln(\lambda_K)\}}.$$

Schabenberger et al. (1999) chose to solve for ψ. Upon substituting the result back into [5.48] one obtains the reparameterized log-logistic equation

$$\mathrm{E}[Y|x] = \delta + \frac{\alpha - \delta}{1 + K/(100 - K)\exp\{\beta \ln(x/\lambda_K)\}}. \qquad [5.49]$$

In the special case where $K = 50$, the term $K/(100 - K)$ in the denominator vanishes.

Two popular ways of expressing the Mitscherlich equation can also be developed by this method. Our [5.44] shows the Mitscherlich equation for crop yield y at nutrient level x as

$$\mathrm{E}[Y|x] = \alpha + (\xi - \alpha)e^{-\kappa x},$$

where α is the yield asymptote and ξ is the yield at nutrient concentration $x = 0$. Suppose we are interested in estimating what Black (1993, p. 273) calls the availability index, the nutrient level x_0 at which the average yield is zero and want to replace ξ in the process. The defining relationship is

$$\mathrm{E}[Y|x_0] = 0 = \alpha + (\xi - \alpha)e^{-\kappa x_0}.$$

Solving for ξ gives $\xi = \alpha(1 - \exp\{\kappa x_0\})$. Substituting back into the original equation and simplifying one obtains the other popular parameterization of the Mitscherlich equation (compare to [5.47]):

$$\begin{aligned} \mathrm{E}[Y] &= \alpha + (\alpha(1 - e^{-\kappa x_0}) - \alpha)e^{-\kappa x} \\ &= \alpha - \alpha e^{\kappa x_0} e^{-\kappa x} \\ &= \alpha\left(1 - e^{-\kappa(x - x_0)}\right). \end{aligned}$$

5.8 Applications

In this section we present applications involving nonlinear statistical models and discuss their implementation with The SAS® System. A good computer program for nonlinear modeling should provide simple commands to generate standard results of any nonlinear analysis such as parameter estimates, their standard errors, hypothesis tests or confidence intervals for the parameters, confidence and prediction intervals for the response, and residuals. It should also allow different fitting methods such as the Gauss-Newton, Newton-Raphson, and Levenberg-Marquardt methods in their appropriately modified forms. An automatic differentiator helps to the user avoid having to code first (Gauss-Newton) and second (Newton-Raphson) derivatives of the mean function with respect to the parameters. The `nlin` procedure of The SAS® System fits these requirements.

In §5.8.1 we analyze a simple data set on sugar cane yields with Mitscherlich's *law of physiological relationships*. The primary purpose is to illustrate a standard nonlinear regression analysis with The SAS® System. But we also provide some additional details on the genesis of this model that is key in agricultural investigations of crop yields as a function of nutrient availability and fit the model in different parameterizations. As discussed in §5.7.1, changing the parameterization of a model changes the statistical properties of the estimators. Ratkowsky's simulation method (Ratkowsky 1983) is implemented for the sugar cane yield data in §5.8.2 to compare the sampling distributions, bias, and excess variance of estimators in different parameterizations of the Mitscherlich equation.

Linear-plateau models play an important role in agronomy. §5.8.3 shows some of the basic operations in fitting a linear-plateau model and its relatives. In §5.8.4 we compare parameters among linear-plateau models corresponding to treatment groups.

Many nonlinear problems require additional programming. A shortcoming of many nonlinear regression packages, `proc nlin` being no exception, is their inability to test linear and nonlinear hypotheses. This is not too surprising since the meaningful set of hypotheses in a nonlinear model depends on the context (model, data, parameterization). For example, the Wald-type F-test of a linear hypotheses H_0:$\mathbf{A}\boldsymbol{\theta} = \mathbf{d}$ in general requires coding the $\mathbf{A}$ matrix and computing the Wald statistic

$$F_{obs}^{(1)} = \left(\mathbf{A}\widehat{\boldsymbol{\theta}} - \mathbf{d}\right)' \left[\mathbf{A}(\mathbf{F}'\mathbf{F})^{-1}\mathbf{A}'\right]^{-1} \left(\mathbf{A}\widehat{\boldsymbol{\theta}} - \mathbf{d}\right) / \left(q\widehat{\sigma}^2\right)$$

in a matrix programming language. Only when the $\mathbf{A}$ matrix has a simple form can the `nlin` procedure be *tricked* in calculating the Walt test directly. The SAS® System provides `proc iml`, an **i**nteractive **m**atrix **l**anguage to perform these tasks as part of the SAS/IML® module. The estimated asymptotic variance covariance matrix $\widehat{\sigma}^2(\mathbf{F}'\mathbf{F})^{-1}$ can be output by `proc nlin` and read into `proc iml`. Fortunately, the `nlmixed` procedure, which was added in Release 8.0 of The SAS® System, has the ability to perform tests of linear and nonlinear combinations of the model parameters, eliminating the need for additional matrix programming. We demonstrate the Wald test for treatment comparisons in §5.8.5 where a nonlinear response is analyzed in a $2 \times 3 \times 6$ factorial design. We analyze the factorial testing for main effects and interactions, and perform pairwise treatment comparisons based on the nonlinear parameters akin to multiple comparisons in a linear analysis of variance model. The `nlmixed` procedure can be used there to formulate contrasts efficiently.

Dose-response models such as the logistic or log-logistic models are among the most frequently used nonlinear equations. Although they offer a great deal of flexibility, they are no panacea for every data set of dose responses. One limitation of the logistic-type models, for example, is that the response monotonically increases or decreases. Hormetic effects, where small dosages of an otherwise toxic substance can have beneficial effects, can throw off a dose-response investigation with a logistic model considerably. In §5.8.6 we provide details on how to construct hormetic models and examine a data set used by Schabenberger et al. (1999) to compare effective dosages among two herbicides, where for a certain weed species one herbicide induces a hormetic response while the other does not.

Yield-density models are a special class of models closely related to linear models. Most yield-density models can be linearized. In §5.8.7 we fit yield-density models to a data set by Mead (1970) on the yield of different onion varieties. Tests of hypotheses comparing the varieties as well as estimation of the genetic and environmental potentials are key in this investigation, which we carry out using the `nlmixed` procedure.

The homogeneous variance assumption is not always tenable in nonlinear regression analyses just as it is not tenable for many linear models. Transformations that stabilize the variance may destroy other desirable properties of the model and transformations that linearize the model do not necessarily stabilize the variance (§5.6). In the case of heterogeneous error variances we prefer to use weighted nonlinear least squares instead of transformations. In §5.8.8 we apply weighted nonlinear least squares to a growth modeling problem and employ a grouping approach to determine appropriate weights.

5.8.1 Basic Nonlinear Analysis with The SAS® System — Mitscherlich's Yield Equation

In this section we analyze a small data set on the yields of sugar cane as a function of nitrogen fertilization. The purpose here is not to draw precise inferences and conclusions about the relationship between crop yield and fertilization, but to demonstrate the steps in fitting a nonlinear model with `proc nlin` of The SAS® System beginning with the derivation of starting values to the actual fitting, the testing of hypotheses, and the examination of the effects of parameterizations. The data for this exercise are shown in Table 5.4.

Table 5.4. Sugar cane yield in randomized complete block design with five blocks and six levels of nitrogen fertilization

Nitrogen	Block					Treatment
(kg/ha)	1	2	3	4	5	sample mean
0	89.49	54.56	74.33	78.20	61.51	71.62
25	108.78	102.01	105.04	105.23	106.52	105.51
50	136.28	129.51	132.54	132.73	134.02	133.01
100	157.63	167.39	155.39	146.85	155.81	156.57
150	185.96	176.66	178.53	195.34	185.56	184.41
200	195.09	190.43	183.52	180.99	205.69	191.14

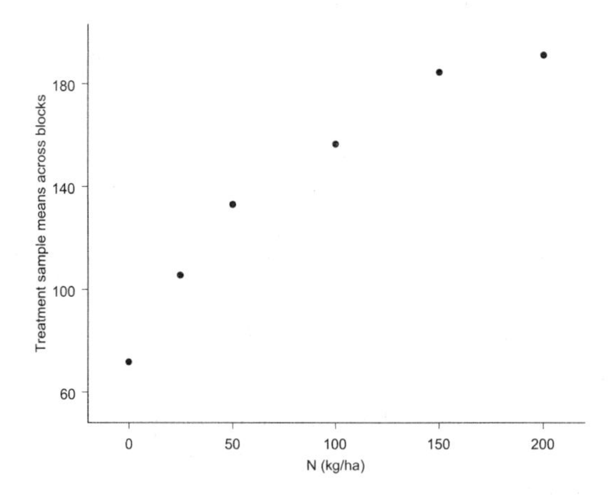

Figure 5.14. Treatment average sugar cane yield vs. nitrogen level applied.

The averages for the nitrogen levels calculated across blocks monotonically increase in the amount of N applied (Figure 5.14). The data do not indicate a decline or a maximum yield at some N level within the range of fertilizer applied nor do they indicate a linear-plateau relationship, since the yield does not appear constant for any level of nitrogen fertilization. Rather, the maximum yield is approached asymptotically. At the control level (0 N), the average yield is of course not 0; it corresponds to the natural fertility of the soil.

Liebig's *law of constant returns*, which implies a linear-plateau relationship (see §5.1), certainly does not apply in this situation. Mitscherlich (1909) noticed by studying experimental data, that the rate of yield (y) increase is often not constant but changes with the amount of fertilizer applied (x). Since the yield relationships he examined appeared to approach some maximum value α asymptotically, he postulated that the rate of change $\partial y/\partial x$ is proportional to the difference between the maximum (α) and the current yield. Consequently, as yield approaches its maximum, the rate of change with fertilizer application approaches zero. Mathematically this relationship between yield y and fertilizer amount x is expressed through the generating equation

$$\partial y/\partial x = \kappa(\alpha - y). \qquad [5.50]$$

The parameter κ is the proportionality constant that Mitscherlich called the effect-factor (*Wirkungsfaktor* in the original publication which appeared in German). The larger κ, the faster yield approaches its asymptote. Solving this generating equation leads to various mathematical forms (parameterizations) of the Mitscherlich equation that are known under different names. Four of them are given in §5.7; a total of eight parameterizations are presented in §A5.9.1. We prefer to call simply the Mitscherlich *equation* what has been termed Mitscherlich's *law of physiological relationships*. Two common forms of the equation are

$$\begin{aligned} y(x) &= \alpha(1 - \exp\{-\kappa(x - x_0)\}) \\ y(x) &= \alpha + (\xi - \alpha)\exp\{-\kappa x\}. \end{aligned} \qquad [5.51]$$

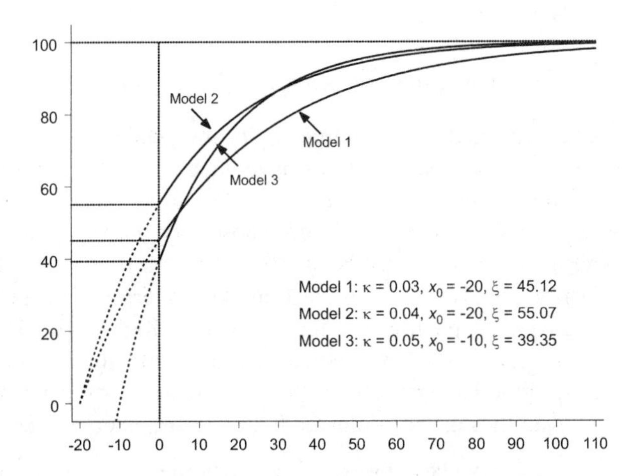

Figure 5.15. Mitscherlich equations for different values of κ, x_0 and ξ, $\alpha = 100$.

Both equations are three parameter models (Figure 5.15). In the first the parameters are α, κ, and x_0; in the second equation the parameters are α, ξ, and κ. α represents the asymptotic yield and x_0 is the fertilizer concentration at which the yield is 0, i.e., $y(x_0) = 0$. Black (1993, p. 273) calls $-x_0$ the availability index of the nutrient in the soil (and seed) when none is added in the fertilizer or as Mead et al. (1993, p. 264) put it, "the amount of fertilizer already in the soil." Since $x = 0$ fertilizer is the minimum that can be *applied*, x_0 is obtained by extrapolating the yield-nutrient relationship below the lowest rate to the point where yield is exactly zero (Figure 5.15). This assumes that the Mitscherlich equation extends past the lowest level applied, which may not be the case. We caution therefore against attaching too much validity to the parameter x_0. The second parameterization replaces the parameter x_0 with the yield that is obtained if no fertilizer is added, $\xi = y(0)$. The relationship between the two parameterizations is

$$\xi = \alpha(1 - \exp\{\kappa x_0\}).$$

In both model formulas, the parameter κ is a scale parameter that **governs** the rate of change. It is not **the** rate of change as is sometimes stated. Figure 5.15 shows three Mitscherlich equations with asymptote $\alpha = 100$ that vary in κ and x_0. With increasing κ, the asymptote is reached more quickly (compare Models 1 and 2 in Figure 5.15). It is also clear from the figure that x_0 is an extrapolated value.

One of the methods for finding starting values in a nonlinear model that was outlined in §5.4.3 relies on the expected value parameterization of the model (Ratkowsky 1990, Ch. 2.3.1). Here we choose values of the regressor variables and rewrite the model in terms of the mean response at those values. We call these expected value parameters since they correspond to the means at the particular values of the regressors that were chosen. For each regressor value for which an expected value parameter is obtained, one parameter of the original model is replaced. An expected value parameterization for the Mitscherlich model due to Schnute and Fournier (1980) is

$$\begin{aligned} y(x) &= \mu^* + (\mu^{**} - \mu^*)(1 - \theta^{m-1})/(1 - \theta^{n-1}) \\ m - 1 &= (n-1)(x - x^*)/(x^{**} - x^*) \\ n &= \text{number of observations.} \end{aligned} \qquad [5.52]$$

Here, μ^* and μ^{**} are the expected value parameters for the yield at nutrient levels x^* and x^{**}, respectively. Expected value parameterizations have advantages and disadvantages. Finding starting values is particularly simple if a model is written in terms of expected value parameters. If, for example, $x^* = 25$ and $x^{**} = 150$ are chosen for the sugar cane data, reasonable starting values (from Fig. 5.14) are identified as $\mu^{*0} = 120$ and $\mu^{**0} = 175$. Models in expected value parameterization are also closer to linear models in terms of the statistical properties of the estimators because of low parameter-effects curvature. Ratkowsky (1990) notes that the Mitscherlich model is notorious for high parameter-effects curvature which gives particular relevance to [5.52]. A disadvantage is that the interpretation of parameters in terms of physical or biological quantities is lost compared to other parameterizations.

Starting values for the standard forms of the Mitscherlich model can also be found relatively easily by using the various devices described in §5.4.3. Consider the model

$$y(x) = \alpha(1 - \exp\{-\kappa(x - x_0)\}).$$

Since α is the upper asymptote, Figure 5.14 would suggest a starting value of $\alpha^0 = 200$. Once α is fixed we can rewrite the model as

$$\ln\{\alpha^0 - y\} = \ln\{\alpha\} + \kappa x_0 - \kappa x.$$

This is a linear regression with response $\ln\{\alpha^0 - y\}$, intercept $\ln\{\alpha\} + \kappa x_0$, and slope κ. For the averages of the sugar cane yield data listed in Table 5.4 and graphed in Figure 5.14 we can use `proc reg` in The SAS® System to find a starting value of κ by fitting the linear regression.

```
data CaneMeans; set CaneMeans;
  y2 = log(200-yield);
run;
proc reg data=CaneMeans; model y2 = nitro; run; quit;
```

From Output 5.8 we gather that a reasonable starting value is $\kappa^0 = 0.01356$. We deliberately ignore all other results from this linear regression since the value $\alpha^0 = 200$ that was substituted to enable a linearization by taking logarithms was only a guess. Finally, we need a starting value for x_0, the nutrient concentration at which the yield is 0. Visually extrapolating the response trend in Figure 5.14, a value of $x_0^0 = -25$ seems not unreasonable as a first guess.

Output 5.8.

```
                         The REG Procedure
                           Model: MODEL1
                      Dependent Variable: y2

                       Analysis of Variance

                               Sum of           Mean
Source               DF       Squares         Square    F Value    Pr > F

Model                 1       5.45931        5.45931     320.02    <.0001
Error                 4       0.06824        0.01706
Corrected Total       5       5.52755

           Root MSE              0.13061    R-Square     0.9877
           Dependent Mean        3.71778    Adj R-Sq     0.9846
           Coeff Var             3.51315

                       Parameter Estimates

                        Parameter        Standard
   Variable      DF      Estimate           Error    t Value    Pr > |t|

   Intercept      1       4.90434         0.08510      57.63      <.0001
   nitro          1      -0.01356      0.00075804     -17.89      <.0001
```

Now that starting values have been assembled we fit the nonlinear regression model with `proc nlin`. The statements that accomplish this in the parameterization for which the starting values were obtained are

```
proc nlin data=CaneMeans method=newton;
  parameters alpha=200 kappa=0.0136 nitro0=-25;
  Mitscherlich = alpha*(1-exp(-kappa*(nitro-nitro0)));
  model yield = Mitscherlich;
run;
```

The `method=newton` option of the `proc nlin` statement selects the Newton-Raphson algorithm. If the `method=` option is omitted, the procedure defaults to the Gauss-Newton method. In either case `proc nlin` implements not the unmodified algorithms but provides internally necessary modifications such as step-halving that stabilize the algorithm. Among other fitting methods that can be chosen are the Marquardt-Levenberg algorithm (`method=Marquardt`) that is appropriate if the columns of the derivative matrix are highly correlated (Marquardt 1963). Prior to Release 6.12 of The SAS® System the user had to specify first derivatives of the mean function with respect to all parameters for the Gauss-Newton method and first and second derivatives for the Newton-Raphson method. To circumvent the specification of derivatives, one could use `method=dud` which invoked a derivative-free method (Ralston and Jennrich 1978). This acronym stands for ***Does** not **U**se **D**erivatives* and the method enjoyed popularity because of this feature. The numerical properties of this algorithm are typically poor and the algorithm is not efficient in terms of computing time. Since The SAS® System calculates derivatives automatically starting with Release 6.12, the DUD method should no longer be used. There is no justification in our opinion for using a method that approximates derivatives over one that determines the actual derivatives. Even in newer releases the user can still enter derivatives through `der.` statements of `proc nlin`. For the Mitscherlich model above one would code, for example,

```
proc nlin data=CaneMeans method=gauss;
  parameters alpha=200 kappa=0.0136 nitro0=-25;
  Mitscherlich = alpha*(1-exp(-kappa*(nitro-nitro0)));
  model yield = Mitscherlich;
  der.alpha = 1 - exp(-kappa * (nitro - nitro0));
  der.kappa = alpha * ((nitro - nitro0) * exp(-kappa * (nitro - nitro0)));
  der.nitro0 = alpha * (-kappa * exp(-kappa * (nitro - nitro0)));
run;
```

to obtain a Gauss-Newton fit of the model. Not only is the added programming not worth the trouble and mistakes in coding the derivatives are costly, the built-in differentiator of `proc nlin` is of such high quality that we recommend allowing The SAS® System to determine the derivatives. If the user wants to examine the derivatives used by SAS®, add the two options `list listder` to the `proc nlin` statement.

Finally, the results of fitting the Mitscherlich model by the Newton-Raphson method in the parameterization

$$y_i = \alpha(1 - \exp\{-\kappa(x_i - x_0)\}) + e_i,\ i = 1,\cdots,6,$$

where the e_i are uncorrelated random errors with mean 0 and variance σ^2, with the statements

```
proc nlin data=CaneMeans method=newton noitprint;
  parameters alpha=200 kappa=0.0136 nitro0=-25;
  Mitscherlich = alpha*(1-exp(-kappa*(nitro-nitro0)));
  model yield = Mitscherlich;
run;
```

shown as Output 5.9. The procedure converges after six iterations with a residual sum of squares of $S(\widehat{\boldsymbol{\theta}}) = 57.2631$. The model fits the data very well as measured by

$$\text{Pseudo-}R^2 = 1 - \frac{57.2631}{10,775.8} = 0.947.$$

The converged iterates (the parameter estimates) are

$$\widehat{\boldsymbol{\theta}} = [\widehat{\alpha}, \widehat{\kappa}, \widehat{x}_0]' = [205.8, 0.0112, -38.7728]',$$

from which a prediction of the mean yield at fertilizer level 60 $kg \times ha^{-1}$, for example, can be obtained as

$$\widehat{y}(60) = 205.8*(1 - \exp\{-0.0112*(60 + 38.7728)\}) = 137.722.$$

Output 5.9.

```
                    The NLIN Procedure
                     Iterative Phase
                Dependent Variable yield
                     Method: Newton

     NOTE: Convergence criterion met.

                    Estimation Summary

              Method                         Newton
              Iterations                          6
              R                            8.93E-10
              PPC                           6.4E-11
              RPC(kappa)                   7.031E-6
              Object                       6.06E-10
              Objective                    57.26315
              Observations Read                   6
              Observations Used                   6
              Observations Missing                0

                                 Sum of        Mean               Approx
Source                   DF     Squares      Square    F Value    Pr > F

Regression                3      128957     42985.6     280.77    0.0004
Residual                  3     57.2631     19.0877
Uncorrected Total         6      129014
Corrected Total           5     10775.8

                                 Approx      Approximate 95% Confidence
     Parameter      Estimate    Std Error             Limits

     alpha             205.8       8.9415        177.3         234.2
     kappa            0.0112      0.00186      0.00529        0.0171
     nitro0         -38.7728       6.5556     -59.6360      -17.9095

               Approximate Correlation Matrix

                     alpha             kappa            nitro0
          alpha   1.0000000        -0.9300182        -0.7512175
          kappa  -0.9300182         1.0000000         0.9124706
          nitro0 -0.7512175         0.9124706         1.0000000
```

For each parameter in the `parameters` statement `proc nlin` lists its estimate, (asymptotic) estimated standard error, and (asymptotic) 95% confidence interval. For example, $\widehat{\alpha} = 205.8$ with $\text{ease}(\widehat{\alpha}) = 8.9415$. The asymptotic 95% confidence interval for α is calculated as

$$\widehat{\alpha} \pm t_{0.025,3} \times \text{ease}(\widehat{\alpha}) = 205.8 \pm 3.182 \times 8.9415 = [177.3, 234.2]\,.$$

Based on this interval one would, for example, reject the hypothesis that the upper yield asymptote is 250 and fail to reject the hypothesis that the asymptote is 200.

The printout of the `Approximate Correlation Matrix` lists the estimated correlation coefficients between the parameter estimates,

$$\widehat{\text{Corr}}\left[\widehat{\theta}_j, \widehat{\theta}_k\right] = \frac{\text{Cov}\left[\widehat{\theta}_j, \widehat{\theta}_k\right]}{\sqrt{\text{Var}\left[\widehat{\theta}_j\right]\text{Var}\left[\widehat{\theta}_k\right]}}.$$

The parameter estimators are fairly highly correlated, $\text{Corr}[\widehat{\alpha}, \widehat{\kappa}] = -0.93$, $\text{Corr}[\widehat{\alpha}, \widehat{x}_0] =$

-0.75, $\text{Corr}[\widehat{\kappa}, \widehat{x}_0] = 0.912$. Studying the derivatives of the Mitscherlich model in this parameterization, this is not surprising. They all involve the same term

$$\exp\{-\kappa(x - x_0)\}.$$

Highly correlated parameter estimators are indicative of poor conditioning of the $\mathbf{F}'\mathbf{F}$ matrix and can cause instabilities during iterations. In the presence of large correlations one should switch to the Marquardt-Levenberg algorithm or change the parameterization. Below we will see how the expected value parameterization leads to considerably smaller correlations.

If the availability index $-x_0$ is of lesser interest than the control yield one can obtain an estimate of $\widehat{y}(0) = \widehat{\xi}$ from the parameter estimates in Output 5.9. Since $\xi = \alpha(1 - \exp\{\kappa x_0\})$, we simply substitute estimates for the unknowns and obtain

$$\widehat{\xi} = \widehat{\alpha}(1 - \exp\{\widehat{\kappa}\widehat{x}_0\}) = 205.8(1 - \exp\{-0.0112*38.7728\}) = 72.5.$$

Although it is easy to obtain the point estimate of ξ, it is not a simple task to calculate the standard error of this estimate needed for a confidence interval, for example. Two possibilities exist to accomplish that. One can refit the model in the parameterization

$$y_i = \alpha + (\xi - \alpha)\exp\{-\kappa x_i\} + e_i,$$

that explicitly involves ξ. `proc nlin` calculates approximate standard errors and 95% confidence intervals for each parameter. The second method uses the capabilities of `proc nlmixed` to estimate the standard error of nonlinear functions of the parameters by the delta method. We demonstrate both approaches.

The statements

```
proc nlin data=CaneMeans method=newton noitprint;
  parameters alpha=200 kappa=0.0136 ycontrol=72;
  Mitscherlich = alpha + (ycontrol - alpha)*(exp(-kappa*nitro));
  model yield = Mitscherlich;
run;
```

fit the model in the new parameterization (Output 5.10). The quality of the model fit has not changed from the first parameterization in terms of x_0. The analysis of variance tables in Outputs 5.9 and 5.10 are identical. Furthermore, the estimates of α and κ and their standard errors have not changed. The parameter labeled `ycontrol` now replaces the term `nitro0` and its estimate agrees with the calculation based on the estimates of the model in the first parameterization. From Output 5.10 we are able to state that with (approximately) 95% confidence the interval $[59.625, 85.440]$ contains the control yield.

Using `proc nlmixed`, the parameterization of the model need not be changed in order to obtain an estimate and a confidence interval of the control yield:

```
proc nlmixed data=CaneMeans df=3 technique=NewRap;
  parameters alpha=200 kappa=0.0136 nitro0=-25;
  s2 = 19.0877;
  Mitscherlich = alpha*(1-exp(-kappa*(nitro-nitro0)));
  model yield ~ normal(Mitscherlich,s2);
  estimate 'ycontrol' alpha*(1-exp(kappa*nitro0));
run;
```

The variance of the error distribution is fixed at the residual mean square from the earlier fits (see Output 5.10). Otherwise `proc nlmixed` will estimate the residual variance by maximum likelihood which would not correspond to an analysis equivalent to what is shown in Outputs 5.9 and 5.10. The `df=3` option in the `proc nlmixed` statement ensures that the procedure uses the same residual degrees of freedom as `proc nlin`. The procedure converged in six iterations to parameter estimates identical to those in Output 5.9 (see Output 5.11). The estimate of the control yield shown under `Additional Estimates` is identical to that in Output 5.10.

Output 5.10.

```
                         The NLIN Procedure
                          Iterative Phase
                      Dependent Variable yield
                          Method: Newton

     NOTE: Convergence criterion met.

                        Estimation Summary

                Method                          Newton
                Iterations                           5
                R                             4.722E-8
                PPC(kappa)                    1.355E-8
                RPC(alpha)                    0.000014
                Object                        2.864E-7
                Objective                     57.26315
                Observations Read                    6
                Observations Used                    6
                Observations Missing                 0

                                  Sum of        Mean               Approx
Source                  DF       Squares      Square    F Value    Pr > F

Regression               3        128957     42985.6     280.77    0.0004
Residual                 3       57.2631     19.0877
Uncorrected Total        6        129014
Corrected Total          5       10775.8

                                   Approx      Approximate 95% Confidence
     Parameter     Estimate     Std Error                Limits

     alpha            205.8        8.9415        177.3         234.2
     kappa           0.0112       0.00186      0.00529        0.0171
     ycontrol       72.5329        4.0558      59.6253       85.4405

                Approximate Correlation Matrix
                        alpha              kappa            initial
         alpha      1.0000000         -0.9300182          0.3861973
         kappa     -0.9300182          1.0000000         -0.5552247
         initial    0.3861973         -0.5552247          1.0000000
```

Finally, we fit the Mitscherlich model in the expected value parameterization [5.52] and choose $x^* = 25$ and $x^{**} = 150$.

```
proc nlin data=CaneMeans method=newton;
  parameters mustar=125 mu2star=175 theta=0.75;
  n       = 6;   xstar  = 25;  x2star = 150;
  m       = (n-1)*(nitro-xstar)/(x2star-xstar) + 1;
  Mitscherlich = mustar + (mu2star-mustar)*(1-theta**(m-1))/
                          (1-theta**(n-1));
  model yield = Mitscherlich;
run;
```

The model fit is identical to the preceding parameterizations (Output 5.12). Notice that the correlations among the parameters are markedly reduced. The estimators of μ^* and μ^{**} are almost orthogonal.

Output 5.11.

```
                         The NLMIXED Procedure

                            Specifications
     Data Set                                       WORK.CANEMEANS
     Dependent Variable                             yield
     Distribution for Dependent Variable            Normal
     Optimization Technique                         Newton-Raphson
     Integration Method                             None

                              Dimensions
                 Observations Used                           6
                 Observations Not Used                       0
                 Total Observations                          6
                 Parameters                                  3
                              Parameters
                 alpha        kappa        nitro0     NegLogLike
                   200       0.0136           -25     22.8642468

                           Iteration History
     Iter      Calls     NegLogLike        Diff      MaxGrad         Slope
        1         10     16.7740728    6.090174     24.52412      -11.0643
        2         15     16.1072166    0.666856     1767.335      -1.48029
        3         20     15.8684503    0.238766     21.15303      -0.46224
        4         25     15.8608402     0.00761      36.2901      -0.01514
        5         30     15.8607649    0.000075     0.008537      -0.00015
        6         35     15.8607649    6.02E-10     0.000046      -1.21E-9

               NOTE: GCONV convergence criterion satisfied.

                            Fit Statistics
                 -2 Log Likelihood                     31.7
                 AIC (smaller is better)               37.7
                 AICC (smaller is better)              49.7
                 BIC (smaller is better)               37.1

                          Parameter Estimates
                         Standard
  Parameter   Estimate      Error   DF   t Value   Pr>|t|       Lower       Upper
  alpha         205.78     8.9496    3     22.99   0.0002      177.30      234.26
  kappa        0.01121   0.001863    3      6.02   0.0092    0.005281     0.01714
  nitro0      -38.7728     6.5613    3     -5.91   0.0097    -59.6539    -17.8917

                         Additional Estimates
                       Standard
  Label       Estimate     Error      DF   t Value   Pr > |t|      Lower   Upper
  ycontrol     72.5329    4.0571       3     17.88     0.0004    59.6215 85.4443
```

To obtain a smooth graph of the response function at a larger number of N concentrations than were applied and approximate 95% confidence limits for the mean predictions, we can use a simple trick. To the data set we append a filler data set containing the concentrations at which the mean sugar cane yield is to be predicted. The response variable is set to missing values in this data set. SAS® will ignore the observations with missing response in fitting the model, but use the observations that have regressor information to calculate predictions. To obtain predictions at nitrogen concentrations between 0 and 200 $kg \times ha^{-1}$ in steps of 2 $kg \times ha^{-1}$ we use the following code.

```
data filler; do nitro=0 to 200 by 2; yield=.; pred=1; output; end; run;
data FitThis; set CaneMeans filler; run;

proc nlin data=FitThis method=newton;
  parameters alpha=200 kappa=0.0136 nitro0=-25;
  Mitscherlich = alpha*(1-exp(-kappa*(nitro-nitro0)));
  model yield = Mitscherlich;
  output out=nlinout predicted=predicted u95m=upperM l95m=lowerM;
run;
proc print data=nlinout(obs=15); run;
```

Output 5.12.

```
                         The NLIN Procedure
                           Iterative Phase
                      Dependent Variable yield
                           Method: Newton
                                                            Sum of
          Iter       mustar       mu2star          theta     Squares
             0        125.0         175.0         0.7500      1659.6
             1        105.3         180.9         0.7500     57.7504
             2        105.1         181.0         0.7556     57.2632
             3        105.1         181.0         0.7556     57.2631
     NOTE: Convergence criterion met.

                         Estimation Summary
                    Method                        Newton
                    Iterations                         3
                    R                           4.211E-8
                    PPC                         3.393E-9
                    RPC(mustar)                 0.000032
                    Object                      1.534E-6
                    Objective                   57.26315
                    Observations Read                  6
                    Observations Used                  6
                    Observations Missing               0

                                 Sum of         Mean                  Approx
Source                  DF      Squares       Square     F Value    Pr > F
Regression               3       128957      42985.6      280.77    0.0004
Residual                 3      57.2631      19.0877
Uncorrected Total        6       129014
Corrected Total          5      10775.8

                                    Approx     Approximate 95% Confidence
     Parameter        Estimate    Std Error               Limits

     mustar              105.1       2.5101      97.1083         113.1
     mu2star             181.0       2.4874        173.1         188.9
     theta              0.7556       0.0352       0.6437        0.8675

                    Approximate Correlation Matrix

                          mustar          mu2star            theta
         mustar        1.0000000        0.0606554       -0.3784466
         mu2star       0.0606554        1.0000000        0.1064661
         theta        -0.3784466        0.1064661        1.0000000
```

The variable `pred` was set to one for observations in the filler data set to distinguish observations from filler data. The `output out=` statement saves predicted values and 95% confidence limits for the mean yield in the data set `nlinout`. The first fifteen observations of the output data set are shown below and a graph of the predictions is illustrated in Figure 5.16. Observations for which `pred=.` are the observations to which the model is fitted, observations with `pred=1` are the filler data.

Output 5.13.

Obs	nitro	yield	pred	PREDICTED	LOWERM	UPPERM
1	0	71.62	.	72.533	59.626	85.440
2	25	105.51	.	105.097	97.108	113.085
3	50	133.01	.	129.702	120.515	138.889
4	100	156.57	.	162.343	153.875	170.812
5	150	184.41	.	180.980	173.064	188.896
6	200	191.14	.	191.621	179.874	203.368
7	0	.	1	72.533	59.626	85.440
8	2	.	1	75.487	63.454	87.520
9	4	.	1	78.375	67.128	89.623
10	6	.	1	81.200	70.647	91.752
11	8	.	1	83.961	74.015	93.908
12	10	.	1	86.662	77.232	96.092
13	12	.	1	89.303	80.302	98.303
14	14	.	1	91.885	83.229	100.540
15	16	.	1	94.410	86.020	102.799

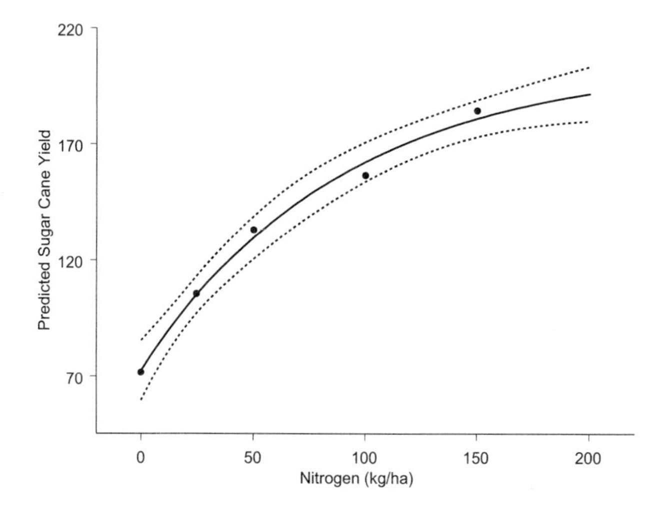

Figure 5.16. Predicted sugar cane yields and approximate 95% confidence limits.

5.8.2 The Sampling Distribution of Nonlinear Estimators — the Mitscherlich Equation Revisited

In §5.7.1 we discussed that strong intrinsic and parameter-effects curvature in a nonlinear model can have serious detrimental effects on parameter estimators, statistical inference, the behavior of residuals, etc. To recall the main results of that section, intrinsic curvature does not affect the fit of the model in the sense that different parameterizations produce the same set of residuals and leave unchanged the analysis of variance decomposition. Changing the parameterization of the model does affect the parameter-effects curvature of the model, however. Parameter estimators in nonlinear models with low curvature behave like their counterparts in the linear model. They are approximately Gaussian-distributed even if the model errors are not, they are asymptotically unbiased and have minimum variance. In nonlinear models some bias will always be incurred due to the nature of the process. Ratkowsky (1983) focuses on the parameter-effects curvature of nonlinear models in particular because (i) he contends that it is the larger of the two curvature components, and (ii) because it can be influenced by the modeler through changing the parameterization. The expected value parameterizations he developed supposedly lead to models with low parameter effects curvature and we expect the estimators in these models to exhibit close-to-linear behavior.

Measures of curvature are not easily computed as they depend on second derivatives of the mean function (Beale 1960, Bates and Watts 1980, Seber and Wild 1989, Ch. 4). A relatively simple approach that allows one to examine the effects of curvature on the distribution of the parameter estimators was given in Ratkowsky (1983). It relies on simulating the sampling distribution of $\widehat{\boldsymbol{\theta}}$ and calculating test statistics to examine bias, excess variability, and non-Gaussianity of the estimators. We now discuss and apply these ideas to the sugar cane yield data of §5.8.1. The two parameterizations we compare are one of the standard equations,

$$\mathrm{E}[Y_i] = \alpha(1 - \exp\{-\kappa(x_i - x_0)\}),$$

and the expected value parameterization [5.52].

To outline the method we focus on the first parameterization, but similar steps are taken for any parameterization. First, the model is fit to the data at hand (Table 5.4, Figure 5.14) and the nonlinear least squares estimates $\widehat{\boldsymbol{\theta}} = [\widehat{\theta}_1, \cdots, \widehat{\theta}_p]'$ and their estimated asymptotic standard errors are obtained along with the residual mean square. These values are shown in Output 5.9 (p. 243). For example, $MSR = 19.0877$. Then simulate K data sets, keeping the regressor values as in Table 5.4, and set the parameters of the model equal to the estimates $\widehat{\boldsymbol{\theta}}$. The error distribution is chosen to be Gaussian with mean zero and variance MSR. Ratkowsky (1983) recommends selecting K fairly large, $K = 1000$, say. For each of these K data sets the model is fit by nonlinear least squares and we obtain parameter vectors $\widehat{\boldsymbol{\theta}}_1, \cdots, \widehat{\boldsymbol{\theta}}_K$. Denote by $\widehat{\theta}_{ij}$ the $j = 1, \cdots, K^{\text{th}}$ estimate of the ith element of $\boldsymbol{\theta}$ and by $\widehat{\theta}_i$ the ith element of $\widehat{\boldsymbol{\theta}}$ from the fit of the model to the original (nonsimulated) data. The relative bias in estimating θ_i is calculated as

$$\text{RelativeBias\%}_i = 100 \times \left(\frac{\widehat{\theta}_{i.} - \widehat{\theta}_i}{\widehat{\theta}_i} \right), \qquad [5.53]$$

where $\widehat{\theta}_{i.}$ is the average of the estimates $\widehat{\theta}_{ij}$ across the K simulations, i.e.,

$$\widehat{\theta}_{i.} = \frac{1}{K} \sum_{j=1}^{K} \widehat{\theta}_{ij}.$$

If the curvature is strong, the estimated variance of the parameter estimators is an underestimate. Similar to the relative bias we calculate the relative excess variance as

$$\text{RelativeExcessVariance\%}_i = 100 \times \left(\frac{s_i^2 - \widehat{\text{Var}}\left[\widehat{\theta}_i\right]}{\widehat{\text{Var}}\left[\widehat{\theta}_i\right]} \right). \qquad [5.54]$$

Here, s_i^2 is the sample variance of the estimates for θ_{ij} in the simulations,

$$s_i^2 = \frac{1}{K-1} \sum_{j=1}^{n} \left(\widehat{\theta}_{ij} - \widehat{\theta}_{i.} \right)^2,$$

and $\widehat{\text{Var}}[\widehat{\theta}_i]$ is the estimated asymptotic variance of $\widehat{\theta}_i$ from the original fit. Whether the relative bias and the excess variance are significant can be tested by calculating test statistics

$$Z^{(1)} = \sqrt{K} \frac{\widehat{\theta}_{i.} - \widehat{\theta}_i}{\widehat{\text{Var}}\left[\widehat{\theta}_i\right]^{1/2}}$$

$$Z^{(2)} = \sqrt{2Ks_i^2 / \widehat{\text{Var}}\left[\widehat{\theta}_i\right]} - \sqrt{2(K-1) - 1}.$$

$Z^{(1)}$ and $Z^{(2)}$ are compared against cutoffs of a standard Gaussian distribution to determine the significance of the bias ($Z^{(1)}$) or the variance excess ($Z^{(2)}$). This seems like a lot of trouble to determine whether model curvature induces bias and excess variability of the coefficients, but it is fairly straightforward to implement the process with The SAS® System. The complete code including tests for Gaussianity, histograms of the parameter estimates in the simulations, and statistical tests for excess variance and bias can be found on the CD-ROM.

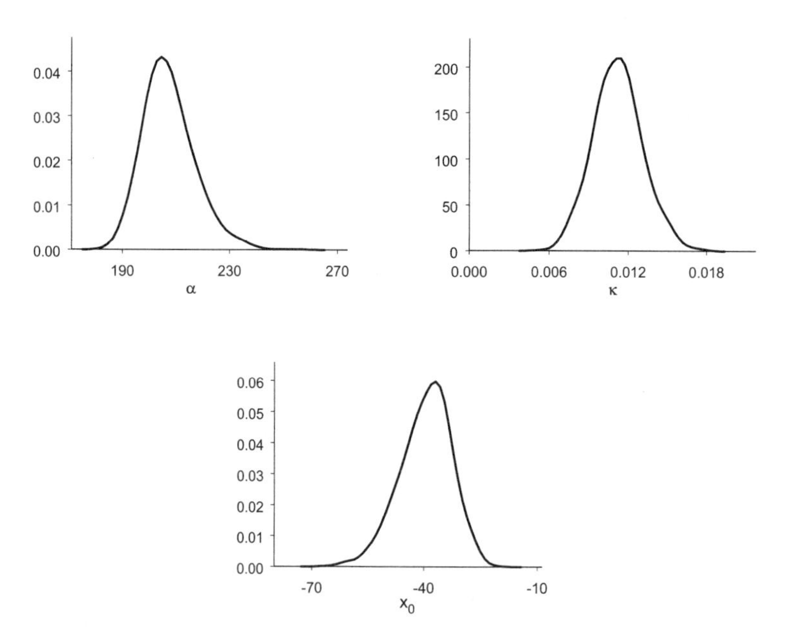

Figure 5.17. Sampling distribution of parameter estimates when fitting Mitscherlich equation in parameterization $\text{E}[Y_i] = \alpha(1 - \exp\{-\kappa(x_i - x_0)\})$ to data in Table 5.4.

We prefer to smooth the sample histograms of the estimates with a nonparametric kernel estimator (§4.7). Figure 5.17 shows the smoothed sampling distributions of the estimates in the standard parameterization of the Mitscherlich equation for $K = 1000$ and Figure 5.18 displays the expected value parameterization. In Figure 5.17 the parameter estimates of α, the yield asymptote, and x_0, the index of availability, appear particularly skewed. The distribution of the former is skewed to the right, that of the latter is skewed to the left. The distribution of the estimates in the expected value parameterization are much less skewed (Figure 5.18). Whether the deviation from a Gaussian distribution is significant can be assessed with a test for normality (Table 5.5). In the standard parameterization all parameter estimates deviate significantly (at the 5% level) from a Gaussian distribution, even $\widehat{\kappa}$, which appears quite symmetric in Figure 5.17. In the expected value parameterization the null hypothesis of a Gaussian distribution cannot be rejected for any of the three parameters.

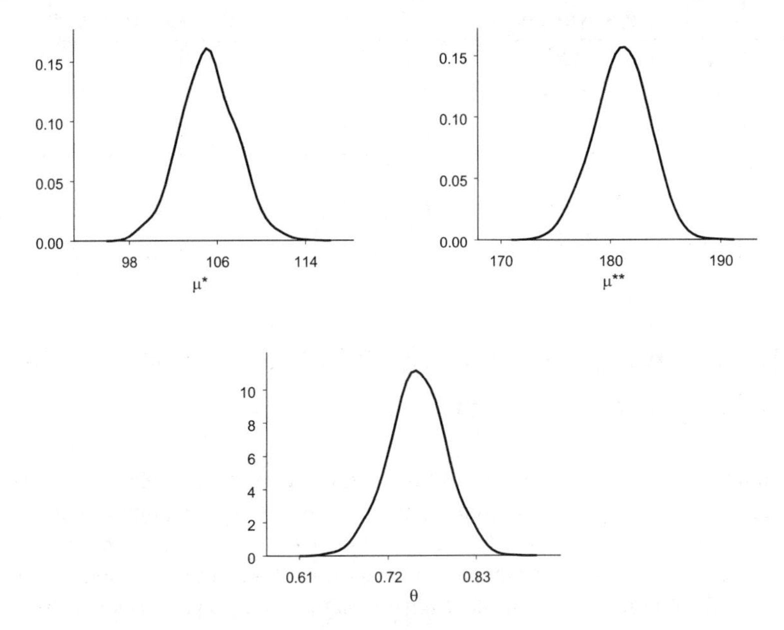

Figure 5.18. Sampling distribution of parameter estimates when fitting Mitscherlich equation in expected value parameterization to data in Table 5.4.

Table 5.5. P-values of tests of Gaussianity for standard and expected value parameterization of Mitscherlich equation fitted to data in Table 5.4

	Parameter Estimate					
	$\widehat{\alpha}$	$\widehat{\kappa}$	$\widehat{x}_0$	$\widehat{\mu}^*$	$\widehat{\mu}^{**}$	θ
***P*-value**	< 0.0001	0.027	< 0.0001	0.432	0.320	0.713

Asymptotic inferences relying on the Gaussian distribution of the estimators seem questionable for this data set based on the normality tests. How about the bias and the variance excesses? From Table 5.6 it is seen that in the standard parameterization the parameters α and x_0 are estimated rather poorly, and their bias is highly significant. Not only are the distributions of these two estimators not Gaussian (Table 5.5), they are also centered at the wrong values. Particularly concerning is the large excess variance in the estimate of the asymptotic yield α. Nonlinear least squares estimation underestimates the variance of this parameter by 15.1%. Significance tests about the yield asymptote should be interpreted with the utmost caution. The expected value parameterization fairs much better. Its relative biases are an order of magnitude smaller than the biases in the standard parameterization and not significant. The variance estimates of $\widehat{\mu}^*$, $\widehat{\mu}^{**}$, and $\widehat{\theta}$ are reliable as shown by the small excess variance and the large p-values.

Table 5.6. Bias [5.53] and variance [5.54] excesses for standard and expected value parameterization of Mitscherlich equation fitted to data in Table 5.4

	Statistic and P-Value				
	$\widehat{\theta}_{i.}$	**RelativeBias%**	P	**RelativeExcessVariance%**	P
$\widehat{\alpha}$	207.4	0.80	< 0.0001	15.10	0.0009
$\widehat{\kappa}$	0.011	– 0.02	0.98	0.48	0.89
$\widehat{x}_0$	– 39.5	1.96	0.0002	5.12	0.24
$\widehat{\mu}^*$	105.2	0.09	0.23	0.10	0.95
$\widehat{\mu}^{**}$	180.9	– 0.002	0.97	– 2.00	0.66
$\widehat{\theta}$	0.756	0.113	0.44	– 0.16	0.99

5.8.3 Linear-Plateau Models and Their Relatives — a Study of Corn Yields from Tennessee

In the introduction to this chapter we mentioned the linear-plateau model (see Figure 5.1) as a manifestation of Liebig's law of the minimum where the rate of change in plant response to changes in the availability of a nutrient is constant until some concentration is reached at which other nutrients become limiting and the response attains a plateau. Plateau-type models are not only applicable in studies of plant nutrition, such relationships can be found in many other situations.

Watts and Bacon (1974) present data from an experiment where sediment was agitated in a tank of fluid. After agitation stopped (time $t = 0$) the height of the clear zone above the sediment was measured for the next 150 minutes (Figure 5.19). The height of the clear zone could be modeled as a single function of the time after agitation or two separate functions could be combined, one describing the initial upward trend, the other flattened behavior to the right. The point at which the switch between the two functions occurs is generally called a change-point. If the two functions connect, it is also termed a **join-point**. Watts and Bacon model the relationship between height of the clear zone and time after agitation ceased with a variation of the hyperbolic model,

$$\mathrm{E}[Y_i] = \beta_0 + \beta_1(t_i - \alpha) + \beta_2(t_i - \alpha)\tanh\left\{\frac{t_i - \alpha}{\gamma}\right\} + e_i.$$

Here, α is the change-point parameter and γ determines the radius of curvature at the change point. The two functions connect smoothly.

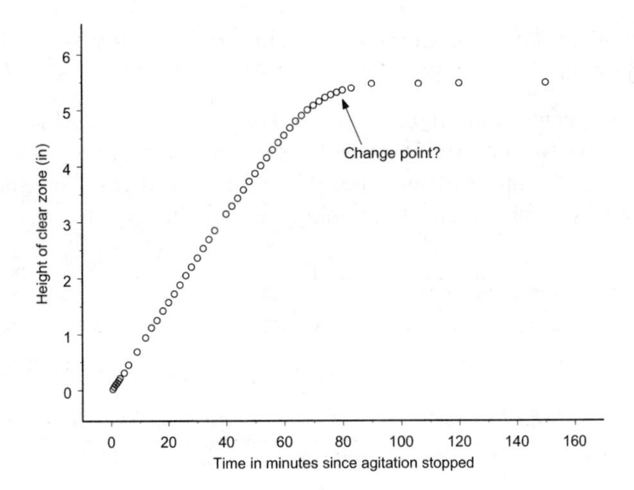

Figure 5.19. Sediment settling data based on Table 2 in Watts and Bacon (1974). Reprinted with permission from *Technometrics*. Copyright © 1974 by the American Statistical Association. All rights reserved.

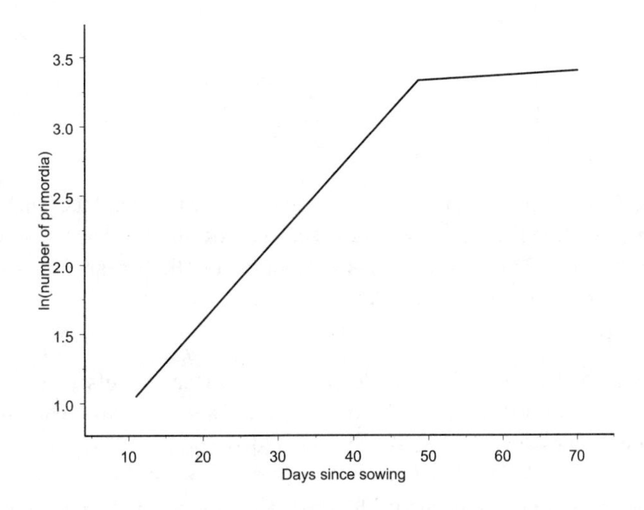

Figure 5.20. Segmented linear trend for Kirby's wheat shoot apex data (Kirby 1974 and Lerman 1980). Adapted with permission from estimates reported by Lerman (1980). Copyright © 1980 by the Royal Statistical Society.

Linear-plateau models are special cases of these segmented models where the transition of the segments is not smooth, there is a kink at the join-point. They are in fact special cases of the **linear-slope** models that connect two linear segments. Kirby (1974) examined the shoot-apex development in wheat where he studied the natural logarithm of the number of primordia as a function of days since sowing (Figure 5.20). Arguing on biological grounds it was believed that the increase in ln(# primordia) slows down sharply (abruptly) at the end of spikelet initiation which can be estimated from mature plants. The kink this creates in the response is obvious in the model graphed in Figure 5.20 which was considered by Lerman (1980) for Kirby's data.

If the linear segment on the right-hand side has zero slope, we obtain the linear-plateau model. Anderson and Nelson (1975) studied various segmented models for crop yield as a function of fertilizer, the linear-plateau model being a special case. We show some of these models in Figure 5.21 along with the terminology used in the sequel.

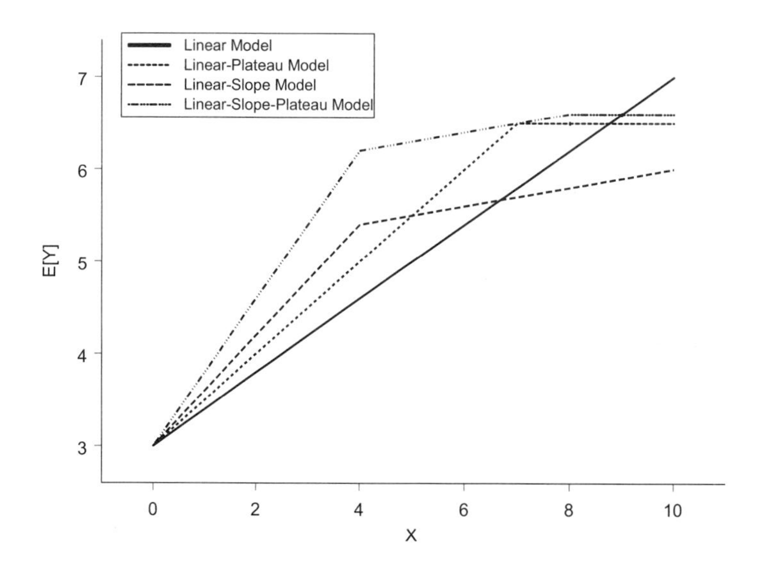

Figure 5.21. Some members of the family of linear segmented models. The linear-slope model (LS) joins two line segments with non-zero slopes, the linear-plateau model (LP) has two line segments, the second of which has zero slope and the linear-slope-plateau model (LSP) has two line segments that connect to a plateau.

Anderson and Nelson (1975) consider the fit of these models to two data sets of corn yields from twenty-two locations in North Carolina and ten site-years in Tennessee. We repeat part of the Tennessee data in Table 5.7 (see also Figure 5.22).

Table 5.7. Tennessee average corn yields for two locations and three years as a function of nitrogen ($\text{kg} \times \text{ha}^{-1}$) based on experiments of Engelstad and Parks (1971)

	Knoxville			**Jackson**		
N ($\text{kg} \times \text{ha}^{-1}$)	**1962**	**1963**	**1964**	**1962**	**1963**	**1964**
0	44.6	45.1	60.9	46.5	29.3	28.8
67	73.0	73.2	75.9	59.0	55.2	37.6
134	75.2	89.3	83.7	71.9	77.3	55.2
201	83.3	91.2	84.3	73.1	88.0	66.8
268	78.4	91.4	81.8	74.5	89.4	67.0
355	80.9	88.0	84.5	75.5	87.0	67.8

Data appeared in Anderson and Nelson (1975). Used with permission of the International Biometric Society.

Graphs of the yields for the two locations and three years are shown in Figure 5.22. Anderson and Nelson (1975) make a very strong case for linear-plateau models and their relatives in studies relating crop yields to nutrients or fertilizers. They show that using a quadratic polynomial when a linear-plateau model is appropriate can lead to seriously flawed inferences (see Figure 5.1, p. 186). The quadratic response model implies a maximum (or minimum) at some concentration while the linear-plateau model asserts that yields remain constant past a critical concentration. Furthermore, the maximum yield achieved under the quadratic model tends to be too large, with positive bias. The quadratic models also tend to fit a larger slope at low concentrations than is supported by the data.

The models distinguished by Anderson and Nelson (1975) are the variations of the linear-plateau models shown in Figure 5.21. Because they do not apply nonlinear fitting techniques they advocate fixing the critical concentrations (N levels of the join-points) and fit the resulting linear model by standard linear regression techniques. The possible values of the join-points are varied only to be among the interior nitrogen concentrations $(67, 134, \cdots, 268)$ or averages of adjacent points $(100.5, 167.5, \cdots, 301.5)$.

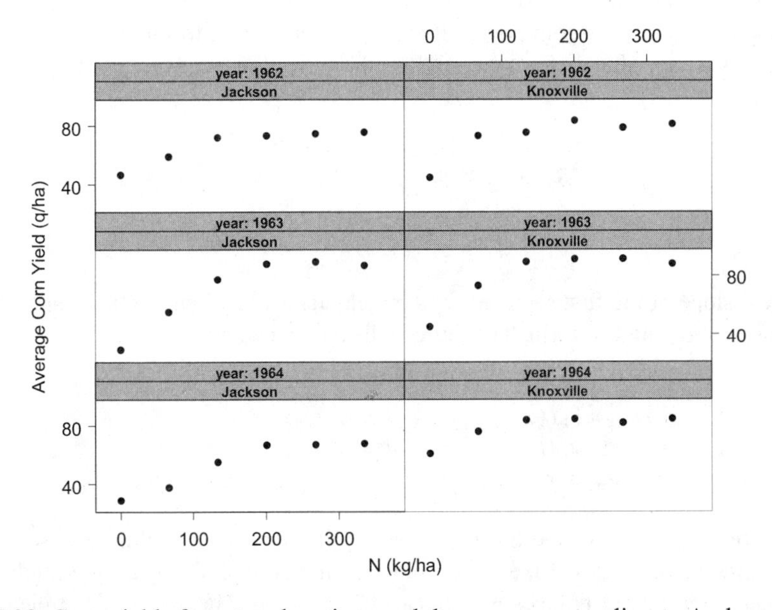

Figure 5.22. Corn yields from two locations and three years, according to Anderson and Nelson (1975).

Table 5.8 shows the models, their residual sums of squares, and the join-points that Anderson and Nelson (1975) determined to best fit the particular subsets of the data. Because we can fit these models as nonlinear models we can estimate the join-points from the data in most cases. As we will see convergence difficulties can be encountered if, for example, a linear-slope-plateau model is fit in nonlinear form to a data set with only six points. The nonlinear version of this model has five parameters and there may not be sufficient information in the data to estimate the parameters.

Table 5.8. Results of fitting the models selected by Anderson and Nelson (1975) for the Tennessee corn yield data of Table 5.7 (The join-points were fixed and the resulting models fit by linear regression)

Location	Year	Model Type	SSR	Join-Point 1	Join-Point 2	k[†]
Knoxville	1962	LSP[‡]	14.26	67	201	3
Knoxville	1963	LP	10.59	100.5		2
Knoxville	1964	LP	4.56	100.5		2
Jackson	1962	LP	11.06	167.5		2
Jackson	1963	LP	7.12	167.5		2
Jackson	1964	LP	13.49	201		2

[†] k = Number of parameters estimated in the linear regression model with fixed join-points
[‡] LSP: Linear-Slope-Plateau model, LP: Linear-Plateau model, LS: Linear-Slope model

Before fitting the models in nonlinear form, we give the mathematical expressions for the LP, LS, and LSP models from which the `model` statements in `proc nlin` will be build. For completeness we include the simple linear regression model too (SLR). Let x denote N concentration and α_1, α_2 the two-join points. Recall that $I(x > \alpha_1)$, for example, is the indicator function that takes on value 1 if $x > \alpha_1$ and 0 otherwise. Furthermore, define the following four quantities

$$\begin{aligned} \theta_1 &= \beta_0 + \beta_1 x \\ \theta_2 &= \beta_0 + \beta_1 \alpha_1 \\ \theta_3 &= \beta_0 + \beta_1 \alpha_1 + \beta_2(x - \alpha_1) \\ \theta_4 &= \beta_0 + \beta_1 \alpha_1 + \beta_2(\alpha_2 - \alpha_1). \end{aligned} \qquad [5.55]$$

θ_1 is the linear slope of the first segment, θ_2 the yield achieved when the first segment reaches concentration $x = \alpha_1$, and so forth. The three models now can be written as

$$\begin{aligned} &\text{SLR:} && \mathrm{E}[Y] = \theta_1 \\ &\text{LP:} && \mathrm{E}[Y] = \theta_1 I(x \le \alpha_1) + \theta_2 I(x > \alpha_1) \\ &\text{LS:} && \mathrm{E}[Y] = \theta_1 I(x \le \alpha_1) + \theta_3 I(x > \alpha_1) \\ &\text{LSP:} && \mathrm{E}[Y] = \theta_1 I(x \le \alpha_1) + \theta_3 I(\alpha_1 < x \le \alpha_2) + \theta_4 I(x > \alpha_2). \end{aligned} \qquad [5.56]$$

We find this representation of the linear-plateau family of models useful because it suggests how to test certain hypotheses. Take the LS model, for example. Comparing θ_3 and θ_2 we see that the linear-slope model reduces to a linear-plateau model if $\beta_2 = 0$ since then $\theta_2 = \theta_3$. Furthermore, if $\beta_1 = \beta_2$, an LS model reduces to the simple linear regression model (SLR). The `proc nlin` statements to fit the various models follow.

```
proc sort data=tennessee; by location year; run;

title 'Linear Plateau (LP) Model';
proc nlin data=tennessee method=newton noitprint;
  parameters b0=45 b1=0.43 a1=67;
  firstterm   = b0+b1*n;
  secondterm  = b0+b1*a1;
  model yield = firstterm*(n <= a1) + secondterm*(n > a1);
  by location year;
run;
```

```
title 'Linear-Slope (LS) Model';
proc nlin data=tennessee method=newton noitprint;
  parameters b0=45 b1=0.43 b2=0 a1=67;
  bounds b2 >= 0;
  firstterm   = b0+b1*n;
  secondterm  = b0+b1*a1+b2*(n-a1);
  model yield = firstterm*(n <= a1) + secondterm*(n > a1);
  by location year;
run;

title 'Linear-Slope-Plateau (LSP) Model for Knoxville 1962';
proc nlin data=tennessee method=newton;
  parameters b0=45 b1=0.43 b2=0 a1=67 a2=150;
  bounds b1 > 0, b2 >= 0;
  firstterm   = b0+b1*n;
  secondterm  = b0+b1*a1+b2*(n-a1);
  thirdterm   = b0+b1*a1+b2*(a2-a1);
  model yield = firstterm*(n <= a1) + secondterm*((n > a1) and (n <= a2)) +
                thirdterm*(n > a2);
run;
```

With `proc nlmixed` we can fit the various models *and* perform the necessary hypothesis tests through the `contrast` or `estimate` statements of the procedure. To fit linear-slope models and compare them to the LP and SLR models use the following statements.

```
proc nlmixed data=tennessee df=3;
  parameters b0=45 b1=0.43 b2=0 a1=67 s=2;
  firstterm   = b0+b1*n;
  secondterm  = b0+b1*a1+b2*(n-a1);
  model yield ~ normal(firstterm*(n <= a1) + secondterm*(n > a1),s*s);
  estimate 'Test against SLR' b1-b2;
  estimate 'Test against LP ' b2;
  contrast 'Test against SLR' b1-b2;
  contrast 'Test against LP ' b2;
  by location year;
run;
```

Table 5.9. Results of fitting the linear-plateau type models to the Tennessee corn yield data of Table 5.7 (The join-points were estimated from the data)

Location	Year	Model Type	SSR	Join-Point 1	$k^{\dagger}$
Knoxville	1962	LP‡	36.10	82.2	3
Knoxville	1963	LP	7.89	107.0	3
Knoxville	1964	LP	4.54	101.3	3
Jackson	1962	LS	0.05	134.8	4
Jackson	1963	LP	5.31	162.5	3
Jackson	1964	LP	13.23	203.9	3

† k = No. of parameters estimated in the nonlinear regression model with estimated join-points
‡ LP: Linear-Plateau model; LS: Linear-Slope model

In Table 5.9 we show the results of the nonlinear models that best fit the six site-years. The linear-slope-plateau model for Knoxville in 1962 did not converge with `proc nlin`. This is not too surprising since this model has 5 parameters ($\beta_0, \beta_1, \beta_2, \alpha_1, \alpha_2$) and only six observations. Not enough information is provided by the data to determine all nonlinear parameters. Instead we determined that a linear-plateau model best fits these data if the join-point is estimated.

Comparing Tables 5.8 and 5.9 several interesting facts emerge. The model selected as *best* by Anderson and Nelson (1975) based on fitting linear regression models with fixed join-points are not necessarily the *best* models selected when the join-points are estimated. For data from Knoxville 1963 and 1964 as well as Jackson 1963 and 1964 both approaches arrive at the same basic model, a linear-plateau relationship. The residual sums of squares between the two approaches then must be close if the join-point in Anderson and Nelson's approach was fixed at a value close to the nonlinear least squares iterate of α_1. This is the case for Knoxville 1964 and Jackson 1964. As the estimated join-point is further removed from the fixed join point (e.g., Knoxville 1963), the residual sum of squares in the nonlinear model is considerably lower than that of the linear model fit.

For the data from the Jackson location in 1962, the nonlinear method selected a different model. Whereas Anderson and Nelson (1975) select an LP model with join-point at 167.5, fitting a series of nonlinear models leads one to a linear-slope (LS) model with join-point at 134.8 kg $\times$ ha^{-1}. The residual sums of squares of the nonlinear model is more than 200 times smaller than that of the linear model. Although the slope (β_2) estimate of the LS model is close to zero (Output 5.14), so is its standard error and the approximate 95% confidence interval for (β_2) does not include zero $([0.0105, 0.0253])$. Not restricting the second segment of the model to be a flat line significantly improves the model fit (not only over a model with fixed join-point, but also over a model with estimated join-point).

Output 5.14.

```
------------------ location=Jackson year=1962 ----------------------

                         The NLIN Procedure
    NOTE: Convergence criterion met.

                         Estimation Summary
                  Method                              Newton
                  Iterations                               7
                  R                                 1.476E-6
                  PPC(b2)                           1.799E-8
                  RPC(b2)                           0.000853
                  Object                            0.000074
                  Objective                         0.053333
                  Observations Read                        6
                  Observations Used                        6
                  Observations Missing                     0

                                 Sum of          Mean                 Approx
Source                    DF    Squares        Square     F Value    Pr > F
Regression                 4    27406.9        6851.7     8419.27    0.0001
Residual                   2     0.0533        0.0267
Uncorrected Total          6    27407.0
Corrected Total            5      673.6

                                    Approx       Approximate 95% Confidence
     Parameter       Estimate    Std Error                Limits
     b0               46.4333       0.1491       45.7919       47.0747
     b1                0.1896      0.00172        0.1821        0.1970
     b2                0.0179      0.00172        0.0105        0.0253
     a1                 134.8       1.6900         127.5         142.0
```

The data from Knoxville in 1962 are a somewhat troubling case. The linear-slope-plateau model that Anderson and Nelson (1975) selected does not converge when the join-points are estimated from the data. Between the LS and LP models the former does not provide a significant improvement over the latter and we are led in the nonlinear analysis to select a linear-plateau model for these data. From the nonlinear predictions shown in Figure 5.23, the linear-plateau model certainly fits the Knoxville 1962 data adequately. Its Pseudo-R^2 is 0.96.

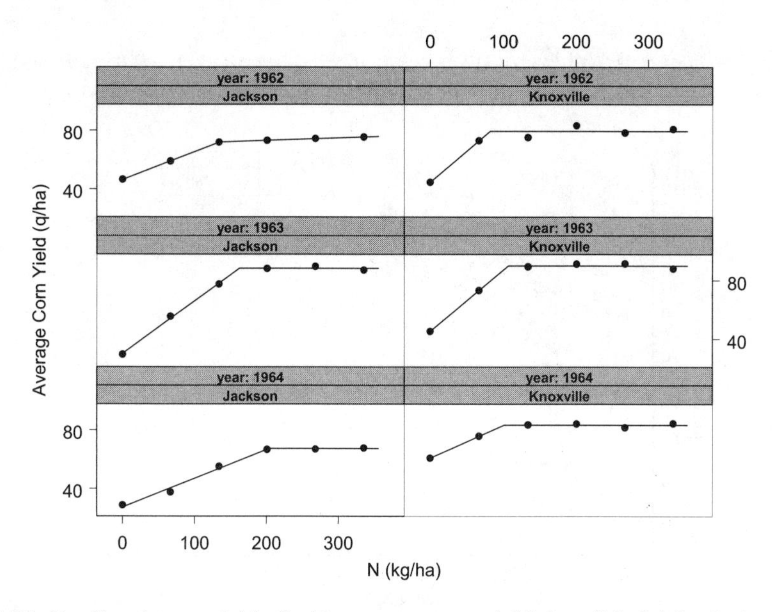

Figure 5.23. Predicted corn yields for Tennessee corn yield data (Model for Jackson 1962 is a linear-slope model; all others are linear-plateau models).

5.8.4 Critical NO_3 Concentrations as a Function of Sampling Depth — Comparing Join-Points in Plateau Models

It has been generally accepted that soil samples for soil NO_3 testing must be collected to depths greater than 30 cm because rainfall moves NO_3 from surface layers quickly to deeper portions of the root zone. Blackmer et al. (1989) suggested that for the late-spring test, nitrate concentrations in the top 30 cm of the soil are indicative of the amounts in the rooting zone because marked dispersion of nitrogen can occur as water moves through macropores (see also Priebe and Blackmer, 1989). Binford, Blackmer, and Cerrato (1992) analyze extensive data from 45 site-years (1346 plot years) collected between 1987 and 1989 in Iowa. When corn plants were 15 to 30 cm tall, samples representing 0 to 30 cm and 30 to 60 cm soil layers were collected. Each site-year included seven to ten rates of N applied before planting. For the N-responsive site years, Figure 5.24 shows relative corn yield for the N rates 0, 112, 224, and 336 kg $\times$ ha^{-1}.

A site was labeled N-responsive, if a plateau model fit the data and relative yield was determined as a percentage of the plateau yield (Cerrato and Blackmer 1990). The data in Figure 5.24 strongly suggest a linear-plateau model with join-point. For a given sampling depth j this linear-response-plateau model becomes

$$\mathrm{E}[Y_{ij}] = \begin{cases} \beta_{0j} + \beta_{1j}NO_{3ij} & NO_{3ij} \leq \alpha_j \\ \beta_{0j} + \beta_{1j}\alpha_j & NO_{3ij} > \alpha_j \end{cases} \qquad [5.57]$$

or

$$\mathrm{E}[Y_{ij}] = (\beta_{0j} + \beta_{1j}NO_{3ij})I\{NO_{3ij} \leq \alpha_j\} + (\beta_{0j} + \beta_{1j}\alpha_j)I\{NO_{3ij} > \alpha_j\}.$$

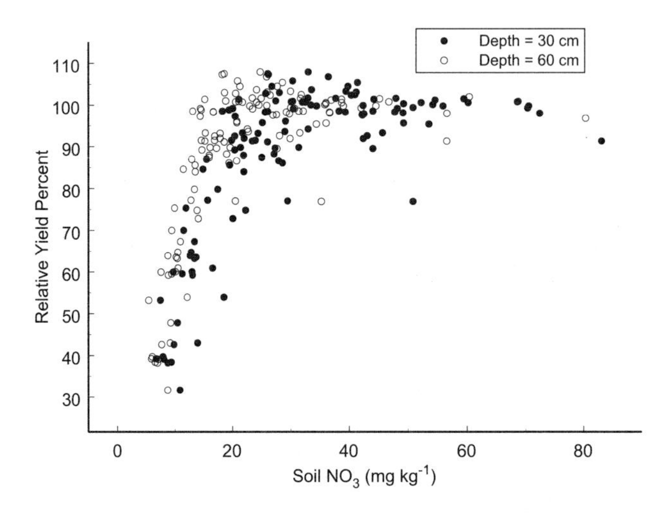

Figure 5.24. Relative yields as a function of soil NO_3 for 30 and 60 cm sampling depths. Data from Binford, Blackmer, and Cerrato (1992, Figures 2c, 3c) containing only N responsive site years on sites that received 0, 112, 224, or 336 kg $\times$ ha^{-1} N. Data kindly made available by Dr. A. Blackmer, Department of Agronomy, Iowa State University. Used with permission.

If soil samples from the top 30 cm ($j = 1$) are indicative of the amount of N in the rooting zone and movement through macropores causes marked dispersion as suggested by Blackmer et al. (1989), one would expect the 0 to 30 cm data to yield a larger intercept ($\beta_{01} > \beta_{02}$), smaller slope ($\beta_{11} < \beta_{12}$) and larger critical concentration ($\alpha_1 > \alpha_2$). The plateaus, however, should not be significantly different. Before testing

$$\begin{array}{lll} H_0: \beta_{01} = \beta_{02} & \text{vs.} & H_1: \beta_{01} > \beta_{02} \\ H_0: \beta_{11} = \beta_{12} & \text{vs.} & H_1: \beta_{11} < \beta_{12} \\ H_0: \alpha_1 = \alpha_2 & \text{vs.} & H_1: \alpha_1 > \alpha_2, \end{array} \qquad [5.58]$$

we examine whether there are any differences in the plateau models between the two sampling depths. To this end we fit the full model [5.57] and compare it to the reduced model

$$\mathrm{E}[Y_{ij}] = \begin{cases} \beta_0 + \beta_1 NO_{3ij} & NO_{3ij} \leq \alpha \\ \beta_0 + \beta_1\alpha & NO_{3ij} > \alpha, \end{cases} \qquad [5.59]$$

that does not vary parameters by sample depth with a sum of squares reduction test.

```
%include 'DriveLetterOfCDROM:\Data\SAS\BlackmerData.txt';
```

```
/* Reduced model [5.59] */
proc nlin data=blackmer noitprint;
  parameters b0=24 b1=24 alpha=25 ;
  model ryp = (b0 + b1*no3)*(no3 <= alpha) + (b0 + b1*alpha)*(no3 > alpha);
run;

/* Full model [5.57] */
proc nlin data=blackmer method=marquardt noitprint;
  parms b01=24 b11=4 alp1=25 del_b0=0 del_b1=0 del_alp=0;
  b02     = b01   + del_b0;
  b12     = b11   + del_b1;
  alp2    = alp1  + del_alp;
  model30 = (b01 + b11*no3)*(no3 <= alp1) + (b01 + b11*alp1)*(no3 > alp1);
  model60 = (b02 + b12*no3)*(no3 <= alp2) + (b02 + b12*alp2)*(no3 > alp2);
  model ryp =  model30*(depth=30) + model60*(depth=60);
run;
```

The full model parameterizes the responses for 60 cm sampling depth $(\beta_{02}, \beta_{12}, \alpha_2)$ as $\beta_{02} = \beta_{01} + \Delta_{\beta_0}$, $\beta_{12} = \beta_{11} + \Delta_{\beta_1}$, $\alpha_2 = \alpha_1 + \Delta_\alpha$ so that differences between the sampling depths in the parameters can be accessed immediately on the output. The reduced model has a residual sum of squares of $SS(\widehat{\boldsymbol{\theta}})_r = 39{,}761.6$ on 477 degrees of freedom (Output 5.15).

Output 5.15.

```
                         The NLIN Procedure

   NOTE: Convergence criterion met.

                         Estimation Summary

               Method                      Gauss-Newton
               Iterations                             8
               R                                      0
               PPC                                    0
               RPC(alpha)                      0.000087
               Object                          4.854E-7
               Objective                       39761.57
               Observations Read                    480
               Observations Used                    480
               Observations Missing                   0

                               Sum of         Mean                   Approx
Source                DF      Squares       Square    F Value      Pr > F

Regression             3      3839878      1279959     774.74      <.0001
Residual             477      39761.6      83.3576
Uncorrected Total    480      3879639
Corrected Total      479       168923

                               Approx      Approximate 95% Confidence
    Parameter      Estimate    Std Error                Limits

    b0               8.7901       2.7688       3.3495      14.2308
    b1               4.8995       0.2207       4.4659       5.3332
    alpha           18.0333       0.3242      17.3963      18.6703
```

The full model's residual sum of squares is $SS(\widehat{\boldsymbol{\theta}})_f = 29{,}236.5$ on 474 degrees of freedom (Output 5.16). The test statistic for the three degree of freedom hypothesis

$$H_0: \begin{bmatrix} \beta_{01} - \beta_{02} \\ \beta_{11} - \beta_{12} \\ \alpha_1 - \alpha_2 \end{bmatrix} = \begin{bmatrix} 0 \\ 0 \\ 0 \end{bmatrix}$$

is

$$F_{obs} = \frac{(39,761.6 - 29,236.5)/3}{29,236.5/474} = 56.879$$

with p-value less than 0.0001.

Since the plateau models are significantly different for the two sampling depths we now proceed to examine the individual hypotheses [5.58]. From Output 5.16 we see that the estimates of the parameters Δ_{β_0}, Δ_{β_1}, and Δ_α have signs consistent with the alternative hypotheses. The asymptotic 95% confidence intervals are two-sided intervals but the alternative hypotheses are one-sided. We thus calculate the one-sided p-values for the three tests with

```
data pvalues;
  tb0    = -9.7424/4.2357; pb0    = ProbT(tb0,474);
  tb1    =  2.1060/0.3205; pb1    = 1-ProbT(tb1,474);
  talpha = -6.8461/0.5691; palpha = ProbT(talpha,474);
run;
proc print data=pvalues; run;
```

Table 5.10. Test statistics and p-values for hypotheses [5.58]

Hypothesis	t_{obs}	p-value
$H_1: \beta_{01} > \beta_{02}$	$-9.7424/4.2357 = -2.300$	0.0109
$H_1: \beta_{11} < \beta_{12}$	$2.1060/0.3203 = 6.575$	< 0.0001
$H_1: \alpha_1 > \alpha_2$	$-6.8461/0.5691 = -12.029$	< 0.0001

Output 5.16.

```
                         The NLIN Procedure

       NOTE: Convergence criterion met.

                            Estimation Summary
                  Method                        Marquardt
                  Iterations                            6
                  R                                     0
                  PPC                                   0
                  RPC(del_alp)                    3.92E-6
                  Object                         1.23E-10
                  Objective                      29236.55
                  Observations Read                   480
                  Observations Used                   480
                  Observations Missing                  0

                                 Sum of          Mean                Approx
Source                  DF      Squares        Square    F Value     Pr > F
Regression               6      3850403        641734     452.94     <.0001
Residual               474      29236.5       61.6805
Uncorrected Total      480      3879639
Corrected Total        479       168923
```

Output 5.16 (continued).

```
                                     Approx        Approximate 95% Confidence
Parameter           Estimate      Std Error                  Limits

b01                  15.1943         2.8322          9.6290       20.7596
b11                   3.5760         0.1762          3.2297        3.9223
alp1                 23.1324         0.4848         22.1797       24.0851
del_b0               -9.7424         4.2357        -18.0657       -1.4191
del_b1                2.1060         0.3203          1.4766        2.7354
del_alp              -6.8461         0.5691         -7.9643       -5.7278
```

Even if a Bonferroni adjustment is made to protect the experimentwise Type-I error rate in this series of three tests, at the experimentwise 5% error level all three tests lead to rejection of their respective null hypotheses. Table 5.11 shows the estimates for the full model. The two plateau values of 97.92 and 97.99 are very close and probably do not warrant a statistical comparison. To demonstrate how a statistical test for $\beta_{01} + \beta_{11}\alpha_1 = \beta_{02} + \beta_{12}\alpha_2$ can be performed we carry it out.

Table 5.11. Parameter estimates in final plateau models

Parameter	Meaning	Estimate
β_{01}	Intercept 30 cm	15.1943
β_{11}	Slope 30 cm	3.5760
α_1	Critical concentration 30 cm	23.1324
$\beta_{01} + \beta_{11}\alpha_1$	Plateau 30 cm	97.9158
β_{02}	Intercept 60 cm	$15.1943 - 9.7424 = 5.4519$
β_{12}	Slope 60 cm	$3.5760 + 2.1060 = 5.6820$
α_2	Critical concentration 60 cm	$23.1324 - 6.8461 = 16.2863$
$\beta_{02} + \beta_{12}\alpha_2$	Plateau 60 cm	97.9900

The first method relies on reparameterizing the model. Let $\beta_{0j} + \beta_{1j}\alpha_j = T_j$ denote the plateau for sampling depth j and notice that the model becomes

$$\begin{aligned}\mathrm{E}[Y_{ij}] &= (T_j + \beta_{1j}(NO_{3ij} - \alpha_j))I\{NO_{3ij} \leq \alpha_j\} + T_j I\{NO_{3ij} > \alpha_j\} \\ &= \beta_{1j}(NO_{3ij} - \alpha_j)I\{NO_{3ij} \leq \alpha_j\} + T_j.\end{aligned}$$

The intercepts β_{01} and β_{02} were eliminated from the model which now contains T_1 and $T_2 = T_1 + \Delta_T$ as parameters. The SAS® statements

```
proc nlin data=blackmer method=marquardt noitprint;
  parms b11=3.56 alp1=23.13 T1=97.91 b12=5.682 alp2=16.28  del_T=0;
  T2       = T1 + del_T;
  model30 = b11*(no3-alp1)*(no3 <= alp1) + T1;
  model60 = b12*(no3-alp2)*(no3 <= alp2) + T2;
  model ryp =  model30*(depth=30) + model60*(depth=60);
run;
```

yield Output 5.17. The approximate 95% confidence interval for Δ_T ($[-1.683, 1.834]$) contains zero and there is insufficient evidence at the 5% significance level to conclude that the relative yield plateaus differ among the sampling depths. The second method of comparing the plateau values relies on the capabilities of `proc nlmixed` to estimate linear and nonlinear functions of the model parameters. Any of the parameterizations of the plateau model will do for this purpose. For example, the statements

```
proc nlmixed data=blackmer df=474;
  parms b01=24 b11=4 alp1=25 del_b0=0 del_b1=0 del_alp=0;
  s2      = 61.6805;
  b02     = b01    + del_b0;
  b12     = b11    + del_b1;
  alp2    = alp1   + del_alp;
  model30 = (b01 + b11*no3)*(no3 <= alp1) + (b01 + b11*alp1)*(no3 > alp1);
  model60 = (b02 + b12*no3)*(no3 <= alp2) + (b02 + b12*alp2)*(no3 > alp2);
  model ryp  ~ normal(model30*(depth=30) + model60*(depth=60),s2);
  estimate 'Difference in Plateaus' b01+b11*alp1 - (b02+b12*alp2);
run;
```

will do the trick (Output 5.18). Since the `nlmixed` procedure approximates a likelihood and estimates all distribution parameters, it would iteratively determine the variance of the Gaussian error distribution. To prevent this we fix the variance with the `s2 = 61.6805;` statement. This is the residual mean square estimate obtained from fitting the full model in `proc nlin` (see Output 5.16 or Output 5.17). Also, because `proc nlmixed` determines residual degrees of freedom by a method different from `proc nlin` we fix the residual degrees of freedom with the `df=` option of the `proc nlmixed` statement.

Output 5.17.

```
                       The NLIN Procedure

    NOTE: Convergence criterion met.

                       Estimation Summary

                 Method                     Marquardt
                 Iterations                         1
                 R                           2.088E-6
                 PPC(alp1)                   4.655E-7
                 RPC(del_T)                  75324.68
                 Object                      0.000106
                 Objective                   29236.55
                 Observations Read                480
                 Observations Used                480
                 Observations Missing               0

                                 Sum of         Mean                  Approx
Source                  DF      Squares       Square    F Value    Pr > F

Regression               6      3850403       641734     452.94    <.0001
Residual               474      29236.5      61.6805
Uncorrected Total      480      3879639
Corrected Total        479       168923

                                  Approx      Approximate 95% Confidence
     Parameter      Estimate    Std Error               Limits

     b11              3.5760       0.1762        3.2297        3.9223
     alp1            23.1324       0.4848       22.1797       24.0851
     T1              97.9156       0.6329       96.6720       99.1592
     b12              5.6820       0.2675        5.1564        6.2076
     alp2            16.2863       0.2980       15.7008       16.8719
     del_T            0.0753       0.8950       -1.6834        1.8340
```

Output 5.18. The NLMIXED Procedure

Specifications

Data Set	WORK.BLACKMER
Dependent Variable	ryp
Distribution for Dependent Variable	Normal
Optimization Technique	Dual Quasi-Newton
Integration Method	None

Dimensions

Observations Used	480
Observations Not Used	0
Total Observations	480
Parameters	6

NOTE: GCONV convergence criterion satisfied.

Fit Statistics

-2 Log Likelihood	3334.7
AIC (smaller is better)	3346.7
AICC (smaller is better)	3346.9
BIC (smaller is better)	3371.8

Parameter Estimates

Parameter	Estimate	Standard Error	DF	t Value	Pr > |t|	Lower	Upper
b01	15.1943	2.8322	474	5.36	<.0001	9.6290	20.7595
b11	3.5760	0.1762	474	20.29	<.0001	3.2297	3.9223
alp1	23.1324	0.4848	474	47.71	<.0001	22.1797	24.0851
del_b0	-9.7424	4.2357	474	-2.30	0.0219	-18.0656	-1.4192
del_b1	2.1060	0.3203	474	6.57	<.0001	1.4766	2.7354
del_alp	-6.8461	0.5691	474	-12.03	<.0001	-7.9643	-5.7278

Additional Estimates

Label	Estimate	Standard Error	DF	t Value	Pr > |t|
Difference in Plateaus	-0.07532	0.8950	474	-0.08	0.9330

Figure 5.25 shows the predicted response functions for the 0 to 30 cm and 0 to 60 cm sampling depths.

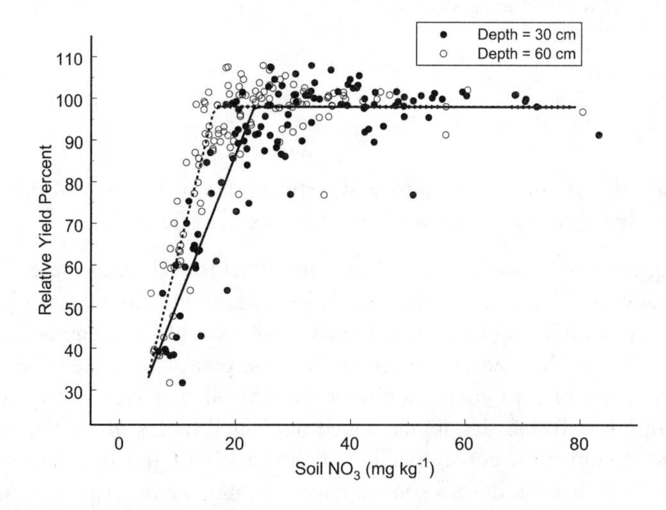

Figure 5.25. Fitted linear-plateau models.

5.8.5 Factorial Treatment Structure With Nonlinear Response

Many agronomic studies involve a treatment structure with more than one factor. Typically, the factors are crossed so that replicates of experimental units are exposed to all possible combinations of the factors. When modeling the mean function of such data with linear combinations of the main effects and interactions of the factors, one naturally arrives at analysis of variance as the tool for statistical inference. If the mean function is nonlinear, it is less clear how to compare treatments, test for main effects, and investigate interactions. As an example we consider a velvetleaf (*Abutilon theophrasti* Medicus) multiple growth stage experiment. Velvetleaf was grown to 5 to 6 cm (2 to 3 leaves), 8 to 13 cm (3 to 4 leaves), and 15 to 20 cm (5 to 6 leaves) in a commercial potting mixture in 1-L plastic pots. Plants were grown in a 16-h photoperiod of natural lighting supplemented with sodium halide lights providing a midday photosynthetic photon flux density of $1,000\ \mu\mathrm{mol/m^2/s}$. Weeds were treated with two herbicides (glufosinate and glyphosate) at six different rates of application. Two separate runs of the experiment were conducted with four replicates each in a randomized complete block design. The above-ground biomass was harvested 14 days after treatment and oven dried. The outcome of interest is the dry weight percentage relative to an untreated control. This and a second multistage growth experiment for common lambsquarter (*Chenopodium album* L.) and several single growth stage experiments are explained in more detail in Tharp, Schabenberger, and Kells (1999).

Considering an analysis of variance model for this experiment one could arrive at

$$\begin{aligned} Y_{ijklm} = \mu + \kappa_i + \nu_{ij} + \alpha_k + \beta_l + \gamma_m + \\ (\alpha\beta)_{kl} + (\alpha\gamma)_{km} + (\beta\gamma)_{lm} + (\alpha\beta\gamma)_{klm} + e_{ijklm}. \end{aligned} \qquad [5.60]$$

where Y_{ijklm} denotes the dry weight percentage, κ_i denotes the ith run $(i = 1, 2)$, ν_{ij} the jth replicate within run i $(j = 1, \cdots, 4)$, α_k the effect of the kth herbicide $(k = 1, 2)$, β_l the effect of the lth size class $(l = 1, \cdots, 3)$, and γ_m the effect of the mth rate $(m = 1, \cdots, 6)$. One can include additional interaction terms in model [5.60] but for expository purposes we will not pursue this issue here. The analysis of variance table (Table 5.12, SAS® output not shown) is produced in SAS® with the statements

```
proc glm data=VelvetFactorial;
  class run rep herb size rate;
  model drywtpct = run rep(run) herb size rate herb*size herb*rate
                   size*rate herb*size*rate;
run; quit;
```

The analysis of variance table shows significant Herb × Rate and Size × Rate interactions (at the 5% level) and significant Rate and Size main effects.

By declaring the rate of application a factor in model [5.60], rates are essentially discretized and the continuity of rates of application is lost. Some information can be recovered by testing for linear, quadratic, up to quintic trends of dry weight percentages. Because of the interactions of rate with the size and herbicide factors, great care should be exercised since these trends may differ for the two herbicides or the three size classes. Unequal spacing of the rates of application is a further hindrance, since published tables of contrast coefficients require a balanced design with equal spacing of the levels of the quantitative factor. From Figure 5.26 it is seen that the dose response curves cannot be described by simple linear or quadratic polynomials, although their general shape appears to vary little by herbicide or size class.

Table 5.12. Analysis of variance for model [5.60]

Effect		**DF**	**SS**	**MS**	F_{obs}	***p*-value**
Run	κ_i	1	541.75	541.75	2.48	0.117
Rep w/in Run	ν_{ij}	6	1453.29	242.22	1.11	0.358
Herbicide	α_k	1	472.78	472.78	2.16	0.143
Size	β_l	2	20391.13	10195.57	46.65	< 0.0001
Rate	γ_m	5	341783.93	68356.79	312.77	< 0.0001
Herb × Size	$(\alpha\beta)_{kl}$	2	285.39	142.69	0.65	0.521
Herb × Rate	$(\alpha\gamma)_{km}$	5	4811.32	962.26	4.40	0.0007
Size × Rate	$(\beta\gamma)_{lm}$	10	6168.70	616.87	2.82	0.003
Herb × Size × Rate	$(\alpha\beta\gamma)_{klm}$	10	3576.70	357.67	1.64	0.097
Error	e_{ijklm}	245	53545.58	218.55		
Total		287	433030.58			

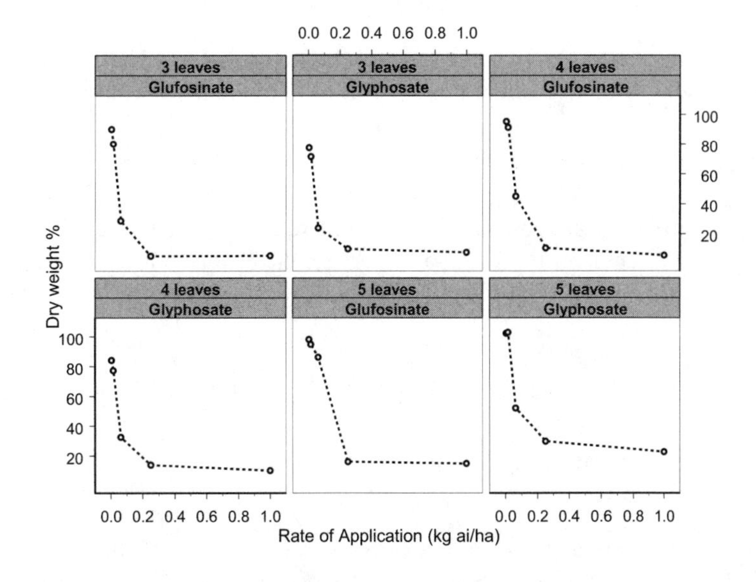

Figure 5.26. Herbicide × Size class (leave stage) sample means as a function of application rate in factorial velvetleaf dose-response experiment. Data made kindly available by Dr. James J. Kells, Department of Crop and Soil Sciences, Michigan State University. Used with permission.

As an alternative approach to the analysis of variance, we consider the data in Figure 5.26 as the raw data for a nonlinear modeling problem. Tharp et al. (1999) analyze these data with a four-parameter log-logistic model. Plotted against log{rate}, the mean dry weight percentages exhibit a definite sigmoidal trend. But graphed against the actual rates, the response appears hyperbolic (Figure 5.26) and for any given Herbicide × Size Class combination the hyperbolic Langmuir model $\alpha\beta x^{\gamma}/(1+\beta x^{\gamma})$, where x is dosage in kg ai/ha appears appropriate. Since the response is expressed as a percentage of the control dry weight, we can fix the parameter α at 100. This reduces the problem to fitting a two-parameter model for each Herbicide × Size combination compared to a three- or four-parameter log-logistic model. Thus, the full model that varies the two Langmuir parameters for the 12 herbicide × size combinations has $2*12 = 24$ parameters (compared to 43 parameters in model [5.60]). The full model we postulate here is

$$\mathrm{E}[Y_{klm}] = 100\frac{\beta_{kl}x_m^{\gamma}}{1+\beta_{kl}x_m^{\gamma}}, \qquad [5.61]$$

where x_m is the mth application rate common to the combination of herbicide k and size class l. For expository purposes we assume here that only the β parameter varies across treatments, so that [5.61] has thirteen parameters. Notice that an additional advantage of model [5.61] over [5.60] is that it entails at most a single two-way interaction compared to three two-way and one three-way interactions in [5.60].

The first hypothesis being tested in [5.61] is that all herbicides and size classes share the same β parameter

$$H_0: \beta_{kl} = \beta_{k'l'} \;\; \forall kk', ll'.$$

This eleven degree of freedom hypothesis is similar to the global ANOVA hypothesis testing equal effects of all treatments. Under this hypothesis [5.61] reduces to a two-parameter model

$$H_0: \text{no treatment effects;} \quad \mathrm{E}[Y_{klm}] = 100\frac{\beta x_m^{\gamma}}{1+\beta x_m^{\gamma}}. \qquad [5.62]$$

Models [5.61] and [5.62] are compared via sum of squares reduction tests. If [5.61] is not a significant improvement over [5.62], stop. Otherwise the next step is to investigate the effects of Herbicide and Size Class separately. This entails two more models, both reduced versions of [5.61]:

$$H_0: \text{no herbicide effects;} \quad \mathrm{E}[Y_{klm}] = 100\frac{\beta_l x_m^{\gamma}}{1+\beta_l x_m^{\gamma}} \qquad [5.63]$$

$$H_0\text{: no size class effects;} \quad \mathrm{E}[Y_{klm}] = 100\frac{\beta_k x_m^{\gamma}}{1+\beta_k x_m^{\gamma}} \qquad [5.64]$$

Models [5.61] through [5.64] are fit in SAS® `proc nlin` with the following series of statements. The full model [5.61] for the 2×3 factorial is fit first (Output 5.19). Size classes are identified with the second subscript corresponding to the 3, 4, and 5 leave stages. For example, `beta_14` is the parameter for herbicide 1 (glufosinate) and size class 2 (4 leave stage).

```
title 'Full Model [5.61]';
proc nlin data=velvet noitprint;
 parameters beta_13=0.05 beta_14=0.05 beta_15=0.05
            beta_23=0.05 beta_24=0.05 beta_25=0.05
            gamma=-1.5;

 alpha = 100;
```

```
 t_1_3  = beta_13*(rate**gamma);
 t_1_4  = beta_14*(rate**gamma);
 t_1_5  = beta_15*(rate**gamma);
 t_2_3  = beta_23*(rate**gamma);
 t_2_4  = beta_24*(rate**gamma);
 t_2_5  = beta_25*(rate**gamma);

 model drywtpct = (alpha*t_1_3/(1+t_1_3))*(herb=1 and size=3) +
                  (alpha*t_1_4/(1+t_1_4))*(herb=1 and size=4) +
                  (alpha*t_1_5/(1+t_1_5))*(herb=1 and size=5) +
                  (alpha*t_2_3/(1+t_2_3))*(herb=2 and size=3) +
                  (alpha*t_2_4/(1+t_2_4))*(herb=2 and size=4) +
                  (alpha*t_2_5/(1+t_2_5))*(herb=2 and size=5);
run; quit;
```

Output 5.19.

```
                         The NLIN Procedure

     NOTE: Convergence criterion met.

                        Estimation Summary

              Method                         Gauss-Newton
              Iterations                               15
              Subiterations                             1
              Average Subiterations              0.066667
              R                                  5.673E-6
              PPC(beta_25)                       0.000012
              RPC(beta_25)                       0.000027
              Object                             2.35E-10
              Objective                          2463.247
              Observations Read                        36
              Observations Used                        36
              Observations Missing                      0

           NOTE: An intercept was not specified for this model.

                                   Sum of        Mean               Approx
Source                    DF      Squares      Square    F Value    Pr > F

Regression                 7       111169     15881.3     186.97    <.0001
Residual                  29       2463.2     84.9396
Uncorrected Total         36       113632
Corrected Total           35      47186.2

                                   Approx      Approximate 95% Confidence
    Parameter       Estimate    Std Error                Limits

    beta_13           0.0179      0.00919     -0.00087        0.0367
    beta_14           0.0329       0.0152      0.00186        0.0639
    beta_15           0.0927       0.0356       0.0199        0.1655
    beta_23           0.0129      0.00698     -0.00141        0.0271
    beta_24           0.0195      0.00984     -0.00066        0.0396
    beta_25           0.0665       0.0272       0.0109        0.1221
    gamma            -1.2007       0.1262      -1.4589       -0.9425
```

The `proc nlin` statements to fit the completely reduced model [5.62] and the models without herbicide ([5.63]) and size class effects ([5.64]) are as follows.

```
title 'Completely Reduced Model [5.62]';
proc nlin data=velvet noitprint;
  parameters beta=0.05 gamma=-1.5;
  alpha = 100;
  term  = beta*(rate**gamma);
  model drywtpct = alpha*term/(1+term);
run; quit;

title 'No Herbicide Effect [5.63]';
proc nlin data=velvet noitprint;
  parameters beta_3=0.05 beta_4=0.05 beta_5=0.05 gamma=-1.5;
  alpha = 100;
  t_3   = beta_3*(rate**gamma);
  t_4   = beta_4*(rate**gamma);
  t_5   = beta_5*(rate**gamma);
  model drywtpct = (alpha*t_3/(1+t_3))*(size=3) +
                   (alpha*t_4/(1+t_4))*(size=4) +
                   (alpha*t_5/(1+t_5))*(size=5);
run; quit;

title 'No Size Effect [5.64]';
proc nlin data=velvet noitprint;
  parameters beta_1=0.05 beta_2=0.05 gamma=-1.5;
  alpha = 100;
  t_1    = beta_1*(rate**gamma);
  t_2    = beta_2*(rate**gamma);
  model drywtpct = (alpha*t_1/(1+t_1))*(herb1=1) +
                   (alpha*t_2/(1+t_2))*(herb1=2);
run; quit;
```

Once the residual sums of squares and degrees of freedom for the models are obtained (Table 5.13, output not shown), the sum of squares reduction tests can be carried out:

$$H_0 : \text{no treatment effects} \quad F_{obs} = \frac{3,176.0/5}{2,463.2/29} = 7.478 \quad p = \Pr(F_{5,29} \geq 7.478) = 0.0001$$

$$H_0 : \text{no herbicide effects} \quad F_{obs} = \frac{247.9/3}{2,463.2/29} = 0.973 \quad p = \Pr(F_{3,29} \geq 0.973) = 0.4188$$

$$H_0\text{: no size class effects} \quad F_{obs} = \frac{2,940.7/4}{2,463.2/29} = 8.655 \quad p = \Pr(F_{4,29} \geq 8.655) = 0.0001$$

Table 5.13. Residual sums of squares for full and various reduced models

Model	Effects in Model	df_{res}	df_{model}	$S(\widehat{\boldsymbol{\theta}})$	$S(\widehat{\boldsymbol{\theta}}) - 2,463.2$
[5.61]	Herbicide, Size	29	7	2,463.2	
[5.62]	none	34	2	5,639.2	3,176.0
[5.63]	Size	32	4	2,711.1	247.9
[5.64]	Herbicide	33	3	5,403.9	2,940.7

The significant treatment effects appear to be due to a size effect alone but comparing models [5.61] through [5.64] is somewhat unsatisfactory. For example, model [5.63] of no Herbicide effects reduces the model degrees of freedom by three although there are only two herbicides. Model [5.63] not only eliminates a Herbicide *main* effect, but also the Herbicide $\times$ Size *interaction* and the resulting model contains a Size Class main effect only. A similar phenomenon can be observed with model [5.64]. There are two degrees of freedom for a Size Class main effect and two degrees of freedom for the interaction; model [5.64] contains four parameters less than the full model. If one is interested in testing for main effects and interactions in a similar fashion as in the linear analysis of variance model, a different method is required to reduce the full model corresponding to the hypotheses

A H_0 : no Herbicide main effect
B H_0 : no Size Class main effect
C H_0 : no Herbicide $\times$ Size Class interaction.

The technique we suggest is an analog of the cell mean representation of a two-way factorial (§4.3.1):

$$\mu_{ij} = \mu + (\mu_{i.} - \mu) + (\mu_{.j} - \mu) + (\mu_{ij} - \mu_{i.} - \mu_{.j} + \mu) = \mu + \alpha_i + \beta_j + (\alpha\beta)_{ij},$$

where the four terms in the sum correspond to the grand mean, factor A main effects, factor B main effects, and A $\times$ B interactions, respectively. The absence of the main effects and interactions in a linear model can be represented by complete sets of contrasts among the cell means as discussed in §4.3.3 (see also Schabenberger, Gregoire, and Kong 2000). With two herbicide levels H_1 and H_2 and three size classes S_3, S_4, and S_5 the cell mean contrasts for the respective effects are given in Table 5.14. Notice that the number of contrasts for each effect equals the number of degrees of freedom for that effect.

Table 5.14. Contrasts for main effects and interactions in unfolded 2×3 factorial design

Effect	**Contrast**	H_1S_3	H_1S_4	H_1S_5	H_2S_3	H_2S_4	H_2S_5
Herbicide Main	H	1	1	1	− 1	− 1	− 1
Size Main	S1	1	− 1	0	1	− 1	0
	S2	1	1	− 2	1	1	− 2
Herb $\times$ Size	(H $\times$ S1)	1	− 1	0	− 1	1	0
	(H $\times$ S2)	1	1	− 2	− 1	− 1	2

To test whether the Herbicide main effect is significant, we fit the full model and test whether the linear combination

$$\mu_{13} + \mu_{14} + \mu_{15} - \mu_{23} - \mu_{24} - \mu_{25}$$

differs significantly from zero. This can be accomplished with the `contrast` statement of the `nlmixed` procedure. As in the previous application we fix the residual degrees of freedom and the error variance estimate to equal those for the full model obtained with `proc nlin`.

```
proc nlmixed data=velvet df=29;
  parameters beta_13=0.05 beta_14=0.05 beta_15=0.05
             beta_23=0.05 beta_24=0.05 beta_25=0.05
             gamma=-1.5;

  s2    = 84.9396;
  alpha = 100;

  mu_13 = alpha*beta_13*(rate**gamma)/(1+beta_13*(rate**gamma));
  mu_14 = alpha*beta_14*(rate**gamma)/(1+beta_14*(rate**gamma));
  mu_15 = alpha*beta_15*(rate**gamma)/(1+beta_15*(rate**gamma));
  mu_23 = alpha*beta_23*(rate**gamma)/(1+beta_23*(rate**gamma));
  mu_24 = alpha*beta_24*(rate**gamma)/(1+beta_24*(rate**gamma));
```

```
  mu_25 = alpha*beta_25*(rate**gamma)/(1+beta_25*(rate**gamma));

  meanfunction =    (mu_13)*(herb=1 and size=3) +
                    (mu_14)*(herb=1 and size=4) +
                    (mu_15)*(herb=1 and size=5) +
                    (mu_23)*(herb=2 and size=3) +
                    (mu_24)*(herb=2 and size=4) +
                    (mu_25)*(herb=2 and size=5);

 model drywtpct ~ normal(meanfunction,s2);

 contrast 'No Herbicide Effect'  mu_13+mu_14+mu_15-mu_23-mu_24-mu_25;
 contrast 'No Size Class Effect' mu_13-mu_14+mu_23-mu_24,
                                 mu_13+mu_14-2*mu_15+mu_23+mu_24-2*mu_25;
 contrast 'No Interaction'       mu_13-mu_14-mu_23+mu_24,
                                 mu_13+mu_14-2*mu_15-mu_23-mu_24+2*mu_25;

run; quit;
```

Output 5.20.

```
                         The NLMIXED Procedure

                            Specifications
     Data Set                                   WORK.VELVET
     Dependent Variable                         drywtpct
     Distribution for Dependent Variable        Normal
     Optimization Technique                     Dual Quasi-Newton
     Integration Method                         None

                              Dimensions
                 Observations Used                        36
                 Observations Not Used                     0
                 Total Observations                       36
                 Parameters                                7

               NOTE: GCONV convergence criterion satisfied.

                            Fit Statistics
                -2 Log Likelihood                  255.1
                AIC (smaller is better)            269.1
                AICC (smaller is better)           273.1
                BIC (smaller is better)            280.2

                          Parameter Estimates
                        Standard
Parameter  Estimate       Error     DF  t Value  Pr > |t|     Lower     Upper
beta_13    0.01792     0.01066     29     1.68    0.1036   -0.00389   0.0397
beta_14    0.03290     0.01769     29     1.86    0.0731   -0.00329   0.0690
beta_15    0.09266     0.04007     29     2.31    0.0281    0.01071   0.1746
beta_23    0.01287    0.008162     29     1.58    0.1257   -0.00382   0.0295
beta_24    0.01947     0.01174     29     1.66    0.1079   -0.00453   0.0435
beta_25    0.06651     0.03578     29     1.86    0.0732   -0.00667   0.1397
gamma      -1.2007      0.1586     29    -7.57    <.0001   -1.5251  -0.8763

                               Contrasts
                                Num     Den
        Label                    DF      DF     F Value      Pr > F
        No Herbicide Effect       1      29        1.93      0.1758
        No Size Class Effect      2      29        4.24      0.0242
        No Interaction            2      29        0.52      0.6010
```

Based on the `Contrasts` table in Output 5.20 we reject the hypothesis of no Size Class effect but fail to reject the hypothesis of no interaction and no Herbicide main effect at the 5% level. Based on these results we could fit a model in which the β parameters vary only by

Size Class, that is [5.63]. Using `proc nlmixed`, pairwise comparisons of the β parameters among size classes are accomplished with

```
proc nlmixed data=velvet df=32;
  parameters beta_3=0.05 beta_4=0.05 beta_5=0.05
             gamma=-1.5;
  s2=84.7219;
  alpha = 100;
  t_3   = beta_3*(rate**gamma);
  t_4   = beta_4*(rate**gamma);
  t_5   = beta_5*(rate**gamma);
  mu_3 = alpha*t_3/(1+t_3);
  mu_4 = alpha*t_4/(1+t_4);
  mu_5 = alpha*t_5/(1+t_5);
  meanfunction = (mu_3)*(size=3) +
                 (mu_4)*(size=4) +
                 (mu_5)*(size=5);
  model drywtpct ~ normal(meanfunction,s2);
  contrast 'beta_3 - beta_4' beta_3 - beta_4;
  contrast 'beta_3 - beta_5' beta_3 - beta_5;
  contrast 'beta_4 - beta_5' beta_4 - beta_5;
run; quit;
```

The `Contrasts` table added to the procedure output reveals that differences between size classes are significant at the 5% level except for the 3 and 4 leaf stages.

Output 5.21.

Contrasts

Label	Num DF	Den DF	F Value	Pr > F
beta_3 - beta_4	1	32	2.03	0.1636
beta_3 - beta_5	1	32	6.11	0.0190
beta_4 - beta_5	1	32	5.30	0.0280

5.8.6 Modeling Hormetic Dose Response through Switching Functions

A model commonly applied in dose-response investigations is the log-logistic model which incorporates sigmoidal behavior symmetric about a point of inflection. A nice feature of this model is its simple reparamaterization in terms of effective dosages, such as LD_{50} or LD_{90} values, which can be estimated directly from empirical data by fitting the model with a nonlinear regression package. In §5.7.2 we reparameterized the log-logistic model

$$\mathrm{E}[Y] = \delta + \frac{\alpha - \delta}{1 + \psi\exp\{\beta\ln(x)\}},$$

where Y is the response to dosage x, in terms of λ_K, the dosage at which the response is K% between the asymptotes δ and α. For example, λ_{50} would be the dosage that achieves a response halfway between the lower and upper asymptotes. The general formula for this reparameterization is

$$\mathrm{E}[Y|x] = \delta + \frac{\alpha - \delta}{1 + K/(100 - K)\exp\{\beta \ln(x/\lambda_K)\}},$$

so that in the special case of $K = 50$ we get

$$\mathrm{E}[Y|x] = \delta + \frac{\alpha - \delta}{1 + \exp\{\beta \ln(x/\lambda_K)\}}.$$

Although the model is popular in dose-response studies it does not necessarily fit all data sets of this type. It assumes, for example, that the trend between δ and α is sigmoidal and monotonically increases or decreases. The frequent application of log-logistic models in herbicide dose-response studies (see, e.g., Streibig 1980, Streibig 1981, Lærke and Streibig 1995, Seefeldt et al. 1995, Hsiao et al. 1996, Sandral et al. 1997) tends to elevate the model in the eyes of some to a law-of-nature to which all data must comply. Whether the relationship between dose and response is best described by a linear, log-logistic, or other model must be re-assessed for every application and every set of empirical data. Figure 5.27 shows the sample mean relative growth percentages of barnyardgrass (*Echinochloa crus-galli* (L.) P. Beauv.) treated with glufosinate [2-amino-4-(hydroxymethylphosphinyl) butanoic acid] + $(NH_4)_2SO_4$ (open circles) and glyphosate [isopropylamine salt of *N*-(phosphonomethyl)glycine] + $(NH_4)_2SO_4$ (closed circles). Growth is expressed relative to an untreated control and the data points shown are sample means calculated across eight replicate values at each concentration. A log-logistic model appears appropriate for the glufosinate response but not for the glyphosate response, which exhibits an effect known as *hormesis*. The term hormesis originates from the Greek for "setting into motion" and the notion that every toxicant is a stimulant at low levels (Schulz 1988, Thiamann 1956) is also known as the Arndt-Schulz law.

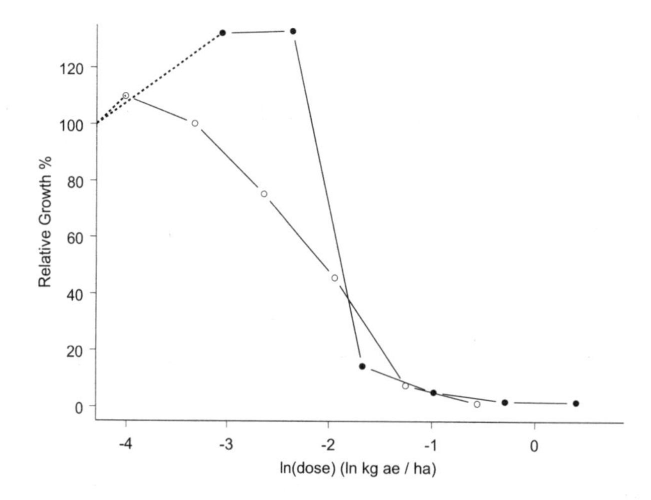

Figure 5.27. Mean relative growth percentages for barnyard grass as a function of log dosage. Open circles represent active ingredient glufosinate, closed circles glyphosate. Data kindly provided by Dr. James J. Kells, Department of Crop and Soil Sciences, Michigan State University. Used with permission.

Hormetic effects can be defined as the failure of a dose-reponse relationship to behave at small dosages as the extrapolation from larger dosages under a theoretical model or otherwise "stylized fact" would lead one to believe. The linear no-threshold hypothesis in radiation studies, for example, invokes a linear extrapolation to zero dose, implying that radiation response is proportional to radiation dosage over the entire range of possible dosages (UNSCEAR 1958). Even in the absence of beneficial effects at low dosages linear trends across a larger dosage range are unlikely in most dose-response investigations. More common are sigmoidal or hyperbolic relationships between dosage and average response. The failure of the linear no-threshold hypothesis in the presence of hormetic effects is twofold: (i) a linear model does not capture the dose-response relationship when there is no hormesis, and (ii) extrapolations to low dosages do not account for hormetic effects.

Several authors have noted that for low herbicide dosages a hormetic effect can occur which raises the average response for low dosages above the control value (Miller et al. 1962, Freney 1965, Wiedman and Appleby 1972). Allender (1997) and Allender et al. (1997) suggested that influx of Ca^{+2} may be involved in the growth stimulation associated with hormesis.The log-logistic function does not accommodate such behavior and Brain and Cousens (1989) suggested a modification to allow for hormesis, namely,

$$\mathrm{E}[Y|x] = \delta + \frac{\alpha - \delta + \gamma x}{1 + \theta\exp\{\beta\ln(x)\}}, \qquad [5.65]$$

where γ measures the initial rate of increase at low dosages. The Brain-Cousens model [5.65] is a simple modification of the log-logistic and it is perhaps somewhat surprising that adding a term γx in the numerator should do the trick. In this parameterization it is straightforward to test the hypothesis of hormetic effects statistically. Fit the model and observe whether the asymptotic confidence interval fo γ includes 0. If the confidence interval fails to include 0 the hypothesis of the absence of a hormetic effect is rejected.

But how can a dose-response model other than the log-logistic be modified if the researcher anticipates hormetic effects or wishes to test their presence? To construct hormetic models we rely on the idea of combining mathematical switching functions. Schabenberger and Birch (2001) proposed hormetic models constructed by this device and the Brain-Cousens model is a special case thereof. In process models for plant growth switching mechanisms are widely used (e.g., Thornley and Johnson 1990), for example, to switch on or off a mathematical function or constant or to switch from one function to another. The switching functions from which we build dose-response models are mathematical functions $S(x)$ that take values between 0 and 1 as dosage x varies. In the log-logistic model

$$Y = \delta + \frac{\alpha - \delta}{1 + \exp\{\beta\ln(x/\lambda_{50})\}}$$

the term $[1 + \exp\{\beta\ln(x/\lambda_{50})\}]^{-1}$ is a switch-off function for $\beta > 0$ (Figure 5.28) and a switch-on function for $\beta < 0$.

With $\beta > 0$, δ is the lower and α the upper asymptote of dose response. The role of the switching function is to determine how the transition between the two extrema takes place. This suggests the following technique to develop dose-response models. Let $S(x, \boldsymbol{\theta})$ be a switch-off function and notice that $R(x, \boldsymbol{\theta}) = 1 - S(x, \boldsymbol{\theta})$ is a switch-on function. Denote the min and max mean dose-response in the absence of any hormetic effects as $\mu_{\min}$ and $\mu_{\max}$. A general dose-response model is then given by

$$Y = \mu_{\min} + (\mu_{\max} - \mu_{\min})S(x,\boldsymbol{\theta}). \quad [5.66]$$

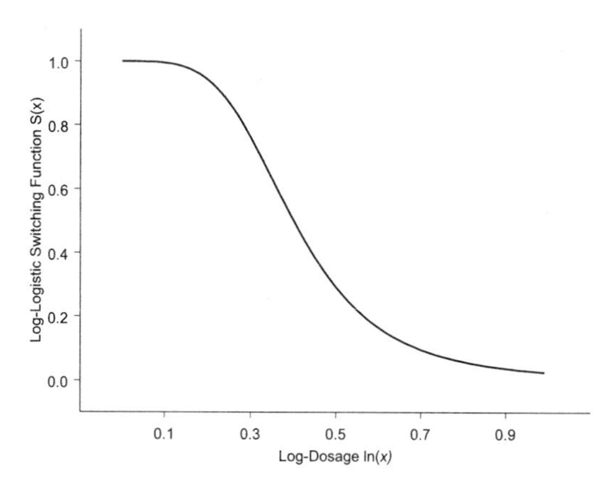

Figure 5.28. Switch-off behavior of the log-logistic term $[1+\exp\{\beta\ln(x/\lambda_{50})\}]^{-1}$ for $\lambda_{50} = 0.4$, $\beta = 4$.

By choosing the switching function from a flexible family of mathematical models, the nonhormetic dose response can take on many shapes, not necessarily sigmoidal and symmetric as implied by the log-logistic model. To identify possible switch-on functions one can choose $R(x,\boldsymbol{\theta})$ as the cumulative distribution function (cdf) of a continuous random variable with unimodal density. Choosing the cdf of a random variable uniformly distributed on the interval (a,b) leads to the switch-off function

$$S(x,\boldsymbol{\theta}) = \frac{b}{b-a} - \frac{1}{b-a}x$$

and a linear interpolation between $\mu_{\min}$ and $\mu_{\max}$. A probably more useful cdf that permits a sigmoidal transition between $\mu_{\min}$ and $\mu_{\max}$ is that of a two-parameter Weibull random variable which leads to the switch-off function (Figure 5.29)

$$S(x,\boldsymbol{\theta}) = \exp\left\{-(x/\alpha)^{\beta}\right\}.$$

Gregoire and Schabenberger (1996b) use a switching function derived from an extreme value distribution in the context of modeling the merchantable volume in a tree bole (see application §8.4.1), namely

$$S(x,\boldsymbol{\theta}) = \exp\{-\alpha x e^{\beta x}\}.$$

This model was derived by considering a switch-off function derived from the Gompertz growth model (Seber and Wild, 1989, p. 330),

$$S(x,\boldsymbol{\theta}) = \exp\{-\exp\{-\alpha(x-\beta)\}\},$$

which is sigmoidal with inflection point at $x = \beta$ but not symmetric about the inflection point. To model the transition in tree profile from neiloid to parabolic to cone-shaped segments, Valentine and Gregoire (2001) use the switch-off function

$$S(x, \boldsymbol{\theta}) = \frac{1}{1 + (x/\alpha)^{\beta}},$$

which is derived from the family of growth models developed for modeling nutritional intake by Morgan et al. (1975). For $\beta > 1$ this switching function has a point of inflection at $x = \alpha\{(\beta - 1)/(\beta + 1)\}^{1/\beta}$ and is hyperbolic for $\beta < 1$. Swinton and Lyford (1996) use the model by Morgan et al. (1975) to test whether crop yield as a function of weed density takes on hyperbolic or sigmoidal shape. Figure 5.29 displays some of the sigmoidal switch-off functions.

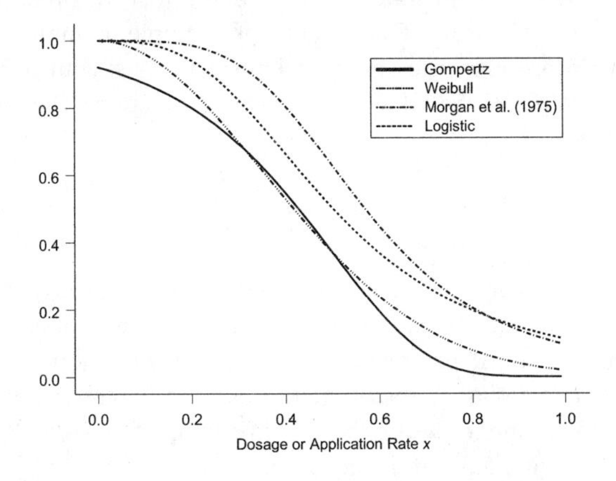

Figure 5.29. Some switch-off functions $S(x, \boldsymbol{\theta})$ discussed in the text. The functions were selected to have inflection points at $x = 0.5$.

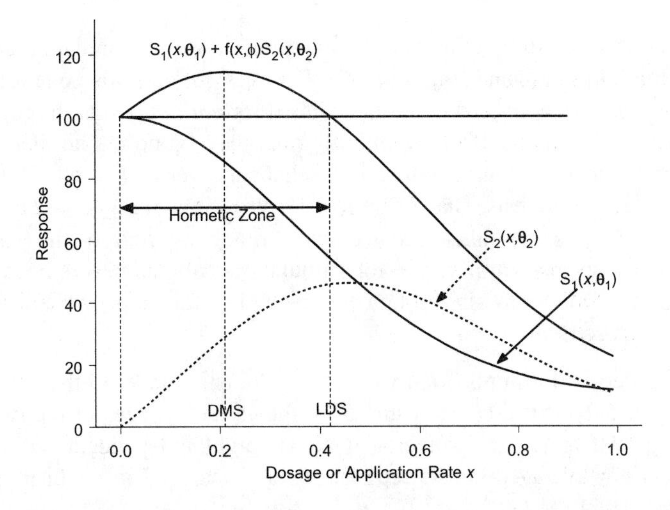

Figure 5.30. Hormetic and nonhormetic dose response. LDS is the limiting dosage for stimulation. DMS is the dosage of maximum stimulation.

Dose-response models without hormetic effect suggest monotonic changes in the response with increasing or decreasing dosage. A hormetic effect is the deviation from this general pattern and in the case of reduced response with increasing dose a beneficial effect is usually observed at low dosages (Figure 5.30).

The method proposed by Schabenberger and Birch (2001) to incorporate hormetic behavior consists of combining a standard model without hormetic effect and a model for the hormetic component. If $S(x,\boldsymbol{\theta})$ is a switch-off function and $f(x,\boldsymbol{\phi})$ is a monotonically increasing function of dosage, then

$$Y = \mu_{\text{min}} + (\mu_{\text{max}} - \mu_{\text{min}})S_1(x,\boldsymbol{\theta}_1) + f(x,\boldsymbol{\phi})S_2(x,\boldsymbol{\theta}_2) \qquad [5.67]$$

is a hormetic model. The switching functions $S_1()$ and $S_2()$ will often be of the same functional form but this is not necessary. One might, for example, combine a Weibull switching function to model the dose-response trend without hormesis with a hormetic component $f(x,\boldsymbol{\phi})S_2(x,\boldsymbol{\theta})$ where $S_2(x,\boldsymbol{\theta})$ is a logistic switching function. The Brain-Cousens model (Brain and Cousens 1989)

$$Y = \delta + \frac{\alpha - \delta + \phi x}{1 + \theta\exp(\beta\ln(x))}$$

is a special case of [5.67] where $S_1(x,\boldsymbol{\theta}) = S_2(x,\boldsymbol{\theta})$ is a log-logistic switch-off function and $f(x,\phi) = \gamma x$. When constructing hormetic models, $f(x,\boldsymbol{\phi})$ should be chosen so that $f(x,\boldsymbol{\phi}) = 0$ for a known set of parameters. The absence of a hormetic effect can then be tested. To prevent a beneficial effect at zero dose we would furthermore require that $f(0,\boldsymbol{\phi}) = 0$. The hormetic model will exhibit a maximum for some dosage (the dosage of maximum stimulation, Figure 5.30) if the equation

$$\frac{\partial f(x,\boldsymbol{\phi})}{\partial x}S_2(x,\boldsymbol{\theta}_2) = -f(x,\boldsymbol{\phi})\frac{\partial S_2(x,\boldsymbol{\theta}_2)}{\partial x}$$

has a solution in x.

The limiting dose for stimulation and the maximum dose of stimulation are only defined for models with a hormetic zone (Figure 5.30). Dosages beyond this zone are interpreted in the same fashion as for a nonhormetic model. This does *not* imply that the researcher can ignore the presence of hormetic effects when only dosages beyond the hormetic zone (such as LD_{50}) are of importance. Through simulation, Schabenberger and Birch (2001) demonstrate the effects of ignoring hormesis. Bias of up to 13% in estimating λ_{20}, λ_{50}, and λ_{75} were observed when hormesis was not taken into account in modeling the growth response. The estimate of the response at the limiting dose for stimulation also had severe negative bias. Once the model accounted for hormesis through the switching function mechanism, these biases were drastically reduced.

In the remainder of this application we fit a log-logistic model to the barnyardgrass data from which Figure 5.27 was created. Table 5.15 shows the dosages of the two active ingredients and the growth percentages averaged across eight independent replications at each dosage. The data not averaged across replications are analyzed in detail in Schabenberger, Tharp, Kells, and Penner (1999).

Table 5.15. Relative growth data for barnyard grass as a function of the concentration of glufosinate and glyphosate† (A control dosage with 100% growth was added)

i	**Glufosinate** ($j=1$) **kg ae / ha**	**Growth %**	**Glyphosate** ($j=2$) **kg ae / ha**	**Growth %**
1	0.018	109.8	0.047	132.3
2	0.036	100.2	0.094	132.9
3	0.071	75.4	0.188	14.5
4	0.143	45.7	0.375	5.3
5	0.286	7.7	0.750	1.9
6	0.572	1.4	1.500	1.8

† Data made kindly available by Dr. James J. Kells, Department of Crop and Soil Sciences, Michigan State University. Used with permission.

To test whether any of the two reponses are hormetic we can fit the Brain-Cousens model

$$\mathrm{E}[Y_{ij}] = \delta_j + \frac{\alpha_j - \delta_j + \gamma_j x_{ij}}{1 + \theta_j \mathrm{exp}\{\beta_j \mathrm{ln}(x_{ij})\}}$$

to the combined data, where Y_{ij} denotes the response for ingredient j at dosage x_{ij}. The main interest in this application is to compare the dosages that lead to 50% reduction in growth response. The Brain-Cousens model is appealing in this regard since it allows fitting a hormetic model to the glyphosate response and a standard log-logistic model to the glufosinate response. It does not incorporate an effective dosage as a parameter in the parameterization [5.65], however. Schabenberger et al. (1999) changed the parameterization of the Brain-Cousens model to enable estimation of λ_K using the method of defining relationships discussed in §5.7.2. The model in which λ_{50}, for example, can be estimated whether the response is hormetic ($\lambda > 0$) or not ($\lambda = 0$), is

$$\mathrm{E}[Y_{ij}] = \delta_j + \frac{\alpha_j - \delta_j + \gamma_j x_{ij}}{1 + \omega_j \mathrm{exp}\{\beta_j \mathrm{ln}\{x_{ij}/\lambda_{50j}\}\}}, \quad \omega_j = 1 + 2\gamma_j \frac{\lambda_{50j}}{\alpha_j - \delta_j}. \qquad [5.68]$$

See Table 1 in Schabenberger et al. (1999) for other parameterizations that allow estimation of general λ_K, LDS, and DMS in the Brain-Cousens model. The `proc nlin` statements to fit model [5.68] are

```
proc nlin data=hormesis method=newton noitprint;
  parameters alpha_glu=100 delta_glu=4 beta_glu=2.0 RD50_glu=0.2
             alpha_gly=100 delta_gly=4 beta_gly=2.0 RD50_gly=0.2
             gamma_glu=300 gamma_gly=300;
  bounds gamma_glu > 0, gamma_gly > 0;
  omega_glu  = 1 + 2*gamma_glu*RD50_glu / (alpha_glu-delta_glu);
  omega_gly  = 1 + 2*gamma_gly*RD50_gly / (alpha_gly-delta_gly);
  term_glu   = 1 + omega_glu * exp(beta_glu*log(rate/RD50_glu));
  term_gly   = 1 + omega_gly * exp(beta_gly*log(rate/RD50_gly));
  model barnyard =
    (delta_glu + (alpha_glu - delta_glu + gamma_glu*rate) / term_glu ) *
                 (Tx = 'glufosinate') +
    (delta_gly + (alpha_gly - delta_gly + gamma_gly*rate) / term_gly ) *
                 (Tx = 'glyphosate') ;
run;
```

The parameters are identified as `*_glu` for glufosinate and `*_gly` for the glyphosate response. The `bounds` statement ensures that `proc nlin` constrains the estimates of the hormesis parameters to be positive. The `omega_*` statements calculate ω_1 and ω_2 and the `term_*` statements the denominator of model [5.68]. The `model` statement uses logical variables to choose between the mean functions for glufosinate and glyphosate.

At this stage we are focusing on parameter estimates for γ_1 and γ_2, as they represent the hormetic component. The parameter are coded as `gamma_glu` and `gamma_gly` (Output 5.22). The asymptotic 95% confidence interval for γ_1 includes 0 ($[-1776.5, 3909.6]$), whereas that for γ_2 does not ($[306.0, 1086.8]$). This confirms the hormetic effect for the glyphosate responses. The p-value for the test of $H_0: \gamma_1 = 0$ vs. $H_1: \gamma_1 > 0$ can be calculated with

```
data pvalue; p=1-ProbT(1066.6/1024.0,4); run; proc print data=pvalue; run;
```

and turns out to be $p = 0.1782$, sufficiently large to dismiss the notion of hormesis for the glufosinate response. The p-value for $H_0: \gamma_2 = 0$ versus $H_1: \gamma_2 > 0$ is $p = 0.0038$ and obtained similarly with the statements

```
data pvalue; p=1-ProbT(696.4/140.6,4); run; proc print data=pvalue; run;
```

Output 5.22.

```
                           The NLIN Procedure

     NOTE: Convergence criterion met.

                          Estimation Summary
                  Method                            Newton
                  Iterations                            11
                  Subiterations                          7
                  Average Subiterations           0.636364
                  R                               6.988E-7
                  PPC(beta_glu)                   4.248E-8
                  RPC(gamma_glu)                   0.00064
                  Object                          1.281E-7
                  Objective                       81.21472
                  Observations Read                     14
                  Observations Used                     14
                  Observations Missing                   0

                                 Sum of        Mean                   Approx
Source                  DF      Squares      Square    F Value    Pr > F
Regression              10      85267.1      8526.7     198.00    <.0001
Residual                 4      81.2147     20.3037
Uncorrected Total       14      85348.3
Corrected Total         13      36262.8

                                   Approx      Approximate 95% Confidence
    Parameter       Estimate    Std Error                Limits

    alpha_glu          100.6       4.5898      87.8883        113.4
    delta_glu       -19.0717      17.9137     -68.8075      30.6642
    beta_glu          1.7568       0.3125       0.8892       2.6245
    RD50_glu          0.1434       0.0343       0.0482       0.2386
    gamma_glu         1066.6       1024.0      -1776.5       3909.6
    alpha_gly        99.9866       4.5055      87.4776        112.5
    delta_gly         2.9241       2.6463      -4.4231      10.2712
    beta_gly          6.2184       0.8018       3.9922       8.4445
    RD50_gly          0.1403      0.00727       0.1201       0.1605
    gamma_gly          696.4        140.6        306.0       1086.8
```

The model we focus on to compare λ_{50} values among the two herbicides thus has a hormetic component for glyphosate ($j = 2$), but not for glufosinate ($j = 1$). Fitting these two apparently different response functions with `proc nlin` is easy; all that is required is to constrain γ_1 to zero in the previous code. This is not the only change we make to the `proc nlin` statements. To obtain a test for H_0:$\lambda_{50(2)} - \lambda_{50(1)} = 0$, we code a new parameter Δ_λ that measures this difference. In other words, instead of coding $\lambda_{50(1)}$ as `RD50_glu` and $\lambda_{50(2)}$ as `RD50_gly` we use

$$\begin{aligned}\texttt{RD50\_glu} &= \lambda_{50(1)}\\ \texttt{RD50\_gly} &= \lambda_{50(1)} + \Delta_\lambda.\end{aligned}$$

We proceed similarly for other parameters, e.g.,

$$\begin{aligned}\texttt{beta\_glu} &= \beta_1\\ \texttt{beta\_gly} &= \beta_1 + \Delta_\beta.\end{aligned}$$

The advantage of this coding method is that parameters for which `proc nlin` gives estimates, estimated asymptotic standard errors, and asymptotic confidence intervals, express differences between the two herbicides. Output 5.23 was generated from the following statements.

```
proc nlin data=hormesis method=newton noitprint;
  parameters alpha_glu=100 delta_glu=4 beta_glu=2.0 RD50_glu=0.2
             alpha_dif=0   delta_dif=0 beta_dif=0   RD50_dif=0
             gamma_gly=300;
  bounds gamma_gly > 0;

  alpha_gly = alpha_glu + alpha_dif;
  delta_gly = delta_glu + delta_dif;
  beta_gly  = beta_glu  + beta_dif;
  RD50_gly  = RD50_glu  + RD50_dif;
  gamma_glu = 0;

  omega_glu = 1 + 2*gamma_glu*RD50_glu / (alpha_glu-delta_glu);
  omega_gly = 1 + 2*gamma_gly*RD50_gly / (alpha_gly-delta_gly);

  term_glu  = 1 + omega_glu * exp(beta_glu*log(rate/RD50_glu));
  term_gly  = 1 + omega_gly * exp(beta_gly*log(rate/RD50_gly));

  model barnyard =
    (delta_glu + (alpha_glu - delta_glu + gamma_glu*rate) / term_glu ) *
                 (Tx = 'glufosinate') +
    (delta_gly + (alpha_gly - delta_gly + gamma_gly*rate) / term_gly ) *
                 (Tx = 'glyphosate') ;
run;
```

Notice that

```
alpha_gly=100 delta_gly=4 beta_gly=2.0 RD50_gly=0.2
```

was removed from the model statement and replaced by

```
alpha_dif=0   delta_dif=0 beta_dif=0   RD50_dif=0.
```

The glyphosate parameters are then reconstructed below the `bounds` statement. A sideeffect of this coding method is the ability to choose zeros as starting values for the Δ parameters assuming initially that the two treatments produce the same response.

Output 5.23.

```
                     The NLIN Procedure

     NOTE: Convergence criterion met.

                       Estimation Summary
              Method                        Newton
              Iterations                        11
              Subiterations                     10
              Average Subiterations       0.909091
              R                           1.452E-8
              PPC                          7.24E-9
              RPC(alpha_dif)              0.000055
              Object                      9.574E-9
              Objective                   125.6643
              Observations Read                 14
              Observations Used                 14
              Observations Missing               0


                                Sum of        Mean               Approx
Source                  DF     Squares      Square    F Value    Pr > F

Regression               9     85222.6      9469.2     179.73    <.0001
Residual                 5       125.7     25.1329
Uncorrected Total       14     85348.3
Corrected Total         13     36262.8


                                 Approx      Approximate 95% Confidence
    Parameter      Estimate    Std Error                Limits

    alpha_glu         105.5       3.5972     96.2174         114.7
    delta_glu       -4.0083       6.3265    -20.2710       12.2543
    beta_glu         2.1651       0.3710      1.2113        3.1188
    RD50_glu         0.1229       0.0128      0.0899        0.1559
    alpha_dif       -5.4777       6.1699    -21.3376       10.3823
    delta_dif        6.9324       6.9781    -11.0050       24.8699
    beta_dif         4.0533       0.9662      1.5697        6.5369
    RD50_dif         0.0174       0.0152     -0.0216        0.0564
    gamma_gly         696.4        156.4       294.3        1098.6
```

The difference in λ_{50} values between the two herbicides is positive (Output 5.23). The λ_{50} estimate for glufosinate is 0.1229 kg ae/ha and that for glyphosate is $0.1229 + 0.0174 = 0.1403$ kg ae/ha. The difference is not statistically significant at the 5% level, since the asymptotic 95% confidence interval for Δ_λ includes zero ($[-.0216, 0.0564]$).

Predicted values for the two herbicides are shown in Figure 5.31. The hormetic effect for glyphosate is very pronounced. The negative estimate $\widehat{\delta}_1 = -4.0083$ suggests that the lower asymptote of relative growth percentages is negative, which is not very meaningful. Fortunately, the growth responses do not achieve that lower asymptote across the range of dosages observed which defuses this issue.

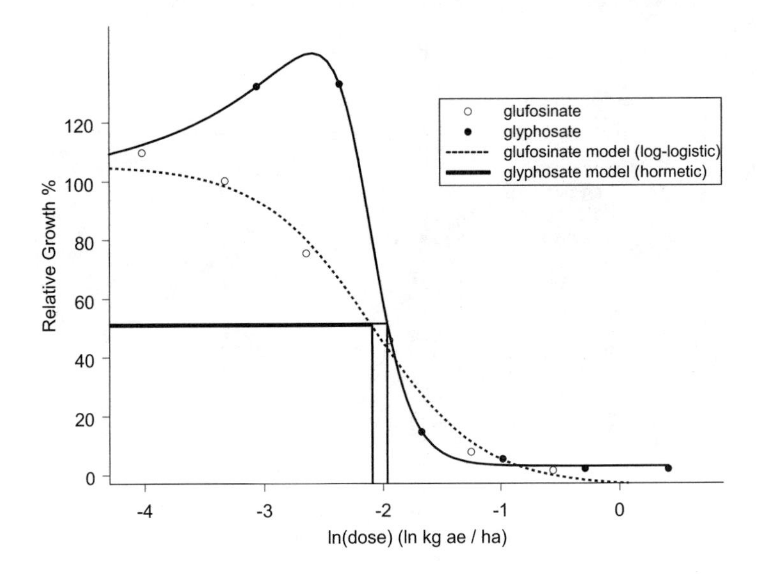

Figure 5.31. Predicted responses for glufosinate and glyphosate for barnyard grass data. Estimated λ_{50} values of 0.1229 kg ae/ha and 0.1403 kg ae/ha are also shown.

What will happen if the hormetic effect for the glyphosate response is being ignored, that is, we fit a log-logistic model

$$\mathrm{E}[Y_{ij}] = \delta_j + \frac{\alpha_j - \delta_j}{1 + \exp\{\beta_j \ln\{x_{ij}/\lambda_{50j}\}\}}?$$

The solid line in Figure 5.31 will be forced to monotonically decrease as the dashed line. Because of the solid circles in excess of 100% the glyphosate model will attempt to stay elevated for as long as possible and then decline sharply toward the solid circles on the right (Figure 5.32). The estimate of β_2 will be large and have very low precision. Also, the residual sum of squares should increase dramatically.

All of these effects are apparent in Output 5.24 which was generated by the statements

```
proc nlin data=hormesis method=newton noitprint;
  parameters alpha_glu=100 delta_glu=4 beta_glu=2.0 RD50_glu=0.122
             alpha_dif=0   delta_dif=0 beta_dif=0   RD50_dif=0;

  alpha_gly = alpha_glu + alpha_dif;
  delta_gly = delta_glu + delta_dif;
  beta_gly  = beta_glu  + beta_dif;
  RD50_gly  = RD50_glu  + RD50_dif;

  term_glu  = 1 + exp(beta_glu*log(rate/RD50_glu));
  term_gly  = 1 + exp(beta_gly*log(rate/RD50_gly));

  model barnyard = (delta_glu + (alpha_glu - delta_glu) / term_glu ) *
                               (Tx = 'glufosinate') +
                   (delta_gly + (alpha_gly - delta_gly) / term_gly ) *
                               (Tx = 'glyphosate') ;
run;
```

Output 5.24.

```
                          The NLIN Procedure

     NOTE: Convergence criterion met.

                         Estimation Summary
                   Method                          Newton
                   Iterations                          77
                   Subiterations                       16
                   Average Subiterations         0.207792
                   R                             9.269E-6
                   PPC(beta_dif)                 0.002645
                   RPC(beta_dif)                 0.005088
                   Object                        8.27E-11
                   Objective                     836.5044
                   Observations Read                   14
                   Observations Used                   14
                   Observations Missing                 0

                             Sum of         Mean                 Approx
Source                DF    Squares       Square    F Value      Pr > F
Regression             8    84511.8      10564.0      36.30      0.0002
Residual               6      836.5        139.4
Uncorrected Total     14    85348.3
Corrected Total       13    36262.8

                                    Approx       Approximate 95% Confidence
    Parameter        Estimate    Std Error                 Limits

    alpha_glu           105.5       8.4724       84.7330         126.2
    delta_glu         -4.0083      14.9006      -40.4688       32.4521
    beta_glu           2.1651       0.8738        0.0268        4.3033
    RD50_glu           0.1229       0.0302        0.0489        0.1969
    alpha_dif         16.2837      10.8745      -10.3252       42.8926
    delta_dif          7.0527      16.3860      -33.0424       47.1477
    beta_dif          31.7039       4595.8      -11213.8       11277.2
    RD50_dif           0.0531       1.5800       -3.8131        3.9193
```

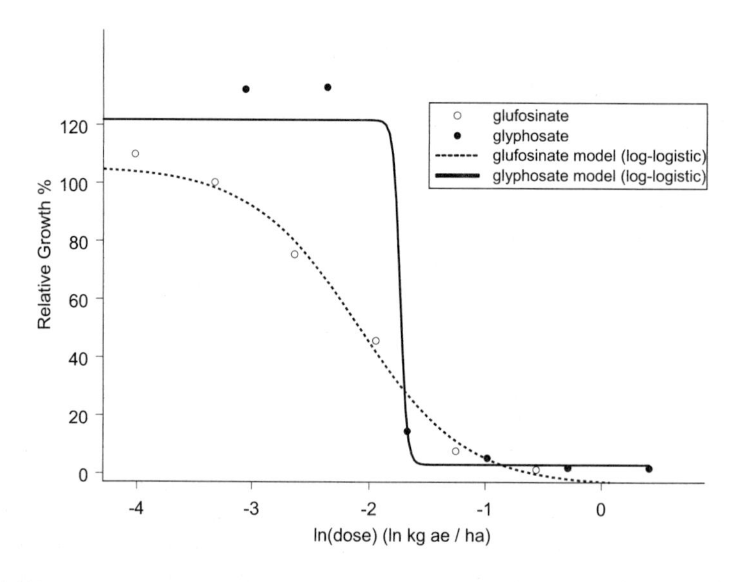

Figure 5.32. Fit of log-logistic model to barnyard grass data.

The parameter Δ_β is estimated as 31.7039 and hence $\widehat{\beta}_2 = 2.1651 + 31.7039 = 33.869$. This is a very unreasonable value and its estimated asymptotic standard error

$$\text{ease}(\widehat{\Delta}_\beta) = 4,595.8$$

is nonsensical. Notice, however, that the fit of the model to the glufosinate data has not changed (compare to Output 5.23, and compare Figures 5.31 and 5.32).

5.8.7 Modeling a Yield-Density Relationship

Mead (1970) notes the long history into examinations of the relationship between crop yield and plant densities. For considerable time the area of most substantial progress and practical importance was the study of competition among plants of a single crop as a function of their density. Mead (1979) classifies experiments in this area into yield-density and genotype competition studies. In this subsection we briefly review the theory behind some common yield-density models and provide a detailed analysis of a data set examined earlier by Mead (1970). There is little to be added to the thorough analysis of these data. This subsection will illustrate the main steps and their implementation with The SAS® System. Additional yield-density models and background material can be found in §A5.9.1 and the references cited there.

To fix ideas let Y denote the **yield per plant** of a species and x the **density per unit area** at which the species grows. The product $U = Yx$ then measures the **yield per unit area**. Two main theories of intraspecies competition are reflected in common yield-density models:

- $\text{E}[U]$ is an increasing function of x that reaches an asymptote for some density x^*. Decreasing the density below x^* will decrease area yield. For densities above x^*, yield per plant decreases at the same rate as the density increases, holding the yield per unit area constant (Figure 5.33a). Yield-density relationships of this kind are termed asymptotic and have been established for peas (Nichols and Nonnecke 1974), tomatoes (Nichols et al. 1973), dwarf beans (Nichols 1974a), and onions (Mead 1970), among other crops. The asymptotic yield can be interpreted as the growth limit for a species in a particular environment.

- $\text{E}[U]$ is a parabolic function without an asymptote, but a maximum that occurs at density $x_{\max}$ (Figure 5.33b). For densities in excess of $x_{\max}$ the yield per plant decreases more quickly than the density increases, reducing the unit area yield. Parabolic relationships were established, for example, for parsnips (Bleasdale and Thompson 1966), sweat corn (Nichols 1974b), and cotton (Hearn 1972).

In either case the yield per plant is a convex function of plant density. Several parameters are of particular interest in the study of yield-density models. If $Y(x)$ tends to a constant c_s as density tends toward 0, c_s reflects the species' potential in the absence of competition from other plants. Similarly, if $U(x)$ tends to a constant c_e as density increases toward infinity, c_e measures the species' potential under increased competition for environmental resources. Ratkowsky (1983, p. 50) terms c_s the *genetic* potential and c_e the *environmental* potential of the species. For asymptotic relationships agronomists are often not only interested in c_s and c_e but also in the density that produces a certain percentage of the asymptotic yield. In Figure

5.33a the density at which 80% of the asymptotic yield ($U(\infty) = 1.25$) is achieved is shown as $x_{0.8}$. If the yield-density relationship is parabolic, the density related to a certain percentage of $U_{\max}$ is usually not of interest, in part because it may not be uniquely determined. The modeler is instead interested in the density at which unit area yield is maximized ($x_{\max}$ in Figure 5.33b).

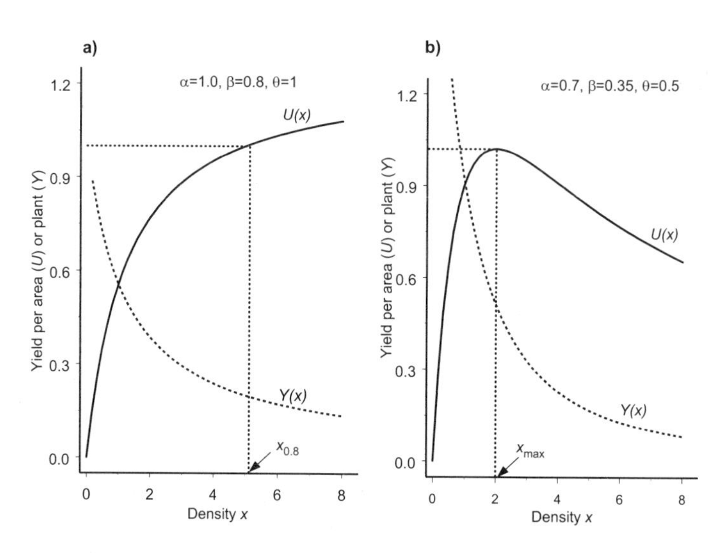

Figure 5.33. Asymptotic (a) and parabolic (b) yield-density relationships. $U(x)$ denotes yield per unit area at density x, $Y(x)$ denotes yield per plant at density x. Both models are based on the Bleasdale-Nelder model $U(x) = x(\alpha + \beta x)^{-1/\theta}$ discussed in the text and in §A5.9.1.

The most basic nonlinear yield-density model is the reciprocal simple linear regression

$$\mathrm{E}[Y] = (\alpha + \beta x)^{-1}$$

due to Shinozaki and Kira (1956). Its area yield function $\mathrm{E}[U] = x(\alpha + \beta x)^{-1}$ is strictly asymptotic with genetic potential $c_s = 1/\alpha$ and asymptotic yield per unit area $U(\infty) = 1/\beta$. In applications one may not want to restrict modeling efforts from the outset to asymptotic relationships and employ a model which allows both asymptotic and parabolic relationships, depending on parameter values. A simple extension of the Shinozaki-Kira model is known as the Bleasdale-Nelder model (Bleasdale and Nelder 1960),

$$\mathrm{E}[Y] = (\alpha + \beta x)^{-1/\theta}$$
$$\mathrm{E}[U] = x(\alpha + \beta x)^{-1/\theta}.$$

This model is extensively discussed in Mead (1970), Gillis and Ratkowsky (1978), Mead (1979), and Ratkowsky (1983). A more general form with four parameters that was originally proposed is discussed in §A5.9.1. For $\theta = 1$ the model is asymptotic and parabolic for $\theta < 1$ (Figure 5.33). A one-sided statistical test of $H_0\colon \theta = 1$ vs. $H_1\colon \theta < 1$ allows testing for asymptotic vs. parabolic structure of the relationship between U and x. Because of its biological relevance, such a test should always be performed when the Bleasdale-Nelder model is considered. Values of θ greater than one are not permissible since $U(x)$ does not have a maximum then. In the Bleasdale-Nelder model with $\theta = 1$ the density at which $K\%$ of the asymptotic yield $(1/\beta)$ are achieved is given by

$$x_{K/100} = \frac{\alpha}{\beta}\left(\frac{K/100}{1 - K/100}\right)$$

and in the parabolic case $(\theta < 1)$ the maximum yield per unit area of

$$U_{\max} = \frac{\theta}{\beta}\left(\frac{1-\theta}{\alpha}\right)^{(1-\theta)/\theta}$$

is obtained at density $x_{\max} = (\alpha/\beta)\{\theta/(1-\theta)\}$.

Ratkowsky (1983, 1990) established rather severe curvature effects in this model, whereas the Shinozaki-Kira model behaves close-to-linear. Through simulation studies Gillis and Ratkowsky (1978) showed that if the true relationship between yield and density is $Y = (\alpha + \beta x)^{-1/\theta}$ but the Bleasdale-Nelder model is fitted, non-negligible bias may be incurred. These authors prefer the model

$$\mathrm{E}[Y] = \left(\alpha + \beta x + \gamma x^2\right)^{-1}$$

due to Holliday (1960) which also allows parabolic and asymptotic relationships and its parameter estimators show close-to-linear behavior.

When fitting yield-density models, care should be exercised because the variance of the plant yield Y typically increases with the yield. Mead (1970) states that the assumption

$$\mathrm{Var}[Y] = \sigma^2 \mathrm{E}[Y]^2$$

is often tenable. Under these circumstances the logarithm of Y has approximately constant variance. If $f(x, \boldsymbol{\theta})$ is the yield-density model, one approach to estimation of the parameters $\boldsymbol{\theta}$ is to fit

$$\mathrm{E}[\ln\{Y\}] = \ln\{f(x, \boldsymbol{\theta})\}$$

assuming that the errors of this model are zero mean Gaussian random variables with constant variance σ^2. For the Bleasdale-Nelder model this leads to

$$\mathrm{E}[\ln\{Y\}] = \ln\left\{(\alpha + \beta x)^{-1/\theta}\right\} = -\frac{1}{\theta}\ln\{\alpha + \beta x\}.$$

Alternatively, one can fit the model $Y = f(x, \boldsymbol{\theta})$ assuming that Y follows a distribution with variance proportional to $\mathrm{E}[Y]^2$. The family of Gamma distributions has this property, for example. Here we will use the logarithmic transformation and revisit the fitting of yield density models in §6.7.3 under the assumption of Gamma distributed yields.

The data in Table 5.16 and Figure 5.34 represent yields per plant of three onion varieties grown at varying densities. There were three replicates of each density, the data values represent their averages. An exploratory graph of $1/Y$ versus density shows that a linear relationship is not unreasonable, confirmed by a loess smooth of $1/Y$ vs. x (Figure 5.34).

Table 5.16. Onion yield-density data of Mead (1970)
(y represents observed yield per plant in grams, x the density in plants $\times$ ft^{-2})

Variety 1		Variety 2		Variety 3	
y	x	y	x	y	x
105.6	3.07	131.6	2.14	116.8	2.48
89.4	3.31	109.1	2.65	91.6	3.53
71.0	5.97	93.7	3.80	72.7	4.45
60.3	6.99	72.2	5.24	52.8	6.23
47.6	8.67	53.1	7.83	48.8	8.23
37.7	13.39	49.7	8.72	39.1	9.59
30.3	17.86	37.8	10.11	30.3	16.87
24.2	21.57	33.3	16.08	24.2	18.69
20.8	28.77	24.5	21.22	20.0	25.74
18.5	31.08	18.3	25.71	16.3	30.33

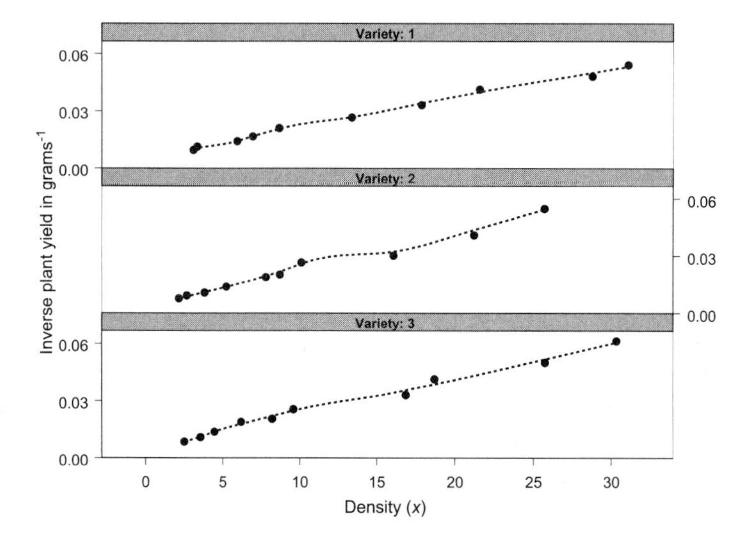

Figure 5.34. Relationships between inverse plant yield and plant densities for three onion varieties. Dashed line is a nonparametric loess fit. Data from Mead (1970).

The relationship between U and density is likely of an asymptotic nature for any of the three varieties. The analysis commences with a fit of the full model

$$\ln\{Y_{ij}\} = \ln\left\{(\alpha_i + \beta_i x_{ij})^{-1/\theta_i}\right\} + e_{ij}, \qquad [5.69]$$

where the subscript $i = 1, \cdots, 3$ denotes the varieties and x_{ij} is the jth density $(j = 1, \cdots, 10)$ at which the yield of variety i was observed. The following hypotheses are to be addressed subsequently:

A $H_0 : \theta_1 = \theta_2 = \theta_3 = 1 \Leftrightarrow$ the three varieties do not differ in the parameter θ and the relationship is asymptotic.

B If the hypothesis in A is rejected then we test separately $H_0{:}\theta_i = 1$ to find out which variety exhibits parabolic behavior. The alternative hypothesis for these tests should be $H_1{:}\theta_i < 1$. If the test in A was not rejected, we proceed with a modified full model

$$\ln\{Y_{ij}\} = -\ln\{\alpha_i + \beta_i x_{ij}\} + e_{ij}.$$

C H_0: $\alpha_1 = \alpha_2 = \alpha_3 \Leftrightarrow$ The varieties do not differ in the parameter α. Depending on the outcomes of A and B the genetic potential is estimated as α_i^{-1}, $\alpha_i^{-1/\theta}$ or α_i^{-1/θ_i}. If the hypothesis H_0: $\alpha_1 = \alpha_2 = \alpha_3$ is rejected one can proceed to test the parameters in pairs.

D H_0: $\beta_1 = \beta_2 = \beta_3 \Leftrightarrow$ The varieties do not differ in the parameter β. Proceed as in C.

Prior to tackling these hypotheses the full model should be tested against the most reduced model

$$\ln\{Y_{ij}\} = \ln\left\{(\alpha + \beta x_{ij})^{-1/\theta}\right\} + e_{ij}$$

or even

$$\ln\{Y_{ij}\} = \ln\{(\alpha + \beta x_{ij})^{-1}\} + e_{ij}$$

to determine whether there are any differences among the three varieties. In addition we are also interested in obtaining confidence intervals for the genetic potential c_s, for the density x_{max} should the relationship be parabolic, for the density $x_{0.9}$ and the asymptote $U(\infty)$ should the relationship be asymptotic. In order to obtain these intervals one could proceed by reparameterizing the model such that c_s, x_{max}, $x_{0.9}$, and $U(\infty)$ are parameters and refit the resulting model(s). Should the relationship be parabolic this proves difficult since c_s, for example, is a function of both α and θ. Instead we fit the final model with the `nlmixed` procedure of The SAS® System that permits the estimation of arbitrary functions of the model parameters and calculates the standard errors of the estimated functions by the delta method.

The full model is fit with the SAS® statements

```
proc nlin data=onions method=marquardt;
  parameters a1=5.4    a2=5.4   a3=5.4
             b1=1.7    b2=1.7   b3=1.7
             t1=1      t2=1     t3=1;
  term1 = (-1/t1)*log(a1 + b1*density);
  term2 = (-1/t2)*log(a2 + b2*density);
  term3 = (-1/t3)*log(a3 + b3*density);
  model logyield = term1*(variety=1)+term2*(variety=2)+term3*(variety=3);
run;
```

Starting values for the α_i and β_i were obtained by assuming $\theta_1 = \theta_2 = \theta_3 = 1$ and fitting the inverse relationship $1/(Y/1000) = \alpha + \beta x$ with a simple linear regression package. For scaling purposes, the plant yield was expressed in kilograms rather than in grams as in Table 5.16. The full model achieves a residual sum of squares of $S(\widehat{\theta}) = 0.1004$ on 21 degrees of freedom. The asymptotic model $\log\{Y_{ij}\} = -\ln\{\alpha + \beta x_{ij}\} + e_{ij}$ with common potentials has a residual sum of square of 0.2140 on 28 degrees of freedom. The initial test for determining whether there are any differences in yield-density response among the three varieties is rejected:

$$F_{obs} = \frac{(0.2140 - 0.1004)/7}{0.1004/21} = 3.394, \; p = 0.0139$$

The model restricted under H_0: $\theta_1 = \theta_2 = \theta_3 = 1$ is fit with the statements

```
proc nlin data=onions method=marquardt;
  parameters a1=5.4   a2=5.4  a3=5.4  b1=1.7   b2=1.7  b3=1.7;
  term1 = -log(a1 + b1*density);  term2 = -log(a2 + b2*density);
  term3 = -log(a3 + b3*density);
  model logyield = term1*(variety=1)+term2*(variety=2)+term3*(variety=3);
run;
```

and achieves $S(\widehat{\boldsymbol{\theta}})_{H_0} = 0.1350$ on 21 degrees of freedom. The F-test has test statistic $F_{obs} = (0.1350 - 0.1004)/(3*0.1004/21) = 2.412$ and p-value $p = \Pr(F_{3,21} > 2.412) = 0.095$. At the 5% significance level H_0 cannot be rejected and the model

$$\ln\{Y_{ij}\} = -\ln\{\alpha_i + \beta_i x_{ij}\} + e_{ij}$$

is used as the full model henceforth. Varying β and fixing α at a common value for the varieties is accomplished with the statements

```
proc nlin data=onions method=marquardt;
  parameters a=4.5  b1=1.65 b2=1.77 b3=1.90;
  term1 = -log(a + b1*density);  term2 = -log(a + b2*density);
  term3 = -log(a + b3*density);
  model logyield = term1*(variety=1)+term2*(variety=2)+term3*(variety=3);
run;
```

This model has $S(\widehat{\boldsymbol{\theta}})_{H_0} = 0.1519$ with 26 residual degrees of freedom. The test for H_0: $\alpha_1 = \alpha_2 = \alpha_3$ leads to $F_{obs} = (0.1519 - 0.1350)/(2*0.1350/24) = 1.50$ $(p = 0.243)$. It is reasonable to assume that the varieties share a common genetic potential. Similarly, the invariance of the β_i can be tested with

```
proc nlin data=onions method=marquardt;
  parameters a1=5.4 a2=5.4 a3=5.4 b=1.7;
  term1 = -log(a1 + b*density);  term2 = -log(a2 + b*density);
  term3 = -log(a3 + b*density);
  model logyield = term1*(variety=1)+term2*(variety=2)+term3*(variety=3);
run;
```

This model achieves $S(\widehat{\boldsymbol{\theta}})_{H_0} = 0.1843$ with 26 residual degrees of freedom. The test for H_0:$\beta_1 = \beta_2 = \beta_3$ leads to $F_{obs} = (0.1843 - 0.1350)/(2*0.1350/24) = 4.38$ $(p = 0.024)$. The notion of invariant β parameters is rejected.

We are now in a position to settle on a final model for the onion yield density data,

$$\ln\{Y_{ij}\} = -\ln\{\alpha + \beta_i x_{ij}\} + e_{ij}. \qquad [5.70]$$

Had we started with the Holliday model instead of the Bleasdale-Nelder model, the initial test for an asymptotic relationship would have yielded $F_{obs} = 1.575$ $(p = 0.225)$ and the final model would have been the same.

We are now interested in calculating confidence intervals for the common genetic potential $(1/\alpha)$, the variety-specific environmental potentials $(1/\beta_i)$, the density that produces 90% of the asymptotic yield ($x_{0.9}$) and the magnitude of 90% of the asymptotic yield ($U_{0.9}$). Furthermore we want to test for significant varietal differences in the environmental potentials and the $x_{0.9}$ values. The quantities of interest are functions of the parameters α and β_i. Care must be exercised to calculate the appropriate standard errors. For example, if ese$(\widehat{\alpha})$ is the estimated standard error of $\widehat{\alpha}$, then ese$(1/\widehat{\alpha}) \neq 1/ese(\widehat{\alpha})$. The correct standard errors of nonlinear functions of the parameters can be obtained by the delta method. The SAS® procedure `nlmixed` enables the estimation of any function of the parameters in a nonlinear model by this method. `Nlmixed` is designed to fit nonlinear models to clustered data where the model contains more than one random effect (§8), but can also be employed to fit regular nonlinear models. A slight difference lies in the estimation of the residual variance Var$[e_{ij}] = \sigma^2$ between the `nlmixed` and `nlin` procedures. To ensure that the two analyses agree, we force `proc nlmixed` to use the residual mean square estimate $\widehat{\sigma}^2 = 0.1519/26 = 0.00584$ of the finally selected model. The complete statements are

```
proc nlmixed data=onions df=26;
  parameters a=4.5 b1=1.65 b2=1.77 b3=1.90;
  s2=0.00584;
  term1 = -log(a + b1*density);
  term2 = -log(a + b2*density);
  term3 = -log(a + b3*density);
  meanmodel = term1*(variety=1)+term2*(variety=2)+term3*(variety=3);
  model logyield ~ Normal(meanmodel,s2);
  estimate 'x_09 (1)' (a/b1)*(0.9/0.1);
  estimate 'x_09 (2)' (a/b2)*(0.9/0.1);
  estimate 'x_09 (3)' (a/b3)*(0.9/0.1);

  estimate 'U_09 (1)' 1000*(0.9/b1);
  estimate 'U_09 (2)' 1000*(0.9/b2);
  estimate 'U_09 (3)' 1000*(0.9/b3);

  estimate 'genetic potential ' 1000/a;  /* common genetic potential */
  estimate 'U(infinity) (1)   ' 1000/b1; /* U asymptote variety 1 */
  estimate 'U(infinity) (2)   ' 1000/b2; /* U asymptote variety 2 */
  estimate 'U(infinity) (3)   ' 1000/b3; /* U asymptote variety 3 */

  /* Test differences among the yield per unit area asymptotes */

  estimate 'U(i)(1) - U(i)(2)' 1000/b1 - 1000/b2;
  estimate 'U(i)(1) - U(i)(3)' 1000/b1 - 1000/b3;
  estimate 'U(i)(2) - U(i)(3)' 1000/b2 - 1000/b3;

/* Test differences among the densities producing 90% of asympt. yield */

  estimate 'x_09(1) - x_09(2) ' (a/b1)*(0.9/0.1) - (a/b2)*(0.9/0.1);
  estimate 'x_09(1) - x_09(3) ' (a/b1)*(0.9/0.1) - (a/b3)*(0.9/0.1);
  estimate 'x_09(2) - x_09(3) ' (a/b2)*(0.9/0.1) - (a/b3)*(0.9/0.1);
run;
```

The final parameter estimates are $\widehat{\alpha} = 4.5364$, $\widehat{\beta}_1 = 1.6611$, $\widehat{\beta}_2 = 1.7866$, and $\widehat{\beta}_3 = 1.9175$ (Output 5.25). The predicted yield per area at density x (Figure 5.35) is thus calculated (in grams $\times$ ft^{-2}) as

$$\begin{aligned} \text{Variety 1:} \quad \widehat{U} &= 1000 \times x(4.5364 + 1.6611x)^{-1} \\ \text{Variety 2:} \quad \widehat{U} &= 1000 \times x(4.5364 + 1.7866x)^{-1} \\ \text{Variety 3:} \quad \widehat{U} &= 1000 \times x(4.5364 + 1.9175x)^{-1}. \end{aligned}$$

Output 5.25.

```
                     The NLMIXED Procedure

                        Specifications
        Data Set                                 WORK.ONIONS
        Dependent Variable                       logyield
        Distribution for Dependent Variable      Normal
        Optimization Technique                   Dual Quasi-Newton
        Integration Method                       None

                       Iteration History

        Iter     Calls    NegLogLike        Diff      MaxGrad        Slope
          1         4     -36.57033     0.173597     1.145899     -90.5784
          2         7    -36.572828     0.002499     0.207954     -1.30563
          3         9    -36.574451     0.001623     0.513823       -0.046
          4        11    -36.576213     0.001763     0.072027     -0.37455
          5        12    -36.576229     0.000016     0.000596     -0.00003
          6        14    -36.576229     1.604E-9     9.455E-7     -3.21E-9
               NOTE: GCONV convergence criterion satisfied.

                      Parameter Estimates

                       Standard
Parameter  Estimate     Error    DF  t Value  Pr > |t|   Lower      Upper
a            4.5364    0.3467    26   13.08    <.0001    3.8237     5.2492
b1           1.6611   0.06212    26   26.74    <.0001    1.5334     1.7888
b2           1.7866   0.07278    26   24.55    <.0001    1.6370     1.9362
b3           1.9175   0.07115    26   26.95    <.0001    1.7713     2.0637

                     Additional Estimates

                              Standard           t
Label                 Estimate     Error   DF   Value  Pr > |t|     Lower     Upper
x_09 (1)                 24.57    2.5076   26    9.80   <.0001    19.4246    29.733
x_09 (2)                 22.85    2.4217   26    9.44   <.0001    17.8741    27.829
x_09 (3)                 21.29    2.1638   26    9.84   <.0001    16.8447    25.740

U_09 (1)                541.81   20.2611   26   26.74   <.0001    500.17     583.46
U_09 (2)                503.74   20.5197   26   24.55   <.0001    461.56     545.92
U_09 (3)                469.36   17.4148   26   26.95   <.0001    433.57     505.16

genetic potential       220.44   16.8493   26   13.08   <.0001    185.80     255.07

U(infinity) (1)         602.01   22.5123   26   26.74   <.0001    555.74     648.29
U(infinity) (2)         559.71   22.7997   26   24.55   <.0001    512.85     606.58
U(infinity) (3)         521.51   19.3498   26   26.95   <.0001    481.74     561.29

U(i)(1) - U(i)(2)        42.30   26.1924   26    1.62   0.1184   -11.5377    96.140
U(i)(1) - U(i)(3)        80.50   24.8330   26    3.24   0.0032    29.4559    131.55
U(i)(2) - U(i)(3)        38.19   24.5919   26    1.55   0.1324   -12.3500    88.748

x_09(1) - x_09(2)         1.72    1.0716   26    1.61   0.1191    -0.4756    3.9298
x_09(1) - x_09(3)         3.28    1.0628   26    3.09   0.0047     1.1021    5.4712
x_09(2) - x_09(3)         1.55    1.0257   26    1.52   0.1404    -0.5487    3.6679
```

The genetic potential is estimated as 220.44 grams $\times$ ft^{-2} with asymptotic 95% confidence interval $[185.8, 255.1]$. The estimated yields per unit area asymptotes for varieties 1, 2, and 3 are 602.01, 559.71, and 521.51 grams $\times$ ft^{-2}, respectively. Only varieties 1 and 3 differ significantly in this parameter $(p = 0.0032)$ and the density that produces 90% of the asymptotic yield $(p = 0.0047)$.

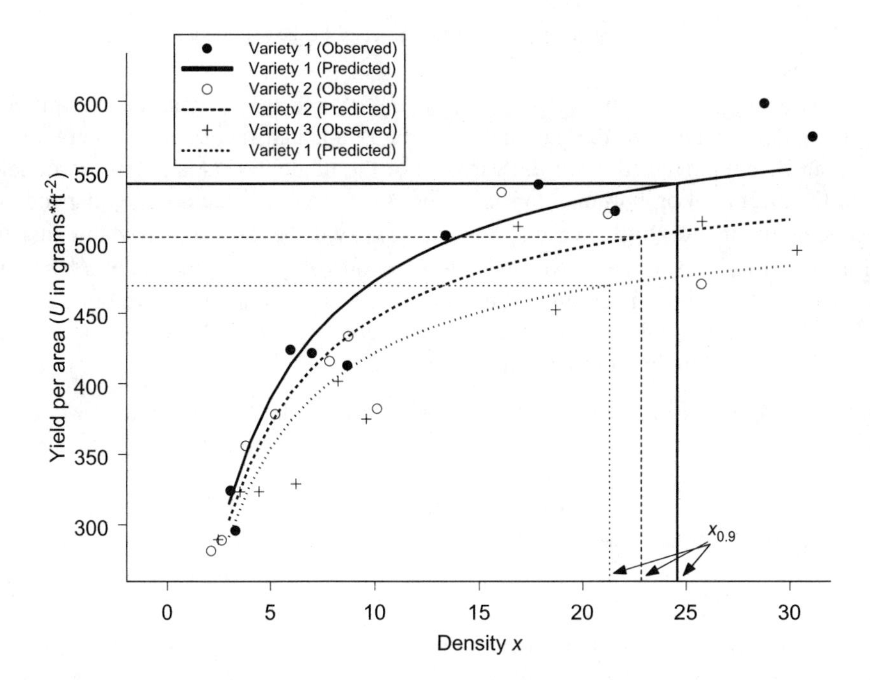

Figure 5.35. Observed and predicted values for onion yield-density data.

5.8.8 Weighted Nonlinear Least Squares Analysis with Heteroscedastic Errors

Heterogeneity of the error variance in nonlinear models presents similar problems as in linear models. In §5.6.2 we discussed various approaches to remedy the heteroscedasticity problem. One relied on applying a Box-Cox type transformation to both sides of the model that stabilizes the variance and to fit the transformed model. The second approach leaves the response unchanged but fits the model by weighted nonlinear least squares where the weights reflect the degree of variance heterogeneity. We generally prefer the second approach over the power-transformation approach because it does not alter the response and the results are interpreted on the original scale of measurement. However, it does require that the relationship between variability and the response can be expressed in mathematical terms so that weights can be defined. In the setting of §5.6.2 the weighted least squares approach relied on the fact that in the model

$$Y_i = f(\mathbf{x}_i, \boldsymbol{\theta}) + e_i$$

the errors e_i are uncorrelated with variance proportional to a power of the mean,

$$\text{Var}[e_i] = \sigma^2 f(\mathbf{x}_i, \boldsymbol{\theta})^\lambda = \sigma^2 \text{E}[Y_i]^\lambda.$$

Consequently, $e_i^* = e_i / f(\mathbf{x}_i, \boldsymbol{\theta})^{\lambda/2}$ will have constant variance σ^2 and a weighted least squares approach would obtain updates of the parameter estimates as

$$\widehat{\theta}^{u+1} = \widehat{\theta}^{u} + \left(\widehat{\mathbf{F}}'\mathbf{W}^{-1}\widehat{\mathbf{F}}\right)^{-1}\widehat{\mathbf{F}}'\mathbf{W}^{-1}\left(\mathbf{y} - \mathbf{f}(\mathbf{x},\widehat{\boldsymbol{\theta}})\right),$$

where $\mathbf{W}$ is a diagonal matrix of weights $f(\mathbf{x}_i, \boldsymbol{\theta})^{\lambda}$. Since the weights depend on $\boldsymbol{\theta}$ they should be updated whenever the parameter vector is updated, i.e., at every iteration. This problem can be circumvented when the variance of the model errors is not proportional to a power of the mean, but proportional to some other function $g(\mathbf{x}_i)$ that does not depend on the parameters of the model. In this case $e_i^* = e_i/\sqrt{g(\mathbf{x}_i)}$ will be the variance stabilizing transform for the errors. But how can we find this function $g(\mathbf{x}_i)$? One approach is by trial and error. Try different weight functions and examine the weighted nonlinear residuals. Settle on the weight function that stabilizes the residual variation. The approach we demonstrate here is also an ad-hoc procedure but it utilizes the data more.

Before going into further details, we take a look at the data and the model we try to fit. The Richards curve (Richards 1959) is a popular model for depicting plant growth owing to its flexibility and simple interpretation. It is not known, however, for excellent statistical properties of its parameter estimates. The Richards model — also known as the Chapman-Richards model (Chapman 1961) — exists in a variety of parameterizations, for example,

$$Y_i = \alpha\left(1 - e^{\beta x_i}\right)^{\gamma} + e_i. \qquad [5.71]$$

Here, α is the maximum growth achievable (upper asymptote), β is the rate of growth, and the parameter γ determines the shape of the curve near the origin. For $\gamma > 1$ the shape is sigmoidal. The covariate x in the Richards model is often the age of an organism, or a measure of its size. Values for γ in the neighborhood of 1.0 are common as are values for β at approximately –0.5.

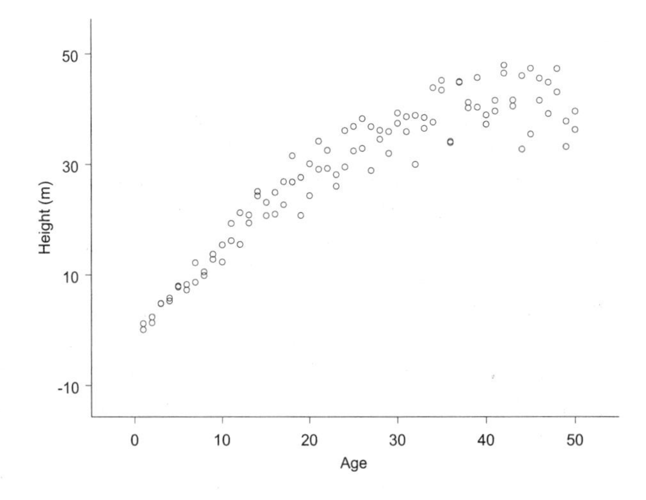

Figure 5.36. Height in meters of 100 Sitka spruce trees as a function of tree age in years. Data generated according to discussion in Rennolls (1993).

Inspired by results in Rennolls (1993) the simulated heights in meters of 100 Sitka spruces (*Picea sitchensis* (Bong.) Carr.) is plotted as a function of their age in years in Figure

5.36. There appears to be an upper height asymptote at approximately 45 meters but it is not obvious whether the model should be sigmoidal or not. Fitting the Richards model [5.71] with $x_i = age_i$, the age of the ith tree , we can let the parameter γ guide whether the shape of the growth response is sigmoidal near the origin. There is clearly variance heterogeneity in these data as the height of older trees varies more than the height of younger trees.

In §4.5.2 we attempted to unlock the relationship between variances and means in an experimental design situation by fitting a linear model between the log sample standard deviation and the log treatment means. This was made possible by having replicate values for each treatment from which the variance in the treatment group could be estimated. In a regression context such as this, there are none or only a few replicate values. One possibility is to group nearby ages and estimate the sample variance within the groups. The groups should be chosen large enough to allow a reasonably stable estimate of the variance and small enough so that the assumption of a constant mean within the group is tenable. An alternative approach is to fit the model by ordinary nonlinear least squares and to plot the squared fitted residuals against the regressor, trying to glean the error variance/regressor relationship from this plot (Figure 5.38 c).

Applying the grouping approach with six points per group, the sample variances in the groups are plotted against the square roots of the average ages in the groups in Figure 5.37. That the variance is proportional to $\sqrt{age}$ is not an unreasonable assumption based on the figure. The diagonal weight matrix **W** for nonlinear weighted least square should have entries $age_i^{1/2}$.

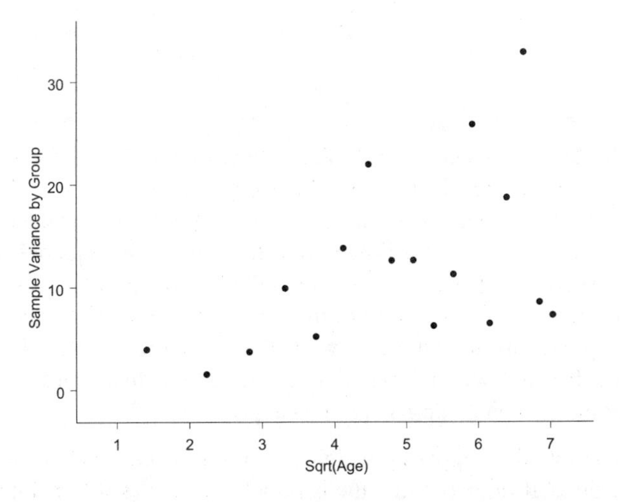

Figure 5.37. Sample standard deviations for groups of six observations against the square root of tree age.

To obtain a weighted nonlinear least squares analysis in SAS®, we call upon the `_weight_` variable in `proc nlin`. The `_weight_` variable can refer to a variable in the data set or to a valid SAS® expression. SAS® calculates it for each observation and assigns the reciprocal values as diagonal elements of the **W** matrix. The statements (Output 5.26) to model the variances of an observation as a multiple of the root ages are

```
proc nlin data=spruce;
  parameters alpha=50 beta=-0.05 gamma=1.0;
  model height = alpha*(1-exp(beta*age))**gamma;
  _weight_ = 1/sqrt(age);
run;
```

Output 5.26.

```
                        The NLIN Procedure

     NOTE: Convergence criterion met.

                          Estimation Summary
                  Method                        Gauss-Newton
                  Iterations                               5
                  R                                 6.582E-7
                  PPC(gamma)                        5.639E-7
                  RPC(gamma)                        0.000015
                  Object                            6.91E-10
                  Objective                         211.0326
                  Observations Read                      100
                  Observations Used                      100
                  Observations Missing                     0

                                Sum of        Mean                   Approx
Source                   DF    Squares      Square    F Value     Pr > F
Regression                3    18008.7      6002.9    2759.19     <.0001
Residual                 97      211.0      2.1756
Uncorrected Total       100    18219.7
Corrected Total          99     5641.2

                                     Approx      Approximate 95% Confidence
  Parameter          Estimate     Std Error                 Limits
  alpha               44.9065        1.5307      41.8684       47.9446
  beta                -0.0682       0.00822      -0.0845       -0.0519
  gamma                1.5222        0.1533       1.2179        1.8265
```

Figure 5.38 displays the residuals from the ordinary nonlinear least squares analyses [(a) and (c)] and the weighted residuals from the weighted analysis (b). The heterogeneous variances in the plot of unweighted residuals is apparent and the weighted residuals show a homogeneous band as residuals of a proper model should. The squared ordinary residuals plotted against the regressor (panel c) do not lend themselves easily to discern how the error variance relates to the regressor. Because of the tightness of the observations for young trees (Figure 5.36) many residuals are close to zero, and a few large residuals for older trees *over-power* the plot. It is because we find this plot of squared residuals hard to interpret for these data that we prefer the grouping approach in this application.

What has been gained by applying weighted nonlinear least squares instead of ordinary nonlinear least squares? It turns out that the parameter estimates differ little between the two analyses and the predicted trends will be almost indistinguishable from each other. The culprit, as so often is the case, is the estimation of the precision of the coefficients and the precision of the predictions. The variance of an observation around the mean trend is assumed to be constant in ordinary nonlinear least squares. From Figure 5.36 it is obvious that this variability is small for young trees and grows with tree age. The estimate of the common variance will then be too small for older trees and too large for younger trees. This effect becomes apparent when we calculate prediction (or confidence) intervals for the weighted and unweighted analyses (Figure 5.39). The 95% prediction intervals are narrower for young trees than the intervals from the unweighted analysis, since they take into account the actual variability around the regression line. The opposite effect can be observed for older trees. The unweighted analysis underestimates variability about the regression and the prediction intervals are not wide enough.

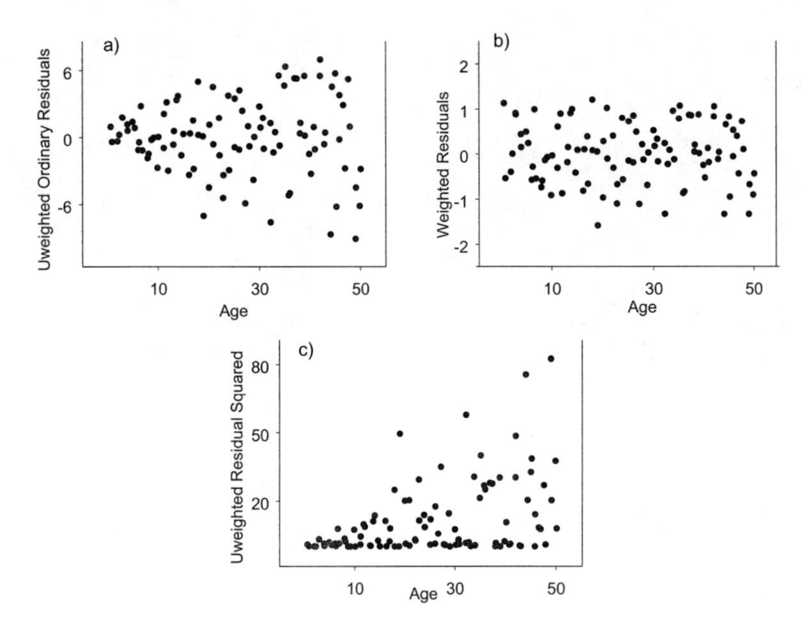

Figure 5.38. Residuals from an ordinary unweighted nonlinear least squares analysis (a), a weighted nonlinear least squares analysis (b), and the square of the ordinary residuals plotted against the regressor (c).

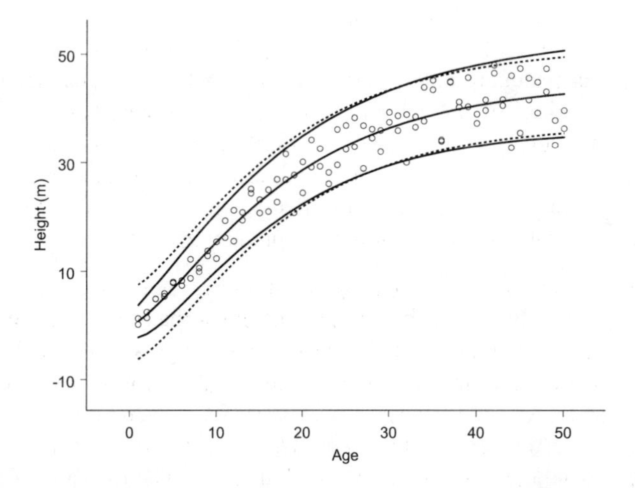

Figure 5.39. Predictions of tree height in weighted nonlinear least squares analysis (solid line in center of point cloud). Upper and lower solid lines are 95% prediction intervals in weighted analysis, dashed lines are 95% prediction intervals from ordinary nonlinear least squares analysis.

Chapter 6

Generalized Linear Models

"The objection is primarily that the theory of errors assumes that errors are purely random, i.e., 1. errors of all magnitudes are possible; 2. smaller errors are more likely to occur than larger ones; 3. positive and negative errors of equal absolute value are equally likely. The theory of Rodewald makes insufficient accommodation of this theory; only two errors are possible, in particular: if K is the germination percentage of seeds, the errors are K and $1-K$." J. C. Kapteyn, objecting to Rodewald's discussion of seed germination counts in terms of binomial probabilities. In Rodewald, H. Zur Methodik der Keimprüfungen, Die Landwirtschaftlichen Versuchs-Stationen, *vol. 49, p. 260. (Quoted in German translated by first author.)*

6.1 Introduction

Box 6.1 Generalized Linear Models

- **Generalized linear models (GLMs) are statistical models that combine elements of linear and nonlinear models.**
- **GLMs apply if responses are distributed independently in the exponential family of distributions, a large family that contains distributions such as the Bernoulli, Binomial, Poisson, Gamma, Beta and Gaussian distributions.**
- **Each GLM has three components: the link function, the linear predictor, and the random component.**
- **The Gaussian linear regression and analysis of variance models are special cases of generalized linear models.**

In the preceding chapters we explored statistical models where the response variable is continuous. Although these models cover a wide range of situations they do not suffice for many data in the plant and soil sciences. For example, the response may not be a continuous variable, but a count or a frequency. The distribution of the errors may have a mean of zero, but may be far from a Gaussian distribution. These data/model breakdowns can be addressed by relying on asymptotic or approximate results, by transforming the data, or by using models specifically designed for the particular response distribution. Poisson-distributed counts, for example, can be approximated by Gaussian random variables if the average count is sufficiently large. In binomial experiments consisting of independent and identical binary random variables the Gaussian approximation can be invoked provided the product of sample size and the smaller of success or failure probability is sufficiently large (≥ 5). When such approximations allow discrete responses to be treated as Gaussian, the temptation to invoke standard analysis of variance or regression analysis is understandable. The analyst must keep in mind, however, that other assumptions may still be violated. Since for Poisson random variables the mean equals the variance, treatments where counts are large on average will also have large variability compared to treatments where counts are small on average. The homoscedasticity assumption in an experiment with count responses is likely to be violated even if a Gaussian approximation to the response distribution holds. When Gaussian approximations fail, transformations of the data can achieve greater symmetry, remove variance heterogeneity, and create a scale on which effects can be modeled additively (Table 6.1). Transforming the data is not without problems, however. The transformation that establishes symmetry may not be the one that homogenizes the variances. Results of statistical analyses are to be interpreted on the transformed scale, which may not be the most meaningful. The square root of weed counts or the arcsine of the proportion of infected plants is not a natural metric for interpretation.

If the probability distribution of the response is known, one should not attempt to force the statistical analysis in a Gaussian framework if tools are available specifically designed for that distribution. Generalized Linear Models (GLMs) extend linear statistical modeling to response distributions that belong to a broad family of distributions, known as the **exponential family**. It contains the Bernoulli, Binomial, Poisson, Negative Binomial, Gamma, Gaussian, Beta, Weibull, and other distributions. GLM theory is based on work by Nelder and Wedderburn (1972) and Wedderburn (1974). It was subsequently popularized in the monograph by McCullagh and Nelder (1989). GLMs combine elements from linear and nonlinear models and we caution the reader on the outset not to confuse the acronym GLM with the `glm` procedure of The SAS® System. The `glm` procedure fits linear models and conducts inference assuming Gaussian errors, a very special case of a generalized linear model. The SAS® acronym stands for *General Linear Model*, the generality being that it can fit regression, analysis of variance, and analysis of covariance models by unweighted or weighted least squares, not that it can fit **generalized** linear models. Some of the procedures in The SAS® System that can fit generalized linear models are `proc genmod`, `proc logistic`, `proc probit`, `proc nlmixed`, and `proc catmod` (see §6.2.3).

Table 6.1. Some common transformations for non-Gaussian data to achieve symmetry and/or stabilize variance

Variable y	**Transformation**
Continuous	$\sqrt{y}$ $\ln\{y\}$ $1/y$
Count (>0)	$\sqrt{y}$ $\sqrt{y+0.375}$ $\ln\{y\}$
Count (≥ 0)	$\sqrt{y}$ $\sqrt{y+0.375}$ $\ln\{y+c\}^{\dagger}$
Proportion	$\arcsin\left(\sqrt{y}\right) = \sin^{-1}\left(\sqrt{y}\right)^{\ddagger}$

†Adding a small value c to count variables that can take on the value 0 enables the logarithmic transformation if $y=0$.

‡The arcsine transformation is useful for binary proportions. If y is a percentage, use $\sin^{-1}\left(\sqrt{y/100}\right)$. The transformation $\sqrt{y+0.375}$ was found to be superior to $\sqrt{y}$ in applications where the average count is small (<3). For binomial data where $n<50$ and the observed proportions are not all between 0.3 and 0.7 it is recommended to apply $\sin^{-1}(\sqrt{(y^{*}+0.375)/(n+0.75)})$ where y^{*} is the binomial count (not the proportion). If $n \geq 50$ or all observed proportions lie between 0.3 and 0.7, use $\sin^{-1}\left(\sqrt{y^{*}/n}\right)$.

Linear models are not well-suited for modeling the effects of experimental factors and covariates on discrete outcomes for a number of reasons. The following example highlights some of the problems encountered when a linear model is applied to a binary response.

Example 6.1. Groundwater pesticide contamination is measured in randomly selected wells under various cropping systems to determine whether wells are above or below acceptable contamination levels. Because of the difficulties in accurately measuring the concentration in parts per million for very low contamination levels, it is only recorded if the concentration at a given well exceeds the acceptable level. An observation from well i thus can take on only two possible values. For purposes of the analysis it is convenient to code the variable as

$$Y_i = \begin{cases} 1 & \text{level exceeded (well contaminated)} \\ 0 & \text{level not exceeded (well not contaminated).} \end{cases}$$

Y_i is a Bernoulli random variable with probability mass function

$$p(y_i) = \pi_i^{y_i}(1-\pi_i)^{1-y_i},$$

where π_i denotes the probability that the ith well is contaminated above threshold level. The expected value of Y_i is easily found as

$$\text{E}[Y_i] = \sum y_i p(y_i) = 1{*}\text{Pr}(Y_i = 1) + 0{*}\text{Pr}(Y_i = 0) = \text{Pr}(Y_i = 1) = \pi_i \, .$$

If we wish to determine whether the probability of well contamination depends on the amount of pesticide applied (X), a traditional linear model

$$Y_i = \beta_0 + \beta_1 x_i + e_i$$

is deficient in a number of ways. The mean response is a probability in the interval $[0, 1]$ but no such restriction is placed on the mean function $\beta_0 + \beta_1 x_i$. Predictions from this model can fall outside the permissible range (Figure 6.1). Also, limits are typically not approached in a linear fashion, but as asymptotes. A reasonable model for the probability of contamination might be sigmoidal in shape. Figure 6.1 shows simulated data for this example highlighting that the response can take on only two values, 0 and 1. The sigmoidal trend stems from a generalized linear model where it is assumed that probabilities approach 0.0 at the same rate with which they approach 1.0.

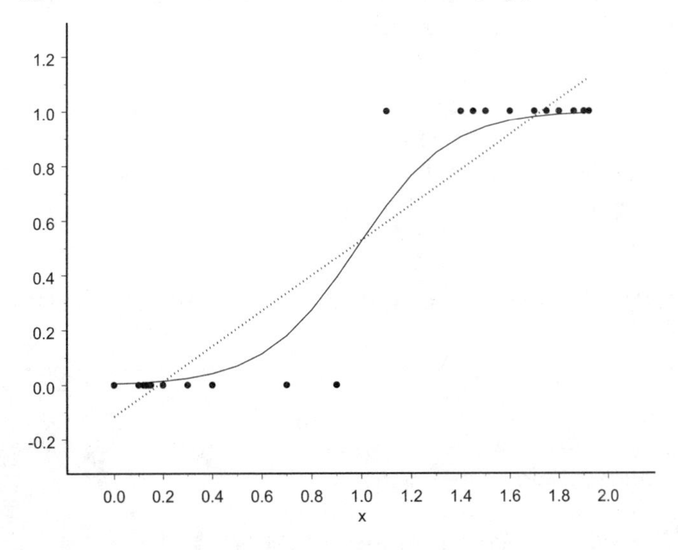

Figure 6.1. Simulated data for well contamination example. Straight line is obtained by fitting the linear regression model $Y_i = \beta_0 + \beta_1 x_i$, the sigmoidal line from a logistic regression model, a generalized linear model.

Generalized linear models inherit from linear models a linear combination of covariates and parameters, $\mathbf{x}_i'\boldsymbol{\beta}$, termed the **linear predictor**. This additive systematic part of the model is not an expression for the mean response $\text{E}[Y_i]$, but for a transformation of $\text{E}[Y_i]$. This transformation, $g(\text{E}[Y_i])$, is called the **link function** of the generalized linear model. It maps the average response on a scale where covariate effects are additive and it ensures range restrictions.

Every distribution in the exponential family suggests a particular link function known as the canonical link function, but the user is at liberty to pair any suitable link function with any distribution in the exponential family. The three components of a GLM, linear predictor, link function, and random component are discussed in §6.2. This section also introduces various procedures of The SAS® System that are capable of fitting generalized linear models to data. How to estimate the parameters of a generalized linear model and how to perform statistical inference is the focus in §6.4. Because of the importance of multinomial, in particular ordinal, responses in agronomy, a separate section is devoted to cumulative link models for ordered outcomes (§6.5).

With few exceptions, distributions in the exponential family have functionally related moments. For example, the mean and variance of a Binomial random variable are $\mathrm{E}[Y] = n\pi$ and $\mathrm{Var}[\pi] = n\pi(1-\pi) = \mathrm{E}[Y](1-\pi)$ where n is the binomial sample size and π is the success probability. If Y has a Poisson distribution then $\mathrm{E}[Y] = \mathrm{Var}[Y]$. Means and variances cannot be determined independently as is the case for Gaussian data. The modeler of discrete data often encounters situation where the data appear more dispersed than is permissible for a particular distribution. This **overdispersion** problem is addressed in §6.6.

Applications of generalized linear models to problems that arise in the plant and soil sciences follow in §6.7. Mathematical details can be found in Appendix A on the CD-ROM as §A6.8.

The models covered in this chapter extend the previously discussed statistical models to non-Gaussian distributions. We assume, however, that data are uncorrelated. The case of correlated, non-Gaussian data in the clustered data setting is discussed in §8 and in the spatial setting in §9.

6.2 Components of a Generalized Linear Model

Box 6.2 Components of a GLM

- **A generalized model consists of a random component, a systematic component, and a link function.**
- **The random component is the distribution of the response chosen from the exponential family.**
- **The systematic component is a linear predictor function $\mathbf{x}'\boldsymbol{\beta}$ that relates a transformation of the mean response to the covariates or effects in $\mathbf{x}$.**
- **The link function is a transformation of the mean response so that covariate effects are additive and range restrictions are ensured. Every distribution in the exponential family has a natural link function.**

6.2.1 Random Component

The random component of a generalized linear model consists of independent observations $Y_1, \cdots, Y_n$ from a distribution that belongs to the **exponential family** (see §A6.8.1 for details). A probability mass or density function is a member of the exponential family if it can be written as

$$f(y) = \exp\{(y\theta - b(\theta))/\psi + c(y, \psi)\} \qquad [6.1]$$

for some functions $b(\bullet)$ and $c(\bullet)$. The parameter θ is called the **natural parameter** and ψ is a dispersion (scale) parameter. Some important members of the exponential family are shown in Table 6.2.

Table 6.2. Important distributions in the exponential family (The Bernoulli, Binomial, Negative Binomial, and Poisson are discrete distributions)

Distribution	$b(\theta)$	$\mathrm{E}[Y] = b'(\theta) = \mu$	$h(\mu)$	ψ	$\theta(\mu)$
Bernoulli, $B(\pi)$	$\ln\{1 + e^\theta\}$	$\pi = \frac{\exp\{\theta\}}{1+\exp\{\theta\}}$	$\pi(1-\pi)$	1	$\ln\{\frac{\pi}{1-\pi}\}$
Binomial, $B(n, \pi)$	$n\ln\{1 + e^\theta\}$	$n\pi = n\frac{\exp\{\theta\}}{1+\exp\{\theta\}}$	$n\pi(1-\pi)$	1	$\ln\{\frac{\pi}{1-\pi}\}$
Negative Binomial, $NB(k, \pi)$	$-k\ln\{1 - e^\theta\}$	$k\frac{1-\pi}{\pi}$ $= k\frac{\exp\{\theta\}}{1-\exp\{\theta\}}$	$k\frac{1-\pi}{\pi^2}$	1	$\ln\{1-\pi\}$
Poisson, $P(\lambda)$	e^θ	$\lambda = \exp\{\theta\}$	λ	1	$\ln\{\lambda\}$
Gaussian, $G(\mu, \sigma^2)$	$\frac{1}{2}\theta^2$	$\mu = \theta$	1	σ^2	μ
Gamma, Gamma(μ, α)	$-\ln\{-\theta\}$	$\mu = -1/\theta$	μ^2	α^{-1}	$-\mu^{-1}$
Exponential, E(μ)	$-\ln\{-\theta\}$	$\mu = -1/\theta$	μ^2	1	$-\mu^{-1}$
Inverse Gaussian, IG(μ, σ^2)	$-(-2\theta)^{0.5}$	$\mu = (-2\theta)^{-0.5}$	μ^3	σ^2	μ^{-2}

The function $b(\theta)$ is important because it relates the natural parameter to the mean and variance of Y. We have $\mathrm{E}[Y] = \mu = b'(\theta)$ and $\mathrm{Var}[Y] = b''(\theta)\psi$ where $b'(\theta)$ and $b''(\theta)$ are the first and second derivatives of $b(\theta)$, respectively. The second derivative $b''(\theta)$ is also termed the **variance function** of the distribution. When the variance function is expressed in terms of the mean μ, instead of the natural parameter θ, it is denoted as $h(\mu)$. Hence, $\mathrm{Var}[Y] = h(\mu)\psi$ (Table 6.2). The variance function $h(\mu)$ depends on the mean for all distributions in Table 6.2, except the Gaussian. An estimate $\widehat{\pi}$ of the success probability π for Bernoulli data thus lends itself directly to a moment estimator of the variance, $\widehat{\pi}(1-\widehat{\pi})$. For

Gaussian data an estimate of the mean does not provide any information about the variability in the data.

The natural parameter θ can be expressed as a function of the mean $\mu = \mathrm{E}[Y]$ by inverting the relationship $b'(\theta) = \mu$. For example, in the Bernoulli case,

$$\mathrm{E}[Y] = \frac{\exp\{\theta\}}{1+\exp\{\theta\}} = \mu \Leftrightarrow \theta = \ln\left\{\frac{\mu}{1-\mu}\right\}.$$

Denoted $\theta(\mu)$ in Table 6.2, this function is called the **natural** or **canonical link function**. It is frequently the link function of choice, but it is not a requirement to retain the canonical link (§6.2.2). We now discuss the important distributions shown in Table 6.2 in more detail.

The **Bernoulli** distribution is a discrete distribution for binary (success/failure) outcomes. If the two possible outcomes are coded

$$Y = \begin{cases} 1 & \text{if outcome is a success} \\ 0 & \text{if outcome is a failure,} \end{cases}$$

the probability mass function (pmf) is

$$p(y) = \pi^y(1-\pi)^{1-y}, \quad y \in \{0,1\}, \qquad [6.2]$$

where π is the success probability $\pi = \Pr(Y=1)$. A **binomial experiment** consists of a series of n independent and identical Bernoulli trials. For example, germinating 100 seeds constitutes a binomial experiment if the seeds germinate independently of each other and are a random sample from a seed lot. Within a binomial experiment, several random variables can be defined. The total number of successes is a **Binomial** random variable with probability mass function

$$p(y) = \binom{n}{y}\pi^y(1-\pi)^{n-y} = \frac{n!}{y!(n-y)!}\pi^y(1-\pi)^{n-y}, \quad y = 0,1,\cdots,n. \qquad [6.3]$$

A Binomial ($B(n,\pi)$) random variable can thus be thought of as the sum of n independent Bernoulli ($B(\pi)$) random variables.

Example 6.2. Seeds are stored at four temperature regimes (T_1 to T_4) and under addition of chemicals at four different concentrations $(0, 0.1, 1.0, 10)$. To study the effects of temperature and chemical concentration a completely randomized experiment is conducted with a 4×4 factorial treatment structure and four replications. For each of the 64 experimental sets, 50 seeds were placed on a dish and the number of seeds that germinated under standard conditions was recorded. The data, taken from Mead, Curnow, and Hasted (1993, p. 325) are shown in Table 6.3.

Let Y_{ijk} denote the number of seeds germinating for the kth replicate of temperature i and chemical concentration j. For example, $y_{121} = 13$ and $y_{122} = 12$ are the realized values for the first and second replication of temperature T_1 and concentration 0.1. These are realizations of $B(50, \pi_{12})$ random variables if the seeds germinated independently in a dish and there are no differences that affect germination between the dishes. Alternatively, one can think of each seed for this treatment combination as a Bernoulli

$B(\pi_{12})$ random variable. Let

$$X_{ijkl} = \begin{cases} 1 & \text{if seed } l \text{ in dish } k \text{ for temperature } i \text{ and concentration } j \text{ germinates} \\ 0 & \text{otherwise} \end{cases}$$

and the counts shown in Table 6.3 are $\sum_{l=1}^{50} x_{ijkl} = y_{ijk}$.

Table 6.3. Germination data from Mead, Curnow, and Hasted (1993, p. 325)†
(Values represent counts out of 50 seeds for four replicates)

	Chemical Concentration			
Temperature	**0** $(j=1)$	**0.1** $(j=2)$	**1.0** $(j=3)$	**10** $(j=4)$
T_1 $(i=1)$	$9,9,3,7$	$13,12,14,15$	$21,23,24,27$	$40,32,43,34$
T_2 $(i=2)$	$19,30,21,29$	$33,32,30,26$	$43,40,37,41$	$48,48,49,48$
T_3 $(i=3)$	$7,7,2,5$	$1,2,4,4,$	$8,10,6,7$	$3,4,8,5$
T_4 $(i=4)$	$4,9,3,7$	$13,6,15,7$	$16,13,18,19$	$13,18,11,16$

†Used with permission.

As for an experiment with continuous response where interest lies in comparing treatment means, we may be interested in similar comparisons of the form

$$\pi_{ij} = \pi_{i'j},$$
$$\pi_{1j} = \pi_{2j} = \pi_{3j} = \pi_{4j},$$
$$\pi_{.1} = \pi_{.2} = \pi_{.3} = \pi_{.4},$$

corresponding to pairwise differences, a slice at concentration j, and the marginal concentration effects. The analysis of these data must recognize, however, that these means are probabilities.

The number of trials in a binomial experiment until the kth success occurs follows the **Negative Binomial** law. One of the many ways in which the probability mass function of a Negative Binomial random variable can be written is

$$p(y) = \binom{k+y-1}{y}(1-\pi)^y \pi^k, \; y = 0, 1, \cdots. \qquad [6.4]$$

See Johnson et al. (1992, Ch. 5) and §A6.8.1 for other parameterizations. A special case of the Negative Binomial distribution is the **Geometric** distribution for which $k = 1$. Notice that the support of the Negative Binomial distribution has no upper bound and the distribution can thus be used to model counts without natural denominator, i.e., counts that cannot be converted to proportions. Examples are the number of weeds per m^2, the number of aflatoxin contaminated peanuts per m^3 and the number of earthworms per ft^3. Traditionally, the **Poisson** distribution is more frequently applied to model such counts than the Negative Binomial distribution. The probability mass function of a Poisson(λ) random variable is

$$p(y) = \frac{\lambda^y}{y!}e^{-\lambda}, \; y = 0, 1, \cdots . \qquad [6.5]$$

A special feature of the Poisson random variable is the identity of mean and variance, $\mathrm{E}[Y] = \mathrm{Var}[Y] = \lambda$. Many count data suggest variation that exceeds the mean count. The Negative Binomial distribution has the same support ($y = 0, 1, \cdots$) as the Poisson distribution but allows greater variability. It is a good alternative model for count data that exhibit excess variation compared to the Poisson model. This connection between Poisson and Negative Binomial distributions can be made more precise if one considers the parameter λ a random Gamma-distributed random variable (see §A6.8.6 for details and §6.7.8 for an application).

The family of **Gamma** distributions encompasses continuous, non-negative, right-skewed probability densities (Figure 6.2). The Gamma distribution has two non-negative parameters, α and β, and density function

$$f(y) = \frac{1}{\Gamma(\alpha)}\beta^{-\alpha}y^{\alpha-1}\exp\{-y/\beta\}, \; y \geq 0, \qquad [6.6]$$

where $\Gamma(\alpha) = \int_0^\infty t^{\alpha-1}e^{-t}dt$ is known as the *gamma* function. The mean and variance of a Gamma random variable are $\mathrm{E}[Y] = \alpha\beta = \mu$ and $\mathrm{Var}[Y] = \alpha\beta^2 = \mu^2/\alpha$. The density function can be rewritten in terms of the mean μ and the scale parameter α as

$$f(y) = \frac{1}{\Gamma(\alpha)y}\left(\frac{y\alpha}{\mu}\right)^{\alpha}\exp\left\{-y\frac{\alpha}{\mu}\right\}, \; y \geq 0, \qquad [6.7]$$

from which the exponential family terms in Table 6.2 were derived. We refer to the parameterization [6.7] when we denote a Gamma(μ, α) random variable. The Exponential distribution for which $\alpha = 1$ and the Chi-squared distribution with ν degrees of freedom for which $\alpha = \nu/2$ and $\beta = 2$ are special cases of Gamma distributions (Figure 6.2).

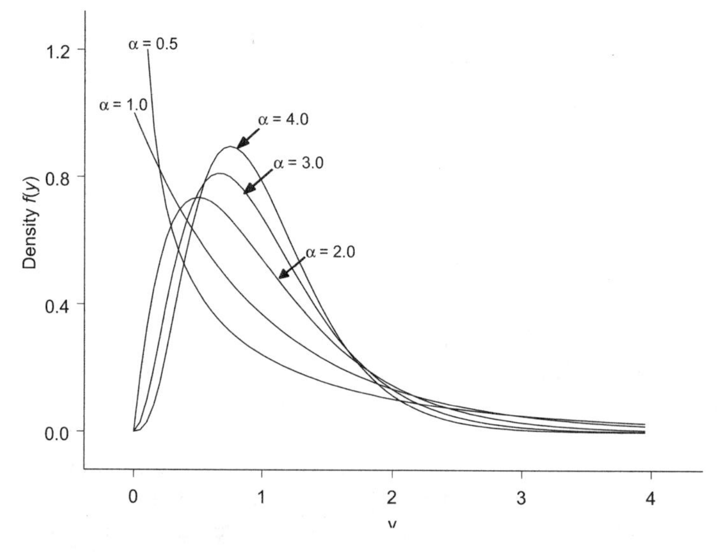

Figure 6.2. Gamma distributions in parameterization [6.7] with $\mu = 1$. As $\alpha \to \infty$, the Gamma distribution tends to a Gaussian.

Because of its skewness, Gamma distributions are useful to model continuous, non-negative, right-skewed outcomes with heterogeneous variances and play an important role in analyzing time-to-event data. If on average α events occur independently in μ/α time units, the time that elapses until the αth event occurs is a Gamma$(\mu/\alpha, \alpha)$ random variable.

The variance of a Gamma random variable is proportional to the square of its mean. Hence, the coefficient of variation of a Gamma-distributed random variable remains constant as its mean changes. This suggests a Gamma model for data in which the standard deviation of the outcome increases linearly with the mean and can be assessed in experiments with replication by calculating standard deviations across replicates.

Example 6.3. McCullagh and Nelder (1989, pp. 317-320) discuss a competition experiment where various seed densities of barley and the weed *Sinapis alba* were grown in a competition experiment with three replications (blocks). We focus here on a subset of their data, the monoculture barley dry weight yields (Table 6.4).

Table 6.4. Monoculture barley yields and seeding densities in three blocks (Experimental units were individual pots)

Seeds sown	Dry weights Block 1	Block 2	Block 3	$\overline{y}$	s	CV
3	2.07	5.32	3.14	3.51	1.65	47.19
5	10.57	13.59	14.69	12.95	2.13	16.47
7	20.87	9.97	5.45	12.09	7.93	65.53
10	6.59	21.40	23.12	17.04	9.09	53.34
15	8.08	11.07	8.28	9.14	1.67	18.28
23	16.70	6.66	19.48	14.28	6.74	47.22
34	21.22	14.25	38.11	24.53	12.27	50.02
51	26.57	39.37	25.53	30.49	7.71	25.28
77	23.71	21.44	19.72	21.62	2.00	9.26
115	20.46	30.92	41.02	30.80	10.28	33.37

Reproduced from McCullagh and Nelder (1989, pp. 317-320) with permission.

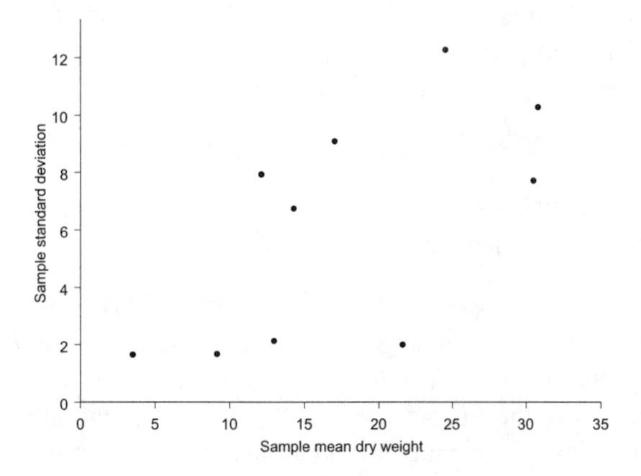

Figure 6.3. Sample standard deviation as a function of sample mean in barley yield monoculture.

Sample standard deviations calculated from only three observations are not reliable. When plotting s against $\overline{y}$, however, a trend between standard deviation and sample mean is obvious and a linear trend $\sqrt{\text{Var}[Y]} = \gamma\text{E}[Y]$ does not seem unreasonable.

In §5.8.7 it was stated that the variability of the yield per plant Y generally increases with Y in yield-density studies and competition experiments. Mead (1970), in a study of the Bleasdale-Nelder yield-density model (Bleasdale and Nelder 1960), concludes that it is reasonable for yield data to assume that $\text{Var}[Y] \propto \text{E}[Y]^2$, precisely the mean-variance relationship implied by the Gamma distribution. Many research workers would choose a logarithmically transformed model

$$\text{E}[\ln\{Y\}] = \ln\{f(x, \boldsymbol{\beta})\},$$

where x denotes plant density, assuming that $\text{Var}[\ln\{Y\}]$ is constant and $\ln\{Y\}$ is Gaussian. As an alternative, one could model the relationship

$$\text{E}[Y] = f(x, \boldsymbol{\beta})$$

assuming that Y is a Gamma random variable. An application of the Gamma distribution to yield-density models is presented in §6.7.3.

Right-skewed distributions should also be expected when the outcome of interest is a dispersion parameter. For example, when modeling variances it is hardly appropriate to assume that these are Gaussian-distributed. The following device can be invoked instead. If $Y_1, \cdots, Y_n$ are a random sample from a Gaussian distribution with variance σ^2 and

$$S^2 = \frac{1}{n-1}\sum_{i=1}^{n}\left(Y_i - \overline{Y}\right)^2$$

denotes the sample variance, then

$$(n-1)S^2/\sigma^2$$

follows a Chi-squared distribution with $n-1$ degrees of freedom. Consequently, S^2 follows a Gamma distribution with parameters $\alpha = (n-1)/2$, $\beta = 2\sigma^2/(n-1)$. In the parameterization [6.7], one obtains

$$S^2 \sim \text{Gamma}\left(\mu = \sigma^2, \alpha = (n-1)/2\right).$$

When modeling sample variances, one should not resort to a Gaussian distribution, but instead draw on a properly scaled Gamma distribution.

Example 6.4. Hart and Schabenberger (1998) studied the variability of the aflatoxin deoxynivalenol (DON) on truckloads of wheat kernels. DON is a toxic secondary metabolite produced by the fungus *Gibberella zeae* during the infection process. Data were gathered in 1996 by selecting at random ten trucks arriving at mill elevators. For each truck ten double-tubed probes were inserted at random from the top in the truck-load and the kernels trapped in the probe extracted, milled, and submitted to enzyme-linked immunosorbent assay (ELISA). Figure 6.4 shows the truck probe-to-probe sample variance as a function of the sample mean toxin concentration per truck.

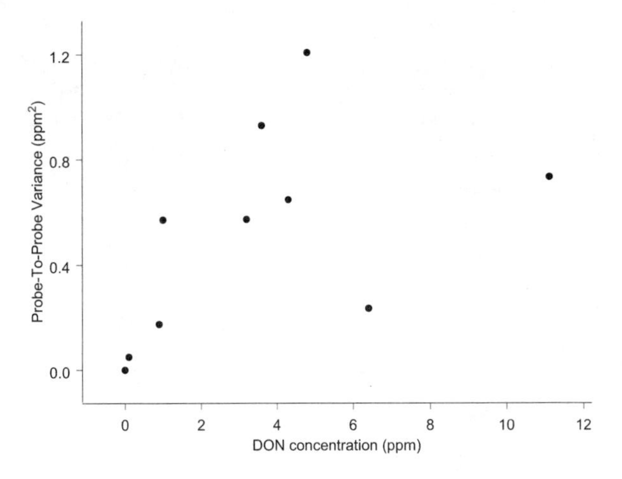

Figure 6.4. Probe-to-probe variances (ppm^2) as a function of average truck toxin concentration. Data kindly provided by Dr. L. Patrick Hart, Department of Crop and Soil Sciences, Michigan State University. Used with permission.

To model the probe-to-probe variability, we choose

$$S^2 \sim \text{Gamma}\left(\mu = g(\mathbf{x}_i'\boldsymbol{\beta}), \alpha = \frac{10-1}{2}\right)$$

where $g(\bullet)$ is a properly chosen link function. These data are analyzed in §6.7.7.

The **Log-Gaussian** distribution is also right-skewed and has variance proportional to the squared expectation. It is popular in modeling bioassay data or growth data where a logarithmic transformation establishes Gaussianity. If $\ln\{Y\}$ is Gaussian with mean μ and variance σ^2, then Y is a Log-Gaussian random variable with mean $\text{E}[Y] = \exp\{\mu + \sigma^2/2\}$ and variance $\text{Var}[Y] = \exp\{2\mu + \sigma^2\}(\exp\{\sigma^2\} - 1)$. It is also a reasonable model for right-skewed data that can be transformed to symmetry by taking logarithms. Unfortunately, it is not a member of the exponential family and does not permit a generalized linear model. In the GLM framework, the Gamma distributions provide an excellent alternative in our opinion. Amemiya (1973) and Firth (1988) discuss testing Log-Gaussian vs. Gamma distributions and vice versa.

The **Inverse Gaussian** distribution is also skew-symmetric with a long right tail and is a member of the exponential family. It is not really related to the Gaussian distribution although there are some parallel developments for the two families. For example, the independence of sample mean and sample variance that is true for the Gaussian is also true for the Inverse Gaussian distribution. The name stems from the inverse relationship between the cumulative generating functions of the two distributions. Also, the formulas of the Gaussian and Inverse Gaussian probability density functions bear some resemblance to each other. Folks and Chhikara (1978) suggested naming it the Tweedie distribution in recognition of fundamental work on this distribution by Tweedie (1945, 1957a, 1957b). A random variable Y is said to have an Inverse Gaussian distribution if its probability density function (Figure 6.5) is given by

$$f(y) = \frac{1}{\sqrt{2\pi y^3 \sigma^2}} \exp\left\{ -\frac{1}{\sigma^2} \frac{(y-\mu)^2}{2\mu^2 y} \right\}, \quad y > 0. \qquad [6.8]$$

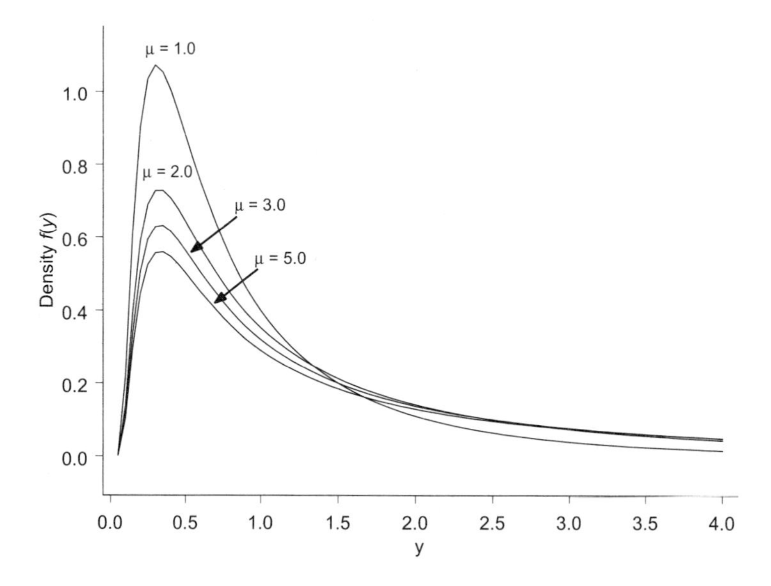

Figure 6.5. Inverse Gaussian distributions with $\sigma^2 = 1$.

The distribution finds application in stochastic processes as the distribution of the first passage time in a Brownian motion, and in analysis of lifetime and reliability data. The relationship with a passage time in Brownian motion suggests its use as the time a tracer remains in an organism. Folks and Chhikara (1978) pointed out that the skewness of the Inverse Gaussian distribution makes it an attractive candidate for right-skewed, non-negative, continuous outcomes whether or not the particular application relates to passage time in a stochastic process. The mean and variance of the Inverse Gaussian are given by $\mathrm{E}[Y] = \mu$ and $\mathrm{Var}[Y] = \mu^3\sigma^2$. Notice that σ^2 is not an independent scale parameter. The variability of Y is determined by μ and σ^2 jointly, whereas for a Gaussian distribution $\mathrm{Var}[Y] = \sigma^2$. The variance of the Inverse Gaussian increases more sharply in the mean than that of the Gamma distribution (for $\sigma^2 = \alpha^{-1}$).

6.2.2 Systematic Component and Link Function

The systematic component of a generalized linear model is a linear combination of covariates (or design effects) and fixed effects parameters. As in linear models we denote the systematic component for the ith observation as

$$\beta_0 + \beta_1 x_{1i} + \beta_2 x_{2i} + \cdots + \beta_k x_{ki} = \mathbf{x}_i'\boldsymbol{\beta}$$

and term it the **linear predictor** η_i. The linear predictor is chosen in generalized linear models in much the same way as the mean function is built in classical linear regression or classification models. For example, if the binomial counts or proportions are analyzed in the completely randomized design of Example 6.2 (p. 306), the linear predictor contains an intercept, temperature, concentration effects, and their interactions. In contrast to the classical linear model where η_i is the mean of an observation the linear predictor in a generalized linear model is set equal to a transformation of the mean,

$$g(\mathrm{E}[Y_i]) = \eta_i = \mathbf{x}_i'\boldsymbol{\beta}. \qquad [6.9]$$

This transformation $g(\bullet)$ is called the **link function** and serves several purposes. It is a transformation of the mean μ_i onto a scale where the covariate effects are additive. In the terminology of §5.6.1, the link function is a linearizing transform and the generalized linear model is intrinsically linear. If one studies a model with mean function

$$\mu_i = \exp\{\beta_0 + \beta_1 x_i\} = \exp\{\beta_0\}\exp\{\beta_1 x_i\},$$

then $\ln\{\mu_i\}$ is a linearizing transformation of the nonlinear mean function. A second purpose of the link function is to confine predictions under the model to a suitable range. If Y_i is a Bernoulli outcome then $\mathrm{E}[Y_i] = \pi_i$ is a success probability which must lie between 0 and 1. Since no restrictions are placed on the parameters in the linear predictor $\mathbf{x}_i'\boldsymbol{\beta}$, the linear predictor can range from $-\infty$ to $+\infty$. To ensure that the predictions are in the proper range one chooses a link function that maps from $(0,1)$ to $(-\infty,\infty)$. One such possibility is the logit transformation

$$\mathrm{logit}(\pi) = \ln\left\{\frac{\pi}{1-\pi}\right\}, \qquad [6.10]$$

which is also the canonical link for the Bernoulli and Binomial distributions (see Table 6.2). Models with logit link are termed **logistic models** and can be expressed as

$$\ln\left\{\frac{\pi}{1-\pi}\right\} = \mathbf{x}'\boldsymbol{\beta}. \qquad [6.11]$$

Link functions must be monotonic and invertible. If $g(\mu_i) = \eta_i$, inversion leads to $\mu_i = g^{-1}(\eta_i)$, and the function $g^{-1}(\bullet)$ is called the **inverse link** function. In the case of the logistic model the inverted relationship between link and linear predictor is

$$\pi = \frac{\exp\{\mathbf{x}'\boldsymbol{\beta}\}}{1+\exp\{\mathbf{x}'\boldsymbol{\beta}\}} = \frac{1}{1+\exp\{-\mathbf{x}'\boldsymbol{\beta}\}}. \qquad [6.12]$$

Once parameter estimates in the generalized linear model have been obtained (§6.4), the mean of the outcome at any value of $\mathbf{x}$ is predicted as

$$\widehat{\mu} = g^{-1}\left(\mathbf{x}'\widehat{\boldsymbol{\beta}}\right).$$

Several link functions can properly restrict the expectation but provide different scales on which the covariate effects are additive. The representation of the probability (mass) density functions in Table 6.2 suggests **canonical link functions** shown there as $\theta(\mu)$. The canonical link for Binomial data is thus the logit link, for Poisson counts the log link, and for Gaussian data the identity link (no transformation). Although relying on the canonical link leads to some simplifications in parameter estimation, these are not of concern to the user of generalized linear models in practice. Functions other than the canonical link may be of interest. We now review popular link functions for binary data and proportions, counts, and continuous variables.

Link Functions for Binary Data and Binomial Proportions

The expected values of binary outcomes and binomial proportions are probabilities confined to the $[0,1]$ interval. Since the linear predictor $\eta_i = \mathbf{x}_i'\boldsymbol{\beta}$ can range from $-\infty$ to ∞, the link function must be a mapping $(0,1) \mapsto (-\infty,\infty)$. Similarly, the inverse link function is a mapping $(-\infty,\infty) \mapsto (0,1)$. A convenient way of deriving such link functions is as follows. Let $F(y) = \Pr(Y \le y)$ denote a cumulative distribution function (cdf) of a random variable Y that ranges over the entire real line (from $-\infty$ to ∞). Since $0 \le F(y) \le 1$, the cdf could be used as an inverse link function. Unfortunately, the inverse cdf (or quantile function) $F^{-1}(\bullet)$ does not exist in closed form for many continuous random variables, although numerically accurate methods exist to calculate $F^{-1}(\bullet)$.

The most popular generalized linear model for Bernoulli and Binomial data is probably the **logistic regression** model. It applies the logit link which is the inverse of the cumulative distribution function of a Logistic random variable. If Y is a Logistic random variable with parameters μ and α, then

$$F(y) = \pi = \frac{\exp\{(y-\mu)/\alpha\}}{1+\exp\{(y-\mu)/\alpha\}}, \quad -\infty < y < \infty, \qquad [6.13]$$

with mean μ and variance $(\alpha\pi)^2/3$. If $\mu = 0$, $\alpha = 1$, the distribution is called the Standard Logistic with cdf $\pi = \exp\{y\}/(1+\exp\{y\})$. Inverting the standard logistic cdf yields the logit function

$$F^{-1}(\pi) = \ln\left\{\frac{\pi}{1-\pi}\right\}.$$

In terms of a generalized linear model for Y_i with $\mathrm{E}[Y_i] = \pi_i$ and linear predictor $\eta_i = \mathbf{x}_i'\boldsymbol{\beta}$ we obtain

$$F^{-1}(\pi_i) = \mathrm{logit}(\pi_i) = \ln\left\{\frac{\pi_i}{1-\pi_i}\right\} = \mathbf{x}_i'\boldsymbol{\beta}. \qquad [6.14]$$

In the logistic model the parameters have a simple interpretation in terms of log odds ratios. Consider a two-group comparison of successes and failures. Define a dummy variable as

$$x_{ij} = \begin{cases} 1 & \text{if } i^{\text{th}} \text{ observation is in the treated group } (j=1) \\ 0 & \text{if } i^{\text{th}} \text{ observation is in the control group } (j=0), \end{cases}$$

and the response as

$$Y_{ij} = \begin{cases} 1 & \text{if } i^{\text{th}} \text{ observation in group } j \text{ results in a success} \\ 0 & \text{if } i^{\text{th}} \text{ observation in group } j \text{ results in a failure.} \end{cases}$$

Because x_{ij} is a dummy variable, the logistic model

$$\text{logit}(\pi_j) = \beta_0 + \beta_1 x_{ij}$$

reduces to

$$\text{logit}(\pi_j) = \begin{cases} \beta_0 + \beta_1 & j = 1 \text{ (treated group)} \\ \beta_0 & j = 0 \text{ (control group).} \end{cases}$$

The gradient β_1 measures the change in the logit between the control and the treated group. In terms of the success and failure probabilities one can construct a 2×2 table.

Table 6.5. Success and failure probabilities in two-group logistic model

	Control Group	**Treated Group**
Success	$\pi_0 = \frac{1}{1+\exp\{-\beta_0\}}$	$\pi_1 = \frac{1}{1+\exp\{-\beta_0-\beta_1\}}$
Failure	$1 - \pi_0 = \frac{\exp\{-\beta_0\}}{1+\exp\{-\beta_0\}}$	$1 - \pi_1 = \frac{\exp\{-\beta_0-\beta_1\}}{1+\exp\{-\beta_0-\beta_1\}}$

The odds O are defined as the ratio of the success and failure probabilities in a particular group,

$$O_{\text{control}} = \frac{\pi_0}{1-\pi_0} = e^{\beta_0}$$
$$O_{\text{treated}} = \frac{\pi_1}{1-\pi_1} = e^{\beta_0+\beta_1}.$$

Successes are $\exp\{\beta_0\}$ times more likely in the control group than failures and $\exp\{\beta_0 + \beta_1\}$ times more likely than failures in the treated group. How much the odds have changed by applying the treatment is expressed by the odds ratio

$$OR = \frac{O_{\text{treated}}}{O_{\text{control}}} = \frac{\exp\{\beta_0 + \beta_1\}}{\exp\{\beta_0\}} = e^{\beta_1},$$

or the log odds ratio $\ln(OR) = \beta_1$. If the log odds ratio is zero, the success/failure ratio is the same in both groups. Successes are then no more likely relative to failures under the treatment than under the control. A test of $H_0{:}\beta_1 = 0$ thus tests for equal success/failure odds in the two groups. From Table 6.5 it is seen that this implies equal success probabilities in the groups. These ideas generalize to comparisons of more than two groups.

Why is the logistic distribution our first choice to develop a link function and not the omnipotent Gaussian distribution? Assume we choose the standard Gaussian cdf

$$\pi = \Phi(\eta) = \int_{-\infty}^{\eta} \frac{1}{\sqrt{2\pi}} \exp\left\{ -\frac{1}{2}t^2 \right\} dt \qquad [6.15]$$

as the inverse link function. The link function is then given by $\Phi^{-1}(\pi) = \eta$ but neither

$\Phi^{-1}(\pi)$ nor $\Phi(\eta)$ exist in closed form and must be evaluated numerically. Models using the inverse Gaussian cdf are termed **probit** models (please note that the inverse Gaussian function refers to the inverse cdf of a Gaussian random variable and is not the same as the **I**nverse Gaussian random variable which is a variable with a right-skewed density). Often, there is little to be gained in practice using a probit over a logistic model. The cumulative distribution functions of the standard logistic and standard Gaussian are very similar (Figure 6.6). Both are sigmoidal and symmetric about $\eta = 0$. The main difference is that the Gaussian tails are less heavy than the Logistic tails and thus approach probability 0 and 1 more quickly. If the distributions are scaled to have the same mean and variance, the cdfs agree even more than is evident in Figure 6.6. Since logit(π) is less cumbersome numerically than $\Phi^{-1}(\pi)$, it is often preferred.

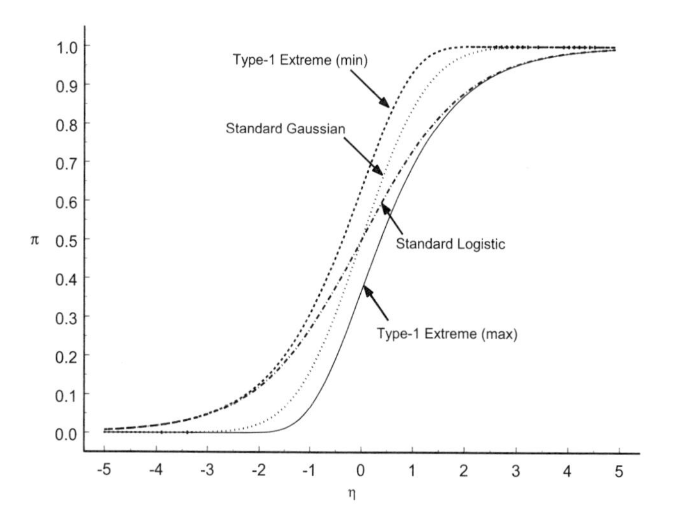

Figure 6.6. Cumulative distribution functions as inverse link functions. The corresponding link functions are the logit for the Standard Logistic, the probit for the Standard Gaussian, the log-log link for the Type-1 Extreme (max) value and complementary log-log link for the Type-1 Extreme (min) value distribution.

The symmetry of the Standard Gaussian and Logistic distributions implies that $\pi = 0$ is approached at the same rate as $\pi = 1$. If π departs from 0 slowly and approaches 1 quickly or vice versa, the probit or logit links are not appropriate. Asymmetric link functions can be derived from appropriate cumulative distribution functions. A Type-I Extreme value distribution (Johnson et al. 1995, Ch. 23) is given by

$$F(y) = \exp\{-\exp\{-(y-\alpha)/\beta\}\} \qquad [6.16]$$

and has a standardized form with $\alpha = 0$, $\beta = 1$. This distribution, also known as the Gumbel or double-exponential distribution arises as the distribution of the largest value in a random sample of size n. By putting $x = -y$ one can obtain the distribution of the smallest value. Consider the standardized form $\pi = F(\eta) = \exp\{-\exp\{\eta\}\}$. Inverting this cdf yields the link function

$$\ln\{-\ln\{\pi\}\} = \mathbf{x}_i'\boldsymbol{\beta}, \qquad [6.17]$$

known as the **log-log link**. Its complement $\ln\{-\ln\{1-\pi\}\}$, obtained by changing successes to failures and vice versa, is known as the **complementary log-log link** derived from $F(\eta) = 1 - \exp\{-\exp\{\eta\}\}$. The complementary log-log link behaves like the logit for π near 0 and has smaller values as π increases. The log-log link behaves like the logit for π near 1 and yields larger values for small π (Figure 6.7).

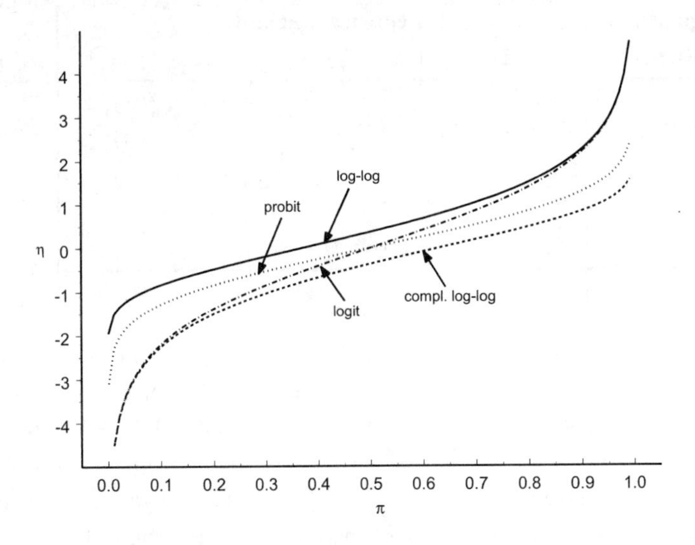

Figure 6.7. Link functions for Bernoulli and Binomial responses.

Link Functions for Counts

Here we consider counts other than Binomial counts, i.e., those without a natural denominator (§2.2). The support of these counts has no upper bound in contrast to the Binomial that represents the number of successes out of n trials. Binomial counts can be converted to proportions by dividing the number of successes by the binomial sample size n. Counts without a natural denominator cannot be expressed as proportions, for example, the number of weeds per m^2. The probability models applied most frequently to such counts are the Poisson and Negative Binomial. A suitable link function for these is an invertible, monotonic mapping $(0,\infty) \mapsto (-\infty,\infty)$. The log link $\ln\{\mathrm{E}[Y_i]\} = \mathbf{x}_i'\boldsymbol{\beta}$ with its inverse $\mathrm{E}[Y_i] = \exp\{\mathbf{x}_i'\boldsymbol{\beta}\}$ has this property and is canonical for both distributions (Table 6.2). The log link leads to parameters interpreted in terms of multiplicative, rather than additive effects. Consider the linear predictor $\eta_i = \beta_0 + \beta_1 x_i$ and a log link function. Then $\exp(\beta_1)$ measures the **relative** increase in the mean if the covariate changes by one unit:

$$\frac{\mu|(x_i+1)}{\mu|x_i} = \frac{\exp\{\beta_0 + \beta_1(x_i+1)\}}{\exp\{\beta_0 + \beta_1 x_i\}} = \exp\{\beta_1\}.$$

Generalized linear models with log link are often called **log-linear models**. They play an important role in regression analysis of counts and in the analysis of contingency tables. Consider the generic layout of a two-way contingency table in Table 6.6. The counts n_{ij} in row i, column j of the table represent the number of times variable X was observed at level j and variable Y simultaneously took on level i.

Table 6.6. Generic layout of a two-way contingency table (n_{ij} denotes the observed count in row $= i$, column j)

Categorical variable Y	**Categorical variable X** $j=1$	$j=2$	$j=3$	$\cdots$	$j=J$	**Row totals**
$i=1$	n_{11}	n_{12}	n_{13}	$\cdots$	n_{1J}	$n_{1.}$
$i=2$	n_{21}	n_{22}	n_{23}	$\cdots$	n_{2J}	$n_{2.}$
$i=3$	n_{31}	n_{32}	n_{33}	$\cdots$	n_{3J}	$n_{3.}$
$\vdots$	$\vdots$	$\vdots$	$\vdots$		$\vdots$	
$i=I$	n_{I1}	n_{I2}	n_{I3}	$\cdots$	n_{IJ}	$n_{I.}$
Column totals	$n_{.1}$	$n_{.2}$	$n_{.3}$		$n_{.J}$	$n_{..}$

If the row and column variables (Y and X) are independent, the cell counts n_{ij} are determined by the marginal row and column totals alone. Under a Poisson sampling model where the count in each cell is the realization of a Poisson(λ_{ij}) random variable, the row and column totals are Poisson$(\lambda_{i.} = \sum_{j=1}^{J}\lambda_{ij})$ and Poisson$(\lambda_{.j} = \sum_{i=1}^{I}\lambda_{ij})$ variables, respectively, and the total sample size is a Poisson$(\lambda_{..} = \sum_{i,j}^{I,J}\lambda_{ij})$ random variable. The expected count λ_{ij} under independence is then related to the marginal expected counts by

$$\lambda_{ij} = \frac{\lambda_{i.}\lambda_{.j}}{\lambda_{..}}$$

Taking logarithms leads to

$$\ln\{\lambda_{ij}\} = -\ln(\lambda_{..}) + \ln\{\lambda_{i.}\} + \ln\{\lambda_{.j}\} = \mu + \alpha_i + \beta_j, \quad [6.18]$$

a generalized linear model with log link for Poisson-distributed random variables and a linear predictor consisting of a grand mean μ, row effects α_i, and column effects β_j. The linear predictor is akin to that in a two-way layout without interactions such as a randomized block design, which is precisely the layout of Table 6.6. We can think of α_i and β_j as main effects of the row and column variables. The Poisson sampling scheme applies if the total number of observations ($n_{..}$) is itself a random variable, i.e., prior to data collection the total number of observations being cross-classified is unknown. If the total sample size is known, one is fortuitously led to the same general decomposition for the expected cell counts as in [6.18]. Conditional on $n_{..}$ the $I \times J$ counts in the table are realizations of a multinomial distribution with cell probabilities π_{ij} and marginal probabilities $\pi_{i.}$ and $\pi_{.j}$. The expected count in cell i, j, if X and Y are independent, is

$$\lambda_{ij} = n_{..}\pi_{i.}\pi_{.j}$$

and taking logarithms leads to

$$\ln\{\lambda_{ij}\} = \ln\{n_{..}\} + \ln\{\pi_{i.}\} + \ln\{\pi_{.j}\} = \mu + \alpha_i + \beta_j\,. \quad [6.19]$$

In §6.7.6 the agreement between two raters of the same experimental material is analyzed by comparing series of log-linear models that structure the interaction between the ratings.

Link Functions for Continuous Data

The identity link is historically the most common link function applied to continuous outcomes. It is the canonical link if data are Gaussian. Hence, classical linear regression and analysis of variance models are special cases of generalized linear models where the random component is Gaussian and the link is the identity function, $\mu = \mathbf{x}'\boldsymbol{\beta}$. Table 6.2 shows that the identity link is not the canonical link in other cases. For Gamma-distributed data the inverse link is the reciprocal link $\eta = 1/\mu$ and for Inverse Gaussian-distributed data $\eta = \mu^{-2}$. The reciprocal link, although canonical, is not necessarily a good choice for skewed, non-negative data. Since the linear predictor ranges over the real line, there is no guarantee that $\mu = 1/\eta$ is positive. When using the reciprocal link for Gamma-distributed outcomes, for example, the requirement $\eta > 0$ requires additional restrictions on the parameters in the model. Constrained estimation may need to be employed to ensure that these restrictions hold. It is simpler to resort to link functions that map $(0, \infty)$ onto $(-\infty, \infty)$ such as the log link. For Gamma and Inverse Gaussian-distributed outcomes, the log link is thus a frequent choice. One should bear in mind, however, that the log link implies multiplicative covariate effects whereas the reciprocal link implies additive effects on the inverse scale. McCullagh and Nelder (1989, p. 293) provide arguments that if the variability in data is small, it is difficult to distinguish between Gaussian models on the logarithmic scale and Gamma models with multiplicative covariate effects, lending additional support for modeling right-skewed, non-negative continuous responses as Gamma variables with logarithmic link function. Whether such data are modeled with a reciprocal or logarithmic link also depends on whether the rate of change or the log rate of change is a more meaningful measure.

In yield density studies it is commonly assumed that yield per plant is inversely related to plant density (see §5.8.7, §5.9.1). If X denotes plant density and a straight-line linear relationship holds between inverse yield and density, the basic Shinozaki and Kira model (Shinozaki and Kira 1956)

$$\mathrm{E}[Y] = (\alpha + \beta x)^{-1} \qquad [6.20]$$

applies. This is a generalized linear model with inverse link and linear predictor $\alpha + \beta x$. Its yield per unit area equation is

$$\mathrm{E}[U] = \frac{x}{\alpha + \beta x}. \qquad [6.21]$$

As for yield per plant we can model yield per unit area as a generalized linear model with inverse link and linear predictor $\beta + \alpha/x$. This is a hyperbolic function of plant density and the reciprocal link is adequate. In terms of the linear predictor η this hyperbolic model gives rise to

$$\eta = \mathrm{E}[U]^{-1} = \beta + \alpha/x. \qquad [6.22]$$

We note in passing that a modification of the hyperbolic function is achieved by including a linear term in plant density known as the **inverse quadratic**,

$$\eta = \beta + \alpha_1/x + \alpha_2 x,$$

a model for unit area yield due to Holliday (1960).

A flexible family of link functions for positive continuous response is given by the family of power transformations made popular by Box and Cox (1964) (see also §5.6.2):

$$\eta = \begin{cases} (\mu^{\lambda} - 1)/\lambda & \lambda \neq 0 \\ \ln\{\mu\} & \lambda = 0. \end{cases} \qquad [6.23]$$

These transformations include as special cases the logarithmic transform, reciprocal transform, and square root transform. If used as link functions, the inverse link functions are

$$\mu = \begin{cases} (\eta\lambda + 1)^{1/\lambda} & \lambda \neq 0 \\ \exp\{\eta\} & \lambda = 0. \end{cases}$$

Figure 6.8 shows the inverse functions for $\lambda = 0.3, 0.5, 1.0, 1.5,$ and 2.0.

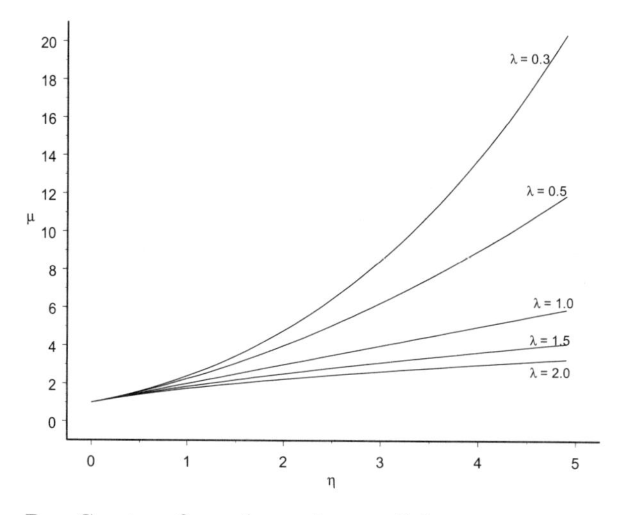

Figure 6.8. Inverse Box-Cox transformation as inverse links.

6.2.3 Generalized Linear Models in The SAS® System

In this subsection we do not fit generalized linear models to data yet. The general approach to parameter estimation and statistical inference in these models must be discussed first (§6.4) before applying the methodology to real data. However, The SAS® System offers various procedural alternatives for GLM analysis and we review here how to specify the various GLM components in some of these procedures. We prefer the powerful procedures `genmod` and `nlmixed` for fitting generalized linear models, but other procedures such as `logistic` offer functionality that cannot be found in `genmod` and `nlmixed`. For example, residual-based logistic regression diagnostics and automated variable selection routines. For cross-classified data the `catmod` procedure can be used. The log-linear models of interest to us can be fit easily with the `genmod` procedure.

The LOGISTIC Procedure

One of the first procedures for generalized linear models in The SAS® System was `proc logistic`. This procedure allows fitting regression type models to Bernoulli, Binomial, and ordinal responses by maximum likelihood (§6.4.1 and §A6.8.2). The possible link functions are the logit, probit, log-log, and complementary log-log links. For ordered responses it fits McCullagh's proportional odds model (see §6.5). The `logistic` procedure offers functionality for logistic regression analysis similar to the `reg` procedure for standard linear regression models. It enables automated variable selection routines and regression diagnostics based on Pearson or deviance residuals (see §6.4.3). Until Release 7.0 of The SAS® System `proc logistic` did not permit a `class` statement; factor and treatment variables had to be coded with a series of dummy variables. An experimental procedure in Release 7.0 (`proc tlogistic`) allowed for the presence of classification variables through a `class` statement. In Release 8.0, the `class` statement was incorporated into `proc logistic` and `proc tlogistic` has disappeared.

Similar to the powerful `genmod` procedure (see below), the response variable can be entered in two different ways in `proc logistic`. The *single-trial* syntax applies to Bernoulli and ordinal responses. The basic model statement then is

```
model response = <covariates and classification effects> / options;
```

Binomial responses are coded in the *events/trial* syntax. This syntax requires two data set variables. The number of Bernoulli trials (= the size of the binomial experiment) is coded as variable `trials` and the number of successes as `events`. Consider the seed germination data in Example 6.2 (Table 6.3, p. 307). The linear predictor of the full model fitted to these data contains temperature and concentration main effects and their interactions. If the data set is entered as shown in Table 6.3, the events/trial syntax would be used.

```
data germination;
  input temp $ conc germnumber;
  trials = 50;
  datalines;
T1 0    9
T1 0    9
T1 0    3
T1 0    7
T1 0.1 13
T1 0.1 12
and so forth
;;
run;
proc logistic data=germination;
  class temp conc / param=glm;
  model germnumber/trials = temp conc temp*conc;
run;
```

The `logistic` procedure in Release 8.0 of The SAS® System inherits from the experimental `tlogistic` procedure in Release 7.0 the ability to use different coding methods for classification variables. The coding method is selected with the `param=` option of the `class` statement. We prefer the coding scheme for classification variables that corresponds to the coding method in the `glm` procedure. This is not the default of `proc logistic` and hence we use the `param=glm` option in the example code above. We prefer glm-type coding because the specification of contrast coefficients in the `contrast` statement of `proc logistic` is then identical to the specification of contrast coefficients in `proc glm` (§4.3.4).

To analyze ordinal responses or Bernoulli variables, the single-trial syntax is used. The next example shows rating data from a factorial experiment with a 4×2 treatment structure arranged in a completely randomized design. The ordered response variable has three levels, *Poor*, *Medium*, and *Good*. Also, a Bernoulli response variable (medresp) is created taking the value 1 if the response was *Medium*, 0 otherwise.

```
data ratings;
  input REP   A   B   RESP $;
  medresp = (resp='Medium');
  datalines;
 1      1      1      Medium
 1      1      2      Medium
 1      2      1      Medium
 1      2      2      Medium
 1      3      1      Good
 1      3      2      Good
 1      4      1      Good
 1      4      2      Good
 2      1      1      Poor
 2      1      2      Medium
 2      2      1      Poor
 2      2      2      Medium
 2      3      1      Good
 2      3      2      Good
 2      4      1      Medium
 2      4      2      Good
 3      1      1      Medium
 3      1      2      Medium
 3      2      1      Poor
 3      2      2      Medium
 3      3      1      Good
 3      3      2      Good
 3      4      1      Good
 3      4      2      Good
 4      1      1      Poor
 4      1      2      Medium
 4      2      1      Poor
 4      2      2      Medium
 4      3      1      Good
 4      3      2      Good
 4      4      1      Medium
 4      4      2      Good
run;
```

The proportional odds model (§6.5) for ordered data is fit with the statements

```
proc logistic data=ratings;
  class A B / param=glm;
  model resp = A B A*B;
run;
```

By default, the values of the response categories are sorted according to their internal format. For a character variable such as RESP, the sort order is alphabetical. This results in the correct order here, since the alphabetical order corresponds to *Good-Medium-Poor*. If, for example, the *Medium* category were renamed *Average*, the internal order of the categories would be *Average-Good-Poor*. To ensure proper category arrangement in this case, one can use the order= option of the proc logistic statement. For example, one can arrange the data such that all responses rated *Good* appear first followed by the *Average* and the *Poor* responses. Then, the correct order is established with the statements

```
proc logistic data=ratings order=data;
  class A B / param=glm;
  model resp = A B A*B;
run;
```

For Bernoulli responses coded 0 and 1, `proc logistic` will also arrange the categories according to internal formatting. For numeric variables this is an ascending order. Consequently, `proc logistic` will model the probability that the variable takes on the value 0. Modeling the probability that the variable takes on the value 1 is usually preferred, since this is the mean of the response. This can be achieved with the `descending` option of the `proc logistic` statement:

```
proc logistic data=ratings descending;
  class A B / param=glm;
  model medresp = A B A*B;
run;
```

By default, `proc logistic` will use a logit link. Different link functions are selected with the `link=` option of the `model` statement. To model the Bernoulli response `medresp` with a complementary log-log link, for example, the statements are

```
proc logistic data=ratings descending;
  class A B / param=glm;
  model medresp = A B A*B / link=cloglog;
run;
```

The `model` statement provides numerous other options, for example the `selection=` option to perform automated covariate selection with backwise, forward, and stepwise methods. The `ctable` option produces a classification table for Bernoulli responses which classifies observed responses depending on whether the predicted responses are above or below some probability threshold, useful to establish the sensitivity and specificity of a logistic model for purposes of classification. The online manuals, help files, and documentation available from SAS Institute discuss additional options and features of the procedure.

The GENMOD Procedure

The `genmod` procedure is the flagship of The SAS® System for fitting generalized linear models. It is more general than the `logistic` procedure in that it allows fitting of models to responses other than Bernoulli, Binomial, and Multinomial. The built-in distributions are the Bernoulli, Binomial, Negative Binomial, Poisson, and Multinomial for discrete data and the Gaussian, Inverse Gaussian, and Gamma distributions for continuous data (the Negative Binomial distribution was added in Release 8.0). The built-in link functions are the identity, log, logit, probit, power, and complementary log-log links. For responses with multinomial distributions (e.g., ordinal responses) cumulative versions of the logit, probit, and complementary log-log links are available (see §6.5). Users who desire to use a link and/or distribution function not in this list can define their own link functions through the `fwdlink` and `invlink` statements of the procedure and the distribution functions through the `deviance` and `variance` statements.

Like `proc logistic` the `genmod` procedure accepts responses coded in single-trial or events/trial syntax. The latter is reserved for grouped Binomial data (see §6.3 on grouped vs. ungrouped data). The `order=` option of the `proc genmod` statement affects the ordering of classification variables as in `proc logistic`, but not the ordering of the response variable. A separate option (`rorder=`) of the `proc genmod` statement is used to determine the ordering of the response.

The code to produce a logistic analysis in the seed germination Example 6.2 (Table 6.3, p. 307) with `proc genmod` is

```
data germination;
  input temp $ conc germnumber;
  trials = 50;
  datalines;
T1 0 9
T1 0 9
T1 0 3
T1 0 7
T1 0.1 13
T1 0.1 12
and so forth
;;
run;
proc genmod data=germination;
  class temp conc;
  model germnumber/trials = temp conc temp*conc link=logit dist=binomial;
run;
```

The link function and distribution are selected with the `link=` and `dist=` options of the `model` statement. The next statements perform a Poisson regression with linear predictor $\eta = \beta_0 + \beta_1 x_1 + \beta_2 x_2$ and log link;

```
proc genmod data=yourdata;
  model count = x1 x2  / link=log dist=poisson;
run;
```

For each distribution, `proc genmod` will apply a default link function if the `link=` option is omitted. These are the canonical links for the distributions in Table 6.2 and the cumulative logit for the multinomial distribution. This does not work the other way around. By specifying a link function but not a distribution function `proc genmod` does not select a distribution for which this is the canonical link. That would be impossible since the canonical link does not identify the distribution. The Negative Binomial and Binomial distributions both have a canonical log link, for example. Since the default distribution of `proc genmod` is the Gaussian distribution (if the response is in single-trial syntax) statements such as

```
proc genmod data=ratings;
  class A B;
  model medresp = A B A*B / link=logit;
run;
```

do not fit a Bernoulli response with a logit link, but a Gaussian response (which is not sensible if `medresp` takes on only values 0 and 1) with a logit link. For the analysis of a Bernoulli random variable in `proc genmod` use instead

```
proc genmod data=ratings;
  class A B;
  model medresp = A B A*B / link=logit dist=binomial;
run;
```

If the events/trials syntax is used the distribution of the response will default to the Binomial and the link to the logit.

The genmod procedure has a contrast statement and an estimate statement akin to the statements of the same name in proc glm. An lsmeans statement is also available except for ordinal responses. The statements

```
proc genmod data=ratings;
  class A B;
  model medresp = A B A*B / dist=binomial link=logit;
  lsmeans A A*B / diff;
run;
```

perform pairwise comparisons for the levels of factor A and the A $\times$ B cell means.

Because there is only one method of coding classification variables in genmod (in contrast to logistic, see above), and this method is identical to the one used in proc glm, coefficients are entered in exactly the same way as in glm. Consider the ratings example above with a 4×2 factorial treatment structure.

```
proc genmod data=ratings;
  class A B;
  model resp = A B A*B / dist=multinomial link=cumlogit;
  contrast 'A1+A2-2A3=0'  A  1 1 -2 0;
run;
```

The contrast statement tests a hypothesis of the form $\mathbf{A}\boldsymbol{\beta} = \mathbf{0}$ based on the asymptotic distribution of the linear combination $\mathbf{A}\widehat{\boldsymbol{\beta}}$ and the estimate statement estimates the linear combination $\mathbf{a}'\boldsymbol{\beta}$. Here, $\boldsymbol{\beta}$ are the parameters in the linear predictor. In other words, the hypothesis is tested on the scale of the linear predictor, not the scale of the mean response. In some instances hypotheses about the mean values can be expressed as linear functions of the βs and the contrast or estimate statement are sufficient. Consider, for example, a logistic regression model for Bernoulli data,

$$\text{logit}(\pi_i) = \ln\left\{\frac{\pi_i}{1-\pi_i}\right\} = \beta_0 + \beta_1 x_{ij},$$

where x_{ij} is a dummy variable,

$$x_{ij} = \begin{cases} 1 & \text{if observation } i \text{ is from treated group } (j=1) \\ 0 & \text{if observation } i \text{ is from control group } (j=0). \end{cases}$$

The success probabilities in the two groups are then

$$\pi_{\text{control}} = \frac{1}{1+\exp\{-\beta_0\}}$$
$$\pi_{\text{treated}} = \frac{1}{1+\exp\{-\beta_0-\beta_1\}}.$$

The hypothesis H_0: $\pi_{\text{control}} = \pi_{\text{treated}}$ can be tested as the simple linear hypothesis H_0:$\beta_1 = 0$. In other instances hypotheses or quantities of interest do not reduce or are equivalent to simple linear functions of the parameter. An estimate of the ratio

$$\frac{\pi_{\text{control}}}{\pi_{\text{treated}}} = \frac{1+\exp\{-\beta_0-\beta_1\}}{1+\exp\{-\beta_0\}},$$

for example, is a nonlinear function of the parameters. The variance of the estimated ratio

$\widehat{\pi}_{\text{control}}/\widehat{\pi}_{\text{treated}}$ must be approximated from a Taylor series expansion, although exact methods for this particular ratio exist (Fieller 1940, see our §A6.8.5). The `nlmixed` procedure, although not specifically designed for generalized linear models, can be used to our advantage in this case.

The NLMIXED Procedure

The `nlmixed` procedure was first introduced into The SAS® System as an experimental procedure in Release 7.0 and has been a full production procedure since Release 8.0. It is designed for models more general than the generalized linear models considered in this chapter. It performs (approximate) maximum likelihood inference in models with multiple random effects (mixed models) and nonlinear mean function (see §8). The procedure can be used, however, even if all model parameters are fixed in the same fashion as linear models can be fit with `proc nlin`. The `nlmixed` procedure allows fitting of nonlinear (fixed and mixed effects) models to data from any distribution, provided that the log-likelihood function can be specified with SAS® programming statements. Several of the common distribution functions are already built into the procedure, among them the Bernoulli, Binomial, Poisson, and Gaussian distribution (the Negative Binomial since Release 8.1). The syntax of the `nlmixed` procedure is akin to that of `proc nlin` with slight differences in the specification of the `model` statement. One of the decisive advantages of `proc nlmixed` is its ability to estimate complicated, nonlinear functions of the parameters and obtain their estimated standard errors by the delta method (Taylor series expansions). We illustrate with a simple example from a dose-response investigation.

Mead et al. (1993, p. 336) provide data on a dose-response experiment with Binomial responses. At each of seven concentrations of an insecticide, twenty larvae were exposed to the insecticide and the number of larvae that did not survive the exposure was recorded. Mead et al. (1993) fit a probit regression model to express the proportion of larvae killed as a function of the $\log_{10}$ concentration of the insecticide. Their model was

$$\text{probit}(\pi) = \Phi^{-1}(\pi) = \beta_0 + \beta_1 \log_{10}(x),$$

where x denotes the insecticide concentration. Alternatively, a logit transformation could be used (see our application in §6.7.1), yielding

$$\text{logit}(\pi) = \ln\left(\frac{\pi}{1-\pi}\right) = \beta_0 + \beta_1 \log_{10}(x).$$

The investigators are interested in estimating the dosage at which the probability that a randomly chosen larva will be killed is 0.5, the so-called LD_{50} (dosage lethal for 50% of the subjects). For any link funtion $g(\pi)$ the LD_{50} is found by solving

$$g(0.5) = \beta_0 + \beta_1 \log_{10}(LD_{50})$$

for LD_{50}. For the probit or logit link we have $g(0.5) = 0$ and thus

$$\begin{aligned} \log_{10}(LD_{50}) &= -\beta_0/\beta_1 \\ LD_{50} &= 10^{-\beta_0/\beta_1}. \end{aligned}$$

Once the parameters of the probit or logistic regression models have been estimated the obvious estimates for the lethal dosages on the logarithmic and original scale of insecticide concentration are

$$-\widehat{\beta}_0/\widehat{\beta}_1 \text{ and } 10^{-\widehat{\beta}_0/\widehat{\beta}_1}.$$

These are nonlinear functions of the parameter estimates and standard errors must be approximated from Taylor series expansions unless one is satisfied with fiducial limits for the LD_{50} (see §A6.8.5). The data set and the `proc nlmixed` code to fit the logistic regression model and to estimate the lethal dosages are as follows.

```
data kills;
  input concentration kills;
  trials = 20;
  logc   = log10(concentration);
  datalines;
 0.375   0
 0.75    1
 1.5     8
 3.0    11
 6.0    16
12.0    18
24.0    20
;;
run;

proc nlmixed data=kills;
  parameters intcpt=-1.7 b=4.0;
  pi = 1/(1+exp(-intcpt - b*logc));
  model kills ~ binomial(trials,pi);
  estimate 'LD50' -intcpt/b;
  estimate 'LD50 original' 10**(-intcpt/b);
run;
```

The `parameters` statement defines the parameters to be estimated and assigns starting values, in the same fashion as the `parameters` statement of `proc nlin`. The statement `pi = 1/(1+exp(-intcpt - b*logc))` calculates the probability of an insect being killed through the inverse link function. This is a regular SAS® programming statement for $\pi = g^{-1}(\eta)$. The `model` statement specifies that observations of the data set variable `kills` are realizations of Binomial random variables with binomial sample size given by the data set variable `trials` and success probability according to `pi`. The `estimate` statements calculate estimates for the LD_{50} values on the logarithmic and original scale of insecticide concentrations.

If a probit analysis is desired as in Mead et al. (1993) the code changes only slightly. Only the statement for $\pi = g^{-1}(\eta)$ must be altered. The inverse of the probit link is the cumulative Standard Gaussian probability density

$$\pi = \Phi(\eta) = \int_{-\infty}^{\eta} \frac{1}{\sqrt{2\pi}} \exp\left\{-\frac{1}{2}t^2\right\} dt,$$

which can be calculated with the `probnorm()` function of The SAS® System (the SAS® function calculating the linked value is not surprisingly called the `probit()` function. A call to `probnorm(1.96)` returns the result 0.975, `probit(0.975)` returns 1.96). Because of the similarities of the logit and probit link functions the same set of starting values can be used for both analyses.

```
proc nlmixed data=kills;
  parameters intcpt=-1.7 b=4.0;
  pi = probnorm(intcpt + b*logc);
```

```
  model kills ~ binomial(trials,pi);
  estimate 'LD50' -intcpt/b;
  estimate 'LD50 original' 10**(-intcpt/b);
run;
```

Like the `genmod` procedure, `nlmixed` allows the user to perform inference for distributions other than the built-in distributions. In `genmod` this is accomplished by programming the deviance (§6.4.3) and variance function (`deviance` and `variance` statements). In `nlmixed` the model statement is altered to

```
model response ~ general(logl);
```

where `logl` is the log-likelihood function (see §6.4.1 and §A6.8.2) of the data constructed with SAS® programming statements. For the Bernoulli distribution the log-likelihood for an individual $(0,1)$ observation is simply $l(\pi;y) = y\ln\{\pi\} + (1-y)\ln\{1-\pi\}$ and for the Binomial the log-likelihood kernel for the binomial count y is $l(\pi;y) = y\ln\{\pi\} + (n-y)\ln\{1-\pi\}$. The following `nlmixed` code also fits the probit model above.

```
proc nlmixed data=kills;
  parameters intcpt=-1.7 b=4.0;
  p    = probnorm(intcpt + b*logc);
  logl = kills*log(p) + (trials-kills)*log(1-p);
  model kills ~ general(logl);
  estimate 'LD50' -intcpt/b;
  estimate 'LD50 original' 10**(-intcpt/b);
run;
```

6.3 Grouped and Ungrouped Data

Box 6.3 Grouped Data

- **Data are grouped if the outcomes represent sums or averages of observations that share the same set of explanatory variables.**
- **Grouping data changes weights in the exponential family models and impacts the asymptotic behavior of parameter estimates.**
- **Binomial counts are always grouped. They are sums of independent Bernoulli variables.**

So far we have implicitly assumed that each data point represents a single observation. This is not necessarily the case. Consider an agronomic field trial in which four varieties of wheat are to be compared with respect to their resistance to infestation with the Hessian fly (*Mayetida destructor*). The varieties are arranged in a randomized block design, and each experimental unit is a 3.7×3.7 m field plot. n_{ij} plants are sampled in the jth block for variety i, z_{ij} of which show damage. If plants on a plot are infected independently of each other the data from each plot can also be considered a set of independent and identically distributed Bernoulli variables

$$Y_{ijk} = \begin{cases} 1 & k^{\text{th}} \text{ plant for variety } i \text{ in block } j \text{ shows damage} \\ 0 & k^{\text{th}} \text{ plant for variety } i \text{ in block } j \text{ shows no damage} \end{cases}, \quad k = 1, \cdots, n_{ij}.$$

A hypothetical data set for the outcomes of such an experiment with two blocks is shown in Table 6.7. The data set contains 56 total observations, 23 of which correspond to damaged plants.

Table 6.7. Hypothetical data for Hessian fly experiment (four varieties in two blocks)

	Block $j = 1$				**Block $j = 2$**			
k	**Entry $i = 1$**	**Entry 2**	**Entry 3**	**Entry 4**	**Entry 1**	**Entry 2**	**Entry 3**	**Entry 4**
1	1	0	0	1	1	0	0	0
2	0	0	1	0	0	0	0	0
3	0	1	0	0	1	1	1	0
4	1	0	0	0	1	0	0	1
5	0	1	0	1	1	0	1	0
6	1	0	1	1	0		0	0
7	1		1	0	0		1	0
8	0			1				1
n_{ij}	8	6	7	8	7	5	7	8
z_{ij}	4	2	3	4	4	1	3	2

One could model the Y_{ijk} with a generalized linear model for 56 binary outcomes. Alternatively, one could model the number of damaged plants (z_{ij}) per plot or the proportion of damaged plants per plot (z_{ij}/n_{ij}). The number of damaged plants corresponds to the sum of the Bernoulli variables

$$z_{ij} = \sum_{k=1}^{n_{ij}} y_{ijk}$$

and the proportion of damaged plants corresponds to their sample average

$$\overline{y}_{ij} = \frac{1}{n_{ij}} z_{ij}.$$

The sums z_{ij} and averages $\overline{y}_{ij}$ are grouped versions of the original data. The sum of independent and identical Bernoulli variables is a Binomial random variable and of course in the exponential family (see Table 6.2 on p. 305). It turns out that the distribution of the average of random variables in the exponential family is also a member of the exponential family. But the number of grouped observations is smaller than the size of the original data set. In Table 6.7 there are 56 ungrouped and 8 grouped observations. A generalized linear model needs to be properly adjusted to reflect this grouping. This adjustment is made either through the variance function or by introducing weights into the analysis. In the Hessian fly example, the variance function $h(\pi_{ij})$ of a Bernoulli observation for variety i in block j is $\pi_{ij}(1 - \pi_{ij})$, but $h(\pi_{ij}) = n_{ij}\pi_{ij}(1 - \pi_{ij})$ for the counts z_{ij} and $h(\pi_{ij}) = n_{ij}^{-1}\pi_{ij}(1 - \pi_{ij})$ for the proportions $\overline{y}_{ij}$. The introduction of weights into the exponential family density or mass function accomplishes the same. Instead of [6.1] we consider the weighted version

$$f(y) = \exp\left\{\frac{(y\theta - b(\theta))}{\psi w} + c(y, \psi, w)\right\}. \qquad [6.24]$$

If y is an individual (ungrouped) observation then $w = 1$, if y represents an average of n observations choose $w = n^{-1}$, and $w = n$ if y is a sum of n observations. If data are grouped, the number of data points is equal to the number of groups, denoted by $n^{(g)}$.

Example 6.5 Earthworm Counts. In 1995 earthworms (*Lubricus terrestris* L.) were counted in four replications of a 2^4 factorial experiment at the W.K. Kellogg Biological Station in Battle Creek, Michigan. The treatment factors and levels were Tillage (chisel-plow and Notill), Input Level (conventional and low), Manure application (yes/no) and Crop (corn and soybean). Of interest was whether the *L. terrestris* density varies under these management protocols and how the various factors act and interact. Table 6.8 displays the total worm count for the 64 $(2^4 \times 4 \text{ replicates})$ experimental units (juvenile and adult worms).

Table 6.8. Ungrouped worm count data ($\#/\text{ft}^2$) in 2^4 factorial design (numbers in each cell of table correspond to counts on replicates)

		Tillage			
		Chisel-Plow		No Tillage	
Crop	**Manure**	***Input Level***		***Input Level***	
		Low	*Conventional*	*Low*	*Conventional*
Corn	Yes	$5,5,4,2$	$5,1,5,0$	$8,4,6,4$	$14,9,9,6$
	No	$3,11,0,0$	$2,0,6,1$	$2,2,11,4$	$15,9,6,4$
Soybean	Yes	$8,6,0,3$	$8,4,2,2$	$2,2,13,7$	$5,3,6,0$
	No	$8,5,3,11$	$2,6,9,4$	$7,5,18,3$	$23,12,17,9$

Unless the replication effects are block effects, the four observations per cell share the same linear predictor consisting of Tillage, Input level, Crop, Manure effects, and their interactions. Grouping to model the averages reduces the 64 observations to $n^{(g)} = 16$ observations.

Table 6.9. Grouped (averaged) worm count data; $w = 1/4$

		Tillage			
		Chisel-Plow		No Tillage	
Crop	**Manure**	***Input Level***		***Input Level***	
		Low	*Conventional*	*Low*	*Conventional*
Corn	Yes	4.00	2.75	5.50	9.50
	No	3.50	2.25	4.75	8.50
Soybean	Yes	4.25	4.00	6.00	3.50
	No	6.75	5.25	8.25	15.25

When grouping data, observations that have the same set of covariates or design effects, i.e., share the same linear predictor $\mathbf{x}'\boldsymbol{\beta}$, are summed or averaged. In the Hessian fly example each block $\times$ entry combination is unique, but n_{ij} observations were collected for each experimental unit. In experiments where treatments are replicated, grouping is often possible even if only a single observation is gathered on each unit. If covariates are continuous and their values unique, grouping is not possible.

In the previous two examples it appears as a matter of convenience whether data are grouped or not. But this choice has subtle implications. Diagnosing the model-data disagreement in generalized linear models based on residuals or goodness-of-fit measures such as Pearson's X^2 statistic or the deviance (§6.4.3) is only meaningful if data are grouped. Grouping is a special case of clustering where the elements of a cluster are reduced to a single observation (the cluster total or average). Asymptotic results for grouped data can be obtained by increasing the number of groups while holding the size of each group constant or by assuming that the number of groups is fixed and the group size grows. The respective asymptotic results are not identical. For ungrouped data, it is only reasonable to consider asymptotic results under the assumption that the sample size n grows to infinity. No distinction between group size and group number is made. Finally, if data are grouped, computations are less timeconsuming. Since generalized linear models are fit by iterative procedures, grouping large data sets as much as possible is recommended.

6.4 Parameter Estimation and Inference

Box 6.4 Estimation and Inference

- **Parameters of GLMs are estimated by maximum likelihood.**
- **The fitting algorithm is a weighted least squares method that is executed iteratively, since the weights change from iteration to iteration. Hence the name iteratively reweighted least squares (IRLS).**
- **Hypotheses are tested with Wald, likelihood-ratio, or score tests. The respective test statistics have large sample (asymptotic) Chi-squared distributions.**

6.4.1 Solving the Likelihood Problem

Estimating the parameters in the linear predictor of a generalized linear model usually proceeds by maximum likelihood methods. Likelihood-based inference requires knowledge of the joint distribution of the n data points (or $n^{(g)}$ groups). For many discrete data such as counts and frequencies, the joint distribution is given by a simple sampling model. If an experiment has only two possible outcomes which occur with probability π and $(1-\pi)$, the distribution of the response is necessarily Bernoulli. Similarly, many count data are analyzed under a Poisson model.

For random variables with distribution in the exponential family the specification of the joint distribution of the data is made simple by the relationship between mean and variance (§6.2.1). If the $i = 1, ..., n$ observations are independent, the likelihood for the complete response vector

$$\mathbf{y} = [y_1, ..., y_n]'$$

becomes

$$\mathfrak{L}(\boldsymbol{\theta}, \psi; \mathbf{y}) = \prod_{i=1}^{n} \mathfrak{L}(\boldsymbol{\theta}, \psi; y_i) = \prod_{i=1}^{n} \exp\{(y_i\theta_i - b(\theta_i))/\psi + c(y_i, \psi)\}, \qquad [6.25]$$

and the log-likelihood function in terms of the natural parameter is

$$l(\boldsymbol{\theta}, \psi; \mathbf{y}) = \sum_{i=1}^{n} (y_i\theta_i - b(\theta_i))/\psi + c(y_i, \psi). \qquad [6.26]$$

Written in terms of the vector of mean parameters, the log-likelihood becomes

$$l(\boldsymbol{\mu}, \psi; \mathbf{y}) = \sum_{i=1}^{n} (y_i\theta(\mu_i) - b(\theta(\mu_i)))/\psi + c(y_i, \psi). \qquad [6.27]$$

Since $\mu_i = g^{-1}(\eta_i) = g^{-1}(\mathbf{x}_i'\boldsymbol{\beta})$, where $g^{-1}(\bullet)$ is the inverse link function, the log-likelihood is a function of the parameter vector $\boldsymbol{\beta}$ and estimates are found as the solutions of

$$\frac{\partial l(\boldsymbol{\mu}, \psi; \mathbf{y})}{\partial \boldsymbol{\beta}} = \mathbf{0}. \qquad [6.28]$$

Details of this maximization problem are found in §A6.8.2.

Since generalized linear models are nonlinear, the estimating equations resemble those for nonlinear models. For a general link function these equations are

$$\mathbf{F}'\mathbf{V}^{-1}(\mathbf{y} - \boldsymbol{\mu}) = \mathbf{0}, \qquad [6.29]$$

and

$$\mathbf{X}'(\mathbf{y} - \boldsymbol{\mu}) = \mathbf{0} \qquad [6.30]$$

if the link is canonical. Here, $\mathbf{V}$ is a diagonal matrix containing the variances of the responses on its diagonal ($\mathbf{V} = \mathrm{Var}[\mathbf{Y}]$) and $\mathbf{F}$ contains derivatives of $\boldsymbol{\mu}$ with respect to $\boldsymbol{\beta}$. Furthermore, if the link is the identity, it follows that a solution to [6.30] is

$$\mathbf{X}'\mathbf{V}^{-1}\mathbf{X}\widehat{\boldsymbol{\beta}} = \mathbf{X}'\mathbf{V}^{-1}\mathbf{y}. \qquad [6.31]$$

This would suggest a generalized least squares estimator $\widehat{\boldsymbol{\beta}}^* = (\mathbf{X}'\mathbf{V}^{-1}\mathbf{X})^{-1}\mathbf{X}'\mathbf{V}^{-1}\mathbf{y}$. The difficulty with generalized linear models is that the variances in $\mathbf{V}$ are functionally dependent on the means. In order to calculate $\widehat{\boldsymbol{\beta}}^*$ which determines the estimate of the mean, $\widehat{\boldsymbol{\mu}}$, $\mathbf{V}$ must be evaluated at some estimate of $\boldsymbol{\mu}$. Once $\widehat{\boldsymbol{\beta}}^*$ is calculated, $\mathbf{V}$ should be updated. The procedure to solve the maximum likelihood problem in generalized linear models is hence iterative where variance estimates are updated after updates of $\widehat{\boldsymbol{\beta}}$, and known as iteratively reweighted least squares (IRLS, §A6.8.3). Upon convergence of the IRLS algorithm, the variance of $\widehat{\boldsymbol{\beta}}$ is estimated as

$$\widehat{\text{Var}}\left[\widehat{\boldsymbol{\beta}}\right] = \left(\widehat{\mathbf{F}}'\widehat{\mathbf{V}}^{-1}\widehat{\mathbf{F}}\right)^{-1}, \qquad [6.32]$$

where the variance and derivative matrix are evaluated at the converged iterate.

The mean response is estimated by evaluating the linear predictor at $\widehat{\boldsymbol{\beta}}$ and substituting the result into the linear predictor,

$$\widehat{\text{E}}[Y] = g^{-1}(\widehat{\eta}) = g^{-1}\left(\mathbf{x}'\widehat{\boldsymbol{\beta}}\right). \qquad [6.33]$$

Because of the nonlinearity of most link functions, $\widehat{\text{E}}[Y]$ is not an unbiased estimator of $\text{E}[Y]$, even if $\widehat{\boldsymbol{\beta}}$ is unbiased for $\boldsymbol{\beta}$, which usually it is not. An exception is the identity link function where $\text{E}[\mathbf{x}'\widehat{\boldsymbol{\beta}}] = \mathbf{x}'\boldsymbol{\beta}$ if the estimator is unbiased and $\text{E}[\mathbf{x}'\widehat{\boldsymbol{\beta}}] = \text{E}[Y]$ provided the model is correct. Estimated standard errors of the predicted mean values are usually derived from Taylor series expansions and are approximate in the following sense. The Taylor series of $g^{-1}(\mathbf{x}'\widehat{\boldsymbol{\beta}})$ around some value $\boldsymbol{\beta}^*$ is an approximate linearization of $g^{-1}(\mathbf{x}'\widehat{\boldsymbol{\beta}})$. The variance of this linearization is a function of the model parameters and is estimated by substituting the parameter estimates without taking into account the uncertainty in these estimates themselves. A Taylor series of $\widehat{\mu} = g^{-1}(\widehat{\eta}) = g^{-1}(\mathbf{x}'\widehat{\boldsymbol{\beta}})$ around η leads to the linearization

$$g^{-1}(\widehat{\eta}) \doteq g^{-1}(\eta) + \frac{\partial g^{-1}(\widehat{\eta})}{\partial \widehat{\eta}_{|\eta}}(\widehat{\eta} - \eta) = \mu + \frac{\partial g^{-1}(\widehat{\eta})}{\partial \widehat{\eta}_{|\eta}}\left(\mathbf{x}'\widehat{\boldsymbol{\beta}} - \mathbf{x}'\boldsymbol{\beta}\right). \qquad [6.34]$$

The approximate variance of the predicted mean is thus

$$\text{Var}[\widehat{\mu}] \doteq \left[\frac{\partial g^{-1}(\eta)}{\partial \eta}\right]^2 \mathbf{x}'\left(\mathbf{F}'\mathbf{V}^{-1}\mathbf{F}\right)^{-1}\mathbf{x}$$

and is estimated as

$$\widehat{\text{Var}}[\widehat{\mu}] \doteq \left[\frac{\partial g^{-1}(\widehat{\eta})}{\partial \widehat{\eta}}\right]^2 \mathbf{x}'\left(\widehat{\mathbf{F}}'\widehat{\mathbf{V}}^{-1}\widehat{\mathbf{F}}\right)^{-1}\mathbf{x} = \left[\frac{\partial g^{-1}(\widehat{\eta})}{\partial \widehat{\eta}}\right]^2 \mathbf{x}'\widehat{\text{Var}}\left[\widehat{\boldsymbol{\beta}}\right]\mathbf{x}. \qquad [6.35]$$

If the link function is canonical, a simplification arises. In that case $\eta = \theta(\mu)$, the derivative $\partial g^{-1}(\eta)/\partial \eta$ can be written as $\partial \mu / \partial \theta(\mu)$ and the standard error of the predicted mean is estimated as

$$\widehat{\text{Var}}[\widehat{\mu}] \doteq h(\widehat{\mu})^2 \mathbf{x}'\widehat{\text{Var}}\left[\widehat{\boldsymbol{\beta}}\right]\mathbf{x}. \qquad [6.36]$$

6.4.2 Testing Hypotheses about Parameters and Their Functions

Under certain regularity conditions which differ for grouped and ungrouped data and are outlined by, e.g., Fahrmeir and Tutz (1994, p.43), the maximum likelihood estimates $\widehat{\boldsymbol{\beta}}$ have an asymptotic Gaussian distribution with variance-covariance matrix $(\mathbf{F}'\mathbf{V}^{-1}\mathbf{F})^{-1}$. Tests of linear hypotheses of the form $H_0{:}\mathbf{A}\boldsymbol{\beta} = \mathbf{d}$ are thus based on standard Gaussian theory. A special case of this linear hypothesis is a test of $H_0{:}\beta_j = 0$ where β_j is the jth element of $\boldsymbol{\beta}$. The standard approach of dividing the estimate of β_j by its estimated standard error is useful for generalized linear models, too.

The statistic

$$W = \frac{\widehat{\beta}_j^2}{\widehat{\text{Var}}\left[\widehat{\beta}_j\right]} = \left(\frac{\widehat{\beta}_j}{\text{ese}\left(\widehat{\beta}_j\right)}\right)^2 \qquad [6.37]$$

has an asymptotic Chi-squared distribution with one degree of freedom. [6.37] is a special case of a Wald test statistic. More generally, to test H_0: $\mathbf{A}\boldsymbol{\beta} = \mathbf{d}$ compare the test statistic

$$W = \left(\mathbf{A}\widehat{\boldsymbol{\beta}} - \mathbf{d}\right)' \left(\mathbf{A}\left(\widehat{\mathbf{F}}'\widehat{\mathbf{V}}^{-1}\widehat{\mathbf{F}}\right)^{-1}\mathbf{A}'\right)^{-1} \left(\mathbf{A}\widehat{\boldsymbol{\beta}} - \mathbf{d}\right) \qquad [6.38]$$

against cutoffs from a Chi-squared distribution with q degrees of freedom (where q is the rank of the matrix $\mathbf{A}$). Such tests are simple to carry out because they only require fitting a single model. The contrast statement in `proc genmod` of The SAS® System implements such linear hypotheses (with $\mathbf{d} = \mathbf{0}$) but does not produce Wald tests by default. Instead it calculates a likelihood ratio test statistic which is computationally more involved but also has better statistical properties (see §1.3). Assume you fit a generalized linear model and obtain the parameter estimates $\widehat{\boldsymbol{\beta}}_f$ from the IRLS algorithm. The subscript f denotes the full model. A reduced model is obtained by invoking the constraint $\mathbf{A}\boldsymbol{\beta} = \mathbf{d}$. Call the estimates obtained under this constraint $\widehat{\boldsymbol{\beta}}_r$. If $l\left(\widehat{\boldsymbol{\mu}}_f, \psi; \mathbf{y}\right)$ is the log-likelihood attained in the full and $l(\widehat{\boldsymbol{\mu}}_r, \psi; \mathbf{y})$ is the log-likelihood in the reduced model, twice their difference

$$\Lambda = 2\left\{l\left(\widehat{\boldsymbol{\mu}}_f, \psi; \mathbf{y}\right) - l(\widehat{\boldsymbol{\mu}}_r, \psi; \mathbf{y})\right\} \qquad [6.39]$$

is asymptotically distributed as a Chi-squared random variable with q degrees of freedom, where q is the rank of $\mathbf{A}$, i.e., q equals the number of constraints imposed in the hypothesis. The log-likelihood is also used to calculate a measure for the goodness-of-fit between model and data, called the **deviance**. In §6.4.3 we define this measure and in §6.4.4 we apply the likelihood ratio test idea as a series of deviance reduction tests.

Hypotheses about nonlinear functions of the parameters can often be expressed in terms of equivalent linear hypotheses. Rather than using approximations based on series expansions to derive standard errors and using a Wald type test, one should test the equivalent linear hypothesis. Consider a two-group comparison of Binomial proportions with a logistic model

$$\text{logit}(\pi_j) = \beta_0 + \beta_1 x_{ij}, \ (j = 1, 2)$$

where

$$x_{ij} = \begin{cases} 1 & \text{if observation } i \text{ is from group 1} \\ 0 & \text{if observation } i \text{ is from group 2.} \end{cases}$$

The ratio

$$\frac{\pi_1}{\pi_2} = \frac{1 + \exp\{-\beta_0 - \beta_1\}}{1 + \exp\{-\beta_0\}}$$

is a nonlinear function of β_0 and β_1 but the hypothesis $H_0: \pi_1/\pi_2 = 1$ is equivalent to the linear hypothesis $H_0: \beta_1 = 0$. In other cases it may not be possible to find an equivalent linear hypothesis. In the logistic dose-response model

$$\text{logit}(\pi) = \beta_0 + \beta_1 x,$$

where x is concentration (dosage) and π the probability of observing a success at that concentration, $LD_{50} = -\beta_0/\beta_1$. A 5% level test whether the LD_{50} is equal to some concentration x_0 could proceed as follows. At convergence of the IRLS algorithm, estimate

$$L\widehat{D}_{50} = -\frac{\widehat{\beta}_0}{\widehat{\beta}_1}$$

and obtain its estimated standard error. Calculate an approximate 95% confidence interval for LD_{50}, relying on the asymptotic Gaussian distribution of $\widehat{\boldsymbol{\beta}}$, as

$$L\widehat{D}_{50} \pm 1.96\,\text{ese}\left(L\widehat{D}_{50}\right).$$

If the confidence interval does not cover the concentration x_0, reject the hypothesis that $LD_{50} = x_0$; otherwise fail to reject. A slightly conservative approach is to replace the standard Gaussian cutoff with a cutoff from a t distribution (which is what `proc nlmixed` does). The key is to derive a good estimate of the standard error of the nonlinear function of the parameters. For general nonlinear functions of the parameters we prefer approximate standard errors calculated from Taylor series expansions. This method is very general and typically produces good approximations. The `estimate` statement of the `nlmixed` procedure in The SAS® System implements the calculation of standard errors by a first-order Taylor series for nonlinear functions of the parameters. For certain functions, such as the LD_{50} above, exact formulas have been developed. Finney (1978, pp. 80-82) gives formulas for fiducial intervals for the LD_{50} based on a theorem by Fieller (1940) and applies the result to test the identity of equipotent dosages for two assay formulations. Fiducial intervals are akin to confidence intervals; the difference between the two approaches is largely philosophical. The interpretation of confidence limits appeals to conceptual, repeated sampling such that the repeatedly calculated intervals include the true parameter value with a specified frequency. Fiducial limits are values of the parameter that would produce an observed statistic such as $L\widehat{D}_{50}$ with a given probability. See Schwertman (1996) and Wang (2000) for further details on the comparison between fiducial and frequentist inference. In §A6.8.5 Fieller's derivation is examined and compared to the expression of the standard error of a ratio of two random variables derived from a first-order Taylor series expansion. As it turns out the Taylor series method results in a very good approximation provided that the slope β_1 in the dose-response model is considerably different from zero. For the approximation to be satisfactory, the standard t statistic for testing $H_0: \beta_1 = 0$,

$$t_{obs} = \frac{\widehat{\beta}_1}{\text{ese}\left(\widehat{\beta}_1\right)},$$

should be in absolute value greater than the t-cutoff divided by $\sqrt{0.05}$,

$$|t_{obs}| \geq t_{\frac{\alpha}{2},v} \Big/ \sqrt{0.05}.$$

Here, v are the degrees of freedom associated with the model deviance and α denotes the significance level. For a 5% significance level this translates into a t_{obs} of about 9 or more in absolute value (Finney 1978, p. 82). It should be noted, however, that the fiducial limits of Fieller (1940) are derived under the assumption of Gaussianity and unbiasedness of the estimators.

6.4.3 Deviance and Pearson's X^2 Statistic

The quality of agreement between model and data in generalized linear models is assessed by two statistics, the Pearson X^2 statistic and the model deviance D. Pearson's statistic is defined as

$$X^2 = \sum_{i=1}^{n^{(g)}} \frac{(y_i - \widehat{\mu}_i)^2}{h(\widehat{\mu}_i)}, \qquad [6.40]$$

where $n^{(g)}$ is the size of the grouped data set and $h(\widehat{\mu}_i)$ is the variance function evaluated at the estimated mean (see §6.2.1 for the definition of the variance function). X^2 thus takes the form of a weighted residual sum of squares. The **deviance** of a generalized linear model is derived from the likelihood principle. It is proportional to twice the difference between the maximized log-likelihood evaluated at the estimated means $\widehat{\mu}_i$ and the largest achievable log likelihood obtained by setting $\widehat{\mu}_i = y_i$. Two versions of the deviance are distinguished, depending on whether the distribution of the response involves a scale parameter ψ or not. Recall from §6.3 the weighted exponential family density

$$f(y_i) = \exp\{(y_i\theta_i - b(\theta_i))/(\psi w_i) + c(y_i, \psi)\}.$$

If $\widehat{\theta}_i = \theta(\widehat{\mu}_i)$ is the estimate of the natural parameter in the model under consideration and $\dot{\theta}_i = \theta(y_i)$ is the canonical link evaluated at the observations, the **scaled deviance** is defined as

$$\begin{aligned} D^*(\mathbf{y};\widehat{\boldsymbol{\mu}}) &= 2\sum_{i=1}^{n^{(g)}} l(\dot{\theta}_i, \psi; y_i) - 2\sum_{i=1}^{n^g} l\left(\widehat{\theta}_i, \psi; y_i\right) \\ &= \frac{2}{\psi}\sum_{i=1}^{n^{(g)}} \left\{ y_i\left(\dot{\theta}_i - \widehat{\theta}_i\right) - b(\dot{\theta}_i) + b\left(\widehat{\theta}_i\right)\right\} \\ &= 2\{l(\mathbf{y}, \psi; \mathbf{y}) - l(\widehat{\boldsymbol{\mu}}, \psi; \mathbf{y})\}. \end{aligned} \qquad [6.41]$$

When $D^*(\mathbf{y};\widehat{\boldsymbol{\mu}})$ is multiplied by the scale parameter ψ, we simply refer to **the deviance** $D(\mathbf{y};\widehat{\boldsymbol{\mu}}) = \psi D^*(\mathbf{y};\widehat{\boldsymbol{\mu}}) = \psi 2\{l(\mathbf{y}, \psi; \mathbf{y}) - l(\widehat{\boldsymbol{\mu}}, \psi; \mathbf{y})\}$. In [6.41], $l(\mathbf{y}, \psi; \mathbf{y})$ refers to the log-likelihood evaluated at $\boldsymbol{\mu} = \mathbf{y}$, and $l(\widehat{\boldsymbol{\mu}}, \psi; \mathbf{y})$ to the log-likelihood obtained from fitting a particular model. If the fitted model is **saturated**, i.e., fits the data perfectly, the (scaled) deviance and X^2 statistics are identically zero.

The utility of X^2 and $D^*(\mathbf{y};\widehat{\mu})$ lies in the fact that, under certain conditions, both have a well-known asymptotic distribution. In particular,

$$X^2/\psi \xrightarrow{d} \chi^2_{n^{(g)}-p}, \quad D^*(\mathbf{y};\widehat{\mu}) \xrightarrow{d} \chi^2_{n^{(g)}-p}, \qquad [6.42]$$

where p is the number of estimated model parameters (the rank of the model matrix $\mathbf{X}$).

Table 6.10 shows deviance functions for various distributions in the exponential family. The deviance in the Gaussian case is simply the residual sum of squares. The deviance-based scale estimate $\widehat{\psi} = D(\mathbf{y};\widehat{\boldsymbol{\mu}})/(n^{(g)} - p)$ in this case is the customary residual mean square error. Furthermore, the deviance and Pearson X^2 statistics are identical then and their scaled versions have exact (rather than approximate) Chi-squared distributions.

Table 6.10. Deviances for some exponential family distributions (in the Binomial case n_i denotes the Binomial sample size and y_i the number of successes)

Distribution	$D(\mathbf{y};\widehat{\boldsymbol{\mu}})$
Bernoulli	$2\sum_{i=1}^{n} - y_i \ln\left\{\frac{\widehat{\mu}_i}{1-\widehat{\mu}_i}\right\} - \ln\{1-\widehat{\mu}_i\}$
Binomial	$2\sum_{i=1}^{n} y_i \ln\left\{\frac{y_i}{\widehat{\mu}_i}\right\} + (n_i - y_i)\ln\left\{\frac{r_i - y_i}{r_i - \widehat{\mu}_i}\right\}$
Negative Binomial	$2\sum_{i=1}^{n} y_i \ln\left\{\frac{y_i}{\widehat{\mu}_i}\right\} - (y_i + 1/k)\ln\left\{\frac{y_i + 1/k}{\widehat{\mu}_i + 1/k}\right\}$
Poisson	$2\sum_{i=1}^{n} y_i \ln\left\{\frac{y_i}{\widehat{\mu}_i}\right\} - (y_i - \widehat{\mu}_i)$
Gaussian	$\sum_{i=1}^{n} (y_i - \widehat{\mu}_i)^2$
Gamma	$2\sum_{i=1}^{n} - \ln\left\{\frac{y_i}{\widehat{\mu}_i}\right\} + (y_i - \widehat{\mu}_i)/\widehat{\mu}_i$
Inverse Gaussian	$\sum_{i=1}^{n} (y_i - \widehat{\mu}_i)^2/(\widehat{\mu}_i^2 y_i)$

As the agreement between data (y_i) and model fit ($\widehat{\mu}_i$) improves, X^2 will decrease in value. On the contrary a model not fitting the data well will result in a large value of X^2 (and a large value for $D^*(\mathbf{y};\widehat{\boldsymbol{\mu}})$). If the conditions for the asymptotic result hold, one can calculate the p-value for H_0: *model fits the data* as

$$\Pr\left(X^2/\psi > \chi^2_{n^{(g)}-p}\right) \quad \text{or} \quad \Pr\left(D^*(\mathbf{y};\widehat{\boldsymbol{\mu}}) > \chi^2_{n^{(g)}-p}\right).$$

If the p-value is sufficiently large, the model is acceptable as a description of the data generating mechanism. Before one can rely on this goodness-of-fit test, the conditions under which the asymptotic result holds, must be met and understood (McCullagh and Nelder 1989, p. 118). The first requirement for the result to hold is independence of the observations. If overdispersion arises from autocorrelation or randomly varying parameters, both of which induce correlations, X^2 and $D^*(\mathbf{y};\widehat{\boldsymbol{\mu}})$ do not have asymptotic Chi-squared distributions. More importantly, it is assumed that data are grouped, the number of groups ($n^{(g)}$) remains fixed, and the sample size in each group tends to infinity, thereby driving the within group variance to zero. If these conditions are not met, large values of X^2 or $D^*(\mathbf{y};\widehat{\boldsymbol{\mu}})$ do not necessarily indicate poor fit and should be interpreted with caution. Therefore, if data are ungrouped (group size is 1, $n^{(g)} = n$), one should not rely on X^2 or $D^*(\mathbf{y}; \widehat{\boldsymbol{\mu}})$ as goodness-of-fit measures.

The asymptotic distributions of X^2/ψ and $D(\mathbf{y}; \widehat{\boldsymbol{\mu}})/\psi$ suggest a simple method to estimate the extra scale parameter ψ in a generalized linear model. Equating X^2/ψ with its asymptotic expectation,

$$\frac{1}{\psi}\mathrm{E}\left[X^2\right] \approx n^{(g)} - p,$$

suggests the estimator $\widehat{\psi} = X^2/(n^{(g)} - p)$. Similarly $D(\mathbf{y}; \widehat{\boldsymbol{\mu}})/(n^{(g)} - p)$ is a deviance-based estimate of the scale parameter ψ.

For distributions where $\psi = 1$ a value of $D(\mathbf{y}; \widehat{\boldsymbol{\mu}})/(n^{(g)} - p)$ or $X^2/(n^{(g)} - p)$ substantially larger than one is indication that the data are more dispersed than is permissible under the assumed probability distribution. This can be used to diagnose an improperly specified model **or** overdispersion (§6.6). It should be noted that if the ratio of deviance or X^2 to its degrees of freedom is large, one should not automatically conclude that the data are overdispersed. Akin to a linear model where omitting an important variable or effect increases the error variance, the deviance will increase if an important effect is unaccounted for.

6.4.4 Testing Hypotheses through Deviance Partitioning

Despite the potential problems of interpreting the deviance for ungrouped data, it has paramount importance in statistical inference for generalized linear models. While the Chi-squared approximation for $D^*(\mathbf{y}; \widehat{\boldsymbol{\mu}})$ may not hold, the distribution of differences of deviances among nested models can well be approximated by Chi-squared distributions, even if data are not grouped. This forms the basis for *Change-In-Deviance* tests, the GLM equivalent of the sum of squares reduction test. Consider a full model M_f and a second (reduced) model M_r obtained by eliminating q parameters of M_f. Usually this means setting one or more parameters in the linear predictor to zero by eliminating treatment effects from M_f. If $\widehat{\boldsymbol{\mu}}_f$ and $\widehat{\boldsymbol{\mu}}_r$ are the respective estimated means, then the increase in deviance incurred by eliminating the q parameters is given by

$$D^*(\mathbf{y}; \widehat{\boldsymbol{\mu}}_r) - D^*\left(\mathbf{y}; \widehat{\boldsymbol{\mu}}_f\right), \qquad [6.43]$$

and has an asymptotic Chi-squared distribution with q degrees of freedom. If [6.43] is significantly large, reject model M_r in favor of M_f. Since $D^*(\mathbf{y}; \widehat{\boldsymbol{\mu}}) = 2\{l(\mathbf{y}, \psi; \mathbf{y}) - l(\widehat{\boldsymbol{\mu}}, \psi; \mathbf{y})\}$ is twice the difference between the maximal and the maximized log-likelihood, [6.43] can be rewritten as

$$D^*(\mathbf{y}; \widehat{\boldsymbol{\mu}}_r) - D^*\left(\mathbf{y}; \widehat{\boldsymbol{\mu}}_f\right) = 2\left\{l\left(\widehat{\boldsymbol{\mu}}_f, \psi; \mathbf{y}\right) - l\left(\widehat{\boldsymbol{\mu}}_r, \psi; \mathbf{y}\right)\right\} = \Lambda, \qquad [6.44]$$

and it is thereby established that this procedure is also the likelihood ratio test (§A6.8.4) for testing M_f versus M_r.

To demonstrate the test of hypotheses through deviance partitioning, we use Example 6.2 (p. 306, data appear in Table 6.3). Recall that the experiment involves a 4×4 factorial treatment structure with factors temperature (T_1, T_2, T_3, T_4) and concentration of a chemical

$(0, 0.1, 1.0, 10)$ and their effect on the germination probability of seeds. For each of the 16 treatment combinations 4 dishes with 50 seeds each are prepared and the number of germinating seeds are counted in each dish. From an experimental design standpoint this is a completely randomized design with a 4×4 treatment structure. Hence, we are interested in determining the significance of the Temperature main effect, the main effect of the chemical Concentration, and the Temperature $\times$ Concentration interaction. If the seeds within a dish and between dishes germinate independently of each other, the germination count in each dish can be modeled as a Binomial$(50, \pi_{ij})$ random variable where π_{ij} denotes the germination probability if temperature i and concentration j are applied. Table 6.11 lists the models successively fit to the data.

Table 6.11. Models successively fit to seed germination data (" $\times$ " in a cell of the table implies that the particular effect is present in the model, grand represents the presence of a grand mean (intercept) in the model)

Model	Grand	Temperature	Concentration	Temp. × Conc.
A	×			
B	×	×		
C	×		×	
D	×	×	×	
E	×	×	×	×

Applying a logit link, model A is fit to the data with the `genmod` procedure statements:

```
proc genmod data=germrate;
  model germ/trials = /link=logit dist=binomial;
run;
```

Output 6.1.

```
                         The GENMOD Procedure

                           Model Information
           Data Set                                WORK.GERMRATE
           Distribution                                 Binomial
           Link Function                                   Logit
           Response Variable (Events)                       germ
           Response Variable (Trials)                     trials
           Observations Used                                  64
           Number Of Events                                 1171
           Number Of Trials                                 3200

                 Criteria For Assessing Goodness Of Fit
       Criterion                  DF          Value           Value/DF
       Deviance                   63      1193.8014            18.9492
       Scaled Deviance            63      1193.8014            18.9492
       Pearson Chi-Square         63      1087.5757            17.2631
       Scaled Pearson X2          63      1087.5757            17.2631
       Log Likelihood                    -2101.6259
Algorithm converged.
                    Analysis Of Parameter Estimates
                        Standard        Wald 95%            Chi-
Parameter   DF  Estimate     Error    Confidence Limits    Square   Pr > ChiSq
Intercept    1   -0.5497  0.0367    -0.6216    -0.4778   224.35       <.0001
Scale        0    1.0000  0.0000     1.0000     1.0000
NOTE: The scale parameter was held fixed.
```

The deviance for this model is 1193.80 on 63 degrees of freedom (Ouput 6.1). It is almost 19 times larger than 1 and this clearly indicates that the model does not account for the variability in the data. This could be due to the seed counts being more dispersed than Binomial random variables and/or the absence of important effects in the model. The intercept estimate $\widehat{\beta}_0 = -0.5497$ translates into an estimated success probability of

$$\widehat{\pi} = \frac{1}{1 + \exp\{0.5497\}} = 0.356.$$

This is the overall proportion of germinating seeds. Tallying all successes (= germinations) in Table 6.3 one obtains $1,171$ germination on 64 dishes containing 50 seeds each and

$$\widehat{\pi} = \frac{1,171}{64*50} = 0.356.$$

Notice that the degrees of freedom for the model denote the number of groups $\left(n^{(g)} = 64\right)$ minus the numbers of estimated parameters.

Models B through E are fit similarly with the SAS® statments (output not shown)

```
/* model B */
proc genmod data=germrate;
  class temp;
  model germ/trials = temp /link=logit dist=binomial;
run;
/* model C */
proc genmod data=germrate;
  class conc;
  model germ/trials = conc /link=logit dist=binomial;
run;
/* model D */
proc genmod data=germrate;
  class temp conc;
  model germ/trials = temp conc /link=logit dist=binomial;
run;
/* model E */
proc genmod data=germrate;
  class temp conc;
  model germ/trials = temp conc temp*conc /link=logit dist=binomial;
run;
```

Their degrees of freedom and deviances are shown in Table 6.12.

Table 6.12. Deviances and X^2 for models A – E in Table 6.11 (DF denotes degrees of freedom of the deviance and X^2 statistic)

Model	DF	Deviance	Deviance/DF	X^2	X^2/DF
A	63	$1,193.80$	18.94	$1,087.58$	17.26
B	60	430.11	7.17	392.58	6.54
C	60	980.09	16.33	897.27	14.95
D	57	148.10	2.59	154.95	2.72
E	48	55.64	1.15	53.95	1.12

The deviance and X^2 statistics for any of the five models are very close. To test hypotheses about the various treatment factors, differences of deviances between two models are compared to cutoffs from Chi-squared distributions with degrees of freedom equal to the difference in DF for the models. For example, comparing the deviances of models $\boxed{A}$ and $\boxed{B}$ yields $1,193.8 - 430.11 = 763.69$ on 3 degrees of freedom. The p-value for this test can be calculated in SAS® with

```
data pvalue; p = 1-probchi(763.69,3); run; proc print; run;
```

The result is a p-value near zero. But what does this test mean? We are comparing a model with Temperature effects only ($\boxed{B}$) against a model without any effects ($\boxed{A}$), that is, in the absence of any concentration effects and/or interactions. We tested the hypothesis that a model with Temperature effects explains the variation in the data as well as a model containing only an intercept. Similarly, the question of a significant Concentration effect in the absence of temperature effects (and the interaction) is addressed by comparing the deviance difference $1,193.8 - 980.09 = 213.71$ against a Chi-squared distribution with 3 degrees of freedom ($p < 0.0001$). The significance of the Concentration effect in a model already containing a Temperature main effect is assessed by the deviance difference $430.11 - 148.10 = 282.01$. This value differs from the deviance reduction of 213.7 which was obtained by adding Concentration effects to the null model. Because of the nonlinearity of the logistic link function the effects in the generalized linear model are not orthogonal. The significance of a particular effect depends on which other effects are present in the model, a feature of sequential tests (§4.3.3) under nonorthogonality. Although either Chi-squared statistic would be significant in this example, it is easy to see that it does make a difference in which order the effects are tested. The most meaningful test that can be derived from Table 6.12 is that of the Temperature $\times$ Concentration interaction by comparing deviances of models $\boxed{D}$ and $\boxed{E}$. Here, the full model includes all possible effects (two main effects and the interaction) and the reduced model ($\boxed{D}$) excludes only the interaction. From this comparison with a deviance difference of $148.10 - 55.64 = 92.46$ on $57 - 48 = 9$ degrees of freedom a p-value of <0.0001 is obtained, sufficient to declare a significant Temperature $\times$ Concentration interaction.

An approach to deviance testing that does not depend on the order in which terms enter the model is to use **partial** deviances, where the contribution of an effect is evaluated as the deviance decrement incurred by adding the effect to a model containing *all other* effects. In `proc genmod` of The SAS® System this is accomplished by adding the `type3` option to the `model` statement. The `type1` option of the `model` statement will conduct a sequential test of model effects. The following statements request sequential (`type1`) and partial (`type3`) likelihood ratio tests in the full model. The `ods` statement preceding the `proc genmod` code excludes the lengthy table of parameter estimates from the output.

```
ods exclude parameterestimates;
proc genmod data=germrate;
  class temp conc;
  model germ/trials = temp conc temp*conc /link=logit
                                          dist=binomial type1 type3;
run;
```

The sequential (`Type1`) and partial (`Type3`) deviance decrements are not identical (Output 6.2). Adding Temperature effects to a model containing no other effects yields a deviance reduction of 763.69 (as also calculated from the data in Table 6.12). Adding Temperature effects to a model containing Concentration effects (and the interactions) yields a deviance reduction of 804.24. The partial and sequential tests for the interaction are the same, because this term entered the model last.

Wald tests for the partial or sequential hypotheses – instead of likelihood ratio tests – are requested with the `wald` option of the `model` statement. In this case both the Chi-squared statistics and the p-values will change because the Wald test statistics do not correspond to a difference in deviances between a full and a reduced model. We prefer likelihood ratio over Wald tests unless data sets are so large that obtaining the computationally more involved likelihood ratio test statistic is prohibitive.

Output 6.2.

```
                         The GENMOD Procedure

                          Model Information

              Data Set                          WORK.GERMRATE
              Distribution                           Binomial
              Link Function                             Logit
              Response Variable (Events)                 germ
              Response Variable (Trials)               trials
              Observations Used                            64
              Number Of Events                           1171
              Number Of Trials                           3200

                    Class Level Information
                Class      Levels    Values

                temp            4    1 2 3 4
                conc            4    0 0.1 1 10

                 Criteria For Assessing Goodness Of Fit

        Criterion                DF            Value          Value/DF

        Deviance                 48          55.6412            1.1592
        Scaled Deviance          48          55.6412            1.1592
        Pearson Chi-Square       48          53.9545            1.1241
        Scaled Pearson X2        48          53.9545            1.1241
        Log Likelihood                    -1532.5458

Algorithm converged.

                  LR Statistics For Type 1 Analysis
                                                 Chi-
         Source          Deviance         DF     Square      Pr > ChiSq

         Intercept      1193.8014
         temp            430.1139          3     763.69          <.0001
         conc            148.1055          3     282.01          <.0001
         temp*conc        55.6412          9      92.46          <.0001

                  LR Statistics For Type 3 Analysis
                                         Chi-
              Source           DF      Square     Pr > ChiSq

              temp              3      804.24         <.0001
              conc              3      198.78         <.0001
              temp*conc         9       92.46         <.0001
```

6.4.5 Generalized R^2 Measures of Goodness-of-Fit

In the nonlinear regression models of Chapter 5 we faced the problem of determining a R^2-type summary measure that expresses the degree to which the model and data agree. The rationale of R^2-type measures is to express the degree of *variation* in the data that is explained or unexplained by a particular model. That has led to the Pseudo-R^2 measure suggested for nonlinear models

$$\text{Pseudo-}R^2 = 1 - \frac{SSR}{SST_m}, \qquad [6.45]$$

where $SST_m = \sum_{i=1}^{n}(y_i - \overline{y})^2$ is the total sum of squares corrected for the mean and $SSR = \sum_{i=1}^{n}(y_i - \widehat{y}_i)^2$ is the residual (error) sum of squares. Even in the absence of an intercept in the model, SST_m is the correct denominator since it is the sample mean $\overline{y}$ that would be used to predict y if the response were unrelated to the covariates in the model. The ratio SSR/SST_m can be interpreted as the proportion of variation unexplained by the model. Generalized linear models are also nonlinear models (unless the link function is the identity link) and a goodness-of-fit measure akin to [6.45] seems reasonable to measure model-data agreement. Instead of sums of squares, the measure should rest on deviances, however. Since for Gaussian data with identity link the deviance is an error sum of squares (Table 6.10), the measure should in this case also reduce to the standard R^2 measure in linear models. It thus seems natural to build a goodness-of-fit measure that involves the deviance of the model that is fit and compares it to the deviance of a null model not containing any explanatory variables. Since differences of scaled deviances are also differences in log likelihoods between full and reduced models we can use

$$l(\widehat{\boldsymbol{\mu}}_f, \psi; \mathbf{y}) - l(\widehat{\boldsymbol{\mu}}_0, \psi; \mathbf{y}),$$

where $l(\widehat{\boldsymbol{\mu}}_f, \psi; \mathbf{y})$ is the log likelihood in the fitted model and $l(\widehat{\boldsymbol{\mu}}_0, \psi; \mathbf{y})$ is the log likelihood in the model containing only an intercept (the null model). For binary response models a generalized R^2 measure was suggested by Maddala (1983). Nagelkerke (1991) points out that it was also proposed for any model fit by the maximum likelihood principle by Cox and Snell (1989, pp. 208-209) and Magee (1990), apparently independently:

$$\begin{aligned} -\ln\{1 - R_*^2\} &= \frac{2}{n}\{l(\widehat{\boldsymbol{\mu}}_f, \psi; \mathbf{y}) - l(\widehat{\boldsymbol{\mu}}_0, \psi; \mathbf{y})\} \\ R_*^2 &= 1 - \exp\left\{-\frac{2}{n}\left(l(\widehat{\boldsymbol{\mu}}_f, \psi; \mathbf{y}) - l(\widehat{\boldsymbol{\mu}}_0, \psi; \mathbf{y})\right)\right\}. \end{aligned} \qquad [6.46]$$

Nagelkerke (1991) discusses that this generalized R^2 measure has several appealing properties. If the covariates in the fitted model have no explanatory power, the log likelihood of the fitted model, $l(\widehat{\boldsymbol{\mu}}_f, \psi; \mathbf{y})$ will be close to the likelihood of the null model and R_*^2 approaches zero. R_*^2 has a direct interpretation in terms of explained *variation* in the sense that it partitions the contributions of covariates in nested models. But unlike the R^2 measure in linear models, [6.46] is not bounded by 1 from above. Its maximum value is

$$\max\{R_*^2\} = 1 - \exp\left\{\frac{2}{n}l(\widehat{\boldsymbol{\mu}}_0, \psi; \mathbf{y})\right\}. \qquad [6.47]$$

Nagelkerke (1991) thus recommends scaling R_*^2 and using the measure

$$\overline{R}^2 = \frac{R_*^2}{\max\{R_*^2\}} \qquad [6.48]$$

instead which is bounded between 0 and 1 and referred to as the rescaled generalized R^2. The `logistic` procedure of The SAS® System calculates both generalized R^2 measures if requested by the `rsquare` option of the `model` statement.

6.5 Modeling an Ordinal Response

Box 6.5 The Proportional Odds Model

- **The proportional odds model (POM) is a statistical model for ordered responses developed by McCullagh (1980). It belongs to the family of cumulative link models and is not a generalized linear model in the narrow sense.**
- **The POM can be thought of as a series of logistic curves and in the two-category case reduces to logistic regression.**
- **The POM can be fit to data with the genmod procedure of The SAS® System that enables statistical inference very much akin to what practitioners expect from an ANOVA-based package.**

Ordinal responses arise frequently in the study of soil and plant data. An ordinal (or ordered) response is a categorical variable whose values are related in a greater/lesser sense. The assessment of turf quality in nine categories from best to worst results in an ordered response variable as does the grouping of annual salaries in income categories. The difference between the two types of ordered responses is that salary categories stem from categorizing an underlying (latent) continuous variable. Anderson (1984) terms this a grouped ordering. The assignment to a category can be made without error and different interpreters will assign salaries to the same income categories provided they use the same grouping. Assessed orderings, on the contrary, involve a more complex process of determining the outcome of an observation. A turf scientist rating the quality of a piece of turf combines information about the time of day, the brightness of the sun, the expectation for the particular grass species, past experience, and the disease and management history of the experimental area. The final assessment of turf quality is a complex aggregate and compilation of these various factors. As a result, there will be variability in the ratings among different interpreters that complicates the analysis of such data. The development of clearcut rules for category assignment helps to reduce the interrater variability in assessed orderings, some room for interpretation of these rules invariably remains.

In this section we are concerned with fitting statistical models to ordinal responses in general and side-step the issue of rater agreement. Log-linear models for contingency tables can be used to describe and infer the degree to which interpreters of the same material rate independently or interact. An application of log-linear modeling of rater association and agreement can be found in §6.7.6. Applications of modeling ordered outcomes with the methods presented in this subsection appear in §6.7.4 and §6.7.5.

The categories of an ordinal variable are frequently coded with numerical values. In the turf sciences, for example, it is customary to rate plant quality or color on a scale between 1 (worst case) and 9 (best case) with integer or even half steps in between. Plant injury is often assessed in 10% (or coarser) categories. Farmers report on a questionnaire whether they perform *low*, *medium*, or *high* input whole-field management and the responses are coded as 1, 2, and 3 in the data file. The obvious temptation is to treat such ordinal outcomes as if they represent measurements of continuous variables. Rating the plant quality of three replications of a particular growth regulator application as 4, 5, and 8 then *naturally* leads to estimates of the mean quality such as $(1/3)*(4+5+8) = 5.66$. If the category values are coded with letters $a, b, c, d, \cdots$ instead, it is obvious that the calculation of an average rating as $(1/3)*(d+e+h)$ is meaningless. Operations like addition or subtraction require that distances between the category values be well-defined. In particular for assessed orderings, this is hardly the case. The difference between a *low* and *medium* response is not the same as that between a *medium* and *high* response, even if the levels are coded 1, 2, and 3. Using numerical values for ordered categories is a mere labeling convenience that does not alter the essential feature of the response as ordinal. The scoring system 1, 2, 3, 4 implies the same ordering as 1, 20, 30, 31 but will lead to different numerical results. Standard analysis of variance followed by standard hypothesis testing procedures requires continuous, Gaussian, univariate, and independent responses. Ordinal response variables violate all of these assumptions. They are categorical rather than continuous, multinomial rather than Gaussian-distributed, multivariate rather than univariate and not independent. If ten responses were observed in three categories, four of which fell into the third category, there are only six responses to be distributed among the remaining categories. Given the responses in the third category, the probability that any of the remaining observations will fall into the first category has changed; hence the counts are not independent. By assigning numerical values to the categories and using standard analysis of variance or regression methods, the user declares the underlying assumptions as immaterial or the procedure as sufficiently robust against their violation.

Snedecor and Cochran (1989, 8th ed., pp. 206-208) conclude that standard analyses such as ANOVA may be appropriate for ordered outcomes if "the [...] classes constructed [...] represent equal gradations on a continuous scale." Unless a latent variable can be identified, verification of this key assumption is impossible and even then may not be tenable. In rating of color, for example, one could construct a latent continuous color variable as a function of red, green, and blue intensity, hue, saturation, and lightness and view the color rating as its categorization in equally spaced intervals. But assessed color rating categories do not necessarily represent equal gradations of this process, even if it can be constructed. In the case of assessed orderings where a latent variable may not exist at all, the assumption of equal gradations on a continuous scale is most questionable. Even if it holds, the ordered response is nevertheless discrete and multivariate rather than continuous and univariate.

Instead of appealing to the restrictive conditions under which analysis of variance *might* be appropriate, we prefer statistical methods that have been specifically developed for ordinal data that take into account the distributional properties of ordered responses, and perhaps most importantly, that do not depend on the actual scoring system in use. The model we rely on most heavily is McCullagh's proportional odds model (POM), which belongs to the class of cumulative link models (McCullagh 1980, McCullagh 1984, McCullagh and Nelder 1989). It is not a bona fide generalized linear model but very closely related to logistic regression models, which justifies its discussion in this chapter (the correspondence of these models to GLMs can be made more precise by using composite link functions, see e.g., Thompson and Baker 1981). For only two ordered categories the POM reduces to a standard GLM for Bernoulli or Binomial outcomes because the Binomial distribution is a special case of the multinomial distribution where one counts the number of outcomes out of N independent Bernoulli experiments that fall into one of J categories. Like other generalized linear models cumulative link models apply a link function to map the parameter of interest onto a scale where effects are linear. Unlike the models for Bernoulli or Binomial data, the link function is not applied to the probability that the response takes on a certain value, but to the cumulative probability that the response occurs in a particular category or below. It is this ingenious construction from which essential simplifications arise. Our focus on the proportional odds model is not only motivated by its elegant formulation, convenient mathematics, and straightforward interpretation. It can furthermore be easily fitted with the `logistic` and `genmod` procedures of The SAS® System and is readily available to those familiar with fitting generalized linear models.

6.5.1 Cumulative Link Models

Assume there exists a continuous variable X with probability density function $f(x)$ and the support of X is divided into categories by a series of cutoff parameters α_j; this establishes a grouped ordering in the sense of Anderson (1984). Rather than the latent variable X we observe the response Y and assign it the value j whenever Y falls between the cutoffs α_{j-1} and α_j (Figure 6.9). If the cutoffs are ordered in the sense that $\alpha_{j-1} < \alpha_j$, the response Y is ordinal. Cumulative link models do not require that a latent continuous variable actually exists, or that the ordering is grouped (rather than assessed). They are simply motivated most easily in the case where Y is a grouping of a continuous, unobserved variable X.

The distribution of the latent variable in Figure 6.9 is $G(4,1)$ and cutoffs were placed at $\alpha_1 = 2.3$, $\alpha_2 = 3.7$, and $\alpha_3 = 5.8$. These cutoffs define a four-category ordinal response Y. The probability to observe $Y = 1$, for example, is obtained as the difference

$$\Pr(X < 2.3) - \Pr(X < -\infty) = \Pr(Z < -1.7) - 0 = 0.045,$$

where Z is a standard Gaussian random variable. Similarly, the probability to observe Y in at most category 2 is

$$\Pr(Y \le 2) = \Pr(X \le 3.7) = \Pr(Z \le -0.3) = 0.382.$$

For a given distribution of the latent variable, the placement of the cutoff parameters determines the probabilities to observe the ordinal variable Y. When fitting a cumulative link model to data, these parameters are estimated along with the effects of covariates and experimental factors. Notice that the number of cutoff parameters that need to be estimated is one less than the number of ordered categories.

To motivate an application consider an experiment conducted in a completely randomized design. If τ_i denotes the effect of the ith treatment and k indexes the replications, we put

$$X_{ik} = \mu + \tau_i + e_{ik}$$

as the model for the latent variable X observed for replicate k of treatment i. The probability that Y_{ik}, the ordinal outcome for replicate k of treatment i, is **at most** in category j is now determined by the distribution of the experimental errors e_{ik} as

$$\Pr(Y_{ik} \leq j) = \Pr(X_{ik} \leq \alpha_j) = \Pr(e_{ik} \leq \alpha_j - \mu - \tau_i) = \Pr\big(e_{ik} \leq \alpha_j^* - \tau_i\big).$$

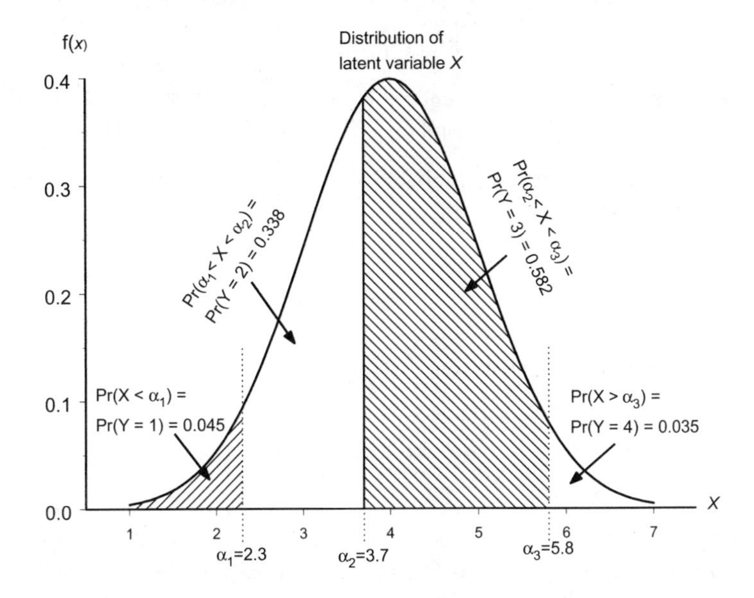

Figure 6.9. Relationship between latent variable $X \sim G(4,1)$ and an ordinal outcome Y. The probability to observe a particular ordered value depends on the distribution of the latent variable and the spacing of the cutoff parameters α_j; $\alpha_0 = -\infty$, $\alpha_4 = \infty$.

The cutoff parameters α_j and the grand mean μ have been combined into a new cutoff $\alpha_j^* = \alpha_j - \mu$ in the last equation. The probability that the ordinal outcome for replicate k of treatment i is at most in category j is a **cumulative** probability, denoted γ_{ikj}.

Choosing a probability distribution for the experimental errors is as easy or difficult as in a standard analysis. The most common choices are to assume that the errors follow a Logistic distribution $\Pr(e \leq t) = 1/(1+e^{-t})$ or a Gaussian distribution. The Logistic error model leads to a model with a logit link function, the Gaussian model results in a probit link function. With a logit link function we are led to

$$\operatorname{logit}(\Pr(Y_{ik} \leq j)) = \operatorname{logit}(\gamma_{ikj}) = \ln\left\{\frac{\Pr(Y_{ik} \leq j)}{\Pr(Y_{ik} > j)}\right\} = \alpha_j^* - \tau_i. \qquad [6.49]$$

The term cumulative link model is now apparent, since the link is applied to the cumulative probabilities γ_{ikj}. This model was first described by McCullagh (1980, 1984) and termed the proportional odds model (see also McCullagh and Nelder, 1989, §5.2.2). The name stems from the fact that γ_{ikj} is a measure of cumulative odds (Agresti 1990, p. 322) and hence the logarithm of the cumulative odds ratio for two treatments is (proportional to) the treatment difference

$$\text{logit}(\gamma_{ikj}) - \text{logit}(\gamma_{i'kj}) = \tau_{i'} - \tau_i.$$

In a regression example where $\text{logit}(\gamma_{ij}) = \alpha_j - \beta x_i$, the cutoff parameters serve as separate intercepts on the logit scale. The slope β measures the change in the cumulative logit if the regressor x changes by one unit. The change in cumulative logits between x_i and $x_{i'}$ is

$$\text{logit}(\gamma_{ij}) - \text{logit}(\gamma_{i'j}) = \beta(x_{i'} - x_i)$$

and proportional to the difference in the regressors. Notice that this effect of the regressors or treatments on the logit scale does not depend on j; it is the same for all categories.

By inverting the logit transform the probability to observe at most category j for replicate k of treatment i is easily calculated as

$$\Pr(Y_{ik} \leq j) = \frac{1}{1 + \exp\left\{-\alpha^*_{jk} + \tau_i\right\}},$$

and category probabilities are obtained from differences:

$$\pi_{ikj} = \Pr(Y_{ik} = j) = \begin{cases} \Pr(Y_{ik} \leq 1) = \gamma_{ik1} & j = 1 \\ \Pr(Y_{ik} \leq j) - \Pr(Y_{ik} \leq j - 1) = \gamma_{ikj} - \gamma_{ikj-1} & 1 < j < J \\ 1 - \Pr(Y_{ik} \leq J - 1) = 1 - \gamma_{ikJ-1} & j = J \end{cases}$$

Here, π_{ikj} is the probability that an outcome for replicate k of treatment i will fall into category j. Notice that the probability to fall into the last category (J) is obtained by subtracting the cumulative probability to fall into the previous category from 1. In fitting the proportional odds model to data this last probability is obtained automatically and is the reason why only $J - 1$ cutoff parameters are needed to model an ordered response with J categories.

The proportional odds model has several important features. It allows one to model ordinal data independently of the scoring system in use. Whether categories are labeled as $a, b, c, \cdots$ or $1, 2, 3, \cdots$ or $1, 20, 34, \cdots$ or *mild, medium, heavy,* $\cdots$ the analysis will be the same. Users are typically more interested in the probabilities that an outcome is in a certain category, rather than cumulative probabilities. The former are easily obtained from the cumulative probabilities by taking differences. The proportional odds model is further invariant under category amalgamations. If the model applies to an ordered outcome with J categories, it also applies if neighboring categories are combined into a new response with $J^* < J$ categories (McCullagh 1980, Greenwood and Farewell 1988). This is an important property since ratings may be collected on a scale finer than that eventually used in the analysis and parameter interpretation should not depend on the number of categories. The development of the proportional odds model was motivated by the existence of a latent variable. It is not a requirement for the validity of this model that such a latent variable exists and it can be used for grouped and assessed orderings alike (McCullagh and Nelder 1989, p.154; Schabenberger 1995).

Other statistical models for ordinal data have been developed. Fienberg's continuation ratio model (Fienberg 1980) models logits of the conditional probabilities to observe category j given that the observation was at least in category j instead of cumulative probabilities (see also Cox 1988, Engel 1988). Continuation ratio models are based on factoring marginal probabilities into a series of conditional probabilities and standard GLMs for binomial outcomes (Nelder and Wedderburn 1972, McCullagh and Nelder 1989) can be applied to the terms in the factorization separately. A disadvantage is that the factorization is not unique. Agresti (1990, p. 318) discusses adjacent category logits where probabilities are modeled relative to a baseline category. Läärä and Matthews (1985) establish an equivalence between continuation ratio and cumulative models if a complementary log-log instead of a logit transform is applied. Studying a biomedical example, Greenwood and Farewell (1988) found that the proportional odds and the continuation ratio models led to the same conclusions regarding the significance of effects.

Cumulative link models are fit by maximum likelihood and estimates are derived by iteratively reweighted least squares as for generalized linear models. The test of hypotheses proceeds along similar lines as discussed in §6.4.2.

6.5.2 Software Implementation and Example

The proportional odds model can be fit to data in various statistical packages. The SAS® System fits cumulative logit models in `proc catmod`, `proc logistic`, and `proc genmod` (starting with Release 7.0). The `logistic` procedure is specifically designed for the proportional odds model. Whenever the response variable has more than two categories, the procedure defaults to the proportional odds model; otherwise it defaults to a logistic regression model for Bernoulli or Binomial responses. Prior to Release 8.0, the `logistic` procedure did not accommodate classification variables in a `class` statement (see §6.2.3 for details). Users of Release 7.0 can access the experimental procedure `proc tlogistic` to fit the proportional odds model with classification variables (treatments, etc.). The `genmod` procedure has been extended in Release 7.0 to fit cumulative logit and other multinomial models. It allows the use of classification variables and performs likelihood ratio and Wald tests. Although `proc genmod` enables the generalized estimating equation (GEE) approach of Liang and Zeger (1986) and Zeger and Liang (1986), only an independence working correlation matrix is permissible for ordinal responses. The proportional odds model is also implemented in the Minitab® package (module OLOGISTIC).

The software implementation and the basic calculations in the proportional odds model are now demonstrated with hypothetical data from an experimental design. Assume four treatments are applied in a completely randomized design with four replications and ordinal ratings (*poor*, *average*, *good*) from each experimental unit are obtained at four occasions $(d = 1, 2, 3, 4)$. The data are shown in Table 6.13. For example, at occasion $d = 1$ all four replicates of treatment A were rated in the *poor* category and at occasion $d = 2$ two replicates were rated *poor*, two replicates were rated *average*.

The model fit to these data is a logistic model for the cumulative probabilities that contains a classification effect for the treatment variable and a continuous covariate for the time effect:

$$\text{logit}\{\gamma_{ijk}\} = \alpha_j^* + \tau_i + \beta t_{ik}. \qquad [6.50]$$

Here, t_{ik} is the time point at which replicate k of treatment i was observed $(i = 1, \cdots, 4; k = 1, \cdots, 4; j = 1, 2)$.

Table 6.13. Observed category frequencies for four treatments at four dates (shown are the total counts across four replicates at each occasion)

	Treatment				
Category	$A\,(i=1)$	$B\,(i=2)$	$C\,(i=3)$	$D\,(i=4)$	$\sum$
poor $(j=1)$	$4,2,4,4$	$4,3,4,4$	$0,0,0,0$	$1,0,0,0$	$9,5,8,8$
average $(j=2)$	$0,2,0,0$	$0,1,0,0$	$1,0,4,4$	$2,2,4,4$	$3,5,8,8$
good	$0,0,0,0$	$0,0,0,0$	$3,4,0,0$	$1,2,0,0$	$4,6,0,0$

The SAS® data step and `proc logistic` statements are:

```
data ordexample;
  input tx $ time rep rating $;
  datalines;
A 1 1 poor
A 1 2 poor
A 1 3 poor
A 1 4 poor
A 2 1 poor
A 2 2 poor
A 2 3 average
A 2 4 average
A 3 1 poor
A 3 2 poor
A 3 3 poor
A 3 4 poor
A 4 1 poor

··· and so forth ···

;;
run;

proc logistic data=ordexample order=data;
  class tx / param=glm;
  model rating = tx time / link=logit rsquare covb;
run;
```

The `order=data` option of `proc logistic` ensures that the categories of the response variable `rating` are internally ordered as they appear in the data set, that is, *poor* before *average* before *good.* If the option were omitted `proc logistic` would sort the levels alphabetically, implying that the category order is *average* – *good* – *poor*, which is not the correct ordination. The `param=glm` option of the class statement asks `proc logistic` to code the classification variable in the same way as `proc glm`. This means that a separate estimate for the last level of the treatment variable `tx` is not estimated to ensure the constraint $\sum_{i=1}^{t=4} \tau_i = 0$. τ_4 will be absorbed into the cutoff parameters and the estimates for the other treatment effects reported by `proc logistic` represent differences with τ_4.

One should always study the `Response Profile` table on the procedure output to make sure that the category ordering used by `proc logistic` (and influenced by the `order=` option) agrees with the intended ordering. The Class Level Information Table shows the levels of all variables listed in the class statement as well as their coding in the design matrix (Output 6.3).

The Score Test for the Proportional Odds Assumption is a test of the assumption that changes in cumulative logits are proportional to changes in the explanatory variables. Two models are compared to calculate this test, a full model in which the slopes and gradients vary by category and a reduced model in which the slopes are the same across categories. Rather than actually fitting the two models, proc logistic performs a score test that requires only the reduced model to be fit (see §A6.8.4). The reduced model in this case is the proportional odds model and rejecting the test leads to the conclusion that it is not appropriate for these data. In this example, the score test cannot be rejected and the p-value of 0.1538 is sufficiently large not to call into doubt the proportionality assumption (Output 6.3).

Output 6.3.

```
                         The LOGISTIC Procedure

                           Model Information
          Data Set                           WORK.ORDEXAMPLE
          Response Variable                  rating
          Number of Response Levels          3
          Number of Observations             64
          Link Function                      Logit
          Optimization Technique             Fisher's scoring

                            Response Profile
                Ordered                                Total
                  Value        rating              Frequency
                      1        poor                       30
                      2        average                    24
                      3        good                       10

                       Class Level Information
                                     Design Variables
           Class       Value        1         2         3         4
           tx          A            1         0         0         0
                       B            0         1         0         0
                       C            0         0         1         0
                       D            0         0         0         1

          Score Test for the Proportional Odds Assumption

                Chi-Square         DF       Pr > ChiSq
                    6.6808          4           0.1538

                         Model Fit Statistics
                                                Intercept
                                 Intercept            and
             Criterion                Only     Covariates
             AIC                   133.667         69.124
             SC                    137.985         82.077
             -2 Log L              129.667         57.124

      R-Square      0.6781     Max-rescaled R-Square      0.7811

                    Type III Analysis of Effects
                                          Wald
             Effect        DF       Chi-Square      Pr > ChiSq
             tx             3          24.5505          <.0001
             time           1           6.2217          0.0126
```

Output 6.3 (continued).

Analysis of Maximum Likelihood Estimates

Parameter		DF	Estimate	Standard Error	Chi-Square	Pr > ChiSq
Intercept		1	-5.6150	1.5459	13.1920	0.0003
Intercept2		1	-0.4865	0.9483	0.2632	0.6079
tx	A	1	5.7068	1.4198	16.1557	<.0001
tx	B	1	6.5190	1.5957	16.6907	<.0001
tx	C	1	-1.4469	0.8560	2.8571	0.0910
tx	D	0	0	.	.	.
time		1	0.8773	0.3517	6.2217	0.0126

The `rsquare` option of the `model` statement in `proc logistic` requests the two generalized R^2 measures discussed in §6.4.5. Denoted as `R-Square` is the generalized measure of Cox and Snell (1989, pp. 208-209) and Magee (1990), and `Max-rescaled R-Square` denotes the generalized measure by Nagelkerke (1991) that ranges between 0 and 1. The log likelihood for the null and fitted models are $l(\widehat{\boldsymbol{\mu}}_0\psi;\mathbf{y}) = -129.667/2 = -64.8335$ and $l(\widehat{\boldsymbol{\mu}}_f,\psi;\mathbf{y}) = -28.562$, respectively, so that

$$R^2_* = 1 - \exp\left\{ -\frac{2}{64}(- 28.562 + 64.8335)\right\} = 0.6781.$$

The `Analysis of Maximum Likelihood Estimates` table shows the parameter estimates and their standard errors as well as Chi-square tests testing each parameter against zero. Notice that the estimate for τ_4 is shown as 0 with no standard error since this effect is absorbed into the cutoffs. The cutoff parameters labeled `Intercept` and `Intercept2` thus are estimates of

$$\alpha_1^* + \tau_4$$
$$\alpha_2^* + \tau_4$$

and the parameters shown as `tx A`, `tx B`, and `tx C` correspond to $\delta_1 = \tau_1 - \tau_4$, $\delta_2 = \tau_2 - \tau_4$, and $\delta_3 = \tau_3 - \tau_4$, respectively. To estimate the probability to observe at most an *average* rating for treatment B at time $t = 1$, for example, calculate

$$\widehat{\alpha}_2^* + \widehat{\tau}_4 + \widehat{\delta}_2 + \widehat{\beta}*1 = -0.4865 + 6.519 + 0.8773*1 = 6.909$$

$$\Pr(Y \leq 2 \text{ at time } 1) = \frac{1}{1 + \exp\{-6.909\}} = 0.999$$

Similarly, for the probability of at most a *poor* rating at time $t = 1$

$$\widehat{\alpha}_1^* + \widehat{\tau}_4 + \widehat{\delta}_2 + \widehat{\beta}*1 = -5.615 + 6.519 + 0.8773*1 = 1.7813$$

$$\Pr(Y \leq 1 \text{ at time } 1) = \frac{1}{1 + \exp\{-1.7813\}} = 0.856$$

For an experimental unit receiving treatment 2, there is an 85.6% chance to receive a *poor* rating and only a 99.9% – 85.6% = 14.3% chance to receive an *average* rating at the first time point (Table 6.14). Each block of three numbers in Table 6.14 is an estimate of the multinomial distribution for a given treatment at a particular time point. A graph of the linear predictors shows the linearity of the model in treatment effects and time on the logit scale and the proportionality assumption which results in parallel lines on that scale (Figure 6.10). Inverting the logit transform to calculate cumulative probabilities from the linear predictors shows the nonlinear dependence of probabilities on treatments and the time covariate (Figure 6.11).

To compare the treatments at a given time point, we formulate linear combinations of the cumulative logits which leads to linear combinations of the parameters. For example, comparing treatments A and B at time $t = 1$ the linear combination is

$$\text{logit}\{\gamma_{1j1}\} - \text{logit}\{\gamma_{2j1}\} = \alpha_j^* + \tau_1 + \beta - \left(\alpha_j^* + \tau_2 + \beta\right) = \tau_1 - \tau_2.$$

The cutoff parameters have no effect on this comparison; the treatment difference has the same magnitude, regardless of the category (Figure 6.10). In terms of the quantities `proc logistic` estimates, the contrast is identical to $\delta_1 - \delta_2$. The variance-covariance matrix (obtained with the `covb` option of the model statement in `proc logistic`, output not given) is shown in Table 6.15. For example, the standard error for $\widehat{\alpha}_1^* + \widehat{\tau}_4$ is $\sqrt{2.389} = 1.545$ as appears on the output in the `Analysis of Maximum Likelihood Estimates` table.

Table 6.14. Predicted category probabilities by treatment and occasion

		Treatment			
	Category	A	B	C	D
Time 1	Poor	0.725	0.856	0.002	0.008
	Average	0.273	0.143	0.256	0.588
	Good	0.002	0.001	0.404	0.742
Time 2	Poor	0.864	0.935	0.004	0.020
	Average	0.135	0.065	0.450	0.759
	Good	0.001	0.000	0.544	0.219
Time 3	Poor	0.938	0.972	0.012	0.048
	Average	0.061	0.028	0.656	0.847
	Good	0.000	0.000	0.105	0.332
Time 4	Poor	0.973	0.988	0.028	0.108
	Average	0.026	0.012	0.800	0.845
	Good	0.000	0.000	0.171	0.046

Table 6.15. Estimated variance-covariance matrix of parameter estimates as obtained from `proc logistic` (`covb` option of the `model` statement)

	$\alpha_1^* + \tau_4$	$\alpha_2^* + \tau_4$	δ_1	δ_2	δ_3	β
$\alpha_1^* + \tau_4$	2.389	0.891	– 1.696	– 1.735	– 0.004	– 0.389
$\alpha_2^* + \tau_4$	0.891	0.899	– 0.462	– 0.486	– 0.313	– 0.241
δ_1	– 1.695	– 0.462	2.015	1.410	0.099	0.169
δ_2	– 1.735	– 0.486	1.410	2.546	0.094	0.182
δ_3	– 0.004	– 0.313	0.099	0.094	0.733	– 0.053
β	– 0.389	– 0.241	0.169	0.182	– 0.053	0.124

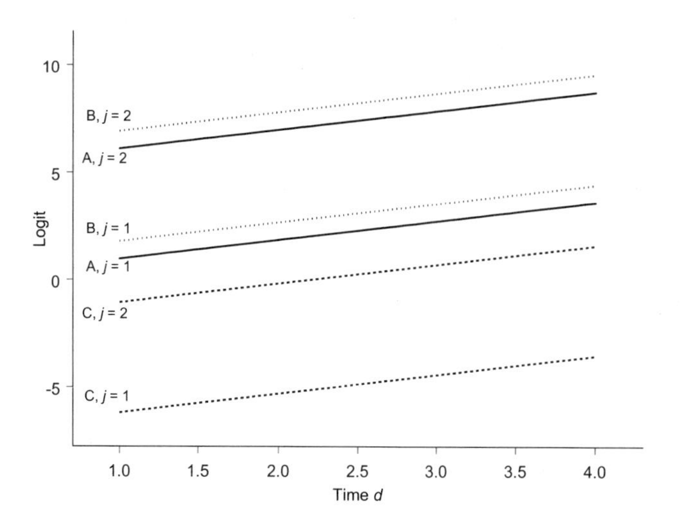

Figure 6.10. Linear predictors for treatments A, B, and C. The vertical difference between the lines for categories $j = 1$ and $j = 2$ is constant for all time points and the same $(5.6150 - 0.4865 = 5.1285)$ for all treatments.

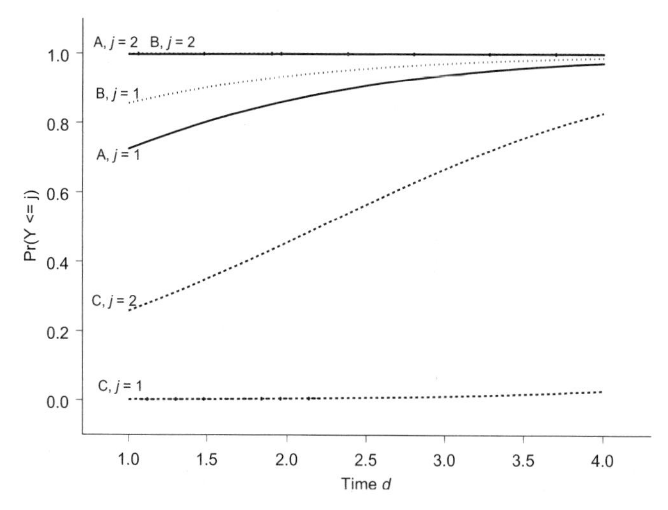

Figure 6.11. Predicted cumulative probabilities for treatments A, B, and C. Because the line for $A, j = 2$ lies completely above the line for $A, j = 1$ in Figure 6.10, the cumulative probability to observe a response in at most category 2 is greater than the cumulative probability to observe a response in at most category 1. Since the cumulative probabilities are ordered, the category probabilities are guaranteed to be non-negative.

The estimated variance of the linear combination $\widehat{\delta}_1 - \widehat{\delta}_2$ is thus

$$\text{Var}\left[\widehat{\delta}_1\right] + \text{Var}\left[\widehat{\delta}_2\right] - 2\text{Cov}\left[\widehat{\delta}_1, \widehat{\delta}_2\right] = 2.015 + 2.546 - 2{*}1.41 = 1.741$$

and the Wald test statistic becomes

$$W = \frac{(5.7068 - 6.519)^2}{1.741} = 0.378.$$

From a Chi-squared distribution with one degree of freedom the p-value 0.538 is obtained. This test can be performed in `proc logistic` by adding the statement

```
contrast 'A vs. B' tx 1 -1 0 0;
```

to the `proc logistic` code above. The test for the treatment main effect can be performed by using a set of orthogonal contrasts, for example let

$$\widehat{\mathbf{c}} = \begin{bmatrix} \widehat{\tau}_1 - \widehat{\tau}_2 \\ \widehat{\tau}_1 + \widehat{\tau}_2 - 2\widehat{\tau}_3 \\ \widehat{\tau}_1 + \widehat{\tau}_2 + \widehat{\tau}_3 - 3\widehat{\tau}_4 \end{bmatrix} = \begin{bmatrix} \widehat{\delta}_1 - \widehat{\delta}_2 \\ \widehat{\delta}_1 + \widehat{\delta}_2 - 2\widehat{\delta}_3 \\ \widehat{\delta}_1 + \widehat{\delta}_2 + \widehat{\delta}_3 \end{bmatrix} = \begin{bmatrix} 1 & -1 & 0 \\ 1 & 1 & -2 \\ 1 & 1 & 1 \end{bmatrix} \begin{bmatrix} \widehat{\delta}_1 \\ \widehat{\delta}_2 \\ \widehat{\delta}_3 \end{bmatrix}$$

which has estimated variance

$$\begin{aligned} \widehat{\mathrm{Var}}[\widehat{\mathbf{c}}] &= \begin{bmatrix} 1 & -1 & 0 \\ 1 & 1 & -2 \\ 1 & 1 & 1 \end{bmatrix} \begin{bmatrix} 2.015 & 1.410 & 0.099 \\ 1.410 & 2.546 & 0.094 \\ 0.099 & 0.094 & 0.733 \end{bmatrix} \begin{bmatrix} 1 & 1 & 1 \\ -1 & 1 & 1 \\ 0 & -2 & 1 \end{bmatrix} \\ &= \begin{bmatrix} 1.741 & -0.541 & -0.526 \\ -0.541 & 9.541 & 5.722 \\ -0.526 & 5.722 & 8.500 \end{bmatrix}. \end{aligned}$$

The Wald statistic for the treatment main effect ($H_0\colon \mathbf{c} = \mathbf{0}$) becomes

$$\begin{aligned} \widehat{\mathbf{c}}'\widehat{\mathrm{Var}}[\widehat{\mathbf{c}}]^{-1}\widehat{\mathbf{c}} &= [-0.812, 15.119, 10.779] \begin{bmatrix} 0.587 & 0.019 & 0.023 \\ 0.019 & 0.176 & -0.118 \\ 0.023 & -0.118 & 0.198 \end{bmatrix} \begin{bmatrix} -0.812 \\ 15.119 \\ 10.779 \end{bmatrix} \\ &= 24.55. \end{aligned}$$

This test is shown on the `proc logistic` output in the `Type III Analysis of Effects` table. The same analysis can be obtained in `proc genmod`. The statements, including all pairwise treatment comparisons follows.

```
proc genmod data=ordexample rorder=data;
  class tx;
  model rating = tx time / link=clogit dist=multinomial type3;
  contrast 'A vs. B' tx 1 -1  0  0;
  contrast 'A vs. C' tx 1  0 -1  0;
  contrast 'A vs. D' tx 1  0  0 -1;
  contrast 'B vs. C' tx 0  1 -1  0;
  contrast 'B vs. D' tx 0  1  0 -1;
  contrast 'C vs. D' tx 0  0  1 -1;
run;
```

More complicated proportional odds models can be fit easily with these procedures. On occasion one may encounter a warning message regarding the separability of data points and a possibly questionable model fit. This can occur, for example, when one treatment's responses are all in the same category, since then there is no variability among the replicates.

This phenomenon is more likely for small data sets and applications where many classification variables are involved in particular interactions. There are several possibilities to correct this: amalgamate adjacent categories to reduce the number of categories; fit main effects and low-order interactions only and exclude high-order interactions; include effects as continuous covariates rather than as classification variables when they relate to some underlying continuous metric such as rates of application or times of measurements.

6.6 Overdispersion

Box 6.6 Overdispersion

- **Overdispersion is the condition by which the variability of the data exceeds the variability expected under a particular probability distribution. Gaussian data are never overdispersed, since the mean and variance can be chosen freely. Overdispersion is an issue for those generalized linear models where the mean and variance are functionally dependent.**

- **Overdispersion can arise from choosing the wrong distributional model, from ignoring important explanatory variables, and from correlations among the observations.**

- **A common remedy for overdispersion is to add a multiplicative overdispersion factor to the variance function. The resulting analysis is no longer maximum likelihood but quasi-likelihood.**

If the variability of a set of data exceeds the variability expected under some reference model we call it **overdispersed** (relative to that reference). Counts, for example, may exhibit more variability than is permissible under a Binomial or Poisson probability model. Overdispersion is a potential problem in statistical models where the first two moments of the response distribution are linked and means and variances are functionally dependent. In Table 6.2 the scale parameter ψ is not present for the discrete distributions and data modeled under these distributions are potentially overdispersed. McCullagh and Nelder (1989, p. 124, p. 193) suggest that overdispersion may be the norm in practice, rather than the exception. In part this is due to the fact that users resort to a small number of probability distributions to model their data. Almost automatically one is led to the Binomial distribution for count variables with a natural denominator and to the Poisson distribution for counts without a natural denominator. One remedy of the overdispersion problem lies in choosing a proper distribution that permits more variability than these standard models such as the Beta-Binomial in place of the Binomial model and the Negative Binomial in place of the Poisson model. Overdispersion can also be caused by an improper choice of covariates and effects to model the data. This effect was obvious for the seed germination data modeled in §6.4.4. When temperature and/or concentration effects were omitted the ratio of the deviance and its degrees of freedom exceeded the benchmark value of one considerably (Table 6.12). Such cases of overdispersion must be addressed by altering the set of effects and covariates, not by postulating a different probability distribution for the data. In what follows we assume that the mean of the responses has been modeled correctly, but that the data nevertheless exhibit variability in excess of our expectation under a certain reference distribution.

Overdispersion is a problem foremost because it affects the estimated precision of the parameter estimates. In §A6.8.2 it is shown that the scale parameter ψ is of no consequence in estimating β and can be dropped from the estimating equations. The (asymptotic) variance-covariance matrix of the maximum likelihood estimates is given by $(\mathbf{F}'\mathbf{V}^{-1}\mathbf{F})^{-1}$ where $\mathbf{V}$ is a diagonal matrix containing the variances $\mathrm{Var}[Y_i] = h(\mu_i)\psi$. Extracting the scale parameter ψ we can simplify:

$$\mathrm{Var}\left[\widehat{\beta}\right] = \psi(\mathbf{F}'\mathrm{Diag}(1/h(\mu_i))\mathbf{F})^{-1}.$$

If under a given probability model the scale parameter ψ is assumed to be 1 but overdispersion exists ($\mathrm{Var}[Y_i] > h(\mu_i)$), the variability of the estimates is larger than what is assumed under the model. The precision of the parameter estimates is overstated, standard error estimates are too small and as a result, test statistics are inflated and p-values are too small. Covariates and effects may be declared significant even when they are not. It is thus important to account for overdispersion present in the data and numerous approaches have been developed to that end. The following four categories are sufficiently broad to cover many overdispersion mechanisms and remedies.

- **Extra scale parameters** are added to the variance function of the generalized linear model. For Binomial data one can assume $\mathrm{Var}[Y] = \phi n\pi(1-\pi)$ instead of the nominal variability $\mathrm{Var}[Y] = n\pi(1-\pi)$. If $\phi > 1$ the model is overdispersed relative to the Binomial and if $\phi < 1$ it is underdispersed. Underdispersion is far less likely and a far less serious problem in data analysis. For count data one can model overdispersion relative to the Poisson(λ) distribution as $\mathrm{Var}[Y] = \lambda\phi$, $\mathrm{Var}[Y] = \lambda(1+\phi)/\phi$, or $\mathrm{Var}[Y] = \lambda + \lambda^2/\phi$. These models have some stochastic foundation in certain mixing models (see below and §A6.8.6). Models with a multiplicative overdispersion parameter such as $\mathrm{Var}[Y] = \phi n\pi(1-\pi)$ for Binomial and $\mathrm{Var}[Y] = \phi\lambda$ for Poisson data can be handled easily with `proc genmod` of The SAS® System. The overdispersion parameter is then estimated from Pearson or deviance residuals by the method discussed in §6.4.3 (`pscale` and `dscale` options of `model` statement).

- **Positive autocorrelation** among observations leads to overdispersion in sums and averages. Let Z_i be an arbitrary random variable with mean μ and variance σ^2 and assume that the Z_i are equicorrelated, $\mathrm{Cov}[Z_i, Z_j] = \rho\sigma^2\ \forall i \neq j$. Assume further that $\rho \geq 0$. We are interested in modeling $Y = \sum_{i=1}^{n} Z_i$, the sum of the Z_i. The mean and variance of this sum follow from first principles as

$$\mathrm{E}[Y] = n\mu$$

and

$$\begin{aligned}\mathrm{Var}[Y] &= n\mathrm{Var}[Z_i] + 2\sum_{i=1}^{n}\sum_{j>i}^{n}\mathrm{Cov}[Z_i, Z_j] = n\sigma^2 + n(n-1)\sigma^2\rho \\ &= n\sigma^2(1+(n-1)\rho) \geq n\sigma^2.\end{aligned}$$

If the elements in the sum were uncorrelated, $\mathrm{Var}[Y] = n\sigma^2$ but the positive autocorrelation thus leads to an overdispersed sum, relative to the model of stochastic independence. This type of overdispersion is accounted for by incorporating the stochastic dependency in the model.

- The parameters of the reference distribution are not assumed to be constant, but random variables and the reference distribution are reckoned conditionally. We term this the **mixing model** approach (not to be confused with the mixed model approach of §7). As a consequence, the unconditional (marginal) distribution is more dispersed than the conditional reference distribution. If the marginal distribution is also in the exponential family, this approach enables maximum likelihood estimation in a genuine generalized linear model. A famous example is the hierarchical model for counts where the average count λ is a Gamma-distributed random variable. If the conditional reference distribution — the distribution of Y for a fixed value of λ — is Poisson, the unconditional distribution of the counts is Negative Binomial. Since the Negative Binomial distribution is a member of the exponential family with canonical link $\log\{\mu\}$, a straightforward generalized linear model for counts emerges that permits more variability than the Poisson(λ) distribution (for an application see §6.7.8).

- Random effects and coefficients are added to the linear predictor. This approach gives rise to **generalized linear mixed models** (GLMM, §8). For example, consider a simple generalized linear regression model with log link,

$$\ln\{\mathrm{E}[Y]\} = \beta_0 + \beta_1 x \Leftrightarrow \mathrm{E}[Y] = \exp\{\beta_0 + \beta_1 x\}$$

  and possible oversdispersion. Adding a random effect e with mean 0 and variance σ^2 in the exponent, turns the linear predictor into a mixed linear predictor. The resulting model can also be reckoned conditionally. $\mathrm{E}[Y|e] = \exp\{\beta_0 + \beta_1 x + e\}$, $e \sim (0, \sigma^2)$. If we assume that $Y|e$ is Poisson-distributed, the resulting unconditional distribution will be overdispersed relative to a Poisson($\exp\{\beta_0 + \beta_1 x\}$) distribution. Unless the distribution of the random effects is chosen carefully, the marginal distribution may not be in the exponential family and maximum likelihood estimation of the parameters may be difficult. Numerical methods for maximizing the marginal (unconditional) log-likelihood function via linearization, quadrature integral approximation, importance sampling and other devices exist, however. The `nlmixed` procedure of The SAS® System is designed to fit such models (§8).

In this chapter we consider a remedy for overdispersion by estimating extra scale parameters (§6.7.2) and by using mixing schemes such as the Poisson/Gamma model. An application of the latter approach is presented in §6.7.8 where poppy counts in a randomized complete block design are more dispersed than is expected under a Poisson model. The poppy count data are revisited in §8.4.2 and modeled with a generalized linear mixed model.

6.7 Applications

The first two applications in this chapter model Binomial outcomes. In §6.7.1 a simple logistic regression model with a single covariate is fit to model the mortality rate of insect larvae exposed to an insecticide. Of particular interest is the estimation of the LD_{50}, the insecticide dosage at which the probability of a randomly chosen larva to succumb to the insecticide exposure is 0.5. In §6.7.2 a field experiment with Binomial outcomes is examined.

Sixteen varieties are arranged in a randomized complete block design and the number of plants infested with the Hessian fly is recorded. Interesting aspects of this experiment are varying binomial sample sizes among the experimental units which invalidates the variance-stabilizing arcsine transformation and possible overdispersion. Yield density models that were examined earlier as nonlinear regression models in §5.8.7 are revisited in §6.7.3. Rather than relying on inverse or logarithmic transformation we treat the yield responses as Gamma-distributed random variables and apply generalized linear model techniques. §6.7.4 and §6.7.5 are dedicated to the analysis of ordinal data. In both cases the treatment structure is a simple two-way factorial. The analysis in §6.7.5 is further complicated by the fact that experimental units were measured repeatedly over time. The analysis of contingency tables is a particularly fertile area for the deployment of generalized linear models. A special class of models, log-linear models for square contingency tables, are discussed in §6.7.6. These models allow estimation of the agreement or disagreement between interpreters of the same material. Generalized linear models can be successfully employed when the outcome of interest is not a mean, but a dispersion parameter, for example a variance. In §6.7.7 we use Gamma regression to model the variability between deoxynivalenol (vomitoxin) probe samples from truckloads of wheat kernels as a function of the toxin load. The final application (§6.7.8) considers count data and demonstrates that the Poisson distribution is not necessarily a suitable model for such data. In the presence of overdispersion, the Negative Binomial distribution is a more reasonable model. We show how to fit models with Negative Binomial responses with the `nlmixed` procedure of The SAS® System.

6.7.1 Dose-Response and LD_{50} Estimation in a Logistic Regression Model

Mead et al. (1993, p. 336) discuss probit analysis for a small data set of the proportion of larvae killed as a function of the concentration of an insecticide. For each of seven concentrations, 20 larvae were exposed to the insecticide and the number of larvae killed was recorded (Table 6.16). Each larva's exposure to the insecticide represents a Bernoulli random variable with outcomes *larva killed* and *larva survived.* If the 20 larvae exposed to the same concentration react independently to the insecticide and if their survival probabilities are the same for a given concentration, then each number in Table 6.16 is a realization of a Binomial$(20, \pi(x))$ random variable, where $\pi(x)$ is the mortality probability if concentration x is applied. If, furthermore, the concentrations are applied independently to the sets of 20 larvae, the experiment consists of seven independent Binomial random variables. Notice that these data are grouped with $n^{(g)} = 7, n = 140$.

Table 6.16. Insecticide concentrations and number of larvae killed out of 20

Concentration x	0.375%	0.75%	1.5%	3%	6%	12%	24%
No. of larvae killed	0	1	8	11	16	18	20

Data from Mead, Curnow, and Hasted (1993, p. 336) and used with permission.

Plots of the logits of the sample proportions against the concentrations and the $\log_{10}$ concentrations are shown in Figure 6.12. The relationship between sample logits and concentrations is clearly not linear. It appears at least quadratic and would suggest a generalized linear model

$$\text{logit}(\pi) = \beta_0 + \beta_1 x + \beta_2 x^2.$$

The quadratic trend does not ensure that the logits are monotonically increasing in x and a more reasonable model posits a linear dependence of the logits on the $\log_{10}$ concentration,

$$\text{logit}(\pi) = \beta_0 + \beta_1 \log_{10}\{x\}. \quad [6.51]$$

Model [6.51] is a classical logistic regression model with a single covariate ($\log_{10}\{x\}$). The analysis by Mead et al. (1993) uses a probit link, and the results are very similar to those from a logistic analysis, because of the similarity of the two link functions.

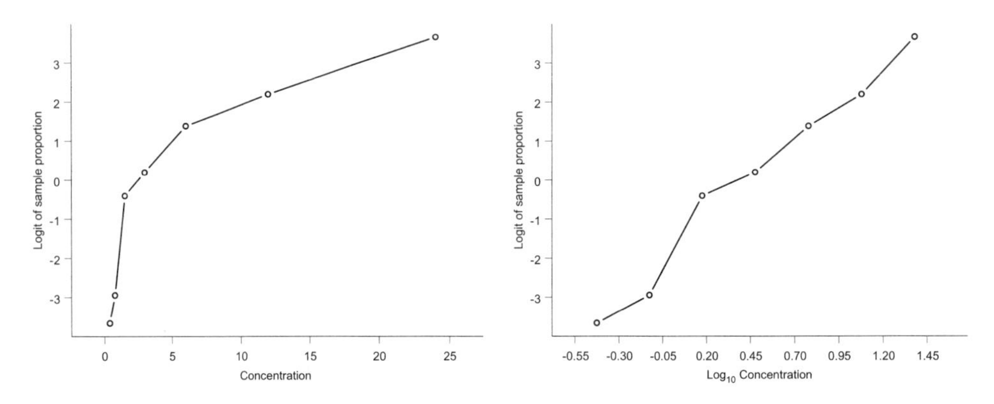

Figure 6.12. Logit of sample proportions against insecticide concentration and logarithm of concentration. A linear relationship between the logit and the $\log_{10}$-concentration is reasonable.

The key relationship in this experiment is the dependence of the probability π that a larva is killed on the insecticide concentration. Once this relationship is modeled other quantities of interest can be estimated. In bioassay and dose-response studies one is often interested in estimating dosages that produce a certain response. For example, the dosage lethal to a randomly selected larva with probability 0.5 (the so-called LD_{50}). In model [6.51] we can establish more generally that if x_α denotes the dosage with mortality rate $0 \leq \alpha \leq 1$, then from

$$\text{logit}(\alpha) = \beta_0 + \beta_1 \log_{10}\{x_\alpha\}$$

the inverse prediction of dosage follows as

$$\begin{aligned} \log_{10}\{x_\alpha\} &= (\text{logit}(\alpha) - \beta_0)/\beta_1 \\ x_\alpha &= 10^{(\text{logit}(\alpha) - \beta_0)/\beta_1}. \end{aligned} \quad [6.52]$$

In the case of LD_{50}, for example, $\alpha = 0.5$, $\text{logit}(0.5) = 0$, and $\log_{10}\{x_{0.5}\} = -\beta_0/\beta_1$. For this particular ratio, fiducial intervals were developed by Finney (1978, pp. 80-82) based on work by Fieller (1940) under the assumption that $\widehat{\beta}_0$ and $\widehat{\beta}_1$ are Gaussian-distributed. These intervals are also developed in our §A6.8.5. For an estimate of the dosage $x_{0.5}$ on the original, rather than the $\log_{10}$ scale, these intervals do not directly apply. We prefer obtaining standard errors for the quantities $\log_{10}\{x_\alpha\}$ and x_α based on Taylor series expansions. If the ratio

$\widehat{\beta}_1/\text{ese}(\widehat{\beta}_1)$ is sufficiently large, the Taylor series based standard errors are very accurate (§6.4.2).

The first step in our analysis is to fit model [6.51] and to determine whether the relationship between mortality probability and insecticide concentration is sufficiently strong. The `data` step and `proc genmod` statements for this logistic regression problem are as follows.

```
data kills;
  input concentration kills;
  trials = 20;
  logc   = log10(concentration);
  datalines;
 0.375   0
 0.75    1
 1.5     8
 3.0    11
 6.0    16
12.0    18
24.0    20
;;
run;

proc genmod data=kills;
  model kills/trials = logc / dist=binomial link=logit;
run;
```

The `proc genmod` output (Output 6.4) indicates a deviance of 4.6206 based on 5 degrees of freedom [$n^{(g)} = 7$ groups minus two estimated parameters (β_0, β_1)]. The deviance/df ratio is close to one and we conclude that overdispersion is not a problem for these data. The parameter estimates are $\widehat{\beta}_0 = -1.7305$ and $\widehat{\beta}_1 = 4.1651$.

Output 6.4.

```
                         The GENMOD Procedure

                          Model Information
              Data Set                              WORK.KILLS
              Distribution                            Binomial
              Link Function                              Logit
              Response Variable (Events)                 kills
              Response Variable (Trials)                trials
              Observations Used                              7
              Number Of Events                              74
              Number Of Trials                             140

                Criteria For Assessing Goodness Of Fit
       Criterion                   DF           Value           Value/DF
       Deviance                     5          4.6206             0.9241
       Scaled Deviance              5          4.6206             0.9241
       Pearson Chi-Square           5          3.8258             0.7652
       Scaled Pearson X2            5          3.8258             0.7652
       Log Likelihood                        -50.0133
Algorithm converged.

                    Analysis Of Parameter Estimates
                        Standard         Wald 95%            Chi-
Parameter  DF  Estimate    Error   Confidence Limits       Square  Pr > ChiSq

Intercept   1   -1.7305   0.3741   -2.4637    -0.9973       21.40      <.0001
logc        1    4.1651   0.6520    2.8872     5.4430       40.81      <.0001
Scale       0    1.0000   0.0000    1.0000     1.0000
NOTE: The scale parameter was held fixed.
```

The Wald test for $H_0\colon \beta_1 = 0$ has test statistic $W = 40.81$ and the hypothesis is clearly rejected. There is a significant relationship between larva mortality and the $\log_{10}$ insecticide concentration. The positive slope estimate indicates that mortality probability increases with the $\log_{10}$ concentration. For example, the probabilities that a randomly selected larva is killed at concentrations $x = 1.5\%$ and $x = 6\%$ are

$$\frac{1}{1+\exp\left\{-\widehat{\beta}_0 - \widehat{\beta}_1\log_{10}\{1.5\}\right\}} = 0.269$$

$$\frac{1}{1+\exp\left\{-\widehat{\beta}_0 - \widehat{\beta}_1\log_{10}\{6\}\right\}} = 0.819.$$

The note that `The scale parameter was held fixed` at the end of the `proc genmod` output indicates that no extra scale parameters were estimated.

How well does this logistic regression model fit the data? To this end we calculate the generalized R^2 measures discussed in §6.4.5. The log likelihood for the full model containing a concentration effect is shown in the output above as $l(\widehat{\boldsymbol{\mu}}_f, \psi; \mathbf{y}) = l(\widehat{\boldsymbol{\mu}}_f, 1; \mathbf{y}) = -50.0133$. The log likelihood for the null model is obtained as $l(\widehat{\boldsymbol{\mu}}_0, 1; \mathbf{y}) = -96.8119$ with the statements (output not shown)

```
proc genmod data=kills;
  model kills/trials =  / dist=binomial link=logit;
run;
```

The generalized R^2 measure

$$R^2_* = 1 - \exp\left\{-\frac{2}{n}\left(l(\widehat{\mu}_f, 1; \mathbf{y}) - l(\widehat{\mu}_0, 1; \mathbf{y})\right)\right\}$$

is then

$$R^2_* = 1 - \exp\left\{-\frac{2}{140}(-50.0133 + 96.8119)\right\} = 0.4875.$$

This value does not appear very large but it should be kept in mind that this measure is not bounded by 1. Also notice that the denominator in the exponent is $n = 140$, the total number of observations, rather than $n^{(g)} = 7$, the number of groups. The rescaled measure $\overline{R}^2$ is obtained by dividing R^2_* with

$$\max\{R^2_*\} = 1 - \exp\left\{\frac{2}{n}l(\widehat{\mu}_0, 1; \mathbf{y})\right\} = 1 - \exp\left\{\frac{-2}{140}96.8119\right\} = 0.749,$$

hence $\overline{R}^2 = 0.4875/0.749 = 0.6508$. With almost 2/3 of the variability in mortality proportions explained by the $\log_{10}$ concentration and a t_{obs} ratio for the slope parameter of $t_{obs} = 4.1651/0.6520 = 6.388$ we are reasonably satisfied with the model fit and proceed to an estimation of the dosages that are lethal to 50% or 80% of the larvae. Based on the estimates of β_0 and β_1 as well as [6.52] we obtain the point estimates

$$\widehat{\log_{10}}\{x_{0.5}\} = 1.7305/4.1651 = 0.4155$$
$$\widehat{x}_{0.5} = 10^{0.4155} = 2.603$$
$$\widehat{\log_{10}}\{x_{0.8}\} = (\text{logit}(0.8) + 1.7305)/4.1651 = 0.7483$$
$$\widehat{x}_{0.8} = 10^{0.7483} = 5.601.$$

To obtain standard errors and confidence intervals for these four quantities `proc nlmixed` is used because of its ability to obtain standard errors for nonlinear functions of parameter estimates by first-order Taylor series. As starting values in the `nlmixed` procedure we use the converged iterates of `proc genmod`. The `df=5` option was added to the `nlmixed` statement to make sure that `proc nlmixed` uses the same degrees of freedom for the determination of p-values as `proc genmod`. The complete SAS® code including the estimation of the lethal dosages on the $\log_{10}$ and the original scale follows.

```
proc nlmixed data=kills df=5;
  parameters intcpt=-1.7305 b=4.165;
  p = 1/(1+exp(-intcpt - b*logc));
  model kills ~ binomial(trials,p);
  estimate 'LD50' -intcpt/b;
  estimate 'LD50 original' 10**(-intcpt/b);
  estimate 'LD80' (log(0.8/0.2)-intcpt)/b;
  estimate 'LD80 original' 10**((log(0.8/0.2)-intcpt)/b);
run;
```

Output 6.5.

```
                    The NLMIXED Procedure

                       Specifications
  Description                                 Value
  Data Set                                    WORK.KILLS
  Dependent Variable                          kills
  Distribution for Dependent Variable         Binomial
  Optimization Technique                      Dual Quasi-Newton
  Integration Method                          None

                         Dimensions
              Description                        Value
              Observations Used                      7
              Observations Not Used                  0
              Total Observations                     7
              Parameters                             2

                         Parameters
                   intcpt            b    NegLogLike
                  -1.7305        4.165    9.50956257

                      Iteration History
  Iter     Calls    NegLogLike        Diff      MaxGrad         Slope
     1         4    9.50956256    8.157E-9      0.00056      -0.00004
     2         7    9.50956255    1.326E-8     0.000373      -6.01E-6
     3         8    9.50956254    4.875E-9     5.963E-7      -9.76E-9

            NOTE: GCONV convergence criterion satisfied.
```

Output 6.5 (continued).

```
                         Fit Statistics
              Description                         Value
              -2 Log Likelihood                    19.0
              AIC (smaller is better)              23.0
              BIC (smaller is better)              22.9
              Log Likelihood                       -9.5
              AIC (larger is better)              -11.5
              BIC (larger is better)              -11.5

                      Parameter Estimates
                   Standard
Parameter  Estimate    Error    DF  t Value  Pr > |t|   Alpha
intcpt      -1.7305   0.3741     5    -4.63    0.0057    0.05
b            4.1651   0.6520     5     6.39    0.0014    0.05

                       Additional Estimates
                          Standard
Label            Estimate    Error   DF  t Value  Pr > |t|  Lower   Upper
LD50               0.4155  0.06085    5     6.83   0.0010  0.2716  0.5594
LD50 original      2.6030   0.3647    5     7.14   0.0008  1.7406  3.4655
LD80               0.7483  0.07944    5     9.42   0.0002  0.5605  0.9362
LD80 original      5.6016   1.0246    5     5.47   0.0028  3.1788  8.0245
```

The `nlmixed` procedure converges after three iterations and reports the same parameter estimates and standard errors as `proc genmod` (Output 6.5). The log likelihood value reported by `proc nlmixed` (-9.5) does not agree with that of the `genmod` procedure (-50.013). The procedures differ with respect to the inclusion/exclusion of constants in the likelihood calculations. Differences in log likelihoods between nested models will be the same for the two procedures. The null model log likelihood reported by `proc nlmixed` (code and output not shown) is -56.3. The log likelihood difference between the two models is thus $-56.3 + 9.5 = -96.81 + 50.01 = -46.8$ in either procedure.

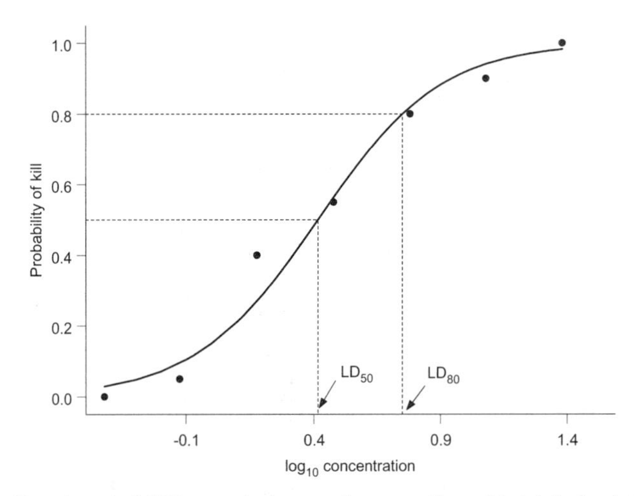

Figure 6.13. Predicted probabilities and observed proportions (dots) in logistic regression model for insecticide kills. Estimated dosages lethal to 50% and 80% of the larvae are also shown.

The table of `Additional Estimates` shows the output for the four `estimate` statements. The point estimates for the lethal dosages agrees with the manual calculation above and the standard errors are obtained from a first-order Taylor series expansion. The values in the columns `Lower` and `Upper` are asymptotic 95% confidence intervals for the estimated quantities.

Figure 6.13 shows the predicted probabilities to kill a randomly selected larva as a function of the $\log_{10}$ concentration. The observed proportions are overlaid and the $\widehat{\log}\{x_{0.5}\}$ and $\widehat{\log}\{x_{0.8}\}$ dosages are shown.

6.7.2 Binomial Proportions in a Randomized Block Design — the Hessian Fly Experiment

Gotway and Stroup (1997) present data from an agronomic field trial in which sixteen varieties of wheat are to be compared with respect to their resistance to infestation with the Hessian fly. The varieties were arranged in a randomized complete block design with four blocks on an 8×8 grid (Figure 6.14). For each of the 64 experimental units the number of plants with insect damage was counted. Let Z_{ij} denote the number of damaged plants for variety i in block j and n_{ij} the number of plants on the experimental unit. The outcome of interest is the sample proportion $Y_{ij} = Z_{ij}/n_{ij}$. If infestations are independent from plant to plant on an experimental unit and the plants are equally likely to become infested, Z_{ij} is a Binomial(n_{ij}, π_{ij}) random variable. By virtue of the random assignment of varieties to grid cells in each block the Z_{ij}'s are also independent of each other.

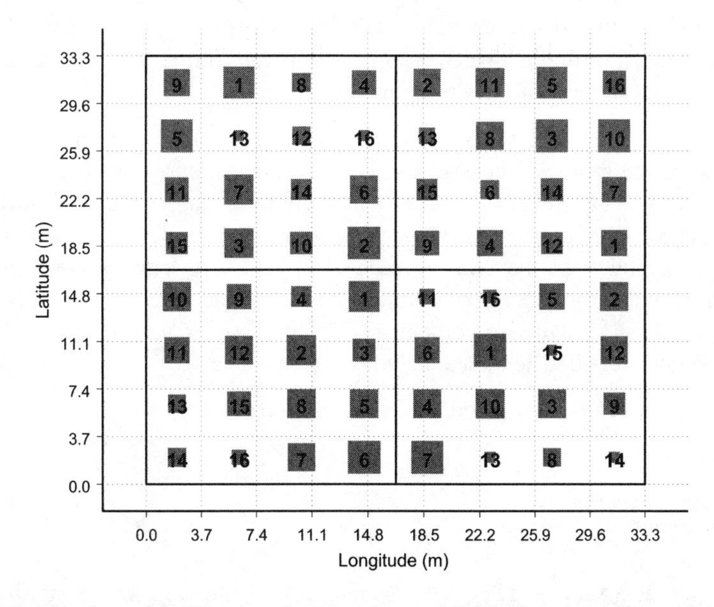

Figure 6.14. Design layout in Hessian fly experiment. Field plots are $3.7m \times 3.7m$. The area of the squares is proportional to the sample proportion of damaged plants; numbers indicate the variety. Block boundaries are shown as solid lines. Data used with permission of the International Biometric Society.

A generalized linear model for this experiment can be set up with a linear predictor that represents the experimental design and a link function for the probability that a randomly selected plant is damaged by Hessian fly infestation. Choosing a logit link function this model becomes

$$\text{logit}(\pi_{ij}) = \ln\left\{\frac{\pi_{ij}}{1-\pi_{ij}}\right\} = \eta_{ij} = \mu + \tau_i + \rho_j, \qquad [6.53]$$

where τ_i is the effect of the ith variety and ρ_j is the effect of the jth block.

Of interest are comparisons of the treatment effects adjusted for the block effects. For example, one may want to test the hypothesis that varieties i and i' have equal probability to be damaged by infestations, i.e, $H_0{:}\pi_i = \pi_{i'}$. These probabilities are not the same as the block-variety specific probabilities π_{ij} in model [6.53]. In a linear model $Y_{ij} = \mu + \tau_i + \rho_j + e_{ij}$ these comparisons are based on the least squares means of the treatment effects. If $\widehat{\mu}$, $\widehat{\tau}_i$, and $\widehat{\rho}_j$ denote the respective least squares estimates and there are $j = 1, \cdots, 4$ blocks as in this example, the least squares mean for treatment i is calculated as

$$\widehat{\mu} + \widehat{\tau}_i + \frac{1}{4}(\widehat{\rho}_1 + \widehat{\rho}_2 + \widehat{\rho}_3 + \widehat{\rho}_4) = \widehat{\mu} + \widehat{\tau}_i + \widehat{\rho}_. . \qquad [6.54]$$

A similar approach can be taken in the generalized linear model. If $\widehat{\mu}$, $\widehat{\tau}_i$, and $\widehat{\rho}_j$ denote the converged IRLS estimates of the parameters in model [6.53]. The treatment specific linear predictor is calculated as the same estimable linear function as in the standard model:

$$\widehat{\eta}_{i.} = \widehat{\mu} + \widehat{\tau}_i + \widehat{\rho}_. \, .$$

The estimate of the marginal probability that variety i is damaged by the Hessian fly is then obtained by inverting the link function, $\widehat{\pi}_{i.} = 1/(1 + \exp\{-\widehat{\eta}_{i.}\})$. Hypothesis tests can be based on a comparison of the $\widehat{\eta}_{i.}$, which are linear functions of the parameter estimates, or the $\widehat{\pi}_{i.}$, which are nonlinear functions of the estimates.

The `proc genmod` code to fit the Binomial proportions with a logit link in a randomized complete block design follows. The `ods exclude` statement suppresses the printing of various default tables. The `lsmeans entry / diff;` statement requests the marginal linear predictors $\widehat{\eta}_{i.}$ for the varieties (entries) as well as all pairwise tests of the form $H_0{:}\,\eta_{i.} = \eta_{i'.}$. Because `lsmeandiffs` is included in the `ods exclude` statement, the lengthy table of differences $\eta_{i.} - \eta_{i'.}$ is not included on the printed output. The ods output `lsmeandiffs=diff;` statement saves the $16{*}15/2 = 120$ pairwise comparisons in the SAS® data set `diff` that is available for post-processing after `proc genmod` concludes.

```
ods exclude ParmInfo ParameterEstimates lsmeandiffs;
proc genmod data=HessFly;
  class block entry;
  model z/n = block entry / link=logit dist=binomial type3;
  lsmeans entry /diff;
  ods output lsmeandiffs=diff;
run;
```

From the `LR Statistics For Type 3 Analysis` table we glean a significant variety (entry) effect ($p < 0.0001$) (Output 6.6). It should not be too surprising that the sixteen varieties do not have the same tendency to be damaged by the Hessian fly. The `Least Squares Means` table lists the marginal linear predictors for the varieties which can be converted into damage probabilities by inverting the logit link function. For variety 1, for example, this estimated marginal probability is $\widehat{\pi}_{1.} = 1/(1 + \exp\{-1.4864\}) = 0.815$ and for variety 8 this probability is only $\widehat{\pi}_{8.} = 1/(1 + \exp\{0.1639\}) = 0.459$.

Output 6.6.

```
                          The GENMOD Procedure

                           Model Information

 Data Set                              WORK.HESSFLY
 Distribution                              Binomial
 Link Function                                Logit
 Response Variable (Events)                       z     No. of damaged plants
 Response Variable (Trials)                       n     No. of plants
 Observations Used                               64
 Number Of Events                               396
 Number Of Trials                               736

                       Class Level Information

       Class       Levels    Values
       block            4    1 2 3 4
       entry           16    1 2 3 4 5 6 7 8 9 10 11 12 13 14 15 16

                 Criteria For Assessing Goodness Of Fit

       Criterion                 DF          Value          Value/DF

       Deviance                  45       123.9550            2.7546
       Scaled Deviance           45       123.9550            2.7546
       Pearson Chi-Square        45       106.7426            2.3721
       Scaled Pearson X2         45       106.7426            2.3721
       Log Likelihood                    -440.6593
Algorithm converged.

                    LR Statistics For Type 3 Analysis

                                           Chi-
               Source            DF      Square     Pr > ChiSq
               block              3        4.27         0.2337
               entry             15      132.62         <.0001

                         Least Squares Means

                                     Standard                 Chi-
  Effect     entry  Estimate          Error       DF        Square      Pr > ChiSq
  entry      1        1.4864         0.3921        1         14.37          0.0002
  entry      2        1.3453         0.3585        1         14.08          0.0002
  entry      3        0.9963         0.3278        1          9.24          0.0024
  entry      4        0.0759         0.2643        1          0.08          0.7740
  entry      5        1.3139         0.3775        1         12.12          0.0005
  entry      6        0.5758         0.3180        1          3.28          0.0701
  entry      7        0.8608         0.3302        1          6.80          0.0091
  entry      8       -0.1639         0.2975        1          0.30          0.5816
  entry      9        0.0960         0.2662        1          0.13          0.7183
  entry      10       0.8413         0.3635        1          5.36          0.0206
  entry      11       0.0313         0.2883        1          0.01          0.9136
  entry      12       0.0423         0.2996        1          0.02          0.8876
  entry      13      -2.0941         0.5330        1         15.44          <.0001
  entry      14      -1.0185         0.3538        1          8.29          0.0040
  entry      15      -0.6303         0.2883        1          4.78          0.0288
  entry      16      -1.4645         0.3713        1         15.56          <.0001
```

To determine which entries differ significantly in the damage probabilities we post-process the data set `diff` by deleting those comparisons which are not significant at a desired significance level and sorting the data set with respect to entries. Choosing significance level $\alpha = 0.05$, the statements

```
data diff; set diff; variety = entry+0; _variety = _entry+0;
  drop entry _entry;
run;
proc sort data=diff(where=(ProbChiSq < 0.05));
  by variety _variety ProbChiSq;
run;
proc print data=diff label;
  var variety _variety Estimate StdErr Df ChiSq ProbChiSq;
run;
```

accomplish that (Output 6.7). The statements `variety = entry+0;` and `_variety = _entry+0;` convert the values for `entry` into numeric format, since they are stored as character variables by `proc genmod`. Entry 1, for example, differs significantly from entries 4, 8, 9, 11, 12, 13, 14, 15, and 16. Notice that the data set variable `entry` was renamed to `variety` to produce Output 6.7. The positive `Estimate` for the comparison of `variety 1` and `_variety 4`, for example, indicates that entry 1 has a higher damage probability than entry 4. Similarly, the negative `Estimate` for the comparison of `variety 4` and `_variety 5` indicates a lower damage probability of variety 4 compared to variety 5.

Output 6.7.

Obs	variety	_variety	Estimate	Std Err	DF	Chi Square	Pr>Chi
1	1	4	1.4104	0.4736	1	8.87	0.0029
2	1	8	1.6503	0.4918	1	11.26	0.0008
3	1	9	1.3903	0.4740	1	8.61	0.0034
4	1	11	1.4551	0.4863	1	8.95	0.0028
5	1	12	1.4440	0.4935	1	8.56	0.0034
6	1	13	3.5805	0.6614	1	29.30	<.0001
7	1	14	2.5048	0.5276	1	22.54	<.0001
8	1	15	2.1166	0.4862	1	18.95	<.0001
9	1	16	2.9509	0.5397	1	29.89	<.0001
10	2	4	1.2694	0.4454	1	8.12	0.0044
11	2	8	1.5092	0.4655	1	10.51	0.0012
12	2	9	1.2492	0.4467	1	7.82	0.0052
13	2	11	1.3140	0.4612	1	8.12	0.0044
14	2	12	1.3030	0.4669	1	7.79	0.0053
15	2	13	3.4394	0.6429	1	28.62	<.0001
16	2	14	2.3638	0.5042	1	21.98	<.0001
17	2	15	1.9756	0.4614	1	18.34	<.0001
18	2	16	2.8098	0.5158	1	29.68	<.0001
19	3	4	0.9204	0.4207	1	4.79	0.0287
20	3	8	1.1602	0.4427	1	6.87	0.0088
21	3	9	0.9003	0.4225	1	4.54	0.0331
22	3	11	0.9651	0.4370	1	4.88	0.0272
23	3	12	0.9540	0.4441	1	4.61	0.0317
24	3	13	3.0904	0.6266	1	24.32	<.0001
25	3	14	2.0148	0.4830	1	17.40	<.0001
26	3	15	1.6266	0.4378	1	13.80	0.0002
27	3	16	2.4608	0.4956	1	24.66	<.0001
28	4	5	-1.2380	0.4607	1	7.22	0.0072
29	4	13	2.1700	0.5954	1	13.28	0.0003

This analysis of the Hessian fly experiment seems *simple enough*. A look at the `Criteria For Assessing Goodness Of Fit` table shows that not all is well, however. The deviance of the fitted model with 123.955 exceeds the degrees of freedom (45) 2.7-fold. In a proper model the deviance is expected to be about as large as its degrees of freedom. We do not advocate formal statistical tests of the deviance/df ratio unless data are grouped. Usually the modeler interprets the ratio subjectively to decide whether a deviation of the ratio from one is reason for concern. First we notice that ratios in excess of one indicate a potential overdispersion problem and then inquire how overdispersion could arise. Omitting important variables from the model leads to excess variability since the linear predictor does not account for important effects. In an experimental design the modeler usually builds the linear predictor from the randomization, treatment, and blocking protocol. In a randomized complete block design, $\eta_{ij} = \mu + \tau_i + \rho_j$ *is* the appropriate linear predictor since all other systematic effects should have been neutralized by randomization. If all necessary effects are included in the model, overdispersion could arise from positive correlations among the observations. Two levels of correlations must be considered here. First, it was assumed that the counts on each experimental unit follow the Binomial law which implies that the n_{ij} Bernoulli(π_{ij}) variables are independent. In other words, the probability of a plant being damaged does not depend on whether neighboring plants on the same experimental unit are infected or not. This seems quite unlikely. We expect infestations to appear in clusters and the Z_{ij} may not be Binomial-distributed. Instead, a probability model that allows for overdispersion relative to the Binomial, for example, the Beta-Binomial model could be used. Second, there are some doubts whether the counts of neighboring units are independent as assumed in the analysis. There may be spatial dependencies among grid cells in the sense that units near each other are more highly correlated than units further apart. This assumption is reasonable if, for example, propensity for infestation is linked to a soil variable that varies spatially. In other cases, such spatial correlations induced by a spatially varying covariate have been confirmed. Randomization of the varieties to experimental units neutralizes such spatial dependencies. On average, each treatment is affected by these effects equally and the overall effect is balanced out. However, the variability due to these spatial effects is not removed from the data. To that end, blocks need to be arranged in such a way that experimental units within a block are homogeneous. Stroup et al. (1994) note that combining adjacent experimental units into blocks in agricultural variety trials can be at variance with an assumption of homogeneity within blocks when more than eight to twelve experimental units are grouped. Spatial trends will then be removed only incompletely and this source of overdispersion prompted Gotway and Stroup (1997) to analyze the Hessian fly data with a model that takes into account the spatial dependence among counts of different experimental units explicitly. We will return to such models and the Hessian fly data in §9.

A quick fix for overdispersed data that does not address the real cause of the overdispersion problem is to estimate a separate scale parameter ϕ in models that would not contain such a parameter otherwise. In the Hessian fly example, this is accomplished by adding the `dscale` or `pscale` option to the `model` statement in `proc genmod`. The former estimates the overdispersion parameter based on the deviance, the latter based on Pearson's statistic (see §6.4.3). The variance of a count Z_{ij} is then modeled as $\text{Var}[Z_{ij}] = \phi n_{ij}\pi_{ij}(1 - \pi_{ij})$ rather than $\text{Var}[Z_{ij}] = n_{ij}\pi_{ij}(1 - \pi_{ij})$. From the statements

```
ods exclude ParmInfo ParameterEstimates lsmeandiffs;
proc genmod data=HessFly;
  class block entry;
  model z/n = block entry / link=logit dist=binomial type3 dscale;
  lsmeans entry /diff;
  ods output lsmeandiffs=diff;
run;
```

a new data set of treatment differences is obtained. After post-processing of the `diff` data set one obtains Output 6.8. Fewer entries are now found significantly different from entry 1 than in the analysis that does not account for overdispersion. Also notice that the estimates of the treatment differences has not changed. The additional overdispersion parameter is a multiplicative parameter that has no effect on the parameter estimates, only on their standard errors. Using the `dscale` estimation method the overdispersion parameter is estimated as $\widehat{\phi} = 123.955/45 = 2.7546$ which is the ratio of deviance and degrees of freedom in the model fitted initially. All standard errors in the preceding partial output are $\sqrt{2.7546}$ larger than the standard errors in the analysis without the overdispersion parameter.

Output 6.8.

Obs	variety	_variety	Estimate	Std Err	DF	Chi Square	Pr>Chi
1	1	8	1.6503	0.8162	1	4.09	0.0432
2	1	13	3.5805	1.0978	1	10.64	0.0011
3	1	14	2.5048	0.8756	1	8.18	0.0042
4	1	15	2.1166	0.8069	1	6.88	0.0087
5	1	16	2.9509	0.8958	1	10.85	0.0010
6	2	13	3.4394	1.0670	1	10.39	0.0013
7	2	14	2.3638	0.8368	1	7.98	0.0047
8	2	15	1.9756	0.7657	1	6.66	0.0099
9	2	16	2.8098	0.8561	1	10.77	0.0010
10	3	13	3.0904	1.0400	1	8.83	0.0030
11	3	14	2.0148	0.8016	1	6.32	0.0120
12	3	15	1.6266	0.7267	1	5.01	0.0252
13	3	16	2.4608	0.8225	1	8.95	0.0028
14	4	13	2.1700	0.9882	1	4.82	0.0281
15	4	16	1.5404	0.7575	1	4.14	0.0420

... and so forth ...

6.7.3 Gamma Regression and Yield Density Models

We now return to the yield density data of McCullagh and Nelder (1989, pp. 317-320) shown in Table 6.4 (p. 309). The data consists of dry weights of barley sown at various seeding densities with three replicates. Recall from §5.8.7 that a customary approach to model a yield-density relationship is to assume that the inverse of yield per plant is a linear function of plant density. The Shinozaki-Kira model (Shinozaki and Kira 1956)

$$\mathrm{E}[Y] = (\beta_0 + \beta_1 x)^{-1}$$

and the Holliday model (Holliday 1960)

$$\mathrm{E}[Y] = \left(\beta_0 + \beta_1 x + \beta_2 x^2\right)^{-1}$$

are representatives of this class of models. Here, x denotes the plant (seeding) density. A standard nonlinear regression approach is then to model, for example, $Y_i =$

$1/(\beta_0 + \beta_1 x_i) + e_i$ where the e_i are independent random errors with mean 0 and variance σ^2. Figure 6.3 (p. 309) suggests that the variability is not homogeneous in these data, however. Whereas one could accommodate variance heterogeneity in the nonlinear model by using weighted nonlinear least squares, Figure 6.3 alerts us to a more subtle problem. The standard deviation of the barley yields seems to be related to the mean yield. Although Figure 6.3 is quite noisy, it is not unreasonable to assume that the standard deviations are proportional to the mean (a regression through the origin of s on $\overline{y}$). Figure 6.15 displays the reciprocal yields and the reciprocal of the sample means across the three blocks against the seeding density. An inverse quadratic relationship

$$\frac{1}{\mathrm{E}[Y]} = \beta_0 + \beta_1 x + \beta_2 x^2$$

as suggested by the Holliday model is reasonable. This model has linear predictor $\eta = \beta_0 + \beta_1 x + \beta_2 x^2$ and reciprocal link function. For the random component we choose Y not to be a Gaussian, but a Gamma random variable. Gamma random variables are non-negative (such as yields), and their standard deviation is proportional to their mean. The Gamma distributions are furthermore not symmetric about the mean but right-skewed (see Figure 6.2, p. 308). The canonical link of a Gamma random variable is the reciprocal link, which provides further support to use this model for yield density investigations where inverse polynomial relationships are common. Unfortunately, the inverse link does not guarantee that the predicted means are non-negative since the linear predictor is not constained to be positive. As an alternative link function for Gamma-distributed random variables, the log link can be used.

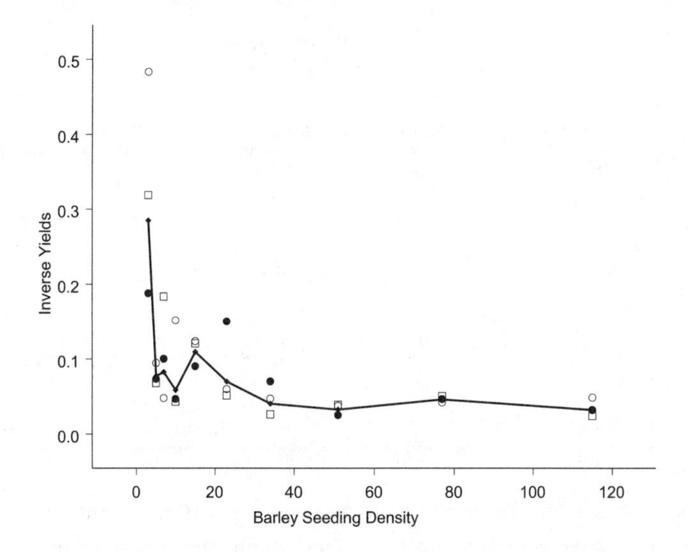

Figure 6.15. Inverse yields and inverse replication averages against seeding density. Disconnected symbols represent observations from blocks 1 to 3, the connected symbols the sample averages. An inverse quadratic relationship is reasonable.

Before fitting a generalized linear model with Gamma errors we must decide whether to fit the model to the 30 observations from the three blocks or to the 10 block averages. In the former case, we must include block effects in the full model and can then test whether it is reasonable that some effects do not vary by blocks. The full model we consider here has a linear predictor

$$\eta_{ij} = \beta_{0j} + \beta_{1j}x_{ij} + \beta_{2j}x_{ij}^2 \qquad [6.55]$$

where the subscript j identifies the blocks. Combined with a reciprocal link function this model will fit a separate inverse quadratic to data from each block. The `proc genmod` code to fit this model with Gamma errors follows. The `noint` option was added to the `model` statement to prevent the addition of an overall intercept term β_0. The `link=power(-1)` option invokes the reciprocal link.

```
ods exclude ParameterEstimates;
proc genmod data=barley;
  class block;
  model bardrwgt = block block*seed block*seed*seed /
                   noint link=power(-1) dist=gamma type3;
run;
```

Output 6.9.

```
                    The GENMOD Procedure

                     Model Information
            Data Set                    WORK.BARLEY
            Distribution                      Gamma
            Link Function                 Power(-1)
            Dependent Variable             BARDRWGT
            Observations Used                    30

                  Class Level Information
               Class        Levels     Values
               BLOCK             3     1 2 3

            Criteria For Assessing Goodness Of Fit
     Criterion                DF           Value          Value/DF
     Deviance                 21          7.8605            0.3743
     Scaled Deviance          21         31.2499            1.4881
     Pearson Chi-Square       21          6.7872            0.3232
     Scaled Pearson X2        21         26.9830            1.2849
     Log Likelihood                     -102.6490
Algorithm converged.

            LR Statistics For Type 3 Analysis
          Source              DF     Chi-Square     Pr > ChiSq
          BLOCK                3          52.39         <.0001
          seed*BLOCK           3           9.19         0.0269
          seed*seed*BLOCK      3           5.66         0.1294
```

The full model has a log likelihood of $l(\widehat{\mu}_f, \widehat{\psi}; \mathbf{y}) = -102.649$ and the `LR Statistics For Type 3 Analysis` table shows that the effect which captures separate quadratic effects for each block is not significant ($p = 0.1294$, Output 6.9). To see whether a common quadratic effect is sufficient, we fit the model as

```
ods exclude obstats;
proc genmod data=barley;
  class block;
  model bardrwgt = block block*seed seed*seed /
                   noint link=power(-1) dist=gamma type3 obstats;
  ods output obstats=stats;
run;
```

and obtain a log likelihood of -102.8819 (Output 6.10). Twice the difference of the log likelihoods, $\Lambda = 2*(-102.649 + 102.8819) = 0.465$, is not significant and we conclude that the quadratic effects need not be varied by blocks ($\Pr(\chi^2_2 \geq 0.465) = 0.793$). The common quadratic effect of seeding density is significant at the 5% level ($p = 0.0227$) and will be retained in the model (Output 6.10). The `obstats` option of the `model` statement requests a table of the linear predictors, predicted values, and various residuals to be calculated for the fitted model.

Output 6.10.

```
                         The GENMOD Procedure

                           Model Information
                  Data Set                  WORK.BARLEY
                  Distribution                    Gamma
                  Link Function               Power(-1)
                  Dependent Variable           BARDRWGT
                  Observations Used                  30

                       Class Level Information
                     Class        Levels     Values
                     BLOCK             3     1 2 3

             Criteria For Assessing Goodness Of Fit
      Criterion                  DF           Value           Value/DF

      Deviance                   23          7.9785             0.3469
      Scaled Deviance            23         31.2678             1.3595
      Pearson Chi-Square         23          6.7327             0.2927
      Scaled Pearson X2          23         26.3856             1.1472
      Log Likelihood                      -102.8819
Algorithm converged.

                  Analysis Of Parameter Estimates
                           Standard  Wald 95% Confidence  Chi-   Pr >
Parameter     DF  Estimate    Error        Limits         Square ChiSq

Intercept       0    0.0000   0.0000    0.0000    0.0000       .      .
BLOCK       1   1    0.1014   0.0193    0.0637    0.1392   27.72 <.0001
BLOCK       2   1    0.1007   0.0189    0.0636    0.1378   28.28 <.0001
BLOCK       3   1    0.0930   0.0177    0.0582    0.1277   27.47 <.0001
seed*BLOCK  1   1   -0.0154   0.0053   -0.0257   -0.0051    8.51 0.0035
seed*BLOCK  2   1   -0.0162   0.0053   -0.0265   -0.0059    9.42 0.0021
seed*BLOCK  3   1   -0.0159   0.0052   -0.0261   -0.0057    9.38 0.0022
seed*seed       1    0.0009   0.0004    0.0002    0.0017    5.69 0.0170
Scale           1    3.9190   0.9719    2.4104    6.3719
NOTE: The scale parameter was estimated by maximum likelihood.

                 LR Statistics For Type 3 Analysis
                                        Chi-
              Source             DF    Square    Pr > ChiSq

              BLOCK               3     51.97        <.0001
              seed*BLOCK          3      8.77        0.0324
              seed*seed           1      5.19        0.0227
```

The `ods exclude obstats;` statement in conjunction with the `ods output obstats=stats;` statement prevents the printing of these statistics to the output window and saves the results in a SAS® data set (named `stats` here). The seeding densities were divided by 10 prior to fiting of this model to allow sufficient significant digits to be displayed in the `Analysis Of Parameter Estimates` table.

The parameterization of the Gamma distribution chosen by `proc genmod` corresponds to our [6.7] (p. 308) and the parameter labeled `Scale` is our α parameter in [6.7]. In this parameterization we have $\mathrm{E}[Y] = \mu$ and $\mathrm{Var}[Y] = \mu^2/\alpha$. We can thus estimate the mean and variance of an observation from block 1 at seeding density 3 as

$$\widehat{\mu} = \frac{1}{0.1014 - 0.0154\frac{3}{10} + 0.0009\frac{9}{100}} = 10.324$$

$$\widehat{\mathrm{Var}}[Y] = 10.324^2/3.919 = 27.197.$$

The fitted barley yields for the three blocks are shown in Figure 6.16. If the model is correct, yields will attain a maximum at a seeding density around 80. While the seeding density of highest yield depends little on the block, the maximum yield attained varies considerably.

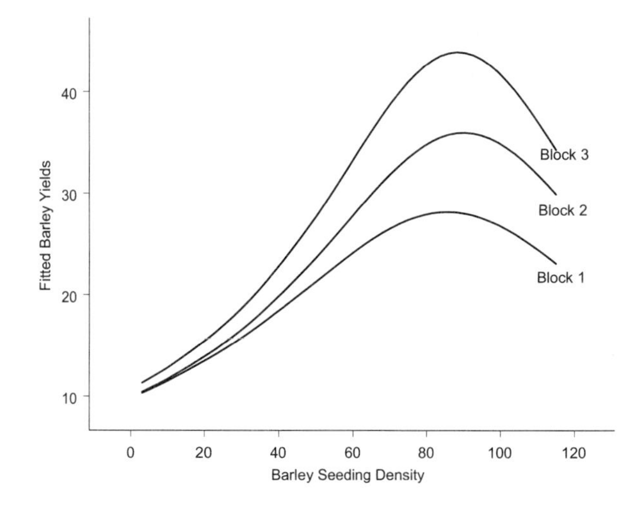

Figure 6.16. Fitted barley yields in the three blocks based on a model with inverse linear effects varied by blocks and a common inverse quadratic effect.

The `stats` data set with the output from the `obstats` option contains several residual diagnostics, such as raw residuals, deviance residuals, Pearson residuals, and their standardized versions. We plotted the Pearson residuals

$$\widehat{r}_{ij} = \frac{y_{ij} - \widehat{\mu}_{ij}}{\sqrt{h(\widehat{\mu}_{ij})}},$$

where $h(\widehat{\mu}_{ij})$ is the variance function evaluated at the fitted mean in Figure 6.17 (open circles). From Table 6.2 (p. 305) the variance function of a Gamma random variable is simply the square of the mean and the Pearson residuals take on the form

$$\widehat{r}_{ij} = \frac{y_{ij} - \widehat{\mu}_{ij}}{\widehat{\mu}_{ij}}. \qquad [6.56]$$

For the observation $y_{11} = 2.07$ from block 1 at seeding density 3 the Pearson residual is

$\widehat{r}_{11} = (2.07 - 10.324)/10.324 = -0.799$. Also shown as closed circles are the studentized residuals from fitting the model

$$Y_{ij} = \frac{1}{\beta_{0j} + \beta_{1j}x_{ij} + \beta_2 x_{ij}^2} + e_{ij}$$

as a nonlinear model with symmetric and homoscedastic errors (in `proc nlin`). If the mean increases with seeding density and the variation of the data is proportional to the mean we expect the variation in the nonlinear regression residuals to increase with seeding density. This effect is obvious in Figure 6.17. The assumption of homoscedastic errors underpinning the nonlinear regression analysis is not tenable; therefore, the Gamma regression is preferred. The tightness of the Pearson residuals in the Gamma regression model at seeding density 77 is due to the fact that these values are close to the density producing the maximum yield (Figure 6.16). Since this critical density is very similar from block to block, but the maximum yields differ greatly, the denominators in [6.56] shrink the raw residuals $y_{ij} - \widehat{\mu}_{ij}$ most for those blocks with high yields.

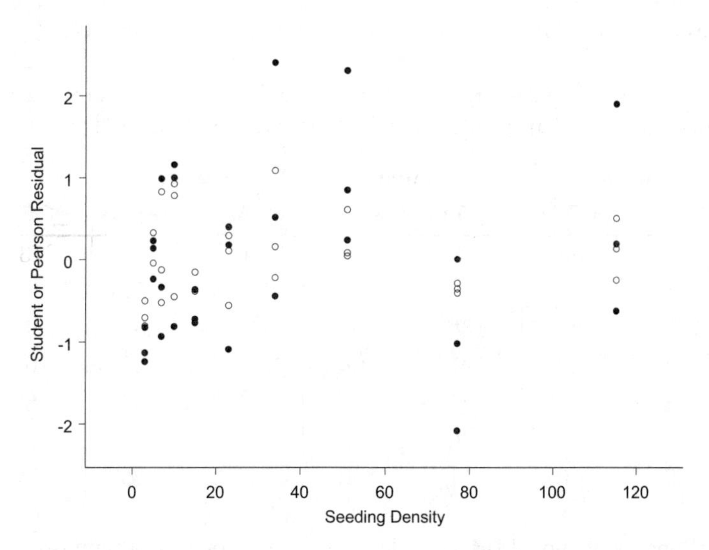

Figure 6.17. Pearson residuals (open circles) from generalized linear Gamma regression model and studentized residuals from nonlinear Gaussian regression model (full circles).

6.7.4 Effects of Judges' Experience on Bean Canning Quality Ratings

Canning quality is one of the most essential traits required in all new dry bean (*Phaseolus vulgaris* L.) varieties, and selection for this trait is a critical part of bean-breeding programs. Advanced lines that are candidates for release as varieties must be evaluated for canning quality for at least three years from samples grown at different locations. Quality is assessed by a panel of judges with varying levels of experience in evaluating breeding lines for visual quality traits. In 1996, 264 bean-breeding lines from four commercial classes were canned according to the procedures described by Walters et al. (1997). These included 62 navy, 65 black, 55 kidney, and 82 pinto bean-breeding lines plus checks and controls. The visual appearance of the processed beans was determined subjectively by a panel of 13 judges on a seven point hedonic scale (1 = very undesirable, 4 = neither desirable nor undesirable, 7 = very desirable). The beans were presented to the panel of judges in a random order at the same time. Prior to evaluating the samples, all judges were shown examples of samples rated as satisfactory (4). Concern exists if certain judges, due to lack of experience, are unable to correctly rate canned samples. From attribute-based product evaluations inferences about the effects of experience can be drawn from the psychology literature. Wallsten and Budescu (1981), for example, report that in the evaluation of a personality profile consisting of fourteen factors, experienced clinical psychologists utilized four to seven factors, whereas psychology graduate students tended to use only the two or three most salient factors. Prior to the bean canning quality rating experiment it was postulated that less experienced judges rate more severely than more experienced judges but also that experience should have little or no effect for navy beans for which the canning procedure was developed. Judges are stratified for purposes of analysis by experience (≤ 5 years, > 5 years). The counts by canning quality, judges experience, and bean-breeding line are listed in Table 6.17.

Table 6.17. Bean rating data. Kindly made available by Dr. Jim Kelly, Department of Crop and Soil Sciences, Michigan State University. Used with permission.

	Black		**Kidney**		**Navies**		**Pinto**	
Score	**≤ 5 ys**	**> 5 ys**	**≤ 5 ys**	**> 5 ys**	**≤ 5 ys**	**> 5 ys**	**≤ 5 ys**	**> 5 ys**
1	13	32	7	10	10	22	13	2
2	91	78	32	31	56	51	29	17
3	123	124	136	96	84	107	91	68
4	72	122	101	104	84	98	109	124
5	24	31	47	71	51	52	60	109
6	2	3	6	18	24	37	25	78
7	0	0	1	0	1	5	1	12

A proportional odds model for the ordered canning scores is fit with `proc genmod` below. The `contrast` statements test the effect of the judges' experience separately for the bean lines. These contrasts correspond to interaction slices by bean lines. The `estimate` statements calculate the linear predictors needed to derive the probabilities to rate each line in category 3 or less and category 4 or less depending on judges experience.

```
ods exclude ParameterEstimates ParmInfo;
proc genmod data=beans;
  class class exper;
  model score = class exper class*exper /
                link=cumlogit dist=multinomial type3;
  contrast 'Experience effect for Black'
                      exper 1 -1 class*exper 1 -1  0  0  0  0  0  0 ;

  contrast 'Experience effect for Kidney'
                      exper 1 -1 class*exper 0  0  1 -1  0  0  0  0;

  contrast 'Experience effect for Navies'
                      exper 1 -1 class*exper 0  0  0  0  1 -1  0  0;
```

```
  contrast 'Experience effect for Pinto'
                     exper 1 -1 class*exper 0  0  0  0  0  0  1 -1;

  estimate 'Black, < 5 years, score < 4'  Intercept 0 0 1
                     class 1 0 0 0 exper 1 0 class*exper 1 0 0 0 0 0 0 0;

  estimate 'Black, > 5 years, score < 4'  Intercept 0 0 1
                     class 1 0 0 0 exper 0 1 class*exper 0 1 0 0 0 0 0 0;

  estimate 'Black, < 5 years, score =< 4' Intercept 0 0 0 1
                     class 1 0 0 0 exper 1 0 class*exper 1 0 0 0 0 0 0 0;

  estimate 'Black, > 5 years, score =< 4' Intercept 0 0 0 1
                     class 1 0 0 0 exper 0 1 class*exper 0 1 0 0 0 0 0 0;
run;
```

There is a significant interaction between bean lines (`class`) and judge experience (Output 6.11). The results of comparing judges with more and less than 5 years experience will depend on the bean line. The contrast slices address this interaction (Output 6.12). The ratings distributions for experienced and less experienced judges is clearly not significantly different for navy beans and at the 5% level not significantly different for black beans ($p = 0.0964$). There are differences betwen the ratings for kidney and pinto beans, however ($p = 0.0051$ and $p < 0.0001$).

Output 6.11.

```
                          The GENMOD Procedure

                          Model Information
              Data Set                             WORK.BEANS
              Distribution                        Multinomial
              Link Function                   Cumulative Logit
              Dependent Variable                        SCORE
              Observations Used                          2795

                       Class Level Information
         Class         Levels     Values
         CLASS              4     Black Kidney Navies Pinto
         EXPER              2     Less than 5 More than 5

                           Response Profile
                  Ordered     Ordered
                    Level     Value            Count
                        1     1                  109
                        2     2                  385
                        3     3                  829
                        4     4                  814
                        5     5                  445
                        6     6                  193
                        7     7                   20

              Criteria For Assessing Goodness Of Fit
   Criterion                    DF          Value          Value/DF
   Log Likelihood                       -4390.5392
Algorithm converged.

                LR Statistics For Type 3 Analysis
          Source                DF      Chi-Square     Pr > ChiSq
          CLASS                  3          262.95         <.0001
          EXPER                  1           37.39         <.0001
          CLASS*EXPER            3           29.40         <.0001
```

To develop an impression of the differences in the rating distributions we calculated the probabilities to obtain ratings below 4, of exactly 4, and above 4. Recall that a score of 4 represents satisfactory quality and that all judges were shown examples thereof prior to the actual canning quality assessment. The probabilities are obtained from the linear predictors calculated with the `estimate` statements (Output 6.11). For black beans and raters with less than 5 years of experience we get

$$\Pr(Score < 4) = \frac{1}{1 + \exp\{-0.7891\}} = 0.69$$

$$\Pr(Score = 4) = \frac{1}{1 + \exp\{-2.1897\}} - 0.69 = 0.21$$

$$\Pr(Score > 4) = 1 - \frac{1}{1 + \exp\{-2.1897\}} = 0.10.$$

Similar calculation for the other groups lead to the probability distributions shown in Figure 6.18.

Well-documented criteria and a canning procedure specifically designed for navy beans explains the absence of differences due to the judges' experience for navy beans ($p = 0.7391$). Black beans in general were of poor quality and low ratings dominated, creating very similar probability distributions for experienced and less experienced judges (Figure 6.18). For kidney and pinto beans experienced and inexperienced judges classify control quality with similar odds, probably because they were shown such quality prior to judging. Experienced judges have a tendency to assign higher quality scores than less experienced judges for these two commercial classes.

Output 6.12.

Contrast Estimate Results

Label	Estimate	Standard Error	Confidence Limits	
Black, < 5 years, score < 4	0.7891	0.1007	0.5917	0.9866
Black, > 5 years, score < 4	0.5677	0.0931	0.3852	0.7501
Black, < 5 years, score =< 4	2.1897	0.1075	1.9789	2.4005
Black, > 5 years, score =< 4	1.9682	0.0996	1.7730	2.1634

Contrast Results

Contrast	DF	Chi-Square	Pr > ChiSq	Type
Experience effect for Black	1	2.76	0.0964	LR
Experience effect for Kidney	1	7.84	0.0051	LR
Experience effect for Navies	1	0.11	0.7391	LR
Experience effect for Pinto	1	58.32	<.0001	LR

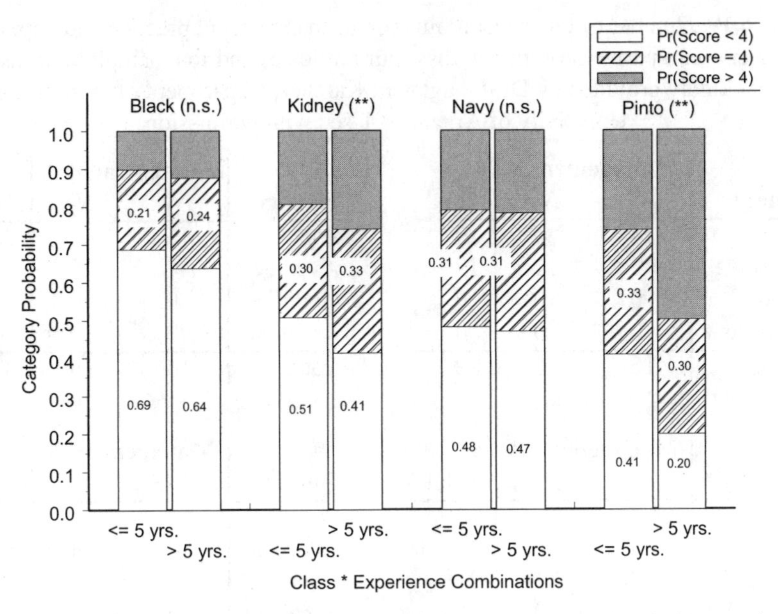

Figure 6.18. Predicted probability distributions for the 4×2 interaction in bean rating experiment. Categories 1 to 3 as well as categories 5 to 7 are amalgamated to emphasize deviations from satisfactory ratings (score = 4).

6.7.5 Ordinal Ratings in a Designed Experiment with Factorial Treatment Structure and Repeated Measures

Turfgrass fertilization traditionally has been accomplished through surface applications. The introduction of the Hydroject® (The Toro Company) has made possible subsurface placement of soluble materials. A study was conducted during the 1997 growing season to compare surface application and subsurface injection of nitrogen on the color of a one-year-old creeping bentgrass (*Agrostis palustris* L. Huds) putting green. The treatment structure comprised a complete 4×2 factorial of factors Management Practice (four levels) and Application Rate (two levels) (Table 6.18). The eight treatment combinations were arranged in a completely randomized design with four replications. Turf color was assessed on each experimental unit in weekly intervals for four weeks as poor, average, good, or excellent (Table 6.19).

Table 6.18. Treatment factors in nitrogen injection study

Management Practice	Rate of Application
1 = N surface applied with no supplemental water injection	$2.5\text{ g} \times \text{m}^{-2}$
2 = N surface applied with supplemental water injection	$5.0\text{ g} \times \text{m}^{-2}$
3 = N injected with #56 nozzle (7.6 cm depth of injection)	
4 = N injected with #53 nozzle (12.7 cm depth of injection)	

Table 6.19. Number of times a combination of management practice and application rate received a particular rating across four replicates and four sampling occasions. Data kindly provided by Dr. Douglas E. Karcher, Department of Horticulture, University of Arkansas. Used with permission.

	Management 1				Management 2		
Quality	N_1	N_2	Total	**Quality**	N_1	N_2	Total
Poor	14	5	19	Poor	15	8	23
Average	2	11	13	Average	1	8	9
Good	0	0	0	Good	0	0	0
Excellent	0	0	0	Excellent	0	0	0
Total	16	16	32	Total	16	16	32

	Management 3				Management 4		
Quality	N_1	N_2	Total	**Quality**	N_1	N_2	Total
Poor	0	0	0	Poor	1	0	1
Average	9	2	11	Average	12	4	16
Good	7	14	21	Good	3	11	14
Excellent	0	0	0	Excellent	0	1	1
Total	16	16	32	Total	16	16	32

Of particular interest were the determination of the water injection effect, the subsurface effect, and the comparison of injection vs. surface applications. These are contrasts among the levels of the factor Management Practice and it first needs to be determined whether the factor interacts with Application Rate.

We fit a proportional odds model to these data containing Management and nitrogen Application Rate effects and their interaction as well as a continuous covariate to model the temporal effects. This is not the most efficient method of accounting for repeated measures; we address repeated measures data structures in more detail in §7. Inclusion of the time variable significantly improves the model fit over a model containing only main effects and interactions of the experimental factors, however. The basic `proc genmod` statements are

```
ods exclude ParameterEstimates ParmInfo;
proc genmod data=mgtN rorder=data;
  class mgt nitro;
  model resp = mgt nitro mgt*nitro date /
               link=cumlogit dist=multinomial type3;
run;
```

The `genmod` output shows a nonsignificant interaction and significant Management, Application Rate, and Time effects (Output 6.13). Since the interaction is not significant, we can proceed to test the contrasts of interest based on the marginal management effects.

Adding the `contrast` statements

```
  contrast 'WIC effect          ' mgt 1 -1       ;
  contrast 'Subsurface effect   ' mgt 0 2 -1 -1  ;
  contrast 'Injected vs surface ' mgt 1 1 -1 -1  ;
```

to the `proc genmod` code produces the additional Output 6.14.

Output 6.13.

```
                  The GENMOD Procedure
                   Model Information
         Data Set                             WORK.MGTN
         Distribution                       Multinomial
         Link Function                 Cumulative Logit
         Dependent Variable                        resp
         Observations Used                          128

               Class Level Information
             Class        Levels      Values
             mgt               4      1 2 3 4
             nitro             2      1 2

                    Response Profile
             Ordered      Ordered
               Level      Value              Count
                   1      Poor                  43
                   2      Average               49
                   3      Good                  35
                   4      Excellen               1

          Criteria For Assessing Goodness Of Fit
   Criterion                  DF           Value          Value/DF
   Log Likelihood                       -64.4218
Algorithm converged.

            LR Statistics For Type 3 Analysis
                                      Chi-
         Source              DF      Square     Pr > ChiSq
         mgt                  3      140.79         <.0001
         nitro                1       43.01         <.0001
         mgt*nitro            3        0.90         0.8262
         date                 1       18.06         <.0001
```

Supplementing nitrogen surface application with water injection does not alter the turf quality significantly. However, the rating distribution of the average of the nitrogen injection treatments is significantly different from the turf quality obtained with nitrogen application and supplemental water injection ($LR = 109.51, p < 0.0001$). Similarly, the average injection treatment leads to a significantly different rating distribution than the average surface application.

Output 6.14.

```
                       Contrast Results

                                         Chi-
 Contrast                      DF      Square    Pr > ChiSq     Type

 WIC effect                     1        1.32        0.2512     LR
 Subsurface effect              1      109.51        <.0001     LR
 Injected vs surface            1      139.65        <.0001     LR
```

To determine these rating distributions for the marginal Management Practice effect and the marginal Rate effect, we obtain the least squares means and convert them into probabilities by inverting the link function. Unfortunately, `proc genmod` does not permit a `lsmeans` statement in combination with the multinomial distribution, i.e., for ordinal response. The marginal treatment means can be constructed with `estimate` statements, however. Adding to the `genmod` code the statements

```
estimate 'nitro 1 mean (<= Poor)' intercept 4 0 0 mgt 1 1 1 1 nitro 4 0
                                  mgt*nitro 1 0 1 0 1 0 1 0 date 10 /
                                  divisor=4;
estimate 'nitro 1 mean (<= Aver)' intercept 0 4 0 mgt 1 1 1 1 nitro 4 0
                                  mgt*nitro 1 0 1 0 1 0 1 0 date 10 /
                                  divisor=4;
estimate 'nitro 1 mean (<= Good)' intercept 0 0 4 mgt 1 1 1 1 nitro 4 0
                                  mgt*nitro 1 0 1 0 1 0 1 0 date 10 /
                                  divisor=4;

estimate 'Mgt 1 mean (<= Poor)'   intercept 2 0 0 mgt 2 0 0 0 nitro 1 1
                                  mgt*nitro 1 1 0 0 0 0 0 0 date 5 /
                                  divisor=2;
estimate 'Mgt 1 mean (<= Average)'intercept 0 2 0 mgt 2 0 0 0 nitro 1 1
                                  mgt*nitro 1 1 0 0 0 0 0 0 date 5 /
                                  divisor=2;
estimate 'Mgt 1 mean (<= Good)'   intercept 0 0 2 mgt 2 0 0 0 nitro 1 1
                                  mgt*nitro 1 1 0 0 0 0 0 0 date 5 /
                                  divisor=2;
```

produces linear predictors from which the marginal probability distributions for rate 2.5 $g \times m^{-2}$ and for surface application with no supplemental water injection can be obtained. Adding similar statements for the second application rate and the other three management practices we obtain the linear predictors in Output 6.15.

Output 6.15.

Contrast Estimate Results

Label	Estimate	Standard Error	Alpha	Confidence Limits	
nitro 1 mean (<= Poor)	-0.9468	0.6192	0.05	-2.1605	0.2669
nitro 1 mean (<= Aver)	4.7147	0.7617	0.05	3.2217	6.2077
nitro 1 mean (<= Good)	10.5043	1.4363	0.05	7.6892	13.3194
nitro 2 mean (<= Poor)	-3.9531	0.6769	0.05	-5.2799	-2.6264
nitro 2 mean (<= Average)	1.7084	0.6085	0.05	0.5156	2.9011
nitro 2 mean (<= Good)	7.4980	1.2335	0.05	5.0803	9.9157
Mgt 1 mean (<= Poor)	0.6986	0.4902	0.05	-0.2621	1.6594
Mgt 1 mean (<= Average)	6.3602	1.1924	0.05	4.0232	8.6971
Mgt 1 mean (<= Good)	12.1498	1.7154	0.05	8.7877	15.5119
Mgt 2 mean (<= Poor)	1.5627	0.6049	0.05	0.3772	2.7483
Mgt 2 mean (<= Average)	7.2243	1.2780	0.05	4.7194	9.7291
Mgt 2 mean (<= Good)	13.0138	1.7854	0.05	9.5145	16.5132
Mgt 3 mean (<= Poor)	-6.5180	1.1934	0.05	-8.8570	-4.1790
Mgt 3 mean (<= Average)	-0.8565	0.4455	0.05	-1.7296	0.0166
Mgt 3 mean (<= Good)	4.9331	1.1077	0.05	2.7621	7.1041
Mgt 4 mean (<= Poor)	-5.5433	1.1078	0.05	-7.7146	-3.3720
Mgt 4 mean (<= Average)	0.1182	0.4710	0.05	-0.8049	1.0414
Mgt 4 mean (<= Good)	5.9078	1.1773	0.05	3.6004	8.2153

A comparison of these marginal distributions is straightforward with `contrast` or `estimate` statements. For example,

```
estimate 'nitro 1 - nitro 2' nitro 1 -1 ;
estimate 'mgt 1   - mgt 2'   mgt 1 -1;
estimate 'mgt 1   - mgt 3'   mgt 1  0 -1;
estimate 'mgt 1   - mgt 4'   mgt 1  0  0 -1;
estimate 'mgt 2   - mgt 3'   mgt 0  1 -1;
estimate 'mgt 2   - mgt 4'   mgt 0  1  0 -1;
estimate 'mgt 3   - mgt 4'   mgt 0  0  1 -1;
```

produces all pairwise marginal comparisons:

Output 6.16.

```
                          Contrast Estimate Results

                                      Standard              Chi-
Label                    Estimate        Error    Alpha    Square     Pr > ChiSq

nitro 1 - nitro 2          3.0063       0.5525     0.05     29.61         <.0001

mgt 1    - mgt 2          -0.8641       0.7726     0.05      1.25         0.2634
mgt 1    - mgt 3           7.2167       1.2735     0.05     32.11         <.0001
mgt 1    - mgt 4           6.2419       1.1947     0.05     27.30         <.0001
mgt 2    - mgt 3           8.0808       1.3561     0.05     35.51         <.0001
mgt 2    - mgt 4           7.1060       1.2785     0.05     30.89         <.0001
mgt 3    - mgt 4          -0.9747       0.6310     0.05      2.39         0.1224
```

Table 6.20 shows the marginal probability distributions and indicates significant differences among the treatment levels. The surface applications lead to *poor* turf quality with high probability. Their ratings are at most *average* in over 90% of the cases. When nitrogen is injected into the soil the rating distributions shift toward higher categories. The #56 nozzle (7.6 cm depth of injection) leads to *good* turf quality in over $2/3$ of the cases. The nozzle that injects nitrogen up to 12.7 cm has a higher probability of *average* ratings, compared to the #56 nozzle, probably because the nitrogen is placed closer to the roots. Although there are no significant differences in the rating distributions between the two surface applications ($p = 0.2634$) and the ratings distributions of the two injection treatments ($p = 0.1224$), the two groups of treatments clearly separate. Based on the results of this analysis one would recommend nitrogen injection of $5 \text{ g} \times \text{m}^{-2}$.

Table 6.20. Predicted marginal probability distributions (by category) for nitrogen injection study

	Management Practice				N Rate	
Quality	1	2	3	4	$2.5 \text{ g} \times \text{m}^{-2}$	$5 \text{ g} \times \text{m}^{-2}$
Poor	0.668	0.827	0.001	0.004	0.279	0.020
Average	0.330	0.172	0.297	0.525	0.712	0.826
Good	0.002	0.001	0.695	0.472	0.008	0.153
Excellent	0+	0+	0.007	0.003	0.001	0.001
	a†	*a*	*b*	*b*	*a*	*b*

†Columns with the same letter are not significantly different in their ratings distributions at the 5% significance level.

6.7.6 Log-Linear Modeling of Rater Agreement

In the applications discussed so far there is a clear distinction between the response and explanatory variables in the model. For example, the response *proportion of Hessian fly-damaged plants* was modeled as a function of (explanatory) block and treatment effects in §6.7.2. This distinction between response and explanatory variables is not possible for certain cross-tabulated data (contingency tables). The data in Table 6.21 represent the results of rating the same 236 experimental units by two different raters on an ordinal scale from 1 to 5. For example, 12 units were rated in category 2 by Rater 2 and in category 1 by Rater 1. An obvious question is whether the ratings of the two interpreters are independent. Should we tackle this by modeling the Rater 2 results as a function of the Rater 1 results or vice versa? There is no response variable or explanatory variable here, only two categorical variables (Rater 1 with five categories and Rater 2 with five categories) and a cross-tabulation of 236 outcomes.

Table 6.21. Observed absolute frequencies in two-rater cross-classification

	Rater 1					
Rater 2	1	2	3	4	5	**Total**
1	10	6	4	2	2	24
2	12	20	16	7	2	57
3	1	12	30	20	6	69
4	4	5	10	25	12	56
5	1	3	3	8	15	30
Total	28	46	63	62	37	236

Table 6.21 is a very special contingency table since the row and column variable have the same categories. We refer to such tables as **matched-pairs** tables. Some of the models discussed and fitted in this subsection are specifically designed for matched-pairs tables, others (such as the independence model) apply to any contingency table. The interested reader can find more details on the fitting of generalized linear models to contingency tables in the monographs by Agresti (1990) and Fienberg (1980).

A closer look at the data table suggests that the ratings are probably not independent. The highest counts appear on the diagonal of the table. If Rater 1 assigns an experimental unit to category i, then there seems to be a high likelihood that Rater 2 also assigns the unit to category i. If we reject the notion of independence, for which we need to develop a statistical test, our interest will shift to determining *how* the two rating schemes depend on each other. Is there more agreement between the ratings in the table than is expected by chance? Is there more disagreement in the table than expected by chance? Is there structure to the disagreement; for example, does Rater 1 systematically assign values to higher categories?

To develop a model for independence of the ratings let X denote the column variable, Y the row variable, and N_{ij} the count observed in row i, column j of the contingency table. Let I and J denote the number of rows and columns. In a square table such as Table 6.21 we necessarily have $I = J$. The independence model does not apply to square or matched-pairs tables alone and we discuss it more generally here. The generic layout of the two-way contingency table we are referring to is shown in Table 6.6 (p. 318). Recall that n_{ij} denotes the observed count in row i and column j of the table, $n_{..}$ denotes the total sample size and $n_{i.}$, $n_{.j}$ are the marginal totals. Under the Poisson sampling model where the count in each cell is the realization of a Poisson(λ_{ij}) random variable, the row and column totals are Poisson$(\lambda_{i.} = \sum_{j=1}^{J} \lambda_{ij})$ and Poisson$(\lambda_{.j} = \sum_{i=1}^{I} \lambda_{ij})$ variables, and the total sample size is a Poisson$(\lambda_{..} = \sum_{i,j}^{I,J} \lambda_{ij})$ random variable. The expected cell count λ_{ij} under independence is then related to the marginal expected counts by

$$\lambda_{ij} = \frac{\lambda_{i.}\lambda_{.j}}{\lambda_{..}}.$$

Taking logarithms leads to a generalized linear model with log link for Poisson random variables and linear predictor

$$\ln\{\lambda_{ij}\} = -\ln\{\lambda_{..}\} + \ln\{\lambda_{i.}\} + \ln\{\lambda_{.j}\} = \mu + \alpha_i + \beta_j. \qquad [6.57]$$

We think of α_i and β_j as the (marginal, main) effects of the row and column variables and independence implies the absence of the $(\alpha\beta)_{ij}$ interaction between the two variables.

There exists, of course, a well-known test for independence of categorical variables in contingency tables based on the Chi-square distribution. It is sometimes referred to as Pearson's Chi-square test. If n_{ij} is the observed count in cell i, j and $e_{ij} = n_{i.}n_{.j}/n_{..}$ is the expected count under independence, then

$$X^2 = \sum_{i=1}^{I}\sum_{j=1}^{J} \frac{(n_{ij} - e_{ij})^2}{e_{ij}} \qquad [6.58]$$

follows asymptotically a Chi-squared distribution with $(I-1)(J-1)$ degrees of freedom. For the approximation to hold we need at least 80% of the expected cell counts to be at least 5 and permit at most one expected cell count of 1 (Cochran 1954). This test and a likelihood ratio test for independence can be calculated in `proc freq` of The SAS® System. The statements

```
proc freq data=rating;
   table rater1*rater2 /chisq nocol norow nopercent expected;
   weight number;
run;
```

perform the Chi-square analysis of independence (option `chisq`). The options `nocol`, `nororow`, and `nopercent` suppress the printing of column, row, and cell percentages. The `expected` option requests a printout of the expected frequencies under independence. None of the expected frequencies is less than 1 and exactly 20 frequencies (80%) exceed 5 (Output 6.17).

The Chi-squared approximation holds and the Pearson test statistic is $X^2 = 103.1089$ $(p < 0.0001)$. There is significant disagreement between the observed counts and the counts expected under an independence model. The hypothesis of independence is rejected. The likelihood ratio test with test statistic $\Lambda = 95.3577$ leads to the same conclusion. The calculation of Λ is discussed below. The expected cell counts on the diagonal of the contingency table reflect the degree of chance agreement and are interpreted as follows: if the counts are distributed completely at random to the cells conditional on preserving the marginal row and column totals, one would expect this degree of agreement between the ratings.

The independence model is rarely the best-fitting model for a contingency table and the modeler needs to consider other models incorporating dependence between the row and column variable. This is certainly the case here since the notion of independence has been clearly rejected. This requires the use of statistical procedures that can fit other models than independence, such as `proc genmod`.

Output 6.17.

```
                    The FREQ Procedure

                 Table of rater1 by rater2

rater1      rater2

Frequency|
Expected |       1|       2|       3|       4|       5|  Total
---------+--------+--------+--------+--------+--------+
       1 |     10 |      6 |      4 |      2 |      2 |     24
         | 2.8475 |  4.678 | 6.4068 | 6.3051 | 3.7627 |
---------+--------+--------+--------+--------+--------+
       2 |     12 |     20 |     16 |      7 |      2 |     57
         | 6.7627 |  11.11 | 15.216 | 14.975 | 8.9364 |
---------+--------+--------+--------+--------+--------+
       3 |      1 |     12 |     30 |     20 |      6 |     69
         | 8.1864 | 13.449 | 18.419 | 18.127 | 10.818 |
---------+--------+--------+--------+--------+--------+
       4 |      4 |      5 |     10 |     25 |     12 |     56
         | 6.6441 | 10.915 | 14.949 | 14.712 | 8.7797 |
---------+--------+--------+--------+--------+--------+
       5 |      1 |      3 |      3 |      8 |     15 |     30
         | 3.5593 | 5.8475 | 8.0085 | 7.8814 | 4.7034 |
---------+--------+--------+--------+--------+--------+
Total          28       46       63       62       37     236

          Statistics for Table of rater1 by rater2

    Statistic                     DF        Value       Prob
    ------------------------------------------------------
    Chi-Square                    16     103.1089     <.0001
    Likelihood Ratio Chi-Square   16      95.3577     <.0001
    Mantel-Haenszel Chi-Square     1      59.2510     <.0001
    Phi Coefficient                        0.6610
    Contingency Coefficient                0.5514
    Cramer's V                             0.3305

                     Sample Size = 236
```

We start by fitting the independence model in `proc genmod` to show the equivalence to the Chi-square analysis in `proc freq`.

```
title1 'Independence model for ratings';
ods exclude ParameterEstimates obstats;
proc genmod data=rating;
  class rater1 rater2;
  model number = rater1 rater2  /link=log error=poisson obstats;
  ods output obstats=stats;
run;

title1 'Predicted cell counts under model of independence';
proc freq data=stats;
  table rater1*rater2 / nocol norow nopercent;
  weight pred;
run;
```

The `ods output obstats=stats;` statement saves the observation statistics table which contains the predicted values (variable `pred`). The `proc freq` code following `proc genmod` tabulates the predicted values for the independence model.

In the `Criteria For Assessing Goodness Of Fit` table we find a model deviance of 95.3577 with 16 degrees of freedom (Output 6.18). This is twice the difference between the log likelihood of a full model containing Rater 1 and Rater 2 main effects *and* their interaction and the (reduced) independence model shown here.

The full model (code and output not shown) has a log likelihood of 368.4737 and the deviance of the independence model becomes $2*(368.4737 - 320.7949) = 95.357$ which is of course the likelihood ratio statistic for testing the absence of Rater 1 $\times$ Rater 2 interactions and identical to the likelihood ratio statistic reported by `proc freq` above. Similarly, the Pearson residual Chi-square statistic of 103.1089 in Output 6.18 is identical to the Chi-Square statistic calculated by `proc freq`. That the independence model fits these data poorly is also conveyed by the "overdispersion" factor of 5.96 (or 6.44). Some important effects are unaccounted for, these are the interactions between the ratings.

Output 6.18.

```
                    Independence model for ratings

                         The GENMOD Procedure

                          Model Information
                  Data Set                      WORK.RATING
                  Distribution                      Poisson
                  Link Function                         Log
                  Dependent Variable                 number
                  Observations Used                      25

                        Class Level Information
                   Class         Levels     Values
                   rater1             5     1 2 3 4 5
                   rater2             5     1 2 3 4 5

                 Criteria For Assessing Goodness Of Fit
     Criterion                  DF           Value           Value/DF
     Deviance                   16         95.3577             5.9599
     Scaled Deviance            16         95.3577             5.9599
     Pearson Chi-Square         16        103.1089             6.4443
     Scaled Pearson X2          16        103.1089             6.4443
     Log Likelihood                       320.7949

Algorithm converged.

                   Analysis Of Parameter Estimates

                                   Standard    Wald 95% Confidence        Chi-
 Parameter        DF    Estimate      Error          Limits             Square

 Intercept         1      1.5483     0.2369      1.0840      2.0126      42.71
 rater1       1    1     -0.2231     0.2739     -0.7599      0.3136       0.66
 rater1       2    1      0.6419     0.2256      0.1998      1.0839       8.10
 rater1       3    1      0.8329     0.2187      0.4043      1.2615      14.51
 rater1       4    1      0.6242     0.2263      0.1807      1.0676       7.61
 rater1       5    0      0.0000     0.0000      0.0000      0.0000        .
 rater2       1    1     -0.2787     0.2505     -0.7696      0.2122       1.24
 rater2       2    1      0.2177     0.2208     -0.2151      0.6505       0.97
 rater2       3    1      0.5322     0.2071      0.1263      0.9382       6.60
 rater2       4    1      0.5162     0.2077      0.1091      0.9234       6.17
 rater2       5    0      0.0000     0.0000      0.0000      0.0000        .
 Scale             0      1.0000     0.0000      1.0000      1.0000
```

From the `Analysis of Parameter Estimates` table the predicted values can be constructed (Output 6.19). The predicted count in cell $1, 1$, for example is obtained from the estimated linear predictor

$$\widehat{\eta}_{11} = 1.5483 - 0.2231 - 0.2787 = 1.0465$$

and the inverse link function

$$\widehat{\mathrm{E}}[N_{11}] = e^{1.0465} = 2.8476.$$

Output 6.19.

```
            Predicted cell counts under model of independence

                              The FREQ Procedure

                          Table of rater1 by rater2
    rater1      rater2

    Frequency|1        |2        |3        |4        |5        |  Total
    ---------+---------+---------+---------+---------+---------+
    1        | 2.8475  |  4.678  | 6.4068  | 6.3051  | 3.7627  |     24
    ---------+---------+---------+---------+---------+---------+
    2        | 6.7627  |  11.11  | 15.216  | 14.975  | 8.9364  |     57
    ---------+---------+---------+---------+---------+---------+
    3        | 8.1864  | 13.449  | 18.419  | 18.127  | 10.818  |     69
    ---------+---------+---------+---------+---------+---------+
    4        | 6.6441  | 10.915  | 14.949  | 14.712  | 8.7797  |     56
    ---------+---------+---------+---------+---------+---------+
    5        | 3.5593  | 5.8475  | 8.0085  | 7.8814  | 4.7034  |     30
    ---------+---------+---------+---------+---------+---------+
    Total          28        46        63        62        37       236
```

Simply adding a general interaction term between row and column variable will not solve the problem because the model

$$\ln\{\lambda_{ij}\} = \mu + \alpha_i + \beta_j + (\alpha\beta)_{ij} \qquad [6.59]$$

is **saturated**, that is, it fits the observed data perfectly. Just as in the case of a general two-way layout (e.g., a randomized block design) adding interactions between the factors depletes the degrees of freedom. The saturated model has a deviance of exactly 0 and $n_{ij} = \exp\left\{\widehat{\mu} + \widehat{\alpha}_i + \widehat{\beta}_j + (\widehat{\alpha\beta})_{ij}\right\}$. The deviation from independence must be structured in some way to preserve degrees of freedom. We distinguish three forms of structured interactions:

- **association** that focuses on structured patterns of dependence between X and Y;
- **agreement** that focuses on the counts on the main diagonal;
- **disagreement** that focuses on the counts in off-diagonal cells.

Modeling association requires that the categories of X and Y are ordered; agreement and disagreement can be modeled with nominal and/or ordered categories. It should be noted that cell counts can show strong association but weak agreement, for example, if one rater consistenly assigns outcomes to higher categories than the other rater.

Linear-By-Linear Association for Ordered Categories

The **linear-by-linear association model** replaces the general interaction term in [6.59] by the term $\gamma u_i v_j$ where γ is an association parameter to be estimated and u_i and v_j are scores assigned to the ordered categories. This model, also known as the uniform association model (Goodman 1979a, 1985; Agresti 1990, pp. 263-265), seeks to detect a particular kind of interaction requiring only one degree of freedom for the estimation of γ beyond the model of independence. It is in spirit akin to Tukey's one degree of freedom test for nonadditivity in two-way layouts without replication (Tukey 1949). The term $\exp\{\gamma u_i v_j\}$ can be thought of as a multiplicative factor that increases or decreases the cell counts away from the independence model. To demonstrate the effect of the linear-by-linear association term we use centered scores. Let $u_i = v_j = 1, 2, \cdots, 4$ and define $\overline{u}$ to be the average of the possible X scores and $\overline{v}$ the corresponding average of the Y scores. Define the linear-by-linear association term as $\gamma u_i^* v_j^*$ where $u_i^* = u_i - \overline{u}$, $v_j^* = v_j - \overline{v}$. For an association parameter of $\gamma = 0.05$ the terms $\exp\{\gamma u_i^* v_j^*\}$ are shown in Table 6.22.

Table 6.22. Multiplicative factors $\exp\{\gamma u_i^* v_j^*\}$ for $\gamma = 0.05$ and centered scores (centered scores u_i^* and v_j^* shown in parentheses)

		v_j			
		1	2	3	4
u_i	u_i^*	(– 1.5)	(– 0.5)	(0.5)	(1.5)
1	(– 1.5)	1.12	1.04	0.96	0.89
2	(– 0.5)	1.04	1.01	0.98	0.96
3	(0.5)	0.96	0.98	1.01	1.03
4	(1.5)	0.89	0.96	1.03	1.12

The multiplicative terms are symmetric about the center of the table and increase along the diagonal toward the corners of the table. At the same time the expected counts are decremented relative to an independence model toward the upper right and lower left corners of the table. The linear-by-linear association model assumes that high (low) values of X pair more frequently with high (low) values of Y than is expected under independence. At the same time high (low) values of X pair less frequently with low (high) values of Y. In cases where it is more difficult to assign outcomes to categories in the middle of the scale than to extreme categories the linear-by-linear association model will tend to fit the data well. For the model fit and the predicted values it does not matter whether the scores are centered or not. Because of the convenient interpretation in terms of multiplicative effects as shown in Table 6.22 we prefer to work with centered scores.

The linear-by-linear association model with centered scores is fit to the rater agreement data using `proc genmod` with the statements

```
title3 'Uniform association model for ratings';
data rating; set rating;
   sc1_centered = rater1-3; sc2_centered = rater2-3;
run;
ods exclude obstats;
proc genmod data=rating;
   class rater1 rater2;
   model number = rater1 rater2 sc1_centered*sc2_centered /
                        link=log error=poisson type3 obstats;
```

```
  ods output obstats=unifassoc;
run;
proc freq data=unifassoc;
  table rater1*rater2 / nocol norow nopercent;
  weight pred;
run;
```

The deviance of this model is much improved over the independence model (Output 6.20). The estimate for the association parameter is $\widehat{\gamma} = 0.4455$ and the likeihood-ratio test for $H_0{:}\gamma = 0$ shows that the addition of the linear-by-linear association significantly improves the model. The likelihood ratio Chi-square statistic of 67.45 equals the difference between the two model deviances $(95.35 - 27.90)$.

Output 6.20.

```
                Uniform association model for ratings

                        The GENMOD Procedure

                         Model Information

                Data Set                     WORK.RATING
                Distribution                     Poisson
                Link Function                        Log
                Dependent Variable                number
                Observations Used                     25

                      Class Level Information

                 Class         Levels     Values
                 rater1             5     1 2 3 4 5
                 rater2             5     1 2 3 4 5

                Criteria For Assessing Goodness Of Fit

       Criterion                  DF            Value          Value/DF
       Deviance                   15          27.9098            1.8607
       Scaled Deviance            15          27.9098            1.8607
       Pearson Chi-Square         15          32.3510            2.1567
       Scaled Pearson X2          15          32.3510            2.1567
       Log Likelihood                        354.5188

Algorithm converged.

                   Analysis Of Parameter Estimates

                                       Standard   Wald 95% Confidence
Parameter             DF    Estimate      Error          Limits
Intercept              1      0.6987     0.2878     0.1347      1.2628
rater1         1       1     -0.0086     0.3022    -0.6009      0.5837
rater1         2       1      1.1680     0.2676     0.6435      1.6924
rater1         3       1      1.4244     0.2621     0.9106      1.9382
rater1         4       1      1.0282     0.2452     0.5475      1.5088
rater1         5       0      0.0000     0.0000     0.0000      0.0000
rater2         1       1     -0.3213     0.2782    -0.8665      0.2239
rater2         2       1      0.5057     0.2477     0.0203      0.9912
rater2         3       1      0.9464     0.2369     0.4820      1.4108
rater2         4       1      0.8313     0.2228     0.3947      1.2680
rater2         5       0      0.0000     0.0000     0.0000      0.0000
sc1_cente*sc2_center   1      0.4455     0.0647     0.3187      0.5722
Scale                  0      1.0000     0.0000     1.0000      1.0000

NOTE: The scale parameter was held fixed.
```

Output 6.20 (continued).

```
                     LR Statistics For Type 3 Analysis

                                               Chi-
        Source                         DF     Square     Pr > ChiSq
        rater1                          4      57.53         <.0001
        rater2                          4      39.18         <.0001
        sc1_cente*sc2_center            1      67.45         <.0001


     Predicted cell counts for linear-by-linear association model

                          The FREQ Procedure

                      Table of rater1 by rater2

rater1      rater2

Frequency|1        |2        |3        |4        |5        |  Total
---------+---------+---------+---------+---------+---------+
1        | 8.5899  | 8.0587  | 5.1371  | 1.8786  | 0.3356  |     24
---------+---------+---------+---------+---------+---------+
2        | 11.431  | 16.742  | 16.662  | 9.5125  | 2.6533  |     57
---------+---------+---------+---------+---------+---------+
3        | 6.0607  | 13.858  | 21.532  | 19.192  | 8.3574  |     69
---------+---------+---------+---------+---------+---------+
4        | 1.6732  | 5.9728  | 14.488  |  20.16  | 13.706  |     56
---------+---------+---------+---------+---------+---------+
5        | 0.2455  | 1.3683  | 5.1816  | 11.257  | 11.948  |     30
---------+---------+---------+---------+---------+---------+
Total          28        46        63        62        37       236
```

Comparing the predicted cell counts to those of the independence model, it is seen how the counts increase in the upper left and lower right corner of the table and decrease toward the upper right and lower left corners.

The fit of the linear-by-linear association model is dramatically improved over the independence model at the cost of only one additional degree of freedom. The model fit is not satisfactory, however. The p-value for the model deviance of $\Pr(\chi^2_{15} \geq 27.90) = 0.022$ indicates that a significant discrepancy between model and data remains. The interactions between Rater 1 and Rater 2 category assignments must be structured further.

Before proceeding with modeling structured interaction as agreement we need to point out that the linear-by-linear association model requires that scores be assigned to the ordered categories. This introduces a subjective element into the analysis, different modelers may assign different sets scores. Log-**multiplicative** models with predictor $\mu + \alpha_i + \beta_j + \gamma\phi_i\omega_j$ have been developed where the category scores ϕ_i and ω_j are themselves parameters to be estimated. For more information about these log-multiplicative models see Becker (1989, 1990a, 1990b) and Goodman (1979b).

Modeling Agreement and Disagreement

Modeling agreement between ratings focuses on the diagonal cells of the table and parameterizes the beyond-chance agreement in the data. It does not require ordered categories as the association models do. The simplest agreement model adds a single parameter δ that models excess counts on the diagonal. Let z_{ij} be a indicator variable such that

$$z_{ij} = \begin{cases} 1 & i = j \\ 0 & \text{otherwise.} \end{cases}$$

The **homogeneous agreement model** is defined as

$$\ln\{\lambda_{ij}\} = \mu + \alpha_i + \beta_j + \delta z_{ij}.$$

Interactions are modeled with a single degree of freedom term. A positive δ indicates that more counts fall on the main diagonal than would be expected under a random assignment of counts to the table (given the marginal totals). The agreement parameter δ can also be made to vary with categories. This **nonhomogeneous agreement model** is defined as

$$\ln\{\lambda_{ij}\} = \mu + \alpha_i + \beta_j + \delta_i z_{ij}.$$

The separate agreement parameters δ_i will saturate the model on the main diagonal, that is predicted and observed counts will agree perfectly for the diagonal cells. In `proc genmod` the homogeneous and nonhomogeneous agreement models are fitted easily by defining an indicator variable for the diagonal (output not shown).

```
data ratings; set ratings; diag = (rater1 = rater2); run;

title1 'Homogeneous Agreement model for ratings';
ods exclude  obstats;
proc genmod data=rating;
  class rater1 rater2;
  model number = rater1 rater2 diag /link=log error=poisson type3;
run;

title1 'Nonhomogeneous Agreement model for ratings';
proc genmod data=rating;
  class rater1 rater2;
  model number = rater1 rater2 diag*rater1 /link=log error=poisson type3;
run;
```

Disagreement models place emphasis on cells off the main diagonal. For example, the model

$$\ln\{\lambda_{ij}\} = \mu + \alpha_i + \beta_j + \delta z_{ij}, \quad z_{ij} = \begin{cases} 1 & |i - j| = 1 \\ 0 & \text{otherwise} \end{cases}$$

adds an additional parameter δ to all cells adjacent to the main diagonal and the model

$$\ln\{\lambda_{ij}\} = \mu + \alpha_i + \beta_j + \delta_+ z_{ij} + \delta_- c_{ij}, \quad z_{ij} = \begin{cases} 1 & i - j = -1 \\ 0 & \text{otherwise} \end{cases}, \quad c_{ij} = \begin{cases} 1 & i - j = 1 \\ 0 & \text{otherwise} \end{cases}$$

adds two separate parameters, one for the first band above the main diagonal (δ_+) and one for the first band below the main diagonal (δ_-). For this and other disagreement structures see Tanner and Young (1985).

When categories are ordered, agreement and association parameters can be combined. A linear-by-linear association model with homogeneous agreement, for example, becomes

$$\ln\{\lambda_{ij}\} = \mu + \alpha_i + \beta_j + \gamma u_i v_j + \delta z_{ij},\ z_{ij} = \begin{cases} 1 & i = j \\ 0 & \text{otherwise.} \end{cases}$$

Table 6.23 lists the deviances and p-values for various models fit to the data in Table 6.21.

Table 6.23. Model deviances for various log-linear models fit to data in Table 6.21

Model	Linear-by-linear assoc.	Homog. agreemt.	Non-homog. agreemt.	Deviance df	D	D/df	p-value
A				16	95.36†	5.96	< 0.0001
B	✓			15	27.90	1.86	0.0222
C		✓		15	43.99	2.93	0.0001
D			✓	11	36.85	3.35	0.0001
E	✓	✓		14	16.61	1.18	0.2776
F	✓		✓	10	15.65	1.56	0.1101

† The independence model

The best-fitting model is the combination of a linear-by-linear association term and a homogeneous agreement term ($p = 0.2776$). The agreement between the two raters is beyond-chance and beyond what the linear association term allows. At the 5% significance level the model with association and nonhomogeneous agreement (F) also does not exhibit a significant deviance. The loss of four degrees of freedom for the extra agreement parameters relative to E is not offset by a sufficient decrease in the deviance, however. The more parsimonious model E is preferred. This model also leads to the deviance/df ratio which is closest to one. It should be noted that the null hypothesis underlying the goodness-of-fit test for model E, for example, is that model E fits the data. A failure to reject this hypothesis is not evidence that model E is the correct one. It simply cannot be ruled out as a possible data-generating mechanism. The p-values for the deviance test should be sufficiently large to rule out an error of the second kind before declaring a model as *the one*.

6.7.7 Modeling the Sample Variance of Scab Infection

In Example 6.4 (p. 310) a data set is displayed showing the variances in deoxynivalenol (DON) concentration among probe samples from truckloads of wheat kernels as a function of the average DON concentration on the truck. The trend in Figure 6.4 suggests that the variability and the mean concentration are not unrelated. If one were to make a sample size recommendation as to how many probe samples are necessary to be within $\pm\Delta$ of the true concentration with a certain level of confidence, an estimate of the variance sensitive to the level of kernel contamination is needed.

A model relating the probe-to-probe variance to the mean concentration is a first step in this direction. As discussed in §6.2.1 it is usually not reasonable to assume that sample variances are Gaussian-distributed. If $Y_1, \cdots, Y_n$ are a random sample from a Gaussian distribution with variance σ^2, then $(n-1)S^2/\sigma^2$ is a Chi-squared random variable on $n-1$ degrees of freedom. Here, S^2 denotes the sample variance,

$$S^2 = \frac{1}{n-1}\sum_{i=1}^{n}\left(Y_i - \overline{Y}\right)^2.$$

Consequently, S^2 is distributed as a Gamma random variable with mean $\mu = \sigma^2$ and scale parameter $\alpha = (n-1)/2$ (in the parameterization [6.7], p. 308). For $\sigma^2 = 2$ and $n = 7$, for example, the skewness of the probability density function of S^2 is apparent in Figure 6.19.

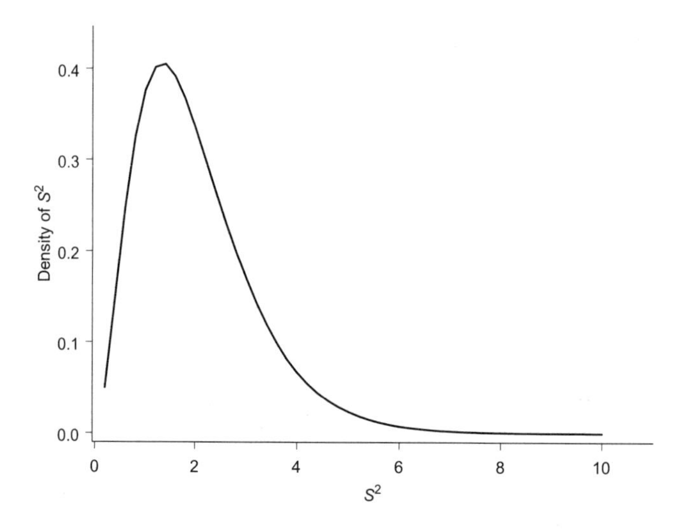

Figure 6.19. Probability density function of S^2 for $\sigma^2 = 2$, $n = 7$ when $(n-1)S^2/\sigma^2$ is distributed as χ^2_6.

Modeling S^2 as a Gamma random variable with $\alpha = (n-1)/2$, we need to find a model for the mean μ. Figure 6.20 shows the logarithm of the truck sample variances plotted against the (natural) logarithm of the deoxynivalenol concentration. From this plot, a model relating the log variance against the log concentration could be considered. That is, we propose to model

$$\ln\{\sigma_i^2\} = \beta_0 + \beta_1 \ln\{x_i\},$$

where x is the mean DON concentration on the ith truck, and $\sigma_i^2 = \exp\{\beta_0 + \beta_1 \ln\{x_i\}\}$ is the mean of a Gamma random variable with scale parameter $\alpha = (n-1)/2 = 4.5$ (since each sample variance is based on $n = 10$ probe samples per truck). Notice that the mean can also be expressed in the form

$$\sigma_i^2 = \exp\{\beta_0 + \beta_1 \ln\{x_i\}\} = e^{\beta_0} x_i^{\beta_1} = \alpha_0 x_i^{\beta_1}.$$

The `proc genmod` statements to fit this model (Output 6.21) are

```
proc genmod data=don;
  model donvar = logmean / link=log dist=gamma scale=4.5 noscale;
run;
```

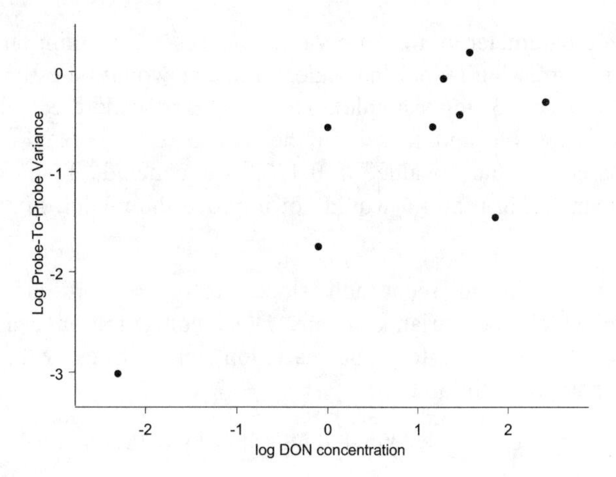

Figure 6.20. Logarithm of probe-to-probe variance against log of average DON concentration.

The `scale=4.5` option sets the Gamma scale parameter α to $(n-1)/2 = 4.5$ and the `noscale` option prevents the scale parameter from being fit iteratively by maximum likelihood. The combination of the two options fixes $\alpha = 4.5$ throughout the estimation process so that only the parameters β_0 and β_1 are being estimated.

Output 6.21.

```
                        The GENMOD Procedure

                         Model Information
                Data Set                      WORK.FITTHIS
                Distribution                         Gamma
                Link Function                          Log
                Dependent Variable                  donvar
                Observations Used                        9
                Missing Values                           1

                 Criteria For Assessing Goodness Of Fit
       Criterion                 DF          Value          Value/DF
       Deviance                   7         2.7712            0.3959
       Scaled Deviance            7        12.4702            1.7815
       Pearson Chi-Square         7         2.4168            0.3453
       Scaled Pearson X2          7        10.8755            1.5536
       Log Likelihood                       0.0675
Algorithm converged.

                    Analysis Of Parameter Estimates

                         Standard       Wald 95%            Chi-
Parameter  DF  Estimate    Error    Confidence Limits     Square   Pr > ChiSq

Intercept   1   -1.2065   0.1938   -1.5864    -0.8266      38.74       <.0001
logmean     1    0.5832   0.1394    0.3100     0.8564      17.50       <.0001
Scale       0    4.5000   0.0000    4.5000     4.5000
NOTE: The scale parameter was held fixed.

                     Lagrange Multiplier Statistics
                  Parameter      Chi-Square     Pr > ChiSq
                  Scale              0.5098         0.4752
```

Fixing the scale parameter at a certain value imposes a constraint on the model since in the regular Gamma regression model the scale parameter would be estimated (see the yield-density application in §6.7.3, for example). The `genmod` procedure calculates a test whether this constraint is reasonable and lists it in the `Lagrange Multiplier Statistics` table (Output 6.21). Based on the p-value of 0.4752 we conclude that estimating the scale parameter rather than fixing it at 4.5 would not improve the model. Fixing the parameter is reasonable.

The estimates for the intercept and slope are $\widehat{\beta}_0 = -1.2065$ and $\widehat{\beta}_1 = 0.5832$, respectively, from which the variance at any DON concentration can be estimated. For example, we expect the probe-to-probe variation on a truck with a deoxynivalenol concentration of 5 parts per million to be

$$\widehat{\sigma}^2 = \exp\{-1.2065 + 0.5832\ln\{5\}\} = 0.765 \text{ ppm}^2.$$

Figure 6.21 displays the predicted variances and approximate 95% confidence bounds for the predicted values. The model does not fit the data perfectly. The DON variance on the truck with an average concentration of 6.4 ppm is considerably off the trend. The model fits rather well for concentrations up to 5 ppm, however. It is noteworthy how the confidence bands widen as the DON concentration increases. Two effects are causing this. First, the probe-to-probe variances are not homoscedastic; according to the Gamma regression model

$$\operatorname{Var}\left[S_i^2\right] = \frac{1}{4.5}\exp\{2\beta_0 + 2\beta_1\ln(x_i)\}$$

increases sharply with x. Second, the data are very sparse for larger concentrations.

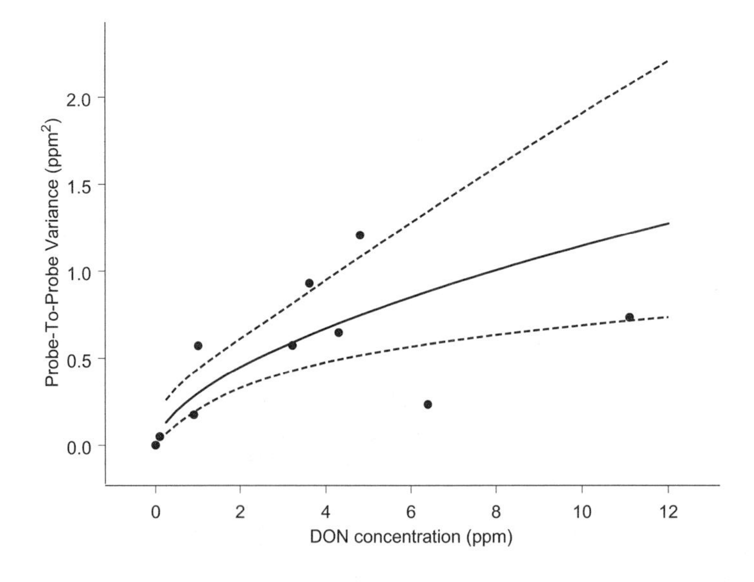

Figure 6.21. Predicted values for probe-to-probe variance from Gamma regression (solid line). Dashed lines are asymptotic 95% confidence intervals for the mean variance at a given DON concentration.

6.7.8 A Poisson/Gamma Mixing Model for Overdispersed Poppy Counts

Mead et al. (1993, p. 144) provide the data in Table 6.24 from a randomized complete block design with six treatments in four blocks in which the response variable was the poppy count on an experimental unit.

Table 6.24. Poppy count data from Mead, Curnow, and Hasted (1993, p. 144)†

Treatment	Block 1	Block 2	Block 3	Block 4
A	538	422	377	315
B	438	442	319	380
C	77	61	157	52
D	115	57	100	45
E	17	31	87	16
F	18	26	77	20

†Used with permission.

Since these are count data, one could assume that the responses are Poisson-distributed and notice that for mean counts greater than $15 - 20$, the Poisson distribution is closely approximated by a Gaussian distribution (Figure 6.22). The temptation to analyze these data by standard analysis of variance assuming Gaussian errors is thus understandable. From Figure 6.22 it can be inferred, however, that the variance of the Gaussian distribution approximating the Poisson mass function is linked to the mean. The distribution approximating the Poisson when the average counts are small will have a smaller variance than the approximating distribution for large counts, thereby violating the homogeneous variance assumption of the standard analysis of variance.

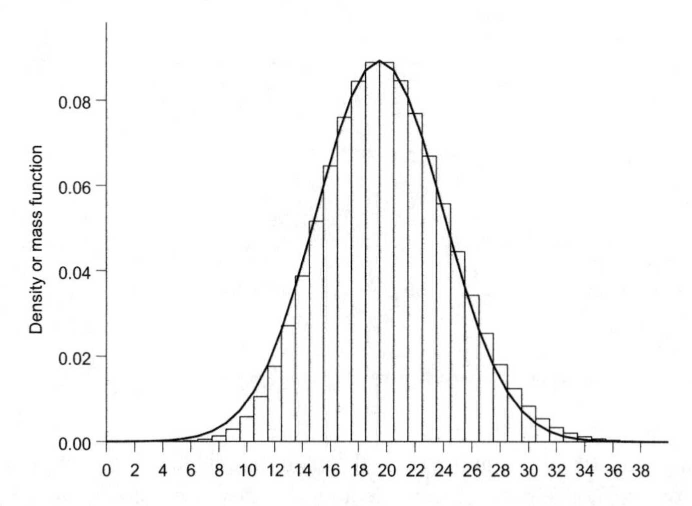

Figure 6.22. Poisson(20) probability mass function (bars) overlaid with probability density function of a Gaussian($\mu = 20, \sigma^2 = 20$) random variable.

Mead et al. (1993, p. 145) highlight this problem by examining the mean square error estimate from the analysis of variance, which is $\widehat{\sigma}^2 = 2,653$. An off-the-cuff confidence interval for the mean poppy count for treatment F leads to

$$\overline{y}_F \pm 2\text{ese}(\overline{y}_F) = 35.25 \pm 2\sqrt{\frac{2,653}{4}} = 35.25 \pm 51.50 = [-16.25, 83.75].$$

Based on this confidence interval one would not reject the idea that the mean poppy count for treatment F could be -10, say. This is a nonsensical result. The variability of the counts is smaller for treatments with small average counts and larger for treatments with large average counts. The analysis of variance mean square error is a pooled estimator of the residual variability being too high for treatments such as F and too small for treatments such as A.

An analysis of these data based on a generalized linear model with Poisson-distributed outcomes and a linear predictor that incorporates block and treatment effects is more reasonable. Using the canonical log link, the following `proc genmod` statements fit this GLM.

```
ods exclude ParameterEstimates;
proc genmod data=poppies;
  class block treatment;
  model count = block treatment / link=log dist=Poisson;
run;
```

Output 6.22.

```
                    The GENMOD Procedure

                     Model Information

             Data Set                      WORK.POPPIES
             Distribution                       Poisson
             Link Function                          Log
             Dependent Variable                   count
             Observations Used                       24

                  Class Level Information

           Class             Levels    Values
           block                  4    1 2 3 4
           treatment              6    A B C D E F

           Criteria For Assessing Goodness Of Fit

  Criterion                  DF           Value        Value/DF
  Deviance                   15        256.2998         17.0867
  Scaled Deviance            15        256.2998         17.0867
  Pearson Chi-Square         15        273.0908         18.2061
  Scaled Pearson X2          15        273.0908         18.2061
  Log Likelihood                     19205.2442

Algorithm converged.
```

The troubling statistic in Output 6.22 is the model deviance of 256.29 compared to its degrees of freedom (15). The data are considerably overdispersed relative to a Poisson distribution. A value of 17.08 should give the modeler pause (since the target ratio is 1). A possible reason for the considerable overdispersion could be an incorrect linear predictor. If these data stem from an experimental design with proper blocking and randomization procedure, no other effects apart from block and treatment effects should be necessary. Ruling out an incorrect linear predictor, (positive) correlations among poppy counts could be the cause of the overdispersion. Mead et al. (1993) explain the overdispersion by the fact that whenever there is one poppy in an experimental unit, there are almost always several, hence poppies are clustered and not distributed completely at random within (and possibly across) experimental units.

The overdispersion problem can be fixed if one assumes that the counts have mean and variance

$$\begin{aligned}\mathrm{E}[Y] &= \lambda \\ \mathrm{Var}[Y] &= \phi\lambda,\end{aligned}$$

where ϕ is a multiplicative overdispersion parameter. In `proc genmod` such an overdispersion parameter can be estimated from either the deviance or Pearson residuals by adding the `dscale` or `pscale` options to the model statement. The statements

```
proc genmod data=poppies;
  class block treatment;
  model count = block treatment / link=log dist=Poisson dscale;
run;
```

fit the same model as above but estimate an extra overdispersion parameter as

$$\widehat{\psi} = \frac{256.29}{15} = 17.08$$

and report its square root as `Scale` in the table of parameter estimates. This approach to accommodating overdispersion has several disadvantages in our opinion.

- It adds a parameter to the variance of the response which is not part of the Poisson distribution. The variance $\mathrm{Var}[Y] = \phi\lambda$ is no longer that of a Poisson random variable with mean λ. The analysis is thus no longer a maximum likelihood analysis for a Poisson model. It is a quasi-likelihood analysis in the sense of McCullagh (1983) and McCullagh and Nelder (1989).

- The parameter estimates are not affected by the inclusion of the extra overdispersion parameter ϕ. Only their standard errors are. For the data set and model considered here, the standard errors of all parameter estimates will be $\sqrt{17.08} = 4.133$ times larger in the model with $\mathrm{Var}[Y] = \phi\lambda$. The predicted mean counts will remain the same, however.

- The addition of an overdispersion parameter does not induce correlations among the outcomes. If data are overdispersed because positive correlations among the observations are ignored, a multiplicative overdispersion parameter is the wrong remedy.

Inducing correlations in random variables is relatively simple if one considers a parameter of the response distribution to be a random variable rather than a constant. For example, in a Binomial(n, π) population one can either assume that the binomial sample size n or the success probability π are random. Reasonable distributions for these parameters could be a Poisson or truncated Poisson for n or the family of Beta distributions for π. In §A6.8.6 we provide details on how these mixing models translate into marginal distributions where the counts are correlated. More importantly, it is easy to show that the marginal variability of a random variable reckoned over the possible values of a randomly varying parameter is larger than the conditional variability one obtains if the parameter is fixed at a certain value.

This mixing approach has particular appeal if the marginal distribution of the response variable remains a member of the exponential family. In this application we can assume that the poppy counts for treatment i in block j are distributed as Poisson(λ_{ij}), but that λ_{ij} is a random variable. Since the mean counts are non-negative we choose a probability distribution for λ_{ij} that has nonzero density on the positive real line. If the mean of a Poisson(λ) variable is distributed as Gamma(μ, α), it turns out that the marginal distribution of the count follows the Negative Binomial distribution which is a member of the exponential family (Table 6.2, p. 305). The details on how the marginal Negative Binomial distribution is derived are provided in §A6.8.6.

Since Release 8.0 of The SAS® System, Negative Binomial outcomes can be modeled with `proc genmod` using the `dist=negbin` option. The canonical link of this distribution is the log link. Unfortunately, there is a bug in Release 8.0 related to the maximum likelihood estimation of the Negative Binomial parameter k (see Table 6.2). This bug has been fixed in Release 8.2 of The SAS® System. Starting with Release 8.1, the Negative Binomial distribution is also an option of `proc nlmixed`. Using Release 8.0 of The SAS® System, we fit count data with a Negative Binomial distribution in `proc nlmixed` to circumvent the problem with `proc genmod`. To this end the log likelihood must be coded with SAS® programming statements. For the poppy count data these statements are as follows.

```
proc nlmixed data=poppies df=14;
   parameters intcpt=3.4
              bl1=0.3 bl2=0.3 bl3=0.3
              tA=1.5 tB=1.5 tC=1.5 tD=1.5 tE=1.5 k=1;
   if block=1 then linp = intcpt + bl1;
   else if block=2 then linp = intcpt + bl2;
   else if block=3 then linp = intcpt + bl3;
   else if block=4 then linp = intcpt;

   if treatment = 'A' then linp = linp + tA;
   else if treatment = 'B' then linp = linp + tB;
   else if treatment = 'C' then linp = linp + tC;
   else if treatment = 'D' then linp = linp + tD;
   else if treatment = 'E' then linp = linp + tE;
   else if treatment = 'F' then linp = linp;

   b = exp(linp)/k;
   ll = lgamma(count+k) - lgamma(k) - lgamma(count + 1)   +
        k*log(1/(b+1)) + count*log(b/(b+1));
   model count ~ general(ll);
run;
```

The `parameters` statement assigns starting values to the model parameters. Although there are four block and six treatment effects, only three block and five treatment effects need to be coded due to the sum-to-zero constraints on block and treatment effects. Because `proc nlmixed` determines residual degrees of freedom by a different method than `proc genmod` we added the `df=14` option to the `proc nlmixed` statement. This will ensure the same degrees of freedom as in the Poisson analysis minus one degree of freedom for the additional parameter of the Negative Binomial that determines the degree of overdispersion relative to the Poisson model. The several lines of `if ... then ...; else ...;` statements that follow the `parameters` statement set up the linear predictor as a function of block effects (`bl1`$\cdots$`bl3`) and treatment effects (`tA`$\cdots$`tE`). The statements

```
b = exp(linp)/k;
ll = lgamma(count+k) - lgamma(k) - lgamma(count + 1)  +
     k*log(1/(b+1)) + count*log(b/(b+1));
```

code the Negative Binomial log likelihood. We have chosen a log likelihood based on a parameterization in A6.8.6. The `model` statement finally instructs `proc nlmixed` to perform a maximum likelihood analysis for the variable `count` where the log likelihood is determined by the variable `ll`. This analysis results in maximum likelihood estimates for Negative Binomial responses with linear predictor

$$\eta_{ij} = \mu + \tau_i + \rho_j.$$

The parameterization used in this `proc nlmixed` code expresses the mean response as simply $\exp\{\eta_{ij}\}$ for ease of comparison with the Poisson analysis. The abridged `nlmixed` output follows.

Output 6.23 (abridged).

```
                      The NLMIXED Procedure

                         Specifications
Description                                      Value
Data Set                                         WORK.POPPIES
Dependent Variable                               count
Distribution for Dependent Variable              General
Optimization Technique                           Dual Quasi-Newton
Integration Method                               None

          NOTE: GCONV convergence criterion satisfied.

                         Fit Statistics
            Description                             Value
            -2 Log Likelihood                       233.1
            AIC (smaller is better)                 253.1
            BIC (smaller is better)                 264.9
            Log Likelihood                         -116.5
            AIC (larger is better)                 -126.5
            BIC (larger is better)                 -132.4

                      Parameter Estimates
                           Standard
Parameter   Estimate       Error     DF   t Value   Pr > |t|

intcpt        3.0433      0.2121     14     14.35     <.0001
bl1           0.3858      0.1856     14      2.08     0.0565
bl2           0.2672      0.1864     14      1.43     0.1737
bl3           0.8431      0.1902     14      4.43     0.0006
tA            2.6304      0.2322     14     11.33     <.0001
tB            2.6185      0.2326     14     11.26     <.0001
tC            0.9618      0.2347     14      4.10     0.0011
tD            0.9173      0.2371     14      3.87     0.0017
tE           0.08057      0.2429     14      0.33     0.7450
k            11.4354      3.7499     14      3.05     0.0087
```

Although it is simple to predict the performance of a treatment in a particular block based on this analysis, this is usually not the final analytical goal. For example, the predicted count for treatment A in block 2 is $\exp\{3.0433 + 0.2672 + 2.6304\} = 380.28$. We are more interested in a comparison of the treatments averaged across the blocks. Before embarking on treatment comparisons we check the significance of the treatment effects. A reduced model with block effects only is fit with the statements

```
proc nlmixed data=poppies df=19;
   parameters intcpt=3.4 bl1=0.3 bl2=0.3 bl3=0.3
                      k=1;
   if block=1 then linp = intcpt + bl1;
   else if block=2 then linp = intcpt + bl2;
   else if block=3 then linp = intcpt + bl3;
   else if block=4 then linp = intcpt;
   b = exp(linp)/k;
   ll = lgamma(count+k) - lgamma(k) - lgamma(count + 1)
           + k*log(1/(b+1)) + count*log(b/(b+1));
   model count ~ general(ll);
run;
```

Minus twice the log likelihood for this model is 295.4 (output not shown) and the likelihood ratio test statistic to test for equal average poppy counts among the treatments is

$$\Lambda = 295.4 - 233.1 = 62.3.$$

The p-value for this statistic is $\Pr(\chi^2_5 > 62.3) < 0.0001$. Hence not all treatments have the same average poppy count and we can proceed with pairwise comparisons based on treatment averages. These averages can be calculated and compared with the `estimate` statement of the `nlmixed` procedure. For example, to compare the predicted counts for treatments A and B, A and C, and D and E we add the statements

```
estimate 'count(A)-count(B)' exp(intcpt+0.25*bl1+0.25*bl2+0.25*bl3+tA) -
                             exp(intcpt+0.25*bl1+0.25*bl2+0.25*bl3+tB);

estimate 'count(A)-count(C)' exp(intcpt+0.25*bl1+0.25*bl2+0.25*bl3+tA) -
                             exp(intcpt+0.25*bl1+0.25*bl2+0.25*bl3+tC);

estimate 'count(D)-count(E)' exp(intcpt+0.25*bl1+0.25*bl2+0.25*bl3+tD) -
                             exp(intcpt+0.25*bl1+0.25*bl2+0.25*bl3+tE);
```

to the `nlmixed` code. The linear predictors for each treatment take averages over the block effects prior to exponentiation. The table added by these three statements to the `proc nlmixed` output shown earlier follows.

Output 6.24.

Additional Estimates

Label	Estimate	Standard Error	DF	t Value	Pr > |t|
count(A)-count(B)	5.0070	89.3979	14	0.06	0.9561
count(A)-count(C)	343.39	65.2337	14	5.26	0.0001
count(D)-count(E)	43.2518	13.5006	14	3.20	0.0064

We conclude that there is no difference between treatments A and B ($p = 0.9561$), but that there are significant differences in average poppy counts between treatments A and C ($p < 0.0001$) and D and E ($p = 0.0064$). Notice that the estimate of 43.2518 for the comparison of predicted counts for treatments D and E is close to the difference in average counts of 44.0 in Table 6.24.

Chapter 7

Linear Mixed Models for Clustered Data

"The new methods occupy an altogether higher plane than that in which ordinary statistics and simple averages move and have their being. Unfortunately, the ideas of which they treat, and still more, the many technical phrases employed in them, are as yet unfamiliar. The arithmetic they require is laborious, and the mathematical investigations on which the arithmetic rests are difficult reading even for experts... This new departure in science makes its appearance under conditions that are unfavourable to its speedy recognition, and those who labour in it must abide for some time in patience before they can receive sympathy from the outside world." Sir Francis Galton.

7.1 Introduction

Box 7.1 Linear Mixed Models

- **A linear mixed effects model contains fixed and random effects and is linear in these effects. Models for subsampling designs or split-plot-type models are mixed models.**
- **Multifactor models where the levels of some factors are predetermined while the levels of other factors are chosen at random are mixed models.**
- **Models in which regression coefficients vary randomly between groups of observations are mixed models.**

A distinction was made in §1.7.5 between fixed, random, and mixed effects models based on the number of random variables and fixed effects involved in the statistical model. A mixed effects model — or mixed model for short — contains fixed effects as well as at least two random variables (one of which is the obligatory model error). Mixed models arise quite frequently in designed experiments. In a completely randomized design with subsampling, for example, t treatments are assigned at random to rt experimental units. A random subsample of n observations is then drawn from every experimental unit. This design is practical if an experimental unit is too large to be measured in its entirety, for example, a field plot contains twelve rows of a particular crop but only three rows per plot can be measured and analyzed. Soil samples are often randomly divided into subsamples prior to laboratory analysis. The statistical model for such a design can be written as

$$\begin{aligned} Y_{ijk} &= \mu + \tau_i + e_{ij} + d_{ijk} \\ i &= 1, \ldots, t;\ j = 1, \ldots, r;\ k = 1, \ldots, n, \end{aligned} \qquad [7.1]$$

where $e_{ijk} \sim (0, \sigma^2), d_{ijk} \sim (0, \sigma_d^2)$ are zero mean random variables representing the experimental and observational errors, respectively. If the treatment effects τ_i are fixed, i.e., the levels of the treatment factor were predetermined and not chosen at random, this is a mixed model.

In the past fifteen years, mixed linear models have risen to great importance in statistical modeling because of their tremendous flexibility. As will be demonstrated shortly, mixed models are more general than standard regression and classification models and contain the latter. In the analysis of designed experiments mixed models have been in use for a long time. They arise naturally through the process of randomization as shown in the introductory example. In agricultural experiments, the most important traditional mixed models are those for subsampling and split-plot type designs (Example 7.1).

Data from subsampling and split-plot designs are clustered structures (§2.4.1). In the former, experimental units are clusters for the subsamples and in the latter whole-plots are clusters for the subplot treatments. In general, mixed models arise very naturally in situations where data are clustered or hierarchically organized. This is by no means restricted to designed experiments with splits or subsampling. Longitudinal studies and repeated measures experiments where observations are gathered repeatedly on subjects or experimental units over time also give rise to clustered data. Here, however, the selections of units within a cluster is not random but ordered along some metric. This metric is usually temporal, but this is not necessary. Lysimeter measurements made at three depths of the same soil profile are also longitudinal in nature with measurements ordered by depth (spatial metric).

Example 7.1. An experiment is planned to investigate different agricultural management strategies and cropping systems. The management alternatives chosen are an organic strategy without use of pesticides (M1), an integrated strategy with low pesticide input (M2), and an integrated strategy with high pesticide input (M3). The crops selected for the experiment are corn, soybeans, and wheat. Since application of a management strategy requires large field units, which can accommodate several different crops, it was decided to assign the management strategies in a randomized complete block design with four blocks (replicates I to IV). Each replicate consists of three fields made homogeneous as far as possible before the experiment by grouping fields according to soil parameters sampled previously. Each field is then subdivided into three plots, and the crops are assigned at random to the fields. This process is repeated in each field. The layout of the design for management types is shown in Figure 7.1. The linear model is $Y_{ij} = \mu + \rho_j + \tau_i + e_{ij}$ $(i = 1, ..., 3; j = 1, .., 4)$, where ρ_j is the effect associated with replicate j, τ_i is the effect of the ith treatment, and $e_{ij} \sim (0, \sigma^2)$ is the experimental error associated with the fields. Figure 7.2 shows the assignment of the three crop types within each field after randomization.

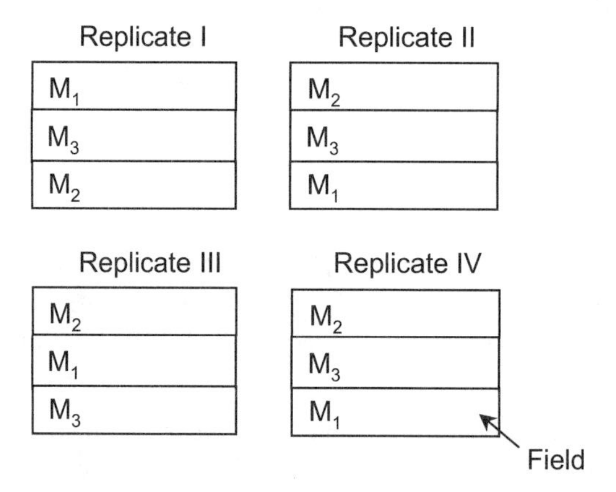

Figure 7.1. Randomized complete block design with four blocks (= replicates) for management strategies M1 to M3.

This experiment involves two separate stages of randomization. The management types are assigned at random to the fields and independently thereof the crops are assigned to the plots within a field. The variability associated with the fields should be independent from the variability associated with the plots. Ignoring the management types and focusing on the crop alone, the design is a randomized complete block with $3*4 = 12$ blocks of size 3 and the model is

$$Y_{ijk} = \mu + \phi_{ij} + \alpha_k + d_{ijk}, \qquad [7.2]$$

$k = 1, .., 3$, where ϕ_{ij} is the effect of the ij^{th} block, α_k is the effect of the kth crop type, and d_{ijk} is the experimental error associated with the plots.

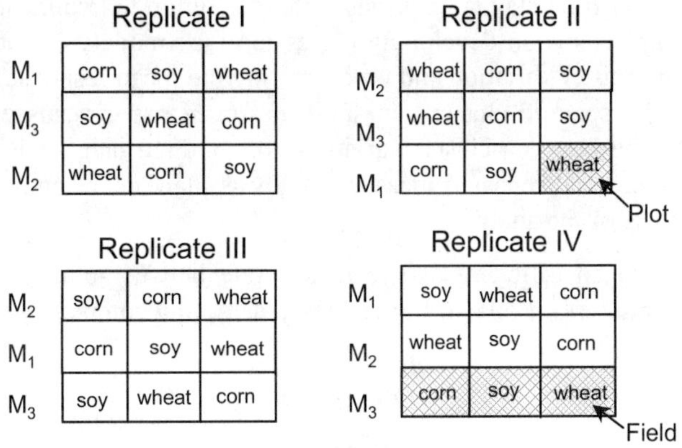

Replicate I

M_1	corn	soy	wheat
M_3	soy	wheat	corn
M_2	wheat	corn	soy

Replicate II

M_2	wheat	corn	soy
M_3	wheat	corn	soy
M_1	corn	soy	wheat

Plot

Replicate III

M_2	soy	corn	wheat
M_1	corn	soy	wheat
M_3	soy	wheat	corn

Replicate IV

M_1	soy	wheat	corn
M_2	wheat	soy	corn
M_3	corn	soy	wheat

Field

Figure 7.2. Split-plot design for management strategies (M1 to M3) and crop types.

Both RCBDs are analysis of variance models with a single error term. The mixed model comes about when the two randomizations are combined into a single model, letting $\phi_{ij} = \rho_j + \tau_i + e_{ij}$,

$$Y_{ijk} = \mu + \rho_j + \tau_i + e_{ij} + \alpha_k + (\tau\alpha)_{ik} + d_{ijk}. \qquad [7.3]$$

The two random variables depicting experimental error on the field and plot level are e_{ij} and d_{ijk}, respectively, and the fixed effects are ρ_j for the jth replicate, τ_i for the ith management strategy, α_k for the kth crop type, and $(\tau\alpha)_{ik}$ for their interaction. This is a classical split-plot design where the whole-plot factor management strategy is arranged in a randomized block design with four replication (blocks) and the subplot factor crop type has three levels.

The distinction between longitudinal and repeated measures data adopted here is as follows: if data are collected repeatedly on experimental material to which treatments were applied initially, the data structure is termed a repeated measure. Data collected repeatedly over time in an observational study are termed longitudinal. A somewhat different distinction between repeated measures and longitudinal data in the literature states that it is assumed that cluster effects do not change with time in repeated measures models, while time as an explanatory variable of within-cluster variation related to growth, development, and aging assumes a center role with longitudinal data (Rao 1965, Hayes 1973). In designed experiments involving time the assumption of time-constant cluster effects is often not tenable. Treatments are applied initially and their effects may wear off over time, changing the relationship among treatments as the experiment progresses. Treatment-time interactions are important aspects of repeated measures experiments and the investigation of trends over time and how they change among treatments can be the focus of the investigation.

The appeal of mixed models lies in their flexibility to handle diverse forms of hierarchically organized data. Depending on application and circumstance the emphasis of data analysis will differ. In designed experiments treatment comparisons, the estimation of treatment effects and contrasts, and the estimation of sources of variability come to the fore. In repeated measures analyses investigating the interactions of treatment and time and modeling the response trends over time play an important role in addition to treatment comparisons. For longitudinal data emphasis is on developing regression-type models that account for cluster-to-cluster and within-cluster variation and on estimating the mean response for the population average and/or for the specific clusters. In short, designed experiments with clustered data structure emphasize the between-cluster variation, longitudinal data analyses emphasize the within-cluster variation, and repeated measures analysis place more emphasis on either one depending on the goals of the analysis.

Mixed models are an efficient vehicle to separate between-cluster and within-cluster variation that is essential in the analysis of clustered data. In a completely randomized design (CRD) with subsampling,

$$Y_{ijk} = \mu + \tau_i + e_{ij} + d_{ijk},$$

where e_{ij} denotes the experimental error (EE) associated with the jth replication of the ith treatment and d_{ijk} is the observational (subsampling) error among the k subsamples from an experimental unit, between-cluster variation of units treated alike is captured by $\text{Var}[e_{ij}]$ and within-cluster variation by $\text{Var}[d_{ijk}]$. To gauge whether treatments are effective the mean square due to treatments should be compared to the magnitude of variation among experimental units treated alike, $MS(EE)$, not the observational error mean square. Analysis based on a model with a single error term, $Y_{ij} = \mu + \tau_i + \epsilon_{ijk}$, assuming the ϵ_{ijk} are independent, would be inappropriate and could lead to erroneous conclusions about treatment performance. We demonstrate this effect with the application in §7.6.3.

Not recognizing variability on the cluster and within-cluster level is dangerous from another point of view. In the CRD with subsampling the experimental errors e_{ij} are uncorrelated as are the observational errors d_{ijk} owing to the random assignment of treatments to experimental units and the random selection of samples from the units. Furthermore, the e_{ij} and d_{ijk} are not correlated with each other. Does that imply that the responses Y_{ij} are not correlated? Some basic covariance operations provide the answer,

$$\text{Cov}[Y_{ijk}, Y_{ijk'}] = \text{Cov}[e_{ij} + d_{ijk}, e_{ij} + d_{ijk'}] = \text{Cov}[e_{ij}, e_{ij}] = \text{Var}[e_{ij}] \neq 0.$$

While observations from different experimental units are uncorrelated, observations from the same cluster are correlated. Ignoring the hierarchical structure of the two error terms by putting $e_{ij} + d_{ijk} = \epsilon_{ijk}$ and assuming the ϵ_{ijk} are independent, ignores correlations among the experimental outcomes. Since $\text{Cov}[Y_{ijk}, Y_{ijk'}] = \text{Var}[e_{ij}] > 0$, the observations are positively correlated. If correlations are ignored, p-values will be too small (§2.5.2), even if the ϵ_{ijk} represented variability of experimental units treated alike, which it does not. In longitudinal and repeated measures data, correlations enter the data more directly. Since measurements are collected repeatedly on subjects or units, these measurements are likely autocorrelated.

Besides correlations among the observations, clustered data provide another challenge for the analyst who must decide whether the emphasis is **cluster-specific** or **population-average** inference. In a longitudinal study, for example, a natural focus of investigation are trends over time. There are two types of trends. One describes the behavior of the population-average (the average trend in the universe of all clusters), the other the behavior of individual clusters. If both types of trends can be modeled, a comparison of an individual cluster to the population behavior allows conclusions about the conformity and similarity of clusters. The overall trend in the population is termed the **population-average** (PA) trend and trends varying by clusters are termed **cluster-specific** (CS) or subject-specific trends (Schabenberger et al. 1995, Schabenberger and Gregoire 1996).

Example 7.2. Gregoire, Schabenberger, and Barrett (1995) analyze data from a longitudinal study of natural grown Douglas fir (*Pseudotsuga menziesii* (Mirb.) Franco) stands (plots) scattered throughout the western Cascades and coastal range of Washington and Oregon in northwestern United States. Plots were visited repeatedly between 1970 and 1982 for a minimum of 6 and a maximum of 10 times. Measurement intervals for a given plot varied between 1 and 3 years. The works of Gregoire (1985, 1987) discuss the data more fully.

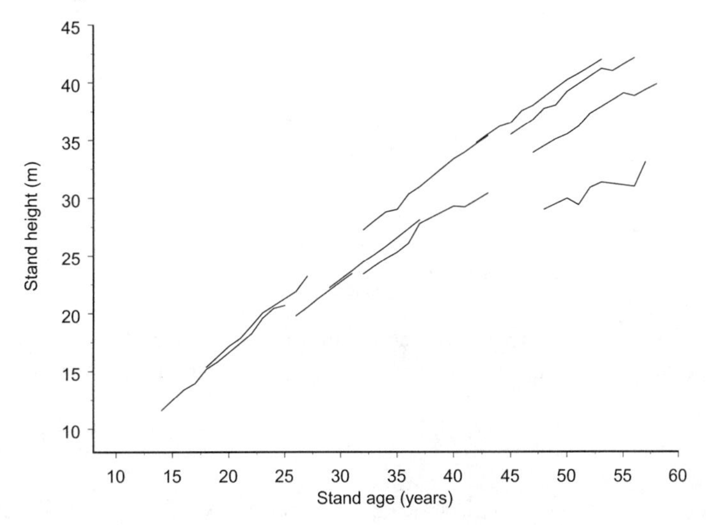

Figure 7.3. Height-age profiles of ten Douglas fir (*Pseudotsuga menziesii*) stands. Data kindly provided by Dr. Timothy G. Gregoire, School of Forestry and Environmental Studies, Yale University. Used with permission.

Figure 7.3 shows the height of the socially dominant trees vs. the stand age for ten of the 65 stands. Each stand depicted represents a cluster of observations. The stands differed in age at the onset of the study. The height development over the range of years during which observations were taken is almost linear for the ten stands. However, the slopes of the trends differ, as well as the maximum heights achieved. This could be due to increased natural variability in height development with age or to differences in micro-site conditions. Growing sites may be more homogeneous among younger than among older stands, a feature often found in man-made forests. Figure 7.4 shows the sample means for each stand (cross-hairs) and the population-averaged trend derived from a simple linear regression model

$$H_{ij} = \beta_0 + \beta_1 age_{ij} + e_{ij},$$

where H_{ij} is the jth height for the ith stand. To predict the height of a stand at a given age not in the data set, the population-average trend would be used. However, to predict the height of a stand for which data was collected, a more precise and accurate prediction should be possible if information about the stand's trend relative to the population trend is utilized. Focusing on the population average only, residuals are measured as deviations between observed values and the population trend. Cluster-specific predictions utilize smaller deviations between observed values and the specific trend.

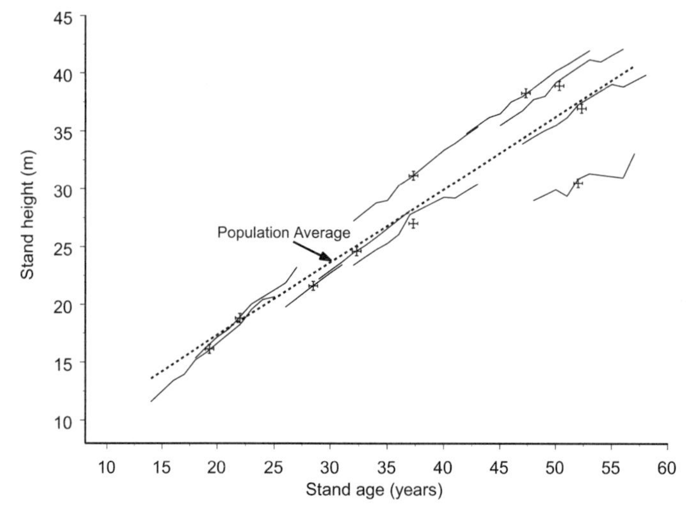

Figure 7.4. Population-average trend and stand- (cluster-) specific means for Douglas fir data. The dotted line represents population-average prediction of heights at a given age.

Cluster-specific inferences exploit an important property of longitudinal data analysis to let each cluster serve as its own control in the separation of age and cohort effects. To demonstrate this concept consider only the first observations of the stands shown in Figure 7.3 at the initial measurement date. These data, too, can be used to derive a height-age relationship (Figure 7.5). The advantage is that by randomly selecting stands the ten observations are independent and standard linear regression methods can be deployed. But do these data represent true growth? The data in Figure 7.5 are **cross-sectional**, referring to a cross-section of individuals at different stages of development. Only if the average height of thirty-year-old stands twenty years from now is identical to the average height of fifty-year-old stands *today*, can cross-sectional data be used for the purpose of modeling growth. The group-to-group differences in development are called **cohort effects**. Longitudinal data allow the unbiased estimation of growth and inference about growth by separating age and cohort effects.

The data in Figure 7.3 could be modeled with a purely fixed effects model. It seems reasonable to assume that the individual trends emanate from the same point and to postulate an intercept β_0 common to all stands. If the trends are linear, only the slopes would need to be varied by stand. The model

$$H_{ij} = \beta_0 + \beta_{1i} age_{ij} + e_{ij} \qquad [7.4]$$

fits a separate slope β_{1i} for each stand. A total of eleven fixed effects have to be estimated: ten slopes and one intercept. If the intercepts are also varied by stands, the fixed effects regression model

$$H_{ij} = \beta_{0i} + \beta_{1i} age_{ij} + e_{ij}$$

could be fitted requiring a total of 20 parameters in the mean function.

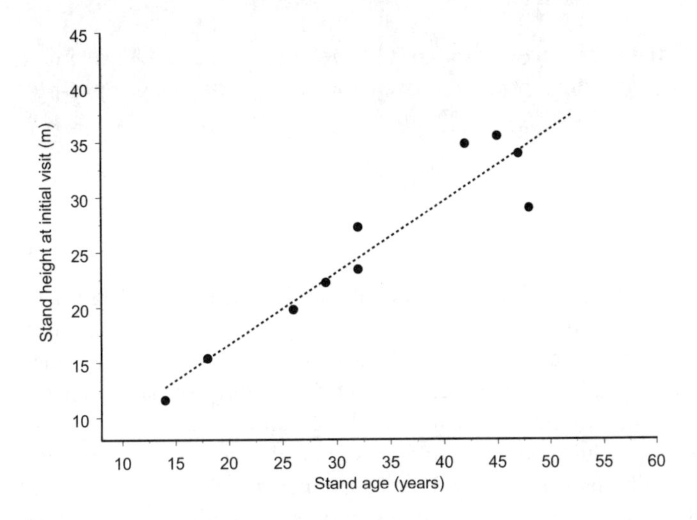

Figure 7.5. Observed stand heights and ages at initial visit and linear trend.

Early approaches to modeling of longitudinal and repeated measures data employed this philosophy: to separate the data into as many subsets as there are clusters and fit a model separately to each cluster. Once the individual estimates $\widehat{\beta}_{01}, \cdots, \widehat{\beta}_{0,10}$ and $\widehat{\beta}_{11}, \cdots, \widehat{\beta}_{1,10}$ were obtained, the population-average estimates were calculated as some weighted average of the cluster-specific intercepts and slopes. This two-step approach is inefficient as it ignores information contributed by other clusters in the estimation process and leads to parameter proliferation. While the data points from cluster 3 do not contribute information about the slope in cluster 1, the information in cluster 3 nevertheless can contribute to the estimation of the variance of observations about the cluster means. As the complexity of the individual trends and the cluster-to-cluster heterogeneity increases, the approach becomes impractical. If the number of observations per cluster is small, it may turn out to be actually impossible. If, for example, the individual trends are quadratic, and only two observations were collected on a particular cluster, the model cannot be fit to that cluster's data. If the mean function is nonlinear, fitting separate regression models to each cluster is plagued with numerical problems as nonlinear models require fairly large amounts of information to produce stable parameter estimates. It behooves us to develop an approach to data analysis that allows cluster-specific *and* population-average inference simultaneously without parameter proliferation. This is achieved by allowing effects in the model to vary at random rather than treating them as fixed. These ideas are cast in the Laird-Ware model.

7.2 The Laird-Ware Model

Box 7.2 Laird-Ware Model

- **The Laird-Ware model is a two-stage model for clustered data where the first stage describes the cluster-specific response and the second stage captures cluster-to-cluster heterogeneity by randomly varying parameters of the first-stage model.**
- **Most linear mixed models can be cast as Laird-Ware models, even if the two-stage concept may not seem natural at first, e.g., split-plot designs.**

7.2.1 Rationale

Although mixed model procedures were developed by Henderson (1950, 1963, 1973), Goldberger (1962), and Harville (1974, 1976a, 1976b) prior to the seminal article of Laird and Ware (1982), it was the Laird and Ware contribution that showed the wide applicability of linear mixed models and provided a convenient framework for parameter estimation and inference. Although their discussion focused on longitudinal data, the applicability of the Laird-Ware model to other clustered data structures is easily recognized. The basic idea is as follows. The probability distribution for thc measurements within a cluster has the same general form for all clusters, but some or all of the parameters defining this distribution vary **randomly** across clusters. First, define $\mathbf{Y}_i$ to be the $(n_i \times 1)$ vector of observations for the ith cluster. In a designed experiment, $\mathbf{Y}_i$ represents the data vector collected from a single experimental unit. Typically, in this case, additional subscripts will be needed to identify replications, treatments, whole-plots, sub-plots, etc. For the time being, it is assumed without loss of generality that the single subscript i identifies an individual cluster. For the leftmost (youngest) douglas fir stand in Figure 7.3, for example, 10 observations were collected at ages 14, 15, 16, 17, 18, 19, 22, 23, 24, 25. The measured heights for this sixth stand in the data set are assembled in the response vector

$$\mathbf{Y}_6 = \begin{bmatrix} Y_{61} \\ Y_{62} \\ Y_{63} \\ Y_{64} \\ Y_{65} \\ Y_{66} \\ Y_{67} \\ Y_{68} \\ Y_{69} \\ Y_{6,10} \end{bmatrix} = \begin{bmatrix} 11.60 \\ 12.50 \\ 13.41 \\ 13.59 \\ 15.19 \\ 15.82 \\ 18.23 \\ 19.57 \\ 20.43 \\ 20.68 \end{bmatrix}.$$

The Laird-Ware model assumes that the average behavior of the clusters is the same for all clusters, varied only by cluster-specific explanatory variables. In matrix notation this is represented for the mean trend as

$$\mathrm{E}[\mathbf{Y}_i] = \mathbf{X}_i\boldsymbol{\beta}, \qquad [7.5]$$

where $\mathbf{X}_i$ is an $(n_i \times p)$ design or regressor matrix and $\boldsymbol{\beta}$ is a $(p \times 1)$ vector of regression coefficients. Observe that clusters share the same parameter vector $\boldsymbol{\beta}$, but clusters can have different values of the regressor variables. If $\mathbf{X}_i$ contains a column of measurement times, for example, these do not have to be the same across clusters. The Laird-Ware model easily accommodates unequal spacing of measurements. Also, the number of cluster elements, n_i, can vary from cluster to cluster. Clusters with the same set of regressors $\mathbf{X}_i$ do not elicit the same response as suggested by the common parameter $\boldsymbol{\beta}$.

To allow clusters to vary in the effect of the explanatory variables on the outcome we can put $\mathrm{E}[\mathbf{Y}_i \mid \mathbf{b}_i] = \mathbf{X}_i(\boldsymbol{\beta} + \mathbf{b}_i) = \mathbf{X}_i\boldsymbol{\beta} + \mathbf{X}_i\mathbf{b}_i$. The $\mathbf{b}_i$ in this expression determine how much the ith cluster population-average response $\mathbf{X}_i\boldsymbol{\beta}$ must be adjusted to capture the cluster-specific behavior $\mathbf{X}_i\boldsymbol{\beta} + \mathbf{X}_i\mathbf{b}_i$. In a practical application not all of the explanatory variables have effects that vary among clusters and we can add generality to the model by putting $\mathrm{E}[\mathbf{Y}_i \mid \mathbf{b}_i] = \mathbf{X}_i\boldsymbol{\beta} + \mathbf{Z}_i\mathbf{b}_i$ where $\mathbf{Z}_i$ is an $(n_i \times k)$ design or regressor matrix. In this formulation not all the columns in $\mathbf{X}_i$ are repeated in $\mathbf{Z}_i$ and on occasion one may place explanatory variables in $\mathbf{Z}_i$ that are not part of $\mathbf{X}_i$ (although this is much less frequent than the opposite case where the columns of $\mathbf{Z}_i$ are a subset of the columns of $\mathbf{X}_i$).

The expectation was reckoned conditionally, because $\mathbf{b}_i$ is a vector of random variables. We assume that $\mathbf{b}_i$ has mean $\mathbf{0}$ and variance-covariance matrix $\mathbf{D}$. Laird and Ware (1982) term this a **two-stage model**. The first stage specifies the conditional distribution of $\mathbf{Y}_i$, given the $\mathbf{b}_i$, as

$$\mathbf{Y}_i|\mathbf{b}_i \sim G(\mathbf{X}_i\boldsymbol{\beta} + \mathbf{Z}_i\mathbf{b}_i, \mathbf{R}_i). \qquad [7.6]$$

Alternatively, we can write this as a linear model

$$\mathbf{Y}_i|\mathbf{b}_i = \mathbf{X}_i\boldsymbol{\beta} + \mathbf{Z}_i\mathbf{b}_i + \mathbf{e}_i,\ \mathbf{e}_i \sim G(\mathbf{0}, \mathbf{R}_i).$$

In the second stage it is assumed that the $\mathbf{b}_i$ have a Gaussian distribution with mean $\mathbf{0}$ and variance matrix $\mathbf{D}$, $\mathbf{b}_i \sim G(\mathbf{0}, \mathbf{D})$. The random effects $\mathbf{b}_i$ are furthermore assumed independent of the errors $\mathbf{e}_i$. The (marginal) distribution of the responses then is also Gaussian:

$$\mathbf{Y}_i \sim G(\mathbf{X}_i\boldsymbol{\beta}, \mathbf{R}_i + \mathbf{Z}_i\mathbf{D}\mathbf{Z}_i'). \qquad [7.7]$$

In the text that follows we will frequently denote $\mathbf{R}_i + \mathbf{Z}_i\mathbf{D}\mathbf{Z}_i'$ as $\mathbf{V}_i$. An alternative expression for the Laird-Ware model is

$$\begin{aligned} &\mathbf{Y}_i = \mathbf{X}_i\boldsymbol{\beta} + \mathbf{Z}_i\mathbf{b}_i + \mathbf{e}_i \\ &\mathbf{e}_i \sim G(\mathbf{0}, \mathbf{R}_i),\ \ \mathbf{b}_i \sim G(\mathbf{0}, \mathbf{D}) \\ &\mathrm{Cov}[\mathbf{e}_i, \mathbf{b}_i] = \mathbf{0}. \end{aligned} \qquad [7.8]$$

Model [7.8] is a classical mixed linear model. It contains a fixed effect mean structure given by $\mathbf{X}_i\boldsymbol{\beta}$ and a random structure given by $\mathbf{Z}_i\mathbf{b}_i + \mathbf{e}_i$. The $\mathbf{b}_i$ are sometimes called the random effects if $\mathbf{Z}_i$ is a design matrix consisting of 0's and 1's, and random coefficients if $\mathbf{Z}_i$ is a regressor matrix. We will refer to $\mathbf{b}_i$ simply as the random effects.

The extent to which clusters vary about the population-average response is expressed by the variability of the $\mathbf{b}_i$. If $\mathbf{D} = \mathbf{0}$, the model reduces to a fixed effects regression or classification model with a single error source $\mathbf{e}_i$. The $\mathbf{e}_i$ are sometimes called the within-cluster errors, since their variance-covariance matrix captures the variability and stochastic dependency within a cluster. The comparison of the conditional distribution $\mathbf{Y}_i|\mathbf{b}_i$ and the unconditional distribution of $\mathbf{Y}_i$ shows how cluster-specific $(\mathbf{X}_i\boldsymbol{\beta} + \mathbf{Z}_i\mathbf{b}_i)$ and population-average inference $(\mathbf{X}_i\boldsymbol{\beta})$ are accommodated in the same modeling framework. To consider a particular cluster's response we condition on $\mathbf{b}_i$. This leaves $\mathbf{e}_i$ as the only random component on the right-hand side of [7.8] and the cluster-specific mean and variance for cluster i are

$$\begin{aligned} \mathrm{E}[\mathbf{Y}_i|\mathbf{b}_i] &= \mathbf{X}_i\boldsymbol{\beta} + \mathbf{Z}_i\mathbf{b}_i \\ \mathrm{Var}[\mathbf{Y}_i|\mathbf{b}_i] &= \mathbf{R}_i. \end{aligned} \qquad [7.9]$$

Taking expectations over the distribution of the random effects, one arrives at

$$\begin{aligned} \mathrm{E}[\mathbf{Y}_i] &= \mathrm{E}[\mathrm{E}[\mathbf{Y}_i|\mathbf{b}_i]] = \mathrm{E}[\mathbf{X}_i\boldsymbol{\beta} + \mathbf{Z}_i\mathbf{b}_i] = \mathbf{X}_i\boldsymbol{\beta} \\ \mathrm{Var}[\mathbf{Y}_i] &= \mathbf{R}_i + \mathbf{Z}_i\mathbf{D}\mathbf{Z}_i' = \mathbf{V}_i. \end{aligned} \qquad [7.10]$$

The marginal variance follows from the standard result by which unconditional variances can be derived from conditional expectations:

$$\mathrm{Var}[Y] = \mathrm{E}[\mathrm{Var}[Y|X]] + \mathrm{Var}[\mathrm{E}[Y|X]].$$

Applying this to the mixed model [7.8] under the assumption that $\mathrm{Cov}[\mathbf{b}_i, \mathbf{e}_i] = \mathbf{0}$ leads to

$$\mathrm{E}[\mathrm{Var}[\mathbf{Y}_i|\mathbf{b}_i]] + \mathrm{Var}[\mathrm{E}[\mathbf{Y}_i|\mathbf{b}_i]] = \mathbf{R}_i + \mathbf{Z}_i\mathbf{D}\mathbf{Z}_i'.$$

In contrast to [7.9], [7.10] expresses the marginal or population-average mean and variance of cluster i.

The Laird-Ware model [7.8] is quite general. If the design matrix for the random effects is absent, $\mathbf{Z}_i = \mathbf{0}$, the Laird-Ware model reduces to the classical linear regression model. Similarly, if the random effects do not vary, i.e.,

$$\mathrm{Var}[\mathbf{b}_i] = \mathbf{D} = \mathbf{0},$$

all random effects must be exactly $\mathbf{b}_i \equiv \mathbf{0}$ since $\mathrm{E}[\mathbf{b}_i] = \mathbf{0}$ and the model reduces to a linear regression model

$$\mathbf{Y}_i = \mathbf{X}_i\boldsymbol{\beta} + \mathbf{e}_i.$$

If the fixed effects coefficient vector $\boldsymbol{\beta}$ is zero, the model becomes a random effects model

$$\mathbf{Y}_i = \mathbf{Z}_i\mathbf{b}_i + \mathbf{e}_i.$$

To motivate the latter consider the following experiment.

Example 7.3. Twenty laboratories are randomly selected from a list of laboratories provided by the Association of Official Seed Analysts (AOSA). Each laboratory receives 4 bags of 100 seeds each, selected at random from a large lot of soybean seeds. The laboratories perform germination tests on the seeds, separately for each of the bags and report the results back to the experimenter. A statistical model to describe the variability of germination test results must accommodate laboratory-to-laboratory differences *and* inhomogeneities in the seed lot. The results from two different laboratories may differ even if they perform exactly the same germination tests with the same precision and accuracy since they received different samples. But even if the samples were exactly the same, the laboratories will not produce exactly the same germination test results, due to differences in technology, seed handling and storage at the facility, experience of personnel, and other sources of variation particular to a specific laboratory. A model for this experiment could be

$$Y_{ij} = \mu + \alpha_i + e_{ij},$$

where Y_{ij} is the germination percentage reported by the ith laboratory for the jth 100 seed sample it received. μ is the overall germination percentage of the seedlot. α_i is a random variable with mean 0 and variance σ^2_α measuring the lab-specific deviation from the overall germination percentage. e_{ij} is a random variable with mean 0 and variance σ^2 measuring intralaboratory variability due to the four samples within a laboratory.

Since apart from the grand mean μ all terms in the model are random, this is a random effects model. In terms of the components of the Laird-Ware model we can define a cluster to consist of the four samples sent to a laboratory and let $\mathbf{Y}_i = [Y_{i1}, \cdots, Y_{i4}]'$. Then our model for the ith laboratory is

$$\mathbf{Y}_i = \mathbf{X}_i\boldsymbol{\beta} + \mathbf{Z}_i\mathbf{b}_i + \mathbf{e}_i = \begin{bmatrix} 1 \\ 1 \\ 1 \\ 1 \end{bmatrix}\mu + \begin{bmatrix} 1 \\ 1 \\ 1 \\ 1 \end{bmatrix}\alpha_i + \begin{bmatrix} e_{i1} \\ e_{i2} \\ e_{i3} \\ e_{i4} \end{bmatrix}.$$

7.2.2 The Two-Stage Concept

Depending on the application it may be more or less natural to appeal to the two-stage concept. In cases where the modeler has in mind a particular class of models that applies in general to different groups, clusters, or treatments, the concept applies immediately. In the first stage we select the population-average model and decide in the second stage which parameters of the model vary at random among the groups. In §5.8.7 nonlinear yield-density models were fit to data from different onion varieties. The basic model investigated there was

$$\ln\{Y_{ij}\} = \ln\left\{(\alpha_i + \beta_i x_{ij})^{-1/\theta_i}\right\} + e_{ij},$$

where Y_{ij} denotes the yield per plant of variety i grown at density x_{ij}. The parameters α, β, and θ were initially assumed to vary among the varieties in a deterministic manner, i.e., were fixed. We could also cast this model in the mixed model framework. Since we are concerned with linear models in this chapter, we concentrate on the inverse plant yield Y_{ij}^{-1} and its relationship to plant density as for the data shown in Table 5.15 and Figure 5.34 (p. 288). It seems reasonable to assume that Y^{-1} is linearly related to density for any of the three varieties. The general model is

$$Y_{ij}^{-1} = \alpha_i + \beta_i x_{ij} + e_{ij}.$$

These data are not longitudinal, since each variety $\times$ density combination was grown independently. The two-stage concept leading to a mixed model nevertheless applies. If the variances are homogeneous, it is reasonable to put $\text{Var}[e_{ij}] = \sigma^2$, $\text{Cov}[e_{ij}, e_{i'j'}] = 0$ whenever $i \neq i'$ or $j \neq j'$. For convenience we identify each variety i (see Table 5.16) as a cluster. The first stage is completed by identifying population parameters, cluster effects, and within-cluster variation in the general model. To this end take $\alpha_i = \alpha + b_{1i}$ and $\beta_i = \beta + b_{2i}$. The model can then be written as

$$Y_{ij}^{-1} = \alpha + b_{1i} + (\beta + b_{2i})x_{ij} + e_{ij}. \qquad [7.11]$$

In this formulation α and β are the population parameters and b_{1i}, b_{2i} measure the degree to which the population-averaged intercept (α) and slope (β) must be modified to accommodate the ith variety's response. These are the cluster effects. The second stage constitutes the assumption that b_{1i} and b_{2i} are randomly drawn from a universe of possible values for the intercept and slope adjustment. In other words, it is assumed that b_{1i} and b_{2i} are random variables with mean zero and variances σ_1^2 and σ_2^2, respectively. Assume $\sigma_2^2 = 0$ for the moment. A random variable whose variance is zero is a constant that takes on its mean value, which in this case is zero. If $\sigma_2^2 = 0$ the model reduces to

$$Y_{ij}^{-1} = \alpha + b_{1i} + \beta x_{ij} + e_{ij},$$

stating that varieties differ in the relationship between inverse plant yield and plant density only in their intercept, not their slope. This is a model with parallel trends among varieties. Imagine there are 30 varieties ($i = 1, \cdots, 30$). The test of slope equality if the β_i are fixed effects is based on the hypothesis $H_0{:}\, \beta_1 = \beta_2 = \cdots = \beta_{30}$, a twenty-nine degree of freedom hypothesis. In the mixed model setup the test of slope equality involves only a single parameter, $H_0{:}\, \sigma_2^2 = 0$. Even in this nonlongitudinal setting, the two-stage concept is immensely appealing if we view varietal differences as random disturbances about a conceptual average variety.

We have identified a population-average model for relating inverse plant yield to density,

$$\text{E}\left[Y_{ij}^{-1}\right] = \alpha + \beta x_{ij},$$

and how to modify the population average with random effects to achieve a cluster-specific (= variety-specific) model parsimoniously. For five hypothetical varieties Figure 7.6 shows the flexibility of the mixed model formulation for model [7.11] under the following assumptions:

- $\sigma_1^2 = \sigma_2^2 = 0$. This is a purely fixed effects model where all varieties share the same dependency on plant density (Figure 7.6 a).
- $\sigma_2^2 = 0$. Varieties vary in intercept (Figure 7.6b).
- $\sigma_1^2 = 0$. Varieties vary in slope (Figure 7.6 c).
- $\sigma_1^2 \neq 0$, $\sigma_2^2 \neq 0$. Varieties differ in slope and intercept (Figure 7.6 d).

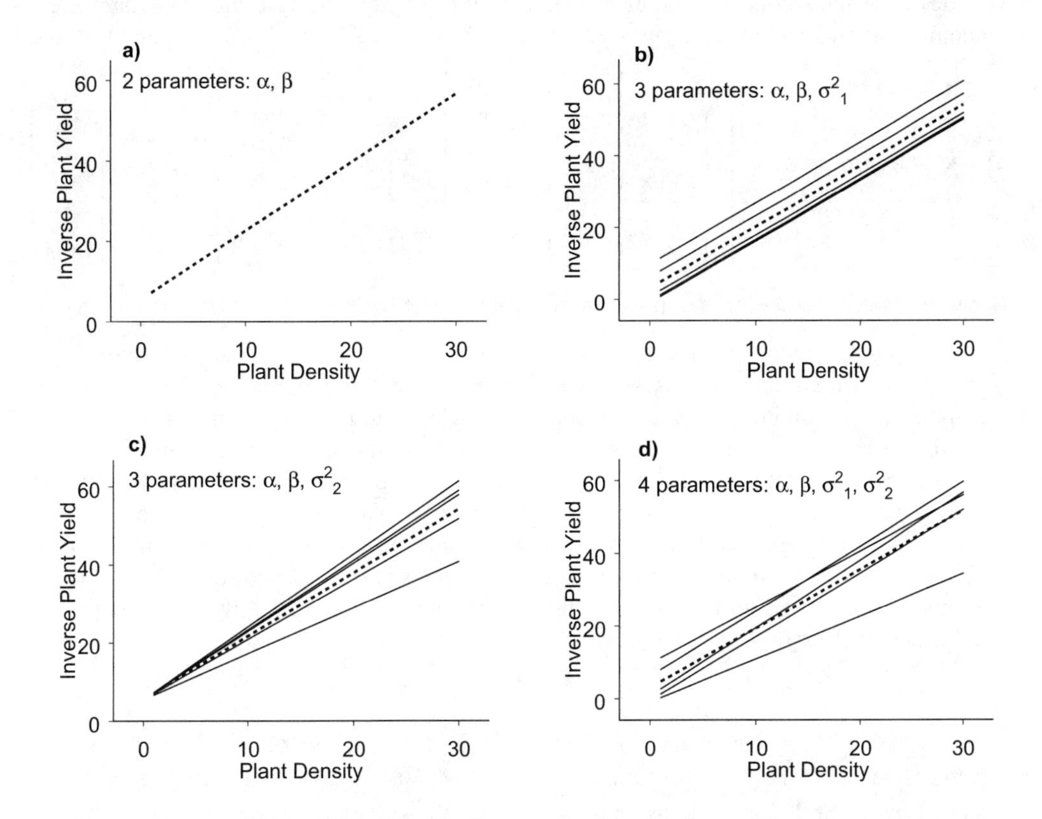

Figure 7.6. Fixed and mixed model trends for five hypothetical varieties. Purely fixed effects model (a), randomly varying intercepts (b), randomly varying slopes (c), randomly varying intercepts and slopes (d). The population-averaged trend is shown as a dashed line in panels (b) to (d). The same differentiation in cluster-specific effects as in (d) with a purely fixed effects model would have required 10 parameters. Number of parameters cited excludes $\text{Var}[e_{ij}] = \sigma^2$.

We complete this example by expressing [7.11] in terms of matrices and vectors as a Laird-Ware model. Collect the ten observations for variety i (see Table 5.16) into the vector $\mathbf{Y}_i$. Collect the ten plant densities for variety i into a two-column matrix $\mathbf{X}_i$, adding an intercept. For variety 1, for example, we have

$$\mathbf{Y}_1 = \begin{bmatrix} 1/105.6 \\ 1/98.4 \\ 1/71.0 \\ 1/60.3 \\ \vdots \\ 1/18.5 \end{bmatrix}, \quad \mathbf{X}_1 = \begin{bmatrix} 1 & 3.07 \\ 1 & 3.31 \\ 1 & 5.97 \\ 1 & 6.99 \\ \vdots & \vdots \\ 1 & 31.08 \end{bmatrix}.$$

If both intercept and slope vary at random among varieties, we have $\mathbf{Z}_i = \mathbf{X}_i$. If only the intercepts vary, $\mathbf{Z}_i$ is the first column of $\mathbf{X}_i$. If only the slopes vary at random among varieties, $\mathbf{Z}_i$ is the second column of $\mathbf{X}_i$. The model [7.11] for variety 1 with both parameters randomly varying, is

$$\mathbf{Y}_1 = \mathbf{X}_1\boldsymbol{\beta} + \mathbf{Z}_1\mathbf{b}_1 + \mathbf{e}_1 = \begin{bmatrix} 1/105.6 \\ 1/98.4 \\ 1/71.0 \\ 1/60.3 \\ \vdots \\ 1/18.5 \end{bmatrix} = \begin{bmatrix} 1 & 3.07 \\ 1 & 3.31 \\ 1 & 5.97 \\ 1 & 6.99 \\ \vdots & \vdots \\ 1 & 31.08 \end{bmatrix} \begin{bmatrix} \alpha \\ \beta \end{bmatrix} + \begin{bmatrix} 1 & 3.07 \\ 1 & 3.31 \\ 1 & 5.97 \\ 1 & 6.99 \\ \vdots & \vdots \\ 1 & 31.08 \end{bmatrix} \begin{bmatrix} b_{11} \\ b_{21} \end{bmatrix} + \begin{bmatrix} e_{11} \\ e_{12} \\ e_{13} \\ e_{14} \\ \vdots \\ e_{1,10} \end{bmatrix}.$$

Because of the experimental setup we can put $\mathrm{Var}[\mathbf{e}_i] = \mathbf{R}_i = \sigma^2\mathbf{I}$.

The two-stage concept may be less immediate when the mixed model structure arises from hierarchical random processes of treatment assignment or sampling as, for example, in split-plot and subsampling designs. A model-based selection of which parameters are random and which are fixed quantities does not occur. Rather, the treatment assignment and/or sampling scheme dictates whether an effect is fixed or random (see §7.2.3). Nevertheless, the resulting models are within the framework of the Laird-Ware model. We illustrate with a simple split-plot design. Consider two whole-plot treatments arranged in a randomized block design with two replications. Each of the four whole-plots is split by two sub-plot treatments. A smaller split-plot design is hardly imaginable. The total design has only eight data points. The linear model for this experiment is

$$Y_{ijk} = \mu + \rho_j + \tau_i + e^*_{ij} + \alpha_k + (\tau\alpha)_{ik} + e_{ijk},$$

where ρ_j $(j = 1, 2)$ are the whole-plot block effects, τ_i $(i = 1, 2)$ are the whole-plot treatment effects, e^*_{ij} are the whole-plot experimental errors, α_k $(k = 1, 2)$ are the sub-plot treatment effects, $(\tau\alpha)_{ik}$ are the interactions, and e_{ijk} denotes the sub-plot experimental errors. Using matrices and vectors the model can be expressed as follows:

$$\begin{bmatrix} Y_{111} \\ Y_{112} \\ \hline Y_{121} \\ Y_{122} \\ \hline Y_{211} \\ Y_{212} \\ \hline Y_{221} \\ Y_{222} \end{bmatrix} = \left[\begin{array}{c|cc|cc|cc|cccc} 1 & 1 & 0 & 1 & 0 & 1 & 0 & 1 & 0 & 0 & 0 \\ 1 & 1 & 0 & 1 & 0 & 0 & 1 & 0 & 1 & 0 & 0 \\ \hline 1 & 0 & 1 & 1 & 0 & 1 & 0 & 1 & 0 & 0 & 0 \\ 1 & 0 & 1 & 1 & 0 & 0 & 1 & 0 & 1 & 0 & 0 \\ \hline 1 & 1 & 0 & 0 & 1 & 1 & 0 & 0 & 0 & 1 & 0 \\ 1 & 1 & 0 & 0 & 1 & 0 & 1 & 0 & 0 & 0 & 1 \\ \hline 1 & 0 & 1 & 0 & 1 & 1 & 0 & 0 & 0 & 1 & 0 \\ 1 & 0 & 1 & 0 & 1 & 0 & 1 & 0 & 0 & 0 & 1 \end{array}\right] \begin{bmatrix} \mu \\ \rho_1 \\ \rho_2 \\ \tau_1 \\ \tau_2 \\ \alpha_1 \\ \alpha_2 \\ (\tau\alpha)_{11} \\ (\tau\alpha)_{12} \\ (\tau\alpha)_{21} \\ (\tau\alpha)_{22} \end{bmatrix} + \left[\begin{array}{cccc} 1 & 0 & 0 & 0 \\ 1 & 0 & 0 & 0 \\ \hline 0 & 1 & 0 & 0 \\ 0 & 1 & 0 & 0 \\ \hline 0 & 0 & 1 & 0 \\ 0 & 0 & 1 & 0 \\ \hline 0 & 0 & 0 & 1 \\ 0 & 0 & 0 & 1 \end{array}\right] \begin{bmatrix} e^*_{11} \\ e^*_{12} \\ e^*_{21} \\ e^*_{22} \end{bmatrix} + \begin{bmatrix} e_{111} \\ e_{112} \\ \hline e_{121} \\ e_{122} \\ \hline e_{211} \\ e_{212} \\ \hline e_{221} \\ e_{222} \end{bmatrix},$$

or

$$\mathbf{Y} = \mathbf{X}\boldsymbol{\beta} + \mathbf{Z}\mathbf{b} + \mathbf{e}.$$

This is a mixed model with four clusters of size two. The horizontal lines delineate observations that belong to the same cluster (whole-plot). In the notation of the Laird-Ware model we identify for the first whole-plot, for example,

$$\mathbf{X}_1 = \begin{bmatrix} 1 & 1 & 0 & 1 & 0 & 1 & 0 & 1 & 0 & 0 & 0 \\ 1 & 1 & 0 & 1 & 0 & 0 & 1 & 0 & 1 & 0 & 0 \end{bmatrix}, \mathbf{Z}_1 = \begin{bmatrix} 1 \\ 1 \end{bmatrix}, \mathbf{b}_1 = e^*_{11}, \mathbf{e} = \begin{bmatrix} e_{111} \\ e_{112} \end{bmatrix},$$

and

$$\boldsymbol{\beta} = [\mu, \rho_1, \rho_2, \tau_1, \tau_2, \alpha_1, \alpha_2, (\tau\alpha)_{11}, (\tau\alpha)_{12}, (\tau\alpha)_{21}, (\tau\alpha)_{22}]'.$$

Observe that the **Z** matrix for the entire data set is

$$\mathbf{Z} = \begin{bmatrix} \mathbf{Z}_1 & \mathbf{0} & \mathbf{0} & \mathbf{0} \\ \mathbf{0} & \mathbf{Z}_2 & \mathbf{0} & \mathbf{0} \\ \mathbf{0} & \mathbf{0} & \mathbf{Z}_3 & \mathbf{0} \\ \mathbf{0} & \mathbf{0} & \mathbf{0} & \mathbf{Z}_4 \end{bmatrix}.$$

This seems like a very tedious exercise. Fortunately, computer software such as `proc mixed` of The SAS® System handles the formulation of the **X** and **Z** matrices. What the user needs to know is which effects of the model are fixed (part of **X**), and which effects are random (part of **Z**).

In the previous examples we focused on casting the models in the mixed model framework by specifying $\mathbf{X}_i$, $\mathbf{Z}_i$, and $\mathbf{b}_i$. Little attention was paid to the variance-covariance matrices **D** and $\mathbf{R}_i$. In split-plot and subsampling designs these matrices are determined by the randomization and sampling protocol. In repeated measures and longitudinal studies the modeler must decide whether random effects/coefficients in $\mathbf{b}_i$ are independent (**D** diagonal) or not and must decide on the structure of $\mathbf{R}_i$. With n_i observations per cluster and if all observations within a cluster are correlated and have unequal variances there are n_i variances and $n_i(n_i - 1)/2$ covariances to be estimated in $\mathbf{R}_i$. To reduce the number of parameters in **D** and $\mathbf{R}_i$ these matrices are usually parameterized and highly structured. In §7.5 we examine popular parsimonious parametric structures. The next example shows how starting from a simple model accommodating the complexities of a real study leads to a mixed model where the modeler makes subsequent adjustments to the fixed and random parts of the model, including the **D** and **R** matrices, always with an eye toward parsimony of the final model.

Example 7.4. Soil nitrate levels and their dependence on the presence or absence of mulch shoot are investigated on bare soils and under alfalfa management. The treatment structure of the experiment is a 2×2 factorial of factors cover (alfalfa/none (bare soil)) and mulch (shoots applied/shoots not applied). Treatments are arranged in three complete blocks each accommodating four plots. Each plot receives one of the four possible treatment combinations. There is *a priori* evidence that the two factors do not interact. The basic linear statistical model for this experiment is given by

$$Y_{ijk} = \mu + \rho_i + \alpha_j + \beta_k + e_{ijk},$$

where $i = 1, ..., 3$ indexes the blocks, α_j are the effects of shoot application, β_k the effects of cover type, and the experimental errors e_{ijk} are independent and identically distributed random variables with mean 0 and variance σ^2. The variance of an individual observation is $\text{Var}[Y_{ijk}] = \sigma^2$.

In order to reduce the costs of the study soil samples are collected on each plot at four randomly chosen locations. The variability of an observation Y_{ijkl}, where $l = 1, ..., 4$

indexes the samples from plot ijk, is increased by the heterogeneity within a plot, σ_p^2 say,

$$\text{Var}[Y_{ijkl}] = \sigma^2 + \sigma_p^2.$$

The revised model must accommodate the two sources of random variation across plots and within plots. This is accomplished by adding another random effect

$$Y_{ijkl} = \mu + \rho_i + \alpha_j + \beta_k + e_{ijk} + f_{ijkl},$$

where $f_{ijkl} \sim (0, \sigma_p^2)$. This is a mixed model with fixed part $\mu + \rho_i + \alpha_j + \beta_k$ and random part $e_{ijk} + f_{ijkl}$. It is reasonable to assume by virtue of randomization that the two random effects are independent and also that

$$\text{Cov}[f_{ijkl}, f_{ijkl'}] = 0.$$

There are no correlations of the measurements within a plot. The **D** matrix of the model in Laird-Ware form will be diagonal.

It is imperative to the investigators to study changes in nitrate levels over time. To this end soil samples at the four randomly chosen locations within a plot are collected in five successive weeks. The data now have an additional repeated measurement structure in addition to a subsampling structure. First, the fixed effects part must be modified to accommodate systematic changes in nitrate levels over time. Treating time as a continuous variable, coded as the number of days t since the initial measurement, the fixed effects part can be revised as

$$\text{E}[Y_{ijklm}] = \mu + \rho_i + \alpha_j + \beta_k + \gamma t_{im},$$

where t_{im} is the time point at which all plots in block i were measured. If the measurement times differ across plots, the variable t would receive subscript ijk instead. The random effects structure is now modified to (a) incorporate the variability of measurements at the same spatial location over time; (b) account for residual temporal autocorrelation among the repeated measurements.

A third random component, $g_{ijklm} \sim (0, \sigma_t^2)$, is added so that the model becomes

$$Y_{ijklm} = \mu + \rho_i + \alpha_j + \beta_k + \gamma t_{im} + e_{ijk} + f_{ijkl} + g_{ijklm}$$

and the variance of an individual observation is

$$\text{Var}[Y_{ijklm}] = \sigma^2 + \sigma_p^2 + \sigma_t^2.$$

With five measurements over time there are 10 unique correlations per sampling location: $\text{Corr}[Y_{ijkl1}, Y_{ijkl2}]$, $\text{Corr}[Y_{ijkl1}, Y_{ijkl3}], \cdots$, $\text{Corr}[Y_{ijkl4}, Y_{ijkl5}]$. Furthermore, it is reasonable that measurements should be more highly correlated the closer together they were taken in time. Choosing a correlation model that depends explicitly on the time of measurement can be accomplished with only a single parameter. The temporal correlation model chosen is

$$\text{Corr}[g_{ijklm}, g_{ijklm'}] = \exp\{-\delta|t_{im} - t_{im'}|\},$$

known as the exponential or continuous AR(1) correlation structure (§7.5). The term $|t_{im} - t_{im'}|$ measures the separation in weeks between two measurements and $\delta > 0$ is a parameter to be estimated from the data. δ determines how quickly the correlations decrease with temporal separation.

The final linear mixed model for analysis is

$$
\begin{aligned}
Y_{ijklm} &= \mu + \rho_i + \alpha_j + \beta_k + \gamma t_{im} + e_{ijk} + f_{ijkl} + g_{ijklm} \\
e_{ijk} &\sim \left(0, \sigma^2\right), f_{ijkl} \sim \left(0, \sigma_p^2\right), g_{ijklm} \sim \left(0, \sigma_t^2\right) \\
\mathrm{Cov}[g_{ijklm}, g_{ijklm'}] &= \sigma_t^2 \exp\{-\delta|t_{im} - t_{im'}|\}.
\end{aligned}
$$

If in the context of repeated measures data each soil sample location within a plot is considered a cluster, σ_t^2 describes the within-cluster heterogeneity and $\sigma^2 + \sigma_p^2$ the between-cluster heterogeneity.

This model can be represented in matrix-vector notation at various levels of clustering. Assuming that a cluster is formed by the repeated measurements at a given location within a plot (index $ijkl$),

$$
\mathbf{Y}_{ijkl} = \mathbf{X}_{ijkl}\boldsymbol{\beta} + \mathbf{Z}_{ijkl}\mathbf{b}_{ijkl} + \mathbf{g}_{ijkl}
$$

where, for example,

$$
\mathbf{X}_{1123} = \begin{bmatrix} 1 & 1 & 0 & 1 & 0 & t_{11} \\ 1 & 1 & 0 & 1 & 0 & t_{12} \\ 1 & 1 & 0 & 1 & 0 & t_{13} \\ 1 & 1 & 0 & 1 & 0 & t_{14} \\ 1 & 1 & 0 & 1 & 0 & t_{15} \end{bmatrix}, \boldsymbol{\beta} = \begin{bmatrix} \mu + \rho_3 + \alpha_2 + \beta_2 \\ \rho_1 - \rho_3 \\ \rho_2 - \rho_3 \\ \alpha_1 - \alpha_2 \\ \beta_1 - \beta_2 \\ \gamma \end{bmatrix}
$$

$$
\mathbf{Z}_{1123} = \begin{bmatrix} 1 & 1 \\ 1 & 1 \\ 1 & 1 \\ 1 & 1 \\ 1 & 1 \end{bmatrix}, \mathbf{b}_{1123} = \begin{bmatrix} e_{112} \\ f_{1123} \end{bmatrix}, \mathbf{g}_{ijkl} = \begin{bmatrix} g_{11231} \\ g_{11232} \\ g_{11233} \\ g_{11234} \\ g_{11235} \end{bmatrix}.
$$

The variance-covariance matrix of the random effects $\mathbf{b}_{ijkl}$ is

$$
\mathrm{Var}[\mathbf{b}_{ijkl}] = \mathbf{D} = \begin{bmatrix} \sigma^2 & 0 \\ 0 & \sigma_p^2 \end{bmatrix}
$$

and of the within-cluster disturbances

$$
\mathrm{Var}[\mathbf{g}_{ijkl}] = \mathbf{R}_{ijkl} = \sigma_t^2 \begin{bmatrix} 1 & e^{-\delta d_{i12}} & e^{-\delta d_{i13}} & e^{-\delta d_{i14}} & e^{-\delta d_{i15}} \\ e^{-\delta d_{i21}} & 1 & e^{-\delta d_{i23}} & e^{-\delta d_{i24}} & e^{-\delta d_{i25}} \\ e^{-\delta d_{i31}} & e^{-\delta d_{i32}} & 1 & e^{-\delta d_{i34}} & e^{-\delta d_{i35}} \\ e^{-\delta d_{i41}} & e^{-\delta d_{i42}} & e^{-\rho \delta_{i43}} & 1 & e^{-\delta d_{i45}} \\ e^{-\delta d_{i51}} & e^{-\delta d_{i52}} & e^{-\delta d_{i53}} & e^{-\delta d_{i54}} & 1 \end{bmatrix},
$$

$d_{imm'} = |t_{im} - t_{im'}|$. With the mixed procedure of The SAS® System this model is analyzed with the following statements.

```
proc mixed data=YourData;
  class block shoot cover location;
  model nitrate = block shoot cover day;
  random block*shoot*cover;           /* random effect e_ijk      */
  random block*shoot*cover*location;  /* random effect f_ijkl     */
                                      /* repeated measures g_ijklm */
  repeated / subject=block*shoot*cover*location type=sp(exp)(day);
run;
```

7.2.3 Fixed or Random Effects

Box 7.3 Fixed or Random?

- **An effect is random if its levels were chosen by some random mechanism from a population (list) of possible levels, or, if the levels were not randomly selected, the effects on the outcome are of a stochastic nature. Otherwise, the effect is considered fixed.**

- **For random effects, inferences can be conducted in three different inference spaces, termed the broad, intermediate, and narrow spaces. Conclusions in fixed effects models apply only to the narrow inference space.**

When appealing to the two-stage concept one assumes that some effects or coefficients of the population-averaged model vary randomly from cluster to cluster. This requires in theory that there is a population or universe of coefficients from which the realizations in the data can be drawn (Longford 1993). In the onion plant density example, it is assumed that intercepts and/or slopes in the universe of varieties vary at random around the average value of α and/or β. Conceptually, this does not cause much difficulty if the varieties were selected at random and stochastic variation between clusters can be reasoned. In many cases the clusters are not selected at random and the question whether an effect is fixed or random is not clear. Imagine, for example, that the same plant density experiment is performed at various locations. If locations were predetermined, rather than randomly selected, can we still attribute differences in variety performance from location to location to stochastic effects, or are these fixed effects? Some modelers would argue that location effects are deterministic, fixed effects because upon repetition of the experiment the same locations would be selected and the same locational effects should operate on the outcome. Others consider locations as surrogates of different environments and consider environmental effects to be stochastic in nature. Repetition of the experiment even at the same locations will produce different outcomes due to changes in the environmental conditions and locational effects should thus be treated as random. A similar discrepancy of opinion applies to the nature of seasonal effects. Are the effects of years considered fixed or random? The years in which an experiment is conducted are most likely not a random sample from a list of possible years. Experiments are conducted when experimental areas can be secured, funds, machinery, and manpower are available. According to the *acid-test* that declares factors as fixed if their levels were pre-determined, years should enter the analysis as fixed effects. But if year effects are viewed as stochastic environmental effects they should enter the model as random effects (see, e.g., Searle 1971, pp. 382-383 and Searle et al. 1992, pp. 15-16).

In a much cited paper, Eisenhart (1947) introduced fixed and random analysis of variance models, which he termed Models I and II, a distinction used frequently to this day. He emphasized two parallel criteria to aid the modeler in the determination whether effects are fixed or random.

(i) If upon repetition of the experiment the "same things" (levels of the factor) would be studied again, the factor is fixed.

(ii) If inferences are to be confined to the factor levels actually employed in the experiment, the factor is fixed. If conclusions are expanded to apply to more general populations, it is random.

The cited test that determines factors as random if their levels are selected by a random mechanism falls under the first criterion. The sampling mechanism itself makes the effect random (Kempthorne 1975). Searle (1971, p. 383) subscribes to the second criterion, that of confining inferences to the levels at hand. If one is interested in conclusions about varietal performance for the specific years and locations in a multiyear, multilocation variety trial, location and year effects would be fixed. If conclusions are to be drawn about the population of locations at which the experiment could have been conducted in particular years, location effects would be random and seasonal effects would be fixed. Finally, if inferences are to pertain to all locations in any season, then both factors would be random. We agree with the notion implied by Searle's (and Eisenhart's second) criterion that it very much depends on the context whether an effect is considered random or not. Robinson (1991) concludes similarly when he states that "The choice of whether a class of effects is to [be] treated as fixed or random may vary with the question which we are trying to answer." His criterion, replacing both (i) and (ii) above, is to ask whether the effects in question come from a probability distribution. If they do, they are random, otherwise they are fixed. Robinson's criterion does not appeal to any sample or inference model and is thus attractive. It is noteworthy that Searle et al. (1992, p. 16) placed more emphasis on the random sampling mechanism than Searle (1971, p. 383). The latter reference reads

> *"In considering these points the important question is that of inference: are inferences going to be drawn from these data about* just *these levels of the factor? "Yes" – then the effects are considered as fixed effects. "No" – then, presumably, inferences will be made not just about the levels occurring in the data but about some population of levels of the factor from which those in the data are presumed to have come; and so the effects are considered as being random."*

Searle et al. (1992, p. 16) state, on the other hand,

> *"In considering these points the important question is that of inference: are the levels of the factor going to be considered a random sample from a population of values? "Yes" – then the effects are going to be considered as random effects. "No" – then, presumably, inferences will be made just about the levels occuring in the data and the effects are considered as fixed effects."*

We emphasize that for the purpose of analysis it is often reasonable to consider effects as random for some questions, and as fixed for others within the same investigation. Assume locations were selected at random from a list of possible locations in a variety trial, so that there is no doubt that they are random effects. One question of interest is which variety is highest yielding across all possible locations. Another question may be whether varieties A and B show significant yield differences at the particular locations used. Under Eisenhart's second criterion one should treat location effects as random for the first analysis and as fixed for the second analysis, which would upset Eisenhart's first criterion. Fortunately, mixed models provide a way out of this dilemma. Within the same analysis we can choose with respect to the random effects different **inference spaces**, depending on the question at hand (see §7.3). Even if an effect is random conclusions can be drawn pertaining only to the factor levels actually used and the effects actually observed (Figure 7.7).

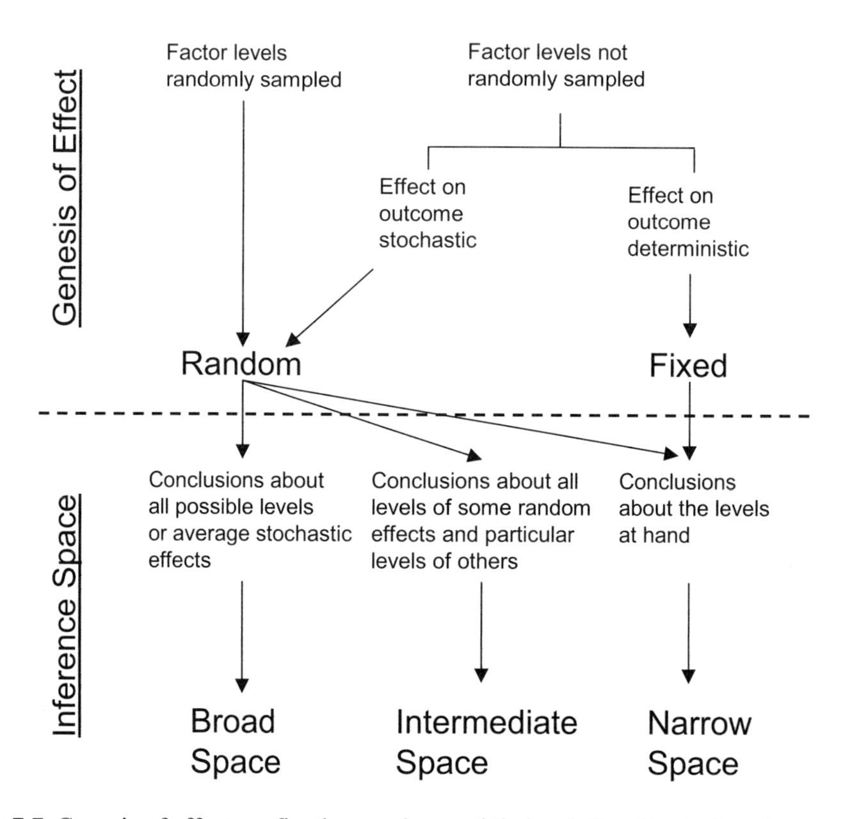

Figure 7.7. Genesis of effects as fixed or random and their relationships to broad, intermediate, and narrow inference spaces (§7.3).

Other arguments have been brought to bear to *solve* the *fixed vs. random* debate more or less successfully. We want to dispense with two of these. The fact that the experimenter does not know with certainty how a particular treatment will perform at a given location does not imply a random location effect. This argument would necessarily lead to all effects being considered random since prior to the experiment none of the effects is known with certainty. Another line of argument considers those effects random that are not under the experimenter's control, such as block and environmental effects. Under this premise the only fixed effects model is that of a completely randomized design and all treatment factors would be fixed. These criteria are neither practical nor sensible. Considering block and other experimental effects (apart from treatment effects) as random even if their selection was deterministic provided their effect on the outcome has a stochastic nature, yields a reasonable middle-ground in our opinion. Treatment factors are obviously random **only** when the treatments are chosen by some random mechanism, for example, when entries are selected at random for a variety trial from a larger list of possible entries. If treatment levels are predetermined, treatment effects are fixed. The interested reader can find a wonderful discourse of these and other issues related to analysis of variance in general and the fixed/random debate in Kempthorne (1975).

7.3 Choosing the Inference Space

Mixed models have an exciting property that enables researchers to perform inferences that apply to different populations of effects. This property can defuse the *fixed vs. random* debate (§7.2.3). Models in which all effects are fixed do not provide this opportunity. To motivate the concept of an **inference space**, we consider a two-factor experiment where the treatment levels were predetermined (are fixed) and the random factor corresponds to environmental effects such as randomly chosen years or locations, or predetermined locations with stochastically varying environment effects. The simple mixed model we have in mind is

$$Y_{ijk} = \mu + \alpha_i + \tau_j + (\alpha\tau)_{ij} + e_{ijk}, \qquad [7.12]$$

where the α_i $(i = 1, \cdots, a)$ are random environmental effects and τ_j $(j = 1, \cdots, t)$ are the fixed treatment effects, e.g., entries (genotypes) in a variety trial. There are k replications of each environment $\times$ entry combination and $(\alpha\tau)_{ij}$ represents genotype $\times$ environment interaction. We observe that because the α_i are random variables, the interaction is also a random quantity. If all effects in [7.12] were fixed, inferences about entry performance would apply to the particular environments (locations) that are selected in the study, but not to other environments that could have been chosen. This inference space is termed the **narrow space** by McLean, Sanders, and Stroup (1991). The narrow inference space can also be chosen in the mixed effects model to evaluate and compare the entries. If genotype performance is of interest in the particular environments in which the experiment was performed and for the particular genotype $\times$ environment interaction, the narrow inference space applies. On other occasions one might be interested in conclusions about the entries that pertain to the universe of all possible environments that could have been chosen for the study. Entry performance is then evaluated relative to potential environmental effects and random genotype $\times$ environment interactions. McLean et al. (1991) term this the **broad inference space** and conclude that it is the appropriate reference for inference if environmental effects are hard to specify. The broad inference space has no counterpart in fixed effects models.

A third inference space, situated between the broad and narrow spaces has been termed the **intermediate space** by McLean et al. (1991). Here, one appeals to specific levels of some random effects, but to the universe of all possible levels with respect to other random effects. In model [7.12] an intermediate inference space applies if one is interested in genotype performance in specific environments but allows the genotype $\times$ environment interaction to vary at random from environment to environment. For purposes of inferences, one would fix α_i and allow $(\alpha\tau)_{ij}$ to vary. When treatment effects τ_j are fixed, the interaction is a random effect since the environmental effects α_i are random. It is our opinion that the intermediate inference space is not meaningful in this particular model. When focusing on a particular environmental effect the interaction should be fixed at the appropriate level too. If, however, the treatment effects were random, too, the intermediate inference space where one focuses on the performance of all genotypes in a particular environment is meaningful.

In terms of testable hypotheses or "estimable" functions in the mixed model we are concerned with linear combinations of the model terms. To demonstrate the distinction between the three inference spaces, we take into account the presence of **b**, not just $\boldsymbol{\beta}$, in specifying these linear combinations. An "estimable" function is now written as

$$\mathbf{A}\boldsymbol{\beta} + \mathbf{M}\mathbf{b} = \mathbf{L}\begin{bmatrix} \boldsymbol{\beta} \\ \mathbf{b} \end{bmatrix},$$

where $\mathbf{L} = [\mathbf{A}, \ \mathbf{M}]$. Since *estimation* of parameters should be distinguished from *prediction* of random variables, we use quotation marks. Setting $\mathbf{M}$ to $\mathbf{0}$, the function becomes $\mathbf{A}\boldsymbol{\beta}$, an estimable function because $\mathbf{A}\boldsymbol{\beta}$ is a matrix of constants. No reference is made to specific random effects and thus the inference is broad. By selecting the entries of $\mathbf{M}$ such that $\mathbf{Mb}$ represents averages over the appropriate random effects, the narrow inference space is chosen and one should refer to $\mathbf{A}\boldsymbol{\beta} + \mathbf{Mb}$ as a predictable function. An intermediate inference space is constructed by averaging some random effects, while setting the coefficients of $\mathbf{M}$ pertaining to other random effects to zero. We illustrate these concepts with an example from Milliken and Johnson (1992, p. 285).

Example 7.8. Machine Productivity. Six employees are randomly selected from the work force of a company that has plans to replace the machines in one of its factories. Three candidate machine types are evaluated. Each employee operates each of the machines in a randomized order. Milliken and Johnson (1992, p. 286) chose the mixed model

$$Y_{ijk} = \mu + \alpha_i + \tau_j + (\alpha\tau)_{ij} + e_{ijk},$$

where τ_j represents the fixed effect of machine type j $(j = 1, \cdots, 3)$, α_i the random effect of employee i $(i = 1, \cdots, 6; \alpha_i \sim G(0, \sigma^2_\alpha))$, $(\alpha\tau)_{ij}$ the machine × employee interaction (a random effect with mean 0 and variance $\sigma^2_{\alpha\tau}$), and e_{ijk} represents experimental errors associated with employee i operating machine j at the kth time. The outcome Y_{ijk} was a productivity score. The data for this experiment appear in Table 23.1 of Milliken and Johnson (1992) and are reproduced on the CD-ROM.

If we want to estimate the mean of machine type 1, for example, we can do this in a broad, narrow, or intermediate inference space. The corresponding expected values are

Broad: $\mathrm{E}[Y_{i1k}] = \mu + \tau_1$

Narrow: $\mathrm{E}[Y_{i1k} | \alpha_i, (\alpha\tau)_{i1}] = \mu + \tau_1 + \frac{1}{6}\sum_{i=1}^{6}\alpha_i + \frac{1}{6}\sum_{i=1}^{6}(\alpha\tau)_{i1}$

Intermediate: $\mathrm{E}[Y_{i1k} | \alpha_i] = \mu + \tau_1 + \frac{1}{6}\sum_{i=1}^{6}\alpha_i$

Estimates of these quantities are obtained by substituting the estimates for $\widehat{\mu}$ and $\widehat{\tau}_1$ and the best linear unbiased predictors (BLUPs) for the random variables α_i and $(\alpha\tau)_{i1}$

from the mixed model analysis. In §7.4 the necessary details of this estimation and prediction process are provided. For now we take for granted that the estimates and BLUPs can be obtained in The SAS® System with the statements

```
proc mixed data=productivity;
  class machine person;
  model score = machine / s;
  random person machine*person / s;
run;
```

The `/s` option on the `model` statement prints the estimates of all fixed effects, the `/s` option on the `random` statement prints the estimated BLUPs for the random effects. The latter are reproduced from SAS® output in Table 7.1.

Table 7.1. Best linear unbiased predictors of α_i and $(\alpha\tau)_{ij}$

Employee (i)	**Best Linear Unbiased Predictors**			
	α_i	$(\alpha\tau)_{i1}$	$(\alpha\tau)_{i2}$	$(\alpha\tau)_{i3}$
1	1.0445	– 0.7501	1.5000	– 0.1142
2	– 1.3759	1.5526	0.6069	– 2.9966
3	5.3608	1.7776	2.2994	– 0.8149
4	– 0.0598	– 1.0394	2.4174	– 1.4144
5	2.5446	– 3.4569	2.1521	2.8532
6	– 7.5143	1.9163	– 8.9757	2.4870
$\sum_{i=1}^{6}$	0	0	0	0

The solutions for the random effects sum to zero across the employees. Hence, when the solutions are substituted into the formulas above for the narrow and intermediate means and averaged, we have, for example,

$$\widehat{\mu} + \widehat{\tau}_1 + \frac{1}{6}\sum_{i=1}^{6}\widehat{\alpha}_i = \widehat{\mu} + \widehat{\tau}_1.$$

The broad, narrow and intermediate estimates of the means will not differ. Provided that the **D** matrix is nonsingular, this will hold in general for linear mixed models. Our prediction of the average production score for machine 1 does not depend on whether we refer to the six employees actually used in the experiment or the population of all company employees from which the six were randomly selected. So wherein lies the difference? Although the point estimates do not differ, the variability of the estimates will differ greatly. Since the mean estimate in the intermediate inference space,

$$\widehat{\mu} + \widehat{\tau}_1 + \frac{1}{6}\sum_{i=1}^{6}\widehat{\alpha}_i,$$

involves the random quantities $\widehat{\alpha}_i$, its variance will exceed that of the mean estimate in the narrow inference space, which is just $\widehat{\mu} + \widehat{\tau}_1$. By the same token the variance of estimates in the broad inference space will exceed that of the estimates in the intermediate space. By appealing to the population of all employees, the additional uncertainty that stems from the random selection of employees must be accounted for. The precision of broad and narrow inference will be identical if the random effects variance σ^2_α is 0, that is, there is no heterogeneity among employees with respect to productivity scores.

Estimates and predictions of various various quantities obtained with `proc mixed` of The SAS® System are shown in the next table from which the impact of choosing the inference space on estimator/predictor precision can be inferred.

Table 7.2. Quantities to be estimated or predicted in machine productivity example (intermediate inference applies to specific employees but randomly varying employee × machine interaction)

Description	ID	Space	Parameter
Machine 1 Mean	(1)	Broad	$\mu + \tau_1$
	(2)	Intermediate	$\mu + \tau_1 + \frac{1}{6}\sum_{i=1}^{6}\alpha_i$
	(3)	Narrow	$\mu + \tau_1 + \frac{1}{6}\sum_{i=1}^{6}\alpha_i + \frac{1}{6}\sum_{i=1}^{6}(\alpha\tau)_{i1}$
Machine 2 Mean	(4)	Broad	$\mu + \tau_2$
	(5)	Intermediate	$\mu + \tau_2 + \frac{1}{6}\sum_{i=1}^{6}\alpha_i$
	(6)	Narrow	$\mu + \tau_2 + \frac{1}{6}\sum_{i=1}^{6}\alpha_i + \frac{1}{6}\sum_{i=1}^{6}(\alpha\tau)_{i2}$
Mach. 1 - Mach. 2	(7)	Broad	$\tau_1 - \tau_2$
	(8)	Narrow	$\tau_1 - \tau_2 + \frac{1}{6}\sum_{i=1}^{6}(\alpha\tau)_{i1} - \frac{1}{6}\sum_{i=1}^{6}(\alpha\tau)_{i2}$
Employee 1 BLUP	(9)		$\mu + \frac{1}{3}\sum_{j=1}^{3}\tau_j + \alpha_1 + \frac{1}{3}\sum_{j=1}^{3}(\alpha\tau)_{1j}$
Employee 2 BLUP	(10)		$\mu + \frac{1}{3}\sum_{j=1}^{3}\tau_j + \alpha_2 + \frac{1}{3}\sum_{j=1}^{3}(\alpha\tau)_{2j}$
Empl. 1 - Empl. 2	(11)		$\alpha_1 - \alpha_2 + \frac{1}{3}\sum_{j=1}^{3}(\alpha\tau)_{1j} - \frac{1}{3}\sum_{j=1}^{3}(\alpha\tau)_{2j}$

The `proc mixed` code to calculate these quantities is as follows.

```
proc mixed data=productivity;
  class machine person;
  model score = machine;
  random person machine*person;

  estimate '(1) Mach. 1 Mean (Broad) ' intercept 1 machine 1 0 0;
  estimate '(2) Mach. 1 Mean (Interm)' intercept 6 machine 6 0 0
              | person 1 1 1 1 1 1  /divisor = 6;
  estimate '(3) Mach. 1 Mean (Narrow)' intercept 6 machine 6 0 0
              | person 1 1 1 1 1 1
       machine*person 1 1 1 1 1 1 0 0 0 0 0 0 0 0 0 0 0 0/divisor = 6;

  estimate '(4) Mach. 2 Mean (Broad) ' intercept 1 machine 0 1 0;
  estimate '(5) Mach. 2 Mean (Interm)' intercept 6 machine 0 6 0
          | person 1 1 1 1 1 1  /divisor = 6;
  estimate '(6) Mach. 2 Mean (Narrow)' intercept 6 machine 0 6 0
          | person 1 1 1 1 1 1
      machine*person 0 0 0 0 0 0 1 1 1 1 1 1 0 0 0 0 0 0/divisor = 6 ;
```

```
  estimate '(7) Mac. 1 vs. Mac. 2 (Broad) ' machine 1 -1;
  estimate '(8) Mac. 1 vs. Mac. 2 (Narrow)' machine 6 -6
        | machine*person 1 1 1 1 1 1 -1 -1 -1 -1 -1 -1 0 0 0 0 0 0/
          divisor = 6 ;

  estimate '(9) Person 1 BLUP' intercept 6 machine 2 2 2
        | person 6 0 0 0 0 0
       machine*person 2 0 0 0 0 0 2 0 0 0 0 0 2 0 0 0 0 0/divisor = 6;
  estimate '(10) Person 2 BLUP' intercept 6 machine 2 2 2
        | person 0 6 0 0 0 0
       machine*person 0 2 0 0 0 0 0 2 0 0 0 0 0 2 0 0 0 0/divisor = 6;
  estimate '(11) Person 1 - Person 2'
        | person 6 -6 0 0 0 0
    machine*person 2 -2 0 0 0 0 2 -2 0 0 0 0 2 -2 0 0 0 0/divisor = 6;
run;
```

When appealing to the narrow or intermediate inference spaces coefficients for the random effects that are being held fixed are added after the | in the `estimate` statements. If no random effects coefficients are specified, the **M** matrix in the linear combination $\mathbf{A}\boldsymbol{\beta} + \mathbf{Mb}$ is set to zero and the inference space is broad. Notice that least squares means calculated with the `lsmeans` statement of the `mixed` procedure are always broad. The abridged output follows.

Output 7.1.

```
                       The Mixed Procedure

                        Model Information
          Data Set                      WORK.PRODUCTIVITY
          Dependent Variable            score
          Covariance Structure          Variance Components
          Estimation Method             REML
          Residual Variance Method      Profile
          Fixed Effects SE Method       Model-Based
          Degrees of Freedom Method     Containment

                     Class Level Information
                  Class       Levels     Values
                  machine          3     1 2 3
                  person           6     1 2 3 4 5 6

                 Covariance Parameter Estimates
                   Cov Parm              Estimate
                   person                 22.8584
                   machine*person         13.9095
                   Residual                0.9246

                            Estimates

                                         Standard
Label                         Estimate      Error    DF   t Value   Pr >|t|
(1) Mach. 1 Mean (Broad)       52.3556     2.4858    10     21.06   <.0001
(2) Mach. 1 Mean (Interm)      52.3556     1.5394    10     34.01   <.0001
(3) Mach. 1 Mean (Narrow)      52.3556     0.2266    10    231.00   <.0001
(4) Mach. 2 Mean (Broad)       60.3222     2.4858    10     24.27   <.0001
(5) Mach. 2 Mean (Interm)      60.3222     1.5394    10     39.19   <.0001
(6) Mach. 2 Mean (Narrow)      60.3222     0.2266    10    266.15   <.0001
(7) Mac. 1 vs. Mac. 2 (Broad)  -7.9667     2.1770    10     -3.66   0.0044
(8) Mac. 1 vs. Mac. 2 (Narrow)-7.9667      0.3205    10    -24.86   <.0001
(9) Person 1 BLUP              60.9064     0.3200    10    190.32   <.0001
(10) Person 2 BLUP             57.9951     0.3200    10    181.22   <.0001
(11) Person 1 - Person 2        2.9113     0.4524    36      6.43   <.0001
```

The table of `Covariance Parameter Estimates` displays the estimates of the variance components of the model; $\widehat{\sigma}^2_\alpha = 22.858$, $\widehat{\sigma}^2_{\alpha\tau} = 13.909$, $\widehat{\sigma}^2 = 0.925$. Because the data are balanced and `proc mixed` estimates variance-covariance parameters by restricted maximum likelihood (by default), these estimates coincide with the method-of-moment estimates derived from expected mean squares and reported in Milliken and Johnson (1992, p. 286).

It is seen from the table of `Estimates` that the means in the broad, intermediate, and narrow inference spaces are identical; for example, 52.3556 is the estimate for the mean production score of machines of type 1, regardless of inference space. The standard errors of the three estimates are largest in the broad inference space and smallest in the narrow inference space. The same holds for differences of the means. Notice that if one would analyze the data as a fixed effects model, the estimates for `(1)` through `(8)` would be identical. Their standard errors would be incorrect, however. The next to the last two estimates are predictions of random effects (`(9)` and `(10)`) and the prediction of the difference of two random effects `(11)`. If one would incorrectly specify the model as a fixed effects model, the estimates *and* their standard errors would be incorrect for `(9) - (11)`.

7.4 Estimation and Inference

The Laird-Ware model

$$\mathbf{Y}_i = \mathbf{X}_i\boldsymbol{\beta} + \mathbf{Z}_i\mathbf{b}_i + \mathbf{e}_i,\ \mathrm{Var}[\mathbf{Y}_i] = \mathbf{Z}_i\mathbf{D}\mathbf{Z}_i' + \mathbf{R}_i$$

contains a fair number of unknown quantities that must be calculated from data. Parameters of the model are $\boldsymbol{\beta}$, $\mathbf{D}$, and $\mathbf{R}_i$; these must be **estimated**. The random effects $\mathbf{b}_i$ are not parameters, but random variables. These must be **predicted** in order to calculate cluster-specific trends and perform cluster-specific inferences. If $\widehat{\boldsymbol{\beta}}$ is an estimator of $\boldsymbol{\beta}$ and $\widehat{\mathbf{b}}_i$ is a predictor of $\mathbf{b}_i$, the population-averaged prediction of $\mathbf{Y}_i$ is

$$\widehat{\mathbf{Y}}_i = \mathbf{X}_i\widehat{\boldsymbol{\beta}}$$

and the cluster-specific prediction is calculated as

$$\widehat{\mathbf{Y}}_i = \mathbf{X}_i\widehat{\boldsymbol{\beta}} + \mathbf{Z}_i\widehat{\mathbf{b}}_i.$$

Henderson (1950) derived estimating equations for $\boldsymbol{\beta}$ and $\mathbf{b}_i$ known as the mixed model equations. We derive the equations in §A7.7.1 and their solutions in §A7.7.2. Briefly, the mixed model equations are

$$\begin{bmatrix}\widehat{\boldsymbol{\beta}} \\ \widehat{\mathbf{b}}\end{bmatrix} = \begin{bmatrix}\mathbf{X}'\mathbf{R}^{-1}\mathbf{X} & \mathbf{X}'\mathbf{R}^{-1}\mathbf{Z} \\ \mathbf{Z}'\mathbf{R}^{-1}\mathbf{X} & \mathbf{Z}'\mathbf{R}^{-1}\mathbf{Z} + \mathbf{B}^{-1}\end{bmatrix}^{-1}\begin{bmatrix}\mathbf{X}'\mathbf{R}^{-1}\mathbf{y} \\ \mathbf{Z}'\mathbf{R}^{-1}\mathbf{y}\end{bmatrix}, \qquad [7.13]$$

where the vectors and matrices of the individual clusters were properly stacked and arranged to eliminate the subscript i and $\mathbf{B}$ is a block-diagonal matrix whose diagonal blocks consist of the matrix $\mathbf{D}$ (see §A7.7.1 for details). The solutions are

$$\widehat{\boldsymbol{\beta}} = \left(\mathbf{X}'\mathbf{V}^{-1}\mathbf{X}\right)^{-1}\mathbf{X}'\mathbf{V}^{-1}\mathbf{y} \qquad [7.14]$$

$$\widehat{\mathbf{b}} = \mathbf{B}\mathbf{Z}'\mathbf{V}^{-1}(\mathbf{y} - \mathbf{X}\widehat{\boldsymbol{\beta}}). \qquad [7.15]$$

The estimate $\widehat{\boldsymbol{\beta}}$ is a generalized least squares estimate. Furthermore, the predictor $\widehat{\mathbf{b}}$ is the **b**est **l**inear **u**nbiased **p**redictor (BLUP) of the random effects $\mathbf{b}$ (§A7.7.2.). Properties of these expressions are easily established. For example,

$$\mathrm{E}\left[\widehat{\boldsymbol{\beta}}\right] = \boldsymbol{\beta}, \quad \mathrm{Var}\left[\widehat{\boldsymbol{\beta}}\right] = \left(\mathbf{X}'\mathbf{V}^{-1}\mathbf{X}\right)^{-1}, \quad \mathrm{E}\left[\widehat{\mathbf{b}}\right] = \mathbf{0}.$$

Since $\mathbf{b}$ is a random variable, more important than evaluating $\mathrm{Var}[\widehat{\mathbf{b}}]$ is the variance of the prediction error $\widehat{\mathbf{b}} - \mathbf{b}$, which can be derived after some tedious calculations (Harville 1976a, Laird and Ware 1982) as

$$\mathrm{Var}\left[\widehat{\mathbf{b}} - \mathbf{b}\right] = \mathbf{B} - \mathbf{B}\mathbf{Z}'\mathbf{V}^{-1}\mathbf{Z}\mathbf{B} + \mathbf{B}\mathbf{Z}'\mathbf{V}^{-1}\mathbf{X}\left(\mathbf{X}'\mathbf{V}^{-1}\mathbf{X}\right)^{-1}\mathbf{X}'\mathbf{V}^{-1}\mathbf{Z}\mathbf{B}. \qquad [7.16]$$

Although not a very illuminating expression, it is [7.16] rather than $\mathrm{Var}[\widehat{\mathbf{b}}]$ that should be reported. When predicting random variables the appropriate measure of uncertainty is the mean square prediction error [7.16], not the variance of the predictor.

Expressions [7.14] and [7.15] assume that the variance-covariance matrix $\mathbf{V}$ and hence $\mathbf{D}$ and $\mathbf{R}$ are known, which almost always they are not. Even in the simplest case of independent and homoscedastic within-cluster errors where $\mathbf{R} = \sigma^2\mathbf{I}$, an estimate of the variance σ^2 is required. It seems reasonable to use instead of [7.14] and [7.15] a substitution estimator/predictor where $\mathbf{V}$ is replaced with an estimate $\widetilde{\mathbf{V}}$,

$$\widehat{\boldsymbol{\beta}} = \left(\mathbf{X}'\widetilde{\mathbf{V}}^{-1}\mathbf{X}\right)^{-1}\mathbf{X}'\widetilde{\mathbf{V}}^{-1}\mathbf{Y}$$

$$\widehat{\mathbf{b}} = \mathbf{B}\mathbf{Z}'\widetilde{\mathbf{V}}^{-1}(\mathbf{Y} - \mathbf{X}\widehat{\boldsymbol{\beta}}).$$

The variance of $\widehat{\boldsymbol{\beta}}$ would then be similarly estimated by substituting $\widetilde{\mathbf{V}}$ into the expression for $\mathrm{Var}[\widehat{\boldsymbol{\beta}}]$,

$$\widehat{\mathrm{Var}}\left[\widehat{\boldsymbol{\beta}}\right] = \left(\mathbf{X}'\widetilde{\mathbf{V}}^{-1}\mathbf{X}\right)^{-1}. \qquad [7.17]$$

$\widehat{\mathrm{Var}}[\widehat{\boldsymbol{\beta}}]$ is a consistent estimator of $\mathrm{Var}[\widehat{\boldsymbol{\beta}}]$ if $\widetilde{\mathbf{V}}$ is consistent for $\mathbf{V}$. It is a biased estimator, however. There are two sources to this bias in that (i) $(\mathbf{X}'\widetilde{\mathbf{V}}^{-1}\mathbf{X})^{-1}$ is a biased estimator of $(\mathbf{X}'\mathbf{V}^{-1}\mathbf{X})^{-1}$ and (ii) the variability that arises from estimating $\mathbf{V}$ by $\widetilde{\mathbf{V}}$ is unaccounted for (Kenward and Roger 1997). Consequently, [7.17] underestimates $\mathrm{Var}[\widehat{\boldsymbol{\beta}}]$. The various approaches to estimation and inference in the linear mixed model to be discussed next depend on how $\mathbf{V}$ is estimated and thus which matrix is used for substitution. The most important principles are maximum likelihood, restricted maximum likelihood, and estimated generalized least squares (generalized estimating equations). Bias corrections and bias corrected estimators of $\mathrm{Var}[\widehat{\boldsymbol{\beta}}]$ for likelihood-type estimation are discussed in Kackar and Harville (1984), Prasad and Rao (1990), Harville and Jeske (1992), and Kenward and Roger (1997).

7.4.1 Maximum and Restricted Maximum Likelihood

We collect all unknown parameters in $\mathbf{D}$ and $\mathbf{R}$ into a parameter vector $\boldsymbol{\theta}$ and call $\boldsymbol{\theta}$ the vector of covariance parameters (although it may also contain variances). The maximum likelihood principle chooses those values $\widehat{\boldsymbol{\theta}}$ and $\widehat{\boldsymbol{\beta}}$ as estimates for $\boldsymbol{\theta}$ and $\boldsymbol{\beta}$ that maximize the joint Gaussian distribution of the $(T \times 1)$ vector $\mathbf{Y}$, i.e., $G(\mathbf{X}\boldsymbol{\beta}, \mathbf{V}(\boldsymbol{\theta}))$. Here, T denotes the total number of observations across all n clusters, $T = \sum_{i=1}^{n} n_i$. Details of this process can be found in §A7.7.3. Briefly, the objective function to be minimized, the negative of twice the Gaussian log-likelihood, is

$$\mathfrak{w}(\boldsymbol{\beta}, \boldsymbol{\theta}; \mathbf{y}) = \ln|\mathbf{V}(\boldsymbol{\theta})| + (\mathbf{y} - \mathbf{X}\boldsymbol{\beta})'\mathbf{V}(\boldsymbol{\theta})^{-1}(\mathbf{y} - \mathbf{X}\boldsymbol{\beta}) + T\ln\{2\pi\}. \qquad [7.18]$$

The problem is solved by first profiling $\boldsymbol{\beta}$ out of the equation. To this end $\boldsymbol{\beta}$ in [7.18] is replaced with $\left(\mathbf{X}'\mathbf{V}(\boldsymbol{\theta})^{-1}\mathbf{X}\right)^{-1}\mathbf{X}'\mathbf{V}(\boldsymbol{\theta})^{-1}\mathbf{y}$ and the resulting expression is minimized with respect to $\boldsymbol{\theta}$. If derivatives of the profiled log-likelihood with respect to one element of $\boldsymbol{\theta}$ depend on other elements of $\boldsymbol{\theta}$, the process is iterative. On occasion, some or all covariance parameters can be estimated in noniterative fashion. For example if $\mathbf{R} = \sigma^2\mathbf{R}^*$, $\mathbf{B} = \sigma^2\mathbf{B}^*$, with $\mathbf{R}^*$ and $\mathbf{B}^*$ known, then

$$\widehat{\sigma}^2 = \frac{1}{T}\left(\mathbf{y} - \mathbf{X}\widehat{\boldsymbol{\beta}}\right)'(\mathbf{Z}\mathbf{B}^*\mathbf{Z}' + \mathbf{R}^*)^{-1}\left(\mathbf{y} - \mathbf{X}\widehat{\boldsymbol{\beta}}\right).$$

Upon convergence of the algorithm, the final iterate $\widehat{\boldsymbol{\theta}}_M$ is the maximum likelihood estimate of the covariance parameters, and

$$\widehat{\boldsymbol{\beta}}_M = \left(\mathbf{X}'\mathbf{V}(\widehat{\boldsymbol{\theta}}_M)^{-1}\mathbf{X}\right)^{-1}\mathbf{X}'\mathbf{V}(\widehat{\boldsymbol{\theta}}_M)^{-1}\mathbf{Y} \qquad [7.19]$$

is the maximum likelihood estimator of $\widehat{\boldsymbol{\beta}}$. Since maximum likelihood estimators (MLE) have certain optimality properties; for example, they are asymptotically the most efficient estimators, substituting the MLE of $\boldsymbol{\theta}$ in the generalized least squares estimate for $\boldsymbol{\beta}$ has much appeal. The predictor for the random effects is calculated as

$$\widehat{\mathbf{b}}_M = \mathbf{B}(\widehat{\boldsymbol{\theta}}_M)\mathbf{Z}'\mathbf{V}(\widehat{\boldsymbol{\theta}}_M)^{-1}\left(\mathbf{y} - \mathbf{X}\widehat{\boldsymbol{\beta}}_M\right). \qquad [7.20]$$

Maximum likelihood estimators of covariance parameters have a shortcoming. They are usually negatively biased, that is, too small on average. The reason for this phenomenon is their not taking into account the number of fixed effects being estimated. A standard example illustrates the problem. In the simple model $Y_i = \beta + e_i$ $(i = 1, \cdots, n)$ where the e_i's are a random sample from a Gaussian distribution with mean 0 and variance σ^2, we wish to find MLEs for β and σ^2. The likelihood for these data is

$$\mathfrak{L}(\beta, \sigma^2; \mathbf{y}) = \prod_{i=1}^{n}\frac{1}{\sqrt{2\pi\sigma^2}}\exp\left\{-\frac{1}{2\sigma^2}(y_i - \beta)^2\right\}$$

and the values $\widehat{\beta}$ and $\widehat{\sigma}^2$that maximize $\mathfrak{L}(\beta, \sigma^2|\mathbf{y})$ necessarily minimize

$$-2l(\beta, \sigma^2; \mathbf{y}) = \mathfrak{w}(\beta, \sigma^2; \mathbf{y}) = \sum_{i=1}^{n}\ln\{2\pi\} + \ln\{\sigma^2\} + \frac{1}{\sigma^2}(y_i - \beta)^2.$$

Setting derivatives with respect to β and σ^2 to zero leads to two equations

$$\boxed{A}: \quad \partial\mathfrak{w}(\beta,\sigma^2;\mathbf{y})/\partial\beta = \frac{1}{\sigma^2}\sum_{i=1}^{n}(y_i-\beta) \equiv 0$$

$$\boxed{B}: \quad \partial\mathfrak{w}(\beta,\sigma^2;\mathbf{y})/\partial\sigma^2 = -\frac{n}{\sigma^2}+\sum_{i=1}^{n}\frac{1}{\sigma^4}(y_i-\beta)^2 \equiv 0.$$

Solving $\boxed{A}$ yields $\widehat{\beta} = n^{-1}\sum_{i=1}^{n}y_i = \overline{y}$. Substituting for β in $\boxed{B}$ and solving yields $\widehat{\sigma}^2_M = n^{-1}\sum_{i=1}^{n}(y_i-\overline{y})^2$. The estimate of β is the familiar sample mean, but the estimate of the variance parameter is *not* the sample variance $s^2 = (n-1)^{-1}\sum_{i=1}^{n}(y_i-\overline{y})^2$. Since the sample variance is an unbiased estimator of σ^2 under random sampling from any distribution, we see that $\widehat{\sigma}^2_M$ has bias

$$\mathrm{E}\left[\widehat{\sigma}^2_M-\sigma^2\right] = \mathrm{E}\left[\frac{n-1}{n}S^2\right]-\sigma^2 = \frac{n-1}{n}\sigma^2-\sigma^2 = -\frac{1}{n}\sigma^2.$$

If β were known there would be only one estimating equation ($\boxed{B}$) and the MLE for σ^2 would be

$$\widehat{\sigma}^2 = \frac{1}{n}\sum_{i=1}^{n}(Y_i-\beta)^2 = \frac{1}{n}\sum_{i=1}^{n}(Y_i-\mathrm{E}[Y_i])^2,$$

which *is* an unbiased estimator of σ^2. The bias of $\widehat{\sigma}^2_M$ originates in not adjusting the divisor of the sum of squares by the number of estimated parameters in the mean function. An alternative estimation method is restricted maximum likelihood (REML), also known as residual maximum likelihood (Patterson and Thompson 1971; Harville 1974, 1977). Here, adjustments are made in the objective function to be minimized that account for the number of estimated mean parameters. Briefly, the idea is as follows (see §A7.7.3 for more details). Rather than maximizing the joint distribution of $\mathbf{Y}$ we focus on the distribution of $\mathbf{KY}$, where $\mathbf{K}$ is a matrix of error contrasts. These contrasts are linear combinations of the observations such that $\mathrm{E}[\mathbf{KY}] = \mathbf{0}$. Hence we require that the inner product of each row of $\mathbf{K}$ with the vector of observations has expectation 0. We illustrate the principle in the simple setting examined above where $Y_i = \beta + e_i$ $(i = 1,\cdots,n)$. Define a new vector

$$\mathbf{U}_{(n-1\times 1)} = \begin{bmatrix} Y_1-\overline{Y} \\ Y_2-\overline{Y} \\ \vdots \\ Y_{n-1}-\overline{Y} \end{bmatrix} \qquad [7.21]$$

and observe that $\mathbf{U}$ has expectation $\mathbf{0}$ and variance-covariance matrix given by

$$\mathrm{Var}[\mathbf{U}] = \sigma^2\begin{bmatrix} 1-\frac{1}{n} & -\frac{1}{n} & \cdots & -\frac{1}{n} \\ -\frac{1}{n} & 1-\frac{1}{n} & \cdots & -\frac{1}{n} \\ \vdots & & \ddots & \vdots \\ -\frac{1}{n} & -\frac{1}{n} & \cdots & 1-\frac{1}{n} \end{bmatrix} = \sigma^2\left(\mathbf{I}_{n-1}-\frac{1}{n}\mathbf{J}_{n-1}\right) = \sigma^2\mathbf{P}.$$

Applying Theorem 8.3.4 in Graybill (1969, p. 190), the inverse of this matrix turns out to have a surprisingly simple form,

$$\text{Var}[\mathbf{U}]^{-1} = \frac{1}{\sigma^2}\begin{bmatrix} 2 & 1 & \cdots & 1 \\ 1 & 2 & \cdots & 1 \\ \vdots & & \ddots & \vdots \\ 1 & 1 & \cdots & 2 \end{bmatrix} = \frac{1}{\sigma^2}(\mathbf{I}_{n-1} + \mathbf{J}_{n-1}).$$

Also, $|\mathbf{P}| = 1/n$ and some algebra shows that $\mathbf{U}'\mathbf{P}^{-1}\mathbf{U} = \sum_{i=1}^n (Y_i - \overline{Y})^2$, the residual sum of squares. The likelihood for $\mathbf{U}$ is called the restricted likelihood of $\mathbf{Y}$ because $\mathbf{U}$ is restricted to have mean $\mathbf{0}$. It can be written as

$$\mathfrak{L}(\sigma^2; \mathbf{u}) = \frac{|\sigma^2\mathbf{P}|^{-1/2}}{(2\pi)^{(n-1)/2}}\exp\left\{ -\frac{1}{2\sigma^2}\mathbf{u}'\mathbf{P}^{-1}\mathbf{u}\right\}$$

and is no longer a function of the mean β. Minus twice the restricted log likelihood becomes

$$\begin{aligned} -2l(\sigma^2; \mathbf{u}) = \mathfrak{w}(\sigma^2; \mathbf{u}) &= (n-1)\ln\{2\pi\} - \ln\{n\} \\ &\quad + (n-1)\ln(\sigma^2) + \sum_{i=1}^n \frac{(y_i - \overline{y})^2}{\sigma^2}. \end{aligned} \qquad [7.22]$$

Setting the derivative of $\mathfrak{w}(\sigma^2; \mathbf{u})$ with respect to σ^2 to zero one obtains the estimating equation that implies the residual maximum likelihood estimate:

$$\frac{\partial \mathfrak{w}(\sigma^2; \mathbf{u})}{\partial \sigma^2} \equiv 0 \Leftrightarrow \frac{(n-1)}{\sigma^2} = \frac{1}{\sigma^4}\sum_{i=1}^n (y_i - \overline{y})^2$$

$$\widehat{\sigma}^2_R = \frac{1}{n-1}\sum_{i=1}^n (y_i - \overline{y})^2.$$

The REML estimator for σ^2 is the sample variance and hence unbiased.

The choice of $\mathbf{U}$ in [7.21] corresponds to a particular matrix $\mathbf{K}$ such that $\mathbf{U} = \mathbf{KY}$. We can express $\mathbf{U}$ formally as $\mathbf{KY}$ where

$$\mathbf{K} = [\,\mathbf{I}_{n-1} \quad \mathbf{0}_{(n-1\times 1)}\,] - [\,\mathbf{J}_{n-1} \quad \mathbf{1}_{(n-1\times 1)}\,].$$

If $\text{E}[\mathbf{Y}] = \mathbf{X}\boldsymbol{\beta}$, $\mathbf{K}$ needs to be chosen such that $\mathbf{KY}$ contains no term in $\boldsymbol{\beta}$. This is equivalent to removing the mean and considering residuals. The alternative name of *residual* maximum likelihood derives from this notion. Fortunately, as long as $\mathbf{K}$ is chosen to be of full row rank and $\mathbf{KX} = \mathbf{0}$, the REML estimates do not depend on the particular choice of error contrasts. In the simple constant mean model $Y_i = \beta + e_i$ we could define an orthogonal contrast matrix

$$\mathbf{C}_{(n-1\times n)} = \begin{bmatrix} 1 & -1 & 0 & \cdots & 0 & 0 \\ 1 & 1 & -2 & \cdots & 0 & 0 \\ \vdots & & & & & \vdots \\ 1 & 1 & 1 & \cdots & 1 & -(n-2) \end{bmatrix}$$

and a diagonal matrix $\mathbf{D}_{(n-1\times n-1)} = \text{Diag}\{(i+i^2)^{-1/2}\}$. Letting $\mathbf{K}^* = \mathbf{DC}$ and $\mathbf{U}^* = \mathbf{KY}$, then $\text{E}[\mathbf{U}^*] = \mathbf{0}$, $\text{Var}[\mathbf{U}^*] = \sigma^2\mathbf{DCC}'\mathbf{D} = \mathbf{I}_{n-1}$, $\mathbf{u}^{*\prime}\mathbf{u}^* = \sum_{i=1}^n (y_i - \overline{y})^2$ and minus twice the log likelihood of $\mathbf{U}^*$ is

$$\mathfrak{w}\left(\sigma^2;\mathbf{u}^*\right) = (n-1)\ln\{2\pi\} + (n-1)\ln\left(\sigma^2\right) + \frac{1}{\sigma^2}\sum_{i=1}^{n}(y_i - \overline{y})^2. \qquad [7.23]$$

Apart from the constant $\ln\{n\}$ this expression is identical to [7.22] and minimization of either function will lead to the same REML estimator of σ^2. For more details on REML estimation see Harville (1974) and Searle et al. (1992, Ch. 6.6). Two generic methods for constructing the **K** matrix are described in §A7.7.3.

REML estimates of variance components and covariance parameters have less bias than maximum likelihood estimates and in certain situations (e.g., certain balanced designs) are unbiased. In a balanced completely randomized design with subsampling, fixed treatment effects and n subsamples per experimental unit, for example, it is well-known that the observational error mean square and experimental error mean square have expectations

$$\begin{aligned}\mathrm{E}[MS(OE)] &= \sigma_o^2 \\ \mathrm{E}[MS(EE)] &= \sigma_o^2 + n\sigma_e^2,\end{aligned}$$

where σ_o^2 and σ_e^2 denote the observational and experimental error variances, respectively. The ANOVA method of estimation (Searle et al. 1992, Ch. 4.4) equates mean squares to their expectations and solves for the variance components. From the above equations we derive the estimators

$$\begin{aligned}\widehat{\sigma}_0^2 &= MS(OE) \\ \widehat{\sigma}_e^2 &= \frac{1}{n}\{MS(EE) - MS(OE)\}.\end{aligned}$$

These estimators are unbiased by construction and identical to the REML estimators in this case (for an application see §7.6.3). A closer look at $\widehat{\sigma}_e^2$ shows that this quantity could possibly be negative. Likelihood estimators must be values in the parameter space. Since $\sigma_e^2 > 0$, a value $\widehat{\sigma}_e^2 < 0$ is considered only a *solution* to the likelihood estimation problem, but not a likelihood *estimate*. Unfortunately, to retain unbiasedness, one has to allow for the possibility of a negative value. One should choose $\max\{\widehat{\sigma}_e^2, 0\}$ as the REML estimator instead. While this introduces some bias, it is the appropriate course of action. Corbeil and Searle (1976) derive solutions for the ML and REML estimates for four standard classification models when data are balanced and examine the properties of the solutions. They call the solutions "ML estimators" or "REML estimators," acknowledging that ignoring the positivity requirement does not produce true likelihood estimators. Lee and Kapadia (1984) examine the bias and variance of ML and REML estimators for one of Corbeil and Searle's models for which the REML solutions are unbiased. This is the balanced two-way mixed model without interaction,

$$Y_{ij} = \mu + \alpha_i + \beta_j + e_{ij},\ (i = 1,\cdots,a; j = 1,\cdots,b).$$

Here, α_i could correspond to the effects of a fixed treatment factor with a levels and β_j to the random effects of a random factor with b levels, $\beta_j \sim G(0, \sigma_b^2)$. Observe that there is only a single observation per combination of the two factors, that is, the design is nonreplicated. The experimental errors are assumed independent Gaussian with mean 0 and variance σ^2. Table 7.3, adapted from Lee and Kapadia (1984), shows the bias, variance, and mean square error of the maximum likelihood and restricted maximum likelihood estimators of σ^2 and σ_b^2 for

$a = 6, b = 10$. ML estimators have the smaller variability throughout, but show non-negligible negative bias, especially if the variability of the random effect is small relative to the error variability. In terms of the mean square error (Variance + Bias^2), REML estimators of σ^2 are superior to ML estimators but the reverse is true for estimates of σ_b^2. Provided σ_b^2 accounts for at least 50% of the response variability, REML estimators are essentially unbiased, since then the probability of obtaining a negative solution for σ_b^2 tends quickly to zero.

Returning to mixed models of the Laird-Ware form, it must be noted that the likelihood for **KY** in REML estimation does not contain any information about the fixed effects $\boldsymbol{\beta}$. REML estimation will produce estimates for $\boldsymbol{\theta}$ only. Once these estimates have been obtained we again put the substitution principle to work. If $\widehat{\boldsymbol{\theta}}_R$ is the REML estimate of $\boldsymbol{\theta}$, the fixed effects are estimated as

$$\widehat{\boldsymbol{\beta}}_R = \left(\mathbf{X}'\mathbf{V}(\widehat{\boldsymbol{\theta}}_R)^{-1}\mathbf{X}\right)^{-1}\mathbf{X}'\mathbf{V}(\widehat{\boldsymbol{\theta}}_R)^{-1}\mathbf{y}, \qquad [7.24]$$

and the random effects are predicted as

$$\widehat{\mathbf{b}}_R = \mathbf{B}(\widehat{\boldsymbol{\theta}}_R)\mathbf{Z}'\mathbf{V}(\widehat{\boldsymbol{\theta}}_R)^{-1}\left(\mathbf{y} - \mathbf{X}\widehat{\boldsymbol{\beta}}_R\right). \qquad [7.25]$$

Because the elements of $\boldsymbol{\theta}$ were estimated, [7.24] is no longer a generalized least squares (GLS) estimate. Because the substituted estimate $\widehat{\boldsymbol{\theta}}_R$ is not a maximum likelihood estimate, [7.24] is also not a maximum likelihood estimate. Instead, it is termed an Estimated GLS (EGLS) estimate.

Table 7.3. Bias (B), variance (Var), and mean square error (MSE) of ML and REML estimates in a balanced, two-way mixed linear model without replication† (fixed factor A has 6, random factor B has 10 levels)

	σ^2					
	ML			**REML**		
$\mathrm{Var}[\beta_j]/\mathrm{Var}[Y_{ij}]$	B	Var	MSE	B	Var	MSE
0.1	– 0.099	0.027	0.037	– 0.010	0.033	0.034
0.3	– 0.071	0.017	0.022	– 0.011	0.021	0.022
0.5	– 0.050	0.009	0.011	– 0.000	0.011	0.011
0.7	– 0.030	0.003	0.004	– 0.000	0.004	0.004
0.9	– 0.010	0.000	0.000	– 0.000	0.000	0.000

	σ_b^2					
	ML			**REML**		
$\mathrm{Var}[\beta_j]/\mathrm{Var}[Y_{ij}]$	B	Var	MSE	B	Var	MSE
0.1	– 0.001	0.010	0.010	0.010	0.012	0.012
0.3	– 0.029	0.031	0.032	0.001	0.039	0.039
0.5	– 0.050	0.062	0.064	0.000	0.076	0.076
0.7	– 0.070	0.101	0.106	0.000	0.106	0.125
0.9	– 0.090	0.151	0.159	0.000	0.187	0.187

†Adapted from Table 1 in Lee and Kapadia (1984). With permission of the International Biometric Society.

Because REML estimation is based on the likelihood principle and REML estimators have lower bias than maximum likelihood estimators, we prefer REML for parameter estimation in mixed models over maximum likelihood estimation and note that it is the default method of the `mixed` procedure in The SAS® System.

7.4.2 Estimated Generalized Least Squares

Maximum likelihood estimation of $(\boldsymbol{\beta}, \boldsymbol{\theta})$ and restricted maximum likelihood estimation of $\boldsymbol{\theta}$ is a numerically expensive process. Also, the estimating equations used in these procedures rely on the distribution of the random effects $\mathbf{b}_i$ and the within-cluster errors $\mathbf{e}_i$ being Gaussian. If the model

$$\begin{aligned}\mathbf{Y}_i &= \mathbf{X}_i\boldsymbol{\beta} + \mathbf{Z}_i\mathbf{b}_i + \mathbf{e}_i \\ \mathbf{e}_i &\sim (\mathbf{0}, \mathbf{R}_i) \\ \mathbf{b}_i &\sim (\mathbf{0}, \mathbf{D})\end{aligned}$$

holds with $\mathrm{Cov}[\mathbf{e}_i, \mathbf{b}_i] = \mathbf{0}$, the generalized least squares estimator

$$\begin{aligned}\widehat{\boldsymbol{\beta}}_{GLS} &= \left(\sum_{i=1}^{n} \mathbf{X}_i'\mathbf{V}_i(\boldsymbol{\theta})^{-1}\mathbf{X}_i\right)^{-1} \sum_{i=1}^{n} \mathbf{X}_i'\mathbf{V}_i(\boldsymbol{\theta})^{-1}\mathbf{Y}_i \\ &= \left(\mathbf{X}'\mathbf{V}(\boldsymbol{\theta})^{-1}\mathbf{X}\right)^{-1}\mathbf{X}'\mathbf{V}(\boldsymbol{\theta})^{-1}\mathbf{Y}\end{aligned} \qquad [7.26]$$

can be derived without any further distributional assumptions such as Gaussianity of $\mathbf{b}_i$ and/or $\mathbf{e}_i$. Only the first two moments of the marginal distribution of $\mathbf{Y}_i$, $\mathrm{E}[\mathbf{Y}_i] = \mathbf{X}_i\boldsymbol{\beta}$ and $\mathrm{Var}[\mathbf{Y}_i] = \mathbf{V}_i = \mathbf{Z}_i\mathbf{D}\mathbf{Z}_i' + \mathbf{R}_i$ are required. The idea of *estimated* generalized least squares (EGLS) is to substitute for $\mathbf{V}(\boldsymbol{\theta})$ a consistent estimator. If $\widehat{\boldsymbol{\theta}}$ is consistent for $\boldsymbol{\theta}$ one can simply substitute $\widehat{\boldsymbol{\theta}}$ for $\boldsymbol{\theta}$ and use $\widehat{\mathbf{V}} = \mathbf{V}(\widehat{\boldsymbol{\theta}})$. The ML and REML estimates [7.19] and [7.24] are of this form. Alternatively, one can estimate $\mathbf{D}$ and $\mathbf{R}$ directly and use $\widehat{\mathbf{V}} = \mathbf{Z}\widehat{\mathbf{B}}\mathbf{Z}' + \widehat{\mathbf{R}}$. In either case the estimator of the fixed effects becomes

$$\widehat{\boldsymbol{\beta}}_{EGLS} = \left(\sum_{i=1}^{n} \mathbf{X}_i'\widehat{\mathbf{V}}_i^{-1}\mathbf{X}_i\right)^{-1} \sum_{i=1}^{n} \mathbf{X}_i'\widehat{\mathbf{V}}_i^{-1}\mathbf{Y}_i = \left(\mathbf{X}'\widehat{\mathbf{V}}^{-1}\mathbf{X}\right)^{-1}\mathbf{X}'\widehat{\mathbf{V}}^{-1}\mathbf{Y}. \qquad [7.27]$$

The ML [7.19], REML [7.24], and EGLS [7.27] estimators of the fixed effects are of the same general form, and they differ only in how $\mathbf{V}$ is estimated. EGLS is appealing when $\mathbf{V}$ can be estimated quickly, preferably with a noniterative method. Vonesh and Chinchilli (1997, Ch. 8.2.4) argue that in applications with a sufficient number of observations and when interest lies primarily in $\boldsymbol{\beta}$, little efficiency is lost. Two basic noniterative methods are outlined in §A7.7.4 for the case where within cluster observations are uncorrelated and homoscedastic, that is, $\mathbf{R}_i = \sigma^2\mathbf{I}$. The first method estimates $\mathbf{D}$ and σ^2 by the method of moments and predicts the random effects with the usual formulas such as [7.15] substituting $\widehat{\mathbf{V}}$ for $\mathbf{V}$. The second method estimates the random effects $\mathbf{b}_i$ by regression methods first and calculates an estimate $\widehat{\mathbf{D}}$ from the $\widehat{\mathbf{b}}_i$.

7.4.3 Hypothesis Testing

Testing of hypotheses in linear mixed models proceeds along very similar lines as in the standard linear model without random effects. Some additional complications arise, however, since the distribution theory of the standard test statistics is not straightforward. To motivate the issues, we distinguish three cases:

(i) $\mathrm{Var}[\mathbf{Y}] = \mathbf{V}$ is completely known;

(ii) $\mathrm{Var}[\mathbf{Y}] = \sigma^2\mathbf{V} = \sigma^2\mathbf{ZBZ}' + \sigma^2\mathbf{R}$ is known up to the scalar constant σ^2;

(iii) $\mathrm{Var}[\mathbf{Y}] = \mathbf{V}(\boldsymbol{\theta})$ or $\mathrm{Var}[\mathbf{Y}] = \sigma^2\mathbf{V}(\boldsymbol{\theta}^*)$ depends on unknown parameters $\boldsymbol{\theta} = [\sigma^2, \boldsymbol{\theta}^*]$.

Of concern is a testable hypothesis of the same form as in the fixed effects linear model, namely $H_0\colon \mathbf{A}\boldsymbol{\beta} = \mathbf{d}$. For cases (i) and (ii) we develop in §A7.7.5 that for Gaussian random effects $\mathbf{b}_i$ and within-cluster errors $\mathbf{e}_i$ exact tests exist. Briefly, in case (i) the statistic

$$W = (\mathbf{A}\widehat{\boldsymbol{\beta}} - \mathbf{d})'\left[\mathbf{A}\left(\mathbf{X}'\mathbf{V}^{-1}\mathbf{X}\right)^{-1}\mathbf{A}'\right]^{-1}(\mathbf{A}\widehat{\boldsymbol{\beta}} - \mathbf{d}) \qquad [7.28]$$

is distributed under the null hypothesis as a Chi-squared variable with $r(\mathbf{A})$ degrees of freedom, where $r(\mathbf{A})$ denotes the rank of the matrix $\mathbf{A}$. Similarly, in case (ii) we have

$$W = (\mathbf{A}\widehat{\boldsymbol{\beta}} - \mathbf{d})'\left[\mathbf{A}\left(\mathbf{X}'\mathbf{V}^{-1}\mathbf{X}\right)^{-1}\mathbf{A}'\right]^{-1}(\mathbf{A}\widehat{\boldsymbol{\beta}} - \mathbf{d})/\sigma^2 \sim \chi^2_{r(\mathbf{A})}.$$

If the unknown σ^2 is replaced with

$$\widehat{\sigma}^2 = \frac{1}{T - r(\mathbf{X})}\left(\mathbf{y} - \mathbf{X}\widehat{\boldsymbol{\beta}}\right)'\mathbf{V}^{-1}\left(\mathbf{y} - \mathbf{X}\widehat{\boldsymbol{\beta}}\right). \qquad [7.29]$$

then

$$F_{obs} = \frac{(\mathbf{A}\widehat{\boldsymbol{\beta}} - \mathbf{d})'\left[\mathbf{A}(\mathbf{X}'\mathbf{V}^{-1}\mathbf{X})^{-1}\mathbf{A}'\right]^{-1}(\mathbf{A}\widehat{\boldsymbol{\beta}} - \mathbf{d})}{r(\mathbf{A})\widehat{\sigma}^2} \qquad [7.30]$$

is distributed as an F variable with $r(\mathbf{A})$ numerator and $T - r(\mathbf{X})$ denominator degrees of freedom. Notice that $\widehat{\sigma}^2$ is not the maximum likelihood estimator of σ^2. A special case of [7.30] is when $\mathbf{A}$ is a vector of zeros with a 1 in the jth position and $\mathbf{d} = \mathbf{0}$. The linear hypothesis $H_0\colon \mathbf{A}\boldsymbol{\beta} = \mathbf{0}$ then becomes $H_0\colon \beta_j = 0$ and F_{obs} can be written as

$$F_{obs} = \frac{\widehat{\beta}_j^2}{\mathrm{ese}\left(\widehat{\beta}_j\right)^2},$$

where $\mathrm{ese}(\widehat{\beta}_j)$ is the estimated standard error of $\widehat{\beta}_j$. This F_{obs} has one numerator degree of freedom and consequently

$$t_{obs} = \mathrm{sign}\left(\widehat{\beta}_j\right) \times \sqrt{F_{obs}} \qquad [7.31]$$

is distributed as a t random variable with $T - r(\mathbf{X})$ degrees of freedom. A $100(1 - \alpha)\%$

confidence interval for β_j is constructed as

$$\widehat{\beta}_j \pm t_{\alpha/2,T-r(\mathbf{X})} \times \text{ese}\left(\widehat{\beta}_j\right).$$

The problematic case is (iii), where the marginal variance-covariance matrix is unknown and more than just a scalar constant must be estimated from the data. The proposal is to replace $\mathbf{V}$ with $\mathbf{V}(\widehat{\boldsymbol{\theta}})$ in [7.28] and $\widehat{\sigma}^2\mathbf{V}(\widehat{\boldsymbol{\theta}}^*)$ in [7.30] and to use as test statistics

$$W^* = (\mathbf{A}\widehat{\boldsymbol{\beta}} - \mathbf{d})'\left[\mathbf{A}\left(\mathbf{X}'\mathbf{V}(\widehat{\boldsymbol{\theta}})^{-1}\mathbf{X}\right)^{-1}\mathbf{A}'\right]^{-1}(\mathbf{A}\widehat{\boldsymbol{\beta}} - \mathbf{d}) \qquad [7.32]$$

if $\text{Var}[\mathbf{Y}] = \mathbf{V}(\boldsymbol{\theta})$ and

$$F^*_{obs} = \frac{(\mathbf{A}\widehat{\boldsymbol{\beta}} - \mathbf{d})'\left[\mathbf{A}\left(\mathbf{X}'\mathbf{V}(\widehat{\boldsymbol{\theta}}^*)^{-1}\mathbf{X}\right)^{-1}\mathbf{A}'\right]^{-1}(\mathbf{A}\widehat{\boldsymbol{\beta}} - \mathbf{d})}{r(\mathbf{A})\widehat{\sigma}^{*2}} \qquad [7.33]$$

if $\text{Var}[\mathbf{Y}] = \sigma^2\mathbf{V}(\boldsymbol{\theta}^*)$. W^* is compared against cutoffs from a Chi-squared distribution with $r(\mathbf{A})$ degrees of freedom and F^*_{obs} against cutoffs from an F distribution. Unfortunately, this substitution has dramatic consequences for the distribution of the resulting statistics. Assume that the estimator of the covariance parameters being substituted is consistent. One can then show that, asymptotically, the distribution of W^* is χ^2 with $r(\mathbf{A})$ degrees of freedom, but the asymptotic distribution of F^*_{obs} is not that of an F random variable. The reason is that $\widehat{\sigma}^{*2}$ converges in distribution to σ^2 and in the limit F^*_{obs} is not the ratio of two independent χ^2 variables divided by their respective degrees of freedom. By comparing F^*_{obs} to cutoffs from an $F_{r(\mathbf{A}),T-r(\mathbf{X})}$ distribution, we do not utilize the correct asymptotic distribution. This argument should lead one to favor [7.32] over [7.33]. It has been established empirically, however, that the Type-I error rates of tests based on [7.33] are closer to the nominal rates than those of [7.32] and p-values of the Chi-square test will be smaller. There is a heuristic explanation for this phenomenon since [7.32] essentially corresponds to using an F distribution with infinitely many denominator degrees of freedom.

In certain balanced cases, for example, in complete split-plot designs, tests based on [7.32] and [7.33] can be exact. In general, one should anticipate, however, that the tests could be distorted. When substituting an estimate of $\boldsymbol{\theta}$, $\mathbf{A}(\mathbf{X}'\mathbf{V}(\widehat{\boldsymbol{\theta}})^{-1}\mathbf{X})^{-1}\mathbf{A}'$ is not an unbiased estimator of the variance of $\mathbf{A}\widehat{\boldsymbol{\beta}}$, which is the centerpiece in [7.28] and [7.30]. Even if $\mathbf{V}(\widehat{\boldsymbol{\theta}})$ were unbiased for $\mathbf{V}(\boldsymbol{\theta})$, $(\mathbf{X}'\mathbf{V}(\widehat{\boldsymbol{\theta}})^{-1}\mathbf{X})^{-1}$ underestimates $\text{Var}[\widehat{\boldsymbol{\beta}}]$ since the uncertainty arising from substituting $\widehat{\boldsymbol{\theta}}$ for $\boldsymbol{\theta}$ is not accounted for. Bias corrections and bias corrected estimators of $\text{Var}[\widehat{\boldsymbol{\beta}}]$ were developed by Kackar and Harville (1984), Prasad and Rao (1990), and Harville and Jeske (1992). Kenward and Roger (1997) combine a bias adjustment with a degree of freedom correction applied to the F test [7.33] that ensures that the actual Type-I error rate is close to the nominal rate in complex mixed models and models with complex error structure. They anticipate the correction to be necessary when sample size is small. But what is a small sample size? To demonstrate the bias in the various tests, we simulated $1,000$ repetitions of a repeated measures experiment under the following conditions. Four treatments are applied in a completely randomized design with three replications and repeated measurements are collected at times $t = 1, 2, 3, 4, 5$ on all experimental units. No observations are missing and all units are measured at the same time intervals. The linear mixed model for this experiment is

$$Y_{ijk} = \mu + \tau_i + e_{ij} + t_k + (\tau t)_{ik} + d_{ijk}, \qquad [7.34]$$

where τ_i measures the effect of treatment i and e_{ij} are independent experimental errors associated with replication j of treatment i. The terms t_k and $(\tau t)_{ik}$ denote time effects and treatment $\times$ time interactions. Finally, d_{ijk} are random disturbances among the serial measurements from an experimental unit. It is assumed that these disturbances are serially correlated according to

$$\mathrm{Corr}[d_{ijk}, d_{ijk'}] = \exp\left\{ -\frac{|t_k - t_{k'}|}{\phi} \right\}. \qquad [7.35]$$

This is known as the exponential correlation model (§7.5). The parameter ϕ determines the strength of the correlation of two disturbances $|t_k - t_{k'}|$ time units apart. Observations from different experimental units were assumed to be uncorrelated in keeping with the random assignment of treatments to experimental units. We simulated the experiments for various values of ϕ (Table 7.4).

Table 7.4. Correlations among the disturbances d_{ijk} in model [7.34] based on the exponential correlation model [7.35]

	Values of ϕ in simulation				
$\lvert t_k - t_{k'}\rvert$	$\frac{1}{3}$	1	2	3	4
1	0.049	0.368	0.606	0.717	0.778
2	0.002	0.135	0.368	0.513	0.607
3	0.000	0.049	0.223	0.368	0.472
4	0.000	0.018	0.135	0.264	0.368

The treatment $\times$ time cell mean structure is shown in Figure 7.8. The treatment means were chosen so that there was no marginal time effect and no marginal treatment effect. Also, there was no difference of the treatments at times 2, 4, or 5. In each of the 1,000 realizations of the experiment we tested the hypotheses

A H_0: no treatment main effect
B H_0: no time main effect
C H_0: no treatment effect at time 2
D H_0: no treatment effect at time 4.

These null hypotheses are true and at the 5% significance level the nominal Type-I error rate of the tests should be 0.05. An appropriate test procedure will be close to this nominal rate when average rejection rates are calculated across the 1,000 repetitions.

The following tests were performed:

- The exact Chi-square test based on [7.28] where the true values of the covariance parameters were used. These values were chosen as $\mathrm{Var}[e_{ij}] = \sigma_e^2 = 1$, $\mathrm{Var}[d_{ijk}] = \sigma_d^2 = 1$, and ϕ according to Table 7.4;

- The asymptotic Chi-square test based on [7.32] where the restricted maximum likelihood estimates of σ_e^2, σ_d^2, and ϕ were substituted;
- The asymptotic F test based on [7.33] where the restricted maximum likelihood estimates of σ_e^2, σ_d^2, and ϕ were substituted;
- The F test based on Kenward and Roger (1997) employing a bias correction in the estimation of Var$[\widehat{\boldsymbol{\beta}}]$ coupled with a degree of freedom adjusted F test.

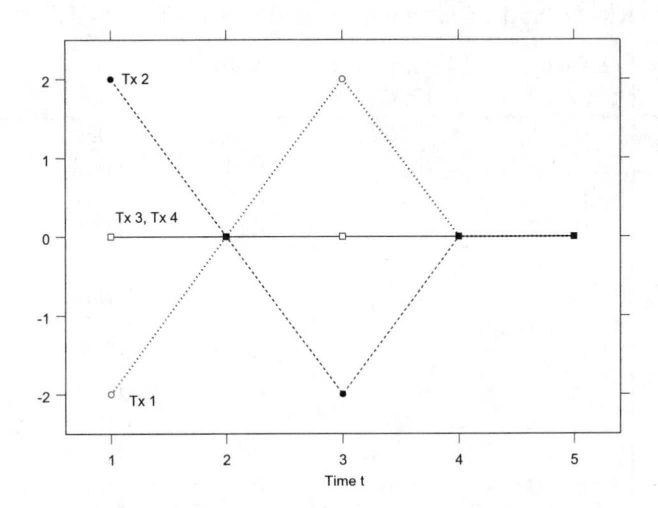

Figure 7.8. Treatment × time cell means in repeated measures simulation based on model [7.34].

The `proc mixed` statements that produce these tests are as follows.

```
/* Analysis with correct covariance parameter estimates */
/* Exact Chi-square test [7.28] */
proc mixed data=sim noprofile;
  class rep tx t;
  model y = tx t tx*t / Chisq ;
  random rep(tx);
  repeated /subject=rep(tx) type=sp(exp)(time);
  /* First parameter is Var[rep(tx)] */
  /* Second parameter is Var[e]       */
  /* Last parameter is range, passed here as a macro variable */
  parms (1) (1) (&phi) / hold=1,2,3;
  by repetition;
run;

/* Asymptotic Chi-square and F tests [7.32] and [7.33] */
proc mixed data=sim;
  class rep tx t;
  model y = tx t tx*t / Chisq ;
  random rep(tx);
  repeated /subject=rep(tx) type=sp(exp)(time);
  by repetition;
run;
```

```
/* Kenward-Roger F Tests */
proc mixed data=sim;
  class rep tx t;
  model y = tx t tx*t / ddfm=KenwardRoger;
  random rep(tx);
  repeated /subject=rep(tx) type=sp(exp)(time);
  by repetition;
run;
```

Table 7.5. Simulated Type-I error rates for exact and asymptotic Chi-square and F tests and Kenward-Roger adjusted F-test (KR-F denotes F^*_{obs} in Kenward-Roger test and KR-df the denominator degrees of freedom for KR-F)

ϕ	H_0	**Exact χ^2** [7.28]	**Asymp. χ^2** [7.32]	**Asymp. F** [7.33]	**KR- F**	**KR-df**
1/3	A	0.051	0.130	0.051	0.055	8.1
	B	0.049	0.092	0.067	0.056	24.9
	C	0.051	0.103	0.077	0.051	12.4
	D	0.056	0.101	0.085	0.055	12.4
1	A	0.055	0.126	0.048	0.051	8.3
	B	0.049	0.091	0.067	0.054	25.6
	C	0.060	0.085	0.069	0.055	18.1
	D	0.055	0.081	0.067	0.053	18.1
2	A	0.054	0.106	0.046	0.055	8.6
	B	0.049	0.091	0.067	0.056	26.2
	C	0.055	0.075	0.059	0.054	22.3
	D	0.049	0.073	0.058	0.055	22.3
3	A	0.053	0.107	0.039	0.049	8.8
	B	0.049	0.092	0.070	0.060	26.6
	C	0.049	0.075	0.053	0.048	24.5
	D	0.048	0.069	0.051	0.048	24.5
4	A	0.056	0.105	0.038	0.050	9.0
	B	0.049	0.090	0.070	0.061	27.1
	C	0.053	0.068	0.052	0.049	25.8
	D	0.042	0.066	0.047	0.044	25.8

The results are displayed in Table 7.5. The exact Chi-square test maintains the nominal Type-I error rate, as it should. The fluctuations around 0.05 are due to simulation variability. Increasing the number of repetitions will decrease this variability. When REML estimators are substituted for $\boldsymbol{\theta}$, the asymptotic Chi-square test performs rather poorly. The Type-I errors are substantially inflated; in many cases they are more than doubled. The asymptotic F test performs better, but the Type-I errors are typically somewhat inflated. Notice that with increasing strength of the serial correlation (increasing ϕ) the inflation is less severe for the asymptotic test, as was also noted by Kenward and Roger (1997). The bias and degree of freedom adjusted Kenward-Roger test performs extremely well. The actual Type-I errors are very close to the nominal error rate of 0.05. Even with sample sizes as small as 60 one should consider this test as a suitable procedure if exact tests of the linear hypothesis do not exist.

The tests discussed so far are based on the exact ($\mathbf{V}(\boldsymbol{\theta})$ known) or asymptotic ($\mathbf{V}(\boldsymbol{\theta})$ unknown) distribution of the fixed effect estimates $\widehat{\boldsymbol{\beta}}$. Tests of any model parameters, $\boldsymbol{\beta}$ or $\boldsymbol{\theta}$, can also be conducted based on the likelihood ratio principle, provided the hypothesis being tested is a simple restriction on $(\boldsymbol{\beta}, \boldsymbol{\theta})$ so that the restricted model is nested within the full model. If $(\widehat{\boldsymbol{\beta}}_M, \widehat{\boldsymbol{\theta}}_M)$ are the maximum likelihood estimate in the full model and $(\widetilde{\boldsymbol{\beta}}_M, \widetilde{\boldsymbol{\theta}}_M)$ are the maximum likelihood estimates under the restricted model, the likelihood-ratio test statistic

$$\Lambda = l\left(\widehat{\boldsymbol{\beta}}_{\boldsymbol{M}}, \widehat{\boldsymbol{\theta}}_M; \mathbf{y}\right) - l\left(\widetilde{\boldsymbol{\beta}}_{\boldsymbol{M}}, \widetilde{\boldsymbol{\theta}}_M; \mathbf{y}\right) \qquad [7.36]$$

has an asymptotic χ^2 distribution with degrees of freedom equal to the number of restrictions imposed. In REML estimation the hypothesis would be imposed on the covariance parameters only (more on this below) and the (residual) likelihood ratio statistic is

$$\Lambda = l\left(\widehat{\boldsymbol{\theta}}_R; \mathbf{y}\right) - l\left(\widetilde{\boldsymbol{\theta}}_R; \mathbf{y}\right).$$

Likelihood ratio tests are our test of choice to test hypotheses about the covariance parameters (provided the two models are nested). For example, consider again model [7.34] with equally spaced repeated measurements. The correlation of the random variables d_{ijk} can also be expressed as

$$\mathrm{Corr}[d_{ijk}, d_{ijk'}] = \rho^{|t_k - t_{k'}|}, \qquad [7.37]$$

where $\rho = \exp\{-1/\phi\}$ in [7.35]. This is known as the first-order autoregressive correlation model (see §7.5.2 for details about the genesis of this model). For a single replicate the correlation matrix now becomes

$$\mathrm{Corr}[\mathbf{d}_{ij}] = \begin{bmatrix} 1 & \rho & \rho^2 & \rho^3 & \rho^4 \\ \rho & 1 & \rho & \rho^2 & \rho^3 \\ \rho^2 & \rho & 1 & \rho & \rho^2 \\ \rho^3 & \rho^2 & \rho & 1 & \rho \\ \rho^4 & \rho^3 & \rho^2 & \rho & 1 \end{bmatrix}.$$

A test of $H_0{:}\rho = 0$ would address the question of whether the within-cluster disturbances (the repeated measures errors) are independent. In this case the mixed effects structure would be identical to that from a standard split-plot design where the temporal measurements comprise the sub-plot treatments. To test $H_0{:}\rho = 0$ with the likelihood ratio test the model is fit first with the autoregressive structure and then with an independence structure $(\rho = 0)$. The difference in their likelihood ratios is then calculated and compared to the cutoffs from a Chi-squared distribution with one degree of freedom. We illustrate the process with one of the repetitions from the simulation experiment. The full model is fit with the `mixed` procedure of The SAS® System:

```
proc mixed data=sim;
  class rep tx t;
  model y = tx t tx*t;
  random rep(tx);
  repeated /subject=rep(tx) type=ar(1);
run;
```

The `repeated` statement indicates that the combinations of replication and treatment variable, which identifies the experimental unit, are the clusters which are considered independent. All observations that share the same replication and treatment values are considered the within-cluster observations and are correlated according to the autoregressive model (`type=ar(1)`, Output 7.2). Notice the table of `Covariance Parameter Estimates` which reports the estimates of the covariance parameters $\widehat{\sigma}_e^2 = 1.276$, $\widehat{\sigma}_d^2 = 0.0658$, and $\widehat{\rho} = 0.2353$. The table of `Fit Statistics` shows the **residual** log likelihood $l(\widehat{\boldsymbol{\theta}}_R; \mathbf{y}) = -29.3$ as `Res Log Likelihood`.

Output 7.2.

```
                    The Mixed Procedure

                    Model Information

Data Set                      WORK.SIM
Dependent Variable            y
Covariance Structures         Variance Components, Autoregressive
Subject Effect                rep(tx)
Estimation Method             REML
Residual Variance Method      Profile
Fixed Effects SE Method       Model-Based
Degrees of Freedom Method     Containment

                  Dimensions
         Covariance Parameters              3
         Columns in X                      30
         Columns in Z                      12
         Subjects                           1
         Max Obs Per Subject               60
         Observations Used                 60
         Observations Not Used              0
         Total Observations                60

          Covariance Parameter Estimates
          Cov Parm      Subject     Estimate
          rep(tx)                     1.2762
          AR(1)         rep(tx)       0.2353
          Residual                   0.06580

                    Fit Statistics
      Res Log Likelihood                          -29.3
      Akaike's Information Criterion              -32.3
      Schwarz's Bayesian Criterion                -33.0
      -2 Res Log Likelihood                        58.6
```

Fitting the model with an independence structure ($\rho = 0$) is accomplished with the statements (Output 7.3)

```
proc mixed data=sim;
  class rep tx t;
  model y = tx t tx*t;
  random rep(tx);
run;
```

Observe that the reduced model has only two covariance parameters, σ_e^2 and σ_d^2. Its **residual** log likelihood is $l(\widetilde{\boldsymbol{\theta}}_R; \mathbf{y}) = -29.6559$. Since one parameter was removed, the likelihood ratio test compares

$$\Lambda = 2(029.3 - (-29.65)) = 0.7$$

against a χ_1^2 distribution. The p-value for this test is $\Pr(\chi_1^2 > 0.7) = 0.4028$ and $H_0{:}\rho = 0$ cannot be rejected. The p-value for this likelihood-ratio test can be conveniently computed with The SAS® System:

```
data test; p = 1-ProbChi(0.7,1); run; proc print data=test; run;
```

Output 7.3.

```
                        The Mixed Procedure

                         Model Information
          Data Set                        WORK.SIM
          Dependent Variable              y
          Covariance Structure            Variance Components
          Estimation Method               REML
          Residual Variance Method        Profile
          Fixed Effects SE Method         Model-Based
          Degrees of Freedom Method       Containment

                            Dimensions
                   Covariance Parameters              2
                   Columns in X                      30
                   Columns in Z                      12
                   Subjects                           1
                   Max Obs Per Subject               60
                   Observations Used                 60
                   Observations Not Used              0
                   Total Observations                60

                     Covariance Parameter Estimates
                          Cov Parm       Estimate
                          rep(tx)          1.2763
                          Residual        0.05798

                           Fit Statistics
                Res Log Likelihood                    -29.6
                Akaike's Information Criterion        -31.6
                Schwarz's Bayesian Criterion          -32.1
                -2 Res Log Likelihood                  59.3
```

In the preceding test we used the restricted maximum likelihood objective function for testing $H_0{:}\rho = 0$. This is justified since the hypothesis was about a covariance parameter. As discussed in §7.4.1, the restricted likelihood contains information about the covariance parameters only, not about the fixed effects $\boldsymbol{\beta}$. To test hypotheses about $\boldsymbol{\beta}$ based on the likelihood ratio principle one should fit the model by maximum likelihood. With the `mixed` procedure this is accomplished by adding the `method=ml` option to the `proc mixed` statement, e.g.,

```
proc mixed data=sim method=ml;
  class rep tx t;
  model y = tx t tx*t;
  random rep(tx);
  repeated /subject=rep(tx) type=ar(1);
run;
```

At times the full and *reduced* models are not nested, that is, the restricted model cannot be obtained from the full model by simply constraining or setting to zero some of its parameters. For example, to compare whether the correlations between the repeated measurements follow the exponential model

$$\text{Corr}[d_{ijk}, d_{ijk'}] = \exp\left\{ -\frac{|t_k - t_{k'}|}{\phi} \right\}$$

or the spherical model

$$\text{Corr}[d_{ijk}, d_{ijk'}] = \left\{1 - \frac{3}{2}\frac{|t_k - t_{k'}|}{\phi} + \frac{1}{2}\left(\frac{|t_k - t_{k'}|}{\phi}\right)^3\right\} I(|t_k - t_{k'}| \leq \phi),$$

one cannot nest one model within the other. A different *test* procedure is needed. The method commonly used relies on comparing overall goodness-of-fit statistics of the competing models (Bozdogan 1987, Wolfinger 1993a). The most important ones are Akaike's information criterion (AIC, Akaike 1974) and Schwarz' criterion (Schwarz 1978). Both are functions of the (restricted) log likelihood with penalty terms added for the number of covariance parameters. In some releases of `proc mixed` a smaller value of the AIC or Schwarz' criterion indicates a better fit. Notice that there are other versions of these two criteria where larger values indicate a better fit. One cannot associate degrees of significance or p-values with these measures. They are interpreted in a *greater/smaller is better* sense only. For the two models fitted above AIC is –32.3 for the model with autoregressive error terms and –31.6 for the model with independent error terms. The AIC criterion leads to the same conclusion as the likelihood-ratio test. The independence model fits this particular set of data better. We recommed to use the AIC or Schwarz criterion for models with *the same fixed effects* terms but different, non-nested covariance structures.

7.5 Correlations in Mixed Models

7.5.1 Induced Correlations and the Direct Approach

A feature of mixed models is to induce correlations even if all the random variables in a model are uncorrelated. While for many applications this is an incidental property of the model structure, the modeler can also purposefully employ this feature in cases where data are known to be correlated, as in longitudinal studies where serial correlation among the repeat measurements is typical. We illustrate the phenomenon of inducing correlations with a simple example.

Consider a linear regression model where the cluster-specific intercepts vary at random (see, for example, Figure 7.6b). Casting this model in the Laird-Ware framework, we write for observation j from cluster i,

$$\begin{aligned} Y_{ij} &= (\beta_0 + b_i) + \beta_1 x_{ij} + e_{ij} \\ \text{Var}[b_i] &= \sigma_b^2,\ \text{Var}[e_{ij}] = \sigma^2 \\ i &= 1, ..., n; j = 1, ..., n_i. \end{aligned} \qquad [7.38]$$

In terms of matrices and vectors, $\mathbf{Y}_i = \mathbf{X}_i\boldsymbol{\beta} + \mathbf{Z}_i\mathbf{b}_i + \mathbf{e}_i$, the model for the ith cluster is written as

$$\begin{bmatrix} Y_{i1} \\ Y_{i2} \\ \vdots \\ Y_{in_i} \end{bmatrix} = \begin{bmatrix} 1 & x_{i1} \\ 1 & x_{i2} \\ \vdots & \vdots \\ 1 & x_{in_i} \end{bmatrix} \begin{bmatrix} \beta_0 \\ \beta_1 \end{bmatrix} + \begin{bmatrix} 1 \\ 1 \\ 1 \\ 1 \end{bmatrix} b_i + \begin{bmatrix} e_{i1} \\ e_{i2} \\ \vdots \\ e_{in_i} \end{bmatrix}.$$

Notice that the $\mathbf{Z}_i$ matrix is the first column of the $\mathbf{X}_i$ matrix (a random intercept). If the error terms e_{ij} are homoscedastic and uncorrelated and $\text{Cov}[b_i, e_{ij}] = 0$, the marginal variance-covariance matrix of the observations from the ith cluster is then

$$\text{Var}[\mathbf{Y}_i] = \sigma_b^2 \mathbf{Z}_i \mathbf{Z}_i' + \sigma^2 \mathbf{I} = \sigma_b^2 \begin{bmatrix} 1 \\ 1 \\ \vdots \\ 1 \end{bmatrix} [1 \quad 1 \quad \cdots \quad 1] + \sigma^2 \begin{bmatrix} 1 & 0 & 0 & 0 \\ 0 & 1 & \cdots & 0 \\ \vdots & \vdots & \ddots & \vdots \\ 0 & 0 & \cdots & 1 \end{bmatrix}$$

$$= \begin{bmatrix} \sigma_b^2 + \sigma^2 & \sigma_b^2 & \cdots & \sigma_b^2 \\ \sigma_b^2 & \sigma_b^2 + \sigma^2 & \cdots & \sigma_b^2 \\ \vdots & \vdots & \ddots & \vdots \\ \sigma_b^2 & \sigma_b^2 & \cdots & \sigma_b^2 + \sigma^2 \end{bmatrix} = \sigma_b^2 \mathbf{J} + \sigma^2 \mathbf{I}.$$

The correlation between the jth and k^{th} observation from a cluster is

$$\text{Corr}[Y_{ij}, Y_{ik}] = \rho = \sigma_b^2 / \left(\sigma_b^2 + \sigma^2\right), \qquad [7.39]$$

the ratio between the variability among clusters and the variability of an observation. $\text{Var}[\mathbf{Y}_i]$ can then also be expressed as

$$\text{Var}[\mathbf{Y}_i] = \left(\sigma_b^2 + \sigma^2\right) \begin{bmatrix} 1 & \rho & \cdots & \rho \\ \rho & 1 & \cdots & \rho \\ \vdots & \vdots & \ddots & \vdots \\ \rho & \rho & \cdots & 1 \end{bmatrix}. \qquad [7.40]$$

The variance-covariance structure [7.40] is known as **compound symmetry** (CS), the exchangeable structure, or the equicorrelation structure and ρ is also called the intracluster correlation coefficient. In this model all observations within a cluster are correlated by the same amount. The structure arises with nested random effects, i.e., $\mathbf{Z}_i$ is a vector of ones and the within-cluster errors are uncorrelated. Hence, subsampling and split-plot designs exhibit this correlation structure. We term correlations that arise from the nested character of the random effects (b_i is a cluster-level random variable, e_{ij} is an observation-specific random variable) **induced** correlations.

In applications where it is known *a priori* that observations are correlated within a cluster one can take a more direct approach. In the model $\mathbf{Y}_i = \mathbf{X}_i \boldsymbol{\beta} + \mathbf{Z}_i \mathbf{b}_i + \mathbf{e}_i$ assume for the moment that $\mathbf{b}_i = \mathbf{0}$. We are then left with a fixed effects linear model where the errors $\mathbf{e}_i$ have variance-covariance matrix $\mathbf{R}_i$. If we let $\text{Var}[Y_{ij}] = \xi$ and

$$\mathbf{R}_i = \xi \begin{bmatrix} 1 & \rho & \cdots & \rho \\ \rho & 1 & \cdots & \rho \\ \vdots & \vdots & \ddots & \vdots \\ \rho & \rho & \cdots & 1 \end{bmatrix}, \qquad [7.41]$$

then the model also has a compound-symmetric correlation structure. Such a model could arise with clusters of size four if one draws blood samples from each leg of a heifer and ρ measures the correlation of serum concentration among the samples from a single animal. No leg takes precedence over any other legs; they are *exchangeable*, with no particular order. In this model the within-cluster variance-covariance matrix was targeted directly to capture the correlations among the observations from the same cluster; we therefore call it **direct modeling** of the correlations. Observe that if $\xi = \sigma_b^2 + \sigma^2$ there is no difference in the marginal variability between [7.40] and [7.41]. In §7.6.3 we examine a subsampling design and show that modeling a random intercept according to [7.40] and modeling the exchangeable structure [7.41] directly leads to the same inference.

There is also a mixture approach where some correlations are induced through random effects *and* the variance-covariance matrix $\mathbf{R}_i$ of the within-cluster errors is also structured. In the linear mixed model with random intercepts above ([7.38]) assume that the measurements from a cluster are repeated observations collected over time. It is then reasonable to assume that the measurements from a given cluster are serially correlated, i.e., $\mathbf{R}_i$ is not a diagonal matrix. If it is furthermore sensible to posit that observations close together in time are more highly correlated than those far apart, one may put

$$\mathbf{R}_i = \sigma^2 \begin{bmatrix} 1 & \rho & \rho^2 & \rho^3 & \cdots & \rho^{n_i-1} \\ \rho & 1 & \rho & \rho^2 & \cdots & \rho^{n_i-2} \\ \rho^2 & \rho & 1 & \rho & \cdots & \rho^{n_i-3} \\ \rho^3 & \rho^2 & \rho & 1 & \cdots & \rho^{n_i-4} \\ \vdots & \vdots & \vdots & \vdots & \ddots & \vdots \\ \rho^{n_i-1} & \rho^{n_i-2} & \rho^{n_i-3} & \rho^{n_i-4} & \cdots & 1 \end{bmatrix}, \rho \geq 0.$$

This correlation structure is known as the first-order autoregressive (AR(1)) structure borrowed from the analysis of time series (see §7.5.2 and §A7.7.6). It is applied frequently if repeated measurements are equally spaced. For example, if repeated measurements on experimental units in a randomized block design are taken in weeks 0, 2, 4, 6, and 8, the correlation between any two measurement two weeks apart is ρ, between any two measurements four weeks apart is ρ^2, and so forth. Combining this within-cluster correlation structure with cluster-specific random intercepts leads to a more complicated marginal variance-covariance structure:

$$\text{Var}[\mathbf{Y}_i] = \sigma_b^2\mathbf{J} + \mathbf{R}_i = \begin{bmatrix} \sigma_b^2 + \sigma^2 & \sigma_b^2 + \sigma^2\rho & \sigma_b^2 + \sigma^2\rho^2 & \cdots & \sigma_b^2 + \sigma^2\rho^{n_i-1} \\ \sigma_b^2 + \sigma^2\rho & \sigma_b^2 + \sigma^2 & \sigma_b^2 + \sigma^2\rho & \cdots & \sigma_b^2 + \sigma^2\rho^{n_i-2} \\ \sigma_b^2 + \sigma^2\rho^2 & \sigma^2\rho & \sigma_b^2 + \sigma^2 & \cdots & \sigma_b^2 + \sigma^2\rho^{n_i-3} \\ \vdots & \vdots & \vdots & \ddots & \vdots \\ \sigma_b^2 + \sigma^2\rho^{n_i-1} & \sigma_b^2 + \sigma^2\rho^{n_i-2} & \cdots & \cdots & \sigma_b^2 + \sigma^2 \end{bmatrix}.$$

Notice that with $\rho > 0$ the within-cluster correlations approach zero with increasing temporal separation and the marginal correlations approach $\sigma_b^2/(\sigma_b^2 + \sigma^2)$. Decaying correlations with increasing separation are typically reasonable. Because mixed models induce correlations through the random effects one can achieve marginal correlations that are functions of a chosen metric quite simply. If correlations are to depend on time, for example, simply include a time variable as a column of $\mathbf{Z}$. The resulting marginal correlation structure may not be meaningful, however. We illustrate with an example.

Example 7.5. The growth pattern of an experimental soybean variety is studied. Eight plots are seeded and the average leaf weight per plot is assessed at weekly intervals following germination. If t measures time since seeding in days, the average growth is assumed to follow a linear model

$$Y_{ij} = \beta_0 + \beta_1 t_{ij} + \beta_2 t_{ij}^2 + e_{ij},$$

where Y_{ij} is the average leaf weight per plant on plot i measured at time t_{ij}. The double subscript for the time variable t allows measurement occasions to differ among plots. Figure 7.9 shows data for the growth of soybeans on the eight plots simulated after Figure 1.2 in Davidian and Giltinan (1995). In this experiment, a plot serves as a cluster and the eight trends differ in their linear gradients (slopes). It is thus reasonable to add random coefficients to β_1. The mixed model becomes

$$Y_{ij} = \beta_0 + (\beta_1 + b_{1i})t_{ij} + \beta_2 t_{ij}^2 + e_{ij}. \qquad [7.42]$$

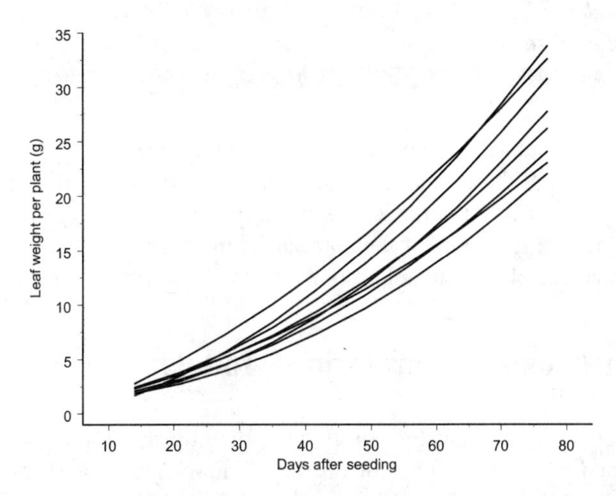

Figure 7.9. Simulated soybean leaf weight profiles fitted to data from eight experimental plots.

If the within-cluster errors are independent, reasonable when the leaf weight is obtained from a random sample of plants from plot i at time t_{ij}, the model in matrix formulation is $\mathbf{Y}_i = \mathbf{X}_i\boldsymbol{\beta} + \mathbf{Z}_i\mathbf{b}_i + \mathbf{e}_i$, $\text{Var}[\mathbf{e}_i] = \sigma^2\mathbf{I}$, with quantities defined as follows:

$$\mathbf{X}_i = \begin{bmatrix} 1 & t_{i1} & t_{i1}^2 \\ 1 & t_{i2} & t_{i2}^2 \\ \vdots & \vdots & \vdots \\ 1 & t_{in_i} & t_{in_i}^2 \end{bmatrix}, \boldsymbol{\beta} = \begin{bmatrix} \beta_0 \\ \beta_1 \\ \beta_2 \end{bmatrix}, \mathbf{Z}_i = \begin{bmatrix} t_{i1} \\ t_{i2} \\ \vdots \\ t_{in_i} \end{bmatrix}, \mathbf{b}_i = b_{1i}, \text{Var}[b_i] = \sigma_b^2.$$

The marginal variance-covariance matrix $\text{Var}[\mathbf{Y}_i] = \mathbf{Z}_i\mathbf{D}\mathbf{Z}_i' + \sigma^2\mathbf{I}$ for [7.42] is now

$$\text{Var}[\mathbf{Y}_i] = \sigma_b^2 \begin{bmatrix} t_{i1}^2 + \sigma^2/\sigma_b^2 & t_{i1}t_{i2} & \cdots & t_{i1}t_{in_i} \\ t_{i2}t_{i1} & t_{i2}^2 + \sigma^2/\sigma_b^2 & \cdots & t_{i2}t_{in_i} \\ \vdots & & \ddots & \vdots \\ t_{in_i}t_{i1} & t_{in_i}t_{i2} & \cdots & t_{in_i}^2 + \sigma^2/\sigma_b^2 \end{bmatrix}.$$

The covariance structure depends on the time variable which is certainly a meaningful metric for the correlations. Whether the correlations are suitable funtions of that metric can be argued. Between any two leaf weight measurements on the same plot we have

$$\text{Corr}[Y_{ij}, Y_{ij'}] = \frac{t_{ij}t_{ij'}}{\sqrt{\left(\sigma_b^2 t_{ij}^2 + \sigma^2\right)\left(\sigma_b^2 t_{ij'}^2 + \sigma^2\right)}}.$$

For illustration let $\sigma^2 = \sigma_b^2 = 1$ and the repeated measurements be coded $t_{i1} = 1$, $t_{i2} = 2$, $t_{i3} = 3$, and so forth. Then $\text{Corr}[Y_{i1}, Y_{i2}] = 2/\sqrt{10} = 0.63$, $\text{Corr}[Y_{i1}, Y_{i3}] = 3/\sqrt{20} = 0.67$, and $\text{Corr}[Y_{i1}, Y_{i4}] = 4/\sqrt{34} = 0.68$. The correlations are not decaying with temporal separation, they increase. Also, two time points equally spaced apart do not have the same correlation. For example, $\text{Corr}[Y_{i2}, Y_{i3}] = 6/\sqrt{50} = 0.85$ which exceeds the correlation between time points 1 and 2. If one would code time as $t_{i1} = 0, t_{i2} = 1, t_{i3} = 2, \cdots$, the correlations of any observations with the first time point would be uniformly zero.

If correlations are subject to modeling, they should be modeled directly on the within-cluster level, i.e., through the $\mathbf{R}_i$ matrix. Interpretation of the correlation pattern should then be confined to cluster-specific inference. Trying to pick up correlations by choosing columns of the $\mathbf{Z}_i$ matrix that are functions of the correlation metameter will not necessarily lead to a meaningful marginal correlation model, or one that can be interpreted with ease.

7.5.2 Within-Cluster Correlation Models

In order to model the within-cluster correlations directly and parsimoniously, we rely on structures that impose a certain behavior on the within-cluster disturbances while requiring only a small number of parameters. In this subsection several of these structures are introduced. The modeler must eventually decide which structure is to be used for a final fit of the model to the data. In our experience it is more important to model the correlation structure in a reasonable and meaningful way rather than to model the correlation structure perfectly. When data are clearly correlated, assuming independence is certainly not a meaningful approach, but one will find that a rather large class of correlation models will provide similar goodness-of-fit to the data. The modeler is encouraged to find one member of this class, rather than to go overboard in an attempt to mold the model too closely to the data at hand. If models are nested, the likelihood ratio test (§7.4.3) can help to decide which correlation structure fits the data best. If models are not nested, we rely on goodness-of-fit criterion such as Akaike's Information Criterion and Schwarz' criterion. If several non-nested models with similar AIC statistic emerge, we remind ourselves of Ockham's razor and choose the most parsimonious one, the one with the fewest parameters. Finding an appropriate function that describes the mean trend, the fixed effects structure, is a somewhat intuitive part of statistical modeling. A plot of the data suggests a particular linear or nonlinear model which can be fitted to data and subsequently improved upon. Gaining insight into the correlation pattern of data over time and/or space is less intuitive. Fortunately, several correlation models have emerged and proven to work well in many applications. Some of these are borrowed from developments in time series analysis or geostatistics, others from the theory of designed experiments.

In what follows we assume that the within-cluster variance-covariance matrix is of the form

$$\text{Var}[\mathbf{e}_i] = \mathbf{R}_i = \mathbf{R}_i(\boldsymbol{\alpha}),$$

where $\boldsymbol{\alpha}$ is a vector of parameters determining the correlations among and the variances of the elements of $\mathbf{e}_i$. In previous notation we labeled $\boldsymbol{\theta}$ the vector of covariance parameters in $\text{Var}[\mathbf{Y}_i] = \mathbf{Z}_i\mathbf{D}\mathbf{Z}_i' + \mathbf{R}_i$. Hence $\boldsymbol{\alpha}$ contains only those covariance parameters that are not contained in $\text{Var}[\mathbf{b}_i] = \mathbf{D}$. In many applications $\boldsymbol{\alpha}$ will be a two-element vector, containing one parameter to model the within-cluster correlations and one parameter to model the within-cluster variances.

We would be remiss not to mention approaches which account for serial correlations but make no assumptions about the structure of $\text{Var}[\mathbf{e}_i]$. One is the multivariate repeated measures approach (Cole and Grizzle 1966, Crowder and Hand 1990, Vonesh and Chinchilli 1997), sometimes labeled multivariate analysis of variance (MANOVA). This approach is restrictive if data are unbalanced or missing or covariates are varying with time. The mixed model approach based on the Laird-Ware model is more general in that it allows clusters of unequal sizes, unequal spacing of observation times or locations and missing observations. The MANOVA approach essentially uses an unstructured variance-covariance matrix, which is one of the structures open to investigation in the Laird-Ware model (Jennrich and Schluchter 1986).

An $(n \times n)$ covariance matrix contains $n(n+1)/2$ unique elements and if a cluster contains 6 measurements, say, up to $6*7/2 = 21$ parameters need to be estimated in addition to the fixed effects parameters and variances of any random effects. Imposing structure on the variance-covariance matrix beyond an unstructured model requires far fewer parameters. In the remainder of this section it is assumed for the sake of simplicity that a cluster contains $n_i = 4$ elements and that measurements are collected in time. Some of these structures will reemerge when we are concerned with spatial data in §9.

The simplest covariance structure is the independence structure,

$$\text{Var}[\mathbf{e}_i] = \sigma^2\mathbf{I} = \sigma^2\begin{bmatrix} 1 & 0 & 0 & 0 \\ 0 & 1 & 0 & 0 \\ 0 & 0 & 1 & 0 \\ 0 & 0 & 0 & 1 \end{bmatrix}. \qquad [7.43]$$

Apart from the scalar σ^2 no additional parameters need to be estimated. The compound-symmetric or exchangeable structure

$$\mathbf{R}_i(\boldsymbol{\alpha}) = \sigma^2\begin{bmatrix} 1 & \rho & \rho & \rho \\ \rho & 1 & \rho & \rho \\ \rho & \rho & 1 & \rho \\ \rho & \rho & \rho & 1 \end{bmatrix}, \boldsymbol{\alpha} = \left[\sigma^2, \rho\right], \qquad [7.44]$$

is suitable if measurements are equicorrelated and exchangeable. If there is no particular ordering among the correlated measurements a compound-symmetric structure may be reasonable. We reiterate that a split-plot or subsampling design also has a compound-symmetric marginal correlation structure. In `proc mixed` one can thus analyze a split-plot experiment in two equivalent ways. The first is to use independent random effects representing whole-plot and sub-plot experimental errors. If the whole-plot design is a randomized completely randomized design, the `proc mixed` statements are

```
proc mixed data=yourdata;
  class rep A B;
  model y = A B A*B;
  random rep(A);
run;
```

It appears that only one random effect has been specified, the whole-plot experimental error `rep(A)`. `Proc mixed` will add a second, residual error term automatically, which corresponds to the sub-plot experimental error. The default containment method of assigning degrees of freedom will ensure that F tests are formulated correctly, that is, whole-plot effects are tested against the whole-plot experimental error variance and sub-plot effects and interactions are tested against the sub-plot experimental error variance. If comparisons of the treatment means are desired, we recommend to add the `ddfm=satterth` option to the `model` statement. This will invoke the Satterthwaite approximation where necessary, for example, when comparing whole-plot treatments at the same level of the sub-plot factor (see §7.6.5 for an example):

```
proc mixed data=yourdata;
  class rep A B;
  model y = rep A B A*B / ddfm=satterth;
  random rep(A);
run;
```

As detailed in §7.5.3 this error structure will give rise to a marginal compound-symmetric structure. The direct approach of modeling a split-plot design is by specifying the compound-symmetric model through the `repeated` statement:

```
proc mixed data=yourdata;
  class rep A B;
  model y = A B A*B / ddfm=satterth;
  repeated / subject=rep(A) type=cs;
run;
```

In longitudinal or repeated measures studies where observations are ordered along a time scale, the compound symmetry structure is often not reasonable. Correlations for pairs of time points are not the same and hence not exchangeable. The analysis of repeated measures data as split-plot-type designs assuming compound symmetry is, however, very common in practice. In section §7.5.3 we examine under which conditions this is an appropriate analysis.

The first, rather crude, modification is not to specify anything about the structure of the correlation matrix and to estimate all unique elements of $\mathbf{R}_i(\boldsymbol{\alpha})$. This unstructured variance-covariance matrix can be expressed as

$$\mathbf{R}_i(\boldsymbol{\alpha}) = \sigma^2 \begin{bmatrix} 1 & \rho_{12} & \rho_{13} & \rho_{14} \\ \rho_{21} & 1 & \rho_{23} & \rho_{24} \\ \rho_{31} & \rho_{32} & 1 & \rho_{34} \\ \rho_{41} & \rho_{42} & \rho_{43} & 1 \end{bmatrix} \qquad [7.45]$$

$$\boldsymbol{\alpha} = \left[\sigma^2, \rho_{12}, \rho_{13}, \cdots, \rho_{34}\right].$$

Here, $\rho_{ij} = \rho_{ji}$ is the correlation between observations i and j within a cluster. This is not a parsimonious structure; there are $n_i \times (n_i - 1)/2$ correlations that need to be estimated. In many repeated measures data sets where the sequence of temporal observations is small, insufficient information is available to estimate all the correlations with satisfactory precision. Furthermore, there is no guarantee that the correlations will decrease with temporal separation. If that is a reasonable stipulation, other models should be employed. The unstructured model is fit by `proc mixed` with the `type=un` option of the `repeated` statement.

The large number of parameters in the unstructured parameterization can be reduced by introducing constraints. For example, one may assume that all c-step correlations are identical,

$$\rho_{jj'} = \rho_j \text{ if } |j - j'| = c.$$

This leads to banded, also called Toeplitz, structures if the diagonal elements are the same. A Toeplitz matrix of order k has $k - 1$ off-diagonals filled with the same element. A 2-banded Toeplitz parameterization is

$$\mathbf{R}_i(\boldsymbol{\alpha}) = \sigma^2 \begin{bmatrix} 1 & \rho_1 & 0 & 0 \\ \rho_1 & 1 & \rho_1 & 0 \\ 0 & \rho_1 & 1 & \rho_1 \\ 0 & 0 & \rho_1 & 1 \end{bmatrix}, \ \boldsymbol{\alpha} = \left[\sigma^2, \rho_1\right], \qquad [7.46]$$

and a 3-banded Toeplitz structure

$$\mathbf{R}_i(\boldsymbol{\alpha}) = \sigma^2 \begin{bmatrix} 1 & \rho_1 & \rho_2 & 0 \\ \rho_1 & 1 & \rho_1 & \rho_2 \\ \rho_2 & \rho_1 & 1 & \rho_1 \\ 0 & \rho_2 & \rho_1 & 1 \end{bmatrix}, \boldsymbol{\alpha} = \left[\sigma^2, \rho_1, \rho_2\right]. \qquad [7.47]$$

A 2-banded Toeplitz structure may be appropriate if, for example, measurements are taken at weekly intervals, but correlations do not extend past a period of seven days. In a turfgrass experiment where mowing clippings are collected weekly but a fast acting growth regulator is applied every ten days, an argument can be made that correlations do not persist over more than two measurement intervals.

Unstructured correlation models can also be banded by setting elements in off-diagonal cells more than $k - 1$ positions from the main diagonal to zero. The 2-banded unstructured parameterization is

$$\mathbf{R}_i(\boldsymbol{\alpha}) = \begin{bmatrix} \sigma_1^2 & \sigma_{12} & 0 & 0 \\ \sigma_{21} & \sigma_2^2 & \sigma_{23} & 0 \\ 0 & \sigma_{32} & \sigma_3^2 & \sigma_{34} \\ 0 & 0 & \sigma_{43} & \sigma_4^2 \end{bmatrix}, \boldsymbol{\alpha} = \left[\sigma_1^2, \cdots, \sigma_4^2, \sigma_{12}, \sigma_{23}, \sigma_{34}\right], \qquad [7.48]$$

and the 3-banded unstructured is

$$\mathbf{R}_i(\boldsymbol{\alpha}) = \begin{bmatrix} \sigma_1^2 & \sigma_{12} & \sigma_{13} & 0 \\ \sigma_{21} & \sigma_2^2 & \sigma_{23} & \sigma_{24} \\ \sigma_{31} & \sigma_{32} & \sigma_3^2 & \sigma_{34} \\ 0 & \sigma_{42} & \sigma_{43} & \sigma_4^2 \end{bmatrix}, \boldsymbol{\alpha} = \left[\sigma_1^2, \cdots, \sigma_4^2, \sigma_{12}, \sigma_{23}, \sigma_{34}, \sigma_{13}, \sigma_{24}\right] \qquad [7.49]$$

The 1-banded unstructured matrix is appropriate for independent observations which differ in their variances, i.e., heteroscedastic data:

$$\mathbf{R}_i(\boldsymbol{\alpha}) = \begin{bmatrix} \sigma_1^2 & 0 & 0 & 0 \\ 0 & \sigma_2^2 & 0 & 0 \\ 0 & 0 & \sigma_3^2 & 0 \\ 0 & 0 & 0 & \sigma_4^2 \end{bmatrix}, \boldsymbol{\alpha} = \left[\sigma_1^2, \cdots, \sigma_4^2\right]. \qquad [7.50]$$

These structures are fit in `proc mixed` with the following options of the `repeated` statement:

```
type=Toep(2) /* 2-banded Toeplitz      */
type=Toep(3) /* 3-banded Toeplitz      */
type=un(2)   /* 2-banded unstructured */
type=un(3)   /* 3-banded unstructured */
type=un(1)   /* 1-banded unstructured */.
```

We now turn to models where the correlations decrease with temporal separation of the measurements. One of the more popular models is borrowed from the analysis of time series data. Assume that a present observation at time t, $Y(t)$, is related to the immediately preceding observation at time $t-1$ through the relationship

$$Y(t) = \rho Y(t-1) + e(t). \qquad [7.51]$$

The $e(t)$'s are uncorrelated, identically distributed random variables with mean 0 and variance σ_e^2 and ρ is the autoregressive parameter of the time-series model. In the vernacular of time series analysis the $e(t)$ are called the random innovations (random shocks) of the process. This model is termed the first-order autoregressive (AR(1)) time series model since an outcome $Y(t)$ is regressed on the immediately preceding observation.

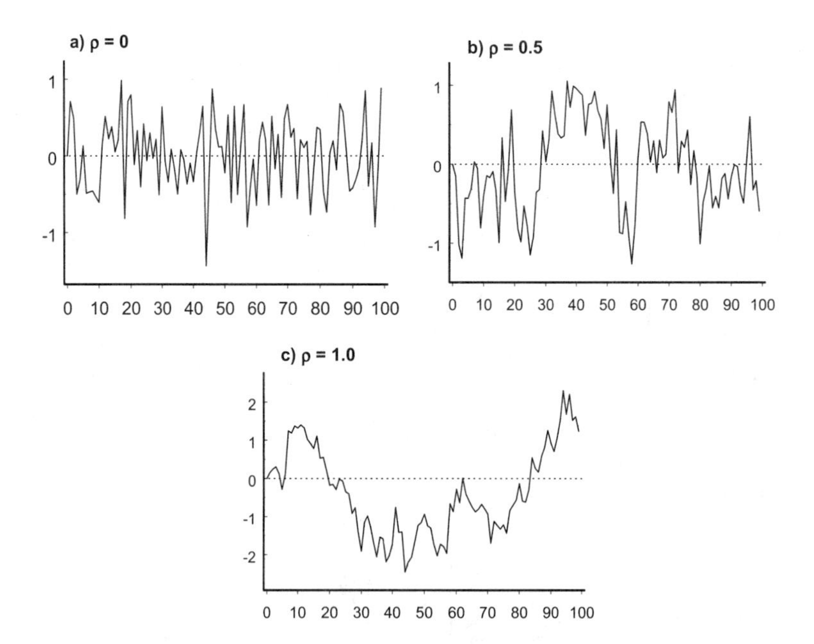

Figure 7.10. Realizations of first-order autoregressive time series with $\sigma_e^2 = 0.2$. a) white noise process; b) $\rho = 0.5$; c) random walk.

Figure 7.10 shows three realizations of first-order autoregressive models with $\rho = 0, 0.5, 1.0$, and mean 0. As ρ increases series of positive and negative deviations from the mean become longer, a sign of positive serial autocorrelation. To study the correlation structure implied by this and other models we examine the covariance and correlation function of the process (see §2.5.1 and §9). For a stationary AR(1) process we have $-1 < \rho < 1$ and the function

$$C(k) = \text{Cov}[Y(t), Y(t-k)]$$

measures the covariance between two observations k time units apart. It is appropriately called the covariance function. Note that $C(0) = \text{Cov}[Y(t), Y(t)] = \text{Var}[Y(t)]$ is the variance of an observation. Under stationarity, the covariances $C(k)$ depend on the temporal separation only, not on the time origin. Time points spaced five units apart are correlated by the same amount, whether the first time point was a Monday or a Wednesday. The **correlation function** $R(k)$ is written simply as $R(k) = C(k)/C(0)$.

The covariance and correlation function of the AR(1) time series [7.51] are derived in §A7.7.6. The elementary recursive relationship is $C(k) = \rho C(k-1) = \rho^k C(0)$ for $k > 0$. Rearranging one obtains the correlation function as $R(k) = C(k)/C(0) = \rho^k$. The autoregressive parameter thus measures the strength of the correlation of observations one time unit apart; the lag-one correlation. For longer temporal separation the correlation is a power of ρ where the exponent equals the number of temporal lags. Since the lags k are discrete, so is the correlation function. For positive ρ the correlations step down every time k increases (Figure 7.11), for negative ρ positive and negative correlations alternate, eventually converging to zero. Negative correlations of adjoining observations are not the norm and are usually indicative of an incorrect model for the mean function. Jones (1993, p. 54) cites as an example a process with circadian rythm. If daily measurements are taken early in the morning and late at night, it is likely that the daily rhythm affects the response and must be accounted for in the mean function. Failure to do so may result in model errors with negative serial correlation.

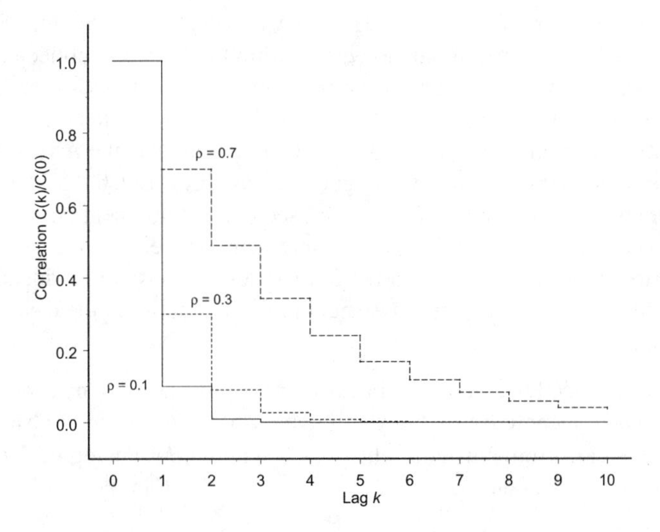

Figure 7.11. Correlation functions for first-order autoregressive processes with $\rho = 0.1, 0.3$, and 0.7.

In terms of the four element cluster i and its variance-covariance matrix $\mathbf{R}_i(\boldsymbol{\alpha})$ the AR(1) process is depicted as

$$\mathbf{R}_i(\boldsymbol{\alpha}) = \sigma^2 \begin{bmatrix} 1 & \rho & \rho^2 & \rho^3 \\ \rho & 1 & \rho & \rho^2 \\ \rho^2 & \rho & 1 & \rho \\ \rho^3 & \rho^2 & \rho & 1 \end{bmatrix}, \quad \boldsymbol{\alpha} = [\sigma^2, \rho], \qquad [7.52]$$

where $C(0) = \sigma^2$. The AR(1) model in longitudinal data analysis dates back to Potthoff and Roy (1964) and is popular for several reasons. The model is parsimonious, the correlation matrix is defined by a single parameter ρ. The model is easy to fit to data. Numerical problems in iterative likelihood or restricted likelihood estimation of $\boldsymbol{\alpha}$ can often be reduced by specifying an AR(1) correlation model rather than some of the more complicated models below. Missing observations within a cluster are not a problem. For example, if it was planned to take measurements at times $1, 2, 3, 4$ but the third measurement was unavailable or destroyed, the row and column associated with the third observation are simply deleted from the correlation matrix:

$$\mathbf{R}_i(\alpha) = \sigma^2 \begin{bmatrix} 1 & \rho & \rho^3 \\ \rho & 1 & \rho^2 \\ \rho^3 & \rho^2 & 1 \end{bmatrix}.$$

The AR(1) model is fit in `proc mixed` with the `type=ar(1)` option of the repeated statement.

The actual measurement times do not enter the correlation matrix in the AR(1) process with discrete lag, only information about whether a measurement occurred after or before another measurement. It is sometimes labeled a discrete autoregressive process for this reason and implicit is the assumption that the measurements are equally spaced. Sometimes there is no basic interval at which observations are taken and observations are unequally spaced. In this case the underlying metric of sampling within a cluster must be continuous. Note that unequal spacing is not a sufficient condition to distinguish continuous from discrete processes. Even if the measurements are equally spaced, the underlying metric may still be continuous. The leaf area of perennial flowers in a multiyear study collected at a few days throughout the years at irregular intervals should not be viewed as discrete daily data with most observations missing, but as unequally spaced observations collected in continuous time with no observations missing. For unequally spaced time intervals the AR(1) model is not the best choice. Assume measurements were gathered at days $1, 2, 6, 11$. The AR(1) model assumes that the correlation between the first and second measurement (spaced one day apart) equals that between the second and third measurement (spaced four days apart). Although there are two pairs of measurements with lag 5, their correlations are not identical. The correlation between the day 1 and day 6 measurement is ρ^2, that between the day 6 and the day 11 measurement is ρ.

With unequally spaced data the actual measurement times should be taken into account in the correlation model. Denote by t_{ij} the jth time at which cluster i was observed. We allow these time points to vary from cluster to cluster. The continuous analog of the discrete AR(1) process has correlation function

$$C(k)/C(0) = \exp\left\{ -\frac{|t_{ij} - t_{ij'}|}{\phi} \right\} \qquad [7.53]$$

and is called the continuous AR(1) or the exponential correlation model (Diggle 1988, 1990; Jones and Boadi-Boateng 1991; Jones 1993; Gregoire et al. 1995). The lag between two observations is measured as the absolute difference of the measurement times t_{ij} and $t_{ij'}, k = t_{ij} - t_{ij'}$. Denoting $C(0) = \sigma^2$ yields the covariance function

$$C(k) = \sigma^2 \exp\left\{ -\frac{|t_{ij} - t_{ij'}|}{\phi} \right\}. \qquad [7.54]$$

For stationarity, it is required that $\phi > 0$ restricting correlations to be positive (Figure 7.12).

The parameter ϕ determines the strength of the correlations, but should *not* be interpreted as a correlation coefficient. Instead it is related to the **practical range** of the temporal process, that is, the time separation at which the correlations have almost vanished. The practical range is usually chosen to be the point at which $C(k)/C(0) = 0.05$. In the continuous AR(1) model [7.54] the practical range equals 3ϕ (Figure 7.12). The **range** of a stationary stochastic process is the lag at which the correlations are exactly zero. The exponential model achieves this value only asymptotically for $|t_{ij} - t_{ij'}| \to \infty$. The within-cluster variance-covariance matrix for the continuous AR(1) model becomes

$$\mathbf{R}_i(\alpha) = \sigma^2 \begin{bmatrix} 1 & e^{-|t_{i1}-t_{i2}|/\phi} & e^{-|t_{i1}-t_{i3}|/\phi} & e^{-|t_{i1}-t_{i4}|/\phi} \\ e^{-|t_{i2}-t_{i1}|/\phi} & 1 & e^{-|t_{i2}-t_{i3}|/\phi} & e^{-|t_{i2}-t_{i4}|/\phi} \\ e^{-|t_{i3}-t_{i1}|/\phi} & e^{-|t_{i3}-t_{i2}|/\phi} & 1 & e^{-|t_{i3}-t_{i4}|/\phi} \\ e^{-|t_{i4}-t_{i1}|/\phi} & e^{-|t_{i4}-t_{i2}|/\phi} & e^{-|t_{i4}-t_{i3}|/\phi} & 1 \end{bmatrix}, \alpha = \left[\sigma^2, \phi\right]. \qquad [7.55]$$

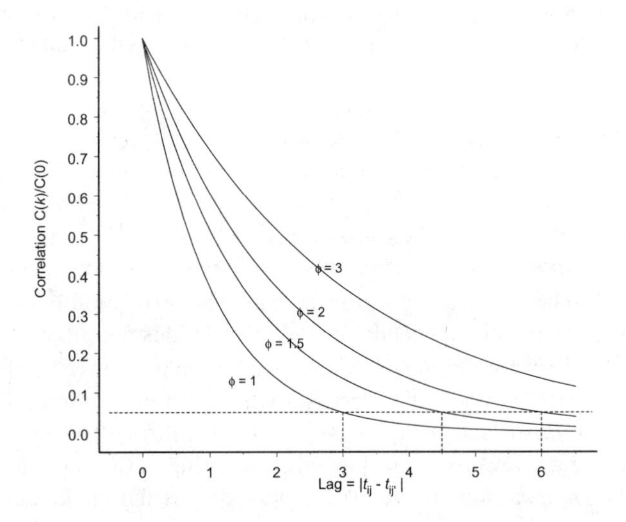

Figure 7.12. Correlation functions of continuous AR(1) processes (exponential models).

The magnitude of ϕ depends on the units in which time was measured. Since the temporal lags and ϕ appear in the exponent of the correlation matrix, numerical overflows or underflows can occur depending on the temporal units used. For example, if time was measured in seconds and the software package fails to report an estimate of ϕ, the iterative estimation algorithms can be helped by rescaling the time variable into minutes or hours. In `proc mixed` of The SAS® System this correlation model is fit with the `type=sp(exp)(time)` option of the `repeated` statement. `sp()` denotes the family of spatial correlation models `proc mixed` provides, `sp(exp)` is the exponential model. In the second set of parentheses are listed the numeric variables in the data set that contain the coordinate information. In the spatial context, one would list the longitude and latitude of the sample locations, e.g., `type=sp(exp)(xcoord ycoord)`. In the temporal setting only one *coordinate* is needed, the time of measurement.

A reparameterization of the exponential model is obtained by setting $\exp\{-1/\phi\} = \rho$:

$$C(k)/C(0) = \rho^{|t_{ij}-t_{ij'}|}. \qquad [7.56]$$

SAS® calls this the *power* model. This terminology is unfortunate because the power model in spatial data analysis is known as a different covariance structure (see §9.2.2). The specification as `type=sp(pow)(time)` in SAS® as a spatial covariance structure suggests that `sp(pow)` refers to the spatial power model. Instead, it refers to [7.56]. It is our experience that numerical difficulties encountered when fitting the exponential model in the form [7.54] can often be overcome by changing to the parameterization [7.56]. From [7.56] the close resemblance of the continuous and discrete AR(1) models is readily established. If observations are equally spaced, that is

$$|t_{ij} - t_{ij'}| = c|j - j'|,$$

for some constant c, the continuous and discrete autoregressive correlation models produce the same correlation function at lag k.

A second model for continuous, equally or unequally spaced observations, is called the gaussian model (Figure 7.13). It differs from the exponential model only in the square of the exponent,

$$C(k)/C(0) = \exp\left\{-\frac{(t_{ij}-t_{ij'})^2}{\phi^2}\right\}. \qquad [7.57]$$

The name must not imply that the gaussian *correlation model* deserves similar veneration as the Gaussian *probability model*. Stein (1999, p. 25) points out that "Nothing could be farther from the truth." The practical range for the gaussian correlation model is $\phi\sqrt{3}$. For the same practical range as in the exponential model, correlations are more persistent over short ranges, they decrease less rapidly (compare the model with $\phi = 6/\sqrt{3}$ in Figure 7.13 to that with $\phi = 2$ in Figure 7.12). Stochastic processes whose autocorrelation follows the gaussian model are highly continuous and smooth (see §9.2.3). It is difficult to imagine physical processes of this kind. We use lowercase spelling when referring to the correlation model to avoid confusion with the Gaussian distribution. From where does the model get its name? A stochastic process with covariance function

$$C(k) = c\exp\{-\beta k^2\},$$

which is a slight reparameterization of [7.57], ($\beta = 1/\phi^2$), has spectral density

$$f(\omega) = \frac{1}{2}\frac{c}{\sqrt{\pi\beta}}\exp\{-\omega^2/4\beta\}$$

which resembles in functional form the Gaussian probability mass function. The gaussian model is fit in `proc mixed` with the `type=sp(gau)(time)` option of the `repeated` statement.

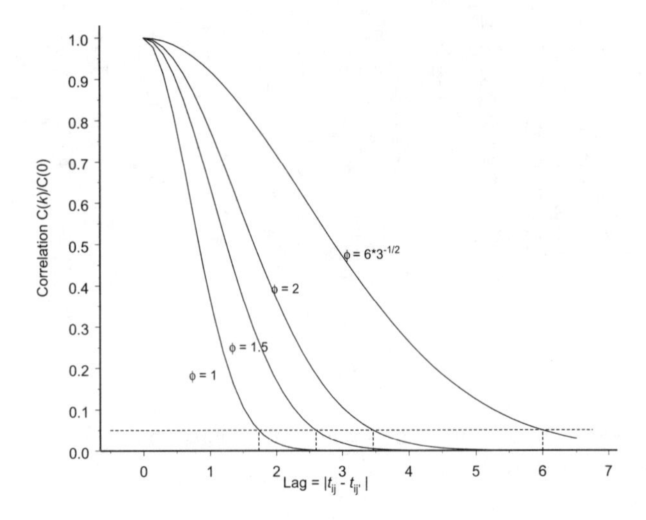

Figure 7.13. Correlation functions in gaussian correlation model. The model with $\phi = 6/\sqrt{3}$ has the same practical range as the model with $\phi = 2$ in Figure 7.12.

A final correlation model for data in continuous time with or without equal spacing is the spherical model

$$C(k)/C(0) = \begin{cases} 1 - \frac{3}{2}\left(\frac{|t_{ij}-t_{ij'}|}{\phi}\right) + \frac{1}{2}\left(\frac{|t_{ij}-t_{ij'}|}{\phi}\right)^3 & |t_{ij} - t_{ij'}| \le \phi \\ 0 & |t_{ij} - t_{ij'}| > \phi. \end{cases} \quad [7.58]$$

In contrast to the exponential and gaussian model the spherical structure has a true range. At lag ϕ the correlation is exactly zero and remains zero thereafter (Figure 7.14).

The spherical model is less smooth than the gaussian correlation model but more so than the exponential model. The spherical model is probably the most popular model for autocorrelated data in geostatistical applications (see §9.2.2). To Stein (1999, p. 52), this popularity is a *mystery* that he attributes to the simple functional form and the "mistaken belief that there is some statistical advantage in having the autocorrelation function being exactly 0 beyond some finite distance." This correlation model is fit in `proc mixed` with the `type=sp(sph)(time)` option of the `repeated` statement.

The exponential, gaussian, and spherical models for processes in continuous time as well as the discrete AR(1) model assume stationarity of the variance of the within-cluster errors. Models that allowed for heterogeneous variances as the unstructured models had many parameters. A class of flexible correlation models which allow for nonstationarity of the within-cluster variances, and changes in the correlations without parameter proliferation was first conceived by Gabriel (1962) and is known as the ante-dependence models. Both continuous and discrete versions of ante-dependence models exist, each in different orders.

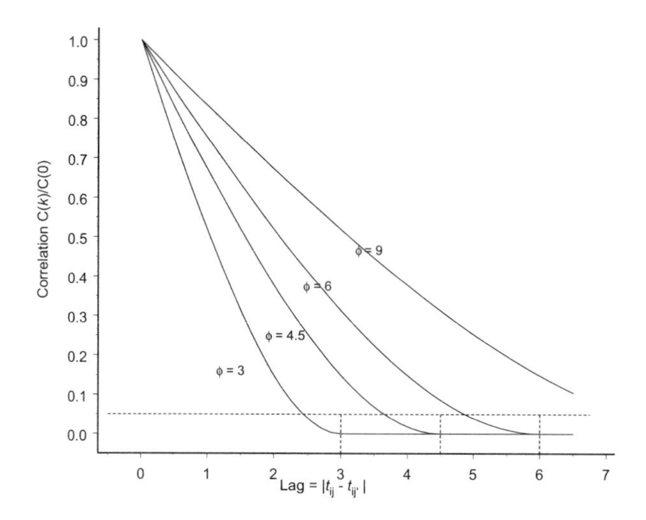

Figure 7.14. Spherical correlation models with ranges equal to the practical ranges for the exponential models in Figure 7.12.

Following Kenward (1987) and Machiavelli and Arnold (1994), the discrete version of a first order ante-dependence model can be expressed as

$$\mathbf{R}_i(\boldsymbol{\alpha}) = \begin{bmatrix} \sigma_1^2 & \sigma_1\sigma_2\rho_1 & \sigma_1\sigma_3\rho_1\rho_2 & \sigma_1\sigma_4\rho_1\rho_2\rho_3 \\ \sigma_2\sigma_1\rho_1 & \sigma_2^2 & \sigma_2\sigma_3\rho_2 & \sigma_2\sigma_4\rho_2\rho_3 \\ \sigma_3\sigma_1\rho_2\rho_1 & \sigma_3\sigma_2\rho_2 & \sigma_3^2 & \sigma_3\sigma_4\rho_3 \\ \sigma_4\sigma_1\rho_3\rho_2\rho_1 & \sigma_4\sigma_2\rho_3\rho_2 & \sigma_4\sigma_3\rho_3 & \sigma_4^2 \end{bmatrix}. \qquad [7.59]$$

Zimmerman and Núñez-Antón (1997) termed it AD(1). For cluster size n_i the discrete AD(1) model contains $2n_i - 1$ parameters, but offers nearly the flexibility of a completely unstructured model with $n_i(n_i+1)/2$ parameters. Zimmerman and Núñez-Antón (1997) discuss extensions to accommodate continuous correlation processes in ante-dependence models. Their continuous first-order model — termed structured ante-dependence model (SAD(1)) — is given by

$$\begin{aligned} \text{Corr}[e_{ij}, e_{ij'}] &= \rho^{f(t_{ij},\boldsymbol{\lambda}) - f(t_{ij'},\boldsymbol{\lambda})}, j' > j \\ \text{Var}[e_{ij}] &= \sigma^2 g(t_{ij}, \boldsymbol{\psi}), j = 1, ..., n_i. \end{aligned} \qquad [7.60]$$

where $f(t_{ij}, \boldsymbol{\lambda})$ and $g(t_{ij}, \boldsymbol{\psi})$ are functions of the measurement times or locations which depend on parameter vectors $\boldsymbol{\lambda}$ and $\boldsymbol{\psi}$. Zimmerman and Núñez-Antón advocate choosing $f(\bullet)$ from the family of Box-Cox transformations (see §5.6.2)

$$f(t_{ij}, \lambda) = \begin{cases} (t_{ij}^\lambda - 1)/\lambda & \lambda \neq 0 \\ \ln\{t_{ij}\} & \lambda = 0 \end{cases}.$$

If $\lambda = 1$ and $g(t_{ij}, \psi) = 1$, the power correlation model results. See Zimmerman and Núñez-

Antón (1997) for higher order ante-dependence models and Machiavelli and Arnold (1994) for variable-order models. The SAD() models alleviate some shortcomings of the stationary continuous models. In growth studies, variability often increases with time. Heteroscedasticity of the within-cluster residuals is already incorporated in ante-dependence models. Also, equidistant observations do not necessarily have the same correlation as is implied by the stationary models discussed above. The discrete first-order ante-dependence model can be fit with `proc mixed` as `type=ante(1)`.

7.5.3 Split-Plots, Repeated Measures, and the Huynh-Feldt Conditions

On the surface repeated measures experiments show similarities to split-plot type designs which frequently prompts research workers to analyze repeated measures data as if it had arisen in a split design. Figure 7.15 shows a single replicate with three treatments (A_1, A_2, A_3) and a four-level split in a genuine split-plot design (Figure 7.15 a) and a repeated measures design (Figure 7.15 b).

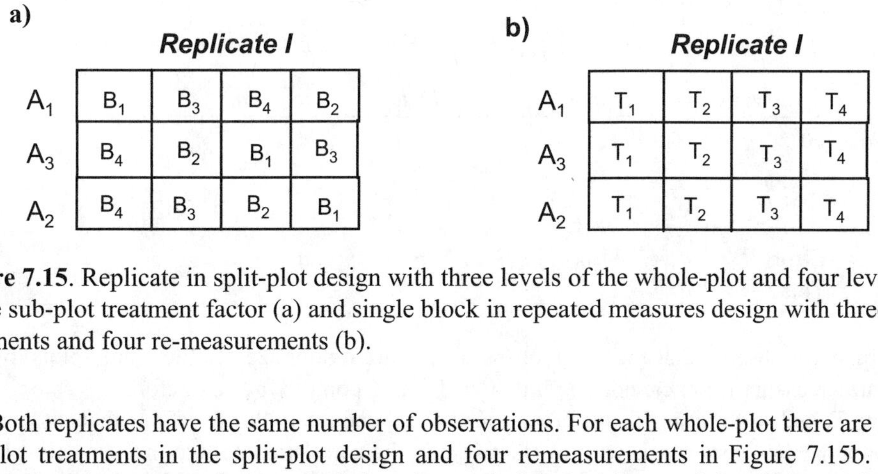

a) *Replicate I*

A_1	B_1	B_3	B_4	B_2
A_3	B_4	B_2	B_1	B_3
A_2	B_4	B_3	B_2	B_1

b) *Replicate I*

A_1	T_1	T_2	T_3	T_4
A_3	T_1	T_2	T_3	T_4
A_2	T_1	T_2	T_3	T_4

Figure 7.15. Replicate in split-plot design with three levels of the whole-plot and four levels of the sub-plot treatment factor (a) and single block in repeated measures design with three treatments and four re-measurements (b).

Both replicates have the same number of observations. For each whole-plot there are four sub-plot treatments in the split-plot design and four remeasurements in Figure 7.15b. The split-plot analysis proceeds by calculating the analysis of variance of the design and then formulating appropriate test statistics to test for factor A main effects, B main effects, and $A \times B$ interactions followed by tests of treatment contrasts or other post-ANOVA procedures. The analysis of variance table for the split-plot design with r replicates is based on the linear mixed model

$$Y_{ijk} = \mu + \rho_j + \alpha_i + e^*_{ij} + \beta_k + (\alpha\beta)_{ik} + e_{ijk},$$

where ρ_j $(j = 1, \cdots, r)$ are the whole-plot replication (block) effects, α_i $(i = 1, \cdots, a)$ are the whole-plot treatment effects, e^*_{ij} is the whole-plot experimental error, β_k $(k = 1, \cdots, b)$ are the sub-plot treatment effects, $(\alpha\beta)_{ik}$ are the interactions, and e_{ijk} denotes the sub-plot experimental errors. Letting σ^2_* denote the whole-plot experimental error variance and σ^2 the sub-plot experimental error variance, Table 7.6 shows the analysis of variance and expected mean squares.

From the expected mean squares it is seen that the appropriate test statistics for main effects and interactions tests are

A	A main effect:	$F_{obs} = MSA/MSE_A$
B	B main effect:	$F_{obs} = MSB/MSE_B$
C	$A \times B$ interaction:	$F_{obs} = MSAB/MSE_B$.

Would it be incorrect to perform a similar analysis in the repeated measures setting, labeling factor B in Table 7.6 as $Time$ and testing

A	A main effect:	$F_{obs} = MSA/MSE_A$
B	$Time$ main effect:	$F_{obs} = MSTime/MSE_{Time}$
C	$A \times Time$ interaction:	$F_{obs} = MSATime/MSE_{Time}$?

Table 7.6. Analysis of variance table of a standard split-plot design (whole-plot factor arranged in a randomized complete block design with r replicates)

Source	df	Mean Squares	Expected Mean Squares
Replicates	$r-1$		
A	$a-1$	MSA	$\sigma^2 + b\sigma_*^2 + \frac{ra}{b-1}\sum_i \alpha_i^2$
Error(A)	$(r-1)(a-1)$	MSE_A	$\sigma^2 + b\sigma_*^2$
B	$b-1$	MSB	$\sigma^2 + \frac{ra}{b-1}\sum_k \beta_k^2$
$A \times B$	$(a-1)(b-1)$	$MSAB$	$\sigma^2 + \frac{r}{(a-1)(b-1)}\sum_{i,k}(\alpha\beta)_{ik}^2$
Error(B)	$a(r-1)(b-1)$	MSE_B	σ^2
Total	$arb-1$		

First we observe that the sub-plot treatments are randomized to the whole-plots, but the repeated measurements cannot be randomized. Time point T_1 occurs before T_2 which occurs before T_3 and so forth. Is this difference substantial enough to throw off the analysis, though? To approach an answer it is worthwhile to study the correlation pattern that the split-plot design implies. Whole-plot errors are independent due to randomization of the whole-plot treatments and so are sub-plot errors. Independent randomizations to whole- and sub-plots also establish that $\text{Cov}[e_{ij}^*, e_{ijk}] = 0$. Since sub-plot errors are nested within whole-plots this is the same setting as in the random intercept model of §7.5.1 and one arrives at a compound-symmetric structure for the observations from the same whole-plot:

$$\begin{aligned}
\text{Cov}[Y_{ijk}, Y_{ijk'}] &= \text{Cov}\left[e_{ij}^* + e_{ijk}, e_{ij}^* + e_{ijk'}\right] = \sigma^2 \\
\text{Corr}[Y_{ijk}, Y_{ijk'}] &= \frac{\text{Cov}[Y_{ijk}, Y_{ijk'}]}{\sqrt{\text{Var}[Y_{ijk}]\text{Var}[Y_{ijk'}]}} = \frac{\sigma_*^2}{\sigma_*^2 + \sigma^2}; \\
\text{Var}[\mathbf{Y}_{ij}] &= \sigma_*^2 \mathbf{J}_b + \sigma^2 \mathbf{I}_b.
\end{aligned}$$

Compound symmetry implies exchangeabiliy of the observations within a whole-plot Exchangeability is appealing since the sub-plot treatments were randomized. No particular ordering or arrangement of treatments within a whole-plot is given preference. This result sheds light on the appropriateness of a split-plot type analysis for repeated measures data. Only if the correlations between any two time points T_k and $T_{k'}$ are identical, will the correlation structure be exchangeable. In that case the ordering of time points is not material and the nonrandomized ordering $T_1, T_2, T_3, \cdots$ will lead to a correct analysis under the split-plot model. There are other correlation structures besides compound symmetry for which repeated measures analysis with a split-plot model is appropriate. These structures are defined through conditions on the variance-covariance matrices known as the Huynh-Feldt conditions (Huynh and Feldt 1970). That compound symmetry satisfies the Huynh-Feldt conditions was established by Geisser and Greenhouse (1958). The Huynh-Feldt conditions are more general than compound symmetry, however. Assume the repeated measures analysis is based on the model

$$Y_{ijk} = \mu + \rho_j + \alpha_i + e^*_{ij} + \tau_k + (\alpha\tau)_{ik} + e_{ijk}, k = 1, \cdots, t,$$

where τ_k represents the time effects and $(\alpha\tau)_{ik}$ the treatment $\times$ time interactions. A test of H_0: $\tau_1 = ... = \tau_t$ via a regular analysis of variance F test of form

$$F_{obs} = \frac{MSTime}{MSE_{Time}}$$

is valid if the variance-covariance matrix of the "sub-plot" errors $\mathbf{e}_{ij} = [e_{ij1}, ..., e_{ijt}]'$ can be expressed in the form

$$\text{Var}[\mathbf{e}_{ij}] = \lambda\mathbf{I}_t + \boldsymbol{\gamma}\mathbf{1}'_t + \mathbf{1}_t\boldsymbol{\gamma}',$$

where $\boldsymbol{\gamma}$ is a vector of parameters and λ is a constant. Similarly, the F test for the whole-plot factor H_0:$\alpha_1 = ... = \alpha_a$ is valid if the whole-plot errors $\mathbf{e}^*_i = [e^*_{i1}, ..., e^*_{ia}]'$ have a variance-covariance matrix which can be expressed as

$$\text{Var}[\mathbf{e}^*_i] = \lambda^*\mathbf{I}_a + \boldsymbol{\gamma}^*\mathbf{1}'_a + \mathbf{1}_a\boldsymbol{\gamma}^{*\prime}.$$

The variance-covariance matrix is then said to meet the Huynh-Feldt conditions. We note in passing that a correct analysis via split-plot ANOVA requires that the condition is met for every random term in the model (Milliken and Johnson 1992, p. 325). Two special cases of variance-covariance matrices that meet the Huynh-Feldt conditions are independence and compound symmetry. The combination of $\lambda = \sigma^2$ and $\boldsymbol{\gamma} = 0$ yields the independence structure $\sigma^2\mathbf{I}$; the combination of $\lambda = \sigma^2$ and $\boldsymbol{\gamma} = \frac{1}{2}[\sigma^2_t, ..., \sigma^2_t]'$ yields a compound symmetric structure $\sigma^2\mathbf{I} + \sigma^2_t\mathbf{J}$.

Analyzing repeated measures data with split-plot models implicitly assumes a compound symmetric or Huynh-Feldt structure which may not be appropriate. If correlations decay over time, for example, the compound symmetric model is not a reasonable correlation model. Two different courses of action can then be taken. One relies on making adjustments to the degrees of freedom for test statistics in the split-plot analysis, the other focuses on modeling the variance-covariance structure of the within-cluster errors. We comment on the adjustment method first.

Box (1954b) developed the measure ϵ for the deviation from the Huynh-Feldt conditions. ϵ is bounded between $(t-1)^{-1}$ and 1, where t is the number of repeated measurements. The degrees of freedom for tests of Time main effects and Treatment $\times$ Time interactions are adjusted by the multiplicative factor ϵ. The standard F statistic

$$MSTime/MSE_{Time}$$

for the Time main effect is not compared against the critical value $F_{\alpha,t-1,a(r-1)(t-1)}$ but against $F_{\alpha,\epsilon(t-1),\epsilon a(r-1)(t-1)}$. Unfortunately, ϵ is unknown. A conservative approach is to set ϵ equal to its lower bound. The critical value then would be $F_{\alpha,1,a(r-1)}$ for the Time main effect test and $F_{\alpha,a-1,a(r-1)}$ for the test of Treatment × Time interactions. The reduction in test power that results from the conservative adjustment is disconcerting. Less power is sacrificed by estimating ϵ from the data. The estimator is discussed in Milliken and Johnson (1992, p. 355). Briefly, let

$$\widehat{\pi}_{kk'} = \frac{1}{a(r-1)}\sum_{i,j}\left(y_{ijk} - \overline{y}_{ij.} - \overline{y}_{i.k} + \overline{y}_{i..}\right)\left(y_{ijk'} - \overline{y}_{ij.} - \overline{y}_{i.k'} + \overline{y}_{i..}\right).$$

Then,

$$\widehat{\epsilon} = (t-1)^{-1}\left(\sum_{k=1}^{t}\widehat{\pi}_{kk}\right)^2 \Big/ \sum_{k,k'}\widehat{\pi}^2_{kk'}. \qquad [7.61]$$

This adjustment factor for the degrees of freedom is less conservative than using the lower bound of ϵ, but more conservative than an adjustment proposed by Huynh and Feldt (1976). The estimated Box epsilon [7.61] and the Huynh-Feldt adjustment are calculated by `proc glm` of The SAS® System if the `repeated` statement of that procedure is used. [7.61] is labeled `Greenhouse-Geisser Epsilon` on the `proc glm` output (Greenhouse and Geisser 1959). To decide whether any adjustment is necessary, i.e., whether the dispersion of the data deviates significantly from the Huynh-Feldt conditions, a test of sphericity can be invoked. This test is available through the `printE` option of the `repeated` statement in `proc glm`. If the sphericity test is rejected, a degree of freedom adjustment is deemed necessary.

This type of repeated measures analysis attempts to coerce the analysis into a split-plot model framework and if that framework does not apply uses *fudge* factors to adjust the end result (critical values or p-values). If the sphericity assumption (i.e., the Huynh-Feldt conditions) are violated, the basic problem from our standpoint is that the statistical model undergirding the analysis is not correct. We prefer a more direct, and hopefully more intuitive, approach to modeling repeated measures data. Compound symmetry is a dispersion/correlation structure that comes about in the split-plot model through nested, independent random components. In a repeated measures setting there is often *a priori* knowledge or theory about the correlation pattern over time. In a two-year study of an annual crop it is reasonable to assume that within a growing season measurements are serially correlated and that correlations wear off with temporal separation. Secondly, it may be reasonable to assume that the measurements at the beginning of the second growing season are independent of the responses at the end of the previous season. Rather than relying on a compound symmetric or Huynh-Feldt correlation structure, we can employ a correlation structure whose behavior is consistent with these assumptions. Of the large number of such correlation models some were discussed in §7.5. Equipped with a statistical package capable of fitting data with the chosen correlation model and a method to distinguish the goodness-of-fit of competing correlation models, the research worker can develop a statistical model that describes more closely the structure of the data without relying on *fudge* factors.

7.6 Applications

The applications of linear mixed models for clustered data we entertain in this section are as varied as the Laird-Ware model. We consider traditional growth studies with linear mixed regression models as well as designed experiments with multiple random effects. §7.6.1 is a study of apple growth over time where we are interested in predicting population-averaged and apple-specific growth trends. The empirical BLUPs will play a key role in estimating the cluster-specfic trends. Because measurements on individual apples were collected repeatedly in time we also pay attention to the possibility of serial correlation in the model residuals. This application is intended to underscore the two-stage concept in mixed modeling. §7.6.2 to §7.6.5 are experimental situations where the statistical model contains multiple random terms for different reasons. In §7.6.2 we analyze data from an on-farm trial where identical experimental designs are laid out on randomly selected farms. The random selection of farms results in a mixed model containing random experimental errors, random farm effects, and random interactions. The estimated BLUPs are again key quantities on which inferences involving particular farms are based.

Subsampling of experimental units also gives rise to clustered data structures and a nesting of experimental and observational error sources. The liabilities of not recognizing the subsampling structure are discussed in §7.6.3 along with the correct analysis based on a linear mixed model. A very special case of an experimental design with a linear mixed model arises when block effects are random effects, for example, when locations are chosen at random. The on-farm experiment in §7.6.2 can be viewed as a design of that nature. If blocks are incomplete in the sense that the size of the block cannot accommodate all treatments, mixed model analyses are more powerful than fixed model analyses because of their ability to recover treatment contrasts from comparisons across blocks. This recovery of interblock information is straightforward with the `mixed` procedure in SAS® and we apply the techniques to a balanced incomplete block design (BIB) in §7.6.4. Finally, a common method for clustering data in experimental designs is the random assignment of treatments to experimental units of different size. The resulting split-type designs that result have a linear mixed model representation. Experimental units can be of different sizes if one group of units is arranged within a larger unit (splitting) or if units are arranged perpendicular to each other (stripping). A combination of both techniques that is quite common in agricultural field experiments is the split-strip-plot design that we analyze in §7.6.5.

Modeling the correlations among repeated observations directly is our preferred method over inducing correlations through random effects or random coefficients. The selection of a suitable correlation structure based on AIC and other fit criteria is the objective of analyzing a factorial treatment structure with repeated measurements in §7.6.6. The model we arrive at is a mixed model because it contains random effects for the experimental errors corresponding to experimental units and serially correlated observational disturbances observed over time within each unit.

Many growth studies involve nonlinear models (§5, §8) or polynomials. How to determine which terms in a growth model are to be made random and which are to be kept fixed is the topic of §7.6.7, concerned with the water usage of horticultural trees. The comparison of treatments in a growth study with complex subsampling design is examined in §7.6.8.

7.6.1 Two-Stage Modeling of Apple Growth over Time

At the Winchester Agricultural Experiment Station of Virginia Polytechnic Institute and State University ten apple trees were randomly selected and twenty-five apples were randomly chosen on each tree. We concentrate the analysis on the apples in the largest size class, those whose initial diameter exceeded 2.75 inches. In total there were eighty apples in that size class. Diameters of the apples were recorded in two-week intervals over a twelve-week period. The observed apple diameters are shown for sixteen of the eighty apples in Figure 7.16. The profiles for the remaining apples are very similar to the ones shown. Only those apples that remained on the tree for the entire three-month period have complete data. Apple 14 on tree 1, for example, was measured only on the first occasion, and only three measurements were available for apple 15 on tree 2 .

Of interest is modeling the apple-specific and the overall growth trends of apples in this size class. One can treat these data as clustered on two levels; trees are clusters of the first size consisting of apples within trees. Apples are clusters of the second size containing the repeated observations over time. It turns out during analysis that tree-to-tree variability is very small, most of the variation in the data is due to apple-specific effects. For this and expository reasons, we focus now on apples as the clusters of interest. At the end of this section we discuss how to modify the analysis to account for tree and apple cluster effects.

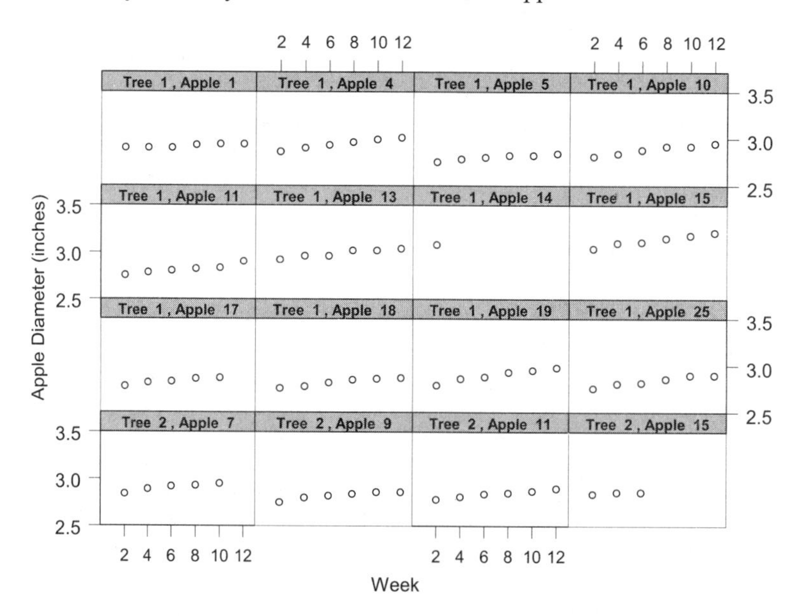

Figure 7.16. Observed diameters over a 12-week period for 16 of the 80 apples. Data kindly provided by Dr. Ross E. Byers, Alson H. Smith, Jr. AREC, Virginia Polytechnic Institute and State University, Winchester, Virginia. Used with permission.

Figure 7.16 suggests that the trends are linear for each apple. A naïve approach to estimating the population-averaged and cluster-specific trends is as follows. Let

$$Y_{ij} = \beta_{0i} + \beta_{1i}t_{ij} + e_{ij}$$

denote a simple linear regression for the data from apple $i = 1, \cdots, 80$. We fit this linear regression separately for each apple and obtain the population-averaged estimates of the overall slope β_0 and intercept β_1 by averaging the apple-specific estimates. In this averaging one can calculate equally weighted averages or take the precision of the apple-specific estimates into account. The equally weighted average approach is implemented in SAS® as follows (Output 7.4).

```
/* variable time is coded as 1,2,...,6 corresponding to weeks 2,4,...,12 */
proc reg data=apples outest=est noprint;
  model diam = time;
  by tree apple;
run;
proc means data=est noprint;
   var intercept time;
   output out=PAestimates mean=beta0 beta1 std=sebeta0 sebeta1;
run;
title 'Naive Apple-Specific Estimates';
proc print data=est(obs=20) label;
  var tree apple _rmse_ intercept time;
run;
title 'Naive PA Estimates';
proc print data=PAEstimates; run;
```

Output 7.4.

Naive Apple-Specific Estimates

Obs	tree	apple	Root mean squared error	Intercept	time
1	1	1	0.009129	2.88333	0.010000
2	1	4	0.005774	2.83667	0.030000
3	1	5	0.008810	2.74267	0.016857
4	1	10	0.012383	2.78867	0.028000
5	1	11	0.016139	2.72133	0.026286
6	1	13	0.014606	2.90267	0.024000
7	1	14	.	3.08000	0.000000
8	1	15	0.010511	3.01867	0.032286
9	1	17	0.008944	2.76600	0.022000
10	1	18	0.011485	2.74133	0.023429
11	1	19	0.014508	2.77133	0.036286
12	1	25	0.013327	2.74067	0.028857
13	2	7	0.013663	2.82800	0.026000
14	2	9	0.014557	2.74667	0.021429
15	2	11	0.006473	2.76267	0.022571
16	2	15	0.008165	2.83333	0.010000
17	2	17	0.015228	2.82867	0.019429
18	2	23	0.016446	2.79267	0.028286
19	2	24	0.010351	2.84000	0.025714
20	2	25	0.013801	2.74333	0.024286

Naive PA Estimates

freq	beta0	beta1	sebeta0	sebeta1
80	2.83467	0.028195	0.092139	.009202401

Notice that for Apple 14 on tree 1, which is the $i = 7^{\text{th}}$ apple in the data set, the coefficient estimate $\widehat{\beta}_{1,7}$ is 0 because only a single observation had been collected on that apple. The apple-specific predictions can be calculated from the coefficients in Output 7.4 as can the population-averaged growth trend. We notice, however, that the ability to do this required estimation of 160 mean parameters, one slope and one intercept for each of eighty apples. Counting the estimation of the residual variance in each model we have a total of 240 estimated parameters. We can contrast the population-average and the apple-specific predictions easily from Output 7.4. The growth of the average apple is predicted as

$$\widetilde{y} = \widetilde{\beta}_0 + \widetilde{\beta}_1 t = 2.83467 + 0.028195 \times t$$

and that for apple 1 on tree 1, for example, as

$$\begin{aligned}\widetilde{y}_1 &= 2.88333 + 0.01 \times t \\ &= \left(\widetilde{\beta}_0 + 0.04866\right) + \left(\widetilde{\beta}_1 - 0.01895\right) \times t.\end{aligned}$$

The quantities 0.04866 and –0.01895 are the adjustments made to the population-averaged estimates of intercept and slope to obtain the predictions for the first apple. We use the $\sim$ notation here, because the averages of the apple-specific fixed effects are not the generalized least squares estimates.

Fitting the apple-specific trends by fitting a model to the data from each apple and then averaging the estimates is a two-stage approach, literally. The cluster-specific estimates are obtained in the first stage and the population-average is determined in the second stage. The two-stage concept in the Laird-Ware model framework leads to the same end result, estimates of the population-average trend and estimates of the cluster-specific trend. It does, however, require fewer parameters. Consider the first-stage model

$$Y_{ij} = (\beta_0 + b_{0i}) + (\beta_1 + b_{1i})t_{ij} + e_{ij}, \qquad [7.62]$$

where Y_{ij} is the measurement taken at time t_{ij} for apple number i. We assume for now that the e_{ij} are uncorrelated Gaussian random errors with zero mean and constant variance σ^2, although the fact that $Y_{i1}, Y_{i2}, \cdots$ are repeated measurements on the same apple should give us pause. We will return to this issue later. To formulate the second stage of the Laird-Ware model we postulate that b_{0i} and b_{1i} are Gaussian random variables with mean 0 and variances σ_0^2 and σ_1^2, respectively. They are assumed not to be correlated with each other and are also not correlated with the error terms e_{ij}. This is obviously a mixed model of Laird-Ware form with 5 parameters $(\beta_0, \beta_1, \sigma_0^2, \sigma_1^2, \sigma^2)$. With the `mixed` procedure of The SAS® System, this is accomplished through the statements

```
proc mixed data=apples;
  class tree apple;
  model diam = time / s;
  random intercept time / subject=tree*apple s;
run;
```

By default the variance components σ_0^2, σ_1^2, and σ^2 will be estimated by restricted maximum likelihood. The `subject=tree*apple` statement identifies the unique combinations of the data set variables `apple` and `tree` as clusters. Both variables are needed here since the apple variable is numbered consecutively starting at 1 for each tree. Technically more correct would be to write `subject=apple(tree)`, since the apple identifiers are nested within trees, but the analysis will be identical to the one above. In writing `subject=` options in `proc mixed`, the user must only provide a variable combination that uniquely identifies the clusters. The `/s` option on the `model` statement requests a printout of the fixed effects estimates $\widehat{\beta}_0$ and $\widehat{\beta}_1$, the same option on the random statement requests a printout of the solutions for the random effects (the estimated BLUPs $\widehat{b}_{0i}, \widehat{b}_{1i}, \cdots \widehat{b}_{0,80}, \widehat{b}_{1,80}$). In Output 7.5 we show only a partial printout of the EBLUPs corresponding to the first ten apples.

Output 7.5.

```
                         The Mixed Procedure

                          Model Information
          Data Set                         WORK.APPLES
          Dependent Variable               diam
          Covariance Structure             Variance Components
          Subject Effect                   tree*apple
          Estimation Method                REML
          Residual Variance Method         Profile
          Fixed Effects SE Method          Model-Based
          Degrees of Freedom Method        Containment

                       Class Level Information
            Class     Levels    Values
            tree          10    1 2 3 4 5 6 7 8 9 10
            apple         24    1 2 3 4 5 7 8 9 10 11 12 13 14
                                15 16 17 18 19 20 21 22 23 24 25

                             Dimensions
                   Covariance Parameters                3
                   Columns in X                         2
                   Columns in Z Per Subject             2
                   Subjects                            80
                   Max Obs Per Subject                  6
                   Observations Used                  451
                   Observations Not Used               29
                   Total Observations                 480

                      Covariance Parameter Estimates
                   Cov Parm       Subject          Estimate
                   Intercept      tree*apple       0.008547
                   time           tree*apple       0.000056
                   Residual                        0.000257

                           Fit Statistics
                  -2 Res Log Likelihood             -1897.7
                  AIC (smaller is better)           -1891.7
                  AICC (smaller is better)          -1891.6
                  BIC (smaller is better)           -1884.5

                       Solution for Fixed Effects
                                    Standard
      Effect          Estimate       Error       DF     t Value     Pr > |t|
      Intercept         2.8345     0.01048       79      270.44       <.0001
      time             0.02849    0.000973       78       29.29       <.0001

                      Solution for Random Effects
                                                Std Err
Effect     tree   apple     Estimate           Pred      DF    t Value     Pr > |t|
Intercept   1       1        0.03475        0.01692     292       2.05       0.0409
time        1       1       -0.01451       0.003472     292      -4.18       <.0001
Intercept   1       4       0.003218        0.01692     292       0.19       0.8493
time        1       4       0.001211       0.003472     292       0.35       0.7274
Intercept   1       5       -0.09808        0.01692     292      -5.80       <.0001
time        1       5       -0.00970       0.003472     292      -2.79       0.0055
Intercept   1      10       -0.04518        0.01692     292      -2.67       0.0080
time        1      10       -0.00061       0.003472     292      -0.17       0.8614
Intercept   1      11        -0.1123        0.01692     292      -6.64       <.0001
```

Output 7.5 (continued).

time	1	11	-0.00229	0.003472	292	-0.66	0.5106
Intercept	1	13	0.06357	0.01692	292	3.76	0.0002
time	1	13	-0.00326	0.003472	292	-0.94	0.3482
Intercept	1	14	0.2094	0.02010	292	10.42	<.0001
time	1	14	0.001382	0.007486	292	0.18	0.8537
Intercept	1	15	0.1830	0.01692	292	10.82	<.0001
time	1	15	0.003882	0.003472	292	1.12	0.2644
Intercept	1	17	-0.07277	0.01759	292	-4.14	<.0001
time	1	17	-0.00491	0.004216	292	-1.16	0.2450
Intercept	1	18	-0.09474	0.01692	292	-5.60	<.0001
time	1	18	-0.00447	0.003472	292	-1.29	0.1989

The estimates of the variance components are fairly small,

$$\widehat{\sigma}_0^2 = 0.0085,\ \widehat{\sigma}_1^2 = 0.000056,\ \widehat{\sigma}^2 = 0.000257,$$

but one should keep in mind that the estimates are scale-dependent. If we model $U_{ij} = 10Y_{ij}$ instead of Y_{ij}, these estimates will be 10^2 times larger without altering the fit of the model.

The fixed effects estimates are $\widehat{\beta}_0 = 2.8345$ and $\widehat{\beta}_1 = 0.02849$, fairly close to the population-averaged estimates obtained by averaging the apple-specific fixed effects coefficients earlier (Output 7.4). The `Solution for Random Effects` table lists the estimated BLUPs. A population averaged prediction is obtained as

$$\widehat{y} = 2.8345 + 0.02849 \times t$$

and the apple-specific predictions for the first apple, for example, as

$$\begin{aligned}\widehat{y}_1 &= \left(\widehat{\beta}_0 + 0.03475\right) + \left(\widehat{\beta}_1 - 0.01451\right) \times t \\ &= 2.86925 + 0.01398 \times t.\end{aligned}$$

In fitting this model it was assumed that the e_{ij} are uncorrelated. This may not be tenable since the measurements from the same apple are taken sequentially in time. To investigate whether there is a significant serial correlation we perform a likelihood ratio test. We fit model [7.62] but assume that the e_{ij} follow a first-order autoregressive model. Since the measurement occasions are equally spaced, this is a reasonable approach. Recall from Output 7.5 that minus twice the restricted (residual) log likelihood of the model with uncorrelated errors is –1897.7. We accomplish fitting a model with AR(1) errors by adding the `repeated` statement as follows:

```
proc mixed data=apples noitprint;
  class tree apple;
  model diam = time / s;
  random intercept time / subject=tree*apple;
  repeated / subject=tree*apple type=ar(1);
run;
```

The estimate of the autocorrelation coefficient, the correlation between diameter measurements on the same apple two weeks apart, is $\widehat{\rho} = 0.3825$ (Output 7.6). It appears fairly substantial, but is adding the autocorrelation to the model a significant improvement? The negative of twice the residual log likelihood in this model is –1910.5 and the likelihood ratio test statistic comparing the models with and without AR(1) correlation is 1910.5 – 1897.7 = 12.8. The p-value for the hypothesis that the autoregressive parameter is zero is thus $\Pr(\chi_1^2 > 12.8) = 0.00035$. Adding the AR(1) serial correlation does significantly improve the model. The impact of adding the AR(1) correlations is primarily on the standard errors of all estimated quantities. The population-averaged estimates as well as the BLUPs for the random effects change very little, hence the impact on the predicted values is minor (not necessarily so the impact on the precision of the predicted values).

Output 7.6.

```
                         The Mixed Procedure

                          Model Information

          Data Set                        WORK.APPLES
          Dependent Variable              diam
          Covariance Structures           Variance Components,
                                          Autoregressive
          Subject Effects                 tree*apple, tree*apple
          Estimation Method               REML
          Residual Variance Method        Profile
          Fixed Effects SE Method         Model-Based
          Degrees of Freedom Method       Containment

                        Class Level Information

            Class     Levels     Values
            tree          10     1 2 3 4 5 6 7 8 9 10
            apple         24     1 2 3 4 5 7 8 9 10 11 12 13 14
                                 15 16 17 18 19 20 21 22 23 24 25

                              Dimensions

                   Covariance Parameters                4
                   Columns in X                         2
                   Columns in Z Per Subject             2
                   Subjects                            80
                   Max Obs Per Subject                  6
                   Observations Used                  451
                   Observations Not Used               29
                   Total Observations                 480

                     Covariance Parameter Estimates

                   Cov Parm        Subject         Estimate
                   Intercept       tree*apple      0.008653
                   time            tree*apple      0.000050
                   AR(1)           tree*apple        0.3825
                   Residual                        0.000365

                            Fit Statistics

                  -2 Res Log Likelihood             -1910.5
                  AIC (smaller is better)           -1902.5
                  AICC (smaller is better)          -1902.4
                  BIC (smaller is better)           -1893.0

                      Solution for Fixed Effects

                                      Standard
      Effect          Estimate           Error       DF     t Value     Pr > |t|
      Intercept         2.8321         0.01068       79      265.30       <.0001
      time             0.02875        0.001017       78       28.28       <.0001
```

Output 7.6 (continued).

Solution for Random Effects

Effect	tree	apple	Estimate	Std Err Pred	DF	t Value	Pr > |t|
Intercept	1	1	0.02936	0.02049	292	1.43	0.1529
time	1	1	-0.01252	0.004189	292	-2.99	0.0030
Intercept	1	4	0.004805	0.02049	292	0.23	0.8147
time	1	4	0.000845	0.004189	292	0.20	0.8403
Intercept	1	5	-0.1024	0.02049	292	-5.00	<.0001
time	1	5	-0.00814	0.004189	292	-1.94	0.0529
Intercept	1	10	-0.04381	0.02049	292	-2.14	0.0333
time	1	10	-0.00080	0.004189	292	-0.19	0.8489

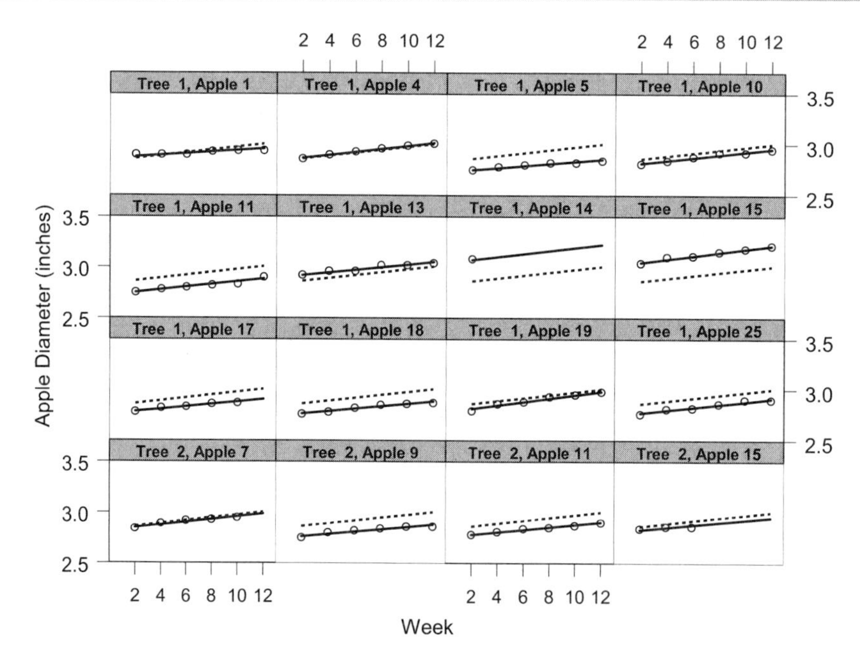

Figure 7.17. Apple-specific predictions (solid lines) from mixed model [7.62] with AR(1) correlated error terms for the same apples shown in Figure 7.16. Dashed lines show population-averaged prediction, circles are raw data.

Once we have settled on the correlation model for these data we should go back and re-evaluate whether the two random effects are in fact needed. Deleting them in turn one obtains a residual log likelihood of 947.75 for the model without random slope, 946.3 for the model without random intercept, and 946.05 for the model without any random effects. Any one likelihood ratio test against the full model in Output 7.6 is significant. Both random effects remain in the model along with the AR(1) serial correlation.

The predictions from this model trace the observed growth profiles very closely (Figure 7.17) and the deviation of the solid from the dashed line in Figure 7.17 is an indication of how strongly a particular apple differs from the average apple. It is clear that most of the apple-to-apple variation is in the actual size of the apple (heterogeneity in intercepts), not the growth rate ($\widehat{\sigma}_0^2$ was larger than $\widehat{\sigma}_1^2$). Notice that population-average and subject-specific predictions have been obtained for apple 14 on tree #1, although only a single diameter had been measured for this apple.

What about the trees in this study? There were after all two levels of random selections. Trees were randomly selected in the orchard and apples were randomly selected from the trees. We can model the data starting with a population average model, adding tree-to-tree heterogeneity, and then apple-to-apple heterogeneity within trees. The code

```
proc mixed data=apples update noitprint;
  class tree apple;
  model diam = time / s;
  random intercept time / subject=tree;
  random intercept time / subject=apple(tree);
  repeated / subject=apple(tree) type=ar(1);
run;
```

uses the `update` option to write to the log window what `proc mixed` is currently doing. Fitting models with many random effects can be time consuming and it is then helpful to find out whether the procedure is still processing.

Notice that there are now two more covariance parameters corresponding to a random intercept and a random slope for the trees (Output 7.7). A likelihood ratio test whether the addition of the two random effects improved the model has test statistic 1917.1 – 1910.5 = 6.6 on two degrees of freedom with p-value $\Pr(\chi_2^2 > 6.6) = 0.037$. Notice, however, that the variance component for the tree-specific random intercept is practically zero. `AIC (smaller is better)` is calculated as minus twice the residual log likelihood plus twice the number of covariance parameters. The AIC adjustment was made for only five, not six covariance parameters. The estimate for the variance of tree-specific random intercepts was set to zero. The data do not support that many random effects. Also, the variance of the random slopes on the tree and the apple level add up to 0.000053, which should be compared to $\widehat{\sigma}_1^2 = 0.00005$ in Output 7.6.

Output 7.7.

```
                    The Mixed Procedure

                     Model Information

   Data Set                      WORK.APPLES
   Dependent Variable            diam
   Covariance Structures         Variance Components,
                                 Autoregressive
   Subject Effects               tree, apple(tree),
                                 apple(tree)
   Estimation Method             REML
   Residual Variance Method      Profile
   Fixed Effects SE Method       Model-Based
   Degrees of Freedom Method     Containment

                 Class Level Information

     Class     Levels    Values
     tree          10    1 2 3 4 5 6 7 8 9 10
     apple         24    1 2 3 4 5 7 8 9 10 11 12 13 14
                         15 16 17 18 19 20 21 22 23 24 25
```

Output 7.7 (continued).

```
                         Dimensions

            Covariance Parameters                 6
            Columns in X                          2
            Columns in Z Per Subject             26
            Subjects                             10
            Max Obs Per Subject                  72
            Observations Used                   451
            Observations Not Used                29
            Total Observations                  480

              Covariance Parameter Estimates

           Cov Parm       Subject           Estimate

           Intercept      tree              3.74E-20
           time           tree              0.000017
           Intercept      apple(tree)       0.008653
           time           apple(tree)       0.000036
           AR(1)          apple(tree)         0.3649
           Residual                         0.000354

                       Fit Statistics

           -2 Res Log Likelihood             -1917.1
           AIC (smaller is better)           -1907.1
           AICC (smaller is better)          -1907.0
           BIC (smaller is better)           -1905.6

                 Solution for Fixed Effects

                             Standard
Effect       Estimate          Error      DF     t Value     Pr > |t|
Intercept      2.8322        0.01066       9      265.58       <.0001
time          0.02870       0.001623       9       17.68       <.0001
```

7.6.2 On-Farm Experimentation with Randomly Selected Farms

On-farm experimentation is part of the technology transfer from the agricultural experiment station to the farm operation. On-farm trials are different from station experiments in several ways. Often only two treatments are being compared, one representing a current standard, the other a treatment in technology transfer from experimental research. Supplementary treatments are sometimes added-on plots smaller than the experimental units to which the main treatments of interest are applied (see, e.g., Petersen 1994). This introduces variance heterogeneity into the data due to differing sizes of experimental units. Expressing outcomes on a per-unit area basis may not remove these effects entirely due to the difference in border effects between large and small units. Replication within a farm is not necessary for a valid experimental design provided that the treatments of interest are applied on more than one farm so that the farms can serve as block effects. Data from such experiments will exhibit large experimental error variability and low power unless the farms are very similar. The average treatment difference is the same on all farms, and a sufficient number of farms are involved in the study. Due to differences in, for example, cropping history, soil types, and lacking control over experimental units and conditions one should anticipate such farm × treatment interaction and allow for replication of the treatments within a farm. In contrast to research station experimentation where locations at which to apply the treatments are often chosen deliberately to reflect certain conditions of interest or because of availability, the farms for on-farm research are often chosen at random to represent the population of farms (conditions) in the region where technology transfer is to take place. If we think of farms as stochastic locational or enviromental effects, these will have to enter any statistical model as random effects (§7.2.3). Treatments, chosen deliberately to reflect current practice and technology to be transferred, are fixed effects on the contrary. As a consequence, statistical models for the analysis of data from on-farm experimentation are typically mixed models.

Consider the (hypothetical) data in Table 7.7 representing wheat yields from eight on-farm block designs, each with three blocks and two treatments. The farms were selected at random from a list of all farms in the area where the new treatment (B) is to be tested.

On each farm we have a randomized block design

$$Y_{ij} = \mu + \tau_i + \rho_j + e_{ij},$$

where τ_i $(i = 1, 2)$ is the effect of the ith treatment and ρ_j $(j = 1, 2, 3)$ denotes the block effects. One could analyze eight separate RCBDs to determine the effectiveness of the treatments by farm. These farm-specific analyses would be powerless since each RCBD has only two degrees of freedom for the experimental error. Also, nothing would be learned about the treatment × farm interaction. A more suitable analysis will combine all the data into a single analysis.

Table 7.7. Data from on-farm trials conducted as randomized block designs with three blocks on each of eight farms

	Block 1		**Block 2**		**Block 3**	
Farm	A	B	A	B	A	B
1	30.86	33.31	30.32	30.94	32.31	35.24
2	31.39	27.87	30.62	25.25	29.93	21.79
3	39.22	41.95	38.96	43.38	35.39	41.09
4	37.19	30.97	36.10	32.55	35.85	33.04
5	24.98	23.39	22.04	24.50	22.93	23.24
6	28.06	28.69	27.98	25.68	25.13	25.88
7	27.82	37.23	25.32	34.45	26.52	32.49
8	29.41	30.98	26.63	30.71	29.60	30.63

The model for this analysis is that of a replicated RCBD with farm effects, treatment effects, block effects nested within farms (since block 1 on farm 1 is a different physical entity than block 1 on farm 2), and treatment × farm interactions. Since the eight farms are a random sample of farms their effects enter the model as random (ϕ_k). The treatment effects (τ_i) are fixed and the interaction between farms and treatments ($(\phi\tau)_{ik}$) is random since it involves the random farm effects. The complete model for analysis is

$$\begin{aligned}
&Y_{ijk} = \mu + \phi_k + \tau_i + \rho_{(j)k} + (\phi\tau)_{ik} + e_{ijk} \qquad [7.63]\\
&\phi_k \sim G(0, \sigma^2_\phi)\\
&\rho_{(j)k} \sim G(0, \sigma^2_\rho)\\
&(\phi\tau)_{ik} \sim G(0, \sigma^2_{\phi\tau})\\
&e_{ijk} \sim G(0, \sigma^2_e)\\
&i = 1,2;\ j = 1,2,3;\ k = 1,\cdots,8.
\end{aligned}$$

Because the farm effects are random it is reasonable to also treat the block effects nested within farms as random variables. To obtain a test of the treatment effects and estimates of all variance components by restricted maximum likelihood we use the `proc mixed` code

```
proc mixed data=onfarm;
  class farm block tx;
  model yield = tx /ddfm=satterth;
  random farm block(farm) farm*tx;
run;
```

Only the treatment effect `tx` is listed in the `model` statement since it is the only fixed effect in the model. The Satterthwaite approximation is invoked here because exact tests may not be available in complex mixed models such as this one.

The largest variance component estimate is $\widehat{\sigma}^2_\phi = 19.9979$ (Output 7.8). The variance between farms is twenty times larger than variation within farms ($\widehat{\sigma}^2_\rho = 0.9848$). The test for differences among the treatments is not significant ($F_{obs} = 0.3, p = 0.6003$). Based on this test one would conclude that the new treatment does not increase or decrease yield over the current standard. It is possible, however, that the marginal treatment effect is masked by an interaction. If, for example, the new treatment (B) outperforms the standard on some farms but performs more poorly than the standard on other farms, it is conceivable that the treatment averages across farms are not very different. To address this question we need to test the significance of the interaction between treatments and farms. Since the interaction terms $(\phi\tau)_{ik}$ are random variables, the `farm*tx` effect appears as a covariance parameter, not as a fixed effect. Fitting a reduced model without the interaction we can calculate a likelihood ratio test statistic to test H_0: $\sigma^2_{\phi\tau} = 0$. The `proc mixed` code

```
proc mixed data=onfarm;
  class farm block tx;
  model yield = tx /ddfm=satterth;
  random farm block(farm) ;
run;
```

produces a `-2 Res Log Likelihood` of 250.4 (output not shown). The difference between this value and the `-2 Res Log Likelihood` of 223.8 is asymptotically the realization of a Chi-square random variable with one degree of freedom. The p-value of the likelihood ratio test of H_0: $\sigma^2_{\phi\tau} = 0$ is thus $\Pr(\chi^2_1 \geq 26.6) < 0.0001$. There is a significant interaction between farms and treatments.

Output 7.8.

```
                    The Mixed Procedure

                     Model Information
   Data Set                       WORK.ONFARM
   Dependent Variable             yield
   Covariance Structure           Variance Components
   Estimation Method              REML
   Residual Variance Method       Profile
   Fixed Effects SE Method        Model-Based
   Degrees of Freedom Method      Satterthwaite

                 Class Level Information
     Class     Levels    Values
     farm           8    1 2 3 4 5 6 7 8
     block          3    1 2 3
     tx             2    A B

                        Dimensions
            Covariance Parameters               4
            Columns in X                        3
            Columns in Z                       48
            Subjects                            1
            Max Obs Per Subject                48
            Observations Used                  48
            Observations Not Used               0
            Total Observations                 48

               Covariance Parameter Estimates
                  Cov Parm          Estimate
                  farm               19.9979
                  block(farm)         0.9848
                  farm*tx             9.3377
                  Residual            1.6121

                        Fit Statistics
         Res Log Likelihood                    -111.9
         Akaike's Information Criterion        -115.9
         Schwarz's Bayesian Criterion          -116.1
         -2 Res Log Likelihood                  223.8

               Type 3 Tests of Fixed Effects
                     Num       Den
       Effect         DF        DF     F Value     Pr > F
       tx              1         7        0.30     0.6003
```

This interaction would normally be investigated with interaction slices by farms producing separate tests of the treatment difference for each farm. Since the interaction term is random this is not possible (slices require fixed effects). However, the best linear unbiased predictors for the treatment means on each farm can be calculated with the procedure. These are the quantities on which treatment comparisons for a given farm should be based. The statements

```
estimate 'Blup Farm 1 tx A' intercept 1 tx 1 0 | farm 1 0 0 0 0 0 0 0
          farm*tx 1 0 0 0 0 0 0 0 0 0 0 0 0 0 0 0;
estimate 'Blup Farm 1 tx B' intercept 1 tx 0 1 | farm 1 0 0 0 0 0 0 0
          farm*tx 0 1 0 0 0 0 0 0 0 0 0 0 0 0 0 0;
```

for example, estimate the BLUPs for the two treatments on farm 1. Notice the vertical slash after `tx 1 0` which narrows the inference space by fixing the farm effects to that of farm 1.

BLUPs for other farms are calculated similarly by shifting the coefficients for `farm` and `farm*tx` effects to the appropriate positions. For example,

```
estimate 'Blup Farm 3 tx A' intercept 1 tx 1 0 | farm 0 0 1 0 0 0 0 0
         farm*tx 0 0 0 0 1 0 0 0 0 0 0 0 0 0 0 0;
estimate 'Blup Farm 3 tx B' intercept 1 tx 0 1 | farm 0 0 1 0 0 0 0 0
         farm*tx 0 0 0 0 0 1 0 0 0 0 0 0 0 0 0 0;
```

estimates the treatment BLUPs for the third farm. The coefficients that were shifted compared to the `estimate` statements for farm 1 are shown in bold. The BLUPs so obtained for the two treatments on all farms follow in Output 7.9. Comparing the `Estimate` values there with the entries in Table 7.7 (p. 475), it is evident that the EBLUPs are the sample means for each treatment calculated across the blocks on a particular farm (apart from roundoff error, the values in Table 7.7 were rounded to two decimal places).

Of interest is of course a comparison of these means by farm. In other words, are there farms where the new treatment outperforms the current standard and farms where the reverse is true? Since the treatment effect was not significant in the analysis of the full model but the likelihood ratio test for the interaction was significant, we almost expect such a relationship. The following `estimate` statements contrast the two treatments for each farm (Output 7.10).

```
estimate 'Tx eff. on Farm 1' tx 1 -1 | farm*tx 1 -1;
estimate 'Tx eff. on Farm 2' tx 1 -1 | farm*tx 0 0 1 -1;
estimate 'Tx eff. on Farm 3' tx 1 -1 | farm*tx 0 0 0 0 1 -1;
estimate 'Tx eff. on Farm 4' tx 1 -1 | farm*tx 0 0 0 0 0 0 1 -1;
estimate 'Tx eff. on Farm 5' tx 1 -1 | farm*tx 0 0 0 0 0 0 0 0 1 -1;
estimate 'Tx eff. on Farm 6' tx 1 -1 | farm*tx 0 0 0 0 0 0 0 0 0 0 1 -1;
estimate 'Tx eff. on Farm 7' tx 1 -1 | farm*tx 0 0 0 0 0 0 0 0 0 0 0 0 1 -1;
estimate 'Tx eff. Farm 8' tx 1 -1 | farm*tx 0 0 0 0 0 0 0 0 0 0 0 0 0 0 1 -1;
```

Output 7.9.

Estimates

Label	Estimate	Standard Error	DF	t Value	Pr > |t|
Blup Farm 1 tx A	31.1595	0.9168	29.2	33.99	<.0001
Blup Farm 1 tx B	33.0960	0.9168	29.2	36.10	<.0001
Blup Farm 2 tx A	30.5384	0.9168	29.2	33.31	<.0001
Blup Farm 2 tx B	25.2136	0.9168	29.2	27.50	<.0001
Blup Farm 3 tx A	37.7253	0.9168	29.2	41.15	<.0001
Blup Farm 3 tx B	41.8261	0.9168	29.2	45.62	<.0001
Blup Farm 4 tx A	36.1593	0.9168	29.2	39.44	<.0001
Blup Farm 4 tx B	32.2394	0.9168	29.2	35.17	<.0001
Blup Farm 5 tx A	23.4730	0.9168	29.2	25.60	<.0001
Blup Farm 5 tx B	23.8930	0.9168	29.2	26.06	<.0001
Blup Farm 6 tx A	27.1144	0.9168	29.2	29.58	<.0001
Blup Farm 6 tx B	26.8735	0.9168	29.2	29.31	<.0001
Blup Farm 7 tx A	26.7533	0.9168	29.2	29.18	<.0001
Blup Farm 7 tx B	34.5251	0.9168	29.2	37.66	<.0001
Blup Farm 8 tx A	28.6076	0.9168	29.2	31.20	<.0001
Blup Farm 8 tx B	30.7610	0.9168	29.2	33.55	<.0001

On farms 2 and 4 the old treatment significantly outperforms the new treatment. On farms 3, 7, and 8 the new treatment significantly outperforms the current standard, however (at the 5% significance level). This reversal of the treatment effects masked the treatment main effect. Whereas the recommendation based on the treatment main effect would have been that one may as well stick with the old treatment and not transfer technology from experiment station research, the analysis of the interaction shows that for farms 3, 7, and 8 (and by implication farms in the target region that are alike) the new treatment holds promise.

Output 7.10.

```
                              Estimates

                                         Standard
Label                    Estimate          Error        DF       t Value     Pr > |t|
Tx eff. on Farm 1         -1.9364         1.0117      17.6         -1.91       0.0720
Tx eff. on Farm 2          5.3249         1.0117      17.6          5.26       <.0001
Tx eff. on Farm 3         -4.1009         1.0117      17.6         -4.05       0.0008
Tx eff. on Farm 4          3.9200         1.0117      17.6          3.87       0.0011
Tx eff. on Farm 5         -0.4200         1.0117      17.6         -0.42       0.6831
Tx eff. on Farm 6          0.2409         1.0117      17.6          0.24       0.8145
Tx eff. on Farm 7         -7.7718         1.0117      17.6         -7.68       <.0001
Tx eff. Farm 8            -2.1535         1.0117      17.6         -2.13       0.0477
```

7.6.3 Nested Errors through Subsampling

Subsampling in designed experiments is the recording of multiple, independent observations on the experimental units. We term the experimental material on which subsamples are collected the observational units to distinguish them from the experimental units to which treatments are assigned. Subsamples are sometimes referred to as pseudo-replications, an unfortunate terminology, because they are not replications in the proper sense. The variation among subsamples from the same experimental unit expresses the homogeneity **within** the unit; we term this source of variation **observational error**. The proper error variation to compare treatment effects is variation among experimental units that received the same treatment, termed **experimental error** variance. In experimental designs with subsampling, care must be exercised to (i) not consider subsamples as replications of the treatments, (ii) separate experimental from observational error, and (iii) to perform tests of hypotheses properly. To demonstrate that subsamples do not substitute for treatment replication consider a 3×2 factorial experiment with six experimental units. Assume that these units are pots containing four plants each. The data structure for such an experiment could be displayed as in Table 7.8.

Table 7.8. Generic data layout of 3×2 factorial without replication and four subsamples per experimental unit

	Factor Combination					
Plant	A_1B_1	A_1B_2	A_2B_1	A_2B_2	A_3B_1	A_3B_2
1	3.5	5.0	5.5	7.0	5.5	6.0
2	4.0	5.5	6.0	9.0	4.5	6.5
3	3.0	4.0	5.0	8.0	8.5	9.5
4	4.5	3.5	5.0	6.5	7.0	7.0

It is tempting to analyze these data with a two-way factorial analysis of variance based on the linear model

$$Y_{ijk} = \mu_{ij} + e^*_{ijk} = \mu + \alpha_i + \beta_j + (\alpha\beta)_{ij} + e^*_{ijk}, \qquad [7.64]$$

assuming that e^*_{ijk} is the "experimental" error for the kth experimental unit receiving level i of factor A and level j of factor B. As will be shown shortly, e^*_{ijk} in [7.64] is the observational error and the experimental error has been confounded with the treatment means. Since The SAS® System cannot know whether repeated values in the data set that share the same treatment assignment represent subsamples or replicates, an analysis of the data in Table 7.8 with model [7.64] will be *successful.* Using `proc glm` significant main effects of factors A and B ($p = 0.0004$ and 0.0144) and a nonsignificant interaction are inferred (Output 7.11).

```
data noreps;
  input A B plant y;
  datalines;
1 1 1 3.5
1 1 2 4.0
1 1 3 3.0
1 1 4 4.5
1 2 1 5.0
1 2 2 5.5
1 2 3 4.0
... and so forth ...
;;
run;

proc glm data=noreps;
  class A B;
  model y = A B A*B;
run; quit;
```

Output 7.11.

```
                            The GLM Procedure

                        Class Level Information
                     Class          Levels     Values
                     A                   3     1 2 3
                     B                   2     1 2
                     Number of observations      24

Dependent Variable: y
                                         Sum of
Source                      DF          Squares    Mean Square  F Value  Pr > F
Model                        5      47.34375000     9.46875000     6.94  0.0009
Error                       18      24.56250000     1.36458333
Corrected Total             23      71.90625000

              R-Square      Coeff Var        Root MSE          y Mean
              0.658409       20.09727        1.168154        5.812500

Source                      DF        Type I SS    Mean Square  F Value  Pr > F
A                            2      34.56250000    17.28125000    12.66  0.0004
B                            1      10.01041667    10.01041667     7.34  0.0144
A*B                          2       2.77083333     1.38541667     1.02  0.3821

Source                      DF      Type III SS    Mean Square  F Value  Pr > F
A                            2      34.56250000    17.28125000    12.66  0.0004
B                            1      10.01041667    10.01041667     7.34  0.0144
A*B                          2       2.77083333     1.38541667     1.02  0.3821
```

Notice that the F statistics on which the p-values are based are obtained by dividing the main effects or interaction mean square by the mean square error of 1.3645. This mean square error is based on 18 degrees of freedom, $4 - 1 = 3$ degrees of freedom for the subsamples in each of 6 experimental units. This analysis is clearly wrong, since the experimental error in this design has $t(r-1)$ degrees of freedom where t denotes the number of treatments and r

the number of replications for each treatment. Since each of the $t = 6$ treatments was assigned to only one pot, we have $r = 1$ and $t(r - 1) = 0$. What SAS® terms the `Error` source in this model is the observational error and $\widehat{\sigma}^2 = 1.36458$ is an estimate of the observational error variance. The correct model for the subsampling design contains separate random terms for experimental and observational error. In the two-factor design we obtain

$$Y_{ijkl} = \mu_{ij} + e_{ijk} + d_{ijkl} \qquad [7.65]$$
$$\text{Var}[e_{ijk}] = \sigma_e^2,\ \text{Var}[d_{ijkl}] = \sigma_d^2\ ,$$

where $k = 1, \cdots, r$ indexes the replications, e_{ijk} is the experimental error as defined above, and d_{ijkl} is the observational (subsampling) error for subample $l = 1, \cdots, n$ on replicate k. σ_e^2 and σ_d^2 are the experimental and observational error variances, respectively. If $k = 1$, as for the data in Table 7.8, the model becomes

$$Y_{ijl} = \mu_{ij} + e_{ij} + d_{ijl} = \mu_{ij}^* + d_{ijl}. \qquad [7.66]$$

and the experimental error is now confounded with the treatments. This is model [7.64] where e_{ijk}^* is replaced with d_{ijl} and μ_{ij} is replaced with μ_{ij}^*. Because μ_{ij} and e_{ij} in [7.66] have the same subscript the two sources of variability are confounded. The only random variation that can be estimated is the variance of d_{ijl}, the observational error. Finally, the observational error mean square is not the correct denominator for F-tests (Table 7.9). The statistic $MS(\text{Treatment})/MS(\text{Obs. Error})$ thus is not a test statistic for the absence of treatment effects $(f(\mu_{ij}^2) = 0)$ but for the simultaneous absence of treatment effects **and** the experimental error, a nonsensical proposition.

Table 7.9. Expected mean squares in subsampling design without treatment replications (model [7.66])

Source of Variation	DF	E[MS]
Treatments + Experimental Error	$t-1$	$\sigma_d^2 + n\sigma_e^2 + f\left(\mu_{ij}^2\right)$
Observational Error	$t(n-1)$	σ_d^2

Table 7.10. Expected mean squares in completely randomized design with subsampling (t denotes number of treatments, r number of replicates, and n number of subsamples)

Source of Variation	DF	E[MS]
Treatments (Tx)	$t-1$	$\sigma_d^2 + n\sigma_e^2 + f(\tau_i^2)$
Experimental Error (EE)	$t(r-1)$	$\sigma_d^2 + n\sigma_e^2$
Observational Error (OE)	$tr(n-1)$	σ_d^2

In subsampling designs *with* treatment replication, experimental and observational error variances are estimable and not confounded with effects. Table 7.10 displays the expected mean squares for t treatments in a completely randomized design with r replications per treatment and n subsamples per experimental unit for the linear model

$$Y_{ijk} = \mu + \tau_i + e_{ij} + d_{ijk}$$
$$i = 1, \cdots, t; j = 1, \cdots, r; k = 1, \cdots, n.$$

Notice that experimental error degrees of freedom are not affected by the number of subsamples, and that $F_{obs} = MS(Tx)/MS(EE)$ is the test statistic for testing treatment effects.

The data in Table 7.11, taken from Steel, Torrie, and Dickey (1997, p. 159), represent a 3×2 factorial treatment structure arranged in completely randomized design with $r = 3$ replicates and $n = 4$ subsamples per experimental unit. From a large group of plants four were randomly assigned to each of 18 pots. Six treatments were then randomly assigned to the pots such that each treatment was replicated three times. The treatments consisted of all possible combinations of three hours of daylight (8, 12, 16 hrs) and two levels of night temperatures (low, high). The outcome of interest was the stem growth of mint plants grown in nutrient solution under the assigned conditions. The experimental units are the pots since treatments were assigned to those. Stem growth was measured for each plant in a pot, hence there are four subsamples per experimental unit.

Table 7.11. One-week stem growth of mint plants data from Steel et al. (1997, p. 159)

Low Night Temperature

	8 hrs			**12 hrs**			**16 hrs**		
Plant	**Pot 1**	**Pot 2**	**Pot 3**	**Pot 1**	**Pot 2**	**Pot 3**	**Pot 1**	**Pot 2**	**Pot 3**
1	3.5	2.5	3.0	5.0	3.5	4.5	5.0	5.5	5.5
2	4.0	4.5	3.0	5.5	3.5	4.0	4.5	6.0	4.5
3	3.0	5.5	2.5	4.0	3.0	4.0	5.0	5.0	6.5
4	4.5	5.0	3.0	3.5	4.0	5.0	4.5	5.0	5.5

High Night Temperature

	8 hrs			**12 hrs**			**16 hrs**		
Plant	**Pot 1**	**Pot 2**	**Pot 3**	**Pot 1**	**Pot 2**	**Pot 3**	**Pot 1**	**Pot 2**	**Pot 3**
1	8.5	6.5	7.0	6.0	6.0	6.5	7.0	6.0	11.0
2	6.0	7.0	7.0	5.5	8.5	6.5	9.0	7.0	7.0
3	9.0	8.0	7.0	3.5	4.5	8.5	8.5	7.0	9.0
4	8.5	6.5	7.0	7.0	7.5	7.5	8.5	7.0	8.0

The linear model for these data is

$$Y_{ijkl} = \mu + \alpha_i + \beta_j + (\alpha\beta)_{ij} + e_{ijk} + d_{ijkl} \quad [7.67]$$
$$i = 1, \cdots, a = 3;\ j = 1, \cdots, b = 2;\ k = 1, \cdots, r = 3; l = 1, \cdots, n = 4$$
$$\text{Var}[e_{ijk}] = \sigma_e^2;\ \text{Var}[d_{ijkl}] = \sigma_d^2,$$

where the e_{ijk} and d_{ijkl} are zero-mean uncorrelated random variables. The analysis of variance is shown in Table 7.12.

Table 7.12. Analysis of variance of mint plants data

Source	**df**
Hours	$a-1=2$
Temperature	$b-1=1$
Hours × Temperature	$(a-1)(b-1)=2$
Experimental Error	$ab(r-1)=6(3-1)=12$
Observational Error	$abr(n-1)=18(4-1)=54$
Total	$abrn-1=71$

The analysis of variance can be obtained with `proc glm` of The SAS® System (Output 7.12):

```
proc glm data=mintstems;
  class hour night pot;
  model growth = hour night hour*night pot(hour*night);
run; quit;
```

The sequential (`Type I`) and partial (`Type III`) sums of squares are identical because the design is orthogonal. Notice that the source denoted `Error` is again the obervational error as can be seen from the associated degrees of freedom and the experimental error is modeled as `pot(hour*night)`. The F statistics calculated by `proc glm` are obtained by dividing the mean square of a source of variability by the mean square for the `Error` source; hence they use the observational error mean square as a denominator and are incorrect. The two error mean square estimates in Output 7.12 are

$$\widehat{\sigma}_d^2 = 0.9340$$
$$\widehat{\sigma}_d^2 + n\widehat{\sigma}_e^2 = 2.1527,$$

hence dividing by the observational mean square error is detrimental in two ways. The F statistic is inflated and the p-value is calculated from an distribution with incorrect (too many) degrees of freedom.

The correct tests can be obtained in two ways with `proc glm`. One can add a `random` statement indicating which terms of the model statement are random variables and The SAS® System will construct the appropriate test statistics based on the formulas of expected mean squares. Alternatively one can use the `test` statement if the correct error term is known. The two methods lead to the following procedure calls (output not shown).

```
proc glm data=mintstems;
  class hour night pot;
  model growth = hour night hour*night pot(hour*night);
  random pot(hour*night) / test;
run; quit;

proc glm data=mintstems;
  class hour night pot;
  model growth = hour night hour*night pot(hour*night);
  test h=hour night hour*night e=pot(hour*night);
run; quit;
```

Output 7.12.

```
                     The GLM Procedure

                  Class Level Information
              Class          Levels    Values
              hour                3     8 12 16
              night               2     Hig Low
              pot                 3     1 2 3

                Number of observations     72

Dependent Variable: growth
                                Sum of
Source                DF       Squares    Mean Square  F Value   Pr > F
Model                 17   205.4756944     12.0868056    12.94   <.0001
Error                 54    50.4375000      0.9340278
Corrected Total       71   255.9131944

           R-Square     Coeff Var      Root MSE    growth Mean
           0.802912      16.70696      0.966451       5.784722

Source                DF     Type I SS    Mean Square  F Value  Pr > F
hour                   2    22.2986111     11.1493056    11.94  <.0001
night                  1   151.6701389    151.6701389   162.38  <.0001
hour*night             2     5.6736111      2.8368056     3.04  0.0562
pot(hour*night)       12    25.8333333      2.1527778     2.30  0.0186

Source                DF   Type III SS    Mean Square  F Value  Pr > F
hour                   2    22.2986111     11.1493056    11.94  <.0001
night                  1   151.6701389    151.6701389   162.38  <.0001
hour*night             2     5.6736111      2.8368056     3.04  0.0562
pot(hour*night)       12    25.8333333      2.1527778     2.30  0.0186
```

A more elegant approach is to use `proc mixed` which is specifically designed for mixed models. The statements to analyze the two-way factorial with subsampling are

```
proc mixed data=mintstems;
  class hour night pot;
  model growth = hour night hour*night;
  random pot(hour*night);
run; quit;
```

or

```
proc mixed data=mintstems;
  class hour night pot;
  model growth = hour night hour*night;
  random intercept / subject=pot(hour*night);
run; quit;
```

The two versions of `proc mixed` code differ only in the form of the `random` statement and yield identical results. The second form explicitly defines the experimental units `pot(hour*night)` as clusters and the columns of the $\mathbf{Z}_i$ matrix as having an intercept only. The residual maximum likelihood estimates of the variance components σ_e^2 and σ_d^2 are $\widehat{\sigma}_e^2 = 0.3047$ and $\widehat{\sigma}_{d,}^2 = 0.9340$, respectively (Output 7.13).

Output 7.13.

```
                    The Mixed Procedure

                    Model Information

Data Set                        WORK.MINTSTEMS
Dependent Variable              growth
Covariance Structure            Variance Components
Estimation Method               REML
Residual Variance Method        Profile
Fixed Effects SE Method         Model-Based
Degrees of Freedom Method       Containment

                  Class Level Information
  Class     Levels    Values
  hour           3    8 12 16
  night          2    Hig Low
  pot            3    1 2 3

                      Dimensions
         Covariance Parameters                 2
         Columns in X                         12
         Columns in Z                         18
         Subjects                              1
         Max Obs Per Subject                  72
         Observations Used                    72
         Observations Not Used                 0
         Total Observations                   72

           Covariance Parameter Estimates

      Cov Parm      Subject                Estimate
      Intercept     pot(hour*night)          0.3047
      Residual                               0.9340
                    Fit Statistics

      Res Log Likelihood                     -103.9
      Akaike's Information Criterion         -105.9
      Schwarz's Bayesian Criterion           -106.8
      -2 Res Log Likelihood                   207.7

            Type 3 Tests of Fixed Effects

                    Num       Den
  Effect             DF        DF    F Value    Pr > F
  hour                2        12       5.18    0.0239
  night               1        12      70.45    <.0001
  hour*night          2        12       1.32    0.3038
```

The latter estimate is labeled as `Residual` in the `Covariance Parameter Estimates` table. Since the data are completely balanced these estimates are identical to the method of moment estimator implied by the analysis of variance. From $\widehat{\sigma}_d^2 + n\widehat{\sigma}_e^2 = 2.1527$ and $\widehat{\sigma}_d^2 = 0.934$ one obtains the moment estimator of the experimental error variance as

$$\widehat{\sigma}_e^2 = (2.1527 - 0.9340)/4 = 0.3047.$$

Results of the main effects and interaction tests are shown in the `Type 3 Tests of Fixed Effects` table. The F statistics are identical to those obtained in `proc glm` **if** one uses the correct mean square error term there. For example from Output 7.12 one obtains

$$F_{obs} = \frac{MS(Hour)}{MS(EE)} = \frac{11.1493}{2.1527} = 5.18$$
$$F_{obs} = \frac{MS(Night)}{MS(EE)} = \frac{151.670}{2.1527} = 70.45$$
$$F_{obs} = \frac{MS(Hour \times Night)}{MS(EE)} = \frac{2.8368}{2.1527} = 1.32.$$

These are the F statistics shown in Output 7.13. Also notice that the denominator degrees of freedom are set to the correct degrees of freedom associated with the experimental error (Table 7.12).

The marginal correlation structure in the subsampling model [7.67] is compound symmetric, observational errors are nested within experimental errors. Adding the `v=list` option to the `random` statement of `proc mixed` requests a printout of the (estimated) marginal variance-covariance matrices of the clusters (subjects) in `list`. For example,

```
random intercept / subject=pot(hour*night) v=1;
```

requests a printout of the variance-covariance matrix for the first cluster (Output 7.14). It is easy to verify that this matrix is of the form

$$\widehat{\sigma}_e^2\mathbf{J}_4 + \widehat{\sigma}_d^2\mathbf{I}_4.$$

Output 7.14.

Estimated V Matrix for pot(hour*night) 1 8 High

Row	Col1	Col2	Col3	Col4
1	1.2387	0.3047	0.3047	0.3047
2	0.3047	1.2387	0.3047	0.3047
3	0.3047	0.3047	1.2387	0.3047
4	0.3047	0.3047	0.3047	1.2387

The same analysis can thus be obtained by modeling the marginal variance-covariance matrix $\text{Var}[\mathbf{Y}_i]$ directly as a compound symmetric matrix. Replacing the `random` statement with a `repeated` statement and choosing the appropriate covariance structure (`type=cs`), the statements

```
proc mixed data=mintstems noitprint;
  class hour night pot;
  model growth = hour night hour*night;
  repeated / sub=pot(hour*night) type=cs r=1;
run; quit;
```

lead to the same results as in Output 7.13 and Output 7.14, only the covariance parameter `Intercept` in Output 7.13 has been renamed to `CS` (Output 7.15).

Output 7.15.

```
                         The Mixed Procedure

                          Model Information

        Data Set                        WORK.MINTSTEMS
        Dependent Variable              growth
        Covariance Structure            Compound Symmetry
        Subject Effect                  pot(hour*night)
        Estimation Method               REML
        Residual Variance Method        Profile
        Fixed Effects SE Method         Model-Based
        Degrees of Freedom Method       Between-Within

                       Class Level Information

          Class     Levels    Values

          hour           3    8 12 16
          night          2    Hig Low
          pot            3    1 2 3

                             Dimensions

               Covariance Parameters                2
               Columns in X                        12
               Columns in Z                         0
               Subjects                            18
               Max Obs Per Subject                  4
               Observations Used                   72
               Observations Not Used                0
               Total Observations                  72

            Estimated R Matrix for pot(hour*night) 1 8 Hig

          Row          Col1          Col2          Col3          Col4

            1        1.2387        0.3047        0.3047        0.3047
            2        0.3047        1.2387        0.3047        0.3047
            3        0.3047        0.3047        1.2387        0.3047
            4        0.3047        0.3047        0.3047        1.2387

                    Covariance Parameter Estimates

             Cov Parm      Subject               Estimate
             CS            pot(hour*night)         0.3047
             Residual                              0.9340

                           Fit Statistics

             Res Log Likelihood                     -103.9
             Akaike's Information Criterion         -105.9
             Schwarz's Bayesian Criterion           -106.8
             -2 Res Log Likelihood                   207.7

                    Type 3 Tests of Fixed Effects
                          Num       Den
           Effect          DF        DF     F Value     Pr > F
           hour             2        12        5.18     0.0239
           night            1        12       70.45     <.0001
           hour*night       2        12        1.32     0.3038
```

7.6.4 Recovery of Inter-Block Information in Incomplete Block Designs

Incompleteness of block designs can have many causes. Some are by design, others by *accident*. Among the accidental causes are destruction or loss of experimental units and discarding of erroneous measurements. Frequently incompleteness is a design feature if the size of the blocks is such that not all treatments can be accommodated. Since calculations for incomplete designs are considerably more involved than for completely balanced designs, experimental plans were developed that ensure some sort of balance in the treatment allocation to either reduce computational burden and/or to ensure a certain precision in treatment comparisons. Incomplete block designs in this category are known as balanced incomplete block designs (BIBs), partially balanced incomplete block designs (PBIBs) and various special cases thereof, such as the lattice designs (see, e.g. Yates 1936, 1940; Bose and Nair 1939; Cochran and Cox 1957 as some of the historically significant references on the subject). We will not discuss the various forms of incomplete block designs here in detail, but the basic issues that come to bear when not all treatments are allocated in every block. Hoshmand (1994, Ch. 4.3) discusses various forms of agronomic lattice designs, which are special cases of BIBs or PBIBs.

To illustrate the problem that arises in incomplete block designs consider the following treatment layout in a BIB with $t = 5$ treatments in $b = 10$ blocks of size $k = 3$.

Table 7.13. A balanced incomplete block design (BIB) (Treatments that appear in a particular block are marked as x)

	Treatment				
Block	A	B	C	D	E
1		x	x		x
2	x			x	x
3		x		x	x
4			x	x	x
5	x	x		x	
6	x	x			x
7	x	x	x		
8		x	x	x	
9	x		x		x
10	x		x	x	

This design is balanced in two ways. Each treatment is replicated the same number of times (6) throughout the experiment and each pair of treatments appears the same number of times (3) within a block. For example, treatments B and C appear in block 1, 7, 8. Treatments A and B appear in blocks 5, 6, and 7. As a result, all treatment comparisons will be made with the same precision in the experiment. However, because of the incompleteness, block and treatment effects are not orthogonal. Whether block effects are removed or not prior to assessing treatment effects is critical. To see this consider a comparison of treatments A and B. The naïve approach is to base this comparison on the two arithmetic averages $\overline{y}_A$ and $\overline{y}_B$. Their difference is not an estimate of the treatment effect; however, since these are averages calculated over different blocks. $\overline{y}_A$ is calculated from information in blocks $2, 5, 6, 7, 9, 10$ and $\overline{y}_B$ from information in blocks $1, 3, 5, 6, 7, 8$. The difference $\overline{y}_A - \overline{y}_B$ carries not only information about differences between the treatments but also about block effects. To obtain a fair comparison of the treatments unaffected by the block effects, the treatment sum of squares must be adjusted for the block effects and treatment means are not estimated as arithmetic averages. A statistical model for the design in Table 7.13 is $Y_{ij} = \mu + \rho_i + \tau_j + e_{ij}$ where the ρ_i are block effects $(i = 1, ..., 10)$, τ_j are the treatment effects $(j = 1, ..., 5)$ and e_{ij} are independent experimental errors with mean 0 and variance σ^2. The only difference between this linear model and one for a randomized *complete* block design is that not all combinations ij are possible. The appropriate estimate of the mean of the jth treatment in the incomplete design is $\widehat{\mu} + \widehat{\tau}_j$ where carets denote the least squares estimate. $\widehat{\mu} + \widehat{\tau}_j$ is also known as the **least squares mean** for treatment j. In fact, these estimates are always appropriate. In a balanced design it turns out that the least squares estimates are identical to the arithmetic averages. The question thus should not be when one should use the least squares means for treatment comparisons, but when one can rely on arithmetic means.

We illustrate the effect of nonorthogonality with data from a balanced incomplete block design reported by Cochran and Cox (1957, p. 448). Thirteen hybrids of corn were arranged in a field experiment in blocks of size $k = 4$ such that each pair of treatments appeared once in a block throughout the experiment and each treatment is replicated four times. This arrangement requires $b = 13$ blocks.

Table 7.14. Experimental layout of BIB in Cochran and Cox (1957, p. 448)†
(showing yield of corn in pounds per plot)

	Hybrid												
Block	1	2	3	4	5	6	7	8	9	10	11	12	13
1			25.3			19.9			29.0		24.6		
2			23.0	19.8				33.3				22.7	
3										16.2	19.3	31.7	26.6
4		27.3			27.0			35.6			17.4		
5							23.4	30.5	30.8	32.4			
6				30.6	32.4	27.2				32.8			
7	34.7				31.1				25.7			30.5	
8			34.4		32.4		33.3						36.9
9	38.2	32.9	37.3							31.3			
10		28.7		30.7					26.9				35.3
11	36.6			31.1			31.1				28.4		
12	31.8					33.7		27.8					41.1
13		30.3				31.5	39.3					26.7	

We obtain the analysis of variance for these data with `proc glm` (Output 7.16).

```
proc glm data=cornyld;
  class block hybrid;
  model yield = block hybrid;
  lsmeans hybrid / stderr;
  means   hybrid;
run; quit;
```

Output 7.16.

```
                         The GLM Procedure

                      Class Level Information

          Class          Levels    Values
          block              13    1 2 3 4 5 6 7 8 9 10 11 12 13
          hybrid             13    1 2 3 4 5 6 7 8 9 10 11 12 13

                    Number of observations    52

Dependent Variable: yield   corn yield in pounds per plot

                                    Sum of
Source                     DF      Squares   Mean Square  F Value   Pr > F
Model                      24  1017.929231     42.413718     2.13   0.0298
Error                      27   538.217500     19.933981
Corrected Total            51  1556.146731

           R-Square     Coeff Var      Root MSE    yield Mean
           0.654134      14.99302      4.464749      29.77885

Source                     DF      Type I SS   Mean Square  F Value   Pr > F
block                      12    689.3842308    57.4486859     2.88   0.0109
hybrid                     12    328.5450000    27.3787500     1.37   0.2378

Source                     DF    Type III SS   Mean Square  F Value   Pr > F
block                      12    475.2650000    39.6054167     1.99   0.0677
hybrid                     12    328.5450000    27.3787500     1.37   0.2378

                        Least Squares Means

                                        Standard
           hybrid    yield LSMEAN          Error    Pr > |t|

           1           33.0019231      2.4586721      <.0001
           2           28.2711538      2.4586721      <.0001
           3           30.2173077      2.4586721      <.0001
           4           28.1019231      2.4586721      <.0001
           5           29.9557692      2.4586721      <.0001
           6           27.1019231      2.4586721      <.0001
           7           29.7250000      2.4586721      <.0001
           8           33.7173077      2.4586721      <.0001
           9           29.0173077      2.4586721      <.0001
           10          28.0250000      2.4586721      <.0001
           11          24.5250000      2.4586721      <.0001
           12          30.0865385      2.4586721      <.0001
           13          35.3788462      2.4586721      <.0001

            Level of            ------------yield------------
            hybrid        N             Mean          Std Dev

            1             4       35.3250000       2.75121185
            2             4       29.8000000       2.40277617
            3             4       30.0000000       6.92194578
            4             4       28.0500000       5.50424079
            5             4       30.7250000       2.55783111
            6             4       28.0750000       6.08187197
            7             4       31.7750000       6.57133929
            8             4       31.8000000       3.38526218
            9             4       28.1000000       2.25831796
            10            4       28.1750000       8.00848508
            11            4       22.4250000       5.01489448
            12            4       27.9000000       4.06939799
            13            4       34.9750000       6.09555849
```

What SAS® terms `Type I SS` and `Type III SS` are sequential and partial sums of squares, respectively. Sequential sums of squares are the sum of squares contributions of sources given that variability of the previously listed sources has been accounted for. The sequential `block` sum of squares of 689.38 is the sum of squares among block averages and the sequential `hybrid` sum of squares of 328.54 is the contribution of the treatment variability after adjusting for block effects. Inferences about treatment effects are to be based on the partial sums of squares. The nonorthogonality of this design is evidenced by the fact that the `Type I SS` and the `Type III SS` differ. In an orthogonal design, the two sets of sums of squares would be identical. Whenever the design is nonorthogonal great care must be exercised to estimate treatment means properly. The list of least squares means shows the estimates $\widehat{\mu}+\widehat{\tau}_j$ that are adjusted for the block effects. Notice that all least squares means have the same standard error, since every treatment is replicated the same number of times. The final part of the output shows the result of the `means` statement. These are the arithmetic sample averages of the observations for a particular treatment which do not estimate treatment means unbiasedly unless every treatment appears in every block. One must not base treatment comparisons on these quantities in a nonorthogonal design. The column `Std Dev` is the standard deviation of the four observations for each treatment. It is not the standard deviation of a treatment based on the analysis of variance.

An analysis of an incomplete block design such as the `proc glm` analysis above is termed an **intra-block** analysis that obtains treatment information by comparing block-adjusted least squares estimates. Yates (1936, 1940) coined the term along with the term **inter-block** analysis that also recovers treatment information contained in the block totals (averages). In incomplete block designs contrasts of block averages also contain contrasts among the treatments. To see this consider blocks 1 and 3 in Table 7.13. The first block contains treatments B, C, and E, the third block contains treatments B, D, and E. If $\overline{Y}_{1.}$ denotes the average in block 1 and $\overline{Y}_{3.}$ the average in block 3, then we have

$$\mathrm{E}\left[\overline{Y}_{1.}\right] = \mu + \rho_1 + \frac{1}{3}(\tau_B + \tau_C + \tau_E)$$
$$\mathrm{E}\left[\overline{Y}_{3.}\right] = \mu + \rho_3 + \frac{1}{3}(\tau_B + \tau_D + \tau_E).$$

The difference of the block averages contains information about the treatments, namely, $\mathrm{E}[\overline{Y}_{1.} - \overline{Y}_{3.}] = \rho_1 - \rho_3 + \frac{1}{3}(\tau_C - \tau_D)$. Unfortunately, this is not just a contrast among treatments, but involves the effects of the two blocks. The solution to uncovering the inter-block information is to let the block effects be random (with mean 0) since then $\mathrm{E}[\overline{Y}_{1.} - \overline{Y}_{3.}] = 0 - 0 + \frac{1}{3}(\tau_C - \tau_D) = \frac{1}{3}(\tau_C - \tau_D)$, a contrast between treatment effects. The linear mixed model for the incomplete block design now becomes

$$Y_{ij} = \mu + \rho_i + \tau_j + e_{ij},\ e_{ij} \sim \left(0, \sigma^2\right), \rho_j \sim \left(0, \sigma^2_\rho\right),$$

where the e_{ij} and ρ_j are independent. The term $\rho_1 - \rho_3 + \frac{1}{3}(\tau_C - \tau_D)$ now represents the conditional (narrow inference space) comparison of the two block means and the unconditional (broad inference space) comparison is $\mathrm{E}[\overline{Y}_{1.} - \overline{Y}_{3.}] = \mathrm{E}[\mathrm{E}[\overline{Y}_{1.} - \overline{Y}_{3.}|\rho]] = \frac{1}{3}(\tau_C - \tau_D)$.

For the corn hybrid experiment of Cochran and Cox (1957, p. 448) the inter-block analysis is carried out with the following `proc mixed` statements.

```
proc mixed data=cornyld;
  class block hybrid;
  model yield = hybrid;
  random block;
  lsmeans hybrid ;
  estimate 'hybrid 1 broad ' intercept 1 hybrid 1;
  estimate 'hybrid 1 narrow' intercept 13 hybrid 13 |
                           block 1 1 1 1 1 1 1 1 1 1 1 1 1 / divisor=13;
  estimate 'hybrid 1 in block 1' intercept 1 hybrid 1 | block 1;
  estimate 'hybrid 1 in block 2' intercept 1 hybrid 1 | block 0 1;
  estimate 'hybrid 1 vs hybrid2' hybrid 1 -1;
run; quit;
```

The inter-block analysis is invoked by moving the `block` term from the model statement to the `random` statement. The `estimate` statements are not necessary unless one wants to estimate treatment means in the narrow or intermediate inference spaces (§7.3). The `lsmeans` statement requests block-adjusted estimates of the hybrid means in the broad inference space. On Output 7.17 we notice that the F statistic for hybrid differences in the mixed analysis $(F_{obs} = 1.67)$ has changed from the intra-block analysis in `proc glm` $(F_{obs} = 1.37)$. This reflects the additional treatment information recovered by the inter-block analysis. Furthermore, the estimates of the treatment means have changed as compared to the least squares means reported by `proc glm`. The additional information recovered from block averages surfaces here again. In the example of §7.3 it was noted that the estimates of factor means would be the same if all factors would have been considered fixed and only the standard errors would differ between the fixed effects and mixed effects analysis. This statement was correct there because the design was completely balanced and hence orthogonal. In a nonorthogonal incomplete block design both the estimates of the treatment means as well as their standard errors differ between the fixed effects and mixed effects analysis.

Output 7.17.

```
                    The Mixed Procedure

                     Model Information
     Data Set                        WORK.CORNYLD
     Dependent Variable              yield
     Covariance Structure            Variance Components
     Estimation Method               REML
     Residual Variance Method        Profile
     Fixed Effects SE Method         Model-Based
     Degrees of Freedom Method       Containment

                  Class Level Information
       Class     Levels    Values
       block         13    1 2 3 4 5 6 7 8 9 10 11 12 13
       hybrid        13    1 2 3 4 5 6 7 8 9 10 11 12 13

                         Dimensions
              Covariance Parameters               2
              Columns in X                       14
              Columns in Z                       13
              Subjects                            1
              Max Obs Per Subject                52
              Observations Used                  52
              Observations Not Used               0
              Total Observations                 52

                 Covariance Parameter Estimates
                     Cov Parm      Estimate
                     block           6.0527
                     Residual       19.9340
```

Output 7.17 (continued).

```
                 Fit Statistics
Res Log Likelihood                        -126.8
Akaike's Information Criterion            -128.8
Schwarz's Bayesian Criterion              -129.4
-2 Res Log Likelihood                      253.6

           Type 3 Tests of Fixed Effects
                 Num     Den
Effect            DF      DF    F Value    Pr > F
hybrid            12      27       1.67    0.1293

                          Estimates
                           Standard
Label                  Estimate     Error     DF    t Value    Pr > |t|
hybrid 1 broad          34.1712    2.4447     27      13.98      <.0001
hybrid 1 narrow         34.1712    2.3475     27      14.56      <.0001
hybrid 1 in block 1     32.6735    2.9651     27      11.02      <.0001
hybrid 1 in block 2     31.2751    2.9651     27      10.55      <.0001
hybrid 1 in block 7     34.1635    2.6960     27      12.67      <.0001
hybrid 1 vs hybrid2      5.1305    3.3331     27       1.54      0.1354

                     Least Squares Means
                           Standard
 Effect    hybrid    Estimate     Error     DF    t Value    Pr > |t|
 hybrid      1        34.1712    2.4447     27      13.98      <.0001
 hybrid      2        29.0406    2.4447     27      11.88      <.0001
 hybrid      3        30.1079    2.4447     27      12.32      <.0001
 hybrid      4        28.0758    2.4447     27      11.48      <.0001
 hybrid      5        30.3429    2.4447     27      12.41      <.0001
 hybrid      6        27.5917    2.4447     27      11.29      <.0001
 hybrid      7        30.7568    2.4447     27      12.58      <.0001
 hybrid      8        32.7523    2.4447     27      13.40      <.0001
 hybrid      9        28.5556    2.4447     27      11.68      <.0001
 hybrid     10        28.1005    2.4447     27      11.49      <.0001
 hybrid     11        23.4680    2.4447     27       9.60      <.0001
 hybrid     12        28.9860    2.4447     27      11.86      <.0001
 hybrid     13        35.1756    2.4447     27      14.39      <.0001
```

The first two `estimate` statements produce the hybrid 1 estimate in the broad and narrow inference space and show that the `lsmeans` statement operates in the broad inference space. The third through fifth `estimate` statements show how to estimate the hybrid mean in a particular plot. Notice that hybrid 1 did not appear in blocks 1 or 2 in the experiment but did in block 7. Nevertheless, we are able to predict how well the hybrid would have done in blocks 1 and 2, although this prediction is less precise than prediction of the hybrid's performance in blocks were the hybrid was observed. In an intra-block analysis where blocks are fixed, it is not possible to differentiate a hybrid's performance by block.

7.6.5 A Split-Strip-Plot Experiment for Soybean Yield

An experiment was conducted at the Tidewater Agricultural Research and Extension Center in Suffolk, Virginia to investigate how soybean yield response depended on soybean cultivar, row spacing, and plant population. The three factors and their levels considered in the experiment were

- Cultivar (AG3601, AG3701, AG4601, AG4701)

• Plant Population (60, 120, 180, 240, 300 thousand per acre)

• Row spacing (9", 18").

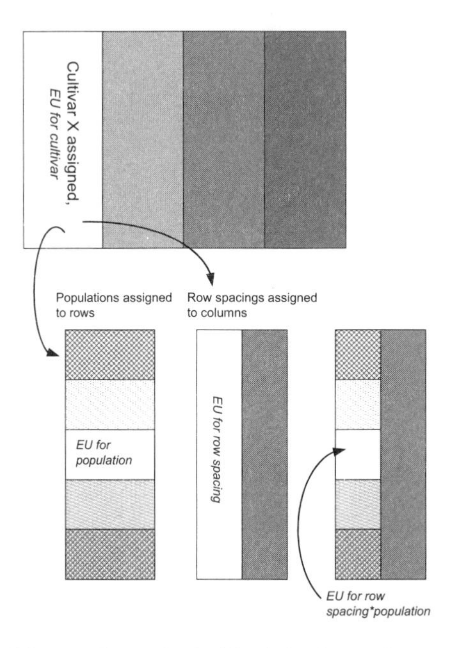

Figure 7.18. Experimental layout for a single block in the soybean yield experiment. Cultivars (varieties) were assigned to the large experimental units (plots), row spacings and population densities to perpendicular strips within the plots.The experiment was brought to our attention and the data were made kindly available by Dr. David Holshouser, Tidewater Agricultural Research and Extension Center, Virginia Polytechnic Institute and State University. Used with permission.

Although the experiment was conducted in four site-years, we consider only a single site-year here. At each site four replications of the cultivars were arranged in a randomized block design. Because of technical limitations, it was decided to apply the row spacing and population densities in strips within a cultivar experimental unit (plot). It was determined at random which side (strip) of the plot received 9" spacing. Then the population densities were assigned randomly to five strips running perpendicular to the row spacing strips. Figure 7.18 displays a schematic layout of one of the four blocks in the experiment.

The factors Row Spacing and Population Density are a split of the experimental unit to which a cultivar is assigned, but are not arranged in a 2×5 factorial structure. Considering the cultivar experimental units, Row Spacing and Population Density form a strip-plot (split-block) design with 16 blocks (replications). Each cultivar experimental unit serves as a replicate for the split-block design of the other two factors. We call this design a split-strip-plot design.

There are experimental units of four different sizes in this experiment, hence the linear model will contain four different experimental error sources of variability associated with the plot, the columns, the rows, and their intersection. Before engaging in an analysis of data from a complex experiment such as this, it is helpful to develop the source of variability and degree of freedom decomposition. Correct specification of the programming statements can then be more easily checked. As is good practice for designs with a split, the whole-plot and subplot design analysis of variance can be developed separately. On the whole-plot (Cultivar) level we have a simple randomized complete block design of four treatments in four blocks (Table 7.15). The sub-plot source and degree of freedom decomposition regards each experimental unit in the whole-plot design as a replicate. Hence, there are $ar - 1 = 15$ replicate degrees of freedom for the sub-plot analysis, which is a strip-plot (split-block) design.

Table 7.15. Whole-plot analysis of variance in soybean yield example

Source	**df**	$=$
Block	$r-1$	3
Cultivar	$a-1$	3
Error(1)	$(r-1)(a-1)$	9
Total	$ar-1$	15

Table 7.16. Sub-plot analysis of variance in soybean yield example

Source	**df**	$=$
Replicate	$ar-1$	15
Row Spacing	$b-1$	1
Error(2^*)	$(ar-1)(b-1)$	15
Population	$c-1$	4
Error(3^*)	$(ar-1)(c-1)$	60
Row Sp. $\times$ Population	$(b-1)(c-1)$	4
Error(4^*)	$(ar-1)(b-1)(c-1)$	60
Total	$arbc-1$	159

Upon combining the whole-plot and sub-plot analysis, the Replicate source in Table 7.16 is replaced with the whole-plot decomposition in Table 7.15. Furthermore, interactions between whole-plot factor Cultivar and all subplot factors are added. The degrees of freedom for the interactions are removed from the corresponding sub-plot errors Error(2^*) through Error(4^*).

The degrees of freedom for Error(2), for example, are obtained as

$$\begin{aligned} df_{Error(2^*)} - df_{Cultivar \times RowSp.} &= (ar-1)(b-1) - (a-1)(b-1) \\ &= (ar-1-a+1)(b-1) \\ &= (ar-a)(b-1) = a(r-1)(b-1), \end{aligned}$$

and similarly for the other sub-plot error terms. The linear model for this experiment has as many terms as there are rows in Table 7.17. In two steps the model can be defined as

$$\begin{aligned} Y_{ijkl} &= \mu_{ijk} + r_l + e^1_{il} + e^2_{ijl} + e^3_{ikl} + e^4_{ijkl} \\ \mu_{ijk} &= \mu + \alpha_i + \beta_j + (\alpha\beta)_{ij} + \gamma_k + (\alpha\gamma)_{ik} + (\beta\gamma)_{jk} + (\alpha\beta\gamma)_{ijk}. \end{aligned} \quad [7.68]$$

μ_{ijk} denotes the mean of the treatment combination of the ith cultivar $(i = 1, \cdots, 4)$, j^{th} row spacing ($j = 1, 2$), and k^{th} population ($k = 1, \cdots, 5$). It is decomposed into a grand mean (μ), main effects of Cultivar (α_i), Row Spacing (β_j), Population (γ_k) and their respective interactions. The first line of model [7.68] expresses the observation Y_{ijkl} as a sum of the mean μ_{ijkl} and various random components. r_l is the random effect of the lth block (whole-plot), assumed $G(0, \sigma_r^2)$. e_{il}^1 is the experimental error on the whole-plot assumed $G(0, \sigma_1^2)$, e_{ijl}^2 is the experimental error on a row spacing strip assumed $G(0, \sigma_2^2)$, e_{ikl}^3 is the experimental error on a population density strip assumed $G(0, \sigma_3^2)$, and finally, e_{ijkl}^4 is the experimental error on the intersection of perpendicular strips assumed $G(0, \sigma_4^2)$ (see Figure 7.18). All random components are independent by virtue of independent randomizations.

Table 7.17. Sources of variability and degrees of freedom in soybean yield example

Source	df	=
Block	$r - 1$	3
Cultivar	$a - 1$	3
Error(1)	$(r - 1)(a - 1)$	9
Row Spacing	$b - 1$	1
Cultivar × Row Sp.	$(a - 1)(b - 1)$	3
Error(2)	$a(r - 1)(b - 1)$	12
Population	$c - 1$	4
Cultivar × Population	$(a - 1)(c - 1)$	12
Error(3)	$a(r - 1)(c - 1)$	48
Row Sp. × Population	$(b - 1)(c - 1)$	4
Cultivar × Row Sp. × Population	$(a - 1)(b - 1)(c - 1)$	12
Error(4)	$a(r - 1)(b - 1)(c - 1)$	48
Total	$arbc - 1$	159

We consider the blocks random in this analysis for two reasons. We posit that the blocks are only a smaller subset of possible conditions over which inferences are to be drawn. Secondly, the Block × Cultivar interaction serves as the experimental error term on the whole-plot. How can this interaction be random if Cultivar and Block factors are fixed? Some research workers adopt the viewpoint that this apparent inconsistency should not be of concern. A `block*cultivar` term will be used in the SAS® code only to generate the necessary error term. We do remind the reader, however, that treating Block × Cultivar interactions as random and blocks as fixed corresponds to choosing an intermediate inference space. As discussed in §7.3 this choice results in smaller standard errrors (and p-values) compared to the broad inference space in which random effects are allowed to vary. Treating the blocks as random in the analysis could be viewed as a somewhat conservative approach.

In a three-factor experiment it is difficult to roadmap the analysis from main effect and interaction tests to contrasts, multiple comparisons, slices, etc. Whether marginal mean comparisons are meaningful depends on which factors interact. The first step in the analysis is thus to produce tests of the main effects and interactions. The `proc mixed` statements

```
proc mixed data=soybeanyield;
  /* rep  = whole-plot replication variable */
  /* tpop = target population density       */
  class rep cultivar tpop rowspace;
  model yield = cultivar
                rowspace rowspace*cultivar
                tpop tpop*cultivar
                rowspace*tpop rowspace*tpop*cultivar / ddfm=satterth;
  random rep
         rep*cultivar
         rep*cultivar*rowspace
         rep*cultivar*tpop;
run;
```

accomplish that (Output 7.18).

Notice that all random terms in model [7.68] are listed in the `random` statement and only fixed effects appear in the `model` statement of the procedure. Furthermore, the error term with the most subscripts (e^4_{ijkl}) and the constant term (μ) do not need to be specified. Altogether there are eleven effects specified between the `model` and the `random` statements (compare to Table 7.17 and model [7.68]). In split-type designs great care must be exercised to ensure that the inference is as accurate as possible. For example, in a regular split-plot design no exact test exists to compare whole-plot treatment means for a given level of the sub-plot treatment, even if the design is balanced. In split-block (strip-plot) designs where two factors A and B are applied perpendicular to each other, no exact test exists to compare two A means at the same level of B and two B means at the same level of A. In these cases we rely on approximate procedures to calculate test statistics, degrees of freedom, and p-values. We choose Satterthwaite's method (Satterthwaite 1946) for split-type designs throughout. If data are balanced and all treatment factors are fixed, the tests for main effects and interactions in split-plot and split-block models are exact and do not require further approximations. When data are unbalanced and/or some treatment factors are fixed while others are random, exact F-tests cannot necessarily be constructed and the Satterthwaite approximation again becomes important. In `proc mixed` this approximation is invoked with the `ddfm=satterth` option of the `model` statement. In the soybean trial, six yield observations were missing, reducing the total number of observations from 160 to 154. If no observations were missing, the Satterthwaite approximation would not be necessary to test main effects and interactions.

Output 7.18.

```
                         The Mixed Procedure

                          Model Information
          Data Set                     WORK.SOYBEANYIELD
          Dependent Variable           YIELD
          Covariance Structure         Variance Components
          Estimation Method            REML
          Residual Variance Method     Profile
          Fixed Effects SE Method      Model-Based
          Degrees of Freedom Method    Satterthwaite

                       Class Level Information
           Class        Levels    Values
           REP               4    1 2 3 4
           CULTIVAR          4    AG3601 AG3701 AG4601 AG4701
           TPOP              5    60 120 180 240 300
           ROWSPACE          2    9 18
```

Output 7.18 (continued).

```
                         Dimensions

          Covariance Parameters                 5
          Columns in X                         90
          Columns in Z                        132
          Subjects                              1
          Max Obs Per Subject                 160
          Observations Used                   154
          Observations Not Used                 6
          Total Observations                  160

            Covariance Parameter Estimates

           Cov Parm                      Estimate

           REP                             3.0368
           REP*CULTIVAR                    0.4524
           REP*CULTIVAR*ROWSPACE           1.2442
           REP*CULTIVAR*TPOP               2.4215
           Residual                        3.9276

                      Fit Statistics

       Res Log Likelihood                     -302.6
       Akaike's Information Criterion         -307.6
       Schwarz's Bayesian Criterion           -306.1
       -2 Res Log Likelihood                   605.3

             Type 3 Tests of Fixed Effects

                             Num      Den
Effect                        DF       DF    F Value    Pr > F

CULTIVAR                       3     9.16      8.77    0.0047
ROWSPACE                       1     11.3      3.72    0.0795
CULTIVAR*ROWSPACE              3     11.3      6.01    0.0108
TPOP                           4     46.9     32.08    <.0001
CULTIVAR*TPOP                 12     46.8      1.26    0.2749
TPOP*ROWSPACE                  4     45.7      1.06    0.3870
CULTIVAR*TPOP*ROWSPAC         12     45.6      2.60    0.0100
```

The estimates of the variance components are shown in the `Covariance Parameter Estimates` table as $\widehat{\sigma}_r^2 = 3.037$, $\widehat{\sigma}_1^2 = 0.452$, $\widehat{\sigma}_2^2 = 1.244$, $\widehat{\sigma}_3^2 = 2.421$, and $\widehat{\sigma}_4^2 = 3.928$. The denominator degrees of freedom for the F statistics in the `Type 3 Tests of Fixed Effects` table were adjusted by the Satterthwaite procedure due to the missingness of the observations. For complete data we would have expected denominator degrees of freedom of 9, 12, 12, 48, 48, 48, and 48. At the 5% significance level the three-way interaction, the Population Density main effect, the Cultivar × Row Spacing interaction, and the Cultivar main effect are significant. Because of the significance of the three-way interaction, the two-way interactions that appear nonsignificant may be masked, and similarly for the Row Spacing main effect.

The next step in the analysis is to investigate the interaction pattern more closely. Because of the significance of the three-way interaction, we start there. Figure 7.19 shows the estimated three-way cell means (least square means) for the Cultivar × Population × Row Spacing combinations. Since the factor Population Density is quantitative, trends of soybean yield in density are investigated later via regression contrasts. The Row Spacing effect for a given population density and cultivar combination and the Cultivar effect for a given density and row spacing can be assessed by slicing the three-way interaction. To this end add the statement

```
lsmeans cultivar*rowspace*tpop / slice=(cultivar*tpop tpop*rowspace);
```

to the `proc mixed` code (Output 7.19). The first block of tests in the `Tests of Effect Slices` table compares 9 inches vs. 18 inches row spacing for cultivar AG3601 at various population densities. The second block for cultivar AG3701 and so forth. The last block compares cultivars for a given combination of population density and row spacing.

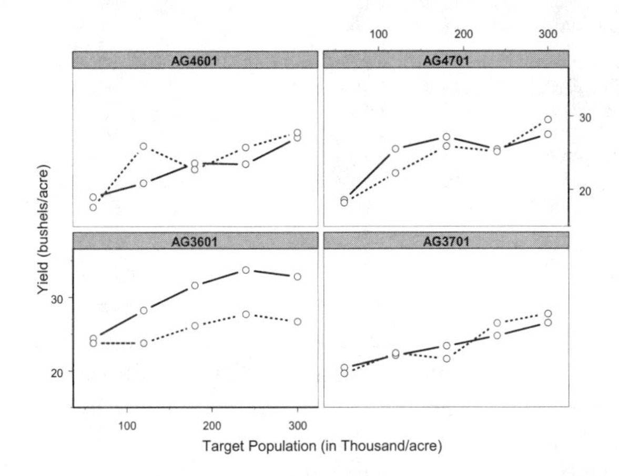

Figure 7.19. Three-way least squares means for factors Cultivar, Row Spacing, and Population Density. Solid line refers to 9-inch spacing, dashed line to 18-inch spacing.

For cultivar AG3601 it is striking that there is no spacing effect at 60,000 plants per acre ($p = 0.8804$), but there are significant spacing effects for all greater population densities. This effect is also visible in Figure 7.19. For the other cultivars the row spacing effects are absent with two exceptions: AG4601 and AG4701 at $120,000$ plants per acre ($p = 0.0030$ and 0.0459, respectively). The last block of tests reveals that at 9-inch spacing there are significant differences among the cultivars at any population density (e.g., $p = 0.0198$ at $60,000\ N/acre$). For the wider row spacing cultivar effects are mostly absent.

Output 7.19.

```
                         Tests of Effect Slices

                                                  Num    Den
Effect              CULTIVAR  TPOP  ROWSPACE   DF     DF  F Value Pr > F

CULTIVA*TPOP*ROWSPAC  AG3601    60                1   54.4     0.02 0.8804
CULTIVA*TPOP*ROWSPAC  AG3601   120                1   45.7     7.66 0.0081
CULTIVA*TPOP*ROWSPAC  AG3601   180                1   45.7    11.59 0.0014
CULTIVA*TPOP*ROWSPAC  AG3601   240                1   51.6    13.19 0.0006
CULTIVA*TPOP*ROWSPAC  AG3601   300                1   45.7    14.51 0.0004

CULTIVA*TPOP*ROWSPAC  AG3701    60                1   45.7     0.23 0.6321
CULTIVA*TPOP*ROWSPAC  AG3701   120                1   54.4     0.04 0.8426
CULTIVA*TPOP*ROWSPAC  AG3701   180                1   51.6     0.85 0.3622
CULTIVA*TPOP*ROWSPAC  AG3701   240                1   45.7     1.12 0.2960
CULTIVA*TPOP*ROWSPAC  AG3701   300                1   45.7     0.58 0.4501
```

Output 7.19 (continued).

```
CULTIVA*TPOP*ROWSPAC  AG4601    60            1  45.7     0.76 0.3885
CULTIVA*TPOP*ROWSPAC  AG4601   120            1  45.7     9.86 0.0030
CULTIVA*TPOP*ROWSPAC  AG4601   180            1  45.7     0.30 0.5890
CULTIVA*TPOP*ROWSPAC  AG4601   240            1  45.7     2.00 0.1639
CULTIVA*TPOP*ROWSPAC  AG4601   300            1  45.7     0.18 0.6766

CULTIVA*TPOP*ROWSPAC  AG4701    60            1  45.7     0.05 0.8166
CULTIVA*TPOP*ROWSPAC  AG4701   120            1  45.7     4.21 0.0459
CULTIVA*TPOP*ROWSPAC  AG4701   180            1  45.7     0.60 0.4410
CULTIVA*TPOP*ROWSPAC  AG4701   240            1  45.7     0.04 0.8407
CULTIVA*TPOP*ROWSPAC  AG4701   300            1  45.7     1.55 0.2199

CULTIVA*TPOP*ROWSPAC            60     9      3  76.5     3.49 0.0198
CULTIVA*TPOP*ROWSPAC           120     9      3  76.5     5.54 0.0017
CULTIVA*TPOP*ROWSPAC           180     9      3  79.2     7.19 0.0003
CULTIVA*TPOP*ROWSPAC           240     9      3  79.1     9.00 <.0001
CULTIVA*TPOP*ROWSPAC           300     9      3  76.5     4.23 0.0081
CULTIVA*TPOP*ROWSPAC            60    18      3  80.1     3.68 0.0154
CULTIVA*TPOP*ROWSPAC           120    18      3  80.3     1.39 0.2530
CULTIVA*TPOP*ROWSPAC           180    18      3  79.2     2.39 0.0750
CULTIVA*TPOP*ROWSPAC           240    18      3  79.1     0.34 0.7982
CULTIVA*TPOP*ROWSPAC           300    18      3  76.5     0.68 0.5660
```

To investigate the nature of the yield dependency on Population Density, the information in the table of slices is very helpful. It suggests that for cultivars AG3701, AG4601, and AG4701 it is not necessary to distinguish trends among row spacings. Determining the nature of the trend averaged across row spacings for these cultivars is sufficient. The levels of the factor Population Density are equally spaced and we use the standard orthogonal polynomial coefficients to test for linear, quadratic, cubic, and quartic trends. The following twelve `contrast` statements are added to the `proc mixed` code to test trends for AG3701, AG4601, and AG4701 across row spacings (Output 7.20).

```
contrast 'AG3701 quartic   ' tpop  1 -4  6 -4 1
                             cultivar*tpop 0 0 0 0 0  1 -4  6 -4 1;

contrast 'AG3701 cubic     ' tpop -1  2  0 -2 1
                             cultivar*tpop 0 0 0 0 0 -1  2  0 -2 1;

contrast 'AG3701 quadratic' tpop  2 -1 -2 -1 2
                             cultivar*tpop 0 0 0 0 0  2 -1 -2 -1 2;

contrast 'AG3701 linear    ' tpop -2 -1  0  1 2
                             cultivar*tpop 0 0 0 0 0 -2 -1  0  1 2;

contrast 'AG4601 quartic   ' tpop  1 -4  6 -4 1
                             cultivar*tpop 0 0 0 0 0 0 0 0 0 0 1 -4  6 -4 1;

contrast 'AG4601 cubic     ' tpop -1  2  0 -2 1
                             cultivar*tpop 0 0 0 0 0 0 0 0 0 0 -1 2  0 -2 1;

contrast 'AG4601 quadratic' tpop  2 -1 -2 -1 2
                             cultivar*tpop 0 0 0 0 0 0 0 0 0 0 2 -1 -2 -1 2;

contrast 'AG4601 linear    ' tpop -2 -1  0  1 2
                             cultivar*tpop 0 0 0 0 0 0 0 0 0 0 -2 -1 0 1 2;

contrast 'AG4701 quartic   ' tpop  1 -4  6 -4 1
                      cultivar*tpop 0 0 0 0 0 0 0 0 0 0 0 0 0 0 0  1 -4  6 -4 1;

contrast 'AG4701 cubic     ' tpop -1  2  0 -2 1
                      cultivar*tpop 0 0 0 0 0 0 0 0 0 0 0 0 0 0 0 -1  2  0 -2 1;
```

```
contrast 'AG4701 quadratic' tpop  2 -1 -2 -1 2
                   cultivar*tpop 0 0 0 0 0 0 0 0 0 0 0 0 0 0 0  2 -1 -2 -1 2;

contrast 'AG4701 linear   ' tpop -2 -1  0  1 2
                   cultivar*tpop 0 0 0 0 0 0 0 0 0 0 0 0 0 0 0 -2 -1  0  1 2;
```

Output 7.20.

Contrasts

Label	Num DF	Den DF	F Value	Pr > F
AG3701 quartic	1	49	1.14	0.2899
AG3701 cubic	1	48.9	0.02	0.9004
AG3701 quadratic	1	46.9	0.18	0.6710
AG3701 linear	1	46.6	27.53	<.0001
AG4601 quartic	1	45.8	0.71	0.4050
AG4601 cubic	1	45.8	4.08	0.0493
AG4601 quadratic	1	45.8	0.55	0.4610
AG4601 linear	1	45.8	34.46	<.0001
AG4701 quartic	1	45.8	1.13	0.2933
AG4701 cubic	1	45.8	4.75	0.0346
AG4701 quadratic	1	45.8	4.61	0.0371
AG4701 linear	1	45.8	42.61	<.0001

Yield is a linearly increasing function of population density for AG3701. Cultivar AG4601 shows a slight cubic effect in addition to a linear term and AG4701 shows polynomial terms up to the third order. To model the yield response (Y) as a function of population (X), one would thus choose the models

$$\begin{aligned} \text{AG3701: } & Y = \beta_0 + \beta_1 x + e \\ \text{AG4601: } & Y = \beta_0 + \beta_1 x + \beta_2 x^3 + e \\ \text{AG4701: } & Y = \beta_0 + \beta_1 x + \beta_2 x^2 + \beta_3 x^3 + e. \end{aligned}$$

The `contrast` statements to discern the row-spacing specific trends for cultivar AG3601 are more involved (Output 7.21):

```
contrast "AG3601 quart., 9inch"       tpop 1 -4  6 -4 1
                            cultivar*tpop 1 -4  6 -4 1
                            tpop*rowspace 1 0 -4 0 6 0 -4 0 1 0
                            cultivar*tpop*rowspace 1 0 -4 0 6 0 -4 0 1 0;

contrast "AG3601 cubic , 9inch"       tpop -1  2  0 -2 1
                            cultivar*tpop -1  2  0 -2 1
                            tpop*rowspace -1 0 2 0 0 0 -2 0 1 0
                            cultivar*tpop*rowspace -1 0 2 0 0 0 -2 0 1 0;

contrast "AG3601 quadr., 9inch"       tpop 2 -1 -2 -1 2
                            cultivar*tpop 2 -1 -2 -1 2
                           tpop*rowspace 2 0 -1 0 -2 0 -1 0 2 0
                           cultivar*tpop*rowspace 2 0 -1 0 -2 0 -1 0 2 0;

contrast "AG3601 linear, 9inch"      tpop -2 -1  0  1 2
                           cultivar*tpop -2 -1  0  1 2
```

```
                        tpop*rowspace -2 0 -1 0 0 0 1 0 2 0
                        cultivar*tpop*rowspace -2 0 -1 0 0 0 1 0 2 0;

contrast "AG3601 quart.,18inch"      tpop 1 -4  6 -4 1
                        cultivar*tpop 1 -4  6 -4 1
                        tpop*rowspace 0 1 0 -4 0 6 0 -4 0 1
                        cultivar*tpop*rowspace 0 1 0 -4 0 6 0 -4 0 1;

contrast "AG3601 cubic ,18inch"      tpop -1  2  0 -2 1
                        cultivar*tpop -1  2  0 -2 1
                        tpop*rowspace 0 -1 0 2 0 0 0 -2 0 1
                        cultivar*tpop*rowspace 0 -1 0 2 0 0 0 -2 0 1;

contrast "AG3601 quadr.,18inch"      tpop 2 -1 -2 -1 2
                        cultivar*tpop 2 -1 -2 -1 2
                        tpop*rowspace 0 2 0 -1 0 -2 0 -1 0 2
                        cultivar*tpop*rowspace 0 2 0 -1 0 -2 0 -1 0 2;

contrast "AG3601 linear,18inch"      tpop -2 -1  0  1 2
                        cultivar*tpop -2 -1  0  1 2
                        tpop*rowspace 0 -2 0 -1 0 0 0 1 0 2
                        cultivar*tpop*rowspace 0 -2 0 -1 0 0 0 1 0 2;
```

For this cultivar at 9-inch row spacing, yield depends on population density in quadratic and linear fashion. At 18-inch row spacing, yield is not responsive to changes in the population density. The slight yield increase at 18-inch spacing (Figure 7.19) is evident in the marginally significant linear trend ($p = 0.0526$).

Output 7.21.

Contrasts

Label	Num DF	Den DF	F Value	Pr > F
AG3601 quart., 9inch	1	80.3	0.01	0.9037
AG3601 cubic , 9inch	1	80.9	0.42	0.5196
AG3601 quadr., 9inch	1	79.6	5.14	0.0261
AG3601 linear, 9inch	1	79.7	30.40	<.0001
AG3601 quart.,18inch	1	80.8	0.12	0.7350
AG3601 cubic ,18inch	1	81.6	0.97	0.3271
AG3601 quadr.,18inch	1	84.2	0.10	0.7519
AG3601 linear,18inch	1	85.9	3.86	0.0526

This analysis of the three-way interaction leads to the overall conclusion that only for cultivar AG3601 is row spacing of importance for a given population density. Does this conclusion prevail when yields are averaged across different population densities? A look at the significant Cultivar × Rowspace interaction confirms this. Figure 7.20 shows the corresponding two-way least square means and the p-values from slicing this interaction by cultivar (bottom margin) and by spacing (right margin). Significant differences exist among varieties for 9-inch spacing ($p < 0.0001$) but not for 18-inch spacing ($p = 0.4204$). Averaged across the population densities, only variety AG3601 shows a significant yield difference among the two row spacings ($p = 0.0008$). The p-values in Figure 7.20 were obtained with the statement (Output 7.22)

```
lsmeans cultivar*rowspace / slice=(cultivar rowspace);
```

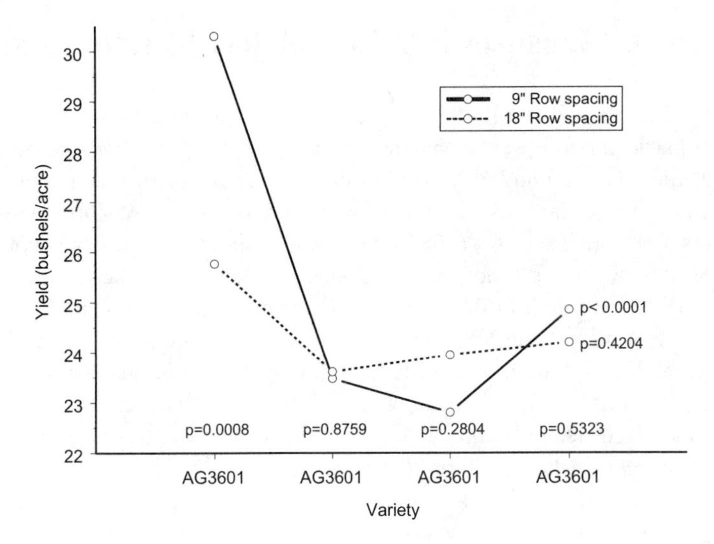

Figure 7.20. Two-way Cultivar × Row Spacing least squares means. *p*-values from slices of the two-way interaction are shown in the margins.

Output 7.22. Tests of Effect Slices

Effect	CULTIVAR	TPOP	ROWSPACE	Num DF	Den DF	F Value	Pr > F
CULTIVAR*ROWSPACE	AG3601			1	11.8	19.80	0.0008
CULTIVAR*ROWSPACE	AG3701			1	11.8	0.03	0.8759
CULTIVAR*ROWSPACE	AG4601			1	10.8	1.29	0.2804
CULTIVAR*ROWSPACE	AG4701			1	10.8	0.42	0.5323
CULTIVAR*ROWSPACE			9	3	17.3	14.79	<.0001
CULTIVAR*ROWSPACE			18	3	17.7	0.99	0.4204

Finally, we can raise the question, which varieties differ significantly from each other at 9-inch row spacing. The previous slice shows that the question is not of interest at 18-inch spacing. The statement

```
lsmeans cultivar*rowspace / diff;
```

compares all Cultivar × Row Spacing combinations in pairs and produces many comparisons that are not of interest. The trimmed output that follows shows only those comparisons where factor spacing was held fixed at 9 inches. Cultivar AG3601 is significantly higher yielding (`Estimates` are positive) than any of the other cultivars.

Output 7.23. Differences of Least Squares Means

CULTIVAR	ROWSPACE	_CULTIVAR	_ROWSPACE	Estimate	Pr > |t|
AG3601	9	AG3701	9	6.7783	<.0001
AG3601	9	AG4601	9	7.3617	<.0001
AG3601	9	AG4701	9	5.3167	0.0004
AG3701	9	AG4601	9	0.5834	0.6406
AG3701	9	AG4701	9	-1.4616	0.2500
AG4601	9	AG4701	9	-2.0450	0.1117

7.6.6 Repeated Measures in a Completely Randomized Design

Water was leached through soil columns to examine secondary minerals formed from weathering, in particular to observe changes in the weathering products of biotite. Eighteen columns were filled with silt and clay sized biotite. 2.5 grams of surface material were added to each column. The surface material was collected from the A horizon of spruce and hardwood forests or consisted of washed quartz sand. The columns were kept at a constant temperature of $4°C$ and three times per week were treated with either 5 or 20 ml of water, simulating two different rainfall rates. Three replicate columns were available for each Rainfall Rate $\times$ Surface Material combination and arranged on a rack in a completely randomized design. At days 5, 21, 40, and 57 water leaching through the columns was sampled. This is a repeated measures study with 4 repeat observations. The basic experimental design is a completely randomized design with a 2×3 factorial treatment structure of factors Rainfall Rate (5 ml, 20 ml) and Surface Material (spruce, hardwood, sand). Table 7.18 shows the pH values of the leachate at the four sampling dates for the eighteen experimental units and Figure 7.21 shows the estimated means over time for the six treatments (obtained from analysis that follows).

Table 7.18. Repeated measures leachate data for 2×3 factorial in a CRD†

		Rainfall Rate 5 ml			**Rainfall Rate 20 ml**		
Day	**Rep**	**Sand**	**Spruce**	**Hardwood**	**Sand**	**Spruce**	**Hardwood**
5	1	6.08	5.76	6.51	6.40	5.69	6.75
	2	6.50	5.18	6.45	6.49	4.97	6.99
	3	6.54	5.52	6.60	6.57	5.29	6.94
21	1	5.26	5.80	6.00	6.05	5.96	6.00
	2	6.24	5.57	6.08	6.16	5.54	6.36
	3	6.02	5.40	6.00	5.96	5.45	6.39
40	1	5.86	6.10	6.34	6.51	6.35	6.76
	2	6.40	6.23	6.16	6.51	6.17	6.66
	3	6.06	5.13	6.02	6.48	5.97	6.56
57	1	6.07	6.24	6.38	6.65	5.93	6.11
	2	6.38	5.74	6.03	6.35	5.84	6.60
	3	6.50	5.39	5.99	6.89	6.09	6.54

†Data kindly provided by Dr. Lucian W. Zelazny and Mr. Ryan Reed, Department of Crop and Soil Environmental Sciences, Virginia Poytechnic Institute and State University (see Reed 2000). Used with permission.

The basic model for the analysis of these data comprises fixed effects for the Surface and Rainfall Rate effects, temporal effects and all possible two-way interactions and one three-way interaction. Random effects are associated with the replicates stemming from random assignment of treatments to the columns and within-column disturbances over time. The model can be expressed as

$$Y = \mu + \alpha_i + \beta_j + (\alpha\beta)_{ij} + e_{ijk} + \tau_l + (\alpha\tau)_{il} + (\beta\tau)_{jl} + (\alpha\beta\tau)_{ijl} + f_{ijkl} \quad [7.69]$$

where

α_i is the effect of the ith surface type $(i = 1, \cdots, 3)$

β_j is the effect of the jth rainfall rate ($j = 1, 2$)

e_{ijk} is the experimental error associated with replicate (column) k ($k = 1, \cdots, 3$), assumed independent Gaussian with mean 0 and variance σ_e^2.

τ_l is the effect of the lth time point ($l = 1, \cdots, 4$)

f_{ijkl} is a random disturbance associated with the lth repeated measurement for the kth replicate.

The remaining terms in model [7.69] denote interactions between the various factors in obvious fashion.

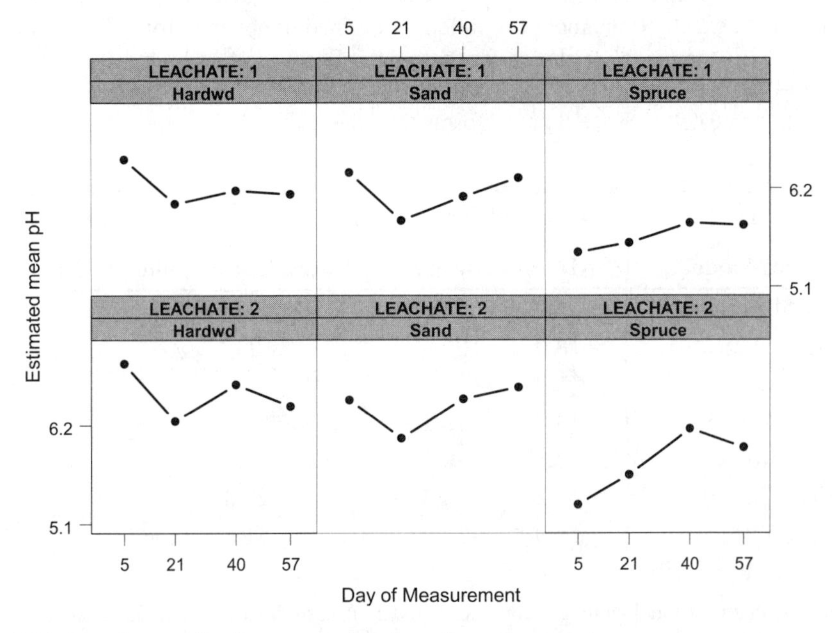

Figure 7.21. Leachate × Surface means by day of measurement.

If this basic model is accepted the first question that needs to be addressed is that of the possible correlation model to be used for the f_{ijkl} from the same column. We fit seven different correlation models and compare their respective AIC and Schwarz criteria (Table 7.19). These criteria point to several correlation models. The AR(1), exponential, gaussian, and compound symmetric models have similar fit statistics. The AIC criterion is calculated as the residual log likelihood minus the number of the covariance parameters ("larger is better" version). The unstructured models appear to fit the data well as judged by the negative of the residual log likelihood (which we try to minimize). Their large number of covariance parameters carries a hefty penalty in the determination of AIC, however.

The power model [7.56] is a reparameterization of the exponential model [7.54]. Hence, their fit statistics are identical. The AR(1) and the exponential model are identical if the data are equally spaced. For the data considered here, the measurement intervals of 16, 19, and 17 days are almost identical which explains the small difference in AIC between the two models. The ante-dependence model is also penalized substantially because of its many covariance parameters. Should we ignore the possibility of serial correlations and continue with a compound symmetry model, akin to a split-plot design where the whole-plot factor is a 2×3 factorial? Fortunately, the power and compound symmetry models are nested. A test of $H_0{:}\rho = 0$ in the power model can be carried out as a likelihood-ratio test. The test statistic is $LR = 24.0 - 21.6 = 2.4$ and the probability that a Chi-squared random variable with one degree of freedom exceeds 2.4 is $\Pr(\chi^2_1 \geq 2.4) = 0.121$. At the 5% level, we cannot reject the hypothesis, however. To continue with a compound symmetry model implies *acceptance* of H_0, and in our opinion the p-value is not large enough to rule out the possibility of a Type-II error. We are not sold on independence of the repeated measurements. For the gaussian model [7.57] the likelihood ratio statistic would be even greater, but the two models are not directly nested. The gaussian model approaches compound symmetry as the parameter α approaches 0. At exactly 0 the model is no longer defined. It appears from Table 7.19 that the gaussian model is the model of choice for these data. Because of its high degree of regularity at short lag distances, we do not recommend it in general for most repeated measures data (see comments in §7.5.2). Since it outperforms the other parsimonious models we use it here.

Table 7.19. Akaike's information criterion and Schwarz' criterion for various covariance models (The last column contains the number of covariance parameters)

Covariance Model	AIC	Schwarz	– 2 Res. Log Likelihood	Parameters
Compound Symmetry	– 14.0	– 14.9	24.0	2
Unstructured	– 18.0	– 22.9	13.9[†]	11
Unstructured(2)	– 15.9	– 19.4	15.7	8
Exponential, [7.54]	– 13.8	– 15.1	21.6	3
Power, [7.56]	– 13.8	– 15.1	21.6	3
Gaussian, [7.57]	– 13.2	– 14.5	20.3	3
AR(1), [7.52]	– 13.9	– 15.2	21.8	3
Antedependence, [7.59]	– 17.5	– 21.0	19.0	8

[†]The unstructured model led to a second derivative matrix of the log likelihood function which was not positive definite. It is not considered further for these data.

The `proc mixed` code to analyze this repeated measures design follows. The `model` statement contains the main effects and interactions of the three factors Rainfall Rate, Surface Material, and Time. The `random` statement identifies the experimental errors e_{ijk} and the `repeated` statement instructs `proc mixed` to treat the observations from the same replicate as correlated according to the gaussian covariance model. The `r=1` and `rcorr=1` options of the repeated statement request a printout of the $\mathbf{R}(\boldsymbol{\alpha})$ and the correlation matrix for the first subject in the data set. The initial sorting of the data set is good practice to ensure that the observations are ordered by increasing time of measurement within each experimental unit. The variable `t` created in the data set is used in the `type=sp(gau)(t)` option to represent temporal distance between repeated observations on the same experimental unit. It is measured in number of weeks since the initial measurement (since `time` is measured in days).

```
proc sort data=Leachate; by surface leachate rep time; run;
data Leachate; set Leachate; t = (time - 5)/7; run;
proc mixed data=Leachate noitprint;
  class rep surface rainfall time;
  model ph = surface rainfall surface*rainfall
```

```
        time time*surface time*rainfall time*surface*rainfall;
  random rep(surface*rainfall);
  repeated / subject=rep(surface*rainfall) type=sp(gau)(t) r=1 rcorr=1;
run;
```

The restricted maximum likelihood algorithm converged to the covariance parameter estimates $\widehat{\sigma}_e^2 = 0.01997$, $\widehat{\sigma}_f^2 = 0.05029$, and $\widehat{\phi} = 2.5870$ (Output 7.24).

Output 7.24.

```
                        The Mixed Procedure

                         Model Information
        Data Set                     WORK.LEACHATE
        Dependent Variable           PH
        Covariance Structures        Variance Components, Spatial Gaussian
        Subject Effect               REP(SURFACE*RAINFAL)
        Estimation Method            REML
        Residual Variance Method     Profile
        Fixed Effects SE Method      Model-Based
        Degrees of Freedom Method    Containment

                      Class Level Information
           Class       Levels    Values
           REP              3    1 2 3
           SURFACE          3    hardwd sand spruce
           RAINFALL         2    1 2
           TIME             4    5 21 40 57

                            Dimensions
                Covariance Parameters                3
                Columns in X                        60
                Columns in Z                        18
                Subjects                             1
                Max Obs Per Subject                 72
                Observations Used                   72
                Observations Not Used                0
                Total Observations                  72


                 Estimated R Matrix for Subject 1
            Row        Col1         Col2         Col3         Col4
              1     0.05029      0.02304     0.001200     0.000013
              2     0.02304      0.05029      0.01673     0.000967
              3    0.001200      0.01673      0.05029      0.02083
              4    0.000013     0.000967      0.02083      0.05029

           Estimated R Correlation Matrix for Subject 1
            Row        Col1         Col2         Col3         Col4
              1      1.0000       0.4581      0.02386     0.000263
              2      0.4581       1.0000       0.3326      0.01922
              3     0.02386       0.3326       1.0000       0.4143
              4    0.000263      0.01922       0.4143       1.0000

                 Covariance Parameter Estimates
        Cov Parm                    Subject                 Estimate
        REP(SURFACE*RAINFAL)                                 0.01997
        SP(GAU)                     REP(SURFACE*RAINFAL)      2.5870
        Residual                                             0.05029
                            Fit Statistics
                Res Log Likelihood                     -10.2
                Akaike's Information Criterion         -13.2
                Schwarz's Bayesian Criterion           -14.5
                -2 Res Log Likelihood                   20.3
```

Output 7.24 (continued).

Type 3 Tests of Fixed Effects

Effect	Num DF	Den DF	F Value	Pr > F
SURFACE	2	12	19.13	0.0002
RAINFALL	1	12	6.09	0.0296
SURFACE*RAINFALL	2	12	0.56	0.5875
TIME	3	36	15.57	<.0001
SURFACE*TIME	6	36	8.84	<.0001
RAINFALL*TIME	3	36	1.79	0.1665
SURFACE*RAINFAL*TIME	6	36	0.48	0.8203

The covariance between the first and second remeasurement disturbances ($t = (21 - 5)/7 = 2.285$ weeks) is estimated as

$$\widehat{\text{Cov}}[f_{ijk1}, f_{ijk2}] = 0.05029 \times \exp\left\{ -\frac{2.285^2}{2.587^2} \right\} = 0.02305$$

and the correlation is $\exp\{ -2.285^2/2.587^2\} = 0.458$. These values can be found in the first row, second column of the `Estimated R Matrix for Subject 1` and the `Estimated R Correlation Matrix for Subject 1`. The continuity of the gaussian covariance model near the origin lets correlations decay rather quickly. The correlation between the first and third measurement ($t = 5.0$ weeks) is only 0.0238. The estimated practical range of the correlation process is $\sqrt{3} \times \widehat{\phi} = 4.48$ weeks. After four and a half weeks leachate pHs from the same soil column are essentially uncorrelated.

The variance component for the experimental error is rather small. If the fixed effects part of the model contains numerous effects and the within-cluster correlations are also modeled, one will frequently find that the algorithm fails to provide estimates for some or all effects in the `random` statement. The likelihood solutions for the covariance parameters of these effects are outside the permissible range. After dropping the `random` statement (or individual effects in the `random` statement), and retaining the `repeated` statement the algorithm then often converges successfully. While this is a reasonable approach, one should be cautioned that this effectively alters the statistical model being fit. Dropping the `random` statement here corresponds to the assumption that $\sigma_e^2 \equiv 0$, i.e., there is no experimental error variability associated with the experimental unit and all stochastic variation arises from the within cluster process.

For some covariance models, combining a `random` and `repeated` statement is impossible, since the effects are confounded. For example, compound symmetry is implied by two nested random effects. In the absence of a `repeated` statement a residual error is always added to the model and hence the statements

```
proc mixed data=whatever;
  class ...;
  model y = ... ;
  random rep(surface*rainfall);
run;
```

and

```
proc mixed data=whatever;
  class ...;
  model y = ... ;
  repeated / subject=rep(surface*rainfall) type=cs;
run;
```

will fit the same model. Combining the `random` and `repeated` statement with `type=cs` will lead to aliasing of one of the variance components.

Returning to the example at hand, we glean from the Table of `Type 3 Tests of Fixed Effects` in Output 7.24 that the Surface $\times$ Time interaction, the Time main effect, the Rainfall main effect, and the Surface main effect are significant at the 5% level.

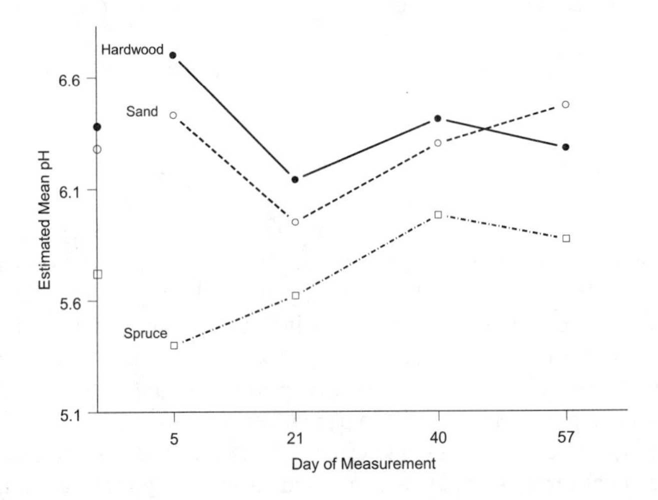

Figure 7.22. Surface least squares means by day of measurement. Estimates of the marginal surface means are shown along the vertical axis with the same symbols as the trends over time.

The next step is to investigate the significant effects, starting with the two-way interaction. The presence of the interaction is not surprising considering the graph of the two-way least square means (Figure 7.22). After studying the interaction pattern we can also conduct comparisons of the marginal surface means since the trends over time do not criss-cross wildly. The `proc mixed` code that follows requests least squares means and their differences for the two rainfall levels (there will be only one difference), slices of the `surface*time` interaction by either factor and differences of the Surface $\times$ Time least squares means that are shown in Figure 7.22.

```
ods listing close;
proc mixed data=Leachate;
  class rep surface rainfall time;
  model ph = surface rainfall surface*rainfall
             time time*surface time*rainfall time*surface*rainfall;
  random rep(surface*rainfall);
  repeated / subject=rep(surface*rainfall) type=sp(gau)(t) r=1 rcorr=1;
  lsmeans rainfall / diff;
  lsmeans surface*time /slice=(time surface) diff;
```

```
   ods output diffs=diffs;
   ods output slices=slices;
run;
ods listing;
title "Tests of Effect Slices";
proc print data=slices; run;

/* process least square means differences */
data diffs; set diffs;
   if upcase(effect)="SURFACE*TIME" then do;
      if (surface ne _surface) and (time ne _time) then delete;
   end;
   LSD05 = tinv(0.975,df)*stderr;
   if probt > 0.05 then sig='   ';
   else if probt > 0.025 then sig = '*';
   else if probt > 0.01  then sig = '**'; else sig = '***';
run;
proc sort data=diffs; by descending rainfall time _time surface _surface; run;
title "Least Square Means Differences";
proc print data=diffs noobs;
   var surface rainfall time _surface _rainfall _time
       estimate LSD05 probt sig;
run;
```

The `ods listing close` statement prior to the `proc mixed` call suppresses printing of procedural output to the screen. Instead, the output of interest (slices and least squares mean differences) is saved to data sets (`diffs` and `slices`) with the `ods output` statement. We do this for two reasons: (1) parts of the `proc mixed` output such as the `Dimensions` table, the `Type 3 Tests of Fixed Effects` have already been studied above. (2) the set of least squares mean differences for interactions contains many comparisons not of interest. For example, a comparison of spruce surface at day 5 vs. sand surface at day 40 is hardly meaningful (it is also not a simple effect!). The data step following the `proc mixed` code processes the data set containing the least squares mean differences. It deletes two-way mean comparisons that do not correspond to simple effects, calculates the least significant differences (LSD) at the 5% level for all comparisons, and indicates the significance of the comparisons at the 5% ($*$), 2.5% ($**$), and 1% level ($***$).

The slices show that at any time point there are significant differences in pH among the three surfaces (Output 7.25). From Figure 7.22 we surmise that these differences are probably not between the sand and hardwood surface, but are between these surfaces and the spruce surface. The least squares mean differences confirm that. The first observation among the least squares mean differences compares the marginal Rainfall means, the next three compare the marginal Surface means that are graphed along the vertical axis of Figure 7.22. Comparisons of the Surface $\times$ Time means start with the fifth observation. For example, the difference between estimated means of hardwood and sand surfaces at day 5 is 0.2767 with an LSD of 0.3103. At any given time point hardwood and sand surfaces are not statistically different at the 5% level, whereas the spruce surface leads to significantly higher acidity of the leachate. Notice that at any point in time the least significant difference (LSD) to compare surfaces is 0.3103. The least significant differences to compare time points for a given surface depend on the temporal separation, however. If the repeated measurements were uncorrelated, there would be only one LSD to compare surfaces at a given time point and one LSD to compare time points for a given surface.

Output 7.25.

Tests of Effect Slices

Obs	Effect	SURFACE	TIME	Num DF	Den DF	FValue	ProbF
1	SURFACE*TIME		5	2	36	40.38	<.0001
2	SURFACE*TIME		21	2	36	5.87	0.0062
3	SURFACE*TIME		40	2	36	4.21	0.0228
4	SURFACE*TIME		57	2	36	8.03	0.0013
5	SURFACE*TIME	hardwd	_	3	36	15.13	<.0001
6	SURFACE*TIME	sand	_	3	36	10.92	<.0001
7	SURFACE*TIME	spruce	_	3	36	7.19	0.0007

Least Square Means Differences

SURFACE	RAINFALL	TIME	_SURFACE	_RAINFALL	_TIME	Estimate	LSD05	Probt	sig
	1	_		2	_	-0.2339	0.20643	0.0296	*
hardwd	_	_	sand	_	_	0.09542	0.25282	0.4269	
hardwd	_	_	spruce	_	_	0.6637	0.25282	<.0001	***
sand	_	_	spruce	_	_	0.5683	0.25282	0.0004	***
hardwd	_	5	sand	_	5	0.2767	0.31038	0.0790	
hardwd	_	5	spruce	_	5	1.3050	0.31038	<.0001	***
sand	_	5	spruce	_	5	1.0283	0.31038	<.0001	***
hardwd	_	5	hardwd	_	21	0.5683	0.19330	<.0001	***
sand	_	5	sand	_	21	0.4817	0.19330	<.0001	***
spruce	_	5	spruce	_	21	-0.2183	0.19330	0.0279	*
hardwd	_	5	hardwd	_	40	0.2900	0.25944	0.0295	*
sand	_	5	sand	_	40	0.1267	0.25944	0.3287	
spruce	_	5	spruce	_	40	-0.5867	0.25944	<.0001	***
hardwd	_	5	hardwd	_	57	0.4317	0.26256	0.0020	***
sand	_	5	sand	_	57	-0.04333	0.26256	0.7398	
spruce	_	5	spruce	_	57	-0.4700	0.26256	0.0009	***
hardwd	_	21	sand	_	21	0.1900	0.31038	0.2225	
hardwd	_	21	spruce	_	21	0.5183	0.31038	0.0017	***
sand	_	21	spruce	_	21	0.3283	0.31038	0.0387	*
hardwd	_	21	hardwd	_	40	-0.2783	0.21452	0.0124	**
sand	_	21	sand	_	40	-0.3550	0.21452	0.0019	***
spruce	_	21	spruce	_	40	-0.3683	0.21452	0.0013	***
hardwd	_	21	hardwd	_	57	-0.1367	0.26005	0.2936	
sand	_	21	sand	_	57	-0.5250	0.26005	0.0002	***
spruce	_	21	spruce	_	57	-0.2517	0.26005	0.0574	
hardwd	_	40	sand	_	40	0.1133	0.31038	0.4638	
hardwd	_	40	spruce	_	40	0.4283	0.31038	0.0082	***
sand	_	40	spruce	_	40	0.3150	0.31038	0.0469	*
hardwd	_	40	hardwd	_	57	0.1417	0.20097	0.1614	
sand	_	40	sand	_	57	-0.1700	0.20097	0.0948	
spruce	_	40	spruce	_	57	0.1167	0.20097	0.2468	
hardwd	_	57	sand	_	57	-0.1983	0.31038	0.2032	
hardwd	_	57	spruce	_	57	0.4033	0.31038	0.0123	**
sand	_	57	spruce	_	57	0.6017	0.31038	0.0004	***

7.6.7 A Longitudinal Study of Water Usage in Horticultural Trees

In this application we fit mixed polynomial models to describe the change over time in water usage during production of landscape trees. Four groups of trees — consisting of two age classes and two species — were of interest to the investigators. Ten trees were randomly selected for each species $\times$ age class combination. We label the groups simply as Age 1, 2 and Species 1, 2. Over a period of approximately 4.5 months the water usage of the trees was assessed regularly, creating a longitudinal time series for each tree. Although it was originally intended to measure water usage on every fifth day, slight variations in the measurement intervals created an unequally spaced longitudinal sequence. For the four groups of trees the water usage averaged across the trees shows distinct trends (Figure 7.23). For species 1 water usage throughout the growing season is quadratic or possibly cubic with a maximum at approximately day 250. There seems to be little difference between older (Age class 2) and younger trees (Age class 1) for species 1. For the second species the trends appear quadratic but with very different maxima. Not only do young trees vary little in their water usage over time, there is a sharp drop-off for older trees of species 2 at about day 250 (Figure 7.23).

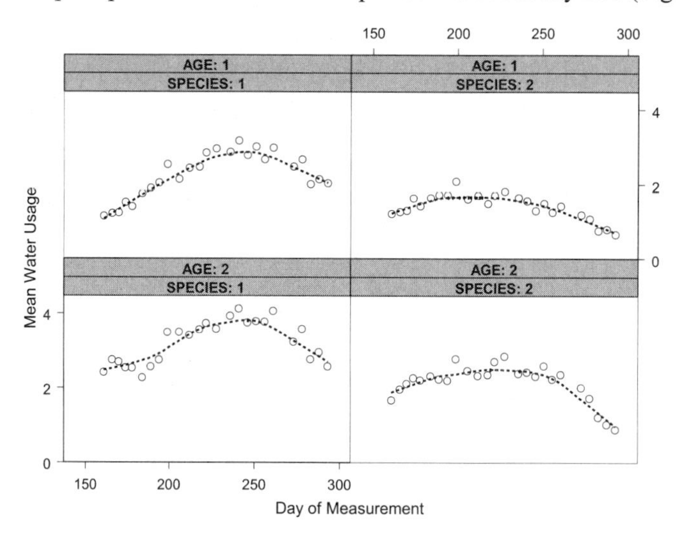

Figure 7.23. Age and species-specific averages for water usage data. The dashed line represents an exploratory loess fit to suggest a parametric polynomial model. Data made kindly available by Dr. Roger Harris and Dr. Robert Witmer, Department of Horticulture, Virginia Polytechnic Institute and State University. Used with permission.

A model for the age and species-specific trends over time based on Figure 7.23 could be the following. Let Y_{ijk} denote the water usage at time t_k of the average tree in age group i $(i = 1, 2)$ for species j $(j = 1, 2)$. Define a_i to be a binary regressor (dummy) variable taking on value 1 if an observation is from age group 1 and 0 otherwise. Similarly define s_j as a dummy variable taking on value 1 for species 1 and value 0 for species 2. Then,

$$\mathrm{E}[Y_{ijk}] = \beta_0 + \beta_1 a_i + \beta_2 s_j + \beta_3 a_i s_j + \beta_4 t_k + \beta_5 t_k^2 + \beta_6 a_i t_k + \beta_7 a_i t_k^2 + \beta_8 s_j t_k + \beta_9 s_j t_k^2 + \beta_{10} a_i s_j t_k + \beta_{11} a_i s_j t_k^2. \quad [7.70]$$

This model appears highly parameterized, but is really not. It allows for separate linear and quadratic trends among the four groups. The intercepts, linear, and quadratic slopes are constructed from β_0 through β_{11} according to Table 7.20.

Table 7.20. Intercepts and Gradients for Age and Species Groups in Model [7.70]

Group	a_i	s_j	Intercept	Linear Gradient	Quadratic Gradient
Age 1, Species 1	1	1	$\beta_0+\beta_1+\beta_2+\beta_3$	$\beta_4+\beta_6+\beta_8+\beta_{10}$	$\beta_5+\beta_7+\beta_9+\beta_{11}$
Age 1, Species 2	1	0	$\beta_0+\beta_1$	$\beta_4+\beta_6$	$\beta_5+\beta_7$
Age 2, Species 1	0	1	$\beta_0+\beta_2$	$\beta_4+\beta_8$	$\beta_5+\beta_9$
Age 2, Species 2	0	0	β_0	β_4	β_5

Model [7.70] is our notion of a population-average model that permits inference about the effects of age, species, and their interaction. It does not accommodate the sequence of measurements for individual trees, however. First, we notice that varying the parameters of model [7.70] on a tree-by-tree basis, the total number of parameters in the mean function alone would be $10 \times 12 = 120$, an unreasonable number. Second, not all of the trees will deviate significantly from the population average, finding out which ones are different is a time-consuming exercise in a model with that many parameters. The hierarchical, two-stage mixed model idea comes to our rescue. Rather than modeling tree-to-tree variability within each group through a large number of fixed effects, we allow one or more of the parameters in model [7.70] to vary at random among trees. This approach is supported by the random selection of trees within each age class and for each species. The BLUPs of these random effects then can be used to (i) assess whether a tree differs in its water usage significantly from the group average and (ii) to produce tree-specific predictions of water usage. To decide which of the polynomial parameters to vary at random, we first focus on a single group, trees of species 2 at age 2. The observed water usage over time for the 10 trees in this group is shown in Figure 7.24. If we adopt a quadratic response model for the average tree in this group we can posit random coefficients for the intercept, the linear, and the quadratic gradient. It is unlikely that the model will support all three parameters being random.

Either a random intercept, random linear slope, or both seem possible. Nevertheless, we commence modeling with the largest possible model

$$Y_{mk} = \left(\beta_0 + b_m^0\right) + \left(\beta_1 + b_m^1\right)t_k + \left(\beta_2 + b_m^2\right)t_k^2 + e_{mk}, \quad [7.71]$$

where m denotes the tree ($m = 1, 2, \cdots, 10$), k the time point and b_m^0 through b_m^2 are random effects/coefficients assumed independent of the model disturbances e_{mk} and independent of each other. The variances of these random effects are denoted σ_0^2 through σ_2^2, respectively. The SAS® `proc mixed` code to fit model [7.71] to the data from age group 2, species 2 is as follows (Output 7.26).

```
data age2sp2; set wateruse(where=((age=2) and (species=2))); t=time/100; run;
proc mixed data=age2sp2 covtest noitprint;
  class treecnt;
  model wu = t t*t / s;
```

```
  random intercept t t*t / subject=treecnt;
run;
```

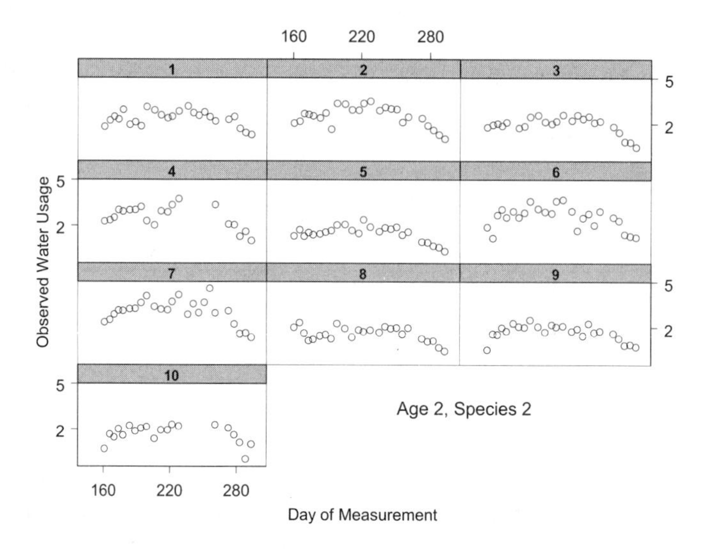

Figure 7.24. Longitudinal observations for ten trees of species 2 in age group 2. The trajectories suggest quadratic effects with randomly varying slope or intercept.

Output 7.26.

```
                        The Mixed Procedure

                         Model Information

          Data Set                     WORK.AGE2SP2
          Dependent Variable           wu
          Covariance Structure         Variance Components
          Subject Effect               treecnt
          Estimation Method            REML
          Residual Variance Method     Profile
          Fixed Effects SE Method      Model-Based
          Degrees of Freedom Method    Containment

                      Class Level Information

           Class       Levels    Values
           treecnt         10    31 32 33 34 35 36 37 38 39 40

                            Dimensions

                 Covariance Parameters               4
                 Columns in X                        3
                 Columns in Z Per Subject            3
                 Subjects                           10
                 Max Obs Per Subject                25
                 Observations Used                 239
                 Observations Not Used               0
                 Total Observations                239
```

Output 7.26 (continued).

```
                Covariance Parameter Estimates

                                         Standard        Z
Cov Parm     Subject     Estimate        Error       Value        Pr Z
Intercept    treecnt       0.2691       0.1298        2.07      0.0190
t            treecnt            0            .           .           .
t*t          treecnt            0            .           .           .
Residual                   0.1472      0.01382       10.65      <.0001

                   Fit Statistics
            -2 Res Log Likelihood          262.5
            AIC (smaller is better)        266.5
            AICC (smaller is better)       266.6
            BIC (smaller is better)        267.1
```

The time variable (measured in days since Jan 01) has been rescaled to allow estimates of the fixed effects and BLUPs to be shown on the output with sufficient decimal places. The estimates for the variance components σ_1^2 and σ_2^2 are zero (Output 7.26). This is an indication that not all three random effects can be supported by these data, as was already anticipated. The `-2 Res Log Likelihood` of 262.5 is the same value one would achieve in a model where only the intercept varies at random. Fitting models with either $\sigma_2^2 = 0$ or $\sigma_1^2 = 0$ also leads to zero estimates for the linear or quadratic variance component (apart from the intercept). We interpret this as evidence that the data support only one of the parameters being random, not that the intercept being random necessarily provides the best fit. Next all models with a single random effect are investigated, as well as the purely fixed effects model:

$$\begin{aligned} Y_{mk} &= \left(\beta_0 + b_m^0\right) + \beta_1 t_k + \beta_2 t_k^2 + e_{mk} \\ Y_{mk} &= \beta_0 + \left(\beta_1 + b_m^1\right) t_k + \beta_2 t_k^2 + e_{mk} \\ Y_{mk} &= \beta_0 + \beta_1 t_k + \left(\beta_2 + b_m^2\right) t_k^2 + e_{mk} \\ Y_{mk} &= \beta_0 + \beta_1 t_k + \beta_2 t_k^2 + e_{mk}. \end{aligned}$$

The `-2 Res Log Likelihoods` of these models are, respectively, 262.5, 280.5, 314.6, and 461.8. The last model is the fixed effects model without any random effects and likelihood ratio tests can be constructed to test the significance of any of the random components.

$$\begin{array}{lll} H_0: \sigma_0^2 = 0 & \Lambda = 461.8 - 262.5 = 199.3 & p < 0.0001 \\ H_0: \sigma_1^2 = 0 & \Lambda = 461.8 - 280.5 = 181.3 & p < 0.0001 \\ H_0: \sigma_2^2 = 0 & \Lambda = 461.8 - 314.6 = 147.2 & p < 0.0001. \end{array}$$

Incorporating any of the random effects provides a significant improvement in fit over a purely fixed effects model. The largest improvement (smallest `-2 Res Log Likelihood`) is obtained with a randomly varying intercept.

So far, it has been assumed that the model disturbances e_{mk} are uncorrelated. Since the data are longitudinal in nature, it is conceivable that residual serial correlation remains even after inclusion of one or more random effects. To check this possibility, we fit a model with an exponential correlation structure (since the measurement times are not quite equally spaced) via

```
proc mixed data=age2sp2 noitprint covtest;
  class treecnt;
  model wu = t t*t;
  random intercept /subject=treecnt s;
  repeated /subject=treecnt type=sp(exp)(time);
run;
```

Another drop in minus twice the residual log likelihood from 262.5 to 251.3 is confirmed with Output 7.27. The difference of $\Lambda = 262.5 - 251.3 = 11.2$ is significant with p-value of $\Pr(\chi^2_1 \geq 11.2) = 0.0008$. Adding an autoregressive correlation structure for the e_{mk} significantly improved the model fit. It should be noted that the p-value from the likelihood ratio test differs from the p-value of 0.0067 reported by `proc mixed` in the `Covariance Parameter Estimates` table. The p-value reported there is for a Wald-type test statistic obtained from comparing a z test statistic against an asymptotic standard Gaussian distribution. The test statistic is simply the estimate of the covariance parameter divided by its standard error. Estimates of variances and covariances are usually far from Gaussian-distributed and the Wald-type tests for covariance parameters produced by the `covtest` option of the `proc mixed` statement are not very reliable. We prefer the likelihood ratio test whenever possible.

Output 7.27.

```
                    The Mixed Procedure

                     Model Information
     Data Set                      WORK.AGE2SP2
     Dependent Variable            wu
     Covariance Structures         Variance Components,
                                   Spatial Exponential
     Subject Effects               treecnt, treecnt
     Estimation Method             REML
     Residual Variance Method      Profile
     Fixed Effects SE Method       Model-Based
     Degrees of Freedom Method     Containment

                 Class Level Information
      Class       Levels     Values
      treecnt         10     1 2 3 4 5 6 7 8 9 10

                      Dimensions
          Covariance Parameters                 3
          Columns in X                          3
          Columns in Z Per Subject              1
          Subjects                             10
          Max Obs Per Subject                  25
          Observations Used                   239
          Observations Not Used                 0
          Total Observations                  239

               Covariance Parameter Estimates
                                      Standard         Z
Cov Parm     Subject     Estimate        Error     Value         Pr Z
Intercept    treecnt       0.2656       0.1301      2.04       0.0206
SP(EXP)      treecnt       3.7945       0.8175      4.64       <.0001
Residual                   0.1541      0.01624      9.49       <.0001

                     Fit Statistics
          -2 Res Log Likelihood              251.3
          AIC (smaller is better)            257.3
          AICC (smaller is better)           257.4
          BIC (smaller is better)            258.2
```

Output 7.27 (continued).

Solution for Fixed Effects

Effect	Estimate	Standard Error	DF	t Value	Pr > |t|
Intercept	-11.2233	1.0989	9	-10.21	<.0001
t	12.7121	0.9794	227	12.98	<.0001
t*t	-2.9137	0.2149	227	-13.56	<.0001

Solution for Random Effects

Effect	treecnt	Estimate	Std Err Pred	DF	t Value	Pr > |t|
Intercept	1	0.1930	0.1877	227	1.03	0.3049
Intercept	2	0.3425	0.1877	227	1.82	0.0694
Intercept	3	-0.1988	0.1881	227	-1.06	0.2917
Intercept	4	0.4539	0.1918	227	2.37	0.0188
Intercept	5	-0.6414	0.1877	227	-3.42	0.0007
Intercept	6	0.3769	0.1877	227	2.01	0.0458
Intercept	7	0.8410	0.1877	227	4.48	<.0001
Intercept	8	-0.5528	0.1877	227	-2.95	0.0036
Intercept	9	-0.4453	0.1877	227	-2.37	0.0185
Intercept	10	-0.3689	0.1918	227	-1.92	0.0556

The `/s` options on the `model` and `random` statements of `proc mixed` yield printouts of the fixed effects estimates and the BLUPs. We obtain $\widehat{\beta}_0 = -11.22$, $\widehat{\beta}_1 = 12.71$, and $\widehat{\beta}_2 = -2.914$ for the fixed effects estimates. Water usage of the average tree in this group is thus predicted as $\widehat{y} = -11.22 + 12.71t - 2.914t^2$. The best linear unbiased predictors of $\mathbf{b} = [b_1^0, b_2^0, \cdots, b_m^0]$ are displayed as `Solutions for Random Effects`. For example the tree-specific prediction for tree #1 is

$$\widehat{y} = -11.22 + 0.1930 + 12.71t - 2.914t^2.$$

The BLUPs are significantly different from zero (at the 5% level) for trees 4, 5, 6, 7, 8, and 9.

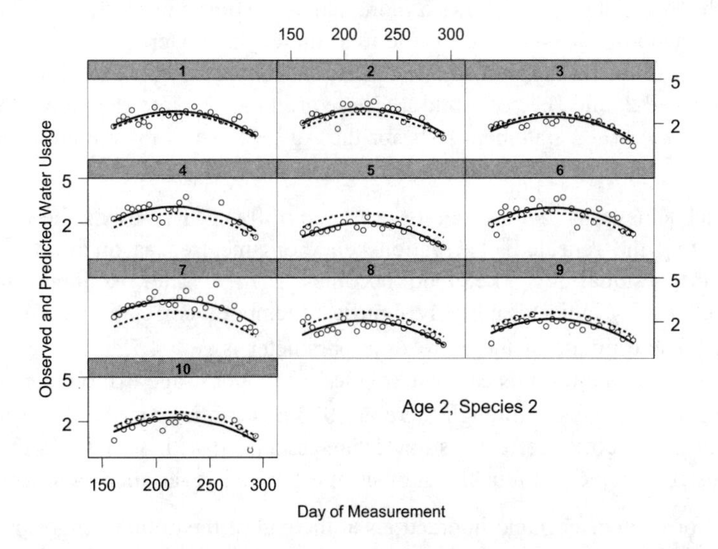

Figure 7.25. Predictions from random intercept model with continuous AR(1) errors for species 2 at age 2. Population average fit shown as a dashed lines, cluster-specific predictions shown as solid lines.

The tree-specific trends show a noticeable deviation from the population-averaged prediction for trees that have a significant BLUP (Figure 7.25). For any of these trees it is obvious that the tree-specific prediction provides a much better fit to the data than the population average.

The previous `proc mixed` runs and inferences were for one of the four groups only. Next we need to combine the data across the age and species groups which entails adding a mixed effects structure to model [7.70]. Based on what was learned from investigating the Age 2, Species 2 group, it is tempting to add a random intercept and an autoregressive structure for the within-tree disturbances and leave it at that. The mixed model analyst will quickly notice when dealing with combined data sets that random effects that may not be estimable for a subset of the data can successfully be estimated based on a larger set of data. In this application it turned out that upon combining the data from the four groups not only a random intercept, but also a random linear slope could be estimated. The statements

```
proc mixed data=wateruse noitprint;
  class age species treecnt;
  model wu = age*species age*species*t age*species*t*t / s;
  estimate 'Intercpt Age1-Age2 = Age Main' age*species 1  1 -1 -1;
  estimate 'Intercpt Sp1 -Sp2  = Sp. Main' age*species 1 -1  1 -1;
  estimate 'Intercpt Age*Sp    = Age*Sp. ' age*species 1 -1 -1  1;
  random intercept t /subject=age*species*treecnt s;
  repeated /subject=age*species*treecnt type=sp(exp)(time);
run;
```

fit the model with random intercept and random (linear) slope and an autoregressive correlation structure for the within-tree errors (Output 7.28). The `subject=` options of the `random` and `repeated` statements identify the units that are to be considered uncorrelated in the analysis. Observations with different values of the variables `age`, `species`, and `treecnt` are considered to be from different clusters and hence uncorrelated. Any set of observations with the same values of these variables is considered correlated. The use of the `age` and `species` variables in the `class` statement allows a more concise expression of the fixed effects part of model [7.70]. The term `age*species` fits the four intercepts, the term `age*species*t` the four linear slopes and so forth. The first two `estimate` statements compare the intercept estimates between ages 1 and 2, and species 1 and 2. These are inquiries into the Age or Species main effect. The third `estimate` statement tests for the Age $\times$ Species interaction averaged across time.

This model achieves a `-2 Res Log Likelihood` of 995.8. Removing the `repeated` statement and treating the repeated observations on the same tree as uncorrelated, twice the negative of the residual log likelihood becomes 1074.2. The likelihood ratio statistic $1074.2 - 995.8 = 78.4$ indicates a highly significant temporal correlation among the repeated measurements. The estimate of the correlation parameter is $\widehat{\phi} = 4.7217$. Since the measurement times are coded in days, this estimate implies that water usage exhibits temporal correlations over $3\widehat{\phi} = 14.16$ days. Although there are 953 observations in the data set, notice that the test statistics for the fixed effects estimates are associated with 36 degrees of freedom, the number of clusters (subjects) minus the number of estimated covariance parameters.

The tests for main effects and interactions at the end of the output suggest no differences in the intercepts between ages 1 and 2, but differences in intercepts between the species. Because of the significant interaction between Age Class and Species we decide to retain all four fixed effect intercepts in model [7.70]. Similar tests can be performed to determine whether the linear and quadratic gradients differ among the species and ages.

Output 7.28. (abridged)

```
                        The Mixed Procedure

                        Model Information

          Data Set                        WORK.WATERUSE
          Dependent Variable              wu
          Covariance Structures           Variance Components,
                                          Spatial Exponential
          Subject Effects                 age*species*treecnt,
                                          age*species*treecnt
          Estimation Method               REML
          Residual Variance Method        Profile
          Fixed Effects SE Method         Model-Based
          Degrees of Freedom Method       Containment

                      Class Level Information

           Class       Levels     Values

           age              2     1 2
           species          2     1 2
           treecnt         40     1 2 3 4 5 6 7 8 9 10 11 12 13
                                  14 15 16 17 18 19 20 21 22 23
                                  24 25 26 27 28 29 30 31 32 33
                                  34 35 36 37 38 39 40

                    Covariance Parameter Estimates

              Cov Parm      Subject                 Estimate
              Intercept     age*species*treecnt       0.1549
              t             age*species*treecnt      0.02785
              SP(EXP)       age*species*treecnt       4.7217
              Residual                                0.1600

                          Fit Statistics

                  -2 Res Log Likelihood            995.8
                  AIC (smaller is better)         1003.8
                  AICC (smaller is better)        1003.8
                  BIC (smaller is better)         1010.5

                      Solution for Fixed Effects

                                                 Standard
Effect            age  species   Estimate        Error      DF   t Value    Pr > |t|

age*species        1    1         -14.5695      1.1780     36    -12.37      <.0001
age*species        1    2          -6.7496      1.1779     36     -5.73      <.0001
age*species        2    1         -12.3060      1.1771     36    -10.45      <.0001
age*species        2    2         -11.3629      1.1771     36     -9.65      <.0001

t*age*species      1    1          14.4083      1.0572     36     13.63      <.0001
t*age*species      1    2           7.9204      1.0571     36      7.49      <.0001
t*age*species      2    1          13.5380      1.0563     36     12.82      <.0001
t*age*species      2    2          12.8370      1.0563     36     12.15      <.0001

t*t*age*species 1       1          -2.9805      0.2316     36    -12.87      <.0001
t*t*age*species 1       2          -1.8449      0.2316     36     -7.97      <.0001
t*t*age*species 2       1          -2.8567      0.2314     36    -12.34      <.0001
t*t*age*species 2       2          -2.9403      0.2314     36    -12.71      <.0001
```

Output 7.28 (continued).

```
                        Solution for Random Effects

                                              Std Err
Effect     age species treecnt Estimate         Pred    DF   t Value   Pr > |t|
Intercept 1    1        1       -0.04041     0.2731    869    -0.15    0.8824
t          1   1        1        0.04677     0.1180    869     0.40    0.6921
Intercept 1    1        2       -0.03392     0.2738    869    -0.12    0.9014
t          1   1        2        0.07863     0.1181    869     0.67    0.5057
Intercept 1    1        3        0.08246     0.2731    869     0.30    0.7628
t          1   1        3       -0.05849     0.1180    869    -0.50    0.6204
Intercept 1    1        4        0.07395     0.2731    869     0.27    0.7866
t          1   1        4       -0.01493     0.1180    869    -0.13    0.8994
Intercept 1    1        5        -0.1568     0.2731    869    -0.57    0.5660
t          1   1        5       -0.07856     0.1180    869    -0.67    0.5059

      (and so forth for all trees and Age × Species combinations)

                                Estimates

                                              Standard
Label                                Estimate     Error    DF t Value   Pr > |t|
Intercpt Age1-Age2 = Age Main          2.3498    2.3551    36    1.00     0.3251
Intercpt Sp1 -Sp2  = Sp. Main         -8.7630    2.3551    36   -3.72     0.0007
Intercpt Age*Sp    = Age*Sp.          -6.8769    2.3551    36   -2.92     0.0060
```

7.6.8 Cumulative Growth of Muskmelons in Subsampling Design

To study the cumulative yield of muskmelons under various irrigation and mulch application strategies an experiment was conducted between 1997 and 1999 at the Tidewater Agricultural Research and Extension Center, Virginia Polytechnic Institute and State University. On plots of size 6 feet by 25 feet, melons were grown under the following four treatments.

- Nonirrigated without mulch or row cover;
- Drip/trickle irrigation without mulch or row cover;
- Drip/trickle irrigation with black plastic mulch;
- Drip/trickle irrigation with red plastic mulch.

There were four replicate plots for each treatment. In 1997, mature melons were harvested on each plot at days 81, 84, 88, and 91. The average yield per plot was converted into yield per hectare and added to the previous yield. Figure 7.26 shows the mean cumulative yields in tons $\times$ ha^{-1} for the four treatments in 1997. The cumulative yield increases over time since it is obtained from adding positive yield figures. In the absence of mulching the cumulative yields are considerably lower compared to the red or black mulch applications. There seems to be little difference in yields between the two plastic mulch types.

These data display several interesting features. The experimental units are the plots to which a particular treatment was applied. The mature melons harvested at a particular day represent observational units. The number of matured melons varies from plot to plot and hence the number of subsamples from which the yield per hectare is calculated differs. If the variability of melon weights is homogeneous across all plots, the variability of observed yields per hectare increases with the number of melons harvested. Also, the cumulative yields which are the focus of the analysis (Figure 7.26) are not independent, even if the muskmelons matured independently on a given plot. The cumulative yield at day 88 is the sum of the cumulative yield at day 84 and the yield observed at day 88. To build a statistical model for these data the two issues of subsampling and correlated responses must be kept in mind.

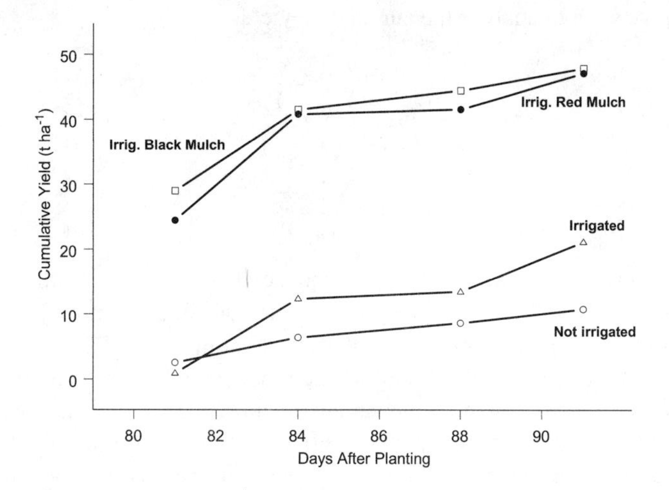

Figure 7.26. Average cumulative yields vs. days after planting in muskmelon study for 1997. The data for this experiment was kindly provided by Dr. Norris L. Powell, Tidewater Agricultural Research and Extension Center, Virginia Polytechnic Institute and State University. Used with permission.

The investigators were particularly interested in the following questions and hypotheses:

[A] In the absence of mulch, is there a benefit of irrigation?

[B] Is there a difference in cumulative yields between red and black plastic mulch?

[C] Is there a benefit of mulching beyond irrigation?

[D] What is the mulch effect?

[E] Can mulching shorten the growing period, i.e., is the yield of mulched plots at the beginning of the observation period comparable to the yield of unmulched plots at the end of the observation period?

These questions have a temporal component; in the case of [A], for example, there may be a beneficial effect late in the season rather than early in the season. In building a statistical model for this experiment we commence by formulating a model for the yield per hectare on a given plot at a given point in time (Y_{ijk}):

$$Y_{ijk} = \mu_{ijk} + e_{ij} + d_{ijk}. \qquad [7.72]$$

In [7.72] μ_{ijk} denotes the mean of the ith treatment ($i = 1, \cdots, 4$) on replicate (plot) j at the kth harvesting day after planting ($k = 1, \cdots, 4$). The e_{ij} are experimental errors associated with replicate (plot) j of treatment i and d_{ijk} is the subsampling error from harvesting n_{ijk} melons on replicate j of treatment i at time k. If σ_m^2 is the variability of melon weights (converted to hectare basis) and $\text{Var}[e_{ij}] = \sigma^2$ is the experimental error variability, then

$$\text{Var}[Y_{ijk}] = \sigma^2 + n_{ijk}\sigma_m^2.$$

Model [7.72] is a relatively straightforward mixed model, but the Y_{ijk} are not the outcome of interest. Rather, we wish to analyze the cumulative yields

$$\begin{aligned} U_{ij1} &= Y_{ij1} \\ U_{ij2} &= Y_{ij1} + Y_{ij2} \\ &\vdots \\ U_{ijk} &= \sum_{p=1}^{k} Y_{ijp}. \end{aligned}$$

That the U_{ijk} are correlated even if the Y_{ijk} are independent is easy to establish. For example, $\text{Cov}[U_{ij1}, U_{ij2}] = \text{Cov}[Y_{ij1}, Y_{ij1} + Y_{ij2}] = \text{Var}[Y_{ij1}]$. The accumulation of yields results in an accumulation of the experimental errors and the subsampling errors. The resulting correlation structure is rather complicated and difficult to code in a statistical computing package. As an alternative, we choose the following route: an unstructured variance-covariance matrix for the accumulated d_{ijk} is combined with a random effect for the experimental errors. The mean model is a cubic response model with different trends for each treatment. This mean model will produce the same estimates of treatment means at days 81, 84, 88, and 91 as a block design analysis but also allows for estimating treatment differences at other time points. The comparisons A – E are coded with the `estimate` statement of `proc mixed`. The variable `t` in the code that follows is 80 days less than time after planting. The first harvesting time thus coincides with $t = 1$. Divisors in estimate statements are used here to ensure that the resulting estimates are properly scaled. For example, a statement such as

```
estimate 'A1+A2 vs. B1+B2' tx 1 1 -1 -1;
```

compares the sum of treatment means for A_1 and A_2 to the sum of B_1 and B_2. Using a divisor has no effect on the significance level of the comparison, and the statement

```
estimate 'Average(A1 A2) vs. Average(B1 B2)' tx 1 1 -1 -1 / divisor=2;
```

will produce the same p-value. The estimate itself will be the difference of sums in the first case and the difference of averages in the second case.

```
proc mixed data=melons97 convh=0.002 noitprint;
  class rep tx;
  model cumyld = tx tx*t tx*t*t tx*t*t*t / noint;
  random rep*tx; /* experimental error */
  repeated /subject=rep*tx type=un;

  /* A at days 81, 84, 88, 91 corresponding to t=1, 4, 8, 11 */
  estimate 'Bare vs. Irrig (81)' tx 1 -1 0 0 tx*t  1  -1 0 0
                                 tx*t*t 1 -1 0 0  tx*t*t*t 1  -1  0 0;
  estimate 'Bare vs. Irrig (84)' tx 1 -1 0 0 tx*t  4  -4 0 0
                                 tx*t*t 16 -16 0 0 tx*t*t*t 64 -64 0 0;
  estimate 'Bare vs. Irrig (88)' tx 1 -1 0 0 tx*t  8  -8 0 0
                                 tx*t*t 64 -64 0 0 tx*t*t*t 512 -512 0 0;
  estimate 'Bare vs. Irrig (91)' tx 1 -1 0 0 tx*t 11 -11 0 0
                                 tx*t*t 121 -121 0 0 tx*t*t*t 1331 -1331 0 0;
```

```
/* [B] at days 81, 84, 88, 91 corresponding to t=1, 4, 8, 11 */
estimate 'IrrBl vs. IrrRed (81)' tx 0 0 1 -1 tx*t 0 0  1  -1
                              tx*t*t 0 0   1   -1  tx*t*t*t 0 0     1  -1 ;
estimate 'IrrBl vs. IrrRed (84)' tx 0 0 1 -1 tx*t 0 0  4  -4
                              tx*t*t 0 0  16  -16 tx*t*t*t 0 0   64   -64;
estimate 'IrrBl vs. IrrRed (88)' tx 0 0 1 -1 tx*t 0 0  8  -8
                              tx*t*t 0 0  64  -64 tx*t*t*t 0 0  512  -512;
estimate 'IrrBl vs. IrrRed (91)' tx 0 0 1 -1 tx*t 0 0 11 -11
                              tx*t*t 0 0 121 -121 tx*t*t*t 0 0 1331 -1331;

/* [C] at days 81, 84, 88, 91 corresponding to t=1, 4, 8, 11 */
estimate 'Irrig vs. Mulch (81)' tx 0 2 -1 -1 tx*t 0 2 -1  -1
               tx*t*t 0 2 -1 -1 tx*t*t*t 0 2 -1 -1 /divisor=2;
estimate 'Irrig vs. Mulch (84)' tx 0 2 -1 -1 tx*t 0 8 -4  -4
               tx*t*t 0 32 -16 -16 tx*t*t*t 0 128 -64 -64 /divisor=2;
estimate 'Irrig vs. Mulch (88)' tx 0 2 -1 -1 tx*t 0 16 -8  -8
               tx*t*t 0 128 -64 -64 tx*t*t*t 0 1024 -512 -512 /divisor=2;
estimate 'Irrig vs. Mulch (91)' tx 0 2 -1 -1 tx*t 0 22 -11 -11
            tx*t*t 0 242 -121 -121 tx*t*t*t 0 2662 -1331 -1331 /divisor=2;

/* [D] at days 81, 84, 88, 91 corresponding to t=1, 4, 8, 11 */
estimate 'Mulch vs. NoMulch (81)' tx 1 1 -1 -1 tx*t  1  1 -1 -1
           tx*t*t   1  1 -1 -1 tx*t*t*t  1  1   -1 -1 /divisor=2;
estimate 'Mulch vs. NoMulch (84)' tx 1 1 -1 -1 tx*t  4  4 -4 -4
           tx*t*t  16  16 -16 -16  tx*t*t*t   64 64 -64 -64 /divisor=2;
estimate 'Mulch vs. NoMulch (88)' tx 1 1 -1 -1 tx*t  8  8 -8 -8
           tx*t*t  64  64 -64 -64  tx*t*t*t  512  512 -512 -512
           /divisor=2;
estimate 'Mulch vs. NoMulch (91)' tx 1 1 -1 -1 tx*t 11 11 -11 -11
           tx*t*t 121 121 -121 -121 tx*t*t*t 1331 1331 -1331 -1331
           /divisor=2;

/* [E]  */
estimate 'Irrig(91) - Mulch(81)' tx 0 2 -1 -1 tx*t 0 22 -1 -1
            tx*t*t 0 242 -1 -1 tx*t*t*t 0 2662 -1 -1 /divisor=2;
run;
```

The parameters of the unstructured variance-covariance matrix of the accumulated subsampling errors reflect the differences in variability (Output 7.29). The variance at the initial measurement occasion (day 81), shown as `UN(1,1)`, is considerably larger than the other variances (`UN(2,2)` through `UN(4,4)`). This reflects the fact that especially on the plastic mulch-treated plots, initially the most melons were harvested at day 81. From the results of the `estimates` statements it can be seen that the separation between nonirrigated and irrigated plots ([A]) increases over time, but the difference in cumulative yields never achieves significance at the 5% level (even if the comparison would be one-sided). Similarly, there is no difference between red and black mulch cumulative yields at any point in time ([B]). Applying either plastic mulch type in addition to irrigation raises cumulative melon yields by 25.8 to 29.5 tons $\times$ ha^{-1}. These differences are highly significant ([C]) and amplified even more strongly if the two no-mulch treatments are combined ([D]). Interestingly, the initial yield of the mulched plots was 5.56 tons per hectare higher than the final yield 91 days after planting on the irrigated plots, although this difference was not significant ($p = 0.458$, [E]).

Output 7.29. (abridged)

```
                    The Mixed Procedure
                     Model Information
Data Set                     WORK.MELONS97
Dependent Variable           CUMYLD
Covariance Structures        Variance Components, Unstructured
Subject Effect               REP*TX
Estimation Method            REML
Residual Variance Method     None
Fixed Effects SE Method      Model-Based
Degrees of Freedom Method    Containment

               Class Level Information
  Class    Levels   Values
  REP           4   1 2 3 4
  TX            4   BareSoil Irrig IrrigBlack IrrigRed

           Covariance Parameter Estimates
           Cov Parm    Subject    Estimate
           REP*TX                  64.2255
           UN(1,1)     REP*TX       146.67
           UN(2,1)     REP*TX      36.3238
           UN(2,2)     REP*TX      38.2103
           UN(3,1)     REP*TX      18.8947
           UN(3,2)     REP*TX      15.9693
           UN(3,3)     REP*TX      19.0087
           UN(4,1)     REP*TX      24.4520
           UN(4,2)     REP*TX      14.1106
           UN(4,3)     REP*TX      29.5125
           UN(4,4)     REP*TX      50.5624

                    Fit Statistics
      Res Log Likelihood                  -190.3
      Akaike's Information Criterion      -201.3
      Schwarz's Bayesian Criterion        -205.5
      -2 Res Log Likelihood                380.5

                             Estimates
                                      Standard
Label                      Estimate      Error    DF   t Value   Pr > |t|
[A]
Bare vs. Irrig (81)          1.6425    10.2688    36      0.16     0.8738
Bare vs. Irrig (84)         -5.9495     7.1567    36     -0.83     0.4113
Bare vs. Irrig (88)         -4.8272     6.4511    36     -0.75     0.4592
Bare vs. Irrig (91)        -10.3727     7.5759    36     -1.37     0.1794

[B]
IrrBl vs. IrrRed (81)        4.5288    10.2688    36      0.44     0.6618
IrrBl vs. IrrRed (84)        0.7070     7.1567    36      0.10     0.9219
IrrBl vs. IrrRed (88)        2.8887     6.4511    36      0.45     0.6570
IrrBl vs. IrrRed (91)        0.7420     7.5759    36      0.10     0.9225

[C]
Irrig vs. Mulch (81)       -25.8126     8.8931    36     -2.90     0.0063
Irrig vs. Mulch (84)       -28.7948     6.1979    36     -4.65     <.0001
Irrig vs. Mulch (88)       -29.5604     5.5868    36     -5.29     <.0001
Irrig vs. Mulch (91)       -26.3408     6.5609    36     -4.01     0.0003

[D]
Mulch vs. NoMulch (81)      -24.9914     7.2611    36     -3.44     0.0015
Mulch vs. NoMulch (84)      -31.7695     5.0605    36     -6.28     <.0001
Mulch vs. NoMulch (88)      -31.9740     4.5616    36     -7.01     <.0001
Mulch vs. NoMulch (91)      -31.5271     5.3570    36     -5.89     <.0001

[E]
Irrig(91) - Mulch(81)       -5.5644     7.4202    36     -0.75     0.4582
```

Chapter 8

Nonlinear Models for Clustered Data

"Although this may seem a paradox, all exact science is dominated by the idea of approximation." Bertrand Russell.

8.1 Introduction

Box 8.1 NLMMs and GLMMs

- **Models for data that are clustered or otherwise call for the inclusion of random effects do not necessarily have a linear mean function.**
- **Nonlinear mixed models (NLMMs) arise when nonlinear mean functions as in §5 are applied to clustered data.**
- **Generalized linear mixed models (GLMMs) arise when clustered data are modeled where the (conditional) response has a distribution in the exponential family.**

Chapters 5 and 6 discussed general nonlinear models and generalized linear models (GLM) for independent data. It was emphasized in §6 that GLMs are special cases of nonlinear models where a linear predictor is placed inside a nonlinear function. Chapter 7 digressed from the nonlinear model theme by introducing linear models for clustered data where the observations within a cluster are possibly correlated. To capture cluster-to-cluster as well as within-cluster variability we appealed to the idea of randomly varying cluster effects which gave rise to the Laird-Ware model

$$\mathbf{Y}_i = \mathbf{X}_i\boldsymbol{\beta} + \mathbf{Z}_i\mathbf{b}_i + \mathbf{e}_i.$$

Recall that the $\mathbf{b}_i$ are random effects or coefficients, with mean $\mathbf{0}$ and variance-covariance matrix $\mathbf{D}$, that vary across clusters and the $\mathbf{e}_i$ are the within-cluster errors. Throughout Chapter 7 it was assumed that the mean function is linear. There are situations, however, where the model calls for the inclusion of random effects and the mean function is nonlinear.

Figure 8.1 shows the cumulative bole volume profiles of three yellow poplar (*Liriodendron tulipifera* L.) trees that are part of a data set of 336 randomly selected trees. The volume of a bole was obtained by felling the tree, delimbing the bole and cutting it into four-foot-long sections. The volume of each section was determined by geometric principles assuming a circular shape of the bole cross-section and accumulated with the volumes of lower sections. This process is repeated to the top of the tree bole if total-bole volume is the desired response variable, or to the point where the bole diameter has tapered to the merchantable diameter, if merchantable volume is the response variable. If d_{ij} is the cross-sectional diameter of the bole of the ith tree at the jth height of measurement, then $r_{ij} = 1 - d_{ij}/\max(d_{ij})$ is termed the complementary diameter. It is zero at the stump and approaches one at the tree tip.

The trees differ considerably in size (total height and breast height diameter), hence their total cumulative volumes differ greatly. The general shapes of the tree profiles are similar, however, suggesting a sigmoidal increase of cumulative volume with increasing complementary diameter. These data are furthermore clustered. Each of the $n = 336$ trees represents a cluster of observations. Since the individual bole segments were cut at equal four-foot intervals the number of observations within a cluster (n_i) varies from tree to tree. A cumulative bole volume model for these data will have a nonlinear mean function that captures the sigmoidal behavior of the response and account for differences in size and shape of the trees through random effects capturing tree-to-tree variability. It will be a nonlinear mixed model combining concepts of §5 and §7. These data are modeled in application §8.4.1.

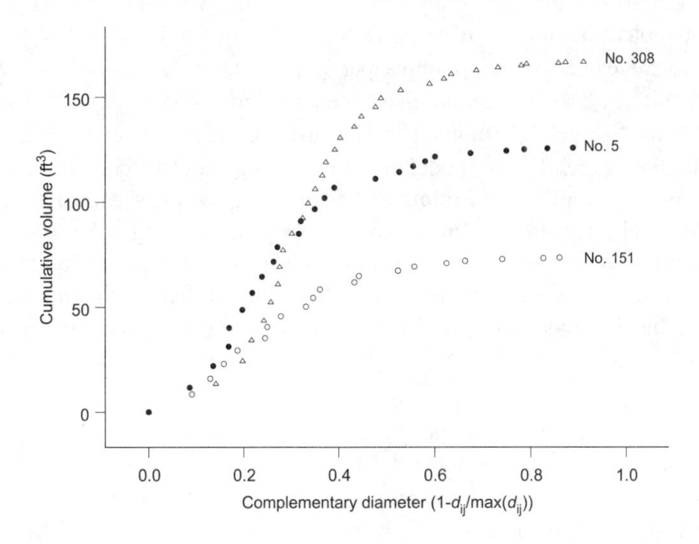

Figure 8.1. Cumulative volume profiles of three yellow poplar (*Liriodendron tulipifera* L.) trees as a function of the complementary bole diameter. Data kindly provided by Dr. David Loftis, USDA Forest Service, originally collected by Dr. Donald E. Beck (see Beck 1963).

Mixed models with nonlinear mean function also come about when generalized linear models are applied to data with clustered structure. Recall the Hessian fly experiment of §6.7.2 where 16 varieties were arranged in a randomized block design with four blocks and the outcome was the proportion of plants on an experimental unit infested with the Hessian fly. The model applied there was

$$\log\left\{\frac{\pi_{ij}}{1-\pi_{ij}}\right\} = \mu + \tau_i + \rho_j,$$

where π_{ij} is the probability of infestation for variety i in block j, and τ_i, ρ_j are the treatment and block effects, respectively. Expressed as a statistical model for the observed data this model can be written as

$$Y_{ij} = \frac{1}{1+\exp\{-\mu-\tau_i-\rho_j\}} + e_{ij}, \qquad [8.1]$$

where Y_{ij} is the proportion of infested plants and e_{ij} is a random variable with mean 0 and variance $\pi_{ij}(1-\pi_{ij})/n_{ij}$ (a so-called shifted binomial random variable). If the blocks were not predetermined but selected at random, then the ρ_j are random effects and model [8.1] is a nonlinear mixed model. It is nonlinear because the linear predictor $\mu+\tau_i+\rho_j$ is inside a nonlinear function, and it is a mixed model because of fixed treatment effects and random block effects.

The problem of overdispersion in generalized linear models was discussed in §6.6. Different strategies for modeling overdispersed data were (i) the addition of extra scale parameters that alter the dispersion but not the mean of the response, (ii) parameter mixing, and (iii) generalized linear mixed models. An extra scale parameter was used to model the Hessian fly data in §6.7.2. For the poppy count data in §6.7.8 parameter mixing was applied. Recall that Y_{ij} denoted the number of poppies for treatment i in block j. It was assumed that given λ_{ij}, the average number of poppies per unit area, the counts $Y_{ij}|\lambda_{ij}$ were Poisson distributed and that λ_{ij} was a Gamma-distributed random variable. This led to a Negative Binomial model for the poppy counts Y_{ij}. Because this distribution is in the exponential family of distributions (§6.2.1), the model could be fit as a standard generalized linear model. The parameter mixing approach is intuitive because a quantity assumed to be fixed in a reference model is allowed to vary at random thereby introducing more uncertainty in the marginal distribution of the response and accounting for the overdispersion in the data. The generalized linear mixed model approach is equally intuitive. Let Y_{ij} denote the poppy count for treatment i in block j and assume that $Y_{ij}|d_{ij}$ are Poisson distributed with mean

$$\lambda_{ij} = \exp\{\mu + \tau_i + \rho_j + d_{ij}\}.$$

The model for the log intensity of poppy counts appears to be a classical model for a randomized block design with error term d_{ij}. In fact, we specify that the d_{ij} are independent Gaussian random variables with mean 0 and variance σ^2. Compare this to the standard Poisson model without overdispersion, $Y_{ij} \sim \text{Poisson}(\exp\{\mu + \tau_i + \rho_j\})$. The uncertainty in d_{ij} increases the uncertainty in Y_{ij}. The model

$$Y_{ij}|d_{ij} \sim \text{Poisson}\left(e^{\mu+\tau_i+\rho_j+d_{ij}}\right) \qquad [8.2]$$

is also a parameter mixing model. If the distribution of the d_{ij} is carefully chosen, one can average over it analytically to derive the marginal distribution of Y_{ij} on which inference is based. In §6.7.8 the distribution of λ_{ij} was chosen as Gamma because the marginal distribution then had a known form. In other situations the marginal distribution may be difficult to obtain or intractable. We then start with a generalized linear mixed model such as [8.2] and approximate the marginal distribution (see §8.3.2). The poppy count data are modeled with a generalized linear mixed model in §8.4.2.

8.2 Nonlinear and Generalized Linear Mixed Models

Denote as Y_{ij} the jth response from cluster i $(i = 1, \cdots, n; j = 1, \cdots, n_i)$. In §5 we used the general notation $Y_i = f(\mathbf{x}_i, \boldsymbol{\theta}) + e_i$ to denote a nonlinear model with mean parameters $\boldsymbol{\theta}$. To reflect the clustered nature of the data and the involvement of random effects we now write

$$Y_{ij} = f(\mathbf{x}_{ij}, \boldsymbol{\theta}, \mathbf{b}_i) + e_{ij}. \qquad [8.3]$$

The vector $\mathbf{x}_{ij}$ is a vector of regressor or design variables and $\boldsymbol{\theta}$ is a vector of fixed effects. The vector $\mathbf{b}_i$ denotes a vector of random effects (or coefficients) modeling the cluster-to-cluster heterogeneity. As in §7, the $\mathbf{b}_i$ are zero mean random variables with variance-covariance matrix $\mathbf{D}$. The e_{ij} are within-cluster errors with mean 0 which might be correlated. Because computational difficulties in fitting nonlinear models are greater than those in fitting linear models it is commonly assumed that the within-cluster errors are uncorrelated, but this is not necessary. The function $f(\cdot)$ can be any nonlinear function. A special case of model [8.3] is a generalized linear mixed model (GLMM) where $f()$ is the inverse of a link function and the variation of the e_{ij} is modeled to reflect the variance of the appropriate conditional distribution in the exponential family (see §6.2.1). For a GLMM we assume that the linear predictor has the form of a Laird-Ware model and write

$$Y_{ij} = g^{-1}\left(\mathbf{x}'_{ij}\boldsymbol{\beta} + \mathbf{z}'_{ij}\mathbf{b}_i\right) + e_{ij}. \qquad [8.4]$$

When combining the responses for a particular cluster in a vector $\mathbf{Y}_i = [Y_{i1}, \cdots, Y_{in_i}]'$ models [8.3] and [8.4] will be written as

$$\mathbf{Y}_i = \mathbf{f}(\mathbf{x}, \boldsymbol{\theta}, \mathbf{b}_i) + \mathbf{e}_i$$
$$\mathbf{Y}_i = \mathbf{g}^{-1}(\mathbf{X}_i\boldsymbol{\beta} + \mathbf{Z}_i\mathbf{b}_i) + \mathbf{e}_i.$$

The variance-covariance matrix of the within-cluster error vector $\mathbf{e}_i$ will be denoted $\mathbf{R}_i$ to keep with the notation in §7.

The involvement of the random effects $\mathbf{b}_i$ inside the inverse link function $g^{-1}()$ or the general nonlinear function $f()$ poses a particular complication for nonlinear mixed models. In the Laird-Ware model $\mathbf{Y}_i = \mathbf{X}_i\boldsymbol{\beta} + \mathbf{Z}_i\mathbf{b}_i + \mathbf{e}_i$ it is easy to obtain the conditional and marginal means and variances because of the linearity of the model:

$$\mathrm{E}[\mathbf{Y}_i|\mathbf{b}_i] = \mathbf{X}_i\boldsymbol{\beta} + \mathbf{Z}_i\mathbf{b}_i \qquad \mathrm{Var}[\mathbf{Y}_i|\mathbf{b}_i] = \mathbf{R}_i$$
$$\mathrm{E}[\mathbf{Y}_i] = \mathbf{X}_i\boldsymbol{\beta} \qquad \mathrm{Var}[\mathbf{Y}_i] = \mathbf{Z}_i\mathbf{B}\mathbf{Z}'_i + \mathbf{R}_i.$$

Maximum (or restricted) maximum likelihood estimation is based on the marginal distribution of $\mathbf{Y}_i$ which turns out to be multivariate Gaussian if the $\mathbf{b}_i$ and $\mathbf{e}_i$ are Gaussian-distributed. Even when $\mathbf{b}_i$ and $\mathbf{e}_i$ are Gaussian, the distribution of $\mathbf{Y}_i$ in the general model $\mathbf{Y}_i = \mathbf{f}(\mathbf{x}, \boldsymbol{\theta}, \mathbf{b}_i) + \mathbf{e}_i$ is not necessarily Gaussian since $\mathbf{Y}_i$ is no longer a linear combination of Gaussian random variables. Even deriving the marginal mean and variance of $\mathbf{Y}_i$ proves to be a difficult undertaking. If the distribution of the within-cluster errors $\mathbf{e}_i$ is known finding the conditional distribution of $\mathbf{Y}_i|\mathbf{b}_i$ is simple. The *marginal* distribution of $\mathbf{Y}_i$, which is key in statistical inference, remains elusive in the nonlinear mixed model. The various approaches put forth in the literature differ in the technique and rationale applied to approximate this marginal distribution.

8.3 Toward an Approximate Objective Function

Box 8.2 Estimation Approaches

- **There are three basic approaches to parameter estimation in a NLMM or GLMM: individual estimates methods, linearization based methods, and integral approximation methods. Only the latter two are considered here.**

- **A linearization method approximates the nonlinear mixed model by a Taylor series to arrive at a pseudo-model which is typically of the Laird-Ware form. Assuming Gaussianity of the pseudo-response estimation proceeds by ML or REML (as in §7).**

- **Integral approximation methods use quadrature or Monte Carlo integration to calculate the marginal distribution of the data and maximize its likelihood.**

Applying first principles the joint distribution of $\mathbf{Y}_i$ and $\mathbf{b}_i$ can be written as

$$f_{y,b}(\mathbf{y}_i, \mathbf{b}_i) = f_{y|b}(\mathbf{y}_i|\mathbf{b}_i) f_b(\mathbf{b}_i),$$

where $f_{y,b}$, $f_{y|b}$, and f_b denote the joint, conditional, and marginal probability density (mass) functions, respectively. The marginal distribution of $\mathbf{Y}_i$ is obtained by integrating this joint density over the distribution of the random effects,

$$f_y(\mathbf{y}_i) = \int f_{y|b}(\mathbf{y}_i|\mathbf{b}_i) f_b(\mathbf{b}_i) d\mathbf{b}_i. \qquad [8.5]$$

If the distribution $f_y(\mathbf{y}_i)$ is known the maximum likelihood principle can be invoked to obtain estimates of the unknown parameters. Unfortunately, this distribution is usually intractable. The relevant approaches to arrive at a solution to this problem can be classified into two broad categories, **linearization** and **integral approximation** methods. Linearization methods replace the nonlinear mixed model with an approximate linear model. They are also called pseudo-data methods since the response being modeled is not $\mathbf{Y}_i$ but a function thereof. Some linearization methods are parametric in the sense that they assume a distribution for the pseudo-response. Other linearization methods are *semi*-parametric in the sense that they estimate the parameters without distributional assumptions beyond the first two moments of the pseudo-response. Methods of integral approximation assume a particular distribution for the random effects $\mathbf{b}_i$ and for the conditional distribution of $\mathbf{Y}_i|\mathbf{b}_i$ and approximate the integral [8.5] by numerical techniques (quadrature methods or Monte Carlo integration).

Before proceeding we need to point out that linearization and integral approximation techniques are not the only possible methods for estimating the parameters of a nonlinear mixed model. For example, if the random effects $\mathbf{b}_i$ enter the model linearly, Vonesh and Carter (1992) apply iteratively reweighted least squares estimation akin to that in a generalized linear model. Parameter estimates in nonlinear mixed models can also be obtained in a two-stage approach known as the **individual estimates** method. In the first stage the nonlinear model is fit to each cluster separately and in the second stage the *cluster-specific* estimates are combined (averaged) to arrive at population-averaged values. Success of inference based on individual estimates depends on having sufficient measurements on each cluster to estimate the nonlinear response reliably and on methods for combining the individual estimates into population average estimates efficiently. One advantage of the linearization methods is not to depend on the ability to fit the model to each cluster separately. Clusters that do not contribute information about the entire response profiles, for example, because of missing values, will nevertheless contribute to estimation in linearization methods. Earlier applications of the individual estimates method suffered from not borrowing strength across clusters in the estimation process (see, e.g., Korn and Whittemore 1997 and Biging 1985). Davidian and Giltinan (1993) improved the method considerably by fitting the nonlinear model separately to each subject but using weight matrices that are estimated across subjects. For more details on two-stage methods that build on individual estimates derived from each cluster the reader is referred to Ch. 5. in Davidian and Giltinan (1995) and references therein.

The distinction between inference based on linearization and inference based on individual estimates goes back to Sheiner and Beal (1980) who pioneered procedures for fitting nonlinear mixed effects models.

In our experience, most nonlinear mixed models were until recently fit by one of the linearization techniques. High-dimensional quadrature methods are computationally demanding and were not readily available in standard statistical software packages. The `%nlinmix` macro distributed by SAS Institute (www.sas.com), for example, implements linearization methods. The `nlmixed` procedure, a recent addition to The SAS® System, implements the integral approximation method.

8.3.1 Three Linearizations

Box 8.3 Linearizations

- **A population-averaged (PA) linearization expands the model $f(x, \theta, b)$ or $g^{-1}(x, \beta, b)$ around an estimate $\widehat{\beta}$ of β and the mean of the random effects.**
- **A subject-specific (SS) or cluster-specific linearization expands the model around an estimate $\widehat{\beta}$ of β and a current predictor $\widehat{b}$ of b.**
- **SS expansions are more accurate but more sensitive to model misspecification.**

The Gauss-Newton method (§A5.10.3) for fitting a nonlinear model (for independent data) starts with a least squares objective function to be minimized, the residual sum of squares $S(\boldsymbol{\theta}) = (\mathbf{y} - \mathbf{f}(\mathbf{x}, \boldsymbol{\theta}))'(\mathbf{y} - \mathbf{f}(\mathbf{x}, \boldsymbol{\theta}))$. It then approximates the mean function $\mathbf{f}(\mathbf{x}, \boldsymbol{\theta})$ by a first-order Taylor series about $\widehat{\boldsymbol{\theta}}$ and substitutes the approximate model back into $S(\boldsymbol{\theta})$. This yields an approximated residual sum of squares to be minimized. In nonlinear mixed models a similar rationale can be applied. Approximate the conditional mean function by a Taylor series about some value chosen for $\boldsymbol{\theta}$ and some value chosen for $\mathbf{b}_i$. This leads to an approximate linear mixed model whose parameters are then estimated by one of the techniques from §7. What are sometimes termed the first- and second-order expansion methods (Littell et al. 1996) differ in whether the mean is expanded about the mean of the $\mathbf{b}_i$, $\mathrm{E}[\mathbf{b}_i] = \mathbf{0}$, or the estimated BLUP $\widehat{\mathbf{b}}_i$.

For the discussion that follows we consider the stacked form of the model to eliminate the cluster subscript i. Let $\mathbf{Y} = [\mathbf{Y}_1', \cdots, \mathbf{Y}_n']'$, $\mathbf{e} = [\mathbf{e}_1', \cdots, \mathbf{e}_n']'$ and denote the model vector as

$$\mathbf{f}(\mathbf{x}, \boldsymbol{\theta}, \mathbf{b}) = \begin{bmatrix} f(\mathbf{x}_{11}, \boldsymbol{\theta}, \mathbf{b}_1) \\ f(\mathbf{x}_{12}, \boldsymbol{\theta}, \mathbf{b}_1) \\ \vdots \\ f(\mathbf{x}_{1n_1}, \boldsymbol{\theta}, \mathbf{b}_1) \\ f(\mathbf{x}_{21}, \boldsymbol{\theta}, \mathbf{b}_2) \\ \vdots \\ f(\mathbf{x}_{nn_n}, \boldsymbol{\theta}, \mathbf{b}_n) \end{bmatrix}.$$

If $\mathbf{b}_i$ has mean $\mathbf{0}$ and variance-covariance matrix $\mathbf{D}$, then we refer to $\mathbf{b} = [\mathbf{b}_1', \cdots, \mathbf{b}_n']'$ as the vector of random effects across all clusters, a random variable with mean $\mathbf{0}$ and variance-covariance matrix $\mathbf{B} = \text{Diag}\{\mathbf{D}\}$. In the sequel we present the rationale behind the various linearization methods. Detailed formulas and derivations can be found in §A8.5.1.

The first linearization method expands $\mathbf{f}(\mathbf{x}, \boldsymbol{\theta}, \mathbf{b})$ about a current estimate $\widehat{\boldsymbol{\theta}}$ of $\boldsymbol{\theta}$ and the mean of $\mathbf{b}$, $\text{E}[\mathbf{b}] = \mathbf{0}$. Littell et al. (1996, p. 463) term this the approximate first-order method. We prefer to call it the population-average (PA) expansion. A subject-specific (SS) expansion is obtained as a linearization about $\widehat{\boldsymbol{\theta}}$ and a predictor of $\widehat{\mathbf{b}}$, commonly chosen to be the estimated BLUP. Littell et al. (1996, p. 463) refer to it as the approximate second-order method. Finally, the generalized estimating equations of Zeger, Liang and Albert (1988) can be adapted for the case of a nonlinear mixed model with continuous response based on an expansion about $\text{E}[\mathbf{b}] = \mathbf{0}$ only (see §A8.5.2). We term this case the GEE expansion. The three linearizations lead to the approximate models shown in Table 8.1.

Table 8.1. Approximate models based on linearizations

Expansion			
Type	**About**	**Pseudo-Response**	**Model**
PA	$\widehat{\boldsymbol{\theta}}$, $\text{E}[\mathbf{b}]$	$\mathbf{Y}^* = \mathbf{Y} - \mathbf{f}(\mathbf{x}, \widehat{\boldsymbol{\theta}}, \mathbf{0}) + \mathbf{X}^*\widehat{\boldsymbol{\theta}}$	$\mathbf{Y}^* = \mathbf{X}^*\boldsymbol{\theta} + \mathbf{Z}^*\mathbf{b} + \mathbf{e}$
SS	$\widehat{\boldsymbol{\theta}}, \widehat{\mathbf{b}}$	$\widetilde{\mathbf{Y}} = \mathbf{Y} - \mathbf{f}(\mathbf{x}, \widehat{\boldsymbol{\theta}}, \widehat{\mathbf{b}}) + \widetilde{\mathbf{X}}\,\widehat{\boldsymbol{\theta}} + \widetilde{\mathbf{Z}}\,\widehat{\mathbf{b}}$	$\widetilde{\mathbf{Y}} = \widetilde{\mathbf{X}}\,\boldsymbol{\theta} + \widetilde{\mathbf{Z}}\,\mathbf{b} + \mathbf{e}$
GEE	$\text{E}[\mathbf{b}]$	not necessary	$\mathbf{Y} = \mathbf{f}(\mathbf{x}, \boldsymbol{\theta}, \mathbf{0}) + \dot{\mathbf{Z}}\mathbf{b} + \mathbf{e}$

The matrices $\mathbf{X}^*$, $\mathbf{Z}^*$, $\widetilde{\mathbf{X}}$, and $\widetilde{\mathbf{Z}}$ are matrices of derivatives of the function $\mathbf{f}(\mathbf{x}, \boldsymbol{\theta}, \mathbf{b})$ defined as follows:

$$\mathbf{X}^* = \frac{\partial \mathbf{f}(\mathbf{x}, \boldsymbol{\theta}, \mathbf{b})}{\partial \boldsymbol{\theta}'}\bigg|_{\widehat{\boldsymbol{\theta}}, \mathbf{0}} \qquad \mathbf{Z}^* = \frac{\partial \mathbf{f}(\mathbf{x}, \boldsymbol{\theta}, \mathbf{b})}{\partial \mathbf{b}'}\bigg|_{\widehat{\boldsymbol{\theta}}, \mathbf{0}}$$

$$\widetilde{\mathbf{X}} = \frac{\partial \mathbf{f}(\mathbf{x}, \boldsymbol{\theta}, \mathbf{b})}{\partial \boldsymbol{\theta}'}\bigg|_{\widehat{\boldsymbol{\theta}}, \widehat{\mathbf{b}}} \qquad \widetilde{\mathbf{Z}} = \frac{\partial \mathbf{f}(\mathbf{x}, \boldsymbol{\theta}, \mathbf{b})}{\partial \mathbf{b}'}\bigg|_{\widehat{\boldsymbol{\theta}}, \widehat{\mathbf{b}}} \qquad \dot{\mathbf{Z}} = \frac{\partial \mathbf{f}(\mathbf{x}, \boldsymbol{\theta}, \mathbf{b})}{\partial \mathbf{b}'}\bigg|_{\mathbf{0}}$$

Starred matrices are evaluated at $\widehat{\boldsymbol{\theta}}$ and $\text{E}[\mathbf{b}] = \mathbf{0}$, matrices with tildes are evaluated at $\widehat{\boldsymbol{\theta}}$ and $\widehat{\mathbf{b}}$. From the last column of Table 8.1 it is seen that for the PA and SS expansion, the linearized model is a linear mixed model of the Laird-Ware form where the matrices $\mathbf{X}$ and $\mathbf{Z}$ have been replaced by the respective derivative matrices.

The correspondence to estimating the parameters in a regular nonlinear model is worth pointing out. There, the model is linearized as

$$\mathbf{Y} \doteq \mathbf{f}(\mathbf{x}, \widehat{\boldsymbol{\theta}}) + \partial \mathbf{f}(\mathbf{x}, \boldsymbol{\theta})/\partial \boldsymbol{\theta}' \times (\boldsymbol{\theta} - \widehat{\boldsymbol{\theta}}) + \mathbf{e} = \mathbf{f}(\mathbf{x}, \widehat{\boldsymbol{\theta}}) + \mathbf{F}(\boldsymbol{\theta} - \widehat{\boldsymbol{\theta}}) + \mathbf{e},$$

which yields an approximate linear model

$$\mathbf{Y}^* = \mathbf{Y} - \mathbf{f}(\mathbf{x}, \widehat{\boldsymbol{\theta}}) + \mathbf{F}\widehat{\boldsymbol{\theta}} = \mathbf{F}\boldsymbol{\theta} + \mathbf{e}.$$

After an estimate of $\boldsymbol{\theta}$ is obtained, the pseudo-response $\mathbf{Y}^*$ and the derivative matrix $\mathbf{F}$ are updated and the linear model for the next iteration is obtained. This process continues until a convergence criterion is met.

In a nonlinear mixed model the approximate model is a linear *mixed* model. Given some starting values for $\boldsymbol{\theta}$ in the PA expansion and $\boldsymbol{\theta}$ and $\mathbf{b}$ in the SS expansion the linear mixed model is fit and new estimates and EBLUPs are obtained. The pseudo-response and the derivative matrices are updated and the next linear mixed model is fit. This process continues until some convergence criterion is met. Notice that with the PA expansion the EBLUPs $\widehat{\mathbf{b}}$ are not needed, whereas in the SS expansion the pseudo-response depends on them. In a PA expansion it is sufficient to obtain the EBLUPs at the end of the iterative process. If the number of clusters is large, this can speed up the estimation process.

If the random effects $\mathbf{b}$ and the within-cluster errors $\mathbf{e}$ are Gaussian distributed the marginal distribution of $\mathbf{Y}^*$ or $\widetilde{\mathbf{Y}}$ is also Gaussian and the linearized models can be fit with the `mixed` procedure of The SAS® System. This is the implementation behind the `%nlinmix` macro available from SAS Institute (www.sas.com). Either the maximum or restricted maximum likelihood principle can then be employed. Fitting a series of nonlinear mixed effects models Gregoire and Schabenberger (1996b) noted that the restricted log likelihood was uniformly larger in the SS expansion than in the PA expansion indicating a better fit in the former. Although some care must be exercised in comparing likelihoods across models that differ in their response, it is clear that the SS expansion provides a better approximation to the nonlinear mixed model problem than the PA expansion. The PA approximation, on the other hand, allows both cluster-specific and population-averaged inference and is relatively robust to model misspecification. Only cluster-specific inference is supported by the SS expansion which also suffers from greater sensitivity to model misspecification and potential convergence problems. The estimates from a PA expansion are similar to those of Sheiner and Beal (1980, 1985). Although these authors employ an expansion about E[$\mathbf{b}$], the estimates are not identical since Sheiner and Beal use an extended least squares criterion. The estimates obtained from an SS expansion are identical to the estimates of Lindstrom and Bates (1990). Their estimation algorithm differs from the one outlined here, however. For more details on these computational alternatives see Lindstrom and Bates (1988, 1990) and Wolfinger (1993b).

The process of fitting nonlinear mixed models based on linearizations and the Gaussian error assumption is doubly iterative. The estimation of the covariance parameters in a Gaussian linear mixed model by ML or REML is an iterative process (see §7) and once estimates have been obtained, the components of the linearized mixed model are updated and the process is repeated. Computation time required for these models can be formidable. Estimation algorithms where estimates of the covariance parameters are obtained in noniterative fashion are thus appealing. Relaxing the assumption of Gaussian distributions for $\mathbf{b}$ and $\mathbf{e}$ is also of great interest. One such method is based on an extension of the generalized estimating equations (GEEs) of Liang and Zeger (1986), Zeger and Liang (1986), and Zeger, Liang, and Albert (1988) to nonlinear mixed models (details in §A8.5.2). The GEE expansion in Table 8.1 gives rise to the model $\mathbf{Y} = \mathbf{f}(\mathbf{x}, \boldsymbol{\theta}, \mathbf{0}) + \dot{\mathbf{Z}}\mathbf{b} + \mathbf{e}$. Assuming only that $\mathbf{b} \sim (\mathbf{0}, \mathbf{B})$ and $\mathbf{e} \sim (\mathbf{0}, \mathbf{R})$, the marginal mean and variance of $\mathbf{Y}$ are approximated as $\mathbf{f}(\mathbf{x}, \boldsymbol{\theta}, \mathbf{0})$ and $\dot{\mathbf{V}} = \dot{\mathbf{Z}}\mathbf{B}\dot{\mathbf{Z}}' + \mathbf{R}$, respectively. Estimates of $\boldsymbol{\theta}$ are obtained by solving the estimating equation

$$U(\boldsymbol{\theta};\mathbf{y},\mathbf{V}) = \frac{\partial \mathbf{f}(\mathbf{x}, \boldsymbol{\theta}, \mathbf{0})}{\partial \boldsymbol{\theta}'}\dot{\mathbf{V}}^{-1}(\mathbf{y} - \mathbf{f}(\mathbf{x}, \boldsymbol{\theta}, \mathbf{0})) \equiv \mathbf{0}. \quad [8.6]$$

Since $\dot{\mathbf{V}}$ contains unknown quantities, it must be estimated from the data before estimates of $\boldsymbol{\theta}$ can be calculated. For the case where $\mathbf{R} = \sigma^2\mathbf{I}$, Schabenberger (1994) derives method of moment estimators for the unknowns in $\dot{\mathbf{V}}$. Because these estimates depend on the current solution $\widehat{\boldsymbol{\theta}}$, the process remains iterative. After an update of $\widehat{\boldsymbol{\theta}}$, given a current estimate of $\dot{\mathbf{V}}$, $\widehat{\dot{\mathbf{V}}}$ is re-estimated followed by another update of $\widehat{\boldsymbol{\theta}}$ and so forth. Because the moment estimators are noniterative this procedure usually converges rather quickly. The performance of the method of moments estimators will improve with increasing number of clusters. The estimates so obtained can be used as starting values for a parametric fit of the nonlinear mixed model based on a PA or SS expansion assuming Gaussian pseudo-response. Gregoire and Schabenberger (1996a, 1996b) note that the predicted cluster-specific profiles from a GEE and REML fit were nearly identical and indistinguishable on a graph of plotted response curves. These authors conclude that the GEE estimation method constitutes a viable estimation approach in its own right and is more than just a vehicle to produce starting values for the other linearization methods.

8.3.2 Linearization in Generalized Linear Mixed Models

With generalized linear mixed models the analyst faces simplifying and complicating circumstances. A simplification arises because of the linearity of the predictor. A complication arises because the variance of the responses for non-normal data can depend on the mean itself and the assumption of Gaussian errors is not tenable. Of the various linearization methods proposed to fit a generalized linear mixed model the most significant in our opinion is the pseudo-likelihood method of Wolfinger and O'Connell (1993) which subsumes other methods such as those proposed by Breslow and Clayton (1993). This approach has been coded in the `%glimmix` macro available from SAS Institute (www.sas.com). It is by far not the only approach, however, and much research has taken place in this important area. Schabenberger and Gregoire (1996), without being exhaustive, describe eight subject-specific and two population-averaged methods for estimating the parameters of a generalized linear mixed models, most of which invoke linearizations at some stage. In this subsection we discuss the pseudo-likelihood approach of Wolfinger and O'Connell (1993) to show the relationship of their linearization method to those of the previous subsection and to discuss the additional approximations that are involved because the response may be non-Gaussian.

We commence by formulating a generalized linear mixed model as a transformation of a Laird-Ware model. Define the linear predictor vector for the observations from cluster i as $\boldsymbol{\eta}_i = \mathbf{X}_i\boldsymbol{\beta} + \mathbf{Z}_i\mathbf{b}_i$. The first stage of the Laird-Ware model reckons the response conditional on the random effects. In the case of a generalized linear mixed model with link function $g()$ we obtain

$$\mathbf{g}(\mathrm{E}[\mathbf{Y}_i|\mathbf{b}_i]) = \mathbf{g}(\boldsymbol{\mu}_i^{\mathrm{b}}) = \mathbf{X}_i\boldsymbol{\beta} + \mathbf{Z}_i\mathbf{b}_i.$$

To incorporate the stochastic nature of the response the observational model is expressed as random deviations of $\mathbf{Y}_i|\mathbf{b}_i$ from its mean,

$$\mathbf{Y}_i|\mathbf{b}_i = \mathbf{g}^{-1}(\mathbf{X}_i\boldsymbol{\beta} + \mathbf{Z}_i\mathbf{b}_i) + \mathbf{e}_i. \qquad [8.7]$$

The conditional distribution of the $Y_{ij}|\mathbf{b}_i$ is chosen as an exponential family member (this can be relaxed) with variance function $\mathrm{Var}[e_{ij}] = \psi h(\mu)$. In many situations ψ will be one (see Table 6.2) although the method of Wolfinger and O'Connell allows for estimation of an extra scale parameter. The variance-covariance matrix of $\mathbf{e}_i$ can be expressed as $\mathrm{Var}[\mathbf{e}_i] =$

$\mathbf{A}^{1/2}(\boldsymbol{\mu}_i^{\mathrm{b}})\mathbf{C}_i\mathbf{A}^{1/2}(\boldsymbol{\mu}_i^{\mathrm{b}})$, where $\mathbf{A}$ is a diagonal matrix containing the variance functions and $\mathbf{C}_i$ is a within-cluster correlation matrix. If, conditional on the random effects, the observations from a cluster are uncorrelated, then $\mathbf{C}_i = \mathbf{I}$ and $\mathrm{Var}[\mathbf{e}_i] = \mathbf{A}(\boldsymbol{\mu}_i^{\mathrm{b}})$.

The problem is to get from [8.7] to the marginal distribution of $\mathbf{Y}_i$. The approach by Wolfinger and O'Connell (1993) considers three separate approximations to accomplish that. The **analytic** approximation expands $\mathbf{g}^{-1}(\mathbf{X}_i\boldsymbol{\beta} + \mathbf{Z}_i\mathbf{b}_i)$ into a Taylor series about $\widehat{\boldsymbol{\beta}}$ and $\widehat{\mathbf{b}}$ (or $\widehat{\boldsymbol{\beta}}$ and $\mathrm{E}[\mathbf{b}]$). Substituting back into [8.7] yields an approximated residual vector $\dot{\mathbf{e}}_i|\mathbf{b}_i$. The **moment** approximation equates the mean and variance of $\dot{\mathbf{e}}_i|\mathbf{b}_i$ with the mean and variance of $\mathbf{e}_i|\mathbf{b}_i$. Finally, the **probabilistic** approximation assumes $\dot{\mathbf{e}}_i|\mathbf{b}_i$ to be Gaussian-distributed (Lindstrom and Bates 1990, Laird and Louis 1982). This yields an approximate linear mixed model which is still conditioned on $\mathbf{b}_i$ (as in the first stage of the Laird-Ware model). Since the distribution of the pseudo-model is Gaussian, however, finding the mean and variance of the marginal distribution (which is also Gaussian) is straightforward. The details of the three-step linearization/approximation can be found in §A8.5.3. From the standpoint of linearizing the model this approach is not very different from the linearizations in §8.3.1 and the approach can be carried out with PA, SS, or GEE expansions (see Schabenberger and Gregoire 1996). Compared to a nonlinear mixed model for continuous response the extra step required is the probabilistic approximation that invokes the Gaussian framework. The end result is a vector of pseudo-responses that follows a Gaussian linear mixed model. If $\boldsymbol{\Delta}_i$ is the diagonal matrix of first derivatives of the mean with respect to the linear predictor the vector of pseudo-responses for cluster i is

$$\boldsymbol{\nu}_i = \boldsymbol{\Delta}_i^{-1}\{\mathbf{Y}_i|\mathbf{b}_i - \mathbf{g}^{-1}(\widehat{\boldsymbol{\eta}}_i)\} + \mathbf{X}_i\widehat{\boldsymbol{\beta}} + \mathbf{Z}_i\widehat{\mathbf{b}}_i. \qquad [8.8]$$

The conditional and marginal distributions of $\boldsymbol{\nu}_i$ are

$$\begin{aligned} \boldsymbol{\nu}_i|\mathbf{b}_i &\sim G(\mathbf{X}_i\boldsymbol{\beta} + \mathbf{Z}_i\mathbf{b}_i, \boldsymbol{\Delta}_i^{-1}\mathbf{A}_i^{1/2}\mathbf{C}_i\mathbf{A}_i^{1/2}\boldsymbol{\Delta}_i^{-1}) \\ \boldsymbol{\nu}_i &\sim G\left(\mathbf{X}_i\boldsymbol{\beta}, \mathbf{Z}_i\mathbf{D}\mathbf{Z}_i' + \boldsymbol{\Delta}_i^{-1}\mathbf{A}_i^{1/2}\mathbf{C}_i\mathbf{A}_i^{1/2}\boldsymbol{\Delta}_i^{-1}\right). \end{aligned} \qquad [8.9]$$

The role of the within-cluster error dispersion matrix $\mathbf{R}$ is now played by $\boldsymbol{\Delta}_i^{-1}\mathbf{A}_i^{1/2}\mathbf{C}_i\mathbf{A}_i^{1/2}\boldsymbol{\Delta}_i^{-1}$. Because of the linearity of the linear predictor the $\mathbf{Z}$ and $\mathbf{X}$ matrices in [8.9] are the same matrices as in model [8.7]. They do not depend on the expansion locus and/or current solutions of the fixed and random effects. The linear predictor $\widehat{\boldsymbol{\eta}}_i$ and the gradient matrix $\boldsymbol{\Delta}_i$ are evaluated at the current solutions, however. The process of fitting a generalized linear mixed model based on this linearization is thus again doubly iterative. The parameters in $\mathbf{D}$ and $\mathbf{C}$ are estimated iteratively by maximum or restricted maximum likelihood as in any Gaussian linear mixed model. Once these estimates and updates of $\widehat{\boldsymbol{\beta}}$ have been obtained the pseudo-mixed model is recalculated for the next iteration and the process is repeated. Noniterative methods of estimating the covariance parameters are available. For details see Schabenberger and Gregoire (1996).

8.3.3 Integral Approximation Methods

The initial problem that led to the various linearization methods was the need to obtain the marginal distribution

$$f_y(\mathbf{y}_i) = \int f_{y|b}(\mathbf{y}_i|\mathbf{b}_i) f_b(\mathbf{b}_i) d\mathbf{b}_i, \qquad [8.10]$$

which is needed to calculate the likelihood for the entire data. Assuming that clusters are independent, this likelihood is simply the product $f_y(\mathbf{y}) = \prod_{i=1}^{n} f_y(\mathbf{y}_i)$. The linearizations combined with a Gaussian assumption for the errors $\mathbf{e}_i$ lead to pseudo-models where $f_{y|b}(\mathbf{y}_i^*|\mathbf{b}_i)$, $f_{y|b}(\widetilde{\mathbf{y}}_i|\mathbf{b}_i)$, or $f_{\nu|b}(\boldsymbol{\nu}_i|\mathbf{b}_i)$ are Gaussian. Consequently, the marginal distributions $f_y(\mathbf{y}_i^*)$, $f_y(\widetilde{\mathbf{y}}_i)$, and $f_\nu(\boldsymbol{\nu}_i)$ are Gaussian if $\mathbf{b}_i \sim G(\mathbf{0}, \mathbf{D})$ and the problem of calculating the high-dimensional integral in [8.10] is defused. The linearization methods rest on the assumption that the values which maximize the approximate likelihoods

$$f_y(\mathbf{y}^*) = \prod_{i=1}^{n} f_y(\mathbf{y}_i^*)$$

$$f_y(\widetilde{\mathbf{y}}) = \prod_{i=1}^{n} f_y(\widetilde{\mathbf{y}}_i)$$

$$f_\nu(\boldsymbol{\nu}) = \prod_{i=1}^{n} f_\nu(\boldsymbol{\nu}_i)$$

are close to the values which maximize $f_y(\mathbf{y}) = \prod_{i=1}^{n} f_y(\mathbf{y}_i)$.

A more direct approach is to avoid linearization altogether and to compute the integral [8.10]. A closed form solution is usually not available if the random effects enter the model in nonlinear form but the integral can be approximated. This approach has many merits. It is the likelihood of the data, not some pseudo-data, that is being maximized. If the conditional distribution $f_{y|b}(\mathbf{y}_i|\mathbf{b}_i)$ is not Gaussian as, for example, in a generalized linear mixed model, the linearization methods coerce the pseudo-data in a Laird-Ware model; the assumption that the pseudo-data follow a Gaussian distribution may be tenuous. After all, these are transformations of residuals whose distribution may be far from Gaussian. If the integral approximation method is independent of $f_{y|b}(\mathbf{y}_i|\mathbf{b}_i)$ any conditional distribution can be accommodated. A loose comparison of linearization vs. integral approximation methods can be drawn by recalling the difference between the Gauss-Newton and Newton-Raphson methods for fitting a nonlinear model (in the nonclustered case, §A5.10.3 and §A5.10.5). The objective there is to minimize the residual sum of squares $S(\boldsymbol{\theta})$ akin to the objective here to maximize the likelihood of the data. The Gauss-Newton (GN) method starts with a linearization of the nonlinear model, substitutes the linear pseudo-model into the sum of squares and maximizes the result. The Newton-Raphson (NR) method approximates the residual sum of squares (the objective function) and seeks the maximum of the approximated $S(\boldsymbol{\theta})$. Since the NR method targets the objective function directly while GN targets the model, the former will usually provide a better approximation. The downsides of the NR method are the need for second derivatives and possibly increased computing time. This is offset somewhat by NR requiring fewer iterations. A similar tradeoff exists between linearizations and integral approximations. The latter target the objective function directly and can be made highly accurate. They may require considerable computing resources, however, especially if the number of random effects is large.

Gaussian quadrature methods are numerical devices for approximating an integral, essentially replacing the integral with a weighted sum. In contrast to simple integral approximation rules (e.g., the extended trapezoidal rule) which evaluate the integrand at equally spaced intervals, quadrature rules evaluate the function at unequally spaced intervals (nodes). Nodes and weights are chosen so that the result is exact for a particular class of polynomial functions. Different variations of quadrature yield exact results for different classes of functions. If the integrand can be expressed as $W(x)f(x)$, where $f(x)$ is a polynomial in x, then Gauss-Legendre quadrature yields exact results for $f(x)$ provided the number of nodes is at least one less than the order of the polynomial. Gauss-Hermite quadrature is exact for functions of the form $\exp\{-x^2\}f(x)$, Gauss-Laguerre quadrature for functions of the form $x^{\alpha}\exp\{-x\}f(x)$, and so forth (see Press et al. 1992, Ch. 4.5 for an excellent discussion and comparison). By choosing the quadrature variant carefully, a high degree of accuracy can be obtained with only a few nodes. Fewer than ten nodes are often sufficient to achieve good results. This is important because the evaluation of a p-dimensional integral requires n^p function evaluations if n is the number of nodes sufficient for a one-dimensional integral. A second complication with quadrature in several dimensions is the specification of the nodes which must lie inside the volume of integration. A computationally efficient method of estimating a high-dimensional integral is through importance sampling, a variation of Monte Carlo integration. Gauss-Hermite quadrature and importance sampling to approximate the integral [8.10] for a nonlinear mixed model are discussed in Pinheiro and Bates (1995) and implemented in the `nlmixed` procedure of The SAS® System. This procedure performs sophisticated adaptive versions of quadrature and importance sampling. The quadrature adaptation consists of centering the nodes at the current estimates of the random effects and by scaling the nodes by their variances. The number of quadrature nodes is also chosen adaptively by `proc nlmixed`. When the relative change between log likelihood calculations is less than 0.0001, the lesser number of quadrature points is used. In the case of a single quadrature point, the procedure performs the Laplace approximation as described in Wolfinger (1993b).

8.4 Applications

We have given linearization methods a fair amount of discussion, although we prefer integral approximations. Until recently, most statistical software capable of fitting nonlinear mixed models relied on linearizations. Only a few specialized packages performed integral approximations. The `%nlinmix` and `%glimmix` macros distributed by SAS Institute (http://ftp.sas.com/techsup/download/stat/) perform the SS and PA linearizations discussed in §8.3.1 and §8.3.2. The text by Littell et al. (1996) gives numerous examples on their usage. With Release 8.0 of The SAS® System the `nlmixed` procedure has become available. Although it has been used in previous chapters to fit various nonmixed models it was specifically designed to fit nonlinear and generalized linear mixed models by integral approximation methods. For models that can be fit by either the `%nlinmix`/`%glimmix` macros and the `nlmixed` procedure, we prefer the latter. Although integral approximations are computationally intensive, the `nlmixed` procedure is highly efficient and converges reliably and faster (in our experience) than the linearization-based macros. It furthermore allows the optimization of a general log-likelihood function which opens up the possibility of modeling mixed models with conditional distributions that are not in the exponential family or not already coded in the procedure. Among the conditional distributions currently (as of Release 8.01) available in `proc nlmixed` are the Gaussian, Bernoulli, Binomial, Gamma, Negative Binomial, and Poisson distribution. Among the limitations of the procedure is the restriction to one level of random effects nesting. Only one level of clustering is possible but multiple random effects at the cluster level are permitted. The random effects distribution is restricted to Gaussian. Since the basic syntax of the procedure resembles that of `proc nlin`, the user has to supply starting values for all parameters and the coding of classification variables can be tedious (see §8.4.2 and §8.4.3 for applications). Furthermore, there is no support for modeling within-cluster correlations in the conditional distributions. Since `proc mixed` provides this possibility through the `repeated` statement and the linearization algorithms essentially call a linear mixed model procedure repeatedly, random effects and within-cluster correlations can be accommodated in linearization approaches. In that case the `%glimmix` and `%nlinmix` macro should be used. It has been our experience, however, that after modeling the heterogeneity across clusters through random effects the data do not support further modeling of within-cluster correlations in many situations. The combination of random effects and a nondiagonal **R** matrix in nonlinear mixed models appears to invite convergence troubles. Finally, it should be noted that the integral [8.10] is that of the marginal likelihood of $\mathbf{y}$, not that of $\mathbf{Ky}$, say. The `nlmixed` procedure performs approximate maximum likelihood inference and no REML alternative is available. A REML approach is conceivable if, for example, one were to approximate the distribution of $\mathbf{u}_i = \mathbf{Ky}_i$,

$$f_y(\mathbf{u}_i) = \int f_{y|b}(\mathbf{u}_i|\mathbf{b}_i) f_b(\mathbf{b}_i) d\mathbf{b}_i.$$

Open questions are the nature of the conditional distribution if $\mathbf{y}_i|\mathbf{b}_i$ is distributed in the exponential family, the transformation $\mathbf{Ky}_i$ that yields $\mathrm{E}[\mathbf{Ky}_i] = \mathbf{0}$, and how to obtain the fixed effects estimates. Another REML approach would be to assume a distribution for $\boldsymbol{\theta}$ and integrate over the distributions of $\mathbf{b}_i$ *and* $\boldsymbol{\theta}$ (Wolfinger 2001, personal communication). Only when the number of fixed effects is large would the difference in bias between REML and ML estimation likely be noticeable. And with a large number of fixed effects quadrature methods are then likely to prove too cumbersome computationally.

For the applications that follow we have chosen a longitudinal study and two designed experiments. The yellow poplar cumulative tree-bole volume data are longitudinal in nature. Instead of a temporal metric, the observations were collected along a spatial metric, the tree bole. In §8.4.1 the cumulative bole volume, a continuous response, is modeled with a nonlinear volume-ratio model and population-averaged vs. tree-specific predictions are compared. The responses for applications §8.4.2 and §8.4.3 are not Gaussian and not continuous. In §8.4.2 the poppy count data is revisited and the overdispersion is modeled with a Poisson/Gaussian mixing model. In contrast to the Poisson/Gamma mixing model of §6.7.8 this model does not permit the analytic derivation of the marginal distribution, and we use a generalized linear mixed model approach with integral approximation. The Poisson/Gaussian model can be thought of as an extension of the Poisson generalized linear model to the mixed model framework and to clustered data. In §8.4.3 we extend the proportional odds model for ordinal response to the clustered data framework. There we analyze the data from a repeated measures experiment where the outcome was a visual rating of plant quality.

8.4.1 A Nonlinear Mixed Model for Cumulative Tree Bole Volume

Since the introduction of regression methods into forestry more than 60 years ago, one of the pressing questions (to forest biometricians) concerns the (merchantable) volume of standing trees. This has commonly been addressed by fitting linear or nonlinear regression models. Because of differences across and within species due to physiographic, climatic, management, and other effects, these volume equations are typically fit separately to sample data from regional populations of a tree species. Although the goal is to predict the woody volume of standing trees, the equations are fitted to measurements conducted on felled trees. A tree is felled and delimbed and cut into short sections measuring at most several feet. The volume of each section is determined by geometric principles and accumulated with the volume of the lower sections. This process is carried out until the tree bole has reached a threshold diameter marking the limits of merchantability or to the top of the bole if total-bole volume is the response of interest. Obviously, the consecutive measurements on sections of a particular tree bole are not independent. Even if they were, the process of accumulating sections of the tree bole would induce correlations among the cumulative volume measurements of a tree.

The strategy to fit a new bole-volume equation when the upper-bole merchantability diameter changes due to changes in milling technology or market conditions is a costly endeavor. Burkhart (1977) suggested a modeling approach where the woody volume V_d up to a bole diameter d is expressed as the product of the total-bole volume (V_0) and the ratio R_d of merchantable volume to total volume,

$$V_d = V_0 R_d. \qquad [8.11]$$

Such models are referred to as volume-ratio models (Newberry and Burk 1985, Avery and Burkhart 1994). Gregoire and Schabenberger (1996b) cite the growing number of applications of these models in forestry (Golden et al. 1982, Knoebel et al. 1983, Van Deusen et al. 1981, Newberry and Burk 1985, Amateis and Burkhart 1987, Bailey 1994). The correlations among measurements collected on a single tree bole were not taken into account until the work by Gregoire and Schabenberger (1996a) who employed nonlinear mixed models. In their models the correlations in the marginal distribution of observations from the same tree bole were induced by random effects that varied across trees. Although a direct approach of modeling the within-cluster correlation structure is more appealing, it is not clear how to model the correlations among accumulated observations. Using random effects to capture tree-to-tree variability furthermore allows modeling the volume of the individual tree bole (cluster-specific predictions) as well as the volume of the average tree (population-averaged predictions).

Figure 8.2 depicts the cumulative volume profiles for $n = 336$ yellow poplar (*Liriodendron tulipifera* L.) trees. The trees were felled and measured for the purpose of developing a bole-volume equation for the Appalachian region of the southeastern United States (Beck 1963). The trees vary greatly in the cumulative volume profiles (Figure 8.2) which is partly due to the differences in tree size. The total volume ranges from 0.02 to 259.8 ft^3, total tree height ranges from 12.0 to 138.0 ft, and diameter at breast height from 0.7 to 30.0 inches (Table 8.2). Breast height, which is typically 4.5 feet above ground, is the customary height at which to measure a tree's reference diameter. Differences in tree bole shape are apparent after adjusting for tree size and plotting the relative cumulative volumes V_d/V_0 (Figure 8.3).

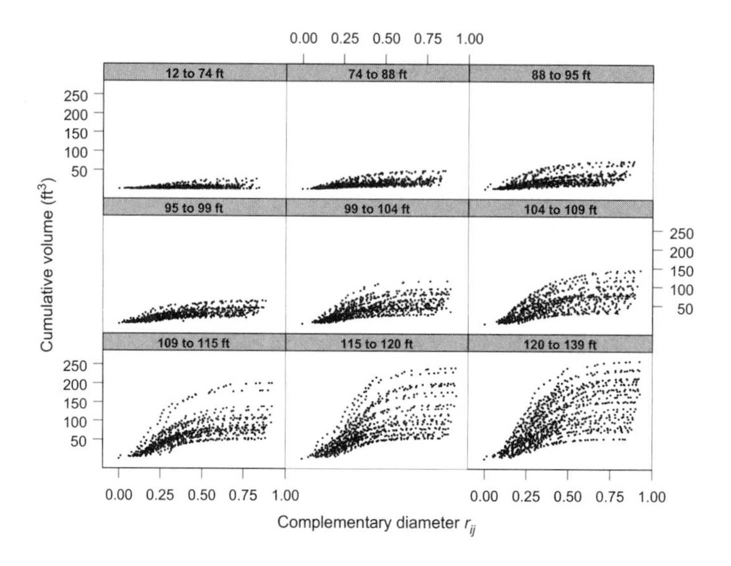

Figure 8.2. Cumulative volume profiles for yellow poplar trees graphed against the complementary diameter $r_{ij} = 1 - d_{ij}/\max(d_{ij})$. d_{ij} denotes the cross-sectional bole diameter of tree i at the jth height of measurement ($i = 1, \cdots, n = 336$). Trees are grouped by total tree height (ft).

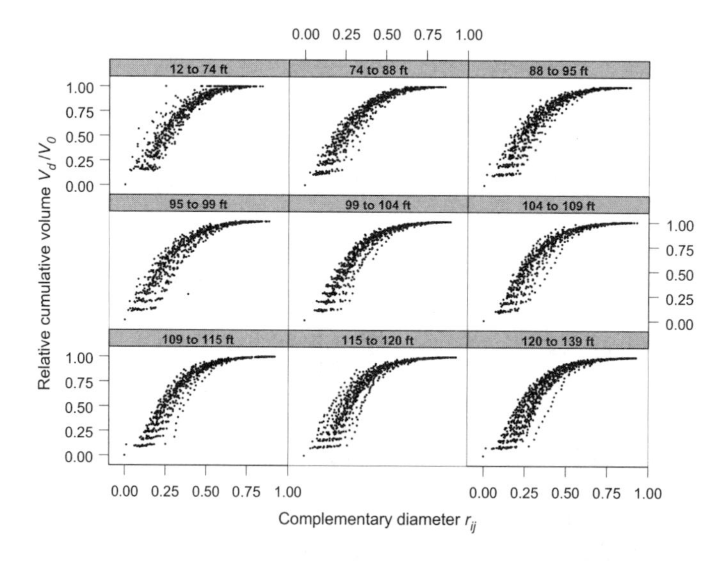

Figure 8.3. Relative cumulative volume V_d/V_0 for 336 yellow poplar trees graphed against complementary diameter. Trees are grouped by total tree height (ft).

Table 8.2. Descriptive statistics for $n = 336$ yellow poplar trees

Variable	Symbol	Sample Mean	Std. Dev.	Min	Max
Diameter at Breast Height (inches)	D_i	13.22	6.51	0.7	30.0
Total Height (feet)	H_i	90.82	26.60	12.0	138.0
Max. Diameter (inches)	$\max\{d_{ij}\}$	15.06	7.66	1.0	35.2
Total Volume (cubic feet)	V_{0i}	54.35	54.04	0.02	259.8
Number of Sections	n_i	19.75		3.0	32.0

To develop models for V_0 and R_d with the intent to fit the two simultaneously while accounting for tree-to-tree differences in size and shape of the volume profiles, we start with a model for the total bole volume V_0. A simple model relating V_0 to easily obtainable tree size variables is

$$V_{i0} = \beta_0 + \beta_1 \frac{D_i^2 H_i}{1000} + e_i.$$

For the yellow poplar data this model fits very well (Figure 8.4) although there is some evidence of heteroscedasticity. The variation in total bole volume for small trees is less than that for larger trees. An ordinary least squares fit yields $\widehat{\beta}_0 = 1.0416$, $\widehat{\beta}_1 = 2.2806$, and $R^2 = 0.99$. The regressor was scaled by the factor $1,000$ so that the estimates are of similar magnitude.

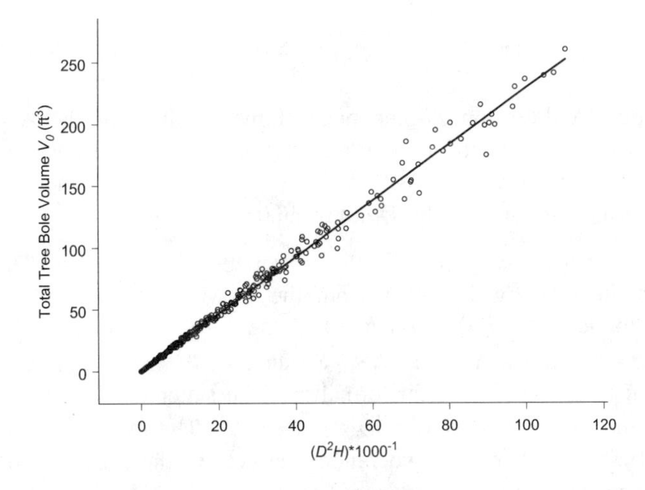

Figure 8.4. Simple linear regression of total tree volume V_0 against D^2H. The data set to which this regression is fit contains $n = 336$ observations, one per tree ($R^2 = 0.99$).

To develop a model for the ratio term R_d, one can think of R_d as a mathematical switch-on function in the terminology of §5.8.6. Since these functions range between 0 and 1 and switch-on behavior is usually nonlinear, a good place to start the search for a ratio model is with the cumulative distribution function (cdf) of a continuous random variable. Gregoire and Schabenberger (1996b) modified the cdf of a Type-I extreme value random variable

(Johnson, Kotz and Balakrishnan 1995, Ch. 22)

$$\Pr(X \le x) = \exp\{-\exp\{-(x-\xi)/\theta\}\}.$$

Letting $t = d/D$, they used

$$R_d = \exp\{-\beta_2 t' \exp\{\beta_3 t\}\}, \qquad [8.12]$$

where $t' = t/1{,}000$. The R_d term is always positive and tends to one as $d \to 0$. The logical constraints ($V_d \ge 0$, $V_d \le V_0$, and $V_{d=0} = V_0$) any reasonable volume-ratio model must obey are thus guaranteed. The fixed effects volume-ratio model for the cumulative volume of the ith tree up to diameter d_j now becomes

$$V_{id_j} = V_{i0}R_{id_j} = \left(\beta_0 + \beta_1 \frac{D_i^2 H_i}{1000}\right)\exp\{-\beta_2 t'_{ij}\exp\{\beta_3 t_{ij}\}\} + e_{ij}. \qquad [8.13]$$

The yellow poplar data are modeled in Gregoire and Schabenberger (1996b) with nonlinear mixed models based on linearization methods and generalized estimating equations. Here we fit the same basic model selected by these authors as superior from a number of models that differ in the type and number of random effects based on quadrature integral approximation methods with `proc nlmixed`. Tree-to-tree heterogeneity is accounted for as variability in size, reflected in variations in total volume, and as variability in shape of the volume profile. The former calls for inclusion of random tree effects in the total volume component V_{i0}, the latter for random effects in the ratio term. The model selected by Gregoire and Schabenberger (1996b) was

$$V_{id_j} = V_{i0}R_{idj} = \left(\beta_0 + \{\beta_1 + b_{1i}\}\frac{D_i^2 H_i}{1000}\right)\exp\{-\{\beta_2 + b_{2i}\}t'_{ij}\exp\{\beta_3 t_{ij}\}\} + e_{ij},$$

where the b_{1i} model random slopes in the total-volume equation and the b_{2i} model the rate of change and point of inflection in the ratio terms. The variances of these random effects are denoted σ_1^2 and σ_2^2, respectively. The within-cluster errors e_{ij} are assumed homoscedastic and uncorrelated Gaussian random variables with mean 0 and variance σ^2.

The model is fit in `proc nlmixed` with the statements that follow. The starting values were chosen as the converged iterates from the REML fit based on linearization. The conditional distribution $f_{y|b}(\mathbf{y}|\mathbf{b})$ is specified in the `model` statement of the procedure. In contrast to other procedures in The SAS® System, `proc nlmixed` uses syntax to denote distributions akin to our mathematical formalism. The statement $V_{id_j}|\mathbf{b} \sim G(V_{i0}R_{id_j}, \sigma^2)$ is translated into `model cumv ~ normal(TotV*R,resvar);`. The `random` statement specifies the distribution of $\mathbf{b}_i$. Since there are two random effects in the model, two means must be specified, two variances and one covariance. The statement

```
random u1 u2 ~ normal([0,0],[varu1,0,varu2]) subject=tn;
```

is the translation of

$$\mathbf{b}_i \sim G\left(\begin{bmatrix}0\\0\end{bmatrix}, \mathbf{D} = \begin{bmatrix}\sigma_1^2 & 0\\ 0 & \sigma_2^2\end{bmatrix}\right); i = 1,\cdots,n; \mathrm{Cov}[\mathbf{b}_i, \mathbf{b}_j] = \mathbf{0}.$$

The `predict` statements calculate predicted values for each observation in the data set. The first of the two statements evaluates the mean without considering the random effects. This is the approximate population-average mean response after taking a Taylor series of the model about E[**b**]. The second `predict` statement calculates the cluster-specific predictions.

```
proc nlmixed data=ypoplar tech=newrap;
  parms beta0=0.25 beta1=2.3 beta2=2.87 beta3=6.7 resvar=4.8
    varu1=0.023 varu2=0.245; /* resvar denotes σ², varu1 σ₁² and varu2 σ₂² */
  X    = dbh*dbh*totht/1000;
  TotV = beta0 + (beta1+u1)*X;
  R    = exp(-(beta2+u2)*(t/1000)*exp(beta3*t));
  model cumv ~ normal(TotV*R,resvar);
  random u1 u2 ~ normal([0,0],[varu1,0,varu2]) subject=tn out=EBlups;
  predict (beta0+beta1*X)*exp(-beta2*t/1000*exp(beta3*t)) out=predPA;
  predict TotV*R out=predB;
run;
```

Output 8.1.

```
                    The NLMIXED Procedure

                       Specifications
     Data Set                                 WORK.YPOPLAR
     Dependent Variable                       cumv
     Distribution for Dependent Variable      Normal
     Random Effects                           u1 u2
     Distribution for Random Effects          Normal
     Subject Variable                         tn
     Optimization Technique                   Newton-Raphson
     Integration Method                       Adaptive Gaussian Quadrature
                          Dimensions
           Observations Used                   6636
           Observations Not Used                  0
           Total Observations                  6636
           Subjects                             336
           Max Obs Per Subject                   32
           Parameters                             7
           Quadrature Points                      1

                          Parameters
  b0        b1        b2        b3    resvar     varu1     varu2     NegLogLike
0.25       2.3      2.87       6.7       4.8     0.023     0.245     15535.9783

                       Iteration History
     Iter     Calls    NegLogLike         Diff      MaxGrad        Slope
        1        18    15532.1097     3.868562      9.49243     -7.48218
        2        27    15532.0946     0.015093     0.021953      -0.0301
        3        36    15532.0946     3.317E-7     2.185E-7     -6.64E-7
          NOTE: GCONV convergence criterion satisfied.

                       Fit Statistics
           -2 Log Likelihood                  31064
           AIC (smaller is better)            31078
           AICC (smaller is better)           31078
           BIC (smaller is better)            31105

                    Parameter Estimates
                    Standard
Parameter Estimate     Error   DF t Value Pr > |t|  Alpha     Lower  Upper
b0           0.2535   0.1292  334    1.96   0.0506   0.05 -0.00070 0.5078
b1           2.2939  0.01272  334  180.38   <.0001   0.05   2.2689 2.3189
b2           2.7529  0.06336  334   43.45   <.0001   0.05   2.6282 2.8775
b3           6.7480  0.02237  334  301.69   <.0001   0.05   6.7040 6.7920
resvar       4.9455  0.08923  334   55.42   <.0001   0.05   4.7700 5.1211
varu1       0.02292  0.00214  334   10.69   <.0001   0.05   0.0187 0.0271
varu2        0.2302  0.02334  334    9.86   <.0001   0.05   0.1843 0.2761
```

Since the starting values are the converged values of a linearization followed by REML estimation, the integral approximation method converges quickly after only three iterations (Output 8.1). The estimates of the fixed effects are $\widehat{\boldsymbol{\beta}} = [0.2535, 2.2939, 2.7529, 6.7480]$ and those of the random effects are $\widehat{\sigma}^2 = 4.9455$, $\widehat{\sigma}_1^2 = 0.02292$, and $\widehat{\sigma}_2^2 = 0.2302$. Notice that the degrees of freedom equal the number of clusters minus the number of random effects in the model (apart from e_{ij}). The asymptotic 95% confidence intervals for the variances of the random effects do not include zero and based on this evidence one would conclude that the inclusion of the random effects improved the model fit. A better test can be obtained by fitting the models without random effects or only one random effect and comparing minus twice the log likelihoods (Table 8.3)

Table 8.3. Minus twice log likelihoods for various models fit to the yellow poplar data (Models differ only in the number of random effects)

Model	Random Effects	-2 Log Likelihood
A	b_1 and b_2	31,064 (Output 8.1)
B	b_1	35,983
C	b_2	39,181
D	none	43,402 (Output 8.2)

The model with two random effects has the smallest value for minus twice the log likelihood and is a significant improvement over any of the other models. Note that D is the purely fixed effect model which fits only a population-average curve and does not take into account any clustering. Fit statistics and parameter estimates for model D are shown in Output 8.2. Since this model (incorrectly) assumes that all observations are independent, its degrees of freedom are no longer equal to the number of clusters minus the number of covariance parameters. The estimate of the residual variation is considerably larger than in the random effects model A. Residuals are measured against the population average in model D and against the tree-specific predictions in model A.

Output 8.2. (abridged)

```
                          The NLMIXED Procedure

                             Fit Statistics

              -2 Log Likelihood                         43402
              AIC (smaller is better)                   43412
              AICC (smaller is better)                  43412
              BIC (smaller is better)                   43446

                          Parameter Estimates

                                  Standard
      Parameter    Estimate          Error      DF    t Value    Pr > |t|

      b0             1.4693         0.1672    6636       8.79      <.0001
      b1             2.2430       0.005112    6636     438.79      <.0001
      b2             4.1712         0.2198    6636      18.98      <.0001
      b3             6.2777        0.05930    6636     105.86      <.0001
      resvar        40.5512         0.7040    6636      57.60      <.0001
```

We selected four trees (5, 151, 279, and 308) from the data set to show the difference between the population-averaged and cluster-specific predictions (Figure 8.5). The trees vary appreciably in size and total volume. The population average fits fairly well to tree #279 and the lower part of the bole of tree #151. For the medium to large sized tree #5 the PA predictions overestimate the cumulative volume in the tree bole. For the large tree #308, the population average overestimates the volume in the lower parts of the tree bole where most of the valuable timber is accrued. Except for the smallest tree, the tree-specific predictions provide an excellent fit to the data. An operator that processes high-grade timber in a sawmill where adjustments of the cutting tools on a tree-by-tree basis are feasible, would use the tree-specific cumulative volume profiles to maximize the output of high-quality lumber. If adjustments to the saws on an individual tree basis are not economically feasible, because the timber is of lesser quality, for example, one can use the population-average profiles to determine the settings.

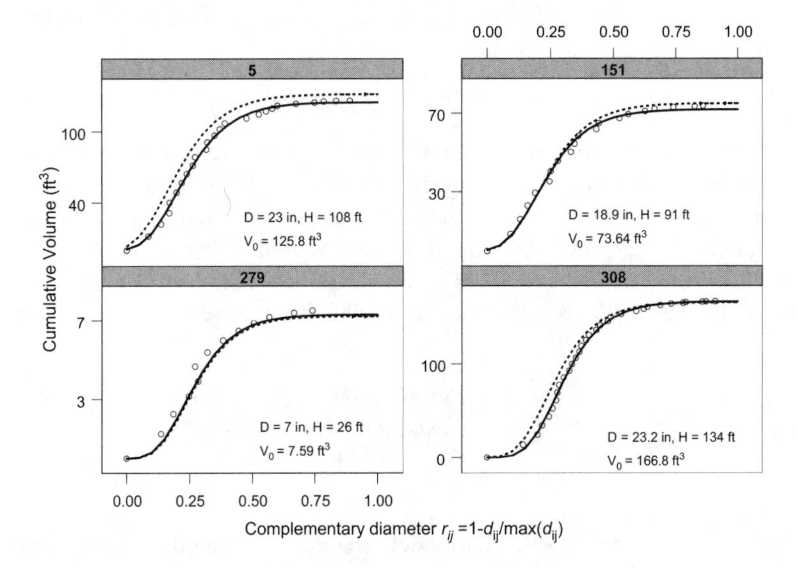

Figure 8.5. Population-averaged (dashed line) and cluster-specific (solid line) predictions for four of the 336 trees. Panel headings are the tree identifiers.

8.4.2 Poppy Counts Revisited — a Generalized Linear Mixed Model for Overdispersed Count Data

In §6.7.8 we analyzed the poppy count data of Mead et al. (1993, p. 144) which represents counts obtained in a randomized block design with six treatments and four blocks (see Table 6.24). An analysis as a generalized linear model for Poisson data with linear predictor

$$\eta_{ij} = \mu + \tau_i + \rho_j;\ i = 1, \cdots, 6; j = 1, \cdots, 4,$$

where τ_i denotes treatments and ρ_j block effects, showed considerable overdispersion. The overdispersion problem was tackled there by assuming that Y_{ij}, the poppy count for treatment

i in block j, was not a Poisson random variable with mean $\lambda_{ij} = \exp\{\eta_{ij}\}$, but that λ_{ij} was a Gamma random variable. The *conditional* distribution $Y_{ij}|\lambda_{ij}$ was modeled as a Poisson(λ_{ij}) random variable. This construction allowed the analytic derivation of the marginal probability mass function

$$p(y) = \int p(y|\lambda)f(\lambda)d\lambda, \qquad [8.14]$$

which turned out to follow the Negative Binomial law. Since this distribution is a member of the exponential family (Table 6.2) the resulting model could be fit as a generalized linear model. We used `proc nlmixed` to estimate the parameters of the model not because this was a mixed model, but because of a problem associated with the `dist=negbin` option of `proc genmod` (in the SAS® release we used that has been subsequently corrected) and in anticipation of fitting the model that follows.

An alternative approach to the Poisson/Gamma mixing procedure is to assume that the linear predictor is a linear mixed model

$$\eta_{ij} = \mu + \tau_i + \rho_j + d_{ij}, \qquad [8.15]$$

where d_{ij} are independent Gaussian random variables with mean 0 and variance σ_d^2. These additional random variables introduce extra variability into the system associated with the experimental units. Conditional on d_{ij}, the poppy counts are again modeled as Poisson random variables. The marginal distribution of the counts in the Poisson/Gaussian mixing model is elusive, however. The integral [8.14] can not be evaluated in closed form. It can be approximated by the methods of §8.3.3, however. This generalized linear mixed model becomes

$$\begin{aligned} Y_{ij}|\lambda_{ij} &= \text{Poisson}(\lambda_{ij}) \\ \lambda_{ij} &= \exp\{\mu + \tau_i + \rho_j + d_{ij}\} \\ d_{ij} &\sim G(0, \sigma^2). \end{aligned}$$

The `proc nlmixed` code to fit this model is somewhat lengthy, because treatment and block effects are classification variables and must be coded inside the procedure. The block of `if .. else ..` statements in the code below sets up the linear predictor for the various combinations of block and treatment effects. The last level of either factor is set to zero and its effect is absorbed into the intercept. This parameterization coincides with that of `proc genmod`. The variance of the d_{ij} is not estimated directly, because σ^2 is bounded by zero from below. Instead, we estimate the logarithm of the standard deviation which can range over the real line (parameter `logsig`).

```
proc nlmixed data=poppies df=14;
   parameters intcpt=3.4 bl1=0.3 bl2=0.3 bl3=0.3
              tA=1.5 tB=1.5 tC=1.5 tD=1.5 tE=1.5
              logsig=0;
   if      block=1 then linp = intcpt + bl1;
   else if block=2 then linp = intcpt + bl2;
   else if block=3 then linp = intcpt + bl3;
   else if block=4 then linp = intcpt;
   if      treatment = 'A' then linp = linp + tA;
   else if treatment = 'B' then linp = linp + tB;
   else if treatment = 'C' then linp = linp + tC;
   else if treatment = 'D' then linp = linp + tD;
```

```
   else if treatment = 'E' then linp = linp + tE;
   else if treatment = 'F' then linp = linp;

   lambda = exp(linp + d);

   estimate 'A Lsmean' intcpt+0.25*bl1+0.25*bl2+0.25*bl3+tA;
   estimate 'B Lsmean' intcpt+0.25*bl1+0.25*bl2+0.25*bl3+tB;
   estimate 'C Lsmean' intcpt+0.25*bl1+0.25*bl2+0.25*bl3+tC;
   estimate 'D Lsmean' intcpt+0.25*bl1+0.25*bl2+0.25*bl3+tD;
   estimate 'E Lsmean' intcpt+0.25*bl1+0.25*bl2+0.25*bl3+tE;
   estimate 'F Lsmean' intcpt+0.25*bl1+0.25*bl2+0.25*bl3;
   estimate 'sigma^2'  exp(2*logsig);

   model count ~ poisson(lambda);
   random d ~ normal(0,exp(2*logsig)) subject=plot;
run;
```

The statement `lambda = exp(linp + d);` calculates the conditional Poisson mean λ_{ij}. Notice that the random effect `d` does not appear in the parameters statement. Only the dispersion of d_{ij} is a parameter of the model. The `model` statement informs the procedure that the counts are modeled (conditionally) as Poisson random variables and the `random` statement determines the distribution of the random effects. Only the `normal()` keyword can be used in the `random` statement. The first argument of `normal()` defines the mean of d_{ij}, the second the variance σ^2. Since we are also interested in the estimate of σ^2 this value is calculated in the last of the `estimate` statements. The other `estimate` statements calculate the "least squares" means of the treatments on the log scale and averaged across the random effects. The `subject=plot` statement identifies the experimental unit as the cluster which yields a model with a single observation per cluster (`Dimensions` table in Output 8.3). The degrees of freedom were set to coincide with the Negative Binomial analysis in §6.7.8. The Poisson model without overdispersion had 15 deviance degrees of freedom. The Negative Binomial model estimated one additional parameter (labeled `k` there). Similarly, the Poisson/Gaussian model adds one parameter, the variance of the d_{ij}.

The initial negative log likelihood of 145.18 calculated from the starting values improved during 21 iterations that followed. The converged negative log likelihood was 116.89. The important question is whether the addition of the random variables d_{ij} in the linear predictor improved the model over the standard Poisson generalized linear model. If H_0: $\sigma^2 = 0$ can be rejected, the Poisson/Gaussian model is superior. From the result of the last `estimate` statement it is seen that the approximate 95% confidence interval for σ^2 is $[0.0260, 0.1582]$ and one would conclude that there is extra variation among the experimental units beyond that accounted for by the Poisson law. A better approach is to fit a Poisson model with linear predictor $\eta_{ij} = \mu + \tau_i + \rho_j$ and to compare the log likelihoods of the two models with a likelihood ratio test. For the reduced model one obtains a negative log-likelihood of 204.978. The likelihood ratio test statistic $\Lambda = 409.95 - 233.8 = 176.15$ is highly significant.

To compare this model against the Negative Binomial model in §6.7.8 a likelihood ratio test can not be employed because the Poisson/Gamma and the Poisson/Gaussian models are not nested. For the comparison among non-nested models AIC can be used. The information criteria for the two models are very close (Poisson/Gamma: AIC $= 253.1$, Poisson/Gaussian: AIC $= 253.8$). Note that since both models have the same number of parameters, the difference of their AIC values equals twice the difference of their log-likelihoods. From a statistical point of view either model may be chosen.

Output 8.3.

```
                         The NLMIXED Procedure

                             Specifications
    Data Set                                     WORK.POPPIES
    Dependent Variable                           count
    Distribution for Dependent Variable          Poisson
    Random Effects                               d
    Distribution for Random Effects              Normal
    Subject Variable                             plot
    Optimization Technique                       Dual Quasi-Newton
    Integration Method                           Adaptive Gaussian
                                                 Quadrature
                               Dimensions
                Observations Used                    24
                Observations Not Used                 0
                Total Observations                   24
                Subjects                             24
                Max Obs Per Subject                   1
                Parameters                           10
                Quadrature Points                     1

                               Parameters
intcpt   bl1   bl2   bl3   tA    tB    tC  tD    tE     logsig    NegLogLike
   3.4   0.3   0.3   0.3  1.5   1.5   1.5 1.5   1.5          0    145.186427

                           Iteration History

    Iter     Calls     NegLogLike        Diff      MaxGrad          Slope
       1         2     141.890476     3.29595     11.45812       -110.573
       2         3     135.687876      6.2026     22.94238       -87.5166
       3         6     131.122483    4.565393     34.73824       -429.497
... and so forth ...
      20        33     116.899994    7.962E-7     0.001561        -1.85E-6
      21        35     116.899994    6.486E-8     0.000319        -1.72E-7

              NOTE: GCONV convergence criterion satisfied.

                             Fit Statistics

                -2 Log Likelihood                   233.8
                AIC (smaller is better)             253.8
                AICC (smaller is better)            270.7
                BIC (smaller is better)             265.6

                          Parameter Estimates

                     Standard
Parameter Estimate    Error    DF  t Value Pr > |t| Alpha   Lower     Upper

intcpt      3.0246   0.2171    14    13.93   <.0001  0.05    2.5589   3.4903
bl1         0.3615   0.1929    14     1.87   0.0820  0.05  -0.05227   0.7753
bl2         0.2754   0.1927    14     1.43   0.1749  0.05   -0.1379   0.6887
bl3         0.7707   0.1897    14     4.06   0.0012  0.05    0.3639   1.1775
tA          2.6272   0.2363    14    11.12   <.0001  0.05    2.1205   3.1339
tB          2.5930   0.2363    14    10.97   <.0001  0.05    2.0862   3.0998
tC          0.9900   0.2418    14     4.09   0.0011  0.05    0.4713   1.5087
tD          0.9256   0.2422    14     3.82   0.0019  0.05    0.4061   1.4452
tE         0.05305   0.2526    14     0.21   0.8367  0.05   -0.4887   0.5948
logsig     -1.1925   0.1673    14    -7.13   <.0001  0.05   -1.5512  -0.8337
```

Output 8.3 (continued).

Additional Estimates

Label	Estimate	Standard Error	DF	t Value	Pr > |t|	Alpha	Lower	Upper
A Lsmean	6.0037	0.1538	14	39.04	<.0001	0.05	5.6739	6.3336
B Lsmean	5.9695	0.1538	14	38.80	<.0001	0.05	5.6396	6.2995
C Lsmean	4.3665	0.1625	14	26.87	<.0001	0.05	4.0180	4.7150
D Lsmean	4.3021	0.1631	14	26.38	<.0001	0.05	3.9524	4.6519
E Lsmean	3.4295	0.1793	14	19.13	<.0001	0.05	3.0451	3.8140
F Lsmean	3.3765	0.1794	14	18.82	<.0001	0.05	2.9918	3.7612
sigma^2	0.09209	0.0308	14	2.99	0.0098	0.05	0.0260	0.1582

Of further interest is a comparison of the parameter estimates and estimates of the treatment means among the various models that have been fit to these data (Output 8.4, SAS® code on CD-ROM). We consider

[A]: a Poisson generalized linear model;

[B]: a Poisson model with multiplicative overdispersion factor;

[C]: a Poisson/Gaussian mixing model;

[D]: a Poisson/Gamma mixing model.

Output 8.4.

effect	Poi Est[1]	Poi/ Gauss Est[2]	Poi/Gam Est[3]	Poi StdErr[4]	Poi/OD StdErr[5]	Poi/Gauss StdErr[6]	Poi/Gam StdErr[7]
intcpt	3.286	3.025	3.043	.0915	.3783	.2171	.2121
bl1	.3736	.3615	.3858	.0452	.1867	.1929	.1856
bl2	.2270	.2754	.2672	.0466	.1926	.1927	.1864
bl3	.2940	.7707	.8431	.0459	.1898	.1897	.1902
tA	2.504	2.627	2.630	.0895	.3700	.2363	.2322
tB	2.459	2.593	2.618	.0897	.3707	.2363	.2326
tC	.9440	.9900	.9618	.1014	.4193	.2418	.2347
tD	.8536	.9256	.9173	.1028	.4248	.2422	.2371
tE	.1120	.0530	.0806	.1184	.4896	.2526	.2429
A Lsmean	6.014	6.004	6.048	.0247	.1021	.1538	.1510
B Lsmean	5.969	5.970	6.036	.0253	.1044	.1538	.1516
C Lsmean	4.454	4.366	4.379	.0537	.2221	.1625	.1585
D Lsmean	4.363	4.302	4.335	.0562	.2323	.1631	.1597
E Lsmean	3.622	3.430	3.498	.0814	.3365	.1793	.1729
F Lsmean	3.510	3.377	3.417	.0861	.3559	.1794	.1742
k	.	.	11.44	.	.	.	3.750
sigma^2	.	.0921	.	.	.	.0308	.

[1]: estimates in Poisson models [A] and [B]. Standard errors in [4] and [5], respectively.
[2]: estimates in Poisson/Gaussian mixing model [C]. Standard errors in [6]
[3]: estimates in Poisson/Gamma mixing model [D]. Standard errors in [7]

A multiplicative overdispersion factor does not alter the estimates, only their precision. Comparing columns 4 and 5 of Output 8.4 the extent to which the regular GLM overstates the precision of estimates is obvious. Estimates of the parameters as well as the treatment means (on the log scale) are close for all methods. The standard errors of the two mixing models are also very close.

An interesting aspect of the generalized linear mixed model is the ability to predict the cluster-specific responses. Since each experimental unit serves as a cluster (of size one) this corresponds to predicting the plot-specific poppy counts. If all effects in the model were fixed, the term d_{ij} would represent the block $\times$ treatment interaction. The model would be saturated with a deviance of zero and predicted counts would coincide with the observed counts. In the generalized mixed linear model the d_{ij} are random variables and only a single degree of freedom is lost to the estimation of its variance (compared to $3 \times 5 = 15$ degrees of freedom for a fixed effect interaction). After calculating the predictors $\widehat{d}_{ij}$ of the random effects, cluster-specific predictions of the counts are obtained as

$$\widehat{Y}_{ij}|d_{ij} = \exp\left\{\widehat{\mu} + \widehat{\tau}_i + \widehat{\rho}_j + \widehat{d}_{ij}\right\}.$$

In `proc nlmixed` this is accomplished by adding the statement

```
predict exp(linp + d) out=sspred;
```

to the code above. Taking a Taylor series of $\exp\{\mu + \tau_i + \rho_j + d_{ij}\}$ about $\mathrm{E}[d_{ij}] = 0$, the marginal average $\mathrm{E}[Y_{ij}]$ can be approximated as $\exp\{\mu + \tau_i + \rho_j\}$ and estimated as $\exp\left\{\widehat{\mu} + \widehat{\tau}_i + \widehat{\rho}_j\right\}$. These *PA* predictions can be obtained with the `nlmixed` statement

```
predict exp(linp)     out=papred;
```

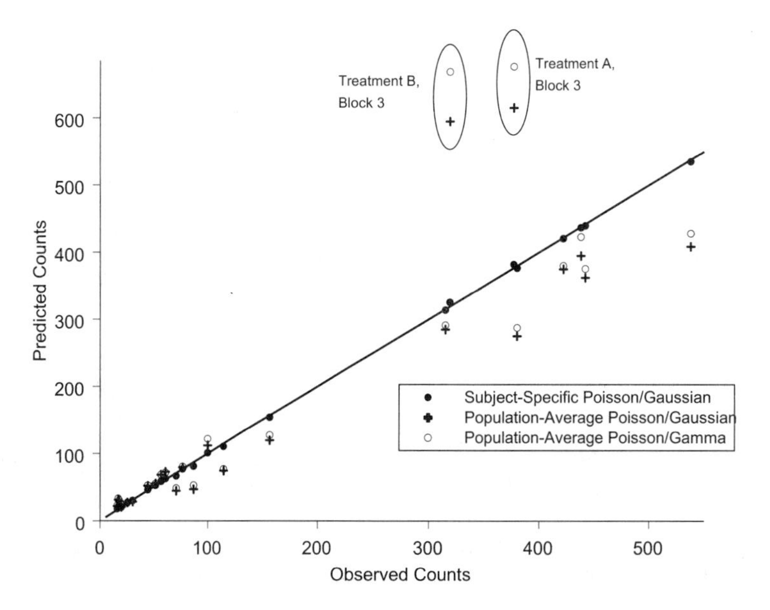

Figure 8.6. Predicted vs. observed poppy counts in Poisson/Gaussian and Poisson/Gamma (= Negative Binomial) mixing models.

The population-averaged and cluster-specific predictions are plotted against the observed poppy counts in Figure 8.6. The cluster-specific predictions for the Poisson/Gaussian model are very close to the 45° line but the model is not saturated; the predictions do not reproduce the data. There are still fourteen degrees of freedom left! The Negative Binomial model based on Poisson/Gamma mixing does not provide the opportunity to predict poppy counts on the plot level because the conditional distribution is not involved at any stage of estimation. The *PA* predictions of the two mixing models are very similar as is expected from the agreement of their parameter estimates (Output 8.4). The predicted values for treatments A and B in block 3 do not concur well with the observed counts, however. There appears to be block × treatment interaction which is even more evident by plotting the *PA* predicted counts by treatments against blocks (Figure 8.7).

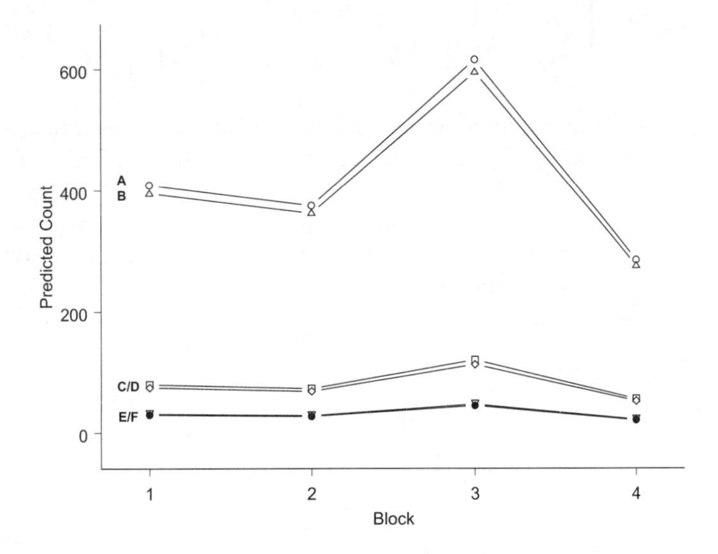

Figure 8.7. Population-averaged predicted cell means in Poisson/Gaussian mixing model.

8.4.3 Repeated Measures with an Ordinal Response

The proportional odds model (POM) of McCullagh (1980) was introduced in §6.5 as an extension of generalized linear models to model an ordinal response. Recall that the POM is a special case of a cumulative link model where the probability that an observation falls into category j or below is modeled. In the case of a logit link with only two categories (a binary response) the POM reduces to a standard logistic regression or classification model. As with any other response, repeated measures are a frequent occurrence in agronomic investigations. They give rise to clustered data structures with correlations among the repeat observations on the same experimental unit that must be accounted for in the analysis, whether the response is univariate or multivariate (as is the case for ordinal data).

The data in Table 8.4 stem from an experiment studying turfgrass quality for five varieties. The varieties were applied independently to seventeen (varieties 2 and 3) or eighteen (varieties 1, 4 and 5) plots. The plots were visited in May, July, and September of the growing season and turf quality was rated on a three-point ordinal scale as *low*, *medium*, or *excellent*.

Table 8.4. Repeated measures of turfgrass ratings in three categories (Numbers represent the number of plots on which a particular rating was observed)

Variety	No. of Plots	May			July			September		
		Low	Med.	Exc.	Low	Med	Exc.	Low	Med	Exc.
1	18	4	10	4	1	9	8	0	12	6
2	17	2	11	4	0	7	10	0	9	8
3	17	2	11	4	2	8	7	2	11	4
4	18	8	7	3	4	8	6	4	13	1
5	18	1	11	6	3	4	11	3	6	9

It appears that the probability to observe a low rating decreases over time and the probability of excellent turf quality appears to be largest in July. Varietal differences in the ratings distributions seem to be minor. To confirm the presence or absence of varietal effects, trends over time, and possibly variety $\times$ time interactions, a proportional odds model containing these effects is fit. A standard model ignoring the possibility of correlations over time can be fit with the `genmod` procedure in The SAS® System (see §§6.5.2, 6.7.4., and 6.7.5 for additional code examples):

```
data counts;
  input rating $ variety month count;
  datalines;
 low          1          5          4
 med          1          5         10
 xce          1          5          4
 low          1          7          1
 med          1          7          9
 xce          1          7          8
 med          1          9         12
 xce          1          9          6
and so forth ...
;;
run;

proc genmod data=counts;
  class variety;
  model rating = variety month month*month variety*month
                / link=cumlogit dist=multinomial type3;
  freq count;
run;
```

This model incorporates both linear and quadratic time effects because the data in Table 8.4 suggest that the rating distributions in July may be different from those in May or September. The term `variety*month` models differences in the linear slopes among the varieties.

The fit achieves a -2 log-likelihood of 483.12 (Output 8.5). From the `LR Statistics For Type 3 Analysis` table it is seen that only the linear and quadratic effects in time appear to be significant. Varietal differences in the rating distributions appear to be absent ($p = 0.7299$) and trends over time appear not to differ among the five varieties ($p = 0.7810$).

Output 8.5.

```
                    The GENMOD Procedure

                     Model Information
     Data Set                                          WORK.COUNTS
     Distribution                                      Multinomial
     Link Function                                Cumulative Logit
     Dependent Variable                                     rating
     Frequency Weight Variable                               count
     Observations Used                                          42
     Sum Of Frequency Weights                                  264
     Probabilities Modeled        Pr( Low Ordered Values of rating )


                   Class Level Information
              Class          Levels    Values
              variety             5    1 2 3 4 5

                      Response Profile
                Ordered    Ordered
                  Level    Value          Count
                      1    low               36
                      2    med              137
                      3    xce               91

             Criteria For Assessing Goodness Of Fit
      Criterion                DF           Value          Value/DF
      Log Likelihood                    -241.5596
Algorithm converged.

                  Analysis Of Parameter Estimates
                                Standard       Wald 95%         Chi-   Pr >
Parameter         DF  Estimate     Error  Confidence Limits   Square   ChiSq

Intercept1         1    7.1408    3.2141    0.8413   13.4403     4.94   0.0263
Intercept2         1    9.9000    3.2451    3.5398   16.2603     9.31   0.0023
variety        1   1    1.5787    1.6553   -1.6657    4.8232     0.91   0.3402
variety        2   1    1.2614    1.6611   -1.9943    4.5171     0.58   0.4476
variety        3   1    0.0813    1.6658   -3.1835    3.3462     0.00   0.9611
variety        4   1    1.7832    1.6690   -1.4879    5.0543     1.14   0.2853
variety        5   0    0.0000    0.0000    0.0000    0.0000      .       .
month              1   -2.8467    0.9331   -4.6756   -1.0178     9.31   0.0023
month*month        1    0.1975    0.0658    0.0686    0.3264     9.02   0.0027
month*variety 1    1   -0.1559    0.2299   -0.6065    0.2947     0.46   0.4977
month*variety 2    1   -0.1773    0.2321   -0.6322    0.2777     0.58   0.4450
month*variety 3    1    0.0810    0.2332   -0.3760    0.5380     0.12   0.7282
month*variety 4    1   -0.0328    0.2312   -0.4860    0.4204     0.02   0.8871
month*variety 5    0    0.0000    0.0000    0.0000    0.0000      .       .
Scale              0    1.0000    0.0000    1.0000    1.0000

NOTE: The scale parameter was held fixed.

                 LR Statistics For Type 3 Analysis
                                       Chi-
             Source               DF     Square    Pr > ChiSq
             variety               4       2.03        0.7299
             month                 1      10.07        0.0015
             month*month           1       9.22        0.0024
             month*variety         4       1.75        0.7810
```

Since this model does not account for correlations over time it is difficult to say whether these findings persist if the temporal correlations are incorporated in a model. In particular, because modeling the correlations through random effects will not only change the standard error estimates but the estimates of the model coefficients themselves. Positive autocorrelation leads to overdispersed data and one approach to remedy the situation is by formulating a *mixed* proportional odds model where, given some random effects, the data follow a POM and to perform maximum likelihood inference based on the marginal distribution of the data. This indirect approach of modeling correlations (see §7.5.1 for the distinction between direct and induced correlation models) is reasonable in models for correlated data where the mean function is nonlinear.

Using the same regressors and fixed effects as in the previous fit, we now add a random effect that models the plot-to-plot variability. This is reasonable since treatments have been assigned at random to plots, because extra variation is likely to be related to excess variation among the experimental units, and the plots have been remeasured (are the clusters). This fit is obtained with the `nlmixed` procedure in SAS®. We note in passing that because `proc nlmixed` uses an integral approximation based on quadrature, this modeling approach is identical to the one put forth by Jansen (1990) for ordinal data with overdispersion and Hedeker and Gibbons (1994) for clustered ordinal data.

The data must be set up differently for the `nlmixed` operation, however. The `counts` data set used with `proc genmod` lists for all varieties and months the number of plots that were assigned a particular rating (variable `count`). The data set `CountProfiles` used to fit the mixed model variety of the POM contains the number of response profiles over time. The first three observations show one unique response profile for variety 1. A low rating in May was followed by two medium ratings in July and September. Two of the 18 plots for this variety exhibited that particular response profile (variable `count`). The remaining triplets of observations in the data set `CountProfiles` give the response profiles for this and the other varieties. The `sub` variable identifies the clusters for this study, corresponding to the plots. It works in conjunction with the `replicate` statement of `proc nlmixed`. The first triplet of observations are identified as belonging to the same plot (cluster) and the value of the `count` variable determines that there are two plots (experimental units) with this response profile.

```
data CountProfiles;
  label rating = '1=low, 2=medium, 3=excellent';
  input rating variety month sub count;
  datalines;
  1         1         5         1         2
  2         1         7         1         2
  2         1         9         1         2
  1         1         5         2         1
  2         1         7         2         1
  3         1         9         2         1
  1         1         5         3         1
  3         1         7         3         1
  2         1         9         3         1
and so forth ...
;;
run;
proc nlmixed data=CountProfiles;
   parms i1=7.14  i2=9.900                         /* cutoffs                  */
         v1=1.57  v2=1.26 v3=0.08  v4=1.783        /* variety effects          */
         m=-2.85 m2=0.197                          /* month and month^2 slope  */
         mv1=-0.15 mv2=-0.17 mv3=0.08 mv4=-0.03    /* Variety spec. slopes */
         sd=1;              /* standard deviation of random plot errors */
```

```
    if variety=1      then linp = v1 + m*month + m2*month*month + mv1*month;
    else if variety=2 then linp = v2 + m*month + m2*month*month + mv2*month;
    else if variety=3 then linp = v3 + m*month + m2*month*month + mv3*month;
    else if variety=4 then linp = v4 + m*month + m2*month*month + mv4*month;
    else linp = m*month + m2*month*month;

    linp = linp + ploterror;

    /* Now build the category probabilities */
    if (rating=1) then do;
        catprob = 1/(1+exp(-i1-linp));
    end; else if (rating=2) then do;
        catprob = 1/(1+exp(-i2-linp)) - 1/(1+exp(-i1-linp));
    end; else catprob = 1- 1/(1+exp(-i2-linp));

    /* Now build the log-likelihood function */
    if (catprob > 1e-8) then ll=log(catprob); else ll=-1e100;
    model rating ~ general(ll);
    random ploterror ~ normal(0,sd*sd) subject=sub;
    replicate count;
run;
```

The block of `if .. else ..` statements sets up the linear predictor apart from the two cutoffs (parameters `i1` and `i2`) needed to model a three-category ordinal response and the random plot effect. The latter is added in the `linp = linp + ploterror;` statement. The second block of `if .. then .. else ..` statements calculates the category probabilities from which the multinomial log-likelihood is built. Should a category probability be very small a log-likelihood contribution of 10^{-100} is assigned to avoid computational inaccuracies when taking the logarithm of a quantity close to zero. The `random` statement models the plot errors as Gaussian random variables with mean zero and variance $\sigma^2 =$ `sd*sd`. In the vernacular of mixing models, this is a Multinomial/Gaussian model. The `replicate` statement identifies the variable in the data set which indicates the number of response profiles for a particular variety. This statement must not be confused with the `repeated` statement of the `mixed` procedure. As starting values for `proc nlmixed` the estimates from Output 8.5 were chosen. The starting value for the standard deviation of the random plot errors was guessed.

The `Dimensions` table shows that the data have been set up properly (Output 8.6). Although there are three observations in each response profile, the `replicate` statement uses only the last observation in each profile to determine the number of plots that have the particular profile. The number of clusters is correctly determined as 88 and the number of repeated measurements as three (`Max Obs Per Subject`). The adaptive quadrature determined that three quadrature points provided sufficient accuracy in the integration problem.

The procedure required thirty-four iterations until further updates did not provide an improvement in the log-likelihood. The -2 log-likelihood at convergence of 456.0 is considerably less than that of the independence model (483.1, Output 8.5). The difference of 27.1 is highly significant ($\Pr(\chi^2_1 > 27.1) < 0.0001$), an improvement over the independence model brought about only by the inclusion of the random plot errors.

From the 95% confidence bounds on the parameters it is seen that the linear and quadratic time effects (`m` and `m2`) are significant, their bounds do not include zero. The confidence interval for the standard deviation of the plot errors also does not include zero, supporting the finding obtained by the likelihood ratio test, that the inclusion of the random plot errors is a significant improvement of the model.

Output 8.6.

```
                         The NLMIXED Procedure

                            Specifications
     Data Set                                   WORK.COUNTPROFILES
     Dependent Variable                         rating
     Distribution for Dependent Variable        General
     Random Effects                             ploterror
     Distribution for Random Effects            Normal
     Subject Variable                           sub
     Replicate Variable                         count
     Optimization Technique                     Dual Quasi-Newton
     Integration Method                         Adaptive Gaussian Quadrature

                              Dimensions
                 Observations Used                        129
                 Observations Not Used                      0
                 Total Observations                       129
                 Subjects                                  88
                 Max Obs Per Subject                        3
                 Parameters                                13
                 Quadrature Points                          3

                              Parameters
     i1         i2          v1          v2          v3          v4          m
   7.14        9.9        1.57        1.26        0.08       1.783      -2.85
     m2        mv1         mv2         mv3         mv4          sd  NegLogLike
  0.197      -0.15       -0.17        0.08       -0.03           1   233.22016

                           Iteration History
     Iter      Calls    NegLogLike          Diff      MaxGrad          Slope
        1          5    233.216559      0.003601     35.97236       -213.078
        2          8    232.974984      0.241575      197.669       -0.05738
        3         10    232.018721      0.956263     11.48055       -2.90352
        :
       33         66    227.995816      0.000395     0.018264       -0.00075
       34         68    227.995816     1.656E-7     0.003131       -3.05E-7

              NOTE: GCONV convergence criterion satisfied.

                            Fit Statistics
                 -2 Log Likelihood                      456.0
                 AIC (smaller is better)                482.0
                 AICC (smaller is better)               485.2
                 BIC (smaller is better)                514.2

                          Parameter Estimates
                     Standard
Parameter Estimate      Error  DF t Value  Pr > |t|  Alpha     Lower    Upper

i1           9.0840    3.6818  87    2.47    0.0156   0.05    1.7660   16.402
i2          12.8787    3.7694  87    3.42    0.0010   0.05    5.3866   20.371
v1           2.2926    1.9212  87    1.19    0.2360   0.05   -1.5260    6.111
v2           1.8926    1.9490  87    0.97    0.3342   0.05   -1.9812    5.766
v3           0.1835    1.9142  87    0.10    0.9239   0.05   -3.6213    3.988
v4           2.2680    1.9586  87    1.16    0.2500   0.05   -1.6249    6.161
m           -3.6865    1.0743  87   -3.43    0.0009   0.05   -5.8217   -1.551
m2           0.2576   0.07556  87    3.41    0.0010   0.05    0.1074    0.408
mv1         -0.2561    0.2572  87   -1.00    0.3221   0.05   -0.7673    0.255
mv2         -0.2951    0.2631  87   -1.12    0.2650   0.05   -0.8179    0.228
mv3         0.08752    0.2562  87    0.34    0.7335   0.05   -0.4218    0.597
mv4        -0.03720    0.2595  87   -0.14    0.8864   0.05   -0.5530    0.479
sd           1.5680    0.2784  87    5.63    <.0001   0.05    1.0146    2.121
```

The confidence bounds for the varieties and the variety × month interaction terms include zero and on first glance one would conclude that there are no varietal effects at work. Because of the coding of the classification variables, `v1` for example, does not measure the intercept for variety 1, rather the difference between the intercepts of varieties 1 and 5. To test the significance of various effects we consider the model whose output is shown in Output 8.6 as the full model and fit various reduced versions of it (Table 8.5).

The likelihood ratio test statistics (Λ) and p-values represent comparisons to the full model [A]. Removing the variety × month interaction from the model does not significantly impair the fit ($p = 0.508$) but removing any other combination of effects in addition to the interaction does worsen the model. Based on these results one could adopt model [B] as the *new* full model. Since variety effects have not been removed by themselves, one can test their significance by comparing the -2 log likelihoods of models [B] and [E]. The p-value is calculated as

$$p = \Pr\left(\chi_4^2 > 471.7 - 459.3\right) = \Pr\left(\chi_4^2 > 12.4\right) = 0.014.$$

Variety effects are significant in model [B].

Table 8.5. -2 Log-likelihoods for various mixed models fitted to the repeated measures turf ratings (All models contain a random plot effect)

Model	Fixed Effects: included	Fixed Effects: dropped	$-2\log L$	df†	Λ	p
[A]	Variety, t, t^2, Variety × t	–	456.0	–		
[B]	Variety, t, t^2	Variety × t	459.3	4	3.3	0.508
[C]	Variety, t	Variety × t, t^2	472.8	5	16.8	0.005
[D]	Variety	Variety × t, t, t^2	475.7	6	19.7	0.003
[E]	t, t^2	Variety, Variety × t	471.7	8	15.7	0.046

†: df denotes the degrees of freedom dropped compared to the full model [A].

Output 8.7.

```
                         Parameter Estimates

                      Standard
Parameter Estimate       Error  DF t Value Pr > |t|  Alpha     Lower    Upper

i1            9.6333    3.5009  87    2.75   0.0072   0.05    2.6748   16.591
i2           13.3522    3.5946  87    3.71   0.0004   0.05    6.2075   20.497

v1            0.4999    0.6644  87    0.75   0.4538   0.05   -0.8206    1.821
v2           -0.1523    0.6778  87   -0.22   0.8227   0.05   -1.4995    1.195
v3            0.7909    0.6720  87    1.18   0.2424   0.05   -0.5448    2.127
v4            1.9887    0.6803  87    2.92   0.0044   0.05    0.6365    3.341

m            -3.7288    1.0575  87   -3.53   0.0007   0.05   -5.8307   -1.627
m2            0.2538   0.07494  87    3.39   0.0011   0.05    0.1048    0.403

sd            1.5205    0.2729  87    5.57   <.0001   0.05    0.9781    2.063
```

The model finally selected is B and its parameter estimates are shown in Output 8.7. From these estimates the variety specific probability distributions over time can be calculated (Figure 8.8). Perhaps surprisingly, the drop in low rating probabilities is less striking than appears in Table 8.4. Except for variety 4 the probability of receiving a low rating remains constant throughout the three-month period. Excellent ratings are most common around July but only for varieties 2 and 5 is excellent turf quality in that period more likely than medium quality.

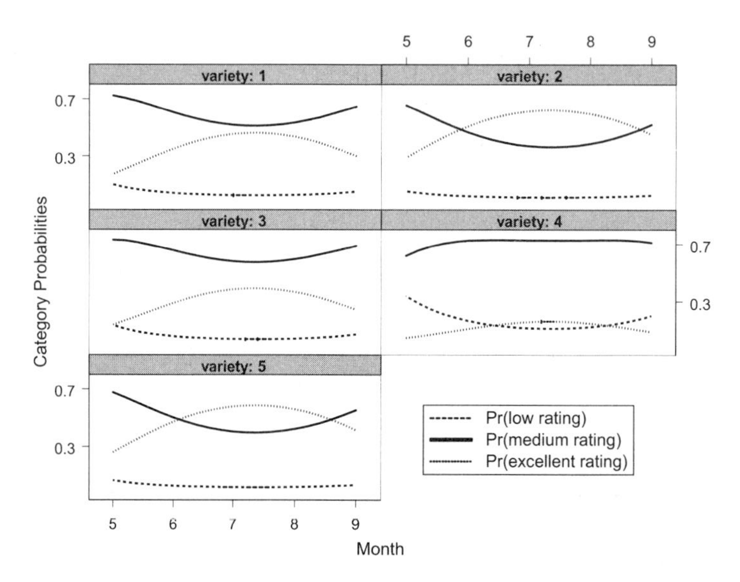

Figure 8.8. Change in category probabilities over time by varieties.

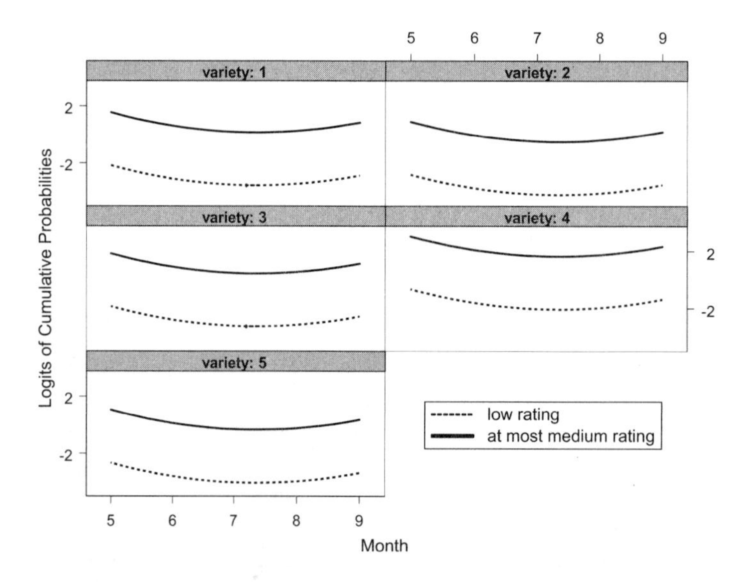

Figure 8.9. Logits of cumulative predicted probabilities.

The linear portion of the model is best interpreted on the logit scale (Figure 8.9). The final model contains varietal differences and linear and quadratic time effects. Especially variety 4 has elevated intercepts compared to the other entries. From Output 8.7 we see that the confidence interval $[0.6365, 3.341]$ for coefficient `v4` does not contain zero. Since variety 5 is the benchmark, this implies that varieties 4 and 5 are significantly different in the elevation of the lines in Figure 8.9. The statements to fit the selected model with `proc nlmixed` and to obtain pairwise comparisons of the variety effects (intercepts) follows below. Results of the pairwise comparisons are shown in Output 8.8. At the 5% significance level variety 4 is significantly different in the rating probability distributions from varieties 1, 2, and 5 ($p = 0.0284$, 0.0027, and 0.0044, respectively).

```
proc nlmixed data=CountProfiles;
   parms i1=5.024 i2=7.793 v1=1.61 v2=0.88 v3=0.337 v4=2.155
         m=-2.34 m2=0.168 sd=1;
   v5 = 0;
   array vv{5} v1-v5;

   linp = vv{variety};
   linp = linp + m*month + m2*month*month+ ploterror;

   if (rating=1)      then catprob= 1/(1+exp(-i1-linp));
   else if (rating=2) then catprob=1/(1+exp(-i2-linp))-1/(1+exp(-i1-linp));
   else                    catprob= 1- 1/(1+exp(-i2-linp));
   if (catprob > 1e-8) then ll=log(catprob); else ll=-1e100;
   model rating ~ general(ll);
   random ploterror ~ normal(0,sd*sd) subject=sub;
   replicate count;
   /* pairwise treatment comparisons */
   estimate 'v1-v2' v1-v2;
   estimate 'v1-v3' v1-v3;
   estimate 'v1-v4' v1-v4;
   estimate 'v1-v5' v1;
   estimate 'v2-v3' v2-v3;
   estimate 'v2-v4' v2-v4;
   estimate 'v2-v5' v2;
   estimate 'v3-v4' v3-v4;
   estimate 'v3-v5' v3;
   estimate 'v4-v5' v4;
run;
```

Output 8.8.

Additional Estimates

Label	Estimate	Standard Error	DF	t Value	Pr > |t|	Alpha	Lower	Upper
v1-v2	0.6522	0.6716	87	0.97	0.3342	0.05	-0.6826	1.9871
v1-v3	-0.2910	0.6630	87	-0.44	0.6618	0.05	-1.6087	1.0268
v1-v4	-1.4887	0.6680	87	-2.23	0.0284	0.05	-2.8164	-0.1610
v1-v5	0.4999	0.6644	87	0.75	0.4538	0.05	-0.8206	1.8205
v2-v3	-0.9432	0.6806	87	-1.39	0.1693	0.05	-2.2959	0.4095
v2-v4	-2.1410	0.6929	87	-3.09	0.0027	0.05	-3.5182	-0.7638
v2-v5	-0.1523	0.6778	87	-0.22	0.8227	0.05	-1.4995	1.1949
v3-v4	-1.1977	0.6698	87	-1.79	0.0772	0.05	-2.5291	0.1336
v3-v5	0.7909	0.6720	87	1.18	0.2424	0.05	-0.5448	2.1267
v4-v5	1.9887	0.6803	87	2.92	0.0044	0.05	0.6365	3.3409

Chapter 9

Statistical Models for Spatial Data

Space. The Frontier. Finally!

9.1 Changing the Mindset

9.1.1 Samples of Size One

We could call this section *Introduction to Statistical Models for Spatial Data*, but the entire chapter is a mere introduction (superficial, at best) to the topic of statistical analysis of spatial data. We will only scratch the surface of many important issues such as the analysis of lattice data, cokriging or the modeling of spatial point patterns. Some of the methods of spatial data analysis are discussed only by way of application in §9.8 without a detailed precursor. On the other hand this section is hopefully more than an introduction to what follows later in this chapter. What we are trying to achieve is a change in mindset, a way of looking at data that differs substantively from any of the viewpoints we have taken in previous chapters. This is necessary not only to convey the special standing statistical techniques for spatial data should be awarded in the research worker's toolbox but also to underline the differences in subject matter origin and mathematical-statistical content from the methods discussed so far. No other area of statistical endeavor promises to impact the plant and soil scientist's approach to data collection and analysis as spatial statistics does. And no other area requires tools that are further removed from the statistical topics to which students, scholars, and research workers have been traditionally exposed.

Observing spatial data entails the recording of an attribute of interest *and* the attribute location. Parting with notation used earlier we denote the attribute of interest being measured by Z and the location at which we observe this attribute as $\mathbf{s}$. A spatial observation is then denoted as $Z(\mathbf{s})$, the observation of attribute Z at location $\mathbf{s}$. The bold-faced vector notation $\mathbf{s}$ is used to emphasize that $\mathbf{s}$ typically contains multidimensional coordinates. The case we will consider throughout this chapter is where $\mathbf{s}$ is a point in $\mathbb{R}^2$, two-dimensional Euclidean space and the elements of $\mathbf{s}$ represent longitude and latitude in the plane. As an example consider yield monitoring a corn field which may give rise to $1,000$ spatially referenced observations. The data consist of $Z(\mathbf{s}_1), \cdots, Z(\mathbf{s}_{1000})$, where $Z(\mathbf{s}_i)$ denotes the corn yield at location $\mathbf{s}_i$. Should we think of these $1,000$ observations as a (random) sample of wheat yields of size $n = 1,000$? First we note that the sample locations were not chosen at random since the combine collects samples at systematic intervals. Second, it is (fairly) obvious that the $1,000$ observations cannot possibly be independent as a random sample would imply. If you were given the information that $Z(\mathbf{s}_5)$ is 135 bushels per acre, how surprised would you be to find out that the next observation, $Z(\mathbf{s}_6)$, collected only a few feet from $Z(\mathbf{s}_5)$, was 142 bushels per acre? We would not be surprised at all. In fact we would be surprised if $Z(\mathbf{s}_6) = 21$ bushels per acre. This phenomenon is sometimes referred to as Tobler's law of geography: "Everything is related to everything else, but near things are more related than distant things." Tobler's law of geography (Tobler 1970) instructs that we should expect relationships between spatially distributed quantities and that the strength of the relationships is a function of their spatial separation. In the sequel we will define and model numerous mathematical forms of this sentiment. But there is a deeper issue to be considered here.

When the biomass of a random sample of fifty plants is observed, fifty realizations from a univariate distribution, the distribution of plant biomass, are obtained. If we measure not only the total plant biomass but the above- and below-ground biomass we obtain fifty realizations from a bivariate distribution. The below- and above-ground biomass of a single plant are a sample of size one from this bivariate distribution. If Y_1 denotes above-ground and Y_2 below-ground biomass, the realized value of this single observation may be $\mathbf{y} = [1.6, 0.5]'$ (in appropriate units).

a) Univariate Random Variable

$$Y \sim G(0,1)$$

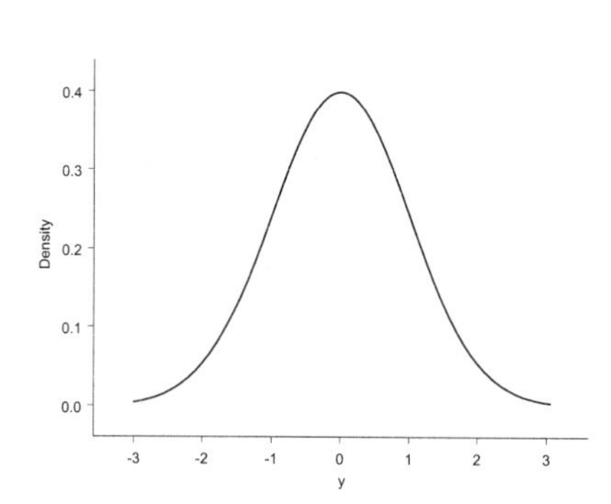

A single realization: $y = 1.43$

b) Bivariate Random Variable

$$\begin{bmatrix} Y_1 \\ Y_2 \end{bmatrix} \sim G\left(\begin{bmatrix} 0 \\ 0 \end{bmatrix}, \begin{bmatrix} 1 & 0.3 \\ 0.3 & 1 \end{bmatrix} \right)$$

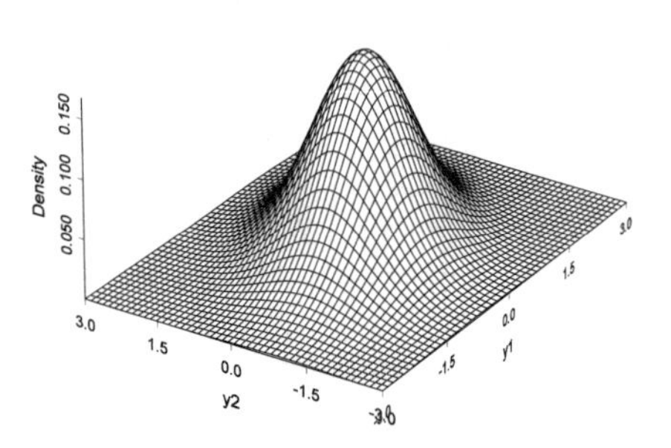

A single realization: $\begin{bmatrix} y_1 \\ y_2 \end{bmatrix} = \begin{bmatrix} 1.6 \\ 0.5 \end{bmatrix}$

c) A Spatial Random Field

$$\{Z(\mathbf{s}) : \mathbf{s} \in D \subset \mathbb{R}^2\}$$

A single realization:

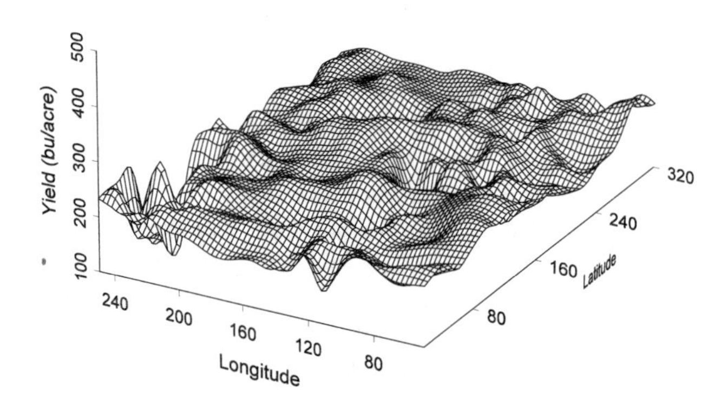

Figure 9.1. Univariate (a) and bivariate (b) random variables and the realization of a stochastic process in $\mathbb{R}^2$ (c). In panels (a) and (b), the graph represents the distribution (process) from which a single realization is drawn. In panel (c) the graph **is** the realization.

What is obtained by measuring crop yield at 200 locations in a wheat field or by recording the locations of a group of red cockaded woodpeckers during the month of July or the locations of craters on the surface of the moon or the proportion of farmers employing site-specific management by county in Virginia? The answer, perhaps surprisingly, is: a *single* realization of an n-dimensional random variable. Notice the difference between one realization of above- and below-ground biomass of a plant and measuring lime requirement at two spatial locations. In the former case two different attributes are observed, whereas in the latter the same attribute is measured at different locations.

To emphasize this viewpoint of spatial data recall the notion of clustering in data from §2.4 and §2.6. Assume data consists of k clusters of size n_i so that the total number of observations is $\sum_{i=1}^{k} n_i = n$. We make the implicit assumptions that observations from different clusters are uncorrelated but that observations from the same cluster may be correlated, for example, because they represent repeated measures. The response vector $\mathbf{Y}_i$ for the ith cluster is an $(n_i \times 1)$ vector and the observed response $\mathbf{y}_i$ is a single realization from an n_i-dimensional distribution. The case of unclustered data is a special case of this structure with $n_i = 1$ and the number of observations is equal to the number of clusters. In that case, y_i is a single realization from a univariate (1-dimensional) distribution. Spatial data is also a special case of a clustered structure ($k = 1$). It represents a single cluster, so to speak.

Figure 9.1 puts these notions into perspective. The graphs shown in panels a and b represent the particular population or distribution from which individual realizations are drawn. In the case of spatial data (panel c), the figure represents the realization itself (the draw) that is obtained. To summarize, spatial observations $Z(\mathbf{s}_1), \cdots, Z(\mathbf{s}_n)$ are not the same variable Z observed n times over, but the variables $Z(\mathbf{s}_1), Z(\mathbf{s}_2), \cdots, Z(\mathbf{s}_n)$ observed once. But what kind of distribution or process do we draw from to generate the realization in Figure 9.1c?

9.1.2 Random Functions and Random Fields

Box 9.1 Random Fields

- **A set of spatial data is considered a realization of a random experiment. For any outcome ω of the experiment, a single realization of $Z(\mathbf{s})$ is obtained. This is the realization of a random field, a stochastic process.**
- **$Z(\mathbf{s}_0)$ is a random variable by considering the distribution of all possible realizations at the location $\mathbf{s}_0$.**
- **When a random field is sampled, samples are drawn from one particular realization of the random experiment.**

Consider $Z(\mathbf{s})$ as a function of the spatial coordinates $\mathbf{s}$ that are elements of a set D, which we call the domain. For now assume that D is a continuous set. To incorporate stochastic behavior (randomness), $Z(\mathbf{s})$ is considered the realization of a random experiment. To make the dependence on the random experiment explicit we use the notation $Z(\mathbf{s}, \omega)$ for the time being, where ω is the outcome of a particular experiment. Hence, we are really concerned with a function of two variables, $\mathbf{s}$ and ω. This is called a random function because the surface obtained depends on a random experiment. Figure 9.2 shows four realizations of a random function where the domain is the rectangle $(0, 5) \times (0, 5)$. Imagine a soil sample is poured from a bucket onto a flat surface and the depth of the soil is measured. This may be the realization in the upper left panel of Figure 9.2. Now put the soil back in the bucket and pour it again. This produces the realization in the upper right panel of the figure and so forth. Every pouring ω constitutes a random experiment, the result is a function $Z(\,\cdot\,, \omega)$.

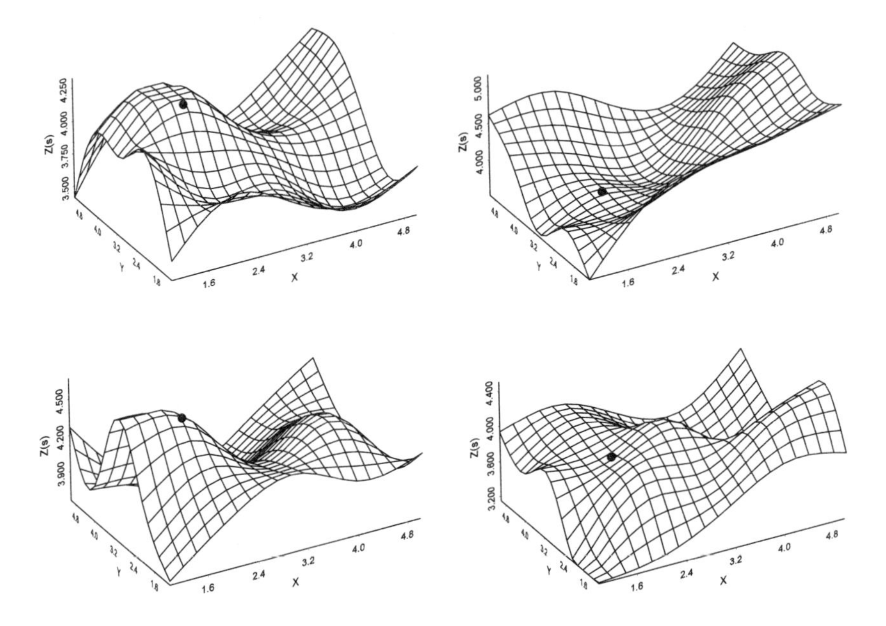

Figure 9.2. Four realizations of a random function in $\mathbb{R}^2$. The domain D is continuous and given by the rectangle $(0, 5) \times (0, 5)$. Realizations at $\mathbf{s}_0 = [2, 3]$ are shown as dots.

The mechanism by which we consider the attribute at a particular location $\mathbf{s}_0$ to be a random variable is to imagine all possible realizations (all possible outcomes ω) of the random experiment at that location. In Figure 9.2 four realizations at $\mathbf{s}_0 = [2, 3]$ are shown as dots. The distribution of the attribute at $\mathbf{s}_0$ over all possible realizations of the random function is the probability distribution of the random variable $Z(\mathbf{s}_0, \,\cdot\,)$.

In what follows we will ignore the explicit dependence of the random function on the random experiment ω and simply refer to $Z(\mathbf{s})$, the attribute at location $\mathbf{s}$, which is a random variable by this mechanism. It is important to note, however, that the randomness of $Z(\mathbf{s})$ stems from a super-population model that is alluded to in Figure 9.2. This has several important ramifications. Whether we sample the spatial attribute by randomly placing sample locations in D or with a systematic grid has no bearing on the randomness of $Z(\mathbf{s})$. A spatial attribute is not considered random because we performed random sampling. Even if we observed $Z(\mathbf{s})$ everywhere in D with an exhaustive sampling procedure (which is impossible if D is continuous), we would be assessing only one realization of the random function $Z(\mathbf{s}, \omega)$. The sample is drawn from a single panel of Figure 9.2. Again, this underlines the notion that a sample $Z(\mathbf{s}_1), \cdots, Z(\mathbf{s}_n)$ is a sample of size one.

In mathematical statistics random functions are known as **stochastic processes**. Those processes where D is two- or more-dimensional are also called **random fields**. This nomenclature has nothing to do with the agricultural notion of a field, although the theory of random fields is fruitfully applied there. The upper-case/lower-case distinction of random variables that we have maintained so far is somewhat difficult to uphold for random functions without making notation too cumbersome. What do we mean by $Z(\mathbf{s})$? The random realizations of the function or the random variable at location $\mathbf{s}$? Similarly, does $z(\mathbf{s})$ represent the realization at location $\mathbf{s}$ or the realization of the function itself? In what follows we suppress the explicit dependence of the random function on ω and use notation that is common in the stochastic process literature:

$$\left\{Z(\mathbf{s}) : \mathbf{s} \in D \subset \mathbb{R}^2\right\} \qquad [9.1]$$

denotes a spatial random field with a two-dimensional domain. The attribute of interest, Z, is a stochastic process with domain (or index set) D which itself is a subset of $\mathbb{R}^2$. When we have in mind the random variable at $\mathbf{s}$, we use $Z(\mathbf{s})$ and denote its realization as $z(\mathbf{s})$. The vector of all observations is denoted $\mathbf{Z}(\mathbf{s}) = [Z(\mathbf{s}_1), \cdots, Z(\mathbf{s}_n)]'$. Definition [9.1] is quite abstract but it can be fleshed out by considering various types of spatial data.

9.1.3 Types of Spatial Data

Box 9.2 Spatial Data Types

- **Three categories of spatial data are distinguished: geostatistical data, lattice data, and spatial point patterns.**
- **The domain D is fixed and continuous for geostatistical data, fixed and discrete for lattice data, and a random set for point data.**
- **The scientific questions raised differ substantially among the three data types and specific tools have been developed to address these questions. Some of the tools are transitive in that they can be applied to any of the data types, others are particular to a specific spatial data structure.**

Many practitioners associate with spatial data analysis terms like geostatistics and methods such as kriging. **Geostatistical data** is only one of many spatial data types which can be defined through the domain D of the random field [9.1]. In the case of geostatistical data the domain is a fixed, continuous set; the number of locations at which observations can be made is not countable. Between any two sample locations $\mathbf{s}_i$ and $\mathbf{s}_j$ an infinite number of additional samples can be placed in theory. Furthermore, there is nothing random about the locations themselves. Examples of geostatistical data are measuring the electrical conductivity of soil, yield monitoring a field, and sampling the ore grade of a rock formation. Figure 9.3 (left panel) shows 72 locations on a shooting range at which the lead concentration was measured. The shooter location is at coordinate $x = 100$, $y = 0$.

Because of the continuity of D, geostatistical data is also referred to as spatial data with continuous variation. This does not imply that the attribute Z is continuous. The nature of the attribute Z as discrete or continuous does not alter the nature of the spatial data type. Whether one is interested in the presence/absence of a microbial species in a series of soil samples (Z is binary), the soil pH (Z is continuous), or the number of macropores (Z is a count), the data are geostatistical unless there is only a countable number of sample locations and/or the domain changes from realization to realization of the random function at random. Whether data are collected on regular grids, irregular grids, or by random sampling of locations also has no bearing on the nature of the spatial data type. The continuity of the fixed domain D is what matters, not how it is sampled. Figure 9.3 (right panel) shows the locations at which wheat was sampled for determination of the deoxinyvalenol concentration in a Michigan field in 2000. The basic layout is systematic, consisting of four transects with equally spaced sample intervals along the transects. At every other transect location a cluster of samples is collected by branching 2, 6, and 8 feet perpendicular to the transect direction. Since samples could have been placed anywhere within the field, the data are geostatistical.

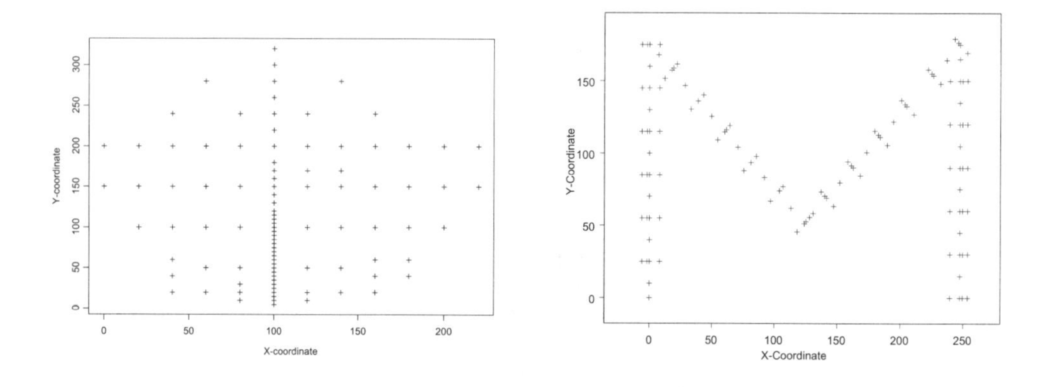

Figure 9.3. Left panel: sampling locations on shooting range at which lead concentrations were measured. Right panel: Sample grid for collecting wheat kernels in a field with wheat scab. Lead data kindly provided by Dr. James R. Craig and Dr. Donald Rimstidt, Dept. of Geological Sciences, Virginia Polytechnic Institute and State University.

Data collected in a systematic sample scheme such as a grid, unaligned grid, or a transect-type sample is sometimes incorrectly termed *lattice data*. Lattice data is a spatial data type where D is a fixed and discrete (and hence countable) set of locations. The number of locations at which measurements can be made might be infinite with lattice data; the key is that the possible sample locations can be enumerated. Examples are attributes recorded by county, city block, or census tract and information obtained from pixel images. When data are collected by census tract or county, there is no space defined between these discrete units at which observations can be made. Whether the units are aligned and shaped regularly as pixels or experimental units in a field experiment, or irregularly shaped such as counties, does not matter for the classification as lattice data. An example of lattice data is shown in Figure 9.4, which depicts the number of sudden infant deaths in North Carolina between 1974 and 1978 relative to the number of live births. This famous data set appears in Cressie (1993) and is analyzed in great depth there. A case of lattice data of particular interest to agronomists arises as data from planned field experiments. The discrete spatial units are the experimental units which are typically arranged in some regular fashion and equally sized (Figure 9.5).

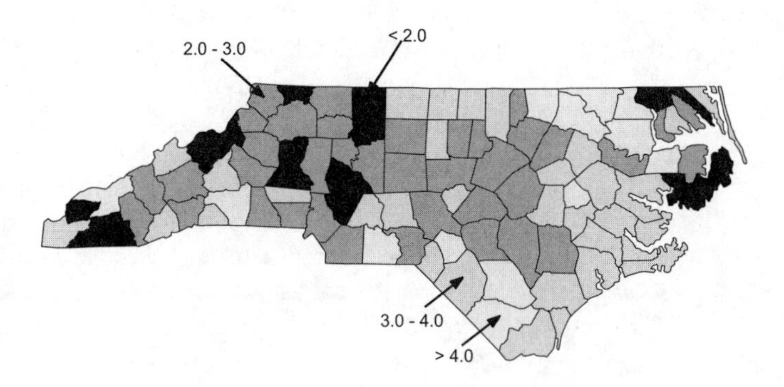

Figure 9.4. Sudden infant deaths (SIDs) in North Carolina 1974-1978. Shown are the number of SIDs relative to the number of live births. These data are included with S+SpatialStats®.

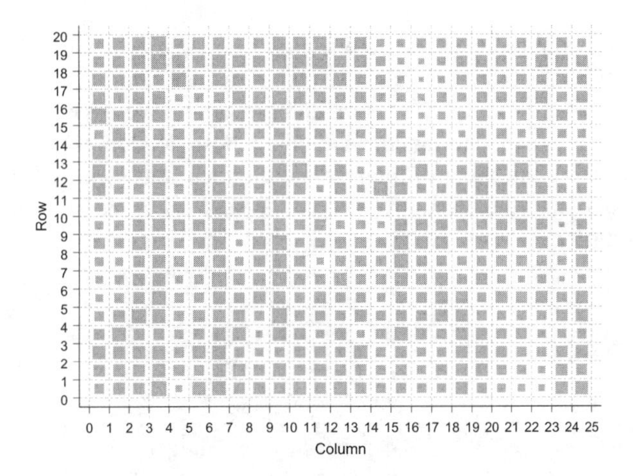

Figure 9.5. Grain yield data of Mercer and Hall (1911) at Rothamsted Experimental Station. The area of the squares in each lattice cell is proportional to the grain yield. From Table 6.1 of Andrews and Herzberg (1985). Used with permission.

Geostatistical and lattice data have in common that the domain is fixed, not random. To develop the idea of a random domain, consider $Z(\mathbf{s})$ to be crop yield and define an indicator variable $U(\mathbf{s})$ which takes on the value 1 if the yield is below some threshold level and 0 otherwise,

$$U(\mathbf{s}) = \begin{cases} 1 & Z(\mathbf{s}) < c \\ 0 & \text{otherwise.} \end{cases}$$

If $Z(\mathbf{s})$ is geostatistical, so is $U(\mathbf{s})$. The random function $U(\mathbf{s})$ now returns the values 1 and 0 instead of the continuous attribute yield.

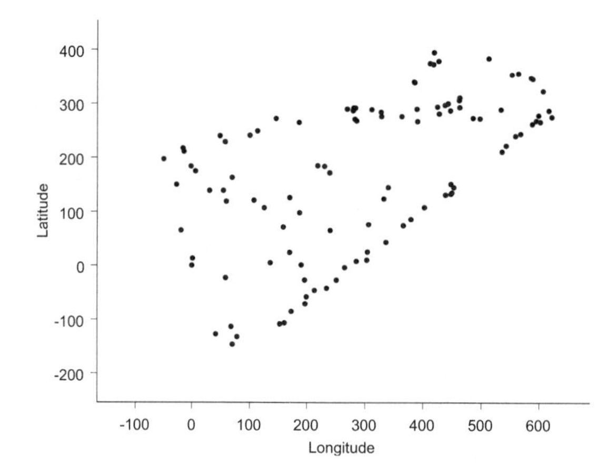

Figure 9.6. Locations in a corn field where yield per unit area is less than some threshold value.

Now imagine throwing away all the points where the yield is above threshold and retaining only those locations for which $U(\mathbf{s}) = 1$. Define a new domain D^* which consists of those points where $Z(\mathbf{s}) < c$. Since $Z(\mathbf{s})$ is random, so is D^*. We have replaced the attribute $Z(\mathbf{s})$ (and $U(\mathbf{s})$) with a degenerate random variable whose domain D^* consists of the locations at which we observe the event of interest and the focus has switched from studying the attribute itself to studying the locations (Figure 9.6). Such processes are termed **point processes** or **point patterns**.

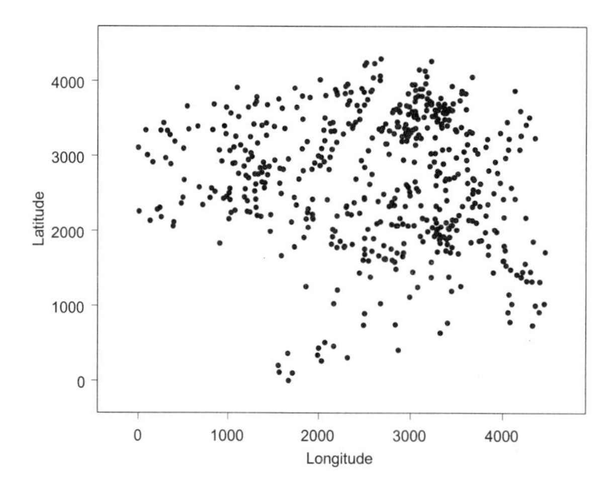

Figure 9.7. Spatial distribution of a group of red cockaded woodpeckers in the Fort Bragg area of North Carolina. Data kindly made available by Dr. Jeff Walters, Dept. of Biology, Virginia Polytechnic Institute and State University. Used with permission.

Most point patterns do not arise through this transformation of an underlying random function, they are observed directly. The emergence of plants, the distribution of seeds, the location of macropores in soil, the location of scab-infected wheat kernels on a truck load of kernels are examples of this spatial data type (Figure 9.7). With geostatistical and lattice data, statistical modeling focuses on the $Z()$ process (since D is not random), whereas with point data, we focus on the D process.

The distinction between the three types of spatial data is not always clear-cut. Geostatistical data can be converted into lattice data by integrating the data over finite regions. By changing the conceptual index set D, one can move from lattice data to geostatistical data (consider D continuous, not discrete). Aggregating a point pattern over regions results in lattice data. Lattice data, in some sense, are not as refined as geostatistical or point data, since they can be obtained by reduction of the other spatial data types (integration of geostatistical data or enumerating events in a point pattern). The questions of interest can vary greatly among the three data types, however, and many statistical tools are specific to a particular data type.

With geostatistical data a frequent application is to produce continuous surfaces (maps) of the attribute Z based on samples taken at some locations. The kriging methods developed for this purpose (§9.4) predict $Z(\mathbf{s}_0)$ at some unsampled location $\mathbf{s}_0$ based on the sample $Z(\mathbf{s}_1), \cdots, Z(\mathbf{s}_n)$. For a mapped point pattern where all events within an area of interest have been located, the issue of predicting at unobserved locations never arises. There are no unobserved locations by definition. Point pattern analyses usually commence by raising the elementary question, "Are the events (points) distributed completely at random," and if they are not whether the events are more aggregated or more regularly distributed as expected under a completely random placement. One approach to addressing this question is to study the distribution of nearest-neighbor distances (§9.7.3) in the observed point pattern. For geostatistical data examining the distance between a sample point and its nearest neighbor is nonsensical since the sample locations are placed according to a sample design. The issue of where events are located never surfaces with geostatistical or lattice data.

The various methods for modeling and analyzing geostatistical, lattice, and point data have in common that they rely heavily on stochastic properties of the spatial random field. If what we observe must be considered a sample of size one, how can we possibly learn anything about the variation of $Z(\mathbf{s})$ or the covariances among $Z(\mathbf{s}_i)$ and $Z(\mathbf{s}_j)$, for example? Certain assumptions, such as having obtained a random sample, are made frequently in classical statistics since they provide the underpinnings of the stochastic structure that enables analysis.

If $Y_1, \cdots, Y_n$ are a random sample from a population with mean μ and variance σ^2, for example, then we automatically know that $\overline{Y}$ and S^2 are unbiased for μ and σ^2, respectively. Random sampling provides the needed replication mechanism without which estimation of variation is difficult. Similar assumptions are made with spatial data but rather than targeting the random mechanism by which the observations are obtained they focus on the (internal) stochastic structure of the random field. Rather than replication of the data, we are looking for replication in the data. These assumptions are summarized under the headings **stationarity** and **isotropy**.

9.1.4 Stationarity and Isotropy — the Built-in Replication Mechanism of Random Fields

Box 9.3 Stationarity and Isotropy

- **Stationarity is a property of self-replication of a stochastic process. It implies the lack of importance of absolute coordinates.**
- **The three important varieties of stationarity are strict, second-order, and intrinsic stationarity.**
- **Whereas in a stationary random field absolute coordinate differences are immaterial, the orientation (angle) of coordinate differences matters. Stationary random fields in which the orientation of coordinate differences is not of consequence are called isotropic.**

Stationarity in simple terms means that the random field looks similar in different parts of the domain D, it replicates itself. Consider two observations at $Z(\mathbf{s})$ and $Z(\mathbf{s}+\mathbf{h})$. The vector $\mathbf{h}$ is a displacement by which we move from location $\mathbf{s}$ to location $\mathbf{u} = \mathbf{s} + \mathbf{h}$; it is referred to as the lag-vector (or lag for short). If the random field is *self-replicating*, the stochastic properties of $Z(\mathbf{s})$ and $Z(\mathbf{s}+\mathbf{h})$ should be similar. For example, to estimate the covariance between locations distance $\mathbf{h}$ apart, we might consider all pairs $(Z(\mathbf{s}_i), Z(\mathbf{s}_i+\mathbf{h}))$ in the estimation process, regardless of where $\mathbf{s}_i$ is located. Stationarity is the absence of an origin, the spatial process has reached a state of equilibrium. Stationarity assumptions are also made for time series data. There it means that it does not matter when a time shift is considered in terms of absolute time, only how large the time shift is. In a stationary time series one can talk about a difference of two days without worrying that the first occasion was a Saturday. In the spatial context stationarity means the lack of importance of absolute coordinates. There are different degrees of stationarity, however, and before we can make the various definitions more precise a few comments are in order.

Since $Z(\mathbf{s})$ is a random variable, it has moments and a distribution. For example, $\mathrm{E}[Z(\mathbf{s})] = \mu(\mathbf{s})$ is the mean of $Z(\mathbf{s})$ at location $\mathbf{s}$ and $\mathrm{Var}[Z(\mathbf{s})]$ is its variance. Analysts and practitioners often refer to Gaussian random fields, but some care is necessary to be clear about what is assumed to follow the Gaussian law. A Gaussian random field is defined as a random function whose finite-dimensional distributions are multivariate Gaussian. That is, the cumulative distribution function

$$\Pr(Z(\mathbf{s}_1) < z_1, \cdots, Z(\mathbf{s}_k) < z_k) \qquad [9.2]$$

is that of a k-variate Gaussian distribution for all k. By the properties of the multivariate Gaussian distribution (§3.7) this implies that any $Z(\mathbf{s}_i)$ is a univariate Gaussian random variable. The reverse is not true. If $Z(\mathbf{s}_i)$ is Gaussian does not imply that [9.2] is a multivariate Gaussian probability. Chilès and Delfiner (1999, p. 17) point out that this leap of faith is sometimes made. The spatial distribution of $Z(\mathbf{s})$ is defined by the multivariate cumulative distribution function [9.2], not the marginal distribution of $Z(\mathbf{s})$.

The first, and most restrictive, definition of stationarity is **strong** (or strict) **stationarity**. It implies that

$$\Pr(Z(\mathbf{s}_1) < z_1, \cdots, Z(\mathbf{s}_k) < z_k) = \Pr(Z(\mathbf{s}_1 + \mathrm{h}) < z_1, \cdots, Z(\mathbf{s}_k + \mathrm{h}) < z_k), \quad [9.3]$$

meaning that the spatial distribution is invariant under translation of the coordinates by the vector **h**. The random field repeats itself throughout the domain. Geometrically, this implies that the spatial distribution is invariant under rotating and stretching of the coordinate system. As the name suggests, strong stationarity is a very strict condition, more restrictive than what is required for many of the statistical methods that follow. Two important versions of stationarity, second-order (weak) and intrinsic stationarity are defined in terms of moments of $Z(\mathbf{s})$. A random field is **second-order stationary** if $\mathrm{E}[Z(\mathbf{s})] = \mu$ and $\mathrm{Cov}[Z(\mathbf{s}), Z(\mathbf{s} + \mathbf{h})] = C^*(\mathbf{h})$. The first assumption states that the mean of the random field is constant and does not depend on locations. The second assumption states that the covariance between two observations is only a function of their spatial separation. The function $C^*(\mathbf{h})$ is called the **covariance function** or the **covariogram** of the spatial process. If a random field is strictly stationary it is also second-order stationary, but the reverse is not necessarily true. The reasons are similar to those that disallow inferring the distribution of a random variable from its mean and variance alone. An exception is the Gaussian random variable where zero covariance does imply independence. Similarly for random fields. If a Gaussian random field is second-order stationary it is also strictly stationary.

Imagine that we wish to estimate the covariance function $C^*(\mathbf{h})$ in a second-order stationary process. For a lag vector $\mathbf{h} = [-35.35, 35.35]'$ all pairs of points that are exactly distance **h** apart can be utilized (Figure 9.8).

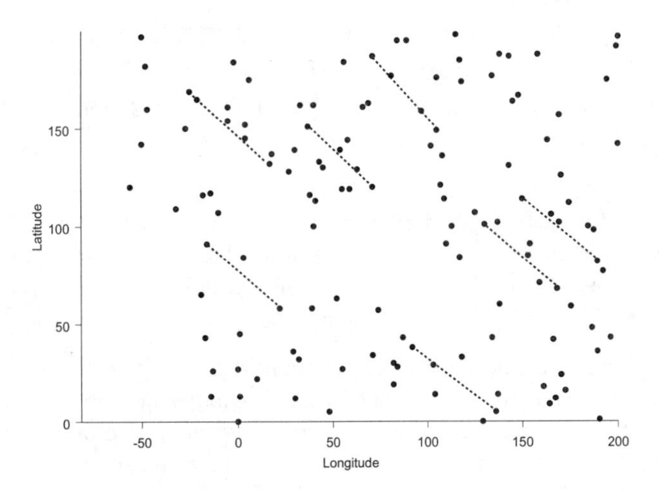

Figure 9.8. The notion of second-order stationarity. Pairs of points separated by the same lag vector (here, $\mathbf{h} = [-35.35, 35.35]'$) provide built-in replication to assess the spatial dependency for the particular choice of **h**.

While stationarity reflects the lack of importance of absolute coordinates, the direction in which the lag vector **h** is assessed still plays an important role. We could not combine the pairs of observations in Figure 9.8 whose lag vector is $\mathbf{h} = [-35.35, -35.35]'$, they are oriented perpendicular to the rays shown in the figure. The condition by which the random field is also invariant under rotation and reflection, is known as **isotropy**. In a second-order stationary random field with isotropic covariogram the covariance between any two points $Z(\mathbf{s})$ and $Z(\mathbf{s}+\mathbf{h})$ is only a function of the Euclidean distance $||\mathbf{h}||$ between the two points,

$$\text{Cov}[Z(\mathbf{s}), Z(\mathbf{s}+\mathbf{h})] = C(||\mathbf{h}||).$$

The Euclidean distance is defined as follows. Let $\mathbf{s} = [x, y]'$ where x and y are the longitude and latitude, respectively. The Euclidean distance (Figure 9.9) between $\mathbf{s}_1$ and $\mathbf{s}_2$ is then $||\mathbf{s}_1 - \mathbf{s}_2|| = \sqrt{(x_1 - x_2)^2 + (y_1 - y_2)^2}$. Random fields that are stationary but not isotropic are called anisotropic.

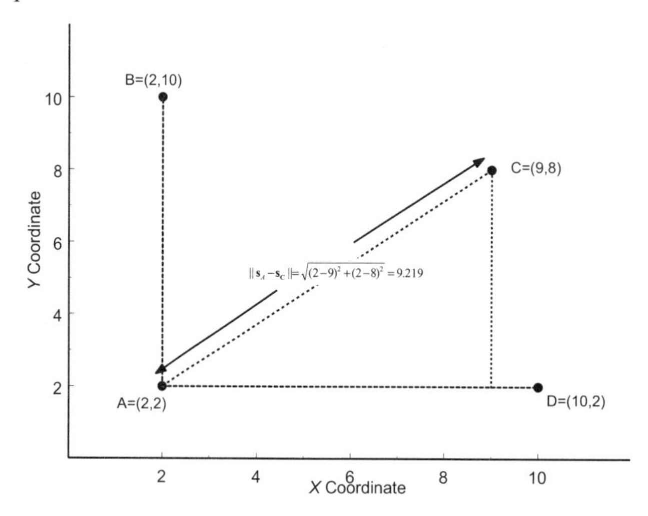

Figure 9.9. Euclidean distance between points $A = (x = 2, y = 2)$ and $C = (9, 8)$. The Euclidean distance between A and C and between A and D is 8.

Note that we have distinguished between the covariogram $C^*(\mathbf{h})$ of the second-order stationary random field and the covariogram $C(||\mathbf{h}||)$ which is also isotropic, since these are different functions. In the sequel we will often refer to only $C()$ and whether the function depends on $\mathbf{h}$ or on $||\mathbf{h}||$ is sufficient to distinguish the general from the isotropic case.

In a process with isotropic covariance function it does not matter how the lag vector between pairs of points is oriented, only that the Euclidean distance between pairs of points is the same (Figure 9.10). To visualize the difference between second-order stationary random fields with isotropic and anisotropic covariance function, realizations were simulated with `proc sim2d` in The SAS® System. The statements

```
proc sim2d outsim=RandomFields;
   grid x=1 to 10 by 1  y=1 to 10 by 1;
   simulate numreal=1 angle=90 range=5 scale=0.75 ratio=0.3 form=gaussian;
   simulate numreal=1 angle= 0 range=5 scale=0.75 ratio=1   form=gaussian;
run;
```

generate two sets of spatial data on a 10×10 grid. The first `simulate` statement generates a random field in which data points are correlated up to Euclidean distance $5\sqrt{3} = 8.66$ in the East-West direction, but correlated up to a much smaller distance in the North-South direction (the genesis of this value called the spatial range is discussed in §9.2.1). The second `simulate` statement generates a random field with isotropic covariance, i.e., the spatial dependence develops similarly in all directions (Figure 9.11).

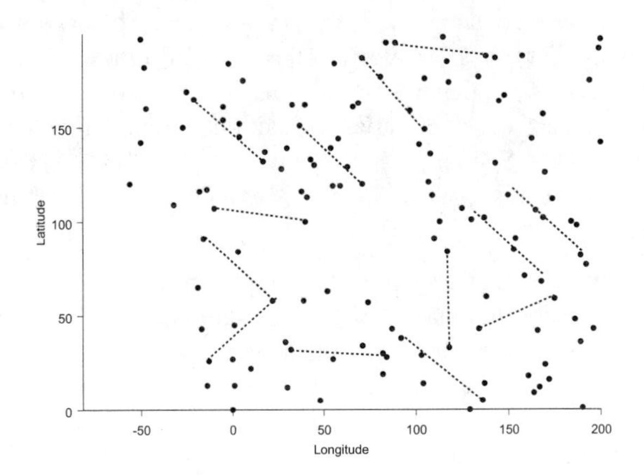

Figure 9.10. The notion of isotropy. Pairs of points separated by the same Euclidean distance (here $||\mathbf{h}|| = 50$) provide built-in replication to assess spatial dependency at that lag. Orientation of the distance vector is immaterial.

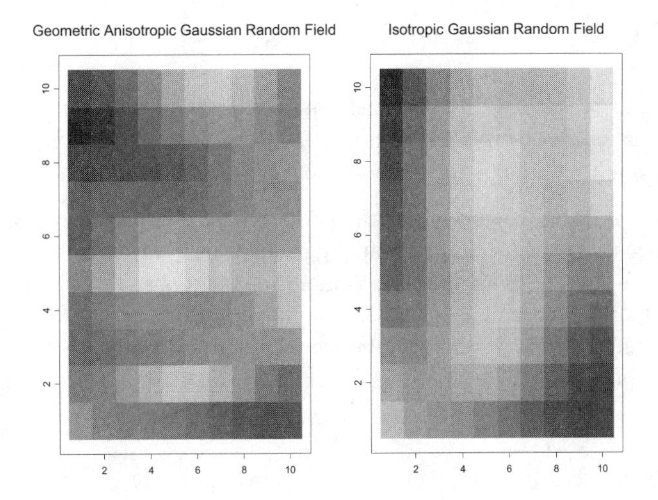

Figure 9.11. Anisotropic (left) and isotropic (right), stationary Gaussian random fields.

The anisotropic field changes its values in the East-West direction more slowly than in the North-South direction. For the isotropic random field the spatial dependency between data points develops in the same fashion in all directions. If one were to estimate the covariance between two points $||\mathbf{h}|| = 3$ distance units apart, for example, the covariance must be estimated separately in the North-South and the East-West direction in the anisotropic case. In the isotropic case any pair of points will do provided their Euclidean distance is 3 units, regardless of the orientation of the points.

From the existence of the covariogram in a second-order stationary random field we can derive an interesting property. Since $\mathrm{Cov}[Z(\mathbf{s}), Z(\mathbf{s}+\mathbf{h})] = C(\mathbf{h})$ does not depend on absolute coordinates and $\mathrm{Cov}[Z(\mathbf{s}), Z(\mathbf{s})] = \mathrm{Var}[Z(\mathbf{s})] = C(\mathbf{0})$, it follows that the variance of the attribute is constant and does not depend on location. A second-order stationary spatial process thus has a constant mean, constant variance, and a covariance function that does not depend on absolute coordinates. Such a process can be thought of as the spatial equivalent of a random sample in classical statistic which gives rise to independent random variables with the same mean and dispersion.

In time series analysis, stationarity is just as important as with spatial data. A frequent device employed to turn a nonstationary series into a stationary one is differencing. Let $Y(t)$ denote an observation in a time series at time t and consider the random walk $Y(t) = Y(t-1) + e(t)$ where the $e(t)$ are independent random variables with mean 0 and variance σ^2. It is easy to show that the random walk is not second-order stationary. We have $\mathrm{E}[Y(t)] = \mathrm{E}[Y(t-k)]$ but the variance is not constant ($\mathrm{Var}[Y(t)] = t\sigma^2$) and the covariance does depend on the origin, $\mathrm{Cov}[Y(t), Y(t-k)] = (t-k)\sigma^2$. While $Y(t)$ is not stationary, the first differences $Y(t) - Y(t-1)$ are second-order stationary. A similar device is used in spatial statistics when the increments $Z(\mathbf{s}) - Z(\mathbf{s}+\mathbf{h})$ are second-order stationary. This form of stationarity is called **intrinsic stationarity**. It is usually defined as follows: if $\mathrm{E}[Z(\mathbf{s})] = \mu$ and

$$\frac{1}{2}\mathrm{Var}[Z(\mathbf{s}) - Z(\mathbf{s}+\mathbf{h})] = \gamma(\mathbf{h}), \qquad [9.4]$$

then $Z(\mathbf{s})$ is said to be intrinsically stationary or said to satisfy the intrinsic hypothesis. The function $\gamma(\mathbf{h})$ is called the **semivariogram** of $Z(\mathbf{s})$. If the semivariogram is a function of the Euclidean distance $||\mathbf{h}||$, then it is furthermore isotropic.

The semivariogram and covariogram are parameters of the spatial process and play a critical role in the geostatistical method of spatial data analysis. Statisticians are used to working with covariances, while the semivariogram is more frequently used in the geostatistical literature. Both are important ingredients of the kriging methods for spatial prediction. In a second-order stationary random field a simple relationship between $\gamma(\mathbf{h})$ and $C(\mathbf{h})$ can be used to derive one from the other,

$$\gamma(\mathbf{h}) = C(\mathbf{0}) - C(\mathbf{h}), \qquad [9.5]$$

and kriging predictors can be written in terms of semivariances or covariances. However, if a process is intrinsically stationary it may not be second-order stationary, the class of intrinsic processes is larger. Care should be exercised when calculating $C(\mathbf{h})$ as $C(\mathbf{0}) - \gamma(\mathbf{h})$. If the process is intrinsic but not second-order stationary, $C(\mathbf{h})$ is not a parameter.

We want to emphasize that it is because of the factor ½ that $\gamma(\mathbf{h})$ is termed the **semi**-variogram and $2\gamma(\mathbf{h})$ is termed the variogram. Statistical packages and the spatial statistics literature are not consistent in this terminology. The variogram procedure in The SAS® System, for example, estimates the semivariogram but denotes it as the variogram on the output. Kaluzny et al. (1998, p. 68) acknowledge in the S+SpatialStats® user manual that $\gamma(\mathbf{h})$ is the semivariogram but refer to it in the manual as the variogram for *conciseness*. Chilès and

Delfiner (1999, p. 32) acknowledge that $\gamma(\mathbf{h})$ is "also called" the semivariogram but refer to $\gamma(\mathbf{h})$ as the variogram for simplicity and because this terminology "tends to become established." Assume a second-order random field whose isotropic covariogram approaches 0 as $||\mathbf{h}|| \to \infty$. Then $\gamma(||\mathbf{h}||)$ approaches the constant $C(\mathbf{0})$ which is the variance of an observation. A graph of the **semi**variogram will thus provide a simple estimate of $\text{Var}[Z(\mathbf{s})]$ as the asymptote of the semivariogram. The asymptote of the variogram estimates twice the variance of an observation and there is nothing concise, simple, or established about missing by a factor of 2. We refer to $\gamma(\mathbf{h})$ as the semivariogram and to $2\gamma(\mathbf{h})$ as the variogram throughout.

The semivariogram is a structural tool that can convey a great deal about the nature and structure of spatial dependency in a random field. It is also a parameter of the process that must be estimated from the data. Estimating a semivariogram is usually a two-step process: (i) derive an empirical estimate of the semivariogram from the data and (ii) fit a theoretical semivariogram model to the empirical estimate. Because of the importance of the semivariogram in spatial statistics the next section is devoted to semivariogram analysis and estimation.

9.2 Semivariogram Analysis and Estimation

Box 9.4 Semivariogram

- **The semivariogram of a spatial process is one half of the variance of the difference between observations.**
- **The semivariogram conveys information about the spatial structure and the degree of continuity of a random field (§9.2.1, §9.2.3).**
- **Theoretical models (§9.2.2) are fit to the data to arrive at an estimated semivariogram that satisfies the properties needed for subsequent analysis (§9.2.4).**

9.2.1 Elements of the Semivariogram

A valid covariance function must satisfy certain properties. It must be even in the sense that $C(\mathbf{h}) = C(-\mathbf{h})$ since we must have $\text{Cov}[Z(\mathbf{s}), Z(\mathbf{s}+\mathbf{h})] = \text{Cov}[Z(\mathbf{s}+\mathbf{h}), Z(\mathbf{s})]$. Furthermore, covariance functions must be non-negative definite, i.e.,

$$\sum_{i=1}^{n}\sum_{j=1}^{n} \alpha_i \alpha_j C(\mathbf{s}_i - \mathbf{s}_j) \geq 0, \qquad [9.6]$$

for all constants $\alpha_1, \cdots, \alpha_n$ and spatial locations. This condition guarantees that the variance of spatial predictions are non-negative. As a consequence it can be shown by the Cauchy-Schwartz inequality that $|C(\mathbf{h})| \leq C(\mathbf{0})$ and we have already established that $C(\mathbf{0}) \geq 0$, since $C(\mathbf{0})$ is the variance of an observation. In practice, we often consider only covariance functions that have the following additional properties: they are positive and decrease monotonically with spatial separation. For some critical distance $||\mathbf{h}_c||$, the covariance function is then either identically zero or it approaches 0 as lag distance increases. Similar conditions and properties arise for valid semivariograms. Semivariograms have the evenness property $\gamma(\mathbf{h}) = \gamma(-\mathbf{h})$ and pass through the origin, $\gamma(\mathbf{0}) = 0$, since $\text{Var}[Z(\mathbf{s}) - Z(\mathbf{s}+\mathbf{0})] = \text{Var}[0] = 0$. Finally, a valid semivariogram must be conditionally negative-definite, i.e.,

$$\sum_{i=1}^{n}\sum_{j=1}^{n}\alpha_i\alpha_j\gamma(\mathbf{s}_i - \mathbf{s}_j) \leq 0, \qquad [9.7]$$

for any number of spatial locations and constants $\alpha_1, \cdots, \alpha_n$ such that $\sum_{i=1}^{n}\alpha_i = 0$ (Cressie 1993, p. 86). When the additional conditions (positive, monotonic decreasing) are imposed on the covariance function the semivariogram of a second-order stationary random field takes on a very characteristic shape (Figure 9.12). It rises from the origin monotonically to an upper asymptote called the **sill** of the semivariogram. The sill corresponds to $\text{Var}[Z(\mathbf{s})] = C(\mathbf{0})$. When the semivariogram meets the asymptote the covariance $C(\mathbf{h})$ is zero since $\gamma(\mathbf{h}) = C(\mathbf{0}) - C(\mathbf{h})$. The distance at which this occurs is called the **range** of the semivariogram. In Figure 9.12 the semivariogram approaches the sill only asymptotically. In this case the **practical range** is defined as the lag distance at which the semivariogram achieves 95% of the sill, here, the practical range is $||\mathbf{h}|| = 15$.

Observations that are spatially separated by more than the range are uncorrelated (or practically uncorrelated when separated by more than the practical range). Spatial autocorrelation exists only for pairs of points separated by less than the (practical) range. The more quickly the semivariogram rises from the origin to the sill the more quickly autocorrelations decline.

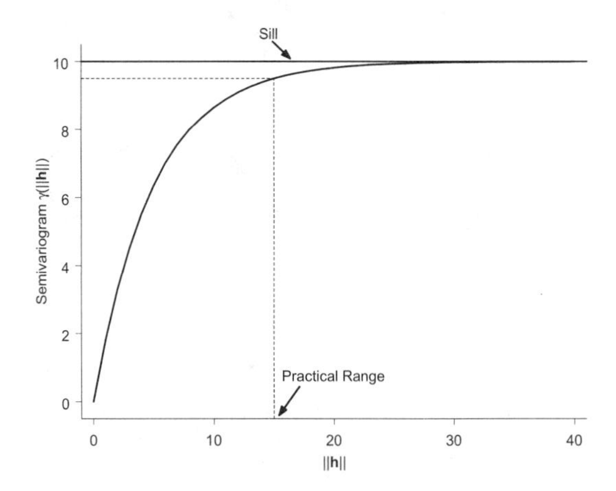

Figure 9.12. Semivariogram of a second-order stationary process with positive covariance function for which $C(||\mathbf{h}||) \to 0$ as $||\mathbf{h}||$ increases. The semivariogram has sill 10 and practical range 15.

An intrinsically but not second-order stationary random field has a semivariogram that does not reach an upper asymptote. The semivariance may increase with spatial separation as in Figure 9.13. Obviously, there is no range defined in this case. The increase of the semivariogram with $||\mathbf{h}||$ can not be arbitrary, however. It must rise less quickly than $2/||\mathbf{h}||^2$ (this check is sometimes referred to as the test of the intrinsic hypothesis) because $2\gamma(||\mathbf{h}||)/||\mathbf{h}||^2$ must approach 0 as $||\mathbf{h}|| \rightarrow \infty$.

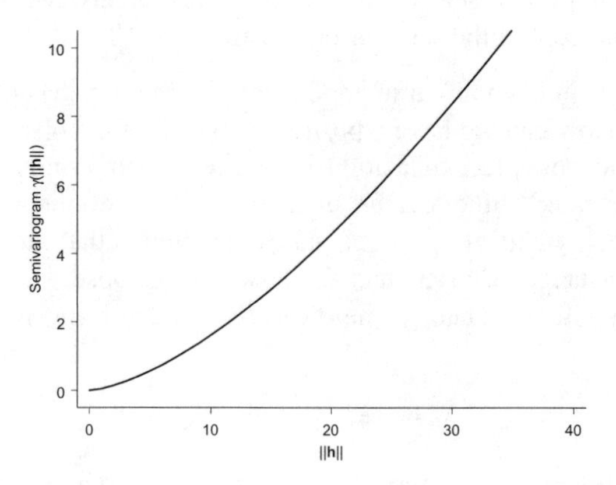

Figure 9.13. Semivariogram of an intrinsically but not second-order stationary process.

By definition we have $\gamma(\mathbf{0}) = 0$ but many data sets do not seem to comply with that property. Figure 9.14 shows an estimate of the semivariogram for the Mercer and Hall grain yield data plotted in Figure 9.5 (p. 569). The estimated semivariogram values are represented by the dots and a nonparametric loess smooth of these values was also added to the graph. The semivariogram appears to reach or at least to approach an asymptote with increasing lag distance, the sill is approximately 0.18.

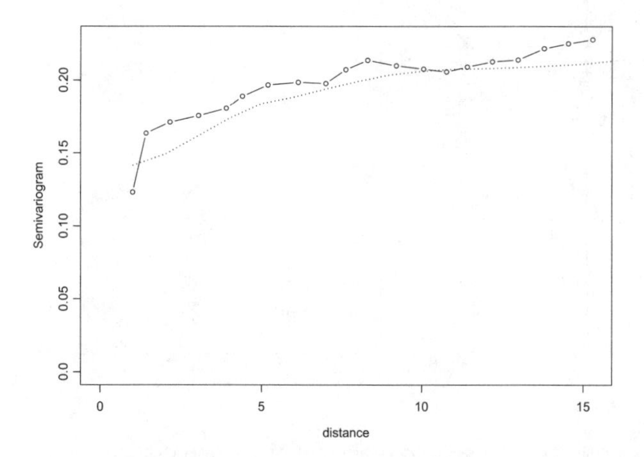

Figure 9.14. Classical semivariogram estimator (see §9.2.4) for Mercer and Hall grain data (connected dots). Dashed line is loess smooth of the semivariogram estimator.

Notice that the empirical semivariogram commences at $||\mathbf{h}|| = 1$, since this is the smallest lag between experimental units. We do not recommend smoothing semivariogram estimates with standard nonparametric procedures because the resulting fit may not have the required properties. It may not be conditionally negative definite (nonparametric semivariogram estimation is discussed in 9.2.4). The loess smooth was added to the figure only to suggest an overall trend in the semivariogram estimate. By connecting the dots of the semivariogram estimates it appears that the trend could pass through the origin as is required. However, the loess smooth of the empirical semivariogram indicates otherwise. Extrapolation below $||\mathbf{h}|| = 1$ suggests an intercept of the semivariogram around 0.13.

This phenomenon is quite common in applications, namely, $\gamma(\mathbf{h}) \to \theta_0 \neq 0$ as $||\mathbf{h}|| \to 0$. How can this happen? How can we have a positive variance of the observation differences at the **same** location? One possible explanation is **measurement error**. If a measurement at location $\mathbf{s}$ cannot be repeated without error, then repeat observations at $\mathbf{s}$ will exhibit variability; call this variance component σ_e^2. A second explanation is that there is a spatial process $\eta(\mathbf{s})$ operating at lag distances shorter than the smallest lag observed in the data set. This **microscale** process has sill σ_η^2. Then, if the measurement error and microscale process are independent,

$$\theta_0 = \sigma_e^2 + \sigma_\eta^2.$$

If any of the two components is not zero, the semivariogram exhibits a discontinuity at the origin. The magnitude of this discontinuity is called the **nugget effect**. The term stems from the idea that ore nuggets are dispersed throughout a larger body of rock but at distances smaller than the smallest sample distance. If a semivariogram has nugget θ_0 and sill $C(\mathbf{0})$, the difference $C(\mathbf{0}) - \theta_0$ is called the **partial sill** of the semivariogram. The practical range is then defined as the lag distance at which the semivariogram has achieved $\theta_0 + 0.95(C(\mathbf{0}) - \theta_0)$ (Figure 9.15).

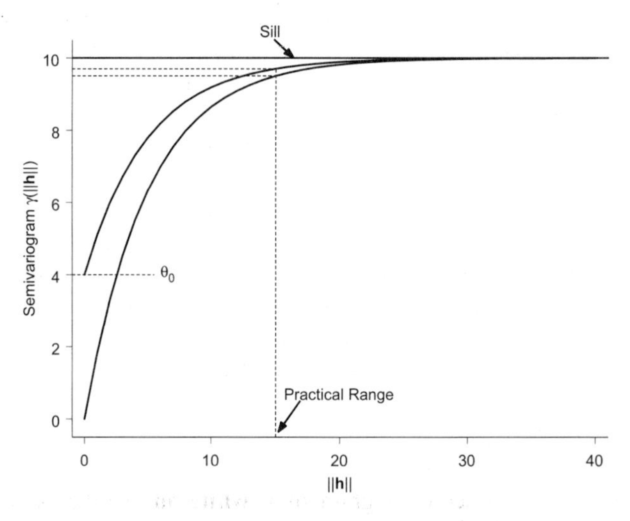

Figure 9.15. Semivariogram of a second-order stationary process with and without nugget effect. The no-nugget semivariogram (lower line) has sill 10 and practical range 15, the nugget semivariogram has $\theta_0 = 4$, partial sill $C(\mathbf{0}) - \theta_0 = 6$, and the same practical range.

In the presence of a nugget effect the relationship between semivariogram and covariogram must be slightly altered. In the no-nugget model we can put $\text{Var}[Z(\mathbf{s}_i)] = C(\mathbf{0}) = \sigma^2$. For the nugget model define $\text{Var}[Z(\mathbf{s}_i)] = C(\mathbf{0}) = \theta_0 + \xi$ where ξ is the partial sill (and $\sigma^2 = \text{Var}[Z(\mathbf{s}_i)] = \theta_0 + \xi$). The semivariogram can now be expressed as $\gamma(\mathbf{h}) = \theta_0 + \xi f(\mathbf{h})$

(see §9.2.2 for some $f(\mathbf{h})$). Then

$$C(\mathbf{h}) = \begin{cases} \xi(1 - f(\mathbf{h})) & ||\mathbf{h}|| > 0 \\ \theta_0 + \xi & \mathbf{h} = \mathbf{0}. \end{cases} \qquad [9.8]$$

In the presence of a nugget effect a useful statistic is the Relative Structured Variability (RSV). It measures the relative elevation of the semivariogram over the nugget effect in percent:

$$RSV = \left(\frac{\xi}{\xi + \theta_0}\right) \times 100\% = \left(1 - \frac{\theta_0}{\xi + \theta_0}\right) \times 100\%. \qquad [9.9]$$

One interpretation of RSV is as the degree to which variability is spatially structured. The unstructured part of the variability is due to measurement error and/or microscale variability. The larger the RSV, the more efficient geostatistical prediction will be compared to methods of prediction that ignore spatial information and the greater the continuity of the spatial process (see §9.2.3)

9.2.2 Parametric Isotropic Semivariogram Models

Estimation of the semivariogram by parametric statistical methods requires the selection of a semivariogram model $\gamma(\mathbf{h}; \boldsymbol{\theta})$ that is fit to data. $\boldsymbol{\theta}$ is a vector of parameters that is estimated from the data by direct or indirect methods. We consider those methods as indirect that process the data $Z(\mathbf{s}_1), \cdots, Z(\mathbf{s}_n)$ in some form, for example by averaging squared differences $\{Z(\mathbf{s}_i) - Z(\mathbf{s}_i + \mathbf{h})\}^2$, and then fit the semivariogram model to these summaries. Functions that serve as semivariogram models must be conditionally negative definite and a relatively small number of such functions is used in practice. We introduce the key models in this subsection; many more semivariogram models can be found in e.g., Journel and Huijbregts (1978), Cressie (1993), Stein (1999) and Chilès and Delfiner (1999). The models that follow are isotropic, θ_0 denotes the nugget and θ_s the sill parameter provided the model is second-order stationary. The models are all valid in $\mathbb{R}^2$, some are valid for higher dimensional problems. Note that a semivariogram that is valid in $\mathbb{R}^d$ is also valid in $\mathbb{R}^s$ where $s < d$. Since we are concerned with two-dimensional random fields in this chapter we do not further comment on the valid dimensions of any of the semivariogram models.

Nugget-Only Model

The nugget-only model is the semivariogram of a **white-noise** process, where the $Z(\mathbf{s}_i)$ behave like a random sample, all having the same mean, variance with no correlations among them. The model is void of spatial structure, the relative structured variability is zero. The nugget-only model is of course second-order stationary and a valid semivariogram in any dimension. Nugget-only models are not that uncommon although analysts keen on applying techniques from the spatial statistics toolbox such as kriging are usually less enthusiastic when a nugget-only model is obtained. A nugget-only model is an appropriate model if the smallest sample distance in the data is greater than the range of the spatial process. Sampling an attribute on a regular grid whose spatial range is unknown may invariably lead to a nugget-only model if grid points are spaced too far apart.

$$\gamma(\mathbf{h};\theta_s) = \begin{cases} 0 & \mathbf{h} = \mathbf{0} \\ \theta_s & \mathbf{h} \neq \mathbf{0} \end{cases}$$

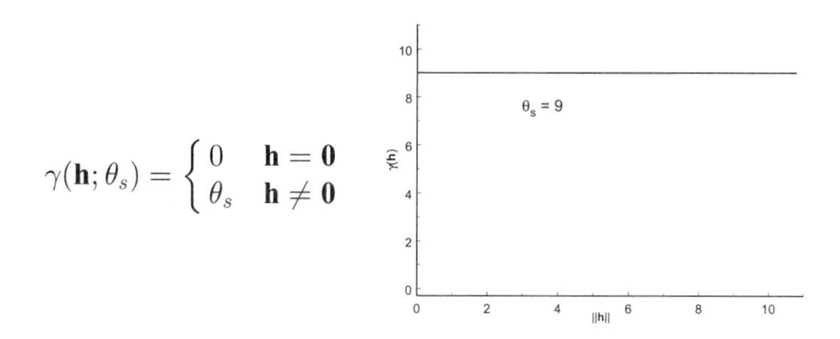

Linear Model

The Linear Model is intrinsically stationary with parameters θ_0 and β, both of which must be positive. Covariances or semivariances usually do not change linearly over a large range, but linear change of the semivariogram near the origin is often reasonable. If a linear semivariogram model is found to fit the data in practice, it is possible that one has observed the initial increase of a second-order stationary model that is linear or close to linear near the origin but failed to collect samples far enough apart to capture the range and sill of the semivariogram. A second-order stationary semivariogram model that behaves linearly near the origin is the spherical model.

$$\gamma(\mathbf{h};\boldsymbol{\theta}) = \begin{cases} 0 & \mathbf{h} = \mathbf{0} \\ \theta_0 + \beta||\mathbf{h}|| & \mathbf{h} \neq \mathbf{0} \end{cases}$$

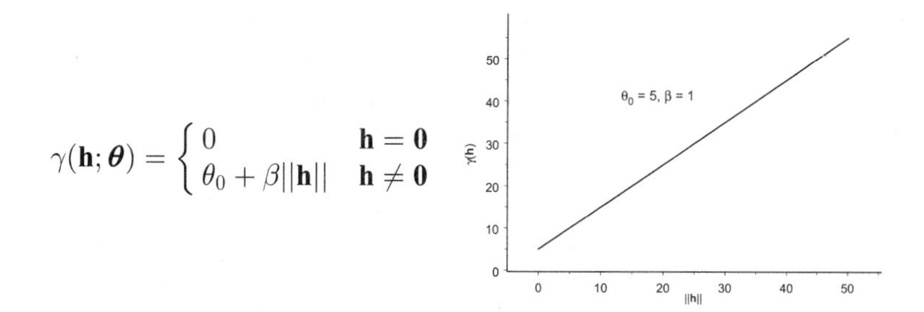

Spherical Model

The spherical model is one of the most popular semivariogram models in applied spatial statistics for second-order stationary random fields. Its two main characteristics are linear behavior near the origin and the fact that at distance α the semivariogram meets the sill and remains flat thereafter. This sometimes creates a visible kink at $||\mathbf{h}|| = \alpha$. The spherical semivariogram thus has a range α, rather than a practical range. The popularity of the spherical model in the geostatistical literature is a mystery to Stein (1999, p. 52) who argues that perhaps "there is a mistaken belief that there is some statistical advantage in having the autocorrelation function being exactly zero beyond some finite distance" (α). The fact that $\gamma(\mathbf{h};\boldsymbol{\theta})$ is only once differentiable at $||\mathbf{h}|| = \alpha$ can lead to problems in likelihood estimation that relies on derivatives. Stein (1999, p. 52) concludes that the spherical model is a poor substitute for the exponential model (see next). He recommends using the square of the spherical model $\gamma(\mathbf{h};\boldsymbol{\theta})^2$ instead of $\gamma(\mathbf{h};\boldsymbol{\theta})$, since the former provides two derivatives on $(0,\infty)$. The behavior of the squared spherical semivariogram near the origin is not linear, however (see §9.2.3 on the effect of the near-origin behavior).

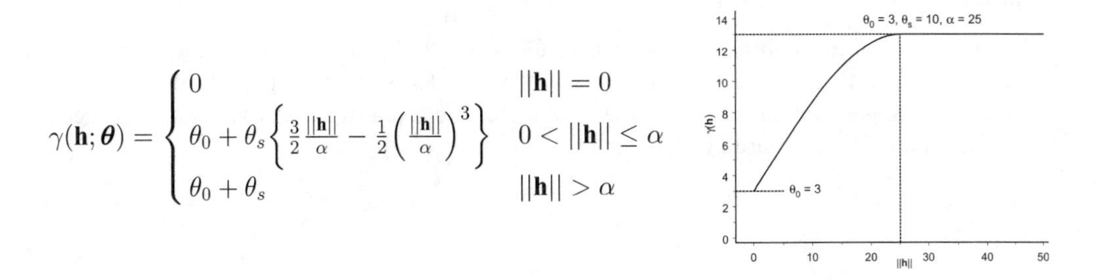

$$\gamma(\mathbf{h};\boldsymbol{\theta}) = \begin{cases} 0 & ||\mathbf{h}|| = 0 \\ \theta_0 + \theta_s\left\{\frac{3}{2}\frac{||\mathbf{h}||}{\alpha} - \frac{1}{2}\left(\frac{||\mathbf{h}||}{\alpha}\right)^3\right\} & 0 < ||\mathbf{h}|| \leq \alpha \\ \theta_0 + \theta_s & ||\mathbf{h}|| > \alpha \end{cases}$$

Exponential Model

The second-order stationary exponential model is a very useful model that has been found to fit spatial data in varied applications well. It approaches the sill θ_s asymptotically as $||\mathbf{h}|| \to \infty$. In the parameterization shown below the parameter α is the practical range of the semivariogram (Figure 9.12 is an exponential semivariogram without nugget). Often the model can be found in a parameterization where the exponent is $-\,||\mathbf{h}||/\alpha$. The practical range then corresponds to 3α. The SAS® System and S+SpatialStats® use this parameterization. For the same range and sill as the spherical model the exponential model rises more quickly from the origin and yields autocorrelations at short lag distances smaller than those of the spherical model. A random field with an exponential semivariogram is less regular (less continuous) on short distances than the spherical model (§9.2.3).

$$\gamma(\mathbf{h};\boldsymbol{\theta}) = \begin{cases} 0 & \mathbf{h} = \mathbf{0} \\ \theta_0 + \theta_s\left\{1 - \exp\left\{-\frac{3||\mathbf{h}||}{\alpha}\right\}\right\} & \mathbf{h} \neq \mathbf{0} \end{cases}$$

The covariance function of the exponential model without nugget effect is

$$C(\mathrm{h}) = \begin{cases} \theta_s(\exp\{-3||\mathbf{h}||/\alpha\}) & ||\mathbf{h}|| > 0 \\ \theta_s & \mathbf{h} = \mathbf{0}. \end{cases}$$

It is easily seen that this is a special case of the covariance model introduced in §7.5.2 for modeling the within-cluster correlations in repeated measures data (see formula [7.54] on p. 457). There it was referred to as the continuous AR(1) model because of its relationship to a continuous first-order regressive time series. The extra constant 3 was not used there, since the temporal range is usually of less interest when modeling clustered repeated measures data than the range for spatial data. The temporal separation $|t_{ij} - t_{ij'}|$ between two observations from the same cluster is now replaced by the Euclidean distance $||\mathbf{h}||$. For the exponential semivariogram to be valid we need to have $\theta_0 \geq 0$, $\theta_s \geq 0$, and $\alpha \geq 0$.

Gaussian Model

This model exhibits quadratic behavior near the origin and produces short-range correlations that are higher than for any of the other second-order stationary models with the same (practical) range. Notice that the *only* difference between the gaussian and exponential semivariogram is the square in the exponent.

$$\gamma(\mathbf{h};\boldsymbol{\theta}) = \begin{cases} 0 & \mathbf{h} = \mathbf{0} \\ \theta_0 + \theta_s\left\{1 - \exp\left\{-3\left(\frac{||\mathbf{h}||}{\alpha}\right)^2\right\}\right\} & \mathbf{h} \neq \mathbf{0} \end{cases}$$

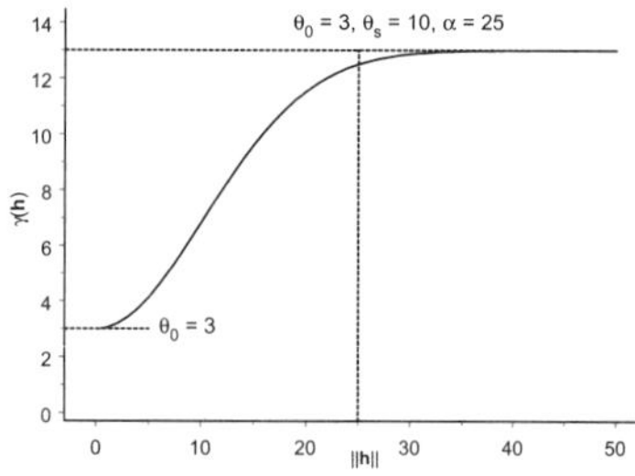

This is a fairly subtle difference that has considerable implications. The gaussian model is the most continuous near the origin of the models considered here. In fact, it is infinitely differentiable near 0. This implies a very smooth, regular spatial process (see §9.2.3). It is so smooth that knowing the value at 0 and the values of all partial derivatives determines the values in the random field at any arbitrary location. Such smoothness is unrealistic for most processes.

The name should not imply that this semivariogram model deserves similar veneration in spatial statistics as is awarded rightfully to the Gaussian distribution in classical statistics. The name stems from the fact that a stochastic process with covariance function

$$C(t) = c\exp\{-\alpha t^2\}$$

has spectral density

$$f(\omega) = \frac{1}{2}\frac{c}{\sqrt{\pi\alpha}}\exp\{-\omega^2/4\alpha\},$$

which resembles in functional form the Gaussian probability mass function. Furthermore, one should not assume that the semivariogram of a Gaussian random field (see §9.1.4 for the definition) is necessarily of this form. It most likely will not be.

As for the exponential model, the parameter α is the practical range and The SAS® System and S+SpatialStats® again drop the factor 3 in the opponent. In their parameterization the practical range is $\sqrt{3}\alpha$.

Power Model

This is an intrinsically stationary model but only for $0 \leq \lambda < 2$. Otherwise the variogram would increase faster than $||\mathbf{h}||^2$, which is in violation with the intrinsic hypothesis. The parameter β furthermore must be positive. The linear semivariogram model is a special case of the power model with $\lambda = 1$. Note that the covariance model `proc mixed` in The SAS®

System terms the power model is a reparameterization of the exponential model and not the power model shown here.

$$\gamma(\mathbf{h};\boldsymbol{\theta}) = \begin{cases} 0 & \mathbf{h} = \mathbf{0} \\ \theta_0 + \beta||\mathbf{h}||^{\lambda} & \mathbf{h} \neq \mathbf{0} \end{cases}$$

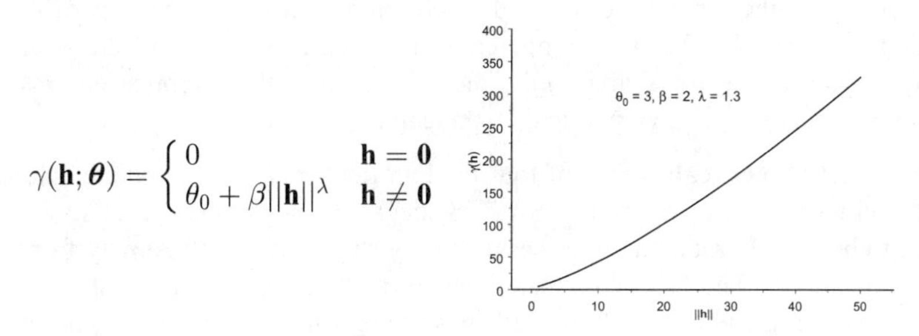

Wave (Cardinal Sine) Model

The semivariogram models discussed thus far permit only positive autocorrelation. This implies that within the range a large (small) value $Z(\mathbf{s})$ is likely to be paired with a large (small) value at $Z(\mathbf{s}+\mathbf{h})$. A semivariogram that permits positive and negative autocorrelation is the wave (or cardinal sine) semivariogram. It fluctuates about the sill θ_s and the fluctuations decrease with increasing lag. At lag distances where the semivariogram is above the sill the spatial correlation is negative; it is positive when the semivariogram drops below the sill. All parameters of the model must be positive.

$$\gamma(\mathbf{h};\boldsymbol{\theta}) = \begin{cases} 0 & \mathbf{h} = \mathbf{0} \\ \theta_0 + \theta_s\left\{1 - \alpha\sin\left(\frac{||\mathbf{h}||}{\alpha}\right)/||\mathbf{h}||\right\} & \mathbf{h} \neq \mathbf{0} \end{cases}$$

The term $||\mathbf{h}||/\alpha$ is best measured in radians. In the Figure above we have chosen $\alpha = 25*\pi/180$. The practical range is the value where the peaks/valleys of the covariogram are no greater than $0.05C(\mathbf{0})$, approximately $\pi*6.5*\alpha$. A process with a wave semivariogram has some form of periodicity.

9.2.3 The Degree of Spatial Continuity (Structure)

The notion of spatial structure of a random field is related to its degree of smoothness or continuity. The slower the increase of the semivariogram near the origin, the more the process is spatially structured, the smoother it is. Prediction of $Z(\mathbf{s})$ at unobserved locations is easier (more precise) if a process is smooth. With increasing irregularity (= lack of structure) less information can be gleaned about the process by considering the values at neighboring locations. The greatest absence of structure is encountered when there is a discontinuity at the origin, a nugget effect. Consequently, the nugget-only model is completely spatially unstructured. For the same sill and range, the exponential semivariogram rises faster than the spherical and hence the former is less spatially structured than the latter. The correctness of statistical inferences, on the other hand, depends increasingly on the correctness of the semivariogram model as processes become smoother. Semivariograms with a quadratic effect near the origin (gaussian model, for example) are more continuous than semivariograms that behave close to linear near the origin (spherical, exponential).

Figure 9.16 shows realizations of four random fields that were simulated along a transect of length 50 with `proc sim2d` of The SAS® System. The degree of spatial structure increases from top to bottom. For a smooth process the gray shades vary only slowly from an observation to its neighbors. The (positive) spatial autocorrelations are strong on short distances. The random field with the nugget-only model shows the greatest degree of irregularity (least continuity) followed by the exponential model that appears considerably smoother. In the nugget-only model a large (dark) observation can be followed by a small (light-colored) observation, whereas the exponential model maintains similar shading over short distances. The spherical model is smoother than the exponential and the gaussian model exhibits the greatest degree of regularity. As mentioned before, a process with gaussian semivariogram is smoother than what one should reasonably expect in practice.

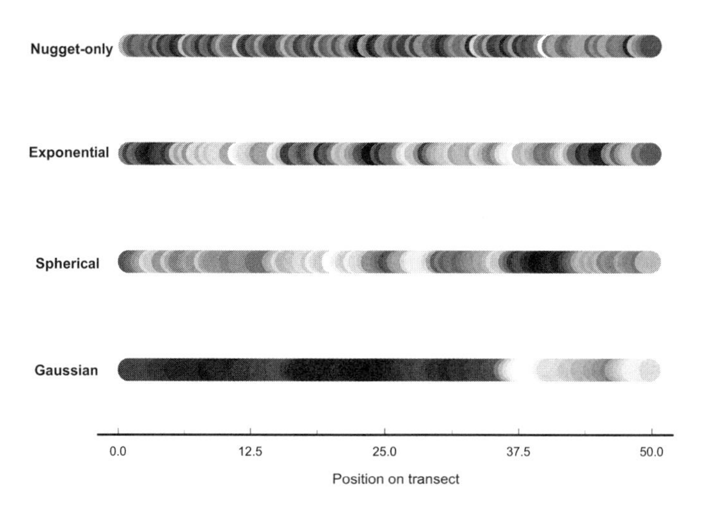

Figure 9.16. Simulated Gaussian random fields that differ in their isotropic semivariogram structure. In all cases the semivariogram has sill 2, the exponential, spherical, and gaussian semivariograms have range (practical range) 5.

Because the (practical) range of the semivariogram has a convenient interpretation as the distance beyond which observations are not spatially autocorrelated it is often interpreted as a zone of influence, or a scale of variability of $Z(\mathbf{s})$, or in terms of the degree of homogeneity of the process. It must be noted that $\text{Var}[Z(\mathbf{s})]$, the variability (scale) of $Z(\mathbf{s})$, is not a function of spatial location in a second-order stationary process. The variances of observations are the same everywhere and the process is homogeneous in this sense, regardless of the magnitude of the variability. The scale of variability the range refers to is the spatial range over which the variance of *differences* $Z(\mathbf{s}) - Z(\mathbf{s}+\mathbf{h})$ changes. For distances exceeding the range $\text{Var}[Z(\mathbf{s}) - Z(\mathbf{s}+\mathbf{h})]$ is constant. From Figure 9.16 it is also seen that processes with the same range can be very different in their respective degrees of continuity. A process with a gaussian semivariogram implies short-range correlation of much greater magnitude than a process with an exponential semivariogram and the same range.

For second-order stationary processes different measures have been proposed to capture the degree of spatial structure. The Relative Structured Variability (RSV [9.9]) measures that aspect of continuity which is influenced by the nugget effect. To incorporate the spatial range *and* the form of the semivariogram model, **integral scales** as defined by Russo and Bresler (1981) and Russo and Jury (1987a) are useful. If $\rho(\mathbf{h}) = C(\mathbf{h})/C(\mathbf{0})$ is the correlation function of the process, then

$$J_1 = \int_0^\infty \rho(h)dh \qquad J_2 = \left\{ 2\int_0^\infty \rho(h)hdh \right\}^{½}$$

are the integral scale measures for one- and two-dimensional processes, respectively, where h denotes Euclidean distance ($h = ||\mathbf{h}||$). Consider a two-dimensional process with an exponential semivariogram, no nugget, and practical range α. Then,

$$J_1 = \int_0^\infty \exp\{-3h/\alpha\}dh = \alpha/3 \qquad J_2 = \left\{ 2\int_0^\infty h\exp\{-3h/\alpha\}dh \right\}^{½} = \alpha\sqrt{2}/3.$$

Integral scales are used to define the distance over which observations are highly related rather than relying on the (practical) range, which is the distance beyond which observations are not related at all. For a process with gaussian semivariogram and practical range α, by comparison, one obtains $J_1 = 0.5{*}\alpha\sqrt{\pi/3} \approx 0.51\alpha$. The more continuous gaussian process has a longer integral scale, correlations wear off more slowly. An alternative measure to define distances over which observations are highly related is obtained by choosing a critical value of the autocorrelation and to solve the correlation function for it. The distance $h(\alpha, c)$ at which an exponential semivariogram with range α (and no nugget) achieves correlation $0 < c \le 1$ is $h(\alpha, c) = \alpha(-\ln\{c\}/3)$. For the gaussian semivariogram this distance is $h(\alpha, c) = \alpha\sqrt{-\ln\{c\}/3}$. The more continuous process maintains autocorrelations over longer distances.

Solie, Raun, and Stone (1999) argue that integral scales provide objective measures for the distance at which soil and plant variables are highly correlated and are useful when this distance cannot be determined based on subject matter alone. The integral scale these authors employ is a modification of J_1 where the autocorrelation function is integrated only to the (practical) range, $J = \int_0^\alpha \rho(h)dh$. For the exponential semivariogram with no nugget effect, this yields $J \approx 0.95{*}\alpha/3$.

9.2.4 Semivariogram Estimation and Fitting

In this section we introduce the most important estimators of the semivariogram and methods of fitting theoretical semivariograms to data. Mathematical background material can be found in §A9.9.1 (on estimation) and §A9.9.2 (on fitting). An application of the estimators and fitting methods is discussed in our §9.8.2.

Empirical Semivariogram Estimators

Estimators in this class are based on summary statistics of functions of the paired differences $Z(\mathbf{s}_i) - Z(\mathbf{s}_j)$. Recall that the semivariogram is defined as $\gamma(\mathbf{h}) = 0.5\mathrm{Var}[Z(\mathbf{s}) - Z(\mathbf{s}+\mathbf{h})]$ and that any kind of stationarity implies at least that the mean $\mathrm{E}[Z(\mathbf{s})]$ is constant. The squared difference $(Z(\mathbf{s}_i) - Z(\mathbf{s}_j))^2$ is then an unbiased estimator of $\mathrm{Var}[Z(\mathbf{s}_i) - Z(\mathbf{s}_j)]$ since $\mathrm{E}[Z(\mathbf{s}_i) - Z(\mathbf{s}_j)] = 0$ and an unbiased estimator of $\gamma(\mathbf{h})$ is obtained by calculating the average of one half of the squared differences of all pairs of observations that are exactly distance $\mathbf{h}$ apart. Mathematically, this estimator is expressed as

$$\widehat{\gamma}(\mathbf{h}) = \frac{1}{2|N(\mathbf{h})|}\sum_{N(\mathbf{h})}(Z(\mathbf{s}_i) - Z(\mathbf{s}_j))^2, \qquad [9.10]$$

where $N(\mathbf{h})$ is the set of location pairs that are separated by the lag vector $\mathbf{h}$ and $|N(\mathbf{h})|$ denotes the number of unique pairs in the set $N(\mathbf{h})$. Notice that $Z(\mathbf{s}_1) - Z(\mathbf{s}_2)$ and $Z(\mathbf{s}_2) - Z(\mathbf{s}_1)$ are the same pair in this calculation and are not counted twice. If the semivariogram of the random field is isotropic, $\mathbf{h}$ is replaced by $||\mathbf{h}||$. [9.10] is known as the classical semivariogram estimator due to Matheron (1962) and also called the Matheron estimator. Its properties are generally appealing. It is an unbiased estimator of $\gamma(\mathbf{h})$ provided the mean of the random field is constant and it behaves similar to the semivariogram: it is an even function $\widehat{\gamma}(\mathbf{h}) = \widehat{\gamma}(-\mathbf{h})$ and $\widehat{\gamma}(\mathbf{0}) = 0$. A disadvantage of the Matheron estimator is its sensitivity to outliers. If $Z(\mathbf{s}_j)$ is an outlying observation the difference $Z(\mathbf{s}_i) - Z(\mathbf{s}_j)$ will be large and squaring the difference amplifies the contribution to the empirical semivariogram estimate at lag $\mathbf{s}_i - \mathbf{s}_j$. In addition, outlying observations contribute to the estimation of $\gamma(\mathbf{h})$ at various lags and exert their influence on more than one $\widehat{\gamma}(\mathbf{h})$ value. Consider the following hypothetical data chosen small to demonstrate the effect. The data represent five locations on a (3×4) grid.

Table 9.1. A spatial data set containing an outlying observation $Z([3,4]) = 20$

	Column (y)			
Row (x)	1	2	3	4
1	1			4
2		2		
3	3			20

The observation in row 3, column 4 is considerably larger than the remaining four observations. What is its effect on the Matheron semivariogram estimator? There are five lag distances in these data, at $||\mathbf{h}|| = \sqrt{2}$, 2, $\sqrt{5}$, 3, and $\sqrt{13}$ distance units. For each lag there are exactly two data pairs. For example, the pairs contributing to the estimation of the semivariogram at $||\mathbf{h}|| = 3$ are $\{Z([1,1]), Z([1,4])\}$ and $\{Z([3,1]), Z([3,4])\}$. The variogram estimates are

$$
\begin{aligned}
2\widehat{\gamma}(\sqrt{2}) &= \frac{1}{2}\{(1-2)^2 + (2-3)^2\} = 1 \\
2\widehat{\gamma}(2) &= \frac{1}{2}\{(1-3)^2 + (4-20)^2\} = 130 \\
2\widehat{\gamma}(\sqrt{5}) &= \frac{1}{2}\{(4-2)^2 + (20-2)^2\} = 164 \\
2\widehat{\gamma}(3) &= \frac{1}{2}\{(4-1)^2 + (20-3)^2\} = 149 \\
2\widehat{\gamma}(\sqrt{13}) &= \frac{1}{2}\{(3-4)^2 + (20-1)^2\} = 181.
\end{aligned}
$$

The outlying observation exerts negative influence at four of the five lags and dominates the sum of squared differences. If the outlying observation is removed the variogram estimates are $2\widehat{\gamma}(2) = 4$, $2\widehat{\gamma}(\sqrt{2}) = 1$, $2\widehat{\gamma}(\sqrt{5}) = 4$, $2\widehat{\gamma}(3) = 9$, and $2\widehat{\gamma}(\sqrt{13}) = 1$.

Cressie and Hawkins (1980) derived an estimator of the semivariogram that is not as susceptible to outliers. Details of the derivation are found in their paper and are reiterated in our §A9.9.1. What has been termed the robust semivariogram estimator is based on absolute differences $|Z(\mathbf{s}_i) - Z(\mathbf{s}_j)|$ rather than squared differences. The estimator has a slightly more complicated form than the Matheron estimator and is given by

$$
\overline{\gamma}(\mathbf{h}) = 0.5\left(\frac{1}{|N(\mathbf{h})|}\sum_{N(\mathbf{h})}|Z(\mathbf{s}_i) - Z(\mathbf{s}_j)|^{1/2}\right)^4 \Big/ \left(0.457 + \frac{0.494}{|N(\mathbf{h})|}\right). \qquad [9.11]
$$

Square roots of absolute differences are averaged first and then raised to the fourth power. The influence of outlying observations is reduced because absolute differences are more stable than squared differences and averaging is carried out before converting into the units of a variance. Note that the attribute *robust* pertains to outlier contamination of the data; it should not imply that [9.11] is robust against other violations, such as nonconstancy of the mean. This estimator is not unbiased for the semivariogram but the term $0.457 + 0.494/|N(\mathbf{h})|$ in the denominator reduces the bias considerably. Calculating the robust estimator for the spatial data set with an outlier one obtains (where $0.457 + 0.494/2 = 0.704$)

$$
\begin{aligned}
2\overline{\gamma}(\sqrt{2}) &= \left\{\frac{1}{2}\left(\sqrt{|1-2|} + \sqrt{|2-3|}\right)\right\}^4 /0.704 = 1.42 \\
2\overline{\gamma}(2) &= \left\{\frac{1}{2}\left(\sqrt{|1-3|} + \sqrt{|4-20|}\right)\right\}^4 /0.704 = 76.3 \\
2\overline{\gamma}(\sqrt{5}) &= \left\{\frac{1}{2}\left(\sqrt{|4-2|} + \sqrt{|20-2|}\right)\right\}^4 /0.704 = 90.9 \\
2\overline{\gamma}(3) &= \left\{\frac{1}{2}\left(\sqrt{|4-1|} + \sqrt{|20-3|}\right)\right\}^4 /0.704 = 104.3 \\
2\overline{\gamma}(\sqrt{13}) &= \left\{\frac{1}{2}\left(\sqrt{|3-4|} + \sqrt{|20-1|}\right)\right\}^4 /0.704 = 73.2.
\end{aligned}
$$

The influence of the outlier is clearly subdued compared to the Matheron estimator.

Other proposals to make semivariogram estimation less susceptible to outliers have been put forth. Considering the median of squared differences instead of the average squared differences is one approach. Armstrong and Delfiner (1980) generalize this idea to using any quantiles of the distribution of $\{Z(\mathbf{s}_i) - Z(\mathbf{s}_j)\}^2$. A median-based estimator is a special case thereof. It is given by

$$\widetilde{\gamma}_{0.5}(\mathbf{h}) = \text{median}\left\{\frac{1}{2}\{Z(\mathbf{s}_i) - Z(\mathbf{s}_j)\}^2\right\}/0.4549. \qquad [9.12]$$

The Matheron estimator [9.10] and the robust estimator [9.11] remain the most important estimators of the empirical semivariogram in practice, however.

The precision of an empirical estimator at a given lag depends on the number of pairs available at that lag that can be averaged or otherwise summarized. The recommendations that at least 50 (Chilès and Delfiner 1999, p. 38) or 30 (Journel and Huijbregts 1978, p. 194) unique pairs should be available for every lag vector $\mathbf{h}$ or distance $||\mathbf{h}||$ are common. Even with 50 pairs the empirical semivariogram can be quite erratic for larger lags and simulation studies suggest that the approximate number of pairs can be considerably larger. Webster and Oliver (1992) conclude through simulation that at least 200 to 300 observations are required to estimate a semivariogram reliably. Cressie (1985) shows that the variance of the Matheron semivariogram estimator can be approximated as

$$\text{Var}[\widehat{\gamma}(\mathbf{h})] \doteq \frac{2\gamma^2(\mathbf{h})}{|N(\mathbf{h})|}. \qquad [9.13]$$

As the semivariogram increases, so does the variance of the estimator. When the semivariogram is intrinsic but not second-order stationary, the variability of $\widehat{\gamma}(\mathbf{h})$ for large lags can make it difficult to recognize the underlying structure unless $N(\mathbf{h})$ is large. We show in §A9.9.1 that [9.13] can be a poor approximation to $\text{Var}[\widehat{\gamma}(\mathbf{h})]$ which also depends on the degree of spatial autocorrelation and the spatial arrangement of the sampling locations. This latter dependence has been employed to determine sample grids and layouts that lead to good properties of the empirical semivariogram estimator without requiring too many observations. For details see, for example, Russo (1984), Warrick and Myers (1987), and Zheng and Silliman (2000).

With irregularly spaced data the number of observations at a given lag may be small, some lags may be even unique. To collect a sufficient number of pairs the set $N(\mathbf{h})$ is then defined as the collection of pairs for which locations are separated by $\mathbf{h} \pm \boldsymbol{\epsilon}$ or $||\mathbf{h}|| \pm \epsilon$, where ϵ is some lag tolerance. In other words, the empirical semivariogram is calculated only for discrete lag classes and all observations within a lag class are considered representing that particular lag. This introduces two potential problems. The term $\{Z(\mathbf{s}_i) - Z(\mathbf{s}_j)\}^2$ is an unbiased estimator of $2\gamma(\mathbf{s}_i - \mathbf{s}_j)$, but not of $2\gamma(\mathbf{s}_i - \mathbf{s}_j + \boldsymbol{\epsilon})$ and grouping lags into lag classes introduces some bias. Furthermore, the empirical semivariogram depends on the width and number of lag classes, which introduces a subjective element into the analysis.

The goal of semivariogram estimation is not to estimate the empirical semivariogram given by [9.10], [9.11], or [9.12] but to estimate the unknown parameters of a theoretical semivariogram model $\gamma(\mathbf{h};\boldsymbol{\theta})$. The least squares and nonparametric approaches fit the semivariogram model to the empirical semivariogram. If [9.10] was calculated at lags $\mathbf{h}_1, \mathbf{h}_2, \cdots, \mathbf{h}_k$, then $\widehat{\gamma}(\mathbf{h}_1), \widehat{\gamma}(\mathbf{h}_2), \cdots, \widehat{\gamma}(\mathbf{h}_k)$ serve as the data to which the semivariogram model is fit

(Figure 9.17). We call this the indirect approach to semivariogram estimation since an empirical estimate is obtained first which then serves as the data. Note that by choosing more lag classes one can apparently increase the size of this *data set*. Of the direct approaches we consider likelihood methods (maximum likelihood and restricted maximum likelihood) as well as a likelihood-type method (composite likelihood).

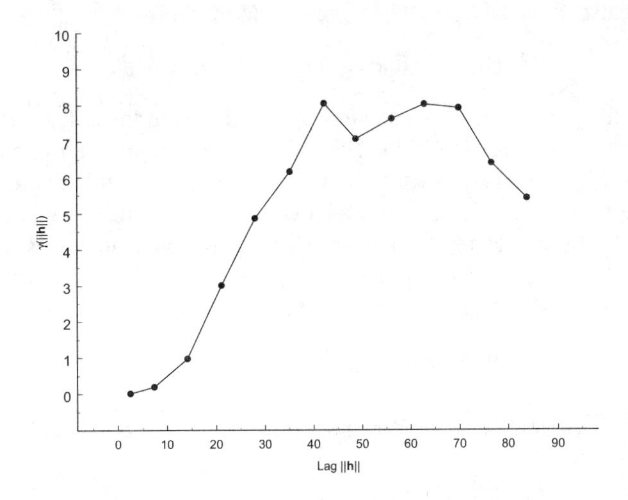

Figure 9.17. A robust empirical estimate of the semivariogram. $\overline{\gamma}(\mathbf{h})$ was calculated at $k = 13$ lag classes of width 7. The semivariogram estimates are plotted at the average lag distance within each class. Connecting the dots does not guarantee that the resulting function is conditionally negative definite.

Fitting the Semivariogram by Least Squares

In least squares methods to estimate the semivariogram the empirical semivariogram values $\widehat{\gamma}(\mathbf{h}_1), \cdots, \widehat{\gamma}(\mathbf{h}_k)$ or $\overline{\gamma}(\mathbf{h}_1), \cdots, \overline{\gamma}(\mathbf{h}_k)$ (or some other empirical estimate of $\gamma(\mathbf{h})$) serve as the responses; k denotes the number of lag classes. We discuss the least squares methods using the Matheron estimator. The robust estimator can be used instead. Ordinary least squares (OLS) estimates of $\boldsymbol{\theta}$ are found by minimizing the sum of squared deviations between the empirical semivariogram and a theoretical semivariogram:

$$\sum_{i=1}^{k} (\widehat{\gamma}(\mathbf{h}_i) - \gamma(\mathbf{h}_i; \boldsymbol{\theta}))^2. \qquad [9.14]$$

OLS requires that the data points are uncorrelated and homoscedastic. Both assumptions are not met. For the Matheron estimator Cressie (1985) showed that its variance is approximately

$$\text{Var}[\widehat{\gamma}(\mathbf{h})] \doteq \frac{2\gamma^2(\mathbf{h})}{|N(\mathbf{h})|}. \qquad [9.15]$$

It depends on the true semivariogram value at lag $\mathbf{h}$ and the number of unique data pairs at that lag. The $\widehat{\gamma}(\mathbf{h}_i)$ are also not uncorrelated. The same data point $Z(\mathbf{s}_i)$ contributes to the estimation at different lags *and* there is spatial autocorrelation among the data points. The same essential problems remain if $\overline{\gamma}(\mathbf{h})$ is used in place of $\widehat{\gamma}(\mathbf{h})$. The robust estimator has an advantage, however; its values are less correlated than those of the Matheron estimator.

One should use a generalized least squares criterion rather than ordinary least squares. Write $\widehat{\boldsymbol{\gamma}}(\mathbf{h}) = [\widehat{\gamma}(\mathbf{h}_1), \cdots, \widehat{\gamma}(\mathbf{h}_k)]'$ and $\boldsymbol{\gamma}(\mathbf{h};\boldsymbol{\theta}) = [\gamma(\mathbf{h}_1;\boldsymbol{\theta}), \cdots, \gamma(\mathbf{h}_k;\boldsymbol{\theta})]'$ and denote the variance-covariance matrix of $\widehat{\boldsymbol{\gamma}}(\mathbf{h})$ by $\mathbf{V}$. Then one should minimize

$$(\widehat{\boldsymbol{\gamma}}(\mathbf{h}) - \boldsymbol{\gamma}(\mathbf{h};\boldsymbol{\theta}))'\mathbf{V}^{-1}(\widehat{\boldsymbol{\gamma}}(\mathbf{h}) - \boldsymbol{\gamma}(\mathbf{h};\boldsymbol{\theta})). \qquad [9.16]$$

The problem with the generalized least squares approach lies in the determination of the variance-covariance matrix $\mathbf{V}$. Cressie (1985, 1993 p. 96) gives expressions from which the off-diagonal entries of $\mathbf{V}$ can be calculated for a Gaussian random field. These are complicated expressions of the true semivariogram and as a simplification one often resorts to weighted least squares (WLS) fitting. Here, $\mathbf{V}$ is replaced by a diagonal matrix $\mathbf{W}$ that contains the variances of the $\widehat{\gamma}(\mathbf{h}_i)$ on the diagonal and the approximation [9.15] is used to calculate the diagonal entries. The weighted least squares estimates of $\boldsymbol{\theta}$ are obtained by minimizing

$$(\widehat{\boldsymbol{\gamma}}(\mathbf{h}) - \boldsymbol{\gamma}(\mathbf{h};\boldsymbol{\theta}))'\mathbf{W}^{-1}(\widehat{\boldsymbol{\gamma}}(\mathbf{h}) - \boldsymbol{\gamma}(\mathbf{h};\boldsymbol{\theta})), \qquad [9.17]$$

where $\mathbf{W} = \text{Diag}\{2\gamma^2(\mathbf{h};\boldsymbol{\theta})/|N(\mathbf{h})|\}$. We show in §A9.9.2 that this is equivalent to minimizing

$$\sum_{i=1}^{k} |N(\mathbf{h}_i)| \left\{ \frac{\widehat{\gamma}(\mathbf{h}_i)}{\gamma(\mathbf{h}_i;\boldsymbol{\theta})} - 1 \right\}^2, \qquad [9.18]$$

which is (2.6.12) in Cressie (1993, p. 96). If the robust estimator is used instead of the Matheron estimator, $\widehat{\gamma}(\mathbf{h}_i)$ in [9.18] is replaced with $\overline{\gamma}(\mathbf{h}_i)$. Note that semivariogram models are typically nonlinear, with the exception of the nugget-only and the linear model, and minimization of these objective functions requires nonlinear methods.

The weighted least squares method for fitting semivariogram models is very common in practice. One must keep in mind that minimizing [9.18] is an approximate method. First, [9.15] is an approximation for the variance of the empirical estimator. Second, $\mathbf{W}$ is a poor approximation for $\mathbf{V}$. The weighted least squares method is a poor substitute for the generalized least squares method that should be used. Delfiner (1976) developed a different weighted least squares method that is implemented in the geostatistical package BLUEPACK (Delfiner, Renard, and Chilès 1978). Zimmerman and Zimmerman (1991) compared various semivariogram fitting methods in an extensive simulation study and concluded that there is little to choose between ordinary and weighted least squares. The Gaussian random fields simulated by Zimmerman and Zimmerman (1991) had a linear semivariogram with nugget effect and a no-nugget exponential structure. Neither the WLS or OLS estimates were uniformly superior in terms of bias for the linear semivariogram. The weighted least squares method due to Delfiner (1976) performed very poorly, however, and was uniformly inferior to all other methods (including the likelihood methods to be discussed next). In case of the exponential semivariogram the least squares estimators of the sill exhibited considerable positive bias, in particular when the spatial dependence was weak.

In WLS and OLS fitting of the semivariogram care should be exercised in the interpretation of the standard errors for the parameter estimates reported by statistical packages.

Neither method uses the correct variance-covariance matrix $\mathbf{V}$. Instead, WLS uses a diagonal matrix where the diagonal entries of $\mathbf{V}$ are approximated, and OLS uses a scaled identity matrix. Also, the size of the data set to which the semivariogram model is fit depends on the number of lag classes k which is chosen by the user. The number of lag classes k to which the semivariogram is fit is also often smaller than the number of lag classes for which the empirical semivariogram estimator was calculated. Especially values at large lags and lag classes for which the number of pairs does not exceed the rule of thumb value $|N(\mathbf{h})| > 30$ (or $|N(\mathbf{h})| > 50$) are removed before fitting the semivariogram by the least squares method. This invariably results in a data set that is slanted towards the chosen semivariogram model because lag classes whose empirical semivariogram values are consistent with the modeled trend are usually retained and those lag classes are removed whose values appear erratic. Journel and Huijbregts (1978, p. 194) recommend using only lags (lag classes) less than half of the maximum lag in the data set.

Fitting the Semivariogram by Maximum Likelihood

The least squares methods did not require that the random field is a Gaussian random field or knowledge of any distributional properties of the random field above the constant mean and the correctness of the semivariogram model. Maximum likelihood estimation requires knowledge of the distribution of the data. Usually it is assumed that the data are Gaussian and for spatial data this implies sampling from a Gaussian random field. $\mathbf{Z}(\mathbf{s}) = [Z(\mathbf{s}_1), \cdots, Z(\mathbf{s}_n)]'$ is then an n-variate Gaussian random variable. Under the assumption of second-order stationarity its mean and covariance matrix can be written as $\mathrm{E}[\mathbf{Z}(\mathbf{s})] = \boldsymbol{\mu} = [\mu_1, \cdots, \mu_n]'$ and $\mathrm{Var}[\mathbf{Z}(\mathbf{s})] = \boldsymbol{\Sigma}(\boldsymbol{\theta})$,

$$\boldsymbol{\Sigma}(\boldsymbol{\theta}) = \begin{bmatrix} C(\mathbf{0};\boldsymbol{\theta}) & C(\mathbf{s}_1-\mathbf{s}_2;\boldsymbol{\theta}) & C(\mathbf{s}_1-\mathbf{s}_3;\boldsymbol{\theta}) & \cdots & C(\mathbf{s}_1-\mathbf{s}_n;\boldsymbol{\theta}) \\ C(\mathbf{s}_2-\mathbf{s}_1;\boldsymbol{\theta}) & C(\mathbf{0};\boldsymbol{\theta}) & C(\mathbf{s}_2-\mathbf{s}_3;\boldsymbol{\theta}) & \cdots & C(\mathbf{s}_2-\mathbf{s}_n;\boldsymbol{\theta}) \\ C(\mathbf{s}_3-\mathbf{s}_1;\boldsymbol{\theta}) & C(\mathbf{s}_3-\mathbf{s}_2;\boldsymbol{\theta}) & C(\mathbf{0};\boldsymbol{\theta}) & \cdots & C(\mathbf{s}_3-\mathbf{s}_n;\boldsymbol{\theta}) \\ \vdots & & & \ddots & \vdots \\ C(\mathbf{s}_n-\mathbf{s}_1;\boldsymbol{\theta}) & C(\mathbf{s}_n-\mathbf{s}_2;\boldsymbol{\theta}) & \cdots & C(\mathbf{s}_n-\mathbf{s}_{n-1};\boldsymbol{\theta}) & C(\mathbf{0};\boldsymbol{\theta}) \end{bmatrix}.$$

Note that instead of the semivariogram we work with the covariance function here, but because the process is second-order stationary, the semivariogram and covariogram are related by

$$\gamma(\mathbf{h};\boldsymbol{\theta}) = C(\mathbf{0};\boldsymbol{\theta}) - C(\mathbf{h};\boldsymbol{\theta}).$$

In short, $\mathbf{Z}(\mathbf{s}) \sim G(\boldsymbol{\mu}, \boldsymbol{\Sigma}(\boldsymbol{\theta}))$ where $\boldsymbol{\theta}$ is the vector containing the parameters of the covariogram. The negative log-likelihood of $\mathbf{Z}(\mathbf{s})$ is

$$l(\boldsymbol{\theta}, \boldsymbol{\mu}; \mathbf{z}(\mathbf{s})) = \frac{n}{2}\ln(2\pi) + \frac{1}{2}\ln|\boldsymbol{\Sigma}(\boldsymbol{\theta})| + \frac{1}{2}(\mathbf{z}(\mathbf{s}) - \boldsymbol{\mu})'\boldsymbol{\Sigma}(\boldsymbol{\theta})^{-1}(\mathbf{z}(\mathbf{s}) - \boldsymbol{\mu}) \qquad [9.19]$$

and maximum likelihood (ML) estimates of $\boldsymbol{\theta}$ (and $\boldsymbol{\mu}$) are obtained as minimizers of this expression. Compare this objective function to that for fitting a linear mixed model by maximum likelihood in §7.4.1 and §A7.7.3. There the objective was to estimate the parameters in the variance-covariance matrix and the unknown mean vector. The same idea applies here. If $\boldsymbol{\mu} = \mathbf{X}\boldsymbol{\beta}$, where $\boldsymbol{\beta}$ are unknown parameters of the mean, maximum likelihood estimation provides simultaneous estimates of the large-scale mean structure (called the drift in the geostatistical literature) and the spatial dependency. This is an advantage over the indirect least squares methods where the assumption of meanstationarity $(\mathrm{E}[Z(\mathbf{s}_i)] = \mu)$ is critical to obtain the empirical semivariogram estimate.

Unbiasedness is not an asset of the ML estimator of the spatial dependence parameters $\boldsymbol{\theta}$. It is well known that maximum likelihood estimators of covariance parameters are negatively biased. In §7 the restricted maximum likelihood (REML) method of Patterson and Thompson (1971) and Harville (1974, 1977) was advocated to reduce this bias. The same ideas apply here. Instead of the likelihood of $\mathbf{Z}(\mathbf{s})$ we consider that of $\mathbf{KZ}(\mathbf{s})$ where $\mathbf{A}$ is a matrix of error contrasts such that $\mathrm{E}[\mathbf{KZ}(\mathbf{s})] = \mathbf{0}$. Instead of [9.19] we minimize

$$l(\boldsymbol{\theta}; \mathbf{Kz}(\mathbf{s})) = (n-1)\ln(2\pi) + \ln|\mathbf{K}\boldsymbol{\Sigma}(\boldsymbol{\theta})\mathbf{K}'| + \mathbf{z}(\mathbf{s})'\mathbf{K}'[\mathbf{K}\boldsymbol{\Sigma}(\boldsymbol{\theta})\mathbf{K}']^{-1}\mathbf{Kz}(\mathbf{s}). \quad [9.20]$$

Although REML estimation is well-established in statistical theory and applications, in the geostatistical arena it appeared first in work by Kitanidis and coworkers in the mid-1980s (Kitanidis 1983, Kitanidis and Vomvoris 1983, Kitanidis and Lane 1985).

An advantage of the ML and REML approaches is their direct dependence on the data $\mathbf{Z}(\mathbf{s})$. No grouping in lag classes is necessary. Because maximum likelihood estimation does not require an empirical semivariogram estimate, Chilès and Delfiner (1999, p. 109) call it a blind method that tends "to be used only when the presence of a strong drift causes the sample variogram to be hopelessly biased." We disagree with this stance, the ML estimators have many appealing properties, e.g., asymptotic efficiency. In their simulation study Zimmerman and Zimmerman (1991) found the ML estimators of the semivariogram sill to be much less variable than any other estimators of that parameter. Also, likelihood-based estimators outperformed the least squares estimators when the spatial dependence was weak. Fitting of a semivariogram by likelihood methods is not a *blind* process. If the random field has large-scale mean $\mathrm{E}[\mathbf{Z}(\mathbf{s})] = \mathbf{X}\boldsymbol{\beta}$, then one can obtain residuals from an initial ordinary least squares fit of the mean $\mathbf{X}\boldsymbol{\beta}$ and use the residuals to calculate an empirical semivariogram which guides the user to the formulation of a theoretical semivariogram or covariogram model. Then, both the mean parameters $\boldsymbol{\beta}$ and the covariance parameters $\boldsymbol{\theta}$ are estimated simultaneously by maximum likelihood or restricted maximum likelihood. To call these methods blind suggests that the models are formulated without examination of the data and without a thoughtful selection of the model.

Fitting the Semivariogram by Composite Likelihood

The idea of composite likelihood estimation is quite simple and dates back to work by Lindsay (1988). Lele (1997) and Curriero and Lele (1999) applied it to semivariogram estimation and Heagerty and Lele (1998) to prediction of spatial binary data. Let $Y_1, \cdots, Y_n$ be random variables whose marginal distribution $f(y_i, \boldsymbol{\theta})$ is known up to a parameter vector $\boldsymbol{\theta} = [\theta_1, \cdots, \theta_p]'$. Then $l(\boldsymbol{\theta}; y_i) = \ln\{f(y_i, \boldsymbol{\theta})\}$ is a true log-likelihood. For maximum likelihood estimation the log-likelihood for the joint distribution of the Y_i is needed. Often this joint distribution is known as in the previous paragraphs. If the Y_i are independent then the full data log-likelihood is particularly simple: it is the sum of the individual terms $l(\boldsymbol{\theta}; y_i)$. There are instances, however, when the complete data log-likelihood is not known or intractable. A case in point are correlated observations. What is lost by using

$$\sum_{i=1}^{n} l(\boldsymbol{\theta}; y_i) \qquad [9.21]$$

as the objective function for maximization? Obviously, [9.21] is not a log-likelihood although the individual terms $l(\boldsymbol{\theta}; y_i)$ are. The estimates obtained by maximizing [9.21] cannot be as efficient as ML estimates, which is easily established from key results in estimating function theory (see Godambe 1960, Heyde 1997, and our §A9.9.2). The function [9.21] is called a composite log-likelihood and its derivative,

$$CS(\theta_k; \mathbf{y}) = \sum_{i=1}^{n} \frac{\partial l(\theta_k; y_i)}{\partial \theta_k} \qquad [9.22]$$

is the composite score function for θ_k. Setting the composite score functions for $\theta_1, \cdots, \theta_p$ to zero and solving the resulting system of equations yields the composite likelihood estimates.

Applying this idea to the problem of estimating the semivariogram we commence by considering the $n(n-1)/2$ unique pairwise differences $T_{ij} = Z(\mathbf{s}_i) - Z(\mathbf{s}_j)$. When the $Z(\mathbf{s}_i)$ are Gaussian with the same mean and the random field is intrinsically stationary (this is a weaker assumption than assuming an intrinsically stationary Gaussian random field), then T_{ij} is a Gaussian random variable with mean 0 and variance $2\gamma(\mathbf{s}_i - \mathbf{s}_j; \boldsymbol{\theta})$. We show in §A9.9.2 that the composite score function for the T_{ij} is

$$CS(\boldsymbol{\theta}, \mathbf{t}) = \sum_{i=1}^{n-1} \sum_{j>i}^{n} \frac{\partial \gamma(\mathbf{s}_i - \mathbf{s}_j; \boldsymbol{\theta})}{\partial \boldsymbol{\theta}} \frac{1}{4\gamma^2(\mathbf{s}_i - \mathbf{s}_j; \boldsymbol{\theta})} \left(t_{ij}^2 - 2\gamma(\mathbf{s}_i - \mathbf{s}_j; \boldsymbol{\theta})\right). \qquad [9.23]$$

Although this is a complicated looking expression, it is really the nonlinear weighted least squares objective function in the model

$$T_{ij}^2 = 2\gamma(\mathbf{s}_i - \mathbf{s}_j; \boldsymbol{\theta}) + e_{ij},$$

where the e_{ij} are independent random variables with mean 0 and variance $8\gamma^2(\mathbf{s}_i - \mathbf{s}_j; \boldsymbol{\theta})$. Note the correspondence of $\mathrm{Var}[T_{ij}^2]$ to Cressie's variance approximation for the Matheron estimator [9.15]. The expressions are the same considering that $\widehat{\gamma}(\mathbf{h})$ is an average of the T_{ij}.

The composite likelihood estimator can be calculated easily with a nonlinear regression package capable of weighted least squares fitting such as `proc nlin` in The SAS® System. Obtaining the (restricted) maximum likelihood estimate requires a procedure that can minimize [9.19] or [9.20] such as `proc mixed`. The minimization problem in ML or REML estimation is numerically much more involved. One of the main problems there is that the matrix $\boldsymbol{\Sigma}(\boldsymbol{\theta})$ must be inverted repeatedly. For clustered data as in §7, where the variance-covariance matrix is block-diagonal, this is not too cumbersome; the matrix can be inverted block by block. In the case of spatial data $\boldsymbol{\Sigma}(\boldsymbol{\theta})$ does not have a block-diagonal structure and in general no shortcuts can be taken. Zimmerman (1989) derives some simplifications when the observations are collected on a rectangular or parallelogram grid. Composite likelihood (CL) estimation on the other hand replaces the inversion of one large matrix with many inversions of small matrices. The largest matrix to be inverted for a semivariogram model with 3 parameters (nugget, sill, range) is a 3×3 matrix. However, CL estimation processes many more data points. With $n = 100$ spatial observations there are $n(n-1)/2 = 4,950$ pairs. That many observations are hardly needed. It is quite reasonable to remove those pairs from estimation whose spatial distance is many times greater than the likely range or even to randomly subsample the $n(n-1)/2$ distances. An advantage over the least squares type methods is the reliance on the data directly without binning pairwise differences into lag classes.

Recently, generalized estimating equations (GEE) have received considerable attention. In the mid-1980s they were mostly employed for the estimation of mean parameters following work by Liang and Zeger (1986) and Zeger and Liang (1986). Later the GEE methodology was extended to the estimation of association parameters, for example, the correlation among repeated measurements (see, e.g., Prentice 1988, Zhao and Prentice 1990). McShane et al. (1997) applied GEE techniques for the estimation of the dependence in spatial data. It turns out that there is a direct connection between generalized estimating equations for dependence parameters and the composite likelihood method. §A9.9.2 contains the details.

Adding Flexibility: Nested Models and Nonparametric Fitting

One of the appealing features of nonparametric methods of statistical modeling is the absence of a rigid mathematical model. In the parametric setting the user chooses a class of models and estimates the unknown parameters of the model based on data to select one member of the class, which is the fitted model. For example, when fitting an exponential semivariogram, we assume that the process has a semivariogram of form

$$\gamma(\mathbf{h};\boldsymbol{\theta}) = \begin{cases} 0 & \mathbf{h} = \mathbf{0} \\ \theta_0 + \theta_s\left\{1 - \exp\left\{-\frac{3\|\mathbf{h}\|}{\alpha}\right\}\right\} & \mathbf{h} \neq \mathbf{0} \end{cases}$$

and the nugget θ_0, sill θ_s, and practical range α are estimated based on data. Our list of isotropic semivariogram models in §9.2.2 is relatively short. Although many more semivariogram models are known, typically users resort to one of the models shown there. In applications one may find that none of these describes the empirical semivariogram well, for example, because the random field does not have constant mean, is anisotropic, or consists of different scales of variation. The latter reason is the idea behind what is termed the linear model of regionalization in the geostatistical literature (see, for example, Goovaerts 1997, Ch. 4.2.3). Statistically, it is based on the facts that (i) if $C_1(\mathbf{h})$ and $C_2(\mathbf{h})$ are valid covariance structures in $\mathbb{R}^2$, then $C_1(\mathbf{h}) + C_2(\mathbf{h})$ is also a valid covariance structure in $\mathbb{R}^2$; (ii) if $C(\mathbf{h})$ is a valid structure, so is $bC(\mathbf{h})$ provided $b > 0$. As a consequence, linear combinations of permissible covariance models lead to an overall permissible model. The coefficients in the linear combination must be positive, however. The same results hold for semivariograms.

The linear model of regionalization assumes that the random function $Z(\mathbf{s})$ is a linear combination of p stationary zero-mean random functions. If $U_j(\mathbf{s})$ is a second-order stationary random function with $\mathrm{E}[U_j(\mathbf{s})] = 0$, $\mathrm{Cov}[U_j(\mathbf{s}), U_j(\mathbf{s}+\mathbf{h})] = C_j(\mathbf{h})$ and $a_1, \cdots, a_p$ are positive constants, then

$$Z(\mathbf{s}) = \sum_{j=1}^{p} a_j U_j(\mathbf{s}) + \mu \qquad [9.24]$$

is a second-order stationary random function with mean μ, covariance function

$$\begin{aligned} C_Z(\mathbf{h}) &= \text{Cov}[Z(\mathbf{s}+\mathbf{h}), Z(\mathbf{s})] \\ &= \sum_{j,k} a_j a_k \text{Cov}[U_j(\mathbf{s}), U_k(\mathbf{s}+\mathbf{h})] \\ &= \sum_{j=1}^{p} a_j^2 C_j(\mathbf{h}), \end{aligned} \quad [9.25]$$

and semivariogram $\gamma_Z(\mathbf{h}) = \sum_{j=1}^{p} a_j^2 \gamma_j(\mathbf{h})$ provided that the individual processes $U_1(\mathbf{s}), \cdots, U_p(\mathbf{s})$ are not correlated. If the individual semivariograms $\gamma_j(\mathbf{h})$ have sill 1, then $\sum_{j=1}^{p} a_j^2$ is the variance of an observation. Covariogram and semivariogram models derived from a regionalization such as [9.24] are called nested models. Every semivariogram model containing a nugget effect is thus a nested model.

The variability of a soil property is related to many causes that have different spatial scales, each scale integrating variability at all smaller scales (Russo and Jury 1987a). If the total variability of an attribute varies with the spatial scale or resolution, nested models can capture this dependency, if properly modeled. Nesting models is thus a convenient way to construct theoretical semivariogram models that offer greater flexibility than the basic models in §9.2.2. Nested models are not universally accepted, however. Stein (1999, p. 13) takes exception to nested models where the individual components are spherical models. A danger of nesting semivariograms is to model the effects of nonconstancy of the mean on the empirical semivariogram through a creative combination of second-order stationary and intrinsically stationary semivariograms. Even if this combination fits the empirical semivariogram well, a critical assumption of variogram analysis has been violated. Furthermore, the assumption of mutual independence of the individual random functions $U_1(\mathbf{s}), \cdots, U_p(\mathbf{s})$ must be evaluated with great care. Nugget effects that are due to measurement errors are reasonably assumed to be independent of the other components. But a component describing smaller scale variability due to soil nutrients may not be independent of a larger scale component due to soil types or geology.

To increase the flexibility in modeling the semivariogram of stationary isotropic processes without violating the condition of positive definiteness of the covariogram (conditional negative definiteness of the semivariogram), nonparametric methods can be employed. The rationale behind the nonparametric estimators (a special topic in §A9.9.3) is akin to the nesting of covariogram models in the linear model of regionalization. The covariogram is expressed as a weighted combination of functions, each of which is a valid covariance function. Instead of combining theoretical covariogram models, however, the nonparametric approach combines positive-definite functions that are derived from a spectral representation. These are termed the basis functions. For data on a transect the basis function is $\cos(h)$, for data in the plane it is the Bessel function of the first kind of order zero, and for data in three dimensions it is $\sin(h)/h$ (Figure 9.18).

The flexibility of the nonparametric approach is demonstrated in Figure 9.19 which shows semivariograms constructed with $m = 5$ equally spaced nodes and a maximum lag of $h = 10$. The functions shown as solid lines have equal weights $a_j^2 = 0.2$. The dashed lines are produced by setting $a_1^2 = a_5^2 = 0.5$ and all other weights to zero. The smoothness of the semivariogram decreases with the unevenness of the weights and the number of sign changes of the basis function.

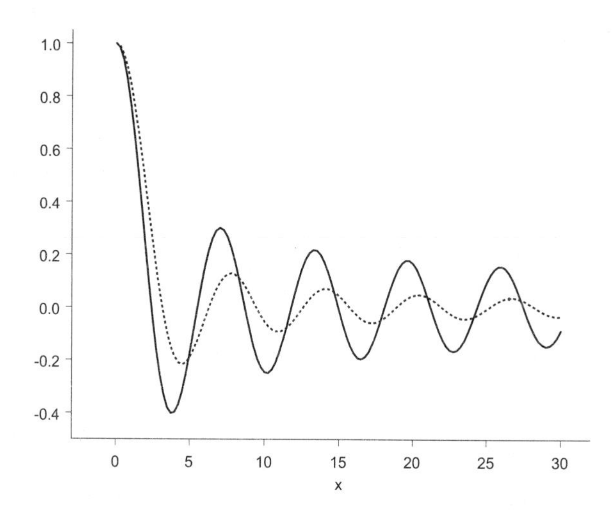

Figure 9.18. Basis functions for two-dimensional data (solid line, Bessel function of the first kind of order 0) and for three-dimensional data (dashed line, $\sin(x)/x$).

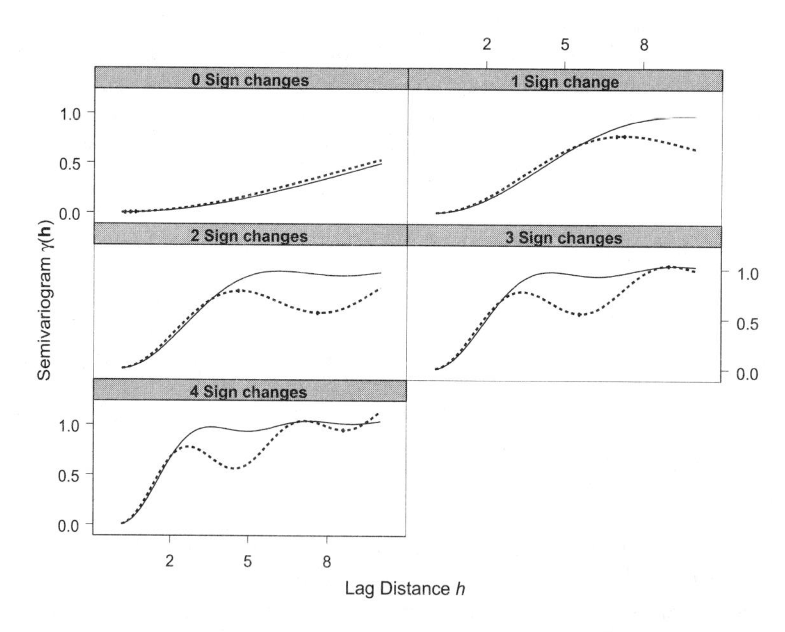

Figure 9.19. Semivariogram models constructed as linear combinations of Bessel functions of the first kind (of order 0) as a function of the weight distribution and the number of sign changes of the basis function. Semivariogram with five nodes and equal weights shown as solid lines, semivariograms with unequal weights shown as dashed lines.

9.3 The Spatial Model

Box 9.5 Spatial Model

- **The Spatial Model is a statistical model decomposing the variability in a random function into a deterministic mean structure and one or more spatial random processes.**

- **The decomposition is not unique and components may be confounded in a particular application.**

- **We distinguish between signal and mean models. In the former interest is primarily in prediction of the signal, whereas in the latter, estimation of the mean structure is more important.**

- **Reactive effects are modeled through the mean structure, interactive effects are modeled through the random structure. For geostatistical data interactive effects are represented through stationary random processes, for lattice data through autoregressive neighborhood structures.**

So far we have been concerned with properties of random fields and the semivariogram or covariogram of a stationary process. Although the fitting of a semivariogram entails modeling, this is only one aspect of representing the structure in spatial data in a manner conducive to a statistical analysis. The constancy of the mean assumption implied by stationarity, for example, is not reasonable in many applications. In a field experiment where treatments are applied to experimental units the variation among units is not just due to spatial variation about a constant mean but also due to the effects of the treatments. Our view of spatial data must be extended to accommodate changes in the mean structure, stationarity, and measurement error. One place to start is to decompose the variability in $Z(\mathbf{s})$ into various sources. Following Cressie (1993, Ch. 3.1), we write for geostatistical data

$$Z(\mathbf{s}) = \mu(\mathbf{s}) + W(\mathbf{s}) + \eta(\mathbf{s}) + e(\mathbf{s}). \qquad [9.26]$$

This decomposition is akin to the breakdown into sources of variability in an analysis of variance model, but it is largely operational. It may be impossible in a particular application to separate the components. It is, however, an excellent starting point to delineate some of the approaches to spatial data analysis. The large-scale variation of $Z(\mathbf{s})$ is expressed through the deterministic mean $\mu(\mathbf{s})$. By implication all other components must have expectation 0. The mean can depend on spatial location and other variables. $W(\mathbf{s})$ is called the **smooth small-scale** variation; it is a stationary process with semivariogram $\gamma_W(\mathbf{h})$ whose range is larger than the smallest lag distance in the sample. The variogram of the smooth-scale variation should thus exhibit some spatial structure and can be modeled by the techniques in §9.2.4. $\eta(\mathbf{s})$ is a spatial process with variogram $\gamma_\eta(\mathbf{h})$ whose range is less than the smallest lag in the data set. Cressie (1993, p. 112) terms it **microscale** variation. The semivariogram of $\eta(\mathbf{s})$ cannot be modeled, no data are available at lags less than its range. The presence of micro-scale variation is reflected in the variogram of $Z(\mathbf{s}) - \mu(\mathbf{s})$ as a nugget effect which measures the sill σ_η^2 of $\gamma_\eta(\mathbf{h})$. $e(\mathbf{s})$, finally, is a white-noise process representing measurement error. The variance of $e(\mathbf{s})$, σ_e^2, also contributes to the nugget effect. There are three random components on the right-hand side of the mixed model [9.26]. These are usually assumed to be independent of each other.

With this decomposition in place, we define two basic types of models by combining one or more components.

1. **Signal Model**. Let $S(\mathbf{s}) = \mu(\mathbf{s}) + W(\mathbf{s}) + \eta(\mathbf{s})$ denote the **signal** of the process. Then, $Z(\mathbf{s}) = S(\mathbf{s}) + e(\mathbf{s})$. If the task is to predict the attribute of interest at unobserved locations it is not reasonable to predict the noisy version $Z(\mathbf{s})$ which is affected by measurement error (unless $\sigma_e^2 = 0$). We are interested in the value of a soil attribute at location $\mathbf{s}_0$, not in the value we would measure in error when a sample at $\mathbf{s}_0$ were taken. Spatial prediction (kriging) should always focus on predicting the signal $S()$. Only if the data are measured without error is prediction of $Z(\mathbf{s})$ the appropriate course of action (in this case $Z(\mathbf{s}) = S(\mathbf{s})$). The *controversy* whether the kriging predictor is a perfect interpolator or not is in large measure explained by considering prediction of $Z(\mathbf{s})$ vs. $S(\mathbf{s})$ in the presence of measurement error (see §A9.9.5) and whether the nugget effect is recognized as originating from measurement error or is part of the signal through micro-scale variation ($\sigma_\eta^2 > 0, \sigma_e^2 = 0$).

2. **Mean Model**. Let $\delta(\mathbf{s}) = W(\mathbf{s}) + \eta(\mathbf{s}) + e(\mathbf{s})$ denote the **error process** (the stochastic part of $Z(\mathbf{s})$). Then, $Z(\mathbf{s}) = \mu(\mathbf{s}) + \delta(\mathbf{s})$. This model is the entry point for spatial regression and analysis of variance where focus is on modeling the mean function $\mu(\mathbf{s})$ as a function of covariates and point location and $\delta(\mathbf{s})$ is assumed to have spatial autocorrelation structure (§9.5). In this formulation $\mu(\mathbf{s})$ is sometimes called the large-scale trend and $\delta(\mathbf{s})$ simply the small-scale trend. All types of spatial data have their specific tools to investigate large- and small-scale properties of a random field. The large-scale trend is captured by the mean function for geostatistical data, the mean vector for lattice data, and the intensity for a spatial point process. The small-scale structure is captured by the semivariogram and covariogram for geostatistical and lattice data, and the K-function for point patterns (§9.7).

The signal and mean models have different focal points. In the former we are primarily interested in the stochastic behavior of the random field and if it is spatially structured employ this fact to predict $Z(\mathbf{s})$ or $S(\mathbf{s})$ at observed and unobserved locations. The mean $\mathrm{E}[Z(\mathbf{s})] = \mu(\mathbf{s})$ is somewhat ancillary in these investigations apart from the fact that if $\mu(\mathbf{s})$ depends on location we must pay it special attention to model the stochastic structure properly since $S(\mathbf{s})$ (and by extension $Z(\mathbf{s})$) will not be stationary. This is the realm of the geostatistical method (§9.4) that calls on kriging methods. It is in the assumptions about $\mu(\mathbf{s})$ that geostatistical methods of spatial prediction are differentiated into simple, ordinary, and universal kriging. In the mean model $Z(\mathbf{s}) = \mu(\mathbf{s}) + \delta(\mathbf{s})$ interest lies primarily in modeling the large-scale trend (mean structure $\mu(\mathbf{s})$) of the process and in turn the stochastic structure that arises from $\delta(\mathbf{s})$ is somewhat ancillary. We must pay attention to the stochastic properties of $\delta(\mathbf{s})$, however, to ensure that the inferences drawn about the mean are appropriate. For example, if spatial autocorrelation exists, the semivariogram of $\delta(\mathbf{s})$ will not be a nugget-only model and taking the correlations among observations into account is critical to obtain reliable inference about the mean $\mu(\mathbf{s})$. An example application where $\mu(\mathbf{s})$ is of primary importance is the execution of a large field experiment where the experimental units are arranged in such a fashion that the effects of spatial autocorrelation are not removed by the blocking scheme. Although randomization neutralizes the spatial effects by balancing them across the units under conceptual repetitions of the basic experiment, we execute the experiment only once and may obtain a layout where spatial dependency among experimental units increases the experimental error variance to a point where meaningful inferences about the treatment effects (information that is captured by $\mu(\mathbf{s})$) are no longer possible. Incorporating the spatial effects in the analysis by modeling $\mu(\mathbf{s})$ as a function of treatment effects and $\delta(\mathbf{s})$ as a spatial random field can assist in recovering vital information about treatment performance.

In particular for mean models the analyst must decide which effects are part of the large-scale structure $\mu(\mathbf{s})$ and which are components of the error structure $\delta(\mathbf{s})$. There is no unanimity between researchers. One modeler's fixed effect is someone else's random effect. This contributes to the nonuniqueness of the decomposition [9.26]. Consider the special case where $\mu(\mathbf{s})$ is linear,

$$Z(\mathbf{s}) = \mathbf{x}'(\mathbf{s})\boldsymbol{\beta} + \delta(\mathbf{s}),$$

and $\mathbf{x}(\mathbf{s})$ is a vector of regressor variables that can depend on spatial coordinates alone or on other explanatory variables and factors. Cliff and Ord (1981, Ch. 6) distinguish between **reaction** and **interaction** models. In a reaction model sites react to outside influences, e.g., plants will react to the availability of nutrients in the root zone. Since this availability varies spatially, plant size or biomass will exhibit a regression-like dependence on nutrient availability. It is then reasonable to include nutrient availability as a covariate in the regressor vector $\mathbf{x}(\mathbf{s})$. In an interaction model, sites react not to outside influences but react with each other. Neighboring plants compete with each other for resources, for example. In general, when the dominant spatial effects are caused by sites reacting to external forces, these effects should be part of the mean function $\mathbf{x}'(\mathbf{s})\boldsymbol{\beta}$. Interactive effects (reaction among sites) call for modeling spatial variability through the spatial autocorrelation structure of the error process.

The distinction between reactive and interactive models is useful, but not cut-and-dried. Significant autocorrelation in the data does not imply an interactive model over a reactive one or vice versa. Spatial autocorrelation can be spurious if caused by large-scale trends or real if caused by cumulative small-scale, spatially varying components. The error structure is thus often thought of as the local structure and the mean is referred to as the global structure. With increasing complexity of the mean model $\mathbf{x}'(\mathbf{s})\boldsymbol{\beta}$, for example, as higher-order terms are added to a response surface, the mean will be more spatially variable and more localized. In a two-way row-column layout (randomized block design) where rows and columns interact one could model the data as

$$Z_{ij} = \mu + \alpha_i + \beta_j + \gamma s_1 s_2 + e_{ij},\ e_{ij} \sim iid\left(0, \sigma^2\right),$$

where α_i denotes row, β_j column effects and s_1, s_2 are the cell coordinates. This model assumes that the term $\gamma s_1 s_2$ removes any residual spatial autocorrelation, hence the errors e_{ij} are uncorrelated. Alternatively, one could invoke the model

$$Z_{ij} = \mu + \alpha_i + \beta_j + \delta_{ij},$$

where the δ_{ij} are autocorrelated. One modeler's reactive effect will be another modeler's interactive effect.

With geostatistical data the spatial dependency between $\delta(\mathbf{s}_i)$ and $\delta(\mathbf{s}_j)$ (the interaction) is modeled through the semivariogram or covariogram of the $\delta()$ process. If the spatial domain is discrete (lattice data), modifications are necessary since $W(\mathbf{s})$ and $\eta(\mathbf{s})$ in decomposition [9.26] are smooth-scale stationary processes with a continuous domain. As before, reactive effects can be modeled as effects on the mean structure through regressor variables in $\mathbf{x}(\mathbf{s})$. Interactions between sites can be incorporated into the model in the following manner. The response $Z(\mathbf{s}_i)$ at location $\mathbf{s}_i$ is decomposed into three components: (i) the mean $\mu(\mathbf{s}_i)$, (ii) a contribution from the neighboring observations, and (iii) random error. Mathematically, we can express the decomposition as

$$\begin{aligned} Z(\mathbf{s}_i) &= \mu(\mathbf{s}_i) + \delta^*(\mathbf{s}_i) \\ &= \mu(\mathbf{s}_i) + \sum_{j=1}^{n} b_{ij}(Z(\mathbf{s}_j) - \mu(\mathbf{s}_j)) + e(\mathbf{s}_i). \end{aligned} \quad [9.27]$$

The contribution to $Z(\mathbf{s}_i)$ made by other sites is a linear combination of residuals at other locations. By convention we put $b_{ii} = 0$ in [9.27]. The $e(\mathbf{s}_i)$'s are uncorrelated random errors with mean 0 and variance σ_i^2. If all $b_{ij} = 0$ and $\mu(\mathbf{s}_i) = \mathbf{x}'(\mathbf{s}_i)\boldsymbol{\beta}$, the model reduces to $Z(\mathbf{s}_i) = \mathbf{x}'(\mathbf{s}_i)\boldsymbol{\beta} + e(\mathbf{s}_i)$, a standard linear regression model. The interaction coefficients b_{ij} contain information about the strength of the dependence between sites $Z(\mathbf{s}_i)$ and $Z(\mathbf{s}_j)$. Since $\sum_{j=1}^{n} b_{ij}(Z(\mathbf{s}_j) - \mu(\mathbf{s}_j))$ is a function of random variables it can be considered part of the *error* process. In a model for lattice data it can be thought of as replacing the smooth-scale random function $W(\mathbf{s})$ in [9.26]. Model [9.27] is the spatial equivalent of an autoregressive time series model where the current value in the series depends on previous values. In the spatial case we potentially let $Z(\mathbf{s}_i)$ depend on all other sites since space is not directed. More precisely, model [9.27] is the spatial equivalent of a simultaneous time series model, hence the denomination as a **S**imultaneous **S**patial **A**uto**r**egressive (SSAR) model. We discuss SSAR and a further class of interaction models for lattice data, the **C**onditional **S**patial **A**uto**r**egressive (CSAR) models, in §9.6.

Depending on whether the spatial process has a continuous or discrete domain, we now have two types of mean models. Let $\mathbf{Z}(\mathbf{s}) = [Z(\mathbf{s}_1), \cdots, Z(\mathbf{s}_n)]'$ be the $(n \times 1)$ vector of the attribute Z at all observed locations, $\mathbf{X}(\mathbf{s})$ be the $(n \times k)$ regressor matrix

$$\mathbf{X}(\mathbf{s}) = \begin{bmatrix} \mathbf{x}'(\mathbf{s}_1) \\ \mathbf{x}'(\mathbf{s}_2) \\ \vdots \\ \mathbf{x}'(\mathbf{s}_n) \end{bmatrix},$$

and $\boldsymbol{\delta}(\mathbf{s}) = [\delta(\mathbf{s}_1), \cdots, \delta(\mathbf{s}_n)]'$ the vector of errors in the mean model for geostatistical data. The model can be written as

$$\mathbf{Z}(\mathbf{s}) = \mathbf{X}(\mathbf{s})\boldsymbol{\beta} + \boldsymbol{\delta}(\mathbf{s}), \quad [9.28]$$

where $\mathrm{E}[\boldsymbol{\delta}(\mathbf{s})] = \mathbf{0}$ and the variance-covariance matrix of $\boldsymbol{\delta}(\mathbf{s})$ contains the covariance function of the $\delta()$ process. If $\mathrm{Cov}[\delta(\mathbf{s}_i), \delta(\mathbf{s}_j)] = C(\mathbf{s}_i - \mathbf{s}_j;\boldsymbol{\theta})$, then

$$\mathrm{Var}[\boldsymbol{\delta}(\mathbf{s})] = \boldsymbol{\Sigma}(\boldsymbol{\theta}) = \begin{bmatrix} C(\mathbf{0};\boldsymbol{\theta}) & C(\mathbf{s}_1 - \mathbf{s}_2;\boldsymbol{\theta}) & \cdots & C(\mathbf{s}_1 - \mathbf{s}_n;\boldsymbol{\theta}) \\ C(\mathbf{s}_2 - \mathbf{s}_1;\boldsymbol{\theta}) & C(\mathbf{0};\boldsymbol{\theta}) & \cdots & C(\mathbf{s}_2 - \mathbf{s}_n;\boldsymbol{\theta}) \\ \vdots & & \ddots & \vdots \\ C(\mathbf{s}_n - \mathbf{s}_1;\boldsymbol{\theta}) & \cdots & C(\mathbf{s}_n - \mathbf{s}_{n-1};\boldsymbol{\theta}) & C(\mathbf{0};\boldsymbol{\theta}) \end{bmatrix}.$$

Note that $\boldsymbol{\Sigma}(\boldsymbol{\theta})$ is also the variance-covariance matrix of $\mathbf{Z}(\mathbf{s})$. Unknown quantities in this model are the vector of fixed effects $\boldsymbol{\beta}$ in the mean function and the vector $\boldsymbol{\theta}$ in the covariance function. Now consider the SSAR model for lattice data ([9.27]). Collecting the autoregressive coefficients b_{ij} into matrix $\mathbf{B}$ and assuming $\boldsymbol{\mu}(\mathbf{s}) = \mathbf{X}(\mathbf{s})\boldsymbol{\beta}$, [9.27] can be written as

$$\begin{aligned} \mathbf{Z}(\mathbf{s}) &= \mathbf{X}(\mathbf{s})\boldsymbol{\beta} + \mathbf{B}(\mathbf{Z}(\mathbf{s}) - \mathbf{X}(\mathbf{s})\boldsymbol{\beta}) + \mathbf{e}(\mathbf{s}) \\ &\Leftrightarrow (\mathbf{I} - \mathbf{B})(\mathbf{Z}(\mathbf{s}) - \mathbf{X}(\mathbf{s})\boldsymbol{\beta}) = \mathbf{e}(\mathbf{s}). \end{aligned} \qquad [9.29]$$

It follows that $\text{Var}[\mathbf{Z}(\mathbf{s})] = (\mathbf{I} - \mathbf{B})^{-1}\text{Var}[\mathbf{e}(\mathbf{s})](\mathbf{I} - \mathbf{B}')^{-1}$. In applications of the SSAR model it is often assumed that the errors are homoscedastic with variance σ^2. Then, $\text{Var}[\mathbf{Z}(\mathbf{s})] = \sigma^2(\mathbf{I} - \mathbf{B})^{-1}(\mathbf{I} - \mathbf{B}')^{-1}$. Parameters of the SSAR model are the vector of fixed effects $\boldsymbol{\beta}$, the residual variance σ^2, and the entries of the matrix $\mathbf{B}$. While the covariance matrix $\boldsymbol{\Sigma}(\boldsymbol{\theta})$ depends on only a few parameters with geostatistical data (nugget, sill, range), the matrix $\mathbf{B}$ can contain many unknowns. It is not even required that $\mathbf{B}$ is symmetric, only that $\mathbf{I} - \mathbf{B}$ is invertible. For purposes of parameter estimation it is thus required to place some structure on $\mathbf{B}$ to reduce the number of unknowns. For example, one can put $\mathbf{B} = \rho\mathbf{W}$, where $\mathbf{W}$ is a matrix selected by the user that identifies which sites are spatially connected and the parameter ρ determines the strength of the spatial dependence. Table 9.2 summarizes the key differences between the mean models for geostatistical and lattice data. How to structure the matrix $\mathbf{B}$ in the SSAR model and the corresponding matrix in the CSAR model is discussed in §9.6.

Table 9.2. Mean models for geostatistical and lattice data (with $\text{Var}[\mathbf{e}(\mathbf{s})] = \sigma^2\mathbf{I}$)

	Geostatistical Data	**Lattice Data**
Model $\mathbf{Z}(\mathbf{s}) =$	$\mathbf{X}(\mathbf{s})\boldsymbol{\beta} + \boldsymbol{\delta}(\mathbf{s})$	$\mathbf{X}(\mathbf{s})\boldsymbol{\beta} + \mathbf{B}(\mathbf{Z}(\mathbf{s}) - \mathbf{X}(\mathbf{s})\boldsymbol{\beta}) + \mathbf{e}(\mathbf{s})$
$\text{E}[\mathbf{Z}(\mathbf{s})]$	$\mathbf{X}(\mathbf{s})\boldsymbol{\beta}$	$\mathbf{X}(\mathbf{s})\boldsymbol{\beta}$
$\text{Var}[\mathbf{Z}(\mathbf{s})]$	$\boldsymbol{\Sigma}(\boldsymbol{\theta})$	$\sigma^2(\mathbf{I} - \mathbf{B})^{-1}(\mathbf{I} - \mathbf{B}')^{-1}$
Mean parameters	$\boldsymbol{\beta}$	$\boldsymbol{\beta}$
Dependency parameters	$\boldsymbol{\theta}$	σ^2, $\mathbf{B}$

9.4 Spatial Prediction and the Kriging Paradigm

9.4.1 Motivation of the Prediction Problem

Box 9.6 Prediction vs. Estimation

- **Prediction is the determination of the value of a random variable. Estimation is the determination of the value of an unknown constant.**

- **If interest lies in the value of a random field at location s then we should predict $Z(\mathbf{s})$ or the signal $S(\mathbf{s})$. If the average value at location s across all realizations of the random experiment is of interest, we should estimate $\text{E}[Z(\mathbf{s})]$.**

- **Kriging methods are solutions to the prediction problem where requirements for the predictor are combined with assumptions about the spatial model.**

A common goal in the analysis of geostatistical data is the mapping of the random function $Z(\mathbf{s})$ in some region of interest. The sampling process produces observations $Z(\mathbf{s}_1),\cdots,Z(\mathbf{s}_n)$ but $Z(\mathbf{s})$ varies continuously through the domain D. To produce a map of $Z(\mathbf{s})$ requires prediction of $Z()$ at unobserved locations $\mathbf{s}_0$. What is commonly referred to as the **geostatistical method** consists of the following steps (at least the first 6).

1. Using exploratory techniques, prior knowledge, and/or anything else, posit a model of possibly nonstationary mean plus second-order or intrinsically stationary error for the $Z(\mathbf{s})$ process that generated the data.
2. Estimate the mean function by ordinary least squares, smoothing, or median polishing to **detrend** the data. If the mean is stationary this step is not necessary. The methods for detrending employed at this step usually do not take autocorrelation into account.
3. Using the residuals obtained in step 2 (or the original data if the mean is stationary), fit a semivariogram model $\gamma(\mathbf{h};\boldsymbol{\theta})$ by one of the methods in §9.2.4.
4. Statistical estimates of the spatial dependence in hand (from step 3) return to step 2 to re-estimate the parameters of the mean function, now taking into account the spatial autocorrelation.
5. Obtain new residuals from step 4 and iterate steps 2 through 4, if necessary.
6. Predict the attribute $Z()$ at unobserved locations and calculate the corresponding mean square prediction errors.

If the mean is stationary or if the steps of detrending the data and subsequent estimation of the semivariogram (or covariogram) are not iterated, the geostatistical method consists of only steps 1, 2, 3, and 6. This section is concerned with the last item in this process, the prediction of the attribute (mapping).

Understanding the difference between **predicting** $Z(\mathbf{s}_0)$, which is a random variable, and **estimation** of the mean of $Z(\mathbf{s}_0)$, which is a constant, is essential to gain an appreciation for the geostatistical methods employed to that end. To motivate the problem of spatial prediction focus first on a classical linear model $\mathbf{Y} = \mathbf{X}\boldsymbol{\beta} + \mathbf{e}$, $\mathbf{e} \sim (\mathbf{0}, \sigma^2\mathbf{I})$. What do we mean by a *predicted* value at a regressor value $\mathbf{x}_0$? We can think of a large number of possible outcomes that share the same set of regressors $\mathbf{x}_0$ and average their response values. This average is an estimate of the mean of Y at $\mathbf{x}_0$, $\mathrm{E}[Y|\mathbf{x}_0]$. Once we have fitted the model to data and obtained estimates $\widehat{\boldsymbol{\beta}}$ the obvious estimate of this quantity is $\mathbf{x}_0'\widehat{\boldsymbol{\beta}}$. What if a predicted value is interpreted as the response of the *next* observation that has regressors $\mathbf{x}_0$? This definition appeals not to infinitely many observations at $\mathbf{x}_0$, but a single one. Rather than predicting the expected value of Y, Y itself is then of interest. In the spatial context imagine that $Z(\mathbf{s})$ is the soil loss potential at location $\mathbf{s}$ in a particular field. If an agronomist is interested in the soil loss potential of a large number of fields with properties similar to the sampled one, the important quantity would be $\mathrm{E}[Z(\mathbf{s})]$. An agronomist interested in the soil loss potential of the sampled field at a location not contained in the sample would want to predict $Z(\mathbf{s}_0)$, the actual soil loss potential at location $\mathbf{s}_0$. Returning to the classical linear model example, it turns out that regardless of whether interest is in a single value or an average, the predictor of $Y|\mathbf{x}_0$ and the estimator of $\mathrm{E}[Y|\mathbf{x}_0]$ are the same:

$$\widehat{\mathrm{E}}[Y|\mathbf{x}_0] = \mathbf{x}_0'\widehat{\boldsymbol{\beta}}$$
$$\widehat{Y}|\mathbf{x}_0 = \mathbf{x}_0'\widehat{\boldsymbol{\beta}}.$$

The difference between the two *predictions* does not lie in the predicted value, but in their precision (standard errors). Standard linear model theory, where $\boldsymbol{\beta}$ is estimated by ordinary least squares, instructs that

$$\mathrm{Var}\left[\widehat{\mathrm{E}}[Y|\mathbf{x}_0]\right] = \mathrm{Var}\left[\mathbf{x}_0'\widehat{\boldsymbol{\beta}}\right] = \sigma^2\mathbf{x}_0'(\mathbf{X}'\mathbf{X})^{-1}\mathbf{x}_0 \qquad [9.30]$$

is the variance of the predicted mean at $\mathbf{x}_0$. When predicting random variables one considers the variance of the **prediction error** $Y|\mathbf{x}_0 - \widehat{Y}|\mathbf{x}_0$ to take into account the variability of the new observation. If the new observation $Y|\mathbf{x}_0$ is uncorrelated with $\mathbf{Y}$, then

$$\mathrm{Var}\left[Y|\mathbf{x}_0 - \widehat{Y}|\mathbf{x}_0\right] = \mathrm{Var}[Y|\mathbf{x}_0] + \sigma^2\mathbf{x}_0'(\mathbf{X}'\mathbf{X})^{-1}\mathbf{x}_0. \qquad [9.31]$$

Although the same formula ($\mathbf{x}_0'\boldsymbol{\beta}$) is used for predictions, the uncertainty associated with predicting a random variable ([9.31]) exceeds the uncertainty in predicting the mean ([9.30]). To consider the variance of the prediction error in one case and the variance of the predictor in the other is not arbitrary. Consider some quantity U is to be predicted. We use some function $f(\mathbf{Y})$ of the data as the predictor. If $\mathrm{E}[U] = \mathrm{E}[f(\mathbf{Y})]$, the mean square prediction error is

$$\begin{aligned} MSE[U, f(\mathbf{Y})] &= \mathrm{E}\left[(U - f(\mathbf{Y}))^2\right] = \mathrm{Var}[U - f(\mathbf{Y})] \\ &= \mathrm{Var}[U] + \mathrm{Var}[f(\mathbf{Y})] - 2\mathrm{Cov}[U, f(\mathbf{Y})]. \end{aligned} \qquad [9.32]$$

In the standard linear model estimating $\mathrm{E}[Y|\mathbf{x}_0]$ and predicting $Y|\mathbf{x}_0$ correspond to the following:

Target U	$f(\mathbf{Y})$	$\mathrm{Var}[U]$	$\mathrm{Var}[f(\mathbf{Y})]$	$\mathrm{Cov}[U, f(\mathbf{Y})]$	$MSE[U, f(\mathbf{Y})]$
$\mathrm{E}[Y\|\mathbf{x}_0]$	$\mathbf{x}_0'\widehat{\boldsymbol{\beta}}$	0	$\sigma^2\mathbf{x}_0'(\mathbf{X}'\mathbf{X})^{-1}\mathbf{x}_0$	0	$\sigma^2\mathbf{x}_0'(\mathbf{X}'\mathbf{X})^{-1}\mathbf{x}_0$
$Y\|\mathbf{x}_0$	$\mathbf{x}_0'\widehat{\boldsymbol{\beta}}$	$\mathrm{Var}[Y\|\mathbf{x}_0]$	$\sigma^2\mathbf{x}_0'(\mathbf{X}'\mathbf{X})^{-1}\mathbf{x}_0$	$0^\dagger$	$\mathrm{Var}[Y\|\mathbf{x}_0] + \sigma^2\mathbf{x}_0'(\mathbf{X}'\mathbf{X})^{-1}\mathbf{x}_0$

† provided $Y|\mathbf{x}_0$ is independent of the observed vector $\mathbf{Y}$ on which the estimate $\widehat{\boldsymbol{\beta}}$ is based.

The variance formulas [9.30] and [9.31] *are* mean square errors and well-known from linear model theory for uncorrelated data. Under these conditions it turns out that $\widehat{Y}|\mathbf{x}_0 = \mathbf{x}_0'\widehat{\boldsymbol{\beta}}$ is the best linear unbiased predictor of $Y|\mathbf{x}_0$ and that $\widehat{\mathrm{E}}[Y|\mathbf{x}_0] = \mathbf{x}_0'\widehat{\boldsymbol{\beta}}$ is the best linear unbiased estimator of $\mathrm{E}[Y|\mathbf{x}_0]$. Expression [9.31] applies only, however, if $Y|\mathbf{x}_0$ **and Y** are not correlated.

Spatial data exhibit spatial autocorrelations which are a function of the proximity of observations. Denote by $Z(\mathbf{s}_1), \cdots, Z(\mathbf{s}_n)$ the attribute at the observed locations $\mathbf{s}_1, \cdots, \mathbf{s}_n$ and as $\mathbf{s}_0$ the target location where prediction is desired. If the observations are spatially correlated, then $Z(\mathbf{s}_0)$ is also correlated with the observations unless the target location $\mathbf{s}_0$ is further removed from the observed locations than the spatial range (Figure 9.20). We must then ask which function of the data best predicts $Z(\mathbf{s}_0)$ and how to measure the mean square prediction error. In order to solve this problem we need to define what *best* means. **Kriging** methods are solutions to the prediction problem where a predictor is best if it (i) minimizes the mean square prediction error, (ii) is linear in the observed values $Z(\mathbf{s}_1), \cdots, Z(\mathbf{s}_n)$, and (iii) is unbiased in the sense that the mean of the predicted value at $\mathbf{s}_0$ equals the mean of $Z(\mathbf{s}_0)$. There are many variants of the kriging method (Table 9.3) and their combinations create a stunning array of techniques (e.g., universal block kriging, lognormal cokriging, $\cdots$).

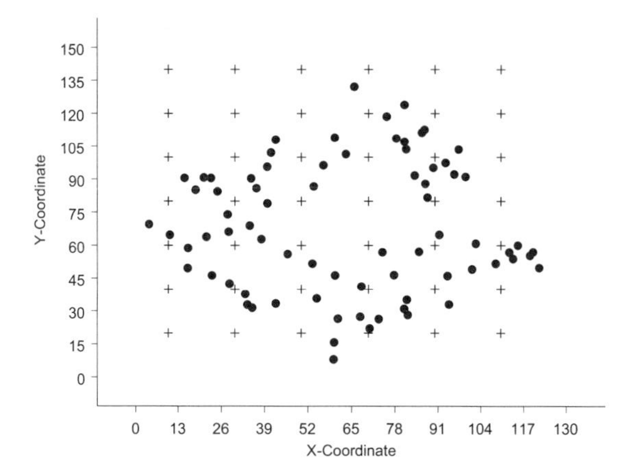

Figure 9.20. Observed sample locations (dots). Crosses denote target locations for prediction. Strength of correlation between observations at target locations and at observed locations depends on the distance between dots and crosses.

Table 9.3. Different kriging methods and commonly encountered names

Distinguished by		**Method known as**
Size of Target		
	Points	(Point) Kriging
	Areas	Block Kriging
What is known		
	μ known	Simple Kriging
	μ unknown but constant	Ordinary Kriging
	$\boldsymbol{\mu} = \mathrm{E}[\mathbf{Z}(\mathbf{s})] = \mathbf{X}\boldsymbol{\beta}$, $\boldsymbol{\beta}$ unknown	Universal Kriging
Distribution		
	$Z(\mathbf{s}) \sim$ Gaussian	Kriging
	$\ln Z(\mathbf{s}) \sim$ Gaussian	Lognormal Kriging
	$\phi(Z(\mathbf{s})) \sim$ Gaussian	Transgaussian Kriging
	$Z(\mathbf{s})$ is an indicator variable	Indicator Kriging
	Gaussian with isolated outliers	Robust Kriging
	Unknown	Median Polish Kriging
	$\mu(\mathbf{s})$ is itself a random process	Bayesian Kriging
Number of attributes		
	Single attribute $Z(\mathbf{s})$	Kriging
	Multiple attributes $Z_1(\mathbf{s}), \cdots, Z_k(\mathbf{s})$	Cokriging
Linearity		
	Predictor linear in $\mathbf{Z}(\mathbf{s})$	Kriging
	Predictor linear in functions of $\mathbf{Z}(\mathbf{s})$	Disjunctive Kriging

The term kriging was coined by Matheron (1963) who named the method after the South African mining engineer, D.G. Krige. Cressie (1993, p. 119) points out that kriging is used both as a noun and a verb. As a noun, kriging implies optimal prediction of $Z(\mathbf{s}_0)$; as a verb it implies optimally predicting $Z(\mathbf{s}_0)$. Ordinary and universal kriging are the most elementary variants and those most frequently applied in agricultural practice. They are discussed in §9.4.3. In the literature kriging is often referred to as an optimal prediction method, implying that a kriging predictor beats any other predictor. This is not necessarily true. If, for example, the discrepancy between the target $Z(\mathbf{s}_0)$ and the predictor $p(\mathbf{Z};\mathbf{s}_0)$ is measured as $|Z(\mathbf{s}_0) - p(\mathbf{Z};\mathbf{s}_0)|$, they are not best in the sense of minimizing the average discrepancy. If $Z(\mathbf{s})$ is not a Gaussian random field the kriging predictor is not optimal unless further restrictions are imposed. It is thus important to understand the conditions under which kriging methods are *best* to avoid overstating their faculties.

9.4.2 The Concept of Optimal Prediction

Box 9.7 The Optimal Predictor

- **To find the optimal predictor $p(\mathbf{Z};\mathbf{s}_0)$ for $Z(\mathbf{s}_0)$ requires a measure for the loss incurred by using $p(\mathbf{Z};\mathbf{s}_0)$ for prediction at $\mathbf{s}_0$. Different loss functions result in different *best* predictors.**

- **The loss function of greatest importance in statistics is squared-error loss, $\{Z(\mathbf{s}_0) - p(\mathbf{Z};\mathbf{s}_0)\}^2$. Its expected value is the mean square prediction error (MSPE).**

- **The predictor that minimizes the mean square prediction error is the conditional mean $\mathrm{E}[Z(\mathbf{s}_0)|\mathbf{Z}(\mathbf{s})]$. If the random field is Gaussian the conditional mean is linear in the observed values.**

When a statistical method is labeled as optimal or best, we need to inquire under which conditions optimality holds; there are few methods that are uniformly best. The famous pooled t-test, for example, is a *uniformly* most powerful test, but only if *uniformly* means among *all* tests for comparing the means of two Gaussian populations with common variance. If $Z(\mathbf{s}_0)$ is the target quantity to be predicted at location $\mathbf{s}_0$ a measure of the **loss** incurred by using some predictor $p(\mathbf{Z};\mathbf{s}_0)$ for $Z(\mathbf{s}_0)$ is required (we use $p(\mathbf{Z};\mathbf{s}_0)$ as a shortcut for $p(\mathbf{Z}(\mathbf{s});\mathbf{s}_0)$). The most common loss function in statistics is squared error loss

$$\{Z(\mathbf{s}_0) - p(\mathbf{Z};\mathbf{s}_0)\}^2 \qquad [9.33]$$

because of its mathematical tractability and simple interpretation. But [9.33] is not directly useful since it is a random quantity that depends on unknowns. Instead, we consider its average, $\mathrm{E}[\{Z(\mathbf{s}_0) - p(\mathbf{Z};\mathbf{s}_0)\}^2]$, the mean square error of using $p(\mathbf{Z};\mathbf{s}_0)$ to predict $Z(\mathbf{s}_0)$. This expected value is also called the Bayes risk under squared error. If squared error loss is accepted as the suitable loss function, among all possible predictors the one that should be chosen is that which minimizes the Bayes risk. This turns out to be the conditional expectation $p^0(\mathbf{Z};\mathbf{s}_0) = \mathrm{E}[Z(\mathbf{s}_0)\,|\,\mathbf{Z}(\mathbf{s})]$. The minimized mean square prediction error (MSPE) then takes on the following, surprising form:

$$\mathrm{E}\left[\{Z(\mathbf{s}_0) - p^0(\mathbf{Z};\mathbf{s}_0)\}^2\right] = \mathrm{Var}[Z(\mathbf{s}_0)] - \mathrm{Var}\left[p^0(\mathbf{Z};\mathbf{s}_0)\right]. \qquad [9.34]$$

This is a stunning result since variances are usually added, not subtracted. From [9.34] it is immediately obvious that the conditional mean must be less variable than the random field at $\mathbf{s}_0$, because the mean square error is a positive quantity. Perhaps even more surprising is the fact that the MSPE is small if the variance of the predictor is large. Consider a time series where the value of the series at time 20 is to be predicted (Figure 9.21). Three different types of predictors are used. The sample average $\overline{y}$ and two nonparametric fits that differ in their smoothness. The sample mean $\overline{y}$ is the smoothest of the three predictors, since it does not change with time. The loess fit with large bandwidth (dashed line) is less smooth than $\overline{y}$ and more smooth than the loess fit with small bandwidth (solid line). The less smooth the predictor, the greater its variability and the more closely it will follow the data. The chance that a smooth predictor is close to the unknown observation at $t = 20$ is smaller than for one of the variable (more jagged) predictors. The most variable predictor is one that interpolates the data points (connects the dots). Such a predictor is said to **honor the data** or to be **a perfect interpolator**. In the absence of measurement error the classical kriging predictors have precisely this property to interpolate the observed data points (see §A9.9.5).

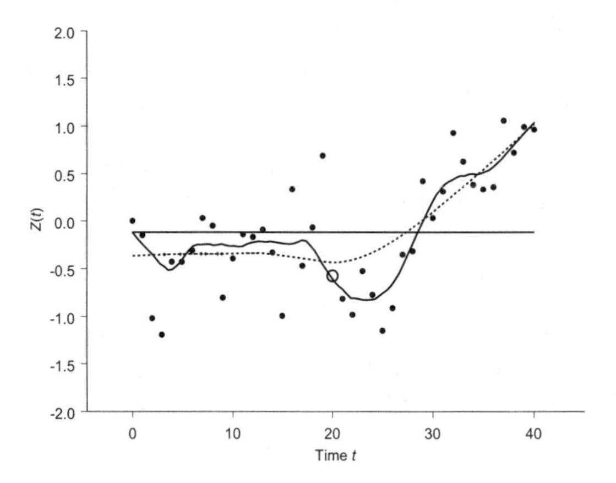

Figure 9.21. Prediction of a target point at $t = 20$ (circle) in a time series of length 40. The predictors are loess smooths with small (irregular solid line) and large bandwidth (dashed line) as well as the arithmetic average $\overline{y}$ (horizontal line).

Although the conditional mean $\mathrm{E}[Z(\mathbf{s}_0)\,|\,\mathbf{Z}(\mathbf{s})]$ is the optimal predictor of $Z(\mathbf{s}_0)$ under squared error loss, it is not the predictor usually applied. $\mathrm{E}[Z(\mathbf{s}_0)\,|\,\mathbf{Z}(\mathbf{s})]$ can be a complicated nonlinear function of the observations. A notable exception occurs when $Z(\mathbf{s})$ is a Gaussian random field and $[Z(\mathbf{s}_0), \mathbf{Z}(\mathbf{s})]$ are jointly multivariate Gaussian distributed. Define the following quantities

$$\begin{array}{ll} \mathrm{E}[\mathbf{Z}(\mathbf{s})] = \boldsymbol{\mu}(\mathbf{s}) & \mathrm{E}[Z(\mathbf{s}_0)] = \mu(\mathbf{s}_0) \\ \mathrm{Var}[\mathbf{Z}(\mathbf{s})] = \boldsymbol{\Sigma} & \mathrm{Var}[Z(\mathbf{s}_0)] = \sigma^2 \\ \mathrm{Cov}[Z(\mathbf{s}_0), \mathbf{Z}(\mathbf{s})] = \mathbf{c} & \end{array} \qquad [9.35]$$

for a Gaussian random field. The joint distribution of $Z(\mathbf{s}_0)$ and $\mathbf{Z}(\mathbf{s})$ then can be written as

$$\begin{bmatrix} Z(\mathbf{s}_0) \\ \mathbf{Z}(\mathbf{s}) \end{bmatrix} \sim G\left(\begin{bmatrix} \mu(\mathbf{s}_0) \\ \boldsymbol{\mu}(\mathbf{s}) \end{bmatrix}, \begin{bmatrix} \sigma^2 & \mathbf{c}' \\ \mathbf{c} & \boldsymbol{\Sigma} \end{bmatrix} \right).$$

Recalling results from §3.7 the conditional distribution of $Z(\mathbf{s}_0)$ given $\mathbf{Z}(\mathbf{s})$ is univariate Gaussian with mean $\mu(\mathbf{s}_0) + \mathbf{c}'\boldsymbol{\Sigma}^{-1}(\mathbf{Z}(\mathbf{s}) - \boldsymbol{\mu}(\mathbf{s}))$ and variance $\sigma^2 - \mathbf{c}'\boldsymbol{\Sigma}^{-1}\mathbf{c}$. The optimal predictor under squared error loss is thus

$$p^0(\mathbf{Z}; \mathbf{s}_0) = \mathrm{E}[Z(\mathbf{s}_0)|\mathbf{Z}(\mathbf{s})] = \mu(\mathbf{s}_0) + \mathbf{c}'\boldsymbol{\Sigma}^{-1}(\mathbf{Z}(\mathbf{s}) - \boldsymbol{\mu}(\mathbf{s})). \qquad [9.36]$$

This important expression is worthy of some comments. First, the conditional mean *is* a linear function of the observed data $\mathbf{Z}(\mathbf{s})$. Evaluating the statistical properties of the predictor such as its mean and variance is thus simple:

$$\mathrm{E}\left[p^0(\mathbf{Z}; \mathbf{s}_0)\right] = \mu(\mathbf{s}_0), \quad \mathrm{Var}\left[p^0(\mathbf{Z}; \mathbf{s}_0)\right] = \mathbf{c}'\boldsymbol{\Sigma}^{-1}\mathbf{c}.$$

Second, the predictor is a perfect interpolator. Assume you wish to predict at all observed locations. To this end, replace in [9.36] $\mu(\mathbf{s}_0)$ with $\boldsymbol{\mu}(\mathbf{s})$ and $\mathbf{c}'$ with $\boldsymbol{\Sigma}$. The predictor becomes

$$p^0(\mathbf{Z}; \mathbf{s}) = \boldsymbol{\mu}(\mathbf{s}) + \boldsymbol{\Sigma}\boldsymbol{\Sigma}^{-1}(\mathbf{Z}(\mathbf{s}) - \boldsymbol{\mu}(\mathbf{s})) = \mathbf{Z}(\mathbf{s}).$$

Third, imagine that $Z(\mathbf{s}_0)$ and $\mathbf{Z}(\mathbf{s})$ are not correlated. Then $\mathbf{c} = \mathbf{0}$ and [9.36] reduces to $p^0(\mathbf{Z}; \mathbf{s}_0) = \mu(\mathbf{s}_0)$, the (unconditional) mean at the unsampled location. One interpretation of the optimal predictor is to consider $\mathbf{c}'\boldsymbol{\Sigma}^{-1}(\mathbf{Z}(\mathbf{s}) - \boldsymbol{\mu}(\mathbf{s}))$ as the adjustment to the unconditional mean that draws on the spatial autocorrelation between attributes at the unsampled and sampled locations. If $Z(\mathbf{s}_0)$ is correlated with observations nearby, then using the information from other locations strengthens our ability to make predictions about the value at the new location (since the MSPE if $\mathbf{c} = \mathbf{0}$ is σ^2). Fourth, the variance of the conditional distribution equals the mean square prediction error.

Two important questions arise. Do the simple form and appealing properties of the *best* predictor prevail if the random field is not Gaussian-distributed? If the means $\boldsymbol{\mu}(\mathbf{s})$ and $\mu(\mathbf{s}_0)$ are unknown are predictors of the form

$$\widehat{\mu}(\mathbf{s}_0) + \mathbf{c}'\boldsymbol{\Sigma}^{-1}(\mathbf{Z}(\mathbf{s}) - \widehat{\boldsymbol{\mu}}(\mathbf{s}))$$

still *best* in some sense? To answer these questions we now relate the decision-theoretic setup in this subsection to the basic kriging methods.

9.4.3 Ordinary and Universal Kriging

Box 9.8 Kriging and Best Prediction

- **Kriging predictors are the best linear unbiased predictors under squared error loss.**
- **Simple, ordinary, and universal kriging differ in their assumptions about the mean structure $\mu(\mathbf{s})$ of the spatial model.**

The classical kriging techniques are methods for predicting $Z(\mathbf{s}_0)$ based on combining assumptions about the spatial model with requirements about the predictor $p(\mathbf{Z};\mathbf{s}_0)$. The usual set of requirements are

(i) $p(\mathbf{Z};\mathbf{s}_0)$ is a linear combination of the observed values $Z(\mathbf{s}_1),\cdots,Z(\mathbf{s}_n)$.

(ii) $p(\mathbf{Z};\mathbf{s}_0)$ is unbiased in the sense that $\mathrm{E}[p(\mathbf{Z};\mathbf{s}_0)] = \mathrm{E}[Z(\mathbf{s}_0)]$.

(iii) $p(\mathbf{Z};\mathbf{s}_0)$ minimizes the mean square prediction error.

Requirement (i) states that the predictors have the general form

$$p(\mathbf{Z};\mathbf{s}_0) = \sum_{i=1}^{n}\lambda(\mathbf{s}_i)Z(\mathbf{s}_i), \qquad [9.37]$$

where $\lambda(\mathbf{s}_i)$ is a weight associated with the observation at location $\mathbf{s}_i$. Relative to other weights, $\lambda(\mathbf{s}_i)$ determines how much the observation $Z(\mathbf{s}_i)$ contributes to the predicted value at location $\mathbf{s}_0$. To satisfy requirements (ii) and (iii) the weights are chosen to minimize

$$\mathrm{E}\left[\left\{Z(\mathbf{s}_0) - \sum_{i=1}^{n}\lambda(\mathbf{s}_i)Z(\mathbf{s}_i)\right\}^2\right]$$

subject to certain constraints that guarantee unbiasedness. These constraints depend on the model assumptions. The three basic kriging methods, simple, ordinary, and universal kriging, are distinguished according to the mean structure of the spatial model

$$Z(\mathbf{s}) = \mu(\mathbf{s}) + \delta(\mathbf{s}).$$

Table 9.4. Simple, ordinary, and universal kriging model assumptions

Method	**Assumption about $\mu(\mathbf{s})$**	$\delta(\mathbf{s})$
Simple Kriging	$\mu(\mathbf{s})$ is known	second-order or intrinsically stationary
Ordinary Kriging	$\mu(\mathbf{s}) = \mu$, μ unknown	second-order or intrinsically stationary
Universal Kriging	$\mu(\mathbf{s}) = \mathbf{x}'(\mathbf{s})\boldsymbol{\beta}$, $\boldsymbol{\beta}$ unknown	second-order or intrinsically stationary

Simple Kriging

The solution to this minimization problem if $\mu(\mathbf{s})$ (and thus $\mu(\mathbf{s}_0)$) is known is called the simple kriging predictor (Matheron 1971)

$$p_{SK}(\mathbf{Z};\mathbf{s}_0) = \mu(\mathbf{s}_0) + \mathbf{c}'\boldsymbol{\Sigma}^{-1}(\mathbf{Z}(\mathbf{s}) - \boldsymbol{\mu}(\mathbf{s})). \qquad [9.38]$$

The details of the derivation can be found in Cressie (1993, p. 109 and our §A9.9.4). Note that $\mu(\mathbf{s}_0)$ in [9.38] is a scalar and $\mathbf{Z}(\mathbf{s})$ and $\boldsymbol{\mu}(\mathbf{s})$ are vectors (see [9.35] on p. 608 for definitions). The simple kriging predictor is unbiased since $\mathrm{E}[p_{SK}(\mathbf{Z};\mathbf{s}_0)] = \mu(\mathbf{s}_0) = \mathrm{E}[Z(\mathbf{s}_0)]$ and bears a striking resemblance to the conditional mean under Gaussianity ([9.36]). Simple kriging is thus **the** optimal method of spatial prediction (under squared error loss) in a Gaussian random field since $p_{SK}(\mathbf{Z};\mathbf{s}_0)$ equals the conditional mean. No other predictor then has a smaller mean square prediction error, not even when nonlinear functions of the data are considered. If the random field is not Gaussian, $p_{SK}(\mathbf{Z}; \mathbf{s}_0)$ is no longer best in that sense, it is best only among all predictors that are linear in the data and unbiased. It is the **b**est **l**inear **u**nbiased **p**redictor (BLUP). The minimized mean square prediction error of an unbiased kriging predictor is often called the **kriging variance** or the **kriging error**. It is easy to establish that the kriging variance for the simple kriging predictor is

$$\sigma^2_{SK}(\mathbf{s}_0) = \sigma^2 - \mathbf{c}'\mathbf{\Sigma}^{-1}\mathbf{c}, \qquad [9.39]$$

where σ^2 is the variance of the random field at $\mathbf{s}_0$. We assume here that the random field is second-order stationary so that $\text{Var}[Z(\mathbf{s})] = \text{Var}[Z(\mathbf{s}_0)] = \sigma^2$ and the covariance function exists (otherwise $C(\mathbf{h})$ is a nonexisting parameter and [9.38], [9.39] should be expressed in terms of the semivariogram).

Simple kriging is useful in that it determines the benchmark for other kriging methods. The assumption that the mean is known everywhere is not tenable for most applications. An exception is the kriging of residuals from a fit of the mean function. If the mean model is correct the residuals will have a known, zero mean. How much is lost by estimating an unknown mean can be inferred by comparing the simple kriging variance [9.39] with similar expressions for the methods that follow.

Universal and Ordinary Kriging

Universal and ordinary kriging have in common that the mean of the random field is not known and can be expressed by a linear model. The more general case is $\mu(\mathbf{s}) = \mathbf{x}'(\mathbf{s})\boldsymbol{\beta}$ where the mean is a linear regression on some regressor variables $\mathbf{x}(\mathbf{s})$. We keep the argument $(\mathbf{s})$ to underline that the regressor variables will often be the spatial coordinates themselves or functions thereof. In many cases $\mathbf{x}(\mathbf{s})$ consists only of spatial coordinates (apart from an intercept). As a special case we can assume that the mean of the random field does not change with spatial locations but is unknown. Then replace $\mathbf{x}'(\mathbf{s})$ with 1 and $\boldsymbol{\beta}$ with μ, the unknown mean. The latter simplification gives rise to the **ordinary kriging** predictor. It is the predictor

$$p_{OK}(\mathbf{Z}; \mathbf{s}_0) = \sum_{i=1}^{n} \lambda_{OK}(\mathbf{s}_i) Z(\mathbf{s}_i)$$

which minimizes the mean square prediction error subject to an unbiasedness constraint. This constraint can be found by noticing that $\text{E}[p_{OK}(\mathbf{Z}; \mathbf{s}_0)] = \text{E}[\sum_{i=1}^{n} \lambda_{OK}(\mathbf{s}_i) Z(\mathbf{s}_i)] = \sum_{i=1}^{n} \lambda_{OK}(\mathbf{s}_i)\mu$ which must equal μ for $p_{OK}(\mathbf{Z}; \mathbf{s}_0)$ to be unbiased. As a consequence the weights must sum to one. This does not imply, by the way, that kriging weights are positive.

If the mean of the random field is $\mu(\mathbf{s}) = \mathbf{x}'(\mathbf{s})\boldsymbol{\beta}$ it is not sufficient to require that the kriging weights sum to one. Instead we need

$$\text{E}\left[\sum_{i=1}^{n} \lambda_{UK}(\mathbf{s}_i) Z(\mathbf{s}_i)\right] = \sum_{i=1}^{n} \lambda_{UK}(\mathbf{s}_i)\mathbf{x}'(\mathbf{s}_i)\boldsymbol{\beta} = \mathbf{x}'(\mathbf{s}_0)\boldsymbol{\beta}.$$

Using matrix/vector notation this constraint can be expressed more elegantly. Write the universal kriging model as $\mathbf{Z}(\mathbf{s}) = \mathbf{X}(\mathbf{s})\boldsymbol{\beta} + \boldsymbol{\delta}(\mathbf{s})$ and the predictor as

$$p_{UK}(\mathbf{Z};\mathbf{s}_0) = \sum_{i=1}^{n} \lambda_{UK}(\mathbf{s}_i) Z(\mathbf{s}_i) = \boldsymbol{\lambda}'\mathbf{X}(\mathbf{s})\boldsymbol{\beta},$$

where $\boldsymbol{\lambda}$ is the vector of (universal) kriging weights. For $p_{UK}(\mathbf{Z};\mathbf{s}_0)$ to be unbiased we need $\boldsymbol{\lambda}'\mathbf{X} = \mathbf{x}'(\mathbf{s}_0)$.

Minimization of

$$\mathrm{E}\left[\left\{Z(\mathbf{s}_0) - \sum_{i=1}^{n} \lambda_{OK}(\mathbf{s}_i) Z(\mathbf{s}_i)\right\}^2\right] \text{ subject to } \boldsymbol{\lambda}'\mathbf{1} = 1$$

to find the ordinary kriging weights and of

$$\mathrm{E}\left[\left\{Z(\mathbf{s}_0) - \sum_{i=1}^{n} \lambda_{OK}(\mathbf{s}_i) Z(\mathbf{s}_i)\right\}^2\right] \text{ subject to } \boldsymbol{\lambda}'\mathbf{X} = \mathbf{x}'(\mathbf{s}_0)$$

to derive the universal kriging weights is a constrained optimization problem. It can be solved as an unconstrained minimization problem using one (OK) or more (UK) Lagrange multipliers (see §A9.9.4 for details and derivations). The resulting predictors can be expressed in numerous ways. We prefer

$$p_{UK}(\mathbf{Z};\mathbf{s}_0) = \mathbf{x}'(\mathbf{s}_0)\widehat{\boldsymbol{\beta}}_{GLS} + \mathbf{c}'\boldsymbol{\Sigma}^{-1}\left(\mathbf{Z}(\mathbf{s}) - \mathbf{X}\widehat{\boldsymbol{\beta}}_{GLS}\right), \quad [9.40]$$

where $\widehat{\boldsymbol{\beta}}_{GLS}$ is the generalized least squares estimator

$$\widehat{\boldsymbol{\beta}}_{GLS} = \left(\mathbf{X}'\boldsymbol{\Sigma}^{-1}\mathbf{X}\right)^{-1}\mathbf{X}'\boldsymbol{\Sigma}^{-1}\mathbf{Z}(\mathbf{s}). \quad [9.41]$$

As a special case of [9.40], where $\mathbf{x} = 1$, the ordinary kriging predictor is obtained:

$$p_{OK}(\mathbf{Z};\mathbf{s}_0) = \widehat{\mu} + \mathbf{c}'\boldsymbol{\Sigma}^{-1}(\mathbf{Z}(\mathbf{s}) - \mathbf{1}\widehat{\mu}) \quad [9.42]$$

Here, $\widehat{\mu}$ is the generalized least squares estimator of the mean,

$$\widehat{\mu} = \left(\mathbf{1}'\boldsymbol{\Sigma}^{-1}\mathbf{1}\right)^{-1}\mathbf{1}'\boldsymbol{\Sigma}^{-1}\mathbf{Z}(\mathbf{s}) = \frac{\mathbf{1}'\boldsymbol{\Sigma}^{-1}\mathbf{Z}(\mathbf{s})}{\mathbf{1}'\boldsymbol{\Sigma}^{-1}\mathbf{1}}. \quad [9.43]$$

Comparing [9.40] with [9.36], the optimal predictor in a Gaussian random field, we again notice a striking resemblance. The question raised at the end of the previous subsection about the effects of substituting an estimate $\widehat{\mu}(\mathbf{s}_0)$ for the unknown mean can now be answered. Substituting an estimate retains certain best properties of the linear predictor. It remains unbiased provided the estimate for the mean is unbiased and the model for the mean is correct. It remains an exact interpolator and the predictor has the form of an estimate of the mean $(\mathbf{x}'(\mathbf{s}_0)\widehat{\boldsymbol{\beta}})$ adjusted by surrounding values with adjustments depending on the strength of the spatial correlation $(\mathbf{c}, \boldsymbol{\Sigma})$. Kriging predictors are obtained *only* if the generalized least squares estimates [9.41] (or [9.43] for OK) are being substituted, however.

In the formulations [9.40] and [9.42] it is not immediately obvious what the kriging weights are. Some algebra leads to

$$p_{UK}(\mathbf{Z};\mathbf{s}_0) = \boldsymbol{\lambda}'_{UK}\mathbf{Z}(\mathbf{s}) = \left[\mathbf{c} + \mathbf{X}\left(\mathbf{X}'\boldsymbol{\Sigma}^{-1}\mathbf{X}\right)^{-1}\left(\mathbf{x}(\mathbf{s}_0) - \mathbf{X}'\boldsymbol{\Sigma}^{-1}\mathbf{c}\right)\right]'\boldsymbol{\Sigma}^{-1}\mathbf{Z}(\mathbf{s})$$
$$p_{OK}(\mathbf{Z};\mathbf{s}_0) = \boldsymbol{\lambda}'_{OK}\mathbf{Z}(\mathbf{s}) = \left[\mathbf{c} + \mathbf{1}\left(\mathbf{1}'\boldsymbol{\Sigma}^{-1}\mathbf{1}\right)^{-1}\left(1 - \mathbf{1}'\boldsymbol{\Sigma}^{-1}\mathbf{c}\right)\right]'\boldsymbol{\Sigma}^{-1}\mathbf{Z}(\mathbf{s}). \qquad [9.44]$$

The kriging variances are calculated as

$$\sigma^2_{UK} = \sigma^2 - \mathbf{c}'\boldsymbol{\Sigma}^{-1}\mathbf{c} + \left(\mathbf{x}(\mathbf{s}_0) - \mathbf{X}'\boldsymbol{\Sigma}^{-1}\mathbf{c}\right)'\left(\mathbf{X}'\boldsymbol{\Sigma}^{-1}\mathbf{X}\right)^{-1}\left(\mathbf{x}(\mathbf{s}_0) - \mathbf{X}'\boldsymbol{\Sigma}^{-1}\mathbf{c}\right)$$
$$\sigma^2_{OK} = \sigma^2 - \mathbf{c}'\boldsymbol{\Sigma}^{-1}\mathbf{c} + \left(1 - \mathbf{1}'\boldsymbol{\Sigma}^{-1}\mathbf{c}\right)^2/\left(\mathbf{1}'\boldsymbol{\Sigma}^{-1}\mathbf{1}\right). \qquad [9.45]$$

Compare these expressions to the kriging error for simple kriging [9.39]. Since

$$\left(\mathbf{x}(\mathbf{s}_0) - \mathbf{X}'\boldsymbol{\Sigma}^{-1}\mathbf{c}\right)'\left(\mathbf{X}'\boldsymbol{\Sigma}^{-1}\mathbf{X}\right)^{-1}\left(\mathbf{x}(\mathbf{s}_0) - \mathbf{X}'\boldsymbol{\Sigma}^{-1}\mathbf{c}\right)$$

is a quadratic form it is positive definite and $\sigma^2_{UK} > \sigma^2_{SK}$. The mean square prediction error increases if the unknown mean is estimated from the data. Expressions in terms of the semi-variances are given in §A9.9.4.

9.4.4 Some Notes on Kriging

Is Kriging Perfect Interpolation?

Consider the decomposition $Z(\mathbf{s}) = S(\mathbf{s}) + e(\mathbf{s})$, where $S(\mathbf{s})$ is the signal of the process. If $e(\mathbf{s})$ is pure measurement error, then one should predict the signal $S()$, rather than the error-contaminated $Z()$ process. If the data are affected by measurement error, one is not interested in predicting the erroneous observation, but the amount that is actually there. The *controversy* whether kriging is a perfect interpolation method (honors the data) is concerned with the nature of the nugget effect as micro-scale variability or measurement error and whether predictions focus on the $Z()$ or the $S()$ process. In §A9.9.5 kriging of the $Z()$ and $S()$ process are compared for semivariograms with and without nugget effect. It is assumed there that $e(\mathbf{s})$ does not contain a micro-scale variability component, i.e., $\text{Var}[e(\mathbf{s})]$ is made up of measurement error in its entirety. The main findings of the comparison in §A9.5.5 are as follows. In the absence of a nugget effect predictions of $Z(\mathbf{s})$ and $S(\mathbf{s})$ agree in value and precision for observed and unobserved locations. This is obvious, since then $Z(\mathbf{s}) = S(\mathbf{s})$. In a model where the nugget effect is measurement error, predictions of $Z(\mathbf{s})$ and $S(\mathbf{s})$ agree in value at unobserved locations but not in precision. Predictions of $Z(\mathbf{s})$ are less precise. At observed locations predictions of $Z(\mathbf{s})$ honor the data even in the presence of a nugget effect. Predictions of $S(\mathbf{s})$ at observed locations honor the data only in the absence of a nugget effect. So, when is kriging not a perfect interpolator? When predicting the signal at observed locations and the nugget effect contains a measurement error component.

The Cat and Mouse Game of Universal Kriging

In §9.4.3 the universal kriging predictor in the spatial model

$$\mathbf{Z}(\mathbf{s}) = \mathbf{X}(\mathbf{s})\boldsymbol{\beta} + \boldsymbol{\delta}(\mathbf{s})$$

was given in two equivalent ways:

$$p_{UK}(\mathbf{Z};\mathbf{s}_0) = \boldsymbol{\lambda}'_{UK}\mathbf{Z}(\mathbf{s}) = \left[\mathbf{c} + \mathbf{X}\left(\mathbf{X}'\boldsymbol{\Sigma}^{-1}\mathbf{X}\right)^{-1}\left(\mathbf{x}(\mathbf{s}_0) - \mathbf{X}'\boldsymbol{\Sigma}^{-1}\mathbf{c}\right)\right]'\boldsymbol{\Sigma}^{-1}\mathbf{Z}(\mathbf{s})$$
$$= \mathbf{x}'(\mathbf{s}_0)\widehat{\boldsymbol{\beta}}_{GLS} + \mathbf{c}'\boldsymbol{\Sigma}^{-1}\left(\mathbf{Z}(\mathbf{s}) - \mathbf{X}\widehat{\boldsymbol{\beta}}_{GLS}\right), \quad [9.46]$$

where $\widehat{\boldsymbol{\beta}}_{GLS}$ is the generalized least squares estimator $\widehat{\boldsymbol{\beta}}_{GLS} = \left(\mathbf{X}'\boldsymbol{\Sigma}^{-1}\mathbf{X}\right)^{-1}\mathbf{X}'\boldsymbol{\Sigma}^{-1}\mathbf{Z}(\mathbf{s})$. The covariance matrix $\boldsymbol{\Sigma}$ is usually constructed from a model of the semivariogram utilizing the simple relationship between covariances and semivariances in second-order stationary random fields. The modeling of the semivariogram requires, however, that the random field is mean stationary, i.e., the absence of large-scale structure. If $\mathrm{E}[Z(\mathbf{s})] = \mathbf{x}'(\mathbf{s})\boldsymbol{\beta}$ the large-scale trend must be removed before the semivariogram can be modeled. Failure to do so can severely distort the semivariogram. Figure 9.22 shows empirical semivariograms (dots) calculated from two sets of deterministic data. The left panel is the semivariogram of $Z(x) = 1 + 0.5x$ where x is a point on a transect. The right-hand panel is the semivariogram of $Z(x) = 1 + 0.22x + 0.022x^2 - 0.0013x^3$. The power model fits the empirical semivariogram in the left panel very well and the gaussian model provides a decent fit to the semivariogram of the cubic polynomial. The shapes of the semivariograms are due to trend only, however. There is nothing stochastic about the data. One must not conclude based on these graphs that the process on the left is intrinsically stationary and that the process on the right is second-order stationary.

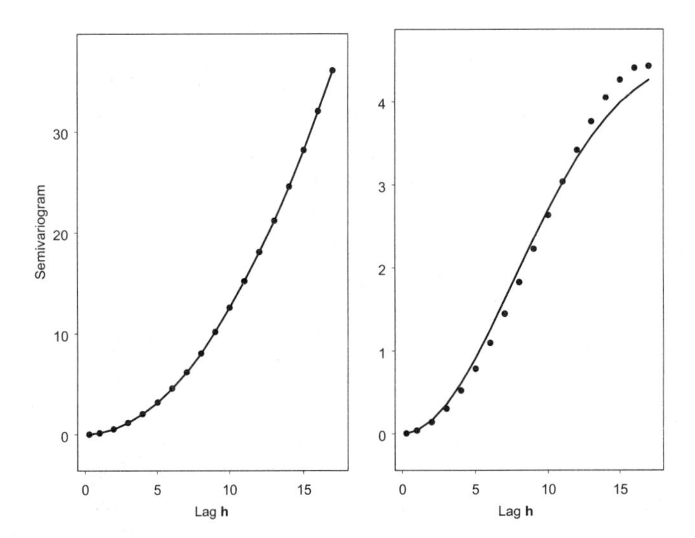

Figure 9.22. Empirical semivariograms (dots) for evaluations of deterministic (trend-only) functions. A power semivariogram was fitted to the semivariogram on the left and a gaussian model to the one on the right.

Certain types of mean nonstationarity (drift) can be inferred from the semivariogram (Neuman and Jacobson 1984, Russo and Jury 1987b). A linear drift, for example, causes the semivariogram to increase as in the left-hand panel of Figure 9.22. Using the semivariogram as a diagnostic procedure for detecting trends in the mean function is dangerous as the drift-contaminated semivariogram may suggest a valid theoretical semivariogram model. Kriging a process with linear drift and exponential semivariogram by ignoring the drift and using a power semivariogram model is not the appropriate course of action. This issue must not be confused with the question whether the semivariogram should be defined as

$$\frac{1}{2}\mathrm{Var}[Z(\mathbf{s}) - Z(\mathbf{s}+\mathbf{h})] \quad \text{or} \quad \frac{1}{2}\mathrm{E}\left[\{Z(\mathbf{s}) - Z(\mathbf{s}+\mathbf{h})\}^2\right].$$

The two definitions are equivalent for stationary processes. Intrinsic or second-order stationary require *first* that the mean is constant. Only if that is the case can the semivariogram (or covariogram) be examined for signs of spatial stochastic structure in the process.

Trend removal is thus critical prior to the estimation of the semivariogram. But efficient estimation of the large-scale trend requires knowledge of the variance-covariance matrix $\boldsymbol{\Sigma}$ as can be seen from the formula of the generalized least squares estimator. In short, we need to know $\boldsymbol{\beta}$ before estimation of $\boldsymbol{\Sigma}$ is possible but efficient estimation of $\boldsymbol{\beta}$ requires knowledge of $\boldsymbol{\Sigma}$.

One approach out of this quandary is to detrend the data by a method that does not require $\boldsymbol{\Sigma}$, for example, ordinary least squares or median polishing. The residuals from this fit are then used to estimate the semivariogram from which $\widehat{\boldsymbol{\Sigma}}$ is being constructed. These steps should be iterated. Once an estimate of $\widehat{\boldsymbol{\Sigma}}$ has been obtained a more efficient estimate of the trend is garnered by estimated generalized least squares (EGLS) as

$$\widehat{\boldsymbol{\beta}}_{EGLS} = \left(\mathbf{X}'\widehat{\boldsymbol{\Sigma}}^{-1}\mathbf{X}\right)^{-1}\mathbf{X}'\widehat{\boldsymbol{\Sigma}}^{-1}\mathbf{Z}(\mathbf{s}).$$

New residuals are obtained and the estimate of the semivariogram is updated. The only downside of this approach is that the residuals obtained from detrending the data do not exactly behave as the random component $\boldsymbol{\delta}(\mathbf{s})$ of the model does. Assume that the model is initially detrended by ordinary least squares,

$$\widehat{\boldsymbol{\beta}}_{OLS} = (\mathbf{X}'\mathbf{X})^{-1}\mathbf{X}'\mathbf{Z}(\mathbf{s}),$$

and the fitted residuals $\widehat{\boldsymbol{\delta}}_{OLS} = \mathbf{Z}(\mathbf{s}) - \mathbf{X}\widehat{\boldsymbol{\beta}}_{OLS}$ are formed. The error process $\boldsymbol{\delta}(\mathbf{s})$ of the model has mean $\mathbf{0}$, variance-covariance matrix $\boldsymbol{\Sigma}$, and semivariogram $\gamma_\delta(\mathbf{h}) = ½\mathrm{Var}[\delta(\mathbf{s}) - \delta(\mathbf{s}+\mathbf{h})]$. The vector of fitted residuals also has mean $\mathbf{0}$, provided the model for the large-scale structure was correct, but it does not have the proper semivariogram or variance-covariance matrix. Since $\widehat{\boldsymbol{\delta}}(\mathbf{s}) = (\mathbf{I} - \mathbf{H})\mathbf{Z}(\mathbf{s})$ where $\mathbf{H}$ is the *hat* matrix $\mathbf{H} = \mathbf{X}(\mathbf{X}'\mathbf{X})^{-1}\mathbf{X}'$, it is established that

$$\mathrm{Var}\left[\widehat{\boldsymbol{\delta}}(\mathbf{s})\right] = (\mathbf{I} - \mathbf{H})\boldsymbol{\Sigma}(\mathbf{I} - \mathbf{H}) \neq \boldsymbol{\Sigma}.$$

The fitted residuals exhibit more negative correlations than the error process and the estimate of the semivariogram based on the residuals will be biased. Furthermore, the residuals do not have the same variance as the $\boldsymbol{\delta}(\mathbf{s})$ process. It should be noted that if $\boldsymbol{\Sigma}$ were known and the semivariogram were estimated based on GLS residuals $\widehat{\boldsymbol{\delta}}_{OLS} = \mathbf{Z}(\mathbf{s}) - \mathbf{X}\widehat{\boldsymbol{\beta}}_{GLS}$, the semivariogram estimator would still be biased (see Cressie 1993, p. 166). The bias comes about because residuals satisfy constraints that are not properties of the error process. For example, the fitted OLS residuals will sum to zero. The degree to which a semivariogram estimate derived from fitted residuals is biased depends on the method used for detrending as well as the method of semivariogram estimation. Since the bias is typically more substantial at large lags, Cressie (1993, p. 168) reasons that weighted least squares fitting of the semivariogram model is to be preferred over ordinary least squares fitting because the former places more weight on small lags (see §9.2.4). In conjunction with choosing the kriging neighborhood (see below) so that the semivariogram must be evaluated only for small lags, the impact of the bias on the kriging predictions can be reduced. If the large-scale trend occurs in only one direction, then the problem of detrending the data can be circumvented by using only pairs in the perpendicular direction of the trend to model the semivariogram.

The following method is termed "universal kriging" on occasion. Fit the large-scale mean by ordinary least squares and obtain the residuals $\widehat{\boldsymbol{\delta}}_{OLS} = \mathbf{Z}(\mathbf{s}) - \mathbf{X}\widehat{\boldsymbol{\beta}}_{OLS}$. Estimate the semivariogram from these residuals and then perform simple kriging (because the residuals are known to have mean zero) on the residuals. To obtain a prediction of $Z(\mathbf{s}_0)$ the OLS estimate of the mean is added to the kriging prediction of the residual. Formally this approach can be expressed as

$$\widetilde{p}\ (\mathbf{Z};\mathbf{s}_0) = \mathbf{x}_0'\widehat{\boldsymbol{\beta}}_{OLS} + p_{SK}(\widehat{\boldsymbol{\delta}};\mathbf{s}_0), \qquad [9.47]$$

where $p_{SK}(\widehat{\boldsymbol{\delta}};\mathbf{s}_0)$ is the simple kriging predictor of the residual at location $\mathbf{s}_0$ which uses the variance-covariance matrix $\widehat{\boldsymbol{\Sigma}}^*$ formed from the semivariogram fitted to the residuals. Hence, $p_{SK}(\widehat{\boldsymbol{\delta}};\mathbf{s}_0) = \mathbf{c}'\widehat{\boldsymbol{\Sigma}}^{*-1}\widehat{\boldsymbol{\delta}}(\mathbf{s})$, making use of the fact that $\mathrm{E}[\widehat{\boldsymbol{\delta}}(\mathbf{s})] = \mathbf{0}$. Comparing the naïve predictor [9.47] with the universal kriging predictor [9.46] it is clear that the naïve approach is not equivalent to universal kriging. $\boldsymbol{\Sigma}$ is not known and the trend is not estimated by GLS. The predictor [9.47] does have some nice properties, however. It is an unbiased predictor in the sense that $\mathrm{E}[\ \widetilde{p}\ (\mathbf{Z};\mathbf{s}_0)] = Z(\mathbf{s}_0)$ and it remains a perfect interpolator. In fact all methods that are perfect interpolators of the residuals are perfect interpolators of $Z(\mathbf{s})$ even if the trend model is incorrectly specified. Furthermore, this approach does not require iterations. The naïve predictor is not a best linear unbiased predictor, however, and the mean square prediction error should not be calculated by the usual formulas for kriging variances (e.g., [9.45]). The naïve approach can be generalized in the sense that one might use any suitable method for trend removal to obtain residuals, krige those, and add the kriged residuals to the estimated trend. This is the rationale behind median polish kriging recommended by Cressie (1986) for random fields with drift to avoid the operational difficulties of universal kriging.

It must be noted that these difficulties do not arise in maximum likelihood (ML) or restricted maximum likelihood (REML) estimation. In §9.2.4 ML and REML for estimating the covariance parameters of a second-order stationary spatial process were discussed. Now consider the spatial model

$$\mathbf{Z}(\mathbf{s}) = \mathbf{X}(\mathbf{s})\boldsymbol{\beta} + \boldsymbol{\delta}(\mathbf{s}),$$

where $\boldsymbol{\delta}(\mathbf{s})$ is a second-order stationary random field with mean $\mathbf{0}$ and variance-covariance matrix $\boldsymbol{\Sigma}(\boldsymbol{\theta})$. Under Gaussianity, $G \sim (\mathbf{X}(\mathbf{s})\boldsymbol{\beta},\boldsymbol{\Sigma}(\boldsymbol{\theta}))$ and estimates of the mean parameters $\boldsymbol{\beta}$ and the spatial dependency parameters $\boldsymbol{\theta}$ can be obtained *simultaneously* by maximizing the likelihood or restricted likelihood of $\mathbf{Z}(\mathbf{s})$. In practice one must choose a parametric covariance model for $\boldsymbol{\Sigma}(\boldsymbol{\theta})$, which seems to open the same Pandora's box as in the cat and mouse game of universal kriging. The operational difficulties are minor in the case of ML or REML estimation, however. Initially, one should estimate $\boldsymbol{\beta}$ by ordinary least squares and calculate the empirical semivariogram of the residuals. From a graph of the empirical semivariogram possible parametric models for the semivariogram (covariogram) can be determined. In contrast to the least squares based methods discussed above one does not estimate $\boldsymbol{\theta}$ based on the empirical semivariogram of the residuals. This is left to the likelihood procedure. The advantage of ML or REML estimation is to provide estimates of the spatial autocorrelation structure and the mean parameters simultaneously where $\widehat{\boldsymbol{\beta}}$ is adjusted for the spatial correlation and $\widehat{\boldsymbol{\theta}}$ is adjusted for the nonconstant mean. Furthermore, such models can be easily fit with the `mixed` procedure of The SAS® System. We present applications in §9.8.4 and §9.8.5.

Local Kriging and the Kriging Neighborhood

The kriging predictors

$$p_{UK}(\mathbf{Z};\mathbf{s}_0) = \mathbf{x}'(\mathbf{s}_0)\widehat{\boldsymbol{\beta}} + \mathbf{c}'\boldsymbol{\Sigma}^{-1}\left(\mathbf{Z}(\mathbf{s}) - \mathbf{X}\widehat{\boldsymbol{\beta}}\right)$$
$$p_{OK}(\mathbf{Z};\mathbf{s}_0) = \widehat{\mu} + \mathbf{c}'\boldsymbol{\Sigma}^{-1}(\mathbf{Z}(\mathbf{s}) - \mathbf{1}\widehat{\mu})$$

must be calculated for each location at which predictions are desired. Although only the vector $\mathbf{c}$ of covariances between $Z(\mathbf{s}_0)$ and $\mathbf{Z}(\mathbf{s})$ must be recalculated every time $\mathbf{s}_0$ changes, even for moderately sized spatial data sets the inversion (and storage) of the matrix $\boldsymbol{\Sigma}$ is a formidable problem. A solution to this problem is to consider for prediction of $Z(\mathbf{s}_0)$ only observed data points within a neighborhood of $\mathbf{s}_0$, called the **kriging neighborhood**. As $\mathbf{s}_0$ changes this is akin to sliding a window across the domain and to exclude all points outside the window in calculating the kriging predictor. If $n(\mathbf{s}_0) = 25$ points are in the neighborhood at $\mathbf{s}_0$, then only a (25×25) matrix must be inverted. Using a kriging neighborhood rather than all the data is sometimes referred to as local kriging. It has its advantages and disadvantages.

Among the advantages of local kriging is not only computational efficiency; it might also be reasonable to assume that the mean is at least locally stationary, even if the mean is globally nonstationary. This justification is akin to the reasoning behind using local linear regressions in the nonparametric estimation of complicated trends (see §4.7). Ordinary kriging performed locally is another approach to avoid the operational difficulties with universal kriging performed globally. Local kriging essentially assigns kriging weight $\lambda(\mathbf{s}_i) = 0$ to all points $\mathbf{s}_i$ outside the kriging neighborhood. Since the best linear unbiased predictor is obtained by allowing all data points to contribute to the prediction of $Z(\mathbf{s}_0)$, local kriging predictors are no longer best. In addition, the user needs to decide on the size and shape of the kriging neighborhood. This is no trivial task. The optimal kriging neighborhood depends in a complex fashion on the parameters of the semivariogram, the large-scale trend, and the spatial configuration of the sampling points.

At first glance it may seem reasonable to define the kriging neighborhood as a circle around $\mathbf{s}_0$ with radius equal to the range of the semivariogram. This is not a good solution, because although points further removed from $\mathbf{s}_0$ than the range are not spatially autocorrelated with $Z(\mathbf{s}_0)$, they are autocorrelated with points that lie within the range from $\mathbf{s}_0$. Chilès and Delfiner (1999, p. 205) refer to this as the **relay effect**. A practical solution in our opinion is to select the radius of the kriging neighborhood as the lag distance up to which the empirical semivariogram was modeled. One half of the maximum lag distance in the data is a frequent recommendation. The shape of the kriging neighborhood also deserves consideration. Rules that define neighborhoods as the n^* closest points will lead to elongated shapes if the sampling intensity along a transect is higher than perpendicular to the transect. We prefer circular kriging neighborhoods in general that can be suitably expanded based on some criteria about the minimum number of points in the neighborhood. The `krige2d` procedure in The SAS® System provides flexibility in determining the kriging neighborhood. The statements

```
proc krige2d data=ThatsYourData outest=krige;
  coordinates xcoord=x ycoord=y;
  grid griddata=predlocs xcoord=x ycoord=y;
  predict var=Z radius=10 minpoints=15 maxpoints=35;
  model form=exponential scale=10 range=7;
run;
```

for example, use a circular kriging neighborhood with a 10-unit radius. If the number of points in this radius is less than 15 the radius is suitably increased to honor the `minpoints=` option. If the neighborhood with radius 10 contains more than 35 observation the radius is similarly decreased. If the neighborhood is defined as the nearest n^* observation, the `predict` statement is replaced by (for $n^* = 20$) `predict var=Z radius=10 numpoints=20;`.

Positivity Constraints

In ordinary kriging, the only constraint placed on the kriging weights $\lambda(\mathbf{s}_i)$ is

$$\sum_{i=1}^{n} \lambda(\mathbf{s}_i) = 1,$$

which guarantees unbiasedness. This does not rule out that individual kriging weights may be negative. For attributes that take on positive values only (yields, weights, probabilities, etc.) a potential problem lurks here since the prediction of a negative quantity is not meaningful. But just because some kriging weights arc negative does not imply that the resulting predictor

$$\sum_{i=1}^{n} \lambda(\mathbf{s}_i) Z(\mathbf{s}_i)$$

is negative and in many applications the predicted values will honor the positivity requirement. To exclude the possibility of negative predicted values additional constraints can be imposed on the kriging weights. For example, rather than minimizing

$$\mathrm{E}\left[\left\{ Z(\mathbf{s}_0) - \sum_{i=1}^{n} \lambda(\mathbf{s}_i) Z(\mathbf{s}_i) \right\}^2\right]$$

subject to $\sum_{i=1}^{n} \lambda(\mathbf{s}_i) = 1$, one can minimize the mean square error subject to $\sum_{i=1}^{n} \lambda(\mathbf{s}_i) = 1$ and $\lambda(\mathbf{s}_i) \geq 0$ for all $\mathbf{s}_i$. Barnes and Johnson (1984) solve this minimization problem through quadratic programming and find that a solution can always be obtained in the case of an unknown but constant mean, thereby providing an extension of ordinary kriging. The positivity and the sum-to-one constraint together also ensure that predicted values lie between the smallest and largest value at observed locations. In our opinion this is actually a drawback. Unless there is a compelling reason to the contrary, one should allow the predictions to extend outside the range of observed values. A case where it is meaningful to restrict the range of the predicted values is indicator kriging (§A9.9.6) where the attribute being predicted is a binary $(0, 1)$ variable. Since the mean of a binary random variable is a probability, predictions outside of the $(0, 1)$ interval are difficult to justify. Cressie (1993, p. 143) calls the extra constraint of positive kriging weights "heavy-handed."

Kriging Variance Overstates Precision

The formulas for the kriging predictors and the corresponding kriging variances ([9.45]) contain the variance-covariance matrix $\boldsymbol{\Sigma}$ which is usually unknown since the semivariogram $\gamma(\mathbf{h})$ is unknown. An estimate $\widehat{\boldsymbol{\Sigma}}$ of $\boldsymbol{\Sigma}$ is substituted in the relevant expressions. The uncertainty associated with the estimation of the semivariances or covariances should be accounted for in the determination of the mean square prediction error. This is typically not done. The kriging predictions are obtained as if $\widehat{\boldsymbol{\Sigma}}$ is the correct variance-covariance matrix of $\mathbf{Z}(\mathbf{s})$. As a consequence, the kriging variance obtained by substituting $\widehat{\boldsymbol{\Sigma}}$ for $\boldsymbol{\Sigma}$ is an underestimate of the mean square prediction error.

9.4.5 Extensions to Multiple Attributes

Box 9.9 Cokriging and Spatial Regression

- **If a spatial data set consists of more than one attribute and stochastic relationships exist among them, these relationships can be exploited to improve predictive ability.**

- **Commonly one attribute, $Z_1(\mathbf{s})$, say, is designated the primary attribute and $Z_2(\mathbf{s}), \cdots, Z_k(\mathbf{s})$ are termed the secondary attributes.**

- **Cokriging is a *multivariate* spatial prediction method that relies on the spatial autocorrelation of the primary and secondary attributes as well as the cross-covariances among the primary and the secondary attributes.**

- **Spatial regression is a *multiple* spatial prediction method where the mean of the primary attribute is modeled as a function of secondary attributes.**

The spatial prediction methods discussed thus far predict a single attribute $Z(\mathbf{s})$ at unobserved locations $\mathbf{s}_0$. In most applications, data collection is not restricted to a single attribute. Other variables are collected at the same or different spatial locations or the same attribute is observed at different time points. Consider the case of two spatially varying attributes Z_1 and Z_2 for the time being. To be general it is not required that Z_1 and Z_2 are observed at the same locations although this will often be the case. The vectors of observations on Z_1 and Z_2 are denoted

$$\mathbf{Z}_1(\mathbf{s}_1) = [Z_1(\mathbf{s}_{11}), \cdots, Z_1(\mathbf{s}_{1n_1})]$$
$$\mathbf{Z}_2(\mathbf{s}_2) = [Z_2(\mathbf{s}_{21}), \cdots, Z_1(\mathbf{s}_{2n_2})].$$

If $\mathbf{s}_{1j} = \mathbf{s}_{2j}$, then Z_1 and Z_2 are said to be colocated, otherwise the attributes are termed noncolocated. Figure 9.23 shows the sampling locations at which soil samples were obtained in a chisel-plowed field and the relationship between soil carbon and total soil nitrogen at the sampled locations. Figure 9.24 shows the relationship between total organic carbon percentage and sand percentage in sediment samples from the Chesapeake Bay collected through the Environmental Monitoring and Assessment Program (EMAP) of the U.S.-EPA. In both cases two colocated attributes (C, N) $(C, Sand\%)$ have been observed. The relationships between C and N is very strong in the field sample and reasonably strong in the aquatic sediment samples.

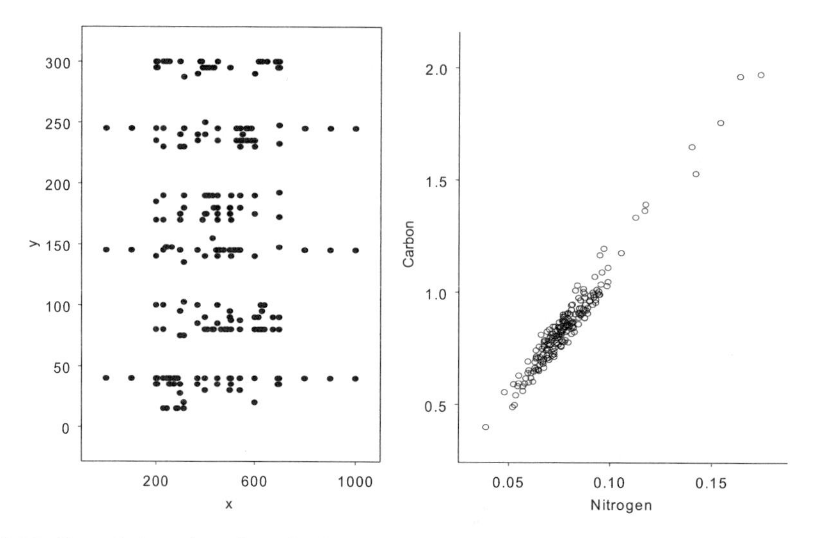

Figure 9.23. Sample locations in a field where carbon and nitrogen were measured in soil samples (left panel). Relationship between soil carbon and total soil nitrogen (right panel). Data kindly provided by Dr. Thomas G. Mueller, Department of Agronomy, University of Kentucky. Used with permission.

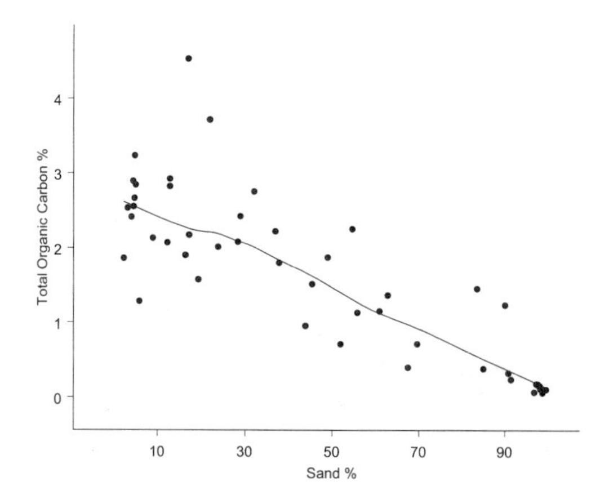

Figure 9.24. Total organic carbon percentage as a function of sand percentage collected at 47 base stations in Chesapeake Bay in 1993 through the US Environmental Protection Agency's Environmental Monitoring and Assessment Program (EMAP).

The attributes are usually not symmetric in that one attribute is designated the *primary* variable of interest and the other attributes are *secondary* or auxiliary variables. Without loss of generality we designate $Z_1(\mathbf{s})$ as the primary attribute and $Z_2(\mathbf{s}), \cdots, Z_k(\mathbf{s})$ as the secondary attributes. For example, $Z_1(\mathbf{s})$ may be the in-situ measurements of the plant canopy temperature on a grassland site and $Z_2(\mathbf{s})$ is the temperature obtained from remotely sensed thermal infrared radiation (see Harris and Johnson, 1996 for an application). The interest lies in predicting the *in-situ* canopy temperature $Z_1(\mathbf{s})$ while utilizing the information collected on the secondary attribute and its relationship with $Z_1(\mathbf{s})$. If primary and secondary attributes are spatially structured and/or the secondary variables are related to the primary variable, predictive ability should be enhanced by utilizing the secondary information.

Focusing on one primary and one secondary attribute in what follows two basic approaches can be distinguished (extensions to more than one secondary attribute are straightforward).

- $Z_1(\mathbf{s})$ and $Z_2(\mathbf{s})$ are stationary random fields with covariance functions $C_1(\mathbf{h})$ and $C_2(\mathbf{h})$, respectively. They also covary spatially giving rise to a cross-attribute covariance function $\text{Cov}[Z_1(\mathbf{s}), Z_2(\mathbf{s}+\mathbf{h})] = C_{12}(\mathbf{h})$. By stationarity it follows that $\text{E}[Z_1(\mathbf{s})] = \mu_1$ and $\text{E}[Z_2(\mathbf{s})] = \mu_2$ and the dependence of $Z_1(\mathbf{s})$ on $Z_2(\mathbf{s})$ is stochastic in nature, captured by $C_{12}(\mathbf{h})$. This approach leads to the **cokriging** methods.

- $\delta_1(\mathbf{s})$ is a stationary random field and $Z_1(\mathbf{s})$ is related to $Z_2(\mathbf{s})$ and $\delta_1(\mathbf{s})$ through a spatial regression model $Z_1(\mathbf{s}) = f(z_2(\mathbf{s}), \boldsymbol{\beta}) + \delta_1(\mathbf{s})$, where the semivariogram of $\delta_1(\mathbf{s})$ is that of the detrended process $Z_1(\mathbf{s}) - f(z_2(\mathbf{s}), \boldsymbol{\beta})$. No assumptions about the stochastic properties of $Z_2(\mathbf{s})$ are made, the observed values $z_2(\mathbf{s}_1), \cdots, z_2(\mathbf{s}_n)$ are considered fixed in the analysis. The relationship between $\text{E}[Z_1(\mathbf{s})]$ and $Z_2(\mathbf{s})$ is deterministic, captured by the mean model $f(z_2(\mathbf{s}), \boldsymbol{\beta})$. This is a special case of a spatial regression model (§9.5).

Ordinary Cokriging

The goal of cokriging is to find a best linear unbiased predictor of $Z_1(\mathbf{s}_0)$, the primary attribute at a new location $\mathbf{s}_0$, based on $\mathbf{Z}_1(\mathbf{s})$ *and* $\mathbf{Z}_2(\mathbf{s})$. Extending the notation from ordinary kriging the predictor can be written as

$$p_1(\mathbf{Z}_1, \mathbf{Z}_2; \mathbf{s}_0) = \sum_{i=1}^{n_1} \lambda_{1i} Z_1(\mathbf{s}_i) + \sum_{j=1}^{n_2} \lambda_{2i} Z_2(\mathbf{s}_j) = \boldsymbol{\lambda}_1' \mathbf{Z}_1(\mathbf{s}) + \boldsymbol{\lambda}_2' \mathbf{Z}_2(\mathbf{s}). \qquad [9.48]$$

It is not required that $Z_1(\mathbf{s})$ and $Z_2(\mathbf{s})$ are colocated. Certain assumptions must be made, however. It is assumed that the means of $Z_1(\mathbf{s})$ and $Z_2(\mathbf{s})$ are constant across the domain D and that $Z_1(\mathbf{s})$ has covariogram $C_1(\mathbf{h})$ and $Z_2(\mathbf{s})$ has covariogram $C_2(\mathbf{h})$. Furthermore, there is a cross-covariance function that expresses the spatial dependency between $Z_1(\mathbf{s})$ and $Z_2(\mathbf{s}+\mathbf{h})$,

$$C_{12}(\mathbf{h}) = \text{Cov}[Z_1(\mathbf{s}), Z_2(\mathbf{s}+\mathbf{h})]. \qquad [9.49]$$

The unbiasedness requirement implies that $\text{E}[p_1(\mathbf{Z}_1, \mathbf{Z}_2; \mathbf{s}_0)] = \text{E}[Z_1(\mathbf{s}_0)] = \mu_1$ which implies in turn that $\mathbf{1}'\boldsymbol{\lambda}_1 = 1$ and $\mathbf{1}'\boldsymbol{\lambda}_2 = 0$. Minimizing the mean square prediction error

$$\text{E}\left[\{Z_1(\mathbf{s}_0) - p_1(\mathbf{Z}_1, \mathbf{Z}_2; \mathbf{s}_0)\}^2\right]$$

subject to these constraints gives rise to the system of cokriging equations

$$\begin{bmatrix} \boldsymbol{\Sigma}_{11} & \boldsymbol{\Sigma}_{12} & \mathbf{1} & \mathbf{0} \\ \boldsymbol{\Sigma}_{21} & \boldsymbol{\Sigma}_{22} & \mathbf{0} & \mathbf{1} \\ \mathbf{1}' & \mathbf{0} & \mathbf{0} & \mathbf{0} \\ \mathbf{0} & \mathbf{1}' & \mathbf{0} & \mathbf{0} \end{bmatrix} \begin{bmatrix} \boldsymbol{\lambda}_1 \\ \boldsymbol{\lambda}_2 \\ m_1 \\ m_2 \end{bmatrix} = \begin{bmatrix} \mathbf{c}_{10} \\ \mathbf{c}_{20} \\ 1 \\ 0 \end{bmatrix}, \qquad [9.50]$$

where $\boldsymbol{\Sigma}_{11} = \mathrm{Var}[\mathbf{Z}_1(\mathbf{s})]$, $\boldsymbol{\Sigma}_{22} = \mathrm{Var}[\mathbf{Z}_2(\mathbf{s})]$, $\boldsymbol{\Sigma}_{12} = \mathrm{Cov}[\mathbf{Z}_1(\mathbf{s}), \mathbf{Z}_2(\mathbf{s})]$, m_1 and m_2 are Lagrange multipliers and $\mathbf{c}_{10} = \mathrm{Cov}[\mathbf{Z}_1(\mathbf{s}), Z_1(\mathbf{s}_0)]$, $\mathbf{c}_{20} = \mathrm{Cov}[\mathbf{Z}_2(\mathbf{s}), Z_1(\mathbf{s}_0)]$. These equations are solved for $\boldsymbol{\lambda}_1$, $\boldsymbol{\lambda}_2$, m_1 and m_2 and the minimized mean square prediction error, the cokriging variance, is calculated as

$$\sigma^2_{CK}(\mathbf{s}_0) = \mathrm{Var}[Z_1(\mathbf{s}_0)] - \boldsymbol{\lambda}_1'\mathbf{c}_{10} - \boldsymbol{\lambda}_2'\mathbf{c}_{20} + m_1.$$

Cokriging utilizes two types of correlations. Spatial autocorrelation due to spatial proximity ($\boldsymbol{\Sigma}_{11}$ and $\boldsymbol{\Sigma}_{22}$) and correlation among the attributes ($\boldsymbol{\Sigma}_{12}$). To get a better understanding of the cokriging system of equations we first consider the special case where the secondary variable is not correlated with the primary variable. Then $\boldsymbol{\Sigma}_{12} = \mathbf{0}$ and $\mathbf{c}_{20} = \mathbf{0}$ and the cokriging equations reduce to

$$\boxed{\text{A}} \quad \boldsymbol{\Sigma}_{11}\boldsymbol{\lambda}_1 - \mathbf{1}m_1 = \mathbf{c}_{10} \qquad \mathbf{1}'m_1 = 1$$

$$\boxed{\text{B}} \quad \boldsymbol{\Sigma}_{22}\boldsymbol{\lambda}_2 - \mathbf{1}m_2 = \mathbf{c}_{20} \equiv \mathbf{0} \quad \mathbf{1}'m_2 = 0.$$

Equations $\boxed{\text{A}}$ are the ordinary kriging equations (see §A9.9.4) and $\boldsymbol{\lambda}_1$ will be identical to the ordinary kriging weights. From $\boxed{\text{B}}$ one obtains $\boldsymbol{\lambda}_2 = \boldsymbol{\Sigma}_{22}^{-1}(\mathbf{c}_{20} - \mathbf{1}\,\mathbf{1}'\boldsymbol{\Sigma}_{22}^{-1}\mathbf{c}_{20}/(\mathbf{1}'\boldsymbol{\Sigma}_{22}^{-1}\mathbf{1})) = \mathbf{0}$. The cokriging predictor reduces to the ordinary kriging predictor,

$$p_1(\mathbf{Z}_1, \mathbf{Z}_2; \mathbf{s}_0) = p_{OK}(\mathbf{Z}_1; \mathbf{s}_0) = \boldsymbol{\lambda}_1^{OK}\mathbf{Z}_1(\mathbf{s}),$$

and $\sigma^2_{CK}(\mathbf{s}_0)$ reduces to $\sigma^2_{OK}(\mathbf{s}_0)$. There is no benefit in using a secondary attribute unless it is correlated with the primary attribute. There is also no harm in doing so.

Now consider the special case where $Z_i(\mathbf{s})$ is the observation of $Z(\mathbf{s})$ at time t_i so that the variance-covariance matrix $\boldsymbol{\Sigma}_{11}$ describes spatial dependencies at time t_1 and $\boldsymbol{\Sigma}_{22}$ spatial dependencies at time t_2. Then $\boldsymbol{\Sigma}_{12}$ contains the covariances of the single attribute across space and time and the kriging system can be used to produce maps of the attribute at future points in time.

Extensions of ordinary cokriging to universal cokriging are relatively straightforward in principle. Chilès and Delfiner (1999, Ch. 5.4) consider several cases. If the mean functions $\mathrm{E}[\mathbf{Z}_1(\mathbf{s})] = \mathbf{X}_1\boldsymbol{\beta}_1$ and $\mathrm{E}[\mathbf{Z}_2(\mathbf{s})] = \mathbf{X}_2\boldsymbol{\beta}_2$ are unrelated, that is, each variable has a mean function on its own and the coefficients are not related, the cokriging system [9.50] is extended to

$$\begin{bmatrix} \boldsymbol{\Sigma}_{11} & \boldsymbol{\Sigma}_{12} & \mathbf{X}_1 & \mathbf{0} \\ \boldsymbol{\Sigma}_{21} & \boldsymbol{\Sigma}_{22} & \mathbf{0} & \mathbf{X}_2 \\ \mathbf{X}_1' & \mathbf{0} & \mathbf{0} & \mathbf{0} \\ \mathbf{0} & \mathbf{X}_2' & \mathbf{0} & \mathbf{0} \end{bmatrix} \begin{bmatrix} \boldsymbol{\lambda}_1 \\ \boldsymbol{\lambda}_2 \\ \mathbf{m}_1 \\ \mathbf{m}_2 \end{bmatrix} = \begin{bmatrix} \mathbf{c}_{10} \\ \mathbf{c}_{20} \\ \mathbf{x}_{10} \\ \mathbf{0} \end{bmatrix}. \qquad [9.51]$$

Universal cokriging has not received much application since it is hard to imagine a situation in which the mean functions of the attributes are unrelated with $\boldsymbol{\Sigma}_{12}$ not being a zero matrix. Chilès and Delfiner (1999, p. 301) argue that cokriging is typically performed as ordinary cokriging for this reason.

The two sets of constraints that the kriging weights for the primary variable sum to one and those for the secondary attributes sum to zero were made to ensure unbiasedness. Goovaerts (1998) points out that most of the secondary weights in $\boldsymbol{\lambda}_2$ tend to be small and because of the sum-to-zero constraint many of them will be negative which increases the risk to obtain negative predicted values for the primary attribute. To enhance the contribution of the secondary data and to limit the possibility of negative predictions Isaaks and Srivastava (1989, p. 416) proposed replacing the two constraints with a single sum-to-one constraint for all weights, primary and secondary. This requires that the secondary attributes be rescaled to have the same mean as the primary variable. An additional advantage of a single constraint comes to bear in what has been termed **collocated cokriging**.

The cokriging predictor [9.48] does not require that $Z_2(\mathbf{s})$ is observed at the prediction location $\mathbf{s}_0$. If $Z_1(\mathbf{s}_0)$ were observed there is of course no point in predicting it unless there is measurement error and prediction of the signal $S_1(\mathbf{s}_0)$ matters. Collocated cokriging (Xu et al. 1992) is a simplification of cokriging where the secondary attribute is available at the prediction location, while the primary attribute is not. The collocated cokriging system thus uses only $n_1 + 1$ instead of $n_1 + n_2$ observations and if separate constraints for the primary and secondary attribute are entertained must be performed as simple cokriging, since λ_2 would be zero in an ordinary cokriging algorithm. Replacing the two constraints with a single sum-to-one constraint avoids this problem and permits collocated cokriging to be performed as an ordinary cokriging method.

Finally, using only a single constraint for the combined weights of all attributes has advantages in the case of colocated data. Part of the problem in implementing cokriging is having to estimate the cross-covariances $\boldsymbol{\Sigma}_{12}$. A model commonly employed is the proportional covariance model where $\boldsymbol{\Sigma}_{12} \propto \boldsymbol{\Sigma}_{11}$. If $Z_1(\mathbf{s})$ and $Z_2(\mathbf{s})$ are colocated the secondary weights will be zero under the proportional covariance model when the constraints $1'\boldsymbol{\lambda}_1 = 1$, $\mathbf{1}'\boldsymbol{\lambda}_2 = 0$ are used. A single constraint avoids this problem and allows the secondary variable to contribute to the prediction of $Z_1(\mathbf{s}_0)$.

Multiple Spatial Regression

The second approach to utilizing information on secondary attributes is to consider the mean of $Z_1(\mathbf{s})$ to be a (linear) function of the secondary attributes which are considered fixed,

$$\mathrm{E}[Z_1(\mathbf{s})] = \beta_0 + \beta_1 Z_2(\mathbf{s}) + \beta_2 Z_3(\mathbf{s}) + \cdots + \beta_{k-1} Z_k(\mathbf{s}) = \mathbf{x}(\mathbf{s})'\boldsymbol{\beta}. \qquad [9.52]$$

The error term of this model is a (second-order) stationary spatial process $\delta_1(\mathbf{s})$ with mean 0 and covariance function $C_1(\mathbf{h})$ (semivariogram $\gamma_1(\mathbf{h})$). The complete model can then be written as

$$\begin{aligned} Z_1(\mathbf{s}) &= \mathbf{x}'(\mathbf{s})\boldsymbol{\beta} + \delta_1(\mathbf{s}) \\ \mathrm{E}[\delta_1(\mathbf{s})] &= 0 \\ \mathrm{Cov}[\delta_1(\mathbf{s}), \delta_1(\mathbf{s}+\mathbf{h})] &= C_1(\mathbf{h}). \end{aligned} \qquad [9.53]$$

This model resembles the universal kriging model but while there $\mathbf{x}(\mathbf{s})$ is a polynomial response surface in the spatial coordinates, here $\mathbf{x}(\mathbf{s})$ is a function of secondary attributes observed at locations $\mathbf{s}$. It is for this reason that the multiple spatial regression model is particularly meaningful in our opinion. In many applications colorful maps of an attribute are attractive but not the most important result of the analysis. That crop yield varies across a field is one thing. The ability to associate this variation with variables that capture soil properties, agricultural management strategies, etc., is more meaningful since it allows the user to interpolate and extrapolate (within reasonable limits) crop yield to other situations. It is a much overlooked fact that geostatistical kriging methods produce predicted surfaces for the primary attribute at hand that are difficult to transfer to other environments and experimental situations. If spatial predictions of soil carbon ($Z_1(\mathbf{s})$) rely on the stochastic spatial variability of soil carbon *and* its relationship to soil nitrogen ($Z_2(\mathbf{s})$), for example, a map of soil carbon can be produced from samples of soil nitrogen collected on a different field provided the relationship between $Z_1(\mathbf{s})$ and $Z_2(\mathbf{s})$ can be transferred to the new location and the stochastic properties (the covariance function $C_1(\mathbf{h})$) are similar in the two instances.

One advantage of the spatial regression model [9.53] over the cokriging system is that it only requires the spatial covariance function of the primary attribute. The covariance function of the secondary attribute and the cross-covariances are not required. Spatial predictions of $Z_1(\mathbf{s}_0)$ can then be performed as modified universal kriging predictions where the polynomial response surface in the spatial coordinates is replaced by the mean function [9.52]. A second advantage is that simultaneous estimation of $\boldsymbol{\beta}$ and the parameters determining the covariogram $C_1(\mathbf{h})$ is possible based on maximum likelihood or restricted maximum likelihood techniques. If one is willing to make the assumption that $\delta_1(\mathbf{s})$ is a Gaussian random field the `mixed` procedure in The SAS® System can be used for this purpose. Imagine that the data in Figure 9.23 are supplemented by similar samples of soil carbon and soil nitrogen on no-till plots (which they actually were, see §9.8.4 for the application). The interest of the researcher shifts from producing maps of soil carbon under no-till and chisel-plow strategies to tests of the hypothesis that soil carbon mean or variability is identical under the two management regimes. This is easily accomplished with spatial regression models by incorporating into the mean function [9.52] not only secondary attributes such as soil nitrogen levels, but classification variables that identify treatments.

The multiple spatial regression approach has some drawbacks as compared to cokriging, however. Only colocated samples of $Z_1(\mathbf{s})$ and all secondary attributes can be used in estimation. If only one of the secondary attributes has not been observed at a particular location the information collected on any attribute at that location will be lost. In the chisel-plow vs. no-till example this does not imply that all locations have to be managed under chisel-plowed and no-till regimes. It implies that for any location where the primary attribute (soil carbon) has been observed it can be identified whether it belonged to the chisel-plow or no-till treatment. If, in addition, soil carbon is modeled as a function of tillage and soil nitrogen, only those records will be retained in the analysis where soil carbon and soil nitrogen have been measured. If the secondary attribute stems from a much more dense sampling scheme, e.g., $Z_2(\mathbf{s})$ has been remotely sensed and $Z_1(\mathbf{s})$ is a more sparse *in-situ* sampling, there is no problem since each primary attribute can be matched with a secondary attribute. If the secondary attribute has been sampled sparsely compared to the primary attribute, e.g., $Z_1(\mathbf{s})$ stems from yield monitoring and $Z_2(\mathbf{s})$ from grid soil sampling, the applicability of multiple spatial regression is limited. The attribute sampled most coarsely determines the spatial resolution of the analysis.

Another "drawback" relates to the prediction of the primary attribute at a new location $\mathbf{s}_0$. It is required that the secondary attributes have been observed at the prediction location $\mathbf{s}_0$. If soil carbon is modeled as a function of tillage regime and soil nitrogen, prediction at a new location cannot be performed unless the tillage regime and soil nitrogen at $\mathbf{s}_0$ are known. This is a feature of all regression methods and should not be construed as a shortcoming. If the modeler determines that Y depends on X and the value of X is unknown the model cannot be used to produce a prediction of Y.

9.5 Spatial Regression and Classification Models

9.5.1 Random Field Linear Models

A spatial regression or classification model is a model for geostatistical data where interest lies primarily in statistical inference about the mean function $\mathrm{E}[Z(\mathbf{s})] = \mu(\mathbf{s})$. The most important application of these models in the crop and soil sciences is the analysis of field experiments where the experimental units exhibit spatial autocorrelation. The agricultural variety trial is probably the most important type of experiment to which the models in this section can be applied, but any situation in which $\mathrm{E}[Z(\mathbf{s})]$ is modeled as a function of other variables in addition to a spatially autocorrelated error process falls under this heading. Variety trials are particularly important here because of their respective size. Randomization of treatments to experimental units neutralizes the effects of spatial correlation among experimental units and provides the framework for statistical inference in which cause-and-effect relationships can be examined. These trials are often conducted as randomized block designs and, because of the large number of varieties involved, the blocks can be substantial in size. Combining adjacent experimental units into blocks in agricultural variety trials can be at variance with an assumption of homogeneity within blocks. Stroup, Baenziger and Mulitze (1994) notice that if more than eight to twelve experimental units are grouped, spatial trends will be removed only incompletely. Although randomization continues to neutralize these effects, it does not eliminate them as a source of experimental error.

Figure 9.25 shows the layout of a randomized complete block design conducted as a field experiment in Alliance, Nebraska. The experiment consisted of 56 wheat cultivars arranged in four blocks and is discussed in Stroup et al. (1994) and Littell et al. (1996).

Analysis of the plot yields in this RCBD reveals a p-value for the hypothesis of no varietal differences of $p = 0.7119$ along with a coefficient of variation of $CV = 27.58\%$. A p-value that large should give the experimenter pause. That there are no yield differences among 56 varieties is very unlikely. The large coefficient of variation conveys the considerable magnitude of the experimental error variance. Blocking as shown in Figure 9.25 did not eliminate the spatial dependencies among experimental units and left any spatial trends to randomization which increased the experimental error. The large p-value is not evidence of an absence of varietal differences, but of an experimental design lacking power to detect these differences.

Instead of the classical RCBD analysis one can adopt a modeling philosophy where the variability in the data from the experiment is decomposed into large-scale trends and smooth-scale spatial variation. Contributing to the large-scale trends are treatment effects, deterministic effects of spatial location, and other explanatory variables. The smooth-scale variation consists of a spatial random field that captures, for example, smooth fertility trends.

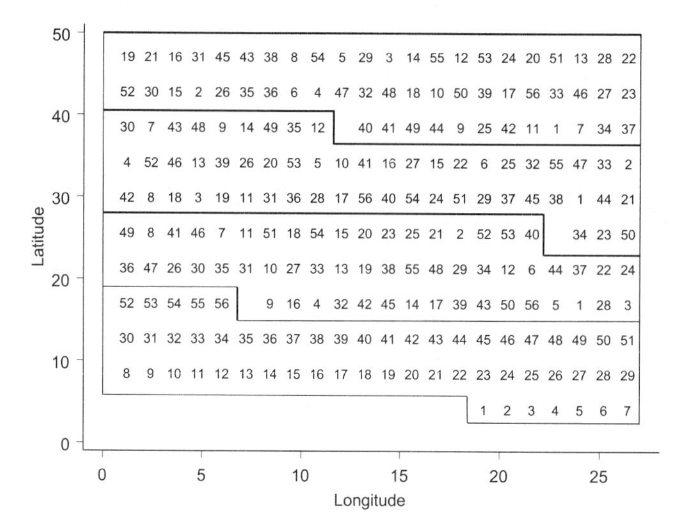

Figure 9.25. Layout of wheat variety trial at Alliance, Nebraska. Lines show block boundaries, numbers identify the placement of varieties within blocks. There are four blocks and 56 varieties. Drawn from data in Littell et al. (1996).

In the notation of §9.4 we are concerned with the spatial mean model

$$Z(\mathbf{s}) = \mu(\mathbf{s}) + \delta(\mathbf{s}),$$

where $\delta(\mathbf{s})$ is assumed to be a second-order stationary spatial process with semivariogram $\gamma(\mathbf{h})$ and covariogram $C(\mathbf{h})$. The mean model $\mu(\mathbf{s})$ is assumed to be linear in the large-scale effects, so that we can write

$$Z(\mathbf{s}) = \mathbf{X}(\mathbf{s})\boldsymbol{\beta} + \delta(\mathbf{s}). \qquad [9.54]$$

We maintain the dependency of the design/regressor matrix $\mathbf{X}(\mathbf{s})$ on the spatial location since $\mathbf{X}(\mathbf{s})$ may contain, apart from design (e.g., block) and treatment effects, other variables that depend on the spatial location of the experimental units, or the coordinates of observations themselves although that is not necessarily so. Zimmerman and Harville (1991) refer to [9.54] as a **random field linear model**. Since the spatial autocorrelation structure of $\delta(\mathbf{s})$ is modeled through a semivariogram or covariogram we take a direct approach to modeling spatial dependence rather than an autoregressive approach (in the vernacular of §9.3). This can be rectified with the earlier observation that data from field experiments are typically lattice data where autoregressive methods are more appropriate by considering each observation as concentrated at the centroid of the experimental unit (see Ripley 1981, p. 94, for a contrasting view that utilizes block averages instead of point observations).

The model for the semivariogram/covariogram is critically important for the quality of spatial predictions in kriging methods. In spatial random field models, where the mean function is of primary importance, it turns out that it is important to do a reasonable job at modeling the second order structure of $\delta(\mathbf{s})$, but as Zimmerman and Harville (1991) note, treatment comparisons are relatively insensitive to the choice of covariance functions (provided the set of functions considered is a reasonable one and that the mean function is properly specified). Besag and Kempton (1986) found that inclusion of a nugget effect also appears to be unnecessary in many field-plot experiments.

Before proceeding further with random field linear models we need to remind the reader of the adage that *one modeler's random effect is another modeler's fixed effect.* Statistical models that incorporate spatial trends in the analysis of field experiments have a long history. In contrast to the random field models, previous attempts of incorporating the spatial structure focused on the mean function $\mu(\mathbf{s})$ rather than the stochastic component of the model. The term *trend analysis* has been used in the literature to describe methods that incorporate covariate terms that are functions of the spatial coordinates. In a standard RCBD analysis where Y_{ij} denotes the observation on treatment i in block j, the statistical model for the analysis of variance is

$$Y_{ij} = \mu + \rho_j + \tau_i + e_{ij},$$

where the experimental errors e_{ij} are uncorrelated random variables with mean 0 and variance σ^2. A trend analysis changes this model to

$$Y_{ij} = \mu + \tau_i + \theta_{kl} + e_{ij}, \qquad [9.55]$$

where θ_{kl} is a polynomial in the row and column indices of the experimental units (Brownie, Bowman, and Burton 1993). If r_k is the kth row and c_l the lth column of the field layout, then one may choose $\theta_{kl} = \beta_1 r_k + \beta_2 c_l + \beta_3 r_k^2 + \beta_4 c_k^2 + \beta_5 r_k c_l$, for example, a second-order response surface in the row and column indices. The difference to a random field linear model is that the deterministic term θ_{kl} is assumed to account for the spatial dependency between experimental units. It is a fixed effect. It does, however, appeal to the notion of a smooth-scale variation in the sense that the spatial trends move smoothly across block boundaries. The block effects have disappeared from model [9.55]. Applications of these trend analysis models can be found in Federer and Schlottfeldt (1954), Kirk, Haynes, and Monroe (1980), and Bowman (1990). Because it is assumed that the error terms e_{ij} remain uncorrelated they are not spatial random field models in our sense and will not be discussed further. For a comparison of trend and random field analyses see Brownie et al. (1993).

A second type of model that maintains independence of the errors are the nearest-neighbor models which are based on differencing observations with each other or by taking differences between plot yields and cultivar averages. The Papadakis nearest-neighbor analysis (Papadakis 1937), for example, calculates residuals between plot yields and arithmetic treatment averages in the East-West and North-South direction and uses these residuals as covariances in the mean model (the θ_{kl} part of the trend analysis model). The Schwarzbach analysis relies on adjusted cultivar means which are arithmetic means corrected for average responses in neighboring plots (Schwarzbach 1984).

In practical applications it may be difficult to choose between these various approaches to model spatial dependencies and to discriminate between different models. For example, changing the fixed effects trend by including or eliminating terms in a trend analysis will change the autocorrelation of the model residuals. Brownie and Gumpertz (1997) conclude that it is necessary to account for major spatial trends as fixed effects in the model but also that random field analyses are surprisingly robust to moderate misspecification of the fixed trend and retain a high degree of validity of tests and estimates of precision. The reason, in our opinion, is that a model which simultaneously models large- and small-scale stochastic trends is able – within limits – to capture omitted trends in the mean model through the spatial dependency structure in the error process. Zimmerman and Harville (1991) refer to this effect as the covariance function "soaking up" spatial heterogeneity that would otherwise be fitted through fixed effects in the mean function. A trend analysis model or nearest-neighbor model that assumes that the mean function is correctly specified and the errors are uncorrelated will be invalid if the mean function is not modeled properly. There is nothing in the error structure that can "soak up" the ill-specification of the fixed effects.

9.5.2 Some Philosophical Considerations

Modeling data from a field experiment with random field methods seems like a win-win situation. The modeler can add or delete terms to the fixed effects part of the model that capture large-scale trends and let the covariance function of the error process $\delta(\mathbf{s})$ pick up any smooth-scale spatial variation of the omitted effects. As always, there is no free lunch and the analyst must be aware of the differences between modeling the data from a designed experiment vs. relying on randomization theory. The classical analysis of an experimental design stems from its underlying linear model which in turn is generated by the particular error-control, treatment, and observational designs. The ability to perform cause-and-effect inferences rests on these design components. Randomization ensures that the unaccounted effects — such as systematic spatial trends among the experimental units — are balanced out. This implies that expectations are reckoned over the randomization distribution of the design. In the Alliance, Nebraska wheat yield variety trial this distribution is formed by all possible arrangements of the 56 treatments to the $56 \times 4 = 224$ experimental units. The observed outcomes are considered fixed in the randomization approach. Assume, for a moment, that the three rightmost columns of experimental units in Figure 9.25 are systematically different from the other units. Should we take this into account in specifying the statistical model for the analysis or appeal to the fact that under randomization such effects are washed (balanced) out? There are three schools of thought:

1. Appeal to the randomization distribution because it allows causal inference. In effect, stick with the randomized complete block analysis. If it does not work out because blocking was carried out incorrectly, learn from the mistake and fix the problem the next time a variety trial with fifty-six treatments is conducted.

2. Do not appeal to the randomization distribution and model the variability and effects for *this* particular set of data. This is a modeling exercise determining which effects are modeled as part of the mean structure $\mathbf{X}(\mathbf{s})$ and which effects are "soaked up" by the error structure.

3. Appeal to randomization but also to the fact that stochastic elements beyond the randomization of treatments to units are at work. In developing the analysis appeal to a model where the errors are no longer independent and take expectation with respect to the joint distribution of randomization *and* the spatial process.

The three approaches differ in what is considered the correct model for analysis and how it is used. In (1) the correct model stems from the error-control, treatment, and observational design components. Treatment comparisons will always be unbiased under this approach, but can be inefficient if the design was not chosen carefully (as is the case in the Alliance-Nebraska case). In (2) the analyst is charged to develop a suitable model. Statistical inference proceeds assuming that the selected model is correct. If a wrong model is used, treatment comparisons will be biased. Since there is never unshakable evidence that the final model is correct one can no longer make causal statements about the effect of treatments on the outcome. Statistical inference is *associative* rather than causal. The third approach is a mixture technique. It recognizes dependencies among the experimental units and the fact that treatments are randomly assigned to the units. Expectations of mean squares are calculated first over the randomization distribution conditional on the spatial process and then over the spatial process (see, for example, Grondona and Cressie 1991).

In spatial analyses the observed data are considered a realization of a random field and modeling the mean and dispersion structure proceeds in an observational manner. Whether a spatial model will provide a more efficient analysis will depend to what extent large-scale and small-scale trends are conducive to modeling. Besag and Kempton (1986) conclude that many agronomic experiments are not carried out in a sophisticated manner. The reasons may be convenience, unfamiliarity of the experimenter with more complex design choices, or tradition. We agree that it is hardly reasonable to conduct a field experiment with 56 treatments in a randomized complete block design. An incomplete block design or a resolvable, cyclic design may have been more appropriate. Nevertheless, many experiments are still conducted in this fashion. Bartlett (1938, 1978a) views analyses that emphasize the spatial context over the design context as ancillary devices to salvage efficiency in experiments that could have been designed more appropriately. Spatial random field models are more than salvage tools. They are statistical models that describe the variation in data, whether the data stem from a designed experiment or an observational study. By switching from a design-based analysis to one based on modeling, the ability to draw causal inferences is sacrificed, however.

9.5.3 Parameter Estimation

In matrix-vector notation model [9.54] can be written as

$$\mathbf{Z}(\mathbf{s}) = \mathbf{X}(\mathbf{s})\boldsymbol{\beta} + \boldsymbol{\delta}(\mathbf{s}), \quad \boldsymbol{\delta}(\mathbf{s}) \sim (\mathbf{0}, \boldsymbol{\Sigma}(\boldsymbol{\theta})) \qquad [9.56]$$

and the parameters of the model to be estimated are $\boldsymbol{\phi} = [\boldsymbol{\beta}, \boldsymbol{\theta}]'$. $\boldsymbol{\theta}$ relates to the spatial dependency structure and $\boldsymbol{\beta}$ to the large-scale trend. As models for $\boldsymbol{\Sigma}(\boldsymbol{\theta})$ we usually consider covariograms that are derived from the isotropic semivariogram models in §9.2.2, keeping the number of parameters in $\boldsymbol{\theta}$ small. Because we work with covariances, it is assumed that the process is second-order stationary so that its covariogram is well-defined. Two general approaches to parameter estimation can be distinguished. Likelihood and likelihood-type methods which estimate $\boldsymbol{\theta}$ and $\boldsymbol{\beta}$ simultaneously and least squares methods that estimate $\boldsymbol{\beta}$ given an externally obtained estimate of the spatial dependency.

Least Squares Methods

If $\boldsymbol{\Sigma}(\boldsymbol{\theta})$ were known parameter estimates for $\boldsymbol{\beta}$ can be obtained by generalized least squares (GLS):

$$\widehat{\boldsymbol{\beta}}_{GLS} = \left(\mathbf{X}'\boldsymbol{\Sigma}(\boldsymbol{\theta})^{-1}\mathbf{X}\right)^{-1}\mathbf{X}'\boldsymbol{\Sigma}(\boldsymbol{\theta})^{-1}\mathbf{Z}(\mathbf{s}).$$

Since $\boldsymbol{\Sigma}(\boldsymbol{\theta})$ is usually unknown we are faced with a similar quandary as in universal kriging.

Estimating $\boldsymbol{\theta}$ through semivariogram analysis requires detrending of the data, that is, an estimate of $\boldsymbol{\beta}$. Efficient estimation of $\boldsymbol{\beta}$ requires knowledge of $\boldsymbol{\theta}$. The usual approach is to

1. Assume $\boldsymbol{\Sigma}(\boldsymbol{\theta}) = \sigma^2\mathbf{I}$ and fit the model by ordinary least squares to obtain $\widehat{\boldsymbol{\beta}}_{OLS}$.
2. Obtain the OLS residuals $\widehat{\mathbf{e}}(\mathbf{s}) = \mathbf{Z}(\mathbf{s}) - \mathbf{X}(\mathbf{s})\widehat{\boldsymbol{\beta}}_{OLS}$.
3. Fit a parametric, second-order stationary semivariogram based on the $\widehat{\mathbf{e}}(\mathbf{s})$ to obtain $\widehat{\boldsymbol{\theta}}$.
4. Use the estimates from the semivariogram fit to construct the $\boldsymbol{\Sigma}(\widehat{\boldsymbol{\theta}})$ matrix.

These steps can (and should) be iterated, replacing OLS residuals in step 2. with GLS residuals after the first iteration. The final estimates of the mean parameters are estimated generalized least square estimates

$$\widehat{\boldsymbol{\beta}}_{EGLS} = \left(\mathbf{X}'\boldsymbol{\Sigma}(\widehat{\boldsymbol{\theta}})^{-1}\mathbf{X}\right)^{-1}\mathbf{X}'\boldsymbol{\Sigma}(\widehat{\boldsymbol{\theta}})^{-1}\mathbf{Z}(\mathbf{s}). \qquad [9.57]$$

The same issues as in §9.4.4 must be raised here. The residuals lead to a biased estimate of the semivariogram of $\delta(\mathbf{s})$ and $\widehat{\boldsymbol{\beta}}_{OLS}$ is an inefficient estimator of the large-scale trend parameters. Since the emphasis in spatial random field linear models is often not on predicting but on estimation and hypothesis testing about $\boldsymbol{\beta}$ these issues are not quite as critical as in the case of universal kriging. If the results of a random field linear model analysis are used to predict $Z(\mathbf{s}_0)$ as a function of covariates and the spatial autocorrelation structure, the issues regain importance.

Likelihood Methods

Likelihood methods circumvent these problems because the mean and covariance parameters are estimated simultaneously. On the other hand they require distributional assumptions about $Z(\mathbf{s})$ or $\delta(\mathbf{s})$. If $\delta(\mathbf{s})$ is a Gaussian random field, then twice the negative log-likelihood of $\mathbf{Z}(\mathbf{s})$ is

$$\mathfrak{w}(\boldsymbol{\beta},\boldsymbol{\theta};\mathbf{z}(\mathbf{s})) = n\ln\{2\pi\} + \ln|\boldsymbol{\Sigma}(\boldsymbol{\theta})| + (\mathbf{z}(\mathbf{s}) - \mathbf{X}(\mathbf{s})\boldsymbol{\beta})'\boldsymbol{\Sigma}(\boldsymbol{\theta})^{-1}(\mathbf{z}(\mathbf{s}) - \mathbf{X}(\mathbf{s})\boldsymbol{\beta}).$$

and the maximum likelihood estimates $\widehat{\boldsymbol{\beta}}_M$, $\widehat{\boldsymbol{\theta}}_M$ minimize this expression. This process is generally iterative and can be simplified by profiling the likelihood. This numerically efficient method can be applied if some parameters have a closed-form solution given the others. First consider $\boldsymbol{\theta}$ fixed and known. Minimizing $\mathfrak{w}(\boldsymbol{\beta},\boldsymbol{\theta};\mathbf{z}(\mathbf{s}))$ is then equivalent to minimizing $(\mathbf{z} - \mathbf{X}(\mathbf{s})\boldsymbol{\beta})'\ \boldsymbol{\Sigma}(\boldsymbol{\theta})^{-1}(\mathbf{z} - \mathbf{X}(\mathbf{s})\boldsymbol{\beta})$. Since this is a generalized residual sum of squares, the maximum likelihood estimate of $\boldsymbol{\beta}$ (given $\boldsymbol{\theta}$) is

$$\widehat{\boldsymbol{\beta}}_{GLS} = \left(\mathbf{X}'\boldsymbol{\Sigma}(\boldsymbol{\theta})^{-1}\mathbf{X}\right)^{-1}\mathbf{X}'\boldsymbol{\Sigma}(\boldsymbol{\theta})^{-1}\mathbf{Z}(\mathbf{s}).$$

The profiled (negative) log likelihood is obtained by substituting this expression back into $\mathfrak{w}(\boldsymbol{\beta},\boldsymbol{\theta};\mathbf{z}(\mathbf{s}))$ which is then only a function of $\boldsymbol{\theta}$ and is minimized with respect to $\boldsymbol{\theta}$. The resulting estimate $\widehat{\boldsymbol{\theta}}_M$ is the maximum likelihood estimate of $\boldsymbol{\theta}$ and the MLE of $\boldsymbol{\beta}$ is

$$\widehat{\boldsymbol{\beta}}_M = \left(\mathbf{X}'\boldsymbol{\Sigma}(\widehat{\boldsymbol{\theta}}_M)^{-1}\mathbf{X}\right)^{-1}\mathbf{X}'\boldsymbol{\Sigma}(\widehat{\boldsymbol{\theta}}_M)^{-1}\mathbf{Z}(\mathbf{s}). \qquad [9.58]$$

The maximum likelihood ([9.58]) and estimated generalized least squares estimates [9.57] are very similar. They differ only in the covariance parameter estimate that is being substituted. To reduce the bias in maximum likelihood estimates of the covariance parameters it is again recommended to perform restricted maximum likelihood estimation. The REML estimates of the large-scale trend parameters are obtained as

$$\widehat{\boldsymbol{\beta}}_R = \left(\mathbf{X}'\boldsymbol{\Sigma}(\widehat{\boldsymbol{\theta}}_R)^{-1}\mathbf{X}\right)^{-1}\mathbf{X}'\boldsymbol{\Sigma}(\widehat{\boldsymbol{\theta}}_R)^{-1}\mathbf{Z}(\mathbf{s}). \qquad [9.59]$$

Software Implementation

The three methods, GLS, ML, and REML, lead to very similar formulas for the $\boldsymbol{\beta}$ estimates. The `mixed` procedure in The SAS® System can be used to obtain any one of the three. The spatial covariance structure of $\delta(\mathbf{s})$ is specified through the `repeated` statement of the procedure. In contrast to clustered data models in §7, all data points are potentially autocorrelated which calls for the `subject=intercept` option of the repeated statement.

Assume that an analysis of OLS residuals leads to an exponential semivariogram with practical range 4.5, partial sill 10.5, and nugget 2.0. The spatial coordinates of the data points are stored in variables `xloc` and `yloc` of the SAS data set. The mean model consists of treatment effects and a linear response surface in the coordinates. The following statements obtain the EGLS estimates [9.57], preventing `proc mixed` from iteratively updating the covariance parameters (`noiter` option of `parms` statement). The `noprofile` option prevents the profiling of an extra scale parameter from $\boldsymbol{\Sigma}(\boldsymbol{\theta})$. The Table of `Covariance Parameter Estimates` will contain three rows entitled `Variance`, `SP(EXP)`, and `Residual`. These correspond to the partial sill, the range, and the nugget effect, respectively. Notice that the parameterization of the exponential covariogram in `proc mixed` considers the range parameter to be one third of the practical range.

```
/* -------------------------------------------------- */
/* Fit the model by EGLS for fixed covariogram estimates */
/* -------------------------------------------------- */
proc mixed data=RFLMExample noprofile ;
  class treatment;
  model Z = treatment xloc yloc xloc*yloc / s;
  parms /* sill   */ ( 10.5 )
        /* range  */ ( 1.5  )
        /* nugget */ ( 2.0  ) / noiter;
  /* The local option of the repeated statement adds the */
  /* nugget effect                                        */
  repeated /subject=intercept local type=sp(exp)(xloc yloc);
run; quit;
```

Restricted maximum likelihood estimates are obtained in `proc mixed` with the statements

```
proc mixed data=RFLMExample noprofile ;
  class treatment;
  model Z = treatment xloc yloc xloc*yloc / s;
  parms /* sill   */ ( 6   to 12 by 2   )
        /* range  */ ( 0.5 to  3 by 1.5 )
        /* nugget */ ( 1   to  4 by 1.0 );
  repeated /subject=intercept local type=sp(exp)(xloc yloc);
run; quit;
```

The `noiter` option was removed from the `parms` statement which prompts the procedure to iteratively update the covariance parameter estimate $\boldsymbol{\theta}$. For each element of $\boldsymbol{\theta}$ a range of starting values is given. This can considerably speed up estimation, which can require formidable resources for large data sets. If the grid of starting values is too fine this is somewhat counterproductive as the procedure then has to evaluate many combinations of possible starting values before settling on the best set. The default estimation procedure for covariance parameter estimation is restricted maximum likelihood and the code example above yields $\widehat{\boldsymbol{\beta}}_R$ as in [9.59]. To obtain maximum likelihood estimates add the `method=ml` option to the `proc mixed` statement.

9.6 Autoregressive Models for Lattice Data

Box 9.10 Lattice Models

- **Models for spatial lattice data are close relatives of time series models.**
- **A lattice model commences with the user's definition of spatial connectivity among sites. This choice is then combined with an appropriate model for the marginal or conditional distribution of the data that is consistent with the neighborhood structure.**
- **Depending on whether the joint or conditional distribution of $Z(\mathbf{s})$ is being modeled, SSAR and CSAR models for lattice data are distinguished.**

9.6.1 The Neighborhood Structure

Lattice data are spatial data where the index set D is a fixed, discrete subset of $\mathbb{R}^2$ of countable points and $Z(\mathbf{s})$ is a random variable at location $\mathbf{s} \in D$. Examples of lattice data are observations made by census tract, county, or city blocks, data from field trials and remotely sensed images. Keeping with the literature on lattice data we call the locations $\mathbf{s} \in D$ the **sites** of the lattice. It is common to enumerate the countable set of sites in a lattice, for example, counties or census tracts can be numbered from 1 to n. Since the numbering in itself does not convey any spatial information it is necessary to define a location feature of each site such as the county center or the seat of the county government. On rectangular lattices (field experiments) the center of the unit is often used or experimental units can be identified by row and column number.

Modeling the spatial dependence among observations via the semivariogram or covariogram requires a smooth-scale spatial structure and a continuous spatial process. With lattice data other means of capturing the spatial dependence are needed. The notion of stationarity is of somewhat questionable value for processes operating on irregularly shaped area units or partitions (census tracts, counties, landscapes, regions, states, etc.). Even if there exists an underlying stationary continuous-space process, variances and covariances will not be the same for all areas if the observations arise from different area integrations. Stationary covariance or semivariogram functions are then not useful to describe the stochastic interarea relationships. For lattice data a different system is needed. This starts with a definition of what is considered the **neighborhood** of site $\mathbf{s}_i$. By choosing the neighborhood structure the modeler determines which sites are spatially connected and the degree of their connectedness. In a regular lattice with nearest neighbor dependence, for example, a site $\mathbf{s}_i$ can be declared as being connected only to its immediate neighbors. Cliff and Ord (1981) and Upton and Fingleton (1985) distinguish the rook, bishop, and queen definition of spatial contiguity drawing on the respective moves on the chess board (Figure 9.26). The rook definition is sometimes identified as *the* nearest-neighbor definition.

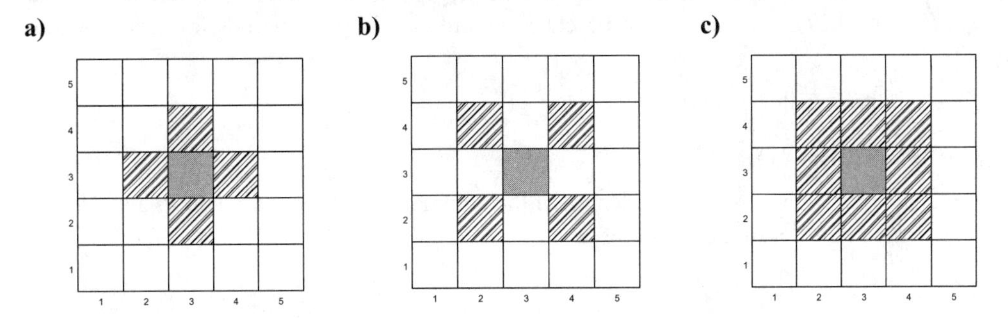

Figure 9.26. Definitions of spatial contiguity (neighborhood, connectedness) on a regular lattice. (a) rook's definition: edges abut; (b) bishop's definition: touching corners; (c) queen's definition: touching corners and edges.

On an irregular lattice neighborhoods are defined differently. Consider a lattice consisting of area units such as counties (Figure 9.27). A neighborhood definition akin to the queen definition on a regular lattice is to collect all sites into the neighborhood of a site that share a common boundary with the county. Alternatively one can identify a single point $\mathbf{s}_i$ with each county, the seat of the county government, for example. The neighborhood then can be defined as all counties within a given distance from the county seat or as all counties whose county seat is within a given distance of $\mathbf{s}_i$.

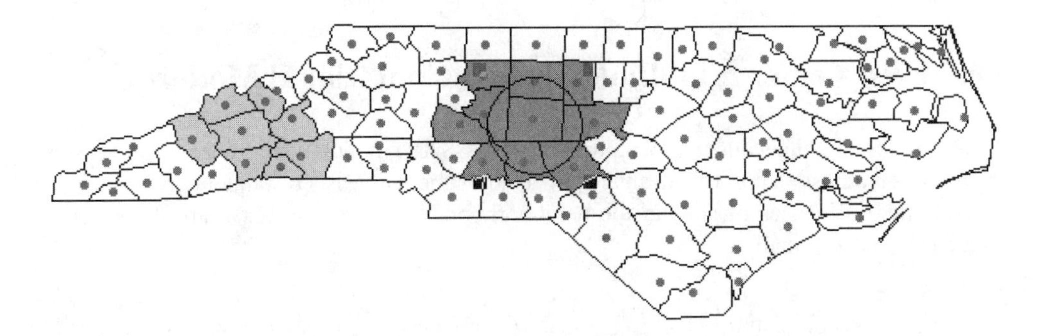

Figure 9.27. Counties of North Carolina and possible neighborhood definitions for two counties: Counties with adjoining borders (left part of map) or counties within 30 miles of the county seat (center part of map). County seats are shown as dots.

Once the neighbors of all sites are identified weights w_{ij} are assigned. If sites $\mathbf{s}_i$ and $\mathbf{s}_j$ are not connected then $w_{ij} = 0$, but what weights are to be assigned to sites that are neighbors? The simplest solution is through binary variables:

$$w_{ij} = \begin{cases} 1 & \text{if } \mathbf{s}_j \text{ is a neighbor of } \mathbf{s}_i \\ 0 & \text{if } \mathbf{s}_j \text{ is not a neighbor of } \mathbf{s}_i \\ 0 & i = j. \end{cases}$$

A binary weighing scheme is reasonable when sites are spaced regularly by uniform distances (Figure 9.26). If sites are arranged irregularly or represent area units of different size and shape (Figure 9.27), weights should be chosen more carefully. Some possibilities (Haining 1990) are

- $w_{ij} = ||\mathbf{s}_i - \mathbf{s}_j||^{-\gamma}, \gamma \geq 0$
- $w_{ij} = \exp\{||\mathbf{s}_i - \mathbf{s}_j||^{-\gamma}\}$
- $w_{ij} = (l_{ij}/l_i)^{\gamma}$ where l_{ij} is the length of the common border between areas i and j and l_i is the perimeter of the border of area i
- $w_{ij} = (l_{ij}/l_i)^{\tau}/||\mathbf{s}_i - \mathbf{s}_j||^{-\gamma}$.

The weights are then collected in an $(n \times n)$ weight matrix $\mathbf{W} = [w_{ij}]$ and the statistical model is parameterized to incorporate the large-scale mean structure as well as the interactive spatial correlation structure. The two approaches that are common lead to the simultaneous and the conditional spatial autoregressive models. These combine the user's choice of neighbors with an appropriate model for the marginal or conditional distribution of the data that is consistent with the neighborhood structure.

The choice of the neighborhood structure is largely subjective. The fact that two sites have nonzero connectivity weights does not imply a causal relationship between the responses at the sites. It is a representation of the local variation due to extraneous conditions. Besag (1975) calls this “"third-party" dependence.” Imagine a locally varying regressor variable on which $Z(\mathbf{s})$ depends has not been observed. The localized neighborhood structure supplants the missing information by defining groups of sites which would have been affected similarly by the unobserved variable because they are in spatial proximity of each other.

9.6.2 First-Order Simultaneous and Conditional Models

In the simplest simultaneous spatial autoregressive (SSAR) model the response at site $\mathbf{s}_i$ is expressed as an adjustment to the mean at $\mathbf{s}_i$. The adjustment consists of random error $e(\mathbf{s}_i)$ and the weighted influences of sites in the neighborhood. If $\mathrm{E}[Z(\mathbf{s}_i)] = \mu(\mathbf{s}_i)$, an SSAR model can be expressed formally as

$$Z(\mathbf{s}_i) = \mu(\mathbf{s}_i) + \rho_s \sum_{j=1}^{n} w_{ij}(Z(\mathbf{s}_j) - \mu(\mathbf{s}_j)) + e(\mathbf{s}_i). \qquad [9.60]$$

In contrast to a spatial regression model, where secondary attributes are used as regressors and their values are considered fixed, the SSAR model regresses $Z(\mathbf{s}_i)$ onto neighboring values of the *same* attribute and these remain random variables. The parameter ρ_s measures the strength of the spatial autocorrelation but in contrast to autoregressive time series models cannot be interpreted as a correlation parameter. For example, the range of ρ_s depends on the structure of $\mathbf{W}$. Using matrix and vector notation and assuming that $\mu(\mathbf{s}_i) = \mathbf{x}'(\mathbf{s})\boldsymbol{\beta}$, [9.60] can be expressed more concisely as

$$\mathbf{Z}(\mathbf{s}) = \mathbf{X}(\mathbf{s})\boldsymbol{\beta} + \rho_s\mathbf{W}(\mathbf{Z}(\mathbf{s}) - \mathbf{X}(\mathbf{s})\boldsymbol{\beta}) + \mathbf{e}(\mathbf{s}).$$

It follows that $\mathrm{E}[\mathbf{Z}(\mathbf{s})] = \mathbf{X}(\mathbf{s})\boldsymbol{\beta}$ and if the $e(\mathbf{s}_i)$ are homoscedastic with variance σ_s^2 that $\mathrm{Var}[\mathbf{Z}(\mathbf{s})] = \sigma_s^2(\mathbf{I} - \rho_s\mathbf{W})^{-1}(\mathbf{I} - \rho_s\mathbf{W}')^{-1}$.

Instead of choosing a semivariogram or covariogram model for the smooth-scale variation as in a spatial regression model this spatial autoregressive model for lattice data requires estimation of only one parameter associated with the spatial autocorrelations, ρ_s. The structure and degree of the autocorrelation is determined jointly by the structure of $\mathbf{W}$ and the magnitude of ρ_s.

The second class of autoregressive models for lattice data, the conditional spatial autoregressive (CSAR) models, commence with the conditional mean and variance of a site's response given the observed values at all other sites. Denote by $\mathbf{z}(\mathbf{s})_{-i}$ the vector of observed values at all sites except the ith one. A CSAR model is then defined through

$$\mathrm{E}[Z(\mathbf{s}_i)|\mathbf{z}(\mathbf{s})_{-i}] = \mu(\mathbf{s}_i) + \rho_c\sum_{j=1}^{n} w_{ij}(\mathbf{z}(\mathbf{s}_j) - \mu(\mathbf{s}_j)), \quad \mathrm{Var}[Z(\mathbf{s}_i)|\mathbf{z}(\mathbf{s})_{-i}] = \sigma_i^2. \quad [9.61]$$

For spatial data the conditional and simultaneous formulations lead to different models. Assume that the conditional variances are identical, $\sigma_i^2 \equiv \sigma_c^2$. The marginal variance in the CSAR model is then given by $\mathrm{Var}[\mathbf{Z}(\mathbf{s})] = \sigma_c^2(\mathbf{I} - \rho_c\mathbf{W})^{-1}$ which is to be compared against $\mathrm{Var}[\mathbf{Z}(\mathbf{s})] = \sigma_s^2(\mathbf{I} - \rho_s\mathbf{W})^{-1}(\mathbf{I} - \rho_s\mathbf{W}')^{-1}$ in the simultaneous scheme. Even if $\sigma_c^2 = \sigma_s^2$ and $\rho_c = \rho_s$, the variance-covariance matrices will differ. Furthermore, since the variance-covariance matrix of $\mathbf{Z}(\mathbf{s})$ is symmetric it is necessary in the CSAR model with constant conditional variance that the weight matrix $\mathbf{W}$ be symmetric. The SSAR model imposes no such restriction on the weights. If a lattice consists of irregularly shaped area units asymmetric weights are often reasonable. Consider a study of urban sprawl with an irregular lattice of counties where a large county containing a metropolitan area is surrounded by smaller rural counties. It is reasonable to assume that what happens in a small county is very much determined by developments in the metropolitan area while the development of the major city will be much less influenced by a rural county. In a regular lattice asymmetric dependency parameters may also be possible. A site located on the edge of the lattice will depend on an interior site differently from how an interior site depends on an edge site. In these cases an asymmetric neighborhood structure is called for which rules out the CSAR model unless the conditional variances are adjusted.

SSAR models have disadvantages, too. The model disturbances $e(\mathbf{s}_i)$ and the responses $Z(\mathbf{s}_j)$ are not uncorrelated in these models which is in contrast to autoregressive time series models. This causes ordinary least squares estimators to be inconsistent. In matrix/vector notation the CSAR model can be written as

$$\mathbf{Z}(\mathbf{s}) - \boldsymbol{\mu}(\mathbf{s}) = \rho_c\mathbf{W}(\mathbf{Z}(\mathbf{s}) - \boldsymbol{\mu}(\mathbf{s})) + \boldsymbol{\nu}(\mathbf{s}),$$

where $\boldsymbol{\nu}(\mathbf{s}) = (\mathbf{I} - \rho_c\mathbf{W})(\mathbf{Z}(\mathbf{s}) - \boldsymbol{\mu}(\mathbf{s}))$ is a vector of pseudo-errors that is uncorrelated with $\mathbf{Z}(\mathbf{s})$ and hence the CSAR model retains this feature of the related time series model. This has advantages in parameter estimation. Cressie (1993) reasons that when the process has achieved stability, symmetric dependencies among sites are in general more natural than asymmetric dependencies and concludes that CSAR models are more natural than SSAR models.

The parameter ρ is called an interaction parameter since autoregressive models are interactive models. This parameter measures the correlation between neighboring time points in a first-order autoregressive time series but not so with spatial data. The matrix $(\mathbf{I} - \rho\mathbf{W})$ must be invertible and hence its determinant $|\mathbf{I} - \rho\mathbf{W}|$ must be nonzero. This places restrictions on the possible values of ρ. If $\{\lambda_i\}$ denotes the set of eigenvalues of $\mathbf{W}$, then, if the smallest eigenvalue is negative and the largest eigenvalue is positive,

$$\frac{1}{\min\{\lambda_i\}} < \rho < \frac{1}{\max\{\lambda_i\}}.$$

For square lattices ρ is restricted to $-0.25 < \rho < 0.25$ as the size of the lattice increases. The range of the interaction parameter can be affected by standardization. If rows of $\mathbf{W}$ are standardized to sum to one then $|\rho| < 1$ (Haining 1990, p. 82). For regular lattices and the rook neighborhood definition without row standardization the permissible ranges for ρ are shown in the next table.

Table 9.5. Limits on interaction parameter in first-order spatial autoregressive models as a function of the size of a regular lattice

Lattice size	$\frac{1}{\min\{\lambda_i\}}$	$\frac{1}{\max\{\lambda_i\}}$
3×3	-0.354	0.354
4×4	-0.309	0.309
5×5	-0.289	0.289
6×6	-0.277	0.277
7×7	-0.271	0.271
8×8	-0.266	0.266
9×9	-0.263	0.263
10×10	-0.261	0.261
20×20	-0.253	0.253

Models [9.60] and [9.61] are termed first-order models since they involve only one set of neighborhood weights and a single interaction parameter. To make $\mathbf{Z}(\mathbf{s})$ a function of two interaction parameters that measure different distance effects, the SSAR model can be modified to a second order model as

$$\mathbf{Z}(\mathbf{s}) = \mathbf{X}(\mathbf{s})\boldsymbol{\beta} + (\rho_1\mathbf{W}_1 + \rho_2\mathbf{W}_2)(\mathbf{Z}(\mathbf{s}) - \mathbf{X}(\mathbf{s})\boldsymbol{\beta}) + \mathbf{e}(\mathbf{s}).$$

For example, $\mathbf{W}_1$ can be a rook neighborhood structure and $\mathbf{W}_2$ a bishop neighborhood structure. The CSAR model can be similarly extended to a higher order scheme (see Whittle 1954, Besag 1974, Haining 1990).

9.6.3 Parameter Estimation

The marginal mean and variance in the (homoscedastic) first order SSAR and CSAR models are

$$\begin{aligned} &\text{SSAR:} \quad \mathrm{E}[\mathbf{Z}(\mathbf{s})] = \mathbf{X}(\mathbf{s})\boldsymbol{\beta} \quad \mathrm{Var}[\mathbf{Z}(\mathbf{s})] = \sigma_s^2(\mathbf{I}-\rho_s\mathbf{W})^{-1}(\mathbf{I}-\rho_s\mathbf{W}')^{-1} \\ &\text{CSAR:} \quad \mathrm{E}[\mathbf{Z}(\mathbf{s})] = \mathbf{X}(\mathbf{s})\boldsymbol{\beta} \quad \mathrm{Var}[\mathbf{Z}(\mathbf{s})] = \sigma_c^2(\mathbf{I}-\rho_c\mathbf{W})^{-1} \end{aligned}$$

and one could estimate the mean parameters by least squares, minimizing

$$\begin{aligned} &\text{SSAR:} \quad \sigma_s^{-2}(\mathbf{Z}(\mathbf{s})-\mathbf{X}(\mathbf{s})\boldsymbol{\beta})'(\mathbf{I}-\rho_s\mathbf{W}')(\mathbf{I}-\rho_s\mathbf{W})(\mathbf{Z}(\mathbf{s})-\mathbf{X}(\mathbf{s})\boldsymbol{\beta}) \\ &\text{CSAR:} \quad \sigma_c^{-2}(\mathbf{Z}(\mathbf{s})-\mathbf{X}(\mathbf{s})\boldsymbol{\beta})'(\mathbf{I}-\rho_c\mathbf{W})(\mathbf{Z}(\mathbf{s})-\mathbf{X}(\mathbf{s})\boldsymbol{\beta}). \end{aligned}$$

Unfortunately, because the errors $e(\mathbf{s}_i)$ and data $Z(\mathbf{s}_j)$ in the SSAR model are not uncorrelated, the least squares estimates in the simultaneous scheme are not consistent (Whittle 1954, Mead 1967, Ord 1975). Ord (1975) devised a modified least squares procedure when $\mathrm{E}[Z(\mathbf{s})] = 0$ that yields consistent estimates but comments on its low efficiency. The CSAR model does not suffer from this shortcoming and least squares estimation is possible there. When the spatial autoregressive model contains reactive effects ($\boldsymbol{\beta}$) in addition to an autoregressive structure it is desirable to obtain estimates of the large-scale mean structure and the interaction parameters simultaneously. The maximum likelihood method seems to be the method of choice. Unless the distribution of $\mathbf{Z}(\mathbf{s})$ is Gaussian, maximum likelihood estimation is numerically cumbersome, however. Ord (1975) adapted an iterative procedure developed by Cochrane and Orcutt (1949) for estimation in simultaneous time series models to obtain maximum likelihood estimates in the SSAR model. Haining (1990, p. 128) notes that no proof exists that this adapted algorithm converges to a local minimum in the spatial case. If the Gaussian assumption is reasonable we prefer maximum likelihood estimation of the parameters in a simultaneous scheme with a profiling algorithm as outlined in §A9.9.7.

9.6.4 Choosing the Neighborhood Structure

Choosing the neighborhood structure and thereby the weight matrix $\mathbf{W}$ is of importance in lattice models but often carried out in an *ad-hoc* manner without clear guidelines. The specification of $\mathbf{W}$ represents *a priori* knowledge of the range and intensity of a spatial effect for a set of area units constituting a geographical system. Forms used in the specification of $\mathbf{W}$ include binary contiguity matrices (rook, queen, bishop's definition), row-standardized forms, length of common boundary, intercentroid distances among others. Of interest are the following three questions:

- Does the choice of $\mathbf{W}$ make any practical difference in the statistical analysis of spatial lattice data?
- In what ways does the misspecification of a geographic weight matrix influence statistical analysis?
- Are there rule-of-thumb directions to guide specification of $\mathbf{W}$ for a given spatial landscape?

Griffith (1996) has addressed these questions and developed some guidelines and rules-of-thumb. We repeat the main results. Assume that the true model is $\mathbf{Z}(\mathbf{s}) = \mathbf{X}\boldsymbol{\beta} + \mathbf{e}(\mathbf{s})$ with $\mathbf{e}(\mathbf{s}) \sim G(\mathbf{0}, \sigma^2\mathbf{V})$ but the model $\mathbf{Z}(\mathbf{s}) = \mathbf{X}\boldsymbol{\beta} + \mathbf{e}^*(\mathbf{s})$ with $\mathbf{e}^*(\mathbf{s}) \sim G(\mathbf{0}, \sigma^2(\mathbf{I} - \rho\mathbf{W})^{-1})$ is fit. That is, instead of $\mathbf{V}$ we are using the matrix $\mathbf{A} = (\mathbf{I} - \rho\mathbf{W})^{-1}$.

It is easy to show that under this misspecification the generalized least squares estimator

$$\widehat{\boldsymbol{\beta}}_A = \left(\mathbf{X}'\mathbf{A}^{-1}\mathbf{X}\right)^{-1}\mathbf{X}'\mathbf{A}^{-1}\mathbf{Z}(\mathbf{s})$$

is unbiased but the residual based estimator

$$\widehat{\sigma}^2 = \left(\mathbf{Z}(\mathbf{s}) - \mathbf{X}\widehat{\boldsymbol{\beta}}\right)'\mathbf{A}^{-1}\left(\mathbf{Z}(\mathbf{s}) - \mathbf{X}\widehat{\boldsymbol{\beta}}\right)/(n - r(\mathbf{X}))$$

is a biased estimator of σ^2.

Misspecification of the geographic weight matrix tends to suppress statistical efficiency. If $\widehat{\boldsymbol{\beta}}_A$ are the estimates obtained under misspecification (using $\mathbf{A}$ instead of $\mathbf{V}$), then $\text{Var}[\mathbf{c}'\widehat{\boldsymbol{\beta}}_A] \geq \text{Var}[\mathbf{c}'\widehat{\boldsymbol{\beta}}_V]$. Moderate to strong positive autocorrelation that is ignored nearly destroys the efficiency of the ordinary least squares estimator.

Griffith (1996) recommends the following:

1. It is better to posit some reasonable geographic weight matrix than to assume independence.
2. A relatively large number of area units should be employed, at least 60.
3. Lower-order spatial models should be given preference over those of higher-order.
4. It is better to employ a somewhat underspecified than a somewhat overspecified geographic weight matrix as long as $\mathbf{W} \neq \mathbf{0}$.

The upshot is that first-order neighbor connectivity definitions are usually sufficient on regular lattices and little is gained by extending to second-order definitions. On irregular lattices complex neighborhood definitions are often not supported and large data sets are necessary to distinguish between competing specifications of the $\mathbf{W}$ matrix. Less is more.

9.7 Analyzing Mapped Spatial Point Patterns

9.7.1. Introduction

The preceding discussion considered the random field $\{Z(\mathbf{s}) : \mathbf{s} \in D \subset \mathbb{R}^2\}$ where D was a fixed set. D was discrete in the case of lattice data and continuous in the case of geostatistical data. A spatial point pattern (SPP) is a random field where D is a random set of locations at which certain events of interest occurred. Unless $Z(\mathbf{s})$ is itself a random variable — a situation we will exclude from the discussion here (see Cressie 1993, Ch. 8.7 on marked point processes) — the focal point of statistical inquiry is the random set D itself. What kind of statistical questions may be associated with studying the set of locations D? For example, one may ask whether the distribution of events in space is completely random or whether events appear more clustered or regular than is expected under a complete random placement. If events have a tendency to group together in space, we may wish to examine what kind of stochastic model can adequately describe the process, that is, we seek a model which can serve as the data generating mechanism for the observed point pattern. Figure 9.28 shows the locations of 514 maple trees in the Lansing Woods of Clinton County, Michigan. Do these points appear to be placed by a mechanism that arranges tree locations independently and uniformly throughout the study region?

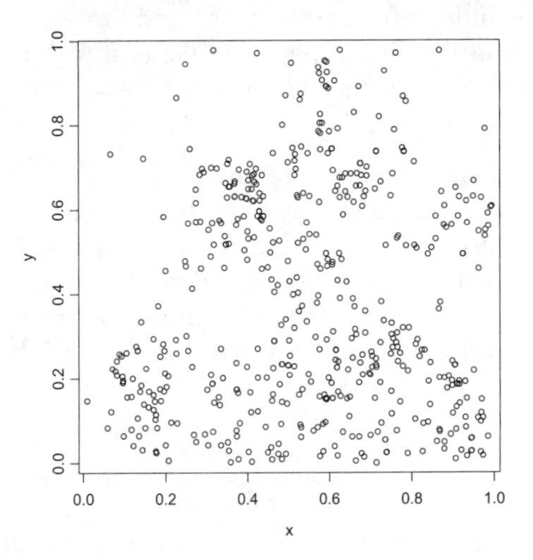

Figure 9.28. Location of 514 maple trees in Lansing Woods, Clinton County, Michigan. Data described by Gerrard (1969), appear in Diggle (1983), and are included in S+SpatialStats®.

The events (location of trees) graphed in Figure 9.28 represent a **mapped** point pattern where all events within the study region have been located. A **sampled** point pattern on the other hand is one where a finite number of sample points is selected. At each point one collects either an area sample by counting the number of events in a sampling area around the point or a distance sample by recording the distance between the sample points and nearby events. This chapter is concerned only with the analysis of mapped patterns. Diggle (1983) is an excellent reference for the analysis of sampled patterns.

A data set containing the results of mapping a spatial point pattern is deceivingly simple. It may contain only the longitude and latitude of the recorded events. Answering such simple questions as *are the points distributed at random* requires tools, however, that are quite different from what the reader has been exposed to so far. For example, little rigorous theory is available to derive the distribution of even simple test statistics and testing hypotheses in spatial point patterns relies heavily on Monte Carlo (computer simulation) methods. It is our opinion that point pattern data is collected quite frequently in agronomic studies but rarely recognized and analyzed as such. This chapter is a brief introduction into spatial point pattern analysis. The interested reader is encouraged to further the limited discussion we provide with resources such as Ripley (1981), Diggle (1983), Ripley (1988), and our §A9.9.10 to A9.9.13.

9.7.2. Random, Aggregated, and Regular Patterns — the Notion of Complete Spatial Randomness

When comparing a set of treatments in an analysis of variance the global hypothesis addressed first is that there are no differences in mean response among the treatments. In many applications rejection of this global hypothesis is not a big surprise and the analyst proceeds with post-ANOVA procedures (contrasts, multiple comparison procedures, etc.) to shed light on how exactly treatments differ in soliciting a response. If the global hypothesis cannot be rejected, however, there is little (no) incentive to proceed further. The global hypothesis for spatial point patterns akin to this initial inquiry in the analysis of variance is whether the events are distributed completely at random. If this hypothesis cannot be rejected there also is little incentive to further inquiries.

Complete spatial randomness (CSR) of events implies that events are uniformly distributed *and* that events are distributed independently of each other. Uniformity means that the expected number of events per unit area is the same throughout the region. Events exhibit no tendency to occupy particular regions in space. Formally, uniformity of events — or the lack thereof — is expressed through the (first-order) **intensity** function of the SPP. Let $N(A)$ denote the number of events that are observed in a region A. The first-order intensity function $\lambda(\mathbf{s})$ is defined as

$$\lambda(\mathbf{s}) \quad = \lim_{|d\mathbf{s}| \to \infty} \frac{\mathrm{E}[N(d\mathbf{s})]}{|d\mathbf{s}|}. \qquad [9.62]$$

In [9.62] $d\mathbf{s}$ is an infinitesimal region (a small disk centered at location $\mathbf{s}$) and $|d\mathbf{s}|$ is its area (volume). As the radius of the disk shrinks toward zero the expected number of events in this area goes to zero but so does the area $|d\mathbf{s}|$. The function $\lambda(\mathbf{s})$ obtained in the limit is the first-order intensity function of the spatial point process. Once $\lambda(\mathbf{s})$ is known, the expected number of events in a region A can be determined by integrating the first-order intensity,

$$\mathrm{E}[N(A)] = \mu(A) = \int_A \lambda(\mathbf{s})d\mathbf{s}. \qquad [9.63]$$

Uniformity of events, one of the conditions of complete spatial randomness, implies that $\lambda(\mathbf{s}) = \lambda$. The average number of events per unit area $(\lambda(\mathbf{s}))$ does not depend on spatial location, it is the same everywhere. A point process with this property is termed **homogeneous** or first-order stationary. The expected number of events in A is then simply $\lambda|A|$, the (constant) expected number of events per unit area times the area. It is now seen that the assumption of a constant first-order intensity is the SPP equivalent to the assumption for geostatistical and lattice processes that the mean $\mathrm{E}[Z(\mathbf{s})]$ is constant. For the latter data types constancy of the mean does not imply the absence of spatial autocorrelations. By the same token spatial point processes where $\lambda(\mathbf{s})$ is independent of location are not necessarily CSR processes.

The first-order intensity conveys no information about the possible interaction of events just as the means of two random variables tell us nothing about their covariance. The CSR hypothesis requires that beyond uniformity the number of events in disjoint regions are independent, $\mathrm{Cov}[N(A), N(B)] = 0$ if $A \cap B = \emptyset$. The spatial point process that embodies CSR is the **homogeneous Poisson process** (HPP). Testing the CSR hypothesis is equivalent to asking whether the observed point pattern could be the realization of an HPP, defined through the following postulates:

(i) the counts in any finite region A have a Poisson distribution with mean $\lambda|A|$, where $|A|$ is the area of A and λ is some positive constant;

(ii) counts in disjoint regions are independent;

(iii) given n events in A, the locations $\mathbf{s}_1, \mathbf{s}_2, \cdots, \mathbf{s}_n$ are a random sample from a uniform distribution on A.

A deviation from CSR implies that points are either not independent or not uniformly distributed. A point pattern is called **aggregated** or **clustered** if events separated by short distances occur more frequently than is expected under CSR, and **regular** if they occur less frequently than in a homogeneous Poisson process (Figure 9.29).

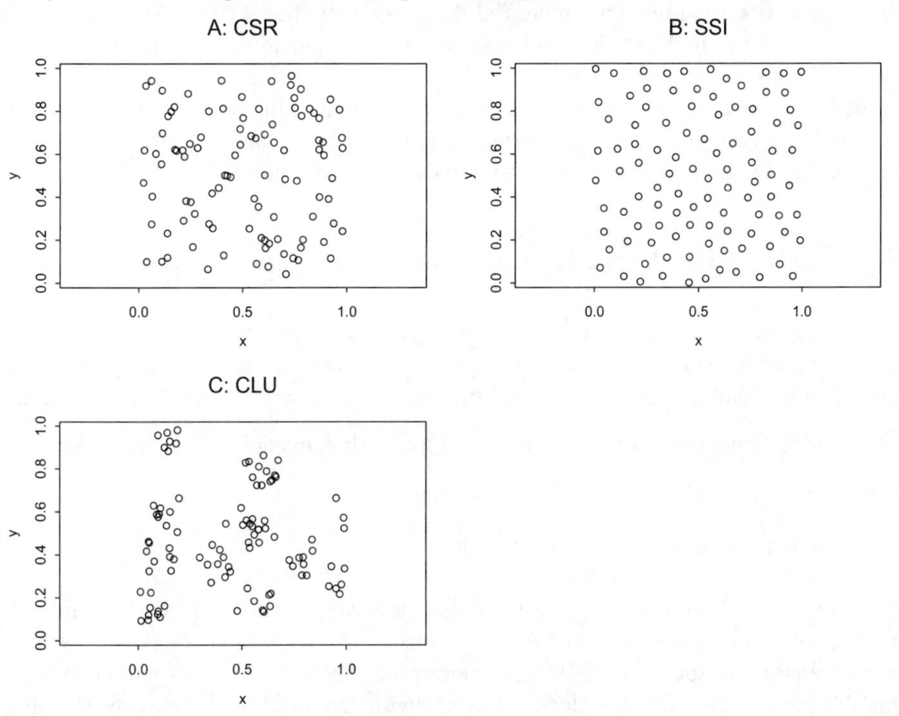

Figure 9.29. Realizations of a CSR (A), regular (B), and clustered (C) process. Each pattern has 100 events on the unit square. SSI is the simple sequential inhibition process (Diggle et al. 1976, Diggle 1983) which does not permit events within a minimum distance of other events (see §A9.9.11).

Aggregated patterns are common and several theories have been developed to explain the formation of clusters in biological applications. One explanation for clustering is through a contagious process where the presence of one or more organisms increases the probability of other organisms occurring in the same sample. In an aggregated process, contagiousness is positive, resulting in an excess of events at small distances and fewer events separated by large distances compared to a CSR process. Aggregation has also been explained in terms of reproductive and dispersal mechanisms. Offspring of plants reproducing vegetatively tend to cluster around parent plants. Plants that reproduce by seed may show a degree of aggregation if seeds do not disperse easily (heavy seeds). Site conditions influence the adaptability and relative abundance of species on a given site. If the success of a species varies with respect to site conditions, the spatial distribution of the species will vary from site to site. Plant density will tend to be higher on preferable sites. A regular spatial distribution is one in which events are more evenly distributed than expected under CSR. It has been speculated that regular spatial patterns could occur in nature when there is a high degree of competition for space between individuals (negative contagiousness). Light sensitivity of trees in older hardwood stands result in a regular distribution of trees to maintain a minimal growing space. The distribution of cell nuclei (or cell centers) in tissue exhibits regularity, since the cell occupies space and can not be deformed arbitrarily.

The distinction between random, aggregated, and regular patterns is made for convenience, there exists a continuum among the three types of spatial patterns. Also, one should keep in mind that spatial patterns evolve over time and undergo an evolution that may take the process through clustered, random, and regular stages. The initial distribution of a regenerated oak stand appears clustered due to the limited radius of dispersion of the heavy seeds. Over the years increasing light sensitivity of the species and intertree competition tend to create a regular pattern. Human intervention can alter this evolution.

9.7.3. Testing the CSR Hypothesis in Mapped Point Patterns

The CSR hypothesis asserts that the number of events in any region A with area $|A|$ is a Poisson random variable with mean $\lambda|A|$ and that given n events $\mathbf{s}_i$ in A, the $\mathbf{s}_i$ are an independent random sample from the uniform distribution on A. The implications are that

(a) the intensity does not vary over the region A (homogeneity of the process);

(b) $\Pr(N(A) = k) = \frac{1}{k!}(\lambda|A|)^k \exp\{-\lambda|A|\}$;

(c) there are no interactions among events.

A goodness-of-fit approach to testing for complete spatial randomness is to count events in nonoverlapping subregions and to compare the observed counts against a Poisson distribution with a standard Chi-square test. While counting events is simple, this approach has some ambiguity because the user must decide how to divide the total area into nonoverlapping subregions. A second approach to testing for CSR is to measure various types of distances in the observed point pattern. For example, the average distance between an event and its nearest neighbor in a clustered pattern is smaller than the same average distance in a random pattern. If the distribution of the average nearest neighbor distance under CSR (the null hypothesis) can be determined a statistical test is possible. Since the sampling distributions of distance-based measures are difficult to ascertain, tests of CSR based on distances often rely on Monte Carlo (simulation) methods.

Quadrat counts

Denote as n the number of events in the observed point pattern which occupies region A. The study region A is partitioned into m nonoverlapping subregions (quadrats) of equal area and the number of events is determined in each quadrat. Commonly, A is taken to be square and divided into a $k \times k$ regular grid, so that $m = k^2$. Let n_i be the event count in grid cell $i = 1, \cdots, m$ and notice that $n = \sum_{i=1}^{m} n_i$. Under CSR the process is homogeneous, the number of events in each cell is estimated as $\overline{n} = n/m$, and the statistic

$$X^2 = \sum_{i=1}^{m} \frac{(n_i - \overline{n})^2}{\overline{n}} \qquad [9.64]$$

has an asymptotic χ^2 distribution with $m - 1$ degrees of freedom. Significantly small values of X^2 indicate regularity and significantly large values of X^2 indicate aggregation. For the χ^2 approximation to hold, counts in each quadrat should exceed 4 (≥ 5) in 80% of the cells and should be greater than 1 everywhere. This rule can be used to find a reasonable grid size to partition A.

The quadrat count statistic is closely related to the **index of dispersion** which is a ratio of two variance estimates obtained with and without making any distributional assumptions. Under CSR the number of events in any one of the m subregions is a Poisson random variable whose mean and variance are estimated by $\overline{n}$. Regardless of the spatial point process that generated the observed data, $S^2 = (m-1)^{-1} \sum_{i=1}^{m} (n_i - \overline{n})^2$ estimates the variance of the quadrat counts. The ratio $I = S^2/\overline{n}$ is called the index of dispersion and is related to [9.64] through

$$I = \frac{1}{\overline{n}} S^2 = \frac{1}{m-1} X^2. \qquad [9.65]$$

If the process is clustered the quadrat counts will vary more than what is expected under CSR, and I will be large. If the process is regular the counts will vary less since all n_i will be similar and similar to the mean count $\overline{n}$. The index of dispersion will be small. For a CSR process the index will be about one on average.

To test the CSR hypothesis with quadrat counts for the point patterns shown in Figure 9.29, $k = 4$ bin classes were used to partition the unit square into $4 \times 4 = 16 = m$ quadrats. The resulting quadrat counts follow.

Table 9.6. Quadrat counts for point patterns of Figure 9.29

	A: CSR				B: SSI				C: CLU			
	1	2	3	4	1	2	3	4	1	2	3	4
1	4	5	8	7	7	5	6	6	10	9	9	7
2	3	7	8	3	6	7	5	8	1	10	1	0
3	10	5	7	6	7	6	4	6	6	7	13	10
4	6	6	6	9	7	6	7	7	2	12	3	0

Notice that $\sum_{i=1}^{n} n_i = n = 100$ for all three patterns and $\overline{n} = 6.25$. The even distribution of the sequential inhibition process (B) and the uneven distribution of the counts for the clustered process (C) are reflected in the sample variances of the quadrat counts: $s^2_{csr} = 3.93$, $s^2_{ssi} = 1.0$, and $s^2_{clu} = 19.93$. The indices of dispersion are $I_{csr} = 0.628$, $I_{ssi} = 0.16$, and $I_{clu} = 3.19$. The CSR hypothesis cannot be rejected for process A and is rejected against the regular alternative for process B and against the clustered alternative for process C (Table 9.7).

Table 9.7. X^2 statistics for quadrat counts of CSR, SSI, and CLU processes

Process	X^2	$\Pr(\chi^2_{15} \leq X^2)$	$\Pr(\chi^2_{15} \geq X^2)$
A: CSR	9.44	0.1466	0.8534
B: SSI	2.40	0.0001	0.9999
C: CLU	47.84	0.9998	0.0002

Using quadrat counts and the Chi-square test is simple, but the method is sensitive to the choice of the subregions (grid size). Statistical tests based on distances between events or sampling points and events avoid this problem.

CSR Tests Based on Distances

Most CSR tests for mapped point patterns utilize the distances between events and between sample points and events rather than quadrat counts. This results in tests for CSR which are slightly more computationally involved, but eliminate a subjective element – how to partition the region – from the analysis. Figure 9.30 shows some of the distances that are commonly employed. In mapped patterns we prefer nearest-neighbor distances because of their greater computational efficiency. In a point pattern with n events there are $n(n-1)/2$ interevent distances but only n nearest-neighbor distances.

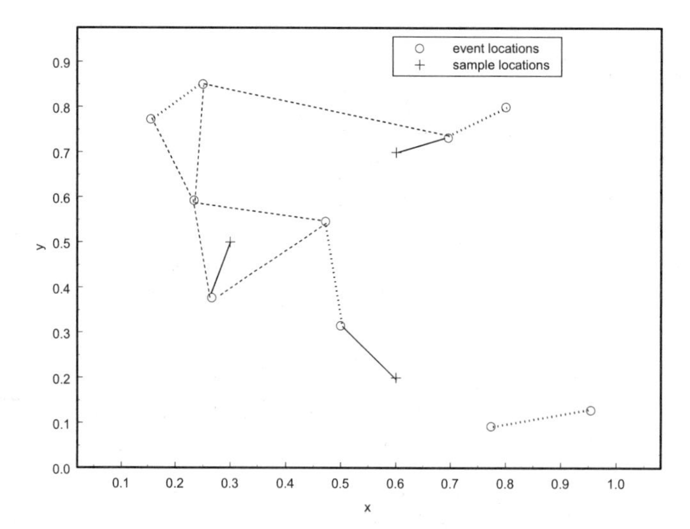

Figure 9.30. Distance measurements used in CSR tests. Solid lines: sample point to nearest event distances. Dashed lines: inter-event distances (also called event-to-event distances). Dotted lines: nearest-neighbor distances (also called event-to-nearest-event distances).

To fix ideas let y_i denote the distance from the event at $\mathbf{s}_i$ to the nearest other event and t_{ij} the Euclidean distance between events at $\mathbf{s}_i$ and $\mathbf{s}_j$. The empirical distribution function of the nearest-neighbor distances is calculated as

$$\widehat{G}_1(y) = \frac{1}{n}\#(y_i \le y) \qquad [9.66]$$

and that of the interevent distances as

$$\widehat{H}_1(t) = 2\frac{\#(t_{ij} \le t)}{n(n-1)}. \qquad [9.67]$$

As a test statistic we may choose the average nearest-neighbor distance $\overline{y}$, the average interevent distance $\overline{t}$, the estimate $\widehat{G}_1(y_0)$ of the probability that the nearest-neighbor distance is at most y_0, or $\widehat{H}_1(t_0)$. It is the user's choice which test statistic to use and in the case of the empirical distribution functions how to select y_0 and/or t_0. It is important that the test statistic constructed from event distances can be interpreted in the context of testing the CSR hypothesis. Compared to the average nearest-neighbor distance expected under CSR, $\overline{y}$ will be smaller in a clustered and larger in a regular pattern. If y_0 is chosen small, $\widehat{G}_1(y_0)$ will be larger than expected under CSR in a clustered pattern and smaller in a regular pattern.

The sampling distributions of distance-based test statistics are usually complicated and elusive, even under the assumption of complete spatial randomness. An exception is the quick test proposed by Ripley and Silverman (1978) that is based on one of the first ordered interevent distances. If $t_1 = \min\{t_{ij}\}$, then t_1^2 has an exponential distribution under CSR and an exact test is possible. Because of possible inaccuracies in determining locations, Ripley and Silverman recommend using the third smallest interevent distance. The asymptotic Chi-squared distribution of these order statistics of interevent distances under CSR is given in their paper. Advances in computing power has made possible to conduct tests of the CSR hypothesis by Monte Carlo (MC) procedures. Among their many advantages is to yield *exact* p-values (exact within simulation variability) and to accommodate irregularly shaped regions.

Recall that the p-value of a statistical test is the probability to obtain a more extreme outcome than what was observed if the null hypothesis is true (§1.6). If the p-value is smaller than the user-selected Type-I error level, the null hypothesis is rejected. An MC test is based on simulating $s-1$ independent sets of data under the null hypothesis and calculating the test statistic for each set. Then the test statistic is obtained from the observed data and combined with the values of the test statistics from simulation to form a set of s values. If the observed value of the test statistic is sufficiently extreme among the s values, the null hypothesis is rejected.

Formally, let u_1 be the value of a statistic U calculated from the observed data and let $u_2, ..., u_s$ be the values of the test statistic generated by independent sampling from the distribution of U under the null hypothesis (H_0). If the null hypothesis is true we have

$$\Pr(u_1 = \max\{u_i, i = 1, ..., s\}) = 1/s.$$

Notice that we consider u_1, the value obtained from the actual (nonsimulated) data, to be part of the sequence of all s values. If we reject H_0 when u_1 ranks k^{th} largest or higher, this is a one-sided test of size k/s. When values of the u_i are tied, one can either randomly sort the u_i's within groups of ties, or choose the least extreme rank for u_1. We prefer the latter method because it is more conservative. Studies have shown that for a 5% level test, $s = 100$ is adequate, whereas $s = 500$ should be used for 1% level tests (Diggle 1983).

To test a point pattern for CSR with a Monte Carlo test we simulate $s - 1$ homogeneous point processes with the same number of events as the observed pattern and calculate the test statistic for each simulated as well as the observed pattern. We prefer to use nearest-neighbor distances and test statistic $\overline{y}$. The SAS® macro `%ghatenv()` contained in file `\SASMacros\NearestNeighbor.sas` accomplishes that for point patterns that are bounded by a rectangular region. The statements

```
%include 'DriveLetterofCDROM:\SASMacros\NearestNeighbor.sas';
%ghatenv(data=maples,xco=x,yco=y,alldist=1,graph=1,sims=20);
proc print data=_ybars; var ybar sim rank; run;
```

perform a nearest-neighbor analysis for the maple data in Figure 9.28. For exposition we use $s - 1 = 20$ which should be increased in real applications. The observed pattern has the smallest average nearest-neighbor distance with $\overline{y}_1 = 0.017828$ (Output 9.1, `sim=0`). This yields a p-value for the hypothesis of complete spatial randomness of $p = 0.0476$ against the clustered alternative and $p = 0.9524$ against the regular alternative (Output 9.2).

Output 9.1.

Obs	ybar	sim	rank
1	0.017828	0	1
2	0.021152	1	2
3	0.021242	1	3
4	0.021348	1	4
5	0.021498	1	5
6	0.021554	1	6
7	0.021596	1	7
8	0.021608	1	8
9	0.021630	1	9
10	0.021674	1	10
11	0.021860	1	11
12	0.022005	1	12
13	0.022013	1	13
14	0.022046	1	14
15	0.022048	1	15
16	0.022070	1	16
17	0.022153	1	17
18	0.022166	1	18
19	0.022172	1	19
20	0.022245	1	20
21	0.023603	1	21

Output 9.2.

Test Statistic	# of MC runs	rank	One Sided Left P	One Sided Right P
0.017828	20	1	0.95238	0.047619

Along with the ranking of the observed test statistic it is useful to prepare a graph of the simulation envelopes for the empirical distribution function of the nearest-neighbor distance (called a G-hat plot). The upper and lower simulation envelopes are defined as

$$U(y) = \max_{i=2,..,s} \left\{ \widehat{G}_i(y) \right\} \text{ and } L(y) = \min_{i=2,...,s} \left\{ \widehat{G}_i(y) \right\}$$

and are plotted against $\overline{G}(y)$, the average empirical distribution at y from the simulations,

$$\overline{G}(y) = \frac{1}{s-1} \sum_{j \neq i} \widehat{G}_j(y).$$

This plot is overlaid with the observed G-function $(\widehat{G}_1(y))$ as shown in Figure 9.31. Clustering is evidenced by the G function rising above the 45-degree line that corresponds to the CSR process (dashed line), regularity by a G function below the dashed line. When the observed G function crosses the upper or lower simulation envelopes the CSR hypothesis is rejected. For the maple data there is very strong evidence that the distribution of maple trees in the particular area is clustered.

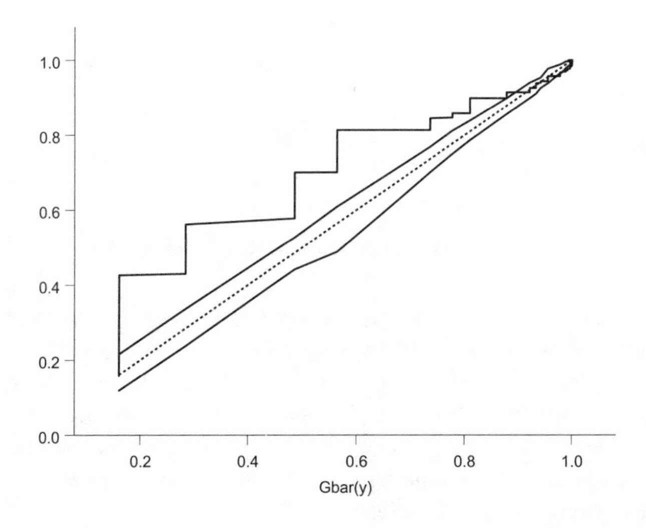

Figure 9.31. Upper $[U(y)]$ and lower $[L(y)]$ simulation envelopes for 20 simulations and observed empirical distribution function (step function) for maple data (Figure 9.28). Dashed line represents G-function for a CSR process.

MC tests require a procedure to simulate the process under the null distribution. To test a point pattern for CSR requires an efficient method to generate a homogeneous Poisson process. The following algorithm simulates this process on the rectangle $(0, 0) \times (a, b)$.

1. Generate a random number from a Poisson(λab) distribution $\rightarrow n$.
2. Order n independent $U(0, a)$ random variables $\rightarrow X_1 < X_2 < \cdots < X_n$.
3. Generate n independent $U(0, b)$ random variables $\rightarrow Y_1, \cdots, Y_n$.
4. Return $(X_1, Y_1), \cdots, (X_n, Y_n)$ as the coordinates of the two-dimensional Poisson process on the rectangle.

This algorithm generates a random number of events in step 1. Typically, simulations are conditioned such that the simulated patterns have the same number of events as the observed pattern (this is called a Binomial process). In this case n is determined by counting the events in the observed pattern and the simulation algorithm consists of steps 2 to 4. If the study region A is not a rectangle, but of irregular shape, create a homogeneous Poisson process on a rectangle which encloses the shape of the study region and generate events until n events fall within the shape of interest.

If it is clear that events are not uniformly distributed, testing the observed pattern against CSR is an important first step but rejection of the CSR hypothesis is not surprising. To test whether the events follow a nonuniform process in which events remain independent one can test the observed pattern against an inhomogeneous Poisson process where $\lambda(\mathbf{s})$ is not a constant but follows a model specified by the user (note that $\lambda(\mathbf{s})$ must be bounded and non-negative). The following algorithm by Lewis and Shedler (1979) simulates an inhomogeneous Poisson process.

1. Simulate a homogeneous Poisson process on A with intensity $\lambda_0 \geq \max\{\lambda(\mathbf{s})\}$ according to the algorithm above $\rightarrow (X_1, Y_1), \cdots, (X_n, Y_n)$.
2. Generate uniform $U(0,1)$ random variables for each event in A that was generated in step 1 $\rightarrow U_1, \cdots, U_n$.
3. If $U_i \leq \lambda(\mathbf{s})/\lambda_0$, retain the event, otherwise discard the event.

9.7.4 Second-Order Properties of Point Patterns

To determine whether events interact (attract or repel each other) it is not sufficient to study the first-order properties of the point pattern. Rejection of the CSR hypothesis may be due to lack of uniformity or lack of independence. With geostatistical data the first-order properties are expressed through the mean function $\mathrm{E}[Z(\mathbf{s})] = \mu(\mathbf{s})$ and the second-order properties through the semivariogram or covariogram of $Z(\mathbf{s})$. For point patterns the first-order intensity $\lambda(\mathbf{s})$ takes the place of the mean function and the second-order properties are expressed through the **second-order intensity function**

$$\lambda_2(\mathbf{s}_1, \mathbf{s}_2) = \lim_{|d\mathbf{s}_1| \rightarrow 0, |d\mathbf{s}_2| \rightarrow 0} \frac{\mathrm{E}[N(d\mathbf{s}_1)N(d\mathbf{s}_2)]}{|d\mathbf{s}_1||d\mathbf{s}_2|}. \qquad [9.68]$$

If the point process is second-order stationary, then $\lambda(\mathbf{s}) = \lambda$ *and* $\lambda_2(\mathbf{s}_1, \mathbf{s}_2) = \lambda_2(\mathbf{s}_1 - \mathbf{s}_2)$; the first-order intensity is constant *and* the second-order intensity depends only on the spatial separation between points. If the process is furthermore isotropic, the second-order intensity does not depend on the direction, only the distance between pairs of points: $\lambda_2(\mathbf{s}_1, \mathbf{s}_2) = \lambda_2(||\mathbf{s}_1 - \mathbf{s}_2||) = \lambda_2(h)$. Notice that any process for which the intensity depends on locations cannot be second-order stationary. The second-order intensity function depends on the expected value of the cross-product of counts in two regions, similar to the covariance between two random variables X and Y, which is a function of their expected cross-product, $\mathrm{Cov}[X,Y] = \mathrm{E}[XY] - \mathrm{E}[X]\mathrm{E}[Y]$. A downside of $\lambda_2(\mathbf{s}_1 - \mathbf{s}_2)$ is its lack of physical interpretability. The remedy is to use interpretable measures that are functions of the second-order intensity or to perform the interaction analysis in the spectral domain (§A9.9.12).

Ripley (1976, 1977) proposed studying the interaction among events in a second-order stationary, isotropic point process through a reduced moment function called the K-function. The K-function at distance h is defined through

$$\lambda K(h) = \text{E}[\#\text{ of extra events within distance } h \text{ from an arbitrary event}], \ \ h > 0.$$

K-function analysis in point patterns takes the place of semivariogram analysis in geostatistical data. The assumption of a constant mean there is replaced with the assumption of a constant intensity function here. The K-function has several advantages over the second-order intensity function [9.68].

- Its definition suggests a method of estimating $K(h)$ from the average number of events less than distance h apart (see §A9.9.10).

- $K(h)$ is easy to interpret. In a clustered pattern a given event is likely to be surrounded by events from the same cluster. The number of extra events within small distances h of an event will be large, and so will be $K(h)$. In an aggregated pattern the number of extra events (and therefore $K(h)$) will be small for small h.

- $K(h)$ is known for important point process models (§A9.9.11). For the homogeneous Poisson process the expected number of events per unit area is λ, the expected number of extra events within distance h is $\lambda\pi h^2$ and $K(h) = \pi h^2$. If a process is first-order stationary comparing a data based estimate $\widehat{K}(h)$ against πh^2 allows testing for interaction among the events. If $K(h) \geq \pi h^2$ for small distances h, the process is clustered whereas $K(h) < \pi h^2$ (for h small) indicates regularity.

- The K-function can be obtained from the second-order intensity of a stationary, isotropic process if $\lambda_2(h)$ is known:

$$K(h) = 2\pi\lambda^{-2}\int_0^h x\lambda_2(x)dx.$$

  Similarly, the second-order intensity can be derived from the K-function.

- If not all events have been mapped and the incompleteness of the data is spatially neutral (events are missing completely at random (MCAR)), the K-function remains an appropriate measure for the second-order properties of the *complete* process. This is known as the invariance of $K(h)$ to random thinning. If the missing data process is MCAR the observed pattern is a realization of a point process whose events are a subset of the complete process generated by retaining or deleting the events in a series of mutually independent Bernoulli trials. Random thinning reduces the intensity λ and the expected number of additional events within distance h of $\mathbf{s}$ by the same factor. Their ratio, which is $K(h)$, remains unchanged.

Ripley's K-function is a useful tool to study second-order properties of stationary, isotropic spatial processes. Just as the semivariogram does not uniquely describe the stochastic properties of a geostatistical random field, the K-function is not a unique descriptor of a point process. Different point processes can have identical K-functions. Baddeley and Silverman (1984) present interesting examples of this phenomenon. Study of second-order properties without requiring isotropy of the point process is possible through spectral tools and periodogram analysis. Because of its relative complexity we discuss spectral analysis of point patterns in §A9.9.12 for the interested reader with a supplementary application in A9.9.13.

Details on the estimation of first- and second-order properties from an observed mapped point pattern can be found in §A9.9.10.

9.8 Applications

The preceding sections hopefully have given the reader an appreciation of the unique stature of statistics for spatial data in the research worker's toolbox and a glimpse of the many types of spatial models and spatial analyses. In the preface to his landmark text, Cressie (1993) notes that "this may be the last time spatial Statistics will be squeezed between two covers." History proved him right. Since the publication of the revised edition of Cressie's *Statistics for Spatial Data*, numerous texts have appeared that deal with primarily one of the three types of spatial data (geostatistical, lattice, point patterns) at length, comparable to this entire volume. The many aspects and methods of this rapidly growing discipline we have failed to address are not countable. Some of the applications that follow are chosen to expose the reader to some topics by way of example.

§9.8.1 reiterates the importance of maintaining the spatial context in data that are georeferenced and the ensuing perils to modeling and data interpretation if this context is overlooked. Global and local versions of Moran's I statistics as a measure of spatial autocorrelation are discussed there. In the analysis of geostatistical data the semivariogram or covariogram plays a central role. Kriging equations depend critically on it and spatial regression/ANOVA models require information about the spatial dependency to estimate coefficients and treatment effects efficiently. §9.8.2 estimates empirical semivariograms and fits theoretical semivariogram models by least squares, (restricted) maximum likelihood, and composite likelihood. Point and block kriging are illustrated in §9.8.3 with an interesting application concerning the amount of lead and its spatial distribution on a shotgun range. Treatment comparisons of random field models and spatial regression models are examined in §9.8.4 and §9.8.5. Most methods for spatial data analysis we presented assume that the response variable is continuous. Many applications with georeferenced data involve discrete or non-Gaussian data. Spatial random field models can be viewed as special cases of mixed models. Extensions of generalized linear mixed models (§8) to the spatial context are discussed in §9.8.6 where the Hessian fly data are tackled with a spatially explicit model. Upon closer inspection many spatial data sets belong in the category of lattice data but are often modeled as if they were geostatistical data. Lattice models can be extremely efficient in explaining spatial variation. In §9.8.7 the spatial structure of wheat yields from a uniformity trial are examined with geostatistical and lattice models. It turns out that a simple lattice model explains the spatial structure more efficiently than the geostatistical approaches. The final application, §9.8.8, demonstrates the basic steps in analyzing a mapped point pattern, estimating its first-order intensity, Ripley's K-function, and Monte-Carlo inferences based on distances to test the hypothesis of complete spatial randomness. A supplementary application concerning the spectral analysis of point patterns can be found in Appendix A1 (§A9.9.13).

While The SAS® System is our computing environment of choice for almost all statistical analyses, its capabilities for spatial data analysis at the time of this writing were limited. The `variogram` procedure is a powerful tool for estimating empirical semivariograms with the classical (Matheron) and robust (Cressie-Hawkins) estimators. The `krige2d` procedure performs ordinary (global and local) kriging admirably. Random field models can be fit with the `mixed` procedure provided the spatial correlation structure can be modeled through the procedure's `repeated` statement. The `kde` procedure efficiently estimates univariate and bivariate densities and can be used to estimate the (first-order) intensity in point patterns. Beyond these procedures the SAS® user must rely on hand-crafted SAS macros and programs tailored to a particular application that fall short of the procedures and functions available in specialized packages. Our preferred computing environment for spatial analyses is the S+SpatialStats® package, available as an add-on module to the S-PLUS® program. For the benefit of the SAS® user we developed several SAS® macros that mimic functions in S+SpatialStats®. The use of the macros is outlined in the applications that follow and the companion CD-ROM.

9.8.1 Exploratory Tools for Spatial Data — Diagnosing Spatial Autocorrelation with Moran's I

The tools for modeling and analyzing spatial data discussed in this chapter have in common the concept that *space matters*. The relationships among georeferenced observations is taken into account, be it through estimating stochastic dependencies (semivariogram modeling for geostatistical data and K-function analysis for point patterns), deterministic structure in the mean function (spatial regression), or expressing observations as functions of neighboring values (lattice models). Just as many of the modeling tools for spatial data are quite different from the techniques applied to model independent data, exploratory spatial data analysis requires additional methods beyond those used in exploratory analysis of independent data. To steer the subsequent analysis in the right direction, exploratory tools for spatial data must allow insight into the spatial structure in the data. Graphical summaries, e.g., stem-and-leaf plots or sample histograms, are pictures of the data, not indications of spatial structure.

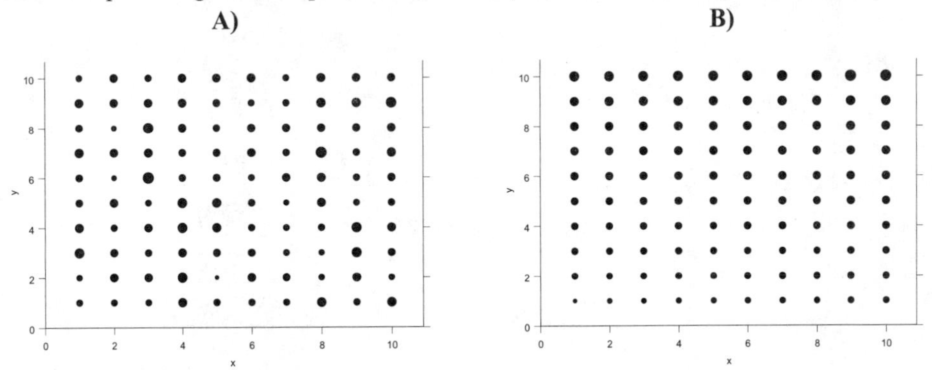

Figure 9.32. Two simulated lattice arrangements with identical frequency distribution of points. Area of dots is proportional to the magnitude of the values.

The importance of retaining spatial information can be demonstrated with the following example. A rectangular 10×10 lattice was filled with 100 observations drawn at random from a $G(0,1)$ distribution. Lattice A is a completely random assignment of observations to lattice positions. Lattice B is an assignment to positions such that a value is surrounded by values similar in magnitude (Figure 9.32).

Histograms of the 100 observed values that do not take into account spatial position will be identical for the two lattices. (Figure 9.33). Plotting observed values against the average value of the nearest neighbors the differences in the spatial distribution between the two lattices emerge (Figure 9.34). The data in lattice A are not spatially correlated, the data in lattice B are very strongly autocorrelated. We note further that the "density" estimate drawn in Figure 9.33 is *not* an estimate of the probability distribution of the data. The probability distribution of a random function is given through [9.3]. Even if the histogram calculated by lumping data across spatial locations appears Gaussian does not imply that the data are a realization of a Gaussian random field.

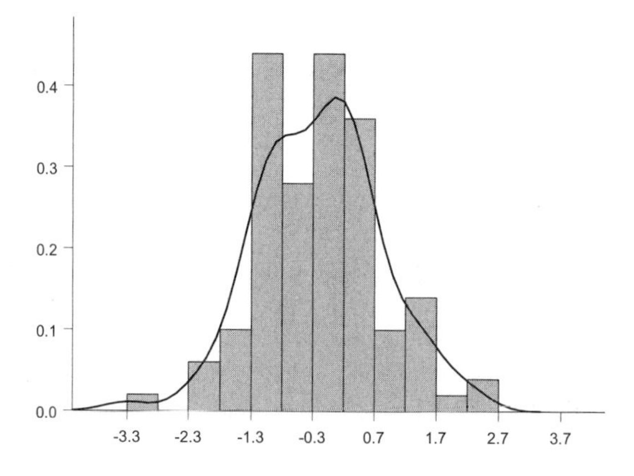

Figure 9.33. Histogram of the 100 realizations in lattices A and B along with kernel density estimate. Both lattices produce identical sample frequencies.

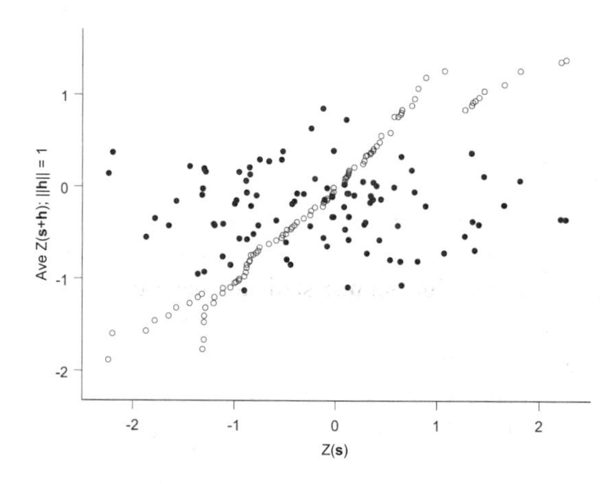

Figure 9.34. Lag-1 plots for lattices A (full circles) and B (open circles) of Figure 9.32. There is no trend between a value and the average value of its immediate neighbors in lattice A but a very strong trend in lattice B.

Distinguishing between the spatial and nonspatial context is also important for outlier detection. An observation that appears unusual in a stem-and-leaf or box-plot is a **distributional** outlier. A **spatial** outlier on the other hand is an observation that is unusual compared to its surrounding values. A data set can have many more spatial than distributional outliers. One method of diagnosing spatial outliers is to median-polish the data (or to remove the large scale trends in the data by some other outlier-resistant method) and to look for outlying observations in a box-plot of the median-polished residuals. Lag-plots such as Figure 9.34 where observations are plotted against averages of surrounding values are also good graphical diagnostics of observations that are unusual spatially.

A first step to incorporate spatial context in describing a set of spatial data is to calculate descriptive statistics and graphical displays separately for sets of spatial coordinates. This is simple if the data are observed on a rectangular lattice such as the Mercer and Hall grain yield data (Figure 9.5, p. 569) where calculations can be performed by rows and columns. Row and column box-plots for these data show a cubic (or even higher) trend in the column medians but no trend in the row medians (Figure 9.35). This finding can be put to use to detrend the data with a parametric model. Without detrending the data semivariogram estimation is possible by considering pairs of data within columns only. The row and column box-plots were calculated in S+SpatialStats® with the statements

```
bwplot(y~grain,data=wheat,ylab="Row",xlab="Grain Yield")
bwplot(x~grain,data=wheat,ylab="Column",xlab="Grain Yield").
```

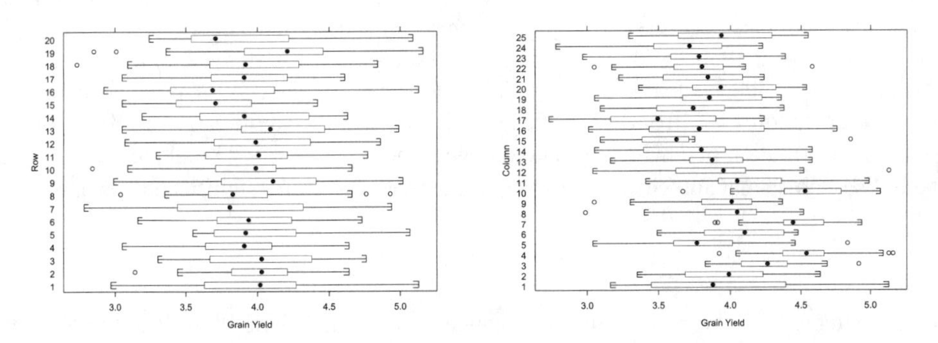

Figure 9.35. Row (left) and column (right) box-plots for Mercer and Hall grain yield data.

Other graphical displays and numerical summary measures were developed specifically for spatial data, for example, to describe, diagnose, and test the degree of spatial autocorrelation. With geostatistical data the empirical semivariogram provides an estimate of the spatial structure. With lattice data **join-count** statistics have been developed for binary and nominal data (see, for example, Moran 1948, Cliff and Ord 1973, and Cliff and Ord 1981). Moran (1950) and Geary (1954) developed autocorrelation coefficients for continuous attributes observed on lattices. These coefficients are known as Moran's I and Geary's c and like many other autocorrelation measures compare an estimate of the covariation among the $Z(\mathbf{s})$ to an estimate of their variation. Since the distribution of I and c tends to a Gaussian distribution with increasing sample size, these summary autocorrelation measures can also be used in confirmatory fashion to test the hypothesis of no (global) spatial autocorrelation in the data. In the remainder of this application we introduce Moran's I, estimation and inference based on this statistic, and a localized form of I that Anselin (1995) termed a LISA (local indicator of spatial association).

Let $Z(\mathbf{s}_i)$, $i = 1, \cdots, n$ denote the attribute Z observed at site $\mathbf{s}_i$ and $U_i = Z(\mathbf{s}_i) - \overline{Z}$ its centered version. Since data are on a lattice let w_{ij} denote the neighborhood connectivity weight between sites $\mathbf{s}_i$ and $\mathbf{s}_j$ with $w_{ii} = 0$. These weights are determined in the same fashion as those for the lattice models in §9.6.1. Moran's I is then defined as

$$I = \frac{n}{\sum_{i,j} w_{ij}} \frac{\sum_{i=1}^{n}\sum_{j=1}^{n} w_{ij} u_i u_j}{\sum_{i=1}^{n} u_i^2}. \qquad [9.69]$$

A more compact expression can be obtained in matrix-vector form by putting $\mathbf{u} = [u_1, \cdots, u_n]'$, $\mathbf{W} = [w_{ij}]$, and $\mathbf{1} = [1, \cdots, 1]'$. Then

$$I = \frac{n}{\mathbf{1}'\mathbf{W}\mathbf{1}} \frac{\mathbf{u}'\mathbf{W}\mathbf{u}}{\mathbf{u}'\mathbf{u}}. \qquad [9.70]$$

In the absence of spatial autocorrelation I has expected value $\mathrm{E}[I] = -1/(n-1)$ and values $I > \mathrm{E}[I]$ indicate positive, values $I < \mathrm{E}[I]$ negative autocorrelation. It should be noted that the Moran test statistic bears a great resemblance to the Durbin-Watson (DW) test statistic used in linear regression analysis to test for serial dependence among residuals (Durbin and Watson 1950, 1951, 1971). The DW test replaces $\mathbf{u}$ with a vector of least squares residuals and considers squared lag-1 serial differences in place of $\mathbf{u}'\mathbf{W}\mathbf{u}$. To determine whether a deviation of I from its expectation is statistically significant one relies on the asymptotic distribution of I which is Gaussian with mean $-1/(n-1)$ and variance σ_I^2. The hypothesis of no spatial autocorrelation is rejected at the $\alpha \times 100\%$ significance level if

$$|Z_{obs}| = \frac{|I - \mathrm{E}[I]|}{\sigma_I}$$

is more extreme than the $z_{\alpha/2}$ cutoff of a standard Gaussian distribution. Right-tailed (left-tailed) tests for positive (negative) autocorrelation compare Z_{obs} to z_α ($z_{1-\alpha}$) cutoffs.

Two approaches are common to derive the variance σ_I^2. One can assume that the $Z(\mathbf{s}_i)$ are Gaussian or adopt a randomization framework. In the Gaussian approach the $Z(\mathbf{s}_i)$ are assumed $G(\mu, \sigma^2)$, so that $U_i \sim (0, \sigma^2(1 - 1/n))$ under the null hypothesis. In the randomization approach the $Z(\mathbf{s}_i)$ are considered fixed and are randomly permuted among the n lattice sites. There are $n!$ equally likely random permutations and σ_I^2 is the variance of the $n!$ Moran I values. A detailed derivation of and formulas for the variances under the two assumptions can be found in Cliff and Ord (1981, Ch. 2.3). If one adopts the randomization framework an empirical p-value for the test of no spatial autocorrelation can be calculated if one ranks the observed value of I among the $n! - 1$ possible remaining permutations. For even medium-sized lattices this is a computationally expensive procedure. The alternatives are to rely on the asymptotic Gaussian distribution to calculate p-values or to compare the observed I against only a random sample of the possible permutations.

The SAS® macro `%MoranI` (contained on CD-ROM) calculates the Z_{obs} statistics and p-values under the Gaussian and randomization assumption. A data set containing the $\mathbf{W}$ matrix is passed to the macro through the `w_data` option. For rectangular lattices the macro `%ContWght` (in file `\SASMacros\ContiguityWeights.sas`) calculates the **W** matrices for classical neighborhood definitions (see Figure 9.26, p. 633). For the Mercer and Hall grain yield data the statements

```
%include 'DriveLetterofCDROM:\Data\SAS\MercerWheatYieldData.sas';
%include 'DriveLetterofCDROM:\SASMacros\ContiguityWeights.sas';
%include 'DriveLetterofCDROM:\SASMacros\MoranI.sas';

title1 "Moran's I for Mercer and Hall Wheat Yield Data, Rook's Move";
%ContWght(rows=20,cols=25,move=rook,out=rook);
%MoranI(data=mercer,y=grain,row=row,col=col,w_data=rook);
```

produce Output 9.3. The observed I value of 0.4066 is clearly greater than the expected value. The standard errors σ_I based on randomization and Gaussianity differ only in the fourth decimal place in this application. In other instances the difference will be more substantial. There is overwhelming evidence that the data exhibit positive autocorrelation.

Moran's I is somewhat sensitive to the choice of the neighborhood matrix $\mathbf{W}$. If the rook definition (edges abut) is replaced by the bishop's move (touching corners),

```
title1 "Moran's I for Mercer and Hall Wheat Yield Data, Bishop's Move";
%ContWght(rows=20,cols=25,move=bishop,out=bishop);
%MoranI(data=mercer,y=grain,row=row,col=col,w_data=bishop);
```

the autocorrelation remains significant but the value of the test statistic is reduced by about 50% (Output 9.4). But it is even more sensitive to large scale trends in the data. For a significant test result based on Moran's I to indicate spatial autocorrelation it is necessary that the mean of $Z(\mathbf{s})$ is stationary. Otherwise subtracting $\overline{Z}$ is not the appropriate shifting of the data that produces zero mean random variables U_i. In fact, the I test may indicate significant "autocorrelation" if data are independent but have not been properly detrended.

Output 9.3.

Moran's I for Mercer and Hall Wheat Yield Data, Rook's Move

Type	Observed I	E[I]	SE[I]	Zobs	Pr(Z > Zobs)
Randomization	0.4066	-.002004	0.0323	12.6508	0
Gaussianity	0.4066	-.002004	0.0322	12.6755	0

Output 9.4.

Moran's I for Mercer and Hall Wheat Yield Data, Bishop's Move

Type	Observed I	E[I]	SE[I]	Zobs	Pr(Z > Zobs)
Randomization	0.19518	-.002004008	0.032378	6.09008	5.6427E-10
Gaussianity	0.19518	-.002004008	0.032315	6.10197	5.2385E-10

This spurious autocorrelation effect can be demonstrated by generating independent observations with a mean structure. On a 10×10 lattice we construct data according to the linear model

$$Z = 1.4 + 0.1x + 0.2y + 0.002x^2 + e, \;\; e \sim iid\, G(0,1),$$

where x and y are the lattice coordinates:

```
data simulate;
  do x = 1 to 10; do y = 1 to 10;
    z = 1.4 + 0.1*x + 0.2*y + 0.002*x*x + rannor(2334);
    output;
  end; end;
run;
title1 "Moran's I for independent data with large-scale trend";
%ContWght(rows=10,cols=10,move=rook,out=rook);
%MoranI(data=simulate,y=z,row=x,col=y,w_data=rook);
```

The test indicates strong positive "autocorrelation" which is an artifact of the changes in $\mathrm{E}[Z]$ rather than stochastic spatial dependency among the sites.

Output 9.5.

Moran's I for independent data with large-scale trend

Type	Observed I	E[I]	SE[I]	Zobs	Pr(Z > Zobs)
Randomization	0.39559	-0.010101	0.073681	5.50604	1.835E-8
Gaussianity	0.39559	-0.010101	0.073104	5.54948	1.4326E-8

If trend contamination distorts inferences about the spatial autocorrelation coefficient, then it seems reasonable to remove the trend and calculate the autocorrelation coefficient from the residuals. If $\widehat{\mathbf{e}} = \mathbf{Y} - \mathbf{X}\widehat{\boldsymbol{\beta}}_{OLS}$ is the residual vector, the I statistic [9.70] is modified to

$$I^* = \frac{n}{\mathbf{1}'\mathbf{W}\mathbf{1}} \frac{\widehat{\mathbf{e}}'\mathbf{W}\widehat{\mathbf{e}}}{\widehat{\mathbf{e}}'\widehat{\mathbf{e}}}. \qquad [9.71]$$

The mean and variance of [9.71] are not the same as those for [9.70]. For example, the mean $\mathrm{E}[I^*]$ now depends on the weights $\mathbf{W}$ and the $\mathbf{X}$ matrix. Expressions for $\mathrm{E}[I^*]$ and $\mathrm{Var}[I^*]$ are found in Cliff and Ord (1981, Ch. 8.3). These were coded in the SAS® macro `%RegressI()` contained in file `\SASMacros\MoranResiduals.sas`. Recall that for the Mercer and Hall grain yield data the exploratory row and column box-plots indicated possible cubic trends in the column medians. To check whether there exists autocorrelation in these data or whether the significant I statistic in Output 9.3 was spurious, the following SAS® code is executed.

```
title1 "Moran's I for Mercer and Hall Wheat Yield Data";
title2 "Calculated for Regression Residuals";
%Include 'DriveLetterofCDROM:\SASMacros\MoranResiduals.sas';
data xmat; set mercer; x1 = col; x2 = col**2; x3 = col**3;
   keep x1 x2 x3;
run;
%RegressI(xmat=xmat,data=mercer,z=grain,weight=rook,local=1);
```

The data set `xmat` contains the regressor variables excluding the intercept. It should not contain any additional variables. This code fits a large-scale mean model with cubic column effects and no row effects (adding higher order terms for column effects leaves the results essentially unchanged). The ordinary least squares estimates are calculated and shown by the macro (Output 9.6). If there is significant autocorrelation in the residuals the standard errors, t-statistics and p-values for the parameter estimates are not reliable, however, and should be disregarded. The value of Z_{obs} is slightly reduced from 12.67 (Output 9.3) to 10.27 indicating that the column trends did add some spurious autocorrelation. The highly significant p-value for I^* shows that further analysis of these data by classical methods for independent data is treacherous. Spatial models and techniques must be used in further inquiry.

Output 9.6.

```
            Moran's I for Mercer and Hall Wheat Yield Data
                 Calculated for Regression Residuals

              Ordinary Least Squares Regression Results

                      OLS Analysis of Variance
                  SS       df       MS        F     Pr>F      R2

Model     13.8261        3   4.6087 25.1656 355E-17   0.1321
Error      90.835      496  0.18314       .       .        .
C.Total   104.661      499        .       .       .        .

                           OLS Estimates
                  Estimate  StdErr     Tobs   Pr>|T|

        Intcpt    3.90872 0.08964 43.6042        0
        x1        0.10664 0.02927 3.64256   0.0003
        x2        -0.0121 0.00259 -4.6896  3.54E-6
        x3        0.00032 0.00007 4.82676  1.85E-6

                         Global Moran's I
          I*       E[I*]      SE[I*]       Zobs    Pr > Zobs
     0.32156     -0.0075     0.03202    10.2773            0
```

The `%RegressI()` and `%MoranI()` macros have an optional parameter `local=`. When set to 1 (default is `local=0`) the macros will not only calculate the global I (or I^*) statistic but local versions thereof. The idea of a local indicator of spatial association (LISA) is due to Anselin (1995). His notion was that although there may be no spatial autocorrelation globally, there may exist local pockets of positive or negative spatial autocorrelation in the data, so called hot-spots. This is only one possible definition of what constitutes a hot-spot. One could also label as hot-spots sites that exceed (or fall short of) a certain threshold level. Hot-spot definitions based on autocorrelation measures designate sites as unusual if the spatial dependency is locally much different from other sites. The LISA version of Moran's I is

$$I_i = \frac{n}{\sum_i^n u_i^2} u_i \sum_j^n w_{ij} u_j, \qquad [9.72]$$

where i indexes the sites in the data set. That is, for each site $\mathbf{s}_i$ we calculate an I statistic based on information from neighboring sites. For a 10×10 lattice there are a total of 101 I statistics. The global I according to [9.69] or [9.71] and 100 local I statistics according to [9.72]. The expected value of I_i is $\mathrm{E}[I_i] = -w_i/(n-1)$ with $w_i = \sum_{j=1}^n w_{ij}$. The interpretation is that if $I_i < \mathrm{E}[I_i]$ then sites connected to $\mathbf{s}_i$ have attribute values dissimilar from $Z(\mathbf{s}_i)$. A high (low) value at $\mathbf{s}_i$ is surrounded by low (high) values. If $I_i > \mathrm{E}[I_i]$, sites connected to $\mathbf{s}_i$ show similar value. A high (low) value at $Z(\mathbf{s}_i)$ is surrounded by high (low) values at connected sites.

The asymptotic Gaussian distribution of I (and I^*) makes tempting the testing of hypotheses based on the LISAs to detect local pockets where spatial autocorrelation is significant. We discourage formal testing procedures based on LISAs. First, there is a serious multiplicity problem. In a data set with n sites one would be testing n hypotheses with grave consequences for Type-I error inflation. Second, the n LISAs are not independent due to spatial dependence *and* shared data points among the LISAs. We prefer a graphical examination of the local I statistics over formal significance tests. The map of LISAs indicates which regions of the domain behave differently from the rest and hopefully identify a spatially variable explanatory variable that can be used in the analysis for adjustments of the large-scale trends.

For the detrended Mercer and Hall grain yield data Figure 9.36 shows sites with positive LISAs. Hot-spots where autocorrelation is locally much greater than for the remainder of the lattice are clearly recognizable (e.g., row 18, col 16).

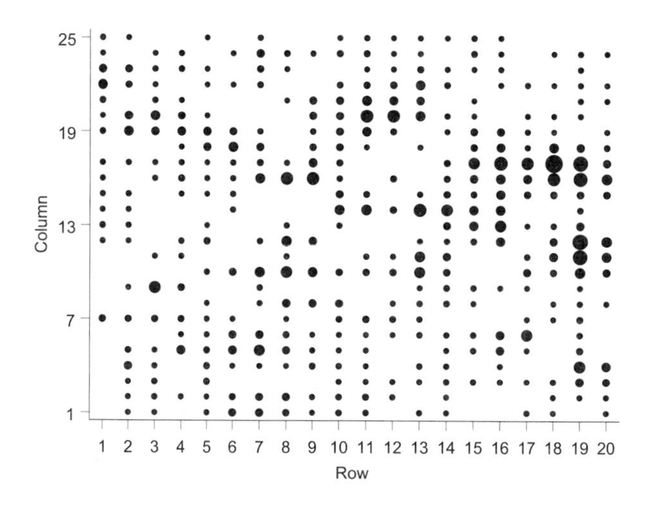

Figure 9.36. Local indicators of positive spatial autocorrelation ($I_i > \mathrm{E}[I_i]$) calculated from regression residuals after removing column trends.

9.8.2 Modeling the Semivariogram of Soil Carbon

In this application we demonstrate the basic steps in modeling the semivariogram of geostatistical data. The data for this application were kindly provided by Dr. Thomas G. Mueller, Department of Agronomy, University of Kentucky. An agricultural field had been in no-till production for more than ten years in a corn-soybean rotation. Along strips total soil carbon percentage was determined at two-hundred sampling sites (Figure 9.37). Eventually, the intermediate strips were chisel-plowed but we concentrate on the no-tillage areas in this application only. A comparison of C/N ratios for chisel-plowed versus no-till treatments can be found in §9.8.4 and a spatial regression application in §9.8.5.

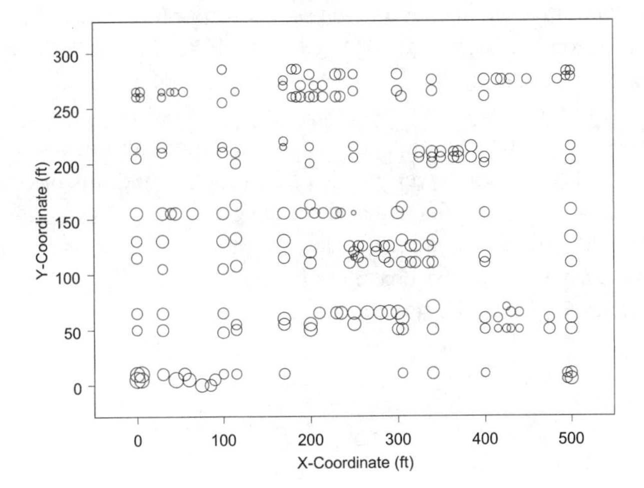

Figure 9.37. Total soil carbon data. Size of dots is proportional to soil carbon percentage. Data kindly provided by Dr. Thomas G. Mueller, Department of Agronomy, University of Kentucky. Used with permission.

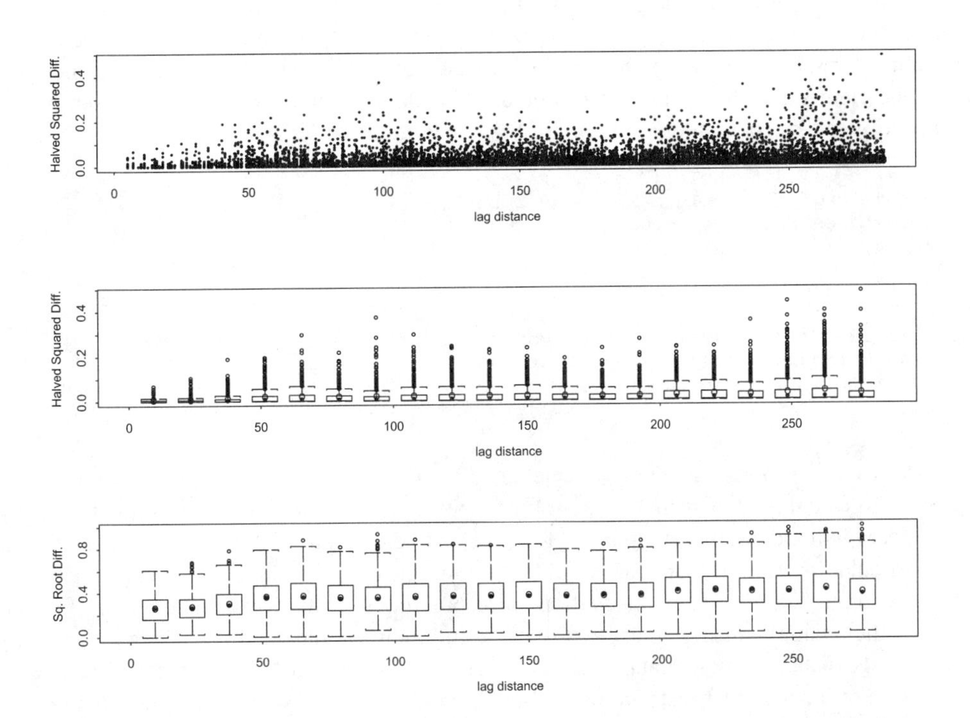

Figure 9.38. Semivariogram cloud (upper panel) and box-plot of halved squared differences (middle panel) for median-polished residuals for total soil carbon percentage. The bottom panel shows box-plots of $\sqrt{|Z(\mathbf{s}_i) - Z(\mathbf{s}_j)|}$. Lag distance in feet.

One of the exploratory tools to examine the second-order properties of geostatistical data is the semivariogram cloud $\{Z(\mathbf{s}_i) - Z(\mathbf{s}_j)\}^2$ (Chauvet 1982, Figure 9.38 upper panel). Because a large number of points share a certain lag distance, this plot is often too busy. A further reduction is possible by calculating summaries such as box-plots of the half squared differences $0.5(Z(\mathbf{s}_i) - Z(\mathbf{s}_j))^2$ or box-plots of the square root differences $\sqrt{|Z(\mathbf{s}_i) - Z(\mathbf{s}_j)|}$. These summaries are shown in the lower panels of Figure 9.38 and it is obvious that the spatial structure can be more easily be discerned from the graph of $\sqrt{|Z(\mathbf{s}_i) - Z(\mathbf{s}_j)|}$ than that of $0.5(Z(\mathbf{s}_i) - Z(\mathbf{s}_j))^2$. The former is more robust to extreme observations which create large deviations that appear as outliers in the box-plots of the $0.5(Z(\mathbf{s}_i) - Z(\mathbf{s}_j))^2$. Figure 9.38 was produced in S+SpatialStats® with the statements

```
par(col=1,mfrow=c(3,1))
vcloud1 <- variogram.cloud(TC ~ loc(x,y),data=notill)
plot(vcloud1,xlab="lag distance",ylab="Halved Squared Diff.",col=1,cex=0.3)
boxplot(vcloud1,mean=T,pch.mean="o",xlab="lag distance",
                ylab="Halved Squared Diff.")
vcloud2 <- variogram.cloud(TCN ~ loc(x,y),data=notill,
                fun=function(zi,zj) sqrt(abs(zi-zj)))
boxplot(vcloud2,mean=T,pch.mean="o",xlab="lag distance",ylab="Sq. Root Diff.")
```

There appears to be spatial structure in these data, the medians increase for small lag distances. The assumption of second-order stationarity is not unreasonable as the medians remain relatively constant for larger lag distances. The square root difference plot is not an estimate of the semivariogram. The square root differences are, however, the basic ingredients of the robust Cressie and Hawkins semivariogram estimator. The halved squared differences are the basic elements of the Matheron estimator. The classical and robust empirical semivariogram estimators are obtained in The SAS® System with the code

```
proc variogram data=NoTillData outvar=svar1;
  compute lagdistance=10 maxlags=40 robust;
  coordinates xcoord=x ycoord=y;
  var TC;
run;
proc print data=svar1; run;

proc variogram data=NoTillData outvar=svar2;
  compute lagdistance=7 maxlags=29 robust;
  coordinates xcoord=x ycoord=y;
  var TC;
run;
```

The first call to `proc variogram` calculates the estimator for forty lags of width 10, the second call calculates the semivariogram for twenty-nine lags of width 7. Thus, the two semivariograms will extend to 400 and 200 feet, respectively (Figure 9.39). The number of pairs in a particular lag class is stored as variable `count` in the output data set of the `variogram` procedure (Output 9.7). The first observation corresponding to `LAG=-1` lists the number of observations, their sample mean (`AVERAGE=0.83672`), and their sample variance (`COVAR=0.025998`). The average distance among data pairs in the first lag class was 8.416 feet, the classical semivariogram estimate at that distance was 0.006533 and the robust semivariogram estimate was 0.005266. The estimate of the covariance function at this lag is 0.023649. Recall that (i) the estimate of the covariogram is biased and (ii) that SAS® reports the **semi**variogram estimates although the columns are labels `VARIOG` and `RVARIO`. The number of pairs at each lag distance are sufficient to produce reliable estimates. Notice that the recommendation is to have at least

30 (50) observations in each lag class. It is our experience that occasionally not even 100 pairs provide reliable estimates of the semivariogram.

Output 9.7.

Obs	LAG	COUNT	DISTANCE	AVERAGE	VARIOG	COVAR	RVARIO
1	-1	200	.	0.83672	.	0.025998	.
2	0	0	.	.	.	.	.
3	1	157	8.416	0.81832	0.006533	0.023649	0.005266
4	2	201	17.836	0.82296	0.006012	0.012518	0.004956
5	3	224	28.631	0.83011	0.009020	0.011840	0.008111
6	4	196	38.941	0.83845	0.012929	0.012672	0.010591
7	5	337	49.221	0.83842	0.020552	0.007283	0.017461
8	6	422	59.246	0.85073	0.022993	0.004784	0.022362
9	7	417	69.361	0.84645	0.018293	0.006607	0.016627
10	8	397	78.999	0.84154	0.019384	0.002179	0.017076
11	9	474	89.184	0.83732	0.019550	0.000790	0.017698
12	10	550	99.689	0.84283	0.020342	0.000327	0.016309
13	11	610	109.367	0.84089	0.021594	-0.001003	0.018939
14	12	550	119.269	0.83355	0.021640	-0.000607	0.019228
15	13	506	129.474	0.82409	0.020955	0.001440	0.019234
16	14	583	139.566	0.83470	0.021357	-0.000521	0.019269
17	15	735	149.374	0.83627	0.021699	0.001650	0.020352
18	16	786	159.507	0.82803	0.019115	0.001175	0.018162
19	17	802	169.781	0.82467	0.019607	0.001251	0.019202
20	18	626	179.443	0.82070	0.018929	0.001839	0.018636
21	19	586	189.454	0.82877	0.017413	0.000579	0.017180
22	20	683	199.384	0.84440	0.021483	0.001346	0.021308
23	21	721	209.227	0.84837	0.027410	-0.001529	0.029724
24	22	733	219.454	0.84424	0.027304	-0.002679	0.029259
25	23	690	229.483	0.83681	0.027045	-0.002245	0.029288
26	24	579	239.428	0.84117	0.026661	-0.001388	0.026804
27	25	583	249.804	0.83451	0.027932	-0.001344	0.025045
28	26	544	259.420	0.83707	0.045123	-0.009930	0.036596
29	27	543	269.434	0.83346	0.035870	-0.004092	0.030603
30	28	479	279.479	0.84181	0.028869	-0.002236	0.025001
31	29	463	289.598	0.83144	0.030523	-0.002682	0.025000
32	30	485	299.997	0.83932	0.028723	0.001142	0.027325
33	31	441	309.639	0.84391	0.031584	-0.001255	0.026534
34	32	405	319.382	0.82694	0.031652	-0.004144	0.028547
35	33	349	329.702	0.83655	0.036750	-0.003053	0.032942
36	34	368	339.701	0.84825	0.035579	-0.002544	0.038313
37	35	230	349.578	0.84224	0.030564	-0.001738	0.030180
38	36	222	359.770	0.84618	0.039159	-0.006564	0.038729
39	37	230	370.194	0.83104	0.033529	-0.007664	0.032587
40	38	162	379.691	0.83849	0.027808	-0.002152	0.022954
41	39	219	389.352	0.84015	0.035187	-0.005041	0.031494
42	40	256	400.117	0.83659	0.032328	-0.003964	0.025213

The empirical semivariogram in the upper panel of Figure 9.39 shows an interesting rise at lag 200 ft. The number of pairs in lag classes 18 – 22 is sufficient to obtain a reliable estimate, so that sparseness of observations cannot be the explanation. The semivariogram appears to have a sill around 0.02 for lags less than 200 feet and a sill of 0.03 for lags greater than 200 feet. Possible explanations are nonstationarity in the mean, and/or a nested stochastic process. The spatial (stochastic) variability may consist of two smooth-scale processes that differ in their range. Whether this feature of the semivariogram is important depends on the intended use of the semivariogram. If the purpose is that of spatial prediction of soil carbon at unobserved locations and kriging is performed within a local neighborhood of 150 feet, say, it is important to capture the spatial structure on that range, since data points more distant than 150 feet from the prediction location are assigned zero weight. The long-range features of the process are of lesser importance then. If the purpose of semivariogram modeling is to partition the spatial process into sub-processes whose sill and range can be linked to physical or biological features or if one performs kriging with a larger kriging radius, then a nested semivariogram model or a nonparametric fit is advised. Large-scale changes in the mean carbon percentages will be revisited in §9.8.5.

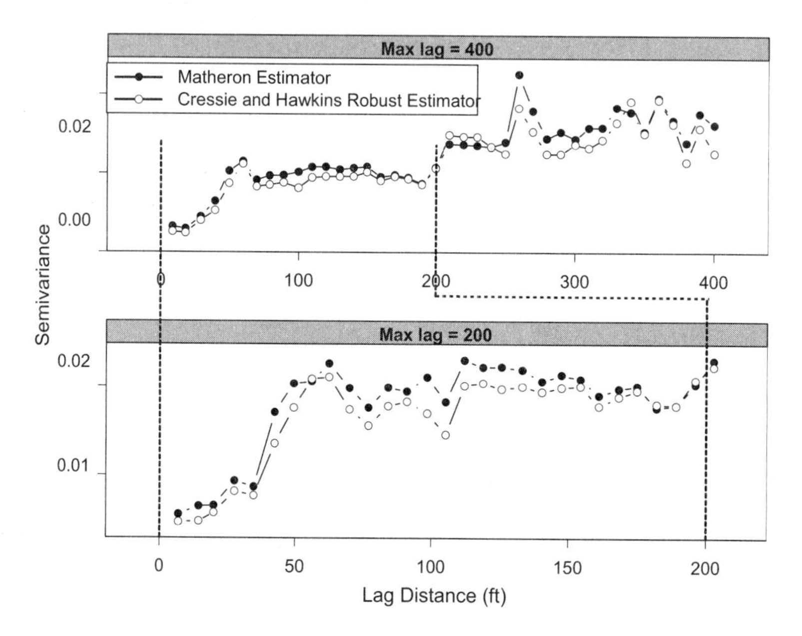

Figure 9.39. Classical and robust empirical semivariogram estimates for soil carbon percentage. Upper panel shows semivariances up to $||\mathbf{h}|| = 400$, lower panel up to $||\mathbf{h}|| = 200$ ft.

Another important feature of the semivariogram in the uper panel of Figure 9.39 is its increasingly erratic nature for lags in excess of 350 feet. There are fewer pairs for large distance classes and the (approximate) variance of the empirical semivariogram increases with the square of $\gamma(\mathbf{h})$ (see [9.13], p. 590). Finally, we note that the robust and classical semivariograms differ little. The robust semivariogram is slightly downward-biased but traces the profile of the classical estimator closely. This suggests that these data are not afflicted with outlying observations. In the case of outliers, the classical estimator often appears shifted upward from the robust estimator by a considerable amount.

In the remainder of this application we fit theoretical semivariogram models to the two semivariograms in Figure 9.39 by ordinary and weighted nonlinear least squares, by (restricted) maximum likelihood and composite likelihood. Recall that maximum and restricted maximum likelihood estimation operates on the *raw* data, not pairwise squared differences, so that one cannot restrict the lag distance. With composite likelihood estimation, this is possible (see §9.2.4 for a comparison of the estimation methods). The semivariogram models investigated are the exponential, spherical, and gaussian models (§9.2.2). We need to decide which semivariogram model best describes the stochastic dependency in the data and whether the model includes a nugget effect or not. The ostensibly simple question about the presence of a nugget effect is not as simple to answer in practice. We illustrate with the fit of the exponential semivariogram to the data in the lower panel of Figure 9.39. The SAS® statements

```
proc nlin data=svar2 noitprint nohalve;
  parameters nugget=0.01 sill=0.02 range=80;
  bounds nugget > 0;
  semivariogram = nugget + (sill-nugget)*(1-exp(-3*distance/range));
  _weight_ = 0.5*count/(semivariogram**2);
  model variog = semivariogram;
run;
```

fit the semivariogram by weighted nonlinear least squares. The `bounds` statement ensures that the estimate of the nugget parameter is positive. Notice that the term `(sill-nugget)` is the partial sill. The asymptotic confidence interval for the nugget parameter includes zero and based on this fact one might be inclined to conclude that a nugget is not needed in the particular model (Output 9.8). The confidence intervals are based on the asymptotic estimated standard errors which are suspect. First, the data points are correlated and the weighted least squares fit takes into account only the heteroscedasticity among the empirical semivariogram values (and that only approximately), not their correlation. Second, the standard errors depend on the number of data points which are the result of a user-driven grouping into lag classes. Because the standard errors are not reliable, the confidence interval should not be trusted. A sum of squares reduction test comparing the fit of a full and a reduced model is also not a viable alternative. The weights depend on the semivariogram being fitted and changing the model (i.e., dropping the nugget) changes the weights. The no-nugget model can be fit by simply removing the nugget from the `parameters` statement in the previous code and fixing its value at zero:

```
proc nlin data=svar2 noitprint nohalve;
  parameters sill=0.02 range=80;
  nugget = 0;
  semivariogram = nugget + (sill-nugget)*(1-exp(-3*distance/range));
  _weight_ = 0.5*count/(semivariogram**2);
  model variog = semivariogram;
run;
```

The sum of squares from two fits with different weights are not comparable (compare Output 9.8 and Output 9.9). For example, the uncorrected total sums of squares are 734.3 and 772.7, respectively.

Output 9.8. (abridged)

The NLIN Procedure

Source	DF	Sum of Squares	Mean Square	F Value	Approx Pr > F
Regression	3	5012.0	1670.7	117.61	<.0001
Residual	26	73.0856	2.8110		
Uncorrected Total	29	5085.1			
Corrected Total	28	734.3			

Parameter	Estimate	Approx Std Error	Approximate 95% Confidence Limits	
nugget	0.000563	0.00169	-0.00291	0.00403
sill	0.0210	0.000680	0.0196	0.0224
range	96.2686	19.7888	55.5926	136.9

Output 9.9. (abridged)

```
                         The NLIN Procedure

     NOTE: Convergence criterion met.

                                      Sum of           Mean                       Approx
Source                       DF      Squares          Square      F Value       Pr > F
Regression                    2       5012.0          2506.0       921.04       <.0001
Residual                     27      73.4622          2.7208
Uncorrected Total            29       5085.5
Corrected Total              28        772.7

                                                Approx          Approximate 95% Confidence
       Parameter        Estimate             Std Error                      Limits
       sill               0.0209              0.000607            0.0197            0.0221
       range             91.4617               11.4371           67.9950             114.9
```

To settle the issue on whether to include a nugget effect with a formal test we can call upon the likelihood ratio principle. Provided we are willing to make the assumption of a Gaussian random field the nugget and no-nugget models can be fit with `proc mixed` of The SAS® System.

```
/* nugget model */
proc mixed data=NoTillData method=ml noprofile;
  model tc = ;
  repeated / subject=intercept type=sp(exp)(x y) local;
  parms /* partial sill   */ ( 0.025 )
        /* range  */ ( 32     )
        /* nugget */ ( 0.005 );
run;

/* no-nugget model */
proc mixed data=NoTillData method=ml;
  model tc = ;
  repeated / subject=intercept type=sp(exp)(x y);
  parms /* range   */ ( 32     )
        /* sill    */ ( 0.025 ) ;
run;
```

To fit a model with nugget effect the `local` option is added to the `repeated` statement and the `noprofile` option is added to the `proc mixed` statement. The latter is necessary to prevent `proc mixed` from estimating an extra scale parameter that it would profile out of the likelihood. The `parms` statement lists starting values for the covariance parameters. In the no-nugget model the `local` and `noprofile` options are removed. Also notice that the order of the covariance parameters in the `parms` statement changes between the nugget and no-nugget models. The correct order in which to enter starting values in the `parms` statement can be gleaned from the `Covariance Parameter Estimates` table of the `proc mixed` output. The `subject=` option of the `repeated` statement informs the procedure which observations are considered correlated in the data. Observations with different values of the subject variable are considered independent. By specifying `subject=intercept` the variable identifying the clusters in the data is a column of ones. Spatial data is treated as if it comprises a single cluster of size n (see §2.6 on the progression of clustering from independent to spatial data).

The converged parameter estimates in the full model (containing a nugget effect) are partial sill $= 0.02299$, nugget $= 0.003557$, and 75.0588 for the *range* parameter. Since `proc mixed` parameterizes the exponential correlation function as $1 - \exp\{-||\mathbf{h}||/\alpha\}$, the estimated practical range is $3\widehat{\alpha} = 3*75.0588 = 225.26$ feet, considerably larger than the estimates from the nonlinear least squares fit. This is not too surprising. The maximum likelihood fit is based on all data, whereas the least squares fit on data pairs with lags up to 200 feet only. If the exponential semivariogram is fit to the data in the upper panel of Figure 9.39, the estimate of the practical range is 840.1 feet.

Minus twice the log likelihood for the full model (containing a nugget effect) and the reduced model without nugget effect are -331.9 and -315.6, respectively (Outputs 9.10 and 9.11). The likelihood ratio statistic of $\Lambda = 331.9 - 315.6 = 16.3$ has p-value $p = 0.00005$. Inclusion of the nugget effect provides a model improvement.

Output 9.10. (abridged)

```
                         The Mixed Procedure

                          Model Information
Data Set                      WORK.NOTILLDATA
Dependent Variable            TC
Covariance Structures         Spatial Exponential, Local Exponential
Subject Effect                Intercept
Estimation Method             ML
Residual Variance Method      None
Fixed Effects SE Method       Model-Based
Degrees of Freedom Method     Between-Within

                Covariance Parameter Estimates

          Cov Parm      Subject        Estimate
          Variance      Intercept       0.02299
          SP(EXP)       Intercept       75.0588
          Residual                     0.003557

                      Fit Statistics
          -2 Log Likelihood                -331.9
          AIC (smaller is better)          -323.9
          AICC (smaller is better)         -323.7
          BIC (smaller is better)          -310.7
```

Output 9.11. (abridged)

```
                         The Mixed Procedure

                          Model Information
Data Set                      WORK.NOTILLDATA
Dependent Variable            TC
Covariance Structure          Spatial Exponential
Subject Effect                Intercept
Estimation Method             ML
Residual Variance Method      Profile
Fixed Effects SE Method       Model-Based
Degrees of Freedom Method     Between-Within

                Covariance Parameter Estimates

          Cov Parm      Subject        Estimate
          SP(EXP)       Intercept       29.5734
          Residual                      0.02402

                      Fit Statistics
          -2 Log Likelihood                -315.6
          AIC (smaller is better)          -309.6
          AICC (smaller is better)         -309.5
          BIC (smaller is better)          -299.7
```

Results of the ordinary and weighted least squares fits for the various semivariogram models are summarized in Table 9.8 and the results for restricted/maximum likelihood in Table 9.9. Missing entries in these tables indicate that the particular model did not converge or that a boundary constraint was violated (nugget estimate less than zero, for example).

Table 9.8. Results of least squares fits. θ_0, ξ, α denote the nugget, sill, and (practical) range, respectively. SSR is residual sum of squares in ordinary least squares fit. Data = 200 refers to lower panel of Figure 9.39 with lags restricted to $\leq$ 200 feet

			WLS			OLS			
Data	**Model**	**Nugget**	θ_0	ξ	α	θ_0	ξ	α	$SSR^{\dagger}$
200	Exponential	No		0.021	91.46		0.021	94.99	149
		Yes	0.0005	0.021	96.27	–	–	–	–
	Gaussian	No		0.020	29.64		0.020	58.09	123
		Yes	0.004	0.021	69.07	0.003	0.020	65.04	102
	Spherical	No		0.020	66.09		0.020	72.23	108
		Yes	0.002	0.020	79.11	–	–	–	–
400	Exponential	No		0.028	202.3		0.034	353.9	907
		Yes	0.009	0.042	840.1	0.008	0.044	867.8	737
	Gaussian	No		0.025	36.88		0.028	100.5	1570
		Yes	0.006	0.027	118.8	0.013	0.035	377.0	799
	Spherical	No		0.026	97.25		0.032	290.0	1210
		Yes	0.011	0.038	534.7	0.009	0.034	411.2	735

$^{\dagger} \times 10^{-4}$

Based on the ordinary least squares fit one would select a gaussian model with nugget effect for the data in the lower panel of Figure 9.39 and a spherical semivariogram with nugget effect for the data in the upper panel. The Pseudo-R^2 measures for these models are 0.87 and 0.75, respectively. While the nugget and sill estimates are fairly stable it is noteworthy that the range estimates can vary widely among different models. Since the range essentially determines the strength of the spatial dependency, kriging predictions from spatial processes that differ greatly in their range can be very different. For the maxlag 400 data, for example, the exponential and spherical models with nugget effects have very similar residual sums of squares (737 and 735), but the range of the exponential model is more than twice that of the spherical model (OLS results). The lesser spatial continuity of the exponential model can be more than offset by a doubling of the range. The weighted least squares estimates of the range appear particularly unstable. An indication of a well-fitting model is good agreement between the ordinary and unweighted least squares estimates. On these grounds the gaussian no-nugget models and the spherical no-nugget model for the second data set can be ruled out. The fitted semivariograms for the gaussian and spherical nugget models are shown in Figure 9.40.

Based on the (restricted) maximum likelihood fits, the spherical nugget model emerges as the best-fitting model (Table 9.9). Nugget and no-nugget models can be compared via likelihood ratio tests which indicates for the exponential and gaussian models that a nugget effect is needed. To compare across semivariogram models the AIC criterion is used since the exponential, gaussian, and spherical models are not nested. This leads to the selection of the spherical nugget model as the best-fitting model in this group. Its range is considerably less than that of the corresponding model fitted by least squares. Notice that the REML estimates are uniformly larger than the ML estimates. REML estimation reduces the negative bias of maximum likelihood estimators of covariance parameters. Also notice that the estimates of the range are less erratic than for the least squares estimates. We could overlay the fitted semivariograms in Figure 9.40 with the semivariogram models implied by the maximum likelihood estimates. This comparison is not fair, however. The least squares methods fit the data to the scatter of points shown in Figure 9.40 whereas the likelihood methods fit a covariance function model to the original data. By virtue of least squares no other fitting method will be closer to the empirical semivariogram cloud. This does not imply that the least squares estimates are the best estimates of the spatial dependency structure.

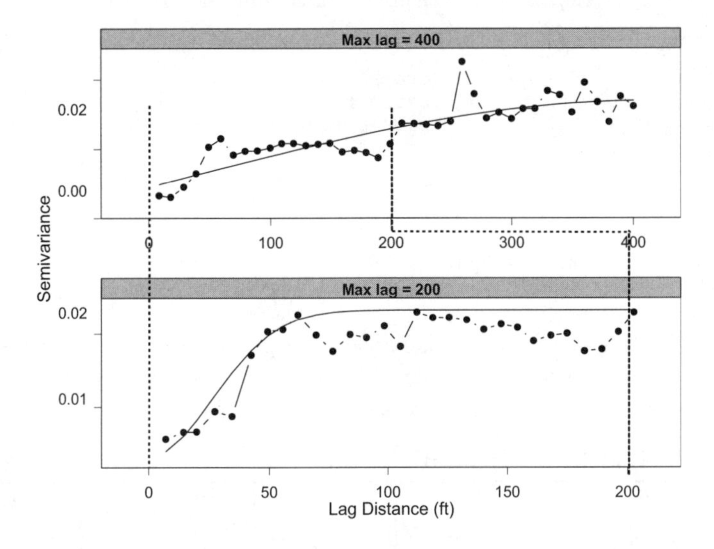

Figure 9.40. Semivariograms fitted by least squares. Spherical nugget semivariogram for max lag = 400 and gaussian nugget semivariogram for max lag = 200. Ordinary least squares fits shown.

Table 9.9. Results of maximum and restricted maximum likelihood estimation. θ_0, ξ, α denote the nugget, sill, and (practical) range, respectively

		ML			REML				
Model	**Nugget**	θ_0	ξ	α	θ_0	ξ	α	**-2logL**†	**AIC**£
Exponential	No		0.024	88.71		0.025	92.52	– 310	– 306
	Yes	0.003	0.026	225.3	0.003	0.031	293.2	– 328	– 322
Gaussian	No		–	–		–	–		
	Yes	0.006	0.024	108.0	0.007	0.026	111.5	– 323	– 317
Spherical	No		0.050	115.6		0.050	115.9	– 296	– 292
	Yes	0.004	0.025	133.2	0.004	0.025	134.6	– 330	– 324

† negative of twice the restricted maximum likelihood
£ Akaike's information criterion (smaller is better variety)

Finally, we conclude this application by fitting the basic semivariogram models by the composite likelihood (CL) principle. This principle is situated between genuine maximum likelihood and least squares. It has features in common with both. As the empirical Matheron semivariogram estimator the principle is based on pairwise squared differences. It does not average the squared differences, however, but fits the semivariogram to a data set of all $n(n-1)/2$ pairwise differences. Composite likelihood estimation has in common with the likelihood principles to stipulate a distribution of the $(Z(\mathbf{s}_i) - Z(\mathbf{s}_j))^2$, thereby indirectly specifying a distribution of the raw data $Z(\mathbf{s}_i)$. It has in common with least squares estimation that restricting estimation of the semivariogram parameters to certain lag distances is simple. The negative impact of erratic squared differences at large lags can thereby be reduced. Furthermore, estimation can be carried out in weighted and unweighted form. The unweighted composite likelihood estimator is in fact a Generalized Estimating Equation estimate as we establish in §A9.9.2. CL estimators are easily obtained with the SAS® macro `%cl()` contained in `\SASMacros\CL.sas`.

```
%include 'DriveLetterofCDROM\SASMacros\CL.sas';
/* Create a data set with the squared differences */
proc variogram data=NoTillData outpair=pairs;
  compute novariogram;
  coordinates xcoord=x ycoord=y;
  var TC;
run;
/* Call the macro */
%cl(nugget=1,covmod=E,maxrng=200,
    nuggetstart=0.005,sillstart=0.03,rangestart=80);
```

Table 9.10. Results of composite likelihood fits. θ_0, ξ, α denote the nugget, sill, and (practical) range, respectively. Data = 200 refers to lower panel of Figure 9.39 with lags restricted to ≤ 200 feet

Data	Model	Nugget	θ_0	ξ	α	$SSR_w^\dagger$
200	Exponential	No		0.021	84.0	12,010
		Yes	0.002	0.021	99.3	11,980
	Gaussian	No		0.020	22.1	12,088
		Yes	0.005	0.020	69.4	11,914
	Spherical	No		0.026	131.3	14,160
		Yes	–	–	–	–
400	Exponential	No		0.028	196.8	24,334
		Yes	0.009	0.048	1078.1	23,361
	Gaussian	No		0.025	29.3	24,793
		Yes	0.016	0.042	528.5	23,193
	Spherical	No		0.036	275.0	31,401
		Yes	0.011	0.048	756.0	23,288

† weighted residual sum of squares

The CL algorithm converged for all but one setting. The estimates are in general close to the least squares estimates in Table 9.8. In situations where the least squares fits to the empirical semivariogram did poorly (e.g., gaussian no-nugget model), so did the composite likelihood fit. The weighted residual sums of squares are not directly comparable among the models as the weights depend on the model. Based on their magnitude alone the spherical model is ruled out for the maxlag 200 data and no-nugget models are ruled out for the maxlag 400 data, however.

9.8.3 Spatial Prediction — Kriging of Lead Concentrations

Recreational shooting ranges are popular as sites to develop and practice sporting firearms activities and, at the same time, are becoming sites of potential concern because of the accumulation of metals, in particular, lead. The data analyzed in this application was collected on a shooting range a few miles west of Blacksburg, Montgomery County, Virginia, operated by the United States Forest Service in the George Washington-Jefferson National Forest. For a more detailed account of this study including analysis of a slightly larger data set see Craig et al. (forthcoming). The range consists of two shooting areas, a rifle range, and a shotgun range. Our data pertain to the shotgun range. The range lies in a second growth mixed hardwood forest on the Devonian Brallier Formation composed primarily of a deeply weathered black shale. The range has been in continuous use since 1993, operated for approximately 350 days a year, and closed periodically for maintenance and general cleaning. The shooting range consists of an open, gently sloping surface, approximately 62 meters in length by about 65 meters in width, that was cleared in the forest. The shooting box is located near the center of the range and is apparently used by most shooters. A clay pigeon launching site is situated approximately seven meters to the right of the shooting box.

Of interest to the investigators was the determination of the area and impact of the shot, the spatial distribution of lead, and an estimate of the total amount of lead on the range. The area of interest was determined to be a rectangle 240 meters wide and 300 meters long with the shooting box located at $x = 100$, $y = 0$ (Figure 9.3, p. 568, left panel). It was believed that much of the shot would occur on the approximately 60×60 surface that had been cleared in front of the shooting box. Accordingly, the initial sampling was carried out at 5-meter intervals along a line extending from $x = 100$ toward the center of the far edge of the cleared area. Additional samples were collected on transects emanating from the central transect at 10- or 20-meter intervals. At each sampling point all material within a 50×50 centimeter square was extracted and sieved through a 6-millimeter metal sieve. After stirring and agitating individually in tap water the samples were transferred to a 14-inch Garrett Gravity trap gold pan. The recovered metal materials were examined under a binocular microscope and all extraneous material was removed. Calibration showed that this procedure had a lead recovery rate of 99.5%. It is recognized that some material could be lost during recovery efforts at any site and that lead lodged in trees is not recovered. The estimates of total lead derived in what follows are thus conservative.

This sampling design does not provide even coverage of the shotgun range; most samples were collected in areas where high lead concentrations were anticipated. An estimate of total lead based on the sample average is thus positively biased, possibly severely so. If $Z(\mathbf{s}_i)$ denotes lead in g/m^2 at sampling site $\mathbf{s}_i$, then $\overline{Z} = 495.21\ g/m^2$ and the total load on the range would be estimated as 35.655 tons. The spatial analysis commences by calculating the empirical semivariogram of the lead concentration (Figure 9.41, top panel). There appears to be little spatial structure in the lead concentrations. Closer examination of the data shows, however, that the lead concentrations in the cleared area of the shotgun range are considerably higher than in other areas creating a right-skewed distribution of lead. Most of the extreme values in the right tail come from this area close to the shooting box. To achieve greater symmetry in the data a log transformation is applied. The empirical semivariogram of the $\ln\{\text{lead}\}$ values shows considerably more spatial structure (Figure 9.41, bottom panel).

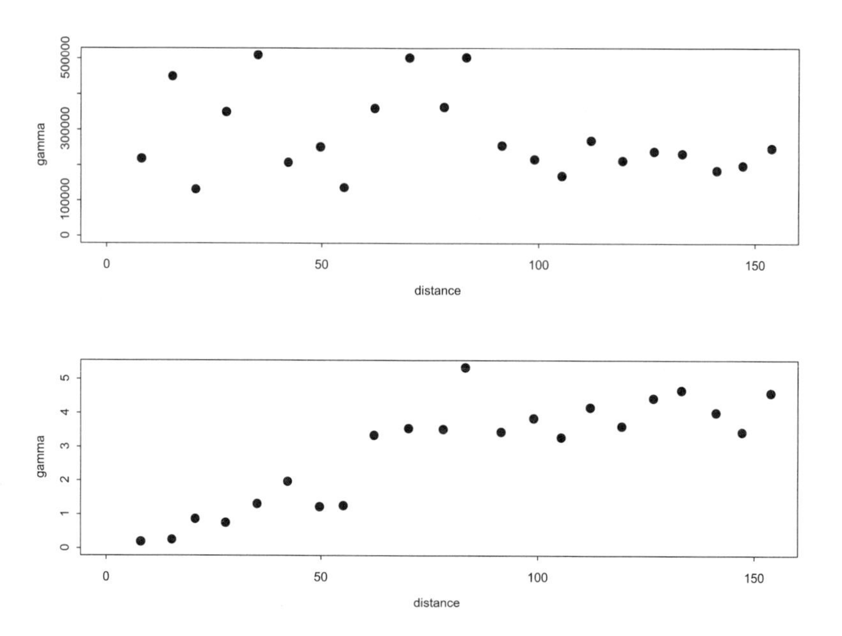

Figure 9.41. Semivariograms of lead (top panel) and ln{lead} (bottom panel) concentrations before detrending.

Fitting a semivariogram by weighted least squares to the log-lead values in S+SpatialStats® with the statements

```
sg.varlglead <- variogram(sg$lglead ~ loc(x,y),data=sg,method="robust",
                lag=7,nlag=22)
SvarFit(data=sg.varlglead,type="spherical",weighted=T,
                start=list(sill=3.5,range=90))
```

yields a sill estimate of 4.31 and a range 128.8 meters. The `SvarFit()` function was developed by the authors and is contained on the CD-ROM.

It is likely that even after the transformation the mean of $\ln\{Z(\mathbf{s})\}$ is not stationary. Removing by ordinary least squares a response surface in the coordinates and fitting a spherical semivariogram by weighted least squares to the residuals leads to Output 9.12 and Figure 9.42. The S+SpatialStats® statements producing the output and figure are

```
sg.lm <- lm(lglead ~ x + y + x*y + x^2,data=sg)
sg.lmres <- sg$lglead - predict(sg.lm)
sg.varlmres <- variogram(sg.lmres ~ loc(x,y),data=sg,method="robust",
               lag=7,nlag=22)
SvarFit(data=sg.varlmres,type="spherical",weighted=T,
               start=list(sill=1.5,range=60))
```

The sill of the semivariogram is drastically reduced as well as the range compared to the nondetrended data. The reduction in sill shows the smaller variability of the model residuals, the reduction in range that spurious autocorrelation caused by a large-scale trend was removed.

Output 9.12.

```
Formula:   ~ spher.wfunnonug(gamma, distance, range, sill, np)

Parameters:
        Value Std. Error  t value
 sill  2.0647   0.124854 16.53680
range 94.9438  11.400800  8.32783

Residual standard error: 2.45191 on 20 degrees of freedom
Residual sum of squares : 120.2369
```

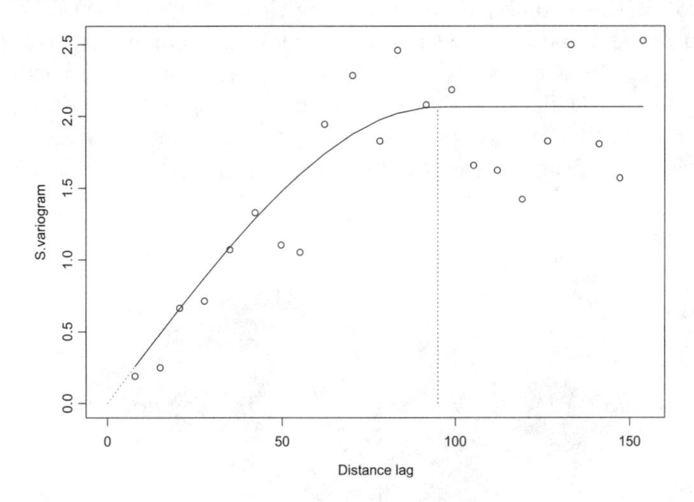

Figure 9.42. Spherical semivariogram fit by weighted least squares to ordinary least squares residuals.

To incorporate the nonconstant mean in spatial predictions of the lead concentrations we perform universal kriging with the same trend model as used in modeling the semivariogram and predict the log-lead concentration on a 240×300 grid. To obtain predictions and 95% prediction intervals on the original scale the predictions and intervals on the logarithmic scale are back-transformed. In this process some transformation bias is incurred. On the upside this procedure guarantees that all predictions are positive without imposing additional constraints on the kriging weights. The S+SpatialStats® statements that solve the universal kriging equations, calculate predictions of log-lead on the grid, and back-transform the results are:

```
sg.ukrige <- krige(lglead ~ loc(x,y) + x + y + x*y +x^2,
                   data=sg,covfun=spher.cov,range=94.94377,
                   sill=2.0469,nugget=0)

grid <- list(x=seq(0,240,4),y=seq(0,300,4))
grid <- expand.grid(grid)
sg.ukp <- predict.krige(sg.ukrige,newdata=grid)

# Confidence intervals and predictions on original scale
sg.ukp$l95 <- sg.ukp$fit - 1.96*sg.ukp$se.fit
sg.ukp$u95 <- sg.ukp$fit + 1.96*sg.ukp$se.fit
sg.ukp$l95 <- exp(sg.ukp$l95)
sg.ukp$u95 <- exp(sg.ukp$u95)
sg.ukp$efit <- exp(sg.ukp$fit)/1000
```

The surface of predicted values shows two large spikes in front of the shooting box (Figure 9.43). The *anomaly* from a homogeneous distribution at 25 to 30 meters results from users mounting targets at this distance. Common targets include clay pigeons, golf balls, plastic jugs, glass bottles, fruits and vegetables. Shooting at sofas has been observed, reports of shooting at computers and toilet bowls are anecdotal, and confirmation exists in the discovery of numerous damaged keyboard keys.

The *anomaly* of lead at approximately 80 meters apparently results from the accumulation of lead fired at elevated trajectories in attempts to hit clay pigeons that have been launched or thrown. This anomaly is wider than the closer one because of the spread of the shot at a greater distance and because shooters are tracking a moving target across the range as they are firing. The peak at approximately 80 meters results from the combined effect of the normal trajectory of much of the shot and from the slowing of some of the shot by leaves and branches. The smaller *anomalies* at approximately 180 meters are more difficult to explain. They may result from a higher trajectory that arcs up over the first line of trees.

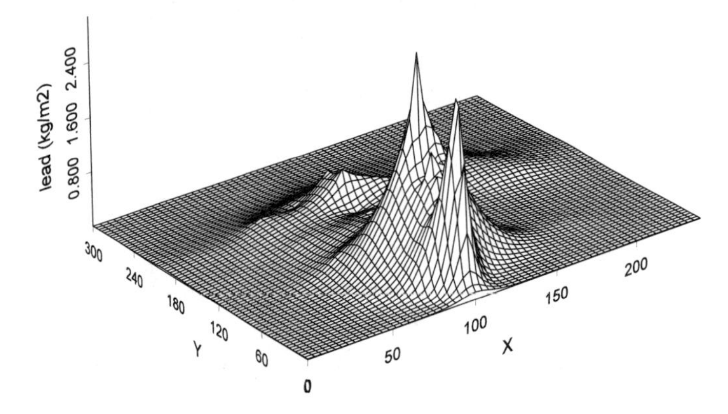

Figure 9.43. Exponentiated $\ln\{\text{lead}\}$ predictions obtained by ordinary kriging.

An estimate of the total lead concentration can be obtained by integrating the surface in Figure 9.43. A quick estimate is calculated as the average ordinate of the surface, which yields 161.229 g/m^2 and a total load of 11.608 tons. Although we believe this estimate of the total lead concentration to be fairly accurate, two problems remain to be resolved. In estimating the total on the logarithmic scale and exponentiating the result some transformation bias is incurred. The total amount of lead will be underestimated. Also, no standard error is available for this estimate. In order to predict the total amount of lead on the original scale without bias we need to consider the block average

$$Z(A) = \int_A Z(\mathbf{u})d\mathbf{u},$$

where A is the rectangle $(0, 240) \times (0, 300)$. Some details on block-kriging appear in §A9.9.6. But there appears to be little spatial structure in the lead concentrations (Figure 9.41, top panel). The solution is to model the large-scale trend allowing the mean of $Z(\mathbf{s})$ to capture the two large spikes in Figure 9.43. The semivariogram can then be developed based on the residuals of this fit. Alternatively, one can exclude the two spikes in calculating the empirical semivariogram. With the former procedure it was determined that a spherical semivariogram with range 88.754 m^2 fits the empirical semivariogram well. With a properly crafted mean model the block total $Z(A)$ can then be obtained by universal block-kriging. The estimate of the total amount of lead so obtained is 13.958 tons with a prediction standard error of 1.68 tons. A 95% prediction interval for the total is thus [10.66 tons, 17.26 tons]. The surface of the universal kriging predictions (Figure 9.44) differs little from back-transformed ordinary kriging predictions on the logarithmic scale (Figure 9.43).

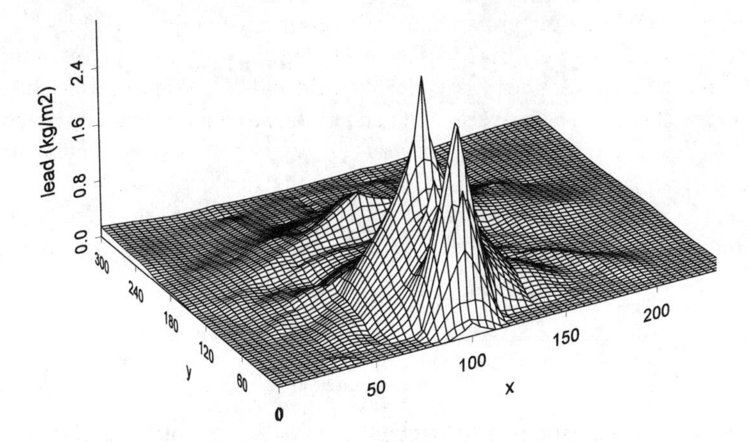

Figure 9.44. Predicted surface of lead in kg/m^2 obtained by universal kriging on the original scale.

9.8.4 Spatial Random Field Models — Comparing C/N Ratios among Tillage Treatments

When data are collected under different conditions, such as treatments, an obvious question is to determine whether the conditions are different from each other, and if so, how the differences manifest themselves. In a classical field experiment contrasts among the treatment means are estimated and tested to formulate statements about the differences among and effects of the experimental conditions. If the data collected under various conditions are autocorrelated, then one needs to rethink what precisely we mean by *differences* in the conditions. We now return to the soil carbon data first introduced in §9.8.2. After ten years of a corn-soybean rotation without tillage, intermediate strips of the field were chisel-plowed. Two months after the soils were first chisel-plowed in the spring samples from 0 to 2 inch depths were collected and total N percentage (TN) and total carbon percentage (CN) were determined. The sampling locations and the strips are shown in Figure 9.45.

Since sampling occurred very soon after tillage we do not anticipate fundamental changes in the TC and TN values or the C/N ratio between the two treatments. Because of the spatial sampling context and the presence of two conditions on the field, however, the data are perfectly suited to demonstrate the basic manipulations and computations involved in a random field analysis that involves treatment structure. We furthermore note from Figure 9.45 that the strips were not randomized. An analysis as a randomized experiment with subsampling of six replications of two treatments is therefore tenuous. Instead, we analyze the data as a spatial random field with a mean structure given by the two treatment conditions and possible spatial autocorrelation among the sampling sites.

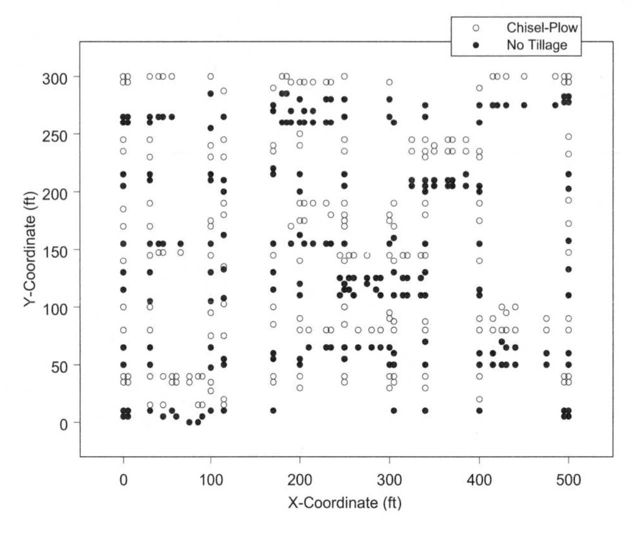

Figure 9.45. Sampling locations at which total soil N (%) and total soil C (%) were observed for two tillage treatments. Treatment strips are oriented in East-West direction.

The target attribute for this application is the C/N ratio and a simplistic pooled t-test comparing the two tillage treatments leads to a p-value of 0.809 from which one would conclude that there are no differences in the average C/N ratios. This test does not account for spatial autocorrelation treating the 195 samples on chisel-plow strips and 200 samples on no-till strips as independent. Furthermore, it does not convey whether there are differences in the spatial structure of the treatments. Even if the means are the same the spatial dependency might develop differently. This, too, would be a difference in the treatments that should be recognized by the analyst. Omnidirectional semivariograms were calculated with the `variogram` procedure in The SAS® System and spherical semivariogram models were fit to the empirical semivariograms (Figure 9.46) with `proc nlin` by weighted least squares:

```
proc sort data=CNRatio; by tillage; run;
proc variogram data=CNRatio outvar=svar;
   compute lagdistance=13.6 maxlag=19 robust;
   coordinates xcoord=x ycoord=y;
   var cn;
   by tillage;
run;
proc nlin data=fitthis nohalve method=newton noitprint;
  parameters sillC=0.093 sillN=0.1414 rangeC=116.6 rangeN=197.2
             nugget=0.1982;
  if tillage='ChiselPlow' then
    sphermodel = nugget + (distance <= rangeC)*sillC*(1.5*(distance/rangeC) -
                   0.5*((distance/rangeC)**3)) + (distance >  rangeC)*sillC;
  else
    sphermodel = nugget + (distance <= rangeN)*sillN*(1.5*(distance/rangeN) -
                    0.5*((distance/rangeN)**3)) + (distance >  rangeN)*sillN;
  model rvario = sphermodel;
  _weight_ = 0.5*count/(sphermodel**2);
run;
```

In anticipation of obtaining generalized least squares and restricted maximum likelihood inferences in `proc mixed` a common nugget effect was fit for both tillage treatments but the sills and ranges of the semivariogram were varied. The sill and range estimates for the chisel-plow treatment were 0.092 and 127.0, respectively. The corresponding estimates for the no-till treatment were 0.1397 and 199.2 (Output 9.13). Notice that to a considerable degree variability in C/N ratios is due to the nugget effect. The relative structured variability is 31% for the chisel-plow and 41% for the no-till treatment. The C/N ratio of the undisturbed no-till sites is more spatially structured, however, as can be seen from the larger range.

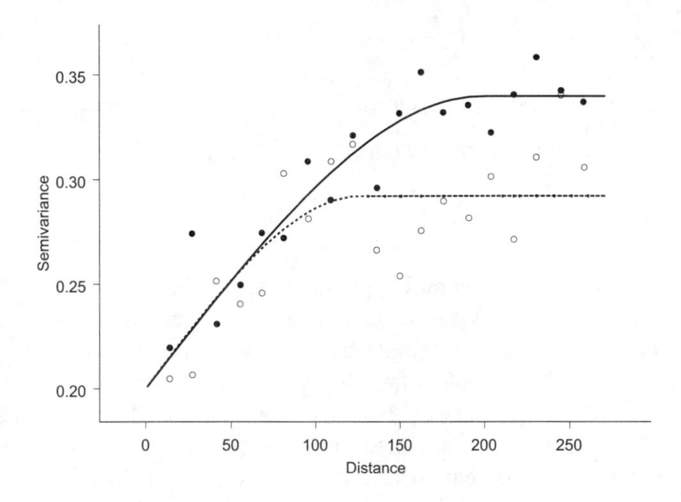

Figure 9.46. Omnidirectional empirical semivariograms for C/N ratio under chisel-plow (open circles) and no-till (full circles) treatments. Weighted least squares fit of spherical semivariograms are shown.

Output 9.13. (abridged)

```
                         The NLIN Procedure

                                Sum of          Mean                 Approx
Source                  DF     Squares        Square    F Value    Pr > F
Regression               5     13117.5        2623.5      23.92    <.0001
Residual                33     53.2243        1.6129
Uncorrected Total       38     13170.7
Corrected Total         37       207.6

                                   Approx      Approximate 95% Confidence
     Parameter      Estimate    Std Error               Limits
     sillC            0.0920       0.0151       0.0612       0.1228
     sillN            0.1397       0.0152       0.1089       0.1706
     rangeC            127.0      19.4203      87.4950        166.5
     rangeN            199.2      29.7131        138.8        259.7
     nugget           0.2000       0.0139       0.1717       0.2284
```

Next we obtain generalized least squares estimates of the treatment effect as well as predictions of the C/N ratio over the entire field with `proc mixed` of The SAS® System. A data set containing the prediction locations for both treatments (data set `filler`) is created and appended to the data set containing the observations. The response variable of the filler data set is set to missing values. This will prevent `proc mixed` from using the information in the prediction data set for estimation. In calculating predicted values these observations can be used, however, since they contain all information apart from the response.

```
data filler;
  do tillage='ChiselPlow','NoTillage';
    do x = 0 to 500 by 10;  do y = 0 to 300 by 10; cn=.; output;  end; end;
  end;
run;
data fitthis; set filler cnratio; run;

proc mixed data=fitthis noprofile;
  class tillage;
  model CN = tillage /ddfm=contain outp=p;
  repeated / subject=intercept type=sp(sph)(x y) local group=tillage;
  parms /* sill  ChiselPlow */    0.0920
        /* range ChiselPlow */  127.0
        /* sill  NoTillage  */    0.1397
        /* range NoTillage  */  199.2
        /* nugget (common)  */    0.2000 /  noiter;
run;
```

The call to `proc mixed` has several important features. The `model` statement describes the mean structure of the model. C/N ratios are assumed to depend on the tillage treatments. The `outp=p` option of the `model` statement produces a data set (named `p`) containing the predicted values. The `repeated` statement identifies the spatial covariance structure to be spherical (`type=sp(sph)(x y)`). The `subject=intercept` option indicates that the data set comprises a single subject, all observations are assumed to be correlated. The `group=tillage` option requests that the spatial covariance parameters are varied by the values of the `tillage` variable. This allows modeling separate covariance structures for the chisel-plow and no-till treatments to reflect the differences in spatial structure evident in Figure 9.46. Finally, the `local` option adds a nugget effect. Since `proc mixed` adds only a single nugget effect, it was important in fitting the semivariograms to ensure that the nugget effect was held the same for the two treatments. The `parms` statement provides starting values for the covariance parameters. The order in which the values are listed equals the order in which the values appear in the `Covariance Parameter Estimates` table of the `proc mixed` output. A trial run is sometimes necessary to determine the correct order. The starting values are set at the converged iterates from the weighted least squares fit of the theoretical semivariogram (Output 9.13). The `noiter` option of the `parms` statement prevents iterations of the covariance parameters and holds them fixed at the starting values provided. To produce restricted maximum likelihood estimates of the covariance parameters, simply remove the `noiter` option. The `noprofile` option of the `proc mixed` statement prevents profiling of the nugget variance. Without this option `proc mixed` would make slight adjustments to the sill and nugget even if the `/noiter` option is specified.

The `Dimensions` table indicates that 395 observations were used in model fitting and 3162 observations were not used (Output 9.14). The latter comprise the `filler` data set of prediction locations for which the `CN` variable was assigned a missing value. The `-2 Res Log Likelihood` of 570.3 in the table of `Fit Statistics` equals minus twice the residual log likelihood in the `Parameter Search` table. The latter table gives the likelihood for all sets of starting values. Here only one set of starting values was used and the equality of the `-2 Res Log Likelihood` values shows that no iterative updates of the covariance parameters took place. The estimates shown in the `Covariance Parameter Estimates` table are identical to the starting values provided in the `parms` statement. Finally, the `Type 3 Tests of Fixed Effects` table shows that there is no significant difference between the mean C/N ratios of the two tillage treatments ($p = 0.799$).

Output 9.14.

```
                        The Mixed Procedure

                         Model Information

          Data Set                     WORK.FITTHIS
          Dependent Variable           cn
          Covariance Structures        Spatial Spherical,
                                       Local Exponential
          Subject Effect               Intercept
          Group Effect                 tillage
          Estimation Method            REML
          Residual Variance Method     None
          Fixed Effects SE Method      Model-Based
          Degrees of Freedom Method    Containment

                      Class Level Information

           Class      Levels    Values
           tillage         2    ChiselPlow NoTillage

                            Dimensions

                    Covariance Parameters             5
                    Columns in X                      3
                    Columns in Z                      0
                    Subjects                          1
                    Max Obs Per Subject             395
                    Observations Used               395
                    Observations Not Used          3162
                    Total Observations             3557

                         Parameter Search

     CovP1      CovP2      CovP3      CovP4      CovP5         -2 Res Log Like
   0.09200     127.00     0.1397     199.20     0.2000                570.2618

                  Covariance Parameter Estimates

         Cov Parm     Subject      Group                      Estimate
         Variance     Intercept    tillage ChiselPlow          0.09200
         SP(SPH)      Intercept    tillage ChiselPlow           127.00
         Variance     Intercept    tillage NoTillage            0.1397
         SP(SPH)      Intercept    tillage NoTillage            199.20
         Residual                                               0.2000

                          Fit Statistics

                   -2 Res Log Likelihood            570.3
                   AIC (smaller is better)          570.3
                   AICC (smaller is better)         570.3
                   BIC (smaller is better)          570.3

                   Type 3 Tests of Fixed Effects

                             Num      Den
              Effect          DF       DF     F Value    Pr > F
              tillage          1      393        0.06    0.7990
```

The predicted C/N surfaces for the two tillage treatments are shown in Figure 9.47. Both surfaces vary about the same mean but the greater spatial continuity (larger range) of the no-till sites is evident in a smoother, less variable surface. Positive autocorrelations are stronger over the same distance under this treatment as compared to the chisel-plow treatment. At this point it is worthwhile to revisit the question raised early in this application. What do we mean by *differences* in experimental conditions if the observations collected from each site have a spatial context? There is no difference in the average C/N values in this study as can be expected when sampling only two months after installment of the treatments. There appear to be differences in the spatial structure of the treatments, however. Fitting a single spherical semivariogram to the empirical semivariograms shown in Figure 9.46 a residual sum of square of 93.09 on 35 degrees of freedom is obtained. A sum of square reduction test leads to

$$F_{obs} = \frac{(93.09 - 53.2)/2}{53.2/33} = 12.37$$

with a p-value of 0.00009. If the semivariogram is estimated by ordinary (instead of weighted) least squares the statistics are $F_{obs} = 11.85$ and $p = 0.0001$. There are significant differences among the treatments in the autocorrelation structure, albeit not in the average C/N ratio. One can argue that after ten years of continuous no-till management there is greater continuity in the C/N ratios compared to what can be observed shortly after a disturbance through plowing.

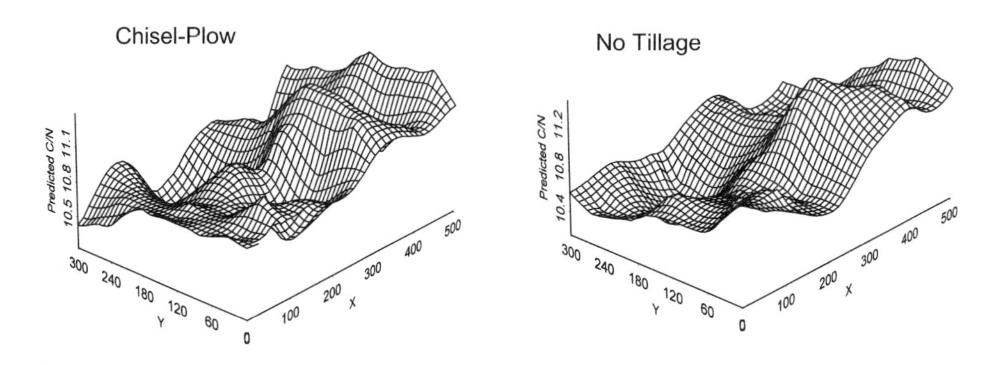

Figure 9.47. Predicted C/N surface under chisel-plow and no-till treatments.

The predicted surfaces in Figure 9.47 were obtained from the generalized least squares fit which assumed that the supplied starting values of the covariance parameters are the true values. This is akin to the assumption in kriging methods that the semivariogram values used in solving the kriging equations are known. Removing the `noiter` option of the `parms` statement in `proc mixed` the spatial covariance parameters are updated iteratively by the method of restricted maximum likelihood. Twice the negative residual log likelihood at convergence can be compared to the same statistic calculated from the starting values. This likelihood ratio test indicates whether the REML estimates are a significant improvement over the starting values. The `mixed` procedure displays the result of this test in the `PARMS Model Likelihood Ratio`

`Test` table (Output 9.15). In this application convergence was achieved after twelve time-consuming iterations with no significant improvement over the starting values ($p = 0.1981$).

Output 9.15. (abridged)

```
                    The Mixed Procedure

                      Fit Statistics
          -2 Res Log Likelihood            562.9
          AIC (smaller is better)          572.9
          AICC (smaller is better)         573.1
          BIC (smaller is better)          592.8


           PARMS Model Likelihood Ratio Test
             DF     Chi-Square      Pr > ChiSq
              5           7.32          0.1981
```

9.8.5 Spatial Random Field Models — Spatial Regression of Soil Carbon on Soil N

In the previous application C/N ratio was modeled directly and compared between the two tillage treatments. In many applications one attribute emerges as the primary variable of interest and other variables are secondary attributes which are to be linked to the primary attribute. This approach is particularly meaningful if the secondary attributes are easy to measure or available in dense coverage (e.g., sensed images, GIS) and the primary attribute is more difficult to determine. If the relationship between primary and secondary attributes can be modeled for a particular data set where both variables have been measured, the model can then be applied to situations where only the secondary attributes are available. Consider that we are interested in predicting soil carbon as a function of soil nitrogen. From Figure 9.48 it is clearly seen that the relationship between TC and TN is very strong ($R^2 = 0.916$), close to linear, and differs not between the two tillage treatments.

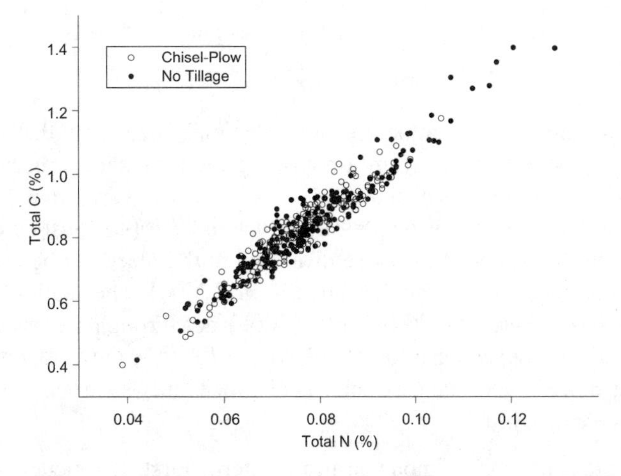

Figure 9.48. Relationship between total C (%) and total N (%) of chisel-plow and no-till areas.

For the time being we disregard the fact that the data are collected under two different tillage regimes. Incorporating additional tillage treatment effects in the models developed subsequently is straightforward. Figure 9.48 belies the fact that both variables are spatially heterogeneous. A side-by-side graph of the TC and TN contours shows more clearly how areas of high (low) TC are associated with areas of high (low) TN (Figure 9.49). Given a sample of TN and TC values we want to model the relationship between TC and TN taking into account that TC observations are spatially autocorrelated and given a sample or map of TN we want to use the modeled relationship to predict total carbon percentage at arbitrary locations.

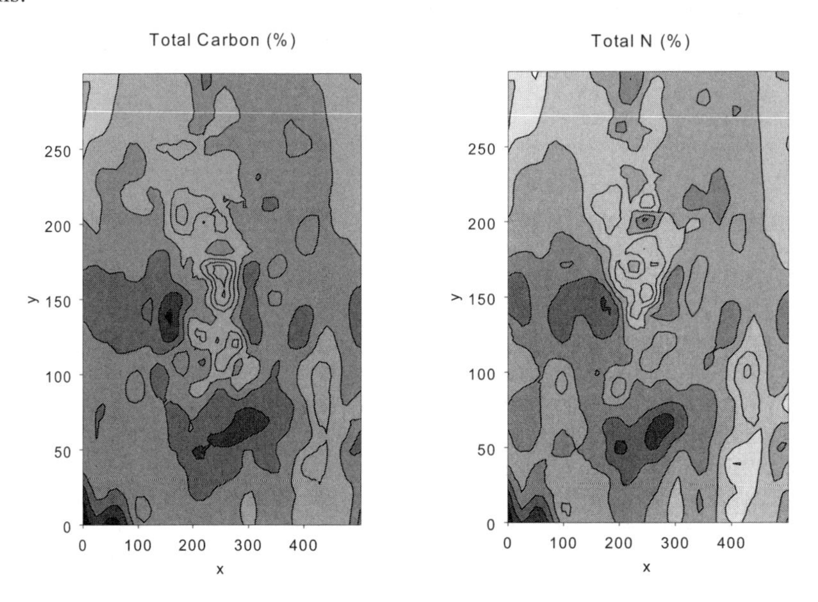

Figure 9.49. Contour plots of ordinary kriging predictions of total C (%) and total N (%) irrespective of tillage treatment.

Based on the relationship in Figure 9.48 an obvious place to start is

$$TC(\mathbf{s}_i) = \beta_0 + \beta_1 TN(\mathbf{s}_i) + e(\mathbf{s}_i), \qquad [9.73]$$

where the errors $e(\mathbf{s}_i)$ are spatially autocorrelated. We emphasize again that such a spatial regression model differs conceptually from a cokriging model where primary and secondary attribute are spatially autocorrelated and models for the semivariogram (covariogram) of $TC(\mathbf{s}_i)$, $TN(\mathbf{s}_i)$ *and* the cross-covariogram of $TC(\mathbf{s}_i)$ and $TN(\mathbf{s}_i)$ must be derived. $TN(\mathbf{s}_i)$ is considered fixed in [9.73] and the only semivariogram that needs to be modeled is that of $TC(\mathbf{s}_i)$ after adjusting its mean for the dependency on $TN(\mathbf{s}_i)$. The spatial regression model expresses the relationship between $TC(\mathbf{s}_i)$ and $TN(\mathbf{s}_i)$ not through a cross-covariogram but models it as deterministic dependency of $\mathrm{E}[TC(\mathbf{s}_i)]$ on $TN(\mathbf{s}_i)$; they are simpler to fit compared to cokriging models and standard statistical procedures such as `proc mixed` of The SAS® System can be employed.

The semivariogram of $e(\mathbf{s}_i)$ is modeled in two steps. First, the model is fit by ordinary least squares and the empirical semivariogram of the OLS residuals is computed to suggest a theoretical semivariogram model. This theoretical model is fit to produce starting values for the semivariogram parameters. Next, the mean and autocorrelation structure (for the suggested theoretical semivariogram model) are estimated simultaneously by (restricted) maximum likelihood. A simple likelihood ratio test can be employed to examine whether the MLEs produce a significantly better fit of the model than the semivariogram parameters derived initially from the OLS residuals. Finally, predictions and their standard errors are calculated for the primary attribute at locations where values of the secondary attribute are available. If mapping of the primary attribute is desired, one can first obtain a surface of the secondary attribute (e.g., the TN surface shown in Figure 9.49) and predict at these locations.

This approach is an obvious generalization of the universal kriging method where now the large-scale trend is not only modeled as functions of the spatial coordinates, but as functions of other, spatially heterogeneous variables. We illustrate the implementation for model [9.73] with The SAS® System. The procedure driving estimation of both the mean function and the autocorrelation structure as well as prediction at unobserved locations is `proc mixed`.

The OLS residuals and their empirical semivariogram are obtained with the statements

```
proc mixed data=CNRatio;
  model TC = TN / outp=OLSresid s;
run;
proc variogram data=OLSResid outvar=svar;
   compute lagdistance=13.5 maxlag=21 robust;
   coordinates xcoord=x ycoord=y;
   var resid;
run;
proc nlin data=svar nohalve noitprint;
  parameters sill=0.0009 Range=40.4 nugget=0.0009;
  expomodel = nugget+sill*(1-exp(-distance/range));
  model variog = expomodel;
  _weight_ = 0.5*count/(expomodel**2);
 run;
```

Output 9.16. (abridged)

The NLIN Procedure

Source	DF	Sum of Squares	Mean Square	F Value	Approx Pr > F
Regression	3	28877.0	9625.7	119.34	<.0001
Residual	19	28.2724	1.4880		
Uncorrected Total	22	28905.3			
Corrected Total	21	383.4			

Parameter	Estimate	Approx Std Error	Approximate 95% Confidence Limits	
sill	0.000864	0.000064	0.000730	0.000999
Range	72.3276	11.9674	47.2796	97.3755
nugget	0.000946	0.000074	0.000790	0.00110

The exponential semivariogram fits the empirical semivariogram well, Pseudo-$R^2 = 1 - 28.27/383.4 = 0.92$ (Output 9.16). We adopt it as the semivariogram model for $e(\mathbf{s}_i)$ in [9.73]. Next, we krige a surface of TN with `proc krige2d` and add the prediction locations to the original data set. In this combined data set (`fitthis` below) the response variable TC will have missing values for all prediction locations. This will prevent `proc mixed` from using the prediction observations in estimating the regression coefficients and the spatial dependency parameters. It will, however, produce predicted values for all observations in the data set that have complete regressor information.

```
data predgrid;
  do x = 0 to 500 by 25; do y = 0 to 300 by 10; TN=.; output; end; end;
run;
data obsdata; set CNRatio(keep=x y tn); run;
proc krige2d data=obsdata outest=krigeEst;
  coordinates xcoord=x ycoord=y;
  grid griddata=predgrid xcoord=x ycoord=y;
  predict var=TN radius=50 maxpoints=50 minpoints=30;
  model form=exponential range=43.063 scale=0.0001775;
run;
data KrigeEst; set KrigeEst;
  if stderr ne .;
  rename gxc=x gyc=y estimate=TN;
run;
data fitthis; set KrigeEst CNRatio; run;
```

The final step is to submit the data `fitthis` to `proc mixed` to fit the spatial regression model [9.73] by (restricted) maximum likelihood:

```
proc mixed data=fitthis noprofile;
  model TC = TN / ddfm=contain s outp=p;
  repeated / subject=intercept type=sp(exp)(x y) local;
  parms /* sill   */  0.000864
        /* range  */  72.3276
        /* nugget */  0.000946 ;
run;
```

The starting values for the autocorrelation parameters are chosen as the converged iterates in Output 9.16. Because the `parms` statement does not have a `/noiter` option `proc mixed` will estimate these parameters iteratively commencing at the starting values. The `outp=p` option of the model statement creates a data set containing the predicted values.

The `Dimensions` table of the output shows that the data are representing a single subject and that 531 of the 926 observations in the data set have not been used in estimation (Output 9.17). These are the observations at the prediction locations for which the response variable `TC` was set to missing. Minus twice the (residual) log likelihood evaluated to -1526.0 at the starting values and was subsequently improved upon during five iterations. At convergence `-2 Res Log Like` $= -1527.76$. The difference between the initial and converged `-2 Res Log Like` can be used to test whether the iterations significantly improved the model fit. The difference is not statistically significant ($p = 0.6253$, see `PARMS Model Likelihood Ratio Test`). The iterated REML estimates of sill, range, and nugget are shown in the `Covariance Parameter Estimates` table.

Output 9.17. (abridged)

```
                    The Mixed Procedure

                         Dimensions
               Covariance Parameters             3
               Columns in X                      2
               Columns in Z                      0
               Subjects                          1
               Max Obs Per Subject             395
               Observations Used               395
               Observations Not Used           531
               Total Observations              926

                      Parameter Search
   CovP1       CovP2       CovP3         Res Log Like   -2 Res Log Like
 0.000946     49.7380    0.000784            763.0047        -1526.0093
                   Convergence criteria met.
```

Output 9.17 (continued).

```
               Covariance Parameter Estimates
             Cov Parm      Subject        Estimate
             Variance      Intercept      0.001193
             SP(EXP)       Intercept       87.7099
             Residual                     0.000801

                       Fit Statistics
           -2 Res Log Likelihood              -1527.8
           AIC (smaller is better)            -1521.8
           AICC (smaller is better)           -1521.7
           BIC (smaller is better)            -1509.8

              PARMS Model Likelihood Ratio Test
                DF     Chi-Square       Pr > ChiSq
                 3           1.75           0.6253

                  Solution for Fixed Effects
                             Standard
Effect          Estimate        Error      DF    t Value     Pr > |t|
Intercept       -0.01542      0.02005     393      -0.77       0.4423
TN               11.1184       0.2066     393      53.83       <.0001
```

The estimates of the regression coefficients are $\widehat{\beta}_0 = -0.01542$ and $\widehat{\beta}_1 = 11.1184$, respectively (`Solution for Fixed Effects` Table). With every additional percent of total N the total C percentage increases by 11.11 units. It is interesting to compare these estimates to the ordinary least squares estimates in the model

$$TC_i = \alpha_0 + \alpha_1 TN_i + e_i, \; e_i \sim iid\left(0, \sigma^2\right),$$

which does not incorporate spatial autocorrelation (Output 9.18). The estimates are slightly different and their standard errors are very optimistic (too small). Furthermore, for a given value of TN, the prediction of TC is the same under the classical regression model, regardless of where the TN observation is located. In the spatial regression model with autocorrelated errors, the best linear unbiased predictions of $TC(\mathbf{s}_i)$ take the spatial correlation structure into account. At two sites with identical TN values the predicted values in the spatial model will differ depending on where the sites are located. Compare, for example the two observations in Output 9.19. Estimates of $\mathrm{E}[TC(\mathbf{s}_i)]$ would be calculated as

$$\widehat{\mathrm{E}}[TC(\mathbf{s}_i)] = \widehat{\beta}_0 + \widehat{\beta}_1 TN(\mathbf{s}_i) = -0.01542 + 11.1184*0.057623 = 0.62523$$

regardless of the location $\mathbf{s}_i$. The values computed by `proc mixed` are predictions of $TC(\mathbf{s}_i)$ and vary by location. If estimates of the mean are desired one can add the statement `outpm=pm` to the code above.

Output 9.18.

```
                  Solution for Fixed Effects

                             Standard
Effect          Estimate        Error      DF    t Value     Pr > |t|

Intercept       -0.03138      0.01323     393      -2.37       0.0181
TN               11.2213       0.1710     393      65.61       <.0001
```

Output 9.19.

```
                x          y            TN              Pred

              420         80      0.057623           0.61429
              420         20      0.057623           0.65172
```

The spatial predictions of the total carbon percentage are shown in the left panel of Figure 9.50 and estimates of the mean in the right panel. Because the underlying TN surface is spatially variable (right panel of Figure 9.49) so are the estimates of $\mathrm{E}[TC(\mathbf{s}_i)]$ which follow the pattern of TN very closely. The predictions of $TC(\mathbf{s}_i)$ follow the same pattern as $\mathrm{E}[TC(\mathbf{s}_i)]$, but exhibit more variability. The left-hand panel of Figure 9.50 is less smooth.

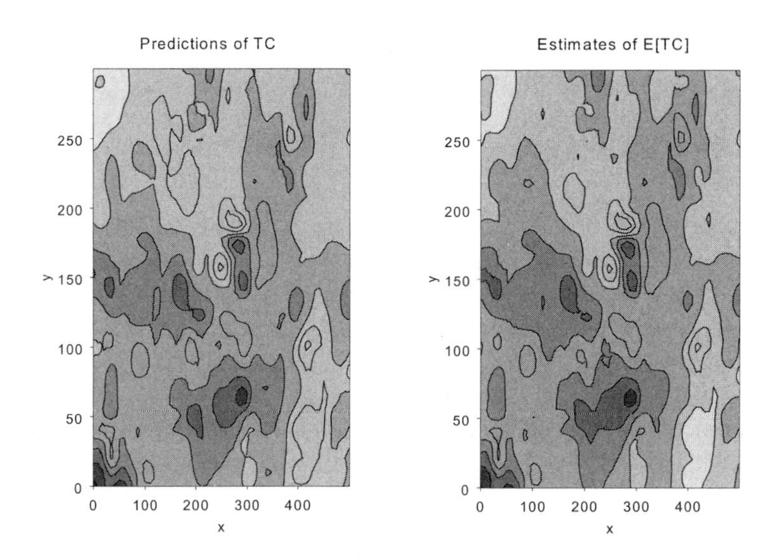

Figure 9.50. Spatial regression predictions and estimates of TC (%).

9.8.6 Spatial Generalized Linear Models — Spatial Trends in the Hessian Fly Experiment

In §6.7.2 we analyzed data from a variety field trial in which sixteen varieties of wheat were compared with respect to their infestation with the Hessian fly. The entries were arranged in a randomized complete block design with four blocks and the outcome of interest was the sample proportion $Y_{ij} = Z_{ij}/n_{ij}$, where Z_{ij} is the number of plants infested with the Hessian fly for entry (= variety) i in block j and n_{ij} is the total number of plants on the experimental unit. Because the data are proportions out of a given total, the Binomial distribution is a natural model and a generalized linear model for Binomial data with a logit link was fit. It was noticed, however, that the data appear overdispersed relative to the Binomial model. One possible reason for overdispersion is positive autocorrelation between the experimental units. If an experimental unit shows a high degree of infestation, it is likely that a nearby unit also is highly infested. This autocorrelation can be linked to spatially varying environmental conditions that inhibit or enhance infestation beyond the varietal differences. One approach to account for overdispersion is to add an additional scale parameter to the variance of the outcome. The Binomial law states that $\text{Var}[Y_{ij}] = \pi_{ij}(1 - \pi_{ij})/n_{ij}$, where π_{ij} is the probability that a plant of variety i in block j is infested. In an overdispersed model one can put $\text{Var}[Y_{ij}] = \phi\pi_{ij}(1 - \pi_{ij})/n_{ij}$ and allow the parameter ϕ to adjust the dispersion. By adding the scale parameter ϕ this model is no longer a Binomial one, but statistical inference can proceed nevertheless along similar lines using quasi-likelihood ideas and estimating ϕ by the method of moments. This was the solution chosen for these data in §6.7.2. If the overdispersion arises from a spatially varying characteristic it is more appropriate to include the spatial variability directly in the model rather than to rely on one multiplicative scale parameter to patch things up. The scale parameter ϕ adjusts only the standard errors of the mean parameter estimates, not the estimates themselves. The estimated probability of infection for a particular variety does not depend on whether ϕ is in the model or not. The models that account for the spatial dependencies among experimental units will adjust the estimates of treatment effects as well as their standard errors.

To motivate statistical models that incorporate spatial autocorrelation and non-normal responses, first recall the case of independent observations. A generalized linear model can be written as

$$Y = g^{-1}(\eta) + e,$$

where $g^{-1}(\eta) = \mu$ is the inverse link function, η is the linear predictor of the form $\mathbf{x}'\boldsymbol{\beta}$ and e is a random error with mean 0 and variance $\text{Var}[e] = h(\mu)\psi$. In the randomized block Hessian fly experiment, if one assumes that Y_{ij} is a Binomial proportion, one obtains $\eta_{ij} = \alpha + \rho_j + \tau_i$, $\text{Var}[e_{ij}] = h(\mu) = g^{-1}(\eta_{ij})(1 - g^{-1}(\eta_{ij}))/n_{ij}$ and $\psi = 1$. Choosing $g()$ as the logit transform is common for such data. In vector/matrix notation the model for the complete data is written as

$$\mathbf{Y} = g^{-1}(\boldsymbol{\eta}) + \mathbf{e}, \quad \mathbf{e} \sim (\mathbf{0}, \text{Diag}\{h(\mu)\}) = (\mathbf{0}, \mathbf{H}(\boldsymbol{\mu})).$$

A spatially varying process that induces autocorrelations can be accommodated in two ways. The **marginal** formulation replaces the error vector **e** with the vector **d** such that

$$\text{Var}[\mathbf{d}] = \mathbf{H}(\boldsymbol{\mu})^{1/2}\mathbf{R}\mathbf{H}(\boldsymbol{\mu})^{1/2}.$$

The matrix **R** is a spatial correlation matrix and the diagonal matrices $\mathbf{H}(\boldsymbol{\mu})$ adjust the correlations to yield the correct variances and covariances. The matrix **R** typically corresponds to the correlation model derived from one of the basic isotropic semivariogram models (§9.2.2). For example, if the spatial dependency between experimental units can be described by an exponential semivariogram, elements of **R** are calculated as $\exp\{-3||\mathbf{s}_k - \mathbf{s}_l||/\alpha\}$, where $\mathbf{s}_k$ and $\mathbf{s}_l$ are the spatial coordinates representing two units. The marginal formulation was chosen by Gotway and Stroup (1997) in modeling the Hessian fly data.

The **conditional** formulation of a spatial generalized linear model assumes that conditionally on the realization of the spatial process the observations are uncorrelated. This formulation is akin to the generalized linear mixed models of §8. In vector notation we can put

$$\mathbf{Y} = g^{-1}(\boldsymbol{\eta} + \mathbf{U}(\mathbf{s})) + \mathbf{e}, \ \mathbf{e} \sim (\mathbf{0}, \mathbf{H}(\boldsymbol{\mu})). \qquad [9.74]$$

$\mathbf{U}(\mathbf{s})$ is a second-order stationary mean zero random field with covariogram $\text{Cov}[U(\mathbf{s}), U(\mathbf{s}+\mathbf{h})] = C(\mathbf{h})$, and semivariogram ½$\text{Var}[U(\mathbf{s}) - U(\mathbf{s}+\mathbf{h})] = \gamma(\mathbf{h})$.

To make estimation practical one chooses a correlation matrix $\mathbf{R}$ in the marginal model or a semivariogram (covariogram) for $U(\mathbf{s})$ in the conditional model from the models in §9.2.2 and replaces $\mathbf{R}$ with $\mathbf{R}(\mathbf{h}, \boldsymbol{\theta})$ and $\gamma(\mathbf{h})$ with $\gamma(\mathbf{h}, \boldsymbol{\theta})$. The estimation process is then doubly iterative. Given an estimate $\widehat{\boldsymbol{\eta}}$ of $\boldsymbol{\eta}$ we estimate $\boldsymbol{\theta}$ and given $\widehat{\boldsymbol{\theta}}$ we update the estimate of $\boldsymbol{\eta}$. Each step is itself iterative and the entire procedure is continued until some overall convergence criterion is met (e.g., track the largest relative change in elements of $\widehat{\boldsymbol{\theta}}$ between iterations). The interested reader can find details of this process in §A9.9.7. It is important to point out, however, that some of the approaches put forth in the literature require the repeated inversion of large matrices which is computationally expensive. The pseudo-likelihood approach of Wolfinger and O'Connell (1993), for example, requires the inversion of an $(n \times n)$ matrix at every iteration that updates $\boldsymbol{\theta}$. Originally designed for clustered data, this is not a big issue there, since $\text{Var}[\mathbf{Y}]$ is block-diagonal and can be inverted in blocks. For spatial data, $\text{Var}[\mathbf{Y}]$ is not block-diagonal and few computational shortcuts are available. Zimmerman (1989) described inversion procedures that exploit the structure of $\text{Var}[\mathbf{Y}]$ when data are collected on a rectangular or parallelogram lattice. For the models considered here, these methods do not apply.

To overcome the possible numerical problems, we consider the following approach. Since the parameter vector $\boldsymbol{\eta}$ is of main interest and the covariance parameter vector $\boldsymbol{\theta}$ is a vector of nuisance parameters, our chief concern is to estimate $\boldsymbol{\theta}$ consistently. A consistent estimate of $\boldsymbol{\theta}$ can be calculated quickly by applying the composite likelihood principle (see §9.2.4). The procedure is as follows. Start by fitting a generalized linear model (§6) for independent data (i.e., assume $U(\mathbf{s}) \equiv 0$). Transform residuals $r(\mathbf{s}_i)$ from the fit in such a way that their mean is (approximately) zero and $\text{Var}[r(\mathbf{s}_i) - r(\mathbf{s}_j)] = 2\gamma(\mathbf{s}_i - \mathbf{s}_j, \boldsymbol{\theta})$. Apply the composite likelihood principle to the transformed residuals to obtain an estimate of $\boldsymbol{\theta}$. Construct an estimate of the marginal variance-covariance matrix $\text{Var}[\mathbf{Y}]$ with $\widehat{\boldsymbol{\theta}}$ and re-estimate the linear predictor $\boldsymbol{\eta}$. Formulate new residuals and continue the process until some convergence criterion is met. For example, continue until the largest relative change in one of the model parameters is less than some critical number ϵ (see §A9.9.7 for details).

The estimation process has been coded in a SAS® macro contained on the CD-ROM (macro `%GlmSpat()` in file `\SASMacros\GLMCompLike.sas`). We demonstrate its usage here for the Hessian fly data. Before fitting the spatial generalized linear model with composite likelihood estimation of the spatial dependence parameters, we compare the estimates of block and treatment effects obtained from a regular generalized linear model with Binomial errors and an overdispersed Binomial model. These models have linear predictor

$$\eta_{ij} = \mu + \tau_i + \rho_j,$$

where τ_i, $i = 1, \cdots, 16$, denotes the entries, and ρ_j, $j = 1, \cdots, 4$, the block effects. The link function was chosen as the logit, and consequently, $\log\{\pi_{ij}/(1 - \pi_{ij})\} = \eta_{ij}$.

```
proc genmod data=HessianFly;
  class block entry;
  model z/n = block entry / link=logit dist=binomial type3;
  ods output ParameterEstimates=GLMEst;
  ods output type3=GLMType3;
run;
```

```
proc genmod data=HessianFly;
  class block entry;
  model z/n = block entry / link=logit dist=binomial type3 dscale;
  ods output ParameterEstimates=ODEst;
  ods output Type3=ODType3;
run;
```

The two `proc genmod` calls fit the regular GLM and the overdispersed GLM (by adding the `dscale` option to the `model` statement) and save estimates as well as the tests for treatment effects in data sets. After processing the output data sets (see code on CD-ROM), we obtain Output 9.20. The estimate of the intercept, block and treatment effects are the same in both models. The standard errors of the overdispersed model are uniformly 1.66 times larger than the standard errors in the regular GLM. This is also reflected in the test of entry effects. The Chi-square statistic in the overdispersed model is $1.66^2 = 2.75$ times *smaller* than the corresponding statistic in the GLM. Not accounting for overdispersion overstates the precision of parameter estimates. Test statistics are too large and p-values too small.

Output 9.20.

Parameter	Level	GLM Estimate	GLM StdErr	Overd. Estimate	Overd. StdErr
Intercept		-1.2936	0.3908	-1.2936	0.6487
block	1	-0.0578	0.2332	-0.0578	0.3870
block	2	-0.1838	0.2303	-0.1838	0.3822
block	3	-0.4420	0.2328	-0.4420	0.3863
entry	1	2.9509	0.5397	2.9509	0.8958
entry	2	2.8098	0.5158	2.8098	0.8561
entry	3	2.4608	0.4956	2.4608	0.8225
entry	4	1.5404	0.4564	1.5404	0.7575
entry	5	2.7784	0.5293	2.7784	0.8785
entry	6	2.0403	0.4889	2.0403	0.8115
entry	7	2.3253	0.4966	2.3253	0.8242
entry	8	1.3006	0.4754	1.3006	0.7890
entry	9	1.5605	0.4569	1.5605	0.7582
entry	10	2.3058	0.5203	2.3058	0.8635
entry	11	1.4957	0.4710	1.4957	0.7818
entry	12	1.5068	0.4767	1.5068	0.7911
entry	13	-0.6296	0.6488	-0.6296	1.0768
entry	14	0.4460	0.5126	0.4460	0.8507
entry	15	0.8342	0.4698	0.8342	0.7798

Source	DF	GLM ChiSq	GLM Pvalue	Overd. Chisq	Overd. Pvalue
block	3	4.27	0.2337	1.55	0.6707
entry	15	132.62	<.0001	48.15	<.0001

The previous analysis maintains that observations from different experimental units are independent; it simply allows the variance of the observations to exceed the variability dictated by the Binomial law. If the data are overdispersed relative to the Binomial model because of positive spatial autocorrelation among the experimental units, the spatial process can be modeled directly. The following code analyzes the Hessian fly experiment with model [9.74], where $U(\mathbf{s})$ has exponential semivariogram without nugget effect (options `Covmod=E`, `nugget=0`). The `sx=` and `sy=` parameters denote the variables of the data set containing longitude and latitude information, the `margin=` parameter specifies the marginal variance function $h(\mu)$. Starting values for the sill and range are set at 1.5 and 5, respectively, and the range parameter is constrained to be at least 2. Setting a minimum value for the range is recommended if numerical problems prevent the macro from converging. This value should not be set, however, without evidence that the range is definitely going to exceed this value.

```
%include 'CDRomDriveLetter:\SASMacros\GLMCompLike.sas';
%glmspat(data=HessianFly,
        procopt=order=data,
        stmts=%str(class block entry;
                   model z/n = block entry / s;
                   lsmeans entry /diff;
                   ),
           sx   = sx,       sy   = sy,
           link=logit,      margin=binomial,
           minrange=2,      CovMod=E,
           nugget=0,        sillstart=1.5,    rangestart=5,
           title=Hessian Fly Data - GLM-CL,
           options=);
```

The `stmts=%str(    )` block of the macro call assembles statements akin to `proc mixed` syntax. The `s` option of the `model` statement requests a printout of the fixed effects estimates (solutions). For predicted values add the `p` option to the `model` statement. The algorithm converged after fourteen iterations, that is, the parameters of the exponential semivariogram were updated fourteen times following an update of the block and entry effects.

The sill and range parameter of the exponential semivariogram (covariogram) are estimated as 0.375 and $9.694\ m$, respectively. Notice that these estimates differ from those of Gotway and Stroup (1997) who estimated the range at $11.6\ m$ and the sill at 3.38. Their model uses a marginal formulation, whereas the model fitted here is a conditional one that incorporates a latent random field inside the link function. Furthermore, in their marginal formulation Gotway and Stroup (1997) settled on a spherical semivariogram. For the conditional model we found an exponential model to fit the semivariogram of the transformed residuals better. Finally, their method does not iterate between updates of the fixed effects and updates of the semivariogram parameters.

The estimates of the fixed effects in the spatial model (Output 9.21) differ from the corresponding estimates in the generalized linear model (compare to Output 9.20), as expected. The overall impact of different estimates is difficult to judge since the treatment effects, which are averaged across the blocks, are of interes. If $\widehat{\eta}_{ij}$ is the estimated linear predictor for entry i in block j, we need to determine $\widehat{\eta}_{i.}$, the effect of the ith entry after removing and averaging over the block effect. The estimates $\widehat{\eta}_{i.}$ are shown in the `Least Squares Means` table (Output 9.22). The least square mean for entry 1, for example, is calculated from the parameter estimates in Output 9.21 as

$$\widehat{\eta}_{1.} = -1.4026 + \frac{1}{4}(-0.1444 - 0.1269 - 0.4431 + 0) + 3.3228 = 1.7416.$$

The probability that a plant of entry 1 is infected with the Hessian Fly is then obtained by applying the inverse link function,

$$\widehat{\pi}_1 = \frac{1}{1 + \exp\{-1.7416\}} = 0.85.$$

The table titled `Differences of Least Square Means` in Output 9.22 can be used to assess differences in the infestation probabilities among pairs of entries. Only part of the lengthy table of least squares mean differences are shown.

Output 9.21.

```
                     Hessian Fly Data - GLM-CL
                 Class Level Information (WORK._CLASS)

        Class     Levels     Values
        block          4     1 2 3 4
        entry         16     1 2 3 4 5 6 7 8 9 10 11 12 13 14 15 16


              Covariance Parameter Estimates (WORK._SOLR)
                          Parameter    Estimate
                             Sill       0.37516
                            Range       9.69449

           Parameter Estimates and Standard Errors (WORK._SOLF)

Effect                              Parameter      Std.      Wald        Pr >
name          block       entry     Estimate      Error     Chi-Sq.     Chi-Sq.
Intercept                            -1.4026     0.5568      6.3453     0.0118
block            1                   -0.1444     0.4283      0.1137     0.7360
block            2                   -0.1269     0.4085      0.0965     0.7560
block            3                   -0.4431     0.4093      1.1721     0.2790
entry                        1        3.3228     0.7004     22.5093     <.0001
entry                        2        3.1181     0.6759     21.2838     <.0001
entry                        3        2.6294     0.6472     16.5047     <.0001
entry                        4        1.8789     0.6083      9.5408     0.0020
entry                        5        2.8513     0.6545     18.9795     <.0001
entry                        6        2.1405     0.6345     11.3826     0.0007
entry                        7        2.4266     0.6428     14.2524     0.0002
entry                        8        1.4999     0.6283      5.6988     0.0170
entry                        9        1.7197     0.6179      7.7469     0.0054
entry                       10        2.3922     0.6601     13.1345     0.0003
entry                       11        1.4721     0.6260      5.5298     0.0187
entry                       12        1.7885     0.6265      8.1487     0.0043
entry                       13       -0.5651     0.7778      0.5278     0.4676
entry                       14        0.6812     0.6494      1.1003     0.2942
entry                       15        0.8458     0.6233      1.8418     0.1747

                    Tests of Fixed Effects (WORK._TESTS)
                               Chi-     Deg. of         Pr >
                  Effect      Square     freed.      Chi-Square
                  BLOCK       1.2556        3          0.73971
                  ENTRY      69.2083       15          0.00000
```

Output 9.22.

```
                        Least Squares Means (WORK._LSM)

                                        Std.Err                     Std.Err of
                                          of        Predicted        Predicted
    Effect     Level       LS Mean      LSMean         Mean             Mean

    ENTRY        1         1.74153      0.52693      0.85088          0.06686
    ENTRY        2         1.53690      0.49153      0.82301          0.07160
    ENTRY        3         1.04815      0.45201      0.74042          0.08688
    ENTRY        4         0.29763      0.40583      0.57386          0.09924
    ENTRY        5         1.27001      0.48065      0.78075          0.08228
    ENTRY        6         0.55929      0.44482      0.63629          0.10294
    ENTRY        7         0.84537      0.44968      0.69960          0.09451
    ENTRY        8        -0.08132      0.42416      0.47968          0.10586
    ENTRY        9         0.13847      0.40264      0.53456          0.10018
    ENTRY       10         0.81100      0.47430      0.69232          0.10103
    ENTRY       11        -0.10911      0.42283      0.47275          0.10539
```

Output 9.22 (continued).

ENTRY	12	0.20723	0.42782	0.55162	0.10581
ENTRY	13	-2.14631	0.63084	0.10468	0.05912
ENTRY	14	-0.90006	0.45888	0.28904	0.09430
ENTRY	15	-0.73541	0.42675	0.32401	0.09347
ENTRY	16	-1.58124	0.49182	0.17062	0.06960

Differences of Least Squares Means (WORK._DIFFS)

Effect	First Level	Sec. Level	LSMeans Difference	SE of Difference	Chi-Square	Deg. of freed.	Pr > Chi-Squ.
ENTRY	1	2	0.2046	0.68852	0.0883	1	0.76631
ENTRY	1	3	0.6934	0.66650	1.0823	1	0.29818
ENTRY	1	4	1.4439	0.63530	5.1656	1	0.02304
ENTRY	1	5	0.4715	0.68719	0.4708	1	0.49262
ENTRY	1	6	1.1822	0.66175	3.1917	1	0.07401
ENTRY	1	7	0.8962	0.66734	1.8034	1	0.17931
ENTRY	1	8	1.8229	0.65175	7.8224	1	0.00516
ENTRY	1	9	1.6031	0.63310	6.4114	1	0.01134
ENTRY	1	10	0.9305	0.68361	1.8528	1	0.17345
ENTRY	1	11	1.8506	0.65423	8.0018	1	0.00467
ENTRY	1	12	1.5343	0.64620	5.6374	1	0.01758
ENTRY	1	13	3.8878	0.80369	23.4015	1	0.00000
ENTRY	1	14	2.6416	0.68019	15.0825	1	0.00010
ENTRY	1	15	2.4769	0.65568	14.2706	1	0.00016
ENTRY	1	16	3.3228	0.70036	22.5093	1	0.00000
ENTRY	2	3	0.4888	0.64143	0.5806	1	0.44608
ENTRY	2	4	1.2393	0.59844	4.2883	1	0.03838
ENTRY	2	5	0.2669	0.65625	0.1654	1	0.68424
ENTRY	2	6	0.9776	0.63584	2.3639	1	0.12417

and so forth ...

To compare the results of the GLM and spatial analysis in terms of infection probabilities, predicted probabilities of infestation with the Hessian fly for the 16 entries in the study were graphed in Figure 9.51. Four methods were employed to calculate these probabilities. The upper left-hand panel shows the probabilities calculated from the entry least squares means in the overdispersed GLM analysis. The predictions in the other three panels are obtained through different techniques of accounting for spatial correlations among experimental units. *Gotway and Stroup* refers to the noniterative technique of Gotway and Stroup (1997), *Pseudo-Likelihood* to the techniques by Wolfinger and O'Connell (1993) that are coded in the `%glimmix()` macro (www.sas.com). It is noteworthy that the predicted probabilities are very similar for the spatial analyses and that the (overdispersed) GLM results in predictions quite similar to the spatial analyses. The standard errors of the predicted probabilities are very homogeneous across entries in the GLM analysis. The dots are of similar size. The spatial analyses show much greater heterogeneity in the standard errors for the predicted infestation probabilities. There is little difference in the standard errors among the three spatial analyses, however.

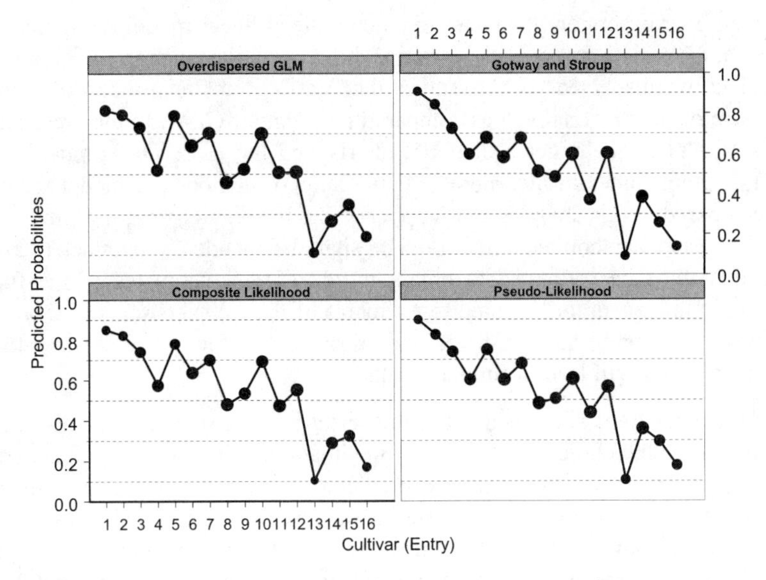

Figure 9.51. Predicted probability of infection by entry for four different methods of incorporating overdispersion or spatial correlations. The size of the dots is proportional to the standard error of the predicted probability.

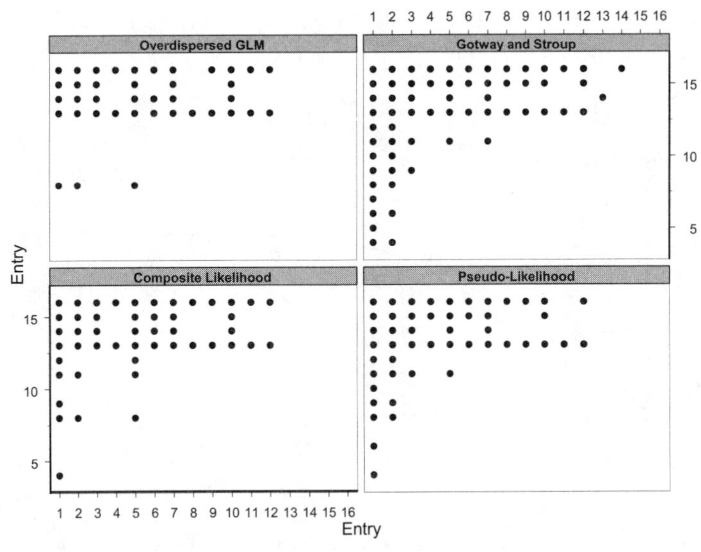

Figure 9.52. Results of pairwise comparisons among entries. A dot indicates a significant difference in infestation probabilities among a pair of entries (at the 5% level).

Differences in predicted probabilities and their standard errors are reflected in multiple comparisons of entries (Figure 9.52). The three spatial analyses produce very similar results. The overdispersed GLM yields fewer significant differences in this application which is due to the large and homogeneous standard errors of the predicted probabilities (Figure 9.51)

Adding an overdispersion parameter to a generalized linear model is simple. It is not the appropriate course of action if overdispersion is due to positive autocorrelation among the observations. The overdispersed GLM assumes that Var[**Y**] remains a diagonal matrix and increases the size of the diagonal values compared to a regular GLM. In the presence of spatial correlations Var[**Y**] is no longer a diagonal matrix and the covariances must be taken into account. Three approaches to incorporate spatial autocorrelation in a model for non-normal data were compared in this analysis and it appears that the results are quite similar. In our experience, this is a rather common finding. The spatial dependency parameters are nuisance parameters that must be estimated to obtain more efficient estimates of the fixed effects parameters and more accurate and precise estimates of their dispersion. The fixed effects are the quantities of primary interest here, and most reasonable methods of estimating the covariance parameters will lead to similar results.

An issue we have not addressed yet is the selection of a semivariogram or covariogram model. Fortunately, this choice is less critical in situations where the mean parameters are of primary interest than in the case of spatial prediction. However, one should still choose the model carefully. We base the initial selection of a semivariogram model on the Pearson residuals from a standard generalized linear model fit. The empirical semivariogram of these residuals will suggest a semivariogram model and starting values for its parameters. Then, the composite likelihood fit is carried out. At convergence we obtain the transformed residuals and calculate their empirical semivariogram (Figure 9.53). The composite likelihood estimate of the semivariogram should fit the empirical semivariogram reasonably well. We do not expect the composite likelihood estimate to fit "too" well. The composite likelihood estimates are obtained by fitting a semivariogram model to pseudo-data $(Z(\mathbf{s}_i) - Z(\mathbf{s}_j))^2$, not by fitting a model to the empirical semivariogram.

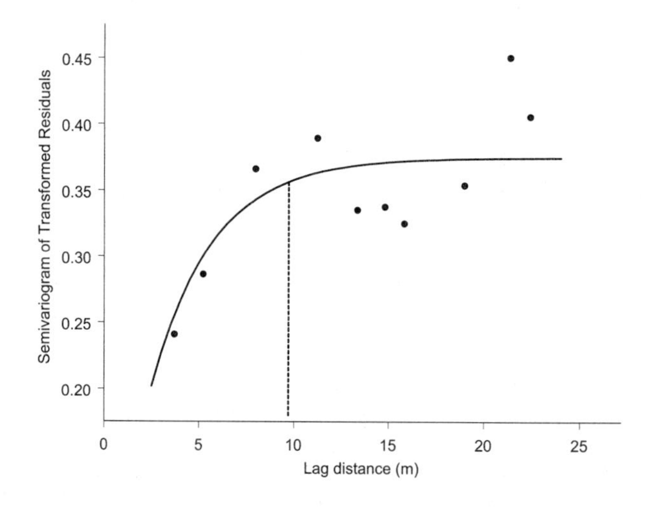

Figure 9.53. Empirical semivariogram of transformed residuals at convergence for the Hessian fly data. Solid line represents composite likelihood estimate of exponential semivariogram.

9.8.7 Simultaneous Spatial Autoregression — Modeling Wiebe's Wheat Yield Data

Correlations among the yields of neighboring field plots can have damaging effect on the experimental error variance of field experiments. Blocking was advocated by R.A. Fisher as a means to eliminate the effects of spatial variability and randomization to neutralize those effects unaccounted for by the blocking scheme. If variability due to blocks does not eliminate the spatial heterogeneity because the shape and size of blocks does not coincide with the pattern of spatial variability, the experimental error will be increased. In the Alliance, Nebraska, wheat variety trial (see §9.5) this was the reason for not finding any significant differences among the 56 varieties. Because of the recognized effect of unaccounted spatial variability among experimental units on the analysis, uniformity trials where a single treatment is applied throughout an experimental area, have received much attention in the past. Limitations in space, time and the economics of operating agricultural experiment stations have almost eliminated uniformity trials as a vehicle to study experimental conditions, although there is much to be learned from past uniformity trials.

Wiebe (1935) discusses the yields of wheat on 1,500 nursery plots grown in the summer of 1927 on the west end of series 100 on the Aberdeen Substation, Aberdeen, Idaho. Each plot consisted of fifteen-foot rows of wheat spaced twelve inches apart. The plots are arranged in a regular rectangular lattice with 125 rows and 12 columns. The data are too voluminous to reproduce here but are contained on the companion CD-ROM and printed in Table 6.2 of Andrews and Herzberg (1985). Griffith and Layne (1999) conduct an analysis of these data with spatial autoregressive lattice models. These authors recommend analyzing a transformation of the grain yields rather than the raw data to achieve greater symmetry in the data and to stabilize the variance. Following their example we analyze the square root of the yields. In our examination of Wiebe's wheat yield data we contrast three different methods to examine, measure, and account for spatial dependencies among field plots. Figure 9.54 shows the sample medians of the square root yields by row and column (series). There are obvious trends in both directions which could be modeled as large-scale trends. A study of the spatial dependency of grain yields should remove these trends and we consider ordinary least squares, median polishing, and a simultaneous spatial autoregressive model to accomplish this task. The trends in the column and row medians appear quadratic or cubic but we choose to remove only linear trends in rows and columns. This appears like a foolish decision at first. What we are interested in, however, is which of the three methods can accommodate the unaccounted large-scale effects and possible spatial autocorrelations among plot yields best in this setting. Recall that a lattice model with autocorrelated spatial errors can be thought of as supplanting the missing information on important covariates with the neighborhood connectivity of the lattice sites. Our expectation is thus that the SSAR model will provide a better fit to the data in light of an ill-specified large-scale trend model compared to the OLS model.

The subsequent analyses were performed with the S+SpatialStats® module. To convert the data from Microsoft® Excel format (CD-ROM) into an S-PLUS® object named `wwy` and to calculate the square root of the wheat yields the statements

```
import.data(FileName = "D:\\...Path to File...\\WiebeWheatYield.xls",
       FileType = "Excel",     TargetStartCol = "1",
       DataFrame = "wwy",      StartCol = "1",
       EndCol = "END",         StartRow = "1",   EndRow = "END")
```

```
wwy$ryield <- sqrt(wwy$yield)
```

are executed. A calculation of Moran's I statistic with the statements

```
wwy.snhbr <- neighbor.grid(nrow=125,ncol=12,neighbor.type="first.order")
spatial.cor(wwy$ryield,neighbor=wwy.snhbr,statistic="moran",
            sampling="free",npermutes=0)
```

shows significant spatial autocorrelation (Output 9.23) among plot yields which may be caused by a nonstationary mean.

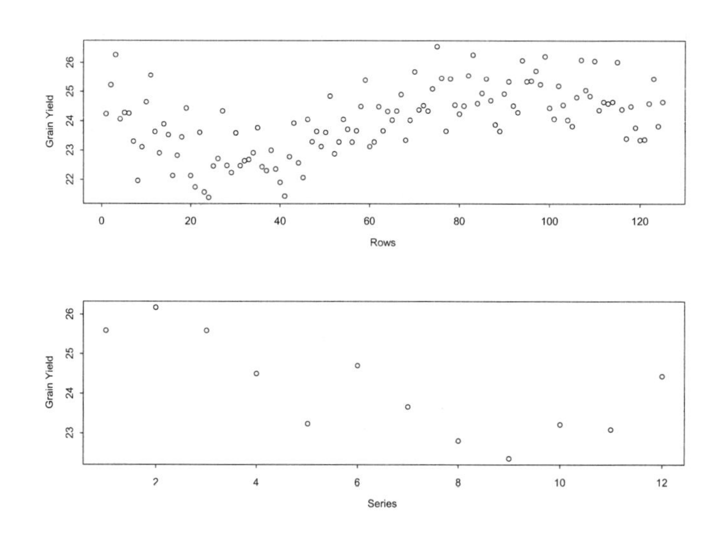

Figure 9.54. Row and column (series) sample medians for Wiebe's wheat yield data.

Output 9.23.

```
        Spatial Correlation Estimate

Statistic = "moran" Sampling = "free"

Correlation =  0.2311
Variance    =  3.484e-4
Std. Error  =  0.01866

Normal statistic =  12.42
Normal p-value (2-sided) =  2.055e-35

Null Hypothesis:  No spatial autocorrelation
```

Before the SSAR model can be fit the neighborhood structure must be defined. We choose a rook definition and standardize the weights to sum to one to mirror the SSAR analysis in Griffith and Layne (1999):

```
wwy.snhbr <- neighbor.grid(nrow=12,ncol=125,neighbor.type="first.order")
n <- wwy.snhbr[length(wwy.snhbr[,1]),1]
for (i in 1:n) {
  wwy.snhbr$weights[wwy.snhbr$row.id==i] <- 1/sum(wwy.snhbr$row.id == i)
```

```
}
wwy.SAR <- slm(ryield ~ row + series ,data=wwy,cov.family=SAR,
              spatial.arglist=list(neighbor=wwy.snhbr),start=0.3)
summary(wwy.SAR)
lrt(wwy.SAR,parameters=c(0))
```

The abbreviated output shows a large estimate of the spatial interaction parameter ($\widehat{\rho} = 0.8469$, Output 9.24) and the likelihood ratio test for $H_0\colon \rho = 0$ is soundly rejected ($p = 0$). There is significant spatial autocorrelation in these data beyond row and column effects.

Output 9.24.

```
Coefficients:
                 Value Std. Error  t value Pr(>|t|)
(Intercept)  24.1736    0.4704    51.3909   0.0000
        row   0.0129    0.0050     2.5732   0.0102
     series  -0.1243    0.0456    -2.7261   0.0065

Residual standard error: 1.0779 on 1496 degrees of freedom

rho =  0.8469

Likelihood Ratio Test

Chisquare statistic = 1182.836, df =1, p.value = 0
```

Note: The estimates for intercept, row and column effects and their standard errors differ from those in Griffith and Layne (1999). These authors standardize the square root yield to have sample mean 0 and sample variance 1 and standardize the row and series effects to have mean 0.

The ordinary least squares analysis assuming independence of the plot yields with the statements

```
wwy.OLS <- lm(ryield ~ row + series, data=wwy)
summary(wwy.OLS)
```

yields parameter estimates that are not too different from the SSAR estimate (Output 9.25) but their standard errors are too optimistic.

Output 9.25.

```
Coefficients:
                 Value Std. Error   t value  Pr(>|t|)
(Intercept)   24.7626    0.1292   191.6848    0.0000
        row    0.0138    0.0013    10.5847    0.0000
     series   -0.2264    0.0136   -16.6716    0.0000

Residual standard error: 1.816 on 1497 degrees of freedom
Multiple R-Squared: 0.2067
F-statistic: 195 on 2 and 1497 degrees of freedom, the p-value is 0
```

Finally, median polishing in S-PLUS® is accomplished with the statements

```
wwy.mp <- twoway(ryield~row+series,data=wwy)
wwy.mp$signal <- wwy$ryield - wwy.mp$residual
```

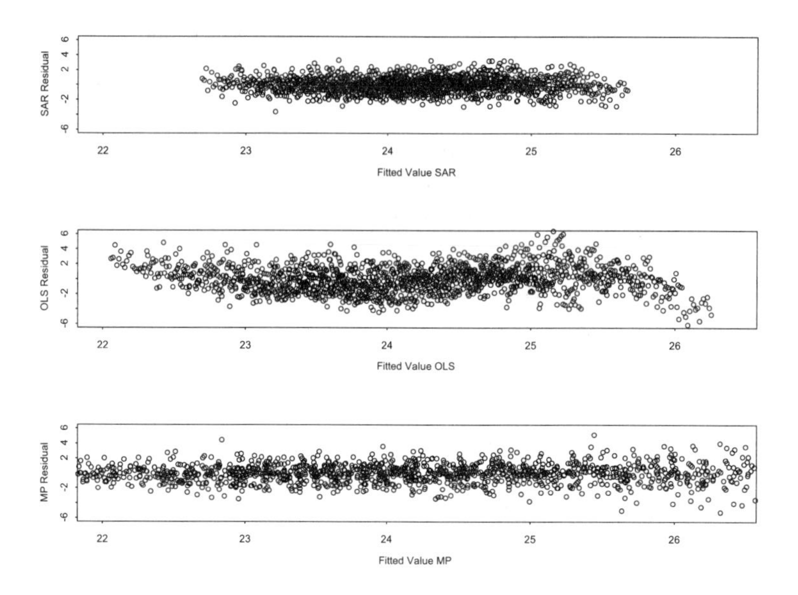

Figure 9.55. Residuals in SSAR model, OLS, and median polished residuals for Wiebe's wheat yield data. Only row and column trends were removed as large-scale trends.

The quality of the fit of the three models is assessed by plotting residuals against the predicted (fitted) yields (Figure 9.55). The spread of the fitted values indicates the variability in the predictions under the respective models. The SSAR model yields the least dispersed fitted values followed by the OLS fit and the median polishing. The OLS residuals exhibit a definite trend which shows the incomplete removal of the large-scale trend. No such trend is apparent in the SSAR residuals. The spatial neighborhood structure has supplanted the missing quadratic and cubic trends in the large-scale model well. Maybe surprisingly, median polishing performs admirably in removing the trend in the data. The residuals exhibit almost no trend. Compared to the SSAR fit the median polished residuals are considerably more dispersed, however. If one compares the sample variance of the various residuals, the incomplete trend removal in the OLS fit and the superior quality of the SSAR fit are evident: $s^2_{OLS} = 3.29$, $s^2_{MP} = 1.84$, $s^2_{SSAR} = 1.16$.

How well the methods accounted for the spatial variability in the plot yields can be studied by calculating the empirical semivariograms of the respective residuals. If spatial variability – both large-scale and small-scale – is accounted for, the empirical semivariogram should resemble a nugget-only model. An assumption of second-order stationarity can be made for the three residual semivariograms but not for the raw data (Figure 9.56). The empirical semivariogram of the SAR residuals shows the small variability (low sill) of these residuals and the complete removal of spatial autocorrelation. Residual spatial dependency remains in the median polished and OLS residuals. The sills of the residual semivariograms agree well with the sample variances. The lattice model clearly outperforms the other two methods of trend removal. It is left to the reader to examine the relative performance of the three approaches if not only linear row and column trends are removed but quadratic or cubic trends.

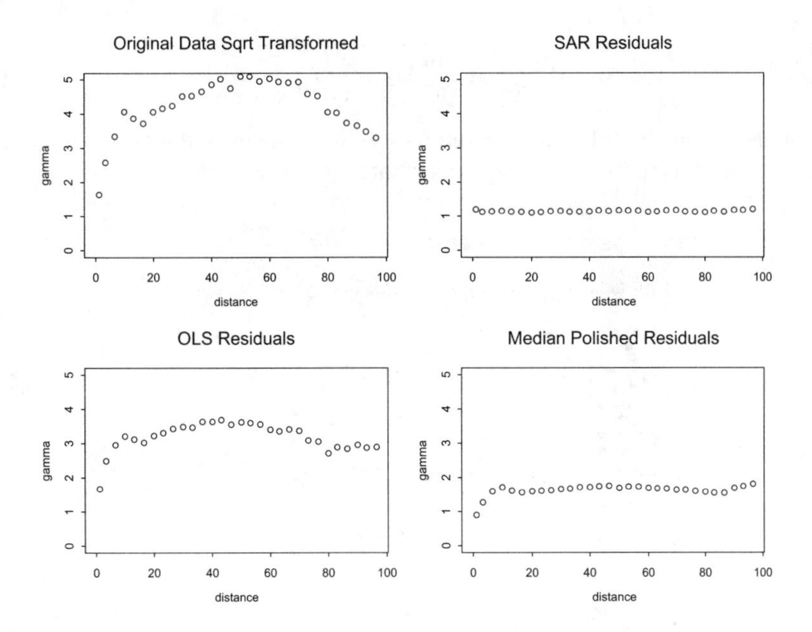

Figure 9.56. Empirical semivariograms of raw data, SSAR, OLS, and median polished residuals for Wiebe's wheat yield data. All semivariograms are scaled identically to highlight the differences in the sill.

9.8.8 Point Patterns — First- and Second-Order Properties of a Mapped Pattern

In this final application we demonstrate the important steps in analyzing a mapped spatial point pattern. We examine a pattern's first- and second-order properties through estimation of the intensity and K-function analysis. The hypothesis of complete spatial randomness is tested by means of Monte Carlo tests based on nearest-neighbor distances. The analysis is carried out with functions of the S+SpatialStats® module and S-PLUS® functions developed by the authors. Nearest-neighbor analyses Monte Carlo tests can also be performed in SAS® with the macro `%Ghatenv()` contained in file `\SASMacros\NearestNeighbor.sas`. The point pattern under consideration is shown in Figure 9.57. It contains $n = 180$ events located on a rectangle with boundary $(0, 400) \times (0, 200)$. This is a simulated point pattern but we shall not reveal the point pattern model that generated it. Rather, we ask the reader to consider Figure 9.57 and query his/her intuition whether the process that produced Figure 9.57 is completely random, clustered, or regular.

If the observed pattern is the realization of a CSR process, then the number of events in nonoverlapping intervals are independent and events are furthermore uniformly distributed. The question of uniformity can be answered by estimating the intensity function

$$\lambda(\mathbf{s}) = \lim_{|d\mathbf{s}| \to 0} \frac{\mathrm{E}[N(d\mathbf{s})]}{|d\mathbf{s}|},$$

that represents the number of events per unit area. If the intensity does not vary with spatial location, the process is first-order stationary (= homogeneous).

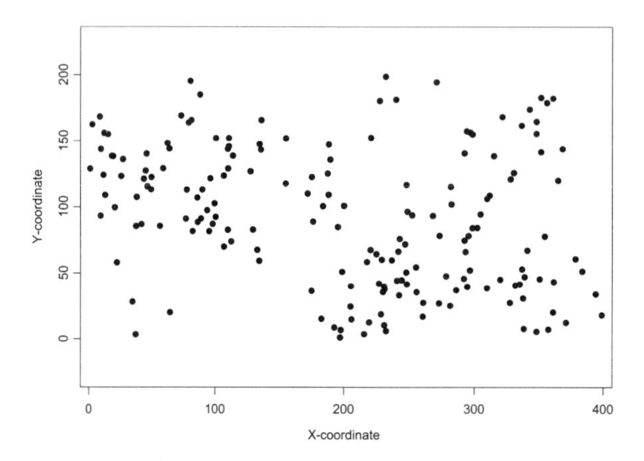

Figure 9.57. Mapped point pattern of $n = 180$ events on the $(0, 400) \times (0, 200)$ rectangle.

The common intensity estimators are discussed in §A9.9.10. The naïve estimator is simply, $\widehat{\lambda} = n/|A|$, where $|A|$ is the area of the domain considered. Here, $n = 180$ and $|A| = 400*200$, hence $\widehat{\lambda} = 0.00225$. This estimator does not vary with spatial location, it is appropriate only *if* the process is homogeneous. Location-dependent estimators can be obtained in a variety of ways. One can grid the domain and count the number of events in a grid cell. This process is usually followed by some type of smoothing of the raw counts. S+SpatialStats® terms this the `binning` estimator. One can also apply nonparametric smoothing techniques such as kernel estimation (§A4.8.7) directly. The smoothness (= spatial resolution) of binning estimators depends on the number of grid cells and the smoothness of kernel estimators on the choice of bandwidth (§4.7.2) The statements below calculate the binning estimator on a 20×10 grid and kernel estimators with gaussian weight function for three different bandwidths (Figure 9.58).

```
par(mfrow=c(2,2))
image(intensity(sppattern,method="binning",nx=20,ny=10))
title(main="20*10 Binning w/ LOESS")
image(intensity(sppattern,method="gauss2d",bw=25))
title(main="Kernel, Bandwidth=25")
image(intensity(sppattern,method="gauss2d",bw=50))
title(main="Kernel, Bandwidth=50")
image(intensity(sppattern,method="gauss2d",bw=100))
title(main="Kernel, Bandwidth=100")
```

With increasing bandwidth the kernel smoother approaches the naïve estimator and the location-dependent features of the intensity can no longer be discerned. The binning estimator as well as the kernel smoothers with bandwidths 25 and 50 show a concentration of events in the southeast and northwest corners of the area.

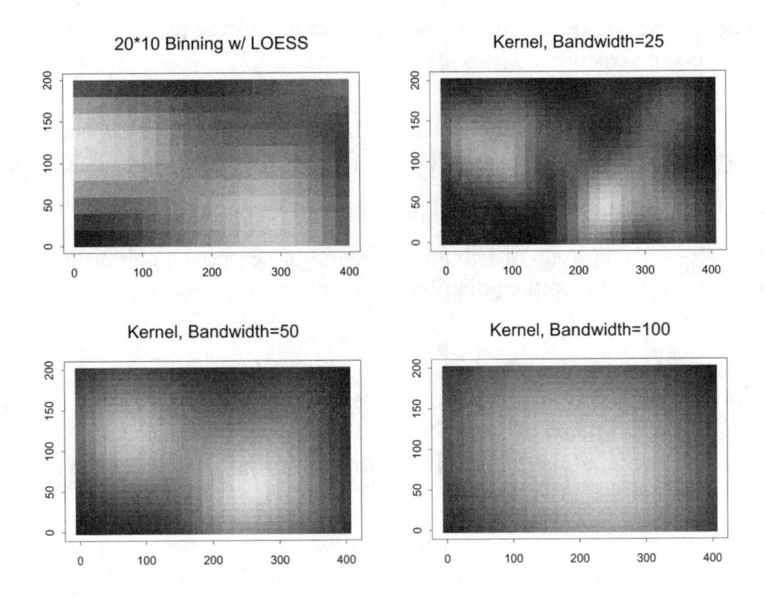

Figure 9.58. Spatially explicit intensity estimators for point pattern in Figure 9.57. Lighter colors correspond to higher intensity.

The first three panels of Figure 9.58 suggest that events tend to group in certain areas, and that the process appears to be clustered. At this point there are three possible explanations, and further progress depends on which is trusted.

1. The number of events in nonoverlapping areas are independent. There is no repulsion or attraction of events. Instead, the first-order intensity (the *mean function*) $\lambda(\mathbf{s})$ is simply a function of spatial location. An inhomogeneous Poisson process is a reasonable model and it remains to estimate the intensity function $\lambda(\mathbf{s})$.

2. The first-order intensity does not depend on the spatial location, i.e., $\lambda(\mathbf{s}) = \lambda$. The grouping of events is due `only` to spatial interaction of events. The second-order properties of the point pattern (the spatial dependency) suffice to explain the nonhomogeneous distribution of events.

3. In addition to interactions among events the first-order intensity is not constant.

The three conditions are roughly equivalent to the following scenarios for geostatistical data. Independent observations with large-scale variation in the mean (1), a constant mean with spatial autocorrelation (2), and large-scale variations combined with spatial autocorrelation (3). While random field models for geostatistical and lattice data allow the separation of large-scale and smooth-scale spatial variation, less constructive theory is available for point pattern analysis. Second-order methods for point pattern analysis require stationarity of the intensity just as semivariogram analysis for geostatistical data requires stationarity of the mean function. There we can either detrend the data or rely on methods that simultaneously estimate the mean and second-order properties (e.g., maximum likelihood). With point patterns, this separation is not straightforward. If we consider explanation 2 we can proceed with examining the second-order properties (K-function) of the process. Following explanations 1 or 2 this is not possible.

We developed an S-PLUS® function that provides a comprehensive analysis of a spatial point pattern (function `GAnalysis()` on CD-ROM). The function performs four specific tasks.

1. It plots the realization of the spatial point pattern (upper left panel of Figure 9.59)
2. It calculates the empirical distribution function of nearest-neighbor distances and compares those to the theoretical distribution function of a CSR process (upper right panel of Figure 9.59).
3. It performs Monte Carlo simulations under the CSR hypothesis and graphs the upper and lower simulation envelopes of nearest neighbor distances against the observed pattern (lower left panel of Figure 9.59). The empirical p-values of the test against the clustered and regular alternative are calculated, the test statistic is the average nearest-neighbor distance.
4. It estimates the K-function from the observed point pattern and plots $\widehat{L}(h) - h = (\widehat{K}/\pi)^{0.5} - h$ against distance h (lower right panel of Figure 9.59). We prefer this graph because a graph of $\widehat{K}(h)$ against the CSR benchmark πh^2 often fails to reveal the subtle deviations from complete spatial randomness. The $\widehat{L} - h$ versus h plot amplifies the deviation from CSR visually. The CSR process is represented by a horizontal line at 0 in this plot. Clustering is indicated when $\widehat{L}(h) - h$ rises above the zero line. Because the variance of $\widehat{K}(h)$ increases sharply with h interpretation of these graphs should be restricted to a distance no greater than one half of the length of the shorter side of the bounding rectangle (here $h = 100$).

After making the `GAnalysis()` function available to S+SpatialStats®, all of these tasks are accomplished by the function call `GAnalysis(sppattern,n=180,sims=100,cluster=" ")`. A descriptive string can be assigned to the `cluster=` argument which will be shown on the output (Figure 9.59, argument was omitted here). Based on the Monte Carlo test of nearest-neighbor distances with 100 simulations we conclude that the observed pattern exhibits clustering. Among all 101 point patterns (one observed, one hundred simulated), the observed pattern had the smallest nearest neighbor distance (rank = 1), leading to a p-value of 0.0099 against the clustered alternative. The observed $\widehat{G}$ function is close to the upper simulation envelope and crosses it repeatedly. The $\widehat{L}(h) - h$ plot shows the elevation above the zero line that corresponds to a CSR process. The expected number of extra events within distance h from an arbitrary event is larger than under CSR. To see a drop of the $\widehat{L}(h) - h$ plot below zero for larger distances in clustered processes is common. This occurs when distances are larger than the cluster diameters and cover a lot of *white space*. Recall the recommendation that $\widehat{L}(h) - h$ not be interpreted for distances in excess of one half of the length of the smaller side of the bounding rectangle. Up to $h = 100$ clustering of the process is implied.

Having concluded that this is a clustered point pattern, based on the $\widehat{G}$ analysis and the $\widehat{L}$ function, we would like to know whether the conclusion is correct. It is indeed. The point pattern in Figure 9.57 was simulated with S+SpatialStats® with the statements

```
set.seed(24)
sppattern <- make.pattern(n=180,process="cluster",radius=35,cpar=25,
                          boundary=bbox(x=c(0,400),y=c(0,200)))
```

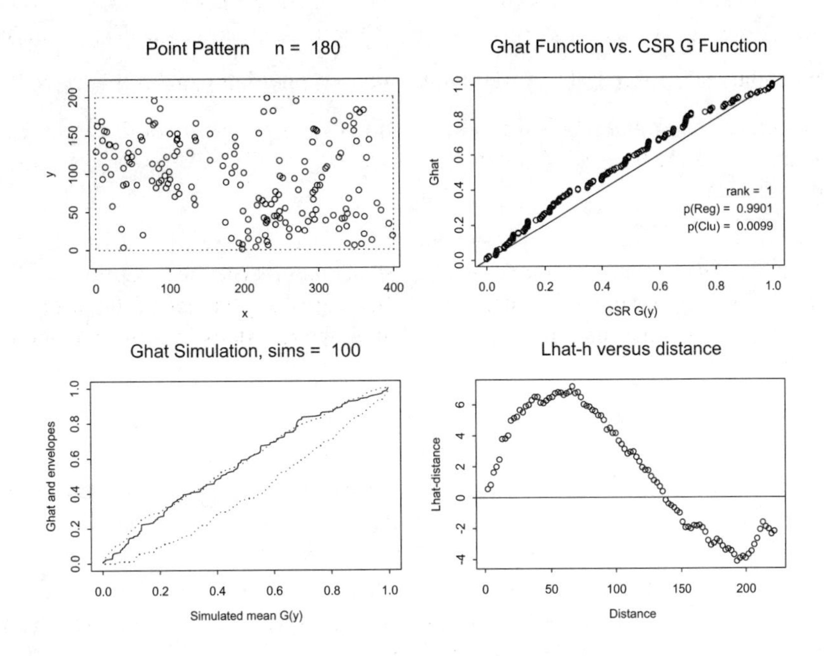

Figure 9.59. Results of analyzing the point pattern in Figure 9.57 with `GAnalysis()`.

The `set.seed()` statement fixes the seed of the random number generator at a given value. Subsequent runs of the program with the same seed will produce identical point patterns. The `make.pattern()` function simulates the realization of a particular point process. Here, a cluster process is chosen with parameters `radius=35` and `cpar=25`. Twenty-five parent events are placed according to a homogeneous Poisson process. Around each parent, offspring events are placed independently of each other within radius 35 of the parent location. Finally, the parent events are deleted and only the offspring locations are retained. This is known as a Poisson Cluster process, special cases of which are the Neyman-Scott processes (see §A9.9.11 and Neyman and Scott 1972). Although this is difficult to discern from Figure 9.57, the process consists of 25 clusters. Furthermore, following explanation 2. above was the correct course of action. This Neyman-Scott process is a stationary process.

Bibliography

Additional references concerning the mathematical details and special topics can be found with Appendix A on the CD-ROM.

Agresti, A. (1990) *Categorical Data Analysis*. John Wiley & Sons, New York

Akaike, H. (1974) A new look at the statistical model identification. *IEEE Transaction on Automatic Control*, AC-19:716-723

Allen, D.M. (1974) The relationship between variable selection and data augmentation and a method of prediction. *Technometrics*, 16:125-127

Allender, W.J. (1997) Effect of trifluoperazine and verapamil on herbicide stimulated growth of cotton. *Journal of Plant Nutrition*, 20(1):69-80

Allender, W.J., Cresswell, G.C., Kaldor, J., and Kennedy, I.R. (1997) Effect of lithium and lanthanum on herbicide induced hormesis in hydroponically-grown cotton and corn. *Journal of Plant Nutrition*, 20:81-95

Amateis, R.L. and Burkhart, H.E. (1987) Cubic-foot volume equations for loblolly pine trees in cutover, site-prepared plantations. *Southern Journal of Applied Forestry*, 11:190-192

Amemiya, T. (1973) Regression analysis when the variance of the dependent variable is proportional to the square of its expectation. *Journal of the American Statistical Association*, 68:928-934

Anderson, J.A. (1984) Regression and ordered categorical variables. *Journal of the Royal Statistical Society (B)*, 46(1):1-30

Anderson, R.L. and Nelson, L.A. (1975) A family of models involving intersecting straight lines and concomitant experimental designs useful in evaluating response to fertilizer nutrients. *Biometrics*, 31:303-318

Anderson, T.W. and Darling, D.A. (1954) A test of goodness of fit. *Journal of the American Statistical Association*, 49:765-769

Andrews, D.F., Bickel, P.J., Hampel, F.R., Huber, P.J., Rogers, W.H., and Tukey, J.W. (1972) *Robust Estimates of Location: Survey and Advances*. Princeton University Press, Princeton, NJ

Andrews, D.F. and Herzberg, A.M. (1985) *Data. A Collection of Problems from Many Fields for the Student and Research Worker*. Springer-Verlag, New York.

Anscombe, F.J. (1948) The transformation of Poisson, binomial, and negative-binomial data. *Biometrika*, 35:246-254

Anscombe, F.J. (1960) Rejection of outliers. *Technometrics*, 2:123-147

Anselin, L. (1995) Local indicators of spatial association — LISA. *Geographical Analysis*, 27:93-115

Armstrong, M. and Delfiner, P. (1980) Towards a more robust variogram: A case study on coal. Technical Report N-671. Centre de Géostatistique, Fontainebleau, France

Baddeley, A.J. and Silverman, B.W. (1984) A cautionary example on the use of second-order methods for analyzing point patterns. *Biometrics*, 40:1089-1093

Bailey, R.L. (1994) A compatible volume-taper model based on the Schumacher and Hall generalized form factor volume equation. *Forest Science*, 40:303-313

Barnes, R.J. and Johnson, T.B. (1984) Positive kriging. In: *Geostatistics for Natural Resource Characterization Part 1* (Verly, G., David, M., Journel, A.G. and Maréchal, A. eEds.) Reidel, Dortrecht, The Netherlands, p. 231-244

Barnett, V. and Lewis, T. (1994) *Outliers in Statistical Data, 3rd ed.* John Wiley & Sons, New York

Bartlett, M.S. (1937a) Properties of sufficiency and statistical tests. *Proceedings of the Royal Statistical Society, Series A*, 160:268-282

Bartlett, M.S. (1937b) Some examples of statistical methods of research in agriculture and applied biology. *Journal of the Royal Statistical Society, Suppl.*, 4:137-183

Bartlett, M.S. (1938) The approximate recovery of information from field experiments with large blocks. *Journal of Agricultural Science*, 28:418-427

Bartlett, M.S. (1978a) Nearest-neighbour models in the analysis of field experiments (with discussion). *Journal of the Royal Statistical Society (B)*, 40:147-174

Bartlett, M.S. (1978b) *Stochastic Processes. Methods and Applications.* Cambridge University Press, London

Bates, D.M. and Watts, D.G. (1980) Relative curvature measures of nonlinearity. *Journal of the Royal Statistical Society (B),* 42:1-25

Bates, D. M., and Watts, D.G. (1981) A relative offset orthogonality convergence criterion for nonlinear least squares. *Technometrics*, 123:179-183.

Beale, E.M.L. (1960) Confidence regions in non-linear estimation. *Journal of the Royal Statistical Society (B)*, 22:41-88

Beaton, A.E. and Tukey, J.W. (1974) The fitting of power series, meaning polynomials, illustrated on band-spectroscopic data. *Technometrics*, 16:147-185

Beck, D.E. (1963) Cubic-foot volume tables for yellow poplar in the southern Appalachians. *USDA Forest Service, Research Note SE-16.*

Becker, M.P. (1989) Square contingency tables having ordered categories and GLIM. *GLIM Newsletter No. 19.* Royal Statistical Society, NAG Group

Becker, M.P. (1990a) Quasisymmetric models for the analysis of square contingency tables. *Journal of the Royal Statistical Society (B)*, 52:369-378

Becker, M.P. (1990b) Algorithm AS 253; Maximum likelihood estimation of the RC(M) association model. *Applied Statistics*, 39:152-167

Beltrami, E. (1998) *Mathematics for Dynamic Modeling. 2nd ed.* Academic Press, San Diego, CA

Berkson, J. (1950) Are there two regressions? *Journal of the American Statistical Association*, 45:164-180

Besag, J.E. (1974) Spatial interaction and the statistical analysis of lattice systems. *Journal of the Royal Statistical Society (B)*, 36:192-236

Besag, J.E. (1975) Statistical analysis of non-lattice data. *The Statistician*, 24:179-195.

Besag, J. and Kempton, R. (1986) Statistical analysis of field experiments using neighboring plots. *Biometrics*, 42(2):231-251

Biging, G.S. (1985) Improved estimates of site index curves using a varying parameter model. *Forest Science*, 31:248-259

Binford, G.D., Blackmer, A.M., and Cerrato, M.E. (1992) Relationship between corn yield and soil nitrate in late spring. *Agronomy Journal*, 84:53-59

Birch, J.B. and Agard, D.B. (1993) Robust inference in regression: a comparative study. *Communications in Statistics – Simulation*, 22(1):217-244

Black, C.A. (1993) *Soil Fertility Evaluation and Control.* Lewis Publishers, Boca Raton, FL

Blackmer, A.M., Pottker, D., Cerrato, M.E., and Webb, J. (1989) Correlations between soil nitrate concentrations in late spring and corn yields in Iowa. *Journal of Production Agriculture*, 2:103-109

Bleasdale, J.K.A. and Nelder, J.A. (1960) Plant population and crop yield. *Nature*, 188:342

Bleasdale, J.K.A. and Thompson, B. (1966) The effects of plant density and the pattern of plant arrangement on the yield of parsnips. *Journal of Horticultural Science* 41:145-153

Bose, R.C. and Nair, K.R. (1939) Partially balanced incomplete block designs. *Sankhya*, 4:337-372

Bowman, D.T. (1990) Trend analysis to improve efficiency of agronomic trials in flue-cured tobacco. *Agronomy Journal*, 82:499-501

Box, G.E.P. (1954a) Some theorems on quadratic forms applied in the study of analysis of variance problems, I. Effects of inequality of variance in the one-way classification. *Annals of Mathematical Statistics*, 25:290-302

Box, G.E.P. (1954b) Some theorems on quadratic forms applied in the study of analysis of variance problems, II. Effects of inequality of variance and of correlations between errors in the two-way classification. *Annals of Mathematical Statistics*, 25:484-498

Box, G.E.P. and Andersen, S.L. (1955) Permutation theory in the derivation of robust criteria and the study of departures from assumption. *Journal of the Royal Statistical Society (B)*, 17:1-26

Box, G.E.P. and Cox, D.R. (1964) The analysis of transformations. *Journal of the Royal Statistical Society (B)*, 26:211-252

Box, G.E.P. Jenkins, G.M., and Reinsel, G.C. (1994) *Time Series Analysis: Forecasting and Control*. Prentice Hall, Englewood Cliffs, NJ

Bozdogan, H. (1987) Model selection and Akaike's information criterion (AIC): the general theory and its analytical extensions. *Psychometrika*, 52:345-370

Brain, P. and Cousens, R. (1989) An equation to describe dose responses where there is stimulation of growth at low doses. *Weed Research*, 29: 93-96

Breslow, N.E. and Clayton, D.G. (1993) Approximate inference in generalized linear mixed models. *Journal of the American Statistical Association*, 88:9-25

Brown, M.B. and Forsythe, A.B. (1974) Robust tests for the equality of variances. *Journal of the American Statistical Association*, 69:364-367

Brown, R.L., Durbin, J., and Evans, J.M. (1975) Techniques for testing the constancy of regression relationships over time. *Journal of the Royal Statistical Society (B)*, 37:149-192

Brownie, C., Bowman, D.T., and Burton, J.W. (1993) Estimating spatial variation in analysis of data from yield trials: a comparison of methods. *Agronomy Journal*, 85:1244-1253

Brownie, C. and Gumpertz, M.L. (1997) Validity of spatial analysis for large field trials. *Journal of Agricultural, Biological, and Environmental Statistics*, 2(1):1-23

Bunke, H. and Bunke, O. (1989) *Nonlinear Regression, Functional Relationships and Robust Methods*. John Wiley & Sons, New York

Burkhart, H.E. (1977) Cubic-foot volume of loblolly pine to any merchantable top limit. *Southern Journal of Applied Forestry*, 1:7-9

Carroll, R.J. and Ruppert, D. (1984) Power transformations when fitting theoretical models to data. *Journal of the American Statistical Association*, 79:321-328

Carroll, R.J. Ruppert, D., and Stefanski, L.A. (1995) *Measurement Error in Nonlinear Models*. Chapman and Hall, New York

Cerrato, M.E. and Blackmer, A.M. (1990) Comparison of models for describing corn yield response to nitrogen fertilizer. *Agronomy Journal*, 82:138-143

Chapman, D.G. (1961) Statistical problems in population dynamics. In: *Proceedings of the Fourth Berkeley Symposium on Mathematical Statistics and Probability*. University of California Press, Berkeley

Chauvet, P. (1982) The variogram cloud. In: *Proceedings of the 17th APCOM International Symposium*. Golden, CO, 757-764

Chilès, J.-P. and Delfiner, P. (1999) *Geostatistics.* John Wiley & Sons, New York

Cleveland, W.S. (1979) Robust locally weighted regression and smoothing scatterplots. *Journal of the American Statistical Association*, 74:829-836

Cleveland, W.S., Devlin, S.J., and Grosse, E. (1988) Regression by local fitting. *Journal of Econometrics*, 37:87-114

Cliff, A.D. and Ord, J.K. (1973) *Spatial Autocorrelation.* Pion, London

Cliff, A.D. and Ord, J.K. (1981) *Spatial Processes; Models and Applications*, Pion, London

Clutter, J.L., Fortson, J.C., Pienaar, L.V. Brister, G.H., and Bailey, R.L. (1992) *Timber Management.* Krieger Publishing, Malabar, FL

Cochran, W.G. (1941) The distribution of the largest of a set of estimated variances as a fraction of their total. *Annals of Eugenics*, 11:47-52

Cochran, W.G. (1954) Some methods for strengthening the common χ^2 tests. *Biometrics*, 10:417-4517

Cochran, W.G. and Cox, G.M. (1957) *Experimental Design 2nd ed.* John Wiley & Sons, New York

Cochrane, D. and Orcutt, G.H. (1949) Applications of least square regression to relationships containing autocorrelated error terms. *Journal of the American Statistical Association*, 44:32-61

Cole, J.W.L. and Grizzle, J.E. (1966) Applications of multivariate analysis of variance to repeated measures experiments. *Biometrics*, 22:810-828

Cole, T.J. (1975) Linear and proportional regression models in the prediction of ventilatory function. *Journal of the Royal Statistics Society (A)*, 138:297-333

Coleman, D., Holland, P., Kaden, N., Klema, V., and Peters, S. C. (1980) A system of subroutines for iteratively re-weighted least-squares computations. *ACM Transactions on Mathematical Software*, 6:327-336.

Colwell, J.D., Suhet, A.R., and Van Raij, B. (1988) Statistical procedures for developing general soil fertility models for variable regions. *Report No. 93*, CSIRO Division of Soils (Australia),

Cook, R.D. (1977) Detection of influential observations in linear regression. *Technometrics*, 19:15-18

Cook, R.D. and Tsai, C.-L. (1985) Residuals in nonlinear regression. *Biometrika*, 72:23-29

Corbeil, R.R. and Searle, S.R. (1976) A comparison of variance component estimators, *Biometrics*, 32:779-791

Courtis, S.A. (1937) What is a growth cycle? *Growth*, 1:247-254

Cousens, R. (1985) A simple model relating yield loss to weed density. *Annals of Applied Biology*, 107:239-252

Cox, C. (1988) Multinomial regression models based on continuation ratios. *Statistics in Medicine*, 7:435-441.

Cox, D.R. and Snell, E.J. (1989) *The Analysis of Binary Data, 2nd ed.* Chapman and Hall, London

Craig, J.R., Edwards, D., Rimstidt, J.D., Scanlon, P.F., Collins, T., Schabenberger, O., and Birch, J.B. Lead Distribution and Loading on a Public Shotgun Range. Submitted to *Environmental Geology*

Craven, P. and Wahba, G. (1979) Smoothing noisy data with spline functions. *Numerical Mathematics*, 31:377-403

Cressie, N. (1985) Fitting variogram models by weighted least squares. *Journal of the International Association for Mathematical Geology*, 17:563-586

Cressie, N.A.C. (1986) Kriging nonstationary data. *Journal of the American Statistical Association*, 81:625-634

Cressie, N.A.C. (1993) *Statistics for Spatial Data. Revised Ed.* John Wiley & Sons, New York

Cressie, N.A.C. and Hawkins, D.M. (1980) Robust estimation of the variogram, I. *Journal of the International Association for Mathematical Geology*, 12:115-125

Crowder, M.J. and Hand, D.J. (1990) *Analysis of Repeated Measures.* Chapman and Hall, New York

Curriero, F.C. and Lele, S. (1999) A composite likelihood approach to semivariogram estimation. *Journal of Agricultural, Biological, and Environmental Statistics*, 4(1):9-28

Davidian, M. and Giltinan, D.M. (1993) Some general estimation methods for nonlinear mixed-effects models. *Journal of Biopharmaceutical Statistics*, 3(1):23-55

Davidian, M. and Giltinan, D.M. (1995) *Nonlinear Models for Repeated Measurement Data*. Chapman and Hall, New York

Delfiner, P. (1976) Linear estimation of nonstationary spatial phenomena. In: *Advanced Geostatistics in the Mining Industry* (M. Guarascio, M. David, C. Huijbregts, eds.) Reidel, Dortrecht, The Netherlands, pp. 49-68

Delfiner, P., Renard D., and Chilès, J.P. (1978) *Bluepack-3D Manual*, Centre de Geostatistique, Fontainebleau, France

Diggle, P. (1983) *Statistical Analysis of Spatial Point Patterns*. Academic Press, London

Diggle, P.J. (1988) An approach to the analysis of repeated measurements. *Biometrics*, 44:959-971

Diggle, P.J. (1990) *Time Series: A Biostatistical Introduction*. Clarendon Press, Oxford, UK

Diggle, P., Besag, J.E. and Gleaves, J.T. (1976) Statistical analysis of spatial patterns by means of distance methods. *Biometrics*, 32:659-667

Diggle, P.J., Liang, K.-Y., and Zeger, S.L. (1994) *Analysis of Longitudinal Data*. Clarendon Press, Oxford, UK

Draper, N.R. and Smith, H. (1981) *Applied Regression Analysis. 2nd ed.* John Wiley & Sons, New York

Dunkl, C.F. and Ramirez, D.E. (2001) Computation of the generalized *F* distribution. *The Australian and New Zealand Journal of Statistics*, 43:21-31

Durbin, J. and Watson, G.S. (1950) Testing for serial correlation in least squares regression. I. *Biometrika*, 37:409-428

Durbin, J. and Watson, G.S. (1951) Testing for serial correlation in least squares regression. II. *Biometrika*, 38:159-178

Durbin, J. and Watson, G.S. (1971) Testing for serial correlation in least squares regression. III. *Biometrika*, 58:1-19

Eisenhart, C. (1947) The assumptions underlying the analysis of variance. *Biometrics*, 3:1-21

Engel, J. (1988) Polytomous logistic regression. *Statistica Neerlandica*, 42(4):233-252.

Emerson, J.D. and Hoaglin, D.C. (1983) Analysis of two-way tables by medians. In: *Understanding Robust and Exploratory Data Analysis* (Hoaglin D.C., Mosteller, F., and Tukey, J.W., eds.), John Wiley & Sons, New York, pp. 166-207

Emerson, J.D. and Wong, G.Y. (1985) Resistant nonadditive fits for two-way tables. In: *Exploring Data Tables, Trends, and Shapes* (Hoaglin, D.C., Mosteller, F., and Tukey, J.W., eds.), John Wiley & Sons, New York, pp. 67-124

Engelstad, O.P. and Parks, W.L. (1971) Variability in optimum N rates for corn. *Agronomy Journal*, 63:21-23

Epanechnikov, V. (1969) Nonparametric estimates of a multivariate probability density. *Theory of Probability and its Applications*, 14:153-158

Eubank, R.L. (1988) *Spline Smoothing and Nonparametric Regression*. Marcel Dekker, New York

Fahrmeir, L. and Tutz, G. (1994) *Multivariate Statistical Modelling Based on Generalized Linear Models*. Springer-Verlag, New York

Federer, W.T. and Schlottfeldt, C.S. (1954) The use of covariance to control gradients in experiments. *Biometrics*, 10:282-290

Fedorov, V.V. (1974) Regression problems with controllable variables subject to error. *Biometrika*, 61:49-56

Fieller, E.C. (1940) The biological standardization of insulin. *Journal of the Royal Statistical Society (Suppl.)*, 7:1-64

Fienberg, S.E. (1980) *The Analysis of Cross-classified Categorical Data*. MIT Press, Cambridge, MA

Finney, D.J. (1978) *Statistical Methods in Biological Assay, 3rd ed.* Macmillan, New York

Firth, D. (1988) Multiplicative errors: log-normal or gamma. *Journal of the Royal Statistical Society (B)*, 50:266-268

Fisher, R.A. (1935) *The Design of Experiments*. Oliver and Boyd, Edinburgh

Fisher, R.A. (1947) *The Design of Experiments, 4th ed.* Oliver and Boyd, Edinburgh

Folks, J.L. and Chhikara, R.S. (1978) The inverse Gaussian distribution and its statistical application: a review. *Journal of the Royal Statistical Society (B)*, 40:263-275

Freney, J.R. (1965) Increased growth and uptake of nutrients by corn plants treated with low levels of simazine. *Australian Journal of Agricultural Research*, 16:257-263

Gabriel, K.R. (1962) Ante-dependence analysis of an ordered set of variables. *Annals of Mathematical Statistics*, 33:201-212

Gallant, A.R. (1975) Nonlinear regression. *The American Statistician*, 29:73-81

Gallant, A.R. (1987) *Nonlinear Statistical Models*. John Wiley & Sons, New York

Gallant, A.R. and Fuller, W.A. (1973) Fitting segmented poynomial regression models whose join points have to be estimated. *Journal of the American Statistical Association*, 68:144-147

Galpin, J.S.and Hawkins, D.M. (1984) The use of recursive residuals in checking model fit in linear regression. *The American Statistician*, 38(2):94-105

Galton, F. (1886) Regression towards mediocrity in hereditary stature. *Journal of the Anthropological Institute*, 15:246-263

Gayen, A.K. (1950) The distribution of the variance ratio in random samples of any size drawn from non-normal universes. *Biometrika*, 37:236-255

Geary, R.C. (1947) Testing for normality. *Biometrika*, 34:209-242

Geary, R.C. (1954) The contiguity ratio and statistical mapping. *The Incorporated Statistician*, 5:115-145

Geisser, S. and Greenhouse, S.W. (1958) An extension of Box's results on the use of the F-distribution in multivariate analysis. *Annals of Mathematical Statistics*, 29:885-891

Gerrard, D.J. (1969) Competition quotient: a new measure of the competition affecting individual forest trees. *Research Bulletin No. 20*, Michigan Agricultural Experiment Station, Michigan State University

Gillis, P.R. and Ratkowsky, D.A. (1978) The behaviour of estimators of the parameters of various yield-density relationships. *Biometrics*, 34:191-198

Gilmour, A.R., Cullis, B.R., and Verbyla, A.P. (1997) Accounting for natural and extraneous variation in the analysis of field experiments. *Journal of Agricultural, Biological, and Environmental Statistics*, 2(3):269-293

Godambe, V.P. (1960) An optimum property of regular maximum likelihood estimation. *Annals of Mathematical Statistics*, 31:1208-1211

Golden, M.S., Knowe, S.A., and Tuttle, C.L. (1982) Cubic-foot volume for yellow-poplar in the hilly coastal plain of Alabama. *Southern Journal of Applied Forestry*, 6:167-171

Goldberg, R.R. (1961) *Fourier Transforms*. Cambridge University Press, Cambridge

Goldberger, A.S. (1962) Best linear unbiased prediction in the generalized linear regression model, *Journal of the American Statistical Association*, 57:369-375

Gompertz, B. (1825) On the nature of the function expressive of the law of human mortality, and on a new method of determining the value of life contingencies. *Phil. Trans. Roy. Soc.*, 513-585

Goodman, L.A. (1979a) Simple models for the analysis of association in cross-classifications having ordered categories. *Journal of the American Statistical Association*, 74:537-552

Goodman, L.A. (1979b) Multiplicative models for square contingency tables with ordered categories. *Biometrika*, 66:413-418

Goodman, L.A. (1985) The analysis of cross-classified data having ordered and/or unordered categories: association models, correlation models, and asymmetry models for contingency tables with or without missing entries. *Annals of Statistics*, 13:10-69

Goovaerts, P. (1997) *Geostatistics for Natural Resources Evaluation.* Oxford University Press, New York

Goovaerts, P. (1998) Ordinary cokriging revisited. *Journal of the International Association of Mathematical Geology*, 30:21-42

Gotway, C.A. and Stroup, W.W. (1997) A generalized linear model approach to spatial data analysis and prediction. *Journal of Agricultural, Biological, and Environmental Statistics*, 2(2):157-178.

Graybill, F.A. (1969) *Matrices with Applications in Statistics. 2nd ed.* Wadsworth International, Belmont, CA.

Greenhouse, S.W. and Geisser, S. (1959) On methods in the analysis of profile data. *Psychometrika*, 32:95-112

Greenwood, C. and Farewell, V. (1988) A comparison of regression models for ordinal data in an analysis of transplanted-kidney function. *Canadian Journal of Statistics*, 16(4):325-335.

Gregoire, T.G. (1985) Generalized error structure for yield models fitted with permanent plot data. Ph.D. dissertation, Yale University, New Haven, CT

Gregoire, T.G. (1987) Generalized error structure for forestry yield models. *Forest Science*, 33:423-444

Gregoire, T.G., Brillinger, D.R., Diggle, P.J., Russek-Cohen, E., Warren, W.G., and Wolfinger, R.D. (eds). (1997) *Modelling Longitudinal and Spatially Correlated Data.* Springer-Verlag, New York, 402 pp.

Gregoire, T.G., Schabenberger, O., and Barrett, J.P. (1995) Linear modelling of irregularly spaced, unbalanced, longitudinal data from permanent plot measurements. *Canadian Journal of Forest Research*, 25(1):137-156

Gregoire, T.G. and Schabenberger, O. (1996a) Nonlinear mixed-effects modeling of cumulative bole volume with spatially correlated within-tree data. *Journal of Agricultural, Biological, and Environmental Statistics*, 1(1):107-119

Gregoire, T.G. and Schabenberger, O. (1996b) A non-linear mixed-effects model to predict cumulative bole volume of standing trees. *Journal of Applied Statistics*, 23(2&3):257-271

Griffith, D.A. (1996) Some guidelines for specifying the geographic weights matrix contained in Spatial statistical models. In: *Practical Handbook of Spatial Statistics* (S.L. Arlinghaus, ed.), CRC Press, Boca Raton, FL, pp. 65-82

Griffith, D.A. and Layne, L.J. (1999) *A Casebook for Spatial Statistical Data Analysis. A Compilation of Analyses of Different Thematic Data Sets.* Oxford University Press, New York

Grondona, M.O. and Cressie, N.A. (1991) Using spatial considerations in the analysis of experiments. *Technometrics*, 33:381-392

Härdle, W. (1990) *Applied Nonparametric Regression.* Cambridge University Press, Cambridge

Haining, R. (1990) *Spatial Data Analysis in the Social and Environmental Sciences.* Cambridge University Press, Cambridge

Hampel, F.R. (1974) The influence curve and its role in robust estimation. *Journal of the American Statistical Association*, 69:383-393

Hampel, F.R., Ronchetti, E.M., Rousseeuw, P.J., and Stahel, W.A. (1986) *Robust Statistics, The Approach Based on Influence Functions.* John Wiley & Sons, New York

Hanks, R.J., Sisson, D.V., Hurst, R.L., and Hubbard, K.G. (1980) Statistical analysis of results from irrigation experiments using the line source sprinkler system. *Journal of the American Soil Science Society*, 44:886-888

Harris, T.R. and Johnson, D.E. (1996) A regression model with spatially correlated errors for comparing remote sensing and in-situ measurements of a grassland site. *Journal of Agricultural, Biological, and Environmental Statistics*, 1:190-204

Hart, L.P. and Schabenberger, O. (1998) Variability of vomitoxin in a wheat scab epidemic. *Plant Disease*, 82:625-630.

Hartley, H.O. (1950) The maximum F-ratio as a short-cut test for heterogeneity of variance. *Biometrika*, 31:249-255

Hartley, H.O. (1961) The modified Gauss-Newton method for the fitting of nonlinear regression functions by least squares. *Technometrics*, 3:269-280

Hartley, H.O. (1964) Exact confidence regions for the parameters in nonlinear regression laws. *Biometrika*, 51:347-353

Hartley, H.O. and Booker, A. (1965) Nonlinear least square estimation. *Annals of Mathematical Statistics*, 36(2):638-650

Harville, D.A. (1974) Bayesian inference for variance components using only error contrasts. *Biometrika*, 61:383-385

Harville, D.A. (1976a) Extension of the Gauss-Markov theorem to include the estimation of random effects. *The Annals of Statistics*, 4:384-395

Harville, D.A. (1976b) Confidence intervals and sets for linear combinations of fixed and random effects. *Biometrics*, 32:320-395

Harville, D.A. (1977) Maximum-likelihood approaches to variance component estimation and to related problems. *Journal of the American Statistical Association*, 72:320-340

Harville, D.A. and Jeske, D.R. (1992) Mean squared error of estimation or prediction under a general linear model. *Journal of the American Statistical Association*, 87:724-731

Hastie, T.J. and Tibshirani, R.J. (1990) *Generalized Additive Models*. Chapman and Hall, New York

Haseman, J.K. and Kupper, L.L. (1979) Analysis of dichotomous response data from certain toxicological experiments. *Biometrics*, 35:281-293

Hayes, W.L. (1973) *Statistics for the Social Sciences*. Holt, Rinehart and Winston, New York

Heagerty, P.J. and Lele, S.R. (1998) A composite likelihood approach to binary spatial data. *Journal of the American Statistical Association*, 93:1099-1111

Healy, M.J.R. (1986) *Matrices for Statistics*. Clarendon Press, Oxford, UK

Hearn, A.B. (1972) Cotton spacing experiments in Uganda. *Journal of Agricultural Science*, 48:19-28

Hedeker, D. and Gibbons, R.D. (1994) A random effects ordinal regression model for multilevel analysis. *Biometrics*, 50:933-944

Henderson, C.R. (1950) The estimation of genetic parameters. *The Annals of Mathematical Statistics*, 21:309-310

Henderson, C.R. (1963) Selection index and expected genetic advance. In: *Statistical Genetics and Plant Breeding* (NRC Publication 982), Washington, D.C. National Academy of Sciences, pp. 141-163

Henderson, C.R. (1973) Sire evaluation and genetic trends. In: *Proceedings of the Animal Breeding and Genetics Symposium in Honor of Dr. J.L. Lush*, Champaign, IL: ASAS and ADSA, pp. 10-41

Heyde, C.C. (1997) *Quasi-likelihood and Its Application: A General Approach to Optimal Parameter Estimation*. Springer-Verlag, New York

Himmelblau, D.M. (1972) A uniform evaluation of unconstrained optimization techniques. *In: Numerical Methods for Nonlinear Optimization* (F.A. Lootsma, ed.), Academic Press, London

Hinkelmann, K. and Kempthorne, O. (1994) *Design and Analysis of Experiments. Volume I. Introduction to Experimental Design*. John Wiley & Sons, New York

Hoerl, A.E. and Kennard, R.W. (1970a) Ridge regression: biased estimation for nonorthogonal problems. *Technometrics*, 12:55-67

Hoerl, A.E. and Kennard, R.W. (1970b) Ridge regression: applications to nonorthogonal problems. *Technometrics*, 12:69-82

Holland, P.W. and Welsch, R.E. (1977) Robust regression using iteratively reweighted least squares. *Communications in Statistics A*, 6:813-888

Holliday, R. (1960) Plant population and crop yield: Part I. *Field Crop Abstracts*, 13:159-167

Hoshmand, A.R. (1994) *Experimental Research Design and Analysis*. CRC Press, Boca Raton, FL

Hsiao, A.I., Liu, S.H. and Quick, W.A. (1996) Effect of ammonium sulfate on the phytotoxicity, foliar uptake, and translocation of imazamethabenz in wild oat. *Journal of Plant Growth Regulation*, 15:115-120

Huber, O. (1981) *Robust Statistics*. John Wiley & Sons, New York

Huber, P.J. (1964) Robust estimation of a location parameter. *Annals of Mathematical Statistics*, 35: 73-101

Huber, P.J. (1973) Robust regression: asymptotics, conjectures, and Monte Carlo. *Annals of Statistics*, 1:799-821

Hurvich, C.M. and Simonoff, J.S. (1998) Smoothing parameter selection in nonparametric regression using an improved Akaike information criterion. *Journal of the Royal Statistical Society (B)*, 60:271-293

Huxley, J.S. (1932) *Problems of Relative Growth*. Dial Press, New York

Huynh, H. and Feldt, L.S. (1970) Conditions under which mean square ratios in repeated measurements designs have exact F-distributions. *Journal of the American Statistical Association*, 65:1582-1589

Huynh, H. and Feldt, L.S. (1976) Estimation of the Box correction for degrees of freedom from sample data in the randomized block and split plot designs. *Journal of Educational Statistics*, 1:69-82

Isaaks, E. and Srivastava, R. (1989) *An Introduction to Applied Geostatistics*. Oxford University Press, New York

Jansen, J. (1990) On the statistical analysis of ordinal data when extravariation is present. *Applied Statistics*, 39:75-84

Jennrich, R.J. and Schluchter, M.D. (1986) Unbalanced repeated-measures models with structured covariance matrices. *Biometrics*, 42:805-820

Jensen, D.R. and Ramirez, D.E. (1998) Some exact properties of Cook's D_I. In: *Handbook of Statistics*, Vol. 16 (Balakrishnan, N. and Rao, C.R. eds)., pp. 387-402 Elsevier Science Publishers, Amsterdam

Jensen, D.R. and Ramirez, D.E. (1999) Recovered errors and normal diagnostics in regression. *Metrica*, 49:107-119

Johnson, N.L., Kotz, S., and Kemp, A.W. (1992) *Univariate Discrete Distributions, 2nd. ed.*, John Wiley & Sons, New York

Johnson, N.L., Kotz, S. and Balakrishnan, N. (1995) *Univariate Continuous Distributions, Vol. 2, 2nd ed.* Wiley and Sons, New York

Jones, R.H. (1993) *Longitudinal Data with Serial Correlation: A State-space Approach*. Chapman and Hall, New York

Jones, R.H. and Boadi-Boateng, F. (1991) Unequally spaced longitudinal data with AR(1) serial correlation. *Biometrics*, 47:161-176

Journel, A.G. and Huijbregts, C.J. (1978) *Mining Geostatistics*. Academic Press, London

Kackar, R.N. and Harville, D.A. (1984) Approximations for standard errors of fixed and random effects in mixed linear models. *Journal of the American Statistical Association*, 79:853-862

Kaluzny, S.P., Vega, S.C., Cardoso, T.P., and Shelly, A.A. (1998) *S+ SpatialStats. User's Manual for Windows® and Unix*. Springer-Verlag, New York

Kempthorne, O. (1952) *Design and Analysis of Experiments*. John Wiley & Sons, New York

Kempthorne, O. (1955) The randomization theory of experimental inference. *Journal of the American Statistical Association*, 50:946-967

Kempthorne, O. (1975) Fixed and mixed model analysis of variance. *Biometrics*, 31:473-486

Kempthorne, O. and Doerfler, T.E. (1969) The behaviour of some significance tests under randomization. *Biometrika*, 56:231-248

Kendall, M.G. and Stuart, A. (1961) *The Advanced Theory of Statistics*, Vol 2. Griffin, London

Kenward, M.G. (1987) A method for comparing profiles of repeated measurements. *Applied Statistics*, 36:296-308

Kenward, M.G. and Roger, J.H. (1997) Small sample inference for fixed effects from restricted maximum likelihood. *Biometrics*, 53:983-997

Kianifard, F. and Swallow, W. H. (1996) A review of the development and application of recursive residuals in linear models. *Journal of the American Statistical Association*, 91:391-400

Kirby, E.J.M. (1974) Ear development in spring wheat. *Journal of the Agricultural Society*, 82:437-447

Kirk, H.J., Haynes, F.L., and Monroe, R.J. (1980) Application of trend analysis to horticultural field trials. *Journal of the American Society of Horticultural Science*, 105:189-193

Kirk, R.E. (1995) *Experimental Design: Procedures for the Behavioral Sciences, 3rd ed.*, Duxbury Press, Belmont, CA

Kitanidis, P.K. (1983) Statistical estimation of polynomial generalized covariance functions and hydrological applications. *Water Resources Research*, 19:909-921

Kitanidis, P.K. and Lane, R.W. (1985) Maximum likelihood parameter estimation of hydrological spatial processes by the Gauss-Newton method. *Journal of Hydrology*, 79:53-71

Kitanidis, P.K. and Vomvoris, E.G. (1983) A geostatistical approach to the inverse problem in groundwater modeling (steady state) and one-dimensional simulations. *Water Resources Research*, 19:677-690

Knoebel, B.R., Burkhart, H.E., and Beck, D.E. (1984) Stem volume and taper functions for yellow-poplar in the southern Appalachians. *Southern Journal of Applied Forestry*, 8:185-188

Korn, E.L. and Whittemore, A.S. (1979) Methods for analyzing panel studies of acute health effects of air pollution. *Biometrics*, 35:795-802

Kvålseth, T.O. (1985) Cautionary note about R^2. *The American Statistician*, 39(4):279-285

Läärä, E. and Matthews, J. N. S. (1985) The equivalence of two models for ordinal data. *Biometrika*, 72:206-207.

Lærke, P.E. and Streibig, J.C. (1995) Foliar absorption of some glyphosate formulations and their efficacy on plants. *Pesticide Science*, 44:107-116

Laird, A.K. (1965) Dynamics of relative growth. *Growth*, 29:249-263

Laird, N.M. (1988) Missing data in longitudinal studies. *Statistics in Medicine*, 7:305-315

Laird, N.M. and Louis, T.A. (1982) Approximate posterior distributions for incomplete data problems. *Journal of the Royal Statistical Society (B)*, 44:190-200

Laird, N.M. and Ware, J.H. (1982) Random-effects models for longitudinal data. *Biometrics*, 38:963-974

Lee, K.R. and Kapadia, C.H. (1984) Variance component estimators for the balanced two-way mixed model. *Biometrics*, 40:507-512

Lele, S. (1997) Estimating functions for semivariogram estimation. In: *Selected Proceedings of the Symposium on Estimating Functions* (I.V. Basawa, V.P. Godambe, and R.L. Taylor, eds.), Hayward, CA: Institute of Mathematical Statistics, pp. 381-396.

Lerman, P.M. (1980) Fitting segmented regression models by grid search. *Applied Statistics*, 29:77-84

Levenberg, K. (1944) A method for the solution of certain problems in least squares. *Quarterly Journal of Applied Mathematics*, 2:164-168

Levene, H. (1960) Robust test for equality of variances. In *Contributions to Probability and Statistics*, I. Olkin (ed.). pp. 278-292. Stanford University Press, Stanford, CA

Lewis, P.A.W. and Shedler, G.S. (1979) Simulation of non-homogeneous Poisson processes by thinning. *Naval Research Logistics Quarterly*, 26:403-413

Liang, K.-Y. and Zeger, S.L. (1986) Longitudinal data analysis using generalized linear models. *Biometrika*, 73:13-22

Liang, K.-Y., Zeger, S.L., and Qaqish, B. (1992) Multivariate regression analysis for categorical data. *Journal of the Royal Statistical Society (B)*, 54:3-40

Lindsay, B.G. (1988), Composite likelihood methods. *Contemporary Mathematics*, 80:221-239

Lindstrom, M.J. and Bates, D.M. (1988) Newton-Raphson and EM algorithms for linear mixed-effects models for repeated measures data. *Journal of the American Statistical Society*, 83:1014-1022

Lindstrom, M.J. and Bates, D.M. (1990) Nonlinear mixed effects models for repeated measures data. *Biometrics*, 46:673-687

Littell, R.C., Milliken, G.A., Stroup, W.W., and Wolfinger, R.D. (1996). *SAS® System for Mixed Models*. SAS Institute Inc., Cary, NC

Little, R.J. and Rubin, D.B. (1987) *Statistical Analysis with Missing Data*. John Wiley & Sons, New York

Longford, N.T. (1993) *Random Coefficient Models*. Clarendon Press, Oxford, UK

Lumer, H. (1937) The consequences of sigmoid growth for relative growth functions. *Growth*, 1:140-154

Machiavelli, R.E. and Arnold, S.F. (1994) Variable order antedependence models. *Communications in Statistics - Theory and Methods*, 23:2683-2699

Maddala, G.S. (1983) *Limited-Dependent and Qualitative Variables in Econometrics*. Cambridge University Press, Cambridge, MA

Magee, L. (1990) R^2 measures based on Wald and likelihood ratio joint significance tests. *The American Statistician*, 44:250-253

Magnus, J.R. (1988) *Matrix Differential Calculus with Applications in Statistics and Econometrics*. John Wiley & Sons, New York

Mallows, C.L. (1973) Some comments on C_p. *Technometrics*, 15:661-675

Marquardt, D.W. (1963) An algorithm for least squares estimation of nonlinear parameters, *Journal of the Society for Industrial and Applied Mathematics*, 2:431-441

Matheron, G. (1962) Traite de Geostatistique Appliquee, Tome I. *Memoires du Bureau de Recherches Geologiques et Minieres*, No. 14. Editions Technip, Paris

Matheron, G. (1963) Principles of geostatistics. *Economic Geology*, 58:1246-1266

Matheron, G. (1971) The theory of regionalized variables and its applications. *Cahiers du Centre de Morphologie Mathematique*, No. 5. Fontainebleau, France

Mays, J., Birch, J.B., and Starnes, B. (2001) Model robust regression: combining parametric, nonparametric, and semiparametric methods. *Journal of Nonparametric Statistics*, 13:245-277

McCullagh, P. (1980) Regression models for ordinal data. *Journal of the Royal Statistical Society (B)*, 42:109-142.

McCullagh, P. (1983) Quasi-likelihood functions. *The Annals of Statistics*, 11:59-67

McCullagh, P. (1984) On the elimination of nuisance parameters in the proportional odds model. *Journal of the Royal Statistical Society (B)*, 46:250-256.

McCullagh, P. and Nelder Frs, J.A. (1989) *Generalized Linear Models. 2nd ed.* Chapman and Hall, New York

McKean, J.W. and Schrader, R.M. (1987) Least absolute errors analysis of variance. In: *Statistical Data Analysis Based on the L_1-Norm and Related Methods* (Dodge, Y., ed.), North-Holland, New York

McLean, R.A., Sanders, W.L., and Stroup, W.W. (1991) A unified approach to mixed linear models. *The American Statistician*, 45:54-64

McPherson, G. (1990) *Statistics in Scientific Investigation*. Springer-Verlag, New York

McShane, L.M., Albert, P.S., and Palmatier, M.A. (1997) A latent process regression model for spatially correlated count data. *Biometrics*, 53:698-706

Mead, R. (1967) A mathematical model for the estimation of inter-plant competition. *Biometrics*, 23:189-205

Mead, R. (1970) Plant density and crop yield. *Applied Statistics*, 19:64-81

Mead, R. (1979) Competition experiments. *Biometrics*, 35:41-54

Mead, R. Curnow, R.N. and Hasted, A.M. (1993) *Statistical Methods in Agriculture and Experimental Biology, 2nd ed.* Chapman and Hall/CRC Press LLC, New York and Boca Raton, FL

Mercer, W.B. and Hall, A.D. (1911) The experimental error of field trials. *Journal of Agricultural Science*, 4:107-132

Miller, M.D., Mikkelsen, D.S., and Huffaker, R.C. (1962) Effects of stimulatory and inhibitory levels of 2,4-D, iron, and chelate supplements on juvenile growth of field beans. *Crop Science*, 2:111-114

Milliken, G.A. and Johnson, D.E. (1992) *Analysis of Messy Data. Volume 1: Designed Experiments*. Chapman and Hall, New York

Minot, C.S. (1908) *The Problem of Age, Growth and Death: A Study of Cytomorphosis*. Knickerbocker Press, New York

Mitscherlich, E.A. (1909) Das Gesetz des Minimums und das Gesetz des Abnehmenden Bodenertrags. *Zeitschrift für Pflanzenernährung, Düngung und Bodenkunde*, 12:273-282

Moore, E.H. (1920) On the reciprocal of the general algebraic matrix. *Bulletin of the American Mathematical Society*, 26:394-395

Moran, P.A.P. (1948) The interpretation of statistical maps. *Journal of the Royal Statistical Society (B)*, 10:243-251

Moran, P.A.P. (1950) Notes on continuous stochastic phenomena. *Biometrika*, 37:17-23

Moran, P.A.P. (1971) Estimating structural and functional relationships. *Journal of Multivariate Analysis*, 1:232-255

Morgan, P.H., Mercer, L.P., and Flodin, N.W. (1975) General model for nutritional responses of higher organisms. *Proceedings of the National Academy of Science*, USA, 72:4327-4331

Morris, G.L. and Odell, P.L. (1968) A characterization for generalized inverses of matrices. *SIAM Review*, 10(2):208-211

Mueller, T.G. (1998) Accuracy of soil property maps for site-specific management. Ph.D. dissertation, Michigan State University, East Lansing, MI (Diss Abstr. 99-22353, Diss Abstr. Int. 60B:0901)

Mueller, T.G., Pierce, F.J., Schabenberger, O., and Warncke, D.D. (2001) Map quality for site-specific fertility management. *Journal of the Soil Science Society of America*, 65: 1547-1558

Myers, R.H. (1990) *Classical and Modern Regression with Applications, 2nd ed.* Duxbury Press, Boston

Nadaraya, E.A. (1964). On estimating regression. *Theory of Probability and its Applications*, 10:186-190

Nagelkerke, N.J.D. (1991) A note on a general definition of the coefficient of determination. *Biometrika*, 78:691-692

Nelder, J.A. and Wedderburn, R.W.M. (1972) Generalized linear models. *Journal of the Royal Statistical Society (A)*, 135:370-384

Neter, J., Wasserman, W., and Kutner, M.H. (1990) *Applied Linear Statistical Models. 3rd ed.*, Irwin, Boston, MA

Neuman, S.P. and Jacobson, E.A. (1984) Analysis of nonintrinsic spatial variability by residual kriging with applications to regional groundwater levels. *Journal of the International Association of Mathematical Geology*, 16:499-521

Newberry, J.D. and Burk, T.E. (1985) S_B distribution-based models for individual tree merchantable volume-total volume ratios, *Forest Science*, 31:389-398

Neyman, J. and Scott, E.L. (1972) Processes of clustering and applications. In: *Stochastic Point Processes* (P.A.W. Lewis, ed.). Wiley and Sons, New York, pp. 646-681

Nichols, M.A. (1974a) Effect of sowing rate and fertilizer application on the yield of dwarf beans. *New Zealand Journal of Experimental Agriculture*, 2:155-158

Nichols, M.A. (1974b) A plant spacing study with sweet corn. *New Zealand Journal of Experimental Agriculture*, 2:377-379

Nichols, M.A. and Nonnecke, I.L. (1974) Plant spacing studies with processing peas in Ontario, Canada. *Scientia Horticulturae* 2:112-122

Nichols, M.A., Nonnecke, I.L., and Pathak, S.C. (1973) Plant density studies with direct seeded tomatoes in Ontario, Canada. *Scientiae Horticulturae*, 1:309-320

Olkin, I., Gleser, L.J., and Derman, C. (1978) *Probability Models and Applications*. Macmillan Publishing, New York

Ord, J.K. (1975) Estimation methods for models of spatial interaction. *Journal of the American Statistical Association*, 70:120-126

Papadakis, J.S. (1937) Méthode statistique pour des expériences sur champ. *Bull. Inst. Amelior. Plant. Thessalonique*, 23

Patterson, H.D. and Thompson, R. (1971) Recovery of inter-block information when block sizes are unequal. *Biometrika*, 58:545-554

Pázman, A. (1993) *Nonlinear Statistical Models*. Kluwer Academic Publishers, London

Pearl, R. and Reed, L.J. (1924) The probable error of certain constraints of the population growth curve. *American Journal of Hygiene*, 4(3):237-240

Pearson, E.S. (1931) The analysis of variance in case of non-normal variation. *Biometrika*, 23:114-133

Penrose, R.A. (1955) A gerneralized inverse for matrices. *Proceedings of the Cambridge Philosophical Society*, 51:406-413

Petersen, R.G. (1994) *Agricultural Field Experiments. Design and Analysis*. Marcel Dekker, New York.

Pierce, F.J., Fortin, M.-C., and Staton, M.J. (1994) Periodic plowing effects on soil properties in a no-till farming system. *Journal of the American Soil Science Society*, 58:1782-1787

Pierce, F.J. and Warncke, D.D. (2000) Soil and crop response to variable-rate liming in two Michigan fields. *Journal of the Soil Science Society of America*, 64:774-780

Pinheiro, J.C. and Bates, D.M. (1995) Approximations to the log-likelihood function in the nonlinear mixed-effects model. *Journal of Computational and Graphical Statistics*, 4:12-35.

Potthoff, R.F. and Roy, S.N. (1964) A generalized mutivariate analysis of variance model useful especially for growth curve problems. *Biometrika*, 51:313-326

Prasad, N.G.N. and Rao, J.N.K. (1990) The estimation of the mean squared error of small-area estimators. *Journal of the American Statistical Association*, 85:163-171

Prentice, R.L. (1988) Correlated binary regression with covariates specific to each binary observation. *Biometrics*, 44:1044-1048

Press, W.H, Teukolsky, S.A., Vetterling, W.T., and Flannery, B.P. (1992) *Numerical Recipes. The Art of Scientific Computing. 2nd ed.* Cambridge University Press, New York

Priebe, D.L. and Blackmer, A.M. (1989) Preferential movement of oxygen-18-labeled water and nitrogen-15-labeled urea through macropores in a Nicollet soil. *Journal of Environmental Quality*, 18:66-72

Quiring, D.P. (1941) The scale of being according to the power formula. *Growth*, 2:335-346

Radosevich, S.R. and Holt, J.S. (1984) *Weed Ecology*. John Wiley & Sons, New York

Ralston, M.L. and Jennrich, R.I. (1978) DUD, a derivative-free algorithm for nonlinear least squares. *Technometrics*, 20:7-14

Rao, C.R. (1965) The theory of least squares when the parameters are stochastic and its application to the analysis of growth curves. *Biometrika*, 58:545-554

Rao, C.R. and Mitra, S.K. (1971) *Generalized Inverse of Matrices and its Applications*. John Wiley & Sons, New York

Rasse, D.P., Smucker, A.J.M., and Schabenberger, O. (1999) Modifications of soil nitrogen pools in response to alfalfa root systems and shoot mulch. *Agronomy Journal*, 91:471-477

Ratkowsky, D.A. (1983) *Nonlinear Regression Modeling*. Marcel Dekker, New York

Ratkowsky, D.A. (1990) *Handbook of Nonlinear Regression Models*. Marcel Dekker, New York

Reed, R.R. (2000) Factors influencing biotite weathering. M.S. Thesis, Department of Crop and Soil Environmental Sciences, Virginia Polytechnic Institute and State University (Available at http://scholar.lib.vt.edu/theses)

Rennolls, K. (1993) Forest height growth modeling. In: *Proceedings from the IUFRO Conference, Copenhagen*, June 14-17, 1993. Forskningsserien Nr. 3, 231-238

Richards, F.J. (1959) A flexible growth function for empirical use. *Journal of Experimental Botany*, 10:290-300

Rigas, A.G. (1991) Spectral analysis of stationary point processes using the fast Fourier transform algorithm. *Journal of Time Series Analysis*. 13:441-450

Ripley, B.D. (1976) The second-order analysis of stationary point processes. *Journal of Applied Probability*, 13:255-266

Ripley, B.D. (1977) Modeling spatial patterns. *Journal of the Royal Statistical Society (B)*, 39:172-192 (with discussion, 192-212)

Ripley, B.D. (1981) *Spatial Statistics*. John Wiley & Sons, New York

Ripley, B.D. (1988) *Statistical Inference for Spatial Processes*. Cambridge University Press, Cambridge

Ripley, B.D. and Silverman, B.W. (1978) Quick tests for spatial interaction. *Biometrika*, 65:641-642

Roberts, H.A., Chancellor, R.J., and Hill, T.A. (1982) The biology of weeds. In: *Weed Control Handbook: Principles*, 7th ed. (H.A. Roberts, ed.). Blackwell Scientific, Oxford, pp. 1-36

Robertson, T.B. (1923) *The Chemical Basis of Growth and Senescence*. J.P. Lippincott Co., Philadelphia and London

Robinson, G.K. (1991) That BLUP is a good thing: the estimation of random effects. *Statistical Science*, 6(1):15-51

Rohde, C.A. (1966) Some results on generalized inverses. *SIAM Review*, 8(2):201-205

Rubin, D.R. (1976) Inference and missing data. *Biometrika*, 63:581-592

Rubinstein, R.Y. (1981) *Simulation and the Monte Carlo Method*. John Wiley & Sons, New York

Russo, D. (1984) Design of an optimal sampling network for estimating the variogram. *Journal of the Soil Science Society of America*, 48:708-716

Russo, D. and Bresler, E. (1981) Soil hydraulic properties as stochastic processes, 1. An analysis of field spatial variability. *Journal of the Soil Science Society of America*, 45:682-687

Russo, D. and Jury, W.A. (1987a) A theoretical study of the estimation of the correlation scale in spatially variable fields. 1. Stationary fields. *Water Resources Research*, 7:1257-1268

Russo, D. and Jury, W.A. (1987b) A theoretical study of the estimation of the correlation scale in spatially variable fields. 2. Nonstationary fields. *Water Resources Research*, 7:1269-1279

Sahai, H. and Ageel, M.I. (2000) *The Analysis of Variance. Fixed, Random and Mixed Models*. Birkhäuser, Boston

Sandland, R.L. (1983) Mathematics and the growth of organisms — some historical impressions. *Mathematical Scientist*, 8:11-30

Sandland, R.L. and McGilchrist, C.A. (1979). Stochastic growth curve analysis. *Biometrics*, 35:255-272

Sandral, G.A., Dear, B.S., Pratley, J.E., and Cullis, B.R. (1997) Herbicide dose rate response curves in subterranean clover determined by a bioassay. *Australian Journal of Experimental Agriculture*, 37:67-74

Satterthwaite, F.E. (1946) An approximate distribution of estimates of variance components. *Biometrics*, 2:110-114

Schabenberger, O. (1994) Nonlinear mixed effects growth models for repeated measures in ecology. In: *Proceedings of the Section on Statstics and the Environment*, Annual Joint Statistical Meetings, Toronto, Canada, Aug. 13-18, 1994, pp. 156-161

Schabenberger, O. (1995) The use of ordinal response methodology in forestry. *Forest Science*, 41(2):321-336.

Schabenberger, O. and Birch, J.B. (2001) Statistical dose-response models with hormetic effects. *International Journal of Human and Ecological Risk Assessment*, 7(4):891-908

Schabenberger, O. and Gregoire, T.G. (1995) A conspectus on estimating function theory and its applicability to recurrent modeling issues in forest biometry. *Silva Fennica*, 29(1):49-70

Schabenberger, O. and Gregoire, T.G. (1996) Population-averaged and subject-specific approaches for clustered categorical data. *Journal of Statistical Computation and Simulation*, 54:231-253

Schabenberger, O., Gregoire, T.G., and Burkhart, H.E. (1995) Commentary: Multi-state models for monitoring individual trees in permanent observation plots by Urfer, W., Schwarzenbach, F.H. Kütting, J., and Müller, P. *Journal of Environmental and Ecological Statistics*, 1(3):171-199

Schabenberger, O., Gregoire, T.G., and Kong, F. (2000) Collections of simple effects and their relationship to main effects and interactions in factorials. *The American Statistician*, 54:210-214

Schabenberger, O., Tharp, B.E., Kells, J.J., and Penner, D. (1999) Statistical tests for hormesis and effective dosages in herbicide dose response. *Agronomy Journal*, 91:713-721

Schnute, J. and Fournier, D. (1980) A new approach to length-frequency analysis: growth structure. *Canadian Journal of Fisheries and Aquatic Science*, 37:1337-1351

Schrader, R.M. and Hettmansberger, T.P. (1980) Robust analysis of variance based on a likelihood criterion. *Biometrika*, 67:93-101

Schrader, R.M. and McKean, J.W. (1977) Robust analysis of variance. *Communications in Statistics A*, 6:979-894

Schulz, H. (1888) Über Hefegifte. *Pflügers Archiv der Gesellschaft für Physiologie*, 42:517-541

Schumacher, F.X. (1939) A new growth curve and its application to timber yield studies. *Journal of Forestry*, 37:819-820

Schwarz, G. (1978) Estimating the dimension of a model. *Annals of Statistics*, 6:461-464

Schwarzbach, W. (1984) A new approach in the evaluation of field trials: The determination of the most likely genetic ranking of varieties. Proceedings EUCARPIA Cer. Sect. Meet., Vortr. Pflanzenzucht, 6:249-259

Schwertman, N.C. (1996) A connection between quadratic-type confidence limits and fiducial limits. *The American Statistician*, 50(3):242-243

Searle, S.R. (1971) *Linear Models*. John Wiley & Sons, New York

Searle, S.R. (1982) *Matrix Algebra Useful for Statisticians*. John Wiley & Sons, New York

Searle, S.R. (1987) *Linear Models for Unbalanced Data*. John Wiley & Sons, New York

Searle, S.R., Casella, G., and McCulloch, C.E. (1992) *Variance Components*. John Wiley & Sons, New York

Seber, G.A.F. and Wild, C.J. (1989) *Nonlinear Regression*. John Wiley & Sons, New York

Seefeldt, S.S., Jensen, J.E., and Fuerst, P. (1995) Log-logistic analysis of herbicide dose-response relationships. *Weed Technology*, 9:218-227

Shapiro, S.S. and Wilk, M.B. (1965) An analysis of variance test for normality (complete samples). *Biometrika*, 52:591-612

Sharples, K. and Breslow, N. (1992) Regression analysis of correlated binary data: some small sample results for the estimating equation approach. *Journal of Statistical Computation and Simulation*, 42:1-20

Sheiner, L.B. and Beal, S.L. (1980) Evaluation of methods for estimating population pharmacokinetic parameters. I. Michaelis-Menten model: routine clinical pharmacokinetic data. *Journal of Pharmacokinetics and Biopharmaceutics*, 8:553-571

Sheiner, L.B. and Beal, S.L. (1985) Pharmacokinetic parameter estimates from several least squares procedures: Superiority of extended least squares. *Journal of Pharmacokinetics and Biopharmaceutics*, 13:185-201

Shinozaki, K. and Kira, T. (1956) Intraspecific competition among higher plants. VII. Logistic theory of the C-D effect. *J. Inst. Polytech. Osaka City University*, D7:35-72

Snedecor, G.W. and Cochran, W.G. (1989) *Statistical Methods, 8th ed.* Iowa State University Press, Ames, Iowa.

Solie, J.B., Raun, W.R., and Stone, M.L. (1999) submeter spatial variability of selected soil and bermudagrass production variables. *Journal of the Soil Science Society of America*, 63:1724-1733

Steel, R.G.D., Torrie, J.H., and Dickey, D.A. (1997) *Principles and Procedures of Statistics. A Biometrical Approach*. McGraw-Hill, New York.

Stein, M.L. (1999) *Interpolation of Spatial Data. Some Theory of Kriging*. Springer-Verlag, New York

Stevens, W.L. (1951) Asymptotic regression. *Biometrics*, 7:247-267

Streibig, J.C. (1980) Models for curve-fitting herbicide dose response data. *Acta Agriculturæ Scandinavica*, 30:59-63

Streibig, J.C. (1981) A method for determining the biological effect of herbicide mixtures. *Weed Science*, 29:469-473

Stroup, W.W., Baenziger, P.S., and Mulitze, D.K. (1994) Removing spatial variation from wheat yield trials: a comparison of methods. *Crop Science*, 86:62-66.

Sweeting, T.J. (1980) Uniform asymptotic normality of the maximum likelihood estimator. *Annals of Statistics*, 8:1375-1381

Swinton, S.M. and Lyford, C.P. (1996) A test for choice between hyperbolic and sigmoidal models of crop yield response to weed density. *Journal of Agricultural, Biological, and Environmental Statistics*, 1:97-106

Tanner, M.A. and Young, M.A. (1985) Modeling ordinal scale disagreement. *Psychological Bulletin*, 98:408-415

Tharp, B.E., Schabenberger, O., and Kells, J.J. (1999) response of annual weed species to glufosinate and glyphosate. *Weed Technology*, 13:542-547

Theil, H. (1971) *Principles of Econometrics*. John Wiley & Sons, New York

Thiamann, K.V. (1956) Promotion and inhibition: twin themes of physiology, *The American Naturalist*, 40:145-162

Thompson, R. and Baker, R.J. (1981) Composite link functions in generalized linear models. *Applied Statistics*, 30:125-131

Thornley, J.H.M. and Johnson, I.R. (1990) *Plant and Crop Models*. Clarendon Press, Oxford, UK

Tobler, W. (1970) A computer movie simulating urban growth in the Detroit region. *Economic Geography*, 46:234-240

Tukey, J.W. (1949) One degree of freedom for nonadditivity. *Biometrics*, 5:232-242

Tukey, J.W. (1977) *Exploratory Data Analysis*. Addison-Wesley, Reading, MA

Tweedie, M.C.K. (1945) Inverse statistical variates. *Nature*, 155:453

Tweedie, M.C.K. (1957a) Statistical properties of inverse Gaussian distributions I. *Annals of Mathematical Statistics*, 28:362-377

Tweedie, M.C.K. (1957b) Statistical properties of inverse Gaussian distributions II. *Annals of Mathematical Statistics*, 28:696-705

UNSCEAR (1958) Report of the United Nations Scientific Committee on the Effects of Atomic Radiation. *Official Records of the General Assembly, 13th Session, Supplement No. 17.*

Upton, G.J.G. and Fingleton, B. (1985) *Spatial Data Analysis by Example, Vol.1: Point Pattern and Quantitative Data.* John Wiley & Sons, New York

Urquhart, N.S. (1968) Computation of generalized inverse matrices which satisfy specified conditions. *SIAM Review*, 10(2):216-218

Utomo, I.H. (1981) Weed competition in upland rice. In: *Proceedings of the 8th Asian-Pacific Weed Science Society Conference*, Vol II: 101-107

Valentine, H.T. and Gregoire, T.G. (2001) A switching model of bole taper. *Canadian Journal of Forest Research.* To appear

Van Deusen, P.C., Sullivan, A.D., and Matney, T.G. (1981) A prediction system for cubic foot volume of loblolly pine applicable through much of its range. *Southern Journal of Applied Forestry*, 5:186-189

Verbeke, G. and Molenberghs, G. (1997) *Linear Mixed Models in Practice: A SAS-oriented Approach.* Springer-Verlag, New York

Verbyla, A.P., Cullis, B.R., Kenward, M.G., and Welham S.J. (1999) The analysis of designed experiments and longitudinal data by using smoothing splines. *Applied Statistics*, 48:269-311

Vitosh, M.L, Johnson, J.W., and Mengel, D.B. (1995) Tri-state fertilizer recommendations for corn, soybeans, wheat and alfalfa. *Michigan State University Extension Bullettin E-2567.*

Von Bertalanffy, L. (1957) Quantitative laws in metabolism and growth. *Quarterly Reviews in Biology*, 32:217-231

Vonesh, E.F. and Carter, R.L. (1992) Mixed-effects nonlinear regression for unbalanced repeated measures. *Biometrics*, 48:1-17

Vonesh, E.F. and Chinchilli, V.M. (1997) *Linear and Nonlinear Models for the Analysis of Repeated Measurements*. Marcel Dekker, New York

Wakeley, J.T. (1949) *Annual Report of the Soils-Weather Project, 1948*. University of North Carolina (Raleigh) Institute of Statistics Mimeo Series, 19

Wallsten, T.S. and Budescu, D.V. (1981) Adaptivity and nonadditivity in judging MMPI profiles. *Journal of Experimental Psychology: Human Perception and Performance*, 7:1096-1109

Walters, K.J., Hosfield, G.L., Uebersax, M.A., and Kelly, J.D. (1997) Navy bean canning quality: correlations, heritability estimates, and randomly mmplified polymorphic DNA markers associated with component traits. *Journal of the American Society for Horticultural Sciences* 122(3): 338-343

Wang, Y.H. (2000) Fiducial intervals: what are they? *The American Statistician*, 54(2):105-111

Warrick, A.W. and Myers, D.E. (1987) Optimization of sampling locations for variogram calculations. *Water Resources Research*, 23:496-500

Watson, G.S. (1964). Smooth regression analysis. *Sankhya (A)*, 26:359-372

Watts, D.G. and Bacon, D.W. (1974) Using an hyperbola as a transition model to fit two-regime straight-line data. *Technometrics*, 16:369-373

Waugh, D.L., Cate Jr., R.B., and Nelson, L.A. (1973) Discontinuous models or rapid correlation, interpretation, and utilization of soil analysis and fertilizer response data. *International Soil Fertility Evaluation and Improvement Program, Technical Bulletin No. 7*, North Carolina State University, Raleigh, NC

Webster, R. and Oliver, M.A. (1992) Sample adequately to estimate variograms for soil properties. *Journal of Soil Science* 43:177-192

Wedderburn, R.W.M. (1974) Quasilikelihood functions, generalized linear models and the Gauss-Newton method. *Biometrika*, 61:439-447

Welch, B.L. (1937) The significance of the difference between two means when the population variances are unequal. *Biometrika*, 29:350-362

White, H. (1980) A heteroskedasticity-consistent covariance matric estimator and a direct test for heteroskedasticity. *Econometrica*, 48:817-838

White, H. (1982) Maximum likelihood estimation of misspecified models. *Econometrics*, 50:1-25

Whittle, P. (1954) On stationary processes in the plane. *Biometrika*, 41:434-449

Wiebe, G.A. (1935) Variation and correlation among 1500 wheat nursery plots. *Journal of Agricultural Research*, 50:331-357

Wiedman, S.J. and Appleby, A.P. (1972) Plant growth stimulation by sublethal concentrations of herbicides. *Weed Research*, 12:65-74

Wilkinson, G.N., Eckert, S.R., Hancock, T.W., and Mayo, O. (1983) Nearest neighbor (NN) analysis of field experiments (with discussion). *Journal of the Royal Statistical Society (B)*, 45:152-212

Winer, B.J. (1971) *Statistical Principles in Experimental Design*. McGraw-Hill, New York

Wishart, J. (1938) Growth rate determinations in nutrition studies with the bacon pig, and their analysis. *Biometrika*, 30:16-28

Wolfinger, R. (1993a) Covariance structure selection in general mixed models. *Communications in Statistics, Simulation and Computation*, 22(4):1079-1106

Wolfinger, R. (1993b) Laplace's approximation for nonlinear mixed models. *Biometrika*, 80:791-795

Wolfinger, R. and O'Connell, M. (1993) Generalized linear mixed models: a pseudo-likelihood approach. *Journal of Statistical Computation and Simulation*, 48:233-243

Wolfinger, R., Tobias, R., and Sall, J. (1994) Computing Gaussian likelihoods and their derivatives for general linear mixed models. *SIAM Journal on Scientific and Statistical Computing*, 15:1294-1310

Xu, W., Tran, T., Srivastava, R., and Journel, A.G. (1992) Integrating seismic data in reservoir modeling: the collocated cokriging alternative. *SPE Paper 24742*, 67th Annual Technical Conference and Exhibition.

Yandell, B.S. (1997) *Practical Data Analysis for Designed Experiments*. Chapman and Hall, New York

Yates, F. (1936) Incomplete randomized blocks. *Annals of Eugenics*, 7:121-140

Yates, F. (1940) The recovery of inter-block information in balanced incomplete block designs. *Annals of Eugenics*, 10:317-325

Zeger, S.L. and Harlow, S.D. (1987) Mathematical models from laws of growth to tools for biological analysis: fifty years of *Growth. Growth*, 51:1-21

Zeger, S.L. and Liang, K.-Y. (1986) Longitudinal data analysis for discrete and continuous outcomes. *Biometrics*, 42:121-130

Zeger, S.L. and Liang, K.-Y. (1992) An overview of methods for the analysis of longitudinal data. *Statistics in Medicine*, 11:1825-1839

Zeger, S.L., Liang, K.-Y., and Albert, P.S. (1988) Models for longitudinal data: a generalized estimating equation approach. *Biometrics*, 44:1049-1060

Zhao, L.P. and Prentice, R.L. (1990) Correlated binary regression using a quadratic exponential model. *Biometrika*, 77:642-648

Zheng, L. and Silliman, S.E. (2000) Estimating the theoretical semivariogram from finite numbers of measurements. *Water Resources Research*, 36:361-366

Zimdahl, R.L. (1980) *Weed-Crop Competition: A Review*. International Plant Protection Center, USA

Zimmerman, D.L. (1989) Computationally exploitable structure of covariance matrices and generalized covariance matrices in spatial models. *Journal of Statistical Computation and Simulation*, 32: 1-15

Zimmerman, D.L. and Harville, D.A. (1991) A random field approach to the analysis of field-plot experiments and other spatial experiments. *Biometrics*, 47:223-239.

Zimmerman, D.L. and Núñez-Antón, V. (1997) Structured antedependence models for longitudinal data. In: *Modelling Longitudinal and Spatially Correlated Data* (Gregoire, T.G., Brillinger, D.R., Diggle, P.J., Russek-Cohen, E., Warren, W.G., and Wolfinger, R.D., eds). Springer-Verlag, New York, pp. 63-76

Zimmerman, D.L. and Zimmerman, M.B. (1991) A comparison of spatial semivariogram estimators and corresponding kriging predictors. *Technometrics*, 33:77-91

Author Index

Main Text only without Appendix

Subject Index